S0-AXJ-869

TERRESTRIAL VEGETATION OF CALIFORNIA

TERRESTRIAL VEGETATION OF CALIFORNIA

NEW EXPANDED EDITION, 1988

Edited by

MICHAEL G. BARBOUR

JACK MAJOR

University of California
Davis

CALIFORNIA NATIVE PLANT SOCIETY

Special Publication Number 9

Second CNPS Printing, 1988.
Third CNPS Printing, 1990.
Fourth CNPS Printing, 1995.

Library of Congress Cataloging-in-Publication Date:

Main entry under title:
Terrestrial vegetation of California.

(Special publication/California Native Plant Society: No. 9)
Includes bibliographies and index.
1. Botany—California—Ecology. I. Barbour, Michael G.
II. Major, Jack, 1917-

QK149.T44 1995 581.9'794 87-71829
ISBN 0-943460-13-1

Printed in the United States of America

PREFACE

In the summer of 1974, for a variety of reasons, we decided that it was time for a technical monograph on the vegetation of California to be written. Many names of potential contributors came to mind, and they in turn suggested others. Their enthusiasm for the project was infectious, and it carried the two year manuscript preparation period through to completion with little discord and many pleasant memories of a sense of community among California ecologists.

Late in 1974, Wiley-Interscience took the young project under its financial wing, based only on a table of contents and a thin book prospectus. Their confidence was impressive, for the sequence of chapters, their titles, and sometimes their authors changed several times. In 1975, Will Küchler joined the group to provide a vegetation map (in back cover pocket) and its description (in the Appendix)—very important contributions toward our presenting a complete story.

The size of the manuscript grew far beyond our early estimate. To keep the size and price of the book within reason, four overview appendices were eliminated: a survey of soils and vegetation, written by Gene Begg and Gordon Huntington of the University of California, Davis; an innovative review of pollination ecology and vegetation by Andy Moldenke of the University of California, Santa Cruz; a catalog of natural areas, compiled by Les Hood for the California Natural Areas Coordinating Council; and an annotated outline of habitat types, prepared by Dan Cheatham (Berkeley) and Bob Haller (Santa Barbara) of the University of California. We miss their contributions very much.

Another concession to size was an early decision to limit the scope of the book to terrestrial vegetation only. However, the absence of chapters on emergent aquatic vegetation, such as tule marsh, or on riparian vegetation is unintentional: we could not find anyone willing to write about them. As pointed out in Chapter 1, the careful reader will soon find other regions or vegetation types also omitted, because so little is known about them. One incidental objective of the book is to point out the limits to our understanding of California vegetation, and in this we have certainly succeeded.

Several ecologists who have made major studies of vegetation types in the state are not contributors to this volume. Some were unable to participate because of other commitments, and some were not asked because their talents duplicated those of contributors who had already agreed to write certain chapters. Their absence here definitely does not imply that we value them less than those who are authors in this book.

Several chapters contain their own acknowledgment sections. From our standpoint as editors, however, we take pleasure in mentioning the following individuals and groups.

The chapter authors deserve special commendation for taking the time and making the hard effort necessary to complete their material within the time and space restrictions imposed on them. All volunteered enthusiastically for the task, and it was a genuine pleasure to work with all of them. They gave us an exciting education in California vegetation. Will Küchler perservered with the map with never a pessimistic statement, despite many crises, financial and other. We are grateful for the monetary support given the mapping by The Ahmanson Foundation, The California Native Plant Society, The Sierra Club, The University of Kansas, The USDA Forest Service, the U.S. Bureau of Reclamation, and Wiley-Interscience. Dr. Küchler's enthusiasm for a first-rate publication was constant, and we thank him for his moral support as well as for this excellent contribution to the book.

Dr. Mary Conway, Life Sciences Editor at Wiley-Interscience, deserves credit for her long-range planning and confidence in our editorial judgment—factors that significantly reduced publication time and the sale price of this volume.

Margaret Lynn Siri drew all the chapter maps and several of the figures. We are grateful for her timely, careful help. The stippled topographic base maps that she used were provided by the Entomology Department of the University of California at Davis. Many of the photographs had to be converted to black and white prints from color transparencies, a difficult process well done by the skilled people in Illustration Services on campus. Timothy Kittel traveled much of the northern part of the state taking pictures of vegetation, and two of the dust jacket pictures (savanna and Mt. Lassen) are his.

The typing of some chapters and half of the Literature Cited sections was done by Wendy Reed; the remaining Literature Cited sections and the tables were typed by Dorothy Benke, Laura Goodman, and Melody Johnson of The Secretariat in Davis, California.

Indexing was a major chore, and we take pleasure thanking the following people who helped lighten that burden: Norma Barbour, Nathan Benedict, Susan Conard, Sandy Chase, Ted DeJong, Rod MacDonald, Mary Major, Bill Overton, Bruce Pavlik, Robert Robichaux, and Dean Taylor. The final copy of the index was typed by the Secretariat. The setting of the text was done by American Graphic Arts.

MICHAEL BARBOUR
JACK MAJOR

Davis, California
February 1977

CONTENTS

INTRODUCTORY OVERVIEW

1 **Introduction** 3
Michael G. Barbour and Jack Major

2 **California climate in relation to vegetation** 11
Jack Major

3 **Research natural areas and related programs in California** 75
Norden H. Cheatham, W. James Barry, and Leslie Hood

4 **The California flora** 109
Peter H. Raven

5 **Outline history of California vegetation** 139
Daniel I. Axelrod

6 **The status of vegetation mapping in California today** 195
Wilmer L. Colwell, Jr.

THE CALIFORNIAN FLORISTIC PROVINCE

7 **Beach and dune** 223
Michael G. Barbour and Ann F. Johnson

8 **Coastal salt marsh** 263
Keith B. Macdonald

9 **The closed-cone pines and cypresses** 295
Richard J. Vogl, Wayne P. Armstrong, Keith L. White, and Kenneth L. Cole

10 **Mixed evergreen forest** 359
John O. Sawyer, Dale A. Thornburgh, and James R. Griffin

11 **Oak woodland** 383
James R. Griffin

12 **Chaparral** 417
Ted L. Hanes

13 **Southern coastal scrub** 471
Harold A. Mooney

14 **Valley grassland** 491
Harold F. Heady

15 **Vernal pools** 515
Robert F. Holland and Subodh K. Jain

THE SIERRAN FLORISTIC PROVINCE

16 **Montane and subalpine forests of the Transverse and Peninsular Ranges** 537
Robert F. Thorne

17 **Montane and subalpine vegetation of the Sierra Nevada and Cascade Ranges** 559
Philip W. Rundel, David J. Parsons, and Donald T. Gordon

18 **Alpine** 601
Jack Major and Dean W. Taylor

THE PACIFIC NORTHWEST FLORISTIC PROVINCE

19 **The redwood forest and associated north coast forests** 679
Paul J. Zinke

20 **Montane and subalpine vegetation of the Klamath Mountains** 699
John O. Sawyer and Dale A. Thornburgh

21 **Coastal prairie and Northern coastal scrub** 733
Harold F. Heady, Theodore C. Foin, Mary M. Hektner, Dean W. Taylor, Michael G. Barbour, and W. James Barry

THE GREAT BASIN FLORISTIC PROVINCE

22 **Sagebrush steppe** 763
James A. Young, Raymond A, Evans, and Jack Major

23 **Transmontane coniferous vegetation** 797
Frank C. Vasek and Robert F. Thorne

THE HOT DESERT FLORISTIC PROVINCE

24 **Mojave desert scrub vegetation** 835
Frank C. Vasek and Michael G. Barbour

25 **Sonoran desert vegetation** 869
Jack H. Burk

THE SOUTHERN CALIFORNIA ISLANDS

26 **The southern California islands** 893
Ralph N. Philbrick and J. Robert Haller

Appendix: The Map of the Natural Vegetation of California 909
A. Will Küchler

Index 939

Table of Conversions Front Endpaper

The California Native Plant Society is an organization of laypeople and professionals united by an interest in the plants of California. It is open to all. Its principal aims are to preserve the native flora and to add to the knowledge of members and the public at large. It seeks to accomplish the former goal in a number of ways: by undertaking a census of rare, endangered, and extinct plants throughout the state; by acting to save endangered areas through publicity, persuasion, and on occasion, legal action; by providing expert testimony to governmental bodies; and by supporting financially and otherwise the establishment of native plant preserves. Its educational work includes: publication of botanical books, a quarterly journal, *Fremontia,* and a quarterly newsletter, *Bulletin;* assistance to teachers and school projects; meetings, field trips, and other activities of local chapters throughout the state. Non-members are welcome to attend meetings and field trips.

The work of the Society is done by volunteers. Money is provided by the dues of members and by funds raised by chapter activities. Additional donations, bequests, and memorial gifts from friends of the Society can assist greatly in carrying forward the work.

California Native Plant Society
1722 J Street, Suite 17
Sacramento, California 94814

INTRODUCTORY OVERVIEW

CHAPTER 1

INTRODUCTION

MICHAEL G. BARBOUR
JACK MAJOR
Botany Department, University of California, Davis

Objectives 3
Format of this book 6
Literature cited 9

OBJECTIVES

California has a rich diversity of vegetation types. About a decade ago, Küchler (1964) compiled a detailed map of potential natural vegetation for the United States, and he showed more than 30 types to exist within the state's boundaries—more than one-fourth of all the types for the entire conterminous United States. Major types on his map, accounting for just over half of California's land surface, included such diverse ones as mixed conifer forest, oakwoods, chaparral, creosote bush, and steppe (Table 1-1). Various classification schemes have identified from 70 plant communities (Knapp 1965), to 29 (Munz and Keck 1959), to as few as 15 (Sochava 1964; see also Major 1963). The text and map of this book have expanded and refined our understanding to the point where more than 50 plant communities can be mapped (see the Appendix). Even so, however, matrices of species versus stands (association tables) are available to document very few of the communities we recognize.

Some California plant communities are endemic to the state, are well known, and have generated considerable lay and scientific interest: coast redwood forest, various oak woodlands, mixed evergreen forest, a variety of chaparral types, mixed conifer forest, red fir forest, and several alpine communities, to name only a few. Many other communities are less well known but deserve further study. A few are nearly extinct, including some that were widespread only 250 yr ago. Several chapters in this book, and a number of other publications as well, have documented dramatic, recent changes in the composition and distribution of redwood and red fir forests, chaparral, grassland, dune scrub, salt marsh, and creosote bush scrub, among others

TABLE 1-1. Area of California dominated by vegetation types mapped by Küchler (1964), and the chapters of this book in which those types are discussed. Planimetry, courtesy of Harold Heady. Vegetation types not mapped by Küchler include the following: beach and dune (Chapter 7), coastal salt marsh (Chapter 8), northern coastal scrub (Chapter 21), vernal pools (Chapter 15), and Channel Island types (Chapter 26). As the total acreage is approximately 100 million, percent land area for any type can be obtained by multiplying acreage by 10^{-6}

Vegetation Type (and Number)	Acres	Hectares	Chapter
Spruce-cedar-hemlock forest (1)	5,009	2,004	19
Cedar-hemlock-Douglas fir forest (2)	2,015,696	806,278	19
Mixed conifer forest (5)	13,641,010	5,522,676	16, 17, 20
Redwood forest (6)	2,320,254	928,102	19
Red fir forest (7)	1,903,490	761,396	17, 20
Lodgepole pine subalpine forest (8)	2,150,944	860,378	16, 18, 20
Pine-cypress forest (9)	123,226	49,290	9
Ponderosa shrub forest (10)	1,695,108	678,043	17, 22, 23
Great Basin pine forest (22)	49,090	19,636	17, 22, 23
Juniper-piñon woodland (23)	2,463,517	985,407	16, 23
Juniper steppe woodland (24)	909,668	363,867	23
California mixed evergreen forest (29)	3,399,232	1,359,693	10
California oakwoods (30)	9,554,518	3,821,807	11
Chaparral (33)	8,500,585	3,400,234	12
Montane chaparral (34)	573,051	229,220	12, 16, 17, 20
Coastal sagebrush (35)	2,473,535	989,414	13
Mosaic of 30 and 35 (36)	641,175	256,470	11, 13
Great Basin sagebrush (38)	1,851,394	740,558	22

TABLE 1-1 [continued]

Vegetation Type (and Number)	Acres	Hectares	Chapter
Saltbush-greasewood (40)	3,104,692	1,241,876	24, 25
Creosote bush (41)	16,355,988	6,542,395	24, 25
Creosote bush-bur sage (42)	5,330,774	2,132,310	25
Paloverde-cactus shrub (43)	1,052,930	421,172	25
Desert sparse vegetation (46)	115,211	46,084	22, 24, 25
Fescue-oatgrass (47)	878,711	351,484	21
California steppe (48)	13,222,242	5,288,897	14, 15
Tule marshes (49)	1,859,409	743,764	...
Alpine meadows (52)	747,370	298,948	18
Sagebrush steppe (55)	3,245,951	1,298,380	22

(see, e.g., Baker 1962; Barry 1972; Dasmann 1965; Mooney and Dunn 1972; Stebbins 1974). If the past is any criterion, only samples of our natural vegetation will be safe from conversion into something else; fortunately, these samples are now being identified (see Chapters 3 and 6).

California also has a rich diversity of field taxonomists and plant ecologists who have been describing the state's vegetation over the past 100 yr. Many names come to mind, but surely a list of those formerly active would include Leroy Abrams, William Brewer, Jens Clausen, Frederic Clements, Alice Eastwood, Edward Greene, Harvey Monroe Hall, Willis Lynn Jepson, John Muir, Philip Munz, Samuel Parish, Arthur Sampson, Forrest Shreve, and A. E. Wieslander. Others, now retired but still in various stages of professional activity, must include Harold Biswell, William Cooper, William Hiesey, John Thomas Howell, Edmund Jaeger, Herbert Mason, Cornelius Muller, G. Ledyard Stebbins, and Ira Wiggins. Clausen, Clements, Hall, Hiesey, Mason, Muller, and Stebbins in particular emphasized autecology, an approach that is properly included in this book.

The diversity and the volume of past and present ecological contributions have created a problem of riches: fragmentation of the literature. The first objective we had in preparing this volume was to summarize and review this literature. A second objective was to build upon that review; to derive conclusions or hypotheses, to formulate models, to reconstruct the past or predict the future, and to highlight the areas that should receive further study.

From the beginning of this project, it was apparent to the editors that meeting these two objectives could not efficiently be accomplished by one, or even by a handful, of authors. Consequently, the result is a series of typically coauthored chapters with a total of nearly 40 contributors. It was also decided at the beginning that the editors would not impose an hierarchical system of classification of communities and vegetation on chapter authors. Indeed, we had none to impose! We *have* attempted, however, to have a uniform approach in each chapter (a statewide outlook, a thorough review, an integration with other chapters, a sampling of tabular data to accompany text descriptions, the inclusion of autecological as well as synecological information, and suggestions for future research), but we have not imposed any bias beyond this. Consequently, we have sacrificed the sort of uniform style found in other recent regional textbooks such as those of Bakker (1971) and Ornduff (1974) for California, Core (1966) for West Virginia, Curtis (1959) for Wisconsin, Franklin and Dyrness (1973) for Oregon and Washington, Lowe and Brown (1973) and Brown and Lowe (1974a,b) for Arizona, and Robichaud and Buell (1973) for New Jersey. Yet we believe that the positive benefits gained from our format are ultimately of more value to the reader.

FORMAT OF THIS BOOK

Statewide review chapters on climate, flora, and the geological history of major vegetation types preface the main portion of the book, Chapters 7–26. These chapters are organized very pragmatically in a format that maximizes the talents of the contributors and minimizes duplication. The 20 chapters are first subdivided into six floristic provinces: Californian, Sierran, Pacific Northwest, Great Basin, Hot Desert, and Channel Islands (see Stebbins and Major 1965). Some chapters are statewide reviews of essentially one vegetation type (e.g., Chapter 8, Coastal salt marsh; Chapter 12, Chaparral), whereas others describe a series of vegetation types within a geographical region of the state (e.g., Chapter 20, Montane and subalpine vegetation of the Klamath Mountains; Chapter 25, Sonoran Desert). As a guide to county names and topographical features mentioned in Chapters 7–26, please refer to Figs. 1-1 and 1-2.

Each vegetation chapter begins with a state map highlighting the major locations of the vegetation types discussed. These maps are diagrammatic, for orientation only. Details are on the 1:1 million map folded in the back pocket of the book. The sketch map is followed by a chapter table of contents. Each chapter closes with suggestions for future research and a bibliography of literature cited.

Although the contributors were not requested to generate new data for the chapters, many do report previously unpublished data, observations, or concepts. In addition, they have made a special effort to review material not readily accessible, such as theses, dissertations, and nonjournal reports issued by state or national agencies. Photographs have been limited to cases where a real sense of the vegetation can be thereby conveyed. Tabular data have been emphasized in preference to photographs or lengthy text summaries. In most cases, common names of taxa have not been used. Nomenclature, unless shown parenthetically, follows Munz (1968) and Munz and Keck (1959). A taxonomic index at the back of the book includes a cross-listing of the synonyms and common names employed.

The editors have refrained from writing a concluding chapter, in part because

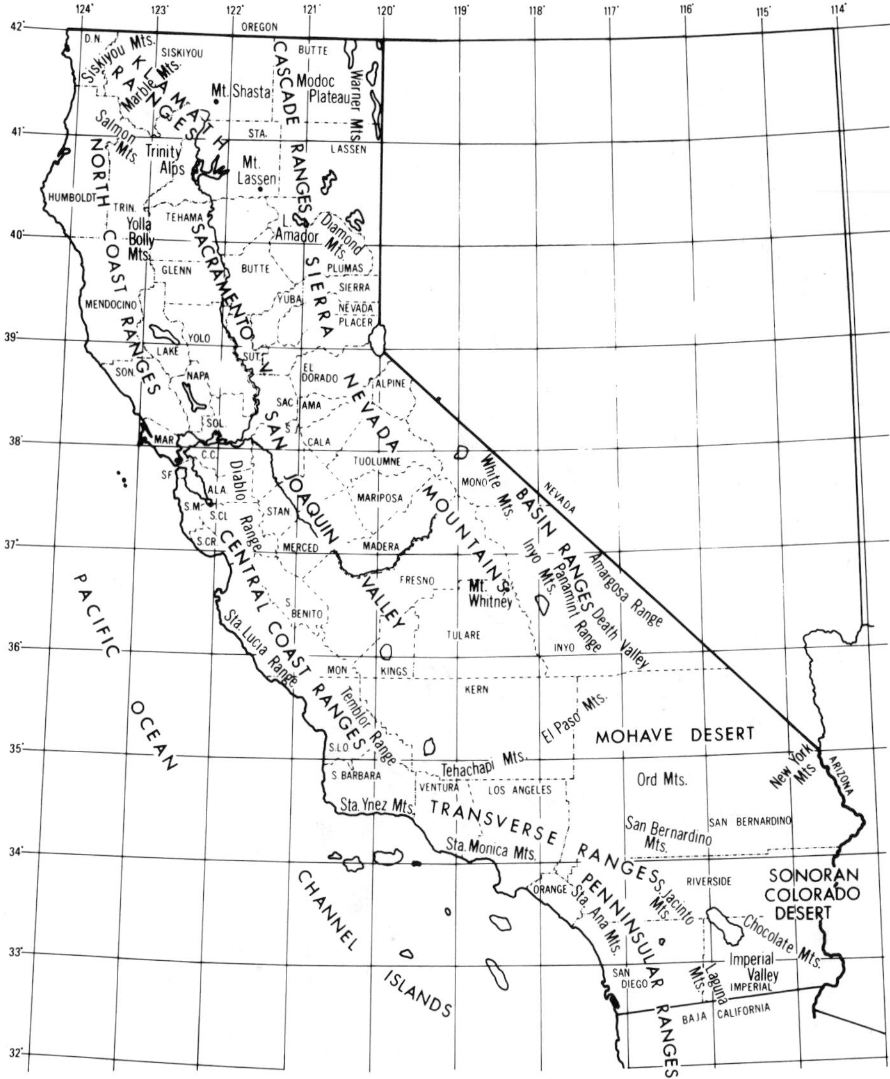

Figure 1-1. Major place names of California: counties and mountain ranges.

such a chapter is bound to be given more attention by the casual reader than it deserves, and we would thus unfairly stamp the book with our own biases. Our only function has been to provide a readable, coherent vehicle in which many contributors summarize our present level of understanding. Perhaps the best concluding chapter will be written 15 yr from now as a preface to someone else's book that will take us to a new level of understanding.

The Appendix is a full explanation to all types shown on the vegetation map. The map, at a scale of 1:1 million, is folded into a pocket on the back cover. A. W. Küchler worked closely with the chapter authors for more than a year in preparing the map and legend. The map has been printed separately by Williams and Heintz of Washington, D.C., and it is copyrighted in Dr. Küchler's name.

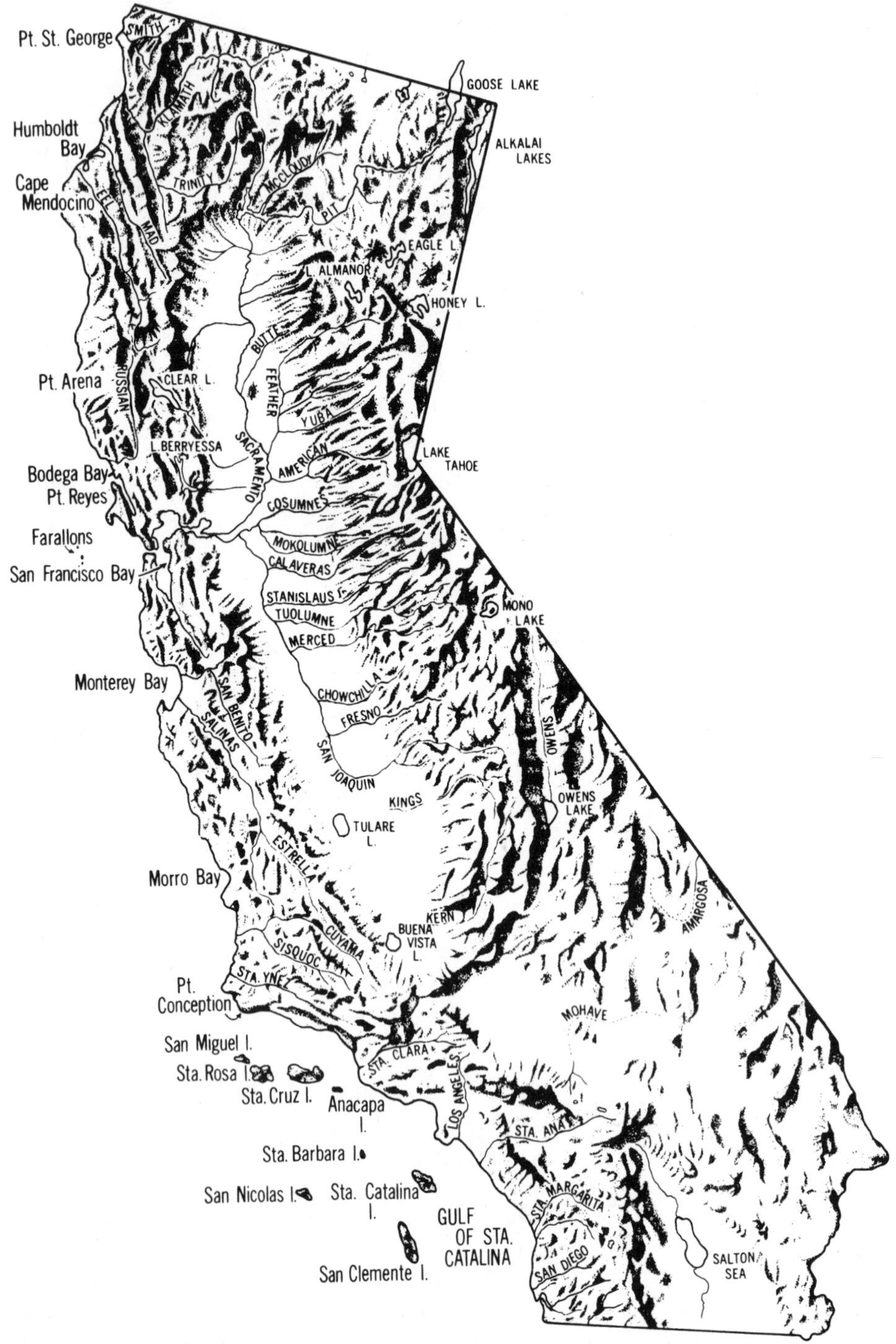

Figure 1-2. Major topographical features of California: rivers, lakes, bays, points, and islands.

LITERATURE CITED

Baker, F. S. 1962. The California region, pp. 460-502. In: J. W. Barrett (ed.). Regional silviculture of the United States. Ronald Press, New York.

Bakker, E. S. 1972. An island called California. Univ. Calif. Press, Berkeley. 357 p.

Barry, W. J. 1972. California prairie ecosystems. Vol. I. The central valley prairie. State of Calif., Resources Agency, Dept. Parks and Recreation, Sacramento. 82 p.

Brown, D. E., and C. H. Lowe. 1974a. A digitized computer-compatible classification for natural and potential vegetation in the southwest with particular reference to Arizona. J. Ariz. Acad. Sci. 9 (Supp. 2). 11 p.

-----. 1974b. The Arizona system for natural and potential vegetation--illustrated summary through the fifth digit for the North American southwest. J. Ariz. Acad. Sci. 9 (Supp. 3). 56 p.

Core, E. L. 1966. Vegetation of West Virginia. McClain, Parsons, Charleston, W. Va. 217 p.

Curtis, J. T. 1959. The vegetation of Wisconsin. Univ. Wisc. Press, Madison. 657 p.

Dasmann, R. F. 1965. The destruction of California. Macmillan, New York.

Franklin, J. F., and C. T. Dyrness. 1973. Natural vegetation of Oregon and Washington. USDA Forest Serv. Gen. Tech. Rep. PNW-8. Washington, D.C. 417 p.

Knapp, R. 1965. Die Vegetation von Nord- und Mittelamerika. Gustav Fischer-Verlag, Stuttgart, Germany. 373 p.

Küchler, A. W. 1964. Potential natural vegetation of the conterminous United States. Amer. Geogr. Soc. Spec. Pub. 36. 116 p. + map.

Lowe, C. H., and D. E. Brown. 1973. The natural vegetation of Arizona. Ariz. Resources Inform. Syst. Pub. 2. Phoenix. 53 p. + map.

Major, J. 1963. Vegetation mapping in California, pp. 195-218. In: Symposium für Vegetations-kartierung, 1959. Verlag J. Cramer, Weinheim, Germany.

Mooney, H. A., and E. L. Dunn. 1972. Land-use history of California and Chile as related to the structure of the sclerophyll scrub vegetations. Madroño 21:305-319.

Munz, P. A. 1968. Supplement to a California flora. Univ. Calif. Press, Berkeley. 224 p.

Munz, P. A., and D. D. Keck. 1959. A California flora. Univ. Calif. Press, Berkeley. 1681 p.

Ornduff, R. 1974. Introduction to California plant life. Univ. Calif. Press, Berkeley. 152 p.

Robichaud, B., and M. F. Buell. 1973. Vegetation of New Jersey. Rutgers Univ. Press, New Brunswick, N.J. 340 p.

Sochava, V. 1964. [Physicogeographical atlas of the world.] Acad. Nauk. USSR, Moscow. 298 p. Translated in: Soviet Geogr., Rev. and Trans. 6(5/6):1-403, 1965.

Stebbins, G. L., and J. Major. 1965. Endemism and speciation in the California flora. Ecol. Monogr. 35:1-35.

Stebbins, R. C. 1974. Off-road vehicles and the fragile desert, Parts 1 and 2. Amer. Biol. Teacher 36:203–208, 220; 36:294–304.

Thorne, R. F. 1976. The vascular plant communities of California. In: J. Latting (ed.). Plant communities of southern California. Calif. Native Plant Soc. Spec. Pub. 2

CHAPTER

2

CALIFORNIA CLIMATE IN RELATION TO VEGETATION

JACK MAJOR

Department of Botany, University of California, Davis

Introduction 11
Climatic data available 12
Energy and water balances 15
An example of a water balance 19
Soil storage of water available to plants 20
Discussion 20
California's climates 21
Horizontal variation 31
Northwest Pacific Coast 31
Mediterranean California 34
Cold Desert 38
Hot Desert 42
Vertical variation 44
Northwest Pacific Coast 45
Mediterranean California 47
Sierra Nevada–Cascade Ranges 53
Northern Sierra Nevade–Cascades 53
Southern Sierra Nevada 60
Cold Desert 64
Hot Desert 66
Literature cited 69

INTRODUCTION

There is no universal correlation between vegetation and climate. This is a general ecological principle that follows immediately from the multifactorial relationship between individual plants and between plant associations or ecological sequences (Jenny 1946, 1941:267–9; Major 1967:94, 109–110) and climates. A single example suffices. *Pinus sabiniana* does not occur at all in a part of the area of foothill woodland climate between roughly the Kings and Kern Rivers in the southern Sierran foothills.

This principle has nothing to do with our arrangement of vegetation into either units or sequences. In fact, the question of whether vegetation does or does not occur in packets (McNaughton 1969:527) is not one for which we can look to vegetation for an answer. It is a question of how the investigator manipulates his impressions (measurements) of the vegetation he studies.

Three levels of climate are differentiated: regional climates, local climates, and microclimates (Major 1951). For definition of regional climates we have the numbers collected and published by official weather bureaus. Efforts are made to have such stations reflect more or less large areas, not local variations in climate. Local climates are functions of relief (Jenny 1941; Geiger 1961). They differ from regional climates because differences in slope, exposure, snow drifting, temperature inversions, cold air drainage, groundwater levels, etc., influence climatic parameters. Microclimates are vegetation or soil properties, dependent on both regional and local climates, on the soil parent material, on the available biota, on the length of time of ecosystem development, and on fire (Jenny 1941; Major 1951; Crocker 1952; Billings 1952).

We can make some generalizations regarding local climates (Geiger 1961; Nash 1963; Hegg 1965; Randall 1972; Rouse 1970), using known relationships between radiation, slope, and exposure. However, in general any discussion of climates as they affect plants must be based in the first instance on regional data. This one is.

CLIMATIC DATA AVAILABLE

The quality of the available data on California's climates can be described in a word: poor. The latest U.S. summary for precipitation and temperature extends up to 1960 (U.S. Weather Bureau 1964), and this is our data base. I have gone back to the 1930 and 1950 summaries for stations no longer recording. In a few cases data from 1974 back into the 1950s or 1960s were compiled.

Temperature data are seemingly straightforward, even though difficult to apply to plants or to plant habitats found in the field (Geiger 1961; Tikhomirov 1963:69; also in Walter 1968:538; Larcher 1973:240). Tikhomirov's arctic tundra *Novosieversia glacialis* (Adams) F. Bolle plant at a soil temperature of 2°C at 20 cm depth and an air shade temperature of 11.7 at 23 cm height had surface temperatures of 15.5 in the flower and 18.9 on the leaves of the basal rosette, and tissue temperatures of 19.4 within the lower part of the stem, 6.3 within the root crown 2 cm below the ground surface, and from 3.9 to 1 within the roots.

On the other hand, measured precipitation is clearly a function of wind, slope (Hamilton 1954), and so on, and phase state of the water.

Measurement of snow precipitation presents problems. Budyko (1974:227–228, 350, 359) and Hare and Hay (1974:117, 120; 1971) both refuse to use precipitation figures in water balance calculations in cases where much of the precipitation occurs as snow. The implication is that, for the mountainous and transmontane parts of California where most of the precipitation does come as snow, water balance calculations may be wrong because of uncertainty regarding the amount of precipitation.

The precipitation data over the Sierra crest at Donner Summit (Fig. 2-11) have suggested a zone of maximum precipitation on the western side of the crest at 1500–

1800 m (Baker 1944:234; Henry 1919; Varney 1920). Uttinger (1951) found no such maximum in the Alps. An alternative explanation is progressive underestimation of precipitation as the proportion of snow in the annual precipitation increases with altitude.

In some coastal areas of California fog drip adds significantly to recorded precipitation. Azevedo and Morgan (1974) recorded more than 200 mm of such water for July–August in coastal forests of Humboldt Co. They quote similar, and higher, figures for the Berkeley Hills (Parsons 1960), the San Francisco Peninsula (Oberlander 1956), and the Santa Ana Mts. (Vogl 1973). This source of water is unrecorded in our climatic records. Its amount is a function of presence and kind of vegetation as well as local topography.

Continentality and maritimity of climate play important roles in determining California's vegetation (Major 1953, etc.). A useful measure of continentality is simply the annual amplitude of mean monthly temperatures (ΔT). However, Khromov (1957) points out that over the open oceans ΔT changes 5.4° with latitude. Therefore $K = \Delta T - 5.4 \sin$ latitude. Conrad (1946) modified ΔT to give a range of continentality from 0 to 100% for the Lofoten Islands to Verkhoyansk. On his scale the coastal area north of San Francisco has negative values.

A different measure, which correlates extremely well with vegetation differences, is Bailey's equability (Eq) (Axelrod 1965). On a diagram of annual mean temperature against ΔT, successive areas denote decreasing equability outward from an annual mean of 14°C and ΔT of 0. Similarly Hintikka (1963) showed quite precise relationships between Scandinavian plant distribution areas and the fields they occupy on a plot of temperatures of the coldest and warmest months.

It is possible to extend short runs of climatic data from a field study locality by using the short run with the overlapping record at a nearby, long-record Weather Bureau station in a regression. In general, temperature records from nearby stations are linearly related as in Fig. 2-1. Obviously monthly means give a suitably wide spread in the data, but daily values, maxima and minima, and so on, can also be used. Campbell (1972) has used the method to predict daily temperatures at an experimental watershed 97 km distant and 377 m higher in elevation than the weather station at Albuquerque, New Mexico. A year's daily (3 hr intervals) temperature readings were used for the correlations. Campbell discusses the accuracy and reliability of the technique. By entering our Fig. 2-1 with the 20 yr mean monthly temperatures for Winters, reasonable monthly means for Monticello Dam were read off.

If the temperatures at the two stations are always equal, the regression line will go through the origin with a slope of 1. If the short-term station is always colder by a constant ΔY than the long-term station, the intercept will be important and negative, although the slope will be 1. If ΔY becomes larger with higher temperatures at the long-term station, the slope will be <1. A combination of these cases is the usual form of the relationship.

Unfortunately precipitation records are not generally linearly related (Fig. 2-2). The problem involved in Fig. 2-2 was to predict the precipitation at Smith Creek (37°19′N, 121°48′W, 641 m altitude), a valley station over the first ridge east of San Jose on the way up Mt. Hamilton (37°21′N, 121°39′W, 1281 m), from the long-term records at Mt. Hamilton (Major 1953). At Smith Creek during some months from 1941 to 1952, mostly during the summer, daily precipitation was

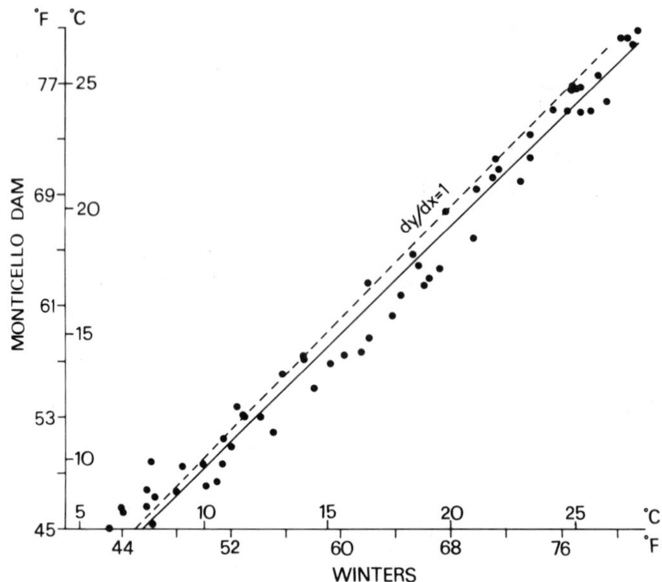

Figure 2-1. Relationship between mean monthly temperatures for Winters (38°32′N, 121°58′W, 41 m elevation), located on the western edge of the Sacramento Valley, and Monticello Dam (38°30′N, 122°07′W, 154 m), located on the eastern slope of the Coast Range about 13 km west of Winters. The monthly mean temperature records are correlated for the 1958–62 period plus December 1957. The line is fitted by eye. See Major and Pyott (1966:260). For n = 61, y = 0.96 + 1.29 in °F, y = 0.96 + 0.076 in °C.

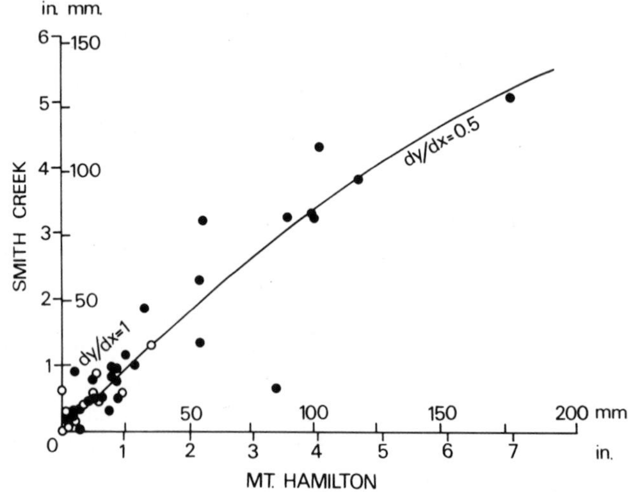

Figure 2-2. Correlation of precipitation from 59 storms at Smith Creek, which has a sporadic record, with the corresponding amounts at Mt. Hamilton, which has a long record. The line is sketched in by eye. See the text.

recorded. For all 59 recorded storms at Smith Creek the precipitation amounts were plotted against the amounts for corresponding storms at Mt. Hamilton 2.9 km eastward and 640 m higher. From the long-term average monthly precipitation figures at the latter, the probable values for Smith Creek were read off the curve.

Radiation data are notably scarce. Radiation is a fundamental feature of climate; net radiation (R_n) is the residual that runs earthly ecosystems. Our potential evapotranspiration approaches R_n as a limit.

California has only eight stations recording (only) total or global radiation in the *Climatic atlas of the U.S.* (U.S. Weather Bureau 1968:70). Annual values in California range from 139 kcal cm^{-2} yr^{-1} at La Jolla on the south coast to 207 at Inyokern in Indian Wells Valley east of Walker Pass at the southern end of the Sierra Nevada. Santa Maria and Los Angeles on the south coast have 175 and 169 kcal; Riverside, somewhat inland, 176. Two central valley stations, Fresno at altitude 101 m and Davis at 16 m, have 164 and 157 kcal, respectively. In the Sierra Nevada at 2060 m radiation at Soda Springs, which is 1° latitude north of Davis, increases to 167 kcal, although cloudiness must increase with altitude, as precipitation increases from 418 mm yr^{-1} at Davis to over 1340 mm yr^{-1} at Soda Springs.

Energy and Water Balances

The elements of climate can be best approached through the energy balance. Radiation into bounded ecosystems is disposed of as shown in Table 2-1. The solar constant is 1360–1396 W m^{-2} (1.94–2.0 cal cm^{-2} min^{-1}) with the lower figure more probable (Monteith 1973:23). The net radiation (R_n) over shorter time periods is disposed of as latent (LE) and sensible (H) heat plus or minus heat from advection, precipitation, photosynthesis–respiration, soil heating–cooling, vegetation heating–cooling, heat equivalent of the mechanical energy of wind, tides, currents, and from water freezing or melting. A plus brings heat into the system; a minus extracts it. The sum of all terms therefore equals zero. Normally LE is $-$, but moisture condensation gives a $+$ figure. Normally H is $-$, but air warmer than the ecosystem surface gives up heat to it ($+$). The latter case is rare, however, and moves little heat since turbulent heat exchange is minimized under conditions of temperature inversion (Budyko 1974:822–3).

Advection can be $+$ when a warm wind blows over the ecosystem, as in an irrigated desert, and $-$ when the wind is cold, like a sea breeze in coastal California. Precipitation can be $+$ as a rain warmer than the surface temperature, and $-$ as colder snow. Respiration can be $+$ and temporarily large in magnitude, as in a forest fire, or permanently significant, as in the combustion of fossil fuels in an urban agglomeration (Myrup 1969). Photosynthesis is $-$, but only a few percent of incoming radiation at maximum. The soil component is $+$ in winter, $-$ in summer, and zero over a year except, for example, where permafrost is melting or forming. The mechanical energy is $+$ if dissipated within the ecosystem, and $-$ if formed in it and carried out. Freezing of water is a $+$ term, and melting $-$. The heat of fusion of water is 333 J cm^{-3} (80 cal g^{-1}). Heat flow from the depths of the earth is negligible, being 0.030–0.070 kcal cm^{-2} yr^{-1} normally, and only 0.150–0.200 in volcanic Iceland (Lamb 1972:14).

In a closed system over a year, then, $R_n = LE + H$, as in Table 2-1. But since

TABLE 2-1. Elements of the heat balance in various landscapes. USSR from Borisov (1967:27); others from Budyko's maps (1963) and diagrams (1958) except the U. S. stations from Sellers (1965:106). Units are kcal cm^{-2} yr^{-1} (x 4180 = J cm^{-2} yr^{-1} or x 41.80 = MJ m^{-2} yr^{-1}). Radia. = radiation, Refl. = reflected, Bal. = balance, Turb. = turbulent, Exchg. = exchange

Landscape	Place	Direct Radia.	Diffuse Radia.	Sum	Refl. Radia.	Absorbed Radia.	Reradi-ated	Net Heat Bal. (R_n)	Turb. Exchg. (H)	Evapotranspiration (LE)	Evapotranspiration (mm H_2O yr^{-1})
					USSR						
Arctic	USSR	10	50	60	42	18	23	-5	...	...	...
Tundra	USSR	25	45	70	40	30	25	5	1	4	70
Turukhansk	USSR	36	39	75	25	50	37	13	0	13	220
Taiga	USSR	35	42	77	37	40	18	22	6	16	270
Mixed forest	USSR	45	40	85	30	55	30	25	7	18	300
Forest steppe	USSR	55	40	95	34	61	31	30	14	16	270
Steppe	USSR	65	35	100	32	68	35	33	18	15	250
Semidesert	USSR	75	35	110	35	75	40	35	25	10	170
Desert	USSR	95	30	125	40	85	43	42	34	8	130

					Outside the USSR						
Atlantic deciduous forest	Paris, France	...	...	105	...	...	...	35	10	25	420
Pacific NW coast	Astoria, Oregon	...	...	128	...	...	...	56	20	36	600
Mediterranean	Lisbon, Portugal	...	...	160	...	...	...	56	25	31	520
	Davis, Calif.	...	...	157	...	...	...	...	...	...	...
Hot desert	Aswan, Egypt	...	...	220	...	...	...	65	65	0.2	3
Colorado desert	Yuma, Arizona	...	...	182	...	...	...	60	56	4	70
Monsoon tropics	Saigon, Viet Nam	...	...	140	...	...	...	78	21	57	950
Tropical rainforest	Manaos, Brazil	...	...	140	...	...	...	74	9	56	1080
Global	Budyko	(1974:226)		126	18	108	36	72	12	60	1000

water has a heat of vaporization (L) that changes only slowly with temperature, 2501 J g^{-1} at 0°C to 2430 at 30°C (596 to 580 cal g^{-1}) or $L = 597 - 0.6T$ in calories per gram, where T is in degrees centigrade, roughly 2500 J (600 cal) is needed to change 1 cm^3 of water from a liquid to a gas, or 2500 J cm^{-2} yr^{-1} or 25 MJ m^{-2} yr^{-1} (600 cal cm^{-2} yr^{-1}) of absorbed radiation will evaporate 1 cm of water or 10 mm yr^{-1}. The connection between the heat and water budgets is therefore direct.

On this basis Budyko defines a radiative index of aridity as the ratio of annual net radiation to the amount of radiation necessary to evaporate the annual precipitation: R_n/LP, where P is mean annual precipitation in centimeters (Budyko 1974:322, 1958:70; Budyko and Efimova 1968:1385). This ratio is 1 when the heat absorbed in an ecosystem is exactly used in evaporating the annual precipitation coming into that system, <1 in humid climates, and >1 in arid climates. Budyko suggests 2 as a steppe limit, 3 for semidesert, and >3 for deserts. As noted above, sensible heat (H) is usually transferred from the earth's surface, leaving R_n as the only nonlateral form of heat income. Therefore R_n is the upper limit of heat used in latent heat transfer.

This is the most rational of such popular climatological ratios, and Budyko (1974:330–5) discusses and disposes of other suggested moisture index ratios. In our terms Budyko's index is essentially the ratio of annual potential evapotranspiration to annual precipitation, PotE/ppt.

Mean yearly indices poorly compare water balances of different kinds of ecosystems. A water balance calculated month by month throughout the year is better. This idea is not new (Contagne 1935; Penck 1910).

Walter's climatic diagrams are pictures of this monthly heat and water balance, assuming 10°C = 20 mm precipitation (Walter and Lieth 1960–67; Walter 1968, 1973; Walter, Harnickell, and Mueller-Dombois 1975). They apply an idea of Gaussen's (1954), which in turn misused Köppen's well-known formulae (Blüthgen 1966:523): Dry climates (B) have precipitation (in cm) < 2 temperature (in °C) for winter rain, ppt < $(2T + 14)$ for well-distributed precipitation, and ppt < $(2T + 28)$ for summer rain. Walter and Lieth's diagrams have been effectively criticized by Carter and Mather (1966) for climatic classification. However, they are not intended as water balance diagrams in any exact sense (Walter et al. 1975), and they are the best graphic representations of climates we have had. They ". . . portray the whole climate of a place in an easily understandable way and show at a glance the most important climatic factors for the growth of plants" (Walter et al. 1975:1). They make possible ready recognition of similarities among and differences between climates. They thus suggest where transfer of data from physiological ecology or synecological studies may have meaning.

Thornthwaite's scheme (1948; Thornthwaite and Mather 1957) is a more realistic estimate of heat income and water use in ecosystems, and we use it here. Mather has a full discussion (1974:100–106). Some applications to California's vegetation have been discussed elsewhere (Major 1963, 1967; Major and Bamberg 1967). For water balance calculations we have used a program developed for a Wang desk computer by Dr. David Randall, who improved an earlier program of Black's (1966).

Thornthwaite (1948) defined potential evapotranspiration (PotE) as the amount of water lost to the atmosphere from a short, green sward of 100% cover and freely supplied with water. Potential evapotranspiration measures monthly water need; precipitation measures water income to the system. A monthly bookkeeping procedure, with the soil assumed to hold 100 mm of water whose availability to plants is inversely proportional to the amount remaining in the soil, gives a total figure for actual evapotranspiration (ActE). Then PotE − ActE = water deficit (*D*), and Precipitation − ActE = water surplus (*S*).

An example of a water balance. At Crescent City in the redwoods the diagram (Fig. 2-5, Table 2-4) shows very high precipitation in winter, with a summer minimum dropping to only 13 mm in August. The temperature curve is flat in this maritime climate with a maximum delayed until August and a monthly range of only 6.6°C. The index of continentality is very small. Monthly need for water is measured by potential evapotranspiration (PotE), which we compute as a function of temperature (Thornthwaite 1948; Thornthwaite et al. 1957; Mather 1974). Potential evapotranspiration is low in winter, relatively large in summer. The total precipitation (1778 mm) cannot be used by the heat supply (PotE, 650 mm water equivalent), nor does the precipitation come when there is heat for plants to use it in evapotranspiration. A water surplus (*S*) for the year of 1234 mm and a deficit (*D*) of 106 mm result. The climate is humid in winter, arid in summer.

From January to May precipitation is greater than potential evapotranspiration, so PotE limits actual water use. In June the current precipitation (42 mm) is exceeded by the current potential evapotranspiration (79 mm). Since the soil is at field capacity after the long winter excess of precipitation over water need, soil water to the extent of 32 mm can be withdrawn and used. June's water deficit of 37 mm is not completely covered by soil-stored water, since we assume that availability of soil water is proportional to the amount left in the soil. A deficit of 5 mm results. The 86 mm of water used from soil storage over June to September is made up by entry of 86 mm of precipitation into the soil after precipitation becomes greater than potential evapotranspiration in October. Further excess precipitation from November to May is surplus, 1234 mm runoff.

Thus the soil storage of 100 mm is not entirely used during the summer drought; 14 mm of available water remains in the soil.

Actual evapotranspiration is low in winter, limited by low heat supply (PotE). It follows the curve of soil water use plus current precipitation until precipitation becomes greater than potential evapotranspiration, when it again is limited by relatively low PotE (Fig. 2-5). The growing season (water use > 10 mm mo^{-1}) is 12 mos. Actually the lowest monthly value is 27 mm of water.

The area under the PotE curve represents maximum water use if water from sources in addition to current precipitation in this climate is available. The water deficit measures possible increase in actual evapotranspiration if more water is made available to the ecosystem in summer, as by irrigation, fog drip, groundwater, or seepage.

The climate of Crescent City is similar in annual heat load to the continental taiga–deciduous forest transition. On the Pacific coast this climate goes north almost through the Alaska Panhandle. With its excess precipitation in all months except summer, only slight summer drought on deep soils, and mild winters, it is a most productive climate, and some of the world's grandest forests are indigenous to it.

Soil storage of water available to plants. The assumed soil storage of water available to plants of 100 mm is a reasonable figure (Russell and Hurlbut 1959:8). Mather (1974:103) assumes 150 mm at Wilmington, Delaware; Birkeland (1974) uses 152 mm. A real figure for a given soil is determinable and can be expressed as percentage depth of the soil profile or, in cases of plants with variable rooting depths, as millimeters per 10 cm of soil depth. As a percentage 5 is too low for all but the sandiest and rockiest soils, and 20 is too high (Russell and Hurlbut 1959:13).

Whether the soil stores 50 or 300 mm water available to plants makes a difference in the water balance (Fig. 2-3, Table 2-2) and therefore in the vegetation.

Discussion. Penman's and other methods of calculating water use are more accurate (Monteith 1973) if only because they (must) use more climatic parameters than mean monthly temperature. A check on the accuracy of the Thornthwaite calculations is the computed water surplus. It should equal runoff, and it does in many cases (cf. Major and Bamberg 1967:101).

"Evaporation" in Weather Bureau records is a poor measure of water loss from a bounded ecosystem. It almost always includes advected heat.

A striking example of discrepancy between small lake (large pan) evaporation and the heat available to ecosystems from radiation is provided by Hare and Hay (1974:116, 118). In the prairies of southern Saskatchewan small lake evaporation is about 102 cm water yr^{-1}. Net radiation in this area is 40 kcal cm^{-2} yr^{-1}, and this amount of heat can evaporate 68 cm water. The discrepancy between small water body evaporation and energy available to a bounded ecosystem for evaporation is 150%. The conclusion follows that the pan records of "evaporation" are a meaningless parameter for regional climate. Walter's criticisms (1967, 1970) of Thornthwaite-type calculations of water balances are therefore incorrect.

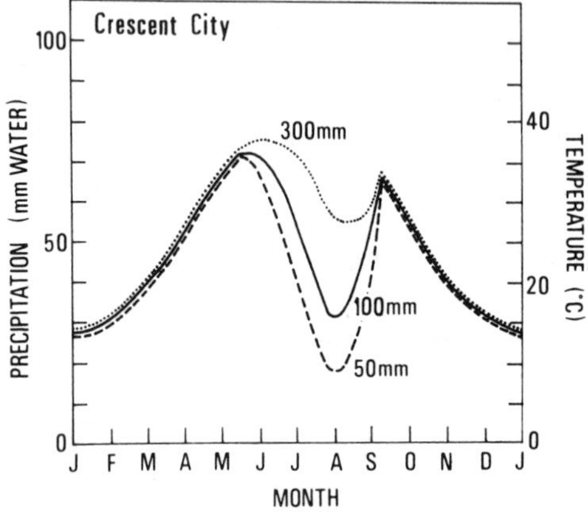

Figure 2-3. Monthly course of actual evapotranspiration at Crescent City, assuming three different storage capacities for soil water available to plants: 50, 100, and 300 mm of water in the rooted profile. See Fig. 2-4 and Tables 2-2 and 2-4.

TABLE 2-2. Changes in elements of the water balance at Crescent City under various amounts of storage of plant-available soil water. See Figs. 2-3 and 2-4 and Table 2-4. All climatic measures are in mm of water yr^{-1}. Annual precipitation is 1777 mm; potential evapotranspiration, 650 mm

Soil Storage (mm)	Actual Evapotranspiration	Water Deficit	Water Surplus
50	506	144	1271
100	544	106	1234
300	597	53	1180

CALIFORNIA'S CLIMATES

The typical Californian climate is zonally subtropical (33–42°N latitude), and it combines the very worst features of arid and humid climates (Table 2-3). It is extremely hot and arid in summer, and extremely cool and humid in winter. The supply of water and the need for water are exactly out of phase. The productivity of natural, zonal vegetation of course reflects these climatic disabilities. The growing season is limited by the cool temperatures of winter as well as the summer drought. Native vegetation is lush in spring, when higher temperatures occur temporarily with adequate water supplies; it is desiccated or fails to grow in summer.

Plant productivity may be high in such a climate under irrigation, using water from local mountains with more precipitation or groundwater from the same source. The tugai vegetation along or in the deltas of such great rivers of the hot deserts of Middle Asia as the Amu-Darya (Oxus) may have a biomass of 11,000 g m^{-2} with an annual productivity of 6000 g m^{-2} yr^{-1}. This is an even higher productivity than that given for tropical rain forest (Rodin and Bazilevich 1967). Native vegetation of the Sacramento–San Joaquin and Colorado River deltas (Leopold 1953), as well as riverine and flood basin vegation, probably originally reflected such optimal conditions (i.e., maximum radiation, ample water, and constant renewal of mineral nutrients) and may have reached comparable productivities. California's irrigated agriculture has replaced native negetation, and it now depends on water imported from mountains as far distant as the eastern slope of the Sierra Nevada or the Rockies—or, even worse, from a very temporary mining of groundwater accumulated over past centuries in the alluvium of desert valleys.

The controls (Trewartha 1961:269ff; Patton 1956) of California's climates are, first, the zonal circulation and, second, the topography. The subtropical high in

TABLE 2-3. Arid versus humid climates (ff. Penck 1910)

Climate			Soil			Vegetation		
	Water Balance							
Kind	Heat	Water	Process	Nutrient Level	Kind	Productivity	Physiognomy	Kinds
Arid	Excess	Deficit	Drought	High	Pedocals	Low	Open	Steppe
Humid	Deficit	Excess	Leaching	Low	Pedalfers	High	Closed	Forest

summer lies off Oregon in the Pacific Ocean. Outbreaks of cold marine air from polar regions (the Aleutian low) sweep eastward with the westerlies north of California. Air coming from the subtropical high is descending and therefore does not precipitate its moisture. Hence California has a summer drought. In winter the low is farther south, and California is in the path of the stormy westerlies. A wet winter results.

The north–south trending coast of California plus the Coast Ranges, and then the Cascade-Sierra axis bordering that coast, turn the zonal circulation southward. Cold water wells up next to the coast. Summer air masses reaching the coast are therefore cooled at the surface and become even more stable. An analogous condition holds in Chile and South Africa but not around the shores of the Mediterranean Sea.

The mountains raise the eastward-moving air; precipitation on them is heavy, and in their lee is a rain shadow, producing steppe to desert climates. The Great Basin is often an area of high pressure in winter, and many cyclones in their passage across the United States are therefore shunted north of California. Except for areas of tule fog, California has surprisingly sunny winters, only shortly broken by stormy periods.

In Fig. 2-4 are portrayed the water balances in three contrasting, but in some respects similar, Californian climates. Crescent City on the northwest coast has one of the wettest, coolest, most maritime, most vegetatively productive climates in California. Palm Springs on the western edge of the Colorado Desert has one of the driest, hottest, most continental, least productive. Davis is intermediate. All three climates have maximum heat supply in summer but maximum water supply in winter.

These kinds of mediterranean climate are typical not only of California but also of the lands surrounding the Old World Mediterranean Basin, parts of Chile, South Africa, and western and south Australia. Five examples of these five climates are diagrammed in Fig. 2-5. Table 2-4 tabulates pertinent climatic data for the eight stations of Figs. 2-4 and 2-5.

The Crescent City climate has been described above. At Palm Springs the water balance is quite different. Precipitation is very low although still concentrated (62%) in the three winter months. Potential evapotranspiration has only 4.7% of its annual value in winter. Actual evapotranspiration is limited to the precipitation amount, 112 mm from precipitation that falls when it can be used plus 27 from soil storage. The growing season (perhaps ActE > 10 mm mo^{-1}) is spread over the five months from December to April, but with an average of only 19 mm mo^{-1}. At a potential leaf water loss of 15 g g^{-1} day^{-1}, this might support gas exchange activity of a biomass of <300 g m^{-2} (15% leaves).

Only a few hot deserts have climates hotter than Palm Springs. Yuma in the Colorado Desert, for example, has PotE of 1327 mm yr^{-1}, 99 mm greater than Palm Springs. This is due to faster warming in spring and slower cooling in fall.

Yuma has only 53 mm less yearly precipitation than Palm Springs, but this means there is no soil storage of moisture at all to give a boost to spring-flowering plants. At Palm Springs average water availability is 24 and 22 mm in February and March for differences between availability and potential losses of only 0 and 25 mm. At Yuma no such difference is less than 20 mm except for January and December, when the possible evapotranspirations are only 16 and 20 mm anyway, even if water were abundantly available. When Yuma has its maximum water use, 16 mm in Sep-

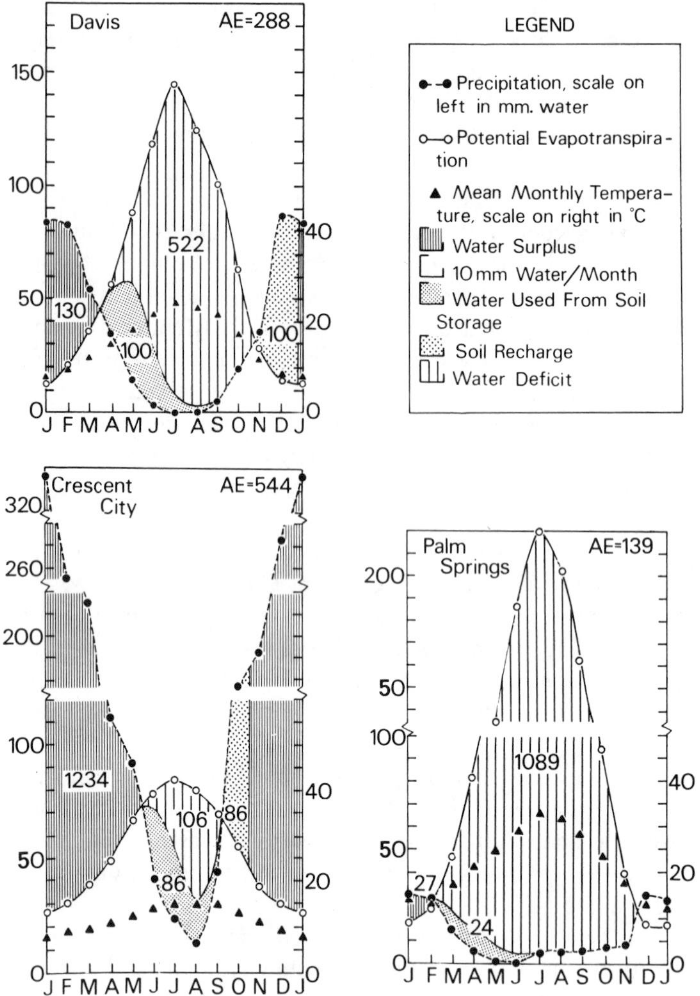

Figure 2-4. Water balance diagrams for three Californian stations with different climates. See Table 2-4.

tember, the difference between this figure and the potential loss of water is 160 mm. The latter may be a measure of drought. Yuma does differ in its late summer, not winter, precipitation maximum. August and September have 34% of the year's total, whereas at Palm Springs two winter months have 40%.

We should note that the low amounts of precipitation recorded for Californian stations in summer are statistical fictions. A 25 mm storm will show an average of 3 mm yr^{-1} after 10 yr of no precipitation at all in that month. Desert precipitation is particularly episodic.

At Davis the climate is not as cool and wet as at Crescent City or as hot and dry as at Palm Springs. It is typically Mediterranean with supply and need for water exactly out of phase. Davis has a more continental climate than San Francisco (Table 2-4) or such coastal Mediterranean stations as Perth, Capetown, and Valparaiso. The San Francisco–Davis increase in continentality is reproduced in Chile from coastal Valparaiso to inland Santiago (ΔT = 12.7, K = 9.7) over approximately the same distance. Davis' climate is less continental than that of Red Bluff (ΔT = 21.1, K = 17.6) or Bakersfield (ΔT = 20.5, K = 17.4) or such eastern Mediterranean stations as Athens.

Since PotE increases along with continentality, the Davis climate is representative in heat availability of a very large area of noncoastal, cismontane California below perhaps 300 m elevation (Arkley and Ulrich 1962:457). This is in general the area of foothill woodland, some kinds of chaparral, and valley grassland vegetation (Chapters 11, 12, and 14). This climate is widespread east of San Francisco Bay, but the heat load is greater to the north and south within the Sacramento and San Joaquin valleys. The interaction of greater precipitation in the Sacramento Valley with the PotE curve makes ActE at Davis representative of the Sacramento Valley rather than the drier San Joaquin, where ActE is lower. Decrease of continentality toward the coast makes Davis' climate similar in ActE also to much of southern California, although the northern California coast has higher values. Increased precipitation in the foothills, along with hotter summers, increases their ActEs above Davis'. Thus Auburn (Fig. 2-8) at 395 m elevation in the Sierra foothills has

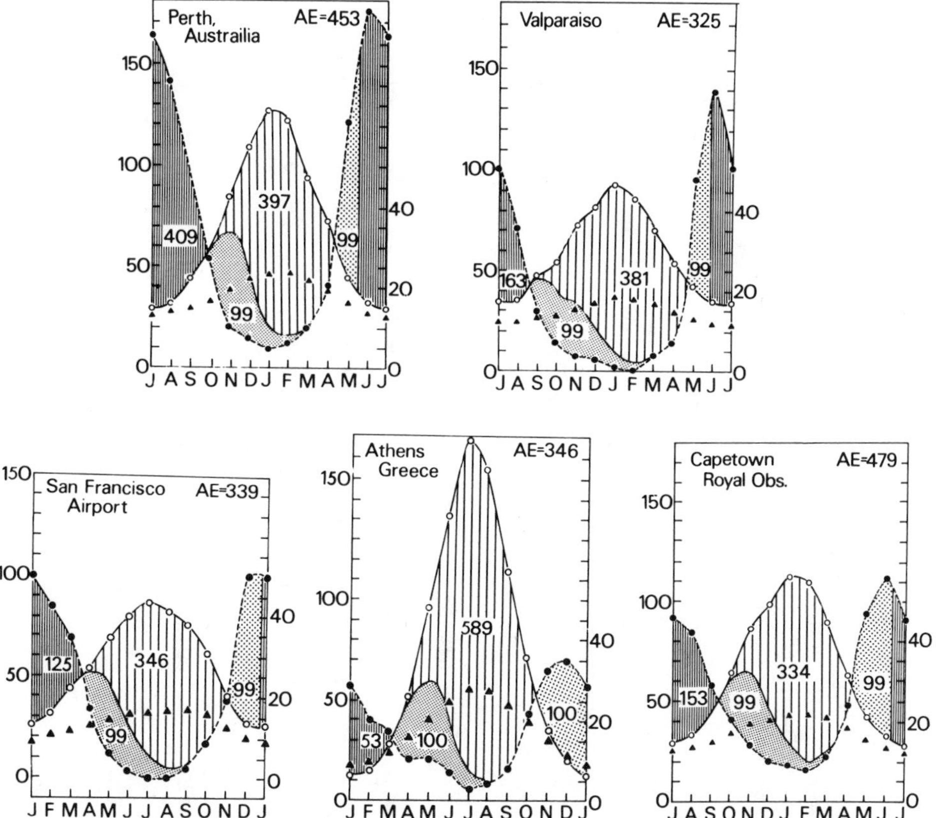

Figure 2-5. Water balance diagrams for five stations in mediterranean climates. See Table 2-4. For legend see Fig. 2-4.

TABLE 2-4. Climatic data (mean annual) for some Mediterranean climates. See Figs. 2-4 and 2-5. *T* = temperature, Ppt = precipitation, PotE = potential evapotranspiration, ActE = actual evapotranspiration, *D* = water deficit, *S* = water surplus, *K* = Khromov's index of continentality. Available soil-stored water is assumed to be 100 mm maximum and unequally available as used

Location	Station	Latitude	Longitude	Elev. (m)	*T* (°C)	Ppt (mm)	PotE (mm)	ActE (mm)	*D* (mm)	*S* (mm)	*K*	ΔT (°C)
California	Crescent City	41°45'N	124°24'W	14	11.4	1777	650	544	106	1234	3.0	6.6
California	Davis	38°32'N	121°45'W	16	15.8	418	810	288	522	130	13.2	16.6
California	Palm Springs	33°49'N	116°31'W	128	22.3	139	1228	139	1089	0	17.7	20.7
Greece	Athens	37°58'N	24°43'E	107	17.9	399	935	346	589	53	15.2	18.5
California	San Francisco	37°37'N	122°23'W	0.3	13.4	463	684	339	346	125	4.6	7.9
Chile	Valparaiso	33°01'S	71°38'W	41	14.5	488	706	325	381	163	3.3	6.2
South Africa	Capetown Observatory	33°56'S	18°29'E	12	17.2	632	812	479	334	153	6.4	9.4
Australia	Perth	31°57'S	115°51'E	60	17.9	862	850	453	397	409	7.6	10.5

both colder springs than Davis and warmer summers by a degree combined with the same annual mean temperature, but over 2× as much precipitation, which increases the Davis ActE by 25%. The effect on soils is probably greater than on vegetation, however, since leaching may be increased 4×.

At Davis deep soils such as the Yolo with perhaps 300 mm of plant-available water storage might boost ActE to the level of total precipitation or by 45%. The growing season would then last clear through September with 19 mm water available in that month (24 mm in October), instead of ending with 29 mm water use in June. On the other hand, shallow hardpan soils (50 mm storage) might reduce ActE by almost a fifth, ending the growing season with 39 mm available in May.

California's climates are variable. Major and Pyott (1966:260) have pointed out that at Winters, just west of Davis, the normal summer drought is broken by some precipitation, usually one storm of 2–26 mm mo^{-1} (average 7 mm), in about half the years of record. Additional data on summer precipitation variability are compiled in Table 2-5. Summer precipitation is usual on the north coast and in the northern Great Basin (Yreka–Ft. Bidwell) and may be abundant. It drops off sharply southward along the coast (Crescent City–Monterey–Los Angeles). It also drops off sharply southward in the Central Valley (Redding–Davis–Bakersfield) with more totally dry summers. Yreka fits this gradient also, and its climate shows similarities to the Great Basin summer climate. South of the latitude of San Francisco Bay dry summers are common with perhaps a drop in frequency southwards. Trewartha (1961:270) states that the latitude of maximum subsidence is 35°N. As one ascends the Sierra Nevada east of Davis (to Placerville and Norden), dry summers decrease and summer precipitation increases. Southward along the crest and then the Great Basin side of the Sierra, Tamarack and Bishop Creek usually have large summer storms, but Bodie to the east in the Great Basin has even fewer dry summers. At a lower elevation Bishop has dry summers and smaller summer storms. Blythe in the very dry Colorado Desert has only a few more dry summers than Bodie and almost as large summer storms. At Blythe most of the rainfall comes at night (Trewartha 1961:265). Finally, the mountains near Los Angeles (Squirrel Inn) have fewer dry summers and larger summer storms.

Table 2-6 shows the lengths of droughts at Davis for a 9 yr period. During the 4 mo wet "winter season" 14% of the 101 recorded storms came at intervals longer than 15 days. This period could use up, on the average, from 10 to 15 mm of water, perhaps all the available water in the surface 10–15 cm of soil. The longest period, 41 days, between these winter storms could use up perhaps 30–40 mm of water, over a third of that available in the soil profile or all the water to 30–40 cm depth. Winter annuals would probably run out of available water during such droughts. According to Table 2-6, over a 20 yr period there would be more than a dozen periods of spring drought up to 1½ mo long, ending with a storm in March or April. In May storms would interrupt a drought period of 3 wk duration in two-thirds of the cases, averaging one storm during the month. A fifth of May storms would interrupt a drought of from 1 to 2 mo. Finally, during the dry "summer season" from June to October droughts are obviously longer with a third of the storms breaking droughts that exceeded 4–5 mo in duration. If October is omitted from the last compilation, two-thirds of the storms break droughts of 2 mo, while well over half break droughts of more than 4–5 mo. On the other hand, "summer season" storms are not unknown. Almost half the May storms come less than 10 days after the preceding

TABLE 2-5. Incidence of summer precipitation for 3 mo at some Californian climatic stations

Station	Latitude (°N)	Altitude (m)	Period (yr)	Precipitation (mm)	
				Minimum	Maximum
Crescent City	41°45'	14	1885-1929	2	190
Yreka	41°43'	802	1871-1930	0 (2x)	101
Ft. Bidwell	41°51'	1370	1866-1930	1	142
Redding	40°35'	176	1875-1930	0 (5x)	119
Davis	38°32'	16	1872-1929	0 (35x)	42
Monterey	36°36'	79	1847-1914	0 (24x)	46
Placerville	38°44'	576	1874-1929	0 (15x)	78
Norden	39°19'	2100	1870-1930	0 (17x)	102
Tamarack	38°36'	2450	1900-1926	0 (3x)	218
Bishop Creek	37°17'	2560	1910-1929	2	127
Bodie	38°14'	2515	1896-1905	0	109
Bishop	37°22'	1251	1844-1917	0 (9x)	44
Los Angeles	34°03'	95	1877-1929	0 (20x)	35
Squirrel Inn	34°18'	1740	1893-1929	0 (12x)	127
Bakersfield	35°25'	151	1889-1929	0 (25x)	15
Blythe	33°37'	81	1910-1929	0 (3x)	77

storm or one every other year, one-fifth of the June–October storms or again one every other year, and one-tenth of the June-September storms or one every 10 yr.

That water balance calculations using only annual values are inadequate is evident from Fig. 2-6 and Table 2-7. The water balances for Ukiah, California, and Asheville, North Carolina, are compared. Climatically the stations are very similar in annual mean temperature and precipitation. They differ in that Ukiah has a mesic Mediterranean climate supporting a mixed evergreen forest (Chapter 10), with extreme contrasts between plant growth possibilities in summer (drought) and winter

TABLE 2-6. Intervals in days between storms at Davis. A storm was defined as precipitation > 0.1 inch day^{-1}. Period 1951-59

Interval (≤ days)	Cumulative Percent Nov.-Feb. (4 mo)	Mar.-Apr. (2 mo)	May (1 mo)	June-Sept. (4 mo)	June-Oct. (5 mo)
5	60	52	22	8	15
10	78	67	44	8	20
15	86	75	67	8	30
20	95	83	67	8	35
25	98	85	78	25	45
41	100	94	89	25	45
47	...	100	89	25	45
63	...	...	100	33	50
120	...	...	...	42	65
162	...	...	...	100	100
Total storms	101	48	9	12	20

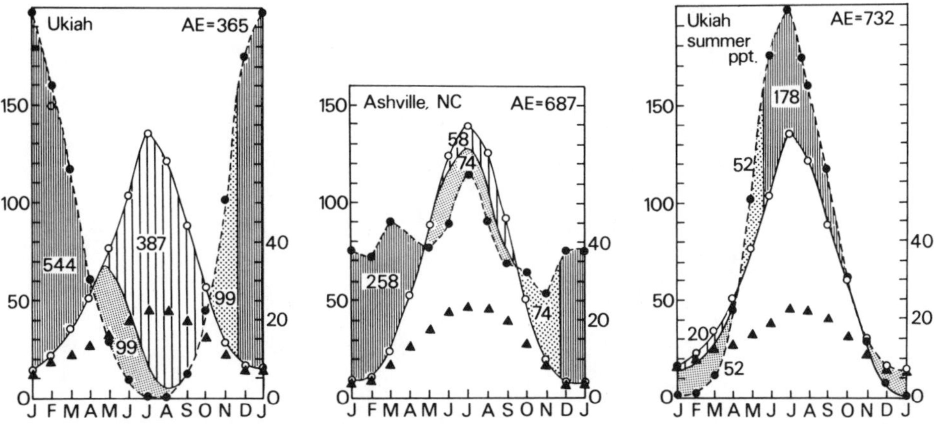

Figure 2-6. A comparison of the water balance diagrams for Asheville, North Carolina, and Ukiah, California—also with the precipitation curve for Ukiah shifted so that need and supply of water are in phase. See Table 2-7. For legend see Fig. 2-4.

TABLE 2-7. Effect of a Mediterranean climate (Ukiah) on the water balance. See Fig. 2-6. Asheville, North Carolina, has similar annual mean temperature, potential evapotranspiration, and precipitation but differs completely from Ukiah in monthly distribution and therefore differs radically in other water balance parameters. If the Ukiah precipitation curve is shifted 6 mo so that the precipitation and temperature distributions are in phase, the Ukiah climate becomes similar to Asheville's

Station	Latitude (°N)	Longitude (°W)	Alt. (m)	Temp. (°C)	PotE (mm)	Ppt (mm)	ActE (mm)	Deficit (mm)	Surplus (mm)	K	ΔT (°C)
Ukiah	39°09'	123°12'	190	14.5	752	909	365	387	544	11.8	15.2
Asheville	35°36'	82°32'	684	13.5	744	945	687	58	258	16.0	19.1
"Ukiah" shifted	...	...	...	14.5	752	909	731	20	178	11.8	15.2

(excessive rain, cool), whereas Asheville has a year-long mesic, temperate climate supporting a deciduous broad-leaved forest. In Fig. 2-6*c* the precipitation curve for Ukiah has been shifted from winter to summer. Both mean temperature and precipitation remain the same. Drought stress decreases from 387 to 20 mm yr^{-1}, runoff or soil leaching decreases also from 544 to 178 mm yr^{-1}, and ActE increases from 365 to 731 mm yr^{-1}, from 48 to 97% of that possible. The corresponding figure for Asheville is 92%.

Horizontal Variation

Within California there is a marvelous variety of climates. Figure 2-7 shows a selection of water balance diagrams over the state. Walter and his co-workers (Walter and Lieth 1960–67; Walter et al. 1975) have similar maps. Maps of PotE and ActE have been printed by Arkley and Ulrich (1962) for California, by Thornthwaite (1948) for the United States, and in the *Fiziko-Geograficheskii Atlas Mira* (1964:22) for the world.

Climatic classifications for California have been discussed by Russell (1926), Kesseli (1942), Walter and Lieth (1960–67), Major (1967), Russell (1946), and Bailey (1966). For our purposes the most useful classification recognizes the differences between floristic regions: Northwest Pacific Coast, Mediterranean California, Sierra Nevada–Cascades, Cold Desert, Hot Desert.

Northwest Pacific Coast. Although annual mean precipitation varies 7× throughout this region in California, from a high of 2780 mm at Monumental at 839 m elevation in the mountains 34 km northeast of Crescent City to only 497 mm at coastal Pt. Reyes and Monterey with 465 mm, ActE varies only from 464 to 412 and 364 mm, but Crescent City has 544 mm (Fig. 2-4, Table 2-4). These are the highest ActE figures in all of California. They are the most productive of vegetation (Rosenzweig 1968). Other high values, going southward along the coast, are Klamath 518 mm, Orick Prairie 496, Eureka 471, Ft. Bragg 452, Pt. Arena 469, and Ft. Ross 474; after Pt. Reyes come San Rafael and wetter, cooler Kentfield on San Francisco Bay with 401 and 426 and drier and cooler San Francisco itself with 339.

Summer temperatures are cool, with highest monthly means in September (14.6, 13.5, and 16.8°C at Crescent City, Pt. Reyes, and Monterey). Winter temperatures are only slightly cooler on the coast with lowest monthly means of 8.0, 9.7, and 9.8°C. At Pt. Reyes the amplitude of mean monthly temperatures is only 3.8°C, only 0.5 more than the amplitude over the open ocean according to Khromov (1957). These even temperatures are reflected in the amplitude of PotE, only 31 mm water at Pt. Reyes. Potential evapotranspiration is only 650 mm at Crescent City, 591 at Monumental, and 634 and 677 at the southern stations.

At Pt. Reyes the month with greatest ActE is April with only 50 mm of water use possible. Crescent City has more favorable conditions for plant growth with its greater summer precipitation, and it has maximum ActE of 73 mm in June. Minimal ActEs are 31 mm at Crescent City but only 9 and 7 mm in August at the two southern stations. Monumental with its cold winter has an ActE of zero, in January, whereas the coastal stations have 27, 35, and 29 mm from north to south.

This cool, equable coastal climate changes quickly inland. Happy Camp is only 66 km inland at 332 m elevation, but it has a mean annual temperature almost as

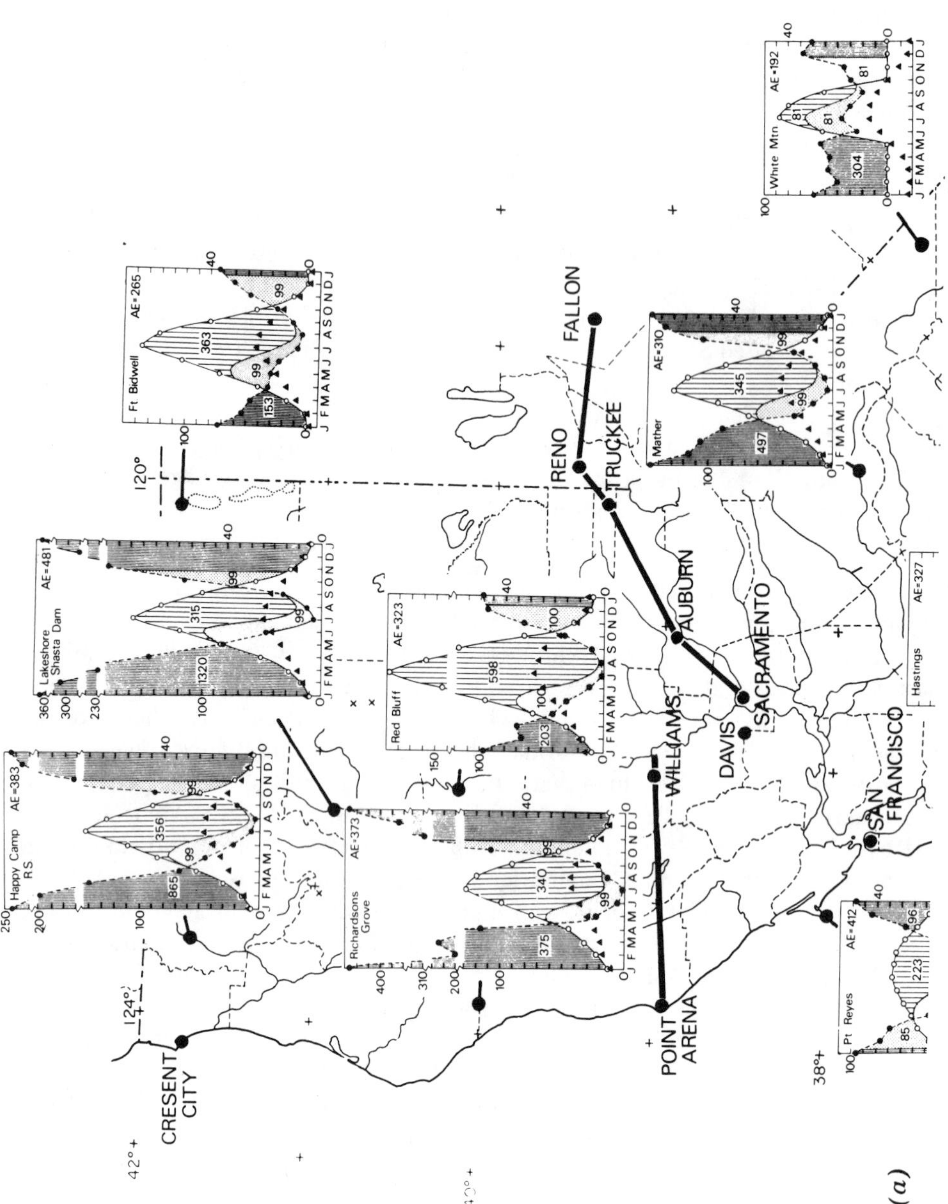
CRESENT CITY
POINT ARENA
WILLIAMS
DAVIS
SACRAMENTO
AUBURN
TRUCKEE
RENO
FALLON
SAN FRANCISCO
Happy Camp RS
AE=383
Richardsons Grove
AE=373
Lakeshore Shasta Dam
AE=481
Red Bluff
AE=323
Ft Bidwell
AE=265
Mather
AE=310
White Mtn
AE=192
Pt Reyes
AE=412
Hastings
AE=327
124°
120°
42°
40°
38°

(a)

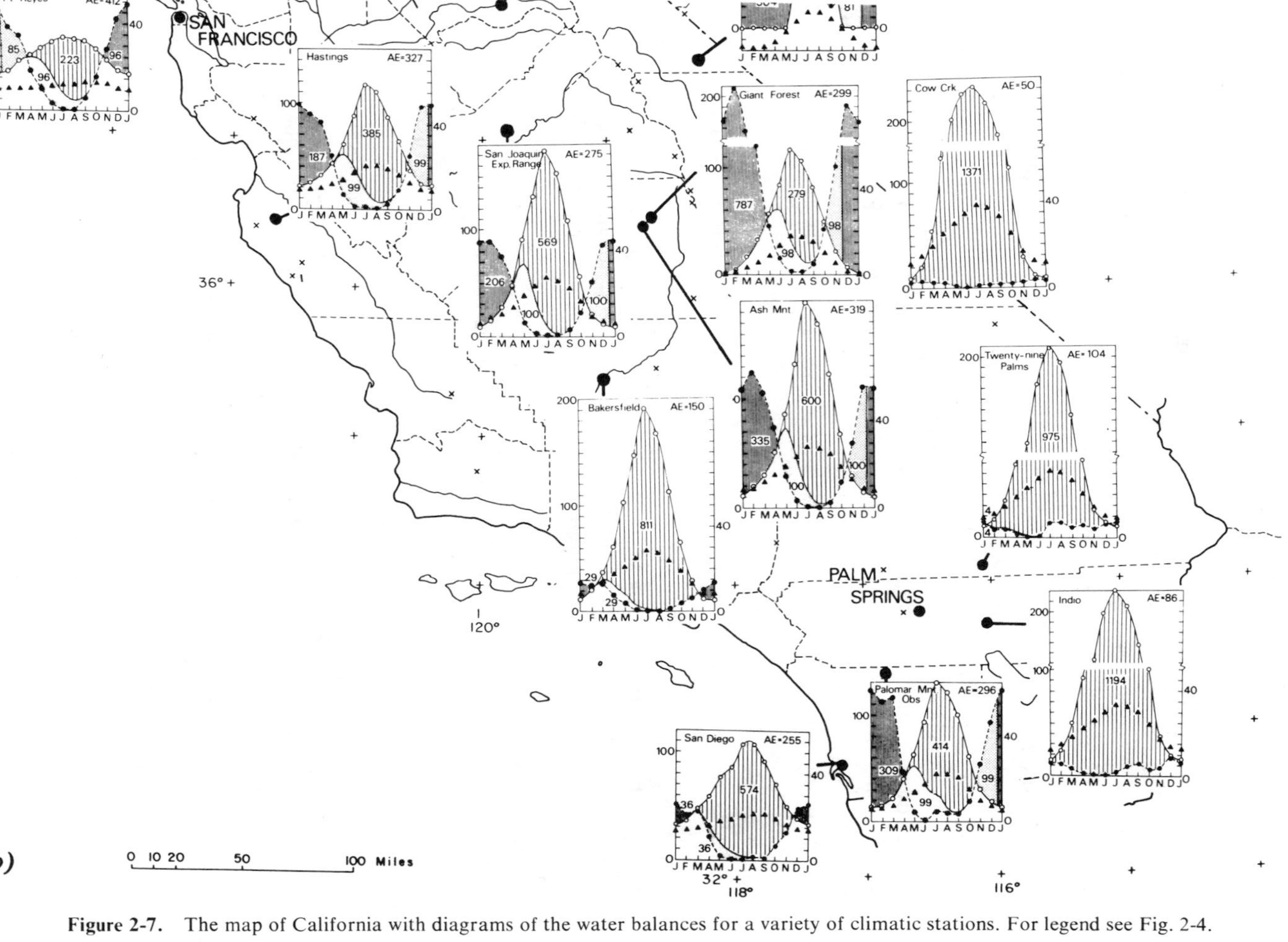

Figure 2-7. The map of California with diagrams of the water balances for a variety of climatic stations. For legend see Fig. 2-4.

high as Monterey's, a higher PotE, a maximum mean monthly temperature 6.2° higher, a minimum 5.0° lower, and an annual ActE 19 mm higher (48 mm greater in summer and 70 mm lower in winter).

San Francisco Bay carries this climate inland. Napa has an ActE of 355 mm since it is wetter than San Francisco but drier than inner Marin Co. Northward in the Napa Valley, St. Helena increases in continentality as much over Napa as Napa does over San Francisco (4.6 to 7.6 to 10.4), but St. Helena's increased precipitation gives an ActE of 360 mm. St. Helena is just out of the redwoods (*Sequoia sempervirens*).

It is these once-magnificent forests that characterize this climatic province in its southern, Californian extension. Actually, much this same climate extends north, seaward of the Coast Ranges, to Anchorage, Alaska, where *Picea sitchensis,* which in California reaches south to 40½°N (disjunctly to 39½°), replaces the rich coniferous flora of more southern latitudes. The inland extension of this climate to the Rockies in southern British Columbia and to Glacier National Park and the Bitterroot Mts. of Montana is also well known.

Fog is common in this coastal climate in California (Byers 1930; Patton 1956), and the redwoods mark the fog belt's reduced evapotranspiration (Byers 1953). Although fog reduces light, photosynthesis may be increased, since leaf water deficits are also reduced and the latter, through their effect on leaf resistances, limit CO_2 assimilation (Hodges 1967).

Mediterranean California. This is "typical" California, similar in climate to the Old World eastern Mediterranean, with no anomalies, no compensating features, a desert in summer, a sodden, dripping landscape in winter, a glory of wildflowers in spring.

The distinction between arid areas (Table 2-3) with so little precipitation, or so much PotE, that the soil is never brought up to storage capacity and humid areas with a surplus of precipitation or too little heat to evapotranspire the available water deserves emphasis. In Mediterranean climates both situations occur through the course of a year. Winters are humid; summers, arid.

This is by no means a uniform region, even in the lowlands. The northern interior valleys have very heavy winter precipitation and do not altogether lack summer rain; in the south the annual precipitation diminishes to desert amounts with no summer rain at all. Coastal stations are cool in summer and maritime in temperature regime; inland summers are hot, and continentality is high. The San Francisco Bay allows cool maritime air to reach inland to the Sierra foothills (Gankin and Major 1964; Patton 1956). River valleys emptying into the Pacific also lead maritime air inland. The higher Coast Ranges have a rain shadow in their lee.

Some examples will be helpful. In the north, Redding (northern Sacramento Valley) annually receives 960 mm of precipitation, Richardson Grove State Park 1826 mm, Garberville (both Eel River) 1446 mm, Ukiah (Russian River) 909 mm. June–September 4 mo precipitations are only 4.9, 2.4, 2.4, and 2.5% of the annual totals but 47, 44, 35, and 23 mm of water. Bakersfield in the southern San Joaquin Valley annually receives only 150 mm of precipitation. Northward Fresno receives 259, Modesto 281, Stockton 339, Davis 418, and Sacramento 430 mm.

Precipitation varies not only spatially but also temporally. Drought has been extreme. In 1850–51 Sacramento received only 119 mm of precipitation during that winter. No rain fell from the 6 mm of May 1850 until January 1851. Soil storage of water reached a calculated total of perhaps 12 mm from January to March 1851. During the entire wet season from November 1850 through May 1851 the need for water was almost consistently greater than the precipitation. Rain was 12 mm greater than PotE in only 1 mo, namely, March 1851. June through August 1851 had no precipitation, and September through November only 84 mm. This drought of 18 mo duration is illustrated in Fig. 2-8.

The period from 1862 to 1864 was as bad with the summer drought of 1862 lasting through November, followed by only 285 mm from December to May (average of 47 mm mo^{-1}), then 5 mo with no rain, and November 1863–May 1864 with only 195 mm (average of 28 mm mo^{-1}); then came another 5 mo long summer drought. In no month during the two winters was more than 70 mm of precipitation received. The period with PotE continuously greater than current precipitation in this drought was 31 mo.

Dale (1959:6) mentions other droughts at 10–15 yr intervals, and in all cases precipitation was insufficient in winter to recharge the soil to field capacity. The drought carried over from one summer dry period to the next, augmenting the effect of the second dry summer. California's summer dry period is a constant; its winter wet period is not—at least in Sacramento and presumably most of the rest of the Central Valley as well.

Average precipitations in June–September at the Central Valley stations mentioned above are Bakersfield 5, 6, 8, and 11 mm and Davis 9 mm—all statistical

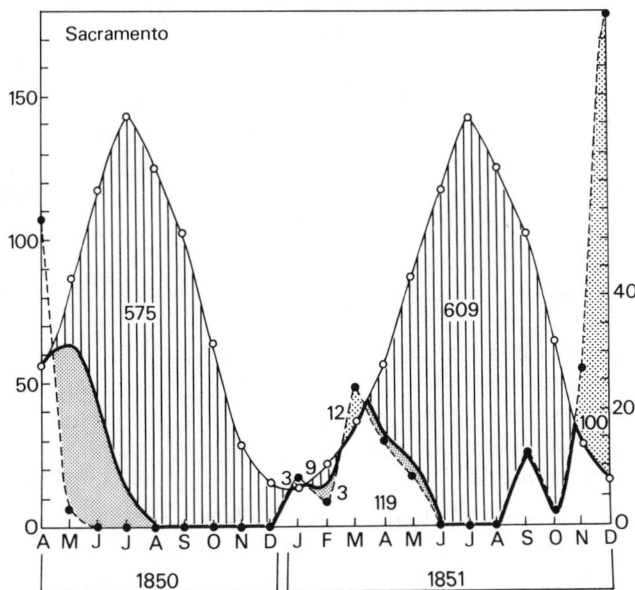

Figure 2-8. The calculated water balance during the 1850–51 drought at Sacramento, California (38°31′N, 121°20′W, 5 m elevation). Annual average PotE = 807 mm, precipitation = 430, ActE = 300, water deficit = 507, surplus = 130 mm, assuming 100 mm available storage. The PotE curve is assumed to be the same in both years, although drought would probably mean higher temperatures than normal and therefore increased PotE. For legend see Fig. 2-4.

artifacts since most summers are dry. Again, ActE is a more revealing parameter: 150 mm in the true desert of the southern San Joaquin Valley, where precipitation is absolutely limiting; 252 mm at Fresno, where precipitation is almost absolutely limiting; and 268, 280, and 288 mm at Davis.

Salinas (lower Salinas Valley) receives 349 mm precipitation, but King City upstream, and in the lee of the Santa Lucia Mts., only 268 mm. Actual evapotranspirations are 327 and 268 mm yr^{-1}.

Napa with only 597 mm precipitation has an ActE of 355 mm, and so does St. Helena (actually 360) up-valley with 834 mm precipitation and almost the same annual temperature. The St. Helena mean for July is 2° hotter and for January 0.8° cooler, however.

Although Ojai, 19 km inland from the coast, has 72 mm more precipitation a year than coastal Santa Barbara, its ActE is less by 23 mm. Ojai's maximum mean monthly temperature is 3.7° higher, its minimum 1.4° cooler. The milder winter temperatures of Santa Barbara result in more water passing through plants, not into runoff. June–October water use is 71 mm at both Ojai and Santa Barbara, but the latter has more December–March ActE by 30 mm: 146 versus 116 mm.

The regular decline of precipitation southward in the Central Valley is interrupted in the rain shadow of the Yolla Bolly Mts., where Williams has only 390 mm yr^{-1}. The western side of the San Joaquin Valley opposite Fresno has only 152 mm precipitation at Panoche Junction, 107 mm less than Fresno. Buttonwillow and Middlewater west of Bakersfield have only 130–129 mm, 21 mm less. At such rainfalls of <150 mm yr^{-1} the soil is annually recharged with water during the winter surplus to the extent of only 30 mm. Only at Fresno does the soil just barely come up to field capacity in winter. Southward at Bakersfield it does not reach this level at all, coming only to 29 mm. This is an absolute desert.

In view of these variations, in this mediterranean climate evapotranspiration is minimal in winter, but not zero, especially in the mild coastal climates. The climate is subtropical, with little frost. In April–May evapotranspiration reaches its maximum and usually exhausts soil-stored water. The amount is fairly constant, 100–120 mm with the higher figures in the north, the lower in the south. June evapotranspiration usually drops to half the preceding month's value. From July until the rains start in October–November ActE is zero. Over a very wide area of California this climate's ActE totals about 300 mm water use a year (Arkley and Ulrich 1962:457).

Annual PotE varies considerably (Arkley and Ulrich 1962:456). Only in coastal stations may it be as low as 700 mm. In the interior it is 800 mm or above, to 900 mm at the northern and to 1000 mm at the southern end of the Central Valley. Ukiah with precipitation of 909 mm yr^{-1}, but with an ActE of only 365 mm, would have an ActE of 731 mm yr^{-1}, 200 mm more than actually occurs at any station in California, if its need for and supply of water were not exactly out of phase (Fig. 2-6, Table 2-7). Such ActEs are found in temperate latitudes only in the southeastern United States and in Europe only in Yugoslavia and the Pontic region of the Black Sea.

On the other hand, in California under former conditions the tule swamps of the Central Valley may have used 800–1000 mm of water a year, resulting in immense annual productivities of vegetation (Rosenzweig 1968). Under irrigation, crops can use similar amounts of water, and productivities are similarly high. Actually irrigated agriculture usually applies more water than the above amounts, but much is not used by plants unless an extreme oasis effect is produced.

Where both PotE and supply of water in summer, as by groundwater, are high, lack of drainage leads to salt accumulation. Saline and alkaline soils have been the result of this natural process not only in the Central Valley, especially southward in the San Joaquin Valley, but also in the transmontane steppe and desert areas. Irrigation has magnified the problem.

In general, in cismontane California winter precipitation is sufficient to fully recharge soil water. Where it is not, the area can properly be termed a desert. South of Fresno with a rather dry, but otherwise typical Californian Mediterranean climate, the southern San Joaquin Valley is such a desert, as at Bakersfield (Fig. 2-7), and so is its west side as far north as Los Banos to Newman.

A variant of the transmontane Hot Desert of southern California reoccurs, isolated from its main, continental area by the more humid coastal mountains, in coastal southern California from Laguna Beach southward. This desert is transitional northward to a maritime mediterranean climate. Los Angeles with 316 mm of precipitation a year barely has a water surplus, and Long Beach with 311 mm does not. Southward at San Diego the soil is recharged only to the extent of 36 mm, and the precipitation of 255 mm absolutely limits ActE. At Chula Vista with a mean annual temperature 1.9° cooler throughout the year than San Diego's 17.3°C, the same situation occurs—no runoff. The PotEs at these stations are 738 and 803 mm yr^{-1}. These are low compared with the values for the transmontane Hot Desert, > 1200 mm yr^{-1} and < 1000 in the Colorado and Mojave deserts, respectively. Equability, on the other hand, is high, 64–68, and no lower northward to Los Angeles out of this coastal desert; it is much lower, however, in the transmontane Hot Desert, 40–50 as extremes but centering around 43 and 47 in the Colorado and Mojave deserts, respectively.

For comparison, Budyko's climatic index of aridity, approximated by the ratio PotE/precipitation, at these stations is as follows: Los Angeles 2.43, Long Beach 2.57, Laguna Beach 2.40, San Diego 3.15, Chula Vista 2.99. In a spring month ActE in these coastal deserts may be fairly high, more than 40 mm, and the annual ActE is 250 mm, so these are strange deserts. But deserts they are (Al-Ani et al. 1972; Pearcy and Harrison 1974).

Even the Channel Islands, at least Avalon Pleasure Pier on Catalina Island, have insufficient precipitation or too much heat to completely recharge their soils of 100 mm storage capacity with water. Precipitation at Avalon is 327 mm yr^{-1}. Potential evapotranspiration is 743 mm yr^{-1}, of which 256 mm or 34% comes in the 6 wet winter months of November–April in this very maritime climate with relatively warm winters (ΔT = 7.2°C, K = 4.1, Eq = 70, PotE/ppt = 2.27). Not enough water is left over for the soil to become completely recharged for the dry season. By comparison Fresno has precipitation of 259 mm yr^{-1}; PotE is 911 mm yr^{-1}, of which 165 mm or only 18% comes in these same 6 wet months. In Fresno's continental climate (ΔT = 19.9, K = 16.7, Eq = 51, PotE/ppt = 3.52) soil water use in winter is low enough so that winter rains do recharge the soil profile with water, assuming that soil storage capacity is 100 mm, and there is 6 mm surplus water left over. Avalon's water balance fits a desert climate, although a maritime one that must be quite different in its effects on vegetation as compared to the continental, transmontane deserts of California. For example, unsatisfied water stress at Avalon (water deficit) is never more than 90 mm mo^{-1}. It reaches 213 mm at Blythe in the true Hot Desert, and even 120 and 99 mm at Reno and Alturas in the Cold Desert at elevations of 1340 and 1410 m (Fig. 2-7). The PotE/ppt values at these stations are as follows: Blythe 11.95, Reno 3.02, Alturas 1.83.

Arid versus humid climates were defined above on the basis of 100 mm possible soil storage of available water. In Mediterranean climates, with an excess of precipitation over PotE in the winter months that is greater than soil storage capacity, a low soil storage capacity would remove the area from a desert classification. This is true for San Diego only if soil storage capacity is less than 36 mm, for Long Beach less than 93, for Los Angeles less than 102, and for Avalon less than 95 mm water. The effects on climatic parameters at Avalon of variations in soil storage capacity are shown in Fig. 2-9 and Table 2-8. Is an area more a desert if its deep soil never reaches field capacity to its full depth of storage (root penetration) but has a higher ActE of 327 mm, or if its shallow soil not only reaches field capacity but also produces a runoff of 46 mm yet has a lower ActE of 282 mm? Obviously a local, perhaps temporary desert occurs in any climate where there is no soil storage of water yet precipitation is not continuous. "Continuous" here means with no interruption longer than the time necessary to evapotranspire perhaps 10 mm of water.

Cold Desert. Only a small area of California is transmontane, and only north of and higher in elevation than Bishop (37½° N) is it a Cold Desert. The Cold Desert is usually taken to coincide with the distribution of sagebrush (*Artemisia tridentata* with its subspecies and close relatives; Beetle 1960) as a characteristic, dominant plant. This distribution extends eastward from the Sierra-Cascades through the Great Basin and Colorado plateaus, through western Colorado, Wyoming, and Montana to the adjoining plains states, northward through eastern Washington and the dry, interior valleys of southern British Columbia, southwestwards through northern Arizona to northwestern New Mexico. Disjunct outliers occur in the southern Californian mountains, in the southern and northern Coast Ranges of California, and in the southern Cascades of Oregon.

The two climates previously discussed are more or less limited in length of grow-

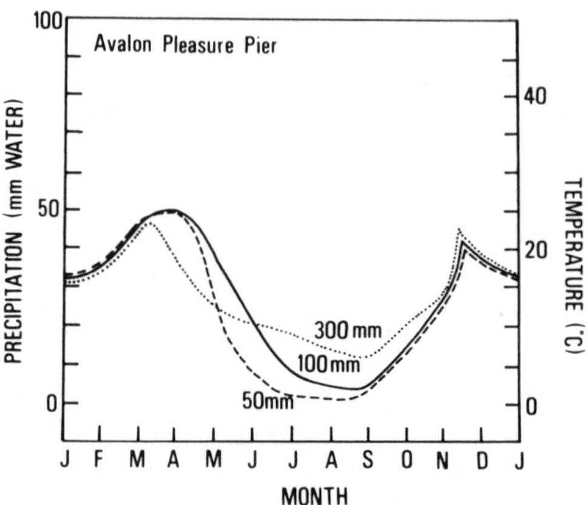

Figure 2-9. The effects on the water balance at Avalon Pleasure Pier, assuming different soil storage capacities for available water. See Fig. 2-3 for Crescent City and Table 2-8.

TABLE 2-8. Differences in climatic parameters at Avalon Pleasure Pier, Catalina Island ($33^{o}21'$N, $118^{o}20'$W, 0 m elevation), for different soil storage capacities for available water. PotE = 743 mm yr^{-1}, precipitation = 327 mm yr^{-1}, PotE/Ppt = 2.27

Soil Water Storage Capacity (mm)	Actual Maximum Soil Storage (mm)	Actual Minimum Soil Storage (mm)	ActE (mm)	Maximum Monthly ActE		Minimum Monthly ActE		Water Deficit (mm)	Water Surplus (mm)
				(mm)	month	(mm)	month		
100	96	1	327	51	Apr.	5	Aug.	416	0
50	50	0	282	49	Apr.	1	Aug.	461	46
300	118	22	327	47	Mar.	11	Sept.	416	0

ing season (ActE > 10 mm mo^{-1}) by a lack of summer-available water. In the Cold Desert both a summer drought and winter temperatures below freezing limit the growing season to a short springtime and a shorter, or even nonexistent, fall period when moisture and heat are both at levels permitting plant activity.

Unfortunately there is even more paucity of climatic records for California's Great Basin area than for cismontane California. Climatic variation is great. As noted by Billings (1954), near Reno even valleys and lower mountain slopes differ very significantly in climates. The piñon–juniper (*Pinus monophylla–Juniperus osteosperma*) belt, for example, lies above the valley pools of cold air. It may also be true, as Hübl and Niklfeld (1973) suggest when comparing forest and alpine belts in the Alps, that in extreme climates, in our case with low precipitation, the partial effects (Major 1951, 1967:94) of soil parent material, local relief, including snow drifting, depth, and persistence, or even differences in available floras between isolated valleys or mountain ranges may be more effective in determining vegetational differences than the (small) changes in climatic parameters.

Summer precipitation occurs in northeasternmost California, or (more accurately) the normal summer drought is shorter with effective amounts of precipitation through June and beginning again in September. The course of ActE is somewhat different therefore from that in cismontaine California, first, because ActE in the winter months is zero and, second, because in the summer months it is greater than 10 mm mo^{-1} in all but the driest places such as Reno, where the soil is not fully recharged by spring snowmelt.

Although the growing season suffers a severe restriction in summer, it at least theoretically includes the summer months at a very low rate of transpiration. The cold winters do limit the growing season to 9 mo or less, even if groundwater is available or summer irrigation is practiced.

The Cold Desert climate has <500 mm yr^{-1} precipitation in the lowlands, most of which comes in winter, but with more in summer than in California's Mediterranean climates. The Sierra–Cascades produce a strong rain shadow, as do the successive ranges eastward. Precipitation minima are <200 mm yr^{-1}, eastward down to 100 mm. Winters are cold, below freezing, with PotE zero for 3 mo with snow. Summers are hot, with maximum PotE in July of 140 mm mo^{-1}. Water stress begins as early as April–May, depending on whether or not soil-stored water was brought up to field capacity by spring snowmelt. Water is deficient until November; then lack of heat brings ActE down to zero in winter. Continentality is extreme, from 17 to >20 eastward. At higher elevations continentality decreases.

Using yearly values, Budyko's radiative index of aridity is >3 for deserts, 2–3 for semideserts, 1–2 for steppe (1974:348). Actually, our Cold Desert is steppe with such more mesic vegetation types as piñon–juniper (*Pinus monophylla-Juniperus osteosperma*), *Agropyron spicatum* bunch grass, or northern juniper woodland (*Juniperus occidentalis*), but mostly semidesert with sagebrush if we let PotE approximate net radiation values in Budyko's index. Values of the index are approximately 1.95, 1.51, and 1.75–2.82 in these three vegetation types according to limited data assembled earlier (Major 1967:121). The woodlands often occur at higher elevations. Sagebrush runs 1.99–2.84 (3.55) in California to western Nevada.

The last figure is for Reno, and we conclude that the western edge of the Great Basin is unique (Billings 1945, 1949). At least as far north as Reno it is lower in elevation, drier, warmer, and less continental than farther east. The ecologically very distinct endemics *Sarcobatus baileyi* Cov., *Abronia crux-maltae* Kell., and *Tetradymia tetrameres* (S. F. Blake) Strother (1974:192–3) occur here. Cronquist et al. (1972:90) add as endemics for the broader Lahontan Basin *Astragalus porrectus* S. Wats., *Eriogonum anemophilum* Greene, *E. rubricaule* Tidestrom, and *Oenothera nevadensis* Kell. In addition *Ephedra nevadensis, Lepidium fremontii, Eriogonum heermanii, Atriplex canescens, Abronia turbinata* (Wilson 1972), *Hermidium alipes, Mirabilis bigelovii, Dalea polydenia, Oryctes nevadensis, Lycium andersonii, Salvia dorrii, Pectocarya setosa, Coldenia nuttallii, Glyptopleura marginata, Hymenoclea monogrya,* and *H. salsola* var. *patula* (A. Nels.) Peterson and Payne (1973, 1974), *Iva nevadensis* Jones, *Bromus rubens, Hilaria jamesii* (West 1972), and *Stipa speciosa* are normally Hot Desert plants, but they extend northward into this area. This "hotspot" is clearly marked on Beeson's map (1974:55) of accumulated cold-stress degree-days below 0°C as a warm extension of the Mojave Desert.

Axelrod (1965:152) shows a nice separation of the three desert areas in California, the Great Basin, Mojave, and Sonoran from north to south, based on means and amplitudes of annual temperatures, although they all center on an amplitude of mean monthly temperatures of 22°C. These areas have equabilities of 50 (52–47), 47 (50–45), and 43 (46–40). Their latitudinal shift makes Khromov's (1957) continentalities 18.3, 18.6, and 19.0. The lowering of mean temperature with latitude northward gives an opposite and larger separation as compared with the latitudinal increase in continentality northward over the oceans, which is used by Khromov to "correct" ΔT. The two measures, equability and continentality, as the names imply have opposite trends but agree in sense.

Billing's studies (1945, 1949, 1954) are most pertinent to our area. He differentiates a drier shadscale (*Atriplex confertifolia*) vegetation from the more mesic sagebrush. Shadscale is usually found nearer the bottoms of the intermontane valleys of the Great Basin than is sagebrush. Being hotter in summer and during the day, and subject to the cold temperature inversions in winter and at night, the continentality of these valleys is greater than that of the sagebrush areas. Budyko's radiative index of aridity, approximated by PotE/ppt, jumps to 5.7 (3.96–7.42) in the shadscale vegetation of western Nevada (Major 1967:121), although it averages only 3.7 eastward through Utah to Colorado. Correspondingly sagebrush figures average 2.4 in the west to 1.5 (1.35–2.10) eastward to Utah and Wyoming (Major 1967:124). Shadscale areas in western Nevada average 114 mm yr^{-1} for both precipitation and ActE, that is, no complete recharge of soil water in the winter (Major 1967:121). This is a truly desert climate according to the definition suggested above. Within sagebrush areas precipitation is over 2× as large (246 vs. 114 mm), and ActE is almost 2× as large (210 vs. 114 mm), according to our limited set of data (Major 1967). Some sagebrush stations with <225 mm precipitation have no soil storage. Potential evapotranspirations are 649 for shadscale versus 604 mm for sagebrush; mean annual temperatures, 11.0 versus 8.8°C. Eastward there is much more variation in the climatic records for both shadscale and sagebrush as noted above, but the distribution of precipitation throughout the year is also quite different eastward, as are available floras.

Axelrod's (1965:153) temperature separation of these deserts suggests that shadscale has an equability of 48 (47–49, decreasing in colder areas), and sagebrush occurs in a colder area with slightly less equability, 47, consonant with the more easterly position of the stations used as representative.

The coldest climates in California outside the highest mountains are in the Cold Desert. Boca at 1688 m near Truckee, in sagebrush and bordering *Pinus ponderosa* forests, recorded −43°C in 1937. Mean annual temperature is low, 5.4°C. Although 134 m higher in elevation, Truckee is 1.0°C warmer annually, with the greatest differences in summer (July 2.2°) and January (1.2°). Even Donner Summit (Norden) just to the west at 2140 m is warmer than Boca (5.6°C mean annual, 2.5° warmer in January, 1.2° in August). In this valley and mountain range physiographical province cold air pools in the valleys must exert a remarkable effect on the vegetation (Billings 1954).

The winter of 1972–73 was cold in California and coldest in the Cold Desert. At Termo in eastern Lassen Co. (40°52′N, 120°27′W, 1614 m) −40° was recorded in December 1972. This was preceded by an October with only 9 days with minima greater than 0°C (extremes +5 to −16°, mean minimum −2.6). November had only 1 day's minimum greater than 0°C (+2 to −13°, mean minimum −4.5). December itself had minima entirely below 0°C, 10 days less than −19° except for two breaks of −12 (mean for the 10 days −22.9). For 11 days the maximum was less than −4°C. Mean minimum for the month was −11.7°; mean maximum, +1.3. January had only 2 days with minima greater than 0°C, 12 with maxima less than 0°. Mean minimum was −11.5°; mean maximum, +1.8. February also had only 2 days with minima greater than 0°C; mean minimum was −5.9 and mean maximum +5.4. Cold returned in March with no minimum above −0.6°C, mean mimimum −5.6, average maximum +6.9. April had no minimum above 0°C, average −3.7, average maximum +14.5. Even May had 10 days with minima less than 0°C.

Bodie (38°14′N, 118°58′W, 2515 m) was consistently much colder than even Termo (Fig. 2-7) or any other place in California in 1972–73. Its mean minima were −5.1°C in October, −13.2 in November, −16.1 in December with the first 14 days averaging −21.9 and the monthly mean maximum +2.4, −16.4 in January, −15.3 in February, −14.6 in March with 7 days below −18.8 and the average maximum +3.3, −9.7 in April, −2.7 in May.

Bishop at 1251 m elevation is not in the Cold Desert, but it is certainly not typical for the Hot Desert either (see the next section). The relatively low PotE at Bishop of 724 mm yr^{-1} may represent a valley receiving cold air drainage. Just above Bishop is a belt of blackbrush (*Coleogyne ramosissima*) vegetation whose Utah occurrence seems to be in a warmer and more continental climate (Major 1967:126) than that of sagebrush. The blackbrush in California is probably in a much less continental climate than that in Utah. Mooney (1973) describes shadscale vegetation with a number of southern plants at 1710 m on the west slope of the White Mts. just above Bishop. *Coleogyne* does not occur in the White Mts. according to Lloyd and Mitchell (cf. Mooney 1973).

Hot Desert. South of the Cold Desert, across the Sierra, and occupying most of southern California is the Hot Desert. If sagebrush typifies the former, creosote bush (*Larrea divaricata*) typifies the latter, and the phylogenetic relationships of these two taxa are most revealing. They relate the Cold Desert to Eurasian continental deserts, and the Hot Desert to subtropical, horse latitude deserts.

The Hot Desert occupies the geomorphologically older of the North American deserts, at lower elevations. It is divided (Shreve 1942) into the northern, cooler (frost present 20–150 days of the year), moister Mojave and the southern, hotter (frost present ca. 10 days a year: U.S. Weather Bureau 1968:28), drier Sonoran Desert, which within California is called the Colorado Desert.

The latter (Fig. 2-7) lies along the Colorado River and the basin of the Salton Sea. It can perhaps be characterized as having at least 200 mm PotE in July, more than 1200 mm PotE a year—to 1327 at Yuma, Arizona, and with the PotE so consistently above the precipitation curve of less than 115 mm yr^{-1} total that soil storage of water is zero as a carryover from a month of more precipitation to one of less precipitation than current PotE. In other words, ActE not only equals precipitation for the year but also more or less equals it for all months. There is usually a water deficit in every month. Elevations here are less than 300 m (1000 feet) and in the Salton Sea Basin to −11 m.

The Mojave Desert occupies a larger area in California. It extends south from Bishop to the Colorado Desert. Maximum monthly PotE is less than 200 mm; yearly, less than 1000 mm. The PotE curve in winter falls sufficiently below the precipitation curve so that storage of a few months precipitation can take place, enough to boost ActE to 20 mm or higher for at least a couple of months. Precipitation and ActE are usually greater than 110 mm yr^{-1} and may be less episodic than in the Colorado Desert, where more of the precipitation comes in the summer than in the winter. In the Mojave Desert winter is the time of maximum precipitation; in the Colorado Desert there are more or less equal peaks in winter and summer.

Death Valley is ecologically unique (Hunt 1966). It combines the heat of the Colorado Desert (PotE of 1400 mm yr^{-1}) with minimal precipitation at all seasons (40–50 mm yr^{-1}). Elevations are down to −86 m, with climatic records from Greenland Ranch at −50 m.

Three of the assumptions made in Thornthwaite's calculations of the water balance are violated in the Hot Desert. 1. The average precipitation is not a very useful measure since precipitation is in fact episodic. See the discussion of the Bagdad precipitation record below. 2. Each episodic rainfall recharges a certain depth of dry soil up to field capacity. A 25 mm rainfall, for example, might bring the 0–25 cm depth of soil up to field capacity; a 15 mm rainfall, only the 0–15 cm depth. Thus soil water is not evenly distributed throughout a 100 mm storage profile and is more episodically available than our diagrams indicate. 3. Either natural or irrigated oases can be expected to have advected heat. They are extraordinary islands of high evapotranspiration. In cases of advected energy the evidence is that Thornthwaite's definition of PotE breaks down, that is, tall vegetation does in fact transpire more than low (Monteny 1969).

Both Shreve (1942:222) and Bailey (1966;25) have called attention to Bagdad's (34°37′N, 115°57′W, 239 m) precipitation record. Within the 24-yr record there are periods of 32, 24, and 36 months with <1 mm precipitation a month. However, the monthly distribution of precipitation is remarkably uniform in one way. If June with 1 and September with only 2 rainy months over the 24 yr record period are omitted, the number of other months with >2 mm averages 5.3 (4–7); the number with >10 mm, 3.9 (3–6). These are remarkably low spreads. Precipitation averages 58 mm, but ranged up to 251 mm in 1905. Only 2 mo average <1.5 mm mo^{-1}; only 3, >7; the other 7, 4–6. In other words, a good storm occurs rarely, but it can occur in almost any month. The 42 monthly totals of >10 mm average a fairly high 30 mm each (10–81 mm mo^{-1}, 10–142 for two consecutive months).

The palm (*Washingtonia filifera*) oases (Vogl and McHargue 1966) have tremendous plant productivity with groundwater supplying water to allow an ActE of more than 1000 mm yr^{-1}. The Imperial Valley is an agricultural equivalent so long as the water of the Colorado River remains available and low enough in salts.

The PotE/ppt ratio in these lowland deserts is 5–10 in the Mojave Desert, 9–18 in the Colorado Desert, and 33.6 in Death Valley.

Vertical Variation

Horizontal variation in climate produces latitudinal zones or, with varying degrees of continentality considered, regions. Vertical variation produces mountain belts.

Reasons for segregating the two kinds of climatic areas are that data are much more abundant for the lowlands, their stations have a better chance of representing regional and not merely local climates, and mountain climates are inherently more ecologically interesting because mountain vegetation is less disturbed. This book is not concerned only with crop or weed vegetation and its ecology. Vertical variations in climate, vertical lapse rates of temperature and precipitation, can be used to approximate mountain climates.

In general, temperature decreases with altitude. The dry adiabatic lapse rate is −0.96°C/100 m rise in elevation; the normal rate is −0.6°C/100 m. Lautensach and Bögel (1956) have generalized the monthly course of lapse rates for different types of climates. In Mediterranean climates summer rates are absolutely smallest; the opposite is true in north temperate climates. In cold, continental climates rates are extreme. Absolute summer rates may be large, and winter rates even positive. Baker (1944) gives graphs of temperatures against altitudes for 27 mountain regions of the western United States. Lapse rates from some of these graphs have been calculated and tabulated by Bamberg and Major (1968:135). July lapse rates center around −0.6°C/100 m, although they are as high as −0.82 in western Wyoming and as low as −0.33 in southern California. Baker observes that the lapse rate in western Wyoming is magnified by the hot Snake River Plains abutting on mountain ranges or plateaus with high valleys having low temperatures under inversions. Conversely, southern California has a cool, maritime lowland abutting on steeply rising mountains with ridge crest or mountaintop climatic stations from which cold air drains freely. January lapse rates are again highest in western Wyoming (−0.86°C/100 m) and lowest in the northern Cascades (−0.33 west slope and −0.38 east slope) or on the western slope of the southern Sierra Nevada (−0.42). The western slope of the northern Cascades has cool lowlands, with clouds over the mountains; the eastern slope graph has excessive scatter. The southern Sierra is warmer than the northern part, and the valley temperatures are not much hotter southward, so lapse rates are reduced.

Precipitation usually increases with altitude. Rates are variable. Baker often shows a zone of maximum precipitation, which is probably related to insufficient snow catch by the rain gages. According to Baker's graphs the eastern slope of the Washington Cascades has a high rate of change of 140 mm/100 m, the seaward slope of the Klamath Mts. has 127, and the western slope of the northern Sierra 110. Low rates of precipitation increase with altitude are found in the Rocky Mts. in continental areas—15–20 mm/100 m or so. Geiger's (1961:465) data from the eastern, continental Alps show 67.5 mm/100 m.

Changes of climate with altitude are discussed further in Brockman-Jerosch (1925–29), Braun-Blanquet (1961), Ellenberg (1963), Bamberg and Major (1968), and others. These discussions of mountain climates need not be repeated here.

A warning is in order. Braun-Blanquet has discussed in detail (1961) the vegetation and ecology of the dry inner valleys of the Alps such as the Durance, Aosta, upper Rhone, upper Rhine, upper Inn, Adda, Isarco, and Adige (Etsch). In all of these valleys increased elevation up-valley does not bring about proportional decrease in temperature; in particular, higher elevations have lower precipitation. The upper valleys are more continental and are in rain shadows of the surrounding peaks. Do such phenomena occur in California? Certainly they do in the upper valleys of the great north coastal rivers such as the Klamath, Trinity, Eel, Russian, and even the Sacramento and San Joaquin. How about the Sierra Nevada? No data are available.

One change of climate with altitude that we need not worry much about from a plant standpoint is the decrease in partial pressure of CO_2 at high altitudes as it affects photosynthesis. It should reduce photosynthesis. Gale (1972a) calculates that the increase in diffusion rate of CO_2 at low pressure partially compensates for the reduced partial pressure. However, the normal decrease of temperature with altitude reduces the diffusion rate. A low mesophyll resistance for CO_2 (r_m) reduces the effect of altitude. Under the high r_m of 6 sec cm^{-1} at 3000 m altitude, with a dry adiabatic lapse rate for temperature Gale calculates that photosynthesis may be reduced to 75% of sea level rate for reasonable values of other environmental and leaf properties.

The effect of increased altitude on transpiration is also complicated (Gale 1972b). Under isothermal conditions increased elevation increases potential transpiration. Increased radiation with increasing altitude enhances this effect, as does increased air humidity. A normal temperature lapse rate changes this effect to a decrease. To correctly partition its heat loss between reradiation, convection, and latent heat, the leaf temperature adjusts so that at high altitudes under reduced air temperature the difference between leaf and air temperatures increases. This result agrees with observations.

Unfortunately for the utility of the above types of calculations plants that grow at high altitudes are often neither morphologically nor physiologically the same as those that grow at low elevations. Hence the calculations should be checked by measurements in the field.

Northwest Pacific Coast. Some climatic changes from coastal Crescent City to Monumental only 34 km inland and 825 m higher have been mentioned above. Precipitation increases at a rapid rate on the seaward slope of the Coast Range as eastward-moving air is forced to rise. The rate of increase is 125 mm/100 m to Monumental and 71.5 mm/100 m to Elk Valley, which is 57 km east of and 508 m higher than Crescent City. The temperature lapse rate to Elk Valley is +45°C/100 m for the annual mean and +0.85 for July but −1.09 for January. The conclusion is that continentality increases inland, not that it increases with elevation.

Since data are lacking up the seaward slope for any Californian coastal mountains in this climate, we turn to Mt. Rainier, Washington. Table 2-9 gives some data. Our three stations extend less than halfway up this great isolated volcanic peak, but Paradise is almost at timberline. Increase of precipitation with altitude is

TABLE 2-9. Characteristics of climate up Mt. Ranier (4390 m), Washington. Lapse rates are in $^{o}C/100$ m for temperatures and in mm/100 m for PotE, precipitation, and ActE

Station	Latitude (N)	Longitude (W)	Elev. (m)	Mean Temp. (^{o}C)			ΔT (^{o}C)	K	Eq	Temp. Lapse Rate from Next Lower Station			PotE		Ppt.		ActE		ActE/PotE
				Year	Jan.	July				Year	Jan.	July	(mm)	lapse rate	(mm)	lapse rate	(mm)	lapse rate	
Paradise	46°47'	121°44'	1692	3.4	-3.3	11.6	15.0	11.1	43	-0.46	-0.29	-0.54							
													471	-13	2637	+64	442	-8	0.94
Longmire	46°45'	121°49'	843	7.3	-0.8	16.2	16.8	12.9	49	-0.40	-0.54	-0.19							
													580	-10	2094	+97	514	+5	0.89
Olympia	46°58'	122°54'	58	10.4	3.4	17.7	14.3	10.4	56				659		1330		477		0.73

most rapid on the lower slopes. Is the decreased rate above due to inefficient snow catch in the gage? Although the midelevation station has maximum ActE, the reasons are complex. Below, there is lack of summer precipitation; above, the growing season is short and cool (7 months vs. 9 and 12 below). Continentality shows the same distribution with a maximum at midaltitude. On the other hand, equability decreases almost uniformly upward at the rate of 0.8 unit/100 m. Temperature lapse rates show some remarkable switches. At low elevations the winter rate is much faster than the summer rate, agreeing with Lautensach and Bögel's generalizations (1956) on seasonal changes in temperature lapse rates in mediterranean climates. In the 850 m below timberline, however, the temperature lapse rate behaves like those in northern temperate mountain climates, namely, faster in summer. Potential evapotranspiration at least decreases fairly uniformly with altitude, somewhat less at low elevations and less than the figure of −16 mm/100 m given for southeastern Alaska by Patric and Black (1968:23).

It is surprising how rapidly some elements of climate change inland. The coastal redwood belt is really very narrow (Griffin and Critchfield 1972:104). Table 2-10 illustrates changes up the Klamath River. Temperature changes are extreme and reflect increased continentality up to Happy Camp but the normal decrease of temperature with altitude farther up-valley. Precipitation drops sharply inland, especially in the valley east of the main crest (Yreka). This kind of valley climatic change as a function of large-scale topography and its effects on vegetation and its ecology have been masterfully described by Braun-Blanquet (1961).

Mediterranean California. Again, coastal stations are anomalous. In Table 2-11 Mt. Tamalpais is compared with Pt. Reyes on the coast to the northwest and with Kentfield and San Rafael to the north on San Francisco Bay. Although Mt. Tamalpais is not very high, its climate is not that of the redwood belt below. Temperature extremes become much more sharply expressed above the fog belt, whose top is about 520 m (Patton 1956:157). Winters are colder. Actual evapotranspiration is consistently reduced as compared with the coastal station values because of lower winter temperatures, and as compared with the inland stations values because of reduced precipitation.

Mount Hamilton (1283 m) provides the opportunity to compare the coastal and Central Valley sides of a Coast Range mountain. Newman to the east is at 28 m, San Jose to the west at 29 m elevation. January temperature lapse rates are −0.25 and −0.39°C/100 m with the more continental Central Valley cooler than the more maritime Coast Range valley. In July roles are reversed with lapse rates of −0.37 and +.14°C/100 m. San Jose often has fog while Mt. Hamilton is sunny. Precipitation increases at a higher rate on the drier Central Valley side: 33 versus 28 mm/100 m, while PotE decreases almost 3× as fast: −16 versus −6 mm/100 m. The ActE increases slightly upward from Newman (+2 mm/100 m) but decreases from San Jose (−1 mm/100 m).

In the inner Coast Ranges two stations on Mt. St. Helena show "normal" relationships to the climate at St. Helena in the upper Napa Valley (Table 2-11). Temperature lapse rates are −0.6°C/100 m. The PotE and ActE uniformly decrease and increase, respectively, at rates of −14 and +8 mm/100 m. Precipitation, however, accelerates its increase at higher altitudes.

Calistoga (39°35′N, 122°35′W, 111 m) and Middletown (38°45′N, 122°37′W,

TABLE 2-10. Some climatic changes inland up the valley of the Klamath River. Lapse rates are in ^{o}C/100 m for temperature and in mm/100 m for PotE, precipitation, and ActE

Station	Distance Inland (km)	Alt. (m)	Mean Temp. (^{o}C)				Eq.	Temp. Lapse Rate from Western Station			PotE		Ppt		ActE	
			Year	Jan.	July	ΔT		Year	Jan.	July	(mm)	lapse rate	(mm)	lapse rate	(mm)	lapse rate
Crescent City	0	14	11.4	8.0	14.3	6.6	69				650		1777		544	
Happy Camp	69	332	13.3	3.9	23.0	19.1	53	+0.60	-1.29	+2.73	739	+28	1247	-148	383	-50
Yreka	126	802	10.9	1.0	22.0	21.0	50	-0.51	-0.62	-0.23	672	-14	444	-171	280	-22
Tulelake	224	1231	8.0	-1.3	18.2	19.5	48	-0.67	-0.53	-0.88	592	-19	259	-43	256	-6

TABLE 2-11. Differences in climates. A. On Mt. Tamalpais, as compared with a nearby seacoast station (Pt. Reyes) and two inland stations on San Francisco Bay. B. On and near Mt. St. Helena. Lapse rates to the highest station in A, from the lowest in B. They are in °C/100 m for temperature and in mm/100 m for PotE, precipitation, and ActE

Station	Lat. (N)	Long. (W)	Alt. (m)	Mean Temp. (°C) Year	Mean Temp. (°C) Jan.	Mean Temp. (°C) July	ΔT (°C)	K	Eq	Temp. Lapse Rate Year	Temp. Lapse Rate Jan.	Temp. Lapse Rate July	PotE (mm)	PotE lapse rate	Ppt (mm)	Ppt lapse rate	ActE (mm)	ActE lapse rate
								A										
Mt. Tamalpais	37°56'	122°35'	747	12.8	6.5	20.6	14.1	10.8	59				690		700		346	
Pt. Reyes	38°00'	123°01'	155	11.4	9.7	12.0	3.8	0.5	74	+0.24	-0.54	+1.45	634	+9	497	+35	412	-11
Kentfield	37°57'	122°33'	3	13.8	7.8	18.7	10.9	7.6	65	-0.13	-0.17	+0.25	709	-3	1184	-65	426	-11
San Rafael	37°58'	122°32'	9	15.0	9.2	19.4	10.4	7.1	65	-0.30	-0.37	+0.16	741	-7	941	-33	401	-7
								B										
Helen Mine	38°42'	122°36'	842	10.2	3.1	16.9	13.8	10.4	51	-0.58	-0.59	-0.57	638	-14	2192	+178	420	+8
Mt. St. Helena	38°40'	122°40'	702	11.0	3.9	17.8	13.9	10.5	52	-0.58	-0.59	-0.56	655	-15	1511	+108	407	+8
St. Helena	38°30'	122°28'	78	14.6	7.6	21.3	13.7	10.3	60				747		834		360	

342 m) are closer to Mt. St. Helena but lack temperature records. Their calculated PotE and ActE fall on the lapse rates given in Table 2-11. Calistoga's precipitation fits, but Middletown lying on the north side of the mountain and protected from the influx of bay air moving up the Napa Valley, is drier and therefore shows a steeper increment of precipitation with altitude. The general rate for all stations averages at least +170 mm/100 m.

The precipitation record at the Helen Mine is simply extraordinary. In January 1909 1814 mm of precipitation was recorded for a U.S. monthly record. As low as 19 mm has been recorded for January over the 22 yr record. February precipitation has varied from 1580 to 20 mm, March 933 to 85, April 340 to 0, May 133 to 0 (5×), June 125 to 0 (10×), July 29 to 0 (19×), August only 4 to 0 (20×), September 232 to 0 (8×), October 401 to 0 (3×), November 704 to 19, and December 878 to 91. Three consecutive months without any precipitation occurred 13 times; the maximum precipitation in a year was 3500 mm.

In southern California Pasadena at 263 m elevation can be compared to Mt. Wilson at 1740 m in the western San Gabriel Range to the north. As usual in coastal California temperatures decrease with altitude at less than the normal lapse rate (−0.41°C/100 mm in January, −0.04 in July). Precipitation here shows a small increase of +20 mm/100 m; PotE, a decrease of −8 mm/100 m. The overall result is a decrease in ActE of only −4 mm/100 m.

Mt. Palomar in the Santa Ana Range to the south at 1691 m is very similar in climate to Mt. Wilson—0.4°C warmer in winter, with less precipitation in the winter half-year (548 vs. 726 mm) and more in the summer (56 vs. 54 mm, 26 vs 7 for July–September).

Squirrel Inn in the Lake Arrowhead area of the San Bernardino Mts. 77 km due east of Mt. Wilson is considerably cooler and wetter (altitude 1730 vs. 1740 m, annual means 11.5 vs. 12.8°C, PotE 660 vs. 694 mm yr^{-1}, precipitation 1041 vs. 780 mm yr^{-1}, ActE 324 vs. 288 mm yr^{-1}). As compared to lowland San Bernardino at 336 m, Squirrel Inn has steeper and more normal temperature lapse rates than the more coastal Mt. Wilson, −0.50°C/100 m in January and −0.34° in July. The increase of precipitation with altitude is doubled to +42 mm/100 m. Whereas ActE is almost constant with altitude, actually −0.5 mm/100 m, PotE decreases at double the Pasadena–Mt. Wilson rate, −15 mm/100 m. San Dimas is intermediate between the western and eastern mountain stations.

Much of the variation in climates eastward across the Coast Ranges is illustrated in the transect of Fig. 2-10. Point Arena is maritime, wet, cool, with ActE high and fairly evenly divided throughout the year or at least not so concentrated in the spring and summer months as is usual in continental, Mediterranean California. The adjacent mountains probably have cooler, still wetter climates. Hopland Largo Station is in the upper Russian River valley, in a rain shadow with a more continental climate, with ActE not only lower in total value but also more seasonally variant. Seasonal distributions of ActE over the transect are given in Table 2-12. Hopland Headquarters is in a north–south fault valley on the west-facing slope above the Russian River. It is wetter and cooler than at river level, although only 100 m higher. For the nearest station with a temperature record, Ukiah (Table 2-7), which is 17 km north and 66 m lower, temperature lapse rates are −1.7°C/100 m in winter and −1.1 in summer. Both are greater than the dry adiabatic, which probably means that the graben where Hopland Headquarters lies is a cold air drainage pocket.

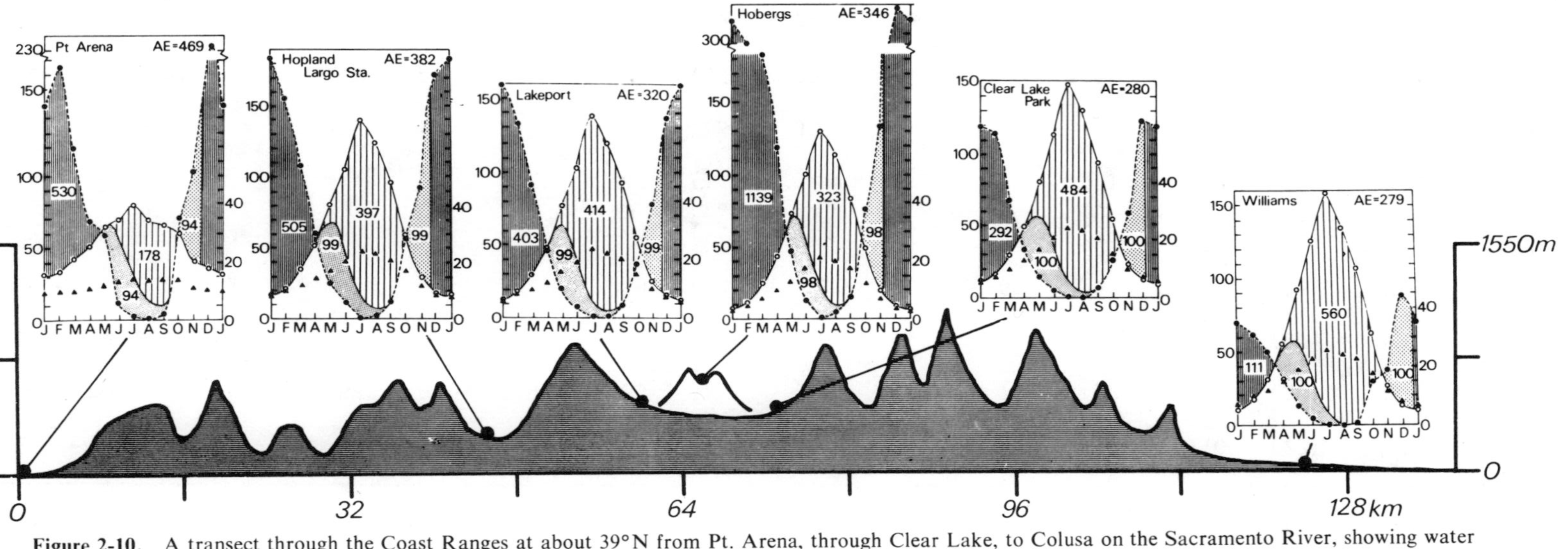

Figure 2-10. A transect through the Coast Ranges at about 39°N from Pt. Arena, through Clear Lake, to Colusa on the Sacramento River, showing water balances. See Table 2-12. For legend see Fig. 4. Vertical exaggeration 14.5×.

TABLE 2-12. Changes in amount and seasonal distribution of actual evapotranspiration on a Coast Range transect from Pt. Arena to Colusa in the Central Valley. The 3 dry months are July-September. See Fig. 2-10

Station	Alt.	Dry Months		Winter		Spring		Summer		Fall		Total	
	(m)	(mm)	(%)	(mm)	(%)	(mm)	(%)	(mm)	(%)	(mm)	(%)	(mm)	(%)
Pt. Arena	72	54	(12)	102	(22)	159	(34)	98	(21)	110	(23)	469	(100)
Hopland Largo	168	36	(9)	55	(14)	156	(41)	71	(19)	100	(26)	382	(100)
Hopland Headquarters	256	38	(11)	50	(14)	137	(40)	67	(19)	92	(27)	346	(100)
Upper Lake 7W	464	41	(11)	48	(13)	146	(39)	85	(23)	93	(25)	372	(100)
Lakeport	410	30	(9)	44	(14)	141	(44)	62	(19)	73	(23)	320	(100)
Upper Lake RS	411	37	(11)	49	(14)	146	(42)	75	(21)	82	(23)	352	(100)
Hobergs	909	51	(15)	27	(8)	139	(40)	93	(27)	86	(25)	345	(100)
Clear Lake Park	398	22	(8)	39	(14)	135	(48)	50	(18)	57	(20)	281	(100)
Williams	37	11	(4)	42	(15)	141	(50)	38	(14)	59	(21)	280	(100)
Colusa	18	12	(4)	44	(16)	148	(53)	40	(14)	48	(17)	280	(100)

Precipitation increases at a rate of +68 mm/100 m from Ukiah, and +77 from Hopland Headquarters. The PotE and ActE decrease at −33 (−56) and −29 (−41) mm/100 m. The mountains separating Hopland from Clear Lake reach 1000 m, and those to the northwest of Clear Lake are still higher (1700 m); therefore the lake, especially its eastern end, is in a pronounced rain shadow. Hobergs at 909 m is somewhat south of the transect and high on the slope of Cobb Mt. (1440 m). Temperatures at Hobergs are lower and precipitation higher as expected. Precipitation probably increases with altitude at a rate of about 100 mm/100 m. Finally, Williams lies across still another range of 850 m altitude and also in the lee of Snow Mt. (2050 m) and the Yolla Bollys (2400 m). It is in the driest part of the Sacramento Valley. The short distance from Williams to Colusa brings an average of 22 mm additional precipitation annually, all in the winter. However, the records may not be strictly comparable.

Sierra Nevada–Cascade Ranges

NORTHERN SIERRA NEVADA–CASCADES. The transect over the central Sierra Nevada from the Central Valley into western Nevada along the original Union Pacific Railroad has a large number of stations that have been much used to describe altitudinal changes in climates (Henry 1919; Varney 1920; Baker 1944) (Figs. 2-11, 2-12). East and west slopes differ. Least-square equations for the four lines of Fig. 2-12 are: West T (°C) = 16.95 − 0.492 altitude (in 100 m), ppt (in mm) = 572 + 60.9 altitude (in 100 m); East T = 17.01 − 0.589 altitude, ppt = −1686 + 140.1 altitude. Thus projected sea level temperatures are remarkably similar on both slopes, and the windward slope has a lower temperature lapse rate than the eastern, which is about normal. Precipitation relationships are very different on the two slopes with great and meaningless differences in projected sea level precipitations and a 2.3× faster rate of increase of precipitation with altitude on the steeper eastern slope.

Squaw Valley can be added to the east side of this transect neatly for its altitude, although its topographical situation is unlike that of any of the other stations. Altitude is thus highly correlated with climate, but the nature of this correlation deserves investigation in quantitative terms related to water balances, and to plant activities if a definition of plant success still eludes us.

Squaw Valley does at least have a climatic record at the base of the ski area. Most of the many ski areas in California and elsewhere have failed to take advantage of their network of mountain transportation to provide climatic records over the often 500–1000 m of vertical drop so readily available. Ski areas must have good weather records for avalanche control, and the climatic interpretations made possible by using such records would benefit not only plant ecologists but also land managers.

Varney (1920) pointed out that the rates of change of precipitation with altitude on the western Sierran slope are absolutely large in winter and low in summer. So precipitation amounts (Fig. 2-11).

This transect shows a decrease in continentality with altitude up the western slope to just short of the crest. This is common in mountains where air drainage is free. Then there is a rapid increase in continentality down the eastern scarp and eastward into the Cold Desert.

On this transect PotE decreases at −15 mm/100 m on the west slope, −19 on the

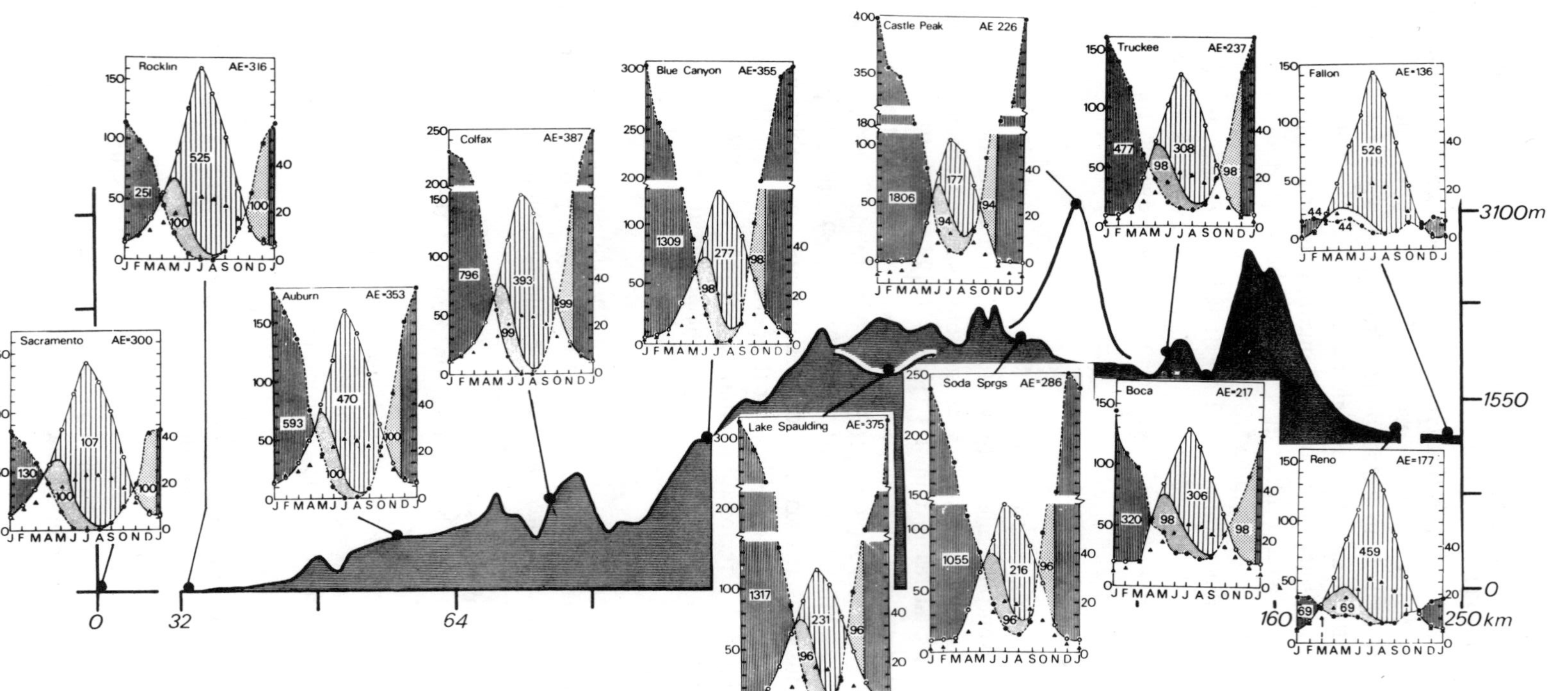

Figure 2-11. A transect over the Sierra Nevada from Sacramento (38.5°N), over Donner Summit (39.3°N), to Reno and Fallon, Nevada (39.5°N), roughly3 along the line of the original Union Pacific Railroad, showing water balances. For legend see Fig. 2-4. Vertical exaggeration 14.5×.

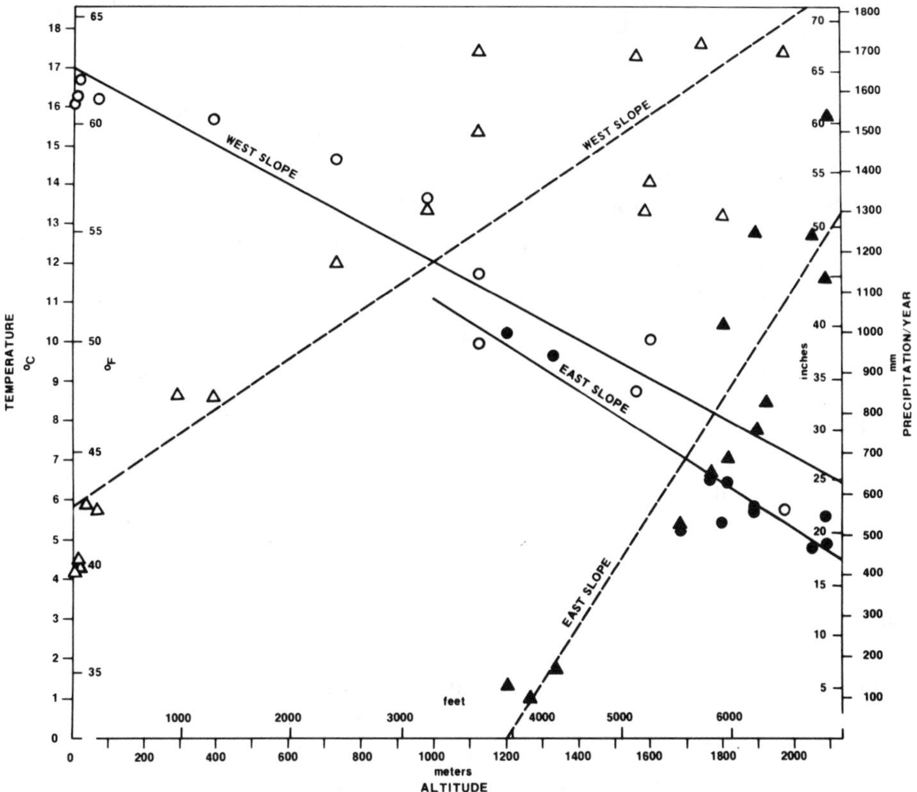

Figure 2-12. Changes with altitude of mean annual temperatures and precipitation over the Sierra Nevada along the transect of Fig. 2-11. Temperatures on west slope ○, on east slope ●. Precipitation on west slope △, on east slope ▲.

east (cf. the discussion of Mt. Rainier). On the west slope PotE drops from 800–850 mm yr^{-1} in the Sacramento Valley, increasing somewhat east of the river, to 500 mm yr^{-1} at 2000 m elevation and down-slope to 650 mm yr^{-1} at Fallon in the shadscale desert of Nevada. An increase of ActE with elevation, in response to increased water and less heat, occurs on both slopes, +9 mm/100 m on the west from 300 mm yr^{-1}, +20 on the east from 140 mm yr^{-1}. On the west slope, however, the positive slope changes to −6 mm/100 m somewhere around 1300 m elevation and 400 mm yr^{-1}. Here increased cold overtakes the advantage of increased precipitation, just as is indicated on Mt. Rainier. The high plant productivity implied by ActE of 400 mm yr^{-1} is worth emphasizing. At 2100 m the east and west slope lines meet at 300 m yr^{-1} with a calculated 226 on Castle Peak (2780 m). The extrapolated data for Castle Peak fit the west slope trends, although the stations from Soda Springs up to Donner Pass do not.

The PotE/ppt ratio is less than 2 in the valley (Davis 1.93, Sacramento 1.87, Citrus Heights 1.43, Rocklin 1.48) but 0.92 at Auburn (395 m). The projected PotE and precipitation lines cross at 375 m. Here the two measures are 800 mm yr^{-1}, the ratio is 1, and this is the arid–humid boundary on the western slope of the Sierra. The index drops to 0.38 at Blue Canyon (1610 m) and to 0.31 at Fordyce Dam (1983

m) but may rise as the crest is approached (0.38 at Soda Springs, 0.44 at Norden), although there is a calculated low of 0.20 on Castle Peak. On the eastern side the ratio decreases upward from Fallon at 1210 m (4.87) to Reno at 1340 m (3.59) and Boca at 1688 m (0.98), then goes to 0.50 at Donner Memorial State Park (1809 m) just below the summit. On this side the PotE and precipitation curves cross at 1610 m with climatic measures at 570 mm yr^{-1}.

Is there a more dramatic expression of the ecological differences between the western and eastern slopes of the Sierra Nevada than the (calculated) change in elevation of the arid–humid boundary from 375 m on the western to 1610 m on the eastern, with PotE and precipitation at levels of 800 versus 570 mm yr^{-1}? The contrast is even greater southward, particularly in the rapid rise of this boundary on the eastern slope.

Much of the precipitation in the higher Sierra Nevada comes as snow. Precipitation that comes as snow can be deposited very irregularly in the landscape. Windblown, bare ridges will alternate with large cornices on the lee of crests. An isolated boulder or tree can pile up a drift in its lee but be surrounded otherwise by a wind-scoured bare area. A depression where snow accumulates may not melt out until August, whereas an even, wind-exposed slope may be bare in April. Overloaded slopes and areas of snow slab formation may produce avalanches that destroy forests or top the trees and remove branches if airborne, scrape slopes clean to the ground, make massive deposits of rock, soil, wood, branches, and conifer needles, and pile up great masses of snow that may last until September even in low elevations.

Precipitation in the Sierra Nevada falls as snow in increasing amounts and proportions at higher elevations. In Table 2-13 some data on snow occurrence are given for the Donner Summit transect above Auburn, Tamarack on the Stockton–Ellery lakes, and Giant Forest and Helm Creek on the Visalia–Sequoia National Park transects to the south, plus Cold Desert stations.

If it is assumed that 1 cm snow gives 1 mm water (i.e., the density of new-fallen snow is 100 kgm^{-3}), the percentage that snow constitutes of total precipitation can be calculated. This density may be too high, however, even for the usually "wet snow" of most Sierran ski resorts since Tamarack then shows 112% snow (89 kgm^{-3}) precipitation during the winter months.

The proportion of snow in annual precipitation and the length of the snowy season both increase rapidly with altitude. Across the summit the proportion of snow and the length of the snowy season both jump higher—and higher than at corresponding altitudes on the western slope. Degree of continentality is nicely correlated with snowiness.

In the southern Sierra, Giant Forest has almost the same total precipitation and total snowfall as Helm Creek at a much higher elevation. Variability is greater at Giant Forest. In its 10 yr record Giant Forest had from 1146 to 197 cm snow annually, even 359 cm in February 1922 (density 82 kgm^{-3}). In April 1923 the snowfall of 288 cm had a putative density of 153 kgm^{-3}! In general, new snow depths in any month vary 10×, although December has had 274 cm (density 97 kgm^{-3}) or zero twice. Plant adaptation to a snow cover must be difficult under such variable circumstances. Probably avalanche destructiveness is increased by this high variability in snowfall, but data are lacking. Mineral King, which has been proposed for skiing and real estate development, is very near Giant Forest.

As the timberline is approached, the increased proportion of snow in total precipitation increases terrain irregularity of available water. Areas usually blown free of snow will therefore develop deeply frozen soil, so any surficial melting during sunny winter days will not penetrate deeply. Freezing will pull water from depth toward the surface. The result during spring snowmelt will be a surficial soil layer that is saturated or more, and no later additions of water to these ridge crest sites.

At this time other areas have deep snow. A 2 m spring layer of corn (firn) snow will contain 1000 mm of water. This is only a little less than the precipitation recorded all year at, for example, Donner Summit.

The alpine landscape in May is a mosaic of bare and still snow-covered areas. The areal form of this mosiac is highly constant from year to year.

Although the windblown site may have only enough water to fill evapotranspiration needs for May, the deep-snow site may have enough for August, although growing seasons may correspond to just these months. Possible expenditures for ActE may be only 40 mm in May but 100 in August.

In addition the deep-snow sites will receive seepage from up-slope. Such lateral movement of water through the soil mantle is probable, given the saturated profile and the positive hydrostatic head. It is made visual by water spurting out of gopher holes below melting snowbanks.

A particularly vivid demonstration of wind movement of snow is provided by recent very extensive clear-cuts supervised by the U.S. Forest Service in old-growth spruce–fir in the Medicine Bow Mts. of Wyoming. These devastated areas are often screened from public view by a narrow band of uncut forest left along roads. These roads are for "recreation," actually built by private timber buyers with proportionate decrease in stumpage prices. Unfortunately the snow is carried by the wind off the clear-cuts, into the fringing strip of trees, and is deposited then quite precisely on the road. In early summer the open clear-cut is bare; the road is neatly buried under 3–5 m of very solid snow. This kind of forestry is not multiple use, but it represents poor management.

In the Cascades east of Red Bluff over the crest to Susanville winter and annual temperature lapse rates are −0.60°C/100 m and −0.80 on western and eastern slopes, while summer rates are −0.70 and −1.10. Potential evapotranspiration drops from 920 mm yr^{-1} in the Central Valley to 550 at 1800 m at the rate of −22 mm/100 m, and over the crest it increases downwards to Susanville with 640 mm yr^{-1} at about the same rate (−21 mm/100 m). On the other hand, ActE figures are irregular and change little with altitude. On the western slope ActE seems to decrease at perhaps −3 mm/100 m to 320 mm yr^{-1}. Eastern slope values are lower, 260 mm yr^{-1}, and show almost no change (−2) with altitude. Precipitation is most irregular. It increases north and east from Red Bluff (526 mm yr^{-1}), and reaches 1308 mm yr^{-1} at Mineral at 1477 m elevation, but Manzanita Lake farther north at 1722 m elevation records less, 1096 mm. Topographical effects in this region are complicated by the bulk of Mt. Lassen (3180 m) and the lowlands north and south of it. Lake Almanor is at only 1373 m altitude, although the passes to east and west are 1750 and 1505 m. There are two trends in precipitation, increasing with elevation and decreasing eastward. Mineral's high precipitation is related not only to its elevation but also to its longitudinal position west of Chester (1379 m and 816 mm yr^{-1}) and Canyon Dam (1389 m and 927 mm yr^{-1}).

The east side, rain shadow stations have low precipitation, 444 mm yr^{-1} at Susan-

TABLE 2-13. Precipitation occurring as snow on a Sierran transect from Sacramento to Fallon, Nevada. See Figs. 2-11 and 2-12 and the text

Station	Altitude (m)	Snow as Percent of Total Ppt	Maximum Percent of Ppt as Snow		Months with Snow Percent >			
			%	Month	2	5	25	75
				Transect Stations				
Colfax	736	5	10	Jan.	Nov.-Apr.	Dec.-Mar.	None	None
Gold Run	988	13	23	Mar.	Nov.-Apr.	Dec.-Apr.	None	None
Towle	1130	23	37	Jan.	Sept.-Apr.	Oct.-Apr.	Jan.-Mar.	None
Emigrant Gap	1593	53	72	Mar.	Sept.-June	Oct.-May	Oct.-May	None
Blue Canyon	1610	33	48	Mar.	Oct.-June	Oct.-May	Dec.-Apr.	None
Cisco	1811	72	90	Mar.-Apr.	Sept.-June	Sept.-June	Oct.-May	Jan.-Apr.
Fordyce Dam	1983	52	73	Feb., Mar.	Sept.-June	Sept.-June	Oct.-May	None

Norden	2140	91	100	Mar.-Apr.	Sept.-June	Sept.-June	Sept.-June	Nov.-May
Truckee	1822	74	87	Feb.	Sept.-June	Sept.-May	Nov.-May	Jan.-Apr.
Boca	1688	75	93	Feb.	Sept.-May	Sept.-May	Oct.-May	Jan.-Apr.
Reno	1340	43	75	Mar.	Oct.May	Oct.-May	Dec.-Apr.	Mar.
Fallon	1210	23	58	Jan.	Oct.-May	Nov.-May	Dec.-Mar.	None
Other Stations								
Tamarack	2440	97	112	Dec.-Feb.	Sept.-June	Sept.-June	Oct.-June	Nov.-May
Ellery Lake	2930	95	117	Mar.	Sept.-July	Sept.-July	Sept.-July	Nov.-Apr.
Giant Forest	1940	66	103	Jan.	Sept.-June	Sept.-June	Oct.-May	Dec.-Mar.
Helm Creek	2445	70	91	Jan.	Sept.-June	Sept.-June	Sept.-June	Jan.-Apr.
Ft. Bidwell	1370	27	43	Dec., Feb.	Oct.-May	Nov.-Apr.	Dec.-Mar.	None
Madeline	1610	51	82	Mar.	Sept.-May	Oct.-May	Nov.-Apr.	None

ville (1262 m) and northward only 338 mm yr^{-1} at Madeline (1611 m). The valley stations have high continentalities in the west, and even higher ones in the east, irregularly less continental with higher elevation. The humid–arid boundary lies only a little above Red Bluff with a PotE/ppt ratio of 1.75, but all east side stations are arid, increasingly so from 1550 m (Westwood 0.97) downwards over the 1750 m pass to Susanville with 1.44, which indicates rain shadow steppe but fits poorly the surrounding *Pinus ponderosa–Quercus kelloggii* vegetation.

Northward from Red Bluff up the Sacramento River precipitation increases very rapidly to Shasta Lake (elevation 330 m) at a rate of about +500 mm/100 m. A rapid drop ensues, −95 mm/100 m up-river to 1000 m elevation and then −290, as Braun-Blanquet (1961) has described, from the large, interior river valleys in the Alps. Temperatures up-river show no such irregularities, although higher precipitation along the lake means somewhat lower temperatures, perhaps −1°C/300–400 mm of precipitation. Lapse rates are large, −1.0°C/100 m in summer, −0.85 in winter and for the annual mean. As temperature decreases, PotE also decreases from 920 mm yr^{-1} in the valley to 640 mm yr^{-1} at 1100 m elevation. The rate of decrease with altitude is about −34 mm/100 m. This is high. Actual evapotranspiration seems to show a maximum where precipitation peaks. It is 320 mm yr^{-1} in the valley and 370 at the head of the Sacramento River, with a maximum of almost 500 mm on Shasta Lake. The latter is very high for California, and the forests are cut over to prove it. The rate of increase is +60 mm/100 m, of decrease −14. Continentality decreases up-river to McCloud, but then increases toward and over the divide from the Klamath River drainage.

The Sacramento–Klamath divide and the upper Klamath region are high in altitude, with the growing season shortened by winter cold, low precipitation of (650)450–300 mm yr^{-1}, low PotE of 560–670 mm yr^{-1}, and low ActE of 250–340 mm yr^{-1}. Both PotE and ActE decrease with altitude, at rates of −19 and −15 mm yr^{-1}. The latter is the same as in the upper Sacramento drainage.

In the 72 m elevation rise over 47 km northward from Red Bluff to Redding the PotE/ppt ratio changes from 1.75 to 0.95. At the arid–humid boundary, where it equals 1, PotE and precipitation are 910 mm yr^{-1} at 170 m elevation. The ratio drops to 0.44 at Lakeshore, where precipitation and ActE are both maximal, and then gradually increases up-river. McCloud is only 0.52, but Mt. Shasta City is 0.69, and Weed only 11 km distant and 11 m lower in elevation (1069 m) is 0.97, with PotE and precipitation northwards equal at 620 mm yr^{-1}. There is forest at Mt. Shasta City, with shrub steppe near Weed. In the upper Klamath, Yreka is 1.51, and both Mt. Hebron Ranger Station and Klamath Falls are 2.00.

SOUTHERN SIERRA NEVADA. South of Donner Summit, from Stockton (7 m elevation) up to Calaveras Big Trees State Park (1430 m) and higher to both Twin (Caples) Lake (2480 m) on the Carson Pass road and Tamarack (2450 m) near Bear Valley on the Ebbetts Pass road, 10 stations, including also Camp Pardee at 200 m, Mokelumne Hill at 426 m, Kennedy Mine at 457 m (precipitation only), Westwood at 834 m (ditto), Salt Springs Reservoir at 1130 m, and Bear River at 1430 m (temperatures from Calaveras Big Trees at 340 m or −2.0°C below), can be assembled in a transect up the western slope of the Sierra Nevada. Temperature decreases from a mean monthly 25°C in July in the valley to 13° at 2500 m, in January from 7° to −4°, for the year from 16° to 4°. An increase of a degree or so occurs just above the valley floor. Overall rates of decrease are $-0.50°C/100$ m in winter to -0.55 in summer, -0.45 annually. Precipitation increases linearly to 1360 mm yr^{-1} at 1400 m at the rate of +70 mm/100 m. A better interpretation is +103 mm/100 m increase to 450 m, then +58 to 1400 m. Precipitation reaches 1380 mm yr^{-1} at 1770 m, and at 2500 m it has ostensibly dropped to 1200 mm yr^{-1}. Potential evapotranspiration decreases at -15 mm/100 m from 850 mm yr^{-1} in the valley to 450 mm yr^{-1} at 2500 m elevation. Actual evapotranspiration increases at +10 mm/100 m to 1100 m elevation for a maximum value of 400 mm yr^{-1}, followed by a decrease at -7 mm/100 m, so that ActE in the valley is about equal to that in subalpine forest (280 mm yr^{-1}). The PotE/ppt ratio indicates semidesert in the valley (2.43), steppe at Camp Pardee (1.66), the arid–humid boundary just below Mokelumne Hill (0.99), crossing below 0.5 just below 1300 m, and down to 0.42 at Twin Lake and 0.38 at Tamarack. At the arid–humid boundary PotE and precipitation are 770 mm yr^{-1}. The ecosystem terms used by Budyko for values of his index on the arid side in Eurasia have no clearly evident meaning in mediterranean climates.

A transect through Yosemite National Park at about latitude 37°40′–38°N is the classical one of Clausen et al. (1940). I have used 13 scattered stations (Table 2-14). Temperature lapse rates cluster closely around $-0.30°C/100$ m for July, 0.37 for January, and -0.35 for the annual mean of western slope stations. Sonora is warmer than expected for its altitude. Dudleys, on the other hand, is cooler than expected by perhaps 1°, probably a reflection of 130 mm yr^{-1} more precipitation. Temperature lapse rates from Mono Lake up to Ellery Lake are much higher than on the western slope, $-0.70°C$ in July and for the annual mean, -0.52 in January. Precipitation increases rapidly (+208 mm/100 m) in the western foothills up to at least 200 m elevation, much less rapidly to 850 mm yr^{-1} at 600 m, and then irregularly to 1600 m. Over this last 1000 m rise precipitation can be as low as 810 or as high as 1140 mm yr^{-1}. There may be an increase at the rate of +24 mm/100 m, but scatter around the line is excessive. On the eastern slope our two points give a rate of +45 mm/100 m.

Compared with conditions in the northern Sierra and Cascades, annual temperatures on the Yosemite transect are no higher at 500 m. They are higher at 1000 m by 0.5° and at 2000 m by 2.5° than on the Donner Summit transect at latitude 39° ± 20′, and higher by 1° at 1000 m and by 4° at 2000 m than on the Red Bluff–Susanville transect at 40°15′N. The latter transects are 140 and 270 km north of Yosemite. The PotE values are highest in the north at lower elevations, and the reverse at higher elevations. At 500 m PotEs are 830, 775, and 750 mm yr^{-1} from north to south, at 1000 m 710, 700, and 700, at 2000 m 460, 550, and 570. Precipitation levels are lower southward from the Donner Summit transect by perhaps 100 mm at 500 m altitude, 250 mm at 1000 m, and 400 mm at 2000 m. The ActE values are consistently 350 mm yr^{-1} at all altitudes on all three transects except for the maximum of 400 mm yr^{-1} at 1000 m on the Donner Summit transect.

On the Yosemite transect PotE is 800 mm yr^{-1} in the valley, down to 600 mm yr^{-1} at 1600 m at a rate of -13 mm/100 m. The rate on the eastern side is steeper, -23 mm yr^{-1} from Mono to Ellery Lake. Our assumed temperature lapse rate above Ellery Lake gives a PotE lapse rate of -16 mm/100 m, the same figure as given for southwestern Alaskan stations by Patric and Black (1968). The ActEs are 350 mm yr^{-1} on the western slope and 230 on the eastern, and they change not at all with altitude above the immediate Central Valley floor (Modesto and Oakdale),

TABLE 2-14. A transect of climatic stations over the Sierra Nevada through Yosemite National Park and vicinity. At Mather monthly mean temperatures were derived from nearby Hetch Hetchy, using a lapse rate of $-0.6^{o}C/100$ m. The asterisks denote stations with temperature and precipitation extrapolated from Ellery Lake, using the above lapse rate for temperature and +100 mm/100 m for precipitation, allocated to the months in proportion to their contributions to total precipitation at Ellery Lake (Major 1967:105)

Station	Latitude (°N)	Longitude (°W)	Alt. (m)	Mean Temp. (°C) July	Jan.	Year	PotE (mm yr^{-1})	Ppt (mm yr^{-1})	ActE (mm yr^{-1})	PotE/Ppt	K
Modesto	37°39'	121°00'	28	24.6	7.3	15.8	818	281	268	2.91	14.0
Oakdale	37°47'	120°41'	66	25.4	7.1	15.9	827	356	279	2.40	15.0
Jacksonville	37°55'	120°21'	214	...	...	...	...	687	...	> 1	...
Sonora	37°59'	120°23'	558	25.4	6.8	15.3	802	828	348	0.97	15.3
Groveland	37°52'	120°13'	863	...	...	...	...	1074	...	< 1	...
Dudleys	37°45'	120°06'	916	20.6	3.7	11.5	669	957	329	0.70	13.6
Hetch Hetchy	37°57'	119°47'	1180	22.0	2.9	12.0	689	867	357	0.80	15.8
Yosemite National Park Headquarters	37°27'	119°35'	1214	21.7	1.8	11.4	675	885	341	0.76	16.6
Mather	37°53'	119°51'	1380	20.8	1.7	10.8	655	807	310	0.81	15.8
Lake Eleanor	37°58'	119°53'	1420	21.8	2.7	11.4	670	1068	343	0.63	15.8
South Entrance Yosemite National Park	37°30'	119°38'	1560	19.2	2.3	9.9	629	1140	343	0.55	13.6
Ellery Lake	37°56'	119°14'	2930	13.0	-5.0	2.2	397	760	236	0.52	14.4
Timberline*	37°55'	119°13'	3230	11.0	-7.0	+0.8	367	994	250	0.37	14.4
Dana Plateau*	37°55'	119°13'	3770	8.0	-10.0	-2.4	272	1534	226	0.18	14.4
Mono Lake	38°01'	119°09'	1985	20.1	-0.3	9.3	615	336	231	1.83	17.1

which is driest and hottest. Oakdale, toward the foothills, repeats the familiar phenomenon of more summer heat and more precipitation above the lowest valley stations, and thus an increase in ActE.

Again the PotE/ppt ratio is large and represents a dry semidesert, according to Budyko, in the valley, but it decreases rapidly with altitude, to cross the arid–humid boundary at about 380 m with PotE and precipitation equal to 780 mm yr^{-1}. The most humid stations are at the highest altitudes, but they are less humid than those on the Donner Summit transect, for example. Mono Lake is again arid, and the arid–humid boundary on the eastern slope may be at 2400 m, 400 m above the valley, with PotE and precipitation at 500 mm yr^{-1}.

Continentalities are low on this transect, without much relationship to altitude.

The next transect southward is at 37° to 37°15′N from Madera (82 m) up through the San Joaquin Experimental Range (448 m), Auberry (612 m), North Fork Ranger Station (913 m), Big Creek Powerhouse (1501 m), Huntington Lake (2140 m), and Florence Lake (2240 m). Here the arid–humid boundary on the western slope is around 750 m elevation with PotE and precipitation at 780 mm yr^{-1}, much higher than only a few km northward. Temperature lapse rates are −0.40°C/100 m in January, −0.48 annual mean, and −0.58 in July, more normal than most. Big Creek Powerhouse is hotter than the regression line by a good 2°C; Madera is somewhat cooler in summer. Potential evapotranspiration falls fairly regularly from 850 mm yr^{-1} in the valley to 630 at 2300 m elevation at a rate of −15 mm/100 m. Precipitation has a maximum somewhat below 1000 m, increasing from 250 mm yr^{-1} in the San Joaquin Valley to 850 mm at 800–900 m elevation at a rate of +102 mm/100 m. Then there is evidently a decrease to 600–700 mm at 2200 m at a rate of −11 mm/100 m. However, a 1921–29 record from Helm Creek (37°02′N, 118°56′W), only 37 km southeast of Huntington Lake at an elevation of 2445 m in the headwaters of the North Fork of the Kings River (cf. Table 2-13), shows precipitation of 919 mm yr^{-1}, so the putative high-altitude decrease may not be real. Actual evapotranspiration, then, has a maximum of 350 at about 1200 m, increasing from 250 in the valley at a rate of +9 mm/100 m and decreasing down to 280 mm yr^{-1} at 2200 m at a rate of −7 mm/100 m.

The southernmost Sierran transect for which data are available extends from Visalia and Lemon Cove in the lower San Joaquin Valley upward into Sequoia National Park to the *Sequoiadendron* groves at 2000 m. Precipitation increases rapidly in the foothills up to 300 m at a rate of 160 mm/100 m, to reach over 500 mm yr^{-1}. The rate of increase above 300 m is evidently +33 mm/100 m, yielding 1100 mm at 2000 m. Whether this increase continues to the crest of the Great Western Divide or further east to the Sierra Crest is unknown. Precipitation at Independence in the Owens Valley at 1200 m is only 133 mm yr^{-1}. Temperatures on this transect decrease with altitude at a rate of −0.6°C/100 m in summer and about −0.45° in winter. The result is a decrease in PotE from 900 mm yr^{-1} at valley elevations to 550 mm at 2000 m elevation, the rate of decrease being about −19 mm/100 m. Although ActE increases very rapidly in the foothills, perhaps at a rate of +42 mm/100 m to 300 m elevation, it increases not at all above there. In the *Sequoiadendron* groves ActE is 300 mm yr^{-1}, the same as in the chaparral and oak woodland at Ash Mt. The latter station, incidentally, is consistently very hot compared with adjacent stations.

The San Joaquin Valley stations are desert by their PotE/ppt ratios (3.62 decreasing to 2.57 with 48 m elevation change, accompanied by 110 mm more precipitation but also 30 mm higher PotE). Three Rivers Powerhouse at 290 m is down to 1.60, and Ash Mt. at 518 m to 1.40. The latter has precipitation up to 654 mm yr^{-1}. Above here the arid–humid boundary is crossed at about 1000 m with PotE and precipitation both 770 mm yr^{-1}. The ratio is 0.5 in the *Sequoiadendron* groves at Giant Forest and Grant Grove at 2000 m.

According to our data, the two southernmost Sierran transects show little or irregular changes from more northern areas. Temperatures are probably higher at 500 m, being 15 ½°C on the Madera–Florence Lake and 17° on the Visalia-Giant Forest transects. At 1000 m temperatures are about 13°; at 2000 m lower, 7½°. Precipitation is lower by 200 mm as compared to the Yosemite transect at 500 m, and 100 mm lower at 1000 m. Potential evapotranspiration increases to 820 and 850 mm yr^{-1} at 500 m, to 740 and 760 at 1000 m, but may be slightly lower at 2000 m. Actual evapotranspiration is lower throughout: at 500 m 280–300 mm yr^{-1}, at 1000 m 340 and 300 mm yr^{-1}, and at 2000 m 300 mm yr^{-1}.

The Kern Canyon might be expected to show the kinds of climatic change the Sacramento River does above Red Bluff. Precipitation does increase at the rate of +94 mm/100 m from Bakersfield to the first Powerhouse at 296 m, but this adds only 130 mm precipitation for a total of 280 mm yr^{-1}. Up-river to 825 m there is no further change in precipitation. This is a nice example of a phenomenon already well described by Braun-Blanquet (1961) in terms not only of climatic factors but also of the effects on vegetation. These are all desert climates, with PotE/ppt ratios for the three stations of 6.41, 3.80, and 3.08. However, the latter two do have runoff, so out of phase are need and supply of water. Up-river from the point of highest precipitation ActE actually decreases at the rate of −4 mm/100 m elevation. Over this stretch PotE decreases at −36 mm/100 m, so ActE is a conservative, integrative measure of climatic effects on the water balance. On the other hand, where no water surplus is generated, as at very dry Bakersfield, ActE increases with altitude faster than PotE, namely, +84 mm/100 m versus +68. Our three points show two trends for temperatures. The first is upward by 1½°C from the valley to 296 m elevation with values of 10° in January, 30° in July, and 20° as a mean annual value. Then temperatures drop at rates of −0.65°C/100 m for January and mean annual, with −0.55 for July. These trends predict 0° for January at 1800 m, 0° for the annual mean at 3300 m, and 10° for July at 3800 m, which may be too high for timberline at this latitude (35½°N), but in this continental climate 12° for July may be a better tree limit than 10°. The latter limit would be at 3500 m.

Cold desert. The western limit of the Great Basin is the Sierra Nevada–Cascade crest. Some changes in climate down this slope are discussed above. Above Bishop (37°22′N, 118°22″W, 1251 m) is a series of three stations up Bishop Creek. The North Fork leads to Piute Pass (Chabot and Billings 1972). Data from these stations allow some interpretation of the southern Sierran east slope climate. The stations up the creek are at 2560–2934 m. Bishop itself is at the northern edge of the Hot Desert, but its surrounding, immediately adjacent mountain slopes are Cold Desert. *Coleogyne ramosissima* vegetation forms a belt just above town on the Sherwin grade and on fans leading up to the Sierra Nevada, whereas the lower slopes of the White Mts. east of town have *Atriplex confertifolia* vegetation.

Chabot and Billings' climatic data extend to 3540 m. It would be useful to have a correlation of their daily summer temperatures with those for Bishop, which has a good long-term record. They give a lapse rate of −0.74°C/100 m for summer temperatures. Our four station plots give −0.60° for July, −0.50 for the annual mean, and −0.35 for January. Their Table 2 of mean weekly maxima for the summer of 1968 fits neatly a lapse rate of −0.60°C/100 m if their 1400 m bajada station is extraordinarily cold and their 3050 m North Lake ridge station is extraordinarily warm. Our annual precipitations increase from 160 mm at Bishop to 430 mm at 3000 m at the rate of +15 mm/100 m. In 1968 and 1970 snow courses along the North Fork of Bishop Creek had less than 400 mm of water on the ground as snow on April 1, but in 1967 and 1969 they had more. Their average at North Lake was slightly more than the total annual precipitation at South Lake, 7½ km to the southeast (record only 1924–30). The increase in the snow courses to 3290 m is moderate, +33 mm/100 m, but to Piute Lake at 3350 m just under Piute Pass to the west at 3485 m it is +660 mm/100 m, so the precipitation on the ground on April 1 at this higher altitude averages 946 mm. This is reasonable, using Ellery Lake as a predictor (Major 1967:119). Summer precipitation increased linearly at rates of 2 and 4 mm/100 m in 1967 and 1968. Totals at 1400 m were 3 and 24 mm, at 3300 m 46 and 90 mm, in these 2 yr. Potential evapotranspiration decreases from 724 mm yr^{-1} in the valley to 450 at 3000 m at the rate of −15 mm/100 m. Actual evapotranspiration increases from 157 mm in the desert valley to 260 mm at 2800 m. It may then decrease. Such a maximum is present but very slight in the Ellery Lake projected data (Major 1967). On the Ellery Lake transect the PotE/ppt ratio is 1 at about 2400 m with values of 500 mm yr^{-1}. On the Bishop Creek transect the valley is more like a desert, with 4.61 for this ratio versus 1.83 at higher Mono Lake. The ratio drops to 1.35 at 2560 m and to 1.17 at 2934 m. At the rate of decrease of the ratio at these upper stations the arid–humid boundary would be about 3200 m. The annual PotE and precipitation curves do cross at about 3100 m with PotE and precipitation at 450 mm yr^{-1}. This is a very high elevation for such a boundary, 700 m higher than near Ellery Lake or a rise of 950 m per degree of latitude southward (110 km).

The White Mts. seem to be the only mountain range within the Great Basin with climatic records. The range extends from the Owens Valley (Bishop at 1251 m) to White Mt. Peak (4342 m). Mooney (1973) gives some climatic data for June–August 1961 and 1962 from vegetation stands that are described floristically and quantitatively. They include *Atriplex confertifolia* vegetation at 1710 m, *Pinus monophylla–Artemisia tridentata* at 2620 m, *Pinus longaeva* (= *P. aristata*) at 3260 m, and alpine (*Trifolium monoense–Koeleria cristata*) at 3870 m. These summer records show temperature lapse rates of weekly maximum temperatures of −0.75°C/100 m up to piñon, −1.00 from there up to bristlecone pine, and −1.30 up to the alpine. Lapse rates for weekly minima are −0.70, −0.70, and −0.95°. Precipitation increased in the summer of 1961 at a rate of +9.5 mm/100 m to the subalpine, with none thereafter. In 1962 the corresponding rate was only 1.5 mm.

The splendid climatic record complied by the White Mt. Research Station (Pace et al. 1974) are for altitudes of 3100 and 3800 m. With Bishop and Deep Springs, in a valley on the east side of the range at 1591 m, these four stations show monthly summer lapse rates of −0.65°C/100 m, in winter −0.45 (decreasing from −0.45 in November to −0.40 in January and then increasing to −0.54 in February). Deep

Springs is warmer by about 2°C than this west-slope transect would indicate in summer and colder by about 1°C in winter. The annual lapse rate is a consistent −0.60°C/100 m. Precipitation increases either at an increased rate above Bishop or at a linear rate from Deep Springs up through the two mountain stations of only +17 mm/100 m. It reaches 496 mm yr^{-1} at 3800 m. From Bishop up through the two White Mt. stations PotE decreases at a linear rate of −18 mm/100 m from 724 to 273 mm yr^{-1} at 3800 m. Upward from the desert valleys ActE increases to a maximum of 220 mm yr^{-1} in the subalpine belt at 3100 m and then decreases into the alpine to <200 mm yr^{-1} at a rate of −4. The PotE/ppt ratio is still 1.15 at 3100 m, although 0.55 at 3800 m. Evidently the arid–humid boundary is at the extreme elevation of 3160 m with PotE and precipitation equaling 380 mm yr^{-1}. The eastern slope of the Sierra Nevada above Bishop has very similar conditions. The change from an arid (steppe) climate to a quite humid one takes place in the White Mts. over a very narrow range of altitude.

Hot desert. Within the Hot Desert the difference between the Mojave and Colorado deserts is shown by Twentynine Palms and Indio, north and south of Joshua Tree National Monument and separated by only 50 km distance but 604 m elevation. January, July, and mean annual temperatures are 8.9 and 12.3°C, 31.3 and 33.4°, 19.7 and 22.8° for lapse rates of −0.56, −0.35, and −0.51°C/100 m. Potential evapotranspirations are 1078 and 1280 mm yr^{-1}, and the lapse rate is −33 mm/100 m; ActEs are the same. The PotE/ppt ratios are 10.3 and 14.9. Soil storage of moisture is 4 mm and 0. Actual evapotranspiration is >9 mm mo^{-1} for 7 and 4 months.

In Saline Valley, separated by high desert ranges from Owens Valley on the west and Death Valley on the east, Randall (1972) derived a lapse rate of minimum temperatures of −0.90°C/100 m between 350 and 1860 m. Hunt (1966) gave a figure of −0.91 for Death Valley near Furnace Creek. The temperature lapse rates Randall used were −0.57°C/100 m for January, −0.97 for July, −0.79 yearly. His calculated PotE figures show lapse rates of −44 and −39 mm/100 m for the intervals from 610 to 1220 m and from 1220 to 1830 m, respectively. Hunt's limited precipitation data show a logarithmic increase of 6.3%/100 m according to Randall (1972:15). This averages out at +6.4 and +9.7 mm/100 m for the two elevation intervals mentioned above. Being equal to precipitation in these truly desert climates, ActE decreases with altitude the same as precipitation. The PotE/ppt ratio changes from 14.20 at the lowest elevation, through 7.26, to 3.50 at 1830 m. Does this kind of rate change mean that a ratio of 1, the arid–humid boundary, would be reached just below 3000 m?

On the western side of the Colorado Desert Mt. Palomar (1691 m) overlooks Borrego Springs (190 m) in the Hot Desert. Latitude is about 33°20′ N. Warner Springs (970 m) between them shows how potentially false it is to use two points only for a lapse rate, and that three points may simply complicate the picture. Temperature lapse rates are much higher immediately above the desert than on the slopes below the summit, namely, for January −0.60°C/100 m versus −0.14, for July −1.10° versus −0.07, annual mean −0.90° versus −0.11. The −0.90°C/100 m showed up at Death and Saline valleys also. Precipitation shows less contrast, +41 mm/100 m at low, +34 at high elevations, with levels of 89, 412, and 605 mm yr^{-1}. Values of PotE and ActE are very high and very low, respectively, in the desert, and they change rapidly above. Potential evapotranspiration drops from 1235 mm yr^{-1} at Borrego to

TABLE 2-15. Two relatively high-altitude Mojave Desert stations compared with nearby stations, Needles in the Colorado Desert and Las Vegas in the Mojave

Station	Latitude (°N)	Longitude (°W)	Elev. (m)	Mean Temp. (°C) Year	July	Jan.	PotE (mm)	Ppt (mm)	ActE (mm)	K	PotE/ Ppt	Maximum Soil Storage (mm)	Maximum Per Month ActE (mm)	month
Mountain Pass	35°28'	115°32'	1440	14.5	27.3	4.4	801	189	189	20.0	4.24	31	35	Aug.
Las Vegas	36°05'	115°10'	661	18.7	32.1	6.2	1058	99	99	22.5	10.69	13	13	July
Mitchell Caverns	34°56'	115°32'	1319	17.2	28.7	7.8	917	179	179	17.6	5.12	23	23	Aug., Nov.
Needles	34°46'	114°37'	278	22.5	35.1	10.7	1263	112	112	21.2	11.28	8	22	Aug.

731 at Warner Springs and to 710 at the top of Mt. Palomar for rates of −65 and −3 mm yr^{-1}. Actual evapotranspiration changes from 89 to 315 to 296, rates of +29 and −3 mm/100 m. Change in the PotE/ppt ratio agrees with the other climatic parameters: rapid change just above the desert, slower on the mountain. Values are 13.87 at Borrego, 1.77 at Warner Springs, 1.18 on Mt. Palomar. The arid–humid boundary must be at perhaps 2500 m if PotE resumes its "normal" lapse rate of −16 mm/100 m. This is very much higher than this boundary on the western slope of the southern Sierra Nevada at latitude 36½°N of 1000 m.

Campo is an elevated (796 m) station also bordering the Colorado Desert on its western side. Its climatic parameters are very similar to those of Warner Springs about 1° latitude farther north. It has the distinction of a "Sonoran" storm which dumped 409 mm of precipitation, 292 mm in 80 min. Differences between the desert and the adjacent mountains may be even more dramatic than our averaged climatic records indicate.

On the eastern side of the Mojave Desert are two relatively high-altitude stations, Mitchell Caverns State Park in the Providence Mts. and Mountain Pass at the south end of the Clark Mts. (peak at 2410 m). Climatic data for these stations and two nearby ones at lower elevations, Las Vegas 76 km NNE of the latter and Needles 88 km ESE of the former, are assembled in Table 2-15. The latter stations are representative of the Mojave and Colorado deserts, respectively. Temperature lapse rates are −0.54°C/100 m for the annual mean up to Mountain Pass from Las Vegas, −0.61 for July, but only −0.23 for January. Corresponding rates from Needles up to Mitchell Caverns are −0.51, −0.61, and −0.28°. The correspondence is striking. The PotE lapse rates are −32 and −33 mm/100 m, double the "normal." Precipitation increases with altitude do differ, +12 and +6 mm/100 m. All the stations are distinguished by a summer peak of precipitation at least as large as the winter peak (Fig. 2-7). In these truly Hot Deserts ActEs are absolutely limited by precipitation. Soil storage of water increases with altitude, and so does maximum ActE in any 1 mo. The PotE/ppt ratios indicate extreme deserts, less extreme at the higher altitudes. By projecting the PotE and precipitation trends we conclude that the arid–humid boundary is at 2800 m with 360 mm yr^{-1} PotE and precipitation for the Las Vegas–Mountain Pass transect and at 3100 m with 310 mm yr^{-1} for the Needles–Mitchell Caverns transect. This is interesting if true. It does say that neither the Clark Mts. nor the Providence Mts. have a zonal belt of humid climates. Plants such as *Abies concolor*, therefore, must exist in these mountains in azonal habitats. Climatically these would probably be relicts of more humid climates of the past unless they occupy sites so odd edaphically that regional, zonal vegetation cannot occupy them (Billings 1950; Gankin and Major 1964).

LITERATURE CITED

Al-Ani, H. A., B. R. Strain, and H. A. Mooney. 1972. The physiological ecology of diverse populations of the desert shrub Simmondsia chinesis. J. Ecol. 60(1):41-57.

Arkley, R. J., and R. Ulrich. 1962. The use of calculated actual and potential evapotranspiration for estimating potential plant growth. Hilgardia 32(1):443-469.

Axelrod, D. I. 1965. A method for determining the altitude of Tertiary floras. Paleobotanist 14:144-171.

Azevedo, J., and D. L. Morgan. 1974. Fog precipitation in coastal California forests. Ecology 55(5):1135-1141.

Bailey, H. P. 1966. The climate of southern California. Calif. Nat. Hist. Guide 17. Univ. Calif. Press, Berkeley. 87 p.

Baker, F. S. 1944. Mountain climates of the western United States. Ecol. Monogr. 14:223-254.

Bamberg, S. A., and J. Major. 1968. Ecology of the vegetation and soils associated with calcareous parent materials in three alpine regions of Montana. Ecol. Monogr. 38:127-167.

Beeson, C. D. 1974. The distribution and synecology of Great Basin pinyon-juniper. M.S. thesis in range management, Univ. Nevada, Reno. 95 p. + map.

Beetle, A. A. 1960. A study of sagebrush, the section Tridentatae of Artemisia. Univ. Wyoming Agric. Exp. Sta. Bull. 368. 83 p.

Billings, W. D. 1945. Plant associations of the Carson Desert region, western Nevada. Butler Univ. Bot. Stud. 7:89-123.

---. 1949. The shadscale vegetation zone of Nevada and eastern California in relation to climate and soils. Amer. Midl. Natur. 42(1):87-109.

---. 1950. Vegetation and plant growth as affected by chemically altered rocks in the western Great Basin. Ecology 31:62-74.

---. 1952. The environmental complex in relation to plant growth and distribution. Quart. Rev. Biol. 27:251-265.

---. 1954. Temperature inversions in the pinyon-juniper zone of a Nevada mountain range. Butler Univ. Bot. Stud. 11:112-118.

Birkeland, P. W. 1974. Pedology, weathering, and geomorphological research. Oxford Univ. Press, New York. 285 p.

Black, P. E. 1966. Thornthwaite's mean annual water balance. Silviculture Gen. Utility Library Program GU-101:1-20. State Univ. Coll. Forestry, Syracuse, N. Y.

Blüthgen, J. 1966. Allgemeine Klimageographie. 2nd ed. Walter de Gruyter, Berlin. 720 p.

Borisov, A. A. 1967. [Climates of the USSR.] 3rd ed. Prosveshchenie, Moscow. 296 p.

Braun-Blanquet, J. 1961. Die inneralpine Trockenvegetation. Gustav Fischer, Stuttgart. 273 p.

Brockmann-Jerosch, H. 1925-29. Die vegetation der Schweiz, Part I. Beitr. geobot. Landesaufn. Schweiz 12. 499 p.

Budyko, M. I. 1958. [The heat balance of the earth's surface.] Translation by N. A. Stepanova of Teplovoi balans zemnoi poverkhnosti, Gidromet. Izd., Leningrad, 1956, 255 p. U.S. Dept. Commerce, Washington, D. C. 259 p.

Budyko, M. I. (ed.). 1963. [Atlas of the heat balance.] Gidromet. Izd, Leningrad.
Budyko, M. I. 1974. Climate and life. Academic Press, New York. 508 p.
Budyko, M. I., and N. A. Efimova. 1968. [Utilization of solar energy by the natural vegetation cover on the territory of the USSR.] Bot. Zhur. 53(10):1384-1389.
Byers, H. R. 1930. Summer sea fogs of the central California coast. Univ. Calif. Pub. Geogr. 3:291-338.
---. 1953. Coast redwoods and fog drip. Ecology 34:192-193.
Campbell, R. E. 1972. Prediction of air temperature at a remote site from official weather station records. USDA Forest Serv., Rocky Mt. Forest and Range Exp. Sta., Res. Note RM-223. 4 p.
Carter, D. G., and J. R. Mather. 1966. Climatic classification for environmental biology. C. W. Thornthwaite Assoc., Lab. of Climatol., Pub. in Climatol. 19:305-395.
Chabot, B. F., and W. D. Billings. 1972. Origins and ecology of the Sierran alpine flora and vegetation. Ecol. Monogr. 42:163-199.
Clausen, J., D. D. Keck, and W. M. Hiesey. 1940. Experimental studies on the nature of species. I. Effect of varied environments on western North American plants. Carnegie Inst. Wash. Pub. 520. 452 p.
Conrad, V. 1946. Usual formulas of continentality and their limits of validity. Amer. Geophys. Union Trans. 27:663-664.
Contagne, M. A. 1935. Comment definir et caracteriser le degree d'aridite d'une region et la variation saisonniere. Meteorologie 9:141-151.
Crocker, R. L. 1952. Soil genesis and the pedogenic factors. Quart. Rev. Biol. 27:139-168.
Cronquist, A., A. H. Holmgren, N. H. Holmgren, and J. L. Reveal. 1972. Intermountain flora. Vol. 1. Hafner, New York. p. 270
Dale, R. F. 1959. Climates of the states: California. Climatography of the U. S. 60-4. U.S. Weather Bur., Washington, D. C. 37 p.
Ellenberg, H. 1963. Vegetation Mitteleuropas mit den Alpen, in kausaler, dynamischer und historischer Sicht. Ulmer, Stuttgart. 945 p.
Fiziko-Geograficheskii Atlas Mira. 1964. Akad. Nauk USSR, Moscow. 298 p.
Gale, J. 1972a. Availability of carbon dioxide for photosynthesis at high altitudes: theoretical considerations. Ecology 53(3):494-497.
---. 1972b. Elevation and transpiration: some theoretical considerations with special reference to mediterranean-type climate. J. Appl. Ecol. 9(3):691-702.
Gankin, R., and J. Major. 1964. Arctostaphylos myrtifolia, its biology and relationship to the problem of endemism. Ecology 45(4):792-808.
Gaussen, H. 1954. Theories et classification des climats et microclimats. Congr. Internat. Bot., 8e, Paris, Sects. 7 and 3:125-130.

Geiger, R. 1961. Das Klima der bodennähen Luftschicht. 4th ed. Die Wissenschaft 78, Vieweg, Braunschweig. 646 p. English Translation: Harvard Univ. Press, Cambridge, Mass., 1965 611 p.

Griffin, J. R., and W. B. Critchfield. 1972. The distribution of forest trees in California. USDA Forest Serv., Pac. SW Forest and Range Exp. Sta., Res. Paper PSW-82. 114 p.

Hamilton, E. L. 1954. Rainfall sampling in rugged terrain. USDA Tech. Bull. 1096. 41 p.

Hare, F. K., and J. E. Hay. 1971. Anomalies in the large-scale annual water balance over northern North America. Canad. Geogr. 15:79-94.

---. 1974. The climate of Canada and Alaska, pp. 49-192. In: R. A. Bryson and F. K. Hare (eds.). World survey of climatology. Vol. 11: Climates of North America. Elsevier, Amsterdam.

Hegg, O. 1965. Untersuchungen zur Pflanzensoziologie und Oekologie im Naturschutzgebiet Hohgant. Beitr. geobot. Landesaufn. Schweiz 46. [Cf. Rev. Ecol. 48(1):172-173, 1967.]

Henry, A. J. 1919. Increase of precipitation with altitude. U.S. Monthly Weather Rev. 47:33-41.

Hintikka, V. 1963. Ueber das Grossklima einiger Pflanzenareale in zwei Klimakoordinatensystemen dargestellt. Ann. Bot. Soc. Zool. Bot. Fenn. Vanamo 34(5):1-64.

Hodges, J. D. 1967. Patterns of photosynthesis under natural environmental conditions. Ecology 48:234-242.

Hübl, E., and H. Niklfeld. 1973. Ueber die regionale Differenzierung von Flora und Vegetation in den Oesterreichischen Alpen. Acta Bot. Hung. 19:147-164.

Hunt, C. B. 1966. Plant ecology of Death Valley, California. U.S. Geol. Survey Prof. Paper 509. 68 p.

Jenny, H. 1941. Factors of soil formation. McGraw-Hill, New York. 281 p.

---. 1946. Arrangement of soil series and types according to functions of soil forming factors. Soil. Sci. 61:375-391.

Kesseli, J. E. 1942. The climates of California according to the Köppen classification. Geogr. Rev. 32:476-480.

Khromov, S. P. 1957. [On the question of continentality of climate.] Izv. Vses. Geogr. Obshch. 89:221-225.

Lamb, H. H. 1972. Climate: present, past and future. Vol. 1. Methuen, London. 613 p.

Larcher, W. 1973. Oekologie der Pflanzen. Universitäts-Taschenbücher 232, E. Ulmer, Stuttgart. 320 p.

Lautensach, H., and R. Bögel. 1956. Der Jahresgang des mittleren geographischen Höhengradienten der Lufttemperatur in den verschiedenen Klimagebieten der Erde. Erdkunde 10:270-282.

Leopold, L. (ed.). 1953. The delta Colorado, pp. 10-30. In Round river, from the journals of Aldo Leopold. Oxford Univ. Press, New York.

McNaughton, S. J. 1969. Climates of organisms. Review of R. H. Shaw (ed.) Ground level climatology, Amer. Assoc. Adv. Sci., 1967, 395 p. Ecology 50(3):526-528.

Major, J. 1951. A functional, factorial approach to plant ecology. Ecology 32:392-412.

---. 1953. The relationship between factors of soil formation and vegetation, with an analysis from the west slope of Mt. Hamilton, California. Ph.D. thesis, Univ. of Calif., Berkeley. 283 p.

---. 1963. A climatic index to vascular plant activity. Ecology 44(3):485-498.

---. 1967. Potential evapotranspiration and plant distribution in Western states with emphasis on California, pp. 93-126. In R. H. Shaw (ed.). Ground level climatology. Amer. Assoc. Adv. Sci., Washington, D. C. 395 p.

Major, J., and S. A. Bamberg. 1967. Comparison of some North American and Eurasian alpine ecosystems, pp. 89-118. In H. E. Wright and W. H. Osborn (eds.). Arctic and alpine environments. Indiana Univ. Press, Bloomington. 308 p.

Major, J., and W. T. Pyott. 1966. Buried, viable seeds in two California bunchgrass sites and their bearing on the definition of a flora. Vegetatio 13(5):253-282.

Mather, J. R. 1974. Climatology: fundamentals and applications. McGraw-Hill, New York. 412 p.

Monteith, J. L. 1973. Principles of environmental physics. Edw. Arnold, London. 241 p.

Monteny, B. 1969. Influence de l'energie advective sur l'evapotranspiration. Oecologia Plant. 4:295-305.

Mooney, H. A. 1973. Plant communities and vegetation, pp. 7-17. In R. M. Lloyd and R. S. Mitchell. A flora of the White Mountains, California and Nevada. Univ. Calif. Press, Berkeley. 208 p.

Myrup, L. O. 1969. A numerical model of the urban heat island. J. Appl. Meteor. 8:908-918. [On p. 912 the additional heat flux from combustion of fossil fuels should be 0.13 cal cm^{-2} min^{-1}.]

Nash, A. J. 1963. A method for evaluating the effects of topography on the soil water balance. Forest Sci. 9:413-422.

Oberlander, G. T. 1956. Summer fog precipitation on the San Francisco Peninsula. Ecology 37:851-852.

Pace, N., D. W. Kiepert, and E. M. Nissen. 1974. Climatological data summary for the Crooked Cr. Laboratory, 1949-1973, and the Barcroft Laboratory, 1953-1973. White Mt. Res. Sta., Univ. Calif., Berkeley.

Parsons, J. J. 1960. Fog drip from coastal stratus, with special reference to California. Weather (London) 15:58-62.

Patric, J. H., and P. E. Black. 1968. Potential evapotranspiration and climate in Alaska by Thornthwaite's classification. U.S. Forest Serv., Pacific NW Forest and Range Exp. Sta. Res. Paper PNW-71:1-28.

Patton, C. P. 1956. Climatology of summer fogs in the San Francisco Bay area. Univ. Calif. Pub. Geogr. 10:113-200.

Pearcy, R. W., and A. T. Harrison. Comparative photosynthetic and respiratory gas exchange characteristics of _Atriplex lentiformis_ (Torr.) Wats. in coastal and desert habitats. Ecology 55:1104-1111.

Penck, A. 1910. Versuch einer Klimaklassifikation auf physiogeographischer Grundlage. Sitzungsber. Preus. Akad. Wiss. Phys. Math. Kl. 12. 236 p.

Peterson, K. M., and W. W. Payne. 1973. The genus Hymenoclea (Compositae: Ambrosieae). Brittonia 25:243-256.

---. 1974. On the correct name for the appressed-winged variety of Hymenoclea salsola (Compositae: Ambrosieae). Brittonia 26:397.

Randall, D. C. 1972. An analysis of some desert shrub vegetation of Saline Valley, California. Ph.D. thesis, Univ. Calif., Davis. 186 p.

Rodin, L. E., and N. I. Bazilevich. 1967. Production and mineral cycling in terrestrial vegetation. Transl. (ed.) G. E. Fogg of [Dynamics of the organic matter and biological cycling of ash elements and nitrogen in the main types of the world's vegetation], Nauka, Moscow, 1965, 253 p. Oliver & Boyd, Edinburgh. 288 p.

Rosenzweig, M. L. 1968. Net primary productivity of terrestrial communities: prediction from climatic data. Amer. Natur. 102(923):67-74.

Rouse, W. R. 1970. Relations between radiant energy supply and evapotranspiration from sloping terrain: an example. Canad. Geogr. 14(1):27-37.

Russell, M. B., and L. W. Hurlbut. 1959. The agricultural water supply. Adv. Agron. 11:6-19.

Russell, R. J. 1926. Climates of California. Univ. Calif. Pub. Geogr. 2(4):73-84.

---. 1946. Climatic transitions and contrasts, pp. 355-379. In R. Peattic (ed.). The Pacific Coast Ranges. Vanguard Press, New York. 402 p.

Sellers, W. D. 1965. Physical climatology. Univ. Chicago Press, Chicago. 272 p.

Shreve, F. 1942. The desert vegetation of North America. Bot. Rev. 8:195-246.

Strother, J. L. 1974. Taxonomy of Tetradymia (Compositae: Senecioneae). Brittonia 26:177-202.

Thornthwaite, C. W. 1948. An approach toward a rational classification of climate. Geogr. Rev. 38:55-94.

Thornthwaite, C. W., and J. R. Mather. 1957. Instructions and tables for computing potential evapotranspiration and the water balance. Drexel Inst. Technol. Pub. Climatol. 10:183-311.

Tikhomirov, B. A. 1963. [Outline of the biology of arctic plants.] Izd. Akad. Nauk USSR, Moscow. 154 p.

Trewartha, G. T. 1961. The earth's problem climates. Univ. Wisc. Press, Madison. 334 p.

U. S. Weather Bureau. Climatic summary of the U.S. through 1930. Sections 15-18. Washington, D. C.

---. Climatic summary of the U.S. --Supplement for 1931 through 1952: California. Washington, D. C.

---. 1964. Climatic summary of the U.S. --Supplement for 1951 through 1960. Washington, D. C.

---. 1968. Climatic atlas of the U.S. Washington, D. C.
Uttinger, H. 1951. Zur Höhenabhängigkeit der Niederschlagsmenge in den Alpen. Arch. Met. Geophys. Biokl., Ser. B, 2:360-382.
Varney, B. M. 1920. Monthly variations of the precipitation-altitude relation in the central Sierra Nevada of California. U. S. Monthly Weather Rev. 48:648-650.
Vogl, R. J. 1973. Ecology of knobcone pine in the Santa Ana Mountains, California. Ecol. Monogr. 43:125-143.
Vogl, R. J., and L. T. McHargue. 1966. Vegetation of California fan palm oases on the San Andreas Fault. Ecology 47:532-540.
Walter, H. 1960. Grundlagen der Pflanzenverbreitung, Part I; Standortslehre, Ulmer, Stuttgart. 566 p.
---. 1967. Das Pampaproblem in vergleichend ökologischer Betrachtung und seine Lösung. Erdkunde 21:181-203.
---. 1968. Die vegetation der Erde. Vol. 2: Die gemässigten und arktischen Zonen. Gustav Fischer-Verlag, Stuttgart. 1001 p.
---. 1970. Ergänzende Betrachtungen zur der im Klimadiagramm-Weltatlas verwendeten Klimadarstellung. Erdkunde 24:145-149.
---. 1973. Vegetation of the earth. Springer, New York. 240 p.
Walter, H., E. Harnickell, and D. Mueller-Dombois. 1975. Climate-diagram maps. Springer, New York. 36 p. + 9 maps.
Walter, H., and H. Lieth. 1960-67. Klimadiagramm Weltatlas. Fischer, Jena.
West, N. E. (ed.). 1972. Galleta: taxonomy, ecology, and management of _Hilaria jamesii_ on western rangelands. Utah Agric. Exp. Sta. Bull. 487. Logan, Utah. 38 p.
Wilson, R. C. 1972. _Abronia_. I. Distribution, ecology and habit of 9 species of _Abronia_ found in California. Aliso 7:421-437.

CHAPTER 3

RESEARCH NATURAL AREAS AND RELATED PROGRAMS IN CALIFORNIA

NORDEN H. (DAN) CHEATHAM
Field Representative, University of California Natural Land and Water Reserves System, Berkeley

LESLIE HOOD
Executive Director, California Natural Areas Coordinating Council, Box 4000 J, Berkeley

W. JAMES BARRY
State Park Plant Ecologist, California Department of Parks and Recreation, Sacramento

Introduction 76
- Definition 76
- Justification for Research Natural Areas 77
- Criteria 77

History 78
Federal research natural area programs 79
- U.S. Forest Service 79
- Bureau of Land Management 80
- National Park Service 81
- Fish and Wildlife Service 81
- Federal Committee on Ecological Reserves 81

State research natural area programs 85
- State Park System 85
- State Department of Fish and Game 86

Educational institutions 90
- University of California Natural Land and Water Reserves System 90
- Other educational institutions 95

Private programs 95
- The Nature Conservancy 95
- Audubon Society Wildlife Sanctuaries 96

Programs related to research natural area programs 96
- Introduction 96
- U.S. Forest Service Special Areas 96
- Other federal programs 101
- State Park System 101
- State Wildlife Areas 101

Programs providing special recognition 101
- Registry of National Natural Landmarks 102
- Programs of professional societies 102

Conclusion **103**
  California Natural Areas Coordinating Council **103**
  Diversity of programs **105**
  Role of the academic and scientific communities **105**
  The dilemma of publicity **106**
Literature cited **107**

INTRODUCTION

This reference book is written for those interested in studying California's diverse vegetation, and it represents an extensive compilation and interpretation of accumulated knowledge indicating what is known and what needs to be studied. Since educators and scientists need sites on which to conduct their field studies, this chapter will discuss various programs designed to accommodate their needs and those of their students, as well as their successors to come.

The emphasis of this chapter will be on areas whose management objectives and supporting administrative authorities are directed primarily at protecting samples of natural ecosystems for scientific study and observation. The discussion will also include related programs that place less emphasis on scientific use but are suitable for many scientific purposes.

Definition

In reviewing various existing research-oriented land management programs, some or all of the following elements are commonly included:

- The areas are part of a network of reserves representing samples typifying the range of natural habitat types and their important variations found within the jurisdiction of the administrative agency. Frequently a unique or an unusual habitat type is represented.
- The areas are officially recognized by the administrative agency as areas managed primarily for scientific study directed toward investigation of the natural ecological processes. Their pristine-like nature is a strong recreational attraction, but they are *not* recreation sites and recreational uses are not permitted by policy or law.
- As far as possible, ordinary physical and biological processes are allowed to operate without human intervention, except that in certain areas deliberate manipulation by approved and proven techniques may be allowed in order to maintain the unique features for which the nature reserve was created or in order to maintain a given transitional stage of natural succession that is a prominent feature of the reserve. However, these are *not* places to conduct extensive manipulative research.
- The areas are suitable sites to conduct long-term studies with the assurance that the plot markers, instrumentation, and so forth, as well as the conditions being studied, will remain undisturbed.

We have chosen to use a definition appearing in *Terminology of Forest Science, Technology, Practice,* and *Products,* edited by Ford-Robertson (1971) and authorized by the Joint FAO/IUFRO Committee on Forestry Bibliography and Terminology, a group whose task was to bring order, on an international basis, to the chaos of diverse terminology in common use. This reference defines a nature reserve (its recognized synonym is "research natural area") as follows: "An area established, by public or private agency, specifically to preserve a representative sample of an ecological community, primarily for scientific and educational purposes."

In this chapter, both terms, "nature reserve" and "research natural area," are used interchangeably.

Justification for Research Natural Areas

There exists a fairly large body of literature on research natural areas, their philosophy, and the justification for establishing them (e.g., Franklin and Trappe 1968; Franklin et al. 1971; Moir 1972; Society for Range Management 1975; The Nature Conservancy 1975). Here again, there is an assortment of special wording to choose from. The common elements are summarized below.

- Research natural areas serve as baselines and controls for measuring the effects of management practices on their counterpart ecosystems subjected to man's direct influence.
- Land use patterns are changing so rapidly that many natural ecosystems are vanishing, or are being severely altered, before they can be described or investigated.
- The gene pools and genetic diversity of the world's biota must be preserved for global ecological and economic stability.
- Scientists and land managers of the future must be trained to understand nature's complexities. Nature reserves contribute to this need.

It must be stressed again that these are not parks or any other forms of recreational area, but places dedicated to preservation and scientific investigation of natural ecosystems. Franklin et al. (1972) compared research natural areas to an "information storage system, or library, of enormous complexity and incomparable significance. The information stored provides research and educational opportunities for the study of natural ecosystems or any of their parts. Like a library, research natural areas are intended for use, but the user must not consume or damage the materials."

Criteria

It is desirable to develop and maintain for educational purposes and scientific study a coordinated network of nature reserves representing California's diversity of natural environment. A site with many habitat types will make a greater contribution than one with only a single habitat type, but there may be occasions when a feature of special interest will override the usually important requirement for habitat diversity. Ecosystems totally free of modern man's influence are no longer to be found. In reality, units within a network of nature reserves may vary from essentially undisturbed ecosystems to ecosystems heavily influenced by modern man. With care and good judgment the emphasis will be on undisturbed ecosystems with selected samples of ecosystems of significant merit elsewhere along the spectrum. This does not preclude a parallel network of reserves dedicated to the manipulation of natural ecosystems or to the study of heavily altered ecosystems.

Ecosystem viability is a prime requisite in establishing a nature reserve. The natural relationships should be essentially intact (i.e., an ecosystem operating as much as possible free of man's influences), and the reserves should be of sufficient size so that the natural balance of the community can be maintained with the survival of all biota assured. Boundary configuration is an important contributor to viability. The boundaries must be chosen so as to encompass the critical landscape features necessary to maintain the ecosystem. A complete watershed is ideal. Whenever possible, a reserve should be buffered from the detrimental impact of adjacent land uses. In some instances a nature reserve will be a remnant ecosystem not meeting the test of viability but having value for study during whatever time is left before the nature reserve values are lost.

It is easy to become enamored with the unusual and overlook the common. Therefore it is important to guard against unbalancing the series of reserves in favor of unusual values, and care should be taken to include typical samples of widely distributed ecological communities. Samples at the limits of their range are important too. In some cases unusual features will be deliberately acquired because they are judged to have special value complementing the existing reserves.

HISTORY

California has played a pioneer role in the efforts to establish research natural areas. On March 25, 1931, Regional Forester S. B. Show appointed a committee to advise him on the establishment of natural areas and experimental forests. The chairman was Assistant Regional Forester L. A. Barrett, and included among the members were Mr. A. E. Wieslander, known for his pioneering work in mapping California's vegetation, and Clarence Dunston, then supervisor of the Lassen National Forest. In April the committee responded with a memorandum of 15 principles to be followed in selecting research natural areas and experimental forests in the California Region. Some of their principles are reflected in the opening paragraphs of this chapter.

While the committee was busy gathering details relative to its task, Dr. Harvey Monroe Hall of the Carnegie Institution of Washington wrote as follows from his office at Stanford University on October 29, 1931, to Major R. Y. Stuart, the Chief of the Forest Service, in Washington, D.C.:

> I desire to place before you a proposal for a scientific reserve of such a nature that it does not appear to fall within any of the classes which the Forest Service has thus far created. In view of this difficulty in classification, and since it is probable that similar reserves will be called for in various national forests as time goes on, I am writing direct to you instead of first taking the matter up with officials of the district in which we are especially interested.

Dr. Hall had in mind an area on the east side of the Sierra Nevada near the town of Lee Vining that was later to become famous as the "Timberline" site for the classic and well-known studies in plant ecology by Clausen et al. (1940). His views on the subject were shaped not only by his research interests but also by his extensive visits to research natural areas in Europe (Hall 1929) and possibly by previous writers (Ashe 1922; Pearson 1922). He was also undoubtedly influenced by the work of the Ecological Society of America and its committee on the preservation of natural conditions. Starting its work in 1917, the committee culminated its efforts in 1926 with the publishing of the *Naturalist's guide to the Americas* (Shelford 1926).

What Dr. Hall did not realize was that his sister institute, the Carnegie Desert Laboratory, in conjunction with the University of Arizona and nature lovers, sportsmen, and public-spirited citizens in Arizona, had formed the Tucson Natural History Society and through the influence of this society had persuaded the Forest Service in 1927 to establish the first federally recognized research natural area in the United States, the Santa Catalina Research Natural Area in the Coronado National Forest. With this precedent set, the Forest Service soon adopted and incorporated Regulation L-20 into the *National forest manual,* as it was then called, thus setting the scene for additional proposals like Dr. Hall's.

While Dr. Hall was documenting his case, so was Barrett's committee, and on June 23, 1932, the Chief of the Forest Service established the Indiana Summit Research Natural Area, a 1936 acre (784 ha) area of Jeffrey pine on the pumice flats of Mono Co. This was followed on January 6, 1933, by establishment of the Harvey Monroe Hall Research Natural Area (Parker 1972), named in honor of Dr. Hall, who had unfortunately died before he could see the results of his efforts. This in turn was soon followed on February 28, 1933, by establishment of the Devils Garden Research Natural Area, an 800 acre (324 ha) natural reserve of Great Basin sagebrush and western juniper located in Modoc Co. These accomplishments were paralleled by similar Forest Service activity in other regions of the United States, thus firmly establishing the concept of officially recognized research natural areas.

FEDERAL RESEARCH NATURAL AREA PROGRAMS

U.S. Forest Service

A direct descendant of Regulation L-20 is Section 251.23 of Title 36 of the Code of Federal Regulations, which specifically states:

> . . . the Chief shall establish a series of research natural areas, sufficient in number and size to illustrate adequately, or typify for research or educational purposes, the important forest and range types in each forest region, as well as other plant communities that have special or unique characteristics of scientific interest and importance. Research natural areas will be retained in a virgin or unmodified condition except where measures are required to maintain a plant community which the area is intended to represent.

The Chief of the Forest Service elaborated on the above mandate in the research section of the *Forest service manual* (FSM 4063) rather than in the recreation management (FSM 2300) or timber management (FSM 2400) sections. This point is important since it clearly establishes the research nature of these areas and their relevance as nature reserves for use by "scientists within and without the Forest Service, and use for certain educational purposes" (FSM 4063).

In addition to defining the policies for use and protection of research natural areas, FSM 4063 defines the process for establishing them. Ultimately, a Forest

Service research natural area is established by signature of the Chief, by virtue of the authority vested in him by the Secretary of Agriculture under Regulation 36 CFR 251.23, cited above. Establishment proposals are forwarded to the Chief jointly by the appropriate Regional Forester and Director of the appropriate forest experiment station. In California, the Regional Forester and the Director of the Pacific Southwest Forest and Range Experiment Station have appointed a joint committee to advise them on the establishment and management of research natural areas.

To date in California, the U.S. Forest Service has established eleven research natural areas (Table 3-1, Part I) totaling approximately 16,170 acres (6550 ha). The first of these was Indiana Summit, established in 1932, and the most recent was Fern Canyon, established in 1972.

Permission to use research natural areas within the National Forests of California may be obtained by writing to the Director, Pacific Southwest Forest and Range Experiment Station, P.O. Box 245, Berkeley, California 94701.

Bureau of Land Management

The relevant authority is Subpart 6225 of Title 43 of the Code of Federal Regulations. In addition to definitions and related matters, it states the following policy:

> Where appropriate the Bureau shall establish and record areas of sufficient number and size to provide adequately for scientific study, research, recreational use and demonstration purposes. These will include:
> (a) The preservation of scenic values, natural wonders and examples of significant natural ecosystems.
> (b) Research and educational areas for scientists to study the ecology, successional trends, and other aspects of the natural environment.
> (c) Preserves for rare and endangered species of plants and animals.

Section 6225.0-5(a) of the Code of Federal Regulations specifically defines Bureau of Land Management research natural areas as areas "established and maintained for the primary purpose of research and education."

Because of the peculiarities of laws governing the public domain, not all land under the Bureau's jurisdiction is eligible for research natural area classification even if natural reserve values are proved to exist. Candidate areas must be on land formally withdrawn from public domain (43 CFR 2300) or classified for retention in federal ownership under the Classification and Multiple Use Act of 1964.

Final establishment of BLM research natural areas is accomplished by administrative action (43 CFR 2072) published in the *Federal register*.

To date 61 areas in California have been inventoried as eligible for establishment as research natural areas. Of these, 38 have been withdrawn or otherwise designated as eligible, but none has been processed to final designation as a research natural area in the full context of CFR 6225.0-5(a).* In the interim these candidate areas can be used for scientific purposes by securing a permit or agreement from the Dis-

* In researching this chapter it was discovered that not all the areas listed in the 1968 *Directory of research natural areas on federal lands* (U.S. Federal Committee on Research Natural Areas 1968) have reached the final designation stage. Hence the discrepancy between the listings in this chapter and those in the 1968 *Directory*. The 1968 *Directory* is in the process of being updated, and the reader should watch for it.

trict Manager or contacting the State Director, U.S. Bureau of Land Management, 2800 Cottage Way, Sacramento. California 95825.

National Park Service

The National Park Service does not have a statutory mandate relating to research natural areas similar to those of the Forest Service or the Bureau of Land Management. However, park superintendents are responsible for proposing candidate research natural areas. After review and recommendation by the Regional Chief Scientist, the Regional Director makes the designation.

Currently there are 13 research natural areas in California on land administered by the National Park Service (Table 3-2). One of these, the Schonchin Lava Tubes Research Natural Area, on the Lava Beds National Monument, is administered by the Pacific Northwest Regional Office in Seattle.

Approval to conduct research within a research natural area is provided by the appropriate Park Superintendent after consultation with professional members of his staff or the regional office, as appropriate.

Fish and Wildlife Service

The Fish and Wildlife Service has a research natural area program with definitions and objectives similar to those of other federal agencies. The program does not have statutory authority, but the *Wildlife refuges handbook* states that the Director of the Service is authorized to establish research natural areas upon screening and forwarding by the Western Regional Director in Portland, Oregon.

The Service's program in California is fairly recent, having had its start in 1972 with the establishment of the San Joaquin Desert Research Natural Area, located in the Kern National Wildlife Refuge. This 2260 acre (915 ha) natural reserve is especially valuable and important in view of the rapid development of the surrounding area after completion of the California Aqueduct. It features vernal pool habitats, alkaline playas, and cottonwood and willow habitats associated with a remnant river bed. This natural reserve can be used by contacting the Refuge Manager, Kern National Wildlife Refuge, P.O. Box 219, Delano, California 93215.

Federal Committee on Ecological Reserves

Originally constituted as the Federal Committee on Research Natural Areas, this committee was established in 1966, in part as an idea whose time had come, and in part as a result of a broad base of activities resulting in the Wilderness Act of 1964, the Land and Water Conservation Fund of 1965, the 1968 report of the President's Scientific Advisory Committee on Environmental Pollution, the National Environmental Policy Act of 1969, the report of the Public Land Law Review Commission, and the creation of the President's Council on Environmental Quality (Romancier 1974). It was truly an interagency committee involving not only the major federal conservation-oriented agencies but also such agencies as USDA's Cooperative State Research Service and Agricultural Research Service, as well as the Atomic Energy Commission Division of Biology and Medicine, Department of Defense Installations and Logistics Office, and the Office of Science and Technology. Such nongovern-

TABLE 3-1. U. S. Forest Service

Name of Area		Size (acres)	(ha)	Cover Type, Species, Features
		I. Research Natural Areas in California		
Backbone Creek	Sierra	430	175	*Carpenteria californica*, related shrubs
Devils Garden	Modoc	800	325	*Juniperus occidentalis*, Great Basin sagebrush
Frenzel Creek	Mendocino	830	335	*Cupressus sargentii*, *C. macnabiana*, serpentine chaparral
Harvey Monroe Hall	Inyo	3,888	1,575	*Pinus contorta* ssp. *murrayana*, *P. albicaulis*, *Tsuga mertensiana*, alpine habitats
Indiana Summit	Inyo	1,170	475	*Pinus jeffreyi*, *P. contorta* ssp. *murrayana*
San Joaquin Experimental Range	Fresno	70	28	*Pinus sabiniana*, *Quercus douglasii*, *Q. wislizenii*
Shasta Mudflow	Shasta-Trinity	3,467	1,405	*Pinus ponderosa*, five accurately dated mudflows yielding data on soil development and associated vegetational changes
White Mt.	Inyo	2,330	944	*Pinus longaeva*, assorted high-elevation habitats of the White Mts.
Fern Canyon	Angeles	1,370	555	*Pseudotsuga macrocarpa*, southern California woodland, scrub oak-chaparral

Blacks Mt.	Blacks Mt. Expl. Forest	521	210	Eastside ponderosa pine, *Juniperus occidentalis*, Great Basin sagebrush
Yurok	Yurok Expl. Forest	150	60	Upland redwood forest, *Chamaecyparis lawsoniana*, *Picea sitchensis*
		II. Botanical Areas in California		
Ancient Bristlecone Pine Forest	Inyo	27,160	11,000	*Pinus longaeva*, assorted high-elevation habitats of the White Mts.
Bodfish Piute Cypress	Sequoia	310	126	*Cupressus nevadensis*
Butterfly Valley	Plumas	400	160	Mixed conifer forest, *Darlingtonia californica* and associated bog species, rich flora of Liliacae and Orchidacae
Carpenteria	Sierra	295	119	*Carpenteria californica*, associated shrubs
Cuesta Ridge	Los Padres	1,334	540	*Cupressus sargentii*, associated species of *Arctostaphylos*, *Ceanothus*, etc., serpentine soils

TABLE 3-2. U. S. National Park Service

Name of Area	Park	Size (acres)	Size (ha)	Features
Carl Inn	Yosemite	488	195	Mixed conifer forest
Castle Rock	Sequoia-Kings Canyon	14,750	5900	Elevation gradient from mixed conifer zone to red fir zone
Esterode Limantour	Pt. Reyes National Seashore	548	219	Estuary
Garfield	Sequoia-Kings Canyon	9,600	3840	Mixed conifer forest, canyon live oak forest
Granite Creek	Yosemite	4,500	1800	Alpine and subalpine habitats
Heather Lake	Sequoia-Kings Canyon	40	16	Western white pine (*Pinus monticola*)
Kaweah Basin	Sequoia-Kings Canyon	13,500	5400	Alpine and subalpine habitats
Little Lost Man Creek	Redwood	2,480	992	Coastal redwood, Douglas fir, western hemlock
Merced River	Yosemite	600	240	Canyon live oak forest
Pt. Reyes Headlands	Pt. Reyes National Seashore	640	256	Exposed coastline with rocky substrate
Schonchin Lava Tubes	Lava Beds National Monument	134	54	Great Basin sagebrush, bunch grass, interesting geology
West Anacapa Island	Channel Islands National Monument	350	140	Rugged rocky coast with coastal bluffs and island chaparral
Whitney Creek	Sequoia-Kings Canyon	75	30	Limber pine (*Pinus flexilis*)

mental agencies as the Smithsonian Institution, The Nature Conservancy, and the American Institute of Biological Sciences were also represented.

One of its accomplishments was to draft a comprehensive statement of standards, guidelines, and policies for research natural areas. This statement was intended for inclusion in administrative manuals, either verbatim or modified in line with individual agency requirements. Although it never reached the level of acceptance that was hoped for, it serves even to this day as a succinct statement of research natural area philosophy and is frequently quoted by nature reserve enthusiasts at all levels of the federal government.

In 1968 the committee published the *Directory of research natural areas on federal lands* (U.S. Federal Committee on Research Natural Areas 1968). This directory had limited practical value for the scientist in the field. Its most useful function was to point out graphically the uneven administration of research natural area programs of the various federal agencies and the importance of updating, codifying, and coordinating administrative procedures.

In 1972 the committee ceased to exist officially. However, a core of dedicated workers maintained the committee's momentum, and it officially surfaced again in 1974 as the Federal Committee on Ecological Reserves, housed in the National Science Foundation, with essentially the same membership, augmented by a host of new agencies and interested observers. On its list of immediate priorities is the updating of the 1968 inventory and the adoption of a standard policy statement for use by the various federal agencies.

STATE RESEARCH NATURAL AREA PROGRAMS

State Park System

The influence and writings of John Muir helped to stimulate the ultimate establishment of the first state park in the nation. In 1864 President Lincoln signed an act deeding Yosemite Valley and the Mariposa Grove to the state of California for park purposes. The land was returned to the federal government in 1905 as part of the establishment of Yosemite National Park. Thus Big Basin State Park, authorized for purchase in 1901, is now the oldest unit in the State Park System. In 1918, a time when commercial logging activities were causing concern to conservation-minded citizens, the Save-the-Redwoods League was organized to protect certain outstanding examples of redwood forest and to aid in preserving them in state ownership for the enjoyment and inspiration of future generations. nine years later, in 1927, the state legislature established the California State Park System as we know it today. The California Department of Parks and Recreation now manages more than 1,000,000 acres (405,000 ha) contained in 200 units encompassing numerous examples of the state's diverse natural ecosystems.

The California Public Resources Code, Division 5, Chapter I, Section 5001.5, states:

> All units which are or shall become a part of the California State Park System . . . shall be classified by the State Park and Recreation Commission into one of the following categories: (a) State Wilderness, (b) State Reserve, (c) State Park, (d) State Recreation Units, (e) Historical Units, (f) Natural Preserves, (g) State Seashore.

Of special interest here are the lands classified as natural preserves and state reserves.

Natural preserves are defined as follows:

> . . . distinct areas of outstanding natural or scientific significance established within the boundaries of other State Park System units. The purpose of natural preserves shall be to preserve such features as rare or endangered plant and animal species and their supporting ecosystems, representative examples of plant or animal communities existing in California prior to the impact of civilization, geological features illustrative of geological processes, significant fossil occurrences or geological features of cultural or economic interest, or unique biogeographical patterns. Areas set aside as natural preserves shall be of sufficient size to allow, where possible, the natural dynamics of ecological interaction to continue without interference, and to provide in all cases, a practicable management unit. Habitat manipulation shall be permitted only in those areas found by scientific analysis to require manipulation to preserve the species or associations which constitute the bases for establishment of the natural preserve."

Table 3-3, Part I, lists the units of the California State Park System classified as natural preserves.

State reserves are defined as:

> . . . embracing outstanding natural or scenic characteristics of statewide significance. The purpose of a state reserve is to preserve the native ecological associations, unique fauna or flora characteristics, geological features and scenic qualities in a condition of undisturbed integrity. Resources manipulation shall be restricted to the minimum required to negate the deleterious influences of man.
>
> Improvements undertaken shall be for the purpose of making the areas available, on a day-use basis, for public enjoyment and education in a manner consistent with the preservation of their natural features. Living and nonliving resources contained within state reserves shall not be disturbed or removed for other than scientific or management purposes.

Table 3-3, Part II, lists the state reserves that are currently available.

Section 5001.65 of the California Public Resources Code states: "Qualified institutions and individuals shall be encouraged to conduct nondestructive forms of scientific investigation within State Park System units, upon receiving prior approval of the Director." Section 4312 of Title 14, California Administrative Code, authorizes the issuance of collecting permits. Because the Department of Parks and Recreation receives a considerable number of requests for scientific collecting permits, it has set up a fairly stringent set of criteria. The applicant must show that the proposed research will have substantial benefit to the State Park System. Benefit may be in the form of data useful to the inventory, management, development, or interpretation of State Park System resources. The department does not issue permits for general classroom collecting. Applications may be obtained by writing to the Chief, Resource Management and Protection Division, California Department of Parks and Recreation, P.O. Box 2390, 1416 Ninth Street, Sacramento, California 94811.

State Department of Fish and Game

The California Fish and Game Code, Division 2, Chapter 5, Article 4, Section 1580 says, "For the purpose of protecting rare or endangered wildlife or aquatic organisms

TABLE 3-3. California State Park System

Name	County	Size (acres)	Size (ha)	Features
		I. Natural Preserves		
Anderson Island	El Dorado	10	4	Blue oak woodland; great blue heron rookery
Doane Valley	San Diego	450	182	Mixed coniferous forest; *Pseudotsuga macrocarpa*, *Pinus ponderosa*, *P. coulteri*
Hagen Canyon	Kern	400	162	Shadscale scrub, Joshua tree woodland; *Hemizonia arida* (endemic to the natural preserves of Red Rock Canyon State Recreation Area)
Heron Rookery	San Luis Obispo	6	2	Bay, coastal sage scrub; great blue heron and black-crowned night heron rookery
Inglenook Fen (proposed)	Mendocino	(400)	(162)	Fen, fen carr, coastal dune, coastal strand, north coastal prairie; *Menyanthes trifoliata*, *Drosera rotundifolia*, *Habenaria* spp., *Campanula californica*
La Jolla Valley	Ventura	600	243	Central Valley prairie; *Stipa pulchra*, *S. lepida*
Least Tern	Orange	5	2	Nesting area of least tern; coastal strand
Los Penasquitos Marsh	San Diego	118	48	Saltwater marsh; *Salicornia* spp.
Mitchell Caverns	San Bernardino	628	254	Creosote bush scrub; piñon woodland; *Yucca schidigera*, *Y. baccata*
Morro Rock	San Luis Obispo	29	12	Coastal bluff and coastal sage scrub; peregrine falcon habitat

TABLE 3-3 [continued]

Name	County	Size (acres)	Size (ha)	Features
Pescadero Marsh	San Mateo	215	87	Freshwater and brackish marsh; lagoon, *Castilleja wrightii*
Pismo Dunes	San Luis Obispo	400	162	Coastal dune; *Artemisia californica*, *Lupinus arboreus*
Red Cliffs	Kern	240	97	Joshua tree woodland; *Yucca brevifolia* var. *herbertii*
Sheep Canyon	Imperial	44,800	18,144	Desert oasis; *Washingtonia filifera*
Sugar Pine Pt.	El Dorado	158	64	Mixed evergreen forest; *Calocedrus decurrens*
Torrey Pines	San Diego	72	29	Torrey pine forest; *Pinus torreyana*
Woodson Bridge	Tehama	260	105	Riparian forest; *Populus*, *Salix*, *Alnus*, *Vitis*
		II. State Reserves		
Año Nuevo	San Mateo	480	194	Pinniped habitat; coastal dune, coastal strand, coastal bluff
Armstrong Redwoods	Sonoma	680	275	Pristine redwood groves, northern oak woodland; mixed evergreen forest
Azalea	Humboldt	30	12	Azalea scrub seral stage of the coast redwood forest
Caspar Headlands	Mendocino	6	2	Sitka spruce krummholz; coastal bluff and north coastal prairie; *Campanula californica*
Indian Creek	Mendocino	15	6	Coast redwood forest

John Little	Monterey	21	9	Coastal sage scrub
Kruse Rhododendron	Sonoma	317	128	Rhododendron-tan oak seral stage of the coast redwood-Douglas fir forest
Los Osos Oaks	San Luis Obispo	85	34	Southern oak woodland; *Quercus agrifolia*, *Rhus diversiloba*
Mailliard Redwoods	Mendocino	242	98	Coast redwood forest
Montgomery Woods	Mendocino	1,142	463	Coast redwood forest
Pygmy Forest Ecological Staircase (proposed)	Mendocino	(800)	(324)	Mendocino cypress pygmy forest
Pt. Lobos	Monterey	1,255	508	Monterey cypress forest, Monterey pine forest, Gowen cypress pygmy forest; mound meadow of north coastal prairie
Smithe Redwoods	Mendocino	462	187	Coast redwood forest; ponderosa pine on serpentine; Sitka spruce forest
Torrey Pines	San Diego	1,030	417	Torrey pine forest; coastal sage scrub; saltwater marsh
Tule Elk	Kern	954	386	Tule elk habitat; alkali flat; saltbush scrub

or specialized habitat types both terrestrial and aquatic, the department, with the approval of the commission, may obtain by purchase, lease, gift or otherwise, land and water for the purpose of establishing ecological reserves." These areas shall be "preserved in a natural condition for the benefit of the general public to observe native flora and fauna and for scientific study." Although the regulations allow for designation of areas where hiking and walking trails may be developed, they go on to say, "Public entry for any or all purposes may be restricted on any or all of the areas at the discretion of the department to protect the wildlife, aquatic life, or habitat of the area."

No collecting is permitted in an ecological reserve except by permit from the Fish and Game Commission, and applicants must concurrently have a valid scientific collecting permit issued by the Department of Fish and Game by its authority under California Administrative Code, Title 14, Division 1, Chapter 11, Section 630.

Presently there are no ecological reserves specifically oriented toward the protection of terrestrial vegetation. Those with significant examples of terrestrial habitat types are listed in Table 3-4 (see also Smith et al. 1974). The Fish and Game Commission may allow commercial exploitation of resources in these reserves by simply issuing a permit or wording the section dealing with a specific reserve in such a way as to protect only certain organisms.

EDUCATIONAL INSTITUTIONS

Although administered by the individual institutions, the nature reserves described below are available to all scientists and students involved in serious study and research.

University of California Natural Land and Water Reserves System

In 1949 zoology graduate student Kenneth S. Norris faced the ultimate frustration of having his unfinished Ph.D. dissertation ruined by the construction of a motel on the site of his field studies. This had happened in spite of his original precaution of selecting what he thought was a remote location. Norris came back from this defeat, and in doing so he not only earned his Ph.D. but also became the motivating force behind the establishment of the University of California Natural Land and Water Reserves System (NLWRS). He wanted to assure that other students would not have to face similar setbacks because of landowners who change their minds about their management goals after promising to cooperate with the scientists, or because of vandals who destroy instrumentation or carefully established field plots. After analyzing the needs and deciding on a course of action (Norris 1968), he and his ad hoc committee convinced then U.C. President Clark Kerr of the need for a system of natural reserves to be administered and managed by an institution of higher education for use by faculty and students in the field-oriented sciences.

At its January 1965 meeting, the Board of Regents formally created the NLWRS and dedicated seven existing university properties as the initial units of the system. Since that time the NLWRS has grown to 22 units (Table 3-5), ranging in size from an 8 acre (3.2 ha) island* serving as a hauling ground and breeding site for marine

* Managed by the university under a use agreement from the California Department of Parks and Recreation.

TABLE 3-4. California Department of Fish and Game Ecological Reserves

Name	County	Size (acres)	(ha)	Features
Fish Slough	Inyo-Mono	104	42	Remnant of pristine Owens Valley aquatic ecosystems
Buena Vista Lagoon	San Diego	71	29	Saltwater marsh, coastal lagoon
Morro Rock	San Luis Obispo	30	12	Nesting habitat of endangered peregrine falcon; coastal bluff, coastal sage scrub
Bolsa Chica	Orange	563	228	Coastal lagoon; saltwater marsh
Tomales Bay	Marin	542	220	Tidal mudflat; saltwater marsh
Hidden Palms	Riverside	143	58	Desert oasis
Santa Cruz Long-Toed Salamander	Santa Cruz	32	13	Salamander habitat
Limestone Salamander	Mariposa	120	49	Salamander habitat

TABLE 3-5. University of California Natural Land and Water Reserves System

Name	County	Size (acres)	Size (ha)	Features
Año Nuevo Island Reserve	San Mateo (20 mi N of Santa Cruz campus)	8	3	Small offshore island; breeding ground for 3 species of pinnipeds; large population of seabirds
Bodega Marine Reserve	Sonoma (on Bodega Head)	326	132	Coastal grassland; rocky coast and intertidal flora and fauna; coastal bluff vegetation; housing and laboratory space
Box Springs Reserve	Riverside (adjacent to Riverside campus)	160	65	Chaparral; coastal sage
Boyd Deep Canyon Desert Research Center	Riverside (near Palm Desert)	14,000	5,670	Desert mountain complex, including an entire drainage system, from montane forest to Sonoran Desert vegetation; limited housing and laboratory space
Burns Piñon Ridge Reserve	San Bernardino (near Yucca Valley)	265	107	Piñon-juniper; rock outcrops; transition between the Mojave and Colorado Deserts and the San Bernardino Mts.
Cheatham Grove Reserve	Humboldt (on the Van Duzen River)	160	65	Mature stand of coastal redwoods on an alluvial flat
Coal Oil Pt. Natural Reserve	Santa Barbara (part of Santa Barbara campus)	49	20	Southern California rocky coast; coastal dunes and bluffs; rocky intertidal reef
Dawson Los Monos Canyon Reserve	San Diego (near Vista)	92	37	Chaparral; permanent stream; riparian woodland

Elliott Chaparral Reserve	San Diego (E of La Jolla and N of Miramar Air Base)	107	43	Chamise chaparral
Hastings Natural History Reservation	Monterey (upper Carmel Valley)	1,600	648	Mixed evergreen forest; foothill woodland; chaparral; limited housing and laboratory space
James San Jacinto Mts. Reserve	Riverside (near Idyllwild)	34	14	Southern California mixed conifer forest; limited housing facilities
Kendall-Frost Mission Bay Marsh Reserve	San Diego (N side of Mission Bay)	20	8	Saltwater marsh with migratory and resident bird population
Pygmy Forest Reserve	Mendocino (near Mendocino)	70	28	Unique dwarfed cypress and pine vegetation on Blacklock soil series
Ryan Oak Glen Reserve	San Diego (near Escondido)	15	6	Chamise chaparral on a north-facing slope; small stand of Engelmann oak; springs and seeps
Sacramento Mts. Reserve	San Bernardino (20 mi W of Needles)	591	293	Unusually dense stand of cholla cactus; 5 other species of cactus also present
Santa Monica Mts. Reserve	Los Angeles (Santa Monica Mts.)	362	147	Chamise chaparral; mixed chaparral; riparian vegetation; complete watershed
Sawyer Trinity Alps Reserve	Siskiyou (near Cecilville)	125	51	Mixed conifer forest
Scripps Shoreline Underwater Reserve	San Diego (fronting Scripps Institution of Oceanography)	254	103	Rocky coast; sandy beach; intertidal zone; coastal bluffs
San Joaquin Freshwater Marsh Reserve	Orange (adjacent to Irvine campus)	202	82	Freshwater marsh; open ponds; numerous aquatic birds

TABLE 3-5 [continued]

Name	County	Size (acres)	Size (ha)	Features
Santa Cruz Island Reserve	Santa Barbara (Channel Islands)	56,000	22,680	Channel Islands woodland; grasslands; chaparral; near-shore marine environments; endemic and relict plant and animal species; limited housing and laboratory space
Unnamed Central Sierra Reserve	Placer (near Donner Summit)	640	259	Alpine lake; alpine meadows; remote location, not presently available for use
Valentine Eastern Sierra Reserve	Mono (near Mammoth Lakes)			
Valentine Camp		136	55	Montane and subalpine habitats of the eastern Sierra Nevada; dry slopes; wet slopes; boggy meadows; terminal moraine; limited housing and laboratory space
Sierra Nevada Aquatic Research Laboratory		51	21	Eastside Sierra trout stream; 4 sections with weirs and other control devices for experimental manipulation; limited housing and laboratory space

mammals to a major desert research center in excess of 14,000 acres (5668 ha). By special arrangement, access to 56,000 acres (22,672 ha) of Santa Cruz Island is also available. As many as 50 reserves are planned to serve the nine campuses of the university, as well as scientific investigators from other institutions.

A university-wide faculty committee, appinted by the President of the University of California, advises on policy matters and screens potential acquisitions. Each campus has a similar faculty committee responsible for managing the individual reserves, each of which, after being added to the NLWRS by action of the Board of Regents, is assigned to one of the campuses. Reserves are acquired by a variety of methods, including gift, lease, use agreement, conservation easement, and purchase. Purchases have been funded through grants from the Ford Foundation and from other nonstate sources.

Additional information can be obtained from the Director, Natural Land and Water Reserves System, University of California, Systemwide Administration, Berkeley, California 94720.

Other Educational Institutions

From time to time various other institutions of higher education have found it desirable to integrate a nature reserve with their educational and research programs. Stanford University has been using the 960 acre (389 ha) Jasper Ridge Nature Reserve since the earliest days of the university. There is an impressive list of graduate degrees and technical publications resulting from studies done on this reserve. Jasper Ridge includes samples of most of the vegetation types typical of the inner Santa Cruz Mts. and presently is the only research natural area containing a sample of serpentine grassland.

The Santa Margarita Ecological Reserve is a 2700 acre (1093 ha) reserve administered by the California State University, San Diego (Cooper et al. 1973). Located on the San Diego–Riverside county line, it is easily accessible to educational institutions throughout southern California.

More recently, California State University, Humboldt, has entered into a management agreement with The Nature Conservancy for the Lanphere-Christensen dunes. This magnificent dune system is located within easy drive of the Humboldt campus. There are also cooperative efforts between The Nature Conservancy and Napa College (Bumpy Camp Reserve) and California State College, Sonoma (Fairfield Osborn Preserve).

PRIVATE PROGRAMS

The Nature Conservancy

The Nature Conservancy is a nationwide, member-governed, nonprofit organization incorporated in 1951 in the District of Columbia for scientific and educational purposes (Goodwin 1968). Its resources are devoted to the preservation of ecologically and environmentally significant land.

The Nature Conservancy works in three ways to protect nature reserves: by purchase of land with funds raised through public subscription, by acceptance of donations of land, and by advance acquisition of land for local, state, and federal governments.

Since its first acquisition in 1954, The Nature Conservancy has acquired some 744,592 acres (301,454 ha) in 1357 projects throughout the United States. Of these, 53 are in California and 29 are managed specifically as nature reserves. In California, the work of The Nature Conservancy is administered by its Western Regional Office, 425 Bush Street, San Francisco, California 94108, and is centered in the efforts of its three California chapters. Table 3-6 lists their accomplishments of special interest to the readers of this chapter.

Audubon Society Wildlife Sanctuaries

The first Audubon Society was formed in 1886 by George Bird Grinnell, a former pupil of Lucy Bakewell Audubon. Grinnell suggested that the society be named in honor of his teacher's husband, John James Audubon (1785–1851), the famous artist, naturalist, and conservationist. The Audubon movement began in the 1880s as a protest against the slaughter of birds for their plumage and as an effort to stop the hunting of birds for sale in the public market, which had brought many species close to extinction. State Audubon societies formed first, beginning with Massachusetts in 1896. The National Audubon Society was incorporated in January 1905 (called, until 1934, the National Association of Audubon Societies). The National Audubon Society is dedicated to helping man understand that conserving our natural resources serves his own best interests. Toward these ends a national system of sanctuaries was established. The system protects threatened species of birds and mammals and their strategic habitats. Visitation and research are considered secondary to the society's basic protective functions.

Some sanctuaries, such as those where nature center activities are conducted, are open to members and the general public on daily or periodic schedules. Because of transportation problems and warden staff limitations, other national sanctuaries can be visited only by prearrangement with the Sanctuary Department. This can be accomplished by writing to the Sanctuary Director, National Audubon Society, West Cornwall Road, Sharon, Connecticutt 06069. Sanctuaries managed by local Audubon societies can be contacted directly. Table 3-7 lists sanctuaries found in California.

PROGRAMS RELATED TO RESEARCH NATURAL AREA PROGRAMS

Introduction

So far this chapter has dealt with programs whose administrative authorities are directed toward meeting the specialized needs of the scientific community. There are, however, programs designed to protect natural ecosystems where public access is permitted in the form of carefully designed nature trails, parking, and picnic areas away from ecologically significant features. Even though public use and enjoyment of natural history features is the primary management objective, by appropriate arrangement these areas can be useful for many scientific investigations.

U.S. Forest Service Special Interest Areas

Chapter 2360 of the recreation management section of the *Forest service manual* provides policies and guidelines applicable to the classification and management of

TABLE 3-6. The Nature Conservancy Projects in California

Name	County	Size (acres)	Size (ha)	Features
Bishop Pine Preserve	Marin	403	163	Complete watershed located on eastern slope of Inverness Ridge, containing California bay and bishop pine; to be transferred to the State of California
Boggs Lake	Lake	104	42	Vernal lake containing rare and endangered flora
Buena Vista Lagoon	San Diego	150	61	Freshwater lagoon and estuary complex; migratory birds; transferred to California State Wildlife Conservation Board
Bumpy Camp	Napa	158	64	Oak woodlands; transferred to Napa College
Cleary Preserve	Napa	400	162	Forest; chaparral and grasslands; streams and waterfall; management authority conveyed to Biological Field Studies Association
Dorland Preserve	Riverside	300	122	Springs and oak woodland
Elkhorn Slough	Monterey	323	131	Undisturbed marine lagoon, salt marsh and mudflats; shore and migratory birds
Fairfield-Osborn Preserve	Sonoma	142	58	Wooded land, streams, ponds, and marsh; leased to Sonoma State College
Forest of Nicene Marks	Santa Cruz	9570	3876	Wilderness area near urban center; forest of second-growth redwood and fir; conveyed to State of California
Martha Alexander Gerbode Preserve	Marin	2138	866	Coastal mountain area, part of the Golden Gate National Recreation Area

TABLE 3-6 [continued]

Name	County	Size (acres)	Size (ha)	Features
Hamilton Sanctuary	Los Angeles	40	16	Piñon-juniper-yucca community adjacent to Angeles National Forest
Hi Mountain Condor Sanctuary	San Luis Obispo	162	66	Transferred to the Los Padres National Forest
Hibbard Property	San Luis Obispo	1500	608	Coastal ridgeline on south extreme of Irish Hills
Holmwood Property	Placer	716	290	Timberland including sections of Ward Creek, principal spawning stream for Tahoe trout; transferred to U. S. Forest Service
Jaeger Sanctuary	Riverside	80	32	Desert flora; important bird study area
Kent Island	Marin	109	44	Sandy island in the midst of the Bolinas Lagoon; abundant shorebird and marine life; transferred to Marin Co.
Lanphere-Christensen Dunes	Humboldt	180	73	Complete and intact coastal dune system; managed by California State University, Humboldt
Lower Tubbs Island	Sonoma	332	134	Marshy island in San Francisco Bay; bird refuge and critical stopover and wintering grounds in the Pacific flyway
Martinez Canyon	Riverside	650	263	Desert bighorn sheep habitat; to be transferred to California Wildlife Conservation Board
McCloud River	Shasta	2330	944	Pristine watershed and trout streams

Morongo Canyon	San Bernardino	80	32	Desert canyon ecosystem; bighorn sheep habitat
Murphy Preserve	Los Angeles	455	180	Chaparral and riparian habitat; uncommon assemblage of biotic communities in the Santa Monica Mts.
Northern California Coast Range Preserve	Mendocino	3890	1575	Old-growth Douglas fir; watershed for Elder Creek
Oasis de los Osos	Riverside	160	65	Desert canyon with live stream; abundant flora and fauna
Paine Wildflower Memorial	Kern	40	16	Remnant of alkali wildflower community; rare desert fauna and flora
Rancho Las Cruces	Santa Barbara	400	162	Typical Transverse Range habitats; oak woodland; habitat for unique species of turtle; watershed of Las Cruces Creek
Sand Ridge Wildflower Preserve	Kern	116	47	Annual wildflowers
Van Duzen Grove	Humboldt	160	65	Redwood grove on the Van Duzen River; managed by the University of California Natural Land and Water Reserves System
Vernal Pools	Tulare	41	17	Annual wildflowers in one of the few remaining remnants of San Joaquin Valley vernal pools

TABLE 3-7. Audubon Society Wildlife Sanctuaries in California

Name	County	Size		Features
		(acres)	(ha)	
Audubon Canyon Ranch	Sonoma	1000	405	Northern oak woodland; coast redwood forest
Bobelaine	Sutter	400	162	Riparian woodland; saltwater marsh
Greco Island	San Mateo	3000	1215	Saltwater marsh
Richardson Bay[a]	Marin	911	369	900-acre mudflats and tidelands; 11-acre north coastal scrub, introduced annual grassland, riparian, interior chaparral
Silverwood	San Diego	212	86	Southern oak woodland, interior chaparral
Starr Ranch[a]	Orange	3000	1215	Sycamore-oak riparian woodland, southern oak woodland, interior chaparral
South San Francisco Bay Sanctuaries[a]	Alameda	2000	810	Saltwater and brackish water marsh

[a]Under National Audubon Society Management

areas with such designations as botanical, zoological, scenic, geological, archaeological, and paleontological areas. These are areas that "possess unusual recreation values and which should be managed in substantially natural condition, but which do not qualify for designation as wilderness, so that the special values are available for public study, use, or enjoyment." Designation of these areas is by action of the Secretary of Agriculture, the Chief of the Forest Service, or the Regional Forester, depending on their size, by the authorities under 36 CFR 294.1(a). All of the botanical areas in California have been established by the Regional Forester and are summarized in Table 3-1, Part II.

Other Federal Programs

National parks, national monuments, national seashores, wild and scenic rivers, wildlife refuges, and wilderness areas are all categories of federal land with highly laudable objectives regarding protection of the natural ecosystems. The level of man's activities in these areas varies, and so does the resulting level of disturbance to ecosystems. Many scientists will find the conditions they need for their studies in these types of areas.

State Park System

Units of the State Park System classified as state wilderness, state seashore, and state park have management objectives conducive to protecting the natural ecosystems. The California Department of Parks and Recreation presently manages 2 state wildernesses, 10 state seashores, and 58 state parks. Natural preserves may be established within these and other units of the system (see an earlier section of this chapter). Scientists are encouraged to investigate State Park System units for unique values worthy of natural preserve status.

State Wildlife Areas

The State Department of Fish and Game manages 19 state wildlife areas. Although these areas preserve and enhance the fish and wildlife resources, they are managed for use and enjoyment, including hunting, by the public. The vegetation is protected except that it "may be cut and used as directed by the Area Manager for the purpose of building blinds on the area during the waterfowl season" (California Administrative Code, Title 14, Division 1, Chapter 8, Section 550).

PROGRAMS PROVIDING SPECIAL RECOGNITION

These programs extend some form of special recognition to a landowner who promises to protect the important nature reserve values of a particular parcel of land. They are especially effective in the case of a farmer, estate owner, or corporation whose land management objectives are not normally oriented toward natural history, but public land management agencies also participate. The main thrust is formalized recognition of the public spirit of the landowner, but the resulting register also serves as an inventory of sites with nature reserve values.

Registry of National Natural Landmarks

This program, administered by the National Park Service, is intended to (1) encourage the preservation of sites illustrating the geological and scientific ecological character of the United States, (2) enhance the educational and scientific value of sites thus preserved, (3) strengthen cultural appreciation of natural history, and (4) foster a wider interest in the conservation of the nation's natural heritage.

In making its inventory of eligible sites, the National Park Service has developed a system of 35 themes broken down into four major categories: landforms of the present, geological history of the earth, land ecosystems, and aquatic ecosystems. If, after study, a site meets certain criteria, it is reviewed by the Advisory Board on National Parks, Historic Sites, Buildings, and Monuments, appointed by the Secretary of the Interior. Upon the recommendation of this board, the Secretary lists the area on the National Registry of Natural Landmarks published annually in the *Federal register.* The owner is then invited to apply for a certificate and a bronze plaque designating the site as a registered natural landmark. Participation is voluntary, and there is no change in ownership. Registration requires the landowner to preserve, insofar as possible, the significant natural values of the site, but he relinguishes none of his rights and privileges to use the land. The federal government gains no possessory interest but will, upon request, provide consultative assistance in protecting and interpreting the natural values of the site. To date in California, 21 sites are listed in the *Federal register.* Of these, 12 have been registered; the remainder are eligible but not yet fully registered. Others are being inventoried (Stebbins and Taylor 1973).

Additional information can be obtained from the Division of Historic Preservation, Western Region, National Park Service, 450 Golden Gate Avenue, Box 36063, San Francisco, California 94102.

Programs of Professional Societies

Many professional societies in the natural sciences have recognized the need for an active and viable nature reserves program as an important contribution to their professions. The oldest of these is the program sponsored by the Society of American Foresters (SAF). In 1947, the SAF created a committee to make an inventory of known natural reserves in the major forest types of the United States and to establish criteria for listing them in the SAF inventory (Buckman and Quintus 1972). The inventory was first published in 1949 and included 68 nature reserves representing 66 forest types. The most recent inventory (Buckman and Quintus 1972) now lists 281 reserves. Of these, 21 are in California. As with the Registry of Natural Landmarks, the SAF has an elaborate qualification procedure, including protection assurance by the landowner. A copy of the inventory can be obtained by writing to the Society of American Foresters, 1010 Sixteenth Street, N.W., Washington, D.C. 20036.

Other professional societies such as the Soil Conservation Society of America (Soil Conservation Society of America 1975), the Wildlife Society, and the Society for Range Management (Laycock 1975) have similar programs. The American Association for the Advancement of Sciences (Peter 1971), Botanical Society of America, Pacific Science Association, Smithsonian Center for Natural Areas, The

Institute of Ecology, and the Western Society of Naturalists have also been active in these matters.

CONCLUSION

California Natural Areas Coordinating Council

In view of the multitude of nature reserve activities in California, there is an obvious need for a coordinated effort in order to minimize competition between programs and to avoid gaps between habitat types in protected status. In the late 1960s such an effort was launched with the creation of the California Natural Areas Coordinating Council (CNACC). The original impetus came from the leadership of Dr. G. Ledyard Stebbins, who persuaded the California Native Plant Society to sponsor a number of meetings that brought together representatives from public agencies and spokespersons from conservation organizations to discuss the problem. There emerged a council where representatives of nature reserves programs could meet to coordinate their activities and discuss matters of mutual concern. The format chosen was a nonprofit corporation consisting of a board of directors to serve as a steering committee, but no formalized membership. This had the advantage of inspiring a freer exchange of information and opinion than might have been possible if participants were obligated to adhere to the official opinions of their agencies or organizations.

One of the most immediate needs was an inventory of the state's areas with nature reserve values, both those with formal status and those not formally protected but worthy of special management attention. An inventory of approximately 1300 areas has now been compiled through massive volunteer participation by the academic community and numerous interested conservationists. The *Inventory of California natural areas,* which includes a brief description of each area, is available for purchase from CNACC, Box 4000J, Berkeley, California 94704.

Selection of an area for inclusion in the list was based on criteria used by the National Park Service in its determination of natural landmarks, with the important proviso that the area need not be of national significance, but rather must have only state or local import. Additionally, a number of areas within or close to metropolitan regions are included, even though they are not as pristine as might be desired. However, they do offer a convenient opportunity for study and research, and their very convenience helps to lessen the impact on similar but less accessible areas.

The natural areas listed fall into two major categories. The first includes unique areas that exhibit a disjunct population, a range limit, an endangered species, a particular formation, a type locality, or some other special feature. The second, and larger, category covers areas that are representative of a particular community, habitat, or formation. It is hoped that, by selecting several such representatives, a valid cross section has been achieved.

Nearly 1300 areas have been listed, ranging in size from 0.4 to 400,000 ha. Approximately three-quarters of these areas are included primarily for their botanical interest. About 20% of the total are areas included for their geological or paleontological features. The remainder are zoological areas. The low percentage of zoological areas is more apparent than real, for a prime botanical area normally supports an abundant fauna.

A list of the names of the areas is on file with the State Office of Planning and Research and in many large libraries; it is in the public domain. However, upon subscription to CNACC, an expanded description of each area can be obtained. As an example, the description for NA 270561, Russian Peak in Siskiyou Co., is given below. The first several lines include a key word description, the county, latitude and longitude, the USGS topographic map on which the area appears, location by township, range, and section, the size, and elevational range. The number assigned to the natural area is followed by two letters (BE in this case) which indicate the state of published documentation for the area and its relative fragility.

RUSSIAN PEAK 471891 BE

Conifer forests, subalpine forest
Siskiyou 41°23′30″ N 122°55′45″ W
Etna 15′ T40N, R9W;R10W S var.
3,200 ha (8,000 a.) 1,304–2,498 m. (4,280–8,196 ft.)
Klamath National Forest; private holdings

A part of a proposed research natural area or botanical area, this area is unique in that it may contain more species of conifers in a localized area than any other area in the world. Plants with Sierran, Cascadian and Great Basin affinities mix here.

Seventeen species are found within 2.6 square kilometers (1 sq. mile). They are whitebark pine, *Pinus albicaulis*, sugar pine, *P. lambertiana*, Western white pine, *P. monticola*, foxtail pine, *P. balfouriana*, lodgepole pine, *P. murrayana*, Jeffrey pine, *P. jeffreyi*, ponderose pine, *P. ponderosa*, white fir, *Abies concolor*, Shasta fir, *A. magnifica* var. *shastensis*, subalpine fir, *A. lasiocarpa*, mountain hemlock, *Tsuga mertensiana*, Douglas fir, *Pseudotsuga menziesii*, Engelmann spruce, *Picea engelmannii*, weeping spruce, *P. breweriana*, incense cedar, *Calocedrus decurrens*, prostrate juniper, *Juniperus communis*, and yew, *Taxus brevifolia*. In addition, there is a variety of montane and subalpine forests, chaparral and meadows, all on granodiorite.

Extensive stands of subalpine fir and Engelmann spruce, both rare in California, are found in the Duck Lake and Sugar Creek drainages.

Weeping spruce is found in many places in this area and is not restricted to isolated stands, as in other parts of the Klamath region. The foxtail pine is found in a single stand on an unglaciated ridge above Duck Lake.

Glacial features, such as lakes, cirques, moraines, and, in some of the northeast-trending creeks, U-shaped valleys, are well-defined here.

Integrity: Low-grade road system in portion.

Use: This area lies within the proposed Russian Peak Wilderness Area. The suggestion has been advanced that this region be designated a Research natural or Botanical Area.

Ref.: Sawyer, J. O. and J. P. Smith, 1973. The Klamath Region. *California Native Plant Society Newsletter*, Vol. VIII, No. 4, p. 3–6.

It is expected that additional areas will be added to the list if they contribute to a more comprehensive coverage, and that some will be eliminated in favor of others because of a change in their status. Many of the areas are on private land, so there is no public control over their present and future use. Thus the inventory should not be considered all inclusive or in its final form. Rather, it is part of a process—the process of creating a full-scale natural area system for California.

Diversity of Programs

Diversity of programs may have weaknesses such as duplication of bureaucracies, resulting in inefficient utilization of budget and manpower resources; lack of uniformity of vocabulary; difficulty in dispersal of information; and competition between programs. But, as in a natural ecosystem, diversity provides for stability of the natural reserves effort. If one bureaucracy suffers a budget cut, others may be able to fill the void. If one program fails to protect a given habitat type, another may be successful. If one program includes a habitat type covered by another program, there is protection in depth against the occurrence of a disaster, such as an epidemic, fire, or flood. If one program suffers from an unsympathetic administration, work still goes on in other programs. The lessons learned by one program can be shared with others.

Each program has its own approach. Some are oriented toward private ownership; others, toward public ownership. Some seek to protect rare and endangered species or unique features. Others are concerned with public enjoyment and interpretation or research and educational use or preservation of "typical" samples of habitat types or the emphasis of botanical versus zoological features, and so on. With such diversity each program can proceed to utilize its strengths with the knowledge that other programs will be developing their respective strengths.

The Role of the Academic and Scientific Communities

Scientists and educators have a multiple role to play. The administrators turn to them for information and advice on which areas to establish and, once established, on the management actions necessary to protect the values, while at the same time utilizing them for research and education. Scientists are also users of the reserves, and frequently they are the managers.

The value of a nature reserve comes from its intrinsic properties, that is, a sample of undisturbed ecosystem, and such value is not a function of the amount of use it receives. This is a subtle point, and it is difficult to convince administrators of its correctness when a reserve recommended by scientists as "valuable" goes unused for scientific investigation. Therefore it is to their own interest that scientists and educators use the reserves. They should establish the baseline studies that they wistfully wish were available (Franklin et al. 1972; Ohman 1973). Their students should use these areas to compile floras, to learn how to make geology, soil, and vegetation maps, or to learn basic field sampling techniques in all disciplines. Only in the gathering and interpretation of these kinds of data will the true "value" of nature reserves be realized.

From the gathering of basic data comes interpretation, not only for scientific understanding but also for guiding management decisions. What basic nature reserve values are present? Which values require special attention? Is a given reserve best allowed to progress through its natural successional changes or, in a specific instance, is it best managed to maintain a particular successional stage through the application of artificial, but planned, management practices? If a field station facility is to be built, where should it be located in order to have a minimum impact on the values of the reserve? The more data that are known, the more sound such decisions are.

The message is clear. The academic and scientific communities have a definite rǫle to play and should do so with diligence and concern for the importance of nature reserves to the field sciences.

The Dilemma of Publicity

To fully utilize the reserves in their educational and research roles, their existence must be communicated to potential users. This leads to the dilemma of attracting users whose objectives or intensity of activities (i.e., overcollecting or trampling) are not compatible with the best interests of the natural reserves. Their pristine-like nature will attract curiosity seekers and recreationists who feel that they too have the right to use these areas in order to enjoy their special character. This becomes a real problem that leads to difficulties for the reserve managers.

The answer to this dilemma lies in a good public relations program. In reserves where the demand is great, scheduled tours and open houses can be utilized to impart two major messages: the value and need of establishing research natural areas in the first place, and the particular features that are important and useful in the reserve in question. This type of activity need only be as elaborate as necessary, but it should be planned as part of the reserve's management activities.

There is another side to the dilemma. The needs and demands of the scientific community will increase as the reputation of research natural areas increases. Administrative procedures should be developed to schedule users, to arbitrate between conflicting projects, to assure orderly activity, and to minimize the impact on the ecosystems. Management plans specifying the kinds of use, zoning, and use levels suitable for a given reserve (Lawrence 1963) should be developed. Depositories of records and collections should be established, as well as a body of scientific literature recording the methods and results of studies utilizing research natural areas.

LITERATURE CITED

Ashe, W. W. 1922. Reserved areas of principal forest types as a guide in developing an American silviculture. J. For. 20:276-283.

Buckman, R. E., and R. L. Quintus. 1972. Natural areas of the Society of American Foresters. Soc. Amer. Forest., Washington, D.C. 38 p.

Clausen, J., D. D. Keck, and W. M. Heisey. 1940. Experimental studies on the nature of species. Carnegie Inst. Wash. Pub. 520.

Cooper, C. F., H. N. Coulombe, R. L. Hays, and P. H. Zedler (eds.). 1973. The Santa Margarita Ecological Reserve: description, present disposition, and recommendations for future use. Calif. State Univ., San Diego. 103 p.

Ford-Robertson, F. C. 1971. Terminology of forest science, technology, practice, and products. Soc. Amer. Forest., Washington, D.C. 349 p.

Franklin, J. F., R. E. Jenkins, and R. M. Romancier. 1972. Research natural areas: contributors to environmental quality programs. J. Environ. Qual. 1:133-139.

Franklin, J. F., and J. M. Trappe. 1968. Natural areas: needs, concepts, and criteria. J. For. 66:456-461.

Goodwin, R. H. 1968. The role of private agencies in natural area preservation. BioScience 18:383-395.

Hall, H. M. 1929. European reservations for the protection of natural conditions. J. For. 27:667-684.

Lawrence, D. B. 1963. Natural areas primarily for research, pp. 3-6. In College natural areas as research and teaching facilities: the proceedings of the first national symposium on college natural areas. The Nature Conservancy, Washington, D.C.

Laycock, W. A. (ed.). 1975. Rangeland reference areas. Soc. Range Manage., Range Sci. Ser. 3, Denver. 66 p.

Moir, W. H. 1972. Natural areas. Science 177:396-400.

Norris, K. S. 1968. California's natural land and water reserve system. BioScience 18:415-417.

Ohman, L. F. 1973. Vegetation data collection in temperate forest research natural areas. USDA Forest Serv., North Central Forest Exp. Sta., Res. Paper NC-92, St. Paul, Minn. 35 p.

Parker, L. 1972. The Harvey Monroe Hall Research Natural Area. Natl. Parks Conserv. 46:25-28.

Pearson, G. A. 1922. Preservation of natural areas in the national forests. Ecology 3:284-287.

Peter, W. G. 1971. AIBS to document U.S. ecosystems. BioScience 21:142-143.

Romancier, R. M. 1974. Natural area programs. J. For. 72:37-42.

Shelford, V. E. (ed.). 1926. Naturalist's guide to the Americas. Ecol. Soc. Amer., Durham, N.C. 761 p.

Smith, E. J. Jr., T. H. Johnson, and D. Zeiner. 1974. The marine life refuges and reserves of California Dept. of Fish and Game. Marine Res. Bull. 1.

Soil Conservation Society of America. 1975. Managed natural areas. Ankeny, Iowa. 24 p.

Stebbins, G. L., and D. W. Taylor. 1973. A survey of the natural history of the South Pacific Border Region, California. Inst. Ecology, Univ. Calif., Davis. 513 p.

The Nature Conservancy. 1975. The preservation of natural diversity: a survey and recommendations. Prepared for U.S. Dept. Interior, Washington, D.C. 212 p.

U.S. Federal Committee on Research Natural Areas. 1968. A directory of research natural areas on federal lands of the United States of America. Super. Docum., Washington, D.C. 129 p.

CHAPTER 4

THE CALIFORNIA FLORA

PETER H. RAVEN

Missouri Botanical Garden, 2345 Tower Grove Avenue, St. Louis

Introduction 110
Numbers of genera and species 110
The California Floristic Province 111
Endemism 111
Endemic genera 111
Endemic species 112
Elements in the flora of California 112
Arcto-Tertiary elements 113
Madro-Tertiary elements 113
Other elements 113
Relationships with South America 114
Relationships with the Mediterranean region 114
Derivation of the flora of California 115
The Transmontane Floristic Provinces 116
Great Basin region 116
Inyo region 117
Mojave Desert region 117
Sonoran Desert region 117
Summary of section 118
Centers of endemism in California 118
Relict areas 119
Other patterns of endemism 120
The endemic plants of California 120
Edaphic endemism in California 121
General 121
Serpentine soils 122
Mode of evolution of California plants 123
Hybridizing complexes 123
Saltational speciation 126
Introduced plants 127
Preservation of the California flora 129
Acknowledgments 129
Literature Cited 131

INTRODUCTION

The flora of California, one the most famous in the world, is a mixture of northern, temperate elements and xeric, southern ones, and is characterized by a large number of species and a very high degree of endemism. That it was a mixture of two distinct floristic elements was recognized clearly by Abrams (1925), was elaborated by Campbell and Wiggins (1947), and was restated on the basis of a more detailed knowledge of the fossil record by Axelrod (1958). This theme will be elaborated in this chapter, and the reasons for the high number of species and the high proportion of endemism in the flora of the state will be considered in detail. Part of the explanation is found in the equable climate that prevailed in California throughout the Tertiary, as pointed out by Abrams (1926), among others. Another important factor has been the late Pliocene to Recent elevation of the Sierra Nevada and adjacent ranges and the concomitant development of a cold offshore current that ultimately resulted in the evolution of a widespread Mediterranean, summer-dry climate during the past million years, as pointed out by Axelrod (1966). The endemics of California are a mixture of relicts and newly produced species, as outlined by Stebbins and Major (1965). It is the latter that have contributed largely to the size of the flora and the overall proportion of endemism in it. A more detailed account of the flora of California will be given elsewhere (Raven and Axelrod report in prep.).

Numbers of Genera and Species

As compared with other temperate regions, California is very rich in species of native vascular plants, with approximately 5057, but not especially rich in native genera, with 875 (Munz and Keck 1959; Smith and Noldecke 1960; Howell 1972; Munz 1968, 1974). California has an area of approximately 411,000 km² (~ 1,000,000 acres), Texas, with 751,000 km², has about 751 native genera and 4196 native species, whereas Japan, with 377,000 km² has about 1098 native genera and 4022 native species (Correll and Johnston 1970; Ohwi 1965). California shares its richness in species with other areas of the world having a Mediterranean climate. The California flora appears even more impressive when the number of genera and species in the California Floristic Province (defined as the part of the state west of the Sierra Nevada–Cascade axis, excluding the deserts; see Fig. 4-1) is considered (Howell 1957). The California Floristic Province, with an area of about 285,000 km² in the state of California, has 743 genera and 4119 species of native vascular plants; there are about 131 genera and 938 species in the state that do not occur within the Californian Floristic Province.

Portions of Oregon and Baja California must also be assigned to the California Floristic Province on the basis of their common flora (Jepson 1925:3; Howell 1957). In Oregon, the boundary runs from Coos Bay southeast to the divide between the Umpqua and Rogue rivers and thence up the arm of the Rogue River valley (in which Ashland is situated) to the California line. This is an area of approximately 25,000 km². According to an analysis of the data in Peck (1961), approximately 90 species in this area are not found in California. In Baja California, the California Floristic Province includes the chaparral and forest belts of the Sierra Juárez and the Sierra San Pedro Mártir, but not their desert eastern flanks, and it extends south along the coast to a point near El Rosario (Howell 1957). Approximately 227

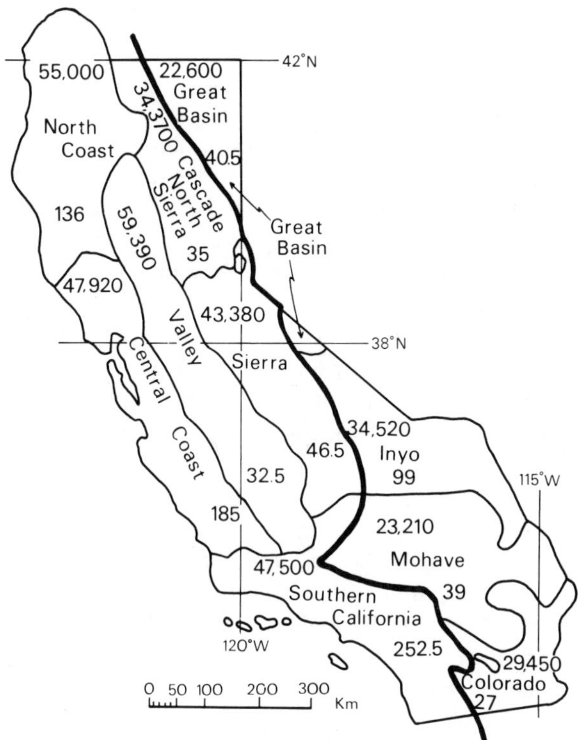

Figure 4-1. The floristic regions of California. In each region, the larger number shows area (in km²) and the smaller shows the number of endemics in 70 large and intermediate genera. The area west of the dark line is in the California Floristic Province, as described in text. Redrawn from Stebbins and Major (1965).

species in this area are not within the borders of California (R. V. Moran, pers. commun.), and the area itself occupies about 14,000 km².

The entire California Floristic Province thus comprises about 324,000 km² and has approximately 770 genera and 4437 species of native vascular plants. This is more than are present in the entire central and northeastern United States and adjacent Canada (Fernald 1950), a region some 10× larger! This region clearly contains the most remarkable assemblage of native plant species in all of temperate and northern North America.

THE CALIFORNIA FLORISTIC PROVINCE

Endemism

Endemic genera. In California 19 of the 863 native genera (2.2%) are endemic. Two of them—*Gilmannia* (Polygonaceae: 1 species) and *Swallenia* (-*Ectosperma,* Poaceae: 1)—are endemic to the Inyo Co. region beyond the boundaries of the California Floristic Province. In the California Floristic Province, there are some 770 genera of native vascular plants, of which 58 (7.5%) are endemic.

As stressed by Howell (1957), the impressive nature of the endemic genera in the California Floristic Province stems not only from their large number, but also from the nature of the genera involved. Many—*Sequoia, Sequoiadendron, Odontostomum, Jepsonia* (Ornduff 1969), *Umbellularia, Romneya, Carpenteria, Sarcodes Hesperelaea,* and *Lyonothamnus* are but a few examples—are outstanding relicts without any close relatives; they doubtless had wider ranges earlier in Tertiary time and are now restricted to relatively mild climates in the California region. Others, such as most of the 23 endemic genera of Asteraceae, probably had a more recent origin and have evolved in response to spreading aridity in Neogene time and more recently.

In addition to the genera that are strictly endemic to the California Floristic Province, many more are nearly restricted to this region. Some range beyond the borders of the Floristic Province onto the Hot Deserts or steppes of the Great Basin or occur in the woodlands and chaparral of Arizona. Others occur beyond the limits of the California Floristic Province in southern Oregon. The entire subtribe Madiinae of Asteraceae, tribe Heliantheae, with 12 genera and 96 species in western North America, is within this region endemic to California except for a few more widely ranging species of *Madia, Layia,* and *Raillardella,* one of *Lagophylla,* and the single species of *Blepharipappus,* all of which also occur within the boundaries of the California Floristic Province. The subtribe also includes three genera and about a dozen species of the Hawaiian Islands, however (Carlquist 1959). A similar pattern is found in the genus *Hesperocnide* (Urticaceae: 2), which has one species in California, a second in Hawaii.

Endemic species. About 1525, or 30.2%, of the vascular plant species of California are endemic. When the California Floristic Province is considered, the percentage of endemism climbs to 48%—some 2133 of the 4427 species. Such a high percentage of endemism is unusual for a continental area, although it may be equaled or exceeded in other regions with a Mediterranean climate. Some of the characteristics of this endemism are a relatively low proportion of monocotyledons (19.2%) and a relatively high proportion of annuals (27.4%).

Moreover, of the approximately 1098 species of annual dicots in the California Floristic Province, some 695 (63.3%) are endemic. Of the 2440 longer-lived species about 1095 (44.9%) are endemic.

Elements in the Flora of California

California has a rich and remarkable flora of native vascular plants for three principal reasons. First, as brought out in more detail in Chapter 5, many Tertiary relicts have survived within the California Floristic Province because of the mild climate of the region and its insulation from the extremes of a continental climate after the mid-Pliocene and subsequent uplift of the Sierra Nevada and associated ranges. Second, the interplay between the mesic elements derived from the Arcto-Tertiary Geoflora with the xeric ones associated with the Madro-Tertiary Geoflora has occurred in a region of high topographic relief and varied soils. Third, outbursts of speciation have occurred in certain woody and herbaceous, especially annual dicot, genera, mainly since the mid-Pliocene, and have been preserved in limited areas in the equable climate of the region. To illustrate these trends, I shall review the various historical elements represented in the California Floristic Province.

Arcto-Tertiary elements. Many of the plants of California have wide distributions in the temperate regions of Eurasia and North America or are clearly derived from taxa that have had such derivations. In other words, California is one of the most important areas of survival and persistence of relicts derived from the great northern temperate forests (Wood 1972; Wolfe 1975), because of equable maritime climate and, since late Pliocene time, the isolation of coastal areas with a moderate climate from spreading aridity from the interior by the uplift of the Sierra–Cascade axis. Examples of California taxa with Arcto-Tertiary affinities include Apiaceae, Caryophyllaceae, most Cyperaceae, Juncaceae, Liliaceae, most Menthaceae, Pinaceae, Ranunculaceae, many Rosaceae, and most Saxifragaceae.

The history of the California region from the Oligocene onward has been one of spreading aridity and progressive interplay between Madro-Tertiary and Arcto-Tertiary elements. A great many plants basically associated with the forests of the north temperate regions, such as *Carex, Delphinium, Iris,* and *Potentilla,* have radiated extensively in California, producing clusters of species. This radiation has generally occurred in the drier vegetation of Madrean derivation, and the dynamic relationships between the two sorts of vegetation, providing numerous opportunities for evolution, have been stressed by many authors.

Plants with Arcto-Tertiary affinities comprose about half of the genera and species found in California at present. There are relatively few annuals (about 12%) among the plants with Arcto-Tertiary affinities, in contrast to 28.6% in the flora of the state as a whole. On the other hand, the monocots amount to about 30% of the total, considerably higher than the percentage for the flora as a whole and more or less consistent with their proportion in other north temperate floras.

Madro-Tertiary elements. A second group of taxa appears characteristically to be associated with drier vegetation types, especially those derived from the Madro-Tertiary Geoflora (Axelrod 1958). These plants are known (or presumed) to have had a long history in association with the sclerophyll vegetation of western North America, and their history is discussed by Axelrod (1975) and in Chapter 5. They comprise about a quarter of the genera and about a third of the species found in California. Included are many Asteraceae and families such as Hydrophyllaceae, Limnanthaceae, and Papaveraceae. Approximately 55% of all these species are annual, representing a strong evolutionary radiation of many genera such as *Clarkia, Cryptantha, Hesperolinon, Lasthenia, Lupinus, Mimulus,* and *Phacelia,* which have been associated with the California Floristic Province probably since their origin. The Mediterranean climate of this region evolved only in latest Tertiary time (Axelrod 1973) and has never been in direct contact with any other region with a Mediterranean climate, this explaining in part the high endemism in California.

A few genera in California, including *Cupressus, Erodium, Lavatera,* the evergreen white oaks, and the annual Inuloid Asteraceae of the western United States, appear to be related to species of the Mediterranean region or the Near East. There is no reason to think that they or their immediate ancestors have ever been associated with the northern temperate forests. Axelrod (1975) has termed such links Madrean-Tethyan.

Other elements. About 15% of the plants of the California Floristic province appear to have radiated into this region from the deserts of other warmer parts of North America. Malvaceae are particularly interesting in this regard, since they seem mainly to have radiated from semiarid tropic-margin situations in the New World. A similar pattern is suggested by Euphorbiaceae. Several groups of plants with what might be considered Arcto-Tertiary or Madrean origins may have radiated first onto the deserts and then back into the California Floristic Province. These include *Astragalus, Cordylanthus, Penstemon, Phacelia,* and some groups of *Allium* and *Galium,* to name just a few. The center of evolution of many of these groups, and others such as Cactaceae, has clearly been in areas of vegetation drier than those now characteristic of the California Floristic Province. They have apparently undergone considerable evolutionary radiation in this area since mid-Pliocene time.

A few groups of plants, mainly water plants, are so widespread that they do not fit easily into any of the groups just mentioned. These cosmopolitan taxa comprise only about 2% of the flora of the state, but 11% of the families (e.g., Alismataceae, Potamogetonaceae). Among them are 2 endemic species of *Callitriche,* 1 of *Elatine,* and 1 of *Isoetes,* out of a total of 92 species, the great majority widespread.

Relationships with South America. Many families of flowering plants seem ultimately to have come from South America to North America (Raven and Axelrod 1974:627–8). Examples of these in the flora of California include Amaryllidaceae–Allieae, Cactaceae, Fabaceae–Mimosoideae, Loasaceae, Nyctaginaceae, and Zygophyllaceae. In addition, there are many other families in which some taxa appear to have come from South America and others from Eurasia; among them are Lauraceae (does *Umbellularia* come from the north or south?), Anacardiaceae, Asteraceae, Euphorbiaceae, Gentianaceae, Rhamnaceae, Solanaceae, and Rubiaceae. The entire order Chenopodiales (Centrospermae) may likewise have originated in West Gondwanaland (= Africa + South America), and thus all of its constituent families are ultimately to be traced there (Raven and Axelrod 1974); but the traces of such ancient events are difficult to interpret in the absence of a fossil record.

Some of the genera in California seem to have originated in South America and to have come to North America and California more recently. Examples are *Acaena, Calandrinia, Gaultheria, Larrea,* and *Pityrogramma.* Taxa such as *Bowlesia, Cardionema, Paronychia,* and *Soliva,* as they exist in California, may be very recent introductions from South America. The others are also recent in a geological sense.

As reviewed by Raven (1963), there are many species or species–pairs common to western North and South America. Most of these are self-compatible or autogamous herbs of open habitats that have achieved their disjunct ranges after the establishment of Mediterranean climates following the late Pleistocene (Moore and Raven 1970; Raven 1972, 1973; Axelrod 1973). The majority of species or species-pairs common to California and Chile have been derived from the north, and the presence of relatively unspecialized groups of species of *Boisduvalia, Chorizanthe,* and *Plagiobothrys* in South America probably reflects more than one time of arrival on that continent rather than origin there.

Relationships with the Mediterranean region. As reviewed by Raven (1973) and mentioned above, a limited number of similarities exist between the floras of the Mediterranean region proper and California. Some appear to have had their origin in differentiation from a common northern flora; examples are *Acer, Alnus, Cercis, Clematis, Crataegus, Fraxinus, Juniperus, Lonicera, Populus, Prunus* subg. *Emplectocladus, Rhamnus, Rosa, Rubus, Smilax, Staphylea, Viburnum,* and *Vitis.* Other genera may have migrated more directly between semiarid regions in Europe and North America via a more southerly route (Madrean-Tethyan: Axelrod 1975). In a few genera (*Galium, Juniperus, Pinus,* and *Quercus* are possible examples) there may be both northern and Madrean-Tethyan links.

A few related species–pairs exist in California and in the Mediterranean region, and these have been reviewed by Raven (1973:217). In view of the many weeds that have originated in the Mediterranean Basin, where agriculture is of such antiquity (Naveh 1967), it should not be surprising that some of the supposed links now seem to have had their origins in human introduction. Among these are *Plantago ovata* Forssk. (Stebbins and Day 1967; Bassett and Baum 1969) and possibly *Oligomeris linifolia,* the only species of Resedaceae in the New World. Others mentioned by Raven (1973) ought to be investigated in detail, as should a few other genera that have ranges disjunct between western North America and the Mediterranean region, including the Inuloid Asteraceae, *Aphanes, Caucalis, Lavatera, Valerianella-Plectritis,* and perhaps *Lupinus* and *Erysimum* (Meusel 1969). The far greater number of disjunct species between areas of Mediterranean climate in California and Chile than between California and the Mediterranean itself may be related to the paths of bird migration between the former areas (Raven 1973).

In several genera, among them *Antirrhinum, Astragalus, Atriplex, Bromus, Juncus* (Hermann 1948), *Poa, Polygonum, Senecio, Silene,* and *Trifolium,* annual species seem to have been independently derived from perennial species in the Old World and in the New. This is an interesting relationship, and one that should be investigated further in a comparative manner.

Derivation of the flora of California. Earlier workers who dealt with the origins of the flora of California (e.g., Abrams 1925, 1926; Jepson 1925; Campbell and Wiggins 1947) tended to stress its unique character. This is not surprising considering that some 58 genera (7.5% of the total) and 48% of the species are endemic in the California Floristic Province. Taking a broader view, however, and looking for ultimate sources, it is clear that the flora of California can be regarded as comprising two elements: a northern, or Arcto-Tertiary, and a southern, or Madro-Tertiary, element (Axelrod 1958), a relationship perceived by Asa Gray (1884) nearly a century ago. Within the area of the broad ecotone between these fundamentally different vegetation types, a Mediterranean climate developed after the mid-Pliocene, and provided a further stimulus for the proliferation of species and probably some herbaceous genera also. Finally, and probably mainly during late Pleistocene and Recent time, plants from the drier parts of the Madro-Tertiary region have entered the California Floristic Province, especially in the drier valleys of cismontane southern California, the San Joaquin Valley, and the inner South Coast Ranges (Axelrod 1966), and evolved a number of endemic species in that region.

More than half of the genera and species in the California Floristic Province have Arcto-Tertiary affinities. Another quarter of the genera and a third of the species are primarily associated with plant formations derived from the Madro-Tertiary Geolora; 20% of the genera and 15% of the species have been derived from the deserts in fairly recent time. The remaining 5% of the genera, with a small number of species, have had miscellaneous origins.

THE TRANSMONTANE FLORISTIC PROVINCES

Thus far, our discussion has been focused on the California Floristic Province, comprising about 75% of the state's area, with some 744 of the 875 genera (85%) and 4119 of the 5057 species (81%) found in the state. In this section, we shall consider the approximately 131 genera and 938 species that occur only east of the mountains and on the deserts, but within the boundaries of California. Here also are our only representatives of the families Acanthaceae, Burseraceae, Elaeagnaceae, Fouquieriaceae, Koeberlinaceae, Krameriaceae, Rafflesiaceae, and Simaroubaceae, 8 of the 154 native families in California.

As brought out clearly by Stebbins and Major (1965), and in Fig. 4-1, reproduced from their paper, there are four distinct transmontane regions in the eastern and southeastern portions of California with affinities outside the state. Each of these units will be discussed in turn.

Great Basin Region

In the northeastern corner of the state, this region comprises all of Modoc Co. and portions of Lassen, Shasta, and Siskiyou Cos., and it also includes northernmost Inyo Co. above the Owens Valley and Mono Co. east of the Sierra Nevada. Although many taxa of the California flora extend northeastward beyond the main Sierra–Cascade axis to the Warner Mts. and other portions of the Modoc Plateau (e.g., Griffin 1966), the flora of this region, which comprises some 22,600 km^2, is essentially that of the Great Basin. Unfortunately, only a portion of its northern, larger part is included by Cronquist et al. (1972) in their *Intermountain Flora.* The Warner Mts., which certainly belong to the Intermountain region, have also been excluded from that work, and they would make an excellent subject for a local floristic study.

Within the northern part of the Great Basin region in California are some 127 species that do not occur in the California Floristic province, including 67 found nowhere else in the state. In the southern part of the same region, there are about 147 species that do not occur in the California Floristic Province, together with 26 found nowhere else in the state, including the genus *Chaetadelpha. Lomatium ravenii* and *Mimulus pygmaeus* are the only endemics known to me to be restricted to the northern part of the Great Basin region within California. There are, however, at least seven endemics within the southern part, together with at least seven others endemic in California and at least five others ranging a short distance into Nevada.

The White Mts., situated at the southwestern corner of the Great Basin region (Stebbins and Major 1965; Cronquist et al. 1972), have been the subject of a valuable flora by Lloyd and Mitchell (1973). Following a detailed analysis of the 763 native species and varieties, the authors concluded that the White Mts. are a Great Basin desert range enriched during the Pleistocene with many boreal elements via the Sierra Nevada; these are especially prevalent in alpine and subalpine areas. Such families as Fagaceae and Ericaceae are absent in the White Mts. In addition, many

Sonoran–Mojavean taxa have come from the south, especially the latest Pleistocene and Recent time. There are only about four endemic species, all alpine. Because there is continuity between the White Mts. and the Inyo Mts., the boundaries of the Inyo and Great Basin regions have been used elastically in compiling the statistics in this section of the present paper.

Inyo Region

The Inyo region is the richest and most interesting in transmontane California. Some 454 species occur in California only within this region, including at least 199 not found elsewhere outside the state. There are at least 43 endemic species within the Inyo region, with 10 other species endemic in California but not found within the California Floristic Province, and more than 25 other species endemic in this area range a short distance into Nevada. The following genera are endemic in the Inyo region, some ranging out of the state of California: *Arctomecon* (Papaveraceae: 2), *Gilmania* (Polygonaceae: 1), *Hecastocleis* (Asteraceae, tribe Mutisieae: 1), *Oxystylis* (Capparaceae: 1), *Scopulophila* (Caryophyllaceae: 1), and *Swallenia* (Poaceae: 1; Soderstrom and Decker 1963). In addition, there are many unique, isolated species, such as *Eriogonum intrafractum, Boerhaavia annulata,* and *Tetracoccus ilicifolius.* Some of the endemics of the Death Valley region may have evolved rapidly, as discussed by Iltis (1957) for *Oxystylis,* whereas others have probably developed over long periods of time on the limestone cliffs or migrated into the region from the south after the last pluvial period. Finally, many genera occur in California only within this region, among them *Buddleia, Chamaesarcha* (excluding *Leucophysalis*), *Cowania, Enceliopsis, Enneapogon, Fallugia, Gaura, Halimolobus, Hedeoma, Laphamia, Maurandya, Monarda, Mortonia, Petradoria, Petrophytum, Physaria, Sanvitalia, Selinocarpus,* and *Tragia.* Endemism on limestone in the deep canyons flanking Death Valley is an important feature of that region.

Mojave Desert Region

In the Mojave Desert there are about 757 species, of which 22 are endemic, 6 others are not found in the California Floristic province but occur in other regions east of the mountains, and at least 7 others are basically endemic but range into Nevada or northwestern Arizona. About 51 of the species in the Mojave Desert are not found in any other part of the state. Approximately half of the species in the Mojave Desert occur in the California Floristic Province, extending to the mountains of southern California, the San Joaquin Valley, or the inner South Coast Ranges, whereas about a third do not. The lines between the Mojave region and the Inyo and Sonoran Desert regions are fairly smooth transitions, and it has been difficult to assign many species to these zones with precision. The preparation of a flora and detailed analysis of the Mojave Desert as a whole, including the portions that lie in Nevada, Utah, and Arizona, would be of great value.

Sonoran Desert Region

The Colorado Desert, the northwestern arm of the Sonoran Desert (Shreve and Wiggins 1964), has about 306 species not found in the California Floristic Province, including about 100 not recurring elsewhere in the state. There are at least eight endemics, together with two other species endemic to the Colorado and Mojave deserts. It is possibly the lack of summer rain over much of the Mojave Desert that restricts a number of desert plants in the California portion of their range to the Colorado Desert, among them the following genera: *Acleisanthes, Ammobroma, Ammoselinum, Beloperone, Bernardia, Bursera, Calliandra, Carnegeia, Colubrina, Condalia, Koeberlinia, Lyrocarpa, Malperia, Olneya, Pilostyles, Schizachyrium, Sesbania,* and *Teucrium.* A number of other genera, including *Achronychia, Cassia, Cercidium, Chilopsis, Dalea, Ditaxis, Fouquieria, Hoffmanseggia, Hyptis, Krameria,* and *Xylorhiza,* are characteristic primarily of the Colorado Desert in California, but range northward to the Mojave Desert or in a few instances to the deserts of Inyo Co. as well.

Summary of Section

Approximately 8 families, 131 genera, and 938 species of vascular plants occur within the boundaries of the state of California but not in the California Floristic Province. Despite the fact that all areas east of the mountains in California intergrade with and form integral parts of larger floristic regions that lie mainly outside the state, there are 2 endemic genera—*Gilmania* and *Swallenia*—and about 85 endemic species in this region, together with at least 4 other genera—*Arctomecon, Hecastocleis, Oxystylis,* and *Scopulophila*—and a minimum of 25 other species endemic to the Death Valley region of the Mojave Desert that extend only a short distance into southern Nevada or northwestern Arizona.

The most important center of endemism east of the mountains in California is the Inyo region, and a detailed floristic account of this region would be of high interest and utility. Two other areas richly deserving more detailed study are the Warner Mts. and the Mojave Desert. In both instances the composition of their floras reflects a mixture of "California" and "extra-Californian" elements that has not been analyzed in detail.

CENTERS OF ENDEMISM IN CALIFORNIA

Stebbins and Major (1965) have provided a useful discussion of endemism in California. Following Favarger and Contandriopoulos (1961), they recognized four classes of endemic species: (1) paleoendemics, (2) schizoendemics, (3) patroendemics, and (4) apoendemics. Paleoendemics are systematically isolated taxa, usually monotypic sections, subgenera, or genera, such as *Sequoia sempervirens, Phacelia dalesiana* (Constance 1952), or *Eriogonum intrafractum.* Schizoendemics are vicarious endemics that have more or less simultaneously diverged from a common ancestor, such as the species of *Ceanothus* sect. *Cerastes.* Patroendemics are diploid or low polyploid endemics that have given rise to more widespread higher polyploids; thus *Poa tenerrima* ($n = 21$) is a highly restricted species related to the widespread *P. scabrella* ($n = 42$). Additional examples are given by Stebbins and Major (1965:17–18). Finally, apoendemics are polyploid endemics that have arisen from widespread diploid or lower polyploid parents; thus *Camissonia hardhamiae* Raven ($n = 21$) of southernmost Monterey and northern San Luis Obispo Co. is derived from the more widespread *C. micrantha* Raven ($n = 7$) and *C. intermedia*

Raven (n = 14; Raven 1969). Other examples are noted by Stebbins and Major (1965:20–22).

Relict Areas

Stebbins and Major (1965) subdivided the state of California into 10 biotic subdivisions, as shown in Fig. 4-1. They then attempted to assess the distribution of relict species within the state, defining "relict species" as monotypic or ditypic genera confined to California or to a part of California and a neighboring region, as well as all species or pairs of species that are the only representatives of their genera in California and are separated from other species of their genera by a considerable distance. The distribution of these relict species is shown in Fig. 4-2, taken from their paper, which shows two areas of maximum concentration of relict species. One is in the Klamath Mt. area of northern California–southwest Oregon, where Arcto-Tertiary relicts predominate. The other is along the northern and western margin of

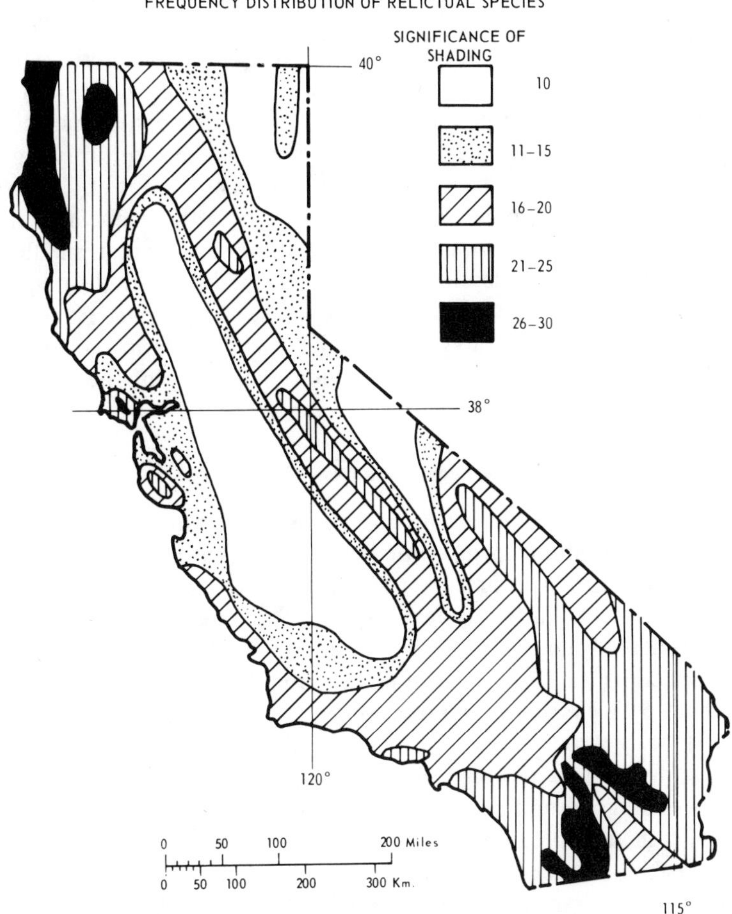

Figure 4-2. The presence of relict species in California. The unshaded areas, for example, contain 10% or fewer of all the relicts within the state; the black areas, 26–30% of the state relicts.

the Colorado Desert, from the Little San Bernardino Mts. along the east slope of the San Jacinto and Santa Rosa Mts., the Borrego Valley area, and southward into Baja California, where Madro-Tertiary relicts abound. As pointed out by Axelrod (1966), these two regions represent areas in which relict species and genera associated, respectively, with the Arcto-Tertiary and Madro-Tertiary geofloras have persisted. It has been the interplay between these elements from the Miocene onward that has given the flora of California its richness. Between them, a region of rather low frequency of relict species, compared with the diversity of the flora as a whole extends from the inner South Coast Ranges to the Tehachapi Mts. This is an area of active speciation but not of relict endemism.

Other Patterns of Endemism

Intermediate habitats in general were found to be richest in neoendemics (Stebbins 1974, Chapter 8), as I shall discuss in more detail in the following pages. In an effort to discover where such endemism in the state is most pronounced, Stebbins and Major (1965) sampled 70 large genera of the California flora. Figure 4-1 shows that the areas richest in endemics in these genera are southern California, the Sierra, and the Central Coast. The poorest are the Central Valley and the deserts. A high degree of endemism occurs in areas that have a great variety of habitats and that, during the Pleistocene at least, were not subjected to such catastrophic changes as glaciation. More endemics are associated with low mountains than with very high ones.

Analyzing the frequency of patroendemics in different regions of California, Stebbins and Major (1965:16) found that they were especially frequent in the Central Coast subdivision of the state, in coastal or maritime sites, and at middle elevations in the Sierra Nevada. Apoendemics, on the other hand, appeared to be less closely confined to coastal areas and to be frequent in the inner Coast Ranges and other areas of recently derived habitats, such as higher elevations in the Sierra Nevada. These results clearly suggest the need for additional study of these phenomena in California.

The Endemic Plants of California

In the late Pliocene, before the elevation of the Sierra Nevada–Cascade axis and the development of a Mediterranean climate in California, the present area of the California Floristic Province presumably was inhabited by about the same number of genera as at present, but many fewer species (perhaps 3400 in place of the 4437 found at present). Monocots probably comprised at least a quarter of the total number of species (vs. 18–19% at present), and the percentage of annuals was probably about 20% (vs. 28% at present). Not more than a third (instead of the present 48%) of the species were probably endemic. These estimates have been obtained by comparison with other regions of similar climate and topography.

Some outstanding examples of groups now seen as strongly associated with, or well developed in, the area at present are Asteraceae–Lactuceae and Madiinae, Hydrophyllaceae, Liliaceae–Calochorteae, Limnanthaceae, Onagraceae–Epilobieae and Onagreae, Papaveraceae–Eschscholzioideae and Platystemonoideae, Polemoniaceae–Gilieae, Polygonaceae–Eriogoneae, Scrophulariaceae–Rhinantheae and perhaps Cheloneae. These groups can reasonably be inferred either to have originated in Madrean vegetation or to have had their primary areas of differentiation there. They, as well as many of the genera endemic to the California Floristic Province, were probably in the region by the Pliocene and provide the distinctive character for the flora of California at present.

With the episodes of mountain building, cooling, and drying of the climate that were accelerated in the late Pliocene and have continued to the present, the numbers of species in many of these genera have been multiplied, in part by the strategies discussed below. Especially among annual dicots of certain families there has been an impressive proliferation of species. The products of such proliferation, existing side by side with relicts surviving in the local areas of equable climate, especially those in which a fair amount of precipitation falls in the summer, are a remarkable part of the flora of California but not an ancient one. They have developed in relation to an ecotone between Arcto-Tertiary and Madro-Tertiary elements in the flora established in the Miocene, and they multiplied during Quaternary time, as discussed in detail by Axelrod (1966).

EDAPHIC ENDEMISM IN CALIFORNIA

General

The restriction of plant taxa to certain soil types is an important feature of the flora in the state of California, just as it is in most arid and semiarid regions of the world. The kinds of patterns found in the state were discussed at some length by Mason (1946,a,b) and then studied experimentally by Kruckeberg (1951, 1954) and others. In general, edaphic endemism is related not to a particular requirement for a certain kind of soil, but rather to the absence of competition on a soil type that is nonzonal for the locality where it occurs (Kruckeberg 1951, 1954; Rune 1954). Particularly at the margins of their ranges, plants tend to grow on soil types that are not occupied by the dominant vegetation of the particular area, a pattern that Raven (1964) suggested would lead to edaphic endemism within a group of plants.

Plants are restricted to many kinds of substrate in California. Some, for example, occur on sands along the coast; examples are *Abronia latifolia, Ambrosia chamissonis, Camissonia cheiranthifolia* (Horn. ex Spreng.) Raimann, *Carpobrotus chilensis* (Mol.) N. E. Br., and *Cirsium rhothophilum.* The beach vegetation of the Pacific coast from Baja California north has been reviewed by Breckon and Barbour (1974). Such species as *Arctostaphylos silvicola* and *Erysimum teretifolium* are restricted to marine sandstones in the Santa Cruz Mts., where until recently an extralimital population of the middle to upper elevation Sierran *Calyptridium monospermum* (Hinton 1975) also occurred.

A particularly interesting edaphically delimited region in coastal northern California is the "pygmy forest" of Mendocino Co. (McMillan 1956, 1959, 1964; Jenny et al. 1969; Westman 1975; see also Chapter 9). Here stunted trees of *Cupressus pygmaea, Pinus muricata,* and *P. contorta* var. *bolanderi* occur within a forest belt consisting of normal-sized trees. They grow on a podsol soil with a white, bleached surface horizon and an iron-rich hardpan layer beneath. No species is known to be endemic to the pygmy forest, but at least two geographical races, *Arc-*

tostaphylos nummularia var. *nummularia* and *Pinus contorta* var. *bolanderi*, have such a distribution (Kruckeberg 1969).

A few species of higher plants are restricted to deposits of seabird guano on islands (Ornduff 1965). One that occurs in California is *Lasthenia minor* ssp. *maritima*, from the Farallon Islands northward.

Various kinds of volcanic soils are likewise important in controlling plant distributions in California; such patterns have been discussed by Mason (1946b). Additional examples are *Madia nutans*, restricted to volcanic ash, and *Polygonum bidwelliae*, *Arctostaphylos elegans*, *Ceanothus purpureus*, and *Eriastrum brandegeae*, endemic to volcanic soils in northern California.

Examples of edaphic endemism on nonvolcanic soils are *Monardella antonia* Hardham and *Eriogonum eastwoodianum* on diatomaceous shales in the inner South Coast Ranges and *Arctostaphylos myrtifolia*, *Helianthemum suffrutescens*, and *Eriogonum apricum* on Eocene laterite and sericitic schists in the Sierra Nevada foothills, especially around Ione, Amador Co. (Gankin and Major 1964).

Restriction to limestone soils is not as common in California as in the Mediterranean region, but there are some significant instances, especially in the desert ranges. Thus *Pinus longaeva* D. K. Bailey, *Cercocarpus intricatus*, *Petrophytum caespitosum*, and *Erigeron clokeyi*, among other species, are almost entirely confined to soils derived from dolomite in the southern White Mts. of eastern California (Lloyd and Mitchell 1973). *Cheilanthes sinuata* var. *cochisensis*, *C. jonesii*, *C. feei*, and *Adiantum capillus-verneris* are among the ferns that occur mainly or exclusively on limestone or other calcareous soils in California. In the mountains around Death Valley, as mentioned above, such restricted endemics as *Eriogonum intrafractum*, *Tetracoccus ilicifolius*, *Maurandya petrophila*, and *Phacelia mustelina* are restricted to limestone cliffs.

One of the most unusual assemblages of plants on limestone in California, that in the Convict Creek basin of Mono Co. on the east side of the Sierra Nevada, has been studied by Major and Bamberg (1963, 1967). Here occur together *Arctostaphylos uva-ursi* at its only station in the Sierra, *Draba nivalis* var. *elongata* (= *D. lonchocarpa* Rydb.), *Kobresia myosuroides*, *Salix brachycarpa*, and *Scirpus rollandii* [= *Trichophorum pumilum* (M. Vahl) Schinz & Thell.], the last four species not known elsewhere in California, but disjunct from the Rocky Mts., the eastern Great Basin, and the Wallowas–North Cascades. *Pedicularis crenulata* occurs at its only California station near the mouth of Convict Creek. This unusual assemblage of plants probably came to the Sierra Nevada from the north and persisted on the unusual soils of the Convict Creek basin because of a lack of competition with the dominant flora of the granodiorite of most of the remainder of the range (Chabot and Billings 1972).

Serpentine Soils

Many species of vascular plants in California are wholly or partly restricted to serpentine and associated ultrabasic rocks. A number of papers have been devoted to them, some discussing the evolutionary processes that account for their origin (Stebbins 1942; Raven 1964), others analyzing soil–plant relationships (e.g., Walker 1954; Kruckeberg 1954), and a third group considering the role of serpentine as an adaphic factor in plant distribution (e.g., Mason 1946a,b; Kruckeberg 1954, 1969; Whittaker 1960; Proctor and Woodell 1975). In his discussion of the serpentine problem, Kruckeberg (1954, 1969) suggested that adaptation to serpentine followed by biotype depletion and development of isolated populations into local endemics provides an explanation of serpentine endemism. This seems acceptable for endemism in local areas, but does not clarify either the problem of discontinuous distribution or the reasons for biotype depletion in nonserpentine areas. Many taxa seem to have originated as serpentine endemics by catastrophic selection in marginal populations (Lewis 1962; Raven 1964); marginal populations often occur on edaphic situations "unusual" for the species as a whole.

On the other hand, such reasoning does not explain the manner by which woody species like *Cupressus sargentii, Quercus durata,* and other taxa attained their present disjunct distributions on ultrabasic rocks. That some of these and other taxa had wide occurrences on nonserpentine sites from the Miocene well into the later Pliocene, that they were confined to serpentine areas as the more widespread ecotypes disappeared when summer rain decreased, and that adaptation to ultrabasic substrates removed them from competition with the adjacent, nonserpentine flora are tenets of a thesis discussed in more detail elsewhere (Raven and Axelrod, rept. in prep.).

MODE OF EVOLUTION OF CALIFORNIA PLANTS

Numerous publications discuss the general outlines of plant evolution. Here the focus is on two of the principal kinds of patterns that are characteristic especially of the California Floristic Province. They seem (in view of the analysis presented earlier in this paper) to be directly responsible for much of the proliferation of species and of endemics in this region. First, the proliferation of species in hybridizing complexes, especially of woody plants, will be discussed, and then the evolution of annual species, mostly dicots, in relation to the hypothesis of saltational speciation.

Hybridizing Complexes

A number of genera of woody plants characteristic of the chaparral and coastal sage vegetation of California are represented by many species. As pointed out by Wells (1969), the ability to sprout from the base after a fire is undoubtedly the primitive condition in such genera, and it has been lost in some species of *Arctostaphylos* and *Ceanothus.* Wells argued convincingly that loss of ability to crown-sprout in about four-fifths of the taxa of these genera tends to be associated with a more rapid adaptation to local conditions and the development of numerous species. In the crown-sprouting taxa, on the other hand, evolution has been more conservative and ecotypic differentiation tends to be the rule. The relationship between crown-sprouting and non-crown-sprouting species of evergreen sclerophylls in the 21 woody genera making up the chaparral vegetation of California and Baja California is shown in Table 4-1. Only 8 of the 50 species of *Arctostaphylos* crown-sprout (P. V. Wells, pers. commun.), and these are the most polymorphic in the genus.

Aside from the special features of the evolution of *Arctostaphylos* and *Ceanothus,* however, it seems to be the ability for differentiated species to hybridize in these

TABLE 4-1. Relation between mode of reproduction in response to fire and extent of differentiation of taxa among the evergreen sclerophylls in the 21 woody genera making up the chaparral vegetation of California and Baja California. Modified from Wells (1969)

Genus	Crown-Sprouting Taxa	Nonsprouting Taxa
Arctostaphylos	16 (21.3%)	59 (78.7%)
Ceanothus	12 (20.7%)	46 (79.3%)
Quercus	12	0
Rhamnus	7	0
Fremontia	6	0
Garrya	6	0
Lonicera	5	0
Cercocarpus	4	0
Adenostoma	3	0
Dendromecon	3	0
Chrysolepis	2	0
Comarostaphylis	2	0
Heteromeles	2	0
Prunus	2	0
Rhus	2	0
Cneoridium	1	0
Malosma	1	0
Ornithostaphylos	1	0
Pickeringia	2	0
Simmondsia	1	0
Xylococcus	1	0

groups that has made possible the relatively rapid reassortment of their genetic material in response to changing climate in a region of changing relief (Epling 1947b; Anderson 1949, 1953; Anderson and Stebbins 1954). They constitute what Grant (e.g., 1971) has termed homogamic complexes, hydrid complexes in which the derived species have originated by processes of hybrid speciation that do not involve drastic changes in the genetic system. Such processes seem to have been very important in California, and there are numerous examples of the proliferation of species by this means. I shall begin with a discussion of *Ceanothus* sect. *Cerastes*, one of the most thoroughly studied.

As pointed out by Johnston (1971), *Ceanothus,* with *Colubrina,* is one of the least specialized genera of Rhamnaceae, and it has no close relatives. There are two sections, *Ceanothus* and *Cerastes.* Section *Ceanothus* is widespread in Mexico and Central America and reaches the eastern United States and southern Canada. It has 32 species, 24 in California, and 21 endemic to the California Floristic Province. Section *Cerastes,* very distinct in its bizarre sunken stomatal crypts and many species with opposite leaves, has 21 species, 20 in California of which 17 are endemic to the California Floristic Province. All of these, as far as is known, have lost the ability to crown-sprout. Two of the species in California range northward to Oregon and Washington, and one, *C. greggii,* occurs south to Oaxaca and east to Texas. The only species not found in California, *C. lanuginosus* (Jones) Rose, occurs in the mountains from Chihuahua to San Luis Potosí.

Although intersectional hybrids are very difficult to obtain, Nobs (1963) demonstrated that all species of sect. *Cerastes* are interfertile. He studied in detail 11 species occurring in the North Coast Ranges of California and found a great diversity in the substrates on which these species occurred. Two were restricted to serpentine, with *C. jepsonii* having a highly discontinuous distribution on serpentine throughout the area; five were limited to the Pliocene Sonoma volcanics; and one was confined to Franciscan marine sandstones of Jurassic age. Within this complex and rapidly changing environmental mosaic, the ability of species of sect. *Cerastes* to hybridize and produce hybrid recombinants better suited to particular combinations of ecological factors than were their parents has doubtless given the group a great advantage. Such disturbed situations must have been very frequent in the North Coast Ranges of California from the Pliocene onward. The ability of hybrids to exploit disturbed situations was early pointed out by Anderson (1948). In turn, this has led to a multiplication of species in this restricted area of a typically Madrean group existing in more generalized vegetation types since at least Miocene time.

A similar pattern has been described for *Mimulus* sect. *Diplacus* by Beeks (1962), in which five species in southern California form a complex much like that described by Nobs for *Ceanothus* sect. *Cerastes.* Observations on species of *Salvia* sect. *Audibertia* in the same region suggest a similar role for hybridization in the evolution of that group (Epling 1938, 1947a,b; Anderson and Anderson 1954; Epling et al. 1962; Webb and Carlquist 1964; Grant and Grant 1964). In the same way, hybridization has probably played a role in the evolution of most genera listed in Table 4-1. As pointed out by Wells (1968,1969) this is especially true of the non-crown-sprouting species of *Arctostaphylos* (e.g., Epling 1947b; Gottlieb 1968), a genus in which all 50 species occur in California, although 1 has attained a circumboreal distribution and is disjunct to Guatemala. That hybridization may be an important factor in the evolution of a relatively small group is suggested by studies of *Adenostoma* by Edgar Anderson (1954). On the other hand, Young (1974), while reporting introgression in *Rhus,* did not regard it as important in the evolution of the group. Patterns suggestive of reticulate evolution at the diploid level also occur in the following additional woody genera of California, in each of which only a single chromosome number is known: *Amelanchier, Berberis, Cornus, Cupressus* (Wolf 1948), *Eriodictyon, Fraxinus* (Miller 1955), *Holodiscus, Keckiella, Lavatera, Lepechinia, Malacothamnus* (Bates 1963), *Opuntia* subg. *Platyopuntia* (Benson and Walkington 1965), *Pinus* (Haller 1962), including subsect. *Oocarpae, Quercus*

(Tucker 1952, 1953; Benson et al. 1967), *Ribes, Sambucus,* and the *Solanum umbelliferum* group. In at least three additional groups, the corresponding patterns probably also occur, but are accompanied by polyploidy: *Eriophyllum* (Mooring 1973, 1975), *Lupinus* (e.g., Nowacki and Dunn 1964), *Salix,* and *Zauschneria.*

As to polyploidy itself as an evolutionary force in California, little can be added to the perceptive discussions of Stebbins and Major (1965) and Stebbins (1971:179–90). Stebbins (1971:185) calculated the percentage of polyploidy for northern coastal California at about 36%, neither strikingly high nor low. As percentages of polyploidy for various habitats within the state are assessed, significant relationships may emerge, but these are certainly not evident at present. A number of excellent studies of polyploid complexes have been made on Californian plants, such as those of Clausen et al. on *Penstemon* (1940) and Madiinae (1945), that of Heckard (1968) on *Castilleja,* and those of Dempster and her co-workers on *Galium* (e.g., Dempster and Stebbins 1968).

Little also can be said about apomixis, which appears to play a very minor role in the flora of California, where it occurs in genera such as *Antennaria, Arabis, Crepis* (Babcock and Stebbins 1938), *Poa,* and *Potentilla* (Clausen et al. 1940), but without any large swarms of species comparable to those in *Crataegus* and *Rubus* in temperate eastern North America. The native Californian species of the last two genera are uniformly sexually reproducing.

Saltational Speciation

The proliferation of annual species in California is associated with polyploidy, autogamy, and saltational speciation. The notion of saltational speciation, the result of catastrophic selection, was first developed by Lewis (1962, 1966, 1969, 1973) as a result of his studies of evolution in *Clarkia.* In a variable and fluctuating climate, occasional extreme reduction of population size may be associated with reorganization of the gene pool and thus give rise to a new population that is at once reproductively and spatially isolated from the parental population and endowed with characteristics whereby it may be uniquely successful in the new habitat. Since marginal populations often grow in edaphic situations unusual for the species as a whole, saltational speciation may frequently give rise to derivative species of narrow edaphic requirements, as discussed by Raven (1964). Furthermore, as pointed out by Bartholomew et al. (1973), saltational speciation may be most important in the annual species of Mediterranean climates, in which seed longevity is usually limited and in which the genetic constitution of the population may therefore change readily if it is reduced to a very low number. This will be especially true if the plants are colonial. In perennial species such as *Delphinium,* or desert annuals such as *Linanthus parryae* (Epling and Dobzhansky 1942; Epling et al. 1960) and *Camissonia* sect. *Chylismia,* dormant genotypes may express themselves in subsequent years to "swamp" the effects of temporary changes due to particular unfavorable or unusual years (Raven 1962). Gottlieb (1973a,b) has stressed the fact that in plants in which seed storage is characteristic the kinds of fluctuation in population size that seem to have led to saltational speciation in *Clarkia* may not be operative, and the limits of the phenomenon should be explored.

At any rate, the derivation of *Clarkia franciscana* from *C. rubicunda* was one of the first instances to be explained by saltational speciation (Lewis and Raven 1958), although the term itself was not coined until several years later. Gottlieb (1973b) has shown that *C. franciscana* is not similar to *C. rubicunda* in the alleles it has fixed at six out of the eight enzyme systems that he studied, and that it is probably not derived from *C. rubicunda* in such a direct fashion as envisioned by Lewis and Raven (1958). On the other hand, the derivation of *C. lingulata* ($n = 9$) from *C. biloba* subsp. *australis* ($n = 8$), as outlined by Lewis and Roberts (1956), is strongly supported by similar electrophoretic data (Gottlieb 1974a). Other examples seem to include the origins of *Clarkia exilis* (Vasek 1958), *C. springvillensis, C. tembloriensis,* and *C.* "Caliente" from *C. unguiculata* (Vasek 1971; Lewis 1973) and the origin of a distinctive, marginal, self-pollinating population of *C. xantiana* studied by Moore and Lewis (1965). All of these evolutionary events in *Clarkia* and the whole pattern of evolution in the genus appear to be correlated with progressive adaptation to aridity (Lewis 1953a,b; Vasek 1964; Vasek and Sauer 1971; Davis 1970), this being accompanied by aneuploid changes in chromosome number, allopolyploidy, and autogamy.

The role of saltational speciation in groups other than *Clarkia* is difficult to assess. It has, however, been invoked by Kyhos (1965) to explain the origin of two annual species of *Chaenactis* (Asteraceae) on the deserts from a third that occurs in more mesic sites in cismontane California. On the other hand, the roles of autogamy and allopolyploidy have been assessed in critical biosystematic studies of many groups of plants found in California, among them some examples discussed earlier in this chapter under edaphic endemism. One of the first of these was the subtribe Madiinae of Asteraceae, studied by Clausen et al. (1945) and discussed by Clausen (1951), and there have been many subsequent detailed studies of other groups.

INTRODUCED PLANTS

The introduction of exotic plants into California began in May 1769, when Father Junipero Serra reached San Diego Bay and founded the first permanent European settlement in Alta California. Hendry (1951) has even inferred that *Rumex crispus, Erodium cicutarium,* and *Sonchus asper* were introduced earlier, but we agree with Frenkel (1970) that this is unlikely. At any event, Frenkel (1970) calculated that at least 16 species of exotic plants were established in California during Spanish colonization, 1769–1824; 63 additional species during Mexican occupation, 1825–48; and 55 more during American pioneer settlement, 1849–60. Thus at least 134 alien species were established in California by 1860. Parish (1920) listed 186 immigrant species, excluding fugitives and waifs, for southern California alone, and Jepson (1925) considered that, at the time he wrote, 292 species of established exotics constituted an integral part of the flora. Robbins et al. (1951) listed 437 introduced species of weeds in California. Munz and Keck (1959) listed 797 species of introduced plants, and Munz added 178 in his supplement (Munz 1968), for a total of 975 species (Howell 1972). Concerning this impressive figure, I agree with Howell (1972) that the representation of species of introduced plants has been greatly inflated by the inclusion of several categories of plants that do not constitute genuine elements in the flora of the state. Many new weeds are being established in California with each passing decade, and some (e.g., *Iris pseudacorus*: Raven and Thomas 1970) are spreading agressively; but others do not appear properly to constitute a part of the naturalized flora of the state.

Because of the equable climate of coastal California and the widespread irrigation during summer dry periods, many species will persist or even seed locally (see discussion by Howell et al. 1958), but they should not be considered members of the flora of the state until it is clear that they are capable of persisting without human interference. The distinction between such species and those accorded full status is by no means always clear, but I believe with Howell (1972) that a conservative interpretation is preferable.

With these factors in mind, the list of introduced plants presented by Munz (1958, 1968) has been scrutinized in the light of various local floras such as Howell (1970) and Howell et al, (1958), and my own observations. I have calculated that the introduced flora of California consists of 2 genera and 2 species of ferns, approximately 235 genera and 487 species of dicots, and about 83 genera and 165 species of monocots, of which 62 genera and 135 species are grasses: a total of 320 genera and 654 species. The monocots comprise about 25.2% of the total, considerably higher than for the native flora of the state (18.1%), entirely because of the rich representation of grasses, which constitute 20.6% of the total, as weeds. In the area covered by Gray's *Manual* (Fernald 1950), monocots comprise 15% of the total introduced plants and 28.2% of the native plants.

Of the 487 species of naturalized dicots, 1 is a tree (*Ailanthus*), 2 are vines (*Lonicera*), 26 are shrubs (16 genera), 163 are biennials and perennials, and 285 (58.5%) are annuals. Of the 165 species of naturalized monocots, 77 are perennials and 88 (53.3%) are annuals. For the naturalized flora of California as a whole, 57% of the species are annuals, a very impressive figure and one that is similar to the values for the weedy floras of the drier parts of the Near East and North America.

Of the 654 species of introduced plants in California, 544 are of Old World and only 110 of New World origin. Among the New World species, some 50 are found in the United States, 23 in tropical America, and 37 in South America. The Old World species include 18 from Australasia, 27 from central and southern Africa, and at least 19 from tropical Asia; the remaining 480 species are from temperate to arid Eurasia and North Africa. Thus nearly three-quarters (73%) of the naturalized flora of California is from Eurasia and North Africa, with very little of that proportion from eastern Asia; 4% is from central and southern Africa; about 2.5% each is from Australasia and tropical Asia; and only 16.8% is from the New World.

Among the native species of California are some that have become weedy during the past century. Robbins et al. (1951) listed 256 native weeds, and Stebbins (1965) provided a useful discussion of 41 that have distinct colonizing tendencies. Some weedy natives, like *Amsinckia, Claytonia, Eschscholzia, Lupinus,* and *Mimulus guttatus,* have become weeds on other continents also. These plants provide interesting opportunities for experimental studies. It is clear, however, that several millenia of cultivation around the Mediterranean, in the Near East, and in North Africa, possibly together with the higher proportion of annual species in the arid parts of the Old World, have resulted in the evolution of the vast majority of the weeds of California.

Because of the special problems concerned with the taxonomic treatment of weedy plants and their economic importance, the preparation of weed flora of California is highly desirable. Standard floras often state the ranges of weeds in vague terms, perhaps because the weeds are expected to attain wider ranges in due course, and collectors, if they do preserve specimens of weeds, often neglect to record the kind of detail that would be most helpful in evaluating their status. A new and critical inventory of the weeds of California would provide an important stimulus to the study of the flora of the state. The considerable advantages of weedy plants for many kinds of evolutionary studies have been stressed by Baker (1974), among others. Some outstanding modern studies have been conducted on the weeds of California, such as those of Iman and Allard (1965), Jain and Marshall (1967) and others on *Avena,* and others reviewed by Baker (1974). In addition, comprehensive investigations like that of Frenkel (1970) on the roadside vegetation of California are badly needed, for, despite the great economic importance of weeds, our ignorance of most aspects of their biology is profound. Agriculture would certainly benefit from such investigations.

PRESERVATION OF THE CALIFORNIA FLORA

No account of the plants of California would be complete without mentioning the threat to many species posed by urbanization, forest destruction, agriculture, and pollution. Many species of *Calochortus, Dudleya, Fritillaria,* and *Lilium* are threatened directly by those who take them from the wild to gardens, in which they may persist for only a short time. Certain cacti, as well as *Agave utahensis* var. *nevadensis* and *Nolina interrata,* are actually exploited commercially. An informed public should bring these practices to a halt at once. Native plants should be introduced into cultivation only by means of seeds and cuttings, and the rare plants of California should be stringently protected in the wild.

Of the endemic native plants of the California Floristic Province, some 28 species and 5 additional infraspecific taxa are apparently extinct already in the wild (Powell 1974, Ripley 1975, Ayensu 1975), although *Cordylanthus palmatus* has been preserved in cultivation and possibly reestablished at a habitat near those where it once grew (Chuang and Heckard 1973). In addition to these, 156 species and 80 additional infraspecific taxa in California are considered endangered, and 279 species and 71 additional subspecies are considered threatened (Ripley 1975, Ayensu 1975). The national list must be reviewed in the light of local lists such as that of Powell (1974), and will doubtless be extended. Thus 459 species and 157 additional infraspecific taxa in California, about 1 out of every 10 species in the total flora, are either extinct already or in danger of extinction. Looking at these figures in another way, about a quarter of all the threatened and endangered plants in the United States occur in California. In view of the extreme specialization and narrow restriction of many taxa in California, this is not surprising, and more and more taxonomic papers will include discussions of the probabilities of extinction of some or all the taxa they are treating, like those of Wilson (1970, 1972) and Chuang and Heckard (1973). It would certainly seem that the citizens of California would want to preserve their irreplaceable heritage of native plants if scientists would take the responsibility of informing them about the problem.

ACKNOWLEDGMENTS

I am especially grateful to H. K. Airy Shaw, D. M. Bates, D. Boufford, T. C. Croat, D. B. Dunn, C. Durrell, A. Gentry, W. F. Grant, L. R. Heckard, M. P. Johnson, M.

C. Johnston, R. M. King, R. M. Lloyd, H. E. Moore, Jr., R. Moran, R. Ornduff, G. L. Stebbins, R. G. Stolze, D. W. Taylor, J. M. Tucker, D. H. Valentine, and P. V. Wells for information they kindly provided during the course of this study. Support by a series of grants from the National Science Foundation is also gratefully acknowledged.

LITERATURE CITED

Abrams, L. 1925. The origin and geographical affinities of the flora of California. Ecology 6:1-6.

---. 1926. Endemism and its significance in the California flora. Proc. Internat. Congr. Plant Sci. 2:1520-1523.

Anderson, E. 1948. Hybridization of the habitat. Evolution 2:1-9.

---. 1949. Introgressive hybridization. John Wiley, New York. 109 p.

---. 1953. Introgressive hybridization. Biol. Rev. 28:280-307.

---. 1954. Introgression in Adenostoma. Ann. Missouri Bot. Gard. 41:339-350.

Anderson, E., and B. R. Anderson. 1954. Introgression of Salvia apiana and S. mellifera. Ann. Missouri Bot. Gard. 41:329-338.

Anderson, E., and G. L. Stebbins. 1954. Hybridization as an evolutionary stimulus. Evolution 8:378-388.

Axelrod, D. I. 1958. Evolution of the Madro-Tertiary Geoflora. Bot. Rev. 24:433-509.

---. 1966. The Pleistocene Soboba flora of southern California. Univ. Calif. Pub. Geol. 60:1-79, pl. 1-14.

---. 1973. History of the mediterranean ecosystem in California, pp. 225-277. In F. di Castri and H. Mooney (eds.). Mediterranean type ecosystems, origin and structure. Springer-Verlag, New York.

---. 1975. Evolution and biogeography of Madrean-Tethyan sclerophyll vegetation. Ann. Missouri Bot. Gard. 62:289-334.

Babcock, E. G., and G. L. Stebbins. 1938. The American species of Crepis. Their interrelationships and distribution as affected by polyploidy and apomixis. Carnegie Inst. Wash. Pub. 504:1-199.

Baker, H. G. 1974. The evolution of weeds. Ann. Rev. Ecol. Syst. 5:1-24.

Bartholomew, B., L. C. Eaton, and P. H. Raven. 1973. Clarkia rubicunda: a model of plant evolution in semiarid regions. Evolution 27:505-517.

Bassett, I. J., and B. R. Baum. 1969. Conspecificity of Plantago fastigiata of North America with P. ovata of the Old World. Canad. J. Bot. 47:1865-1868.

Bates, D. M. 1963. The genus Malacothamnus. Ph.D. dissertation, Univ. Calif., Los Angeles. 171 p.

Beeks, R. M. 1962. Variation and hybridization in southern California populations of Diplacus (Scrophulariaceae). Aliso 5:83-122.

Benson, L., E. A. Phillips, P. A. Wilder, et al. 1967. Evolutionary sorting of characters in a hybrid swarm. I. Direction of slope. Amer. J. Bot. 54:1017-1026.

Benson, L., and D. L. Walkington. 1965. The southern California prickly pears-invasion, adulteration, and trial-by-fire. Ann. Missouri Bot. Gard. 52:262-273.

Breckon, G. J., and M. G. Barbour. 1974. Review of North American Pacific Coast beach vegetation. Madroño 22:333-360.

Campbell, D. H., and I. L. Wiggins. 1947. Origins of the flora of California. Stanford Univ. Pub. Biol. Sci. 10:1-20.

Carlquist, S. 1959. Studies on Madinae: anatomy, cytology, and evolutionary relationships. Aliso 4:171-236.

Chabot, B. F., and W. D. Billings. 1972. Origins and ecology of the Sierran alpine flora and vegetation. Ecol. Monogr. 42:163-199.

Chuang, T.-I., and L. R. Heckard. 1973. Taxonomy of Cordylanthus subgenus Hemistegia (Scrophulariaceae). Brittonia 25:135-158.

Clausen, J. 1951. Stages in the evolution of plant species. Cornell Univ. Press, Ithaca, N.Y. 206 p.

Clausen, J., D. D. Keck, and W. M. Hiesey. 1940. Experimental studies on the nature of species. I. Effect of varied environments on western North American plants. Carnegie Inst. Wash. Pub. 520:1-452.

---. 1945. Experimental studies on the nature of species. II. Plant evolution through amphiploidy and autoploidy, with examples from the Madiinae. Carnegie Inst. Wash. Pub. 546:1-174.

Constance, L. 1952. Howellianthus, a new subgenus of Phacelia. Madroño 11:198-203.

Correll, D. S., and M. C. Johnston. 1970. Manual of the vascular plants of Texas. Texas Res. Found., Renner. 1881 p.

Cronquist, A., A. H. Holmgren, N. J. Holmgren, and J. L. Reveal. 1972. Intermountain flora. Vol. 1. Hafner, New York. 370 p.

Davis, W. S. 1970. The systematics of Clarkia bottae, C. cylindrica, and a related new species, C. rostrata. Brittonia 22:27-284.

Dempster, L. T., and G. L. Stebbins. 1968. A cytotaxonomic revision of the fleshy-fruited Galium species of the Californias and southern Oregon (Rubiaceae). Univ. Calif. Pub. Bot. 46:1-52.

Epling, C. 1938. The Californian salvias. A review of Salvia, section Audibertia. Ann. Missouri Bot. Gard. 25:95-188.

---. 1947a. Natural hybridization of Salvia apiana and S. mellifera. Evolution 1:69-78.

---. 1947b. Actual and potential gene flow in natural populations. Amer. Natur. 81:104-113.

Epling, C., and T. Dobzhansky. 1942. Genetics of natural populations. VI. Microgeographical races in Linanthus parryae. Genetics 27:317-332.

Epling, C., H. Lewis, and F. M. Ball. 1960. The breeding group and seed storage: a study in population dynamics. Evolution 14:238-255.

Epling, C., H. Lewis, and P. H. Raven. 1962. Chromosomes of Salvia: section Audibertia. Aliso 5:217-221.

Favarger, C., and J. Contandriopoulos. 1961. Essai sur l'endémisme. Bul. Soc. Suisse 71:384-408.

Fernald, M. L. 1950. Gray's manual of botany, 8th ed. American Book Co., New York. 1632 p.

Frenkel, R. E. 1970. Ruderal vegetation along some California roadsides. Univ. Calif. Pub. Geogr. 20:1-163.

Gankin, R., and J. Major. 1964. Arctostaphylos myrtifolia, its biology and relationship to the problem of endemism. Ecology 45:792-808.

Gottlieb, L. D. 1968. Hybridization between Arctostaphylos viscida and A. canescens in Oregon. Brittonia 20:83-93.

---. 1973a. Genetic differentiation, sympatric speciation and the origin of a diploid species of Stephanomeria. Amer. J. Bot. 60:545-554.

---. 1973b. Enzyme differentiation and phylogeny in Clarkia franciscana, C. rubicunda and C. amoena. Evolution 27:204-214.

---. 1974a. Genetic confirmation of the origin of Clarkia lingulata. Evolution 28:244-250.

---. 1974b. Genetic stability in a peripheral isolate of Stephanomeria exigua ssp. coronaria that fluctuates in population size. Genetics 76:551-556.

Grant, K. A., and V. Grant. 1964. Mechanical isolation of Salvia apiana and Salvia mellifera (Labiatae). Evolution 18:196-212.

Grant, V. 1971. Plant speciation. Columbia Univ. Press, New York. 435 p.

Gray, A. 1884. Characteristics of the North American flora. Amer. J. Sci. Arts, III, 28:323-340.

Griffin, J. R. 1966. Notes on disjunct foothill species near Burney, California. Leafl. West. Bot. 10:296-298.

Haller, J. R. 1962. Variation and hybridization in ponderosa and Jeffrey pines. Univ. Calif. Pub. Bot. 34:123-166.

Heckard, L. R. 1968. Chromosome numbers and polyploidy in Castilleja (Scrophulariaceae). Brittonia 20:212-226.

Hendry, G. W. 1931. The adobe brick as an historical source. Agric. Hist. 5:110-127.

Hermann, F. J. 1948. The Juncus triformis group in North America. Leafl. West. Bot. 5:109-124.

Hinton, W. F. 1975. Systematics of the Calyptridium umbellatum complex (Portulacaceae). Brittonia 27:197-208.

Howell, J. T. 1957. The California flora and its province. Leafl. West. Bot. 8:133-138.

---. 1972. A statistical estimate of Munz' Supplement to a California flora. Wasmann J. Biol. 30:93-96.

Howell, J. T., P. H. Raven, and P. Rubtzoff. 1958. A flora of San Francisco, California. Wasmann J. Biol. 16:1-157.

Iltis, H. H. 1957. Studies in the Capparidaceae. III. Evolution and phylogeny of the western North American Cleomoideae. Ann. Missouri Bot. Gard. 44:77-119.

Imam, A. G., and R. W. Allard. 1965. Population studies in predominantly self-pollinated species. VI. Genetic variability between and within natural populations of wild oats, Avena fatua L., from differing habitats in California. Genetics 51:49-62.

Jain, S. K., and D. R. Marshall. 1967. Population studies in predominantly self-pollinating species. X. Variation in natural populations of Avena fatua and A. barbata. Amer. Natur. 101:19-33.

Jenny, H., R. J. Arkley, and A. M. Schultz. 1969. The pygmy forest-podsol ecosystem and its dune associates of the Mendocino coast. Madroño 20:60-74.

Jepson, W. L. 1925. A manual of the flowering plants of California. Univ. Calif. Press, Berkeley. 1238 p.

Johnston, M. C. 1971. Revision of Colubrina (Rhamnaceae). Brittonia 23:2-53.

Kruckeberg, A. R. 1951. Intraspecific variation in the response of certain native plant species to serpentine soil. Amer. J. Bot. 38:408-419.

---. 1954. The plant species in relation to serpentine soils. Ecology 35:267-274.

---. 1969. Soil diversity and the distribution of plants, with examples from western North America. Madroño 20:129-154.

Kyhos, D. W. 1965. The independent aneuploid origin of two species of Chaenactis (Compositae) from a common ancestor. Evolution 19:26-43.

Lewis, H. 1953a. The mechanism of evolution in the genus Clarkia. Evolution 7:1-20.

---. 1953b. Chromosome phylogeny and habitat preference in Clarkia. Evolution 7:102-109.

---. 1962. Catastrophic selection as a factor in speciation. Evolution 16:257-271.

---. 1966. Speciation in flowering plants. Science 152:167-172.

---. 1969. Evolutionary processes in the ecosystem. BioScience 19:223-227.

---. 1973. The origin of diploid neospecies in Clarkia. Amer. Natur. 107:161-170.

Lewis, H., and P. H. Raven. 1958. Rapid evolution in Clarkia. Evolution 12:319-326.

Lewis, H., and M. R. Roberts. 1956. The origin of Clarkia lingulata. Evolution 10:126-138.

Lloyd, R. M., and R. S. Mitchell. 1973. A flora of the White Mountains of California and Nevada. Univ. Calif. Press, Berkeley. 202 p.

Major, J., and S. A. Bamberg. 1963. Some cordilleran plant species new for the Sierra Nevada of California. Madroño 17:93-109.

---. 1967. Some cordilleran plants disjunct in the Sierra Nevada of California, and their bearing on Pleistocene ecological conditions, pp. 171-189. In H. E. Wright and W. H. Osborn (eds.). Arctic and alpine environments. Indiana Univ. Press, Bloomington.

Mason, H. L. 1946a. The edaphic factor in narrow endemism. I. The nature of environmental influences. Madroño 8:209-226.

---. 1946b. The edaphic factor in narrow endemism. II. The geographic occurrence of plants of highly restricted patterns of distribution. Madroño 8:241-257.

McMillan, C. 1956. The edaphic restriction of Cupressus and Pinus in the Coast Ranges of central California. Ecol. Monogr. 26:177-212.

---. 1959. Survival of transplanted Cupressus in the pygmy forests of Mendocino County, California. Madroño 15:1-4.

---. 1964. Survival of transplanted Cupressus and Pinus after thirteen years in Mendocino County, California. Madroño 17:250-253.

Meusel, H. 1969. Beziehungen in der florendifferenzierung ven Eurasien und Nordamerika. Flora, Abt. B, 158:537-564.

Miller, G. N. 1955. The genus *Fraxinus*, the ashes, in North America, north of Mexico. Cornell Univ. Agric. Sta. Mem. 335:1-64.
Moore, D. M., and H. Lewis. 1965. The evolution of self-pollination in *Clarkia xantiana*. Evolution 19:104-114.
Moore, D. M., and P. H. Raven. 1970. Cytogenetics, distribution, and amphitropical affinities of South American *Camissonia* (Onagraceae). Evolution 24:816-823.
Mooring, J. S. 1973. Chromosome counts in *Eriophyllum* and other Helenieae (Compositae). Madroño 22:95-97.
---. 1975. A cytogeographic study of *Eriophyllum lanatum* (Compositae, Helenieae). Amer. J. Bot. 62:1027-1037.
Munz, P. A. 1968. Supplement to a California flora. Univ. Calif. Press, Berkeley. 224 p.
---. 1974. A flora of southern California. Univ. Calif. Press, Berkeley. 1086 p.
Munz, P. A., and D. D. Keck. 1959. A California flora. Univ. Calif. Press, Berkeley. 1681 p.
Navah, Z. 1967. Mediterranean ecosystems and vegetation types in California and Israel. Ecology 48:445-459.
Nobs, M. A. 1963. Experimental studies on species relationships in *Ceanothus*. Carnegie Inst. Wash. Pub. 623:1-94.
Nowacki, E., and D. B. Dunn. 1964. Shrubby California lupines and relationships suggested by alkaloid content. Genet. Polon. 5:47-56.
Ohwi, J. 1965. Flora of Japan. Smithsonian Inst., Washington, D.C. 1067 p.
Ornduff, R. 1965. Ornithocoprophilous endemism in Pacific Basin angiosperms. Ecology 46:864-867.
---. 1969. Ecology, morphology and systematics of *Jepsonia* (Saxifragaceae). Brittonia 21:286-298.
Parish, S. B. 1920. The immigrant plants of southern California. Bull. South. Calif. Acad. Sci. 14(4):3-30.
Peck, M. E. 1961. A manual of the higher plants of Oregon, 2nd ed. Bindfords and Mort, with Oregon State Univ. Press, Corvallis. 936 p.
Powell, W. R. 1974. Inventory of rare and endangered vascular plants of California. Calif. Native Plant Soc., Spec. Pub. 1. Berkeley. 56 p.
Proctor, J., and S. R. J. Woodell. 1975. The ecology of serpentine soils. Adv. Ecol. Res. 9:255-366.
Raven, P. H. 1962. The systematics of *Oenothera* subgenus *Chylismia*. Univ. Calif. Pub. Bot. 34:1-122.
---. 1963. A flora of San Clemente Island, California. Aliso 5:289-347.
---. 1964. Catastrophic selection and edaphic endemism. Evolution 18:336-338.
---. 1969. A revision of the genus *Camissonia* (Onagraceae). Contrib. U. S. Natl. Herb. 37:161-396.
---. 1972. Plant species disjunctions: a summary. Ann. Missouri Bot. Gard. 59:234-246.

---. 1973. The evolution of "mediterranean" floras, pp. 213-224. In H. Mooney and F. di Castri (eds.). Evolution of Mediterranean ecosystems. Springer-Verlag, New York.

Raven, P. H., and D. I. Axelrod. 1974. Angiosperm biogeography and past continental movements. Ann. Missouri Bot. Gard. 61:539-673.

Raven, P. H., and J. H. Thomas. 1970. Iris pseudacorus in western North America. Madroño 20:390-391.

Ripley, S. D. 1975. Report on endangered and threatened plant species of the United States. Serial No. 94-A: i-iv, 1-200. U. S. Govt. Printing Office, Washington, D. C.

Robbins, W. W., M. K. Bellue, and W. S. Ball. 1951. Weeds of California. State of Calif., Dept. Agric. 547 p.

Rune, O. 1954. Notes on the flora of the Gaspé Peninsula. Svensk Bot. Tidskr. 48:117-136.

Shreve, F., and I. M. Wiggins. 1964. Vegetation and flora of the Sonoran Desert. 2 vols. Stanford Univ. Press, Stanford, Calif. 1740 p.

Smith, G. L., and A. M. Noldecke. 1960. A statistical report on a California flora. Leafl. West. Bot. 9:117-132.

Soderstrom, T. R., and H. F. Decker. 1963. Swallenia, a new name for the California genus Ectosperma (Gramineae). Madroño 17: 88.

Stebbins, G. L. 1942. The genetic approach to problems of rare and endemic species. Madroño 6:241-258.

---. 1965. Colonizing species of the native California flora, pp. 173-195. In H. G. Baker and G. L. Stebbins (eds.). The genetics of colonizing species. Academic Press, New York.

---. 1971. Chromosomal evolution in higher plants. Addison-Wesley, Reading, Mass. 216 p.

---. 1974. Flowering plants, evolution above the species level. Harvard Univ. Press, Cambridge, Mass. 399 p.

Stebbins, G. L., and A. Day. 1967. Cytogenetic evidence for long continued stability in the genus Plantago. Evolution 21:409-428.

Stebbins, G. L., and D. Major. 1965. Endemism and speciation in the California flora. Ecol. Monogr. 35:1-35.

Tucker, J. M. 1952. Taxonomic interrelationships in the Quercus dumosa complex. Madroño 11:234-251.

---. 1953. The relationship between Quercus dumosa and Quercus turbinella. Madroño 12:49-60.

Vasek, F. C. 1958. The relationship of Clarkia exilis to Clarkia unguiculata. Amer. J. Bot. 45:150-162.

---. 1964. The evolution of Clarkia unguiculata derivatives adapted to relatively xeric environments. Evolution 18:26-42.

---. 1971. Variation of marginal populations of Clarkia. Ecology 53:1046-1051.

Vasek, F. C., and R. H. Sauer. 1971. Seasonal progression of flowering Clarkia. Ecology 52:1038-1045.

Walker, R. B. 1954. Factors affecting plant growth on serpentine soils. Ecology 35:259-266.

Webb, A., and S. Carlquist. 1964. Leaf anatomy as an indicator of Salvia apiana-mellifera introgression. Aliso 5:437-449.

Wells, P. V. 1968. New taxa, combinations, and chromosome numbers in Arctostaphylos (Ericaceae). Madroño 19:193-244.

---. 1969. The relation between mode of reproduction and extent of speciation in woody genera of the California chaparral. Evolution 23:246-267.

Westman, W. E. 1975. Edaphic climax pattern of the pygmy forest region of California. Ecol. Monogr. 45:109-159.

Whittaker, R. H. 1960. Vegetation of the Siskiyou Mountains, Oregon and California. Ecol. Monogr. 30:279-338.

Wilson, R. C. 1970. The rediscovery of _Abronia alpina_, a rare specialized endemic of sandy meadows in the southern Sierra Nevada, California. Aliso 7:201-205.

---. 1972. _Abronia_. I. Distribution, ecology and habit of nine species of _Abronia_ found in California. Aliso 7:421-437.

Wolf, C. B. 1948. Taxonomic and distributional studies of the New World cypresses. Aliso 1:1-250.

Wolfe, J. A. 1975. Some aspects of plant geography of the Northern Hemisphere during the Late Cretaceous and Tertiary. Ann. Missouri Bot. Gard. 62:264-279.

Wood, C. E., Jr. 1972. Morphology and phytogeography: the classical approach to the study of disjunctions. Ann. Missouri Bot. Gard. 59:107-124.

Young, D. A. 1974. Comparative wood anatomy of _Malosma_ and related genera (Anacardiaceae). Aliso 8:133-146.

CHAPTER

5

OUTLINE HISTORY OF CALIFORNIA VEGETATION

DANIEL I. AXELROD

Botany Department, University of California, Davis

Introduction 140
Conifer forests 141
North Coast Forest 145
Mixed Conifer Forest 147
Subalpine Forest 154
Summary 157
Sclerophyll vegetation 157
Mixed Evergreen Forest 158
Oak–Laurel Forest 162
Oak Woodland–Savanna 164
Quercus agrifolia woodland 166
Quercus engelmannii woodland 166
Juglans californica woodland 167
Insular woodland 168
Digger pine-oak woodland 169
Piñon–Juniper Woodland 169
Closed–Cone Pine Forest 171
Subtropic Relicts 172
Summary 172
Scrub vegetation 173
Chaparral 173
Time–space relations 173
Origin and age 175
Coastal Sage 178
Desert 179
Great Basin 179
Mojave–Sonoran 181
Age and origin 182
Late Cretaceous (or older?) 183
Paleocene–Eocene 183
Oligocene–Miocene 184
Pliocene 184
Quaternary 184
Desert disjuncts 184
Summary 185
Conclusions 186
Literature cited 188

INTRODUCTION

The taxa that contribute to the distinctive vegetation of California have come from diverse sources and at different times during the Cenozoic. To understand its historical development, it is necessary to recall that early in the Tertiary the region differed greatly in geography, climate, and vegetation from that of today. The area west of the San Andreas Fault was then situated approximately 300 miles (480 km) farther south, with Monterey near the present Mexican border. The high mountains that now dominate the landscape were not yet in existence. The Sierra Nevada was the site of a low plain in the north, across which flowed a 3 mile (4.8 km) wide river, the remnants of which may still be seen on the scarp above Susanville. To the south, smaller rivers drained across most of the present Sierran axis, which was chiefly an area of low, rolling terrain, increasing gradually in altitude to a low summit region in the Mt. Whitney area. Much of the Klamath Mt. region was also a lowland, the present Coast Ranges were largely submerged, and so were the Transverse Ranges. A shallow, tropical sea lapped against the low Sierran rise, extending southward along the front of the present Peninsular Ranges, which were not then appreciably elevated, and swinging west near Riverside and then south to Oceanside (Nilsen and Clarke 1975).

As judged from the early Eocene Chalk Bluffs flora of the central Sierra Nevada (MacGinitie 1941), the Ione from the Central Coast Ranges at Corral Hollow (MacGinitie 1941), undescribed small Paleocene floras in the Alberhill Formation in the Santa Ana Mts. (UC Mus. Pal.*), the Goler Formation in the El Paso Mts. (UC Mus. Pal.), and the Eocene Moonlight flora in the northern Sierra (UC Mus. Pal.), the region was covered with temperate rainforest grading southward into subtropical forest. Tree ferns, palms, cycads, and numerous large-leaved evergreen dicots of tropical and subtropical families, notably Anacardiaceae, Burseraceae, Caesalpiniaceae, Lauraceae, Meliaceae, Moraceae, Simaroubaceae, Sabiaceae, and Sapindaceae, were dominant. They imply a climate like that now in southeastern Mexico or in southeastern China, where there is heavy rainfall during the warm season and winters are relatively dry and frost free. There are few records of late Eocene and Oligocene floras in California, chiefly because much of the preserved sedimentary record represents marine deposition. The early Oligocene La Porte flora (Potbury 1935) from the northern Sierra Nevada is an evergreen subtropical forest that lived under a drier climate than the Chalk Bluffs, as is evident from its smaller leaves.

The evergreen rainforest reached north into Washington, where members of temperate alliances increase in number, as shown by records of *Metasequoia, Sequoia, Alnus, Carya, Castanopsis, Cladastris, Pterocarya, Quercus, Smilax, Ulmus,* and *Viburnum* (Wolfe 1968). The forest extended up broad valleys into the interior, as indicated by the Lower Clarno and other floras of central Oregon, and by the Roslyn flora of central Washington (Chaney 1938a). Vegetation rapidly changed in composition commencing at the edge of the cordilleran upland. This region included the area of the present Great Basin, stretching northward into British Columbia, with its axis generally through central Nevada and Idaho. From the Jurassic and on down into Paleocene time, this region served as the major source for sediment carried west to the Pacific and east to the great central seaway (Gilluly

* University of California (Berkeley) Museum of Paleontology.

1963). Although the major mountain axes and basins of the Rocky Mts. were outlined by the Eocene, the ranges were still low and regional uplift of a mile or more (1.6 km) did not occur until late in the Tertiary (King 1958; Curtis 1975). It was in this cordilleran axis that forerunners of taxa which contributed to vegetation allied to that now in California are first recorded in abundance during the middle and late Eocene.

Vegetation in the cordilleran region changed with increasing altitude. Lowland subtropical rainforest gave way successively to a rich mixed deciduous hardwood–evergreen dicot forest, to conifer–deciduous hardwood forest, and then to subalpine conifer forest (Axelrod 1968). Local dry areas in the lee of scattered low ranges in the central and southern cordilleran region supported taxa representing sclerophyllous scrub, oak–piñon woodland, and thorn forest (Axelrod 1958b; Becker 1961, 1969; MacGinitie 1953, 1969). The species are related to the near-modern taxa that had appeared by the early Miocene. Because of a continuing cooling and drying trend, they shifted farther west during the Miocene and Pliocene, displacing more mesic forests along the coast that included numerous taxa now in distant regions with summer rainfall. Thus, to interpret the history of California vegetation, it is necessary to follow the coastward movement of forest, sclerophyll, and scrub vegetation, noting important changes in composition and distribution especially during Miocene and Pliocene times, when essentially modern (or living) species were in existence.

To aid in the presentation of vegetation history, Figs. 5-1 to 5-4, 5-6, and 5-7 summarize the geographical occurrence and age of the representative floras to be discussed. Rather than using fossil names unfamiliar to the present readers, fossil species are referred to by the use of parentheses which indicate their modern affinities. Thus *Acer* (*macrophyllum*) refers to a fossil species very similar to the living big-leaf maple. Genera and species in the fossil floras that are allied to taxa now in areas with summer rain are termed *exotics*. These are in the eastern United States (e.g., *Acer, Carya, Fagus, Taxodium, Ulmus*), in eastern Asia (*Betula, Cercidiphyllum, Ginkgo, Metasequoia, Zelkova*), or in the southwestern United States and northern Mexico (e.g., *Quercus, Robinia, Sabal, Sapindus, Thouinia*).

This chapter includes many data that have been gathered with the generous support of the National Science Foundation.

CONIFER FORESTS

By late Paleogene time, two major types of conifer forests were established along the cordilleran axis, stretching from southern British Columbia (Princeton flora: Arnold 1955; Rouse and Mathews 1961) southward for 1500 miles (2400 km) into south-central New Mexico (Hillsboro, Winston floras: UC Mus. Pal.). One type from moderate altitudes (2500–4500 feet; 750–1350 m) represents forests in which both montane conifers and deciduous hardwoods contribute importantly to the community. A second, higher-elevation type includes those in which subalpine conifers are the principal taxa, and deciduous hardwoods are wholly absent or nearly so. Both forest types showed a gradual change in composition southward. The rich northern montane conifer and conifer–hardwood forests with 10–12+ conifers gave way first to those in the central cordilleran area, in which conifers and associated hardwoods were less common, then to forests in the south dominated by one or two conifers with few associates.

In the north, taxa suggest a humid climate with precipitation well distributed through the year, and with low to moderate ranges of temperature. In the transitional area of the central Rocky Mts., precipitation evidently was lower and more seasonal with less winter rain, and summer temperatures were higher. Farther south, effective precipitation decreased, and there probably was considerable drought stress as compared with regions farther north.

Taxa in the northern forests seem directly ancestral to many of those which contributed to the Neogene record in Washington, Oregon, and Idaho, and to those which today make up the rich conifer forests of the humid Northwest. Taxa in the upland Paleogene floras of the central cordilleran area appear ancestral to the Neogene floras of the Coloradan region, as well as the Great Basin, the Sierra Nevada, and border areas. Some taxa of the more impoverished forests of the southern Rocky Mts. persisted over that area and the Great Basin. The forests shifted coastward during the Miocene, arriving in the present coastal strip only in the later Pliocene as climate and terrain became suited for them (Figs. 5-1 and 5-2).

Thus the present conifer forests of California were derived directly from Neogene

Figure 5-1. Some Tertiary floras with taxa representing mixed conifer–deciduous hardwood forest and mixed conifer forest.

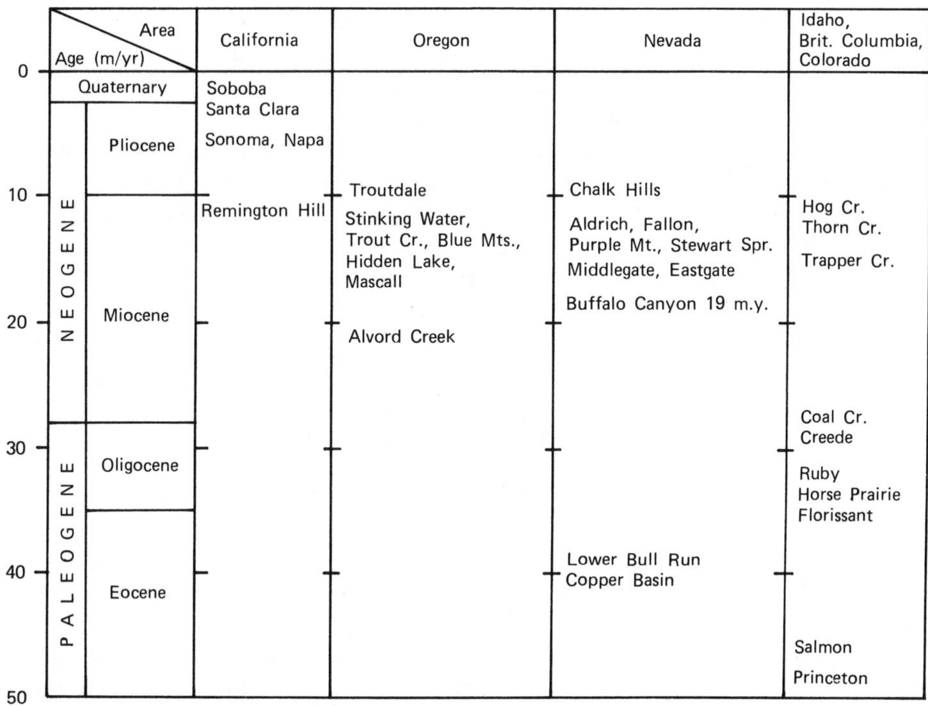

Figure 5-2. Relative ages of Tertiary floras representing mixed conifer–deciduous hardwood forest or, in the late Miocene and Pliocene, chiefly mixed conifer forest (see Fig. 5-1).

forests whose taxa characterized two different floristic and, by inference, climatic regions. Miocene floras from Oregon and Idaho (Chaney and Axelrod 1959), like Blue Mts., Thorn Creek, and Trout Creek, have more mesic taxa than those to the south (Axelrod 1956). Among the conifers, *Metasequoia, Sequoia,* and *Thuja* were in the north, whereas *Sequoiadendron* and its frequent associates, notably *Abies* (*concolor*), and *Pinus* (*ponderosa*) were chiefly to the south. *Chamaecyparis, Picea, Pseudotsuga,* and *Tsuga* were in both areas. The northern floras have numerous deciduous hardwoods represented now by taxa in areas with ample summer rain (*Betula, Carya, Fagus, Ostrya, Quercus, Sassafras, Tilia,* and *Ulmus*). They are less frequent in floras of similar age farther south, floras in which species representing sclerophyllous evergreen vegetation are more abundant. The relations suggest that the northern region had higher precipitation, much summer rainfall, and moderate temperatures.

That the Miocene forests lived under a climate with ample rain in summer is indicated by the occurrence in them of numerous taxa that live only in such regions today. Apart from genera that are now exotic to the West, including *Carya, Diospyros, Fagus, Liquidambar, Pterocarya, Ulmus,* and *Zelkova,* these floras also have species of *Acer, Alnus, Betula, Fraxinus, Populus, Quercus,* and *Rhus* whose nearest descendants occur only in summer rain climates. As summer rainfall gradually decreased after the middle Miocene (15–16 m.y.), these taxa, which were regular members of conifer–deciduous hardwood forests, and also of mixed deciduous mesophytic forests at lower altitudes (e.g., Succor Creek, Spokane, Grand Coulee, Lower Ellensburg, Neroly floras), gradually disappeared from the region.

Total precipitation had decreased sufficiently by the middle Pliocene so that very few forests remained over the lowlands of the Great Basin. It was now largely a region of oak–juniper woodland and savanna, with open grasslands in which riparian woodlands of *Acer, Betula, Populus,* and *Salix* were common. Rare records of small-leaved ecotypes of *Populus, Rhus,* and *Ulmus,* whose descendants now occur in summer rain areas, imply a general deficiency of summer rain over the interior at this time. By contrast, contemporary floras from the windward side of the Sierra–Cascade axis reveal widespread forest and woodland vegetation that indicates higher rainfall, and taxa that require summer rain are recorded there in larger numbers. The floristic evidence suggests that the Sierra-Cascade axis now formed a more effective rain shadow than earlier. This is consistent with structural evidence which shows that major uplift of the Sierra-Cascade-Peninsular range axis occurred during the later Pliocene and Quaternary, and that the Colorado Plateau and bordering ranges were also elevated to their present heights at about the same time. These events created a *double* rainshadow over the Great Basin. Ingress of moist tropical air from the Gulf of Mexico was reduced, and major cyclonic disturbances from the north Pacific could no longer effectively add moisture to the lowlands in the lee of the Sierra-Cascade axis. Furthermore, to judge from climatic principles, it seems clear that, as the oceans became progressively colder, convectional summer showers along the coastward slope would be reduced. This resulted in a gradual decrease in exotic taxa which evidently required summer rain. As thermal contrasts increased, the Pacific high pressure system gradually strengthened and became more stable in summer. This not only tended to block cyclonic disturbances from the north, but also blocked tropical storms from penetrating up the west coast during summer. In this manner, long, dry summers gradually developed over the region, with a full Mediterranean climate appearing only during the Quaternary.

A changing thermal regime is the second factor that accounts for differences between Neogene forests and present ones derived from them. The global temperature gradient was lower during the Miocene because ice caps were only commencing to appear and oceans at middle and higher latitudes were warmer than at present. Furthermore, since the major topographical features that now control provincial climates had not yet appeared, mild temperature characterized the region. Winter and summer extremes were moderated.

As the thermal gradient increased, there was a trend to more extreme summer temperatures and higher evaporation (= drought stress). Together with decreased precipitation in general and a shift in its seasonal occurrence, environments were progressively less favorable for rich mixed conifer–deciduous hardwood and montane conifer forests. As they were restricted to more mesic mountain areas over the interior and toward the coast, their composition and diversity were altered appreciably. Conifer–deciduous hardwood forests were converted into the surviving mixed conifer forests. The rich montane conifer (or subalpine) forests lost numerous taxa and were confined to more local provinces.

Inasmuch as Miocene taxa of the diverse conifer forests closely resembled living species, their morphology and anatomy, even though gained under other climates and at earlier times, appear to have preadapted them to the emerging Mediterranean climate of long summer drought. Gradual adaptation to the progressively longer period of summer drought probably was functional, by a shift in the time of germination into the earlier, moister part of the year.

North Coast Forest

This modern forest stretches from southern Alaska into coast-central California. The forest reaches its greatest diversity on the coastward slopes of the Siskiyou Mts. near the California border. Dominants include *Abies grandis, Chamaecyparis lawsoniana, Picea sitchensis, Pseudotsuga menziesii, Sequoia sempervirens, Thuja plicata,* and *Tsuga heterophylla.* To the south, the forest becomes progressively more attenuated and confined to the coastal strip. *Sequoia* and *Pseudotsuga* are the sole dominants south of Pt. Arena.

Fossil species allied to present members of the north coast forest occur in the Miocene of eastern Oregon and central Idaho. A number of species in the Blue Mts. flora (Chaney and Axelrod 1959) notably *Pseudotsuga* (*menziesii*), *Sequoia* (*sempervirens*), *Thuja* (*plicata*), *Tsuga* (*heterophylla*), *Acer* (*macrophyllum*), *Alnus* (*rhombifolia*), *Lithocarpus* (*densiflora*), *Mahonia* (*aquifolium*), *Populus* (*trichocarpa*), *Prunus* (*demissa*), *Rhamnus* (*californica*), and *Vaccinium* (*parviflorum*), are similar to (or identical with) those in the forest east of Crescent City. In addition, the flora includes species related to those typical of the mixed conifer forest that occurs at the inner margin of the north coast forest or at altitudes above it in mountainous areas. Among these are *Abies* (*concolor*), *A.* (*magnifica*), *Alnus* (*tenuifolia*), *Betula* (*papyrifera*), and *Picea* (*engelmannii*). Other taxa in the flora include a number of exotics, distributed in *Acer* (*pictum, saccharinum*), *Ailanthus, Betula* (*lenta*), *Crataegus* (*hupehensis*), *Fagus, Hydrangea* (*aspera*). *Ilex, Liquidambar, Nyssa, Ostrya, Pterocarya, Sassafras,* and *Viburnum.*

Similar relations are shown by the late Miocene Thorn Creek flora (Smith 1941), situated 50 miles (80 km) north of Boise, Idaho. Species allied to those in the coast forest include *Abies* (*grandis*), *Pseudotsuga* (*menziesii*), *Sequoia* (*sempervirens*), *Thuya* (*plicata*), *Tsuga* (*heterophylla*), *Acer* (*negundo*), *A.* (*rhombifolia*), *Alnus* (*sinuata*), *Amelanchier* (*alnifolia*), *Populus* (*trichocarpa*), *Quercus* (*chrysolepis*), *Salix* (*hookeriana*), *S.* (*lasiandra*), and *S.* (*scouleriana*). Taxa whose nearest descendants now contribute to the mixed conifer forest are *Betula* (*papyrifera*), *Holodiscus* (*dumosus*), *Picea* (*breweriana*), *Pinus* (*monticola*), *Populus* (*angustifolia*), *Quercus* (*garryana*), *Q.* (*kelloggii*), *Salix* (*caudata*), and *Symphoricarpos* (*racemosus*). Exotics are represented by *Acer* (*pictum, saccharinum*), *Betula* (*lenta*), *Carpinus, Carya, Fagus, Populus* (*grandidentata*), *Quercus* (*montana*), and *Ulmus.* The Thorn Creek flora, which appears to be somewhat younger than that of the Blue Mts. and occurs 220 miles (350 km) east-southeast, has a definitely drier aspects. Oaks are especially abundant, and many of the species have smaller leaves than related taxa in somewhat older floras in the nearby region (Axelrod 1964).

It is likely that species of the mixed conifer–deciduous hardwood forest in these and other floras were members of a richer Neogene conifer–deciduous hardwood forest made up of taxa which have since been segregated into two distinct forests—north coast conifer and mixed conifer. Many floras display similar mixtures of taxa, notably the late Eocene Copper Basin (Axelrod 1966a), the transitional Eo-Oligocene Florissant (MacGinitie 1953), and the Oligocene Horse Prairie, Medicine Lodge, and Ruby in southwestern Montana (Becker 1961, 1969). They show a gradual decrease in members of the north coast forest taxa in the interior and a replacement there by taxa ancestral to the Petran mixed conifer forest. By the early Miocene, the Creede flora of Colorado (UC Mus. Pal.) had most of its taxa allied to species that live within 100 miles (160 km) of the site today. These include *Picea* (*pungens*), *Pinus* (*aristata*), *Pseudotsuga* (*glauca*), and diverse dicots distributed among *Acer* (*neomexicanum*), *Cercocarpus* (aff. *paucidentatus*), *Chamaebataria, Jamesia, Peraphyllum, Philadelphus, Physocarpus, Populus* (aff. *balsamifera*), *Ribes* (aff. *cereum, aureum*), *Robinia*, and *Salix*. Exotics are very rare, including a *Mahonia* and an extinct fir (*A. longirostris*) of Asian affinity. Inasmuch as mesic forest taxa such as *Sequoia* and *Chamacyparis* were common in the nearby region in the early Oligocene (Florissant flora), and the Creede flora has no significant group of exotic taxa, modernization had largely been completed in this drier, sunnier region by the end of the Oligocene, though continued evolution of some of its taxa (e.g., *Cercocarpus*) produced modern, living species. This relation was foreshadowed in the Eocene of Nevada and Idaho, for there, taxa of conifer-deciduous hardwood and subalpine forests already had a near-modern morphology. This is clearly exhibited by species of *Acer, Aesculus, Alnus, Berberis, Betula, Amelanchier, Crataegus, Populus, Prunus, Quercus, Rosa, Sassafras, Ulmus* and *Zelkova*, as well as the montane conifers distributed in *Abies, Picea, Larix, Pinus, Pseudotsuga, Tsuga, Chamaecyparis* and *Sequoia*. By contrast, forests in the subtropical lowlands to the west supported evergreen dicots, many of which are either extinct, or differ considerably from modern species. Rates of evolution in these areas were quite different, but for reasons that are still obscure (Axelrod 1966a: 53–57).

The preceding evidence indicates that Paleogene forests were composed of taxa whose descendants now contribute to both coast forest and mixed conifer forest, associated with alliances now in the eastern United States, eastern Asia, and also in central Mexico. The floras reveal a general change from rich northern forests thriving under humid climate to poorer southern ones living under subhumid climate. These forests were segregated into forerunners of modern forests as early as Oligocene time in the interior. Whereas derived members of the mixed conifer forest persisted over the interior (as the Petran forest), and in the higher mountains toward the coast (as the Sierran forest), coast forest taxa were restricted to the moister, more oceanic climate of the Pacific Northwest and to the coastal strip southward into central California. Taxa allied to those in the coast and mixed conifer forests were on the leeward slopes of the western Cascades in the late Oligocene (Brown and Wolfe, *in* Peck *et al.* 1964). The ancestral forest did not appear on the windward slopes until the Pliocene because climate there was too mild and warm for it in the late Miocene, as shown by the numerous mesic conifers (*Cephalotaxus, Fokienia, Metasequoia, Taxodium*) and humid dicots, (laurels, *Exbucklandia, Magnolia*, and others) in the Mollala (UC Mus. Pal.) and Neroly floras (Condit 1938).

Coast conifers first appear in numbers in the coastal strip of Oregon and California in the Pliocene. The Troutdale flora of western Oregon (Chaney 1944) has several coast forest taxa, including *Chrysolepis* (=*Castanopsis*) *chrysophylla*: not *Salix*), *Chamecyparis, Cornus, Sequoia*, and *Tsuga* (not *Abies*). They were associated with numerous exotic hardwoods, notably species of *Diospyros, Liquidambar, Prunus* (*davidiana*), *Pterocarya, Quercus* (*aliena-prinus*), *Sassafras, Ulmus*, and *Zelkova*.

The late Pliocene (3.5 m.y.) Sonoma flora near Santa Rosa represents a mesic phase of the coast forest like that now on the coast from Ft. Bragg northward (Axelrod 1944d). Accumulating at the head of a broad marine embayment that lay to the west, it includes *Abies* (*grandis*) *Chamaecyparis* (*lawsoniana*), *Pseudotsuga* (*menziesii*), *Sequoia*, and *Tsuga* (*heterophylla*). Their associates were *Acer, Arbutus, Lithocarpus, Myrica, Rhododendron,* and *Vaccinium*. Significantly, large, winged seeds of spruce similar to those of *Picea breweriana* occur in the flora, as do leaves of aspen. This implies that segregation of the coast and mixed conifer forests was not yet complete in this region. Sclerophylls occupied nearby warmer sites, apparently lingering here from the middle Pliocene. This flora has the last significant group of exotics in the Late Pliocene of California, including species of *Ilex, Mahonia, Persea, Trapa* and *Ulmus*.

The Napa flora from the Sonoma Formation, 35 miles (55 km) southeast of Santa Rosa, represents vegetation near the inner margin of north coast forest in the late Pliocene (Axelrod 1950a). Apart from *Sequoia* and *Pseudotsuga*, it has a pine related to *Pinus ponderosa*, as well as aspen, *Populus tremuloides*. They imply that segregation of the north coast and mixed conifer forests had not yet been completed in this somewhat drier, interior region. Numerous sclerophylls are present, notably *Cercocarpus* (*montanus*), *Heteromeles* (*arbutifolia*), and *Quercus* (*wislizenii*). Exotics include abundant leaves of *Persea* and a leaflet of *Pterocarya*. The assemblage resembles the modern forest at Los Posadas State Park on the east slope of the Napa Mts., east of St. Helena. As compared with the Sonoma flora at Santa Rosa, the Napa has an interior aspect.

The Quaternary history of the north coast forest is scarcely known, because of a lack of fossil evidence. A significant record is provided by woods of *Sequoia* and *Pseudotsuga* in the Carpinteria flora near Santa Barbara. A geological study (Putnam 1942) shows that the woods, reported as sea drift (Chaney and Mason 1933), occur in a stream deposit inland from the shore (Axelrod 1967b:295). The wood chips probably floated downstream from canyons in nearby hills to the east where the trees were somewhat removed from the fog-bound, closed-cone pine forest near the shore. The present *Arbutus-Lithocarpus-Quercus-Umbellularia* mixed evergreen forest at San Marcos Pass above Santa Barbara, which includes *Mahonia pinnata* and other northern taxa, appears to be a remnant of associates of the north coast forest that shifted southward during the later Quaternary. The stand of *Pseudotsuga* on the north slope of the Purisima Hills (Griffin 1964), 85 miles (137 km) southeast of the southernmost stand on Salmon Creek in the Santa Lucia Range, may also be a remnant of that retreat.

Mixed Conifer Forest

The mixed conifer forest of the Sierra Nevada extends southward into northern Baja California, north into the southern Cascades and Klamath Mts., and down the Coast Ranges. The present forest was derived chiefly from a richer, ancestral one that covered the Great Basin and Columbia Plateau regions in the middle Miocene.

Recall that during the middle Miocene the northern half of the Sierra was quite low (Axelrod 1957). Climate was warm and moist, and the area supported a rich deciduous hardwood forest with numerous evergreen dicots. This is shown by the Miocene Carson Pass and Gold Lake floras (UC Mus. Pal.), with *Actinodaphne, Carya, Liquidambar, Magnolia, Persea, Platanus, Pterocarya, Quercus* (evergreen and deciduous), and many other dicots, but no montane conifers. Closely similar vegetation covered the seaward slopes of the western Cascades, as shown by floral lists for that region (Brown and Wolfe, *in* Peck et al. 1964). A few mesic conifers are recorded in those floras, which are situated fully 450 miles (725 km) farther north, where the climate was moister and cooler. It was in areas to the lee of the Miocene Sierra–Cascade axis that montane conifers and temperate hardwoods occurred in large numbers.

Miocene floras east of the Sierra Nevada are composed of species whose nearest descendants contribute to both the modern Sierran mixed conifer and the fir-subalpine forests (Axelrod 1956, 1959, 1976b). Most of the adequately sampled Miocene floras have species allied to *Abies* (*concolor*), *Chamaecyparis* (*lawsoniana*), *Pinus* (*ponderosa*), *Sequoiadendron* (*giganteum*), and their frequent associates, together with species of *Arbutus, Lithocarpus, Quercus* (*chrysolepis, wislizenii*) and others from the bordering broad-leaved sclerophyll forest. In addition, each flora commonly has species related to those that are confined chiefly to subalpine forests today, notably *Abies* (*magnifica, shastensis*), *Picea* (*breweriana*), *Pinus* (*monticola*), *Populus* (*tremuloides*), *Sorbus* (*scopulina*), and *Tsuga* (*mertensiana*). Under a Miocene climate of mild temperature and summer rainfall, they were regular members of a rich mixed forest. Even today descendants of the fossil species enter the mixed conifer forest in the Siskiyou Mts. There, temperatures are more moderate, the wet season is somewhat longer, some rain occurs in summer, drought stress is lower, and its period is briefer than in the Sierra. While these taxa no doubt ranged up to subalpine levels wherever mountains were sufficiently high (~ 2,000–2,500 feet, or 1,800–1,900 m above the basins), the actual fossils evidently were derived from trees living near the sites of accumulation. An altitude of about 5,000 feet (1,500 m) estimated for the lower margin of Miocene subalpine forest in the western Great Basin parallels its altitude in the western Siskiyous today.

Forests had largely disappeared from the lowlands of the western Great Basin by the early Pliocene. This is indicated by the Truckee (US Mus. Pal.) and Esmeralda (Axelrod 1940b) floras, which are dominated by live oak woodland vegetation. Montane conifers are absent or are represented by only an occasional specimen, frequently a winged seed. Precipitation increased toward the west margin of the area (as it does today), and mixed conifer forest was present there into the early Pliocene. For example, the Mansfield Ranch flora (UC Mus. Pal.) in Hungry Valley west of Pyramid Lake has *Abies* (*concolor*), *Pinus* (*ponderosa*), and *Pseudotsuga* (*menziesii*), with associates of *Amorpha, Cercocarpus,* and *Mahonia.* In addition, the Chalk Hills flora near Virginia City shows that a *Sequoiadendron* forest persisted there into the early Pliocene (Axelrod 1962). Associates include *Abies* (*concolor*), *Chamaecyparis* (*lawsoniana*), *Pinus* (*monticola*), and *Picea* (*breweriana*), with an understory of *Amelanchier, Arbutus, Azalea, Ceanothus* (*intergerrimus, velutinus*), *Lithocarpus, Populus, Prunus,* and *Salix.* By the early Pliocene precipitation was nearly 20–25 inches (50–65 cm) over the interior, as compared with 30–35 inches (70–80 cm) in the late Miocene at the lower edge of *Sequoiadendron* forest. Exotic taxa were now largely absent from the region.

Taxa representing the middle Pliocene Alturas (Axelrod 1944e) and Red Rock (UC Mus. Pal.) floras of northeastern California, and the Deschutes (Chaney 1938b) and Pendleton (UC Mus. Pal.) floras of eastern Oregon, are mainly riparian, implying that forests probably were confined chiefly to the mountains, where precipitation was higher. At the east edge of the Sierra, the Verdi flora (Axelrod 1958a) is also dominated by riparian species. In addition, it contains *Abies* (*concolor*), *Pinus* (*attenuata, lambertiana, ponderosa*), with associates of *Arctostaphylos* (*nevadensis*), *Ceanothus* (*cuneatus*), *Prunus* (*emarginata*), and *Ribes* (*roezelii*), suggesting that the forest reached out onto the floodplain from the low ridge to the west. Mild winters are implied by species of *Quercus* (*lobata, wislizenii*) that contributed to oak woodland and savanna on well-drained warmer slopes, and by the dominant small, ovate to oval leaves of *Populus* that are like those of *P. trichocarpa* in the mild winter lowlands along the coast from west–central California southward. Climate probably was too arid in summer for the more mesic montane conifers (e.g., *Abies shastensis, Picea breweriana, Pinus monticola*) in the nearby mixed conifer forest.

In the basin ranges of eastern California, as well as most of those in the Great Basin, the altitudinal belt that mixed conifer forest might be expected to occupy is now an open sage–meadowland–aspen zone (Billings 1951). It is situated above the piñon–juniper belt and below scattered subalpine stands of *Pinus flexilis* and *P. longeava* ("*aristata*"). This treeless zone has numerous perennials and herbs that are regular associates of the mixed conifer forests of the Wasatch and adjacent ranges or of the Sierra. The absence of mixed conifer forests from most of the basin ranges today, as compared with their rich representation there in the Neocene, is due partly to low precipitation in the mountains, but chiefly to high evaporation, which militates against easy seedling establishment (see Kramer 1969; Zahner 1968). Hence it is understandable that in the eastern (Ruby, Egan, Schell, Creek, Snake) and the southern (Spring, New York, Clark) ranges, where 15–20% of the total precipitation falls in summer, restricted mixed forests of *Abies concolor, Pinus ponderosa,* and *Pseudotsuga* occur, with the last-named taxon confined to the east.

Mixed conifer forest was present in the Sierra during the Pliocene, as shown by the Mt. Reba flora from a site now at 8600 feet (2615 m) (UC Mus. Pal.). It is dominated by leaves of *Quercus* (*chrysolepis*) and *Lithocarpus* (*densiflora*), taxa that today make up evergreen sclerophyllous forest and also contribute to the understory of mixed conifer forest in the northern part of the range and in the Coast Ranges. *Pseudotsuga* (*menziesii*) is common in the flora, as is an extinct *Cupressus* allied to the Asian *funebris* and *torulosa.* Rare remains (two specimens each) of *Abies* (*concolor*) and *Sequoiadendron* (*giganteum*) indicate that they lived nearby in cooler sites. Some summer rain is implied by *Carya* and *Cupressus.* It is estimated that the flora was at an altitude near 2500 feet (750 m), implying that uplift of about 6000 feet (1800 m) has since taken place, and that the present subalpine forest at the fossil locality invaded the area much later. Subalpine taxa that were in nearby Nevada in the Miocene may have occupied higher ridges or upper slopes of small volcanos in the summit section by the early Pliocene, especially in the area south of Carson Pass.

The development of montane Mediterranean climate largely accounts for the composition and distribution of taxa that make up the present mixed conifer forest. As the range of summer temperature increased and as summer drought became more severe, numerous species that had been part of the Neogene mixed conifer forest into the Pliocene (and early Quaternary) were restricted northward. They now occur in northwestern California or farther north in Washington–Idaho, in areas with a longer precipitation season, some summer rain, and milder temperatures in summer.

In addition to range contraction, other taxa now have a narrower altitudinal range. For example, *Abies* (*concolor*) regularly occurred with *Chamaecyparis* (*lawsoniana*), *Pinus* (*ponderosa*), *P.* (*lambertiana*), *Pseudotsuga* (*menziesii*), *Sequoiadendron* (*giganteum*), and their usual associates, as well as with some of the subalpine taxa noted above. In addition, these floras also have many sclerophylls, including *Arbutus, Cercocarpus, Lithocarpus, Quercus* (*chrysolepis, wislizenii*), and others that occur today altitudinally well below *Abies concolor.* Thus, in the Pliocene Verdi flora at the east front of the Sierra, *A. concoloroides* (*concolor*) mingles with live oaks similar to *Q. wislizenii* and *Q. lobata.* Today it is found chiefly in the upper part of the mixed conifer forest belt, mostly at levels above 3,500 feet (1,000 m) in the central Sierra Nevada, and well above descendants of the sclerophyllous taxa that ocurred with it in the Pliocene and earlier. Its shift to higher levels presumably reflects its inability to establish at lower altitudes where summers are now too hot and dry for establishment of its seedings. Fowells (1965) has noted ". . . early drought often hampers seeding development in the southwest, but not to the north where there is more reliable summer rain."

A similar explanation may account for the troublesome problem raised by aspen. Aspen is associated with coast redwood in the late Pliocene Sonoma flora, with live oaks at the lower edge of mixed conifer forest in the Pliocene Verdi flora of western Nevada, and with chaparral and woodland vegetation at the lower margin of mixed conifer forest in the Soboba flora ($\sim$1.5 m.y.) in interior southern California. All these vegetation types are widely removed from aspen today. Since germinating seeds of aspen are especially susceptible to heat and drought (Fowells 1965), climates with summer rain and mild temperature would be more suited to them. In the Rocky Mountains, and also in the Spring (Charleston) Mountains, southern Nevada, aspen and white fir both occur at the lower forest margin, presumably because of adequate summer moisture. Hence, their restriction to higher altitudes in California as summer rains decreased becomes understandable.

Sequoiadendron also reached well down into the sclerophyll zone during the late Miocene, but today is usually situated well above it. It too may have been confined to higher altitudes, to the more equable and moister part of the mixed forest zone, because of increased drought and higher summer temperature at lower levels during the interglacial ages, and especially during the later Xerothermic periods.

A number of taxa that find optimum development in the Coast Ranges and in the Siskiyou–Klamath region are scattered southward into the central Sierra, the area farther south evidently being too dry and hot for them in summer. Among the taxa in this group are *Arbutus menziesii,* **Castanopsis chrysophylla, Ceanothus velutinus,* **Corylus rostrata, Gaultheria ovatifolia, Leucothoe davisiae,* **Lithocarpus densiflora, Rhamnus purshiana,* **Taxus breviflora,* and *Vaccinium parviflorum.* Those marked by an asterisk have been reported from Quaternary pollen floras from the Kern Plateau (Axelrod and Ting 1961). They probably were eliminated there during the Xerothermic, being restricted northward to moister climates with lower drought stress.

A number of species that were in the late Miocene mixed conifer forest in the western Great Basin are now represented by descendants in areas well removed from the more mesic parts of the derived, modern forest. *Acer* (*negundo*), *Arbutus* (*menziesii*), *Castanopsis* (*chrysophylla*), *Fraxinus* (*oregona*), *Lithocarpus* (*densiflora*), *Quercus* (*chrysolepis, wislizenii*), and *Torreya* (*californica*) now occur in the lower, warmer part of the Sierra and in bordering sclerophyll vegetation. Segregation seems related to the development of colder winters, for heavy snows occur regularly today in areas where mixed forest attains best development. All of these broad-leaved sclerophyll taxa were severely injured (broken tops and limbs) by heavy snowfall and ice in the Santa Lucia Range in January 1974, with heavy damage extending down into the lower oak savanna (*Quercus agrifolia, Q. douglasii, Q. lobata*) zone. The occurrence of fossil relatives of sclerophyll taxa with *Abies* (*concolor*), *A.* (*magnifica*), *Picea* (*breweriana*), and *Sequoiadendron* (*giganteum*) in the late Miocene supports the suggestion that they were gradually confined to the lower part of mixed conifer forest as colder winters and heavy snow developed during the Quaternary.

Fossils similar to *Abies bracteata* are in the middle and late Miocene of western Nevada (Axelrod 1976b). They are associated with taxa allied to those in the ecotone between mixed conifer forest and evergreen sclerophyll vegetation. Santa Lucia fir is now in a near-coastal region, associated with broad-leaved sclerophyll vegetation and the lower mixed conifer forest. Many of its regular tree associates are similar to (or identical with) those in the Miocene of Nevada, notably *Acer macrophyllum, A. negundo, Arbutus menziesii, Lithocarpus densiflora, Pinus ponderosa, Platanus racemosa, Pseudotsuga menziesii, Quercus chryselopis,* and *Q. wislizenii,* as are a number of shrubs, including *Ceanothus cuneatus, C. intergerrimus, Cercocarpus betuloides,* and *Heteromeles arbutifolia. Abies bracteata* now inhabits an area where temperatures are higher in winter and lower in summer than in the Sierra, conditions more nearly like those postulated for the Miocene. It presumably was eliminated from the Sierra Nevada by colder and snowier climates of the glacial ages and by hotter and drier climates of the interglacials.

Mixed conifer forest has several codominants in most of its area. However, in the Central Coast Ranges the forest is scattered and the taxa are unequally represented. Thus *Pinus ponderosa* forms nearly pure stands in the Napa Range, in the Santa Cruz Mts., the Mt. Hamilton Range, and on Figueroa Mt. north of Santa Barbara. *Pinus ponderosa, P. lambertiana,* and *Calocedrus* are in the Santa Lucia Range, but mostly as single dominants, not in a mixed forest. Furthermore, the mesic *Abies concolor* is not in the region. Unequal representation of forest taxa in the Central Coast Ranges probably results from the comparatively low altitude. More mesic forest species that were present there during the moister phases of the Quaternary, like *Pinus lambertiana* and *Calocedrus decurrens* in the Santa Cruz Mts. (Dorf 1930), evidently could not retreat upward into montane sites sufficiently cool and moist as the climate became warmer and drier, and were therefore eliminated.

Sierra redwood (*Sequoiadendron*) occurs as widely discontinuous small groves north of the Kings River, and as nearly continuous large ones to the south (Jepson 1910; Griffin and Critchfield 1972). The restricted northern groves may be due to a wetter climate there that supports a dense forest and deep, dry litter in summer, conditions which are inimical to seedling establishment. The large southern groves are confined to uplands between canyons down which the ice streams poured (Muir 1876). Cold air drainage and the ice itself may have separated a previously continuous forest into isolated patches. The drier, warmer postglacial Xerothermic no doubt eliminated or restricted small outliers in both areas.

The occurrence of *Pinus jeffreyi* in the New Idria region represents its only known station in the Coast Ranges south of Hull Mt., northern Lake Co. It may have entered the area by northward migration from the Transverse Ranges during the Wisconsin. In this regard, several herbaceous species, as well as *Artemisia tridentata,* that occur with it are common members of the Jeffrey pine forest in the Sierra Nevada (Griffin 1971). In the Sierra, *P. jeffreyi* lives in areas where the mean January temperature is 30°F (–1°C) or lower. It can be calculated from temperature data at New Idria (alt. 2650 feet, 800 m) that Jeffrey pine at 4800 feet (1485 m) on San Carlos Peak to the south now lives under a mean winter temperature of approximately 42°F (5.5°C), or 8°F (4.5°C) warmer than that at the lower edge of the forest in the Sierra. This appears to represent the amount of postglacial warming in the New Idria area.

No doubt *P. jeffreyi* reached much lower levels in the Sierra Nevada and Klamath Mt. region during the ice ages. Its present occurrences at low altitudes in these areas on diverse ultrabasic rocks also appear to be relict from the Wisconsin (see Raven 1964 for discussion of evolutionary significance). The sites noted by Griffin and Critchfield (1972) and by Whittaker (1960) may be compensatory, in which *P. jeffreyi* escapes competition from taxa in the surrounding mixed conifer forest. In the Siskiyou Mts., other taxa that usually occur in subalpine environments also extend to lower levels on ultrabasic rocks, including *Abies magnifica, Juniperus sibirica, Pinus contorta,* and *P. monticola* (Whittaker 1960). In Plumas Co. west of Quincy, there is a belt of serpentine and associated ultrabasic rocks on which *P. jeffreyi* is disjunct from the main stand of the forest to the east (Griffin and Critchfield 1972). Other taxa occur with it, notably *Quercus vaccinifolia,* that are also generally at higher, cooler altitudes in the forest to the east (J. L. Jenkinson, written commun., June 1975).

Significantly, in all three areas there are additional associated taxa on serpentine that occur also in the Jeffrey pine or higher (subalpine) environments. In the San Benito Mountain area south of New Idria this only includes *Allium burlewii,* probably due to its present warmer, drier climate. However, in the Siskiyous there are *Castilleja miniata, Linnaea borealis,* and *Thlaspi alpestre* (Whittaker 1960); and in the Sierra Nevada near Quincy Jenkinsom records *Castilleja miniata, Claytonia lanceolata, Eriogonum marifolium, Geum macrophyllum, Gilia aggregata, Penstemon newberryi, Spiraea densiflora, Thalictrum fendleri* and *Veratrum californicum* and others. A lowering of mean temperature 6–7°F (3.5–4°C) would be adequate to displace the lower margin of Jeffrey pine–red fir forest to the altitude of the present relict areas where these taxa occur on serpentine and associated ultrabasic rocks. A return to warmer climate would leave them stranded as hardy relicts together with Jeffrey pine.

All three areas also include a number of taxa that are mainly representative of the Great Basin and of a drier climate. The moister, cooler western Siskiyou region has fewer Great Basin taxa, but some are present, notably *Balsamorhiza platylepis* (= *B. hookeri*), *Fritillaria atropurpurea, Monardella odoratisiima,* and *Phlox diffusa.* Detling (1968) recorded additional taxa of this alliance in the Oregon Coast Ranges and in the western Cascades as well. Presumably the taxa in all these areas (and elsewhere) reached maximum distribution during the height of the Xerothermic about 6000 yr ago.

The relative aridity of the mountains of southern California probably accounts for the absence (*Pseudotsuga menziesii, Sequoiadendron, Taxus, Torreya*) or rarity (*Arbutus, Cornus nuttallii, Lithocarpus, Populus tremuloides*) there of plants whose fossil relatives had a wider occurrence in the ancestral mixed conifer forest. This is consistent with the rapid impoverishment of subalpine forest in the southern Sierra and with the absence of several Sierran subalpine trees (*Abies magnifica, Pinus monticola, Tsuga mertensiana*) from the high mountains of southern California. In both areas *Pinus flexilis* contributes to mixed conifer and subalpine forests, and reaches down almost to *Cercocarpus ledifolius, Pinus monophylla, Purshia tridentata,* and other taxa that also have affinities in the Great Basin and southern Rocky Mts.

Taxa typical of the southern Rocky Mts. that have relict occurrences in the mixed conifer forest of southern California include *Arctostaphylos pringlei, Holodiscus microphyllus, Lewisia brachycalyx, Populus angustifolia,* and a number of herbs (Munz 1935). They are not known from the Sierra and apparently are remnants of a montane flora that earlier ranged across the southern Great Basin. Some are in higher basin-ranges in the southern Mojave region. A number of additional taxa that do not reach California, including *Acer glabrum diffusum, Pinus ponderosa scopulorum,* and *Populus tremuloides aurea,* are in the Charleston (= Spring) Mts. (Clokey 1951). Other common Rocky Mountain species in that area—where there is considerable summer rain—are either absent from southern California (*Ceanothus martinii, Quercus gambelii, Rhamnus betulifolia*) or are rare there (*Jamesia americana, Mahonia fremontii, M. repens, Populus angustifolia*).

These links predate the late Pleistocene, for climate then enabled the lower piñon-juniper belt to reach the present desert floor (Wells and Berger 1967). Late Pliocene and early Quaternary pollen floras from the Coso Mts., Panamint Valley, and the adjacent region to the north suggest a forested region (Axelrod and Ting 1960), at which time the links may have extended across the area. They certainly were present in the late Miocene of central Nevada (Axelrod 1956). Those floras show that forest stringers reached down to the margins of lakes, interfingering with broad-leaved sclerophyll vegetation. The forest had several taxa of present Rocky Mt. affinity, notably *Acer* (*grandidentatum*), *Cercocarpus* (*breviflorus*), *Fraxinus* (*anomala*), and *Paxistima* (*myrsinities*), as well as *Fraxinus velutina, Populus* (*angustifolia*), and *Rhus* (*glabra*), which are rare in southern California. The Rocky Mt. *Acer brachypterum* persisted into the early Quaternary in southern California (Axelrod 1966b). It probably was eliminated there as effective summer rain ceased and as summer drought and high evaporation increased later in the Quaternary. The warm dry Xerothermic probably played a crucial role in eliminating many other taxa of present Rocky Mt. distribution from southern California.

In this regard, *P. flexilis* is scattered southward along the east front of the Sierra from near Bridgeport (Twin Lakes) to the Kern Plateau (Monache Mt.) as a regular associate of the *Pinus jeffreyi–Abies concolor–A. magnifica* forest. It is then discontinuous to Mt. Pinos and the high mountains of southern California, where it is in the Jeffrey pine forest. There it is associated with *A. concolor, Pinus contorta, Juniperus occidentalis,* and an understory of east-side shrubs, notably *Artemisia tridentata, Cercocarpus ledifolius,* and *Chrysothamus nauseosus*. Its present discontinuous distribution at high altitudes becomes explicable when it is realized that *P. flexilis* is not confined to the subalpine zone. In the Rocky Mts. it is a regular member of the mixed conifer forest and lives under summer rain like that inferred for the Miocene and Pliocene in California and Nevada. In southern California it also contributes to mixed conifer forest in areas where summer thundershowers are concentrated by favorable topography, as in the upper Santa Ana River drainage. Summer rains were present into the early Quaternary (late Blancan = 1.5 m.y.), and under such conditions *P. flexilis* would be expected to be a more frequent member of the mixed forest, ranging southward into the Peninsular Ranges. The subsequent spread of drier summer climate would have confined it to its present discontinuous areas at higher levels.

Significant forest evolution during the later Quaternary involved the appearance of monoclimax stands, notably the pure Douglas fir forest of northwest California–Oregon. Its recency is inferred from pollen evidence in Oregon (Hansen 1942, 1947) which indicates that pure Douglas fir forests rose to dominance about 6000 B.P. in areas where mixed forests dominated earlier. The pure stands of *Pinus ponderosa* at the lower edge of mixed conifer forest in the Sierra and in the interior valleys of the North Coast Range and Klamath Mts. also appear to be a new forest type. Similarly, pure stands of *P. jeffreyi* at the lower margin of mixed conifer forest on the east slope of the Sierra Nevada appear to be recent. Their area coincides closely with a mean January temperature no warmer than 30°F (–1°C), an environment unsuited to *P. ponderosa*, which may cover slopes just above it (i.e., near Sierraville, Blairsden) where winter temperatures are slightly higher.

Subalpine Forest

Many taxa that contributed to the matrix of conifer–deciduous hardwood forest during the Miocene and into the early Pliocene also reached up into the fir–subalpine zone (Axelrod 1976a). As summer precipitation decreased and extremes of summer temperature with accompanying drought stress increased, these taxa were eliminated gradually from the mixed conifer forest in the Sierra and bordering regions and were restricted to cooler and moister higher elevations.

A number of typical "subalpine indicators" occur in the mixed conifer forest in the Siskiyou Mts. today, though they are not in the forest in the Sierra. Their presence appears to reflect the gradual shift southward to a shorter precipitation season, less summer rain, and greater ranges of temperature, which increase drought stress. This may also explain the absence of several subalpine taxa south of Tulare Co., notably *Abies magnifica, Pinus albicaulis, P. monticola,* and *Tsuga mertensiana,* as well as their absence from the mountains of southern California. Similarly, farther north the subalpine forest gains taxa that are chiefly Cascade or Rocky Mts., as shown by *Abies amabilis, A. procera, A. lasiocarpa,* and *Picea engelmannii,* and by the occurrence there of the Neogene relict, *Picea breweriana*.

The record suggests that in the Sierra Nevada taxa of the subalpine forest were confined to higher evaluations during the latter half of the Quaternary. Warm season rainfall was effective well into the Quaternary, as shown by the record of *Picea* in the forest at Tahoe City (Adam 1973) and by *Magnolia* and *Acer* (*brachypterum*) in the late Blancan (~ 1.5 m.y.) Soboba flora (Axelrod 1966b). In addition, pollen floras from the southern Sierra have species that are now far to the north and indicate moister summers than those there at present (Axelrod and Ting 1960, 1961). These include *Lithocarpus densiflora, Pseudotsuga menziesii, Taxus brevifolia,* and *Tsuga mertensiana,* which are in the Sierra farther north, *Abies grandis, Chrysolepis* (= *Castanopsis*) *chrysophylla, Chamaecyparis lawsoniana,* and *Tsuga heterophylla,* which are typical in the north coast sector, and *Pinus balfouriana,* which is now disjunct from the southern high Sierra to the Klamath Mt.

region. Some annuals and perennials that occur with *Pinus balfouriana* are disjunct with it, including species or species-pairs of *Dicentra, Erythronium, Haplopappus Lewisia,* and *Raillardella* recorded by Jepson (1925:11). In addition, Howell (1945) has noted that *Phacelia orogenes* of the southern high Sierra is related to taxa of the *P. pringlei* group of the Klamath Mt. region. The disjunction is certainly Quaternary. The persistence of "Klamath" taxa in the southern high Sierra Nevada seems related to the fact that there is considerably more warm reason precipitation there than in the central Sierra. Farther north the Sierran axis is so low that subalpine environments are highly restricted. Sites there were strongly affected by the Xerothermic, which may have eliminated possible intermediate geographical links between the Klamath–southern High Sierra regions.

The subalpine forests of the basin ranges in eastern California are often monoclimax stands, chiefly *Pinus flexilis* or *P. longaeva* ("*aristata*"). However, in the eastern Great Basin, notably in the Ruby and Jarbidge Mts., there is a strong Rocky Mt. subalpine element (Loope 1969). *Abies lasiocarpa* occurs there with *Pinus flexilis, Picea engelmannii* is local in the region, and *Pinus albicaulis* is at several sites, mingling with *P. flexilis* and *A. lasiocarpa.* These are the richest subalpine forests in the Great Basin today, but they are highly impoverished as compared with those of the Eocene (Figs. 5-3 and 5-4). The late Eocene Bull Run flora north of Elko has over 12 subalpine species, distributed in *Abies* (3 spp.), *Picea* (3), *Pinus* (3),

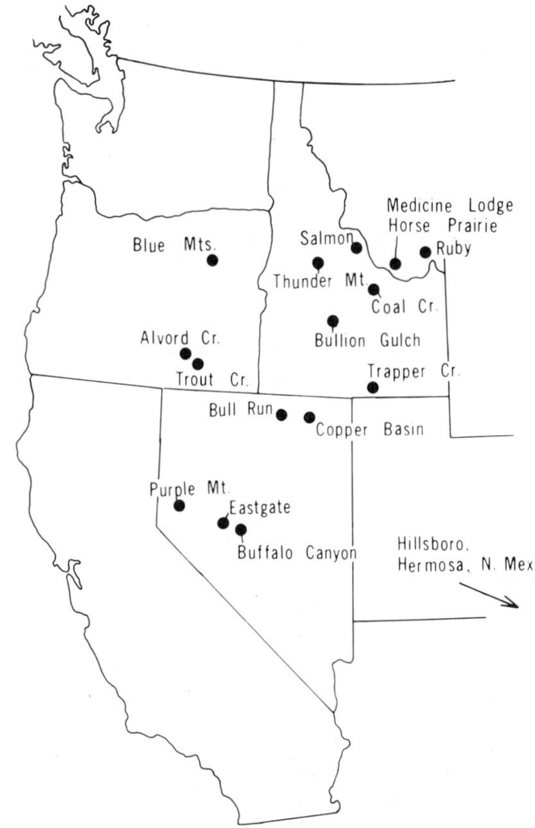

Figure 5-3. Some Tertiary floras with taxa representing subalpine forest.

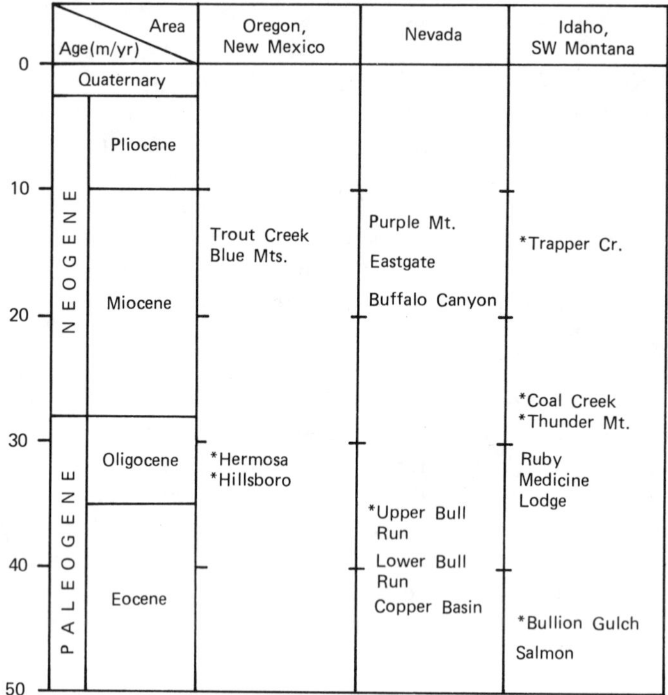

Figure 5-4. Relative ages of Tertiary floras that chiefly represent subalpine forest (*), or that contain numerous taxa that contributed to subalpine forests in bordering mountainous tracts.

Chamaecyparis (2), *Thuja, Tsuga, Larix,* and *Pseudotsuga.* The Miocene Trapper Creek flora in nearby Idaho has 8–9 subalpine species, distributed in *Abies* (2), *Pinus, Picea* (3), *Pseudotsuga, Larix,* and *Tsuga,* as well as a rich conifer–hardwood forest with 5–6 conifers [*Pinus* (2), *Abies, Pseudotsuga, Sequoiadendron, Chamaecyparis*].

The richer development of subalpine Rocky Mt. taxa in northeastern Nevada today is paralleled by the occurrence there of numerous circumpolar arctic–alpine plants (Loope 1969). This is not surprising since the area has some summer rain and it was much moister and cooler during the Quaternary, as shown by heavy glaciation. As drier and hotter summers spread during the Xerothermic, many taxa of the subalpine, as well as the arctic-alpine zone, probably were eliminated there and elsewhere in the Great Basin. A few relicts have survived in cool sites in some of the central basin ranges, and also on the east slope of the central Sierra on Convict Creek (Major and Bamberg 1967), where a limestone substrate removes species of *Arctostaphylos uva-ursi, Draba, Kobresia, Pedicularis, Salix,* and *Scirpus* from competition with the native flora (cf. Raven 1964).

Clearly, the increased diversity of both subalpine and mixed conifer forests east and west of the Great Basin today is related to more adequate precipitation, whether in summer (to the east) or in winter (to the west), and to sufficiently mild temperature so that evaporation is not too high for the establishment of seedlings. It is therefore understandable that the high basin ranges (9000–12,000 feet, 2700–3600 m) with little or no summer rainfall and with high summer temperatures either have subalpine forests of very restricted area and low diversity, or have none at all.

Summary

The western conifer forests were derived from ancestral, generalized communities that appeared first along the cordilleran axis in Eocene time. Richer in conifers and deciduous hardwoods than any surviving forests, they were made up of taxa whose nearest relatives occur in the mixed conifer forests of the western United States and in the conifer–deciduous hardwood forests of the eastern United States and eastern Asia. The ancestral cordilleran forest was differentiated regionally by the close of the Eocene into mesic northern forests that graded through a transitional central region into subhumid southern forests of lower diversity. Their taxa are directly ancestral to those in the Neogene and in surviving north coast, Sierra–Cascade, Petran, and Great Basin forests.

Modernization commenced in the interior and spread westward. Taxa now exotic to the region were eliminated gradually as summer rain decreased and as colder winters developed. By late Pliocene only a few exotics remained in the coastal strip, and some persisted into the earliest Quaternary.

Increasing drought spread progressively over the Great Basin during the Pliocene as mountain axes were elevated to the east and west, creating a double rain shadow. This largely eliminated forests from that region except in favorable upland sites. Segregation of the Petran, Sierran, and coast forests into their present regional subclimates is recent, as judged from the association of their taxa into the Quaternary. Our present forests are highly impoverished in comparison to those of the Neogene, and they are persisting under a wholly new climate to which they gradually adjusted physiologically during the Pliocene.

The sequential changes are depicted in Fig. 5-5 in terms of the opening of new subzones of a montane zone (Simpson 1953). Selection has been for taxa adapted to moister, more equable climates, at one extreme, as compared with dry ones of low equability that scarcely (or do not) support impoverished monoclimax stands.

SCLEROPHYLL VEGETATION

Evergreen sclerophyll vegetation, comprising mixed evergreen forest, live oak woodland–savanna, piñon–juniper woodland, chaparral, and other vegetation types, appeared over southwestern North America during the Eocene (Axelrod 1939, 1958b). Their taxa were derived from ancestral alliances in evergreen forests that had earlier inhabited the region. They appear to have diversified in response to expanding dry climate and to have spread with it during post-Eocene time (Axelrod 1938, 1939, 1958b, 1973, 1975). By the early Miocene, sclerophyll vegetation covered much of southern California, reaching northward along the coastal strip and into central Nevada. To the north, it appears to have covered warmer slopes adjacent to mixed deciduous–hardwood forest and conifer-hardwood forest, with some of the sclerophyll taxa no doubt contributing to those communities as well.

Figures 5-6 and 5-7 summarize the locations and ages of floras reviewed in this section.

It is imperative to realize that evergreen sclerophyll vegetation is not confined to the Mediterranean climate of California. It occurs also from Arizona to western Texas and south into Mexico, all regions with summer rainfall. In those areas, vegetation is similar in physiognomy to that in California, the taxa have similar

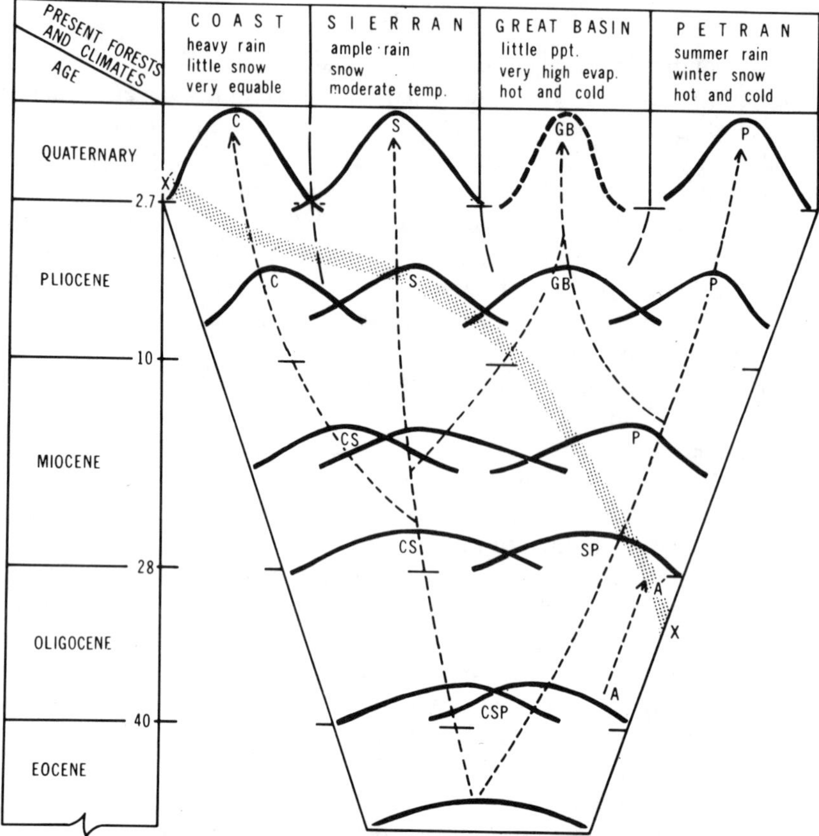

Figure 5-5. Opening of new subzones of the montane forest zone during Tertiary time. A . . . . A′ is the time when many near-modern woody Petran taxa were being established. X . . . . X′ represents approximately the time–space boundary of the change from mixed conifer-deciduous hardwood forest (below) to mixed conifer forest (above). Simplified from Axelrod (1976a).

structural adaptations, and the floras have a number of identical sclerophyllous species or closely related ones. Taxa that are now only in Arizona and areas to the east and south had close allies in California, Nevada, and border areas in Neogene time. Hence it is inferred that in the latter areas sclerophyll vegetation then lived under summer as well as winter precipitation. The taxa that now characterize the sclerophyll vegetation of California are therefore descendants of a richer Neocene flora that was progressively impoverished as summer rain decreased. Taxa of the present sclerophyll vegetation of California were thus preadapted structurally to live under Mediterranean climate, which only appeared here after the Pliocene. Functionally, adaptation largely involved a gradual shift in the period of establishment to the cooler, moister part of the year (Axelrod 1975), a process that presumably commenced in the late Miocene as summer rain decreased.

Mixed Evergreen Forest

This community now extends from the Coast Ranges northward into southern Oregon, southward into coastal southern California, and eastward into the northern

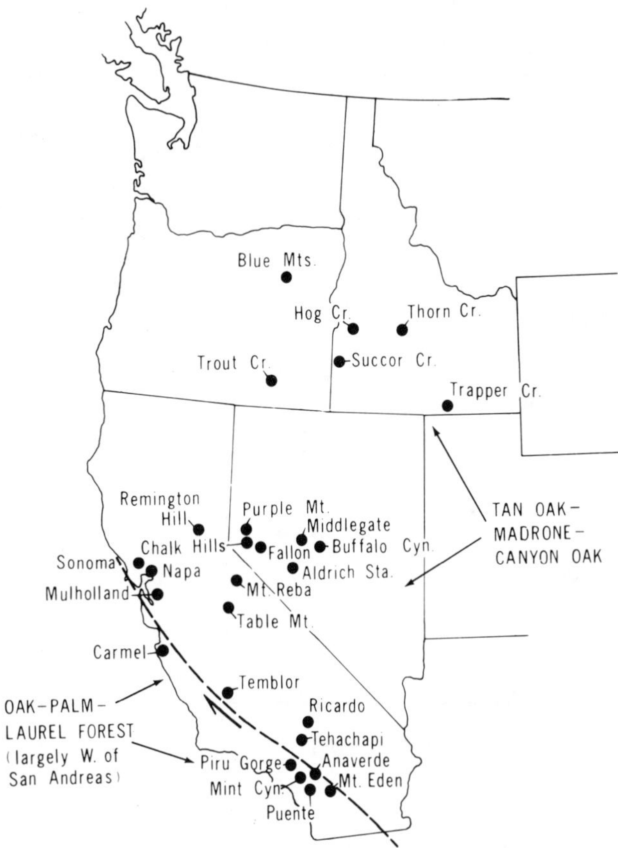

Figure 5-6. Occurrence of some Neogene floras with taxa that contributed to broad-leaved sclerophyll vegetation. All floras west of the San Andreas Fault (and also the Ricardo and Tehachapi in southeastern California) represent oak–palm–laurel forest. All other floras have tan oak–madrone–canyon live oak sclerophyll vegetation.

and central Sierra Nevada. It grades into Douglas fir–coast redwood forest in wetter parts of the coastal strip, and into mixed conifer forest in the higher, cooler parts of the Coast Ranges, Sierra Nevada, and Transverse Ranges. In drier areas, it is replaced by oak woodland, savanna, or chaparral, as in the foothills of the Sierra, the Coast Ranges, and southern California. It represents transitional vegetation between Arcto-Tertiary conifer–hardwood forests and Madro-Tertiary live oak woodland–savanna, a relation already well established by the Miocene.

Mixed evergreen forest is represented in Miocene floras of Oregon, Washington, western Idaho, and Nevada, where it merged into a mesic phase of the rich conifer–deciduous hardwood forests, many of whose taxa find their nearest allies in the southern Cascade–Klamath-Siskiyou region. This is apparent from the composition of the Blue Mt., Trout Creek, Thorn Creek, and other floras of the region where species of *Arbutus, Lithocarpus* (*densiflora*), and *Quercus* (*chrysolepis*) are recorded (Chaney and Axelrod 1959). Their regular associates include *Acer* (*macrophyllum*), *Alnus* (*rhombifolia*), *Amelanchier* (*alnifolia*), *Prunus* (*demissa, emarginata*), *Quercus* (*kelloggii*), *Rhamnus* (*purshiana*), and others that live with the evergreens today. Additional evergreens now exotic to the region include *Cedrela, Ilex, Persea,*

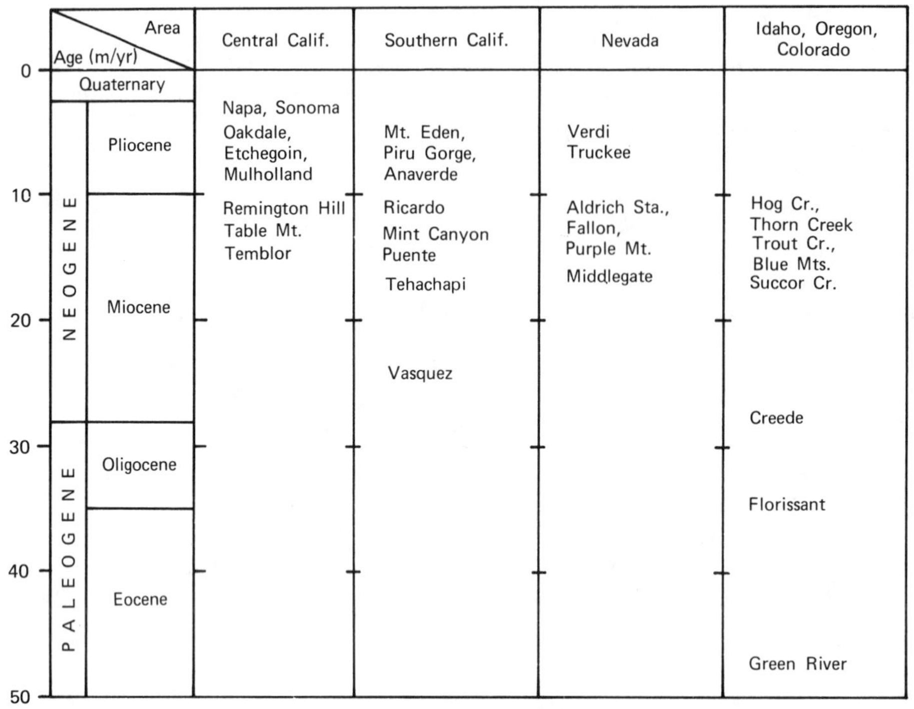

Age (m/yr)		Area: Central Calif.	Southern Calif.	Nevada	Idaho, Oregon, Colorado
0	Quaternary				
NEOGENE	Pliocene	Napa, Sonoma Oakdale, Etchegoin, Mulholland	Mt. Eden, Piru Gorge, Anaverde	Verdi Truckee	
10–20 (NEOGENE)	Miocene	Remington Hill Table Mt. Temblor	Ricardo Mint Canyon Puente Tehachapi Vasquez	Aldrich Sta., Fallon, Purple Mt. Middlegate	Hog Cr., Thorn Creek Trout Cr., Blue Mts. Succor Cr. Creede
30 (PALEOGENE)	Oligocene				Florissant
40–50 (PALEOGENE)	Eocene				Green River

Figure 5-7. Some Tertiary floras with taxa representing broad-leaved sclerophyll vegetation (see Fig. 5-6).

and *Oreopanax,* as well as hardwoods similar to taxa now in the eastern United States or eastern Asia. Apart from species of *Acer, Alnus, Betula, Prunus,* and *Quercus,* the exotics notably include *Ailanthus, Carpinus, Castanea, Carya, Diospyros, Fagus, Liquidambar,* and *Sassafras.* Some of them no doubt entered into the matrix of Miocene mixed evergreen forest in warmer sites. These and other floras in the region also have numerous conifers, including *Abies, Calocedrus, Chamaecyparis, Picea, Pinus, Pseudotsuga, Sequoia,* and *Tsuga,* that towered over the hardwoods and evergreens. They occupied cooler, moister valleys, with the sclerophylls presumably concentrated chiefly on warmer, south-facing slopes. Apart from the exotics, many of the broad-leaved and conifer taxa have survived in scarcely modified form on the mesic, coastward slopes of the Klamath–Siskiyou Mts.

To the south, Miocene floras in the central Great Basin show that sclerophylls were more prominent, with fossils similar to *Arbutus menziesii, Chrysolepis* (= *Castanopsis*) *chrysophylla, Lithocarpus densiflorus,* and *Quercus chrysolepis* represented. Their associates, whose nearest descendants are found in mixed evergreen forest, include *Acer* (*macrophyllum*), *Amelanchier* (*alnifolia*), *Cercocarpus* (*betuloides*), *Heteromeles* (*arbutifolia*), and *Prunus* (*demissa*). In addition, indubitable remains of *Abies bracteata* occur in the later Miocene of western Nevada (Axelrod 1976b), occupying an ecotone between broad-leaved evergreen forest and mixed conifer forest, much as it does today. As noted above, the mixed conifer forest in nearby moister and cooler valleys had a typical Sierran aspect as compared with the more mesic forests to the north. *Sequoiadendron* is prominent in most of the Nevada floras. Among its associates were *Abies* (*concolor, magnifica*), *Chamaecyparis* (*lawsoniana*), *Picea* (*breweriana*), *Pinus* (*ponderosa, monticola*), and *Pseudotsuga* (*menziesii*). Exotic taxa, including *Acer* (*saccharinum*), *Betula* (*papyrifera*), *Diospryos* (*virginiana*), *Gymnocladus* (*dioica*), *Populus* (*heterophylla*), *Ulmus* (*alata*), and *Zelkova* (*serrata*), are less abundant in both a quantitative and a qualitative sense than in floras to the north.

Late Miocene floras reveal a richer mixed evergreen forest on the west Sierran slope. The Remington Hill flora from the northern Sierra Nevada (Condit 1944a) has *Arbutus* (*menziesii*), *Lithocarpus* (*densiflora*), *Persea* (*borbonia*), *Quercus* (*wislizenii*), and *Umbellularia* (*californica*). They probably also formed an understory in a forest characterized by *Sequoia* (*sempervirens*) and *Chamaecyparis* (*lawsoniana*). The forest was confined chiefly to sheltered sites, with the sclerophylls assuming dominance on warmer, drier slopes. Drier sites in the nearby area supported *Arctostaphylos* (*manzanita*), *Ceanothus* (*cuneatus*), *Quercus* (*douglasii*), and *Ungnadia* (*speciosa*). Exotic taxa were more numerous here than east of the low Sierran divide, for this small flora includes *Liquidambar, Mahonia, Persea, Populus* (*grandidentata*), and *Ulmus* (*americana*).

The middle Miocene Temblor flora near Coalinga is rich in mixed evergreen forest taxa (Renney 1972), including *Arbutus* (*menziesii*), *Lithocarpus* (*densiflora*), and *Quercus* (*chrysolepis*). Their associates included oaks no longer in the region, notably *Quercus* (*virginiana*) and *Q.* (*reticulata*), as well as *Magnolia* and *Persea.* Other exotics are *Glyptostrobus, Keteleeria, Carya, Castanea, Diospyros, Nyssa, Photinia,* and *Zelkova,* thus approaching the northern floras in diversity and abundance. They probably formed a mixed forest together with evergreens, and in cooler nearby hills to the west were associated with *Pinus* (five-needled), deciduous oaks (*Q. pseudolyrata*), *Crataegus, Mahonia, Platanus,* and others.

By early Pliocene time, mixed evergreen forest had largely been eliminated from the lowlands over western Nevada, which were dominated then by hardier juniper–oak woodland, with conifer forests evidently confined to higher, moister hills. The Sierra was still quite low, and much of its northern to central section evidently was covered by mixed evergreen forest at altitudes above ~ 1500 feet (450 m). As noted above, this is suggested by the middle Pliocene Mt. Reba flora, dominated by *Lithocarpus* (*densiflorus*) and *Quercus* (*chrysolepis*). Associates included abundant *Pseudotsuga* (*menziesii*) and *Cupressus* (Asian aff.). Rare records of *Abies* and *Sequoiadendron* indicate mixed conifer forest in nearby cooler canyons. The site is now only 10 miles (16 km) from the present crest, at an altitude of 8600 feet (2615 m) at wind-timberline above red fir–white pine–mountain hemlock forest. The flora apparently lived near 2500 feet (750 m) in the middle Pliocene, implying that the Sierra has been elevated on the order of 6000 feet (1800 m) since then (Axelrod, unpub. data).

Broad-leaved sclerophyll vegetation appeared in the San Francisco Bay region in the middle Pliocene, the area earlier having been dominated by deciduous hardwood forest. The Mulholland flora near Moraga has *Arbutus, Heteromeles, Lithocarpus, Lyonothamnus, Quercus,* and *Umbellularia* (Axelrod 1944b). Associated species of *Acer* (*macrophyllum*), *Alnus* (*rhombifolia*), *Amelanchier* (*alnifolia*), *Arctostaphylos* (*glandulosa*), *Cornus* (*californica*), *Platanus* (*racemosa*), *Populus* (*fremontii*), *Rhamnus* (*californica*), and others also contributed to the oak woodland and savanna that bordered this flora in drier sites. The Petaluma flora shows similar vegetation and suggests affinity with the living mixed evergreen forest in the southern Santa Cruz Mts. Some summer rain is indicated by a few exotic taxa in these floras, notably species of *Nyssa, Populus* (*grandidentata*), *Quercus* (*vaseyana*), *Sapindus,* and *Ulmus* (*americana*).

Late Pliocene floras reflect increased precipitation that heralded the build-up of the continental ice sheets. They show that mixed evergreen forest and north coast forest invaded lowland areas where live oak woodland and savanna had dominated earlier. Their subsequent restriction is due to the spread of more recent drier climates. An early Pleistocene flora from the Santa Cruz Mts. (Dorf 1930) shows that mixed conifer forest and mixed evergreen sclerophyll vegetation covered lowlands now dominated by live oak woodland. In southern California, mixed evergreen forest of *Arbutus menziesii, Quercus chrysolepis,* and *Q. wislizenii* inhabited the present San Jacinto Valley, where it was marginal to mixed conifer forest. Mixed evergreen forest has since been confined to more equable coastal slopes of the Santa Ana Mts. and northward, and conifer forests to levels fully 3000 feet (900 m) higher in the nearby San Jacinto Mts. Semidesert sage and chaparral now dominate the fossil site (Axelrod 1966b).

Oak–Laurel Forest

Oak–laurel forest characterizes several undescribed Miocene floras at sites in coastal southern (Puente, Modelo, Topanga floras) and central California (Carmel flora). Situated west of the San Andreas Fault, they have been displaced northwest approximately 180 miles (290 km) since the late Miocene (13 m.y.). Hence, when they were living, the southern California floras were near the present latitude of Ensenada, and the Carmel flora was near Santa Paula. They indicate a subtropical climate well removed from frost.

Typical Arcto-Tertiary deciduous hardwoods (*Ailanthus, Betula, Carya, Castanea, Diospyros, Fagus, Liquidambar, Nyssa, Sassafras, Ulmus, Zelkova*) and conifers (*Abies, Chamaecyparis, Keteleeria, Metasequoia, Picea, Pinus, Pseudotsuga, Sequoia, Thuja, Tsuga*) are wholly absent. The floras are dominated by evergreen oaks [*Quercus* (*chrysolepis, eduardi, potosiana, virginiana*)] and laurels (*Nectandra, Ocotea, Persea, Umbellularia*). Other evergreens include *Annona, Arbutus, Ficus, Ilex, Lyonothamnus, Pistacia, Sabal,* and sclerophyllous shrubs, notably *Cercocarpus, Laurocerasus, Quercus,* and *Rhus*. The Miocene oak–laurel forest shows a relationship to vegetation from southern Sonora southward and in the eastern Sierra Madre. In physiognomy, it recalls the hammock flora of Florida, also dominated by palms, oaks, and laurels.

Oak–laurel forest evidently ranged into the moister interior uplands. The Tehachapi flora has live oaks [*Quercus* (*arizonica, brandegeei, chrysolepis emoryi*)], laurels (*Persea, Umbellularia*), palms (*Erythea, Sabal*), and other evergreens, notably *Arbutus* (*xalapensis*), *Cedrela, Clethra* (*lanata*), *Laurocerasus* (*lyonii*), *Lyonothamnus,* and *Myrica* (*mexicana*) (Axelrod 1939). The Mint Canyon flora (UC Mus. Pal.), then situated near the present site of the Salton Sea, has taxa that represent oak-laurel forest as judged from numerous oaks (5 spp.), laurel, palm, and evergreen cherry (*Laurocerasus*). It covered the moister, cooler, north end of the San Gabriel Mts. which was then much lower in altitude and situated directly south of the locality. In both areas oak-laurel forest evidently was marginal to live oak woodland and savanna of the lowlands, the drier parts of which supported a thorn forest of *Acacia, Bursera, Colubrina, Ficus, Randia,* and others. At the southwest front of the Sierra Nevada northeast of Bakersfield, the Miocene marine sequence has yielded a log of *Diöon,* leaves of *Quercus* (*virginiana*), and cotyledons of *Juglans*. They probably were numbers of an oak-laurel forest that lived in the nearby hills to the east, in moist, warm, sheltered valleys.

The Miocene Temblor flora near Coalinga represents an ecotone between oak-laurel forest and mixed deciduous hardwood forest (Renney 1972). Taxa assigned to oak-laurel forest include *Arbutus* (*menziesii*), *Heteromeles, Ilex, Persea* (*borbonia*), and *Quercus* (*chrysolepis, rugosa = reticulata, virginiana*). Their associates were typical Arcto-Tertiary deciduous hardwoods and conifers like those in middle Miocene floras to the north, including *Castanea* (*americana*), *Keteleeria* (*davidiana*), *Nyssa* (*aquatica*), *Pinus* (five-needle spp.) and *Quercus* (*borealis, falcata*). They probably lived in nearby hills to the west, where the climate was more temperate.

Taxa that contributed to oak-laurel forest were becoming restricted by Pliocene time as summer rainfall diminished and as winter temperatures decreased. Some of the increased cold no doubt resulted from the northward movement of the Pacific plate west of the San Andreas Fault. Records that suggest an impoverished oak-laurel forest occur in the Anaverde flora from the margin of the Mojave Desert, with *Persea* (*podadenia*), *Quercus* (*wislizenii*), and a palm (Axelrod 1950c). It probably occupied nearby moister slopes, above oak savanna–woodland and thorn forest which dominated drier lowlands. Summer rain indicators include *Bumelia, Colubrina, Dodonaea,* and *Eysenhardtia,* as well as palm and *Populus* (*brandegeei*). The Piru Gorge flora to the west has a relict oak–laurel–palm forest (Axelrod 1950d), as represented by *Laurocerasus* (*lyonii*), *Persea* (*podadenia*), *Quercus* (*engelmannii, wislizenii*), and *Sabal* (*uresana*). It blanketed moist, sheltered slopes at the west border of the basin. The absence of thorn forest there probably results from its position farther west under moister climate. Summer rain indicators include *Persea* and *Sabal,* as well as *Acer* (*brachypterum*) and *Populus* (*brandegeei*). The Mount Eden flora near Beaumont also appears to have had a relict oak–laurel forest, as judged from its species of *Arbutus* (*glandulosa*), *Persea* (*podadenia*), and *Quercus* (*agrifolia, chrysolepis, engelmannii*) (Axelrod 1937, 1950b). Woodland probably occupied moist, sheltered valleys with mild temperatures. Thorn forest was present over the drier lowlands, as indicated by *Cercidium, Dodonaea, Eysenhardtia,* and *Ficus*.

A highly modified descendant of the Neocene oak–laurel forest has survived in California. This is represented by the *Umbellularia* woodland on the seaward slopes of mountains at low altitudes and under equable temperatures. It is commonly associated with *Quercus agrifolia* in the coastal strip, and more frequently with *Q. chrysolepis* and *Q. wislizenii* in the interior and at higher levels. Associated stream-border trees regularly include taxa similar to those of the Neocene, probably *Platanus* (*racemosa*) and *Populus* (*fremontii*). To the north and at higher altitudes, the *Umbellularia* woodland is in ecotone with mixed evergreen (tan oak–madrone–canyon live oak) forest, but typically occurs below it, for it requires milder winters. In this regard, reference is made to the remarkable *Umbellularia* savanna [80 foot (24 m) tall trees, 6 foot (1.9 m) dbh] in Marin Co. west of Novato and near Olema (also see Jepson 1910). In the North Coast Ranges, pure stands occupy floodplains and rolling slopes below mixed evergreen forest or redwood–Douglas fir forest, under a climate of high equability (*M* 65–70+).

Oak Woodland–Savanna

Vegetation dominated by live oaks and associated sclerophyllous shrubs, and sometimes with piñon and juniper or cypress, characterized the Miocene of interior southern California. Floras like the Tehachapi (Axelrod 1939) and the Mint Canyon (Axelrod 1940a) include taxa similar to *Cupressus* (*arizonica*), *Pinus* (*monophylla*), and *Quercus* (*arizonica, chrysolepis, emoryi*), together with *Amorpha* (*californica*), *Arctostaphylos* (*pungens*), *Ceanothus* (*crassifolius, cuneatus, insularis, verrucosus*), *Fremontodendron, Quercus* (*turbinella*), and others that contributed to a rich understory. Moist swales and riverbanks supported gallery forests of *Celtis, Erythea, Juglans, Lyonothamnus, Platanus, Populus, Sabal,* and *Salix.* At higher, moister levels, oak woodland graded up into a rich oak–laurel forest (see preceding section). At lower levels, dense woodlands gave way to an open oak savanna with grasslands locally prominent on drier plains. Warmer lowland sites supported a rich thorn scrub of *Acacia, Bursera, Celtis, Condalia, Dodonaea, Ficus, Jatropha, Karwinskia, Lysiloma, Pithecolobium,* and *Randia,* all implying mild, frostless winters and summer rainfall.

Rather similar vegetation persisted into the Pliocene, as shown by the Anaverde (Axelrod 1950c) and Ricardo (Webber 1933) floras, both dominated by live oaks and taxa that evidently represent thorn scrub. Mild winters are implied by palms in these floras, which are in the Mojave Desert, a region now too cold for them. The Mt. Eden flora near Beaumont has a rich live oak woodland as well as thorn scrub vegetation, showing that mild winters persisted there into the middle Pliocene (Axelrod 1950b). With increasing aridity, oak woodland was gradually eliminated from the area of the present desert proper, though oaks (*Q. chrysolepis*) still inhabit the upper desert slopes at the west edge of the province. Higher desert ranges, however, now support a conifer woodland (piñon–juniper), elements of which were already in the region in the Miocene (see below).

It was only at the end of the Miocene that sclerophyllous oak–juniper woodland commenced to spread north into central Nevada to occupy sites marginal to the retreating mixed conifer forest and evergreen sclerophyll vegetation. The oak–conifer woodland is represented chiefly by *Quercus* (*arizonica, wislizenii*) and locally by juniper (*Juniperus osteosperma*) and piñon (*Pinus monophylla*). Scattered through the woodland and savanna were *Aesculus* (*parryi*), *Amelanchier* (*utahensis*), *Amorpha* (*californica*), *Cercis* (*occidentalis*), *Cercocarpus* (*betuloides, breviflorus*), *Peraphyllum* (*ramosissimum*), *Rhamnus* (*californica*), *Rhus* (*ovata*), and *Robinia* (*neomexicana*). Moist stream borders supported *Bumelia* (*lanuginosa*), *Juglans* (*major*), *Platanus* (*racemosa*), *Populus* (*trichocarpa*), *Salix* (*exigua, gooddingii, lasiandra, melanopsis*), and *Sapindus* (*drummondii*). Cooler nearby slopes had an impoverished mixed conifer forest (*Abies, Picea, Pinus, Sequoiadendron*). As judged from the very rare occurrence of exotic taxa in these floras, summer rainfall had decreased appreciably in comparison with the middle Miocene.

By Pliocene time, live oak woodland-savanna with scattered juniper dominated the lowlands of the Great Basin. Common live oaks were *Quercus* (*arizonica, chrysolepis, wislizenii*), and the juniper was *Juniperus* (*californica-osteosperma*).

Their associates were chiefly stream- and lake-border taxa, notably *Amorpha, Celtis, Juglans, Populus, Ribes,* and *Salix.* They imply a wide restriction of woodland and forest, with broad grasslands on the interfluves. Scattered shrubs in the woodland included *Arctostaphylos* (*glauca*), *Cercocarpus* (*betuloides*), *Purshia* (*tridentata*), and *Rhus* (*lanceolata*). Precipitation over the lowlands was nearly 20–25 inches (50–64 cm), as compared with 30–35+ inches (76–89 cm) in the Miocene, when forests lived at the sites of plant accumulation. In the middle Pliocene a relict woodland-savanna covered the east front of the Carson Range near Verdi, Nevada, where several oaks (*Quercus engelmannii, lobata, wislizenii*) are recorded at the lower margin of an impoverished mixed conifer forest (Axelrod 1958a). By this late date, exotic taxa had largely been eliminated from the region. The late Pliocene pollen record from the area presently desert (Axelrod and Ting 1960) provides an indication, not of semidesert communities, but of Sierran mixed conifer forest. Hence it is inferred that live oak woodland was highly restricted by the wetter, colder climate of the late Pliocene, when precipitation increased fully 10–15 inches (25–38 cm) above that of the middle Pliocene.

Oak woodland–savanna penetrated northward along the west flank of the Sierra Nevada as aridity increased during the Miocene. By the close of the epoch it was in the north–central part of the range, as shown by the Remington Hill flora (Condit 1944a). It includes *Quercus* (*douglasii, lobata, wislizenii*) associated with *Arctostaphylos* (*manzanita*), *Ceanothus* (*cuneatus*), and *Prunus* (*ilicifolia*), with *Acer* (*negundo*), *Platanus* (*racemosa*), and *Umbellularia* (*californica*) along streambanks. Summer rainfall was more effective here than on the east side of the range, with species of *Carya, Liquidambar, Magnolia, Persea,* and *Ulmus* recorded. As noted above, they contributed to evergreen sclerophyll forest that graded into a *Sequoia-Chamaecyparis* forest in cooler sites nearby.

The small middle Pliocene Oakdale (Axelrod 1944c) and Turlock Lake floras (UC Mus. Pal.) from the lowest foothill belt include species of *Quercus* (*douglasii, lobata, wislizenii*), implying that an oak savanna covered the broad interfluves and formed dense woodlands that blanketed the moister valley slopes. Associates included *Arctostaphylos* (*manzanita*), *Ceanothus* (*cuneatus*), *Heteromeles* (*arbutifolia*), *Quercus* (*dumosa*), and *Rhamnus* (*californica*). Moist floodplain sites supported *Celtis* (*reticulata*), *Persea* (*podadenia*), *Populus* (*deltoides, trichocarpa*), *Robinia* (*neomexicana*), *Salix* (*lasiolepis*), *Sapindus* (*drummondii*), and *Umbellularia* (*californica*). Rainfall was now greatly reduced in summer, but some still occurred as shown by *Persea, Robinia,* and *Sapindus.* This contrasts markedly with the effectively drier region east of the Sierra.

Live oak woodland did not become dominant in west–central California until the middle Pliocene. Earlier, the area was sufficiently low and well watered so that it supported mixed deciduous hardwood forest on the floodplains and swamp forest (*Taxodium-Nyssa*) in oxbow lakes and swamps (Condit 1938). The first hints of oak woodland are found in the small early Pliocene Black Hawk Ranch flora near Mt. Diablo (Axelrod 1944a). Live oaks wholly dominate the middle Pliocene Mulholland flora (Axelrod 1944b), notably *Quercus* (*agrifolia, lobata, tomentella, vaseyana, wislizenii*). Associates along the stream margins in the oak savanna region were *Alnus* (*rhombifolia*), *Celtis* (*reticulata*), *Lyonothamnus* (*floribundus*), *Platanus* (*racemosa*), *Populus* (*fremontii, wislizenii*), *Salix* (*lasiandra*), *Sapindus* (*drummondii*), and *Toxicodendron* (*diversiloba*), with moister nearby slopes supporting mixed evergreen (tan oak–madrone) forest. A few species, notably *Acer* (*grandidentatum*), *Karwinskia* (*humboldtiana*), *Nyssa* (*sylvatica*), and *Quercus* (*vaseyana*), imply summer rainfall in small amount. Mild winters are suggested by plants whose nearest relatives are now in southern California, notably *Ceanothus* (*spinosus*), *Lyonothamnus, Malosma* (*laurina*), *Quercus* (*tomentella*), and *Schmaltzia* (*ovata*).

After the middle Pliocene, which was the driest part of the Tertiary in California and bordering regions, precipitation increased and climate became colder. Mixed evergreen forest and conifer forest now expanded to assume dominance over lowland areas earlier covered by live oak woodland–savanna. This is shown by the appearance in the late Pliocene of a rich coast conifer forest at Santa Rosa (Axelrod 1944*d*), and by mixed conifer forest and broad-leaved sclerophyll vegetation in the Santa Clara flora of transitional Plio-Pleistocene age (Dorf 1930).

Oak woodland-savanna now occupies a smaller area than the ancestral Neogene vegetation, which ranged across the present northern Sonoran and Mojave deserts and reached north into central Nevada. It also covered parts of the Sierra Nevada and Coast Ranges that are presently forested. The composition of the surviving woodlands varies over the region (Fig. 5-8) and reflects differences in warmth and temperateness of climate (Bailey 1960, 1964, 1971). The segregated modern communities occupy narrower areas than the more generalized, widely distributed Pliocene type of vegetation.

***Quercus agrifolia* woodland.** California live oak is not common in the fossil record. It has been recorded only from western California, generally near the coast, in the Petaluma (Dorf 1930), Mulholland (Axelrod 1944b), Etchegoin (Dorf 1930), and Mt. Eden floras (Axelrod 1950b). All these floras have taxa whose closest analogues are now in mild-winter climates, notably *Ceanothus* (*spinosus*), *Lyonothamnus* (*floribundus*), *Malosma* (*laurina*), and *Pinus* (*radiata*). Inasmuch as there are no known pure communities of *Quercus* (*agrifolia*) in the record, even in the coastal area, the modern association may well be a new one. In this regard, it would be similar to the pure forests of *Pinus ponderosa* and *P. jeffreyi,* and to the pure chaparral of *Adenostoma fasciculatum,* which also appears to be late Pleistocene in age.

***Quercus engelmannii* woodland.** Remains of a fossil oak with leaves similar to those of living *Quercus engelmannii* occur in the Mt. Eden (Axelrod 1950b), Piru Gorge (Axelrod 1950d), and Tehachapi floras (Axelrod 1939) of southern California. It then lived with other oak woodland taxa, with numerous sclerophylls, and with thorn scrub vegetation on the interior. The climate was characterized by summer rain, as implied by the numerous exotics in these floras, and by mild winters, as judged from the presence of taxa that apparently require nearly frost-free conditions (*Ficus, Laurocerasus, Lyonothamnus, Persea*).

The present community dominated by *Q. engelmannii* forms a nearly pure woodland-savanna, or often occurs with *Q. agrifolia.* It inhabits interior valleys of southern California, notably in San Diego, Orange, and Riverside Cos. Small stands disjunct well to the north, as at Claremont, Glendora, and Pasadena, are ecologically out of place, for they occur there with walnut woodland and coastal sage, which indicate a strong maritime influence. The present occurrence of Englemann oak in these areas may be relict from the Xerothermic, having been isolated by the return of more equable conditions (see Fig. 5-8).

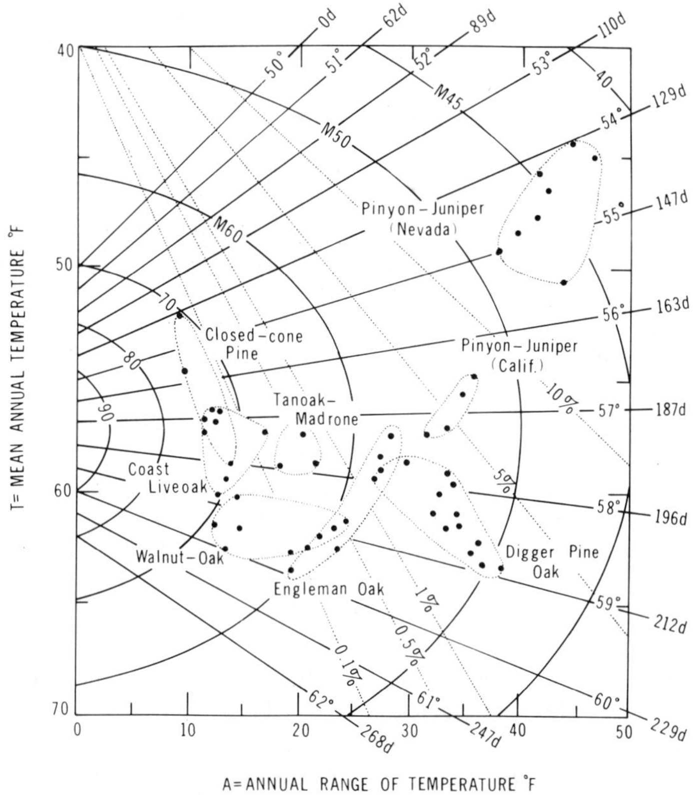

Figure 5-8. Thermal relations of some present sclerophyllous woodland vegetation. Blank areas between communities represent a lack of meteorological stations for data. Radii are lines of warmth, or effective temperature, that depict the number of days in which mean temperature rises above the particular thermal level indicated (Bailey 1960). Arcs nearly normal to the radii represent the temperateness rating, or the equability of climate (Bailey 1960, 1964), which decreases in all directions from a theoretical ideal of 100 at $T = 57.2°F$ and $A = 0°F$. The descending dotted lines give the percent of all hours of the year with freezing temperatures.

***Juglans californica* woodland.** This community extends along the southern California coast from near Santa Barbara southward into Orange Co., reaching inland up the major river valleys where the climate is moderated by a low cloud deck in summer and by warmer maritime air in winter. Walnuts similar to those of *Juglans californica* occur in the Miocene Puente Formation at Puddingstone Dam near Claremont (UC Mus. Pal.), an area where walnut savanna is well developed today. In the Miocene it was associated with a rich oak–palm–laurel forest with *Persea, Quercus* (*virginiana, potosiana*), and *Sabal,* as well as *Cercocarpus, Karwinskia, Lyonothamnus,* and *Quercus* (*turbinella*), which contributed to the understory. Walnuts occur also in the Pliocene Mt. Eden flora near Beaumont (Axelrod 1937), where they formed part of an oak savanna rich in sclerophyllous shrubs, and with a big-cone spruce forest in the nearby hills. Many of its Pliocene associates are represented by similar species that live with California walnut today. Walnut is represented also in the Rancho La Brea flora (UC. Mus. Pal.), though its stratigraphic position is uncertain. It may represent part of a flora that lived there during a drier phase, after the *Pinus radiata–P. muricata* forest disappeared from the region (Axelrod 1967b). In this regard, its present occurrence in the lower Santa Ynez River valley and near Jalama, well north of the main area of the community, has been attributed to northward movement during the Xerothermic (Axelrod 1966b).

Insular woodland. Some of the distinctive woody dicots that contribute to insular woodland–savanna have closely allied species in Neogene rocks on the adjacent continent, notably *Laurocerasus* (*lyonii*), *Lyonothamnus* (*asplenifolius, floribundus*), and *Quercus* (*tomentella*), as do the unique maritime pines allied to those now in Mexico (see below). The fossil record suggests that these taxa have largely been preserved under mild climate, having been eliminated from the nearby continent by more extreme conditions during the Quaternary (Le Conte 1887; Axelrod 1939, 1958b, 1967b).

During the Miocene and Pliocene, their fossil relatives were members of an oak–pine–laurel forest rich in taxa that no longer occur in the region. These include *Clethra, Persea, Quercus* (*potosiana*), and *Sabal,* related to species now in Mexico, and species of *Dodonaea, Eysenhardtia, Quercus* (*emoryi*), *Robinia,* and *Sapindus* now in the southwestern United States. In addition, the floras include taxa that represent thorn forest, notably species of *Acacia, Bursera, Colubrina, Eysenhardtia, Ficus, Pithecolobium,* and *Randia* that also live in areas with summer rainfall and under winters that are frostless or nearly so.

Lyonothamnus occurs at a number of Miocene and Pliocene sites in Nevada. There it was a member of a broad-leaved sclerophyll vegetation, composed of *Arbutus, Cercocarpus, Heteromeles,* and *Quercus* (*chrysolepis, wislizenii*), that covered warmer slopes bordering sites of deposition. On nearby cooler, effectively moister slopes was a mixed conifer–hardwood forest. At least two species allied to *L. asplenifolius* are represented in these floras. The older one has smaller pinnate-pinnatified leaves, fewer divisions to the lamina, and smaller lobes to the pinnae. Both lived contemporaneously, and the one with smaller leaves persisted longest. As now known, its latest occurrence is in the middle Pliocene Truckee flora near Hazen, associated with *Cercocarpus* (cf. *blancheae*), *Juglans* (*californica*), *Populus* (*trichocarpa*), *Quercus* (*wislizenii*), and *Rhus* (*lanceolata*).

These records support the suggestion (Axelrod 1958b) that the living *L. asplenifolius* may have gradually evolved larger leaves as the alliance was confined to the insular environment of mild temperature (*M* 65+) and low evaporation, where water stress is greatly reduced. Under such conditions, larger leaves and increased stature are both favored, a feature shown by many woody plants in insular regions. In this regard, the large size of many of the shrubs in the insular area is not due to evolution in isolation, but reflects a more effective moisture regime resulting from the mild oceanic climate with low evaporation. Some of them (e.g., *Heteromeles*) also reach tree size (40 feet, 12 m) in coastal sites on the mainland, as at Inverness. On this basis, it seems probable that *Lyonothamnus* may have been a shrub to small tree in the understory of broad-leaved sclerophyllous woodland, especially in interior regions during the Miocene and Pliocene.

Fossil species allied to present insular woodland taxa ranged northward into central California as the climate became drier in the Pliocene. In the middle Pliocene Mulholland flora *Lyonothamnus* (*floribundus*) and *Quercus* (*tomentella*) are accompanied by species represented now in southern California, notably *Ceanothus* (*spinosus*), *Malosma* (*laurina*), and *Schmaltzia* (*ovata*), or, in areas farther south, as

Karwinskia (*humboldtiana*), *Populus* (*mexicana*), and *Sapindus* (*drummondii*). All of them disappeared from central California as colder climates appeared in the late Pliocene and as summer rains decreased.

A number of typically insular species have relict occurrences in coastal southern California or Baja California. These include *Cercocarpus blancheae* and *Comarostaphylis diversifolia* in the Santa Ynez and Santa Monica Mts.; *Laurocerasus* (*Prunus*) *lyonii* in San Julio Canyon near Comondú, Baja California; *Xylococcus bicolor* and *Pinus torreyana* at Torrey Pines Park; *Coreopsis gigantea* along the immediate coast in the western Santa Monica and Santa Ynez Mts.; and *Comarostaphylis diversiloba, Pinus muricata, P. remorata, Rhamnus insula,* and *Ribes viburnifolia* in Canyon de las Pinitos southwest of San Vicente, Baja California. The last-named area represents, in essence, an insular flora that survived on the mainland where hilly terrain along the coast has about 20–25 inches (51–64 cm) of rainfall and mild, fog-shrouded summers. Such occurrences suggest that other species now occurring only in the insular region may have been on the mainland in the late Pleistocene, disappearing there as hotter, drier climates spread coastward during the Xerothermic.

Digger pine–oak woodland. There are records of Digger pine woodland in southern California, a region now south of its present area. The records include the middle Pliocene Mt. Eden flora (Axelrod 1937), the late Pliocene Pico flora (Dorf 1930), and a Plio-Pleistocene locality near Santa Paula (Wiggins 1951). In the Pliocene, Digger pine lived with a typical southern California sclerophyllous flora which was then enriched by thorn forest (*Acacia, Bursera, Ficus*) and woodland taxa (*Arbutus, Persea, Sapindus*) now exotic to the region.

Axelrod (1937) earlier suggested that Digger pine, *Quercus douglasii,* and other associates were eliminated from southern California by the development of a warmer winter climate. However, it now seems more probable that the community disappeared there in response to desiccation during the Xerothermic. This is consistent with the occurrence of *Pinus sabiniana* together with *Arctostaphylos glauca* and *Juniperus californica* in the late Pleistocene Carpinteria flora (Chaney and Mason 1933).

Several patches of Digger pine–oak woodland are now disjunct from the main area of the community that encircles the Great Valley and occurs also in the Coast Ranges. These include (1) a stand on Figueroa Mt. north of Santa Barbara; (2) a stand along the seaward slopes of the Santa Lucia Mts. between Salmon Creek and Gorda, close to *Pseudotsuga* and *Sequoia*; (3) large stands north of Mt. Lassen on Hot Creek and near Fall River; (4) populations in the central Sierra, isolated from the woodland of the Great Valley by broad tracts of mixed conifer forest, as at Hetch Hetchy; and (5) stands in the North Coast Ranges, as in the lower Trinity River gorge west of Weaverville, which are isolated from the main community in the Great Valley to the east. They appear to be relict distributions from the Xerothermic, their present isolation due to a return of moister climate that favored the spread of forest.

Piñon-Juniper Woodland

This woodland covers drier, lower, eastern slopes of the Sierra, reaching southward along upper desert slopes of the Transverse and Peninsular ranges into Baja California. It penetrates westward from Antelope Valley across the Mt. Pinos region into upper Cuyama Valley. Juniper ranges farther north in the inner Coast Ranges, reaching Mt. Diablo. At moister, higher levels it is replaced by an impoverished mixed conifer forest. It gives way to sage and desert scrub at lower, drier elevations. To the east, it inhabits high desert ranges in California and Nevada.

During the late Miocene and Pliocene in Nevada, piñon and juniper were associated with taxa that usually are found with them today, notably *Amelanchier* (*utahensis*), *Cercocarpus* (*ledifolius*), *Peraphyllum* (*ramosissimum*), *Purshia* (*tridentata*), and *Ribes* (*cereum*). However, they are all that remain of a once much-richer mixed oak–conifer woodland, not unlike the one that survives on the desert slopes of southern California or in the Mt. Pinos–Cuyama region.

Some of the taxa disappeared from the Great Basin as summer rainfall decreased, and their nearest relatives are now in the southwestern United States, notably *Arbutus* (*arizonica*), *Bumelia* (*lanuginosa*), *Cercocarpus* (*breviflorus*), *Juglans* (*major*), *Quercus* (*arizonica*), *Rhus* (*lanceolata*), and *Sapindus* (*drummondii*). Others appear to have succumbed to lowered winter temperatures in the late Pliocene and Quaternary, for they now occur chiefly in areas of milder climate west of the Sierra or in southern California, as *Ceanothus* (*cuneatus*), *Cercocarpus* (*betuloides*), *Juglans* (*californica*), *Lyonothamnus* (*asplenifolius*), *Quercus* (*chrysolepis, engelmannii, lobata, wislizenii*), and *Rhus* (*ovata*). The surviving conifer woodland is clearly an impoverished, derivative community that has adapted to colder winters and drier summers since the Pliocene (see Fig. 5–8).

Pinus monophylla and *Juniperus californica,* together with *Atriplex, Arctostaphylos glauca, A. pungens, Marah* (=*Echinocystis*), and *Laurocerasus* (*Prunus*) *ilicifolia,* are recorded at McKittrick (Mason 1944), an area now desert. Dated at 38,000 ± 2500 B.P. (Berger and Libby 1966), they indicate that rainfall in the area was fully 10–15 inches (25–38 cm) more then than at present. A similar flora now lives at the southwest corner of Cuyama Valley, where piñon-juniper woodland reaches its northern limit of distribution.

Juniper is abundant in the Rancho La Brea flora (Frost 1927), though its stratigraphic position has not been established. Its occurrence there implies temperatures like those now in the interior, as near San Bernardino or Redlands. This is consistent with the presence of desert shrew (*Notisorex*), yucca night lizard (*Xantusia*), and others in the tar pits which indicate a hotter, drier, interior environment. It appears that temperatures in summer and winter were more extreme than those at Rancho La Brea when *Pinus radiata* lived there in the late Pleistocene, or those in the area today.

Piñon-juniper woodland occurs at several sites in the Mojave Desert in wood rat middens that range from ~ 8,000 to ~ 29,000 B.P. (Wells and Jorgensen 1964; Wells and Berger 1967). The middens are in areas presently desert, indicating that when the woodland existed precipitation was 10–15 inches (25–38 cm) higher, and that the vegetation zone was 1500–3000 feet (456–900 m) lower, depending on slope aspect. The lists of associated taxa appear to indicate no important floristic changes since the Wisconsin, though *Quercus dunnii* (= *palmeri*) has been reported from the Newberry Mts. in the southern tip of Nevada, associated there with a piñon-juniper community. It is no longer in the region, being found in areas to the east and west, in California and Arizona, respectively (Leskinen 1975).

Juniperus is rather widely spread in the cismontane valleys of southern California, ranging north along the front of the San Gabriel Mts. to Claremont. There it overlaps coastal sage (*Malosma laurina, Schmaltzia integrifolia, Salvia leucophylla*) and California walnut (see Fig. 5-8). They appear to have entered this area as climate moderated after the Xerothermic, when juniper evidently spread there from the interior.

Closed-Cone Pine Forest

Closed-cone pine forest ranges discontinuously along the outer coastal strip of California and northern Baja California and also inhabits Santa Cruz and Santa Rosa Islands off the Santa Barbara coast, as well as Cedros and Guadalupe Islands off Baja California. The area it occupies has been displaced approximately 180–200 miles (~ 300 km) northwest since the late Miocene (~ 13 m.y.).

The fossil record does not support the notion (Mason 1934) that *Pinus radiata, remorata,* and *muricata* originated in insular isolation during the Pleistocene. All of them have been recorded from middle and early Pliocene floras, some of them well in the interior (Axelrod 1967a,b). In addition, closed-cone pines are abundantly represented by pollen in the Miocene of coastal southern California (Martin and Gray 1962), associated there with subtropical plants (see the section on Oak–Laurel Forest). Available evidence suggests that they are survivors of an ancient evergreen oak–laurel-pine-palm forest that persisted under mild, equable climate in the coastal strip, and that they originated in the southwestern interior.

Closed-cone pines are members of the subsection Oocarpeae, the other members of the group being *P. patula, greggii, oocarpa,* and *pringlei* of central and southern Mexico (Little and Critchfield 1969). Such a disjunction between related taxa implies ancient ties with the vegetation of Mexico. Hence it is noteworthy that other taxa in the closed-cone pine forest also have their closest allies in the evergreen oak-pine-palm forests of central to southern Mexico, including the following:

Californian Taxa	Allied Mexican Taxa
Arbutus menziesii	*A. xalapensis*
Arctostaphylos pungens	*A. pungens*
Ceanothus arboreus	*C. coeruleus*
Cercocarpus traskiae	*C. mojadensis*
Comarostaphylis diversifolia	*C.* (12 spp.)
Garrya elliptica	*G. ovata*
Gaultheria shallon	*G.* (8 spp.)
Heteromeles arbutifolia	*H. mexicana*
Laurocerasus lyonii	*L. prionophylla*
Myrica californica	*M. mexicana*
Vaccinium ovatum	*V. confertum*

The list becomes more impressive when we recall that others now in the oak–palm-madrone-laurel forests of Mexico had similar taxa in the Miocene and Pliocene of southern California, including species of *Arbutus, Clethra, Myrica, Nectandra, Persea, Quercus,* and *Sabal.*

The fossil record shows that an essentially continuous closed-cone pine forest inhabited the outer coastal strip and islands during the late Pleistocene. For example, *Pinus radiata* has been recorded at Tomales Bay, Thornton Beach, Little Sur, Pt. Sal, Carpinteria, Santa Cruz Island, and Rancho La Brea, all sites well removed from the present groves. *Pinus muricata* also occurs at late Pleistocene localities now distant from its present areas, including Thornton Beach, Little Sur, and Pt. Sal. *Pinus remorata* is recorded in the late Pleistocene at Carpinteria and at Santa Monica during the Plio-Pleistocene transition. This wider distribution is understandable on the basis of moister climates during pluvial times and is implied also by wood of *Sequoia* in stream deposits at Carpinteria (Axelrod 1967b). Furthermore, Munz (1935) showed that numerous species extend from central California into coastal southern California and southward into northern Baja California, or are disjunct from central California (mostly from Monterey Co. north) to Santa Cruz and adjacent islands, as well as to San Clemente and Guadalupe Islands (Raven 1963). The breakup of the more continuous forest and its restriction to localized sites have been ascribed to the coastward spread of drier, hotter climate during the Xerothermic 8000–4000 B.P. (Axelrod 1966b, 1967b, fig. 6).

Subtropic Relicts

Finally, it must be noted that some taxa in the sclerophyll vegetation, and also in the bordering scrub, are the sole members of their families in California, families that are preponderantly tropical to subtropical in optimum development. Among these are *Bursera* (Burseraceae), *Chilopsis* (Bignoniaceae), *Fremontodendron* (Sterculiaceae), *Holocantha* (Simaroubaceae), *Styrax* (Styracaceae), *Umbellularia* (Lauraceae), and *Washingtonia* (Arecaceae). Several of them (*Bursera, Chilopsis, Holocantha, Washingtonia*) are in the desert in southeastern California, but most also enter woodland vegetation in areas of summer rain to the south and east. The taxa are relicts of earlier times, when subtropical vegetation was richly represented here. Many others of similar alliance disappeared in the Pliocene as summer rains decreased and as winter temperatures were lowered, including *Ficus* (Moraceae), *Ilex* (Aquifoliaceae), *Sapindus,* and *Thouinia* (Sapindaceae). As noted above, other survivors include a number of sclerophylls in the closed-cone pine forest whose near allies also contribute to sclerophyll vegetation in the subtropical parts of Mexico.

Summary

The varied kinds of sclerophyll vegetation in California were derived chiefly from an ancestral one that appeared initially in the drier central and southern cordilleran region in Eocene time. Rich in taxa, it comprised a mixture of alliances whose nearest descendants are in the sclerophyll vegetation of northern Mexico, the southwestern United States, and California.

Modernization commenced in the interior and gradually spread coastward with drier climate. As with the conifer forests, relict taxa now exotic to the region persisted in the coastal strip into the early Quaternary under a climate of equable temperature and some summer rain.

As increasing drought spread over the Great Basin in the late Miocene, oak–piñon-juniper woodland with a rich understory of evergreen shrubs became dominant there, replacing mixed conifer forest that shifted into the mountains and westward. However, increasing cold and decreasing effective summer rainfall eliminated many sclerophyll taxa and accounts for the impoverished piñon-juniper woodland that has survived.

Sclerophyll vegetation was rich in taxa in central California into the middle Pliocene, at which time taxa allied to those now in coastal southern California were present there. As summer rainfall decreased and as winter temperature lowered, they disappeared in the central region, together with exotics now found far to the south.

Segregation of sclerophyll vegetation into localized communities continued into the Quaternary, and they now inhabit narrower climatic parameters (Fig. 5-8). Like taxa in the montane forests, by the Miocene sclerophylls were already adapted structurally to the increasingly drier summer climate. They appear to have adjusted physiologically by shifting the period of establishment into the earlier, moister half of the year.

The changes that appear to have taken place in terms of community evolution are depicted in a general way in Fig. 5-9. It shows deployment into narrower, more specialized subzones of warmth and temperateness in the gradually emerging Mediterranean climate.

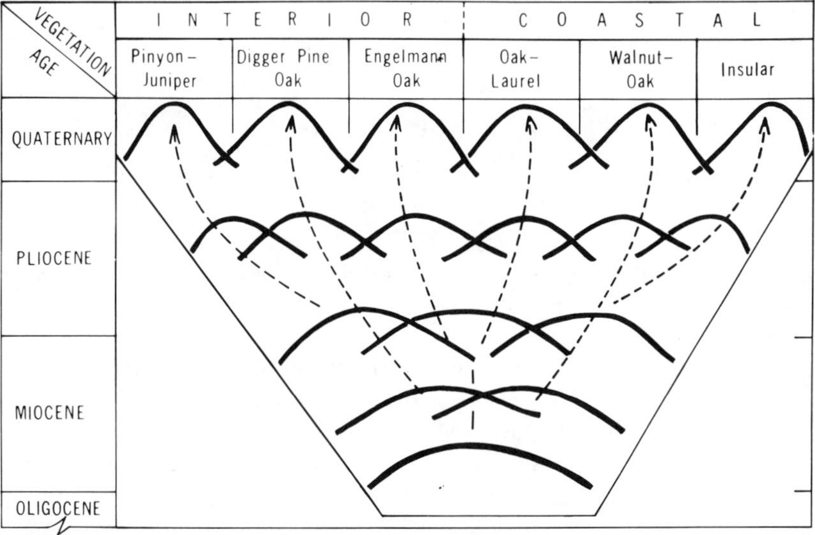

Figure 5-9. Inferred zonal evolution of sclerophyll vegetation during Miocene and later times (see Fig. 5-8).

SCRUB VEGETATION

Figure 5-10 summarizes the locations of floras referred to in this section, and preceding figures indicate their ages.

Chaparral

Time–space relations. Sclerophyllous shrubs that have often been regarded as "chaparral indicators" first appeared in the interior in the Eocene. The Green River

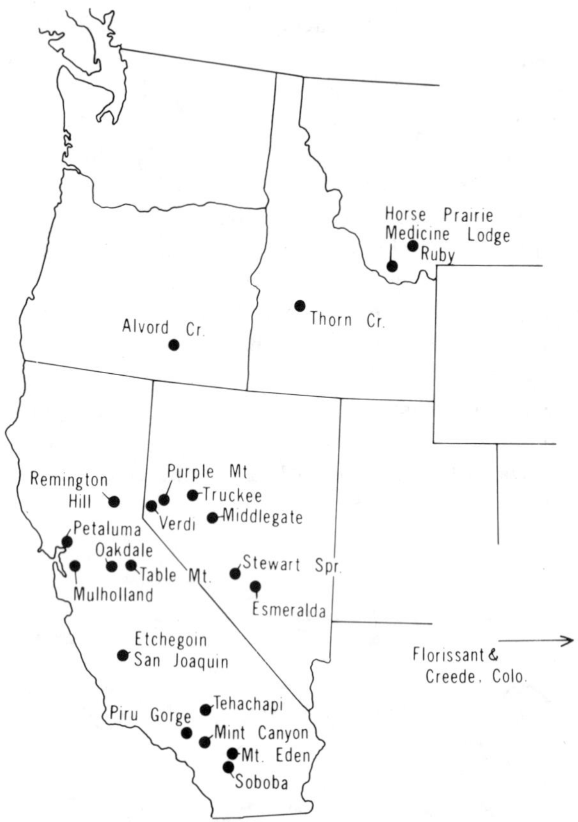

Figure 5-10. Some Tertiary floras with taxa related to alliances now in diverse types of scrub vegetation. Ages are shown in Figs. 5-2 and 5-4.

flora of Colorado–Utah (MacGinitie 1969) has species of *Ptelea, Rhus, Quercus, Vauquelinia,* and others that probably formed an understory in sclerophyllous oak-laurel forest. The Florissant flora (MacGinitie 1953) has *Cercocarpus, Dodonaea, Mahonia, Morus, Philadelphus, Ptelea, Quercus, Rhamnus, Rhus, Vauquelinia,* and *Zizyphus,* which also appear to have contributed to the sclerophyllous woodland in that area, much as they do today in northern Mexico and border areas. A few sclerophyllous shrubs are recorded also in the Oligocene floras of southwestern Montana (Becker 1961, 1969), notably the Ruby, Horse Prairie, and related floras. They were also with sclerophyllous oaks there, living marginal to mixed conifer–deciduous hardwood forest.

It is in the middle Miocene of interior southern California that sclerophyllous shrubs are first recorded in large numbers. The Tehachapi flora (Axelrod 1939) has *Arctostaphylos* (*pungens*), *Ceanothus* (*cuneatus, crassifolius*), *Cercocarpus* (*betuloides, ledifolius*), *Mahonia* (*fremontii*), *Quercus* (*turbinella*), *Rhamnus* (*californica*), and *Rhus* (*ovata, virens*). The Mint Canyon flora of late Miocene age (Axelrod, 1940a), then living near the present site of the Salton Sea, is more diverse, with *Ceanothus* (*cuneatus, crassifolius, perplexans, tomentosus*), *Cercocarpus* (*betuloides*), *Fremontodendron* (*californica*), *Laurocerasus* (*ilicifolia*), *Mahonia* (*fremontii*), *Quercus* (*dunnii, turbinella*), *Rhamnus* (*crocea*), and *Schmaltzia* (*ovata*). In each flora, the shrubs contributed to the understory of live oak woodland and oak-laurel forest. In drier sites with thin soil they may have formed local brushy patches of seral nature. The time of their initial appearance in the region is not now known, but it is probably Oligocene (at least), as judged from the semiarid nature of taxa in the small Vasquez flora (UC Mus. Pal.).

Sclerophyllous shrubs occur in smaller numbers in the Miocene floras of central Nevada because this moister region supported mixed conifer forest and mixed evergreen forest. As judged from evidence provided by the Middlegate flora with *Ceanothus* (*cuneatus*), *Cercocarpus* (*betuloides*), *Fraxinus* (*anomala*), *Heteromeles* (*arbutifolia*), and *Styrax* (*occidentalis*), shrubs were confined to warmer, south-facing slopes, together with oak woodland and mixed evergreen forest. The slightly younger (12–14 m.y.) late Miocene Aldrich Station, Fallon, Chloropagus, and Purple Mt. floras from the nearby region similarly have few sclerophyllous shrubs (Axelrod 1956).

The Pliocene floras of Nevada have few taxa, but sclerophyllous shrubs were present, as shown by *Ceanothus* (*cuneatus*) in the Verdi flora and *Ceanothus* (*cuneatus*) and *Cercocarpus* (*betuloides*) in the Truckee floras. They are generally rare to absent in Miocene and Pliocene floras farther north where forests dominated under cooler and moister climate.

In the Miocene, sclerophyllous shrubs moved northward along the west Sierran flank as an understory to live oak woodland and oak-laurel forest. They are represented by *Cercocarpus* (*betuloides*) and *Quercus* (*turbinella*) in the Table Mt. flora (Condit 1944b) and by *Arctostaphylos* (*manzanita*), *Ceanothus* (*cuneatus*), *Laurocerasus* (*ilicifolia*), and *Mahonia* (*haematocarpa*), in the Remington Hill flora (Condit 1944a). The woodland–sclerophyll shrub community was marginal to a deciduous hardwood floodplain forest at Table Mt., and to a *Sequoia-Chamaecyparis* deciduous hardwood forest at Remington Hill. A few sclerophyllous shrubs are recorded in the Middle Pliocene Oakdale (Axelrod 1944c) and Turlock Lake floras (UC Mus. Pal.) from the live oak savanna belt in the lowest foothills of the Sierra. These include *Arctostapylos* (*mariposa*), *Ceanothus* (*cuneatus*), *Heteromeles* (*arbutifolia*), *Quercus* (*dumosa*), and *Rhamnus* (*ilicifolia*). Taxa (*Persea, Robinia, Sapindus*) that are now found only in areas with summer rainfall are also present.

Sclerophyllous shrubs attained greatest diversity in west–central California in the middle Pliocene, the time of maximum dry climate and optimum development of live oak woodland and mixed evergreen forest in that area. The middle Pliocene Mulholland flora (Axelrod 1944b) has species of *Arctostaphylos, Ceanothus, Cercocarpus, Malosma, Quercus,* and *Schmaltzia.* Some of them are represented now in southern California or in the southwestern United States, a feature shown also by woodland taxa (Axelrod 1944a). As colder climates developed in the late Pliocene and during the glacial–pluvial ages, many sclerophyllous taxa were restricted southward.

Origin and age. Taxa that make up chaparral have a number of structural adaptations, including relatively small, thick, sclerophyllous leaves, often evergreen, with thick cuticle and sunken stomata. They also have hard wood, deep root systems that can tap water in the dry season, and many of them stump-sprout after fire. These adaptations are not unique to taxa of the Mediterranean climate of California.

Many California chaparral species occur in the chaparral of central Arizona, an area with ample summer rain. Some of them range farther east and south into regions where sclerophyll woodland, with many evergreen shrubs in the understory, grades into mesic forest in which other sclerophyllous taxa also occur (Axelrod 1973, 1975).

Taxa with these structures are present in divergent families and in vegetation ranging from rainforest to evergreen scrub at timberline, implying that the evolution of the sclerophyllous "chaparral habit" has been recurrent in angiosperm evolution and is an ancient feature. This is consistent with the presence of sclerophyllous shrubs in Miocene, Oligocene, Eocene, and older floras. The relations suggest sclerophyllous chaparral shrubs are not specialized, but may best be regarded as "generalists". Their structural adaptations are sufficiently fundamental and broad that they have been able—with but minor modification—to become members of many different kinds of vegetation, and in very different climatic regions. Hence, it is inferred that the sclerophylls were largely preadapted structurally to live under spreading climates of lower rainfall, a shorter rainy period, and greater ranges and extremes of temperature. Only minor changes in structure were necessary to adapt, notably decreased stature (small tree to shrub), somewhat smaller leaves, deeper internal crypts, thicker cuticles, or waxy or villous coverings. Although many shrubs sprout immediately after fire, there is no evidence to show that the habit is pyrogenic. Sprouting is frequent in *most* woody plants and is especially common in taxa that make up evergreen subtropical forests, taxa ancestral to those that contribute to woodland and chaparral today. For these reasons, the structural adaptations of sclerophyllous taxa now occurring in areas of Mediterranean-type climate do not represent an evolutionary response to that climate, as many have implied. Mediterranean climate appeared only after the early Pleistocene, fully 15–17 m.y. (at least) after sclerophyllous shrubs very similar to (or identical with) living species were in existence (Axelrod 1973, 1975). Response to the developing mediterranean climate was chiefly functional, by shifting the time of germination and seedling establishment into the moister, cooler part of the year. This is implied also by the occurrence of typical California chaparral species in regions with ample summer rainfall, as in Arizona or eastern Mexico, where they also make up chaparral. These summer-rain climates parallel those inferred for the Neogene of California, and the Mediterranean (Tethyan) region as well. Clearly, the taxa that occur in both regions have adapted to summer-dry climate more recently.

The 12 or so sclerophyllous California chaparral species now in the chaparral of Arizona are operating at peak efficiency under high rainfall and temperature in July–August, much as they are presumed to have done in the Neogene of California. Hence they may be regarded as more ancient ecotypes than the ones that adapted gradually to decreasing summer rain in California during the Pliocene.

Although Clements (1920, 1936) believed chaparral to be an ancient type of climax vegetation, much evidence suggests that chaparral is of comparatively recent origin as a widespread vegetation type (Axelrod 1975). The fossil floras that are presumed to have had dominant chaparral vegetation occur in areas where geological evidence shows topographical diversity insufficient to support it and thorn scrub, oak savanna-woodland, and big-cone forest, or thorn forest, oak-woodland, and mixed conifer forest, all of which also contributed importantly to the record. Furthermore, sclerophyllous shrubs regularly form the understory of pine–oak-laurel woodland, with which they regularly alternate on adjacent slopes. Hence the occurrence of numerous sclerophyllous shrubs in a fossil flora with oak, pine, and madrone need only mean that they formed understory shrubs in the woodland, not necessarily chaparral.

Throughout the Madrean region there are numerous transitions from woodland with a rich understory of shrubs and small trees, through scattered woodland trees in sclerophyllous shrublands, to pure chaparral-covered slopes. These stages occur in proximity and are gradational. Hence, shrublands are only successional, whether in regions of severe summer drought or ample summer rain. During most of the Tertiary, chaparral probably formed only local seral shrublands, though restricted edaphic sites may have been temporarily favorable for small patches of chaparral.

The spread of chaparral was due to several factors. First, commencing chiefly in the later Pliocene, the rapid rise of steep mountain slopes greatly favored the spread of shrubs over trees. Because of natural updrafts, fires from lightning or volcanism would be more effective in eliminating trees on slopes than in earlier, gentle terrains. Most slopes probably would soon revert to woodland with an understory of shrubs, as judged from relict areas today.

Second, the increasingly drier climates of the later interglacial ages favored the spread (temporarily) of chaparral at the expense of woodland. This involved the progressive strengthening of the hot, dry, high-velocity "Santa Ana" winds that rage across the area during autumn. At these times of low humidity, fires would be more effective than earlier, especially on steep slopes. The most favorable time for widespread expansion of chaparral probably was the Xerothermic.

In this regard it is recalled that a number of taxa typical of southern California occur at isolated sites in the Central Coast Ranges and southern Sierra. These seem explicable by the spread of warmer, drier climate after the last glacial, followed by more recent contraction of range as cooler, moister climate returned. Among these species are *Adenostema sparsifolium, Arctostaphylos glauca, Ceanothus divaricatus, C. spinosus, C. oliganthus,* and *Quercus dunnii* (= *palmeri*). Others common in interior southern California are also recorded from the coastal strip at isolated sites (Axelrod 1966b). Some chaparral species typical in southern California penetrate well northward in the inner Coast Ranges. Reaching as far north as the Mt. Hamilton range and/or Mt. Diablo are *Arctostaphylos glauca, A. pungens, Ceanothus divaricatus* (= *leucodermis*), *Laurocerasus ilicifolia, Ribes malvaceum,* and *R. speciosum.*

Chaparral species occur also in the Rogue River basin, Oregon (Jepson 1925; Detling 1953, 1968), notably *Arctostaphylos viscida, Ceanothus cuneatus, Cercocarpus betuloides, Eriodictyon californicum, Garrya fremontii,* and *Rhamnus crocea.* To judge from the pollen record in western Oregon (Hansen 1942, 1947; Hansen and Packard 1947; Heusser 1960), they invaded the area during the Xerothermic. A rise in mean July temperature of 5°F (2°C) and a lowering of precipitation by 10 inches (25 cm) probably would be sufficient to restrict forest on the low Siskiyou divide [4000 ft (1,210 m)] and enable taxa from the adjacent Shasta Valley in California to surmount the divide, invade the Rogue River basin (Detling 1953; Axelrod 1976a), and penetrate northward wherever conditions were suitable, some reaching to the Columbia gorge (Detling 1968).

Segregation of the present communities seems to have occurred in response to the emergence of new subclimates during the Quaternary. Like pure forests of *Pinus ponderosa* and *P. jeffreyi,* the pure *Adenostema fasciculatum* association appears to be a wholly new community, segregated from a richer chaparral in response to hotter and drier climate at lower altitude.

Coastal Sage

The history of coastal sage vegetation must largely be inferential. This is due partly to the fact that many of its semiwoody plants have small, soft leaves that tend to wither on the stems and hence are not likely to contribute to a recognizable fossil record. In addition, scrub communities occur in regions where rainfall is low, and where lakes and rivers are generally rare. Hence, burial of plant remains potentially suitable for preservation in fine-grained sediments is much less likely to occur than in humid, forested terrains. Nonetheless, a few woody plants that contribute to sage vegetation do have fossil records and provide a basis for suggesting its history.

Such inferences may be supplemented by range disjunctions, for they provide evidence for inferring the times of connection, and hence the age and probable history of vegetation to which they contribute. Although the Tertiary pollen record may contain records of sage genera (*Atriplex, Artemisia, Salvia,* etc.), their presence need not demonstrate the existence of sage vegetation. Most sage taxa occur also in the drier parts of forest and woodland, from which they probably have been derived (see below). Furthermore, determination of source for pollen raises insurmountable problems.

Sage is dominated by semiwoody plants with brittle stems, soft wood, and rich branching, as exemplified by species of *Artemisia, Baccharis, Encelia, Ericameria, Eriodictyon, Eriogonum, Hazardia, Helianthemum* and *Salvia.* Ecologists have not yet described the very diverse types of sage that cover the California coastal strip and reach into the interior. Among the clearly recognizable coastal communities are the Diegan sage and Venturan sage of coastal southern California, the distinctive Morroan sage of the Lompoc–Morro Bay region, the Franciscan sage of the central California coast, and the communities in the interior, which are much less diverse in composition, including Coalingan sage.

Some of the woody plants have Neogene records. Among these are *Malosma laurina* and *Schmaltzia integrifolia,* both of which contribute to coastal sage in southern California. They occur in the middle Miocene Tehachapi flora at the southeast corner of the Sierra Nevada, in the late Miocene Mint Canyon flora (then near the site of the Salton Sea), in the Pliocene Anaverde flora near Palmdale at the edge of the Mojave Desert, and in the Pliocene Mt. Eden flora south of Beaumont, which also has *Baccharis* (*sergiloides*). These floras have numerous semiarid taxa that contributed to live oak woodland. Since sage vegetation regularly occurs marginal to and interfingers with oak woodland–savanna vegetation today, the existence of local sage communities on dry sites in the Miocene and Pliocene of southern California seems probable. The distribution of *Malosma* (*laurina*) and *Schmaltzia* (*integrifolia*) in interior areas into the Pliocene shows that they have been restricted coastward to a more maritime climate. It probably was at this time that they were confined to the matrix of coastal sage, which otherwise is composed chiefly of nonwoody ("soft chaparral") species. A few woody endemics in *Arctostaphylos* (e.g., *A. morroensis, pumila, rudis*) and *Ceanothus* (e.g., *C. impressus, maritimus, rigidus*) that occur in coastal sage in the coastal strip appear to be relicts of earlier closed-cone pine forest or live oak woodland communities that survived under a drier but equable coastal climate.

As for its antiquity as a climax formation, sage occupies areas drier than chaparral or woodland-savana, chiefly tracts where precipitation is below 15 inches (38 cm). On this basis, it may be inferred that coastal sage came into existence as a widespread formation only when large areas of relatively low rainfall appeared. This occurred chiefly after the Wisconsin, for forest and woodland dominated areas presently covered with sage during the last glacial–pluvial stage, both on the coast and in the interior.

Desert

Shreve (1942, 1964) outlined the nature of the three major desert floras of California. Some topographic and climatic relationships are shown in Fig. 5–11.

Great Basin. The Great Basin Desert flora is young, as judged from evidence reviewed above with respect to the history of forest and sclerophyll vegetation from the area now desert. During the warm, dry middle Pliocene, small semidesert (sage brush) patches probably were present locally. The shrubs presumably were confined to drier sites in valleys in the lee of ranges, though they were covered chiefly by grassland and scattered stands of juniper–oak-piñon woodland. Semidesert taxa of sage habit no doubt entered the matrix of woodland vegetation, as well as the lower

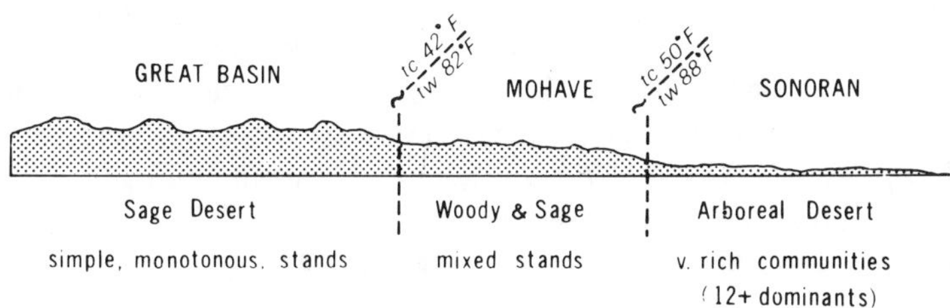

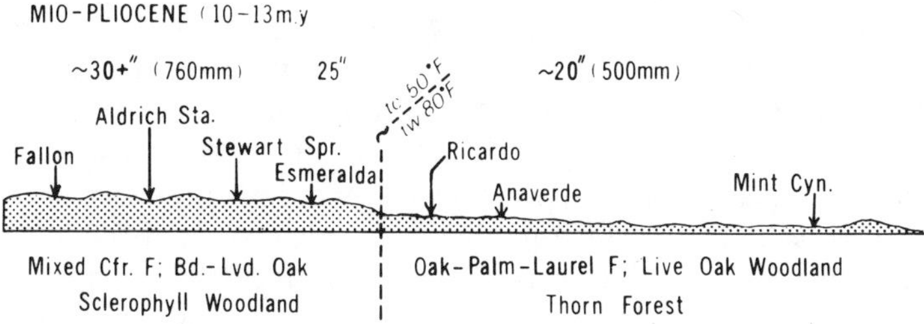

Figure 5-11. Comparison of topography, vegetation, and climate of the present desert region with that at the close of the Miocene. Transects run from west-central Nevada (Fallon), on the left, south to the Salton Sea (Mint Canyon), on the right. The Mojave is an ecotonal desert, separated structurally from the Sonoran province by faulting which elevated it into a colder climate during the late Pliocene and Quaternary. The abbreviations "tc" and "tw" refer to mean temperatures of the coldest and warmest months, respectively.

forest margin, much as they do today. Since the climate was wetter during the late Pliocene and first glacial, areas of semidesert vegetation largely disappeared, or were at least highly restricted to local sites in the driest parts of the conifer woodland belt. Since extremes of temperature were more moderate early in the Quaternary, conifer woodland covered more of the lowlands then than later in the period. During the hotter and drier interglacials it is inferred that conditions were more suited to semidesert communities than to conifer woodland, and the former spread more widely.

The taxa appear to have had two major sources. One group is related to those in the east Asian deserts and semideserts, including species of *Artemisia, Atriplex, Emplectocladus, Ephedra, Kochia, Suaeda,* and numerous grasses and herbs as well. There appears to have been interchange across Beringia during the middle Pliocene, and also during the warmer interglacials when climate was more favorable (Axelrod 1950e). This inference is supported by the continuity of the *Artemisia frigida–Juniperus sibirica* community that links the Yukon with the interior basins of northeast Siberia, and is consistent also with new distributional evidence presented by Yurtsev (1972). These taxa, or closely related (possibly ancestral) ones, enter the drier margins of taiga and adjacent conifer-hardwood forests. Under a climatic trend to drier conditions, shrubs would spread as forest slowly retreated.

Records of *Emplectocladus* (= *Prunus*) (*andersonii*) and *Purshia* (*tridentata*) in the Pliocene of Nevada imply that Great Basin sage probably occupied local drier slopes over the lowlands. There it presumably was marginal to grassland, which is inferred to have covered wide areas because there was still some summer rainfall over the province. *Emplectocladus andersonii* is related to *E. mongolica* of the central Asian desert and steppe. Furthermore, a small-leaved poplar in the middle Pliocene floras of Oregon and Nevada closely resembles *P. tremula* of Eurasia, which ranges out into open grassland and steppe country along watercourses. There are numerous other links between these areas of steppe and desert vegetation, for instance, *Atriplex, Ephedra, Eurotia, Lycium,* and *Suaeda* among the woody plants, and herbaceous alliances as well (see Wang 1961), notably in *Agropyron, Allium, Bromus, Hordeum, Koeleria, Lepidium, Lespedeza, Melica, Stipa, Tanacetum,* and *Thermopsis*.

We may infer that interchange between the interior steppe (sagebrush) and bordering grasslands of western North America and eastern Asia has been an intermittent process. As suggested earlier (Axelrod 1950e), the older semiarid links, such as *Emplectocladus* and *Populus* (*tremula*), may be middle Pliocene chiefly. This was the driest part of the Tertiary and also a time of considerable interchange of mammalian faunas rich in grazers (Repenning 1967). Since they required semiarid open tracts and cold steppe climate, conditions probably were suited for steppe and sagebrush plants, which no doubt provided food for the grazers. Earlier interchange seems less likely because dense, mixed deciduous hardwood forests inhabited Beringia into the Middle and Late Miocene. That further interchange occurred during the warmer phases of the Quaternary is apparent from the critical data of Yurtsev (1972).

The second group of taxa that contributes to the Great Basin Desert and semidesert occurs also in the Mojave and Sonoran regions. These typically soft-stemmed, highly branched shrubs of sage habit, distributed in *Chrysopsis, Chrysothamnus, Ericameria, Eriogonum, Haplopappus, Hazardia, Salvia,*

Stentopsis, Tetradymia, and many others, appear to have originated over the drier parts of southwestern North America. Not only do they occur in all three deserts, but also a number of them form the shrubby sage matrix of diverse woodlands marginal to all the western deserts, and some extend southward into thorn scrub as well. Since most of the genera are concentrated in the drier parts of southwestern North America, they probably originated there. These autochthonous taxa appear to have initially formed part of the understory and herbaceous ground cover on the drier borders of Madro-Tertiary woodland–savanna vegetation, and to have ranged north with it during the Miocene and Pliocene. They persisted over the interior lowlands when the more mesic woodland and then grassland retreated as climate became drier. In this manner, shrubs and herbs on the margins of Arcto-Tertiary forests and Madro-Tertiary woodlands have both contributed taxa to the Great Basin Desert flora.

Mojave-Sonoran During the Miocene and Pliocene there was rapid change in vegetation near the present boundary between the Great Basin and Mojave deserts (Axelrod 1950e). In the late Miocene the change was from a northern area in which forests were dominant to one at the south in which woodland and thorn scrub were characteristic (Fig. 5–11). As judged from modern vegetation, the northern area was humid to subhumid, the southern semiarid. These environmental differences appear to reflect a major change in topography, with the area of the present Great Basin standing near 2000–2500 feet (600–750 m) as an average, and with the Mojave province closer to sea level (Axelrod 1968).

The Miocene and Pliocene floras of the Mojave region, as represented by the Tehachapi, Ricardo, and Anaverde floras, are very similar to the late Miocene Mint Canyon flora, which was then situated near the present site of the Salton Sea. All these floras represent live oak–palm-laurel woodland–savanna with a rich understory of sclerophyllous shrubs ("chaparral") and with a good representation of thorn scrub, which covered drier sites. Their similarity in general composition demonstrates that the present Mojave and Sonoran regions were a floristic unit into the Pliocene. More recent differentiation is related to uplift and rotation of the Mojave block, the movement reflecting jostling between the American and Pacific plates (Garfunkel 1974). Uplift during the late Pliocene and Quaternary transformed the Mojave into a transitional region, intermediate in climate and altitude between the colder Great Basin province to the north and the warmer Sonoran region to the south, and floristically related to both.

The Sonoran Desert flora now has its northern limits in three low basins, the Coahuila ("Colorado Desert"), Blythe, and Needles troughs which project as tongues into the Mojave province. The depressions are mostly below 1500 feet (450 m) and are surrounded at higher levels by the Mojave Desert flora. Climate in the troughs is milder in winter, with January average and minium temperatures fully 10°F (5°C) higher than those in the Mojave area to the north. Among the woody plants in these basins that are typical of the Sonoran Desert are **Acacia greggii*, **Acalypha california, Agave deserti,* **Baccharis sarothroides,* **Bursera microphylla, Callianda eriophylla, Canotia holocantha,* **Cercidium floridum,* **Colubrina californica,* **Condalia lycioides,* **C. parryi, Dalea spinosa, Ephedra trifurca, Fouquieria splendens, Hoffmanseggia microphylla,* **Lycium torreyi, Nolina bigloveii, Olynea tesota,* **Prosopis juliflora,* and **Washingtonia filifera.* They are within a few kilometers of the higher, colder Mojave province but are not in the area today, though those marked (*) have closely similar (or the same) species in the Miocene and Pliocene of the Mojave and its border areas. More numerous species in this group, distributed in *Acacia, Bursera, Euphorbia, Eysenhardtia, Ficus, Lysiloma, Pachycormus, Piscidia, Pithecolobium,* and *Xanthoxylum,* that were in the area now occur to the south and east in the Sonoran Desert, where summer rain is more frequent and winters are frostless.

These plants, which give to the Sonoran Desert its arboreal character, were derived chiefly from thorn forest by gradual adaptation to drier climate during the later Tertiary. They evidently were eliminated from the Mojave area when summer rain ceased and temperatures were lowered as the Mojave block was uplifted. This was the chief factor that differentiated the Mojave from the Sonoran Desert flora. A few of the plants still occur at isolated sites in the Mojave Desert, including *Crossosoma bigelovii, Dalea emoryi, Lycium pallidum,* and *Prosopis juliflora.* Whether they are Tertiary relicts, or invaded the region during one or more of the drier interglacials, or represent an incursion during the Xerothermic (8000-4000 B.P.) is not known.

Age and origin. Johnson (1968) reviewed the history of the Mojave and Sonoran desert floras in California and concluded that, because many desert plants appear to be ancient taxa, desert environment has been a persistent feature of this region since the Late Cretaceous. This is unlikely, however, because fossil floras from the present desert region show that rainfall was greater there from Late Cretaceous into Pliocene time (Axelrod 1950e). During the Late Cretaceous and Paleocene, the entire region of the three western deserts graded from a perhumid subtropical rainforest in the north to dry tropical forest in the south. The Goler flora from the western Mojave Desert (UC Mus. Pal.) which is interbedded with coal deposits, is represented by diverse, large-leaved (notophyll, mesophyll) evergreen dicots and ferns (UC Mus. Pal.). The Late Cretaceous Holz Shale and the Paleocene Alberhill floras of the Santa Ana Mts., which are situated west of the San Andreas Fault represent mesic forests as shown by the large-leaved evergreen dicots and ferns (UC Mus. Pal.); They have been displaced some 300 miles (480 km) northwest from near the latitude of Rosario in Baja California. In addition, *Araucaria* is in the Late Cretaceous near La Mission and near Rosario, Baja California, sites that were then at the present latitude of the Viscaino Desert some 300 miles (480 km) farther south. *Araucaria* species require mild temperatures and ample rainfall. Some live in temperate rainforests, but others inhabit savannas (Brazil), an environment probably similar to that in the Cretaceous of Baja California and border areas. Furthermore, Late Cretaceous plants from near Las Vegas, Nevada, and Winkleman, Arizona, also associated with coal beds, provide clear evidence of a well watered region as does the occurrence of large herbivorous and carnivorous dinosaurs near Rosario in Baja California and in southern Arizona and Sonora.

Most modern desert species appear to have originated from taxa that contributed to the drier margins of the more mesic Tertiary vegetation, which was displaced from the area of the present desert as progressively drier climate developed (Axelrod 1950e). As aridity increased, taxa that earlier were restricted to dry sites were able to spread out into the widening, ever-drier regions, much like a series of ripples on a lake, each ripple representing a group of new taxa better adapted to the ever-increasing drought (Axelrod 1967c, 1972, 1975).

It is critical to realize that all desert growth forms are also found in semiarid to subhumid environments, contributing to oak woodland, grassland, sage, thorn scrub, and other nondesert vegetation. Some of these adaptive types were already present in the Eocene, as shown by the Green River, Puryear, and other floras in which there are numerous small-leaved drought-deciduous taxa (Fabaceae), evergreen shrubs (*Vauquelinia, Quercus*), palms, members of the Crassulaceae, and other taxa. As judged from the known record, these adaptive types originated under climates with a dry (monsoon) season. For many taxa, only a change in the tempo of living—more rapid seed germination, growth, flowering, pollination, seed setting, etc.—would be the chief requirement for becoming a successful desert plant. Such environments probably had only a local spotwise distribution early in the Tertiary, in the lee of low ranges, and on dry edaphic sites. Many desert taxa may thus have been derived from more mesic ancestors that were already structurally adapted (i.e. preadapted) to survive in areas of decreasing moisture.

With regard to the origin of the present Sonoran desert flora, forerunners of some of its taxa probably inhabited local drier tracts in the region during the Cretaceous. Thus, they have been perfecting both their structural and functional adaptations to hotter, drier environments since that time. Through time, as new major alliances (genera, families) originated in the area, or entered the region from other sources (see Axelrod 1975; Raven and Axelrod 1974), increasing numbers of taxa were added to the expanding area of dry climate. The presumed relative ages of some of the typical alliances as listed below are either known, or inferred from fossil evidence.

LATE CRETACEOUS (OR OLDER?). A number of unique desert plants represent essentially monotypic families, often of unique adaptive type (Engler 1914; Axelrod 1950e, 1952, 1970). Their affinities are obscure or unknown but probably lie with ancient, now-extinct, border-tropical stocks. Among these families are Crossosomataceae, Fouquieriaceae, Koeberliniaceae, Krameriaceae (aff. Polygalaceae?), Loasaceae, and Simmondsiaceae (aff. Buxaceae?).

In addition, a number of families that were established in the Late Cretaceous have genera that are now in the desert, or range from the desert into bordering thorn scrub and nearby woodland. Some of them are unique adaptive types that may well be as old as Late Cretaceous, to judge from their occurrence in Eocene floras, where they contributed to thorn forest and arid subtropical forest. These are the kinds of plants that probably were in the Mojave–Sonoran region from mid-Eocene into the Pliocene, when it was covered with dry tropical forest, thorn forest, and oak savanna. The latest species of these genera have adapted to desert environment more recently. Included here are the following: Anacardiaceae (*Pachycormus*), Arecaceae (*Erythea, Washingtonia, Sabal*), Berberidaceae (*Mahonia*), Bignoniaceae (*Chilopsis*), Cactaceae (many), Caesalpiniaceae (*Haematoxylon*), Capparidaceae (*Atamisquaea, Forchammeria, Isomeris*), Fabaceae (*Olynea*), Liliaceae (*Nolina, Yucca, Dasylirion*), Mimosaceae (*Acacia, Prosopis*, etc.), Rosaceae (*Cowania, Vauquelinia*), Rutaceae (*Choisya*), Simaroubaceae (*Castela, Holocantha*), and Zygophyllaceae (*Viscanoa*).

PALEOCENE–EOCENE. Many genera of subtropical to tropical families in the Paleocene–Eocene floras of the Atlantic and Gulf states and central Rocky Mts. lived in subhumid environments, as shown by their occurrence in the middle Eocene (~ 50 m.y.) Green River and Puryear floras. Today most of these genera are represented by species that form phylads extending from tropical savannas into dry tropical forest, thorn forest, and tropical desert. Inasmuch as many of the desert species are unique, they may well have appeared at an early date. Among the genera now found in the Sonoran Desert that were already adapted to subhumid climate by the middle Eocene, and probably earlier to local drier areas, are the following: Bombacaceae (*Bombax*), Boraginaceae (*Cordia*), Caesalpinaceae (*Bauhinia, Cassia*), Euphorbiaceae (*Acalypha, Euphorbia*), Fabaceae (*Erythrina*), Mimosaceae (*Acacia, Leucanea, Pithecollobium*), Moraceae (*Ficus, Morus*), Rhamnaceae (*Colubrina, Zizyphus*), Sapindaceae (*Cardiospermum, Serjanea, Sapindus*), and Sapotaceae (*Bumelia, Sideroxylon*).

Related to this group are a number of taxa, still unrecorded from the early Tertiary, that appear to have evolved at an early date, as judged from the fact that they are unique adaptive types or have obscure affinities. Here may be mentioned *Dalea* in the Fabaceae, *Cercidium* and *Parkinsonia* in the Caesalpiniacae, and *Sageretia* and *Microrhamnus* in the Rhamnaceae.

OLIGOCENE–MIOCENE. This was the time of rapid spread (not necessarily of origin) of many typical western alliances, notably scores of genera of Asteraceae, Poaceae, and probably also Apiaceae, Brassicaceae, Hydrophyllaceae, Menthaceae, Polemoniaceae, and many others that are certainly autochthonous.

Distinctive species of many older genera, in *Abutilon, Agave, Atriplex, Bursera, Dalea, Dudleya, Eriogonum, Euphorbia, Ipomoea, Jatropha, Lycium, Prosopis, Yucca,* and others, are inferred to have originated at this time because some Miocene and Oligocene species are closely similar to them.

PLIOCENE. Scores of species no doubt came into existence as drought spread. Especially active speciation is postulated for the middle Pliocene, the driest part of the Tertiary (Axelrod 1948).

QUATERNARY. Many herbaceous species range from woodland, chaparral, sage, and grassland communities far out into the Sonoran Desert today. They evidently spread into the area during the pluvials, and have been able to persist there as drought has increased. Many of them are not regionally distinct over their wide areas. Others show minor differences (races, subspecies, "weak" species) and may represent taxa that have survived from invasions during earlier pluvial periods. Included among these diverse alliances are numerous species of *Astragalus, Calochortus, Cryptantha, Delphinium, Eriogonum, Lupinus, Lyrocarpa, Mentzelia, Oenothera, Penstemon, Phacelia, Senecio, Sphaeralcea, Streptanthus,* and many others.

Clearly, the flora of the dry regions of western North America increased in diversity as new taxa were added progressively to more ancient ones during Late Cretaceous, Paleocene-Eocene, Oligo-Miocene, Pliocene, and Quaternary times as dry climate continued to expand in area and to increase in severity, culminating in the wide, regional extreme deserts of post-Wisconsin time.

Desert disjuncts. Many taxa typical of the upper desert slopes, or of the desert itself, have outposts in areas well removed from the desert (Axelrod 1966b). They appear to have spread coastward with warmer and drier climate during the Xerothermic.

Semidesert species that occur on the coastal bluffs in San Diego Co., as at Torrey Pines State Park, include *Euphorbia misera, Isomeris arborea, Simmondsia chinensis,* and *Yucca schidigera.* The interior valleys south of Riverside have numerous typically desert shrubs, including *Acacia greggii, Larrea, Nolina, Prosopis,* and many annuals (see Norland 1931), associated with woodland–savanna vegetation. Among the typical desert species in Santa Ana River canyon west of Corona are *Atriplex canescens, Baccharis emoryi, Bebbia Juncea, Haplopappus pinifolia, Lepidospartum squamatum, Pluchea sericea,* and *Tetradymia comosa* (Howell 1929). A comparable list has been assembled for the dry sites on the north slope of the Santa Monica Mts., as near Cornell Corners, where a number of typically interior taxa occur (Axelrod 1966b).

In his study of the flora of San Clemente Island, Raven (1963) noted that a number of taxa, chiefly herbaceous, are mostly of desert occurrence or are found typically in the interior. These include species of *Achillea, Brodiaea, Cheilanthes, Eriastrum, Filago, Lycium, Sida,* and *Tropidocarpum.* They also occur on the southern Catalinan islands, but are largely absent from the northern Channel Islands (Santa Cruz, Santa Rosa, etc.). In that area climate probably was too moist and cool for most of them, and others no doubt disappeared there after the return of cooler, moister conditions.

Taxa typical of the Mojave region and its borders have penetrated into the southern San Joaquin Valley via the Tejon, Tehachapi, and Walker passes, and thence spread northward in the inner Coast Ranges, some as far as Mount Diablo. Sharsmith (1945) lists many typically desert-border species in the Mt. Hamilton Range. Notable among the woody plants are *Forestieria neomexicana, Gutierrezia californica, Juniperus californica, Lepidospartum squamatum, Prosopis juliflora,* and *Stenotopsis linearifolius.* In this connection, it is recalled that the kit fox (*Vulpes macrotis*), a typical denizen of the desert, occurs in the southern San Joaquin Valley, as do the yucca night lizard (*Xantusia vigilis*), desert scaly lizard (*Sceloporous magister*), and leopard lizard (*Crotaphytus wislizenii*). They inhabit saline Kern Desert in Kern and southern Kings Cos., an undrained to poorly drained area largely dammed by alluvial fans from the bordering mountains, except at times of exceptionally wet winters. Dominated by *Prosopis* in the wet Buena Vista basin, and by *Atriplex* elsewhere, it probably originated during the Xerothermic as evaporation increased and the series of lakes disappeared. A rise in mean temperature of 5°–6°F (~ 3°C) and a lowering of effective rainfall by 4–5 inches (10–12.7 cm) during the Xerothermic would have enabled desert taxa to cross Tehachapi Pass and invade Kern Desert and the drier South Coast Ranges.

Summary

Although chaparral, coastal sage, and desert are comparatively recent as regional types of vegetation, many of the taxa that contribute to them are Miocene and older.

Chaparral taxa have been associated with sclerophyllous oak-laurel woodland vegetation since the Eocene. Apart from local drier sites, chaparral did not spread widely until late in the Cenozoic, after mountains had been elevated significantly and the summer dry Mediterranean climate had appeared. Chaparral taxa are not pyrogenic, but were preadapted to survive under such conditions. The spread of chaparral was aided by natural fires, especially during the Xerothermic, but its greatest expansion has occurred under man's influence. It is seral to woodland, not a natural climax.

Taxa that contribute to coastal sage are regular members of the drier parts of woodland–savanna vegetation. Some of the woody taxa in it have records in the middle Miocene. Sage apparently expanded under the influence of drier climate during the later Pleistocene as woodland-savanna was eliminated from the drier areas it now inhabits as a new climax.

Many of the woody taxa that contribute to desert vegetation have similar (or identical) species in the woodland-chaparral and thorn forest vegetation that covered the area presently desert into the Miocene and Pliocene. The taxa appear to have been preadapted structurally to conditions of greater drought, and to have survived as aridity increased and as woodland and thorn forest were eliminated from the area. Taxa now in the desert have diverse ages; monotypic families possibly are Cretaceous, unique genera are probably Eocene, and the others are younger. Many of the herbaceous alliances entered the region only during the last glacial, for they also range well outside the desert and contribute to vegetation bordering it today.

CONCLUSIONS

Many woody plants that make up the varied vegetation zones of California are represented by similar, or the same, species in middle Miocene and younger floras of the western United States. With summer as well as winter precipitation and more equable yearly temperatures, these vegetation zones were more diverse in taxa than are the descendant ones, and they were composed of taxa that now contribute to different segregated communities.

Commencing in the late Miocene and continuing into the Quaternary, many taxa disappeared from the region as summer rainfall gradually decreased. Eliminated first from the interior, their nearest descendants are in the eastern United States and eastern Asia and the southwestern United States and Mexico.

As interior regions became drier and colder, most mesic conifer forests and sclerophyll vegetation shifted coastward. Impoverished forests and piñon–juniper woodland persisted in the interior, and grassland, sagebrush, and desert scrub then spread widely over the lowlands there. As Mediterranean climate appeared, and temperature contrasts increased from coast to interior, the surviving vegetation zones were segregated into communities of more local distribution and lower diversity.

Most woody taxa represented in Miocene and Pliocene floras can scarcely be separated from living species. Hence the structural adaptations of forest trees, sclerophyllous woodland, and chaparral taxa, as well as a number of the diverse forms now in the arboreal Sonoran Desert, were already established 15 million and more years ago. Thus it is inferred that, as summer rains gradually decreased, adaptation to the emerging Mediterranean climate was chiefly physiological. This probably was accomplished by shifting germination and growth into the moister and cooler part of the year, though some Sonoran Desert taxa in California are still triggered by light summer showers and warm nights.

It is inferred that the herbaceous and perennial flora diversified greatly during the Quaternary as intense mountain building created many new subzones with novel topographic–climatic-biotic relations, and as fluctuating glacial–pluvial and interglacial climates repeatedly brought together and then isolated interbreeding populations that created new taxa.

The Xerothermic of 8000–4000 B.P. appears to account for the present discontinuous distributions of many xeric taxa that require warm summers, and for the disjunct occurrence of forests and woodlands in both the Coast Ranges and Sierra Nevada.

LITERATURE CITED

Adam, D. P. 1973. Early Pleistocene (?) pollen spectra from near Lake Tahoe, California. U.S. Geol. Survey J. Res. 1:691-693.

Arnold, C. A. 1955. Tertiary conifers from the Princeton coal field, British Columbia. Univ. Mich. Mus. Paleont. Contrib. 12:245-258.

Axelrod, D. I. 1937. A Pliocene flora from the Mount Eden beds, southern California. Carnegie Inst. Wash. Pub. 476:125-183.

---. 1938. The stratigraphic significance of a southern element in later Tertiary floras of western America. Wash. Acad. Sci. J. 28:314-322.

---. 1939. A Miocene flora from the western border of the Mohave Desert. Carnegie Inst. Wash. Pub. 516:1-128.

---. 1940a. The Mint Canyon flora of southern California: a preliminary statement. Amer. J. Sci. 238:577-585.

---. 1940b. The Pliocene Esmeralda flora of west-central Nevada. Wash. Acad. Sci. 30:163-174.

---. 1944a. The Black Hawk Ranch flora. Carnegie Inst. Wash. Pub. 553:91-101.

---. 1944b. The Mulholland flora. Carnegie Inst. Wash. Pub. 553:103-146.

---. 1944c. The Oakdale flora. Carnegie Inst. Wash. Pub. 553:147-166.

---. 1944d. The Sonoma flora. Carnegie Inst. Wash. Pub. 553:167-206.

---. 1944e. The Alturas flora. Carnegie Inst. Wash. Pub. 553:263-284.

---. 1948. Climate and evolution in western North America during Middle Pliocene time. Evolution 2:127-144.

---. 1950a. A Sonoma florule from Napa, California. Carnegie Inst. Wash. Pub. 590:23-71.

---. 1950b. Further studies of the Mount Eden flora, southern California. Carnegie Inst. Wash. Pub. 590:73-117.

---. 1950c. The Anaverde flora of southern California. Carnegie Inst. Wash. Pub. 590:119-158.

---. 1950d. The Piru Gorge flora of southern California. Carnegie Inst. Wash. Pub. 590:159-214.

---. 1950e. Evolution of desert vegetation. Carnegie Inst. Wash. Pub. 590:215-306.

---. 1952. A theory of angiosperm evolution. Evolution 6:29-60.

---. 1956. Mio-Pliocene floras from west-central Nevada. Univ. Calif. Pub. Geol. Sci. 33:1-316.

---. 1957. Late Tertiary floras and the Sierra Nevadan uplift. Geol. Soc. Amer. Bull. 68:19-46.

---. 1958a. The Pliocene Verdi flora of western Nevada. Univ. Calif. Pub. Geol. Sci. 34:61-160.

---. 1958b. Evolution of the Madro-Tertiary Geoflora. Bot. Rev. 24:433-509.

---. 1959. Late Cenozoic evolution of the Sierran Bigtree forest. Evolution 13:9-23.

---. 1962. A Pliocene *Sequoiadendron* forest from western Nevada. Univ. Calif. Pub. Geol. Sci. 39:195-268.

---. 1964. The Miocene Trapper Creek flora of southern Idaho. Univ. Calif. Pub. Geol. Sci. 51:1-161.
---. 1966a. The Eocene Copper Basin flora of northeastern Nevada. Univ. Calif. Pub. Geol. Sci. 59:1-119.
---. 1966b. The early Pleistocene Soboba flora of southern California. Univ. Calif. Pub. Geol. Sci. 60:1-109.
---. 1967a. Evolution of the California closed-pine forest, pp. 93-150. In R. N. Philbrick (ed.). Proc. symp. on biology of the California Islands, Santa Barbara Bot. Gard., Santa Barbara, Calif.
---. 1967b. Geologic history of the California insular flora, pp. 267-316. In R. N. Philbrick (ed.). Proc. symp. on biology of the California Islands, Santa Barbara Bot. Gard., Santa Barbara, Calif.
---. 1967c. Drought, diastrophism and quantum evolution. Evolution 21:201-209.
---. 1968. Tertiary floras and topographic history of the Snake River Basin, Idaho. Geol. Soc. Amer. Bull. 79:713-734.
---. 1970. Mesozoic paleogeography and early angiosperm history. Bot. Rev. 36:277-319.
---. 1972. Edaphic aridity as a factor in angiosperm evolution. Amer. Natur. 106:311-320.
---. 1973. History of the Mediterranean ecosystem in California, pp. 225-277. In F. di Castri and H. A. Mooney (eds.). Mediterranean type ecosystem: origin and structure. Springer-Verlag.
---. 1975. Evolution and biogeography of Madrean-Tethyan sclerophyll vegetation. Missouri Bot. Gard. Ann. 62:280-334.
---. 1976a. History of the coniferous forests, California and Nevada. Univ. Calif. Pub. Bot. 70:1-62.
---. 1976b. Evolution of the Santa Lucia fir (Abies bracteata) ecosystem. Ann. Missouri Bot. Gard. 63:24-41.
Axelrod, D. I., and W. S. Ting. 1960. Late Pliocene floras east of the Sierra Nevada. Univ. Calif. Pub. Geol. Sci. 39:1-118.
Axelrod, D. I., and W. S. Ting. 1961. Early Pleistocene floras from the Chagoopa surface, southern Sierra Nevada. Univ. Calif. Pub. Geol. Sci. 39:119-194.
Bailey, H. P. 1960. A method of determining the warmth and temperateness of climate. Geograf. Annal. 42:1-16.
---. 1964. Toward a unified concept of the temperate climate. Geogr. Rev. 54:516-545.
---. 1971. On recognition of climatic similarities in global perspective. Origin and structure of ecosystems. Integrated Res. Program, Internat. Biol. Program, Tech. Rept. 71-1:1-12.
Becker, H. F. 1961. Oligocene plants from the upper Ruby River basin, southwestern Montana. Geol. Soc. Amer. Mem. 82. 127 p.
---. 1969. Fossil plants of the Tertiary Beaverhead basins in southwestern Montana. Paleontographica B127:1-142.
Berger, R., and W. F. Libby. 1966. UCLA radiocarbon dates. Radiocarbon 8:467-497.
Billings, W. D. 1951. Vegetational zonation in the Great Basin in western North America, pp. 101-122. In Comp. rend. du colloque sur les bases ecologiques de regeneration de la vegetation des zones arides, Union Internat. Soc. Biol., Paris.

Chaney, R. W. 1938a. Ancient forests of Oregon: a study of earth history in western America. Carnegie Inst. Wash. Pub. 501:631-648.

---. 1938b. The Deschutes flora of eastern Oregon. Carnegie Inst. Wash. Pub. 476:185-216.

---. 1944. The Troutdale flora. Carnegie Inst. Wash. Pub. 476:323-351.

Chaney, R. W., and D. I. Axelrod. 1959. Miocene floras of the Columbia Plateau. Carnegie Inst. Wash. Pub. 617. 237 p.

Chaney, R. W., and H. L. Mason. 1933. A Pleistocene flora from the asphalt deposits at Carpinteria, California. Carnegie Inst. Wash. Pub. 415:45-79.

Clements, F. E. 1920. Plant indicators. Carnegie Inst. Wash. Pub. 290. 388 p.

---. 1936. The origin of the desert climax and climate, pp. 87-140. In T. H. Goodspeed (ed.). Essays in geobotany in honor of William Albert Setchell. Univ. Calif. Press, Berkeley.

Clokey, I. W. 1951. Flora of the Charleston Mountains, Clark County, Nevada. Univ. Calif. Pub. Bot. 24. 274 p.

Condit, C. 1938. The San Pablo flora of west central California. Carnegie Inst. Wash. Pub. 476:217-268.

---. 1944a. The Remington Hill flora. Carnegie Inst. Wash. Pub. 553:21-55.

---. 1944b. The Table Mountain flora. Carnegie Inst. Wash. Pub. 553:57-90.

Curtis, B. F. (ed.) 1975. Cenozoic history of the southern Rocky Mountains. Geol. Soc. Amer. Mem. 144. 279 pp.

Detling, Leroy E. 1953. The chaparral formation of southeastern Oregon with consideration of its post-glacial history. Ecology 42:348-357.

---. 1968. Historical background of the flora of the Pacific Northwest. Univ. Oregon Mus. Nat. Hist. Bull. 13. 57 p.

Dorf, E. 1930. Pliocene floras of California. Carnegie Inst. Wash. Pub. 412:1-112.

Engler, A. 1914. Ueber Herfunft, Alter and Verbreitung extremer xerothermer Pflanzen. K. Preuss. Akad. Wissen. 20:564-621.

Fowells, H. A. 1965. Silvics of forest trees of the United States. USDA Forest Serv. Agric. Handbook 271. 762 pp.

Frost, F. H. 1927. The Pleistocene flora of Rancho La Brea. Univ. Calif. Pub. Bot. 14:73-98.

Garfunkel, Z. 1974. Model for the Late Cenozoic tectonic history of the Mohave Desert, California, and for its relation to adjacent regions. Geol. Soc. Amer. Bull. 85:1931-1944.

Gilluly, J. 1963. The tectonic evolution of the western United States. Quart. J. Geol. Soc. London 119:133-174.

Griffin, J. R. 1964. A new Douglas fir locality in southern California. Forest Sci. 10:317-319.

---. 1971. Vascular plants of the San Benito Mountain natural area. Bur. Land Manage., Folsom, Calif. 9 p. (mimeo.).

Griffin, J. R., and W. B. Critchfield. 1972. The distribution of forest trees in California. USDA Forest Serv. Res. Paper PSW-82/1972. 114 p.

Hansen, H. P. 1942. A pollen study of lake sediments in the low Willamette Valley of western Oregon. Bull. Torrey Bot. Club 69:262-280.

---. 1947. Postglacial forest succession, climate, and chronology in the Pacific Northwest. Amer. Philos. Soc. Trans. 37:1-130.

Hansen, H. P., and E. L. Packard. 1947. Pollen and analysis and the age of proboscidian bones near Silverton, Oregon. Ecology 30:461-468.

Heusser, C. J. 1960. Late Pleistocene environments of North Pacific North America. Amer. Geogr. Soc. Spec. Pub. 35.

Howell, J. T. 1929. The flora of Santa Ana cañon region. Madroño 1:243-253.

---. 1945. Studies in Phacelia: Revision of species related to P. douglasii, P. linearis and P. pringlei. Amer. Midl. Natur. 33:460-494.

Jepson, W. L. 1910. The Silva of California. Univ. Calif. Mem. 2. Univ. Calif. Press, Berkeley. 480 pp.

---. 1925. A manual of the flowering plants of California. Univ. Calif. Press, Berkeley. 1238 pp.

Johnson, A. W. 1968. The evolution of desert vegetation in western North America, pp. 101-140. In G. W. Brown, Jr. (ed.). Desert biology. Academic Press, New York.

King, P. B. 1958. Evolution and the modern surface features of western North America. Amer. Assoc. Adv. Sci. Pub. 51:3-60.

Kramer, P. J. 1969. Plant and soil-water relationships: a modern synthesis. McGraw-Hill, New York. 482 p.

Le Conte, J. 1887. The flora of the coast islands of California in relation to recent changes of physical geography. Amer. J. Sci., Ser. 3, 34:457-480.

Leskinen, P. H. 1975. Occurrence of oaks in Late Pleistocene vegetation in the Mohave Desert of Nevada. Madroño 23:234-235.

Little, E. L., and W. B. Critchfield. 1969. Subdivisions of the genus Pinus (pines). USDA Forest Serv. Misc. Pub. 1144:1-51.

Loope, L. L. 1969. Subalpine and alpine vegetation of northeastern Nevada. Ph. D. dissertation, Duke Univ., Durham, N.C. 292 p.

MacGinitie, H. D. 1941. A middle Eocene flora from the central Sierra Nevada. Carnegie Inst. Wash. Pub. 534. 178 p.

---. 1953. Fossil plants of the Florissant beds, Colorado. Carnegie Inst. Wash. Pub. 599. 188 p.

---. 1969. The Eocene Green River flora of northwestern Colorado and northeastern Utah. Univ. Calif. Pub. Geol. Sci. 83:1-202.

Major, J., and S. A. Bamberg. 1967. Some Cordilleran plants disjunct in the Sierra Nevada of California and their bearing on Pleistocene ecological conditions, pp. 171-188. In H. E. Wright, Jr., and W. H. Osburn (eds.). Arctic and alpine environments. Indiana Univ. Press, Bloomington.

Martin, P. S. and J. Gray. 1962. Pollen analysis and the Cenozoic. Science 137:103-111.

Mason, H. L. 1934. Pleistocene flora of the Tomales Formation. Carnegie Inst. Wash. Pub. 415:81-179.

---. 1944. A Pleistocene flora from the McKittrick asphalt deposits of California. Calif. Acad. Sci. Proc. 25:221-234.

Muir, J. 1876. On the post-glacial history of Sequoia gigantea. Amer. Assoc. Adv. Sci. Proc. 25:242-253.

Munz, P. A. 1935. Manual of southern California botany. Claremont Colls., Claremont, Calif. 642 p.

Nilsen, T. H., and S. H. Clarke, Jr. 1975. Sedimentation and tectonics in the early Tertiary continental borderland of central California. U. S. Geol. Survey Prof. Paper 925. 64 p.

Norland, E. D. 1931. A study of the distribution of southern California plants. M. A. thesis, Pomona Coll., Claremont, Calif. 107 p.

Peck, D., et. al. 1964. Geology of the central and northern parts of the Western Cascade Range in Oregon. U.S. Geol. Survey Prof. Paper 449. 59 pp.

Potbury, S. S. 1935. The La Porte flora of Plumas County, California. Carnegie Inst. Wash. Pub. 465:29-81.

Putnam, W. C. 1942. Geomorphology of the Ventura region, California. Geol. Soc. Amer. Bull. 53:691-754.

Raven, P. H. 1963. A flora of San Clemente Island, California. Aliso 5:289-347.

---. 1964. Catastrophic selection and edaphic endemism. Evolution 64:336-338.

Raven, P. H., and D. I. Axelrod. 1974. Angiosperm biogeography and past continental movements. Ann. Missouri Bot. Gard. 61:539-673.

Renney, K. M. 1972. The Miocene Temblor flora of west central California. M.A. thesis, Univ. Calif., Davis. 106 p.

Repenning, C. A. 1967. Palearctic-Nearctic mammalian dispersal in the late Cenozoic, pp. 288-311. In D. M. Hopkins (ed.). The Bering land bridge. Stanford Univ. Press, Stanford, Calif.

Rouse, G. E., and W. H. Mathews. 1961. Radioactive dating of Tertiary plant-bearing deposits. Science 133:1079-1080.

Sharsmith, H. K. 1945. Flora of the Mount Hamilton Range of California. Amer. Midl. Natur. 34:289-367.

Shreve, F. 1942. The desert vegetation of North America. Bot. Rev. 8:195-246.

---. 1964. Vegetation of the Sonoran Desert, pp. 9-186. In F. Shreve and I. L. Wiggins. Vegetation and flora of the Sonoran Desert. Stanford Univ. Press, Stanford, Calif.

Simpson, G. G. 1953. The major features of evolution. Columbia Univ. Press, New York. 434 p.

Smith, H. V. 1941. A Miocene flora from Thorn Creek, Idaho. Amer. Midl. Natur. 25:473-522.

Wang, Chi-Wu. 1961. The forests of China with a survey of grassland and desert vegetation. Maria Moors Cabot Found. Pub. 5. Harvard Univ., Cambridge, Mass. 313 p.

Webber, I. E. 1933. Woods from the Ricardo Pliocene of Last Chance Gulch, California. Carnegie Inst. Wash. Pub. 412:113-134.

Wells, P. V., and R. Berger. 1967. Late Pleistocene history of coniferous woodland in the Mohave Desert. Science 155:1640-1647.

Wells, P. V., and C. D. Jorgensen. 1964. Pleistocene wood rat middens and climatic change in the Mohave Desert: a record of juniper woodlands. Science 143:1171-1174.

Whittaker, R. H. 1960. Vegetation of Siskiyou Mountains, Oregon and California. Ecol. Monogr. 30:279-338.

Wiggins, I. L. 1951. An additional specimen of *Pinus piperi* Dorf from Ventura County, California. Amer. J. Bot. 38:211-213.

Wolfe, J. A. 1968. Paleogene biostratigraphy of nonmarine rocks in King County, Washington. U. S. Geol. Survey Prof. Paper 571. 33 p.

Yurtsev, B. A. 1972. Phytogeography of northeastern Asia and the problem of Transberingian floristic interrelations, pp. 19-54. In A. Graham (ed.). Floristics and paleofloristics of Asia and eastern North America. Elsevier, New York.

Zahner, R. 1968. Water deficits and growth of trees, pp. 191-254. In T. T. Kozlowski (ed.). Water deficits and plant growth. Vol. 2: Plant water consumption and response. Academic Press, New York.

CHAPTER

6 THE STATUS OF VEGETATION MAPPING IN CALIFORNIA TODAY

WILMER L. COLWELL, JR.

Research Forester, Pacific Southwest Forest and Range Experiment Station, Forest Service, U.S. Department of Agriculture, Berkeley, California

Introduction	196
Vegetation mapping in California	196
The Vegetation Type Map Survey	197
Vegetation type map field sheets	197
Published vegetation type maps (colored prints)	198
Blue-line vegetation type and forest condition maps	198
State maps of vegetation type and timber cropland	200
State Cooperative Soil–Vegetation Survey	201
Timber stand-vegetation cover maps	201
Soil–vegetation maps	205
Upland soils maps	206
Vegetation mapping by other agencies	207
U.S. Forest Service, California Region	207
Soil-vegetation surveys	207
Soil resource inventories (soil surveys)	207
Forest type and classification system	207
Range surveys	208
Bureau of Land Management	208
National Park Service	208
University of California Remote Sensing Research Program	209
Methods used by the Vegetation Type Map and State Cooperative Soil–Vegetation Surveys	210
Vegetation Type Map Survey	210
Basic classification system	210
Supplementary information	211
Tree and shrub maps	211
Sample plots	215
Other information	215
State Cooperative Soil–Vegetation Survey	215
Basic classification system	215
Supplementary information	216
Conclusion	218
Literature cited	219

INTRODUCTION

"The most important act to be done by the board should be the preparation and publication of a forest map of the entire state, showing the amount and kind of timber standing in the different counties, and its commercial use and value." That statement was made by the Hon. Abbot Kinney, Chairman of the State Board of Forestry, during a board meeting on May 19, 1886 (California State Board of Forestry 1886).

The preparation of these maps may well have been the first organized effort to portray natural vegetation in California for its economic importance on an areawide basis. Much of the early vegetation mapping was confined to the routine recording of prominent vegetation along new trails or routes by explorers, and of survey lines by topographical engineering field parties. In the course of reconnaissance geological mapping in the 1870s, the U.S. Geological Survey recorded broad groups of vegetation.

Beginning in 1887, forest mapping was done on small-scale county maps (1:506,880 or 1 inch = 8 miles, 1 cm = 5 km) and covered the more heavily forested portions of the counties in northern California. Later, a system of symbols for commercial conifers and certain hardwood stands was used to indicate dominant species in township maps at a larger scale (1 inch = 1 mile, 1 cm = 0.6 km). Densities of the stands were shown by stippling and shading of a green color on the map.

Today, we are nearing the fulfillment of that early request made nearly 100 yr ago. Most commercial timberlands under federal and state jurisdiction and most of the large lumber company lands are mapped and inventoried to various degrees of intensity as to stand composition, age class, cover density, and volume and growth of commercial tree species. However, detailed vegetation information is still needed for about one-fourth of the state and privately owned commercial forest lands of California.

VEGETATION MAPPING IN CALIFORNIA

Not until 1926 did vegetation mapping in California begin on a large scale. The purpose was to provide data to aid in the development of statewide land use and fire protection policies. The mapping of natural vegetation was carried out by the U.S. Forest Service's California (now Pacific Southwest) Forest and Range Experiment Station in Berkeley (Wieslander 1935b). First priority was to collect enough data to prepare an extensive generalized vegetation map of the chaparral belt in southern California. About 280,000 ha were mapped in winter and spring of 1927 and 1928 (Wieslander 1961).

In 1928, the nationwide Forest Survey was authorized by the McSweeney–McNary Research Act. Thus vegetation mapping became part of the beginning forest survey in California. Again, southern California was the survey site. Needed was more detailed information for the management and protection of chaparral-covered watersheds. Along with the mapping, plots were sampled to obtain additional data about the natural vegetation associations, such as plant size, cover density, and understory species. In 1931, mapping expanded into northern California, beginning with the Trinity National Forest and later extended into Eldorado and Tahoe National Forests. As a prelude to the timber aspects of the Forest Survey, commercial forests were mapped as to virgin, residual, second-growth, or unstocked stands, and as to site index. A. E. Wieslander, who headed this Vegetation Type Map Survey, had the foresight to realize the value of vegetation data over and beyond any immediate purpose. He understood that information would be needed for many purposes in the future. Küchler (1967) reported that Wieslander showed a "particularly keen insight into the problems of vegetation mapping and used a scientific approach to their solution."

From these initial efforts, a format was developed for mapping the natural vegetation of the state. The mapping continued statewide, excluding the desert and agricultural lands of the large valleys, until the work was stopped during World War II. By then, about 16 million ha had been mapped (Wieslander 1961).

In discussing various kinds of classification systems for vegetation mapping, Küchler (1967) concluded that there are five basic types of vegetation maps: ecological, floristic, physiognomical, regional, and combination physiognomical–floristic. He considered the California vegetation mapping programs conducted since 1926 to be of the combination type. Basically, the program recognizes vegetation in the physiognomical form, such as grass, shrubs, hardwoods, and conifers. For each form, the floristic attributes are described and plant species listed.

In 1947, another large vegetation mapping program began—the Cooperative Soil-Vegetation Survey at the California (now Pacific Southwest) Forest and Range Experiment Station. This survey has the unique distinction of mapping and classifying both soils and vegetation, providing more complete resource information for the management of wildlands. The work is done primarily on privately owned and state-owned lands, and the survey is financed by appropriations of the California Legislature to the State Division of Forestry. In addition to the Experiment Station and the Division of Forestry, other organizations cooperating in the survey are the Departments of Agronomy and Range Science and of Soils and Plant Nutrition at the University of California, Davis, and the Department of Forestry and Conservation at the University of California, Berkeley.

To date, more than 4.6 million ha of wildland have been mapped, including about 0.8 million ha on some national forest lands.

The Vegetation Type Map Survey

The original goal of the Vegetation Type Map Survey of mapping vegetation embracing 28 million ha in all of California, excluding the desert and agricultural lands, was never reached (Wieslander 1935a). However, much valuable information is available for the areas completed, and most of it is provided on three kinds of vegetation map formats: (1) the original vegetation type map field sheets, (2) map sheets overprinted in color on U.S. Geological Survey (USGS) 15-min and 30-min topographical quadrangles, and (3) simplified, uncolored blue-line print sheets at a scale of 1:62,500 of two kinds—broad vegetation types, and forest conditions.

Vegetation type map field sheets. The actual field mapping of the vegetation was done on 15-min topographical quadrangle sheets at a scale of 1:62,500. For convenience in mapping, the map sheets were cut into quarters, mounted on a cloth backing, and folded.

The major vegetation types are shown in different colors and are separated by ink lines. The types are further divided into units of different species composition in which the dominant (20% cover or more) species of trees, shrubs, and some herbs and grasses are indicated by inked letter symbols.

A total of 331 map sheets—320 in California and 11 in western Nevada—were completed and are on file (together with a quadrangle checklist) at the Pacific Southwest Forest and Range Experiment Station in Berkeley. The maps cover an area of about 16 million ha. An index map of California (Fig. 6-1) shows the location of the field maps according to a number and letter index of USGS quadrangles. A key to the index system is shown in the left-hand corner.

Published vegetation type maps (colored prints). The Vegetation Type Map Survey issued 23 sheets covering about 2.8 million ha, including most of the chaparral lands in southern California (Fig. 6-2). The sheets are printed in color as overprints on regular USGS 15-min and 30-min topographical quadrangles at scales of 1:62,500 and 1:125,000, respectively. Eight of the maps are on 30-min quadrangle bases. They are on file (together with a quadrangle checklist) at the Pacific Southwest Forest and Range Experiment Station in Berkeley.

As on the basic field maps, the colors show the general or broad vegetation types having similar use, economic importance, or fire hazard characteristics. For each separately colored type area, there may be one or more subtypes within which the dominant species symbols are listed. Map margins contain (1) color legends of the types mapped in the quadrangle, (2) a list of plant species and symbols, (3) brief descriptions of the type classification methods and of the types mapped in the quadrangle, (4) a table summarizing type areas by counties, national forests, and parks, and (5) elevational profiles illustrating the local relation of vegetation types to elevation and slope exposure. Additional profiles have been published separately (Critchfield 1971).

Blue-line vegetation type and forest condition maps. Since only 23 vegetation type maps were published, it was important that vegetation information recorded on the original field sheets for the remaining 13 million ha be made available in some published form. As a result, 265 blue-line prints were compiled and published.

These blue-line vegetation type maps are generalized versions of the original vegetation type field sheets and are printed in 15-min USGS quadrangle units at a scale of 1:62,500. The areas covered are shown in Fig. 6-2. They are available (together with a quadrangle checklist) for purchase from the California Region, Forest Service, 630 Sansome Street, San Francisco, California 94111.

The vegetation units are simplified to the extent of omitting species composition, all areas under 8 ha in size, and others under 16 ha where the vegetation is not markedly different from that of surrounding areas. Each unit has a number corresponding to 23 major vegetation types and other land units. Two special legend sheets define the vegetation types: one for central Sierra and western Nevada, and one for South Coast Ranges. The 23 vegetation types and land units are as follows: grassland, sagebrush, chamise–chaparral, chaparral, semidesert chaparral, woodland–grass, woodland, piñon and juniper, miscellaneous conifers, redwood belt, Douglas fir belt, pine belt, pine–Douglas fir belt, pine–Douglas fir–fir belt, pine–fir belt, fir belt, lodgepole-white pine belt, whitebark-foxtail pine belt, barren, cultivated and urban, desert, plantation, and badlands.

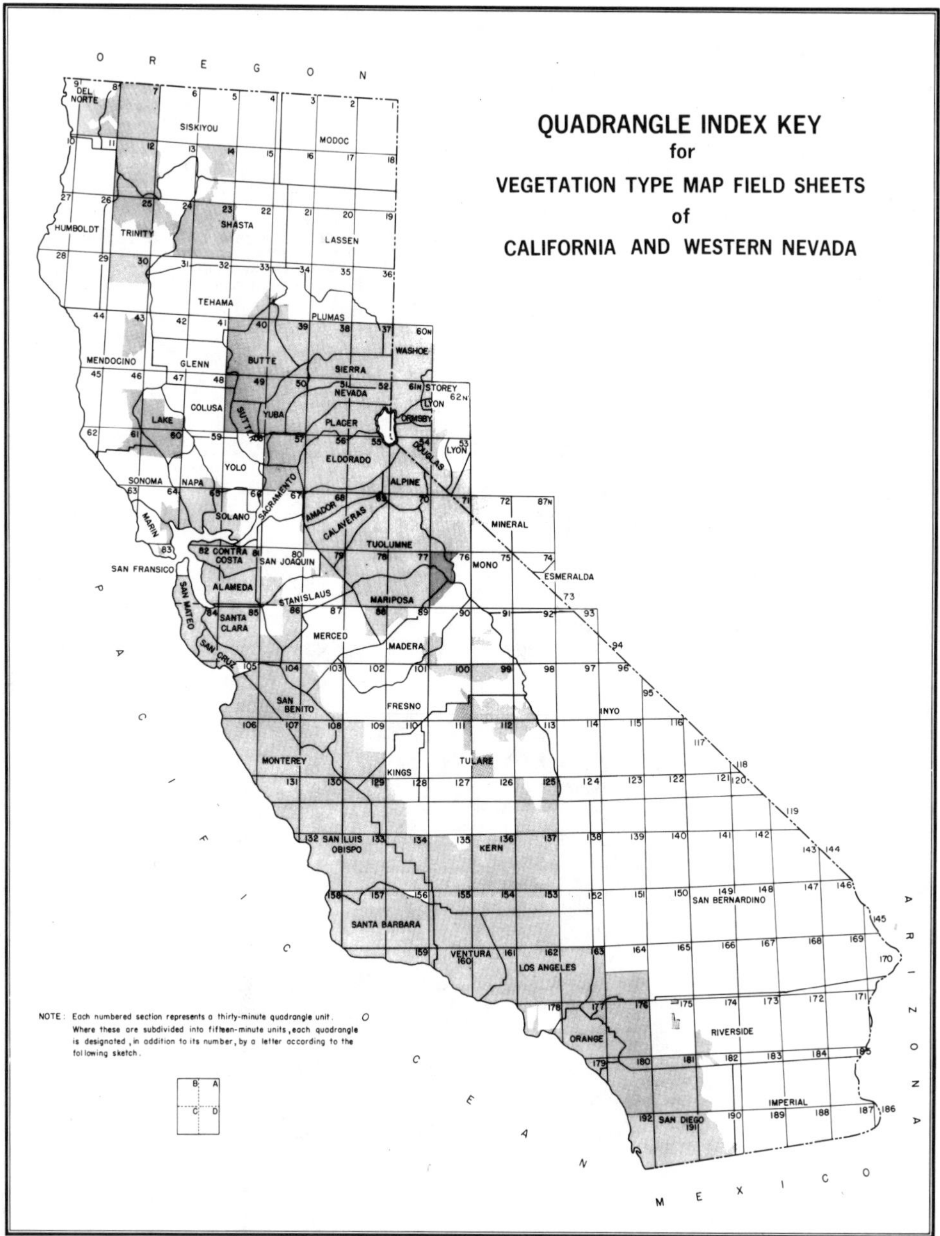

Figure 6-1. Quadrangle index key for vegetation type map field sheets of California and western Nevada. Shaded areas represent quadrangle maps available.

Supplementing these maps are 76 blue-line forest condition maps (Fig. 6-2). They provide additional information on forest conditions collected simultaneously with the mapping of vegetation, but only for lands capable of producing commercial timber crops. Basically, the maps show forest land classified as to virgin timber, residual stands, second-growth timber segregated by age and stocking classes,

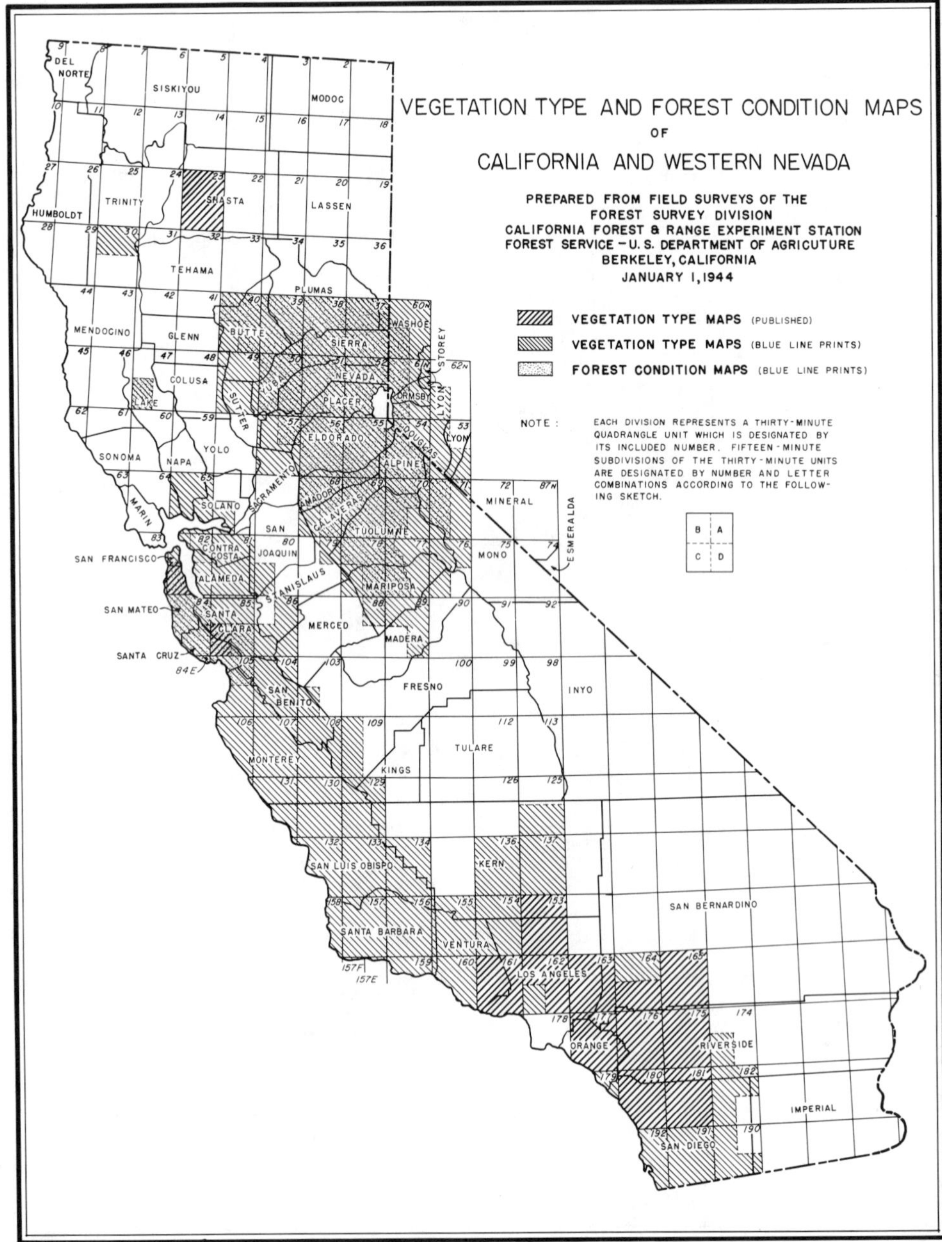

Figure 6-2. Quadrangle index key for published vegetation type and forest condition maps of California and western Nevada.

deforested areas, and other lands. A site index is also indicated in the mapping units of commercial forest land. Forest condition maps, a quadrangle checklist, and explanatory legends are also available for purchase from the U.S. Forest Service in San Francisco.

State maps of vegetation type and timber cropland. Two maps, "Vegetation Types of California" and "Timber Croplands of California," at a scale of 1:1,000,000 (1 inch = 16 miles, 1 cm = 10 km) were published in 1945 as part of the Forest Survey Reappraisal Project (Wieslander and Jensen 1946). The maps are available for purchase from the U.S. Forest Service in San Francisco.

The vegetation type map of the state is a blue-line print suitable for coloring that shows the distribution of 18 different vegetation types and land status units by letter symbols. It was compiled from all available data, including vegetative type maps; National Park Service vegetation type maps; Forest Service, California Region, timber type survey maps; national forest type maps; aerial photos; and ground observation.

The timber croplands map of California is also in blue-line print, showing the division of commercial forest land by three timber age classes for five major timber types. For areas that are poorly stocked or unstocked with commercial conifer timber, the present temporary cover, such as grass, chaparral, sagebrush, woodland, or woodland–grass, is indicated by appropriate symbols. Legends for the vegetation type and timber cropland maps are shown in Table 6-1.

State Cooperative Soil–Vegetation Survey

Three kinds of maps are published: timber stand–vegetation cover maps, soil–vegetation maps, and generalized county maps of upland soils (State Cooperative Soil–Vegetation Survey 1958). Both the soil–vegetation and timber stand–vegetation cover maps are published on a standard 7.5-min quadrangle unit at a scale of 1:31,680 (2 inches = 1 mile, 1 cm = 0.3 km). The generalized county maps are at a scale of 0.5 inch to 1 mile (1 cm = 2 km) and show upland soils grouped by color according to normally associated broad vegetation types. They have been published for only three counties: Mendocino, Glenn, and Lake.

Timber stand–vegetation cover maps. A total of 467 map sheets covering about 5.8 million ha were published during the period from 1948 to 1962. They include much of the woodland, chaparral, and commercial forestlands of the state and privately owned lands outside national forest boundaries in northern California (Fig. 6-3). Portions of some national forests have timber stand maps (Sierra, Mendocino, Shasta–Trinity, and Lassen). The maps, with legends, quadrangle checklist, and tables, are available for purchase from the U.S. Forest Service in San Francisco.

In general, the timber stand-vegetation cover maps show information obtained mainly from study and interpretation of vegetation patterns on stereopairs of vertical aerial photographs and supplementary field observations. The information is shown on the map in three categories: (1) broad vegetation types and other land units in order of abundance, (2) age class of commercial timber, and (3) "density" (actually, cover) of commercial timber by two size classes and total "density" (cover) of all woody vegetation.

The broad vegetation groups (grass, shrubs, trees, and miscellaneous), previously subdivided into major types for the Vegetation Type Map Survey, are now called elements as shown below:

C	Commercial conifers	G	Grass
K	Noncommercial conifers	M	Marsh

TABLE 6-1. Legends for 1 : 1,000,000 scale maps; vegetation types of California and timber croplands of California

Vegetation Types			Timber Croplands	
Tree types			Age class	
Timber	P	Pine	1.	Old-growth stands
	R	Redwood	2.	Young-growth-old-growth stands
	D	Douglas fir	3.	Young-growth stands
	F	Fir	4.	Poorly stocked and unstocked areas. Present cover indicated as follows:
	M	Pine-Douglas fir-Fir		
Subalpine	L	Lodgepole-white bark pine	C	Chaparral
Miscellaneous conifers	N	Piñon pine	G	Grass
	J	Juniper	S	Great Basin sagebrush
	E	Minor conifers	T	Coastal sagebrush
	E-1	Knobcone pine	V	Woodland-grass
	E-2	Monterey pine	W	Woodland (hardwoods)
	E-3	Bishop pine	Timber types	
	E-4	Coulter pine	D	Douglas fir
	E-5	Big-cone spruce	F	Fir
	E-6	Cypress	M	Pine-Douglas fir-Fir
Woodlands	W	Woodland (hardwoods)	P	Pine
	V	Woodland-grass	R	Redwood

Type	Code	Description	Code	Description
Shrub types				
	C	Chaparral	O	Unsuitable for timber production
	S	Great Basin sagebrush		
	T	Coastal sagebrush		
Herbaceous types				
	G	Grassland		
Miscellaneous types				
	B	Barren		
	Z	Desert		
	A	Cultivated urban and industrial		

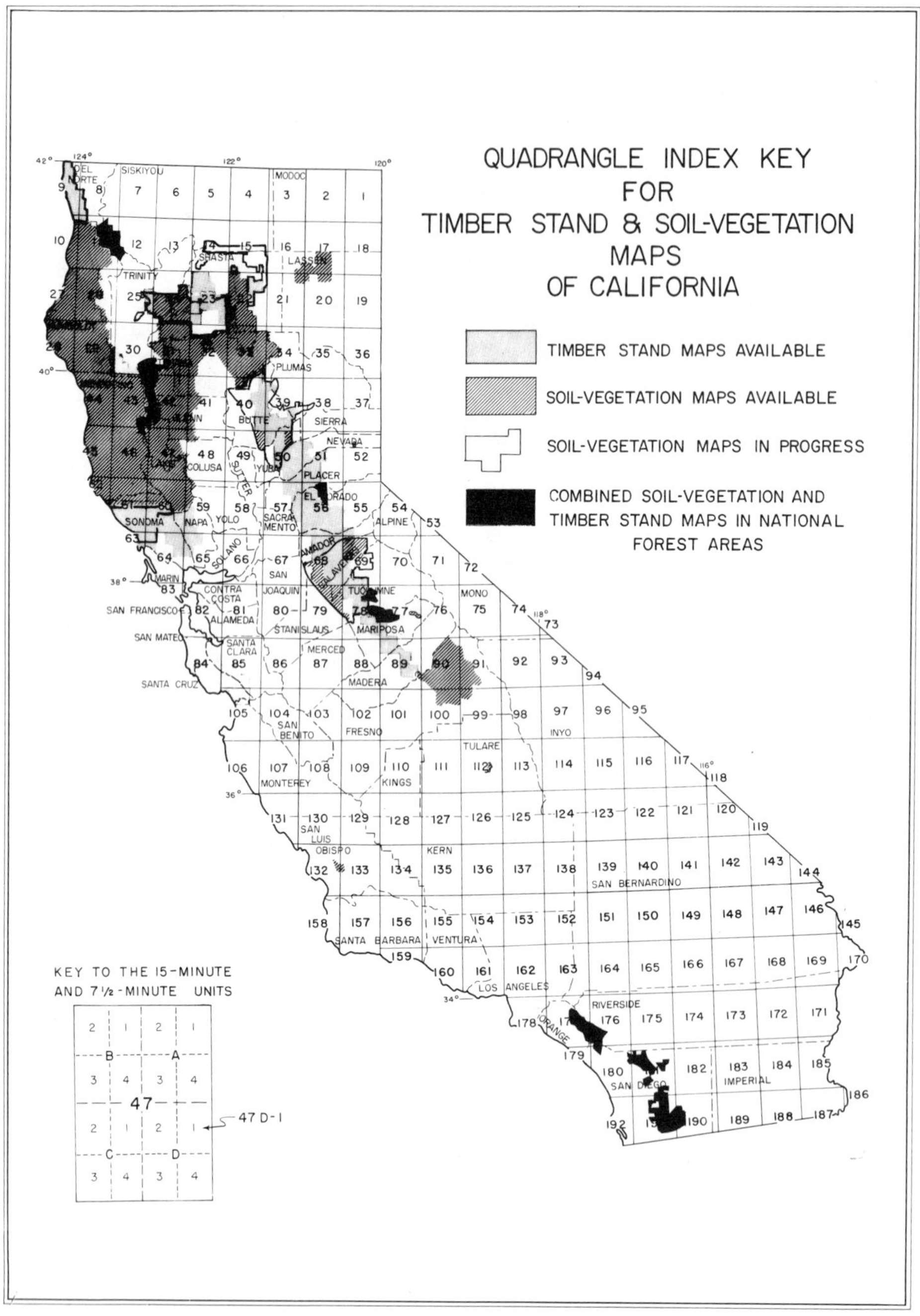

Figure 6-3. Quadrangle index key for published timber stand and soil–vegetation maps of California.

Hy	Hardwoods (young)	B	Bare ground
Ho	Hardwoods (old)	R	Rock
S	Chaparral	A	Cultivated
T	Sagebrush	U	Urban
F	Bushy herbs		

With this basic group of symbols, any land area could be easily classified and delineated on vertical aerial photographs according to the elements present and their order of abundance (Jensen 1947).

The second category gives more detail on the age structure of the commercial conifer tree element (C) on areas considered suitable for growing commercial conifer trees. The age structure is based on the age–size class of individuals that makes up the stands: mature (old) and immature (young). Mature trees are those considered as having passed their optimum growth, while immature trees are those not yet past their optimum. Four age classes are defined: old growth, old growth–young growth, young growth–old growth, and young growth (U.S. Forest Service, California Forest and Range Experiment Station 1954).

The third category is the designation of the "density" or percent cover of the canopies occupied by the various elements. Symbols range from 1 (crowns > 80% cover) to 5 (< 5% cover). "Density" is determined for both sawlog size trees over 27 cm dbh and all size timber trees, and for all woody vegetation.

The symbols for all three categories are shown as fractions on the map with age class and "density" in the numerator, and the elements in the denominator. In areas considered unsuitable for growing commercial timber crops, an "N" replaces the age and timber "density" symbols in the numerator (U.S. Forest Service, California Forest and Range Experiment Station 1949).

Certain requirements have been established for recognition of elements in the aerial photo delineation process. They are very similar to those used in the earlier Vegetation Type Map Survey, but differ in that the elements shown in the timber stand map are always in order of decreasing abundance. However, the tree elements (C, H, and K) are emphasized under certain "density" requirements.

To be recognized, an individual element can occupy from 5 to 100% of the ground area, depending on the kind of element and combinations. Crowns of commercial conifers (C) must occupy at least 5% or more of the ground space. Hardwoods (H) and minor conifers (K) each must occupy at least 5%, or 20% when in combination with 20% or more of commercial conifers. Shrubs (S), sagebrush (T), and bushy herbs (F) each must also occupy 5% to be indicated, but must occupy 20% when in combination with 20% or more of any tree element (C, K, or H). All other elements must occupy at least 20% to be indicated in a delineated area.

The timber stand–vegetation cover maps were used by the Forest Survey in 1946 to provide an age class and "density" stratification of commercial timber areas to a minimum of 16 ha for the purposes of sampling and locating growth and volume plots and of estimating timber volume. When the Cooperative Soil–Vegetation Survey was started in 1947, the timber stand maps were intensified to a 4 ha minimum and provided a framework for vegetation and soil mapping and for determining forest productivity.

Soil–vegetation maps. These maps give detailed information about the soil and vegetation in a 7.5-min quadrangle unit and are accompanied by legends and tables.

As of July 1, 1975, 333 soil-vegetation maps had been published by the Pacific Southwest Forest and Range Experiment Station. They represent an area of about 4.6 million ha in the uplands of California (Fig. 6-3). The soil-vegetation maps and legends, as well as a quadrangle checklist, are available for sale from the U.S. Forest Service in San Francisco.

The soil and vegetation information for each quadrangle map is shown by coded symbols on the map with additional information provided on the tables accompanying each map. In general the maps show the following:

1. Significant woody plant species in order of decreasing dominance within each element present, such as shrub, hardwoods, and conifers, and occurrence of grass, wet meadow, marsh, and other land elements or classes.
2. Woody vegetation classified by cover "density."
3. Timber site quality in areas suitable for growing commercial conifer timber.
4. Soils classified as to soil series and phases (depth class, slope class, etc.).
5. Unclassified areas, such as agricultural land or areas with little or no soil.
6. Locations of soil–vegetation plots.

The tables and legends explain the map symbols and give additional information on the following:

1. Common and scientific names of species mapped.
2. Growth habit, sprouting or nonsprouting characteristics, and browse value of woody and bushy herb vegetation for livestock and wildlife.
3. List of all plant species recorded for the quadrangle area.
4. Estimated suitability of the soils for timber and extensive range use.
5. Some of the general soil characteristics, such as depth, color, texture, acidity or alkalinity, kind of parent material, drainage, permeability of soil to water, and erosion hazard.

The first maps issued by the Cooperative Soil–Vegetation Survey were called vegetation-soil maps. Later the name was changed to soil-vegetation maps; otherwise the format has changed very little, although tabular data have increased. A supplementary publication (Colwell 1974) provides information about the symbols and tables used in the maps.

Upland soils maps. Upland soils maps (scale 1:125,000) are compiled from the detailed soil–vegetation maps, and no areas smaller than 16 ha are shown on the generalized maps. Colors represent different groups of soils that are normally associated with certain broad vegetation types and land conditions. For example, the upland soils map of Mendocino Co. (Gardner et al. 1964; State of California 1951) includes mapping units related to commercial forest types and site, noncommercial conifer forest, grass, oak–grass, chaparral, and unclassified lands, among others.

Separate tables accompany the maps and show some of the characteristics of soils and the estimated suitabilities of the soils for timber production and extensive range use. They are for sale from the State Printing Division in Sacramento.

Vegetation Mapping by Other Agencies

U.S. Forest Service, California Region

SOIL–VEGETATION SURVEYS. In 1958, the U.S. Forest Service's California Region started soil–vegetation surveys on national forests. About 0.5 million ha were mapped on portions of the Eldorado, Mendocino, Six Rivers, and Stanislaus National Forests, and all of the Cleveland National Forest.

The soil-vegetation and timber stand maps were combined into a single 7.5-min quadrangle and titled "Soil-Vegetation and Timber Stand Map." The maps were published during the period 1962–1973. The geographical locations are shown in Fig. 6-3. Although these detailed soil–vegetation surveys have been discontinued, the maps and a quadrangle checklist are available at the California Region headquarters in San Francisco.

The maps are similar in scale and in procedures of mapping to the blue-line soil-vegetation map sheets issued by the State Cooperative Soil-Vegetation Survey. The California Region maps differ by having the symbols and lines overprinted on USGS topographical contour maps, making it easy to locate mapping units in the field. The symbols for age and density of conifer stands, the soil and slope phases, vegetation species, and site class are shown in each mapping unit. Vegetation boundary lines, species symbols, and site quality are indicated in green; the soil symbols and boundaries are shown in red.

Some soil–vegetation surveys were conducted in three watershed areas (Fallen Leaf, Upper Truckee, and Trout Creek) of the U.S. Forest Service's Lake Tahoe Basin Management Unit, but were never published in quadrangle format. These surveys are on file at the Lake Tahoe Basin Management Unit headquarters at South Lake Tahoe.

SOIL RESOURCE INVENTORIES (SOIL SURVEYS). To cover the national forests more rapidly for soil resource information essential to broad land use planning, the soil resource inventory system was developed in 1972 (U.S. Forest Service 1974).

On most forests, the inventory is just beginning. The Shasta–Trinity National Forest completed soil resource inventories at the reconnaissance level for several planning units in the forst, and reports are on file at its headquarters in Redding. As for the vegetation information in these completed soil resource inventory reports, two of them list the symbols for woody vegetation and the other land status elements in order of dominance for each mapping unit. Other reports indicate only the number symbols of a list of vegetation cover types. Most national forests are planning to record only broad, vegetation cover type symbols for each response unit mapped during the inventory phase.

FOREST TYPE AND CLASSIFICATION SYSTEM. This system is the first step in the stratification of timberland as part of the inventory phase of forest management operations in the California Region. It evolved from a long series of many kinds of inventory systems.

When national forests in California were established in the early 1900s, one of the most important tasks undertaken was the inventory of the natural resources in the forests. Timber stand and type maps of the forest areas were compiled from data from many sources: (1) earlier vegetation maps from the state, county, and other government sources (Küchler and McCormick 1965), (2) field maps from land surveyors, (3) old geological exploration notes, and (4) timber survey cruises and reconnaissance mapping by U.S. Forest Service personnel. More detail was obtained in the timber areas as to acreage distribution of forest types, stands size and density, growth, and volume. Information was put on township maps of usually 36 square miles (9300 ha).

Further refinements were made on existing maps when the nationwide Forest Survey was authorized by the McSweeney-McNary Research Act of 1926. Several national forests in the 1930s used vegetation type map procedures and prepared generalized vegetation maps on USGS quadrangles enlarged to a scale of 1:62,500. Since that time, various systems of timber classifications have evolved and have been supplemented by various statistical sampling designs of sample plots to obtain more accurate growth and volume data.

The forest type and classification system is presently being adopted for every national forest in California (U.S. Forest Service 1968). Most of the information is put on plastic map sheets of a township base at 2 inches to the mile (1 cm = 0.16 km). Eventually maps will be prepared on standard 7.5-min topographical quadrangle format at a scale of 1:24,000.

The basic system involves six classification categories, each having several subdivisions: (1) productive status of the land (commercial or noncommercial forest land, or nonforest), (2) name of broad forest type, (3) three most abundant tree species, (4) stand size class, (5) density of stocking, and (6) decadence rating of overstory tree groups. By a system of symbols, the various categories are delineated on aerial photographs down to a minimum of 4 ha and transferred to plastic township map sheets.

Unfortunately, little information is available for nonforest areas of the national forests other than that the land is agricultural, grass, brush, urban, or water. The maps are not published but are on file at national forest headquarters.

RANGE SURVEYS. Some national forests have a large portion of their land in range. Range inventory maps and survey notes that show the different forage and browse types on various range allotment areas are available at forest headquarters.

Bureau of Land Management. The Bureau of Land Management (BLM) of the U.S. Department of Interior has widely scattered parcels of lands all over California, making integrated resource management of its lands difficult. Probably the most consolidated ownerships of BLM lands are in the desert and northeastern parts of the state.

Much of the BLM land is dominated by rangeland and chaparral vegetation types, and these have been inventoried by ocular reconnaissance surveys. Small field maps are made during the reconnaissance, and delineations are made of the vegetation type, species present and their density, the average density of all vegetation within the type, and estimates of productivity.

This information is available in the files of the various district offices throughout the state.

National Park Service. Of the seven U.S. national parks and national recreation areas in California, four were mapped during the Vegetation Type Map Survey from 1930 to 1940.

Vegetation type maps to a scale of 1:62,500 were prepared by National Park

Service personnel under the direction of the California (now Pacific Southwest) Forest and Range Experiment Station for Sequoia National Park and part of Yosemite National Park. Vegetation maps for Lassen National Park were done to a scale of 1:31,680. Kings Canyon National Park was formerly part of Sierra National Forest, and the vegetation there was mapped by the U.S. Forest Service.

The newer parks and some of the older ones have had more recent vegetation surveys. Redwood National Park was mapped by the State Cooperative Soil–Vegetation Survey in 1961, before the park was established. The Whiskeytown portion of the Whiskeytown–Shasta–Trinity National Recreation Area was included in a soil–vegetation survey of the French Gulch quadrangle (Mallory et al. 1973). In 1962, a very detailed soil–vegetation survey was conducted in Yosemite Valley. A timber stand and soil–vegetation map at a scale of 1:24,000 and a report were prepared by Zinke and Alexander (1963). Point Reyes National Seashore was mapped by high-altitude aerial photography in 1973. Ten broad vegetation types and their densities were mapped and supplemented by ground plots to determine species composition. The Sequoia, Kings Canyon, Redwood, and Whiskeytown parks were mapped at a scale of 4 inches to the mile (1 cm = 0.16 km). Considerable detail of tree distribution and density was included with lesser emphasis on shrub vegetation.

Of the national monuments, Pinnacles and Devils Postpile were both mapped by the Vegetation Type Map Survey in the 1930s.

University of California Remote Sensing Research Program. In 1972, ERTS-1 (Earth Resources Technology Satellite) was successfully launched to become the first of a series of unmanned earth-orbiting satellites especially designed to obtain remotely sensed data for use in the inventory and monitoring of natural resources. Because California has a wide range of climatic and topographical conditions, it has proved to be an ideal testing site for conducting integrated remote sensing research in water supply, timber, range, chaparral fuels, agriculture, and recreation (Colwell 1975). Many such studies were conducted by the Remote Sensing Research Program of the Department of Forestry and Conservation, and by the Space Sciences Laboratory, both of the University of California, Berkeley, under National Aeronautic and Space Administration (NASA) grants, and cooperative agreements with the Forest Service and with the Bureaus of Land Management, Reclamation, and Outdoor Recreation.

In studies concerning the inventory of wildland vegetation, ERTS-1 (now called LANDSAT-1) imagery serves as the first stage of stratification in multilevel sampling designs. In conjunction with this process, aerial photography (both high-altitude color-infrared and low-altitude color) and systematically collected ground data are utilized to provide additional sampling levels. The information collected at these levels is used in equations to estimate the condition and volume of a vegetative resource. Depending on given objectives, these estimates can, for example, be made for timber volume, range productivity, or fuel hazard ratings.

In the application of remote sensing to range inventory, LANDSAT-1 imagery enhanced from magnetic and color composite processes detected differences in the conditions of California's annual range on a regional basis. These amounted to differences in production between grazing regions, length of the green feed period, and changes in growth stages (Carneggie et al. 1974).

In another study in the northern desert shrub type in eastern Lassen Co.,

LANDSAT-1 digital data in conjunction with additional sampling levels provided a computer-generated in-place map of 35 vegetative cover types and summary statistics within seven broad vegetative cover classes: conifers, mountain shrub, sagebrush, grass, annual meadow, barren, and waste. Further detail was obtained from aerial photography, supported by ground sampling, to complete the multilevel sampling process. Also, a color-coded map composed of eight fire hazard ratings was prepared from the 35 vegetation cover types, based on relative species compositions, cover densities, stand structures, site aridity, flammability, and resistance to control parameters (DeGloria et al. 1975).

In a related study (Nichols 1974), wildland fuel characteristics in chaparral northwest of Los Angeles were mapped. The LANDSAT-1 multispectral scanner detected differences among trees, shrubs, grass, and bare ground. Some distinctions could also be made of shrub maturity classes, such as pioneer, immature, and mature. In some homogeneous areas, types could be separated by species composition.

Tests were also made in forested areas. On the Plumas National Forest, studies determined the utility of LANDSAT-1 imagery in a total resource inventory application. As in the range studies, multispectral data from the satellite provided a stratification base for the first step in multilevel sampling designs (Gialdini et al. 1975). Major forest types and some broad timber volume classes could be separated.

Imagery from spacecraft has been used mainly as the first stage of stratification for various sampling designs requiring additional photography for ground truth and more detailed sampling. Satellite imagery is useful in the separation of broad vegetation types, condition classes, and other contrasting land conditions.

METHODS USED BY THE VEGETATION TYPE MAP AND STATE COOPERATIVE SOIL-VEGETATION SURVEYS

Since the Vegetation Type Map Survey and the currently operating State Cooperative Soil-Vegetation Survey provide the most extensive detailed map coverage of vegetation in California, the methods and procedures of these surveys are discussed here in some detail.

Vegetation Type Map Survey

Basic classification system. Wieslander (1935b) developed a flexible scheme of type classification simple enough to allow a reasonable rate of mapping progress with accuracy and consistency. From peaks, ridges, and other vantage points, the mapper could locate distinct units of vegetation and sketch the boundaries on the 1:62,500 topographic contour maps. Plant communities were mapped to a minimum of 16 ha. Later on, when aerial photos became available, they were used to substitute for much ground observation.

Because of the great abundance and apparent heterogeneity of plant communities, the concept of dominant species was adopted. Dominant species were defined as those occupying 20% or more of the canopy of trees as a group or shrubs as a group (Wieslander 1961). Understory vegetation was not a major consideration, although sample plots did provide quantitative characterization of both overstory and understory vegetation. Two major vegetation categories were established: single-element type, and mosaic or composite type having more than one element.

The single-element types consist of areas where the dominant vegetation is purely shrubby, arborescent, or herbaceous and occupies 80% or more of the vegetation cover to be designated. Mosaic types consist of two or more elements, each of which forms 20–80% of the overall vegetation cover.

The single-element types were further subdivided into pure and mixed stands. The pure-stand subtype is a unit in which two or more species each occupy at least 20% of the cover. Species are listed in order of decreasing abundance.

In the mosaic types, the percentage of species occurrence is applied in the same manner to each class or element of vegetation separately. Thus in a tree–shrub mosaic the tree component can have tree species recorded in order of abundance if each occupies more than 20% of the tree canopy. The same applies to the shrub component.

When the Vegetation Type Map Survey was discontinued in 1941, it had segregated vegetation into 21 major and 3 miscellaneous types (Table 6-2). Of the major types, 18 were considered dominant natural plant associations made up of one element, such as herbaceous, shrub, or tree type. The other 3 major vegetation types represented the mosaic category of two or more elements: woodland–grass, woodland–sagebrush, and woodland–chaparral. As other mosaics occurred, they were designated as subtypes of the nearest related major type. For example, a shrub–herb mosaic was recorded as a subtype of the nearest related shrub type. These major types are designated on the map by a color legend. All of the vegetation type field maps and the 23 published maps have this standard color legend.

Since the preparation of type maps was part of the Forest Survey, information was needed to show the location and extent of major vegetation types having similar fire hazard characteristics and uses or qualities of economic importance. Priority was given to elements or species that had these characteristics. Within a vegetative classification unit, the dominant plant species were ordinarily listed in decreasing order of abundance, but there were exceptions. In certain mosaic types in which combinations of trees, shrubs, or herbs were mapped, tree species were given first priority in the symbol designation. In tree types, "commercial" conifers were listed before other tree types included in the mosaic unit, regardless of their relative abundance in the type. Thus trees were always listed before shrubs, and commercial trees before other trees. For types having more than one commercial conifer tree, the trees were listed in order of decreasing economic importance rather than in order of abundance. For example, in the redwood belt, redwood was listed first, regardless of the abundance of other conifers present. In mixed conifer types, dominant conifers were listed in order of commercial economic importance, rather than in order of relative abundance of the species, as follows: sugar pine, ponderosa pine, Jeffery pine, Douglas fir, white fir, red fir, and incense cedar.

Supplementary information.

TREE AND SHRUB MAPS. Many important plant species are not indicated on the standard vegetation type map because of their low densities. To record these species, separate topographical quadrangle maps were used to show the occurrence of (1) tree species, either as individuals or groups of trees considered to be ecologically

TABLE 6-2. Vegetation and miscellaneous type classification of the Vegetation Type Map Survey

Major Vegetation Classes and Types	Definition
Herbaceous	
Marshland	Marshes and perpetual wet meadows
Grassland and meadowland	Grasses and other herbaceous vegetation not under cultivation
Shrub	
Sagebrush	Low-growing, slenderly branched shrubs (*Artemisia*, *Salvia*, *Eriogonum*, *Baccharis*, etc.)
Chamise chaparral	Chamise is one of the dominants in chaparral association
Chaparral	Dense, tall, heavily branched shrubs (*Ceanothus*, *Arctostaphylos*, shrub *Quercus*, etc.) exclusive of sagebrush, chamise, and semidesert chaparral
Semidesert chaparral	Open chaparral associations on slopes bordering desert or within range of desert climatic influence
Timberland chaparral	Transition or Canadian life zone shrubs in areas capable of growing commercial conifers removed by fire, logging, or chemicals
Tree-woodland	
Woodland	Dense stands of broad-leaved trees comprising the only dominants; may be in association with Digger pine; includes steam bottom hardwoods

Woodland-chaparral	Mosaic types of both woodland and chaparral species; may include Digger pine or some herbaceous constituents
Woodland-sagebrush	Mosaic types of woodland and sagebrush species; may include herbaceous species
Woodland-grass	Mosaic of open stands of woodland species or Digger pine with herbaceous species
Piñon and juniper	Stands of piñon or juniper or mixtures of both
Miscellaneous conifers	Conifers other than piñon or juniper, or subalpine tree types; includes Monterey, bishop, knobcone, and Coulter pines, big-cone spruce, and similar conifers having little to no value for timber purposes
Tree-commercial conifer	
Redwood belt	Redwood or giant *Sequoia* species as a dominant in stands of coniferous trees to the exclusion of 20% or more of Douglas fir; includes Redwood-Douglas fir subtype if both species occupy 20% or more of coniferous tree species
Douglas fir belt	Douglas fir species occupy 20% or more of coniferous tree species, excluding 20% or more of *Sequoia* or commercial *Pinus* species
Pine belt	One or more pines (sugar, ponderosa, or Jeffrey pine) and incense cedar are dominant to exclusion of true firs; includes pine-Douglas fir subtype
Pine-fir belt	Pines associated in joint dominance with true firs; includes pine-Douglas fir-fir subtype when all occupy 20% or more of coniferous tree species

TABLE 6-2 [continued]

Major Vegetation Classes and Types	Definition
Fir belt	True firs either individually or mixed occupying over 80% of coniferous tree species
Spruce belt	Sitka spruce is dominant in stands of coniferous trees to the exclusion of 20% or more of redwood or Douglas fir
Tree-subalpine	
Lodgepole pine-white pine belt	Lodgepole pine, western white pine, or mountain hemlock are dominant in stands of coniferous tree species excluding commercial trees (other than true firs) or whitebark-foxtail pine type
Whitebark-foxtail pine	Whitebark pine, foxtail pine, limber pine, or bristlecone pine are dominant in a stand of coniferous tree species
Miscellaneous	
Barren	Areas practically devoid of vegetation: beaches, rocks, etc.; semibarren areas occur within any of above vegetative types where ground is less than 50% covered: rocky or eroded slopes, badlands, etc.; desert areas are indicated by dominance of distinctive desert species
Cultivated and urban	Cu-Cultivated areas such as cropland, orchards, natural hay lands Res-Residential, business, and industrial
Plantations	Areas artifically planted with trees other than horticultural crops, usually eucalyptus, Monterey pine, or Monterey cypress

important, (2) shrub species having unique or special importance but not abundant enough to be indicated on the vegetation field map sheets, and (3) small mountain meadows. Recent burned-over areas were also shown on these maps.

SAMPLE PLOTS. Detailed information about species composition, plant size, stand density of shrubs and trees, thickness of leaf litter, soil character, and type of parent material was obtained by means of sample plots. For tree and shrub types, 0.04 ha rectangular plots were sampled to record the dominant species. At the same time, trees ≥ 10 cm dbh were tallied by diameter classes within a 0.08 ha plot. For herbaceous areas, such as burned-over shrub types and natural grassland, all species intercepted along a 101 m line transect were listed.

The plot sites were carefully selected to be representative of the major types and subtypes mapped in a geographical area. Generally, one plot was set up for every 800 ha mapped, and it was located and numbered on a topographical quadrangle map. Data from 18,000 sample plots are on file at the Pacific Southwest Forest and Range Experiment Station, Berkeley.

OTHER INFORMATION. During the Vegetation Type Map Survey, plant specimens were collected of all species shown on the quadrangle or sample plots. Some 30,000 mounted plant specimens were deposited at the Herbarium, Department of Botany, University of California, Berkeley (Wieslander 1961). Representative types were photographed and were recorded on a topographical quadrangle map. The data are on file at the Pacific Southwest Forest and Range Experiment Station.

State Cooperative Soil–Vegetation Survey

Basic classification system. The timber stand–vegetation cover map provides the basic mapping units for the more detailed mapping of (1) vegetation species, in order of abundance, that comprise each vegetation element, (2) site class for "commercial" timber areas, and (3) soil series, soil depth, rockiness and stoniness, erosion condition, and slope class. These are all shown by symbols on the soil–vegetation map (Fig. 6-4).

Detailed explanations of the mapping techniques are given in field manuals, legends, and tables that accompany the maps published by the Soil–Vegetation Survey and other publications (Wieslander and Storie 1953; U.S. Forest Service, California Forest and Range Experiment Station 1954; Mallory et al. 1973; Colwell 1974).

As was done in the Vegetation Type Map Survey, plant species are represented by letter symbols. For each vegetation element (except grass) shown in the denominator of the map symbol on timber stand maps, one or more dominant species are indicated in order of relative abundance within that element. A species is considered dominant if it occupies 20% or more of the canopy of the element to which it belongs on a delineated mapping unit. If five species symbols appear for one element, the relative abundance of the species is variable within the delineated area. Species of herbaceous plants are not indicated for the grass element.

For soil–vegetation maps published after 1963, the percent of canopy cover of total woody vegetation is shown as a number symbol in the numerator of the fraction above the vegetation species. In the enlarged mapping unit of the example in Fig. 6-4 there is a greater proportion of shrubs than hardwoods, commercial conifers, and bare ground, and more hardwoods than conifers and bare ground. In the same manner, species within each element are listed in order of abundance. Thus there is more Cf (*Chamaebatia foliolosa*) than Ama (*Arctostaphylos mariposa*) in the shrub canopy, and more B (*Quercus kelloggi*) than C (*Q. chrysolepis*) or W (*Q. wislizenii*) in the hardwood group, and more I (*Calocedrus decurrens*) than Ŵ (*Abies concolor*) for the conifers (Colwell 1974). Species symbols in parentheses indicate species of commercial conifers reduced in cover to less than 5% because of logging, burning, or clearing.

Although the minimum size of a homogeneous delineated area is 4 ha, some areas have distinct small vegetation groups so intimately mixed that they cannot be shown separately at the scale of mapping. In such cases, two or more groups of cover class and species symbols are shown with a vertical line separating them (Fig. 6-4).

The mapping unit of each delineated commercial timber area also includes the predominant site quality class. Several different site classification systems are used in California, depending on the forest type. For the redwood and Douglas fir types, the Cooperative Soil–Vegetation Survey uses curves by McArdle et al. (1949, revised in 1961). For all other types, except lodgepole and spruce, the site classification curves developed by Arvanitis et al. (1964) and Dunning (1942) are used.

Supplementary information. More than 3000 circular 0.4 ha plots have been sampled during the course of the State Cooperative Soil–Vegetation Survey, and the data obtained are on file at the Pacific Southwest Forest and Range Experiment Station at Berkeley. These plots were chosen as representative of over 400 wildland soils and many plant associations in the grass, chaparral, woodland, and commercial conifer types.

On each plot, complete site characteristics, such as slope, aspect, elevation, erosion, drainage, and surface rockiness, are recorded. A soil pit is dug, and the soil profile is described. For certain modal representative soils, samples are collected and sent to the University of California at Davis for complete physical and chemical analyses. On plots predominantly woody in cover, ocular estimates are made of the percentage of ground occupied by woody plant species and the range of height in the understory and overstory. Herbaceous species are identified and listed. In commercial forest areas, height and age measurements are taken on suitable conifers to determine site index.

For areas mainly in grass or other herbaceous cover, species composition is obtained by the "step-point" sampling method (Evans and Love 1957). At each of 100 uniformally spaced points, the species at or nearest the point is recorded. Plant cover is obtained by estimates in a 30 cm square wire frame for every tenth point step (U.S. Forest Service, California Division of Forestry 1959).

During the course of mapping, prominent species other than those recorded on plots or mapped are noted. The result is a floristic list in the tables accompanying each quadrangle map that is fairly complete—at least for woody plants in upland habitats.

Plant specimens are collected for each species mapped in a county area and are submitted to University of California herbariums at both Davis and Berkeley for identification and storage.

In the past, the Department of Agronomy of the University of California at Davis has conducted greenhouse pot tests on certain key soils to determine their responses

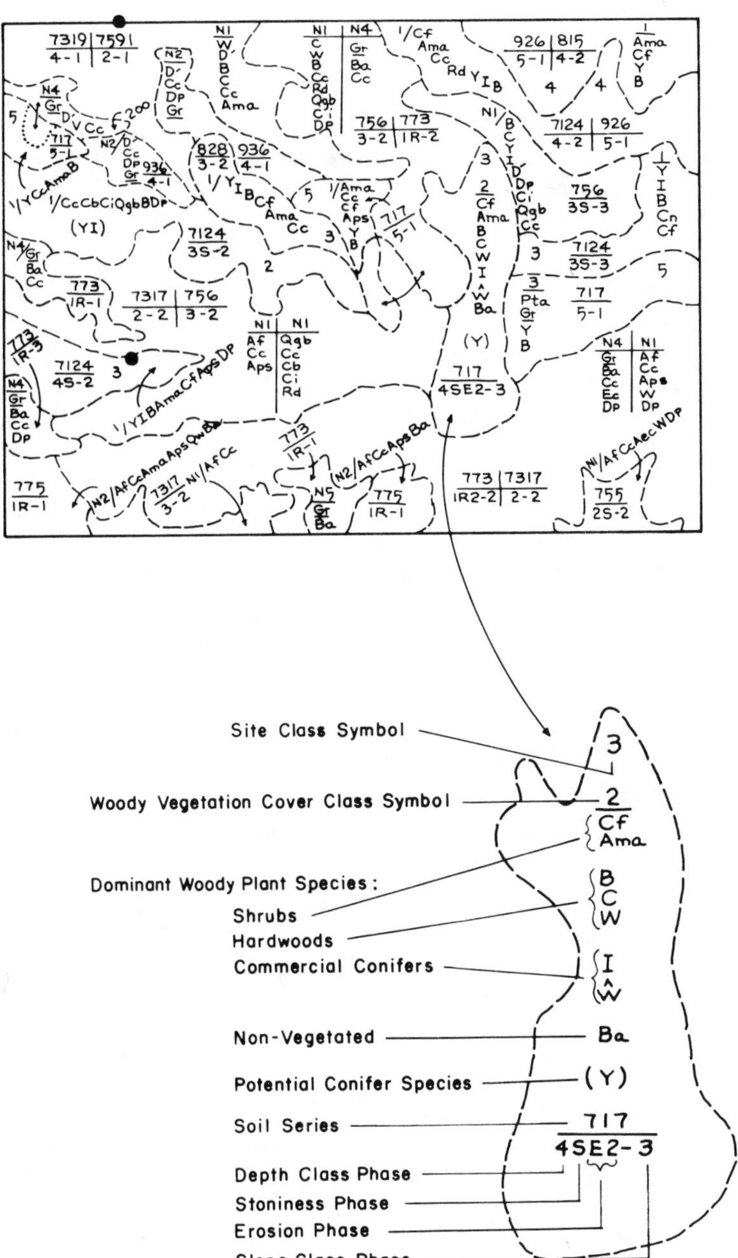

Figure 6-4. Portion of a soil–vegetation map (format after 1964) at scale of 2 inches equals 1 mile (1 cm = 0.3 km). Enlarged mapping unit shows explanation of the symbols.

to fertilizers. Fertilizer field trials on 0.1 ha plots are being conducted on the range soils that cover large acreages. Various kinds and rates of nutrients are added to determine responses in terms of forage yield and species composition. This information can aid the range manager in deciding whether or not to increase the intensity of rangeland use (Powell 1964).

CONCLUSION

Fifty years' mapping of vegetation in California at different levels of classification and intensity by various agencies has resulted in a variety of information that is not fully compatible for uniform application throughout the state. Historically, the U.S. Forest Service, in cooperation with other agencies, has provided most of the leadership for the systematic development of mapping methods in inventorying the vegetation of California.

Lacking is a uniform statewide land classification system in which all types of resource information are organized into an integrated system of land units. The method used by the State Cooperative Soil–Vegetation Survey closely approaches such a system—it stratifies land into basic homogeneous units containing components of vegetation, topography, soil, geology, and potential productivity. A hierarchical framework can then be developed to provide resource data for any defined level of land use planning and land management. A uniform land classification system would strengthen the process of arriving at wise land use decisions.

Finally, I will comment briefly on Dr. Küchler's 1977 map of California vegetation, folded into the back of this book. Clearly, his objectives differ from those of Wieslander, the Vegetation Type Map Survey, and the State Cooperative Soil-Vegetation Survey. Whereas the maps of the latter two show actual vegetation, Küchler's map gives potential or virgin vegetation. For example, in his map montane brushfields are shown as yellow pine or montane conifer forest; chaparral occurring at the western edge of the Sierra Nevada often appears as blue oak-digger pine forest; and presently cultivated or pasture land in the central valley is indicated as riparian forest, tule marsh, or *Stipa* prairie.

Küchler aptly expresses his rationale in the Appendix:

> A vegetation map that shows man-made phytocenoses quickly becomes a historical document . . . [but] a vegetation map gains in value if it . . . portrays the vegetation types as indicators of environmental conditions, that is above all the natural vegetation . . . [T]he new vegetation map of California portrays the natural vegetation. However, it is not practical to distinguish on the map between what is actual and what is potential.

His map will undoubtedly draw sharp debate during the next decade or two. Most montane brushfields can be seral to forests, but chaparral may not be seral to woodland (see Chapter 12), and *Stipa* grassland has probably forever been replaced by "new natives" (see Chapter 14). Furthermore, some of man's more drastic disturbances have caused excessive soil loss, resulting in the irreversible replacement of original vegetation by other plant communities. Each reader may find other points of debate.

Nonetheless, the approach in the new map of showing natural vegetation and more detail in urban-agricultural, desert, and some other areas makes it a useful complement to the last statewide map published—Wieslander's map, which was issued more than 30 years ago. Our understanding of nature advances in steps, and Küchler's map is a useful and welcome step in that continuing process.

LITERATURE CITED

Arvanitis, L. G., J. Lindquist, and M. Palley. 1964. Site index curves for even-aged young growth ponderosa pine of the west-side Sierra Nevada. Calif. For. and Forest Prod. No. 35, Calif. Agric. Exp. Sta., Berkeley, Calif. 8 p.

California State Board of Forestry. 1886. First biennial report of the California State Board of Forestry for the years 1885-86, to Governor George Stoneman. State Printing, Sacramento. 238 p.

Carneggie, D. M., S. D. DeGloria, and R. N. Colwell. 1974. Usefulness of ERTS-1 and supporting aircraft data for monitoring plant development and range conditions in California's annual grassland. Final rep., Contract 53500-CT3-266(N), BLM/USDI, Remote Sensing Res. Program, Univ. Calif., Berkeley. 53 p.

Colwell, R. N. 1975. An integrated study of earth resources in the State of California using remote sensing techniques. Ann. progr. rept., NASA Contract NGL 05-003-404, Space Sci. Ser. 16, Issue 34.

Colwell, W. L., Jr. 1974. Soil-vegetation maps of California. USDA Forest Serv., Pac. SW Forest and Range Exp. Sta., Resources Bull. PSW-13. Berkeley, Calif. 6 p.

Critchfield, W. B. 1971. Profiles of California vegetation. USDA Forest Serv. Res. Paper PSW-76. Pac. SW Forest and Range Exp. Sta., Berkeley, Ca. 54 p., illus.

DeGloria, S. D., S. J. Daus, and R. W. Thomas. 1975. The utilization of remote sensing data for multidisciplinary resource inventory and analysis within a rangeland environment, pp. 640-659. In Proc. ASP-ACSM Fall Convention.

Dunning, D. 1942. A site classification for the mixed-conifer selection forests of the Sierra Nevada. USDA Forest Serv., Calif. Forest and Range Exp. Sta., Res. Note 28. 21 p.

Evans, R. A., and R. M. Love. 1957. The step-point method for sampling-a practical tool in range management. J. Range Manage. 10:205-212.

Gardner, R. A., A. E. Wieslander, R. E. Storie, and K. E. Brawshaw. 1964. Wildland soils and associated vegetation of Mendocino County, California. Div. Forestry, Sacramento, Calif. 113 p.

Gialdini, M. J., S. Titus, J. P. Nichols, and R. W. Thomas. 1975. The integration of manual and automatic image analysis techniques with supporting ground data in multistage sampling framework for timber resource inventories. Three examples. Proc. NASA Earth Resources Survey Symp. 11 p.

Jensen, H. A. 1947. A system for classifying vegetation in California. Calif. Fish and Game 33(4):199-266.

Küchler, A. W. 1967. Vegetation mapping. Ronald Press, New York. 472 p.

Küchler, A. W., and J. McCormick (compilers). 1965. Vegetation maps of North America. In A. W. Küchler (ed.). International bibliography of vegetation maps. Univ. of Kansas Libraries, Lawrence. 435 p.

Mallory, J. I., W. L. Colwell, Jr., and W. R. Powell. 1973. Soils and vegetation of the French Gulch quadrangle (24D-1, 2, 3, 4), Trinity and Shasta Counties, California. USDA Fornst Serv., Pac. SW Forest and Range Exp. Sta., Resources Bull. PSW-12. Berkeley, Calif. 42 p.

McArdle, R. W., W. H. Meyer, and D. Bruce. 1930. The yield of Douglas fir in the Pacific Northwest. (Rev. 1961). USDA Tech. Bull. 201. 64 p.

Nichols, J. D. 1974. Mapping of wildland fuel characteristics of the Santa Monica Mountains of Southern California. Fiscal rep., Cooperative Agreement 21-274, U.S. Forest Serv. Remote Sensing Res. Program, Univ. Calif., Berkeley. 26 p.

Powell, W. R. 1964. Procedures used for range land soil fertility studies. Div. Forestry, Sacramento, Calif. 15 p.

State of California. 1951. Upland soils map of Mendocino County with legends and tables. Div. Forestry, Sacramento, Calif.

State Cooperative Soil-Vegetation Survey. 1958, rev. 1969. Soil-vegetation surveys in California. Div. Forestry, Sacramento, Calif. 31 p.

U.S. Forest Service. 1968. Forest types and classification system, pp. 23-35. In Region 5, Supplement 65. Forest Serv. manual Section 2441 (2441.0--2441.5). San Francisco, Calif.

---. 1974. Region 5, Supplement 12. Forest Service manual, Title 2500-Chapter 2550:Soil survey interpretations and management. San Francisco, Calif. 9 p.

U.S. Forest Service, California Division of Forestry, and University of California. 1959. Field manual: grassland sampling. Supplement 1: Field manual: soil-vegetation surveys in California October 1954.

U.S. Forest Service, California Forest and Range Experiment Station. 1949. The timber stand and vegetation soil maps of California. Berkeley, Calif. 15 p.

---. 1954. Field manual: soil-vegetation surveys in California. Berkeley, Calif. 95 p.

Wieslander, A. E. 1935a. A vegetation type map of California. Madroño 3:140-144.

---. 1935b. First steps of the forest survey in California. J. For. 33:877-884.

---. 1961. California's vegetation maps: Recent advances in botany. Univ. Toronto Press. 4 p.

Wieslander, A. E., and H. A. Jensen. 1946. Forest areas, timber volumes and vegetation types in California. Calif. Forest and Range Exp. Sta., Berkeley, Calif. 66 p.

Wieslander, A. E., and E. Storie. 1953. Vegetational approach to soil surveys in wild-land area. Soil Sci. Proc. 17(2):153-147.

Zinke, P. J., and E. Alexander. 1963. The soil and vegetation of Yosemite Valley. Univ. Calif., Berkeley, Ca. 86 p.

THE CALIFORNIAN FLORISTIC PROVINCE

CHAPTER 7

BEACH AND DUNE

MICHAEL G. BARBOUR
ANN F. JOHNSON
Botany Department, University of California, Davis

Introduction **224**
  Introduced taxa **226**
Beach vegetation **228**
  Topography **228**
  General north–south trends **228**
  Zonation back from tide line **229**
    Overview **229**
    Regional studies **230**
Dune vegetation **241**
  Topography **241**
  Vegetation and succession **242**
    North of Bodega Bay **242**
    South of Bodega Bay **244**
Autecological studies **253**
Areas for future research **253**
Literature cited **258**

INTRODUCTION

The California coastline is more renowned for spectacular cliffs and rocky tidepools than for sandy beaches and extensive dunes. Only 23% of the 1326 km long coast is occupied by beach and dune (Cooper 1967). Major dune localities include eight north of San Francisco, including San Francisco itself, Pt. Reyes, Dillon Beach, Bodega Beach, Pt. Arena, Ft. Bragg, Humbolt Bay, and Crescent City; and five south of San Francisco, including Monterey, Morro Bay, the Santa Maria River complex, Los Angeles, and San Diego Bay. Of these the least altered are Humbolt Bay, Pt. Reyes, Morro Bay, and parts of the Santa Maria complex. Monterey and San Diego Bay areas have been used for military installations; the remaining seven have been extensively altered by grazing, lumbering, or development.

The present location and configuration of these beaches and dunes are largely a result of eustatic sea level rise during the past 20,000 yr (Flint 1971). At maximum glaciation in Late Wisconsin, sea level stood perhaps 100 m lower than at present. Sea level rose linearly with deglaciation until 7000 yr ago, when it stood about 10 m below the present level. It may have continued to rise slowly, to the present, or have been constant for the past 3000 yr (Emery and Milliman 1968). The chemical and physical nature of sand on California beaches and its rate of deposition during Holocene time have depended on erosional processes in coastal watersheds some distance to the north (see, e.g., Minard's 1971 treatment of the area between the Russian River and Drake's Bay). However, recent coastline changes have blocked southerly transport to some beaches. Sediments, such as beach and dune sand, related to marine transgression since Late Winconsin are referred to as Flandrian by some authors (Cooper 1967; Curray 1964). A few areas of extensive dunes are manifestly of greater age; these pre-Flandrian deposits may date to the Sangamon interglacial, 120,000 yr ago, or earlier (Orme 1973, among others).

Davies (1973) summarized the California coast environment as exhibiting (1) a low frequently of high winds of Beaufort force 4 or above; (2) a low frequency of high waves, 2.5 m or above; (3) a mixed tide (lower low, lower high, higher low, higher high tides all within approximately a 24 hr period) of moderate 2–4 m spring tidal range; (4) frost some years, if not every year; and (5) about 150 kcal cm^{-2} yr^{-1} solar radiation. Annual precipitation declines from about 1000 mm in the north to 250 mm at San Diego, but mean annual air temperature increases only from 11 to 17°C along the same gradient (Breckon and Barbour 1974). The difference between mean air temperature for the warmest and coldest months is only 6–9°C, and the range for surface water temperature is even less, 2–5°C. Mean annual surface water temperature rises from about 12°C in the north to 16°C in the south. According to Köppen's scheme, climate north of Pt. Conception is mesothermal (Cs); to the south it is semiarid (BS).

The flora, vegetation, and microenvironment of beach and dune are sufficiently different to warrant their separate treatment in this chapter. Beach is defined here as the expanse of sandy substrate between mean tide and the foredune or, in the absence of a foredune, to the furthest inland reach of storm waves. Because of occasional but recurring inundation by seawater, the beach is a logical extension of the intertidal zone. "Beach" is equivalent to "strand" as used by some authors, but Munz and Keck (1959), among others, lump beach and dune together as strand; thus the term will not be used in this chapter. Another synonym for "beach" is "littoral strip," and the "foredune" becomes the "littoral dune." The beach is characterized by a maritime climate, high exposure to salt spray and sand blast, and a shifting, sandy substrate with low water-holding capacity and low organic matter content.

There has been some debate about the level of soil salinity on the beach and, consequently, whether beach plants are halophytes. Kearney (1904) and Olsson-Seffer (1909) pointed out that the percent of salt by weight in beach sand is very low, but Barbour (1970) concluded that soil salinity in the root zone may reach 10,000 ppm (1%) if one considers the low water-holding capacity of sand and assumes that the salts are dissolved in the remaining water. Kearney's own data for California beaches in midsummer show the same level, based on these assumptions. The source of salts in the sand may be airborne spray, inundation by waves, or a saline water table.

Dunes are here defined to include the sandy, open habitat which extends from the foredune to typically inland vegetation on stabilized substrate. Major differences between beach and dune in salt spray, soil salinity, and air and soil temperatures have been demonstrated at several places along the coast by Johnson (1963), Barbour et al. (1973), Martin and Clements (1939), and Purer (1933, 1936a). Floristic surveys (Barbour et al. 1975; Breckon and Barbour 1974; Cooper 1936; Macdonald and Barbour 1974) reveal that the beach flora, with the exception of *Atriplex leucophylla,* is not confined to the beach and foredune, but can also colonize inland dune areas, where the sand is actively moving.

Foredunes created by the sand-stilling abilities of the rhizomatous grasses *Ammophila arenaria* and *Elymus mollis* are characteristic of beaches north of Monterey. These foredunes generally present a continuous front to shore, rise abruptly if dominated by *Ammophila,* and are more than several meters high. South of Monterey, foredunes are created by the less spectacular sand-stilling qualities of several perennial herbs with woody taproots; these foredunes are broken up into separate hillocks, but their height may still approximate that of continuous foredunes to the north (Fig. 7-1).

Introduced Taxa

Introduced European beach grass, or marram grass (*Ammophila arenaria*), has been widely planted along the Pacific coast during the past 100 yr. It appears to be an escape and fully naturalized today north of San Francisco. It occurs as far south as San Diego Co., but it is not aggressive there. Cooper (1936, 1967) believed that this species has changed the topography of dunes in central and northern California. Before its introduction, foredunes were low, rose gradually, and were dominated by *Elymus;* they led to a series of dunes alternating with swales which all were oriented roughly perpendicular to the coast, that is, aligned with prevailing onshore winds. Although such a virgin dune system is preserved today just south of Trinidad Head (Parker 1974), most systems have been replaced by a steep, *Ammophila*-dominated foredune which gives way inland to a series of dunes and swales oriented parallel to the coast.

Management of beach and dune erosion in Europe and North America typically involves the planting of *Ammophila arenaria* or *A. breviligulata* culms according to standard techniques described by Dolan et al. (1973), McLaughlin and Brown (1952), Ranwell (1972), Wiedemann et al. (1969), and Woodard (1972). In Oregon, Brown and Hafenrichter (1948) demonstrated the sand-stilling superiority of *A. arenaria* over the native *Elymus mollis*.

Although *Ammophila* stabilizes the sand, and its rhizosphere may harbor nitrogen-fixing bacteria (Hassouna and Wareing 1964), it forms a dense cover that appears to exclude many native taxa. Species richness may be halved, in comparison to adjacent *Elymus*-dominated foredunes (Barbour et al. 1976; Parker 1974; Fig. 7-1*c*).

It is likely that the need for erosion management has heightened in this century because of human-related activities which trample or denude dune vegetation. Certainly, near the town of Bodega Bay (NA 490265), wind erosion from the dunes and

Figure 7-1. Typical grass-dominated, continuous foredune (*a*) and forb-dominated, hillock foredune (*b*). In (*c*) the openness of *Elymus*-dominated foredune contrasts with the dense *Ammophila* foredune in the background. Pt. Reyes National Seashore.

Figure 7-1 (Continued)

consequent filling of Bodega Harbor are recent phenomena that began within the memory of area residents still living. Grazing by livestock and attempts at cultivation initiated a cycle of erosion there, but recreational activities have done so elsewhere. Dr. Wayne Williams (pers. commun.) has documented an increase in bare sand and changes in species composition at Morro Bay (NA 401375) due to off-the-road vehicles, and Stone and McBride (1969) have implicated human foot traffic in dune erosion at Asilomar State Park on the Monterey Peninsula (NA

270180). The California Department of Parks and Recreation is experimenting imaginatively at Asilomar to reestablish a native plant cover for erosion control (Barry 1973; Cowan 1975), but such use of native taxa is rare.

Other introduced taxa, now naturalized, which occur on beach and dune include *Cakile edentula, C. maritima, Mesembryanthemum chilense,** *M. edule,* and a number of annual grasses and forbs more restricted to localized wet depressions. The range of *Lupinus arboreus* may have been extended to Humboldt and Del Norte Cos. since 1900 in dune stabilization projects; the record of its natural range is unclear (Parker 1974). *Cakile edentula* spread so rapidly north and south since its introduction into the San Francisco Bay area in the late 1800s that it has mistakenly been listed as a native in some floras (Barbour and Rodman 1970). Rodman (1974) has determined that *C. edentula* ssp. *californica,* as listed in Munz and Keck (1959), is indistinct from *C. edentula* var. *edentula,* which is native to the east coast.

BEACH VEGETATION

Topography

Beach steepness, height, and width are affected by wave height, tidal range, sand grain size, and sand supply (Davies 1973; Perkins 1974). Given California's moderate tidal range, wave height, and fine-grained sand supply, beaches tend to be relatively low and narrow, with a gentle inland rise. The lower half of the beach is bare of plants. The upper half is thinly populated with herbaceous perennials, for the most part, which form low hillocks. The density and species richness increase back to the foredune. Driftwood is abundant on the upper beach along central and northern California.

An ephemeral concave ridge called the berm marks the last higher high water and, consequently, the seaward edge of the beach, just as prominently as the foredune marks the inland edge. The beaches tend to build seaward in summer and to be cut back in winter. Most appear to be stable in width from year to year except in parts of southern California where sand supply has been diminished by dammed and otherwise modified coastal watersheds to their north or by breakwaters.

General North–South Trends

With the exception of *Cakile* species, beach taxa are perennials. Many, but not all, share the following traits: herbaceous, evergreen, succulent, leaves entire, habit prostrate, leading to nearly complete burial in hillocks of sand. Such forbs occupy the entire coast, but north of Santa Cruz (37°N) they are subdominant to two or more perennial grasses.

Beach vegetation is low in species richness and plant cover. Generally, about five taxa compose the vegetation at any one beach, and only one or two contribute a significant amount of cover—usually *Ammophila arenaria* north of 37°N, and *Abronia maritima, Ambrosia chamissonis* (= *Franseria c.*), *Cakile maritima,* or *Mesembryanthemum chilense* south of that latitude. For the entire coast, only 15–30

* There is still debate as to the status, native or introduced, of this species.

taxa are common to the beach (Barbour et al. 1976; Breckon and Barbour 1974; Cooper 1936; Macdonald and Barbour 1974). Cover for any given 1 m^2 can reach 100%, but on the average it is 10–25%, rising to more than 50% on the seaward face and top of the foredune (Barbour et al. 1973, 1976; Johnson 1963; Parker 1974; see also Table 7-2).

If only the 15–30 common taxa are examined, their latitudinal ranges overlap and extend in such a way as to form two "ecofloristic" zones which meet near Pt. Conception, about 34°30′ (Breckon and Barbour 1974, and as amended by Barbour et al. 1975). The zone to the south is characterized by relatively few species, many of these having endemic maritime distributions. The zone to the north is characterized by many more species, some having endemic maritime and others circumarctic distributions. Point Conception falls in a region of climatic shift recognized by Köppen, Russell (1926), Durrenberger (1974), and Thornthwaite (1941). In contrast, the vegetation shift from forb to grass dominance, at 37°N, does not correspond to any obvious climatic shift. In fact, Barbour et al. (1976) concluded that latitude correlates only roughly with beach vegetation composition throughout California. They suggested that this implies that gradients of rainfall, frost, fog, sunshine, air and water temperatures, and general degree of seasonality are less significant in molding beach vegetation than are more local, microenvironmental factors, such as incidence of salt spray, mobility of substrate, presence of introduced taxa, and protection from storm waves.

In brief, *Ammophila arenaria, Elymus mollis, Ambrosia chamissonis,* and *Cakile maritima,* in that order, characterize beaches north of Pt. Reyes (38°N). Any one of the four may be dominant, but all are typically present. *Abronia latifolia* and *Calystegia soldanella* (= *Convolvulus s.*) are common subdominant associates. *Ambrosia chamissonis* is sometimes present as ssp. *bipinnatisecta,* but this subspecies is more common to the south.

Between Pt. Reyes and Morro Bay (38–35°30′N), the two grasses remain but no longer dominate, and we have a *Cakile maritima–Ambrosia chamissonis–Mesembryanthemum chilense* community. *Abronia latifolia, Atriplex leucophylla,* and *Lathyrus littoralis* are common subdominant associates.

South of Morro Bay, the grasses are lost altogether except where planted, and *Abronia maritima, Cakile maritima, Mesembryanthemum chilense,* and *Ambrosia chamissonis* characterize the community. More detail on community composition is given in the next section.

Above-ground phytomass, whether of individual species or of the entire beach community, shows no latitudinal pattern (Barbour and Robichaux 1976). Maximum community phytomass ranges from a low of 26 gm^{-2} to a high of 348 gm^{-2}, depending on the beach's dominant species. This range corresponds to arid steppe and desert and demonstrates the azonal nature of the habitat.

Zonation Back from Tide Line

Overview. Table 7-1 and Fig. 7-2 abstract the zonation of 11 taxa encountered by Barbour et al. (1976) in their transect sampling of 19 California beaches. Most of the taxa (9 out of 14) were found at least once in the leading (seaward) edge of vegetation, but by far the most frequent species in this zone was *Cakile maritima,* present in the leading edge on 53% of the beaches where it occurred at all.

However, the species present in the leading edge seldom peaked in cover near the leading edge. If the upper beach is arbitrarily divided into thirds—a first (seaward) third, a middle third, and a last third—and the percent ground cover for each species is analyzed by beach thirds, then only *Atriplex leucophylla* peaked in cover in the first third on a majority of the beaches. All other species peaked in cover, on most beaches, in the middle or last thirds, or showed no peak at all. *Abronia maritima, Ammophila arenaria, Camissonia cheiranthifolia* (= *Oenothera c.*), and *Lathyrus littoralis,* for example, never peaked in the first third. There is seen, in Table 7-1 and Fig. 7-2, a general attrition of species and plant cover as one moves from foredune to the leading edge.

The ability of most beach taxa to occasionally locate in the leading edge may be due to a suitably short life cycle, favorable microsites, or genetically based tolerance to such factors as inundation. Autecological investigations are necessary to determine which explanation is correct.

Regional studies. A number of studies, some superficial and some in depth, will be highlighted here in a north-to-south sequence. Cooper (1967) briefly discussed beach and dune vegetation at Pt. St. George, near Crescent City. Pioneer hillocks, in front of the foredune, were made by *Abronia latifolia, Ambrosia chamissonis, Ammophila arenaria,* and *Calystegia soldanella.*

Beaches between Trinidad Head and Arcata have been well documented by Johnson (1963) and Parker (1974). Parker recognized an *Elymus–Cakile* community at the seaward extreme of vegetation (Table 7-2), consisting of *Abronia umbellata* var. *breviflora* (which he incorrectly identified as *A. maritima*), *A. latifolia, Ambrosia chamissonis, Cakile maritima,* and *Elymus mollis.* He found the community to be very homogeneous, with constancy above 80% for all taxa except *A. umbellata.* Only *A. umbellata* was restricted to this community, the others being found further inland. The next community inland was on the foredune and was dominated either by beach grass (the *Ammophila–Erechtites* type) or by natives (the *Calystegia-Elymus* type). Some 15 species are characteristic of the former, with *Ammophila* contributing 50–75% of the total 77% cover. The latter community is characterized by an equal number of species and total cover, but dominance is shared by several species instead of being restricted to one (see Table 7-2). Still further inland is a *Poa–Lathyrus* community, which is discussed in the dune section of this chapter.

Since Johnson sampled vegetation with line transects, rather than by the relevé method used by Parker, he was able to document zonation in more detail. The species listed in Table 7-3 belong to the *Elymus-Cakile* community of Parker, but it is obvious that the community is not spatially homogeneous: *Cakile* species completely dominate the first third of the vegetated portion of the beach, while *Abronia, Ambrosia,* and *Elymus* dominate the last third. *Cakile maritima* seems slightly more tolerant of beach extremes than *C. edentula.* This is one of the few California beaches on which both are present; *C. edentula* is absent from beaches to the south, even though at one time its range extended to San Diego. Johnson additionally ran a 330 m long transect parallel to shore, some 6 m in front of the foredune, and recorded plant cover. *Ambrosia* and *Elymus* contributed by far the most cover, but

TABLE 7-1. Summary of species zonation for the 19 beaches illustrated in Fig. 7-1. The first column tabulates the number of beaches on which each taxon was found; the next column, the percentage of those beaches on which each taxon was encountered in the leading (seaward) edge of vegetation; the next four columns, the percentage of those beaches that showed maximum cover for each taxon by beach thirds above the leading edge. Maxima are underlined. From Barbour et al. (1976)

Taxon and Symbol	Number of Beaches	Leading Edge	Cover Peaks in			No Peak
			First Third	Middle Third	Last Third	
Abronia latifolia (ABLA)	11	9	18	27	36	19
A. maritima (ABMA)	5	40		40	60	
Ambrosia chamissonis (and ssp. *bipinnatisecta*) (AMCH)	15	20	13	40	40	7
Ammophila arenaria (AMAR)	7	29		28	72	
Atriplex leucophylla (ATLE)	6	17	50	33	17	
Cakile maritima (CAMA)	17	53	19	41	19	21
Calystegia soldanella (CASO)	9	11	22	11	55	12
Camissonia cheiranthifolia (CACH)	3			33	33	33
Distichlis spicata var. *stolonifera* (DISP)	1					100
Elymus mollis (ELMO)	7	29	15	43	15	27
Fragaria chiloensis (FRCH)	1					100
Lathyrus littoralis (LALI)	4			25	50	25
Mesembryanthemum chilense (MECH)	13	7	7	23	46	13
Poa douglasii (PODO)	1				100	

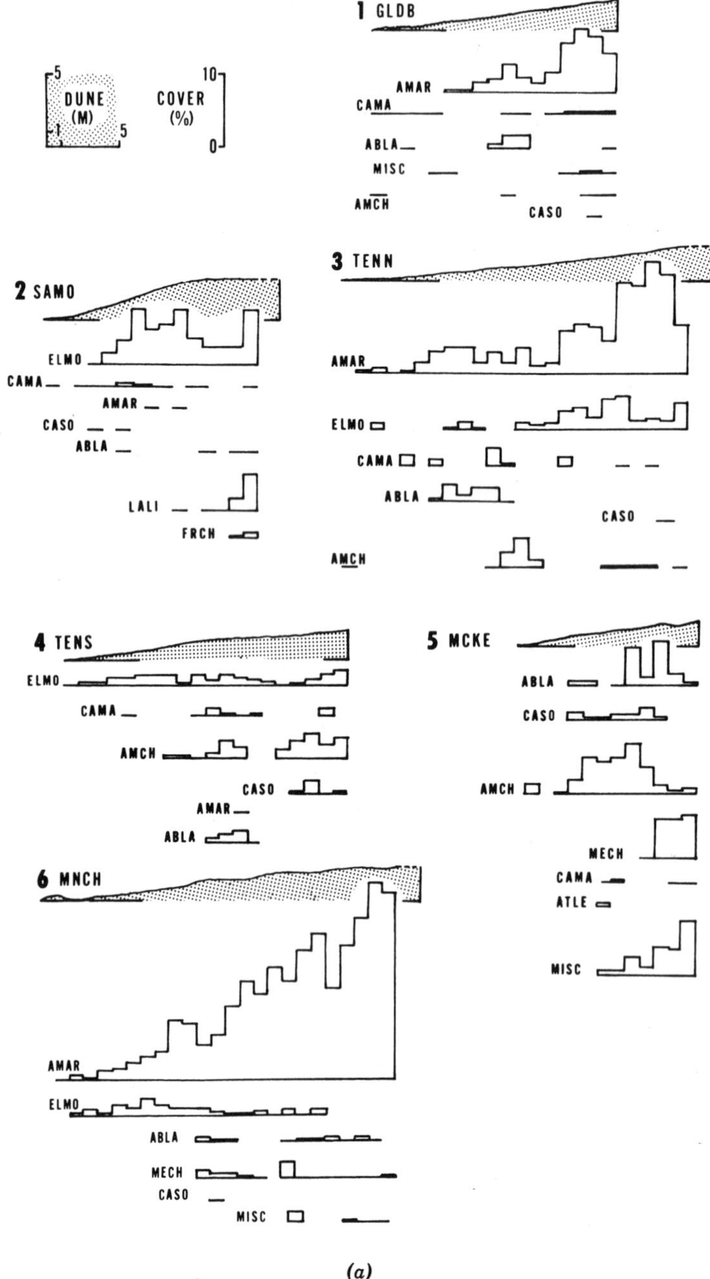

(a)

Figure 7-2. Beach profile diagrams and plant cover summaries for 19 sites. Most data represent an average of 10 transects, 1 m wide, which ran from the leading edge of vegetation to the foredune. Beaches are arranged from north to south: GLDB = Gold Bluffs, SAMO = Samoa, TENN and TENS = Ten Mile River, MCKE = MacKerricher, MNCH = Manchester, DILL = Dillon, REYE = Pt. Reyes, HMOB = Half-Moon Bay, PESC = Pescadero, ZMUD = Zmudowski, SALI = Salinas River, MORB = Morro Bay, CALL = Callander, VAFB = Vandenberg Air Force Base, MCGR = McGrath, TRAN = Trancas, SIST = Silver Strand. From Barbour et al. (1976). For species identification, see Table 7-1, except MISC = non-beach taxa.

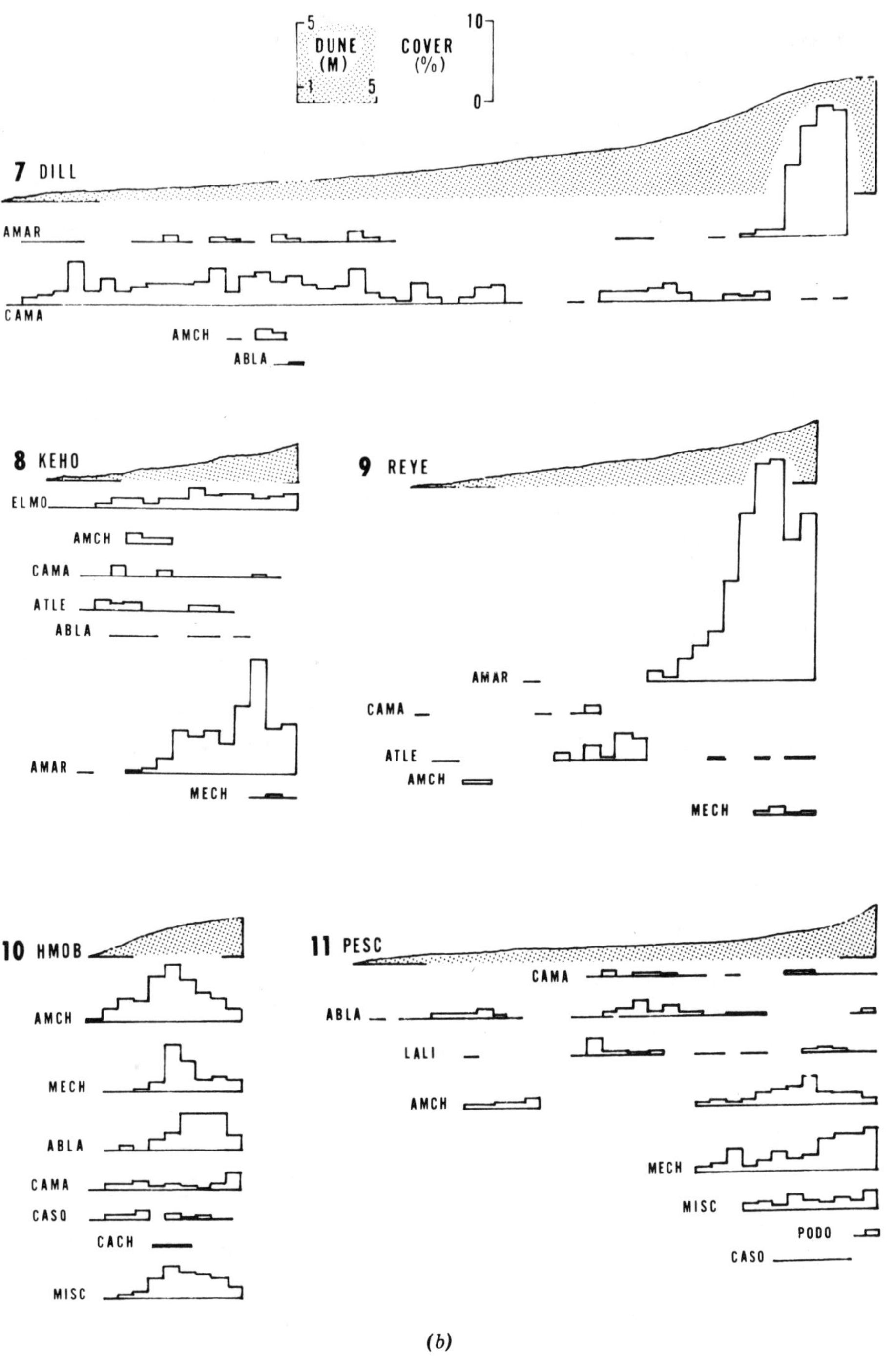

(b)

Figure 7-2 (Continued)

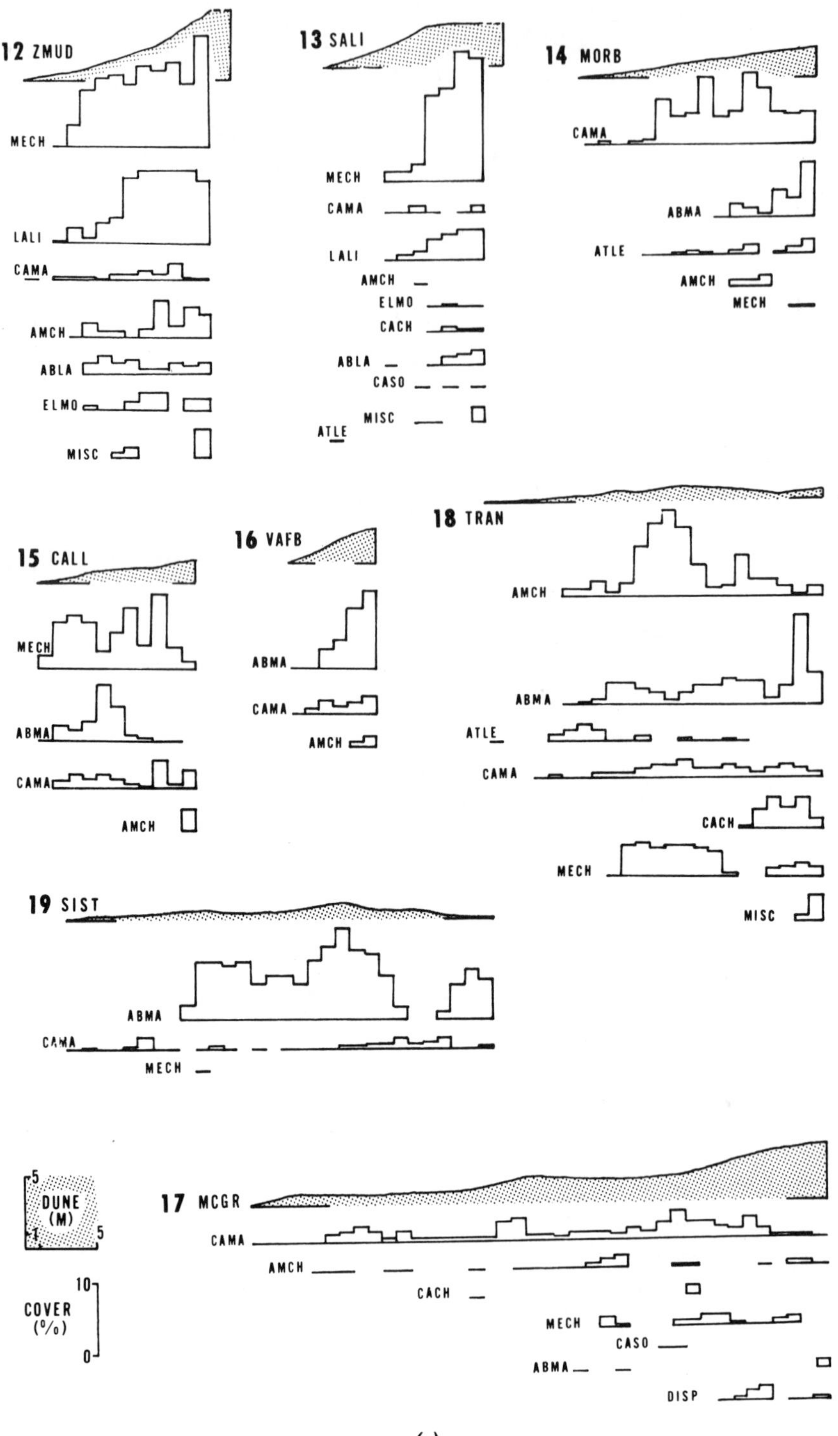

(c)

Figure 7-2 (Continued)

total cover, including overlapping foliage and driftwood, was only 23%. Four beach species (both *Cakile* species, *Calystegia,* and *Elymus*) were restricted to the beach and foredune, but seven others also occurred widely further inland.

Johnson's next community in, on the foredune, was dominated by natives, and was similar to Parker's *Calystegia–Elymus* type except that dominance shifted to *Fragaria chiloensis, Lathyrus littoralis,* and *Poa douglasii.* Eight annual species were present, but they contributed minimal cover. The total was 51%. Johnson found significant changes in community composition as his transects crossed foredune tops and dropped down their lee faces. *Erigeron glaucus* joined *Fragaria* and *Lathyrus* as dominants on foredune tops; also, the species richness and cover of annuals were much greater than on the seaward face. Additional shifts in the presence and importance of annual and perennial species occurred on the lee face.

Howell (1970) wrote of beach vegetation zones in Marin Co. as being just as distinct as those in salt marshes. *Cakile* species are in the most seaward zone. The next zone in is dominated by *Abronia latifolia* and *Ambrosia chamissonis,* among others. A third (foredune) zone supports *Ammophila, Elymus,* and *Poa douglasii.* Our own observations near the town of Bodega Bay show a similar zonation. *Atriplex leucophylla* evidently reaches its northern limit here, but nevertheless still is present in the leading edge, just as it commonly is to the south. *Cakile maritima* may also be in the leading edge or just behind *Atriplex.* Plant cover is very low between Salmon Creek and Bodega Head: one may find only one large hillock of *Cakile* for every 60 m of beachfront. *Mesembryanthemum chilense* and *Ambrosia chamissonis* (and ssp. *bipinnatisecta*) make their appearance next, and finally *Ammophila* enters at the base of the foredune and extends up the seaward face almost as a monoculture, with about 70–80% cover (see Table 7-5).

Beach and dune vegetation near Golden Gate Park on the San Francisco Peninsula was examined by Ramaley (1918) and Reynoldson (1938) before the area had been greatly modified, but their natural history style is unfortunately without quantitative measures. The leading edge was dominated by *Cakile edentula,* with scattered *Ammophila* subdominant. These continued onto the uppermost portion of the beach and shared dominance there with *Abronia latifolia, Agoseris apargioides, Ambrosia chamissonis* (and its ssp.), *Camissonia cheiranthifolia, Calystegia soldanella, Mesembryanthemum chilense,* and *Poa douglasii.* Of these, *Agoseris* was the least abundant. Finally, the foredune was thoroughly dominated by *Ammophila,* but also supported some *Ambrosia, Artemisia pycnocephala, Lupinus arboreus,* and *L. chamissonis.* This is the only reference we have seen showing *Lupinus* on the foredune.

In the Monterey Bay area, *Elymus* may dominate the beach and foredune in places; in other places *Ambrosia, Fragaria chiloensis,* and/or *Mesembryanthemum chilense* dominate (Cooper 1919, 1922, 1967). Cooper added that a fifth species, *Abronia latifolia,* is a pioneer on the beach, forming the first hillocks closest to the tide line, but it does not dominate. He equated its role and zonation with that of *A. maritima* further south.

In San Luis Obispo Co., Hoover (1970) listed eight taxa as occurring on the beach, with *Cakile maritima* the one most frequently encountered, but he did not describe their zonation. The other seven listed were *Abronia latifolia, A. maritima, A. umbellata, Ambrosia chamissonis* ssp. *bipinnatisecta, Atriplex leucophylla, Camissonia cheiranthifolia,* and *Mesembryanthemum chilense.*

TABLE 7-2. Table analysis of beach and dune communities between Trinidad Head and Arcata. Constancy (CoN) is percent occurrence of a taxon for all the relevés placed within that community type. Cover and abundance (CoV/Ab) is the mode, for those relevés, according to this scale: r = erratic occurrence, less than 1% cover; + = occasional and less than 5% cover; 1 = common, but less than 5% cover; 2 = 5-25%; 3 = 25-50%; 4 = 50-75%; 5 = 75-100%. From Parker (1974)

Community Types and Taxa	Beach		Clam Beach Foredune		Vista Pt. Foredune		Transverse Ridge		Open Dune		Longitudinal Ridge		Deflation	
	CoN	CoV/Ab	CoN	CoV/Ab	CoN	CoV/Ab	CoN	CoV/Ab	CoN	CoV/Ab	CoN	CoV/Ab	CoN	CoV/Ab
Elymus-Cakile														
Abronia umbrellata	15	r												
Elymus mollis	100	1	5	+	59	2					9	r		
Cakile maritima	95	+	13	+	12	+					2	r		
Abronia latifolia	89	+	1	+	64	1	5	+	15	1	21	+		
Ambrosia chamissonis	84	+	16	+	53	1	22	+	27	1	19	1		
Ammophila-Erechtities														
Ammophila arenaria	28	+	97	4	27	2	98	3	23	2				
Anaphalis margaritacea			75	+	29	1	56	1	21	1	6	+	15	+
Gnaphalium chilense			43	+	18	1	33	+						
Erechtites prenanthoides			93	1			58	+	43	1				
Convolvulus-Elymus														
Convolvulus soldanella	13	+			43	2					19	+		
Erechtites arguta					34	2	33	+						

Mesembryanthemum chilense					10	2					2	+		
Camissonia cheiranthifolia	7	+			32	1	22	+	4	1	6	+		
Ammophila-Baccharis														
Glehnia leiocarpa							13	+			6	+		
Gnaphalium ramosissimum					15	+			47	+				
Polystichum munitum							13	+						
Baccharis pilularis			35	+	32	2	69	2	65	2	10	+	53	1
Angelica hendersonii							23	1	18	1				
Baccharis-Scrophularia														
Scrophularia californica			2	+	13	1	20	1	87	1			21	1
Gnaphalium purpureum					16	+			17	+				
Heracleum lanatum					17	1			21	2				
Baccharis douglasii									8	+				
Solanum nodiflorum									14	+				
Tanacetum douglasii	8	+	5	+	33	1	33	1	35	2	5	+		
Poa-Lathyrus														
Solidago spathulata					41	1	24	1	23	1	59	2	11	1
Lathyrus littoralis	15	+			25	+	4	+	2	1	61	2		
Erigeron glaucus									16	+	37	1		
Eriogonum latifolium									15	1	33	1		

TABLE 7-2 [continued]

Community Types and Taxa	Beach		Clam Beach Foredune		Vista Pt. Foredune		Transverse Ridge		Open Dune		Longitudinal Ridge		Deflation	
	CoN	CoV/Ab	CoN	CoV/Ab	CoN	CoV/Ab	CoN	CoV/Ab	CoN	CoV/Ab	CoN	CoV/Ab	CoN	CoV/Ab
Fragaria chiloensis									21	1	36	1		
Poa douglasii									46	1	54	2		
P. confinis									14	+	19	+		
Sanicula arctopoides											11	+	6	+
Artemisia pycnocephala	3	r									25	2		
Cardionema ramosissimum							6	+			11	1		
Arctostaphylos uva-ursi											7	1		
Polygonum paronychia											36	1		
Armeria maritima											20	+		
Lupinus biocolor											12	+		
L. littoralis											17	1		
Orthocarpus purpurascens											8	+		
Plectritis samolifolia											7	+		
Plantago hookeriana											9	+		
Lotus micranthus											8	+		
Cryptantha leiocarpa											5	+		
Layia carnosa											13	+		
Polypodium californicium											16	+		

Salix-Rubus														
Lonicera involucrata									7	2	8	2	17	1
Juncus lesueurii									4	1	7	1	40	3
Myrica californica							4	1					11	1
Rubus vitifolius							7	1	8	2			72	2
Alnus oregona													32	2
Salix hookeriana													81	5
Carex obnupta													45	1
Cotula coronopifolia													8	+
Potentilla egedii var. grandis													6	1
Number of relevés	102		90		104		81		86		79		61	
Number of species per relevé (av.)	5		9		14		12		13		16		6	
Bare sand (av. %)	94		23		26		4		7		18		0	

TABLE 7-3. Species presence in each 10' interval along line transects run from the base of foredunes (interval 0-10) toward tide line for beaches near Arcata. Presence is expressed as percent occurrence in the 20 replicate transects. From Johnson (1963)

10 ft (3 m) Intervals	Species						
	Cakile maritima	*C. edentula*	*Elymus mollis*	*Abronia latifolia*	*Ambrosia chamissonis*	*Lathyrus littoralis*	*Eriogonum latifolium*
0-10	20		85	10	15	30	
10-20	5		75	25	30		
20-30	15		60	15	5	5	5
30-40	30	5	50	5	15		
40-50	10		20	5			
50-60	15		15	5			
60-70	15	15	10	5			
70-80	10	10	10				
80-90	40	20					
90-100	30	30	10				
100-110	45	5					
110-120	45	15					
120-130	20	5					
130-140	5						
140-150							

Williams (1974; Williams and Potter 1972; unpub. data) reported that *Cakile maritima* and *Abronia maritima* are the only species consistently found in the leading edge at Morro Bay, but that the "growth and health of these plants increase as distance from the sea increases. It appears that . . . interspecific competition elsewhere may be restrictive further away from the sea, as these two species have been observed in all habitats through the peninsula, but only seasonally, and always in diminutive numbers." Once these pioneers form hillocks, *Ambrosia* and *Mesembryanthemum* can become established in the same zone.

Near Santa Barbara, the leading edge consists of *Abronia maritima, Ambrosia,* and *Atriplex* (Martin and Clements 1939; Smith 1952; Whitfield 1932). Referring to all California beaches south of Pt. Conception, Cooper (1967) stated that the most seaward line of hillocks is dominated by *Atriplex* and a second, more inland, line by *Abronia maritima.*

Couch (1914) roamed Los Angeles beaches and dunes when they were "practically undisturbed," but his sampling scale was too coarse for us to separate beach from dune. Pierce and Pool (1938) clearly stated that *Atriplex* was "found only on the strand at the edge of high water line," but beyond that fact their terminology is too general to extract much more beach information. It is possible that *Abronia maritima* and *Calystegia soldanella* may have dominated the hillock zone just back of *Atriplex.*

Purer (1933, 1936a) was much more exact in her description of zonation of seven beach species between Los Angeles and San Diego. *Abronia maritima* and *Atriplex leucophylla* were the most consistent embryonic hillock formers, occurring closest to the tide line. Secondary invaders, appearing further inland on the beach or on hillocks established by the first two species, were *Ambrosia chamissonis* ssp. *bipinnatisecta, Calystegia soldanella,* and *Mesembryanthemum chilense.* A third, more inland zone supported *Abronia umbellata* and *Camissonia cheiranthifolia,* among others.

Sampling by Barbour et al. (1976; see Fig. 7-2) generally supports earlier southern California work described above, with one exception: *Cakile maritima* is now a prominent part of the beach vegetation and often occurs in the leading edge. The omission of *C. edentula* in the early work is anomalous, because it is clear from herbarium records that it was common to occasional all the way south to San Diego by the 1930s (Barbour and Rodman 1970). Perhaps those earlier botanists recognized it as an introduced "weed" and ignored it in their tabulations.

DUNE VEGETATION

Topography

A small vocabulary of specialized terms describing dune topography has grown through the years, together with regional terms and synonyms. Behind the foredune, or scattered throughout the dunes, may be areas eroded down to or near the water table. Such deflation areas are referred to as plains, if extensive, or as dune hollows, dune slacks, or vernal ponds if more localized. They can be invaded by mesophytic or even hydrophytic trees, shrubs, or herbs (e.g., species of *Alnus, Carex, Juncus, Rubus, Salix, Scirpus*).

Open dunes, if bare of vegetation and receiving sand from the beach as well as losing it by wind erosion, are mobile. They may be oriented with their ridges parallel to shore (transverse dunes), or roughly perpendicular to shore, oriented parallel to prevailing onshore winds (longitudinal dunes). Such open, mobile dunes are equivalent to white or yellow dunes in the European literature. Plants are not entirely absent from open dunes; they may colonize lateral dune faces (as opposed to the windward or lee faces) and they can create hillocks elsewhere (Parker 1974).

Older, more inland dunes are stabilized by a nearly continuous plant cover; these are referred to as stable dunes, gray dunes, or fixed dunes. Localized openings in the plant cover which permit wind erosion are called blowouts, but they are not deep enough to allow invasion by mesophytes. In northern California especially, the sand supply may be so extensive that active dunes invade stable, nondune vegetation. The innermost ridge of sand is generally high and is called a precipitation ridge; sand is blown over the ridge and down the slipface, continuing the process of dune advance. More details on dune terminology can be found in Parker (1974), Salisbury (1952), Ranwell (1972), Wiedemann (1966), and Wiedemann *et al.* (1969).

One microenvironmental feature of dunes is large diurnal soil surface and subsurface temperature fluctuation. Salisbury (1952) made some careful measurements of dune temperature and moisture fluctuations in English dunes and concluded that the temperature fluctuations were sufficient to condense water vapor in the sand (internal dew), and that the amount of water so provided compensated for the amount lost by plants through transpiration. A somewhat related phenomenon has been reported for desert sands (Syvertsen et al. 1975), but its existence in California dunes has not been substantiated.

Both vegetation and presumed successional pathways will be examined in the next section. For ecological reasons, that section is subdivided geographically at about 38°30′N (the latitude of Bodega Bay and the northern tip of Pt. Reyes Peninsula).

With regard to succession, one should not have the idea that a beach community is seral to an open dune community, and open dune to a scrub, chaparral, or dune forest climax community, all on the same plot of ground. Williams and Potter (1972 and unpub. data) suggest instead that each community is zoned back from the tide line, and each has reached its present state of development by succession only if sea level has fluctuated. Climax species will never invade the beach, for example, no matter how much time goes by, unless the tide line regresses some distance out to sea, giving more protection from salinity and wind. Thus, on the beach, pioneers are also climax species. Limited succession may occur on stabilized dunes, in blowouts, on protected slopes, or in deflation areas in the open dunes. A dune succession diagram, then, should be viewed more as a picture of community zonation over space than of change through time at any one location. It is still not clear whether time, or distance from the sea, is most critical to climax on stabilized dunes.

Vegetation and Succession

North of Bodega Bay. Dune vegetation between Trinidad Head and Arcata has been extensively surveyed by Johnson (1963) and Parker (1974) recognized four plant communities, which he characterized by habitat: moving dune, stabilized ridge, vernal pool hollow, and dune forest. He sampled plant cover in all of these except dune forest by a line intercept method.

Much of the surface in open, moving dunes was barren or supported scattered *Abronia latifolia, Ambrosia chamissonis, Lathyrus littoralis,* and *Poa douglasii,* but the lateral slopes to either side of the slipface were partially stabilized (77% cover) by a *Lathyrus* community, with *Fragaria chiloensis, Abronia, Poa,* and *Ambrosia* subdominant. Eleven species of annuals contributed relatively little cover.

Stabilized ridges bore a species-rich herb–shrub community dominated by *Solidago spathulata* and *Lupinus arboreus*. Although total cover approximated that of the *Lathyrus* community on dune slopes, 20 species of annuals here contributed significant cover (20%), and much of the bare sand was covered by litter. Three other species that contributed more than 5% cover were *Achillea borealis* ssp. *arenicola, Poa douglasii* ssp. *macrantha,* and *Polygonum paronychia.*

The vernal pond community (deflation area) lies just behind the foredune. It is flooded during winter and spring, and during the dry season the water table is less than 75 cm below the surface. Dominants were *Salix* (probably *hookeriana*), *Carex obnupta,* and *Juncus lesueurii,* but the presence of some dune forest species (e.g., *Pinus contorta*) indicated to Johnson an eventual succession to forest. Cover was nearly 100%.

Johnson visualized both the *Solidago–Lupinus* ridge community and the *Salix–Carex* hollow community as being seral and leading to a dune scrub with some *Pinus contorta* and *Picea stitchensis,* and ultimately to the dune forest (Fig. 7-3); however, he showed no evidence of species replacement for the xerosere (ridge) path. Parker hypothesized succession to forest as taking place only on a hydrosere path, through

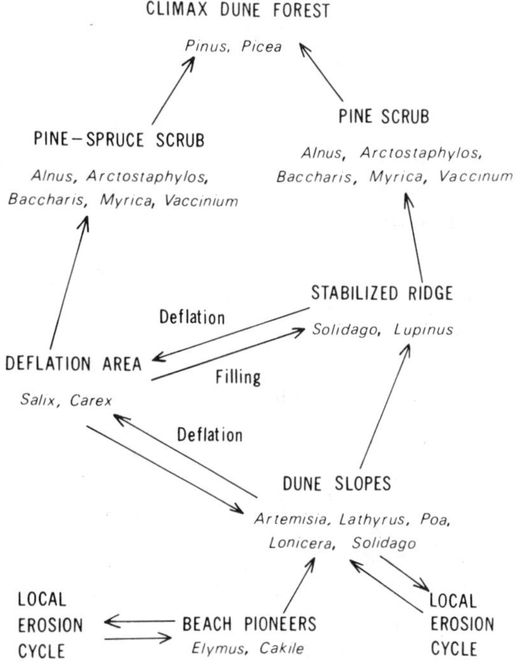

Figure 7-3. Possible pattern of plant succession on dunes between Trinidad Head and Arcata. Adapted from Johnson (1963) and Parker (1974).

Salix–Carex. Both viewed the climax dune forest as being dominated in the overstory by *Picea* and *Pinus*, with some *Abies grandis* and *Psuedotsuga menziesii*. There is a rich shrub layer of *Arctostaphylos uva ursi* var. *coactilis*, *Baccharis pilularis*, *Garrya elliptica*, *Gaultheria shallon*, *Lonicera involucrata*, *Myrica california*, and *Vaccinium ovatum*, among others. An herb stratum is poorly represented. This forest also occurs north of Crescent City (Cooper 1967), but it differs from the dune climax in Oregon by its inclusion of *Pinus*, exclusion of *Tsuga*, and richness of shrubs (Franklin and Dyrness 1973). Table 7-4 illustrates the type of soil changes which accompany this succession.

Parker defined his communities on a floristic, rather than habitat, basis (Table 7-2), so it is not surprising that his summary of dune vegetation is slightly different He visualized two major dune communities, one dominated by *Baccharis pilularis* and *Scrophularia californica*, with considerable *Lupinus arboreus*, and the other dominated by *Lathyrus littoralis* and *Poa douglasii*. He considered the *Poa–Lathyrus* type to belong to a "native dune system" representative of minimal disturbance by human beings or introduced taxa. It has greater cover and species richness than the *Baccharis-Scrophularia* type, and Parker hypothesized that succession to forest climax proceeded through it. The *Poa-Lathyrus* type is well represented on the Lanphere Dunes (NA 121212).

South of Arcata, at Ft. Bragg and Pt. Arena, climax dune forest is dominated by *Pinus muricata*, and just north of the town of Bodega Bay the climax shifts to a dune scrub. The scrub is dominated by low, evergreen, short-lived shrubs, most typically *Haplopappus ericoides* and *Lupinus chamissonis* (Cooper 1967). The northern limit of both is the dunes just north of Bodega Head (NA 490265), and they contribute very little cover there. These dunes have a history of disturbance and have been planted with *Ammophila*. The flora is quite depauperate. Dune scrub is much better developed on Pt. Reyes (Table 7-6, Fig. 7-5*b*).

The dune scrub mentioned above continues with the same dominants, but with a complete change in subdominants, south to the vicinity of Los Angeles. Munz and Keck (1959) did not recognize it as one of their 29 distinctive plant communities. Some of its dominants appear in their list of characteristic species for "coastal strand" and "coastal sage scrub."

South of Bodega Bay. Ramaley (1918) and Reynoldson (1938) viewed the "sand dune formation" along the San Francisco Peninsula as being composed of pioneer, willow thicket, and mixed scrub associations. These are equivalent to open dune, deflation areas, and dune scrub communities described in the preceding section. The most common open dune species, which formed tall hillocks, were *Abronia latifolia* and *Ambrosia chamissonis*. Less common species, which may invade the hillocks above or form smaller hillocks on their own, included *Camissonia cheiranthifolia*, *Fragaria chiloensis*, *Marah fabaceus*, and *Tanacetum camphoratum*. Deflation areas were dominated by *Salix lasiolepis*, but some shrubs of the dune scrub were also present. Herbs included *Eleocharis acicularis*, *Juncus lesueurii*, *J. phaeocephalus*, and *Mimulus guttatus*. Dune scrub occurred on stabilized dunes or in locally protected places. Dominants were 1–2 m tall and included *Lupinus arboreus*, *L. chamissonis*, *Rhamnus californica*, *Baccharis pilularis*, and *Haplopappus ericoides*, apparently in that order. Subshrubs, about 60 cm tall, were mainly *Artemisia pycnocephala* and *Eriophyllum staechadifolium*.

TABLE 7-4. Chemical analysis of beach, moving dune, stabilized ridge, vernal pond (deflation area), and dune forest soils near Arcata. Samples were taken from a 30 cm depth. Data are in parts per million except for pH and humus. From Johnson (1963)

Site	Substance												
	pH	Humus (%)	NH_3	NO_3	P_2O_5	K_2O	Ca	Mg	Na	SO_4	HCO_3	Cl	Total N
Beach	8.2	0.27	1.7	11	4.6	16	37	12	67	8	146	63	112
Foredune	7.1	0.74	1.4	9	7.8	12	24	12	35	8	98	28	350
Moving dune	6.8	0.35	0.9	3	0.3	12	24	11	20	8	85	21	126
Stable ridge	6.5	2.27	1.9	25	7.7	18	41	9	26	12	122	28	1134
Deflation	5.9	2.08	5.9	4	1.7	62	45	28	89	136	171	105	684
Dune forest	5.6	7.48	3.0	15	7.3	31	53	22	57	8	134	48	1988

According to a brief paper by Cooper (1919), which gave no hint of the sort of evidence relied upon, succession in the Monterey Bay region progresses from a pioneer community to dune scrub, then to chaparral, and finally to oak or pine forest. A forest climax in this region is the only exception to his dictum that forest climax is restricted to north of Bodega Bay. His pioneer community includes *Abronia latifolia, A. umbellata, Artemisia pycnocephala, Calystegia soldanella, Camissonia cheiranthifolia, Croton californicum, Eriogonum latifolium,* and *Poa douglasii.* His dune scrub is a nearly closed canopy community dominated by *Haplopappus ericoides,* with *Eriogonum parvifolium, Eriophyllum staechadifolium,* and *Lupinus chamissonis* as important associates. The chaparral seral stage is largely made up of *Adenostoma fasciculatum, Arctostaphylos pumila, A. vestita, Ceanothus ridigus,* and *C. dentatus.* The climax oak forest contains an overstory of *Pinus radiata* and *Quercus agrifolia;* on the peninsula only *Pinus* remains in the overstory.

More recently, Stone and McBride (1969) elaborated on a presumed succesional pathway at Asilomar State Park on the Monterey Peninsula (NA 270180). Open dunes are colonized by a pioneer *Carex pansa* community in swales and by *Artemisia pycnocephala* and/or *Haplopappus ericoides* on slopes. In either case, the next seral stage is a dense dune scrub with scattered *Pinus* or *Quercus* or both. Shrub dominants include *Baccharis pilularis, Eriophyllum staechadifolium, Rhamnus californica, Rhus diversiloba,* and *Rubus vitifolius. Pinus radiata* may invade first, especially in exposed sites, but if fire frequency and deer population are low, *Quercus agrifolia* ultimately replaces it with a completely closed canopy. The shrub stratum then becomes depauperate. A chaparral stage is missing in the Stone-McBride scheme. Figure 7-4 attempts to combine the views of Cooper and of Stone and McBride.

Later studies by Williams and Potter (1972) and Williams (1974) at Morro Bay (NA 401375) are valuable because repeated transect sampling was done for 3 yr; thus variation in time was illustrated, in addition to variation in space. Open, moving dunes are invaded by the pioneers *Ambrosia chamissonis* ssp. *bipinnatisecta* and *Mesembryanthemum chilense,* which form hillocks that can be secondarily invaded by *Lupinus chamissonis;* the hillocks may then reach heights of 10 m or more. Deflation areas contain *Achillea millefolium, Abronia umbellata, Elymus pacificus, Juncus acutus* var. *sphaerocarpus,* and an unidentified species of *Salix.* A third community characterizes stable slopes; it is dominated by *Haplopappus ericoides, Corethrogyne* (*leucophylla*?), *Dudleya caespitosa,* and *Lotus scoparius.* Williams and Potter likened this to the coastal sage scrub community of Munz and Keck (1959), for lack of anything better, but clearly it is dune scrub as described earlier in this chapter.

Williams and Potter also described a variation of dune scrub found on windward, more exposed slopes. The shrub element is poorly represented on such slopes, with the exception of *Lupinus chamissonis,* which is restricted to windward slopes. Common herbs include *Ambrosia chamissonis* ssp. *bipinnatisecta, Cakile maritima, Camissonia cheiranthifolia,* and *Malacothrix incana,* restricted to or more common on windward slopes, and more widely distributed *Abronia maritima, Croton californicus,* and *Mesembryanthemum chilense.* The authors called the dune scrub community on lee slopes an *Haplopappus–Dudleya* type, but they did not name the

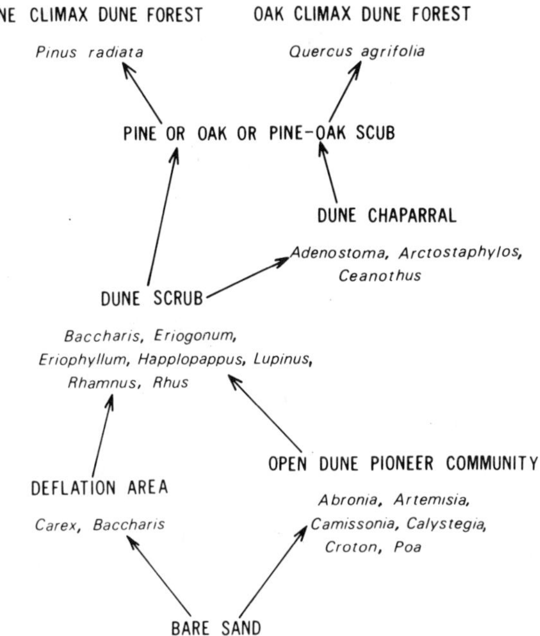

Figure 7-4. Possible pattern of plant succession on dunes in the Monterey Bay region. Adapted from Cooper (1919) and Stone and McBride (1969).

windward variant. Table 7-5 shows that sociability patterns also change, for almost every species, from windward to lee slopes.

Seasonal sampling by Williams and Potter revealed different patterns of growth, and these patterns usually did not correlate with life form. For example, *Ambrosia, Dudleya,* and *Mesembryanthemum* exhibited constant growth from December to June; *Cakile, Erysimum, Haplopappus,* and *Lotus* showed an immediate growth flush in response to winter rains and then a leveling of growth afterward; *Abronia, Achillea,* and *Malacothrix* manifested an increased rate of growth toward the end of the December–June period.

Just as in the Monterey area, dune scrub may itself be successional to chaparral. At Vandenberg Air Force Base, about 90 km south of Morro Bay, we have seen slightly more inland dunes (still with little soil development) dominated by a taller chaparral with scattered *Quercus agrifolia.* Common shrubs include *Adenostoma fasciculatum, Artemisia californica, Arctostaphylos* ssp. (possibly *A. pechoensis* and *A. rudis*), *Ceanothus cuneatus, C. impressus, Leptodactylon californicum,* and *Salvia mellifera.* Wells (1962) found these same species dominant at two other inland dune sites sampled near the towns of Baywood Park and Nipomo, San Luis Obispo Co. They were also the main constituents of the natural vegetation that he sampled on the other types of geological substrata found within the county.

We sampled dune scrub at eight widely scattered locations between Morro Bay and Vandenberg Air Force Base, a region with possibly more hectares of this type in its best condition than any other region within its range. One of the relevés is summarized in Table 7-6 and is photographed in Fig. 7-5*a*. It can be seen that, although

TABLE 7-5. Sociability of dune scrub species on Morro Bay dunes as affected by topography. Unpublished data, courtesy of Wayne T. Williams

Taxon	Lee Slope	Windward Slope
Abronia maritima	Singly	Pure colonies
Achillea millefolium	Small patches	Pure colonies
Ambrosia chamissonis ssp. *binpinnatisecta*	Singly	Extensive patches
Cakile maritima	Singly	Pure colonies
Camissonia cheiranthifolia	Singly	Singly
Corethrogyne leucophylla	Small patches	Singly
Croton californicus	Small patches	Pure colonies
Dudleya caespitosa	Grouped	Singly
Eriogonum parvifolium	Small patches	Not present
Erysimum insulare	Small patches	Singly
Eschscholzia californica	Small patches	Singly
Haplopappus ericoides	Extensive patches	Singly
Lotus scoparius	Singly	Singly
Lupinus chamissonis	Not present	Pure colonies
Malacothrix incana	Not present	Grouped, tufted
Mesembryanthemum chilense	Extensive patches	Pure colonies

the dominant shrubs, *Haplopappus ericoides* and *Lupinus chamissonis*, are present at both Pt. Reyes and Vandenberg, few other species occur in common (Sorensen coefficient of community = 38). Species-pairs are striking, however (in *Artemisia, Eriogonum, Erysimum, Erigeron, Dudleya*). Species characteristic of the south and absent in the north include *Corethrogyne filaginifolia, Croton californicus, Eriogonum parvifolium, Lotus scoparius*, and *Senecio blochmanae*.

Dune vegetation at Manhattan Beach, near Los Angeles, was sampled early this century by Couch (1914), using 100 m^2 quadrats and recording plant density only. On open dunes, *Ambrosia chamissonis* ssp. *bipinnatisecta* was by far the most abundant; *Abronia umbellata* was a common associate. More stabilized ridges and slopes supported a dense dune scrub community with *Adenostoma fasciculatum, Lupinus chamissonis, Eriogonum parvifolium, Lotus junceus* (?), and *Salvia mellifera*, in that order, in the shrub stratum, and *Abronia umbellata, Camissonia cheiranthifolia* var. *suffruticosa, Erysimum suffrutescens*, and *Phacelia douglasii* in the herb stratum.

Pierce and Pool (1938) described three communities on the nearby El Segundo

TABLE 7-6. Summary of cover (%) in dune scrub communities on stable dunes at Vandenberg Air Force Base and Pt. Reyes National Seashore. Vandenberg species marked with an asterisk were found in half or more of similar guadrats placed widely in San Luis Obispo and Santa Barbara Cos. Point Reyes species so marked were found in half or more of quadrats placed along a 15 km segment of the Pt. Reyes coast in Marin Co. The quadrats were 20-40 m^2 in area. The Vandenberg community is shown in Fig. 7-4. P = present, but outside the quadrat

Taxon	Cover	
	Vandenberg	Pt. Reyes
Shrubs		
*Baccharis piluaris ssp. consanguinea	...	10
*Halpopappus ericoides	15	P
*Lupinus chamissonis	<1	30
Subshrubs		
Artemisia californica	15	...
A. pycnocephala	...	P
*Corethrogyne filaginifolia	1	...
*Croton californicus	3	...
Eriastrum densifolium	<1	...
*Eriogonum latifolium	...	10
*E. parvifolium	3	...
*Lotus scoparius	10	...
Polygonum paronychia	...	<1
*Senecio blochmanae	1	...
Perennial herbs		
Abronia latifolia	<1	1
Agoseris apargioides	...	<1
Bromus sp.	<1	...
*Bromus carinatus	...	<1
Camissonia cheiranthifolia	...	<1
*Cardionema ramosissima	...	<1
Carex sp.	10	...
Dudleya farinosa	...	P
*D. lanceolata	2	...

Table 7-6 [continued]

Taxon	Cover Vandenberg	Cover Pt. Reyes
Erigeron blochmanae	1	...
E. glaucus	...	2
**Erysimum concinnum*	...	<1
E. insulare	P	...
**Eschscholzia californica*	1	2
Galium sp.	1	...
Mesembryanthemum chilense	3	P
Monardella undulata (annual)	P	<1
Poa douglasii	...	25
Total number of species:	17	14
Total cover:	68	80

dunes. A pioneer community included *Abronia maritima, A. umbellata, Ambrosia chamissonis* ssp. *bipinnatisecta, Calystegia soldanella, Dithyrea californica* var. *maritima,* and *Salsola kali.* A semiestablished dune community appears to have been dune scrub, for it was dominated by *Haplopappus ericoides* and *Lupinus chamissonis.* An additional, established dune community was quite distinctive: shrubs *Isomeris arborea* and *Rhus integrifolia* over herbaceous *Mesembryanthemum chilense, M. edule, M. crystallinum, Atriplex parishii, Chenopodium album, Cenchrus pauciflora, Mirabilis laevis, Solanum nigrum,* and the cactus *Opuntia occidentalis* var. *littoralis.*

Unfortunately, we have no records of undisturbed dune vegetation in the San Diego region, so it is difficult to know how extensive the established dune community of Pierce and Pool was. Cooper (1967) mentioned that cacti plus *Ephedra californica* and *Rhus integrifolia* were present on dunes at the international border, species also found on dunes on northwestern Baja California (Johnson 1973; Orme 1973).

At Ensenada, approximately 100 km south of the Mexico–California border, Johnson (1973) described the most seaward dunes as supporting pioneer hillocks of *Abronia maritima,* a subspecies of *Haplopappus venetus,* and *Mesembryanthemum chilense.* More protected habitats in this zone exhibited a scrub dominated by *Atriplex canescens, Ephedra californica,* a subspecies of *Haplopappus venetus, Lycium brevipes, Rhus integrifolia,* and *Simmondsia chinensis.* Dunes further back were stabilized with a prostrate herb–low shrub community of 70% cover. It included *Astragalus leucopsis, Cakile maritima, Camissonia cheiranthifolia, Ephedra californica, Helianthus niveus, Lotus distichus, Mesembryanthemum crystallinum, Phacelia distans,* and *Simmondsia chinensis*.

At El Socorro, 100 km further south, Orme employed a "stratified random walk," sampling 25 times with a 12.6 m^2 hoop in each of three dune phases (ages). The data, summarized in Table 7-7, show the same species as were present at Ensenada with the exception of *Atriplex julacea,* a peninsular endemic that does not extend to San Diego. The vegetation of the oldest (phase 1) dunes resembles the inland vegetation of this desert–chaparral region (mean annual precipitation less than half that at Ensenada) and would not be expected to occur at San Diego. Thus the San Diego stabilized dunes probably supported vegetation now extant in northwestern Baja California, dominated by *Ephedra californica, Haplopappus venetus, Lycium brevipes,* and *Rhus integrifolia*.

Figure 7-5. (*a*) *Haplopappus* dune scrub community on stable dunes at Vandenberg Air Force Base. (*b*) *Lupinus chamissonis* dune scrub at Pt. Reyes; *Haplopappus* flowering in foreground. A quadrat summary for these communities appears in Table 7-7.

TABLE 7-7. Plant density (per 12.6 m^2) and cover (%) on El Socorro dunes, Baja California. Phase I dunes are oldest (paleodunes) and are stabilized by a desert scrub community; phases II and III are younger (Flandrian), mobile, and nearer the coast. P = present, but with cover and density less than 0.01. From Orme (1973)

Taxon	Phase III Dunes		Phase II Dunes		Phase I Dunes	
	Density	Cover	Density	Cover	Density	Cover
Abronia maritima	1.96					
Lycium brevipes	0.92	1.36	0.28	0.16	0.96	5.12
Ephedra californica	0.12	0.01	0.16	0.37	0.04	0.04
Atriplex julacea			1.48	2.10		
Helianthus niveus[a]			4.44	1.64		
Haplopappus venetus			4.68	4.56		
Cactaceae			0.08	0.01	3.44	3.65
Simmondsia chinensis			P	P	1.80	2.08
Euphorbia misera			P	P	0.76	3.05
Franseria chenopodifolia			P	P	1.56	4.10
Dudleya spp.					1.28	0.39
Agave shawii					1.16	3.03
Gramineae					13.88	3.68
Other spp.	0.08	0.03	4.04	3.69	2.32	4.28
Total	3.08	4.53	15.16	12.53	27.20	29.42

[a]Possibly misidentified as *Encelia californica* in the original table.

AUTECOLOGICAL STUDIES

Table 7-8 lists the California beach and dune taxa that have been examined autecologically. Most of the studies cited have not gone further than passive field observation of distribution patterns, root and shoot morphology and anatomy, phenology, sensitivity to salt spray and sand burial, and osmotic potential of the sap. Although many of the authors include microenvironmental data sufficient to program growth chambers and there conduct manipulative experiments, an experimental approach has rarely been taken; hence these data go unutilized.

Martin and Clements (1939) did field manipulation with phytometer transplants, and Purer (1936b) manipulated the rhizomes of *Calystegia*, but these examples stand out as exceptions in the literature. More recently, Barbour (1970b,c, 1972) experimented in the field and in growth chambers with *Cakile;* it may be tenuous, however, to extrapolate his findings from this introduced species to native taxa. *Lupinus arboreus* has been examined in a very modern, complete fashion (Davidson 1975; Pitelka 1974) that touches on demography, energy allocation to growth and reproduction, water relations, photosynthesis, and the impact of herbivores; but these studies have been done on grassland populations, and it is not clear whether the results apply to dune populations. Although DeJong (1976) is in the process of measuring the effect of several environmental factors on the photosynthetic rate of the only C_4 beach species, *Atriplex leucophylla,* his study had not been completed at the time this chapter was written.

Some of the taxa occur on other shores (e.g., *Ammophila, Elymus, Cakile*), and autecological work has gone on there. Such studies have been ignored in Table 7-8, however, because of the possibility that ecotypic variation and climatic differences are so significant that the results have little application to California genotypes.

AREAS FOR FUTURE RESEARCH

More quantitative documentation of beach and dune vegetation is needed. Many studies cited in this chapter contain little more than a listing of species in order of importance without any sampling data. This is especially true for climax communities, such as *Pinus–Picea* forest, *Pinus muricata* forest, *Haplopappus* dune scrub, *Adenostoma* dune chaparral, and desert dune scrub. The relationship of such communities to adjacent inland, nondune communities has not been examined.

More geological information about which beaches are prograding and which are retrograding, the volume of sand supply (to beach and ultimately to dune), the sources of sand supply, and the rate of dune movement is needed. Such data are important to basic science and for management purposes.

More information about management options for beach and dune erosion control is also needed. *Ammophila* continues to be used because it has been tried successfully in the past, nursery stock is available, and field planting technology is well known. However, negative aspects of its monoculture are recognized. Perhaps proper planting and management of a mixture of native taxa, together with the provision of walkways for pedestrian traffic and the elimination of horse traffic, cattle grazing, and off-the-road vehicles, would result in stabilization every bit as effective

TABLE 7-8. Summary of autecological studies of Pacific coast beach and dune taxa. An asterisk indicates a non-California or nonbeach study. Ordinarily nonbeach or nondune taxa are included if they were studied in a beach or dune habitat. Not all dune forest trees are included; however, a general reference for them is Fowells (1965)

Taxon	References
Abronia latifolia[1]	Johnson 1963; Kumler 1969; Reynoldson 1938
A. maritima	Johnson 1976; Martin and Clements 1939; Purer 1933, 1936a; Whitfield 1932; Williams and Potter 1972
A. umbellata	Martin and Clements 1939; Purer 1933, 1936a
Achillea borealis	Johnson 1963
A. millefolium	Williams and Potter 1972
Aira praecox	Kumler 1969*
Alnus oregona	Johnson 1963
Ambrosia chamissonis and ssp. *bipinnatisecta*	Johnson 1963; Kumler 1969*; Martin and Clements 1939; Purer 1933, 1936a; Reynoldson 1938; Whitfield 1932; Williams and Potter 1972
A. psilostachya	Whitfield 1932
Ammophila arenaria	Brown and Hafenrichter 1948*; McLaughlin and Brown 1942*; Purer 1933; Reynoldson 1938
A. breviligulata	Brown and Hafenrichter 1948*; Johnson 1963
Amsinckia intermedia	Whitfield 1932
Anaphalis margaritacea	Johnson 1963
Arctostaphylos uva-ursi var *coactilis*	Johnson 1963
Armeria maritima ssp. *californica*	Drysdale 1971*; Johnson 1963
Artemisia pycnocephala	Johnson 1963; Reynoldson 1938
A. vulgare	Whitfield 1932
Atriplex leucophylla	De Jong 1976; Martin and Clements 1939; Purer, 1933, 1936a

[1] See also page 261 for additional information.

TABLE 7-8 [continued]

Taxon	References
Baccharis pilularis and ssp. consanguinea	Johnson 1963; Reynoldson 1938; Whitfield 1932
Brassica campestris	Whitfield 1932
Cakile edentula	Barbour 1970c; Barbour and Rodman 1970; Johnson 1963; Rodman 1974
C. maritima	Barbour 1970b,c, 1972; Barbour et al. 1973; Barbour and Rodman 1970; Johnson 1963; Reynoldson 1938; Rodman 1974; Williams and Potter 1972
Calystegia soldanella	Johnson 1963; Kumler 1969*, Martin and Clements 1939; Purer 1933, 1936a,b; Reynoldson 1938; Whitfield 1932
Camissonia cheiranthifolia	Johnson 1963; Martin and Clements 1939; Purer 1933, 1936a; Williams and Potter 1972
Carex macrocephala	Kumler 1969*
C. obnupta	Johnson 1963
Cerastium holosteoides	Kumler 1969
Cirsium californicum	Whitfield 1932
Corethrogyne leucophylla	Williams and Potter 1972
Croton californicus	Williams and Potter 1972
Dudleya caespitosa	Williams and Potter 1972
Elymus mollis	Brown and Hafenrichter 1948*; Johnson 1963
Erigeron glaucus	Johnson 1963
Erigonum latifolium	Johnson 1963
E. parvifolium	Purer 1934, 1936a; Williams and Potter 1972
Erysimum insulare	Williams and Potter 1972
Eschscholzia californica	Whitfield 1932; Williams and Potter 1972
Fragaria chiloensis	Johnson 1963; Reynoldson 1938
Glehnia leiocarpa	Kumler 1969*

TABLE 7-8 [continued]

Taxon	References
Haplopappus ericoides	Purer 1933, 1936a; Reynoldson 1938; Williams and Potter 1972
H. squarrosa	Whitfield 1932
H. venetus ssp. vernonioides	Purer 1934; Reynoldson 1938
Heteromeles arbutifolia	Purer 1934
Heterotheca grandiflora	Purer 1934; Whitfield 1932
Hypochoeris radicata	Kumler 1969*
Juncus lesueurii	Johnson 1963
Lathyrus littoralis	Johnoon 1963
Lotus scoparius	Williams and Potter 1972
Lupinus arboreus	Barbour et al. 1973; Davidson 1975*; Pitelka 1974*; Reynoldson 1938
L. chamissonis	Martin and Clements 1939; Purer 1933, 1934, 1936a,b; Reynoldson 1938; Whitfield 1932; Williams and Potter 1972
L. littoralis	Kumler 1969*
Malacothrix incana	Williams and Potter 1972
M. saxatilis var. tenuifolia	Whitfield 1932
Mesembryanthemum chilense	Johnson 1963; Purer 1933, 1934, 1936a; Williams and Potter 1972
M. crystallinum	Vivrette 1973
Myrica californica	Johnson 1963
Nicotiana glauca	Purer 1934
Phacelia ramosissima	Reynoldson 1938; Whitfield 1932
Picea sitchensis	Johnson 1963
Pinus contorta	Johnson 1963
Poa confinis	Johnson 1963

TABLE 7-8 [continued]

Taxon	References
P. douglasii	Johnson 1963; Reynoldson 1938
P. macrantha	Kumler 1969*
Polygonum paronychia	Johnson 1963; Kumler 1969*
Potentilla egedii var. *grandis*	Johnson 1963
Rhus integrifolia	Purer 1933, 1934, 1936a
R. laurina	Purer 1934
Rumex crassus	Johnson 1963
Salix lasiolepis	Johnson 1963; Whitfield 1932
Solanum douglasii	Whitfield 1932
Solidago spathulata	Johnson 1963
Suaeda depressa	Whitfield 1932
Tanacetum douglasii	Johnson 1963
Vaccinium ovatum	Johnson 1963

as that achieved with *Ammophila;* yet there has been minimal experimentation along this line.

Finally, an autecological understanding of dominant species needs to be generated now from a more experimental, physiological approach. We have at present little more than correlations between species distribution or behavior and environmental factors; cause and effect relationships have seldom been demonstrated. The relationship between macroclimate and the north–south range of species should be investigated. Of prominent beach taxa, only *Malacothrix incana* has a distribution restricted to California. The others range through the entire state (*Cakile maritima, Calystegia soldanella, Mesembryanthemum chilense*), or at least through much of the state and far to the north (e.g., *Abronia latifolia, Ambrosia chamissonis, Elymus mollis*) or far to the south (*Abronia maritima, A. umbellata, Atriplex leucophylla*). Beach taxa show a relative insensitivity to macroclimatic gradients—a trait not shared by major dune taxa.

LITERATURE CITED

Barbour, M. G. 1970a. Is any angiosperm an obligate halophyte? Amer. Midl. Natur. 84:105-120.

---. 1970b. Germination and early growth of the strand plant *Cakile maritima*. Bull. Torrey Bot. Club 97:13-22.

---. 1970c. Seedling ecology of *Cakile maritima* along the California coast. Bull. Torrey Bot. Club 97:280-289.

---. 1972. Seedling establishment of *Cakile maritima* at Bodega Head, California. Bull. Torrey Bot. Club 99:11-16.

Barbour, M. G., R. B. Craig, F. R. Drysdale, and M. T. Ghiselin. 1973. Coastal ecology: Bodega Head. Univ. Calif. Press, Berkeley. 338 p.

Barbour, M. G., T. M. De Jong, and A. F. Johnson. 1975. Additions and corrections to a review of North American Pacific Coast beach vegetation. Madroño 23:130-134.

---. 1976. Synecology of beach vegetation along the Pacific Coast of the United States: a first approximation. J. Biogeogr. 3:55-69.

Barbour, M. G., and R. H. Robichaux. 1976. Beach phytomass along the California Coast. Bull. Torrey Bot. Club 103:16-20.

Barbour, M. G., and J. E. Rodman. 1970. Saga of the west coast sea-rockets: *Cakile edentula* ssp. *californica* and *C. maritima*. Rhodora 72:370-386.

Barry, W. J. 1973. Problems in dune management. Speech to the Natl. Sci. Found. workshop on ecological succession and land management, held in Ukiah, April 14, 1973. Mimeo. provided by Calif. Dept. Parks and Recreation, Sacramento.

Breckon, G. J., and M. G. Barbour. 1974. Review of North American Pacific Coast beach vegetation. Madroño 22:333-360.

Brown, R. A., and A. L. Hafenrichter. 1948. Factors influencing the production and use of dunegrass clones for erosion control. J. Amer. Soc. Agron. 40:articles beginning on pp. 512, 603, 677.

Cooper, W. S. 1919. Ecology of the strand vegetation of the Pacific coast of North America. Carnegie Inst. Wash. Yearbook 18:96-99.

---. 1922. Strand vegetation of the Pacific coast. Carnegie Inst. Wash. Yearbook 21:74-75.

---. 1936. The strand and dune flora of the Pacific coast of North America: a geographic study, pp. 141-187. In T. H. Goodspeed (ed.). Essays in geobotany. Univ. Calif. Press, Berkeley.

---. 1967. Coastal dunes of California. Mem. Geol. Soc. Amer. 104. Denver, Color. 131 p.

Couch, E. B. 1914. Notes on the ecology of sand dune plants. Plant World 17:93-97.

Cowan, B. 1975. Protecting and restoring native dune plants. Fremontia 3(2):3-7.

Curray, J. R. 1964. Transgressions and regressions, pp. 175-203. In Papers in marine geology: Shepard commemorative volume. Macmillan, New York.

Davidson, E. 1975. Demography of *Lupinus arboreus* at Bodega Head, California. Ph.D. dissertation, Univ. Calif., Davis. 114 p.

Davies, J. L. 1973. Geographical variation in coastal development. Hafner, New York. 204 p.

De Jong, T. M. 1976. Autecology of Atriplex leucophylla on California beaches. Ph.D. dissertation, Univ. Calif., Davis (in prep.).

Dolan, R., P. J. Godfrey, and W. E. Odum. 1973. Man's impact on the barrier islands of North Carolina. Amer. Sci. 61:152-162.

Drysdale, F. R. 1971. Studies in the biology of Armeria maritima with emphasis upon the variety californica as it occurs at Bodega Head, California. Ph.D. dissertation, Univ. Calif., Davis. 141 p.

Durrenberger, R. W. 1974. Patterns on the land. 4th ed. National Press Books, Palo Alta, Calif. 102 p.

Emery, K. O., and J. P. Milliman. 1968. Sea levels in the past 35,000 years. Science 162:1121-1122.

Flint, R. F. 1971. Glacial and Quaternary geology. John Wiley, New York. 892 p.

Fowells, H. A. 1965. Silvics of forest trees of the United States. USDA Agric. Handbook 271. Washington, D.C. 762 p.

Franklin, J. F., and C. T . Dyrness. 1973. Natural vegetation of Oregon and Washington. USDA Forest Serv. Gen. Tech. Rept. PNW-8. Portland, Ore. 417 p.

Hassouna, M. G., and P. F. Wareing. 1964. Possible role of rhizosphere bacteria in the nitrogen nutrition of Ammophila arenaria. Nature 202:467-469.

Hoover, R. F. 1970. The vascular plants of San Luis Obispo County, California. Univ. Calif. Press, Berkeley. 350 p.

Howell, J. T. 1970. Marine flora. 2nd ed. Univ. Calif. Press, Berkeley. 366 p.

Johnson, A. F. 1973. A survey of the strand and dune vegetation along the Pacific coast of Baja California, Mexico. MS. thesis, Univ. Calif., Davis. 125 p.

---. 1976. Ecology of Abronia maritima. Ph.D. dissertation, Univ. Calif., Davis (in prep.).

Johnson, J. W. 1963. Ecological study of dune flora, Humboldt Bay. M.S. thesis, Humboldt State Coll., Arcata, Calif. 447 p.

Kearney, R. H. 1904. Are plants of sea-beaches and dunes true halophytes? Bot. Gaz. 37:424-436.

Kumler, M. L. 1963. Succession and certain adaptive features of plants native to the sand dunes of the Oregon coast. Ph.D. dissertation, Oregon State Univ., Corvallis. 158 p.

---. 1969. Plant succession on the sand dunes of the Oregon coast. Ecology 50:695-704.

Macdonald, K. B., and M. G. Barbour. 1974. Beach and salt marsh vegetation along the Pacific Coast, pp. 175-234. In R. J. Reimold and W. H. Queen (eds.). Ecology of halophytes. Academic Press, New York.

Martin, E. V., and F. E. Clements. 1939. Adaptation and origin in the plant world. I. Factors and functions in coastal dunes. Carnegie Inst. Wash. Pub. 521. 107 p.

McBride, J. R., and E. C. Stone. 1976. Plant succession on the sand dunes of the Monterey Peninsula, California. The Amer. Midl. Natur. 96:118-132.

McLaughlin, W. T., and R. L. Brown. 1942. Controlling coastal sand dunes in the Pacific Northwest. USDA Circ. 660:1-46.

Minard, C. R. 1971. Quaternary beaches and coasts between the Russian River and Drake's Bay, California. Hydraulic Eng. Lab., Coll. Eng., Univ. Calif. Berkeley, HEl 2-35.

Munz, P. A., and D. D. Keck. 1959. A California flora. Univ. Calif. Press, Berkeley. 1681 p.

Olsson-Seffer, P. 1909. Relation of soil and vegetation on sandy sea shores. Bot. Gaz. 47:85-126.

Orme, A. R. 1973. Coastal dune systems of northwest Baja California, Mexico. Office Naval Res. Tech. Rept. 0-73-1. 43 p.

Parker, J. 1974. Coastal dune systems between Mad River and Little River, Humboldt County, California. M.A. thesis, Humboldt State Coll., Arcata, Calif. 62 p.

Perkins, E. J. 1974. The biology of estuaries and coastal waters. Academic Press, New York. 678 p.

Pierce, W. D., and D. Pool. 1938. The fauna and flora of the El Segundo sand dunes. I. General ecology of the dunes. Bull. South. Calif. Acad. Sci. 37:93-97.

Pitelka, L. F. 1974. Energy allocation in annual and perennial lupines (*Lupinus*: Leguminosae). Ph.D. dissertation, Stanford Univ., Stanford, Calif. 126 p.

Purer, E. A. 1933. Studies of certain coastal sand dune plants of southern California. Ph.D. dissertation, Univ. South. Calif. Los Angeles. 207 p.

---. 1934. Foliar differences in eight dune and chaparral species. Ecology 15:197-203.

---. 1936a. Studies of certain coastal and sand dune plants of southern California. Ecol. Monogr. 6:1-87.

---. 1936b. Growth behavior in *Convolvulus soldanella* L. Ecology 17:541-550.

Ramaley, F. 1918. Notes on dune vegetation at San Francisco, California. Plant World 21:191-201.

Ranwell, D. 1972. Ecology of salt marshes and sand dunes. Chapman and Hall, London. 258 p.

Reynoldson, F. 1938. The flora of San Francisco sand-dunes, its composition and adaptations. M.S. thesis, Univ. Calif., Berkeley. 57 p.

Rodman, J. E. 1974. Systematics and evolution of the genus *Cakile* (Cruciferae). Contrib. Grey Herb. 205. Harvard Univ., Cambridge, Mass. 146 p.

Russell, R. J. 1926. Climates of California. Univ. Calif. Pub. Geogr. 2:73-84.

Salisbury, E. 1952. Downs and dunes. G. Bell and Sons, London. 328 p.

Smith, C. F. 1952. A flora of Santa Barbara. Santa Barbara Bot. Gard., Santa Barbara, Calif. 100 p.

Syvertsen, J. P., G. L. Cunningham, and T. P. Feather. 1975. Anomalous diurnal patterns of stem xylem water potential in *Larrea tridentata*. Ecology (in press).

Thornthwaite, C. W. 1941. USDA atlas of climatic types of the United States, 1900-1939. Misc. Pub. 421, U.S. Govt. Printing Office, Washington, D.C. 7 p.

Vivrette, N. J. 1973. Mechanism of invasion and dominance of coastal grasslands by *Mesembryanthemum crystallinum* L. Ph.D. dissertation, Univ. Calif., Santa Barbara. 78 p.

Wells, P. V. 1962. Vegetation in relation to geological substratum and fire in the San Luis Obispo quadrangle, California. Ecol. Monogr. 32:79-102.

Whitfield, C. J. 1932. Osmotic concentrations of chaparral, coastal sagebrush, and dune species of southern California. Ecology 13:279-285.

Wiedemann, A. 1966. Contributions to the plant ecology of the Oregon coastal sand dunes. Ph.D. dissertation, Oregon State Univ., Corvallis. 255 p.

Wiedemann, A. M., L. J. Dennis, and F. H. Smith. 1969. Plants of the Oregon coastal dunes. Oregon State Univ. Book Stores, Corvallis. 117 p.

Williams, W. T. 1974. Species dynamism in the coastal strand plant community at Morro Bay, California. Bull. Torrey Bot. Club 101-83-89.

Williams, W. T., and J. R. Potter. 1972. The coastal strand community at Morro Bay State Park, California. Bull. Torrey Bot. Club 99:163-171.

Woodard, D. W. 1972. San Francisco beach dune stabilization program feasibility, evaluation, and recommendations. USDA, Coastal Eng. Res. Center, Memo. for Records, Washington, D.C. 7 p.

Note added in proof: The following autecological information was inadvertently omitted from Table 7-8 and the literature cited.

Taxa	References
Abronia latifolia, *A. maritima*, *Ambrosia chamissonis*, *A. c.* ssp. *bipinnatisecta*, *Atriplex leucophylla*, *Cakile edentula*, *C. maritima*, *Calystegia soldanella*, *Camissonia cheiranthifolia*, *Lathyrus japonicus*, *L. littoralis*	Barbour, M. G., and T. M. DeJong. 1977. Response of west coast beach taxa to salt spray, seawater inundation, and soil salinity. Bull. Torrey Bot. Club 104 (in press).
Abronia latifolia, *A. maritima*, *Ambrosia chamissonis*, *Ammophila arenaria*, *Amsinckia spectabilis*, *Calystegia soldanella*, *Erysimum concinnum*, *Eschscholzia californica*, *Lathyrus littoralis*, *Layia carnosa*, *Lupinus arboreus*, *Marah fabaceus*	Pitts, W. D. 1976. Plant/animal interaction in the beach and dunes of Point Reyes. Ph.D. dissertation, Univ. Calif., Davis. 246 p.

CHAPTER 8 COASTAL SALT MARSH

KEITH B. MACDONALD
Department of Geological Sciences, University of California, Santa Barbara

Currently, Senior Scientist, Science Applications Inc., Boulder, Colorado

Introduction **264**
  Pleistocene origins **264**
  Historical perspective **266**
  The literature **266**
Environmental variables **268**
  Sedimentation, drainage, and soils **268**
  Marsh topography and tidal flooding **269**
  Salinity relations **271**
  Nutrient relations **272**
Floristics and vegetation **272**
  Floristics **272**
  Vegetation **273**
    San Francisco Bay Area marshes **273**
      Brackish marshes **274**
    Northern California coastal marshes **275**
    Southern California coastal marshes **275**
Autecology **279**
  Spartina foliosa **279**
  Salicornia virginica **280**
  Batis maritima **281**
  Salicornia bigelovii **282**
  Distichlis spicata **282**
  Monanthochloë littoralis **282**
  Salicornia subterminalis **282**
Vertical zonation **283**
Salt marsh development models **286**
Future research **287**
Acknowledgments **288**
Literature cited **289**

INTRODUCTION

Coastal salt marshes are restricted to the upper intertidal zone of protected shallow bays, estuaries, and coastal lagoons. Physical conditions are dominated by the tides, and pronounced environmental gradients are established in response to elevational changes in the frequency and duration of tidal flooding. These gradients are frequently reflected in the vertical zonation of different halophytes across the salt marsh. Studies of this species zonation and its underlying causes have long been a central theme of salt marsh research.

This chapter begins with brief accounts of the Pleistocene origins and historical destruction of California coastal salt marshes. Descriptions of the major physical factors operating in salt marsh environments, and of the species composition and general appearance of their vegetation, follow. Autecological summaries are presented for several prominent, better-known species, and the nature and reality of zonation are discussed. A summary of regional succession patterns and research topics for the future concludes the chapter.

Pleistocene Origins

With the notable exception of the San Francisco Bay complex (Atwater and Hedel 1976), the narrow continental shelf and mountainous coast of California provide few lowland sites suitable for extensive salt marsh development. Salt marshes are thus restricted to relatively small areas fringing the few bays and protected river mouths scattered along the coast (Fig. 8-1). Many of these sites originated in late Pleistocene time, perhaps 15,000 yr ago, when geological evidence indicates that sea level was about 130 m below its present elevation (Milliman and Emery 1968). At that time, many coastal streams had eroded significant valley systems across the present shoreline and flowed into the ocean far seaward of their present positions. The bays

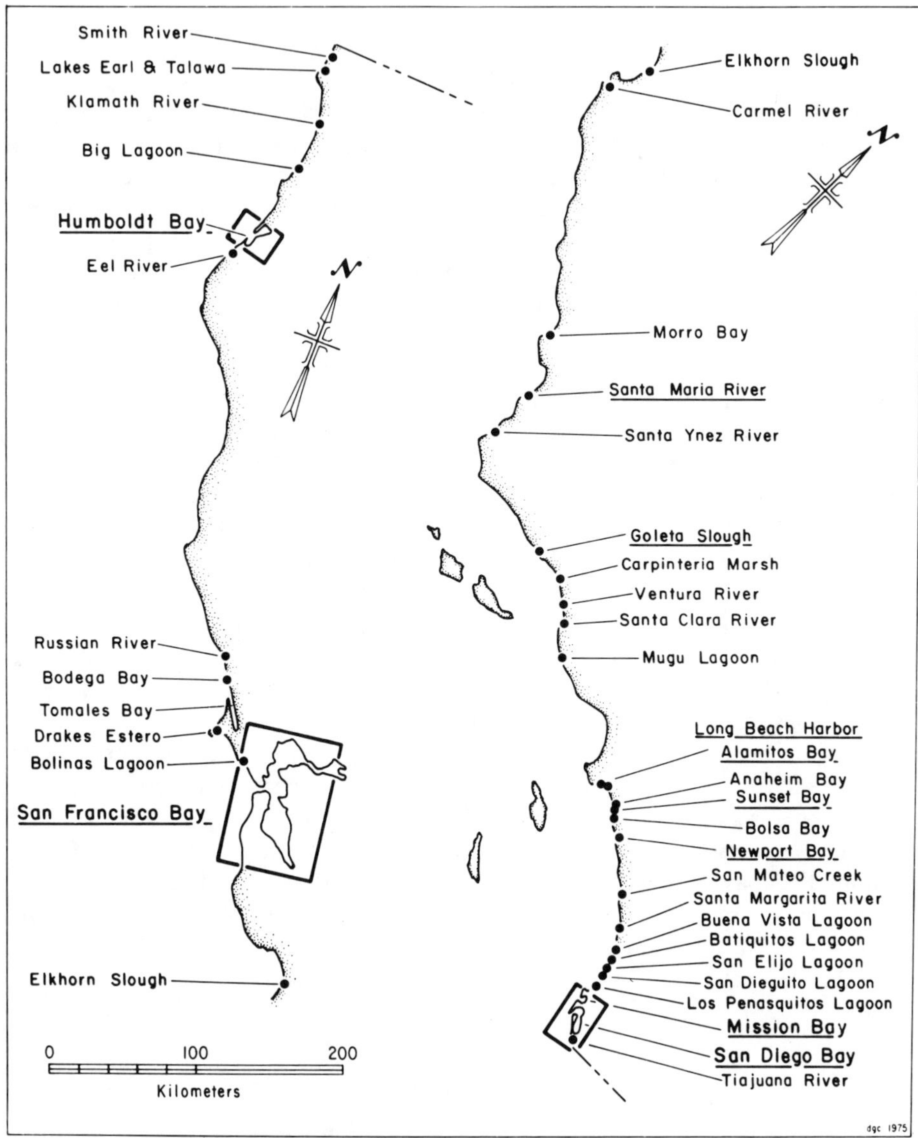

Figure 8-1. California coastal salt marsh sites. Underlined sites have suffered significant declines in salt marsh acreage.

and estuaries that we know today were created as sea level rose to its present elevation and flooded these coastal river valleys. Deposition of marine and alluvial sediments resulted in rapid coastal shallowing, while transport and reworking of beach sands by ocean currents and wave action created spits and barrier beaches that protected the coastal embayments from open ocean waves. The resulting reduction in turbulence and concomitant deposition of finer-grained sediments created intertidal flats ideal for the establishment and rapid expansion of salt marsh vegetation (Upson 1949; Stevenson and Emery 1958; California Department of Water Resources 1968; Warme 1971; Pestrong 1972; Scott et al. 1976).

Historical Perspective

Historically, California's coastal salt marshes were much more extensive than is the case today. Numerous archaelogical middens indicate that these coastal sites were the focus of aboriginal food-gathering cultures that had persisted for several thousand years before European settlement. Early settlers also depended heavily on wetland wildlife resources (Gill and Buckman 1974), but uncontrolled harvest and depletion forced them to turn increasingly to wetland reclamation for livestock and agriculture (Nesbit 1885). By the 1930s much of California's salt marsh had been reclaimed for salt ponds, agriculture, and expanding urbanization. More recently, the increasing needs of commercial shipping and the popularity of recreational and residential marinas have also taken their toll.

The reductions in salt marsh acreage are impressive. Nichols and Wright (1971) documented a 60% reduction—from 811 to 324 km^2—in the peripheral marshlands of San Francisco Bay complex between 1850 (Fig. 8-2) and 1968. Brackish Suisun Marsh (223 km^2: Gill and Buckman 1974) accounts for the majority of the remaining wetlands. Dredging began in San Diego Bay (Fig. 8-2) in the 1850s, and some marshes were used for salt production as early as 1871. Mudie (1970b) documented an 85% reduction in the bay's salt marshes—from 10 to 1.4 km^2—between 1919 and 1970. Reclamation began in Humboldt Bay in 1881; of 28 km^2 of tidal marsh present in 1897, less than 9% (2.4 km^2) remained in 1973 (Fig. 8-3). Smaller marshes along the coast have suffered similar declines.

Current estimates (1975) indicate that 365–370 km^2 (somewhat more than 90,000 acres) of salt marsh still remain along the California coast. The great majority, some 88–89%, is still included within the San Francisco Bay complex. Another 4–5% (16–20 km^2) occurs at sites north of San Francisco, while the remaining 7% (25 km^2) lies along the coast between San Francisco and the Mexican border. The fact that 1 million and 3.5 million acres of salt marsh remain along the Atlantic and Gulf coasts, respectively (Chapman 1974), emphasizes the scarcity of Californian salt marsh. Unfortunately, very little of what remains is presently guaranteed future preservation.

The Literature

Most available studies of California coastal salt marshes are cited where appropriate in the text. Halophyte research has been expanding rapidly in recent years, and several valuable summary volumes are now available. Chapman's classic *Salt*

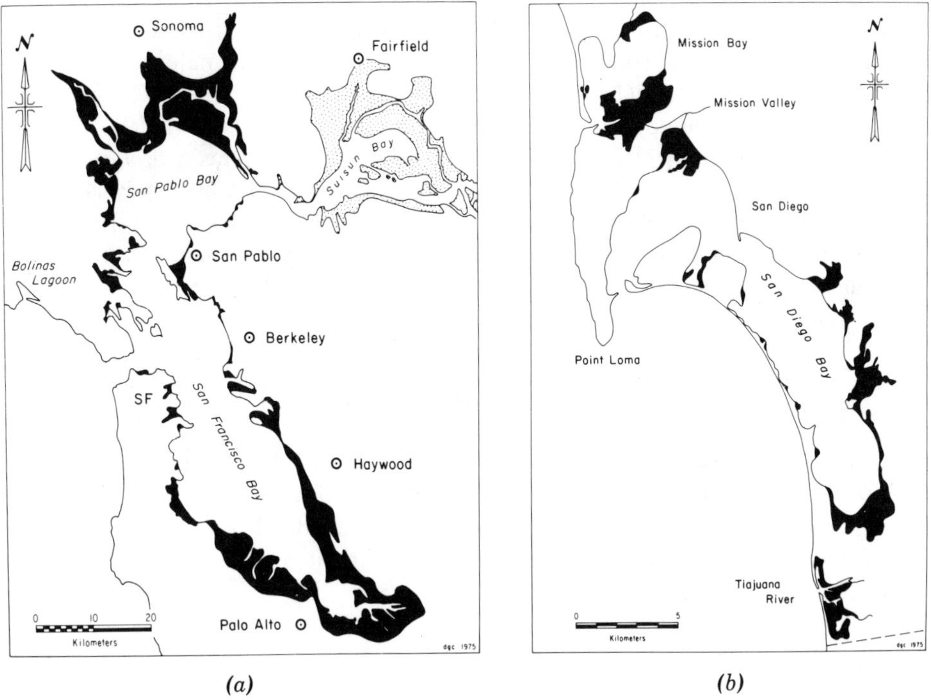

Figure 8-2. Salt marsh distribution (*a*) in San Francisco Bay around 1850 and (*b*) in San Diego Bay around 1890. Coastal salt marsh in black, brackish marshes stippled. After U.S. Coast and Geological Survey maps.

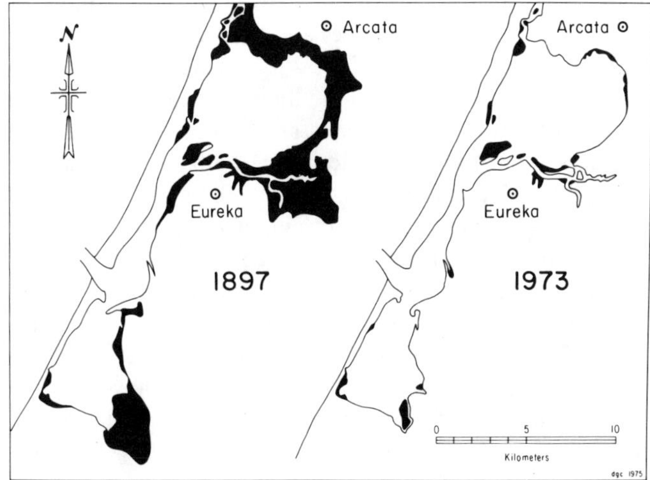

Figure 8-3. Salt marsh distribution in Humboldt Bay. After U.S. Coast and Geological Survey maps.

marshes and salt deserts of the world (1960, 1974) has been revised, and he has an additional volume, *Wet coastal formations,* in press. Other major publications include those of Strogonov (1964), Ranwell (1972), Waisel (1972), and Reimold and Queen (1974). Valuable additional perspectives are provided by Lauff (1967), Green (1968), Castañares and Phleger (1969), the Teals (1969), and the *National estuary study* (1970). Gosselink et al. (1974) have presented a stimulating analysis of the monetary values of natural tidal marshes. About a dozen reports, in the California Dept. of Fish and Game-Coastal Wetlands Series, have been produced for lagoons and estuaries along the coast. Not all of these have been cited in the text.

ENVIRONMENTAL VARIABLES

The distinctive environmental characteristics of coastal salt marshes result principally from four interacting groups of physical variables: (1) sedimentation rates, drainage patterns, and the progressive development of marsh soils; (2) topography and tidal flooding; (3) salinity relations; and 4) nutrient relations. The major roles of these variables are summarized below.

Sedimentation, Drainage, and Soils

Salt marsh extension begins with the first establishment of intertidal vascular halophytes. Plant growth baffles wave and current action, accelerating deposition, while continued root growth and the addition of dead plant material increasingly stabilize and modify the newly deposited sediments. Pestrong (1972) found that flood tides carried offshore sediment onto the upper tidal flats where it was joined by sediment carried out of the salt marsh by ebbing tides. Increases in marsh area and elevation were reflected in growth of the marsh drainage system, which in turn led to accelerated ebb-tide deposition. Large marshes thus expand more rapidly than small ones. Equilibrium is finally reached when marsh expansion reduces the tidal prism (volume of water moved by tidal influence) and thus decreases flood-tide sedimentation. Although ebb-tide deposition continues, it is increasingly offset by tidal scour as the expanding marshes encroach upon the main bay channels.

California salt marsh sedimentation rates are becoming increasingly well known. Stevenson and Emery (1958) and Warme (1971) both cited rates of 1 cm yr^{-1} for tideflats, 0.5 cm yr^{-1} for *Spartina* marsh, and 0.1–0.3 cm yr^{-1} for higher *Salicornia virginica* marsh. Zedler (1975a) cited a high marsh value of 1 cm yr^{-1}, and Pestrong (1965) noted an average of 2 cm yr^{-1} for San Francisco Bay Area marshes, the higher figure probably reflecting their larger size and consequent more rapid growth. Studies using dates from C^{14} analyses and first occurrences of introduced *Eucalyptus* pollen (Mudie 1975, 1976) indicate that American settlement of California accelerated sedimentation rates: values of 0.2–0.3 cm yr^{-1} have prevailed for the last century, whereas depositional rates of <0.1 cm yr^{-1} were more typical earlier.

Several workers have described California salt marsh soils. Pestrong (1965, 1969, 1972) found sediments in Bay Area marshes to be very uniform, suggesting a common depositional history; 65% of the sediment is $<4\ \mu$ in diameter, and of that fraction, the ratio of the clay minerals—montmorillonite, kaolinite, and illite—is

47:31:22. He also confirmed the general observation (Bouma 1963) that salt marshes contain the finest-grained and least well-sorted sediments of any intertidal environment. Soil moisture decreases with increasing elevation (from 130 to 100% of dry weight sediment), while soil organic content increases (13% in *Spartina* soils to 16.4% in *Salicornia* soils). The binding effects of *in situ* root systems progressively increase the shear strength and firmness of marsh soils at higher elevations.

Further south, Stevenson and Emery (1958), Mudie (1970b), and Zedler (1975a) all noted local trends of increasing grain size at higher marsh elevations. Although general trends of decreasing soil moisture and increasing shear strength were confirmed at all sites, soil organic matter patterns were more variable. Warme (1971) noted that relatively little of the abundant high marsh plant litter became incorporated into marsh soils; and, in contrast to Pestrong, Mudie (1970b) and Zedler (1975a) noted decreasing organic content at higher elevations in salt marshes at San Diego and Tijuana (16% in *Spartina* soils to 4% in *Salicornia* soils).

Marsh Topography and Tidal Flooding

Tidal effects dominate the physical environment of coastal salt marshes. Tidal scour can disrupt the substrate, increase turbidity, and uproot seedlings. Tidal submergence may rapidly change ambient temperatures, reduce oxygen and carbon dioxide availability, and decrease incident light, thus reducing both the photosynthetic and the respiration rates of the marsh plants. Tidal flushing also clearly influences soil–water salinities and controls dissolved nutrient and trace element exchange between the marsh plants, substrate, and water. Since the frequency and duration of periods of tidal flooding change with elevation across the intertidal zone, it follows that other tide-related environmental variables also exhibit varying gradients.

Chapman (1960) noted that at many widely separated sites significant differences in tidal flooding regimes could be used to distinguish distinctive *upper* and *lower* salt marsh environments. Ranwell (1972) described similar distinctions between *emergence* and *submergence* marshes and noted that, although the demarcation between them (~ MHW) was difficult to define, it was nonetheless real. Hinde (1954) was unable to distinguish upper and lower marsh environments in south San Francisco Bay. He used continuous tidal records to determine the average *total* submergence and emergence times at 3 cm elevational increments across Palo Alto salt marsh. A recent recalculation of Hinde's data by Zedler (1975a), plotting total hours submerged, as a percentage of the year, versus elevation, yielded a nearly perfect linear relationship [84% at 1.5 m above mean lower low water (MLLW) to 45% at 2.5 m]. Like Hinde, Zedler concluded that the environment "appears to change relatively uniformly with elevation." Both analyses largely miss the point, however, for it is more likely that the marsh plants respond to changes in the maximum duration of *continuous* submersion or exposure (Doty 1946; Macdonald 1969) than to average annual *total* submergence times.

The California coast experiences mixed tides with a pronounced daily inequality. Data collected within Mission Bay marsh with a continuous tide recorder (Bradshaw 1968b; Macdonald 1969) demonstrated the effect of increasing elevation on the maximum duration of both continuous submersion and exposure (Fig. 8-4). A marked reduction in the duration of continuous submersion occurs around mean

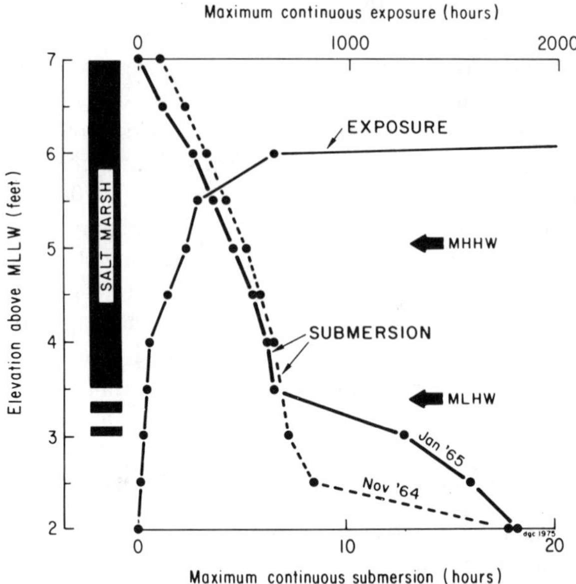

Figure 8-4. Mission Bay Marsh, San Diego: duration of maximum tidal submersion and exposure at successive elevations (Macdonald 1969).

lower high water (MLHW), while the maximum duration of continuous exposure increases significantly above mean higher high water (MHHW). These tidal flooding patterns are not unique to Mission Bay; rather they are characteristic of most Pacific coast marshes (Doty 1946; Macdonald 1967; Jefferson 1974; Eilers 1975). The lower limit of halophytes approximately corresponds with the level at which the intertidal zone is exposed once every 24 hr (MLHW). Two major salt marsh environments can be distinguished above MLHW. The first extends from the lower limit of salt marsh vegetation to ~ MHHW (low marsh); the second includes the zone between MHHW and EHW levels (high marsh). In the former, tidal submergences frequently last more than 6 hr and are never separated by more than 15 days of continuous exposure. In the latter, short tidal submergences (< 6 hr) on several consecutive days are followed by several weeks or even months of continuous exposure. The differences between these zones parallel those recognized by Chapman (1960) and Ranwell (1972) and support their view that such a division is both fundamental and widespread.

At any particular site the upper and lower limits of salt marsh development, relative to MLLW level, are functions of the maximum tidal range. Within this predetermined vertical range the relative extent of high or low marsh environments is controlled by the topography of the marsh surface. Only rarely are both environments equally represented, and a variety of local factors apparently control the dominance of high or low marsh habitats. High sedimentation rates and rapid expansion favor low marsh environments, whereas high marshes are more typical of stable settings with minimal tideflat sedimentation and expansion, or even some landward erosion (Marshall 1948; Hinde 1954; Macdonald 1969; Mahall and Park 1976a).

Salinity Relations

The principal contributions concerning California salt marsh soil salinity regimes are those of Mall (1969), Mudie (1970b), and Mahall and Park (1976b). Additional useful data have been presented by Purer (1942), Phleger and Bradshaw (1966), Bradshaw (1968b), Carpelan (1969), Lane (1969), Barbour et al. (1973), Rollins (1973), Jefferson (1974), and Zedler (1975a). Despite real differences in site characteristics and investigative techniques, apparent conflicts between these various studies can be satisfactorily resolved to fit a single soil salinity model. The three basic elements of this model are as follows:

1. Soil salinities are largely determined by the frequency and duration of periods of tidal flooding (which both introduce and remove salts), evaporation, and leaching of salts by rainfall and freshwater runoff. The net effect of these factors is that soil salinities increase landward to a maximum around MHW (i.e., the low marsh–high marsh ecotone) and then gradually decline (Chapman 1960; Mudie 1970b; Ranwell 1972; Mahall and Park 1976b).
2. Subsurface (>1 m) marsh sediments provide a reservoir for salts that exhibit a cyclic rise and fall within the soil profile in response to seasonal changes in soil moisture (Mall 1969; Rollins 1973). Salinity concentrations increase downward during the wet season but decrease downward in the dry season. A deep-rooted marsh species such as *Spartina foliosa* will thus experience lower root salinities during the spring and summer growing season than a shallow-rooted species such as *Salicornia virginica* (Mahall and Park 1976b).
3. The concurrence between regional climate trends of rising air temperatures and declining precipitation increases net evaporation at lower latitudes. From about San Francisco south, annual evaporation increasingly exceeds precipitation; high marsh soil moistures decline, surface soil salinities increase, and highly saline salt pan habitats with minimal plant cover become more common (Purer 1942; Macdonald 1967; Bradshaw 1968a; Mudie 1970b; cf. Clarke and Hannon 1969–71). Soil salinities fall sharply at the landward limit of maximum tidal flooding or saltwater subirrigation, even though freshwater sources may be minimal (Mudie 1970b; Neuenschwander 1972). North from San Francisco, increasing net precipitation and runoff produce a steady soil salinity decline at higher elevations and a more gradual transition to upland environments (Barbour et al. 1973; Jefferson 1974).

Local deviations from this model result from differences in proximity to tidal creeks or freshwater sources, microtopography and soil aeration, sediment grain size, and organic content. More significant deviations described by Barbour et al. (1973) and Zedler (1975a) can also be satisfactorily explained. Barbour et al. noted a steady landward decline in soil salinities across a narrow fringing marsh at Bodega Bay. Sandy sediments and considerable freshwater seepage both contributed to the lower salinities. More significantly, however, the vegetation suggests that only high marsh environments are represented; thus the model satisfactorily predicts the salinity decline inland. Zedler (1975a), at Tijuana estuary, also found evidence of a steady decline in near-surface (<10 cm) soil paste conductivities from low to high elevations. Her data do not extend much below the *Spartina–Salicornia* ecotone (Zedler, pers. commun.) and thus again satisfactorily fit the model. Regrettably, Zedler cites no soil moisture data; thus the relationship between the soil paste conductivities and *in situ* soil salinities remains unclear.

Nutrient Relations

Several workers have examined the abundance, availability, and ecological significance of both macro- and micronutrients in salt marsh environments (Chapman 1960; Adams 1963; Jefferies 1972; Ranwell 1972). Physiological adaptations of marsh plants to changing nutrient levels have been documented (Stewart et al. 1972, 1973; Chapin 1974), and nutrient enrichment studies have been conducted (Pigott 1969; Valiela and Teal 1974; Newcombe and Pride 1975). It appears that the availability of certain nutrients, particularly nitrogen, may be limiting in some high marsh environments and that some species are better adapted to handle this deficiency than others.

Preliminary nutrient concentration data from California estuaries and lagoons are only now becoming available (Mudie 1970b; U.S. Geological Survey 1970; Bradshaw and Mudie 1972; Macdonald, unpub. data), and California salt marsh studies along these lines await future research. The possible role of marsh clay cation exchange capabilities in influencing nutrient supply (Ranwell 1972) also awaits investigation (cf. Mall 1969; Newcombe and Pride 1975).

FLORISTICS AND VEGETATION

Although both the floristics and the vegetation of California coastal marshes are reasonably well known, the published data are very scattered. The accounts outlined below are based on the references cited, my own field studies (Macdonald 1967, 1969, 1971a,b,c), and two recently prepared literature reviews (Macdonald and Barbour 1974; Macdonald 1976).

Floristics

Nearly 70% of the plants that characterize Pacific coast (23–71°N) salt marshes extend into California (Macdonald and Barbour 1974). The latitudinal ranges of the more frequently encountered and abundant of these are shown in Fig. 8-5. Elsewhere (Macdonald 1976) I have used species range data as a basis for dividing Pacific coast salt marshes into a number of distinct floristic groups. Figure 8-5 includes a portion (32–50°N) of the Q-mode dendrogram, developed from Jaccard coefficients based on 1°N sampling of range data for 88 species, produced during this analysis. Three groups of California coastal marshes can be distinguished. North from Pt. Arena (39°N) the marshes can contain increasing numbers of northerly distributed species, and their potential flora is most closely allied with that of marshes in Oregon and Washington. Central California marshes, including those of the San Francisco Bay and Monterey regions, and marshes from Morro Bay (~35°N) southward represent two district but related groups. Marsh floras from Washington, Oregon, and California all appear more closely similar to the arid Mediterranean

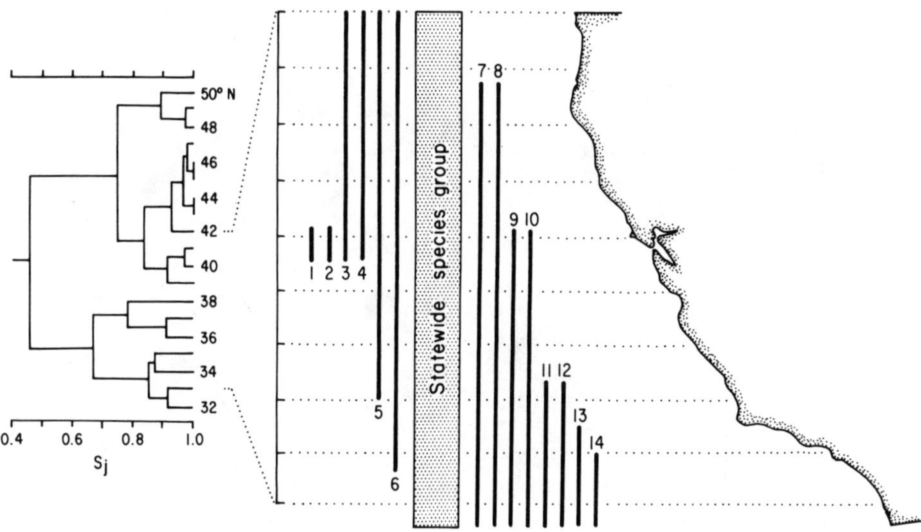

Figure 8-5. California salt marsh floristics. *Left: Q*-mode dendrogram summarizing similarity (Jaccard coefficients) among theoretical marsh sites spaced at 1° intervals along the coast (Macdonald 1976). *Right:* Latitudinal species distributions, modified after Munz (1973). 1. *Grindelia humilis.* 2. *G. paludosa.* 3. *G. stricta.* 4. *Plantago maritima.* 5. *Glaux maritima.* 6. *Potentilla pacifica.* 7. *Limonium californicum.* 8. *Spartina foliosa.* 9. *Frankenia grandifolia.* 10. *Suaeda californica.* 11. *Salicornia subterminalis.* 12. *Batis maritima.* 13. *Monanthochloë littoralis.* 14. *Salicornia bigelovii.* Statewide species group includes *Atriplex patula, Cordylanthus maritimus, Cotula coronopifolia, Cuscuta salina, Distichlis spicata, Jaumea carnosa, Myosurus minimus, Salicornia europaea, S. virginica, Spergularia macrotheca, S. marina, Trifolium wormskjoldii, Triglochin concinna, T. maritima,* and several species of *Scirpus.*

marsh floras of Baja California than to those of cooler, wetter Canadian and Alaskan sites.

Vegetation

The following sections outline the species composition, general appearance, and zonation of the vegetation from the three groups of California salt marshes distinguished above.

San Francisco Bay Area Marshes. Cameron (1972) and Mahall and Park (1976a) described the actively accreting, rapidly expanding salt marshes of north San Pablo Bay (Fig. 8-2). *Spartina foliosa* is the primary colonist on broad tidal mudflats, and virtually pure stands of *Spartina,* a meter or more tall, dominate low marsh environments (Fig. 8-6). A dense bushy cover of *Salicornia virginica* replaces the *Spartina* at about MHW and dominates the high marsh. The *Spartina–Salicornia* ecotone is abrupt, an elevational rise of as little as 7 cm being sufficient for pure stands of *Spartina* to be completely replaced by pure stands of *Salicornia. Distichlis spicata* and *Frankenia grandifolia* are widespread associates within the *Salicornia* zone. At undisturbed sites the salt marshes are succeeded by grassland, brackish marsh, or scrub communities (Marshall 1948; Howell 1937, 1970).

A similar zonation pattern is characteristic of most other San Francisco Bay Area

Figure 8-6. General aspect of *Spartina foliosa* low marsh (high tide, Mission Bay).

marshes, although both the relative extent of the *Spartina* and *Salicornia* communities and the abruptness of their ecotone vary considerably between sites (Marshall 1948; Hinde 1954; Atwater and Hedel 1976). Although *Spartina* and *Salicornia* usually predominate, additional species are also represented. *Distichlis spicata* becomes increasingly common at higher marsh elevations and remains a prominent member of the succeeding maritime grassland belt (Filice 1954; Lane 1969; Howell 1970). *Grindelia* bushes colonize the raised natural levees of high marsh tidal creek banks (Marshall 1948; Hinde 1954), and other high marsh species include *Frankenia grandifolia, Jaumea carnosa, Limonium californicum, Salicornia europaea, Triglochin concinna, T. maritima,* and *Cuscuta salina,* a common parasite of several high marsh species. None of these species occurs in abundance, however. *Suaeda californica* is present but rare in the bay marshes (Munz 1973), gaining prominence further south. An increasing variety of plants occur toward the upper edges of the salt marshes (*Atriplex patula* ssp. *hastata, Cordylanthus* spp., *Cressa truxillensis, Glaux maritima, Lasthenia glabrata, Myosurus minimus, Plantago maritima* ssp. *juncoides, Potentilla pacifica*), and opportunistic introduced species are common in disturbed areas (*Atriplex semibaccata, Chenopodium ambrosioides, Cotula coronopifolia, Monerma cylindrica, Parapholis incurva, Spergularia marina*).

BRACKISH MARSHES. These are also better developed within the San Francisco Bay complex than elsewhere along the California coast. Marshall (1948) provided an excellent general account of their vegetation. *Scirpus acutus* and *Typha latifolia* predominate; associated with them are *Scirpus olneyi, S. robustus,* and *S. californicus.* The vegetation is homogeneous on gentle slopes but becomes more zonal on steeper channel banks. Pure stands of *S. acutus* grow in deep water and are followed by a fairly continuous band of *Typha* higher up the bank. *Typha* is in turn succeeded by alternating monospecific stands of *S. olneyi* and *S. robustus.* Brackish marshes of this type were once the prevailing natural habitat of Suisun Bay (Fig. 8-2; Mall 1969; Gill and Buckman 1974); more recently they have been modified by diking to provide waterfowl habitat.

Marshall (1948) noted a marked difference between the salt marsh–freshwater transition in San Francisco Bay and that in northern San Pablo Bay. The large streams that enter northern San Pablo Bay form brackish marshes at their mouths, and the transition from riparian to salt vegetation is quite gradual. The smaller streams entering San Francisco Bay contain no comparable brackish water areas. Growing low on the banks, saltwater vegetation extends upstream as far as the tides, while freshwater vegetation grows above and beyond the stream banks and thins out toward the marshes.

Northern California coastal marshes. Humboldt Bay is the principal site of coastal salt marshes in northern California (Fig. 8-3; Monroe 1973; Macdonald 1967; Macdonald and Barbour 1974). *Spartina foliosa,* now at its northern limit, remains the usual primary colonist of the tideflats but often shares dominance of the low marsh with *Salicornia virginica. Salicornia* may become a primary invader in sheltered situations, and at Arcata, well within the bay, Macdonald (1967) recorded a reversal of the usual succession. Here firm mudflats were being colonized by circular patches of *Salicornia* up to a meter or more in diameter. The lower marsh, apparently formed through coalescence of such stands, was composed almost entirely of this species. *Spartina foliosa* appeared at intermediate elevations. Instead of occurring in its usual form, it grew in compact, closely spaced tussocks (<1 m tall), occupying irregular areas of the middle marsh and completely surrounded by *Salicornia.*

Although *Salicornia* remains abundant in the high marsh, it often occurs as a codominant with *Distichlis spicata* and *Jaumea carnosa.* Additional, less common species include *Cordylanthus martimus, Cuscuta salina, Limonium californicum* (now at its northern limit), *Myosurus minimus,* and *Triglochin maritima. Atriplex patula, Grindelia stricta, Juncus lesueurii,* and *Potentilla pacifica* occur within the *Distichlis*-dominated salt marsh–grassland ecotone. *Frankenia grandifolia* and *Suaeda californica* are both absent, for neither species extends north of San Francisco Bay (Fig. 8-5).

Narrow fringes of coastal marsh are developed along the estuaries of the Eel, Klamath, and Smith rivers (CDFG 1975; Monroe et al. 1975); however, freshwater influences predominate. Several brackish coastal lagoons—Lakes Earl and Talawa, Big Lagoon, and Stone Lagoon—cut off from ocean circulation by barrier beaches, also contain stands of salt marsh species, predominantly *Distichlis spicata* and *Salicornia virginica.*

Brackish marshes at these sites are similar in appearance to those of Suisun Marsh, but have a rather different species composition. *Scirpus americanus, S. acutus,* and *Typha latifolia,* respectively, are the dominant species in progressively less saline habitats. Additional species include *Carex lyngbyei, C. obnupta, Distichlis spicata, Juncus balticus, J. effusus, Potentilla* spp., and *Scirpus robustus* (Hehnke 1969; Monroe et al. 1975). Several of these species become increasingly important in the coastal salt marsh vegetation of Oregon and Washington (Jefferson 1974; Macdonald and Barbour 1974; Eilers 1975). Salt marshes at Bodega Harbor and Bolinas have been described by Barbour et al. (1973), Standing (1976), and Giguere (1970).

Southern California coastal marshes. Salt marshes are scarce along the 400 km stretch of coast between San Francisco and Pt. Conception ($\sim 34.5°$N), and *Spartina foliosa* is absent until Mugu Lagoon ($\sim 34°$N). *Salicornia virginica* is the dominant species at all of these sites (Fig. 8-7). *Atriplex patula, Cuscuta salina, Distichlis spicata, Frankenia grandifolia, Jaumea carnosa, Limonium californicum,* and *Suaeda californica* are widespread, but rarely abundant. *Batis maritima, Salicornia subterminalis,* and *Monanthochloë littoralis* all occur for the first time (Fig. 8-5) and represent significant additions to the high marsh flora (Smith 1952; Hoover 1970; Browning 1972; Gerdes et al. 1974; Macdonald and Barbour 1974; Mudie 1976).

Two distinct groups of estuaries and coastal lagoons can be recognized in southern California. The first of these includes several extensive, relatively deeper-water sites that have sufficiently large tidal prisms to maintain permanent ocean inlets and continuous tidal action. Mugu Lagoon (Macdonald 1967; Warme 1971), Anaheim Bay, Newport Bay (Stevenson and Emery 1958; Vogl 1966), Mission Bay (Macdonald 1967), San Diego Bay (Mudie 1970b), and the Tijuana River estuary (McIlwee 1970; Zedler 1975a,b) belong to this group. These sites all share closely similar salt marsh floras and consistently exhibit a more complex zonation pattern than that described for marshes in the San Francisco Bay region.

Spartina foliosa remains the usual primary tideflat colonist and dominates low marsh environments (Fig. 8-6). Pure stands of *Spartina* at lower elevations grade into *Spartina-Salicornia* mixtures higher up, where the upper limit of *Spartina foliosa* overlaps the lower limit of *Salicornia virginica.* Open areas within the *Spartina* low marsh zone are often colonized by vigorous individuals of *Salicornia bigelovii,* an annual. *Batis maritima* also appears in this zone but is rarely abundant.

In the San Francisco Bay region, *Spartina*-dominated low marsh is replaced above MHW by a dense *Salicornia virginica*-dominated high marsh assemblage. In southern California however, the *Spartina* assemblage is succeeded by a low ground cover 15–40 cm tall, dominated by *Salicornia bigelovii* and *Batis maritima* (Fig. 8-8). *Batis* becomes more prominent in winter when the *S. bigelovii* dies down. Limited to a narrow zone around MHHW, this is the most vertically restricted of

Figure 8-7. General aspect of *Salicornia virginica* stands (low tide, Carpinteria Marsh).

Figure 8-8. General aspect of *Batis maritima–Salicornia bigelovii* assemblage (low tide, Mugu Lagoon).

the salt marsh vegetation types. Where steeper marsh topography minimizes its extent, stunted specimens of *Spartina* and *Salicornia virginica* persist within the assemblage. On gentle slopes where the zone is best developed, *Spartina* disappears completely and *S. virginica,* when present, appears as a straggling, prostrate ground cover rather than the usual upright bushes. Occasional short specimens of *Frankenia grandifolia* and *Triglochin maritima* appear scattered throughout the zone.

The *Batis maritima–Salicornia bigelovii* assemblage is succeeded by a dense *Salicornia virginica*-dominated high marsh assemblage, similar to that developed at San Francisco Bay (Fig. 8-9). Plant cover is taller (30-50 cm) and more dense (70-

Figure 8-9. General aspect of *Salicornia virginica* high marsh assemblage (low tide, Mugu Lagoon).

100%) than in the middle marsh. Although pure stands of *Salicornia virginica* are not uncommon, the high marsh usually consists of a complex mosaic of *Salicornia* mixed with several other species. *Limonium californicum* and *Suaeda californica* are both conspicuous and widespread species. *Distichlis spicata, Monanthochloë littoralis,* and *Salicornia subterminalis* become increasingly prominent at higher elevations, particularly on sandier soils, and all can form dense clonal stands several meters in diameter. Other common species include *Batis maritima, Cuscuta salina, Frankenia grandifolia, Jaumea carnosa,* and *Triglochin maritima.* Stunted specimens of *Salicornia bigelovii* and *Spartina foliosa* occasionally occur in the lower parts of the high marsh.

At Mugu Lagoon, the largest remaining relatively undisturbed salt marsh in southern California, both plant cover and species diversity decline sharply at higher elevations as the diverse high marsh assemblage gives way to barren salt flats (cf. playas). Sinuous stands of *Salicornia subterminalis* and *Monanthochloë littoralis* separate the flats into individual salt pans, some of which are rimmed by extensive mats of *Batis maritima.* In earlier times the salt flats apparently gave way to maritime grassland, coastal sage scrub, and chaparral—both plant cover and species diversity rapidly rising again beyond the salt marsh–upland ecotone. At other locations where less extensive salt marshes abut steep bluffs or dunes, salt flats are absent and the transition from high marsh to upland vegetation is more abrupt (Vogl 1966; Macdonald 1967; Mudie 1970b; Zedler 1975a).

The second major group of southern California coastal marshes includes most of the remaining sites indicated in Fig. 8-1. These marshes are developed in relatively smaller, shallower estuaries and coastal lagoons whose diminished tidal prisms are inadequate to prevent periodic closure (sometimes seasonal, sometimes long-term) of their ocean inlets. Environmental conditions at these sites are both more variable and more extreme. When under tidal influence, conditions approximate those of the larger lagoons; when closed, however, local freshwater runoff patterns become the dominant influence on changing water levels and salinity regimes (Purer 1942; Carpelan 1969; Macdonald 1971c; Bradshaw and Mudie 1972; Mudie 1976).

Studies of the marsh floras developed at several of these sites (Purer 1942; Bradshaw 1968; Mudie 1970a, 1976; Macdonald et al. 1971) reveal significant differences from the more typical succession outlined above. Three characteristics stand out. First, the dominant low and middle marsh species of other southern California sites (*Spartina foliosa, Batis maritima, Salicornia bigelovii*) are invariably absent. Second, most of the characteristic high marsh species are well represented. Third, large numbers of opportunistic weedy species are usually present. Los Peñasquitos Lagoon, for example (Bradshaw 1968; Mudie et al. 1974; Mudie 1976), contained three salt marsh associations. The lowest levels of the marsh, subject to seasonal daily tidal flooding, were occupied by *Salicornia virginica* pure stands succeeded by *S. virginica–Frankenia grandifolia* mixtures. Most of the remaining intertidal marsh comprised a variable assemblage dominated by *Distichlis spicata. Salicornia virginica* and *Jaumea carnosa* commonly co-occurred with *Distichlis* in clayey soils but were replaced by *Mesembryanthemum chilense* in sandy areas. *Cressa truxillensis, Limonium californicum,* and *Monanthochloë littoralis* were also present. The bluffs around the marsh and an extensive area of salt flats beyond the *Distichlis* association were dominated by *Salicornia subterminalis* and *Suaeda californica.* A variety of small annuals (*Cotula coronopifolia, Lasthenia glabrata,*

Medicago hispida, Melilotus indicus, Mesembryanthemum nodiflorum, Parafolis incurva, Spergularia marina) were recorded from shallow depressions in the salt flats that partially filled with rain each spring. Although a total flora of 106 species was recorded from Los Peñasquitas, several common marsh species—*Atriplex watsonii, Batis maritima, Salicornia bigelovii, Spartina foliosa,* and *Triglochin maritima*—were absent.

Santa Margarita (Fig. 8-1; Mudie 1970a) salt marsh flora is still more depauperate than that of Los Peñasquitas. In addition to the low and middle marsh dominants, *Limonium, Monanthochloë*, and *Suaeda* are also absent.

Because of the low regional rainfall and largely intermittent river runoff, extensive brackish coastal marshes are not developed in southern California. Mudie (1970b) noted that the dominant brackish marsh species in south San Diego Bay were *Scirpus californicus* and *Typha latifolia,* with *Cotula coronopifolia, Polypogon monspeliensis,* and *Rumex crispus* forming a low undercover. Mudie (1970a, 1976) also described brackish marsh development at the mouths of the Santa Margarita, Las Flores, and San Mateo rivers in northern San Diego Co. A *Jaumea*-rush association occupied the salt marsh–freshwater marsh transition. This association was dominated by *Jaumea carnosa* and *Distichlis spicata,* with *Juncus acutus* var. *sphaerocarpus, J. mexicanus,* or *Anemopsis californica* becoming locally abundant. *Typha* (*T. domingensis* and *T. latifolia*)-*Scirpus* (chiefly *S. californicus* and *S. olneyi*) associations dominated less brackish habitats and graded upstream into either willow thickets (*Salix hindsiana, S. lasiolepis*), riparian woodland, or upland brush communities.

AUTECOLOGY

Despite research advantages of a limited flora and pronounced environmental gradients, the ecological strategies of most California halophytes remain poorly known. Indeed, only very recently have carefully coordinated field and laboratory experiments (Mahall and Park 1976; Zedler 1975b) begun to seek the causes behind the field correlations and distributional generalizations noted by earlier workers. Autecological data on several better-known species are summarized below. Additional data appear in several of the review volumes noted earlier in this chapter, and in papers by Purer (1942), Binet (1961), Webb (1966), Barbour (1970), Woodell and Mooney (1970), and Mudie (1976).

Spartina foliosa

Spartina foliosa is a tall (> 1 m) perennial grass (Fig. 8-6) that spreads principally by deep-buried (> 15 cm) rhizomes. It tolerates longer periods of tidal submergence than other salt marsh species and typically dominates low marsh environments (below MHHW). Parnell (1975) was unable to find evidence of different polyploid races ($2n = 60$) despite pronounced variations in culm height. *Spartina* is winter dormant, the stems dying back each year; it grows actively from March through October and flowers from August through October (Purer 1942; Cameron 1972; Mahall and Park 1976a,b,c).

The anatomy of *S. foliosa* has been described by Purer (1942), Mobberley (1956), Levering and Thompson (1971), and Kasapligil (1974). The xerophytic rolled leaves possess well-developed salt glands. Aerenchyma tissue is prominent in leaves, stems, and rhizomes, and Wong (1975) confirmed oxygen transport from leaves to roots. Vegetative reproduction predominates; Mahall and Park (1976a) described mats of seedlings from San Pablo Bay, and prolonged freshwater flooding may stimulate germination (Newcombe and Pride 1975).

Net productivity measurements from stands in San Diego (Mudie 1970b) yeilded an average of 0.8 kg dry wt m^{-2}/yr^{-1}. Cameron (1972) cited much higher values of 1.4–1.7 kg from San Pablo Bay stands, while Mahall and Park (1976a) obtained lower values, 0.3–0.7 kg dry wt m^{-2}/yr^{-1}, from marshes only a few kilometers away. Mahall and Park found that root–rhizome biomass exceeded (1.3–5.5×) shoot biomass and that both figures declined inland across the *Spartina–Salicornia* ecotone in response to increasing soil salinity concentrations. *Spartina* rhizomes penetrated further into the ecotone than above-ground shoots and could produce new shoots if heavy rains reduced salinity levels. Culture experiments confirmed the negative influence of salinity on *Spartina* growth (Phleger 1971; Mahall and Park 1976b). The following conclusions appear warranted:

1. Aggressive vegetative reproduction, under conditions unsuitable for seedling development, allows *Spartina foliosa* to penetrate and stabilize lower intertidal habitats than do other marsh halophytes. An efficient oxygen transport system minimizes possible waterlogging problems.
2. The lower limit of its vertical distribution is largely determined by its inability to withstand longer periods of tidal submergence. The fact that *S. foliosa* extends to lower elevations at more sheltered salt marsh sites (Macdonald 1967) indicates that submergence alone is not the limiting factor (cf. Ranwell 1972). Growth inhibition through direct mechanical effects or the leaching of substances important to plant growth may play a minor role (Mahall and Park 1976c), and inhibition of photosynthesis through increased water turbidity may be significant (Hubbard 1969).
3. The upper limit of *S. foliosa* is marked by growth inhibition and stunting that apparently reflects intolerance of increasing soil salinities (Mahall and Park 1976b), rather than competition or nutrient unavailability (cf. Adams 1963; Seneca 1974; Newcombe and Pride 1975).

Salicornia virginica

Salicornia virginica is a shallow-rooted (< 15 cm) suffrutescent, herbaceous perennial with succulent joined stems and obscure, scalelike leaves (Fig. 8-7). Exhibiting several growth forms and occurring in low marsh through maritime bluff habitats, it is by far the most widespread and abundant California salt marsh halophyte. *Salicornia* breaks seasonal dormancy in March, flowers from June through September, and sets seed (wind pollinated) by October (Purer 1942; Cameron 1972). Reproductive strategies include seedlings, underground rhizomes, and rooting of decumbent branches. Although salt glands are lacking, aerenchyma tissue is present in the stems, and large intercellular spaces develop in the roots (Purer 1942; Anderson, 1974). Mahall and Park (1976c) confirmed that oxygen moves down into the roots; they also presented data discounting soil aeration as a limiting factor in *Salicornia–Spartina* distribution patterns.

Annual net production figures for *S. virginica* include 0.3–0.7 kg dry wt m^{-2}/yr^{-1} in Suisun Marsh (Mall 1969), 1.0–1.2 (Cameron 1972) and 0.03–1.0 kg (Mahall and Park 1976a) in San Pablo Bay, and 1.5–2.5 kg in San Diego (Mudie 1970b). Unlike *Spartina*, Mahall and Park (1976a) noted a close correspondence between above- and below-ground biomass distribution; root–rhizome biomass was 0.3–0.8× above-ground biomass. Although the total abundance of *S. virginica* declined seaward into the *Spartina* zone, the individual plants exhibited better growth (i.e., higher annual production per stem) than others at higher elevations. In San Francisco Bay Area marshes, *Salicornia* biomass rises sharply above the *Spartina–Salicornia* ecotone, but in southern California both its frequency and its cover decline in the *Batis maritima–Salicornia bigelovii*-dominated middle marsh. Whereas it occurs as erect bushes at both lower and higher elevations, it is reduced to a prostrate matlike cover in the middle marsh (Macdonald 1967; Mudie 1970b). Pielou (1974) has theorized that competition could produce such a biomodal abundance distribution, and Zedler (1975a) cited quantitative data supporting this hypothesis.

Salicornia species are highly salt tolerant (Chapman 1960; Barbour 1970), and *S. virginica* is significantly more tolerant of both higher and more variable soil salinities than *Spartina foliosa* (Queen 1974; Mahall and Park 1976b). Frequent tidal submergence inhibits seedling lodgement, germination, and early growth (Stevenson and Emery 1958), and growth of established plants is significantly (30–40%) reduced by water passing across shoot surfaces (Mahall and Park 1976b). The factors controlling *Salicornia's* landward limit remain unresolved. Competition (Zedler 1975a), disturbance by man and animals (Hinde 1954; Daiber 1974; Shanholtzer 1974), and a variety of physical factors—decreasing soil moisture (cf. *S. rubra*: Ungar 1974), changing salinities, or nutrient concentrations (cf. Adams 1963; Valiela and Teal 1974)—could all play a role. The following conclusions appear justified:

1. The lower limit of *Salicornia virginica* primarily reflects poor seedling establishment and early growth, caused by intolerance of increasingly frequent and prolonged tidal submergence.
2. In southern California, changes in *Salicornia* abundance around MHHW reflect competition with *Batis maritima* and *Salicornia bigelovii.*
3. Landward distributional controls remain unresolved but probably include disturbance, competition, and various physical factors. Competition may predominate in northern California, where soil salinities decline steadily inland. Further south, the high salinities and low soil moistures of salt pan habitats may provide physical controls that override competitive factors.

Batis maritima

Batis maritima is a prostrate succulent, whose long (>1 m) horizontal stems and fleshy leaves form a low, open cover (Fig. 8-8). Intercellular air spaces and root lacunae are well developed (Purer 1942). It is most abundant around MHHW but also co-occurs with *Spartina* in the low marsh and rims many high marsh salt pans. Maximum growth occurs from April through December, considerably later than in *Salicornia bigelovii* (March–June), with which it often co-occurs (Purer 1942; Stevenson and Emery 1958; Warme 1971; Zedler 1975a,b).

Salicornia bigelovii

Salicornia bigelovii is the only widespread California salt marsh annual. Low salinities promote germination, but higher salinities stimulate seedling growth and reduce mortality (Purer 1942; Bradshaw 1968; Ungar 1974). Seedlings are most abundant and mortality is lowest on open ground around MHHW; tidal disturbance and dense plant cover limit seedling success at lower and higher elevations, respectively (Zedler 1975a,b). Rapid March through June growth is followed by July through November flowering, seed release, and death.

It seems particularly significant that *S. bigelovii* and *Batis maritima,* two subtropical floral elements with very different ecological strategies, both flourish around MHW. Further north, the *Spartina–S. virginica* ecotone occupies this zone and low plant cover suggests lack of space utilization. In southern California, *S. bigelovii* and *Batis* have successfully invaded this "unoccupied" space, but their spread northward is presumably checked by temperature requirements.

Distichlis spicata

Distichlis spicata is a low-growing (<30 cm) perennial xerophytic grass that spreads principally by buried (~ 6 cm) rhizomes. Salt glands and intercellular air spaces are present (Purer 1942; Anderson 1974). *Distichlis* tolerates a wide range of soil moisture and salinity levels (Purer 1942; Barbour 1970; Barbour et al. 1973). Scattered plants are common in high marsh habitats. Dense clonal mats, tens of meters in diameter, that exclude all other species are not unusual (Stevenson and Emery 1958; Lane 1969; Warme 1971).

Monanthochloë littoralis

Monanthochloë littoralis, which appears only in southern California, is another low-growing perennial grass much like *Distichlis* (Purer 1942; Stephenson 1971). Winter dormant, with vigorous growth from February through June, *Monanthochloë* is restricted to high marsh habitats, where its dense, rhizomatous, creeping growth habit produces extensive clonal mats (Vogl 1966; Zedler 1975a). Purer (1942) suggested that its lower limit reflected intolerance of tidal submergence, while its upper limit was controlled by competition (cf. Stevenson and Emery 1958).

Salicornia subterminalis

Salicornia (= *Arthrocnemum*) *subterminalis* is a perennial succulent, much like *S. virginica* in appearance, that flourishes on inland alkali flats (San Joaquin Valley) as well as in southern California salt marsh and maritime bluff habitats (Fig. 8-5). Removal of Bolsa Bay *Spartina–Salicornia* marsh (6 km^2; Fig. 8-1) from tidal influence in the 1890s provided a natural experiment in the relative environment tolerances of *S. subterminalis* and *S. virginica* (Dillingham Environmental Co. 1971; Macdonald et al. 1971). *Salicornia virginica* flourished under a variety of soil salinity regimes but declined in abundance and died back earlier in the season at progressively drier sites. *Salicornia subterminalis* co-occurred with *S. virginica* at seasonally moist sites where it continued to flourish as soil moistures declined and *S. virginica* died back. It also flourished with *Mesembryanthemum nodiflorum* in highly saline, very dry playa-like habitats where *S. virginica* was absent (Bradshaw

1968; Warme 1971). The vertical range of *S. subterminalis* may well be limited by competition with other high marsh species (particularly *S. virginica*: Zedler 1975a) at the lower end, and with coastal sage scrub species higher up (Vogl 1966; Bradshaw 1968; Munz 1973; cf. Clarke and Hannon 1967–71).

VERTICAL ZONATION

There is no doubt that vertical zonation is one of the most striking features of worldwide tidal marshes (Chapman 1960; Ranwell 1972). A casual observer would have no difficulty in picking out obvious vegetation "zonation" at many California salt marsh sites, yet a field ecologist examining line intercepts in these same marshes may be hard pressed to separate adjacent vegetational zones (Vogl 1966; Zedler 1975a). What, then, is the real nature of California salt marsh zonation?

As is commonly the case in zonation studies of intertidal rocky-shore algae, salt marsh communities are usually separated and named on the basis of one or two dominant species. Since these species achieve their maximum abundance at different intertidal elevations, salt marsh vegetation is conveniently described in terms of successive vertical "zones." Statewide distributional relationships among these dominant species and vertical zones described earlier are summarized in Fig. 8-10. Although such zonal descriptions *could* suggest a succession of uniform communities separated by abrupt ecotones (cf. Mahall and Park 1976a), most salt marsh workers have been careful to point out that zone dominants may also occur, though in lesser abundance, in different zones. Other, less common species may also be distributed with little regard to zonal boundaries between dominants.

Despite the scarcity of quantitative studies of California salt marsh zonation, the data that are available (Vogl 1966; Macdonald 1967; Mudie 1970a; Barbour et al. 1973; Zedler 1975a) support the qualitative impressions outlined above. Vogl (1966) obtained species frequency and cover data from several subjectively defined vegetation "zones" in Newport Bay marshes (Table 8-1). Although different key species

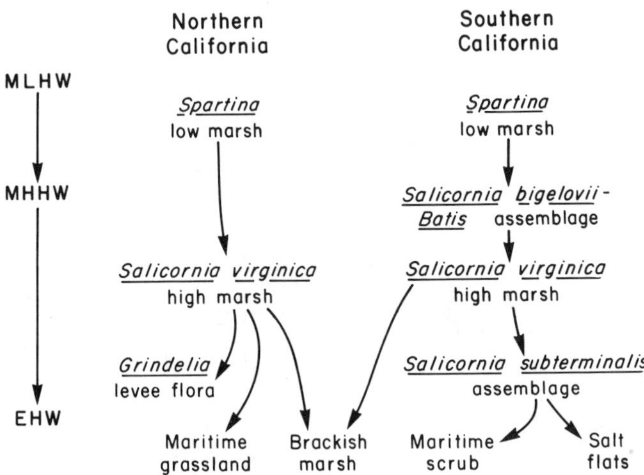

Figure 8-10. Distributional relationships among principal Californian salt marsh plant communities (see text for additional explanation).

TABLE 8-1. Upper Newport Bay: Average percentage species cover, by zone. After Vogl (1966). Each value is an average based on 240 625 cm^2 quadrats in six stands; ranges for six stands are in parentheses. Asterisks indicate presence of additional species

Species	Salt Marsh			Maritime Bluffs
	Lower	Middle	Upper	
Spartina foliosa	38 (4-87)	1 (0-2)	*	
Batis maritima	4 (0-23)	15 (3-42)	1 (1-6)	
Salicornia virginica	4 (1-10)	23 (6-67)	40 (2-65)	11 (0-29)
Suaeda californica	*	*	2 (1-7)	19 (9-25)
Frankenia grandifolia		3 (0-11)	2 (2-12)	*
Distichlis spicata		2 (0-10)	5 (0-29)	
Triglochin maritima		11 (0-42)	1 (0-8)	*
Limonium californicum		1 (0-4)	1 (0-10)	
Monanthochloë littoralis			15 (0-66)	
Cuscuta salina			2 (0-5)	
Juncus acutus			2 (0-14)	
Scirpus californicus			14 (0-90)	
Salicornia subterminalis				24 (0-54)
Mesembryanthemum crystallinum				13 (0-26)
Encelia californica				4 (0-14)

achieved maximum average cover in different zones, most species were recorded over a broad elevational range. Vogl noted that zonal boundaries were indistinct and concluded that floristic composition in fact changed gradually with elevation to form a vegetational continuum. Zedler (1975a) collected similar data from the Tijuana River estuary. Unlike Vogl (1966), who sampled and analyzed his data by zones, she sampled at all elevations and plotted the results within 10 cm elevational classes, thus providing a clearer view of zonal "boundaries." Again, broad species overlaps were common; however, individual species or small groups of species reached peak abundances at different elevations: *Spartina foliosa* dominated the first 40 cm of a 1 m elevational gradient that extended down to about MSL (MHHW ~ 80 cm above MSL). *Spartina* declined abruptly 50–60 cm above MSL, and *Batis maritima* and *Salicornia bigelovii* dominated elevations 50–80 cm above MSL. *Jaumea carnosa* reached its peak frequency and abundance at 70–80 cm, while *Suaeda californica, Frankenia grandifolia,* and *Monanthochloë littoralis* all increased in abundance to dominate elevations 90–100 cm above MSL. *Salicornia virginica* accounted for 15–25% of the plant cover at all but the lowest and highest elevations, where it was replaced by *Spartina* and *Salicornia subterminalis,* respectively. These data confirmed that <10 cm elevational changes can result in abrupt changes in floral composition and dominance relationships (cf. Mahall and Park 1976a), yet Zedler noted that only one of these changes—the replacement of tall *Spartina* grass by low-growing succulents—is readily visible in the field. A second "visual boundary" can be created by the change from prostrate (middle marsh) to erect (high marsh) growth in *Salicornia virginica* (Macdonald 1967; Mudie 1970b).

Even from these few data it is clear that the existence or nonexistence of vegetation zones is largely a matter of definition. The "obvious zonation" noted by the casual observer is most likely to represent changes in plant growth habit, which may or may not reflect real changes in species distributions and abundance. Most salt marsh halophytes have broad distributional ranges; thus species presence–absence analyses rarely identify zoned communities. Zonal separation is more readily achieved through analysis of relative abundance data (Macdonald 1967, 1971a; Zedler 1975a).

The abruptness of zone ecotones depends largely on the nature of distributional overlaps between adjacent dominant species. The effectiveness of specific limiting factors (see the section entitled "Autecology"), the steepness of topographical and environmental gradients (cf. Beals 1969), and the intensity of competition are all important. The specific accretionary and successional history of each marsh also appears to be more critical than is usually assumed (see the next section and Ranwell 1972).

Rigorous low marsh environments contain the least diverse and most uniform (both within and between sites) vegetation of any zone. The landward boundary of this zone is usually abrupt (Vogl 1966; Mahall and Park 1976a; Zedler 1975a)—both visually and in terms of species composition and abundance. This reflects two different tide-related limiting factors that come into effect around MHW, increasing soil salinity on *Spartina* and increasing submergence on *Salicornia virginica.* Most high marsh settings yield increasingly diverse floras that exhibit considerable intra- and intersite variability. Physical controls remain important, but competition becomes increasingly significant and the simple parallel zones characteristic of lower elevations are replaced by a complex multispecies mosaic.

Salt marsh development models

Sufficient geological and ecological data are now available for simple but integrated models of regional salt marsh development and succession (cf. Redfield 1972) to be proposed. A preliminary model appropriate for San Francisco Bay Area marshes is outlined below.

Salt marsh development is initiated by the establishment of *Spartina foliosa* along the landward margin of sheltered intertidal flats. Tide flat sediments are stabilized by *Spartina's* expanding subsurface rhizome mat, and they increase in elevation as its aerial stems trap plant detritus and suspended sediments. The vegetation remains simple, for adjacent upland glycophytes cannot tolerate the saline marsh soils and other halophytes cannot tolerate the degree of tidal submergence that *Spartina foliosa* withstands. As the predominantly *Spartina* marsh increases in area and elevation, the frequency and duration of tidal flooding diminish landward. Longer periods of exposure promote evaporation, and soil salinities rise. *Salicornia virginica* becomes increasingly prominent along the inner marsh margins, for although frequent submergence restricts its establishment, it exhibits a higher salt tolerance than most other local halophytes.

Headward erosion in the rapidly evolving marsh drainage system increases sedimentation at the low marsh perimeter, and marsh expansion continues to accelerate (Pestrong 1972). A narrow ecotone now separates an extensive *Spartina* low marsh community from an increasingly prominent *Salicornia* high marsh assemblage. The landward spread of *Spartina* is limited by high soil salinities around MHW, while seaward penetration of *Salicornia virginica* is restricted by increasing tidal submergence (cf. San Pablo Bay; Cameron 1972; Mahall and Park 1976a,b,c). Above MHW, declining tidal submergence reduces salt influx and increasing exposure allows rainfall and runoff to reduce soil salinities. The high marsh assemblage now becomes increasingly diverse. Physical factors may still be limiting, particularly to seedling establishment; vigorous colonal vegetative expansion is prominent, and interspecific competition is clearly increasingly important.

Encroachment of the expanding low marsh on major tidal channels now begins to reduce the tidal prism and slows net sedimentation (Pestrong 1972). Sedimentation and plant litter accumulation continue in high marsh environments, which now expand at the expense of the low marsh. The *Spartina* community becomes increasingly restricted to a narrow fringe fronting the expanding, almost level high marsh, which by now is occupied by an increasingly diverse and complex vegetation mosaic (cf. Palo Alto: Hinde 1954). The elevated landward marsh margins are increasingly disturbed by animals and man; only storm tides now reach these elevations, and a gradual transition to upland vegetation ensues.

With minor modifications this same model also fits the development of southern California salt marshes. The higher net annual evaporation that characterizes these sites increases high marsh soil salinities and reduces soil moisture levels. Since low marsh environments are little affected by these changes, *Spartina foliosa* behaves much as before. *Salicornia virginica* does less well, however, and its lack of space utilization around MHHW (possibly aided by competition) allows *Batis maritima* and the opportunistic annual *Salicornia bigelovii* to establish a distinctive middle marsh association (cf. Mission Bay: Macdonald 1967). Increasingly dry salty soils toward the landward margins of the high marsh reduce species diversity and encourage the development of barren salt flats—a marked contrast to the gradual high marsh–upland transition seen further north. This ecotone has also proved susceptible to invasion by species not seen further north: *Monanthochloë littoralis* and *Salicornia* (*Arthrocnenum*) *subterminalis* (cf. Mugu Lagoon: Warme 1971).

Both models outlined above suggest that the absence of low marsh *Spartina* communities from small coastal embayments that contain extensive *Salicornia virginica* stands (cf. Warme 1971) can be attributed to one of the following: (1) *Spartina* was originally present but was excluded as high marsh development and expansion encroached upon major tidal channels; (2) suitable *Spartina* habitats are present, but either the species never reached the site or it was unsuccessful in establishing itself; (3) because of the topography and tidal flushing characteristics of the site, only high marsh physical environments are represented.

FUTURE RESEARCH

Recent advances in our understanding of coastal salt marshes have inevitably opened a wealth of new approaches and new leads and raised new questions that demand attention; a few of these are listed below. Publicity, politics, and preservation must not be forgotten, however, for California's marshes represent an important, highly productive, and very vulnerable natural resource (Gosselink et al. 1974; Mudie 1974; Woodhouse et al. 1974; Newcombe and Pride 1975). Perhaps 70% or more of the marshlands present a century ago are gone, and it is critical that the remainder be preserved.

Several types of environmental data are badly needed. Are the tidal regimes of sites with and without low marsh communities really different? To what extent do salt marshes act as nutrient reservoirs? Does the proposed soil salinity model really work, and do soil moisture and salinity regimes exhibit latitudinal trends as suggested? Quantitative vegetational data like those of Zedler (1975a), preferably related to both elevation and local tidal regimes, are also needed.

The major thrust of future research should be directed toward autecological studies of key species, unraveling vegetation dynamics (competition, succession, etc.), and examining physiological tolerance mechanisms. What controls the changing growth form of *Salicornia virginica*? How do *S. virginica* and *S. subterminalis* differ in drought resistance? Does *Batis maritima* actively compete with *S. virginica* or simply fill a "vacant" niche? How do the distributions of established marsh plants and seedlings compare? How does strong clonal growth affect the vegetation mosaic? How stable is the high marsh community? How long do salt marsh plants live? Are nutrients limiting? What mechanisms control nutrient availability and uptake? Thoughtful integration of field observations and manipulative experiments with laboratory culture studies—an approach elegantly demonstrated by Clarke and Hannon (1967–71) and Mahall and Park (1976a,b,c)—should prove particularly rewarding. Finally, the practical aspects of salt marsh reclamation, as so well begun by the U.S. Army Corps of Engineers (1976) for San Francisco Bay, are well worth continuing.

ACKNOWLEDGMENTS

It is a particular pleasure to acknowledge the following individuals with whom I have exchanged reprints and discussed California salt marsh problems: J. Bradshaw, V. Chapman, J. Henrickson, C. Jefferson, B. Mahall, P. Mudie, R. Pestrong, J. Speth, R. Vogl, and J. Zedler. A. Küchler and the editors of this book, Jack Major and Michael Barbour, deserve special thanks for their patient encouragement, as does my family, who kept the pressure on so I wouldn't miss Christmas. Dave Crouch drew most of the figures, Ed Ellison photographed them, and Doris Faust typed the manuscript. Thank you.

LITERATURE CITED

Adams, D. A. 1963. Factors affecting vascular plant zonation in North Carolina salt marshes. Ecology 44:445-456.

Anderson, C. E. 1974. A review of structure in several North Carolina salt marsh plants, pp. 307-344. In R. J. Reimold and W. H. Queen (eds.). Ecology of halophytes. Academic Press, New York.

Atwater, B. F., and C. W. Hedel. 1976. Distribution of seed plants with respect to tide levels and water salinity in the natural tidal marshes of northern San Francisco Bay estuary, California. U.S. Geol. Survey Open File Rept. 76-389. 41 p.

Barbour, M. G. 1970. Is any angiosperm an obligate halophyte? Amer. Midl. Natur. 84:105-120.

Barbour, M. G., R. B. Craig, F. R. Drysdale, and M. T. Ghiselin. 1973. Bodega Head: coastal ecology. Univ. Calif. Press, Berkeley. 338 p.

Beals, E. 1969. Vegetational change along altitudinal gradients. Science 165:981-985.

Binet, P. 1961. Rapports entre le'eau de mar de la germination des semances de Triglochin maritimum L. Bull. Soc. Linn. Normandie, 10th Ser. I:117-132.

Bouma, A. H. 1963. A graphic presentation of the facies model of salt marsh deposits. Sedimentology 2:122-129.

Bradshaw, J. S. 1968a. Report on the biological and ecological relationships in the Los Penasquitos Lagoon and salt marsh area of the Torrey Pines State Reserve. Calif. State Div. Beaches and Parks, Contract 4-05094-033. 113 p.

---. 1968b. Environmental parameters and marsh foraminifera. Limnol. Oceanogr. 13:26-28.

Bradshaw, J. S., and P. J. Mudie. 1972. Some aspects of pollution in San Diego County lagoons. Calif. Mar. Resources Comm., Cal COFI Rept. 16:84-94.

Browning, B. M. 1970. The natural resources of Elkhorn Slough, their present and future use. Calif. State Dept. Fish and Game, Coastland Wetlands Ser. 4 . 105 pp.

California Department of Fish and Game. 1975. Natural resources of the Eel River Delta. Calif. State Dept. Fish and Game, Coastal Wetland Ser. 9. 108 pp.

California Department of Water Resources. 1968. Seawater intrusion: Bolsa-Sunset area, Orange County. Bull. 63-2. 167 pp.

Cameron, G. N. 1972. Analysis of insect trophic diversity in two salt marsh communities. Ecology 53:58-73.

Castañares, A. A., and F. B. Phleger. 1969. Coastal lagoons, a symposium (UNAM-UNESCO). Univ. Nac. Autonoma de Mexico, Mexico. 687 pp.

Carpelan, L. H. 1969. Physical characteristics of southern California coastal lagoons, pp. 319-334. In A. A. Castañares and F. B. Phleger (eds.). Coastal lagoons, a symposium (UNAM-UNESCO). Univ. Nac. Autonoma de Mexico.

Chapin, F. S. 1974. Phosphate absorption capacity and acclimation potential in plants along a latitudinal gradient. Science 183:521-523.

Chapman, V. J. 1960. Salt marshes and salt deserts of the world. Interscience, New York. 392 pp.

---. 1974. Salt marshes and salt deserts of the world. 2nd, supplemented reprint ed.). V. V. J. Cramer, Germany. 494 pp.

---. 1976. Wet coastal formations. Elsevier, Amsterdam (in press).

Clark, L. D., and N. J. Hannon. 1967-71. The mangrove swamp and salt marsh communities of the Sydney district, Parts I-IV. J. Ecol. 55:753-771; 57:213-234; 58:351-369; 59:535-553.

Daiber, F. C. 1974. Salt marsh plants and future salt marshes in relation to animals, pp. 475-510. In R. J. Reimold and W. H. Queen (eds.). Ecology of halophytes. Academic Press, New York.

Dillingham Environmental Co., 1971. An environmental evaluation of the Bolsa Chica Area. Vols. I-III. Dillingham Corp., La Jolla, Calif. 400 pp.

Doty, M. S. 1946. Critical tide factors that are correlated with the vertical distribution of marine algae and other organisms along the Pacific Coast. Ecology 27:315-328.

Eilers, H. P. 1975. Plants, plant communities, net production and tide levels: the ecological biogeography of the Nehalem salt marshes, Tillamook County, Oregon. Ph.D. dissertation, Oregon State Univ., Corvallis. 368 p.

Filice, F. P. 1954. An ecological survey of the Castro Creek area in San Pablo Bay. Wasmann J. Biol. 12:1-24.

Gerdes, G. L., E. R. J. Primbs, and B. M. Browning. 1974. The natural resources of Morro Bay, their status and future. Calif. Dept. Fish and Game, Coastal Wetlands Ser. 8. 138 pp.

Giguere, P. E. 1970. The natural resources of Bolinas Lagoon: their status and future. Calif. Dept. Fish and Game.

Gill, R., and A. R. Buckman. 1974. The natural resources of Suisun Marsh, their status and future. Calif. Dept. Fish and Game, Coastal Wetlands Ser. 9. 152 pp.

Gosselink, J. G., E. P. Odum, and R. M. Pope. 1974. The value of the tidal marsh. Center for Wetland Resources, Louisiana State Univ., Baton Rouge, LSU-SG-74-03. 30 pp.

Green, J. 1968. The biology of estuarine animals. Sidgwick and Jackson, London. 401 p.

Hehnke, M. 1969. Bay and estuary report: Big Lagoon, Stone Lagoon, Freshwater Lagoon, Lake Earl, Lake Talawa. Calif. Dept. Fish and Game Rept.

Hinde, H. P. 1954. Vertical distribution of salt marsh phanerogams in relation to tide levels. Ecol. Monogr. 24:209-225.

Hoover, R. F. 1970. The vascular plants of San Luis Obispo County, California. Univ. Calif. Press, Berkeley. 350 pp.

Howell, J. T. 1937. Marin flora. Univ. Calif. Press, Berkeley. 322 p.

Howell, J. T. 1970. Marin flora. 2nd ed. with supplement. Univ. Calif. Press, Berkeley. 366 pp.

Hubbard, J. C. E. 1969. Light in relation to tidal immersion and the growth of *Spartina townsendii* (s.l.). J. Ecol. 57:795-804.

Jefferies, R. L. 1972. Aspects of salt-marsh ecology with particular reference to inorganic plant nutrition, pp. 61-85. In R. S. K. Barnes and J. Green (eds.). The estuarine environment. Applied Science Pub., London.

Jefferson, C. A. 1974. Plant communities and succession in Oregon coastal marshes. Ph.D. dissertation, Oregon State Univ., Corvallis. 192 p.

Kasapligil, B. 1974. A synoptic report on the morphology and ecological anatomy of Spartina foliosa Trin., pp. 112-127. In C. L. Newcombe and C. R. Pride (eds.). Marsh studies. San Francisco Bay Marine Res. Center, Richmond, Calif.

Lane, R. S. 1969. The insect fauna of a coastal salt marsh. M.A. thesis, San Francisco State College, Calif. 78 p.

Lauff, G. H. 1967. Estuaries. Amer. Assoc. Adv. Sci. Pub., Washington, D.C. 83. 757 p.

Levering, C. A., and W. W. Thompson. 1971. The ultrastructure of the salt gland of Spartina foliosa. Planta 97:183-196.

Macdonald, K. B. 1967. Quantitative studies of salt marsh mollusc faunas from the North American Pacific Coast. Ph.D. dissertation, Univ. Calif., San Diego. 316 p.

---. 1969. Quantitative studies of salt marsh mollusc faunas from the North American Pacific Coast. Ecol. Monogr. 39:33-60.

---. 1971a. Species distribution patterns in Pacific Coast salt marshes. Bull. Ecol. Soc. Amer. 52:20.

---. 1971b. Nutrient cycling in Goleta Slouth, California. Progr. Abs., Amer. Soc. Limnol. Oceanogr. (Pac. Section), San Diego, June 21-25.

---. 1971c. Variations in the physical environment of a coastal slough subject to seasonal closure, p. 141. In Abs. Vol., Second Coastal and Shallow Water Research Conference. Univ. South. Calif., Los Angeles.

---. 1976. Plant and animal communities of Pacific North American salt marshes, pp. 167-191. In V. J. Chapman (ed.). Wet coastal formations. Elsevier, Amsterdam (in press).

Macdonald, K. B., and M. G. Barbour. 1974. Beach and salt marsh vegetation of the North American Pacific Coast, pp. 175-233. In R. J. Reimold and W. H. Queen (eds.). Ecology of halophytes. Academic Press, New York.

Macdonald, K. B., J. Henrickson, R. Feldmeth, G. Collier, and R. Dingman. 1971. Changes in the ecology of a southern California wetland removed from tidal action, p. 144. In Abs. Vol., Second Coastal and Shallow Water Research Conference. Univ. South. Calif., Los Anagels.

Mahall, B. E., and R. B. Park. 1976a,b,c. The ecotone between Spartina foliosa Trin. and Salicornia virginica L. in salt marshes of northern San Francisco Bay. I. Biomass and productivity (a). II. Soil water and salinity (b). III. Soil aeration and tidal immersion (c). J. Ecol. (in press).

Mall, R. E. 1969. Soil-water-salt relationships of waterfowl food plants in the Suisun Marsh of California. Calif. Dept. Fish and Game Wildlife Bull. 1:1-59.

Marshall, J. T. 1948. Ecologic races of song sparrows in the San Francisco Bay region. I-Habitat and abundance. Condor 50:193-215.

McIlwee, W. R. 1970. San Diego County, coastal wetlands inventory: Tijuana Slough. Calif. State Dept. Fish and Game, Prelim. Rept. 62 pp.

Milliman, J. D., and K. O. Emery. 1968. Sea levels during the past 35,000 years. Science 162:1121-1123.

Mobberley, D. G. 1956. Taxonomy and distribution of the genus Spartina. Iowa State Coll. J. Sci. 30:471-574.

Monroe, G. W. 1973. The natural resources of Humboldt Bay. Calif. Dept. Fish and Game, Coastal Wetland Ser. 6. 160 pp.

Monroe, G. W., B. J. Mapes, and P. L. McLoughlin. 1975. The natural resources of Lake Earl and the Smith River Delta. Calif. Dept. Fish and Game, Coastal Wetlands Ser. 10. 131 pp.

Mudie, P. J. 1970a. A survey of the coastal wetland vegetation of north San Diego County. Calif. State Resources Agency, Wildlife Manage. Admin. Rept 70-4. 18 pp.

---. 1970b. A survey of the coastal wetland vegetation of San Diego Bay. Calif. Dept. Fish and Game Contract W26. D25-51. 79 pp.

---. 1974. The potential economic uses of halophytes, pp. 565-598. In R. J. Reinold and W. H. Queen (eds.). Ecology of halophytes. Academic Press, New York.

---. 1975. Palynology of recent coastal lagoon sediments in central and southern California. Bot. Soc. Amer. (abstr.):22.

---. 1976. Ecological and taxonomic diversity in the modern salt marsh floras of southern California and western Baja California. In W. W. Wright (ed.). Plant diversity in aquatic habitats. 2nd South. Calif. Bot. Symp., Los Angeles (in press).

Mudie, P. J., B. Browning, and J. Speth. 1974. The natural resources of Los Penasquitos Lagoon, recommendations for use and development. Calif. Dept. Fish and Game, Coastal Wetlands Ser. 7. 95 pp.

Munz, P. A. 1973. A California flora, with supplement. Univ. Calif. Press, Berkeley. 1905 pp.

National estuary study. 1970. U. S. Dept. Interior, Fish and Wildlife Service. 7 vols. Washington, D. C.

Nesbit, D. M. 1885. Tide marshes of the United States. USDA Misc. Spec. Rept. 7. Washington, D.C. 259 pp.

Neuenschwander, L. F. 1972. A phytosociological study of the transition between salt marsh and terrestrial vegetation of Bahia de San Quintin. M.A. thesis, Los Angeles State Coll., Calif. 60 p.

Newcombe, C. L., and C. R. Pride. 1975. Marsh studies: the establishment of intertidal marsh plants on dredge material substrate. San Francisco Bay Marine Res. Center, Richmond, Calif. 141 pp.

Nichols, D. R., and N. A. Wright. 1971. Preliminary map of historic margins of marshland, San Francisco Bay, California. U. S. Geol. Survey Open File Map, San Francisco Bay Region Environment and Resources Planning Study. Basic Data Contrib. 9 (map and 10 p. report).

Parnell, D. R. 1975. Chromosome numbers in growth forms of *Spartina foliosa* Trin., pp. 128-131. In C. L. Newcombe and C. R. Pride (eds.). Marsh studies. San Francisco Bay Marine Res. Center, Richmond, Calif.

Pestrong, R. 1965. The development of drainage patterns on tidal marshes. Stanford Univ. Pub. Geol. Sci. 10:1-87.

---. 1969. The shear strength of tidal marsh sediments. J. Sed. Pet. 39:322-394.

---. 1972. San Francisco Bay tidelands. Calif. Geol. 25:27-40.

Phleger, C. F. 1971. Effect of salinity on growth of a salt marsh grass. Ecology 52:908-911.

Phleger, F. B., and J. S. Bradshaw. 1966. Sedimentary environments in a marine marsh. Science 154:1551-1553.

Pielou, E. C. 1974. Vegetation zones: repetition of zones on a monotonic environmental gradient. J. Theor. Biol. 47(2):485-489.

Pigott, C. D. 1969. Influence of mineral nutrition on the zonation of flowering plants in coastal salt marshes, pp. 25-35. In I. H. Rorison (ed.). Ecological aspects of the mineral nutrition of plants. Blackwell, Oxford.

Purer, E. A. 1942. Plant ecology of the coastal salt marshlands of San Diego County, California. Ecol. Monogr. 12:81-111.

Queen, W. H. 1974. Physiology of coastal halophytes, pp. 345-353. In R. J. Reimond and W. H. Queen (eds.). Ecology of halophytes. Academic Press, New York.

Ranwell, D. S. 1972. Ecology of salt marshes and sand dunes. Chapman and Hall, London. 258 pp.

Redfield, A. C. 1972. Development of a New England salt marsh. Ecol. Monogr. 42:201-237.

Reimold, R. J., and W. H. Queen. 1974. Ecology of halophytes. Academic Press, New York. 605 p.

Rollins, G. L. 1973. Relationships between soil salinity and the salinity of applied water in the Suisun Marsh of California. Calif. Dept. Fish and Game 59:5-35.

Scott, D. B., P. J. Mudie, and J. S. Bradshaw. 1976. Bethonic foraminifera of three southern California lagoons: modern ecological studies and interpretation of recent stratigraphy. J. Foram. Res. 6:59-75.

Seneca, E. D. 1974. Germination and seedling response of Atlantic and Gulf coast populations of *Spartina alterniflora*. Amer. J. Bot. 61:947-956.

Shanholtzer, A. F. 1974. Relationship of vertebrates to salt marsh plants, pp. 463-474. In R. J. Reimold and W. H. Queen (eds.). Ecology of halophytes. Academic Press, New York.

Smith, C. F. 1952. A flora of Santa Barbara. Santa Barbara Bot. Gard., Santa Barbara, Calif. 100 p.

Standing, J. 1976. The natural resources of Bodega Bay. Calif. Dept. Fish and Game, Coastal Wetlands Ser. (in press).

Stephenson, S. N. 1971. A putative *Distichlis-Monanthochloë* (Poaceae) hybrid from Baja California, Mexico. Madroño 21:125-127.

Stevenson, R. E., and K. O. Emery. 1958. Marshlands at Newport Bay, California. Occas. Papers, Allan Hancock Found. 20:109.

Stewart, G. R., J. A. Lee, and T. O. Orebamjo. 1972. Nitrogen metabolism of halophytes. I. Nitrate-reduction activity in Suaeda maritima. New Phytol. 71:263-268.

---. 1973. Nitrogen metabolism of halophytes. II. Nitrate availability and utilization. New Phytol. 72:539-546.

Strogonov, B. P. 1964. [Physiological basis of salt tolerance in plants.] Akad. Nauk.. SSSR. Translated from Russian. Israel Progr. Sci. Transl., Jerusalem.

Teal, J., and M. Teal. 1969. Life and death of the salt marsh. Little, Brown & Co., Boston. 278 pp.

Ungar, I. A. 1974. Inland halophytes of the United States, pp. 235-305. In R. J. Reimold and W. H. Queen (eds.). Ecology of halophytes. Academic Press, New York.

Upson, J. E. 1949. Late Pleistocene and recent changes of sea level along the coast of Santa Barbara County, California. Amer. J. Sci. 247:94-115.

U. S. Army Crops of Engineers. 1976. Dredge disposal study, San Francisco Bay and Estuary. Appendix K, marshland development. Open File Report, San Francisco, Calif. 61 p. + several enclosures and appendices totalling another 300 p.

U. S. Geological Survey. 1970. A preliminary study of the effects of water circulation in the San Francisco Bay estuary. U.S. Geol. Survey Circ. 637-A,B. 35 p.

Valiela, I., and J. M. Teal. 1974. Nutrient limitation in salt marsh vegetation, pp. 547-563. In R. J. Reimold and W. H. Queen (eds.). Ecology of halophytes. Academic Press, New York.

Vogl, R. J. 1966. Salt-marsh vegegetation of Upper Newport Bay, California. Ecology 47:80-87.

Waisel, Y. 1972. The biology of halophytes. Academic Press, New York. 395 p.

Warme, J. E. 1971. Paleoecological aspects of a modern coastal lagoon. Univ. Calif. Pub. Geol. Sci. 87. 131 pp.

Webb, K. L. 1966. NaCl effects on growth and transpiration in Salicornia bigelovii, a salt marsh halophyte. Plant and Soil 24:261-268.

Wong, G. 1975. Oxygen transport in Spartina foliosa Trin., pp. 132-138. In C. L. Newcombe and C. R. Pride (eds.). Marsh studies. San Francisco Bay Res. Center, Richmond, Calif.

Woodell, S. R. J., and H. A. Mooney. 1970. The effect of sea water on carbon dioxide exchange by the halophyte Limonium californicum (Bois) Hellei. Ann. Bot., New Ser. 34:117-121.

Woodhouse, W. W., E. D. Seneca, and S. W. Broome. 1974. Propagation of Spartina alterniflora for substrate stabilization and salt marsh development. U. S. Army Crops of Engineers, Coastal Eng. Res. Center, Tech. Memo. 46. 155 pp.

Zedler, J. B. 1975a. Salt marsh community structure in the Tijuana Estuary, California. San Diego State Univ., Center for Marine Studies, Contrib. 7. 32 pp.

---. 1975b. Salt marsh community structure along an elevational gradient. Abstracts, AIBS Meeting, Corvallis, Ore. (August 1975).

CHAPTER 9

THE CLOSED-CONE PINES AND CYPRESS

RICHARD J. VOGL
Biology Department, California State University, Los Angeles

WAYNE P. ARMSTRONG
Department of Life Science, Palomar College, San Marcos, California

KEITH L. WHITE
College of Environmental Sciences, University of Wisconsin, Green Bay

KENNETH L. COLE
Department of Geosciences, University of Arizona, Tucson

General introduction 297
Cypress 297
Introduction 297
Taxonomy of *Cupressus* 297
Cupressus communities 298
Cupressus forbesii 298
Cupressus stephensonii 300
Cupressus macrocarpa 303
Cupressus goveniana 303
Cupressus pygmaea 303
Cupressus abramsiana 307
Cupressus sargentii 307
Cupressus macnabiana 307
Cupressus nevadensis 311
Cupressus bakeri 311
Ecological relationships 311
Soil requirements 311
Other site requirements 316
Role of fire: autecology 316
Role of fire: synecology 320
Closed-cone pines 325
Pinus attenuata 325
General biology 325
Vegetation 326
Soil/parent material–climate relationships 327
Role of fire 328
Pinus muricata (including *P. remorata*) 332
Taxonomy and distribution 332
Vegetation 332
Substrate preferences 337
Role of fire 337
Pinus contorta ssp. *contorta* 338
Taxonomy and distribution 338
General description 339
Pinus contorta ssp. *bolanderi* 340
Distribution 340
General description and vegetation 340
Pinus torreyana 341
Distribution 341
Vegetation 341
Pinus radiata 342
Distribution 342
Vegetation 342
Range restrictions 346
Role of fire 247
Summary 347
Impact of animals 347
Impact of man 348
Areas for future research 349
Literature cited 351

GENERAL INTRODUCTION

The closed-cone pines and cypress are unique, disjunct plant communities scattered the length of California's coast, mountains, and islands. These relict species occur on infertile and sometimes unusual substrates. Most stands are influenced by a maritime climate. A number of endemic species are associated with these communities, and general plant diversities and densities tend to be reduced on these impoverished sites.

The characteristic pine species are the closely related knobcone pine (*Pinus attenuata*), bishop pine (*P. muricata*), Monterey pine (*P. radiata*), Santa Cruz Island pine (*P. remorata*), and possibly beach pine (*P. contorta* ssp. *contorta*) and pygmy pine (*P. contorta* ssp. *bolanderi*). Torrey pine (*Pinus torreyana*) is also sometimes included in the group (Knapp 1965). The cypresses include 10 species, 8 of which are endemic to the state. They are Santa Cruz cypress (*Cupressus abramsiana*), Modoc cypress (*C. bakeri*), Tecate cypress (*C. forbesii*), Gowen cypress (*C. goveniana*), MacNab cypress (*C. macnabiana*), Monterey cypress (*C. macrocarpa*), Piute cypress (*C. nevadensis*), Mendocino cypress (*C. pygmaea*), Sargent cypress (*C. sargentii*), Cuyamaca cypress (*C. stephensonii*), and one subspecies, Siskiyou cypress (*C. bakeri* ssp. *matthewsii*). The major species are intimately related to fire, characterized by a closed-cone habit or by serotinous cones, whereby the ovulate cones remain sealed after maturity, usually accumulating on the tree until opened by fire. "Cypress" will be used as both singular and plural in this chapter.

CYPRESS

Introduction

According to Griffin and Critchfield (1972), there are 25 recognized species of *Cupressus* in the northern hemisphere, distributed from the Mediterranean region to China, and from the western United States to Central America. The majority of species occur in western North America in isolated groves from southern Oregon to Baja California, eastward into Arizona, New Mexico, and Texas, and south to Honduras (Wolf 1948; Little 1970). Unlike other coniferous genera, such as *Juniperus* and *Pinus*, *Cupressus* consists of species whose native habitats are restricted to well-defined areas, appropriately termed "arboreal islands" (Bowers 1961). Ten distinguishable species and one subspecies of *Cupressus* are native to California (Munz and Keck 1959).

Most southern California groves occur within chaparral communities. Central and northern California groves are commonly associated with chaparral, closed-cone pine forest, mixed evergreen forest, foothill woodland, yellow pine forest, and north coastal coniferous forest.

Taxonomy of *Cupressus*

Difficulties in distinguishing *Cupressus* species are caused by morphological variability, similarities exhibited by all species, and variation among different populations of the same species. Various species exhibit remarkably similar cones (e.g., Posey and Goggans 1968), and vegetative features such as foliage and bark must be used for their separation. Table 9-1 shows three classification schemes. Eckenwalder (1975) believed that, until more critical evaluations are made, Wolf's (1948) system of separate species is perhaps preferable to Little's (1966; 1970) groupings. Future studies on *Cupressus* may reveal significant genetic and biochemical differences.

Evidence of natural interspecific hybridization in California cypress is limited and inconclusive. It is doubtful that natural hybridization occurs between southern California species because of their disjunct distributions. *Cupressus macnabiana* and *C. sargentii* form associations in the North Coast Ranges, particularly in Napa, Lake, and Mendocino Cos. Wolf (1948) and Griffin and Stone (1967) did not find hybrids, but field observations in northern Napa Co. by Griffin and Critchfield (1972) suggest possible hybridization. Monoterpene studies indicate natural hybridization between *C. macnabiana* and *C. sargentii* (Lawrence et al. 1975).

The present restricted and widely scattered distribution of cypress species suggests relicts of former widespread occurrence (Little 1950; Twisselmann 1962; McMillan 1952; Langenheim and Durham 1963; see also Chapter 5). During glacial times, the temperate coastal climate was more favorable for cypress and closed-cone pines (Griffin 1972). Increased aridity probably accounts for the restriction of many California species to coastal environments (Axelrod 1967). According to Little (1970), environmental conditions changed faster than the species could evolve, and they therefore became restricted to a few locations still suitable for growth and reproduction. Small populations, geographic isolation, and different selection pressures such as drought, fire, and competition have produced enough variation that some groves are now classified as different species. Perhaps species of glandular and eglandular cypress groups originated from separate gene pools that extended through the coastal and inland mountains, respectively.

Cupressus Communities

Detailed locations of cypress groves and associated vegetation have been described by Wolf (1948), McMillan (1952), Twisselmann (1962, 1967), Wagener and Quick (1963), Stone (1965), Armstrong (1966), Griffin and Stone (1967), Posey and Goggans (1967), Little (1966, 1970, 1975), and Griffin and Critchfield (1972).

Cypress species occur in isolated groves the length of California and extend inland up to 281 km. They range in elevation from near sea level (*C. macrocarpa*) to almost 2134 m (*C. bakeri*). Cypress are associated with at least 10 different plant communities. *Adenostoma fasciculatum* is the most frequently associated shrub, occurring with at least 6 of the 10 species. Several endemic species are found in cypress groves. No native cypress occur on the offshore California islands or in the Central Valley, Mojave Desert, Colorado Desert, and the Basin and Range Province east of the Sierra Nevada. Communities dominated by each of the 10 species are reviewed in the following pages.

Cupressus forbesii. Four isolated groves of *C. forbesii* occur in the Peninsular Ranges: on Guatay Mt. (NA 370790), Otay Mt. (NA 371580), and Tecate Peak (NA 372030) in San Diego Co., and on Sierra Peak in the northern Santa Ana Mts. of Orange Co. The cypress range in elevation from 305 m on Otay Mt. to about 1372 m on Guatay Mt. Isolated groves also extend about 241 km south into Baja California (Little 1971). The groves are associated with chaparral, except for one grove in El Cañon de Pinitos, Baja California, associated with *Pinus muricata*

TABLE 9-1. Taxonomic subdivisions of *Cupressus*. Species marked with an asterisk occur in California

Species of *Cupressus* recognized by Wolf (1948) and Munz and Keck (1959)	Suggested Reduction of Species (Wolf 1948)	Revisions by Little (1966, 1970)
1. *C. macrocarpa**	1. *C. macrocarpa*	1. *C. macrocarpa*
2. *C. goveniana**	2. *C. goveniana*	2. *C. goveniana* var. *g*
3. *C. abramsiana**	ssp. *abramsiana*	var. *abramsiana*
4. *C. pygmaea**	ssp. *pygmaea*	var. *pygmaea*
5. *C. sargentii**	ssp. *sargentii*	3. *C. sargentii*
6. *C. guadalupensis*	3. *C. guadalupensis*	4. *C. guadalupensis* var. *g*
7. *C. forbesii**	ssp. *forbesii*	var. *forbesii*
8. *C. macnabiana**	4. *C. macnabiana*	5. *C. macnabiana*
9. *C. bakeri** (and ssp. *matthewsii*	5. *C. bakeri*	6. *C. bakeri*
10. *C. arizonica*	6. *C. arizonica*	7. *C. arizonica* var. *a*
11. *C. montana*	ssp. *montana*	var. *montana*
12. *C. nevadensis**	ssp. *nevadensis*	var. *nevadensis*
13. *C. glabra*	ssp. *glabra*	var. *glabra*
14. *C. stephensonii**	ssp. *stephensonii*	var. *stephensonii*

(Robinson and Epling 1940). Early grove descriptions were reported by Parish (1914) and Howell (1929).

Common associated species are listed in Table 9-2. Endemics in the Otay Mt. grove include *Arctostaphylos otayensis, Ceanothus otayensis,* and *Lepechinia ganderi.* Other species that are rather local in California but extend into Baja California are *Chamaebatia australis* and *Fremontodendron mexicanum. Lathyrus splendens* is confined to southern San Diego Co. and adjacent Baja California, while *Pickerengia montana* is found as far north as Mendocino Co. but is rare in Baja California. Although it occurs elsewhere in southern California, *Nolina parryi* is locally confined to the Sierra Peak grove (Fig. 9-1). *Lepechinia cardiophylla,* a local endemic, also occurs in the grove. The oldest *C. forbesii* tree was found on Sierra Peak and was about 209 yr old.

Cupressus stephensonii. This species is restricted to the southwest slopes of Cuyamaca Peak, San Diego Co. (NA 370390). Wolf (1948) reported trees up to 14.6 m high, but these were apparently killed in a 1950 fire, since the tallest trees found now are about 9.14 m. According to R. V. Moran (pers. commun.), *C. stephensonii* also occurs in the Sierra Juarez, Baja California, scattered along an arroyo east of Santa Catarina; this species was formerly thought to be endemic to Cuyamaca Peak (Wolf 1948; Little 1970). The cypress extend to approximately 1676 m elevation. The elevations of this and the Guatay Mt. *C. forbesii* grove are high enough for snow. In addition to dense chaparral, there are scattered individuals of *Pinus coulteri* among the cypress (Table 9-2), with the upper slopes supporting mixed conifer forest (Armstrong 1966). With the exception of *C. sargentii* in Santa Bar-

Figure 9-1. *Cupressus forbesii* at the crest of a south-facing escarpment in the Sierra Peak grove, Santa Ana Mts. Old trees on these exposed, rocky sites have escaped recurrent fires. Note *Nolina parryi* on right.

TABLE 9-2. Species associated with southern California cypress groves. Compiled from field data taken by W. P. Armstrong

Associated Taxon	Dominant Cypress	
	C. forbesii	*C. stephensonii*
Adenostoma fasciculatum	x	x
Arctostaphylos glandulosa	x	x
A. otayensis	x	
Artemisia californica	x	
Ceanothus crassifolius	x	x
C. foliosus		x
C. leucodermis		x
C. otayensis	x	
C. tomentosus olivaceus	x	
Cercocarpus betuloides	x	x
C. minutiflorus	x	
Chamaebatia australis	x	
Cneoridium dumosum	x	
Dendromecon rigida	x	
Eriodictyon crassifolium	x	
E. trichocalyx	x	
Eriogonum fasciculatum	x	x
Eriophyllum confertiflorum	x	
Fremontodendron mexicanum	x	
Garrya flavescens pallida		x
Haplopappus squarrosus	x	
Helianthemum scoparium	x	
Heteromeles arbutifolia	x	X
Keckiella antirrhinoides	x	
K. cordifolia	x	
Lathyrus laetiflorus alefeldii	x	x
L. splendens	x	
Lepechinia cardiophylla	x	
L. ganderi	x	
Lotus scoparius	x	
Malacothamnus fasciculatus	x	

TABLE 9-2 [continued]

Associated Taxon	Dominant Cypress	
	C. forbesii	*C. stephensonii*
Marah macrocarpus	x	
Mimulus aurantiacus australis	x	
M. clevelandii	x	
M. longiflorus	x	
Nolina parryi	x	
Pickeringia montana	x	
Pinus coulteri		x
Prunus ilicifolia	x	x
Quercus agrifolia	x	x
Q. chrysolepis		x
Q. dumosa	x	x
Rhamnus crocea	x	
R. ilicifolia	x	
Rhus laurina	x	
R. ovata	x	x
Romneya trichocalyx	x	
Rosa californica		x
Salvia apiana	x	
S. clevelandii	x	x
S. mellifera	x	
S. munzii	x	
Symphoricarpos mollis	x	x
Trichostema lanatum	x	
T. parishii	x	
Xylococcus bicolor	x	
Yucca schidigera	x	
Y. whipplei	x	x

bara Co., this is the only southern California cypress associated with pines. It apparently is not associated with rare or endemic shrubs.

Cupressus macrocarpa. Monterey cypress is known from only two localities along the rocky Monterey Co. seacoast: between Pt. Cypress and Pescadero Pt., and at Pt. Lobos (NA 271664, 271665). A detailed account of the Pt. Cypress population before extensive residential development is given by Greene (1929). Some trees receive considerable salt spray. A green alga (*Trentepohlia aurea* var. *polycarpa*) and a lichen (*Ramalina reticulata*) grow on many trees. Point Lobos State Reserve naturalists estimate 200–300 yr as the maximum age for *C. macrocarpa. Cupressus macrocarpa* has been planted extensively along the coast and has become naturalized in some locations (Hoover 1970; Howell 1970). The cypress are associated with dominants of several plant communities, including closed-cone pine forest, northern coastal scrub, and coastal sage scrub (Barry 1974; Table 9-3).

According to Carlquist (1965), the nonwaxy, green foliage of *C. macrocarpa* is less xerophytic than the gray, waxy foliage of species such as *C. sargentii*, a feature that may restrict them to moist coastal exposures. Salt spray may explain the resistance of *C. macrocarpa* on the Monterey Peninsula to cypress canker (*Coryneum cardinale*), since fungus spores exposed to salt water have a low viability (Carlquist 1965). *Cupressus macrocarpa* is extremely susceptible to *Coryneum* in inland areas. Inland cypress species, such as *C. forbesii*, are rather resistant. Parasitic fungi of *Cupressus* have been discussed by Wagener (1948) and Peterson (1967).

Cupressus goveniana. The two known groves of *C. goveniana* are located in Monterey Co. a few kilometers inland from the *C. macrocarpa* groves. One grove occurs on the western slopes of Huckleberry Hill inland from Pt. Cypress; the other, between Gibson and San Jose creeks, inland from Pt. Lobos. The Gibson Creek population contains dwarfed cypress on an exposed ridge and larger cypress on deeper soils (McMillan 1956). The stunted thickets of cypress and pine on Huckleberry Hill are strikingly similar to *C. pygmaea* groves on the pine barrens of Mendocino Co. (Jenny et al. 1969). The cypress receive moisture from fog drip (Wolf 1948). In both groves, the cypress are dominant members of the closed-cone pine forest (Seeman 1970; Table 9-3). The Huckleberry Hill grove (NA 270890) is unique because it is the only place where *Pinus radiata* and *P. muricata* occur together. Several northern Pacific coastal species are associated with *C. goveniana*, including *Gaultheria shallon, Rhododendron macrophyllum, Myrica californica*, and *Vaccinium ovatum*. Two local endemics confined to sandy areas along the Monterey Peninsula are *Haplopappus eastwoodae* and *Arctostaphylos hookeri. Xerophyllum tenax* approaches the southern limit of its range on Huckleberry Hill. After the last fire on Huckleberry Hill, many *Xerophyllum* bloomed vigorously (Griffin 1972). Two herbaceous species, *Navarretia atractyloides* and *N. squarrosa*, occur on locally disturbed sites in both groves (McMillan 1956).

Cupressus pygmaea. This cypress is mainly confined to a narrow discontinuous strip up to several kilometers wide along the Mendocino Co. coast (NA 231665–231698). The strip lies approximately 2.4–3.2 km inland. The range for *C. pygmaea* described by Wolf (1948) was revised by Griffin and Critchfield (1972). The largest population occurs between Ft. Bragg (Virgin Creek) and Albion (Little Salmon

TABLE 9-3. Species associated with cypress groves that occur along the immediate coast of central California. Compiled from field data taken by W. P. Armstrong and all available references

Associated Taxon	Dominant Cypress *C. macrocarpa*	*C. goveniana*	*C. pygmaea*
Adenostoma fasciculatum	x	x	
Arctostaphylos hookeri	x	x	
A. nummularia			x
A. tomentosa	x	x	
A. columbiana			x
A. glandulosa			x
Artemisia californica	x		
Baccharis pilularis	x		
B. pilularsis consanguinea	x		
Ceanothus gloriosus exaltatus			x
C. thyrsiflorus	x		
C. foliosus			x
Chrysolepsis chrysophylla			x
C. chrysophylla minor		x	x
Dudleya farinosa	x		
Erigeron glaucus	x		
Eriophyllum confertiflorum	x		
Gaultheria shallon	x	x	x
Haplopappus eastwoodae		x	
Ledum glandulosum columbianum			x
Lithocarpus densiflorus			x
Mimulus aurantiacus	x	x	
Myrica california	x	x	x
Navarretia atractyloides		x	
N. squarrosa	x	x	x
Picea sitchensis			x
Pinus contorta contorta			x
P. contorta bolanderi			x
P. muricata		x	x
P. radiata	x	x	

TABLE 9-3 [continued]

Associated Taxon	Dominant Cypress		
	C. macrocarpa	*C. goveniana*	*C. pygmaea*
Pseudostuga menziesii			x
Ramalina reticulata (lichen)	x		
Rhododendron macrophyllum		x	x
Rhus diversiloba	x		
Sequoia sempervirens			x
Trentepohlia aurea (green alga)	x		
Tsuga heterophylla			x
Vaccinium ovatum	x	x	x
Viola sempervirens			x
Xerophyllum tenax		x	x

Creek). Several populations are scattered south of Pt. Arena, including Galloway Creek, Slick Rock Creek, and Roseman Creek. Wolf (1948) cited the area inland from Anchor Bay as the southernmost limit, but a grove occurs in Sonoma Co. in Miller Gulch, south of Plantation. This grove is on a partially podzolized soil, 14.5 km west of a large population of *C. sargentii* confined to serpentinite (Powell 1968). As on Huckleberry Hill, the cypress form a dwarf forest on the white sands. On more favorable sites, *C. pygmaea* may reach 45.7+ m in height (Mathews 1929). Ironically, this is the tallest of all cypress species, although its specific name refers to the stunted form. Cypress trees in the Ft. Bragg and Mendocino areas receive the greatest rainfall of any cypress and have the mildest and most continuously damp climate (Wolf 1948). Groves of *C. macrocarpa* and *C. goveniana* also have mild, damp climates because of the ocean proximity.

Throughout its range, *C. pygmaea* is a dominant member of the closed-cone pine forest (Table 9-3). However, surrounding the pine barrens it is associated with dominant species of the north coastal coniferous forest, Douglas fir forest, and redwood forest. Within the pine barrens, stunted individuals of *C. pygmaea* and *Pinus contorta* var. *bolanderi,* varying from 1.8 to 2.4 m in height, grow in profusion (Fig. 9-2). *Pinus muricata* is also interspersed with *C. pygmaea.* A floristic study of the Mendocino coastal plateau at Pt. Arena was made by Hardham and True (1972). Many ericaceous shrubs are associated with *C. pygmaea,* although none is completely restricted to these areas. Common examples are *Arctostaphylos nummularia, A. columbiana, Vaccinium ovatum, Gaultheria shallon, Ledum glandulosum* spp. *columbianum,* and *Rhododendron macrophyllum. Viola sempervirens* blooms practically the entire year. *Navarretia squarrosa* occurs in dry, disturbed places.

Figure 9-2. Pygmy forest near the town of Mendocino, Mendocino Co. (*a*) Typical summer aspect, forest dominated by *Cupressus pygmaea, Pinus contorta* ssp. *bolanderi,* and dwarf form of *P. muricata.* The normal form of bishop pine is in the background, on a different soil type. (*b*) Same area after winter rains, showing flooding. Photograph courtesy of H. Jenny.

Cupressus abramsiana. Three groves occur in the Santa Cruz Mts. of Santa Cruz Co., and a fourth on Butano Ridge (NA 410295), San Mateo Co. Detailed descriptions are given in Wolf (1948), McMillan (1952), and Thomas (1961). The cypress are associated chiefly with chaparral, although some groves contain yellow pine and closed-cone pine forest elements (Table 9-4). Elevations range from approximately 305 to 762 m. The groves in the Santa Cruz Mts. are above the summer fog belt of adjacent redwood and mixed evergreen forests (Wolf 1948). In a geographical sense, Griffin and Critchfield (1972) considered *C. abramsiana* a nonserpentine form of *C. sargentii*, since there are serpentinite outcrops in the Santa Cruz Mts. without cypress. In the Bonnie Doon grove (NA 440272), *C. abramsiana* and *Pinus attenuata* occur as dwarfed individuals. On the fringes, where the sandy soils are deeper, larger individuals of both species occur with *P. ponderosa*. At the Eagle Rock grove (NA 440510), *P. attenuata* occurs principally along the crest of a sandy ridge above the cypress stand. Associated *Arctostaphylos silvicola* is restricted to sandy soils in the Santa Cruz Mts. *Arctostaphylos tomentosa* var. *tomentosiformis* and *A. nummularia* var. *sensitiva* also have rather narrow distributions in the Central Coast Ranges. The annuals *Navarretia atractyloides* and *N. mellita* are frequent on barren sandy sites.

Cupressus sargentii. With the exception of *C. macnabiana*, there are probably more documented locations for *C. sargentii* than for any other California cypress. This species is confined to the inner and outer Coast Ranges, from Red Mt. in extreme northern Mendocino Co. (NA 231831, 231832) to Zaca Peak in Santa Barbara Co. (NA 422615): a 643 km range. In the North Coast Ranges it is commonly found along creeks and lower canyon slopes, most frequently associated with chaparral and foothill woodland communities (Table 9-4). In some locations, cypress occur with dominant members of the mixed evergreen, yellow pine, and closed-cone pine forests. In Sonoma Co., *C. sargentii* is associated with redwood forest components and occupies the same summer fog belt (Wolf 1948). The cypress form dense thickets in burned areas. On more favorable mesic sites, such as Cypress Mt., San Luis Obispo Co. (NA 400398), some trees are estimated to be over 27 m tall. Larger trees have trunks over 90 cm in basal diameter and are without lateral branches for the first 9–12 m.

Throughout most of its range, *C. sargentii* is frequently associated with *Pinus sabiniana, Ceanothus cuneatus, Quercus dumosa*, and *Q. durata*. With few exceptions, *Q. durata* is restricted to serpentine soils. *Ceanothus jepsonii* and *Arctostaphylos pungens* var. *montana* are largely, if not entirely, restricted to serpentine soils (McMillan 1956). *Pinus sabiniana* is also a common associate on serpentinite. Several populations of *C. sargentii* in the inner North Coast Ranges are sympatric with isolated populations of *C. macnabiana*, making these the only two cypress species that occur together. In these associations, *C. macnabiana* tends to be shrubbier and to occupy the upper slopes (Griffin and Critchfield 1972).

Cupressus macnabiana. Numerous scattered groves occur in the inner North Coast Ranges bordering the Sacramento Valley, Sierra Nevada foothills, and Cascade Range. More than 30 groves occur in at least 12 counties, including Sonoma, Napa, Yolo, Mendocino, Lake, Colusa, Tehama, Shasta, Butte, Nevada, Yuba, and

TABLE 9-4. Species associated with cypress groves that occur in the Coast Ranges of central and northern California. Compiled from field data taken by W. P. Armstrong and all available references

Associated Taxon	Dominant Cypress		
	C. abramsiana	*C. sargentii*	*C. macnabiana*
Acer macrophyllum		x	
Adenostoma fasciculatum	x	x	x
Arctostaphylos canescens		x	
A. glauca		x	
A. nummularia sensitiva	x		
A. pungens montana		x	
A. silvicola	x		
A. tomentosa tomentosiformis	x		
A. viscida		x	
Arbutus menziesii		x	
Calocedrus decurrens		x	
Ceanothus cuneatus	x	x	
Chrysolepsis chrysophylla minor	x		
Cupressus macnabiana		x	
C. sargentii			x
Dendromecon rigida	x	x	
Eriodictyon californicum	x		

Garrya buxifolia		x	
G. congdoni			x
Haplopappus ericoides blakei	x		
Heteromeles arbutifolia	x		
Lepechinia calycina	x		
Navarretia atractyloides	x		
N. mellita	x		
Pickeringia montana	x		
Pinus attenuata	x	x	x
P. coulteri		x	
P. lambertiana		x	
P. ponderosa	x		
P. sabiniana		x	x
Pseudotsuga macrocarpa		x	
P. menziesii		x	
Ptelea crenulata			x
Quercus chrysolepis	x	x	
Q. dumosa		x	x
Q. durata		x	x
Q. lobata			x
Q. vaccinifolia		x	

TABLE 9-4 [continued]

Associated Taxon	Dominant Cypress		
	C. abramsiana	*C. sargentii*	*C. macnabiana*
Q. wislizenii frutescens	x		
Rhamnus californica crassifolia			x
Sequoia sempervirens		x	
Styrax officinalis californica			x
Umbellularia californica		x	
Xerophyllum tenax	x		
Yucca whipplei		x	

Amador (Wolf 1948; Griffin and Stone 1967; True 1970; Griffin and Critchfield 1972), making *C. macnabiana* the most widely distributed cypress in California (Little 1975).

Cupressus macnabiana is commonly associated with chaparral and foothill woodland species (Tables 9-4 and 9-5). Several groves contain closed-cone pine forest components, including *Pinus attenuata.* The Magalia grove, Butte Co., is confined to a serpentinite outcrop surrounded by yellow pine forest. In addition to *Pinus sabiniana,* groves of *C. macnabiana* frequently contain *Ceanothus cuneatus, Arctostaphylos viscida,* and *Eriodictyon californicum.* Associated shrubs largely restricted to serpentinite outcrops are *Quercus durata, Ceanothus jepsonii,* and *Garrya congdoni.* In Shasta–Tehama locations, *Salvia sonomensis* covers the ground (Griffin and Stone 1967). This species is rather local in San Diego Co. mountains but does occur near the summit of Cuyamaca Peak, above the grove of *C. stephensonii.* A disjunct population of *Adenostoma fasciculatum* occurs at the Lack Creek grove in Shasta Co. Although it is the dominant shrub in the hills west of the Sacramento Valley, *Adenostoma* is rare in the volcanic region east of the valley (Griffin and Stone 1967).

Cupressus nevadensis. At least nine groves are scattered along 72 km of the central Kern River drainage (Twisselmann 1967) in Kern and Tulare Cos. The cypress groves range in elevation from about 1219 to 1829 m, the second highest elevations for California cypress. *Cupressus nevadensis* is associated with foothill woodland and chaparral (Table 9-5). Some groves contain dominant members of the piñon–juniper woodland, including *Pinus monophylla* and *Juniperus californica.*

***Cupressus bakeri* (including ssp. *matthewsii*).** Several isolated groves occur in the Sierra Nevada, Cascade Range, and Siskiyou Mts. (Wolf 1948; Wagener and Quick 1963; Stone 1965; Little 1970, 1975; Griffin and Critchfield 1972). The northernmost location for *C. bakeri* is near Prospect, Oregon, in the Cascade Range (Little 1970). The two southernmost groves occur in Plumas Co. at 1980 and 2100 m, the highest elevations of any California cypress groves. The largest stand covers around 2800 ha near Timbered Crater (Stone 1965) in Modoc, Shasta, and Siskiyou Cos. Another grove occurs on Goosenest Mt., about 88 km north of Timbered Crater. Wolf (1948) segregated the Goosenest Mt. (NA 470707) and Siskiyou Mt. groves as ssp. *matthewsii.*

Cupressus bakeri is commonly associated with yellow pine forest (Table 9-5). According to Stone (1965), the Timbered Crater grove suggests a transition zone between several plant communities, including northern juniper woodland, yellow pine forest, and sagebrush scrub. Several groves contain *Pinus attenuata.* The high-elevation groves of Plumas Co. are associated with red fir forest dominants.

Wagener and Quick (1963) believed that the Plumas Co. groves introduced enough departures from previous descriptions of the species and subspecies to warrant restudy. Griffin and Critchfield (1972) suggested that the entire subspecies problem be reviewed.

Ecological Relationships

Soil requirements. The distribution of some cypress species correlates with certain soil types. Throughout its range, *Cupressus sargentii* is restricted to rocky outcrops

TABLE 9-5. Species associated with cypress groves that occur in Sierra Nevada foothills, Cascade Range, and Modoc Plateau. Compiled from all available references

Associated Taxon	Dominant Cypress		
	C. nevadensis	*C. macnabiana*	*C. bakeri*
Abies concolor			x
A. magnifica			x
A. magnifica shastensis			x
Adenostoma fasciculatum		x	
Amelanchier pallida			x
Arctostaphylos mariposa	x		
A. patula			x
A. viscida		x	
Artemisia tridentata	x		x
Calocedrus decurrens			x
Ceanothus cordulatus			x
C. cuneatus		x	x
C. greggii vestitus	x		
C. integerrimus macrothyrsus?			x
C. jepsonii		x	
C. lemmonii			x
C. prostratus			x
C. velutinus			x
Cercis occidentalis			x
Cercocarpus betuloides	x		x
C. ledifolius			x
Chamaebatiaria mellefolium			x

TABLE 9-5 [continued]

Associated Taxon	Dominant Cypress		
	C. nevadensis	*C. macnabiana*	*C. bakeri*
Chrysolepsis sempervirens			x
Chrysothamnus viscidiflorus			x
Ephedra viridis	x		
Eriodictyon californicum	x	x	
Fraxinus dipetala		x	
Fremontodendron californicum	x	x	
Garrya congdoni		x	
G. flavescens pallida	x		
G. fremontii			x
Haplopappus arborescens		x	
Juniperus californica	x		
J. occidentalis			x
Phoradendron bolleanun densum	x	x	x
Pickeringia montana		x	
Pinus attenuata		x	x
P. jeffreyi			x
P. lambertiana			x
P. monophylla	x		
P. contorta murrayana			x
P. ponderosa		x	x
P. sabiniana	x	x	
Prunus emarginata			x
P. subcordata			x

TABLE 9-5 [continued]

Associated Taxon	Dominant Cypress		
	C. nevadensis	*C. macnabiana*	*C. bakeri*
P. virginiana demissa			x
Pseudotsuga menziesii			x
Ptelea crenulata		x	
Purshia tridentata			x
Quercus douglasii	x		
Q. durata		x	
Q. garryana		x	x
Q. garryana breweri			x
Q. kelloggii			x
Q. sadleriana			x
Q. wislizenii		x	
Rhamnus californica crassifolia		x	
R. californica tomentella		x	
R. rubra			x
R. rubra obtusissima		x	
Rhus trilobata			x
Rosa gymnocarpa			x
Salvia sonomensis		x	
Symphoricarpos vaccinoides			x
Yucca whipplei	x		

or formations of serpentine (Wolf 1948; Hardham 1962). Although *C. macnabiana* is found on serpentinite outcrops, particularly in the coast ranges, it also occurs on clay and other soils (Wolf 1948). In the Sierra Nevada foothills, widely disjunct populations of *C. macnabiana* occur on serpentinite and on granitic soils (McMillan 1956). Griffin and Stone (1967) reported a variety of substrates for *C. macnabiana,* including soils derived from greenstone, serpentine, gabbro, basalt, tuff, and sandstone. Physical properties vary from those of clay loam more than 1.5 m deep to those of silty loam, which may be only a few centimeters deep. *Cupressus bakeri,* including ssp. *matthewsii,* is largely confined to volcanic lava outcrops (Stone 1965). Groves in the Siskiyou Mts. tend to grow on serpentine (Griffin and Critchfield 1972). Dense *C. pygmaea* thickets occur along a strip of white sandstone soil in Mendocino Co. known as the Mendocino White Plains or Pine Barrens. According to Wolf (1948), the soils are sterile beach terraces that form a hardpan through which water percolates very slowly. The soil is highly acid and podzolic with very shallow A_1 and A_2 horizons above the hardpan, and with possible aluminum toxicities (McMillan 1956; Jenny et al. 1969; Westman 1975). Cypress roots seldom penetrate the hardpan, but form extensive, shallow systems. The shallowest and most highly acid soils support the most dwarfed individuals, and on deeper organic soils individuals of *C. pygmaea* attain larger sizes (McMillan 1956; Westman 1975). The soil infertility of Anchor Bay locations is not as extreme as that of Ft. Bragg (McMillan 1956).

Cupressus goveniana occurs in two small, disjunct populations in Monterey Co. on sandy and podzolic soils. The acid, hardpan soil on Huckleberry Hill represents old beach deposits remarkably similar to those under Mendocino Co. stunted forests of *C. pygmaea* (Griffin 1972); *C. goveniana* soils are more infertile than *C. macrocarpa* soils (Wolf 1948). *Cupressus macrocarpa* is chiefly found on thin residual soils of coarse-grained Santa Lucia granodiorite on Monterey Peninsula. The soils have a 4.5–5.5 pH (Bowen 1966; Carlquist 1965; Williams 1970). *Cupressus abramsiana* is confined to soils principally derived from sandstone. In some locations the sandstone bedrock is exposed or lies only a few centimeters below the soil surface.

McMillan (1956, 1964) studied edaphic restrictions of *C. goveniana, C. macrocarpa, C. abramsiana, C. pygmaea, C. macnabiana,* and *C. sargentii.* These species are commonly found on serpentinite and highly acidic sandy soils derived from sandstone and weathered granodiorites. The soils are typically podzolic, poorly drained, and often less than 30 cm in depth. With the exception of *C. sargentii,* which grew best on serpentine, seedlings of the five species differed only slightly in their responses to serpentine, acidic, and control soils.

Cupressus nevadensis occurs on a variety of soils, including black and red clays, decomposed granite, and fractured rock. Twisselmann (1962) observed that specialized soil preferences limited the distribution of *C. nevadensis* in the Piute Mts. The best growth of *C. nevadensis* occurs in the Bodfish Peak grove in Kern Co. (NA 150215) on red clays covered with a well-developed topsoil. Although the substrate has not been analyzed for deficiencies, toxicities, or unusual physical properties, Twisselmann believed that such stands indicate that aridity, rather than soil, limits distribution.

Cupressus forbesii grows on various soil types, including sedimentary clay, sandstone, and conglomerate, and soils derived from granitic and extrusive metavolcanic rocks. Pequegnat (1951) stated that the Sierra Peak grove of *C. forbesii* had an affinity for soils rich in clay. Cypress trees in the grove were found to be largely confined to clay deposits, but many of the largest and oldest individuals occur on coarse sandstone (Armstrong 1966). Cypress groves in southern California had an average pH of 6.4, ranging from 5.8 on Otay Mt. to 7.5 on Tecate Peak. With the exception of Guatay Mt., *C. forbesii* groves are confined to rocky and infertile soils. The Guatay Mt. grove occurs on soil of granitic origin with obvious organic soil horizons, which also support a dense, vigorous growth of *Arctostaphylos glandulosa* (Armstrong 1966). Abundant organic matter may have accumulated because this highest-elevation stand receives precipitation in the form of snow (Wilson and Vogl 1965) or because a long time has lapsed since the last fire. The chaparral is sparse and stunted on all other *C. forbesii* sites. This was especially evident in the Santa Ana Mts., where the chaparral was dense except in the vicinity of Sierra Peak (Vogl 1976a). The foliage of *Rhus laurina* was also chlorotic in this grove. Pequegnat (1951) considered "Santa Ana" winds responsible for the dwarfed chaparral. This view cannot be considered seriously, however, because these hot, dry winds have an effect on stunted as well as normal chaparral in the northern Santa Ana Mts., and poor soils appear to offer a better explanation. Although *C. forbesii* is able to survive, it is not particularly adapted to thrive on infertile soils. It is restricted to these sites because these shade-intolerant pioneers appear to be unable to compete with typically dense chaparral. Further substrate studies need to be made.

Other site requirements. Site requirements for cypress seedlings are typical of pioneer conifers. These shade-intolerant species survive best in full sunlight on bare mineral soils lacking litter (Roy 1953; Wilde 1958). Seedling mortality is greater in shaded situations with abundant litter because of fungal infections, which cause damping-off (Vaartaja 1952, 1962). Wolf (1948) found poorly drained soils detrimental to cultivated cypress trees, and water-logged soils may account for dwarfed cypress in northern California. Cypress germination is ideally suited to the Mediterranean climate: fires during late summer and fall, followed by winter rains, ensure seed dissemination, bare mineral substrates, full sunlight, and adequate soil moisture. Seed dispersal is accomplished primarily by wind and rain runoff. Dispersal of other coastal cypress may also have occurred because of the buoyant ovulate cones.

With the exception of *C. stephensonii,* which is confined to a southwest exposure, the major portions of all cypress groves in southern California are located on the north sides of mountains, indicating a preference for more mesic sites.

Wolf (1948) reported heavy fog condensation on cypress along the immediate California coast and on Guadalupe Island (Libby et al. 1968). Fog condensation was also reported on cypress in the Santa Lucia Mts. (Hardham 1962), but fog drip does not appear to be a significant source of moisture in the southern California groves (Armstrong 1966).

Role of fire: autecology. The occurrence of even-aged cypress stands that date back to known fires throughout southern California indicates that these trees reproduce after fire. With the exception of Guatay Mt., every cypress grove in Orange and San Diego Cos. has burned at least once between 1944 and 1966

(Armstrong 1966). Fires may have been even more frequent before fire protection. A September 1970 fire narrowly missed the Guatay Mt. grove. Fire records obtained in 1965 correlated with even-aged stands of cypress in several groves (Table 9-6). A 5600 ha fire in November 1914 correlated with a large number of fire-killed trees of uniform age and height on Sierra Peak, still standing in 1966. Another fire in November 1948 related to 17 yr old thickets in the same grove (Fig. 9-3). A 1950 fire burned 4000 ha of Cuyamaca Rancho State Park, killing the larger *C. stephensonii.* Thickets of 15 yr old *C. stephensonii* originated after that fire. Other even-aged stands include a 19 yr old one on Tecate Peak and a 22 yr old one on Otay Mt. (Armstrong 1966). Wolf (1948) cited similar fire correlations in *C. sargentii* groves. Frequent chaparral fires are responsible for cypress thickets, abundant *Ceanothus* shrubs, and other fire-followers. The cypress thickets are conducive to crown fires, which usually consume the entire stand. *C. nevadensis,* occurring in pinyon-juniper woodland, is probably burned less frequently than cypress groves located within chaparral. The coastal *C. macrocarpa* may also burn less frequently.

Figure 9-3. A fire-killed parental tree in a 17 yr old thicket of *Cupressus forbesii* dating back to a November 1948 fire. Excessive competition is apparently responsible for the short stature of the trees. Sierra Peak, Santa Ana Mts.

TABLE 9-6. Ages of individual *Cupressus* trees occurring in various southern California groves. Ages taken in 1965 by W. P. Armstrong. Figures marked with an asterisk were obtained from several samples

Grove Location	Tree Aspect	Height (m)	Diameter at Coring (cm)	Age (yr)
Sierra Peak	Steep south-facing slope	5.2	35.6	209
	Steep south-facing slope	1.8	8.9	46
	Steep south-facing slope	3.7	25.4	67
	Northwest slope (killed by fire)	7.6	22.9	125
	Northwest slope (killed by fire)	6.1	17.8	98
	East-facing cliff edge	7.6	22.9	53
	East-facing cliff edge	9.2	20.3	138
	Northwest slope (killed by 1948 fire)	3.1	6.4	33*
	Northwest slope (germination after 1948 fire)	1.5	1.8	17*
Cuyamaca Peak	Southwest slope along King Creek (germination after 1950 fire)	3.1	6.4	15*
Guatay Mt.	North-facing slope	6.1	38.1	65-70*
	North-facing slope	1.5	3.8	65*
Otay Mt.	North-facing slope	1.5	2.5	22*
	North-facing slope	0.8	1.3	22*
Tecate Peak	North-facing slope	1.5	2.5	19*
	North-facing slope	3.7	3.8	19*
	North-facing slope (killed by fire)	3.7	6.4	63*
	North-facing slope (killed by fire)	6.1	15.2	110*

Groves occurring on rocky outcrops or thinly vegetated sites escape some fires because of discontinuous fuels. *C. Bakeri* associated with yellow pine forest may have been subject to frequent light surface fires prior to fire protection.

Wolf (1948) reported a 38 day germination period for *C. forbesii.* Prodigious numbers of cypress seedlings were growing on Sierra Peak 5 mo after the November 1948 fire (Pequegnat 1951). Cypress seedlings were observed on Tecate Peak 3 mo after the October 1965 fire. The short germination period ensures maximum utilization of moisture during the winter wet season.

The average interval between fires in *C. forbesii* groves is approximately 25 yr (Armstrong 1966). The shortest interval was 15 yr on Sierra Peak, and ranged from 10 to 63 yr on the northeast side of Tecate Peak. Cypress trees generally reach cone-bearing age before another fire occurs. Cones have been observed on 14 yr old *C. forbesii* and *C. stephensonii,* and the presence of mature ovulate cones on lateral branches indicated that the cones were produced at a younger age. Small trees of *C. sargentii* 5–6 yr old produce ovulate cones, apparently within a period less than the minimum fire interval (Wolf 1948).

The number of ovulate cones is variable. Mature trees 9 m tall often contain several thousand cones, a single branch bearing numerous clusters of 20–30 cones. Growth competition was observed to affect cone production. Trees in extremely dense stands generally had fewer cones than widely spaced trees of the same age. Dwarf *C. forbesii* on Otay Mt., 90 cm tall and 22 yr old, had only two or three cones per tree. *Cupressus pygmaea* in Mendocino Co. also had relatively few ovulate cones in comparison to larger open-grown specimens of the same age. A *C. pygmaea* 43 cm tall and 12 yr old had only 5 cones. Tecate Peak 19 yr old *C. forbesii* trees 3.05–4.57 m tall had an average of 50 cones per tree. *Cupressus stephensonii* 14 yr old trees were observed with over 100 mature ovulate cones.

Almost all cones remain unopened on the trees until burned. Heat generated by fire breaks the tight resinous seal between ovuliferous scales, allowing them to open. Seeds are gradually shed from the cones for several months after fire. Charred cones still retained 50% of their seed 5 mo after the 1965 Tecate Peak fire. The abundant production of persistent (for the life of a tree or 25+ yr), unopened ovulate cones effectively ensures reproduction after fire and accounts for the common thickets of stunted trees. In addition, fire is the only effective reproductive agent in opening cones and creating conditions for reestablishment.

Laboratory tests were conducted by W. P. Armstrong (unpub. data) to determine the effects of fire on *C. nevadensis* cones and the rate of seed shed after fire. Ignited cones gradually open as resin melts and boils. Resin pockets within the scales burst because of expanding gases from boiling resin. Rapid charring of the thick scales extinguishes the flames. Seeds are unburned and begin falling as the scales continue to open. Over a 5 mo period, approximately 5000 seeds were extracted from 50 burned cones, an average of 100 seeds per cone. *Cupressus forbesii* was found to have a similar average. Wolf (1948) reported an average of 60–150 seeds per cone for other species.

Unlike other closed-cone conifers, such as *Pinus attenuata* (Vogl 1973), cypress cones usually open when mechanically detached from the tree, the resinous seals breaking as the cones dry. But dried cones do not always open as widely as burned cones, and often retain clusters of exposed seeds lodged between scales. In 1964, 167 unopened *C. forbesii* cones were collected from Sierra Peak; 2 yr later, 58% of the cones had opened and shed seeds, while 42% remained unopened. Most of the unopened cones, however, had slightly separated scales with trapped seeds which had probably lost their viability because of desiccation. Cones attached to broken branches dry out more slowly and remain closed longer. Separate cones and cones attached to branches were collected from cypress species throughout California by W. P. Armstrong (unpub. data). Individual cones usually opened in 6–8 wk, whereas those on branches remained closed up to 6 mo, depending on temperature and humidity. Some cone clusters of *C. forbesii, C. stephensonii, C. nevadensis, C. sargentii,* and *C. abramsiana* have remained closed for over 8 yr. Sierra Peak *C. forbesii* closed cones, some of them estimated to be 25–30 yr old, were seen partially enveloped by exfoliating bark (Armstrong 1966). Although detached cones or cones attached to broken branches do open, seeds shed from them rarely, if ever, result in cypress establishment. These cones and seeds are eventually destroyed by fire, fungi, or animals.

Most southern California groves contain old trees that have survived fires by growing on steep, barren slopes or rocky outcrops where fires have been unable to spread (Fig. 9-1). Tree ages from southern California groves are summarized in Table 9-6. The fire resistance of mature cypress trees is poor, and old trees are generally restricted to unburned sites. Species such as *C. forbesii, C. stephensonii,* and *C. bakeri* have thin exfoliating bark which offers little fire protection. Other species with thick, persistent, fibrous trunks are somewhat fire resistant (Wolf 1948), but their small size, characteristic low and sometimes dead branches, and flammable foliage render them susceptible to crown fires, just as do dense stands. Large individuals of *C. sargentii* and *C. pygmaea* could probably survive surface fires.

Cypress groves are not always completely burned, resulting in a mosaic of differently aged stands of uniform height and density (Armstrong 1966). In the Mayacamas Range of southeastern Mendocino Co., *C. sargentii* exhibits a stratification of different age groups within one stand (Wolf 1948). Dunning (1916) reported two groups of 50 yr old *C. goveniana* mixed with 15 yr old reproduction. Similarly, the 1965 Tecate Peak fire resulted in a marked transition between a new generation of cypress seedlings and surviving 19 yr old thickets (Armstrong 1966).

Without fires, some cones on broken branches or dead trees may eventually release some seeds. But these seeds will probably land on hostile sites with dense chaparral, heavy soil litter, and closed canopies. Even if optimum sites are available, seed dissemination may not coincide with ideal germination conditions; the thin-coated cypress seeds readily lose viability because of desiccation. Stone (1965) reported seedlings of *C. bakeri* in areas that did not show signs of recent fires, but the seedlings were usually in the immediate vicinity of fallen cypress trees and along skid roads resulting from the logging of scattered *Pinus ponderosa.* Successful cypress reproduction, in general, is restricted to burned sites, and other chance reproduction cannot be expected to perpetuate groves and species.

Role of fire: synecology. Armstrong (1966) compared cover, density, and frequency of taxa in several *C. forbesii* burned and unburned stands (Table 9-7). Central and northern California cypress groves may not show the same patterns. On unburned sites, *C. forbesii* (35.1% cover) was the dominant species. Only widely spaced (80 trees per hectare) old trees with irregular, spreading branches occurred on the ridges and steep slopes (35.1% bare ground). The second dominant, *Adenostoma fasciculatum* (19.9% cover), had the highest density (832 stunted shrubs per hectare) and frequency (50%).

On burned sites *C. forbesii* was still dominant (21.4% cover) and had the greatest density (8408 trees per hectare) and frequency (68.8%). Heaviest concentrations of reproduction occurred along erosion channels; apparently seeds were washed and blown into these channels after the last fire. *Adenostoma fasciculatum* was second in dominance (15.9% cover), density (672 trees per hectare) and frequency (25%). *Ceanothus tomentosus* var. *olivaceus* was equally frequent and almost as dense (608 shrubs per hectare). Bare ground constituted 39.8% of the slope, particularly on small ridges and hillocks. Seedling establishment was apparently inhibited on these resistant sandstone areas. *Nolina parryi* was common on these hillocks, reproducing vegetatively from basal rosettes. The greater number of shrub species on the burned sites is a reflection of the more favorable substrates, compared to those of the steep, unburned slopes, rather than increases produced by burning. There were areas of exceptionally dense and stunted cypress within the burned sites. Factors responsible for these extremely dense thickets include the production of a prolific number of seeds by parental trees, localization of seeds in erosion channels, clusters of seeds adhering together, optimum germination and establishment conditions, and an apparently high seed viability. Dense cypress thickets also occur with other species, particularly *C. sargentii, C. goveniana,* and *C. pygmaea.*

Burned and unburned sites in the Tecate Peak grove of *C. forbesii* were sampled 5 mo after an October 1965 fire (Armstrong 1966). On the unburned sites *C. forbesii* was the most abundant species, and *Xylococcus bicolor* was second. Other dominant shrubs were *Ceanothus tomentosus* var. *olivaceus* and *Adenostoma fasciculatum.* There were no seedlings of *C. forbesii* or *Ceanothus* spp. and almost no herbaceous species. Density quadrats on burned sites in the Tecate Peak grove revealed 4019 fire-killed *C. forbesii* per hectare. Seedlings of *C. forbesii* and *Ceanothus tomentosus* var. *olivaceus* were present with densities of 230 and 77 per hectare, respectively. Resprouting individuals of *Xylococcus bicolor* had the highest shrub density (102 per hectare). *Phacelia brachyloba* was the most abundant species. This annual herbaceous species frequently appears on disturbed places, particularly burns. When the same area was resampled in 1970, *Eriophyllum confertiflorum* had the highest density, with no *Phacelia brachyloba* sampled. There were substantial increases in seedlings of *Ceanothus tomentosus* var. *olivaceous* (358 per hectare) and *Xylococcus bicolor* (998 per hectare).

Some burned cypress groves occupy the same acreage as they did before the last fire, whereas others vary. Otay Mt. grove has recently expanded; newly established stands were found on slopes and in canyons below the main grove. An example of *Cupressus* tenacity is the Greenhorn Mt. grove of *C. nevadensis.* In September 1961, this grove was reduced by fire from three mature trees to one, but responded by producing approximately 40 seedlings the next year (Howell 1962). The lone parental tree and seedlings occupy roughly the same area as the original grove. The Sierra Peak cypress densities increased after the 1948 fire, but data from 1970 showed a significant decrease as a result of the fall 1967 fire (Table 9-7). Cypress density in 1966 was 8408 trees per hectare, while the density of fire-killed trees in 1970 was 4928 per hectare. The latter figure is an underestimate because many stunted cypress

TABLE 9-7. Cover (%), density per hectare, and frequency (%) for burned and unburned Sierra Peak cypress groves, Santa Ana Mts. The dead tree counts, marked with an asterisk, are underestimates because many tree stems were consumed in 1967 fire. From Armstrong (1966) and unpublished data obtained by W. P. Armstrong and R. J. Vogl. The 1966 sampling consisted of 40 quadrats, 20 in the unburned and 20 in the burn, along with 40 line intercepts. The 1970 sampling consisted of 20 quadrats. Each quadrat was 13 m^2 and each line intercept was 8 m

Species	Unburned Sites, Sampled in 1966			Burned Sites			
				Sampled 1966, Burned 1948			Sampled 1970, burned 1967
	Cover	Density	Freq.	Cover	Density	Freq.	(dead trees)*
Cupressus forbesii	35.1	80	0.0	21.4	8408	68.8	4928
C. forbesii (seedlings)	0.0	0	0.0	0.0	0	0.0	6
Adenostoma fasciculatum	19.9	832	50.0	15.9	672	25.0	826
Ceanothus crassifolius	1.0	0	0.0	0.0	208	0.0	0
C. tomentosus var. *olivaceus*	0.0	0	0.0	6.8	608	25.0	0
C. tomentosus (seedlings) var. *olivaceus*	0.0	0	0.0	0.0	0	0.0	160
Eriodictyon crassifolium	1.1	48	0.0	1.0	128	6.3	333
Salvia mellifera	6.3	448	12.5	1.6	48	6.3	13
Rhus laurina	7.2	0	0.0	0.0	0	0.0	0
Eriogonum fasciculatum	6.5	256	37.5	9.1	72	12.5	0
Yucca whipplei	0.0	64	0.0	0.0	0	0.0	13
Nolina parryi	0.0	16	0.0	8.5	320	6.3	19
Eriophyllum confertiflorum	0.0	480	0.0	0.3	48	12.5	38
Arctostaphylos glandulosa	0.0	0	0.0	1.7	168	12.5	6

TABLE 9-7 [continued]

Heteromeles arbutifolia	0.0	0	0.0	1.5	0	0.0	6
Lepechinia cardiophylla	0.0	0	0.0	1.9	424	18.8	179
Helianthemum scoparium	0.0	0	0.0	0.4	128	0.0	3994
Mimulus longiflorus	0.0	0	0.0	0.0	32	0.0	0
Artemisia californica	0.0	0	0.0	0.0	16	0.0	0
Elymus condensatus	0.0	0	0.0	0.1	0	0.0	0
Helianthus gracilentus	0.0	0	0.0	0.0	0	0.0	3142
Calystegia macrostegia ssp. *arida*	0.0	0	0.0	0.0	0	0.0	3827
Lotus scoparius	0.0	0	0.0	0.0	0	0.0	454
Gnaphalium californicum	0.0	0	0.0	0.0	0	0.0	6
Brassica geniculata	0.0	0	0.0	0.0	0	0.0	256
Calochortus splendens	0.0	0	0.0	0.0	0	0.0	45
annual grass	0.0	...	12.5	3.0	...	12.5	...
bare ground	35.1	...	...	39.8	...	...	...

stems or trunks were consumed by fire. Cypress seedlings in 1970 were extremely sparse with only 6 per hectare. The highest densities in 1970 were attained by two herbaceous fire-followers, *Helianthemum scoparium* (3994 per hectare) and *Calystegia macrostegia* ssp. *arida* (3827 per hectare). The very low density of cypress seedlings is probably related to unfavorable environmental conditions at the time of seedling establishment.

The Cuyamaca Peak *C. stephensonii* grove was described by Wolf (1948). Observations in 1966 indicated that a 1950 fire had reduced this grove substantially. Very few large individuals were left, and practically all cypress reproduction was confined to the King Creek drainage. The main grove narrowly escaped the September 1970 fire, although several cypress along the eastern fringes were burned. Reproduction from these trees may help to again expand the grove. Coleman (1905) and Wolf (1948) cited locations in central and northern California where fires had also diminished the size of groves.

Preliminary investigations made on Tecate Peak since the October 1965 fire indicate that this grove may also be diminishing in size (Armstrong 1966). Burned slopes that previously supported cypress contained large blackened areas devoid of cypress seedlings. Other areas showed a marked reduction in cypress density. The 1970 density of cypress seedlings on burned sites was 384 per hectare compared to unburned thickets with an average density of 6093 trees per hectare. In 1966, burned slopes were randomly sampled using 1 m^2 quadrats. Preburn densities determined from fire-killed parental trees and cypress seedlings were similar: 3872 and 4646 per hectare, respectively. In 1970, the number of seedlings from the same area was only 40 per hectare. As in the Sierra Peak 1967 fire, unfavorable environmental conditions may have affected cypress reproduction. Another factor appears to be intense competition with resprouting shrubs and very high densities of herbaceous postburn species. Over 100 different native and naturalized species of forbs and grasses were recorded within 5 ha of a burn (W. P. Armstrong, unpub. data). The density of *Phacelia brachyloba* on Tecate Peak 5 mo after an October 1965 fire was 8883 individuals per hectare. In May 1967, *Dicentra chrysantha* formed almost impenetrable thickets on the same burned slopes. Each year different species appeared and flourished. In 1970, *Eriophyllum confertiflorum* dominated with 9357 individuals per hectare. Competition created by these fire-following herbs, along with that from resprouting shrubs, may produce unfavorable conditions for cypress reestablishment.

Fire frequencies and intensities have also changed since fire protection has been established, resulting in unusual fires that may be adversely affecting cypress groves while favoring other species, just as they have reduced tree distributions in the San Bernardino Mts. (Wright 1966; Minnich 1974). Or perhaps the cypress are still unable to adapt, and are continuing to decline by becoming more and more restricted as they suffer repeated setbacks. Another possibility is that fluctuations in grove size and in tree densities occurred naturally in the past, and will continue to occur without jeopardizing cypress populations. Coleman (1905), for example, reported that the Huckleberry Hill *C. goveniana* grove was reduced from over 40 to only a few hectares by the 1901 fire. Dunning (1916) reported reproduction from that fire, and Wolf (1948) described the grove as extending for not over 1 km in any direction, perhaps approximating the 1900 grove size. Fire, apparently, must be considered in long-range terms before its role can be properly evaluated.

CLOSED-CONE PINES

Pinus attenuata

General biology. *Pinus attenuata* (knobcone pine) has the most widespread distribution of the California closed-cone pines and cypress (Sudworth 1908; Cain 1944). It occurs nearly the length of California, growing commonly on the slopes of the North Coast Ranges and Sierra Nevada to above 1524 m, and becoming more discontinuous but still common in central California along the lower slopes of the Sierras south to Yosemite National Park, and in Central Coast Ranges south to the Santa Cruz (Thomas 1961; Taylor 1965) and Santa Lucia Mts. There are only three natural occurrences of knobcone pine in southern California: one in San Luis Obispo Co. near Cuesta Pass (Wells 1962), a widely scattered stand in the western San Bernardino Mts. (Horton 1960; Newcomb 1962; Wright 1966, 1968), and a location in the Santa Ana Mts. (Vogl 1973). It occurs at one location in Baja California, Mexico (Mason 1927; Newcomb 1959), and is abundant in southern Oregon (Sudworth 1908; Whittaker 1960; Critchfield and Little 1966; Waring 1969). Knobcone pine extends inland over 241 km from the coast, occurring from almost sea level to over 1676 m elevation (Sudworth 1908; Badran 1949; Stebbins 1950; Griffin and Critchfield 1972; Newcomb 1962). It is the only important closed-cone pine on the southern California mainland. The exact extent of many knobcone pine stands is unknown, particularly in northern California, and the locations of a few groves are uncertain or unrecorded. Range records have been complicated by apparent confusion with *Pinus muricata* and by the widespread planting and naturalization of *P. attenuata* (Howell 1959; Griffin and Critchfield 1972).

Little is known about the vegetation of *Pinus attenuata* groves or forests. The only known quantitative vegetational surveys of knobcone pine are those of Whittaker (1960) in the Siskiyou Mts. of Oregon and California, Wright (1966) in the San Bernardino Mts., and Vogl (1973) in the Santa Ana Mts., the most northern and southern locations in California. Most references are floristic or descriptive in nature, or only mention the occurrence of knobcone pine or a knobcone pine community (Sudworth 1908; Jepson 1909; Jensen 1939; Peattie 1953; Abrams 1923; Griffin 1958; Munz and Keck 1959; Horton 1960; Gillett et al. 1961; Thomas 1961; Sawyer and Thornburgh 1969; Hoover 1970).

Because of the generally small stature of the trees, their discontinuous distribution, and their common occurrence outside commercial forest areas, knobcone pine has received little silvicultural attention (Fowells 1965). But because of the ability of knobcone pine to hybridize with commercially important *Pinus radiata* (Stockwell and Righter 1946; Bannister 1958a,b; Thomas 1961; Critchfield 1967) and to reseed naturally after fire, the species has come into use for roadside and afforestation plantings, particularly in areas subject to repeated fires (Peattie 1953).

Studies have been conducted to evaluate the genetic (Stebbins 1942, 1950), ecotypic (McMillan 1956; Kruckeberg 1967; Wright 1971), and geographical (Duffield 1951; Newcomb 1962) variations of knobcone pine. Its taxonomic (Jepson 1909; Mason 1930; Newcomb 1962), evolutionary, and past widespread distribution (Mason 1932, 1934; Cain 1944; Benson 1962; Axelrod 1967) and edaphic relationships (Whittaker 1954a,b, 1960; Mason 1946a,b; McMillan 1956; Griffin 1957;

Kruckeberg 1969; Mallory et al. 1973) have also been investigated. Bynum and Miller (1964) and Hawksworth and Wiens (1972) reported on a disease and the dwarf mistletoes of knobcone pine, but the species is apparently not subject to heavy insect or disease attacks (Peattie 1953). Other studies have focused attention on various ecological aspects of the pine (Cowles 1948; Badran 1949; Everett 1957; Wright 1966, 1974; Conkle 1971; Y. B. Linhart, unpub. rept.).

Vegetation. As noted above, detailed descriptions of knobcone pine communities are too few to permit many generalizations. Knobcone pine usually occupies a transitional position between lower chaparral and woodland and higher-elevational coniferous forests, and can form a low elevation treeline (Wright 1966, 1971, 1974). Because of its patchy distribution, it is usually surrounded by another community. It is most commonly associated and mixed with chaparral and woodland (Newcomb 1962).

The associated chaparral is usually dominated by *Adenostoma fasciculatum* and contains *Arctostaphylos* and *Quercus* spp. and other common shrubs (Munz and Keck 1959; Horton 1960; Thomas 1961; Newcomb 1962; Wright 1966; Vogl 1973). *Vaccinium* spp. shrubs occur in some central and northern stands (Whittaker 1960; Thomas 1961). Whittaker (1960) found that *P. attenuata* tended to be concentrated with shrubs, perhaps because these were the only hospitable sites on the serpentinite. The associated woodland is generally dominated by *Quercus* spp., *Pinus coulteri*, or *P. sabiniana* (Munz and Keck 1959; Wright 1966; Vogl 1973).

In the Central Coast Ranges, knobcone pine is found with *Cupressus abramsiana* and *Pinus radiata* (Wolf 1948; Griffin 1958; Munz and Keck 1959; Thomas 1961), and natural hybrids with *P. radiata*, called *P. attenuradiata* Stockw. & Right., occur (Stockwell and Righter 1946; Thomas 1961). This is the only place where *P. attenuata* is found with another closed-cone pine (Stebbins 1950; Critchfield 1967). *Cupressus macnabiana* occurs with *P. attenuata* in several northern California locations (Wolf 1948).

Because of its wide distribution in northern California, knobcone pine is associated with a variety of communities, including north coastal coniferous forest, Douglas fir forest, mixed evergreen forest, and Sierran mixed conifer forest (Fig. 9-4). Common associated conifers are *Libocedrus decurrens, Pseudotsuga menziesii, Pinus jeffreyi, P. monticola,* and *P. lambertiana* (Whittaker 1960).

Despite these associations, knobcone pine often grows alone, particularly on better sites (Sudworth 1908), where it frequently occurs in dense stands or dwarfed thickets (Whittaker 1954b, 1960). It mixes with other communities most commonly on open sites, but the adjacent communities are invariably reduced in species, cover, and size on the knobcone sites. Herbaceous species are usually rare, except for fire-followers and endemics (Whittaker 1954b; Wright 1966; Vogl 1973). Chaparral shrubs occur individually or in small patches among widely spaced pines, with shrub cover often exceeding tree cover (Whittaker 1960; Vogl 1973). Sometimes a mosaic of chaparral, woodland, and pine forest occurs as the various communities respond to topographical and substrate differences.

Detailed floristic investigation will probably reveal localized species restrictions and endemism on other knobcone pine sites, just as have been found for the Santa Ana Mts. location (Vogl 1976a) and other closed-cone species (Mason 1934; Cain 1944; Stebbins 1950; Proctor and Woodell 1975).

Figure 9-4. A stand of *Pinus attenuata* in the Sierra Nevada, about 1100 m elevation, in Placer Co. The short stature of the trees, their high density, and small diameter at breast height are typical.

Soil/parent material–climate relationships. Throughout its range, *Pinus attenuata* occurs on gentle to steep slopes and ridges in dry, shallow, raw, and rocky soils (Sudworth 1908; Jepson 1909; Peattie 1953; Munz and Keck 1959; Thomas 1961; Newcomb 1962; Fowells 1965). Serpentinite is the most common substrate (Whittaker 1954b, 1960; Newcomb 1962; Vogl 1973), but knobcone pine also occurs on nonserpentine soils (McMillan 1956), including sandy (Griffin 1958; Thomas 1961) and granitic types (Wright 1966). Substrate specificity may assume importance as the species become more discontinuous in central and southern California (McMillan 1956; Vogl 1973). This species appears to be less edaphically controlled than the other closed-cone pines and cypress, but this will not be confirmed until more stands are studied. Climatic conditions and plant growth in northern stands, for example, may mask the effects of serpentine (Whittaker 1954b), and substrate specificity may be overlooked. The substrates, whether serpentinite or not, are apparently infertile, acid, and dry or contain toxic elements (Walker 1954; Whittaker 1954a,b; Newcomb 1962; Vogl 1973; Proctor and Woodell 1975; Westman 1975).

These infertile and undeveloped substrates not only restrict knobcone distributions, but also support limited shrub or other plant growth, whereas adjacent nonspecialized or more fertile soils support dense chaparral or woodland without the pines (Whittaker 1954b; Horton 1960; Wright 1966; Vogl 1973). Serpentinite pine stands in the Santa Ana Mts., for example, possessed about 4× more bare ground on the average than the surrounding chaparral and had only 3 common species, whereas the adjacent chaparral supported about 10 common species (Table 9-8; Vogl 1973).

Knobcone pine apparently can tolerate the existing impoverished and possibly toxic edaphic conditions of serpentinite substrates. It is often slow-growing, stunted, and chlorotic on raw serpentine and only begins to grow better on more weathered and deeper soils (Newcomb 1962; Vogl 1973). This conifer is apparently restricted to serpentinite or other infertile substrates by the degree of plant competition or allelopathy found on nonspecialized soils (Whittaker 1954b; Horton 1960; Christensen and Muller 1975). Like the other closed-cone pines and cypress, it is not restricted to poor substrates by its inability to grow on normal soils, for it has been successfully cultivated on a variety of soils in coastal and montane locations in California (Cowles 1948; Everett 1957; Howell 1959). Although it appears to be restricted by its inability to compete with vegetation on normal substrates, particularly during reestablishment after fire (Horton 1960), the exact nature of this competition is unknown, and proof of this hypothesis has not been produced (Kruckeberg 1954; Proctor and Woodell 1975).

Knobcone pine groves and forests generally occur in locations exposed to various amounts of marine air. The extent to which low clouds and fog modify their environments is unknown or unrecorded, but must be less than for the more coastal *Pinus radiata* and *P. muricata.* In the Santa Ana Mts., the characteristic multiple-trunked trees, spreading crowns, medium-length needles, steep slope and ridgetop locations, and sparse stands of *P. attenuata* promote moisture condensation and fog drip. The Santa Ana Mts. stands are enhanced by summer and winter fogs and the high water-retaining capacity of the serpentinite (Vogl 1973). Precipitation in San Luis Obispo, Santa Cruz, Mendocino, and Del Norte Cos. is also augmented by moisture from year-round fogs (Azevedo and Morgan 1974; Westman 1975). Some higher-elevational locations in the Coast Ranges are above the summer fog belt but are affected by frequent fogs the rest of the year. The San Bernardino Mts. stands are inland from the coast but are, nevertheless, influenced by coastal clouds and fog in the winter and spring, and similar conditions may temper the environments of other inland knobcone pine locations.

Climatic conditions favorable for *P. attenuata* apparently occur in northwestern California and southwestern Oregon, where the most extensive stands grow over the widest topographical conditions (Whittaker 1960). Precipitation is the greatest in the north, and snow is common in stands above 914 m and even occurs occasionally in the southern California locations.

Role of fire. The closed-cone habit is more pronounced in *Pinus attenuata* than in the other closed-cone pines and cypress (Benson 1962). Although the persistence of closed cones varies among knobcone populations (Badran 1949; Newcomb 1962; Vogl 1973), cones usually remain unopened and firmly attached to trunks and branches for extensive periods, commonly for the life of a tree (Sudworth 1908; Jepson 1909; Peattie 1953). Detached cones, cones on fallen branches, and cones on

TABLE 9-8. Average cover (%) of common plant species on a Santa Ana Mt. knobcone pine-serpentinite site and on a surrounding chaparral-granitic site. Averages were obtained from eight 30 m line intercepts in each type. Taken from Vogl (1973)

Species	Average Cover	
	Knobcone Pine	Chaparral
Adenostoma fasciculatum	10.3	25.3
Arctostaphylos glandulosa	38.9	5.1
Ceanothus crassifolius	1.2	15.0
Dendromecon rigida	0.1	...
Lotus crassifolius	0.2	0.1
Pinus attenuata	12.5	...
Quercus dumosa	0.2	20.1
Bare ground	48.7	12.8
Ceanothus tomentosus var. *olivaceus*	...	0.3
Cercocarpus betuloides	...	2.8
Chlorogalum pomeridianum	0.7	...
Elymus triticoides	...	0.4
Eriodictyon crassifolium	...	0.1
Eriogonum fasciculatum	...	0.4
Eriophyllum confertiflorum	0.2	...
Fraxinus dipetala	...	0.6
Galium angustifolium	...	0.1
Garrya fremontii	...	1.2
Bromus spp.	...	2.6
Heteromeles arbutifolia	...	6.6
Lepechinia cardiophylla	0.5	...
Marah macrocarpus	...	2.3
Penstemon cordifolius	0.2	...
Rhus diversiloba	...	0.1
R. ovata	...	2.6
Salvia melifera	...	4.0
Solanum douglasii	...	0.4
Symphoricarpos mollis	...	0.6
Trichostema lanatum	0.1	...
Total average cover	65.1	90.7

dead and fallen trees also remain closed (Vogl 1973). Only rarely do the heavy, wooden cones open without fire, and those are usually partially opened by rodents, insects, weathering, or rot. Even when tree squirrels cut off and dismantle cones, or when an occasional cone has fully opened, germination and establishment may not occur. Apparently, the thin-coated and fragile seeds quickly perish if receptive sites and proper climatic conditions are not available. Reproduction is also absent in decadent stands where the majority of trees are senescent or dying, conditions sometimes created by fire prevention. Fire is the usual and necessary cone opener, simultaneously preparing a favorable seedbed. Cones are seldom consumed and seeds rarely damaged in a fire, and the opened cones shed seeds for a considerable time (Vogl 1973). Wet weather followed by dry days may actually cause the cone scales to open wider, shedding additional seeds on an ideal seedbed of wetted ash and soils. The fallen seeds germinate rapidly (Wright 1966).

Trees begin producing cones when 5–12 yr old (Peattie 1953; Newcomb 1962; Vogl 1973) and continue to accumulate cones until a fire occurs or the trees die. Cones are generally produced in pairs on young and stunted trees, but older trees annually produce cones in circles of four, sometimes five, on all major branches. Thickets of stunted pines produce a low number of cones per tree, whereas widely spaced trees tend to have multiple trunks and high numbers of cones (Newcomb 1962), so that in either situation an adequate seed supply ensures reproduction after fire.

Fire creates the pioneer conditions of bare mineral soils and full sunlight necessary for seedling establishment. It temporarily reduces the cover of the dominant and resprouting shrubs and increases the cover of temporary fire-following herbs (Table 9-9; Vogl 1973). Fire releases mineral nutrients from the standing vegetation and litter, thus making more available for the new trees and thereby helping to counter the otherwise slow decomposition rates and general infertility of knobcone sites. Fire removes the established plant cover and litter accumulations and accentuates the continual erosion common to knobcone sites with their steep slopes, thin soils, and friable substrates. The steepness of many serpentinite sites may be related to location on active faults where they are continually rising because of expansion or increases in rock volume (Barnes et al. 1966; Vogl 1973). Fire and erosion counter soil and litter accumulation; the latter result in higher pH, improved nutrient content, and reduction in soil toxicity—all leading eventually to normal vegetation with which the pines cannot compete.

Pinus attenuata occurs in even-aged stands that originate after fire (Newcomb 1962) and often forms mosaics of different-aged stands (Vogl 1973). With the exception of some locations in northern California, the pine grows as sparse stands of scraggly trees, spotty thickets of stunted individuals, or dense stands that quickly stagnate (Jepson 1909; Peattie 1953; Whittaker 1960). Trees on better sites grow rapidly (Newcomb 1962) and attain heights of 15–26 m, but trees on most sites are commonly only 2–15 m high (Peattie 1953; Munz and Keck 1959; Horton 1960; Vogl 1973). The faded and thin foliage and common multiple trunks add to the pine's unimpressive appearance.

Trees have a short potential life span (Griffin 1964; Newcomb 1962; Wright 1966; Vogl 1973), usually reduced even more by recurring fires. Trees that escape fire begin dying when they are about 50 yr old; only a rare tree lives 100 yr. Pines over

TABLE 9-9. Average cover (%) and density per hectare for common species on unburned and recently burned knobcone pine stands in Santa Ana Mts. Averages were obtained from six sample sites. Taken from Vogl (1973)

Species	Cover		Density	
	Unburned	Burned	Unburned	Burned
Adenostoma fasciculatum	18.1	1.7	97	31
Arctostaphylos glandulosa	21.4	18.3	628	532
Bromus spp.	0.1	22.3	Not counted	
Ceanothus papillosus var. *roweanus*	...	1.1	...	115
Dendromecon rigida	...	4.0	3	199
Dryopteris arguta	0.4	...	19	...
Helianthemum scoparium var. *aldersonii*	...	...	10	72
Heteromeles arbutifolia	...	...	6	5
Lepechinia cardiophylla	...	0.4	...	19
Lonicera subspicata var. *johnstonii*	1.2	...	52	...
Lotus crassifolius	...	0.5	6	12
Mimulus moschatus var. *longiflorus*	4.1	...	256	5
Monardella lanceolata	...	...	...	18
Pinus attenuata	15.3	7.8	75	313
Quercus wislizenii var. *frutescens*	5.4	4.5	36	70
Rhamnus crocea	...	0.2	...	97
Symphoricarpos mollis	...	...	3	4
Tauschia arguta	...	0.2	...	133
Bare ground	36.6	48.5		

50 yr old in the Santa Ana Mts. showed abundant signs of deterioration (Vogl 1973), and stands of trees over 75 yr old are practically nonexistent (Horton 1960). Knobcone sites are subject to frequent fires, perhaps once every 33–50 yr, because of their positional relationships with other fire type communities, edaphically dry sites, and early and widespread senescence leading to favorable fuel conditions. Higher fire frequencies usually do not occur, even when adjacent chaparral and yellow pine forests burn more frequently, because of the slower growth, sparser cover, and lighter fuel accumulations on knobcone sites.

P. muricata (including *P. remorata*)

Taxonomy and distribution. *Pinus muricata* (bishop pine) is distributed discontinuously along the coast from Humboldt to Santa Barbara Cos., and on Santa Cruz and Santa Rosa Islands (Duffield 1951; Griffin and Critchfield 1972). The largest and greatest numbers of stands occur in Mendocino, Sonoma, San Luis Obispo, and Santa Barbara Cos. The species also occurs outside of California near San Vicento (Robinson and Epling 1940) and on Cedros Island, Baja California. This island population has been designated *P. muricata* var. *cedroensis* (Howell 1941; Libby et al. 1968) and appears to be similar to the Guadalupe Island Monterey pine variety (*P. radiata* var. *binata*: Griffin and Critchfield 1972).

Pinus muricata stands, like those of other closed-cone species, are believed to be relicts of a more widespread late Tertiary forest (Axelrod 1967 and Chapter 5). *Pinus muricata* is remarkably variable from one disjunct location to another and even varies within one stand. Westman (1975) distinguished between a xeric and a normal form of bishop pine on the Mendocino barrens. This morphological variation exceeds that of the other closed-cone pines and has led to speculation regarding its taxonomic relationships (Duffield 1951; Mirov et al. 1966). Mason (1930) described the Channel Island pines with smooth, symmetrical, nonreflexed cones as *P. remorata* (Santa Cruz Island pine). More recent evidence indicates that these island pines are similar, if not identical, to central California mainland stands of *P. muricata* (Mason 1949; Duffield 1951; Linhart et al. 1967), including their phytochemistry (Forde and Blight 1964) and crossability with other pines (Critchfield 1967). As a result, we are including *P. remorata* as a variation of *P. muricata*. Another variant in northern California, *P. muricata* var. *borealis* (Duffield 1951), differs from pines to the south in color, crossability (Critchfield 1967), and terpene composition (Forde and Blight 1964).

The species exists in a maritime climate, occupying headlands and low hills from near sea level to 400 m elevation, usually within 12 km of the ocean. Rainfall ranges from 63 cm in the north to 37 cm in the south. Additional moisture is supplied by frequent fogs and fog drip, which may be particularly critical during the usually dry summers (Libby et al. 1968). In the La Purisima Hills, Santa Barbara Co., for example, 73% of the summer days are foggy (Cole 1974). Temperatures and evaporation rates are ameliorated by the persistent cloud cover, fog, and moist air.

Vegetation. It is difficult to summarize the vegetation of *Pinus muricata* forests without listing the species common to each location, since the forests have a wide latitudinal spread and vegetational differences. Southern stands are surrounded by chaparral, coastal sage scrub, and coastal grassland; there common chaparral species such as *Adenostoma fasciculatum, Arctostaphylos* spp., *Quercus* spp., and

Heteromeles arbutifolia occur with the pines, but with the exception of *Quercus agrifolia* usually account for only a small part of the plant cover (Peattie 1953; Linhart et al. 1967; Cole 1974; Yeaton 1974). Typical stands of *P. muricata* are dense (60–90% cover) and occupy northern-exposed slopes (Table 9-10). These north-facing slopes often have more northern species such as *Pseudotsuga menziesii, Polystrichum munitum, Arbutus menziesii, Vaccinium ovatum,* and *Lonicera hispidula* var. *vacillans* (Cole 1974).

Northern pine stands occur in Douglas fir forest, mixed evergreen forest, redwood forest, coastal grassland, and pygmy forest. These forests are rich in ericads such as *Arctostaphylos* spp., *Vaccinium ovatum, V. parvifolium, Gaultheria shallon, Arbutus menziesii, Rododendron macrophyllum,* and *Ledum glandulosum* ssp. *columbianum* (Westman 1975; Fig. 9-5).

In the La Purisima Hills 3.8% of the plants (Cole 1974), and on the Mendocino Plains 8% of the plants, are endemic (Westman 1975). Other associated species include *Myrica california, Berberis nervosa, Pteridium aquilinum, Rubus spectabilis, Goodyera oblongifolia,* and *Rhamnus purshiana* (Westman 1975).

Figure 9-5. A *Pinus muricata* forest near Salt Point, Sonoma Co. The upper canopy is not closed. Typical understory trees and shrubs include *Lithocarpus densiflora, Myrica californica, Gaultheria shallon, Vaccinium ovatum,* and *Ledum glandulosum.* Note figure in foreground, for scale.

TABLE 9-10. Vegetation cover of a *Pinus muricata* stand on diatomaceous mudstone, and adjacent coastal sage scrub on clayey mudstone and coastal sage scrub-grassland on sand. All stands occurred on north-facing slopes and were in the same vicinity within the La Purisima Hills, Santa Barbara Co. Each cover figure (%) is an average of five 30 m line intercepts. From Cole (1974)

Species	Cover		
	Diatomaceous Mudstone	Clayey Mudstone	Sand
Pinus muricata	53.9	...	...
Artemisia californica	...	45.9	20.1
Arctostaphylos crustacea	19.3	...	...
Elymus condensatus	...	13.3	14.7
Quercus agrifolia	12.3	...	10.3
Lotus scoparius	7.0	8.5	...
Mimulus aurantiacus ssp. *lompocensis*	...	7.9	6.8
Rhus diversiloba	1.5	4.5	7.6
Baccharis pilularis ssp. *consanguinea*	0.4	5.4	5.0
Stachys chamissonis	...	3.5	...
Eriophyllum confertiflorum	...	1.0	4.1
Cercocarpus betuloides var. *blancheae*	...	...	3.9
Penstemon cordifolius	...	...	3.3
Horkelia cuneata	...	...	3.0
Rhamnus crocea	...	...	2.9
Ribes speciosum	...	...	2.8

Artemisia dracunculus	...	...	2.4
Heteromeles arbutifolia	2.3	...	...
Vaccinium ovatum	2.3	...	...
Eriogonum parvifolium	0.4	1.9	...
Haplopappus ericoides	...	...	1.3
Sanicula crassicaulis	...	1.1	...
Achillea borealis ssp. *californica*	...	...	0.8
Solanum umbelliferum var. *incanum*	...	0.7	...
Salvia spathacea	...	...	0.6
Corethrogyne filaginifolia var. *bernardina*	...	0.6	...
Salvia spathacea	0.3	...	...
Scrophularia californica	...	0.3	...
Dryopteris arguta	0.2	0.3	0.3
Marah fabaceus var. *agrostis*	...	0.6	0.3
Phacelia ramosissima var. *austrolitoralis*	...	...	0.3
Dentaria californica	0.2	...	...
Juncus patens	...	...	0.2
Verbena lasiostachys	...	...	0.2
Orthocarpus purpurascens var. *pallidus*	...	...	0.2
Odiantium jordanii	...	0.1	0.1
Cirsium occidentale	...	...	0.1

TABLE 9-10 [continued]

Species	Cover		
	Diatomaceous Mudstone	Clayey Mudstone	Sand
Pityrogramma triangularis	...	...	0.1
Gnaphalium microcephalum	...	0.4	< 0.1
Stachys bullata	...	...	< 0.1
Oenothera cruciata	...	...	< 0.1
Grasses	...	15.7	25.2
Bare ground	10.0	3.2	6.0
Total average cover	101.1	110.0	118.0

Substrate preferences. *Pinus muricata* has definite soil–parent material preferences. It occurs on a variety of substrates, but they are all generally shallow, acid, and poorly drained (Duffield 1951; McMillan 1956).

In Mendocino Co. it grows on, and adjacent to, the podzolized soils of the pygmy forest. These white sands are acid (pH 4.3–4.7), have possible toxicities and nutrient deficiencies (Westman 1975), are flooded by perched water on top of an iron hardpan in the B horizon, and have a high, persistent water table between the hardpan and sandstone terraces below. Pines on these sites are small and stunted (Westman and Whittaker 1975), but those on adjacent sands or those that have penetrated the hardpan to permanent water supplies (Jenny et al. 1969) attain the largest sizes for this species.

The Monterey Peninsula bishop pine forest occurs on similar infertile, podzolized, and swampy soils (Griffin 1972; Griffin and Critchfield 1972). *Pinus muricata* mixes with *P. radiata* at this location, but differing flowering times prevent hybridization (Duffield 1951; Critchfield 1967).

In Santa Barbara and San Luis Obispo Cos., bishop pine has a definite substrate preference, with the densest forests restricted to diatomaceous shale. In several localities the bishop pine forest stops abruptly along geological contacts (Fig. 9-6). This chalklike substrate produces shallow, highly acid (pH 4.7) soils with a high water-retaining capacity (Cole 1974). In the La Purisima Hills, large areas of coastal sage scrub or grassland appear to occur on diatomaceous shale, but these treeless communities actually grow on a remarkably similar clayey mudstone which contains some diatom frustules (Cole 1974). A small number of isolated trees or small groves (2% of the La Purisima Hills population) occur on Careaga sandstone and Orcutt sand and apparently constitute the basis of previous reports (Griffin 1964; Axelrod 1967) that *P. muricata* is not edaphically restricted. These deep, white, quartz sand pockets also support reduced plant growth and apparently mimic the edaphic conditions produced on the diatomaceous shale. We have found no evidence to support Axelrod's (1967) speculation that "*muricata*" type trees grow on a variety of substrates, whereas "*remorata*" types are restricted to sandstone.

The Santa Cruz Island populations occur on sandy soils from diatomaceous parent material, weathered basalt, and granitic materials, and the Santa Rosa Island population grows on a siliceous soil (Linhart et al. 1967). Further soil–parent material generalizations cannot be made until the island and northern location substrates have been studied. It is probable that *P. muricata* has survived climatic changes since Pleistocene because of reduced plant competition on the poor, acid, and often swampy soils, the tempering effect of the prevalent fogs (Libby et al. 1968), and possible high water tables.

Role of fire. Stands of *Pinus muricata* are characteristically even aged, originating after fires (Linhart et al. 1967; Cole 1974). Whorls of firmly attached cones are produced on the main leaders each year, and the persistent cones remain sealed for a number of years (Peattie 1953). Cones are usually opened by fire, but on rare occasions old cones open on hot days. In some locations, gray squirrels (*Sciurus griesus*) remove cones and cut them open to feed on the seeds. Trees may be dwarfed as on the Mendocino Plains, occur in stagnated thickets as in the La Purisima Hills or on the west end of Santa Cruz Island, or occur as open grown, twisted, sheared, and topped by wind and salt spray as on Inverness Ridge. Stand sizes of these apparently short-lived pines sometimes fluctuate from one fire to another (Linhart et al. 1967;

Figure 9-6. The role of soil type in the distribution of ***Pinus muricata*** in the La Purisima Hills, Santa Barbara Co. In the upper photograph, bishop pine forest grows on diatomaceous mudstone in lower left, coastal sage scrub and grassland grow on clayey mudstone in center, and oak–grassland grows on Orcutt sand at top. In the lower photograph, bishop pine torest grows on diatomaceous mudstone, with chaparral on Careaga sandstone and isolated pine groves on deep sand pockets.

Cole 1974), and groves may have been eliminated from a few locations by unknown natural circumstances (Griffin and Critchfield 1972). It is possible that all the trees in a grove are stripped of their cones by squirrels, leaving no chance for recovery after fire. Squirrels have been also known to damage pines by stripping bark (Cole 1974). A fire-free period of 80+ yr would also allow trees to succumb to diseases and die without reproducing. Bishop pine is susceptible to rust gall infection (*Peridermium harknessii*) and to secondary infection by a fungus (*Nectria fuckeliana*: Byler and Cobb 1972; Byler and Platt 1972). Fire appears to be a critical factor in the continuance of *P. muricata,* as it is with other closed-cone species, and may influence its distribution and abundance as well (Libby et al. 1968; Wells 1962).

P. contorta spp. *contorta*

Taxonomy and distribution. Critchfield (1957) proposed four subspecies of *Pinus contorta,* which is a taxon of subsection Contortae, a morphologically similar group of small-cone pines (Critchfield 1975). One of these is *Pinus contorta* ssp. *contorta* (beach pine), which is found along the Pacific coast from Mendocino Co., California, to southern Alaska. Although the southern extent is at Manchester, Mendocino Co., the subspecies is uncommon in California, becoming common in Oregon and further north (Griffin and Critchfield 1972).

Although we are including beach pine among the closed-cone pines, there is a question as to whether this subspecies possesses a distinct closed-cone habit such as that found in *P. contorta* ssp. *latifolia* or *P. banksiana.* In this respect, it may be most closely related to *P. contorta* ssp. *murrayana* of the Sierra Nevada–Cascade Ranges (Fowells 1965). However, we include it in this chapter for the sake of completeness.

General description. *Pinus contorta* ssp. *contorta* is confined to coastal dune (Fig. 9-7) and seaside bluff habitats in California, but farther north it commonly occurs to about 150 m elevation, sometimes to about 900 m (Sudworth 1908; Franklin and Dyrness 1973).

Synecological descriptions are not quantitative (Clark 1937; Cooper 1967;

Figure 9-7. A dense *Pinus contorta* ssp. *contorta* forest on stabilized dunes near Arcata, Humboldt Co. The trees appear dense and even aged, with an occasional *Picea sitchensis* present, and there is a dense Ericaceae- and lichen-dominated understory. *Pteridium aquilinum* is also shown in the foreground.

Wiedemann 1966). This very small to medium-sized tree appears to withstand saline, xeric, and even hydric conditions since it occurs on the leeward sides of active sand dunes, stabilized dunes, the marshy edges of brackish and freshwater sloughs, and exposed rocky headlands and bluffs, where it is subject to heavy wind and salt spray. It also occurs in and around sphagnum bogs farther north. Fog and fog drip may ameliorate the pine's maritime environment.

According to Fowells (1965), closed or serotinous cones are uncommon in most populations of *P. contorta* ssp. *contorta.* McMillan (1956) and Munz and Keck (1959) stated that cones open at maturity, but Sudworth (1908), Sargent (1922), Taylor and Little (1950), and Bowers (1961) reported serotinous, or both serotinous and nonserotinous, cones. Perhaps the possession of serotinous cones varies with local populations because of genetic variations and the hybridization that freely occurs among these pines (Critchfield 1975). Since coastal dune and bluff communities are usually considered to be fire free (Vogl 1976b), the possession of closed cones would seem to be of little advantage; but some fires along the Pacific coast do burn out to the shoreline, and beach pines with serotinous cones would benefit. The role of fire will not be resolved until stand structure, age composition, and reproductive requirements are known.

Soils may be highly acid and wet or may be dry sands and gravels. The common distorted and scrubby growth forms on exposed sites appear to be due to salt spray and prevailing winds. Stunting may be caused by poor drainage and acid soils. Trees only 12 m tall may be over 200 yr old (Peattie 1953). Trees often occur as open-grown individuals, scrubby thickets, or sparse to dense groves. The inland edges of beach pine stands often grade into north coastal coniferous or redwood forests. In these inland areas the subspecies grows with straight, clean trunks 12-16 m high in typical forest stands (Peattie 1953; Munz and Keck 1959).

P. contorta ssp. *bolanderi*

Distribution. *Pinus contorta* ssp. *bolanderi* (pygmy pine) is endemic to the white plains or barrens of coastal Mendocino Co. (see Fig. 9-2). It has the most limited distribution of the four subspecies of *Pinus contorta,* a distribution that generally coincides with that of *Cupressus pygmaea,* although it is more restricted. Its distribution may be controlled edaphically, since it occurs exclusively on the poorly drained soils of the barrens (Westman 1975).

General description and vegetation. The narrow coastal strip of closed-cone pine forest in Mendocino Co. is dominated by *Pinus contorta* ssp. *bolanderi, P. muricata,* and *Cupressus pygmaea.* Pygmy pine usually forms a dwarfed forest with thickets of pygmy cypress and several endemic shrub species, including *Arctostaphylos nummularia.* Associated vegetation and the general environment are discussed in the section on *Cupressus pygmaea.* Westman (1975) and Westman and Whittaker (1975) conducted a detailed soil and general vegetational analysis of this community. They noted the abundance of lichen encrustations on soil and plants in the pygmy conifer forest. The forest is relatively open with light on the forest floor about 25% of full sun. On the most acidic, shallow, podzolic soils, pygmy pine may only reach 0.3–0.6 m in height in 50–60 yr (Ornduff 1974), or 2.4–3.0 m in 122 yr (McMillan 1956). The stunted growth form is less pronounced when individuals occur on more favorable sites. Westman (1975) stated that frequently flooded areas produced a shorter pygmy forest with an abundance of *Ledum glandulosum* and *Lilium maritimum*.

The few small cones of *Pinus contorta* ssp. *bolanderi* are persistent on the trees and remain closed after reaching maturity (Griffin and Critchfield 1972), although trees with open cones can be found. Bowers (1961) stated that cones may remain closed for 15–20 yr. Thickets of stunted trees are often apparently even aged, relating to the closed-cone habit and fire origin. Study is needed to determine fire frequencies, stand structure, and average length of life of this unique species.

Pinus torreyana

Distribution. *Pinus torreyana* (Torrey pine) has been included as part of the California closed-cone pine and cypress forests by Knapp (1965) and Griffin and Critchfield (1972), but not by Munz and Keck (1959) and others. Even though we are including it here, its large, stalked female cones are not comparable to the closed or serotinous cones of the other pines and cypress discussed in this chapter, but are most similar to those of *Pinus sabiniana* (Digger pine) and possibly *P. coulteri* (Coulter pine, Haller 1967). *Pinus torreyana* is included in this chapter largely because of its disjunct coastal and insular distribution, a distribution comparable to that of the closed-cone species.

Torrey pine is the most restricted pine species in California, occurring in only two localities. The mainland location is north of San Diego near Del Mar, where it is scattered for about 8 km along the eroded coastal bluffs or marine terraces on both sides of Soledad Valley (Munz and Keck 1959; Griffin and Critchfield 1972). It also occurs 180 km to the northwest on Santa Rosa Island, where it occupies a coastal area about 0.8 km long and 0.4 km wide (Haller 1967). The pines occur from near sea level to about 150 m elevation and extend inland only short distances from the sea. No fossil record exists for *P. torreyana*, but it is believed to be a relict of the Pleistocene flora of California.

Vegetation. The mainland stand of Torrey pine consists of trees 10–15 m in height that are usually open grown and scraggly in appearance. The trees are often widely dispersed or occur in scattered groves, occupying ridgetops, slopes, and gullies. Most trees are apparently less than 100 yr old. Some of the existing groupings apparently originated after fire (Haller 1967).

The associated vegetation is a sparse coastal sage scrub and chaparral. Sea slopes often contain dune scrub elements, and some valleys grade into coastal salt marsh. The dominant shrub associates include the common *Adenostoma fasciculatum*, *Artemisia californica*, *Heteromeles arbutifolia*, *Rhus integrifolia*, *R. laurina*, *Quercus dumosa*, *Baccharis viminea*, *Dendromecon rigida*, *Encelia californica*, *Salvia apiana*, *S. mellifera*, *Haplopappus squarrosus*, and *Arctostaphylos glandulosa* var. *crassifolia*. Less common species include *Cercocarpus minutiflorus*, *Xylococcus bicolor*, *Simmondsia chinensis*, *Yucca schidegera*, *Echinocactus viridescens*, *Mammillaria dioica*, and *Dudleya* spp. An unpublished 1970 flora checklist by Fleming, Michael, and Irwin for Torrey Pines Reserve lists 117 species. A number of the associated species found in Torrey Pines Reserve are more characteristic of regions to the south and at the desert edge.

The island stand is dominated by *Adenostoma fasciculatum* and *Heteromeles arbutifolia,* but the shrubs do not form typical dense chaparral (Haller 1967; see also Chapter 26). All the trees on exposed ridges are wind pruned, and none is more than 11 m tall. R. N. Philbrick (pers. commun.) did not find trees over 75 yr old. There is a lower number of associated coastal sage scrub and chaparral species in the island grove.

Although the pines in the two localities have been compared for their morphological and genetic differences (Haller 1967), including interesting three- and four-needle fascicle variations, and terpene compositions have been studied (Zavarin et al. 1967), quantitative vegetation studies do not exist.

The two locations have similar maritime climates with low winter rainfalls and frequent fogs. The island population is probably subject to stronger winds than the mainland stand, and possibly to more salt spray. The pines grow in shallow, open sandy soils at both locations—on sandstone on the mainland, and on sandstone and diatomaceous shale on Santa Rosa Island. Little is known of the life history and ecological relationships of the species in its natural state, even though Torrey pine may be the rarest North American pine.

Pinus radiata

Distribution. At first discovery in 1602, *Pinus radiata* (Monterey pine) occurred in only three small areas along the California coast—primarily near Monterey, but also to the northwest near Año Nuevo Pt. and to the southeast near Cambria. Monterey pine has since been planted around the world because of its rapid growth and desirable lumber and pulpwood quality. It is now the leading exotic plantation species in Spain, Chile, Australia, and New Zealand (Scott 1960), and is of lesser but still significant importance in the Union of South Africa, Kenya, Argentina, and Uruguay (Scott 1960; Roy 1966). Consequently, much of the literature on Monterey pine has been published by foresters working on overseas plantations (Pert, 1963).

Native stands of Monterey pine have been studied by Coleman (1905), Dunning (1916), Larsen (1915), Lindsay (1932), Mason (1934), Stoddard (1947), Moulds (n.d., 1950?), McDonald (1959), Offord (1964), Forde (1962, 1966), and Roy (1966). Summary treatments of Monterey pine in its California environment are found in Forde (1966), Roy (1966), and Scott (1960). Excellent photographs of Monterey pine stands at Año Nuevo Pt., Monterey, and Cambria are found in Offord (1964).

The natural range of Monterey pine embraces some 5400 ha (Roy 1966), including about 400 ha near Año Nuevo Pt. 65 km northwest of Monterey, about 4000 ha near Monterey, and about 1000 ha near Cambria 135 km southeast of Monterey. Offord (1964), Forde (1966), and Roy (1966) presented detailed range maps of Monterey pine at these three localities. The pines occur within 8 km of the coast on all exposures, from sea level to 300 m. Nearly all of the pine stands occur on private land and thus have been or remain subject to logging, grazing, and urbanization.

Vegetation. Surprisingly few quantitative data are available on the composition of the native Monterey pine forests. Coleman (1905), Dunning (1916), Mason (1934), Moulds, and Forde (1962) listed associated tree, shrub, and herbaceous species. Roy (1966) provided a compendium of these species lists.

More recently, the tree, sapling, and understory vegetation in 48 stands of Monterey pine was sampled in 1965 and 1966 by Keith White. Of these stands, 4 were near Año Nuevo Pt., 37 near Monterey, and 6 near Cambria. Each stand sampled was about 1 ha in size. Homogeneity of structure and composition within any stand was maintained by remaining on the same ownership parcel (to avoid encompassing different land use histories) and on uniform topography. Although diversity of sites was a goal, the stands sampled were not intended to represent all possible site conditions. It was not possible to sample undisturbed stands since all of the Monterey pine forest has at some time been grazed, logged, or burned. Cattle had access to 24 of the 48 stands sampled.

In each stand a presence list was made for all vascular species observed. Beatrice F. Howitt was especially helpful in identifying plants. Herbs, shrubs, and tree seedlings were tallied in 40 1,000 cm^2 quadrats. Trees (stems greater than 10.1 cm dbh) and saplings (2.5–10.1 cm dbh) were sampled for frequency, basal area, and density at 10 points with the point-centered quarter method.

In selecting stands for sampling, the initial choice was usually based on observed dominance by Monterey pine trees. Indeed, Monterey pine made up 80% or more of the trees in most stands. But in 6 of the 48 stands, Monterey pine constituted less than 50%, and in 1 stand only 12%, of the 40 trees sampled. The latter stand, near Año Nuevo Pt., is probably the northernmost occurrence for Monterey pine; 29 of the 40 trees sampled were *Quercus agrifolia,* 2 were *Pseudotsuga menziesii,* 1 was *Pinus attenuata,* and 3 were *Arbutus mensiesii.* Four of the six stands appeared to have been cut over some 15 yr earlier, when pines may well have been cut in preference to live oaks.

Of the 1920 trees sampled in the 48 stands, 85% were Monterey pine and 15% were coast live oak. The percentage of Monterey pine as saplings was lower (69%), and as seedlings much lower (49%). There appears to be no clear explanation for the decline in Monterey pine from seedling to tree stages. Live oak seedlings should be more successful than Monterey pine seedlings in the pine litter for two reasons: more stored food in the oak acorns, and greater shade tolerance by oak seedlings. The shade-tolerant saplings and trees of live oak should also be more successful than pines under a pine canopy, yet they are not. In addition, previous cutting should have favored the live oak trees because of the preference for Monterey pine lumber. Perhaps the oaks are subject to greater mortality as seedlings because of browsing by cattle, although Monterey pine appears to suffer greater browsing damage. Oak seeds may be eliminated more often by birds and mammals than are pine seeds. Surface fires may also favor pines over oaks, environmental conditions may favor pine establishment and dominance, or it may be that the prevailing conditions are such that Monterey pine is replaced by itself in a cyclic fashion. Further investigation is needed to explain this.

The average number of trees in all 48 stands, calculated from the distances measured at the quarter points, was 93.0 per hectare. The average tree diameter was 35.4 cm, and the largest Monterey pine observed measured 124 cm. Only 14 Monterey pines out of 1624 measured exceeded 81.3 cm in diameter, but this may not reflect the real potential because ubiquitous cutting has undoubtedly depleted the larger size classes.

The most common (present in 10 or more stands) herb and shrub species found in the 48 stands are listed in Table 9-11 in rank order according to an index of commonness (species presence percentage times quadrat frequency). Note that only 9 of the 136 herb and shrub species were present in more than 50% of the stands. From cursory observation of the Monterey pine forests a high presence percentage would

TABLE 9-11. Common herb and shrub species in 48 stands of Pinus muricata forest. All species listed were present in 10 or more stands. Arctostaphylos tomentosa and Rhamnus californicus (not listed) were present in 10 or more stands, but not within the sample quadrats. Species marked with an asterisk are shrubs. Data collected by K. L. White

Species	Presence (%)	Average Quadrat Frequency (%)
Galium californicum	95.8	47.6
Rhus diversiloba*	93.7	29.3
Agrostis diegoensis	77.0	27.4
Vicia americana	54.1	35.7
Symphoricarpos mollis*	43.7	34.0
Rubus ursinus*	60.4	20.6
Elymus glaucus	70.8	16.1
Vaccinium ovatum*	29.1	35.5
Stachys bullata	54.1	14.2
Satureja douglasii	52.0	14.6
Bromus rigidus	31.2	19.1
Mimulus aurantiacus*	89.5	6.0
Carex globosa	39.5	11.3
Dryopteris arguta	43.7	9.3
Sanicula laciniata	20.8	18.7
Bromus carinatus	35.4	9.7
Lathyrus vestitus	27.0	12.7
Fragaria californica	33.3	9.5
Stipa lepida	35.4	7.2
Galium aparine	27.0	8.6
Sanicula crassicaulis	39.5	4.7
Iris douglasiana	33.3	5.0
Achillea borealis	31.2	5.3
Pteridium aquilinum var. lanuginosum	25.0	6.5
Melica imperfecta	22.9	6.7
Ceonothus thrysiflorus*	22.9	2.5

be expected for the introduced shrub *Cytissus monspessulanus,* but it was recorded in only one stand (near Monterey) since it occurs primarily on disturbed soil, such as roadsides, excluded from this study.

Shrub cover, measured along 91 m line transects in the 48 stands, ranged from 0 to 71%; the average for all 48 stands was 34.4%. Only eight shrubs were found in more than 10 stands. *Vaccinium ovatum* provided the most cover, about 7%.

Attempts to make successional predictions along classical lines from the limited field data may be misleading, but a three-dimensional ordination of the 48 stands suggested the following temporal relationships of shrub understory to Monterey pine canopy. After fire an even-aged stand of young pines develops. Four such stands were sampled in this study: tree diameter at breast height was 15–25 cm, tree age was 45 yr, tree density was 160–200 per hectare, and there was high cover (ca. 50%) of the shrubs *Arctostaphylos tomentosa* and *Vaccinium ovatum* (Fig. 9-8).

With competition and uneven growth, some pines overtop others and mortality occurs. Seven such stands were sampled: tree diameter at breast height was 20–40 cm, tree age was 65 yr, tree density was 80–120 per hectare, and shrub cover remained high (ca. 40%), with *Rhus diversiloba* and *Mimulus aurantiacus* in addition to the two shrubs named above. If these stands were disturbed lightly, cover by *Rubus ursinus* increased.

Figure 9-8. A closed-canopy *Pinus radiata* forest near Monterey, Monterey Co. The trees average 25 yr old, 18 m tall, and 143 per hectare. Sapling and seedling densities are 23 and 19 per hectare, respectively. Shrub cover is 91%, dominated by *Arctostaphylos tomentosa* and *Vaccinium ovatum.* Herbs are rare. This stand appears to be regrowth from a previous clear cutting.

As the pines grow older, tree density and shrub cover remain about the same, but the tree canopy begins to open. Younger age classes of pines occur as a result of enhanced growth after surface fires. Nine such stands were sampled: tree diameter at breast height was 15–70 cm, tree density was 80–120 per hectare, and shrub cover (around 40%) was dominated by *Symphoricarpos mollis* with some *Rhus diversiloba.*

When the initial pine age class reaches diameters of 100 cm, *Quercus agrifolia* sometimes makes up 25–50% of the trees. One or more younger age classes of pines occur in these "old" stands. Twelve such stands were sampled. The number of shrub species was low, and the shrub cover was very low (about 10%), dominated by *Mimulus aurantiacus* and *Rhus diversiloba.* With disturbance, such as selective logging, the number of shrub species and the cover increased.

Range restrictions. The restricted range of Monterey pine has always been intriguing. The exact original limits of the stands in the 1600s are not known, and cannot be deduced accurately from contemporary limits because numerous plantings have been made over the years and now appear natural. The contemporary limits are probably not very different from the original limits at Año Nuevo Pt. and Monterey but may be reduced at Cambria (Forde 1966). At least a half-dozen limiting factors have been suggested as setting the distribution limits. The most common is that the inland reach of summer fog sets the range limit (Larsen 1915; Mason 1934; Stoddard 1947; Moulds 1950?; McDonald 1959; Forde 1966). Mason (1934) noted that natural reproduction of Monterey pine in planted areas of the Berkeley Hills was occurring in only one small area where summer fog was more frequent and longer lasting. McDonald (1959) measured fog drip under Monterey pine stands during 1 wk in August and found larger values at higher elevations and at sites closer to the ocean. It may be that summer fog does affect the inland range limit, but its influence should be negligible on the similarly abrupt limits north and south along the coast.

Another factor suggested as limiting the range is soil type (Lindsay 1932; Mason 1934; McDonald 1959; Roy 1966). Within its range, Monterey pine grows on several different sandy loam soils developed above a variety of parent materials but derived mainly from overlying shallow marine sediments (Roy 1966). Lindsay (1932) observed fading out of the Monterey pine type inland from the coast at Monterey coincidentally with a change from deep sandy loams to heavier clay soils. Offord (1964) suggested that poor drainage and aeration in heavy clay soils favored the development of root diseases and reduced the vigor of Monterey pines. McDonald (1959) considered soil depth the main factor limiting distribution, finding no growth of pine on the fringe areas of the forest at Monterey where soil depth became less than 23 cm. Roy (1966) concluded that Monterey pine is an edaphic climax at both Monterey and Cambria.

Other factors have also been suggested as limiting the range: less than 38 mm annual precepitation (Larsen 1915), nondependability of rainfall (Lindsay 1932), and frost at higher elevations (Scott 1960). Factors suggested as limiting for good growth are amount of soil moisture, amount of total soil nitrogen, amount of water-soluble soil phosphate (McDonald 1959), and root diseases (Offord 1964). It has also been suggested that Monterey pine is a senescent species whose range is shrinking in the process of extinction (Mason 1932), although present observations along Monterey pine boundaries do not reveal active advance or retreat.

Grazing does not appear to limit seedling establishment. White's data showed that seedling densities were both high and low in grazed as well as ungrazed stands: the average number of seedlings of all trees was 59 per hectare, and density varied from 0 to 524, without apparent relation to exposure, distance from the coast, elevation, slope, or soil type.

There is no clear understanding of what it is that restricts the range of Monterey pine, and so Mason's (1934) conclusion still holds: "So far as climate, topography, and soil are concerned, there is no apparent reason why the forest does not extend itself farther."

Role of fire. Monterey pine cones remain attached to the trees for years, but they open and close several times during this period; thus a constant, though meager, seed rain results. Small numbers of pine seedlings were found in 34 of the 48 stands of White's study. The seedlings were often only a few centimeters tall, with thin stems and sparse needles, and very probably were several years old with growth suppressed because of the accumulated litter.

On the one small recent surface burn White sampled, there were some 490 seedlings per hectare in January after a previous spring burn, but they were 30–56 cm tall and had robust stems and profuse, dark green needle growth. Hence, even though the accumulating cones open without fire and modest recruitment takes place, optimum conditions for reestablishment occur with fire, whereby maximum numbers of cones are opened and a receptive seedbed is prepared. Monterey pine is exceptional among the closed-cone pines and cypress in that fires that produce optimum reproduction are not as often the catastrophic types common to the other species, but are more frequently surface fires in which parent trees survive (as indicated by basal wounds or fire scars on trees). The role of fire in Monterey pine, including possible distributional relationships, needs study before sound management plans can be made. Fire frequencies and intensities may be particularly critical in the continuation of this conifer.

SUMMARY

Impact of Animals

Most southern and central California cypress and closed-cone pine forests were part of Spanish and California ranches and were subject to grazing by domestic livestock. Cypress and closed-cone pine trees are generally considered to be undesirable forage for livestock, although young plants are vulnerable, particularly on overgrazed ranges. The soft, succulent seedlings of the pines are more susceptible to herbivore utilization than the cypress with their sharp juvenile foliage.

Some island populations are still under heavy grazing by cattle, feral sheep, feral pigs, and goats. As a result, some stands such as those on Cedros Island are endangered, others are without reproduction, and some show a reduction in associated species or have had serious erosion and soil losses. The systematic removal and exclusion of feral sheep on Santa Cruz Island, for example, has been an essential step in assuring the continuance of bishop pine. Some stands in northern California have been subject to grazing, and current grazing is affecting many Monterey pine and some cypress groves. The intensive grazing and frequent burning associated with the Spanish and early California ranching era might have eliminated some cypress or pine populations without such events being recorded. Although most of the southern California locations are presently without livestock, the past impact may have permanently altered vegetational composition, since grazing promoted non-palatable species at the expense of palatable species.

Rabbit and deer browsing takes place in the mainland populations and some of the island stands, but the impacts of these herbivores have not been recorded. Squirrels have been removing the accumulated cones from a number of closed-cone pine stands and, along with bark stripping, have been exerting negative impacts on the pines. Porcupines may also be causing some damage to pine stands in northern California by stripping bark, destroying shoot leaders, and girdling trees. Ground-dwelling rodents destroy seeds released when squirrels open cones, or when an occasional cone opens without fire, since reproduction without fire appears to be rare or nonexistent for most of these conifers. Protected deer populations in such places as Point Lobos Reserve are increasing and may build up until they seriously damage the conifer stands.

Impact of Man

Although current logging operations are generally having little effect on closed-cone pine and cypress forests, past logging has seriously affected some areas. Monterey and bishop pine stands in the vicinity of Monterey Bay were altered or reduced by early logging operations. Wolf (1948) cited several examples of cypress groves being diminished as a result of cutting for fence posts. Logging may have also changed the composition of some closed-cone conifer forests on the Mendocino Co. plains (Westman 1975). Scattered cutting for firewood has occurred in a number of Monterey, bishop, beach, and knobcone pine stands. Cutting still continues in pine and cypress stands in central and northern California.

Greater threats to these conifer stands are road building and sand, gravel, and mining operations. Most coastal populations have been dissected and reduced by roads. In a few places, the narrow strip of conifers remaining between the road and shoreline has been further reduced by beach and shoreline erosion. Several bishop pine groves in the vicinity of Lompoc, Santa Barbara Co., have been severely reduced or eliminated by diatomaceous earth strip-mining operations. A large portion of the Sierra Peak Tecate cypress grove in the Santa Ana Mts. has been destroyed by strip mining for clay deposits. Some conifer stands have been disturbed by the removal of sand or gravel. A large sand pit, for example, has endangered Gowen cypress on Huckelberry Hill, Monterey Co.

The La Purisima Hills bishop pine groves are studded with oil pumps and storage tanks and have been cut up by numerous service roads. Urbanization, including golf courses and resorts, and the building of seaside homes and mountain cabins have also made inroads into some conifer stands. This is particularly serious on the periphery of the mainland Torrey pine stand (Bedford 1965) and in the groves of Monterey pine and associated conifers. Little (1975) believed that the proximity of Torrey pines to densely populated areas may lead to destruction by air pollution. This is unlikely because of its coastal location, but the two southernmost stands of knobcone pine and the Santa Ana Mt. Tecate cypress are subject to heavy smog, which may be weakening trees (McBride et al. 1975).

Although a few groves have been lost and others disturbed because of man's perturbations (Wolf 1948), many of the groves or forests are presently secure. A summary of the status of common California closed-cone pines and cypress, including species that are protected in preserves and those that are endangered, has been made by Little (1975). Efforts should be made to include more of these local or rare conifers in natural areas or preserves.

Many of the groves have been visually scarred, and trees and associated plants destroyed, by the construction of fire breaks or fuel breaks. Wide firebreaks, for example, have destroyed Tecate cypress in the Guatay Mt. and Otay Mt. groves, another has eliminated several *Cupressus abramsiana* in the Bonnie Doon grove, and knobcone pines have been destroyed in the Santa Ana Mts. Firebreaks could have been constructed by staying outside these unique groves. Fire protection personnel have apparently been unaware of the unique status of these conifers or have had little regard for them because of their low commercial value.

A more serious effect of fire protection has been the disruption of natural fire frequencies and intensities, and in some places the nearly complete elimination of fires. Total exclusion has been facilitated by overgrazing, road building, urban encroachment, and the establishment of state parks and preserves. Since most of these conifers are obligate fire types, fire cannot be eliminated or replaced by any other process. Many groves are presently stagnated, are in a state of decline, or are suffering from unusual competition (Vogl 1973) because of fire suppression. If fire is not reintroduced, these groves and forests will eventually be eliminated.

AREAS FOR FUTURE RESEARCH

The closed-cone pine and cypress forests contribute to the uniqueness of California's flora and vegetation. These conifers, as well as some associated species, are either endemic to California or restricted to California and adjacent Pacific coast locations.

Although most of these species and forests have been well known by biologists, there is a paucity of quantitative and comprehensive vegetation studies. As a result, it is difficult to generalize about the vegetation. In some instances, attention has been focused on the unusual and unique features of the group, rather than the common and general, thus further complicating attempts at generalization.

All of the species apparently display some degree of edaphic or soil–parent material preferences. The vegetation is generally more sparse and smaller in stature than that on the surrounding soils. In some locations substrate relationships may have been overlooked or oversimplified (Whittaker 1960), perhaps even misidentified (Cole 1974). Substrate effects may operate only in concert with a multiplicity of other ecological factors (Vogl 1973). Careful geological surveys, detailed substrate analyses, and transplant and growth experiments are needed to determine more conclusively the substrate role.

Further work is needed also on the taxonomic relationship of the cypress and closed-cone pines, particularly by ecological, chemical, and anatomical comparisons. Some of the edemism present in these conifers and associated species, particularly at the varietal level, may be nothing more than phenotypic variation caused by unusual substrates.

Investigation into the evolutionary history of these species should also be expanded. What other factors, in addition to increasing aridity, could account for the present disjunct distributions? As California became more arid, perhaps lightning fires occurred more often or burned larger areas. Fire frequencies might also have increased with the arrival of early man. Contemporary closed-cone pines and cypress may represent species adapted to, and survivors of, a trial by fires that accompanied aridity, rather than representing the vestiges of species that could not adapt and change to drought alone.

The life cycles of these conifers and associated species are intimately related to fire, in that they are fire adapted and fire dependent. Further research is needed to understand these fire relationships, which vary between species with their variable closed-cone habits and differing fire frequencies. The rarer species and forests are receiving protection or are in need of it, but protection often takes the form of trying to undo nature by attempting to reduce or eliminate natural and necessary extremes or catastrophes. The continuance of these conifers, however, can be assured only by ecological management, which will have to include the reintroduction of fire. The distributional studies of these conifers must now be followed by synecological and autecological studies so that sound management practices can be established.

LITERATURE CITED

Abrams, L. 1923. Illustrated flora of the Pacific States. Vol. 1. Stanford Univ. Press, Stanford, Calif. 538 p.

Armstrong, W. P. 1966. Ecological and taxonomic relationships of Cupressus in southern California. M.A. thesis, Calif. State Univ., Los Angeles. 124 p.

Axelrod, D. I. 1958. Evolution of the Madro-Tertiary geoflora. Bot. Rev. 24:434-509.

---. 1967. Evolution of the California closed-cone pine forest, pp. 93-149. In R. N. Philbrick (ed.). Proc. Symp. on the biology of the California Islands. Santa Barbara Bot. Gard., Santa Barbara, Calif.

Azavedo, J., and D. L. Morgan. 1974. Fog precipitation in coastal California forest. Ecology 55:1135-1141.

Badran, O. A. 1949. Maintenance of seed viability in closed-cone pines. M.S. thesis, Univ. Calif., Berkeley. 63 p.

Bannister, M. H. 1958a. Evidence of hypridization between Pinus attenuata and P. radiata in New Zealand. Trans. Roy. Soc. New Zealand 85:217-225.

---. 1958b. Variation in samples of two-year-old Pinus attenuata, P. radiata and their hybrids. Trans. Roy. Soc. New Zealand 85:227-236.

Barnes, I., V. C. LaMarche, Jr., and G. Himmelberg. 1966. Geochemical evidence of present-day serpentinization. Science 156:830-832.

Barry, J. W. 1974. Plant succession at Point Lobos State Reserve. Calif. Native Plant Soc., Monterey Bay Chapter. 13 p.

Bedford, H. 1965. A rare evergreen defies extinction. Audubon 67(3):182-183.

Benson, L. 1962. Plant taxonomy: methods and principles. Ronald Press, New York. 494 p.

Bowen, O. E. 1966. Stratigraphy, structure and oil possibilities in Monterey and Salinas quadrangles, California, pp. 48-67. In Symp. Pac. Section, Amer. Assoc. Petroleum Engineers. Bakersfield, Calif.

Bowers, N. A. 1961. Cone-bearing trees of the Pacific Coast. Pacific Books, Palo Alto, Calif. 169 p.

Byler, J. W., and F. W. Cobb, Jr. 1972. Effects of secondary fungi on the epidemiology of western gall rust. Canad. J. Bot. 50:1061-1066.

Byler, J. W., and W. D. Platt. 1972. Cone infection by Peridermium harknessii. Canad. J. Bot. 50:1429-1430.

Bynum, H. H., and D. R. Miller. 1964. Elytoderma deformans found on knobcone pine (Pinus attenuata) in California. Plant Disease Reporter 48:828.

Cain, S. A. 1944. Foundations of plant geography. Harper and Brothers, New York. 556 p.

Carlquist, S. 1965. Rare cypress clings to coast habitat. Nat. Hist. 74(8):38-43.

Chaney, R. W., and H. L. Mason. 1930. A Pleistocene flora from Santa Cruz Island, California. Carnegie Inst. Wash. Pub. 415:1-24.

Christensen, N. L., and C. H. Muller. 1975. Effects of fire on factors controlling plant growth in Adenostoma chaparral. Ecol. Monogr. 45:29-55.

Clark, H. W. 1937. Association types in the North Coast Ranges of California. Ecology 18:214-230.

Cole, K. L. 1974. Edaphic restrictions in the La Purisima Hills with special reference to Pinus muricata D. Don. M.S. thesis, Calif. State Univ., Los Angeles. 67 p.

Coleman, G. A. 1905. Report upon Monterey pine, made for the Pacific Improvement Company. Unpub. rept., U. S. Forest Serv., Pac. S W Forest and Range Exp. Sta., Berkeley, Calif. 75 p.

Conkle, M. T. 1971. Isozyme specificity during germination and early growth of knobcone pine. Forest Sci. 17:494-498.

Cooper, W. S. 1967. Coastal dunes of California. Geol. Soc. Amer. Mem. 104. 131 p.

Cowles, F. H. 1948. Handy knob cone pines thrive on tough going. Santa Barbara News Press 85:1-7 (May 9, 1948).

Critchfield, W. B. 1957. Geographic variation in Pinus contorta. Maria Moors Cabot Found. Pub. Harvard Univ., Cambridge, Mass. 118 p.

---. 1967. Crossability and relationships of the closed-cone pines. Silvae Genet. 16:89-97.

---. 1975. Interspecific hybridization in Pinus: a summary review, pp. 99-105. In D. P. Fowler, and C. W. Yeatman (eds.). Symp. on interspecific and interprovenance hybridization in forest trees. Proc. 14th Meeting, Canad. Tree Improv. Assoc. Part 2.

Critchfield, W. B., and E. L. Little, Jr. 1966. Geographic distribution of the pines of the world. USDA Misc. Pub. 991. 97 p.

Duffield, J. W. 1951. Interrelationships of the California closed-cone pines with special reference to Pinus muricata D. Don. Ph.D. dissertation, Univ. Calif., Berkeley. 108 p.

Dunning, D. 1916. A working plan for the Del Monte Forest of the Pacific Improvement Company. M.S. thesis, School of Forestry, Univ. Calif., Berkeley. 78 p.

Eckenwalder, J. E. 1975. Book reviews of Profiles in California vegetation, by W. B. Critchfield, and The distribution of forest trees in California, by J. R. Griffin and W. B. Critchfield. Madroño 23:103-104.

Everett, P. C. 1957. A summary of the culture of California plants at the Rancho Santa Ana Botanic Garden, 1927-1950. Rancho Santa Ana Bot. Gard., Claremont, Calif. 223 p.

Forde, M. B. 1962. Variation in the natural population of Monterey pine (Pinus radiata D. Don) in California. Ph.D. dissertation, Univ. Calif., Berkeley. 328 p.

---. 1964. Variations in natural populations of Pinus radiata in California, Parts I-IV. New Zealand J. Bot. 2:213-257, 459-501.

---. 1966. Pinus radiata in California. New Zealand J. For. 11:20-42.

Forde, M. B., and M. M. Blight. 1964. Geographical variation in the turpentine of bishop pine. New Zealand J. Bot. 2:44-52.

Fowells, H. A. 1965. Silvics of forest trees of the United States. USDA Agric. Handbook 271. 762 p.

Franklin, J. F., and C. T. Dyrness. 1973. Natural vegetation of Oregon and Washington. USDA Forest Serv., Pacific NW Forest and Range Exp. Sta., Gen. Tech. Rept. PNW-8. 417 p.

Gillett, G. W., J. T. Howell, and H. Leschke. 1961. A flora of Lassen Volcanic National Park, California. Wasmann J. Biol. 19:1-185.

Green, H. A. 1929. Historical note on the Monterey cypress at Cypress Point. Madroño 1:197-198.

Griffin, J. R. 1958. A study of the distribution of Pinus ponderosa Laws. and Pinus attenuata Lemm. on sandy soils in Santa Cruz County, California. M.S. thesis, School of Forestry, Univ. Calif., Berkeley. 40 p.

---. 1964. Isolated Pinus ponderosa forests on sandy soils near Santa Cruz, California. Ecology 45:410-412.

---. 1972. What's so special about Huckleberry Hill on Monterey Peninsula? pp. 3-7. In Forest Heritage, a natural history of the Del Monte Forest. Calif. Native Plant Soc., Berkeley, Calif.

Griffin, J. R., and W. B. Critchfield. 1972. The distribution of forest trees in California. USDA Forest Serv. Res. Paper PSW-82. 114 p.

Griffin, J. R., and C. O. Stone. 1967. MacNab cypress in northern California: a geographic review. Madroño 19:19-27.

Haller, R. J. 1967. A comparison of the mainland and island populations of Torrey pine, pp. 79-88. In R. N. Philbrick (ed.). Proc. Symp. on the biology of the California Islands. Santa Barbara Bot. Gard., Santa Barbara, Calif.

Hardham, C. B. 1962. The Santa Lucia Cupressus sargentii groves and their associated northern hydrophilous and endemic species. Madroño 16:173-179.

Hardham, C. B., and G. H. True, Jr. 1972. A floristic study of Point Arena, Mendocino County, California. Madroño 21:499-504.

Hawksworth, F. G., and D. Wiens. 1972. Biology and classification of dwarf mistletoes (Arceuthobium). USDA Agric. Handbook. 401. 234 p.

Hoover, R. F. 1970. The vascular plants of San Luis Obispo County, California. Univ. Calif. Press, Berkeley. 350 p.

Horton, J. S. 1960. Vegetation types of the San Bernardino Mountains. USDA For. Serv., Pac. SW Forest and Range Exp. Sta., Tech. Paper 44. 29 p.

Howell, J. T. 1929. The flora of the Santa Ana Canyon region. Madroño 1:243-253.

---. 1941. The closed-cone pines of insular California. Leafl. West. Bot. 3:1-8.

---. 1959. After ten years: names and notes for the Marin flora. Leafl. West. Bot. 9:53-66.

---. 1962. A Piute cypress postscript. Leafl. West. Bot. 9:253-254.

---. 1970. Marin flora. 2nd ed. Univ. Calif. Press, Berkeley. 366 p.

Howitt, B. F., and J. T. Howell. 1964. The vascular plants of Monterey County, California. Wasmann J. Biol. 22:1-184.

Jenny, H., R. J. Arkley, and A. M. Schultz. 1969. The pygmy forest-podsol ecosystem and its dune associates of the Mendocino coast. Madroño 20:60-74.

Jensen, H. A. 1939. Vegetation types and forest conditions of the Santa Cruz Mountains unit of California. USDA Forest Serv., Calif. Forest Range Exp. Sta., Forest Survey Release 1. 55 p.

Jepson, W. L. 1909. The trees of California. Cunningham, Curtis, and Welch, San Francisco. 228 p.

---. 1923. A manual of the flowering plants of California. Univ. Calif. Press, Berkeley. 1238 p.

King, R. C. 1962. Genetics. Oxford Univ. Press, New York. 347 p.

Knapp, R. 1965. Die vegetation von Nord- und Mittelamerika und der Hawaii-Inseln Gustav Fischer-Verlag, Stuttgart. 373 p.

Komarek, E. V., Sr. 1965. Fire ecology-grasslands and man, pp. 169-220. In Proc. Fourth Ann. Tall Timbers Fire Ecol. Conf.

Kruckeberg, A. R. 1954. The ecology of serpentine soils. III. Plant species in relation to serpentine soils. Ecology 35:267-274.

---. 1967. Ecotypic response to ultramafic soils by some plant species of northwestern United States. Brittonia 19:133-151.

---. 1969. Soil diversity and the distribution of plants, with examples from western North America. Madroño 20:129-154.

Langenheim, J. H., and J. W. Durham. 1963. Quaternary closed-cone pine flora from travertine near Little Sur, California. Madroño 17:33-51.

Larsen, L. T. 1915. Monterey pine. Proc. Soc. Amer. Forest. 10:68-74.

Lawrence, L., R. Bartschot, E. Zavarin, and J. R. Griffin. 1975. Natural hybridization of _Cupressus sargentii_ and _C. macnabiana_ and the composition of the derived essential oils. Biochem. Syst. Ecol. 2:113-119.

Libby, W. J., M. H. Bannister, and Y. B. Linhart. 1968. The pines of Cedros and Guadalupe Islands. J. For. 66:846-853.

Lindsay, A. D. 1932. Monterey pine (_Pinus radiata_ D. Don) in its native habitats. (Australian) Commonwealth For. Bur. Bull. 10. (1937 reprint, 57 p.).

Linhart, Y. B., B. Burr, and M. T. Conkle. 1967. The closed-cone pines of the northern Channel Islands, pp. 151-177. In R. N. Philbrick (ed.). Proc. symp. on the biology of the California Islands. Santa Barbara Bot. Gard., Santa Barbara, Calif.

Little, E. L., Jr. 1950. Southwestern trees: a guide to the native species of New Mexico and Arizona. USDA Agric. Handbook 9. 109 p.

---. 1953. Check list of native and naturalized trees of the United States (including Alaska). USDA Agric. Handbook 41. 472 p.

---. 1966. Varietal transfers in _Cupressus_ and _Chamaecyparis_. Madroño 18:161-167.

---. 1970. Names of New World cypresses (_Cupressus_). Phytologia 20:429-445.

---. 1971. Atlas of United States trees. Vol. I. Conifers and important hardwoods. USDA Misc. Pub. 1146. 9 p., 200 maps.

---. 1975. Rare and local conifers in the United States. USDA Conserv. Res. Rept. 19. 25 p.

Mallory, J. I., W. L. Colwell, Jr., and W. R. Powell. 1973. Soils and vegetation of the French Gulch quadrangle, Trinity and Shasta Counties, California. USDA Forest Serv. Resources Bull. PSW-12. 42 p.

Mason, H. L. 1927. Fossil records of some west American conifers. Carnegie Inst. Wash. Pub. 346:139-158.

---. 1930. The Santa Cruz Island pine. Madroño 2:8-10.

---. 1932. A phylogenetic series of the California closed-cone pines suggested by the fossil record. Madroño 2:49-56.

---. 1934. Contributions to paleontology. IV. A Pleistocene flora of the Tomales formation. Carnegie Inst. Wash. Pub. 415-81-179.

---. 1946a. The edaphic factor in narrow endemism. I. The nature of environmental influences. Madroño 8:209-226.

---. 1946b. The edaphic factor in narrow endemism. II. The geographical occurrence of plants of highly restricted patterns of distribution. Madroño 8:241-257.

---. 1949. Evidence for the genetic submergence of _Pinus remorata_, pp. 356-362. In G. L. Jepson, G. G. Simpson, and E. Mayr (eds.). Genetics, paleontology, and evolution. Princeton Univ. Press, Princeton, N. J.

Mathews, W. C. 1929. Measurements of _Cupressus pygmaea_ Sarg. on the Mendocino "Pine Barrens" or "White Plains." Madroño 1:216-218.

McBride, J. R., V. P. Semion, and P. R. Miller. 1975. Impact of air pollution on the growth of ponderosa pine. Calif. Agric. 29(12):8-9.

McDonald, J. B. 1959. An ecological study of Monterey pine in Monterey County, California. M.S. thesis, School of Forestry, Univ. Calif., Berkeley. 58 p.

McMillan, C. 1952. The third locality for _Cupressus abramsiana_ Wolf. Madroño 11:189-194.

---. 1956. The edaphic restriction of _Cupressus_ and _Pinus_ in the Coast Ranges of central California. Ecol. Monogr. 26:177-212.

---. 1964. Survival of transplanted _Cupressus_ and _Pinus_ after thirteen years in Mendocino County, California. Madroño 17:250-254.

McMinn, H. E., and E. Maino. 1937. An illustrated manual of Pacific Coast trees. Univ. Calif. Press, Berkeley. 409 p.

Minnich, R. A. 1974. The impact of fire suppression on southern California conifer forests: a case study of the Big Bear fire, November 13 to 16, 1970, pp. 45-74. In Proc. symp. on living with the chaparral. Sierra Club, Calif.

Mirov, N. T., E. Zavarin, K. Snajberg, and K. Costello. 1966. Further studies of turpentine composition of _Pinus muricata_ in relation to its taxonomy. Phytochemistry 5:343-355.

Moulds, F. R. n.d. (1950?). Ecology and silvicluture of _Pinus radiata_ D. Don in California and southern Australia. Ph.D. dissertation, Yale Univ., New Haven, Conn.

Munz, P. A. and D. D. Keck. 1959. A California flora. Univ. Calif. Press, Berkeley. 1681 p.

Newcomb, G. B. 1959. The relationships of the pines of insular Baja California. Proc. Ninth Internat. Bot. Congr., Montereal 2:281 (abstr.).

---. 1962. Geographic variation in Pinus attenauta Lemm. Ph.D. dissertation, Univ. Calif., Berkeley. 191 p.

Offord, H. R. 1964. Diseases of Monterey pine in native stands of California and in plantations of western North America. USDA Forest. Serv. Res. Paper PSW-14. 37 p.

Ornduff, R. 1974. An introduction to California plant life. Univ. Calif. Press, Berkeley. 152 p.

Parish, S. B. 1914. The Tecate cypress. Bull. South. Calif. Acad. Sci. 13:11-13.

Peattie, D. C. 1953. A natural history of western trees. Houghton Mifflin, Boston. 751 p.

Pequegnat, W. E. 1951. The biota of the Santa Ana Mountains. J. Entomol. Zool. 42:1-84.

Pert, M. 1963. Pinus radiata - a bibliography to 1963 (Australian). Forestry and Timber Bureau, Canberra. 145 p.

Peterson, R. S. 1967. Cypress rusts in California and Baja California. Madroño 19:47-54.

Posey, C. E., and J. F. Goggans. 1967. Observations on species of cypress indigenous to the United States. Auburn Univ. Agric. Exp. Sta. Circ. 153. 190 p.

---. 1968. Variation in seeds and ovulate cones of some species and varieties of Cupresses. Auburn Univ. Agric. Exp. Sta. Circ. 160. 23 p.

Powell, W. R. 1968. Some vascular plants in Sonoma County. State Cooperative Soil-Vegetation Survey, Calif. Div. Forestry, Sacramento. 18 p.

Proctor, J., and S. R. J. Woodell. 1975. The ecology of serpentine soils. Adv. Ecol. Res. 9:263-349.

Robinson, W., and C. Epling. 1940. Pinus muricata and Cupressus forbesii in Baja California. Madroño 5:248-250.

Roy, D. F. 1953. Effects of ground cover and class of planting stock on survival of transplants in the eastside pine type of California. USDA Forest Serv. Res. Notes, Calif. Forest and Range Exp. Sta., Bull. 87. 6 p.

---. 1966. Silvical characteristics of Monterey pine (Pinus radiata D. Don). USDA Forest Serv., Pac. SW Forest and Range Exp. Sta., Res. Paper 31. 21 p.

Sargent, C. S. 1922. Manual of the trees of North America. Houghton Mifflin, Boston. 910 p.

Sawyer, J. O., Jr., and D. A. Thornburgh. 1969. Ecological reconnaissance of relict conifers in the Klamath region. Unpub. rept. on file at the Pac. SW Forest and Range Exp. Sta., Berkeley, Calif. 97 p.

Scott, C. W. 1960. Pinus radiata. UN, FAO Forest and Forest Prod. Stud. 14. 328 p.

Seeman, E. L. 1970. Vegetation type map of Huckleberry Hill, Monterey Peninsula. Prepared by J. A. Roberts Assoc., Tustin, Calif., for Del Monte Properties Co., Pebble Beach, Calif.

Stebbins, G. L., Jr. 1942. The genetic approach to problems of rare and endemic species. Madroño 6:241-272.

---. 1950. Variation and evolution in plants. Columbia Univ. Press, New York. 643 p.

Stockwell, P., and F. I. Righter. 1946. Pinus: the fertile species hybrid between knobcone and Monterey pines. Madroño 8:157-160.

Stoddard, C. H. 1947. Forestry in the Monterey pines. Amer. Forest 53:251, 275, 277.

Stone, C. O. 1965. Modoc cypress, Cupressus bakeri Jeps., does occur in Modoc County. Aliso 6:77-87.

Sudworth, G. B. 1908. Forest trees of the Pacific slope. U. S. Dept. Agric., Washington, D. C. 441 p.

Taylor, B. 1965. The canyon recovered. Westways 57(11):22-24.

Taylor, R. F., and E. L. Little, Jr. 1950. Pocket guide to Alaska trees. USDA Agric. Handbook 5. 63 p.

Thomas, J. H. 1961. Flora of the Santa Cruz Mountains of California. Stanford Univ. Press, Stanford, Calif. 434 p.

True, G. H., Jr. 1970. A preliminary checklist of the plants of Nevada County (unpub. rept.). 20 p.

Twisselmann, E. C. 1962. The Piute cypress. Leafl. West. Bot. 9:248-253.

---. 1967. A flora of Kern County, California. Wasmann J. Biol. 25:1-395.

Vaartaja, O. 1952. Forest humus quality and light conditions as factors influencing damping-off. Phytopathology 42:501-506.

---. 1962. The relationship of fungi to survival of shaded tree seedlings. Ecology 43:547-549.

Vogl, R. J. 1973. Ecology of knobcone pine in the Santa Ana Mountains, California. Ecol. Monogr. 43:125-143.

---. 1976a. An introduction to the vegetation of the Santa Ana and San Jacinto Mts. In J. Latting (ed.). Plant communities of southern California. Calif. Native Plant Soc., Spec. Pub. 2 p. 77-98.

---. 1976b. Fire: a destructive menacn or a natural process. In J. Cairns (ed.). Symposium on recovery and restoration of damaged ecosystems. Univ. Press of Virginia, Charlottesville. (in press).

Wagener, W. W. 1948. The New World cypresses. II. Dieases of American cypresses. Aliso 1:257-321.

Wagnner, W. W., and C. R. Quick. 1963. Cupressus bakeri - an extension of the known botanical range. Aliso 5:351-352.

Walker, R. B. 1954. The ecology of serpentine soils. II. Factors affecting plant growth on serpentine soils. Ecology 35:259-266.

Waring, R. H. 1969. Forest plants of the eastern Siskiyous: their environmental and vegetational distribution. Northwest Sci. 43:1-17.

Wells, P. V. 1962. Vegetation in relation to geological substratum and fire in the San Luis Obispo quadrangle, California. Ecol. Monogr. 32:79-103.

Westman, W. E. 1975. Edaphic climax pattern of the pygmy forest region of California. Ecol. Monogr. 42:109-135.

Westman, W. E., and R. H. Whittaker. 1975. The pygmy forest region of northern California: studies on biomass and primary productivity. J. Ecol. 63:493-520.

Whittaker, R. H. 1954a. The ecology of serpentine soils. I. Introduction. Ecology 35:258-259.

---. 1954b. The ecology of serpentine soils. IV. The vegetational response to serpentine soils. Ecology 35:275-288.

---. 1960. Vegetation of the Siskiyou Mountains, Oregon and California. Ecol. Monogr. 30:279-308.

Wiedemann, A. M. 1966. Contributions to the plant ecology of the Oregon coastal sand dunes. Ph.D. dissertation, Oregon State Univ., Corvallis. 255 p.

Wiggins, I. L. 1933. New plants from Baja California. Contrib. Dudley Herb. 1:161.

Wilde, S. A. 1958. Forest soils. Ronald Press, New York. 537 p.

Williams, J. W. 1970. Geomorphic history of Carmel Valley and Monterey Peninsula, California. Ph.D. dissertation, Stanford Univ., Stanford, Calif. 106 p.

Wilson, R. C., and R. J. Vogl. 1965. Manzanita chaparral in the Santa Ana Mountains, California. Madroño 18:47-62.

Wolf, C. B. 1948. The New World cypresses. I. Taxonomic and distributional studies of the New World cypresses. Aliso 1:1-250.

Wright, R. D. 1966. Lower elevational limits of montane trees. I. Vegetational and environmental survey in the San Bernardino Mountains of California. Bot. Gaz. 127:184-193.

---. 1971. Local phytosynthetic ecotypes of *Pinus attenuata* as related to altitude. Madroño 21:254-257.

---. 1974. Rising atmospheric CO_2 and photosynthesis of San Bernardino Mountain plants. Amer. Midl. Natur. 91:360-370.

Yeaton, R. I. 1974. An ecological analysis of chaparral and pine forest bird communities on Santa Cruz Island and mainland California. Ecology 55:959-973.

Zavarin, E., W. Hathaway, T. Reichert, and Y. Linhart. 1967. Chemotaxonomic study of *Pinus torreyana* Parry turpentine. Phytochemistry 6:1019-1023.

Zavarin, E., L. Lawerence, and M. C. Thomas. 1971. Compositional variations of leaf monoterpenes in *Cupressus macrocarpa*, *C. pygmaea*, *C. goveniana*, *C. ambramsiana*, and *C. sargentii*. Phytochemistry 10:379-393.

CHAPTER

10

MIXED EVERGREEN FOREST

JOHN O. SAWYER
DALE A. THORNBURGH
Humboldt State University, Arcata, California

JAMES R. GRIFFIN
Hastings Natural History Reservation,
Carmel Valley, California

Introduction **360**
***Pseudotsuga*-hardwood forests** **361**
  Klamath region forests **362**
    Forests on granitic and metamorphic parent materials **362**
    Forests on ultrabasic parent materials **365**
    Summary **366**
  North Coast Range forests **366**
    The mosaic of vegetation types **366**
    Forest composition **367**
    Areas for future research **369**
  Successional status of some tree species **369**
Mixed hardwood forests **371**
  ***Quercus agrifolia–Arbutus* forest** **371**
  ***Lithocarpus–Arbutus–Quercus* forest** **373**
  ***Quercus chrysolepis-Pinus coulteri* forest** **374**
  Minor forest types **376**
    ***Abies bracteata* forest** **376**
    ***Pseudotsuga macrocarpa* forest** **376**
Summary **377**
Literature cited **379**

INTRODUCTION

The term "mixed evergreen forest" describes a characteristic set of coastal California mountain communities. Many species assemblages and several physiognomical classes are included within this broad category. The closed stands and the broad-leaved, sclerophyllous nature of the dominants typify these forests, which may also contain a minor to significant conifer component. In terms of physiognomical–ecological classes (Mueller-Dombois and Ellenberg 1974), this type includes temperate evergreen coniferous forests, evergeeen giant conifer forests with evergreen sclerophyllous understories, evergreen broad-leaved sclerophyllous forests, and mainly evergreen broad-leaved woodlands. *Acer macrophyllum, Quercus chrysolepis,* and *Umbellularia californica* more or less range throughout the type. Characteristic dominants, at least in some phases, are *Arbutus menziesii, Lithocarpus densiflora, Quercus agrifolia, Q. chrysolepis,* and *Pseudotsuga menziesii.*

The mixed evergreen forest, as discussed here, is extensive in California from Del Norte and Siskiyou Cos. south in varying forms to San Diego Co. From the Santa Lucia Range north, the type is more or less continuous, but to the south it becomes increasingly restricted to mesic slopes. Additionally, forests in southwestern Oregon (Franklin and Dyrness 1973) are included in our definition.

The mixed evergreen forest is bounded by a series of vegetation types. The mesic border is with coniferous forests of the Pacific Northwest (hemlock zone and Sitka spruce zone forests of Franklin and Dyrness 1973), redwood forests (Chapter 19), and montane coniferous forests (Chapter 20). The xeric border is with chaparral, oak woodland, and grassland communities (Chapters 11, 12, 14). In northern California the mixed evergreen forest forms a complicated mosaic with northern oak woodland (Chapter 11) and coastal prairie (Chapter 21). Central coastal mountain phases continue to form this characteristic mosaic with oak woodland, grassland, and coastal sage scrub (Chapter 13), as well as with chaparral. In the southern mountains this type is more montane in character.

Recognition of the mixed evergreen forest as a major vegetation type in California developed slowly. Early descriptions (Harshberger 1911; Shelford et al. 1926) are sketchy. Clements (1920) did not mention such important trees as *Lithocarpus* and *Quercus agrifolia.* Shreve (1927) described Santa Lucia Mt. forests of mixed hardwood character. Clark (1937) only briefly detailed Mendocino Co. forests with *Pseudotsuga* dominance.

Clear recognition of the mixed evergreen forest began with Cooper's (1922) monograph of the "broad-sclerophyll forest formation." He described stands dominated by mixtures of *Lithocarpus, Quercus agrifolia, Q. chrysolepis, Arbutus,* and *Umbellularia,* along with *Q. kelloggii, Acer macrophyllum,* and minor numbers of conifers. Although use of Cooper's "broad-sclerophyll" name has declined, the concept of mixed hardwood communities as part of the mixed evergreen forest has continued (Knapp 1965; Küchler 1964; Munz and Keck 1949, 1950, 1959; Talley 1974; Whittaker 1960).

Munz and Keck's description included *Pseudotsuga* as a dominant along with *Lithocarpus, Arbutus, Chrysolepis chrysophylla, Umbellularia, Acer macrophyllum, Quercus chrysolepis, Q. kelloggii,* and shrubs. Küchler followed their lead in his description of California mixed evergreen. Whittaker described mixed evergreen forests of southwestern Oregon and detailed especially the role played by parent material differences in determining composition, physiognomy, and pattern. His stress on the "central significance" of the Klamath region forests (Whittaker 1961) may be of wide phytogeographical implication, but these descriptions referred to the northernmost phases, which include even *Tsuga heterophylla* (Franklin and Dyrness 1973).

Obviously Munz, Whittaker, and others somewhat expanded Cooper's original concept to include these northern California and Oregon forests with *Pseudotsuga* dominance. Cooper described some "transitional phases to humid coastal forests of Douglas fir of Oregon and Washington" in western Trinity Co. and near Garberville, Humboldt Co. In view of the nature of these *Pseudotsuga* forests in Humboldt, Del Norte, Trinity, and Siskiyou Cos., they must be included in a definition of mixed evergreen forest rather than be considered as southern extensions of the north coast coniferous forest of Oregon and Washington. The Douglas fir forest of Munz and Keck (1959) and Küchler (1964) in California is therefore included here.

The mixed evergreen forest can be divided into the *Pseudotsuga*-hardwood forests of the Klamath Mts. and North Coast Ranges and the mixed hardwood forests of southern California and the South Coast Ranges. Within the *Pseudotsuga*-hardwood category it is convenient to differentiate the forests of the Klamath Mts. from the forests of the coastal mountains, which grow mainly on Franciscian Formation parent materials. Within the mixed hardwood category it is convenient to recognize a low-elevation *Quercus agrifolia–Arbutus* forest, a redwood border *Lithocarpus–Arbutus–Quercus* forest, and a montane *Quercus chrysolepis–Pinus coulteri* forest.

PSEUDOTSUGA–HARDWOOD FORESTS

North from Sonoma Co., *Pseudotsuga menziesii* assumes an increasingly important role in the mixed evergreen forest. It dominates many phases and can be considered the major species of modal communities of the North Coast Range and Klamath

Mts. Within these two provinces, forest patterns differ to such an extent that further subdivision is necessary in spite of floristic similarity. North Coast Range *Pseudotsuga*-hardwood forests form an extensive mosaic with northern oak woodland and coastal prairie communities, whereas Klamath region forests are more or less continuous. Additionally, in the Klamath region *Quercus garryana* and *Q. kelloggii* may form closed forests rather than open woodlands, and grassland openings are minor.

Klamath Region Forests

Forests on granitic and metamorphic parent materials. These forests are best known from Whittaker's (1960) monograph on Siskiyou Mt. vegetation. From sampling done mainly in Oregon, he described three general forest types segregated by topographical positions below 1000 m on diorite. *Pseudotsuga menziesii* and *Chamaecyparis lawsoniana* codominate dense ravine forests with well-developed shrub layers of *Corylus cornuta, Acer circinatum, Gaultheria shallon, Berberis nervosa,* and few herbs. Submesic slope forests of *Pseudotsuga* and sclerophylls (*Lithocarpus densiflora, Quercus chrysolepis, Arbutus menziesii, Chrysolepis chrysophylla*) form a two-storied canopy over a moderate shrub layer of mostly *Rhus diversiloba* and a sparse herb layer. These are described as grading into sclerophyll-dominated forests of more xeric slopes. Shrub and herb coverage declines with dryness, but there is little compositional change. Low-elevation forests on gabbro are described as similar in composition but more open in character. These Oregon descriptions generally apply to Del Norte and Siskiyou Co. vegetation patterns.

Rockey et al. (1966) described and mapped dominant forest vegetation of the Orleans District, Six Rivers National Forest, Humboldt Co. Below 1000 m on Neuns and Boomer soils, developed from metamorphosed basic igneous rocks of the Lower Galice Formation (Irwin 1960), the dominant trees are *Lithocarpus* and *Arbutus* with minor growth of *Pseudotsuga* and *Pinus lambertiana. Quercus kelloggii, Q. chrysolepis,* and *Chrysolepis* occasionally occur in these stands. On the same soils at higher elevations *Pseudotsuga* and *Pinus lambertiana* dominate the hardwoods. The principal understory species is *Gaultheria shallon.* Surprisingly, there is little evidence of aspect differentiation.

On Hugo, Sheetiron, and Josephine soils, developed from metamorphosed sedimentary rocks of the Upper Galice Formation, *Pseudotsuga, Lithocarpus,* and *Arbutus* form a mixed forest (Fig. 10-1). *Chamaecyparis* becomes a component on the lower slopes with deeper soil phases. The principal understory shrubs mapped are *Vaccinium ovatum, Ceanothus velutinus, Gaultheria shallon,* and *Berberis nervosa.*

Steep slope phases of Neuns and Boomer soils support *Pseudotsuga* and *Pinus lambertiana* in fairly dense stands with occasional hardwoods. On very steep (>70%) colluvial slopes with little soil development, *Quercus chrysolepis* is the dominant tree with occasional *Lithocarpus, Arbutus,* and *Pseudotsuga.* Dominant shrubs are *Arctostaphylos patula* and *Q. vaccinifolia.*

Lower-elevation forests (< 1000 m) on granitic parent materials were studied by Mize (1973) in the Wooley Creek and Irving Creek drainages near Somes Bar, Siskiyou Co. They show generally similar patterns to those on metasediments in the

Figure 10-1. Mixed evergreen forest, *Pseudotsuga*-hardwood phase. *Pseudotsuga, Lithocarpus,* and *Arbutus* present in background; *Arbutus* in foreground. Trinity National Forest, Humboldt Co.

Orleans District. Mize recognized four vegetation types from his detailed survey, which included descriptions of understory composition. Available soil moisture, resulting from differences in topographical position, aspect, and percent slope, was judged to be the important environmental determinant. Mize described more differentiated forests in his area of stronger relief than did Whittaker or Rockey. His four types are summarized below.

A *Pseudotsuga menziesii–Taxus brevifolia* type is located above 310 m in steep-sided ravines and draws with permanent streams. *Pseudotsuga* dominates the canopy, and *Taxus* and *Lithocarpus* the understory, with an occasional *Acer macrophyllum. Gaultheria shallon, Berberis nervosa, Acer circinatum,* and *Corylus cornuta* form a well-developed shrub layer interspersed with *Polystichum munitum, Vancouveria planipetala,* and *Achlys triphylla*

A *Pseudotsuga menziesii–Achlys triphylla* type forms the most productive stands above 500 m on northeasterly aspects with moderate (< 60%) slopes. *Pseudotsuga* again dominates with some *Pinus lambertiana* and *P. ponderosa.* The understory is composed of a scattering of shrubs over a well-developed herb layer including *Achlys triphylla, Chimaphila umbellata, Whipplea modesta, Adenocaulon bicolor, Pyrola picta, Goodyera oblongifolia, Disporum hookeri* var. *trachyandrum,* and *Vancouveria planipetala.* Taylor (unpub. data) sampled such forests on the South Fork of the Trinity River (Table 10-1).

A *Pseudotsuga menziesii–Pinus lambertiana* type covers moderate southerly facing slopes between 340 and 750 m. *Pseudotsuga* and *Pinus lambertiana* share dominance over a secondary canopy of *Lithocarpus.* Herb and shrub layers are poorly developed, although *Rhus diversiloba* is common.

A *Quercus chrysolepis–Pseudotsuga menziesii* type occupies steep (>85%)

TABLE 10-1. Basal area (m^2 ha^{-1}) and density per hectare for all trees over 1.5 m tall in mixed evergreen forest, northern (Douglas fir) phase. Data are means of 18 plots on generally north-facing slopes along the South Fork of the Trinity River, Siskiyou Co., ranging in elevation from 900 to 1370 m. Unpublished data, courtesy of Dean Taylor

Taxa	Basal area	Density
Psueodtsuga menziesii	25.7	297.6
Pinus lambertiana	2.9	13.4
Pinus ponderosa	2.5	5.9
Abies concolor	2.2	81.5
Arbutus menziesii	1.1	23.0
Chrysolepsis chrysophylla	1.0	29.4
Quercus kelloggii	0.9	18.0
Calocedrus decurrens	0.5	9.2
Quercus chrysolepis	0.1	2.6
Cornus nuttallii	0.1	10.1
Acer macrophyllum	0.1	1.3
Totals	37.1	474.6

southwesterly slopes with shallow rocky soils. *Quercus chrysolepis* dominates the canopy, along with *Pseudotsuga* and *Arbutus*. *Lithocarpus* grows only as a shrub in the very poorly developed shrub and herb layers, where the understory is typified by moss-covered rocks.

Emmingham (1973) conducted a forest inventory in the Siskiyou National Forest of Oregon as part of the Illinois River Wild River Study. The study determined timber volumes and potential productivities of some 36,425 ha, which include some of the sampling sites used by Whittaker (1960). Emmingham tentatively recognized 12 forest types on sedimentary, metasedimentary, volcanic, granitic, and ultrabasic parent materials.

Excluding those on ultrabasics, his forests may be categorized as (1) those on steep slopes with rock mulch (colluvial) soils and (2) those on granitic parent materials or sediments with steep slopes that are drier than (3) those on the deeper soils over various sediments. The rock mulch group is dominated by *Quercus chrysolepis*. Individuals of *Pseudotsuga* are scattered on the drier phases but become more important, as does *Lithocarpus*, on the relatively more moist areas with thick mulch. *Rhus* is the common understory associate.

Rapidly weathering hornblende, diorite, and granodiorite produce soils that tend to be droughty in comparison to those on sediments. *Pseudotsuga* dominates these forests with *Pinus lambertiana* and *P. ponderosa*, *Lithocarpus*, and *Q. chrysolepis* as secondary associates. *Vaccinium ovatum* is a common understory shrub, especially on north-facing slopes or along rocky streamside terraces.

Such forests are generally found at elevations below the *Pseudotsuga* forests on deep soils (100–125 cm) of more gentle slopes and ridges, or on silty river terraces. These are the most productive habitats, with forests dominated by *Pseudotsuga* with *Pinus lambertiana*, *Chrysolepis*, *Lithocarpus*, and *Quercus kelloggii* associated in varying degrees. The understory is composed of *Lithocarpus* and *Q. chrysolepis* as small shrubs, *Berberis nervosa*, *Rosa gymnocarpa*, *Rubus vitifolius*, *Pteridium aquilinum*, *Goodyera oblongifolia*, *Chimaphila umbellata*, *Disporum hookerii* var. *trachyandrum*, and *Xerophyllum tenax*. These forests are characterized by a well-developed cover ground layer, in comparison to those on drier, less productive habitats. Emmingham's descriptions represent Del Norte and Siskiyou Co. vegetation patterns better than do those of Whittaker.

Forests on ultrabasic parent materials. Forest patterns on perioditie, dunite, and serpentine show strikingly different patterns from those on adjacent metamorphic and granitic parent materials (Kruckeberg 1954; 1969a,b; Whittaker 1954, 1960). Again Whittaker's work forms the most easily accessible reference for "serpentine effects" in the mixed evergreen forest. He stressed that species composition and physiognomy are quite different for comparable habitats on these parent materials from those of more "normal" mineral composition. He described ravine forests dominated by *Chamaecyparis*, *Pinus monticola*, and *Pseudotsuga*, a forest–shrub complex on mesic slopes, and *Pinus jeffreyi* woodland on xeric slopes. In all types *Pseudotsuga* attains less dominance than on diorites and gabbros, and deciduous trees are absent. Additionally, the sclerophyllous trees common on adjacent parent materials are replaced by a rather continuous layer of shrubs (*Quercus vaccinifolia*, *Lithocarpus densiflora* var. *echinioides*, a shrubby form of *Umbellularia californica*, *Chrysolepis chrysophylla* var. *nana*, *Rhamnus californica* var. *occidentalis*, and *Q. garryana* var. *breweri*).

On mesic sites, several conifers (*Calocedrus decurrens*, *Pseudotsuga*, *Pinus jeffreyi*, *P. monticola*, *P. lambertiana*, and *P. attenuata*) occur in varying degrees of dominance in rather open stands. On slightly drier sites, shrubs become the dominant layer and trees decrease in density. Also, grassy patches with *Poa rhizomata*, *Trisetum canescens*, *Melica geyeri*, *Luzula campestris*, and *Calamagrostis koeleroides* increase in number and extent. This alternation of patches of grassland and shrubland was described by Whittaker as the "two-phase effect." The grassy phase increases in extent as the most xeric stands of *P. jeffreyi* are reached. These woodlands include scattered individuals of *Arctostaphylos viscida* and grasses (*Stipa lemmoni*, *Sitanion jubatum*, *Melica geyeri*, *Elymus glauca*, *Festuca ovina*, *Bromus breviaristanus*, *Danthonia californica*, *Agrostis hallii*, and *Koeleria cristata*).

Rockey et al. (1966) described forests growing on ultrabasics in the Orleans District, Six Rivers National Forest. Forests on Dubakella and Weitchpec soils are very open. Conifers in decreasing order of importance are *Pseudotsuga, Pinus lambertiana, P. jeffreyi,* and *Calocedrus decurrens.* Occasional lower canopy hardwoods (*Lithocarpus, Arbutus, Chrysolepis,* and *Quercus chrysolepis*) are also found on soils developed from less serpentinized rock. The heavy shrub layer is dominated by *Q. vaccinifolia, Arctostaphylos canescens, Rhamnus californica* var. *occidentalis, Ceanothus cuneatus,* and *Q. sadleriana.*

Emmingham (1973) described a series of forest communities in the Siskiyou National Forest, Oregon, which apply to California. At low elevations his *P. jeffreyi* woodland description is much like Whittaker's. A more mesic type dominated by *P. lambertiana* is also included. This open pine forest includes a mixture of conifers with *Pseudotsuga, Calocedrus,* and *P. jeffreyi* well represented over a diverse shrub layer dominated by *Q. vaccinifolia,* the shrub form of *Umbellularia,* and *Rhamnus californica* var. *occidentalis.* In Del Norte Co. these *P. lambertiana* forests are more common than *P. jeffreyi* woodlands, and they additionally include *P. attenuata* and *P. contorta* as successional associates. Even *P. monticola* can be found in such communities to elevations below 150 m.

Summary. The descriptions by Whittaker, Rockey, Mize, and Emmingham present a regional pattern. On metamorphic and granitic parent materials, the driest, discontinuous, rocky soils of steep canyon slopes are dominated by *Quercus chrysolepis.* Dry, continuous soils are characterized by *Pseudotsuga menziesii, Lithocarpus densiflora,* and *Arbutus menziesii,* with *Pinus ponderosa, P. lambertiana,* and *Chrysolepis chrysophylla* as secondary associates. On deep, mesic soils, *Pseudotsuga* dominates over an herb-rich ground layer. The wettest sites include *Chamaecyparis lawsoniana, Taxus brevifolia,* and various nonsclerophyllous shrubs. On ultrabasics, xeric *Pinus jeffreyi* woodland gives way to mesic open forests with understories of sclerophyllous shrubs; additionally, canopy dominance changes to *P. lambertiana* and even *Pseudotsuga* with increasing habitat favorability. The resulting forest mosaic is inherently complex because of regional geological and topographical diversity, and is further complicated by successional patterns from fire and other disturbances.

North Coast Range Forests

The mosaic of vegetation types. *Pseudotsuga*-hardwood forests form part of an extensive mosaic with northern oak woodland and coastal prairie in the north-central portion of the mixed evergreen forest. As in the Klamath region, these forests show various combinations of *Pseudotsuga, Lithocarpus,* and *Arbutus* dominance on deeper, well-watered soils. In southeastern portions of Humboldt and Mendocino Cos., *Pinus ponderosa* becomes a major codominant in forests and woodlands. At higher elevations *P. lambertiana* is of secondary importance, and to the south *Quercus agrifolia* becomes an increasingly common associate.

This region has a long history of intensive grazing and more recent extensive logging. Presettlement patterns, no doubt, have been greatly modified in comparison to those of Klamath region forests. It is therefore more difficult to assess the role of land use practices in the extremely complicated vegetation patterns seen today.

Much of the area in Humboldt and Mendocino Cos. has been surveyed with both timber stand–vegetation maps and soil–vegetation maps published since 1950 (State Cooperative Soil–Vegetation Survey 1969). These maps document the complicated nature of the soil–vegetation mosaic, with some 50 different forest soil series and about 20 different grassland soil series in Humboldt Co. alone (Cooper and Krohn 1968). The nature of the grassland–woodland–forest mosaic and the causal relations responsible for it constitute major unanswered questions.

Soils in Humboldt Co. are readily classifiable into those which support forest, grassland, woodland, or brush on soft sedimentary, hard sedimentary and metamorphic, or igneous (including ultrabasic) parent materials. Soft sediments are more coastal and generally support *Sequoia sempervirens* forests. The most common soil series to support *Pseudotsuga*–hardwoods is Hugo, which develops from hard sedimentary Franciscian rocks. The Tyson series over hard sediments and metamorphics is characteristic for oak woodlands. The unstable, slumping Yorkville soils over metasedimentary, glaucophane, and sericite shists support oak woodland and grassland (DeLapp 1960; Black 1964). Extensive stable grassland soils include Laughlin and Kneeland series over hard sediments (Storie and Wieslander 1952). Supposedly, even within the *Pseudotsuga*–hardwood forests, soil differences are reflected in forest composition. Cooper and Krohn (1968) suggested that the order of dominant hardwoods may be reversed by soil type; on Melborne soils *Lithocarpus* dominates *Arbutus*, while on Hugo soils *Arbutus* is more common than *Lithocarpus*.

Waring and Major (1964), drawing on soil–vegetation map information along with extensive field and laboratory analyses, concluded that, of gradients in moisture, nutrients, light, and temperature, soil moisture availability is primary in determining forest composition in southern Humboldt Co. Additionally, they supplied autecological information for *Pseudotsuga*–hardwood associates. For example, the role of *Acer macrophyllum* as a riparian species was seen even in these mesic forests. The sensitivity of *Sequoia* to soil moisture stress, in comparison to mixed evergreen dominants, explained its restriction to well-watered soils in the study area. Their observation that oak dominance increases on drier, upland soils (*Quercus garryana, Q. kelloggii*) and canyon slopes (*Q. chrysolepis*) seems to represent a regional phenomenon.

Cooper (1972) documented annual soil moisture cycles for three soil series (Hugo, Tyson, Yorkville) in central Humboldt Co. Annual (May 1971–April 1972) soil moisture–soil stress cycles were documented by gravimetric and pressure plate–membrane analyses. There was rapid drying of surface layers of Yorkville soils, probably because of strong water use by grassland plants with shallow, fibrous root systems. The forest and woodland soils were more depleted of moisture at >1 m during the drying phase with slower reduction of surface moisture after cessation of spring rains. Finer-textured, deeper Hugo soils maintained higher water availability at greater depths in soil profiles throughout summer drought. These soils support *Pseudotsuga*–hardwood forests.

Forest composition. The tree layer of the mature modal community consists of both *Pseudotsuga* and *Lithocarpus*. The canopy is of two heights with *Pseudotsuga* forming an irregular upper stratum as tall as 65 m. *Lithocarpus* forms a more continuous lower canopy at heights to 35 m. *Arbutus*, on south slopes, and *Chryso-*

lepis, on north slopes, are secondary components in the lower-level canopy. The shrub layer is very dense and is composed of shrubby *Lithocarpus* (not var. *echinioides*) and a mixture, on north slopes of *Cornus nuttallii, Gaultheria shallon, Vaccinium ovatum, Berberis nervosa, Rhododendron macrophyllum*, and *Corylus cornuta*, and on south slopes of *Rosa gymnocarpa* and *Rhus*. The very sparse herb layer on more moist slopes includes *Oxalis oregana, Vancouveria hexandra, V. planipetala, Polystichum munitum, Asarum caudatum, Disporum hookerii*, and *Goodyera oblongifolia*. These herbs gradually change to a layer dominated by *Whipplea modesta* and *Pteridium aquilinum* on drier slopes.

In some areas, *Pseudotsuga* dominates forests on both north and south slopes (the *Pseudotsuga* phase). It produces a continuous canopy, with *Lithocarpus* occurring only as a shrub. *Arbutus* and *Chrysolepis* rarely grow in such stands. Few shrubs occur, but the less-developed herb layer is similar to the modal type. It might appear that these stands grow on more productive sites; however, their habitat completely overlaps that of the modal phase. These stands tend to be even aged, suggesting a successional nature. More mature, uneven-aged stands on similar sites approach the composition of the modal community. In other areas of comparable habitat, *Lithocarpus* is a dominant of early successional stages. As *Lithocarpus* matures, *Pseudotsuga* invades to form a mature forest characteristic of the modal community.

Tsuga heterophylla extends into California immediately east of the *Sequoia* forests and contributes to a *Tsuga* phase of mixed evergreen forest. Stands dominated by *Tsuga* can be found in Goose Creek, Blue Creek, Redwood Creek, and Mad River drainages of Del Norte and Humboldt Cos. In locations that have been free of fire for a considerable period of time, *Tsuga* occurs as a codominant with *Pseudotsuga* and *Lithocarpus*. On lower moist slopes *Tsuga* is a major species in the seedling and sapling strata, whereas on upper dry slopes many *Pseudotsuga* and *Lithocarpus* seedlings and saplings are present, but only an occasional *Tsuga*. In some forests, *Tsuga* completely dominates stands to the exclusion of *Pseudotsuga* and *Lithocarpus*. In other areas, *Chamaecyparis* is a codominant. *Arbutus* and *Chrysolepis* are minor components in some upper slope stands. A well-developed shrub layer is dominated by *Gaultheria shallon* and *Rhododendron macrophyllum*, with minor amounts of *Berberis nervosa, Vaccinium ovatum*, and *Rosa gymnocarpa*. Clumps of *Polystichum munitum* and *Pteridium aquilinum* grow throughout these forests. Underneath the dense layer of shrubs and ferns is a scattering of *Oxalis oregana* and *Viola sempervirens* on wetter sites, and *Xerophyllum* and *Achlys triphylla* on drier sites.

Variants of this *Tsuga* phase can be found in the lower Mad River drainage. *Abies grandis* and *Thuja plicata* are major local dominants along with *Tsuga* and *Pseudotsuga*. Associated with these conifers are *Lithocarpus* and *Umbellularia*. All species are reproducing in undisturbed stands. *Polystichum* is normally the dominant understory plant. These habitats are close to the *Sequoia* type and are within the tolerance limits of *Sequoia*. Perhaps *Sequoia* has been excluded by past events and has not reinvaded.

Pinus ponderosa increases in importance in *Pseudotsuga*-hardwood forests on drier slopes in southeastern Humboldt and eastern Mendocino Cos. (the *Pinus ponderosa* phase). In numerous locations *P. ponderosa* and *Pseudotsuga* completely dominate the canopy. Other communities in this area contain components of the modal type, as *Lithocarpus, Arbutus,* and *Quercus chrysolepis* are occasionally of some importance.

Areas for future research. In the north coastal mountains, *Pseudotsuga*-hardwood forests presently form a complicated mosaic of early and late successional communities resulting from a long history of fire, grazing, and logging. The larger characteristic pattern of forest–woodland–prairie, so typical of the north coastal region, is little understood and needs further study. Unfortunately the effects of disturbance of these communities in this area are so pervasive that the natural patterns will be difficult to determine.

Successional Status of Some Tree Species

Pseudotsuga menziesii, Lithocarpus densiflora, and *Arbutus menziesii* are the most important trees in the *Pseudotsuga*-hardwood forests of northern California. They interact to create several different successional sequences toward a typical modal climax community.

Pseudotsuga menziesii. This species can be considered moderately shade tolerant in northern California. Seedlings can become established under various shade levels of 20% full sun with optimum survival at about 50% shade; optimum growth occurs in up to 75% full sun (Strothman 1972). *Pseudotsuga* is long lived, reaching ages of 400–600 yr in these forests. Height growth is fairly rapid to maturity at 100–150 yr. Thereafter, growth rates slow considerably with mature trees reaching maximum heights of 50–75 m. Young seedlings and saplings can grow at a slow rate for a considerable time under mature trees and will then increase rapidly in height as openings in the canopy occur.

Lithocarpus densiflora. This species can regenerate as sprouts or seedlings. Vigorous sprouts, which develop from dormant buds at the base, are produced when aerial parts are removed by cutting or fire. They grow rapidly either from tree stumps or shrub bases. *Lithocarpus* sprouts quickly, and often provides severe competition to *Pseudotsuga* seedlings (Roy 1957). It can also regenerate from seed under most light conditions, with best success in full sun. Regeneration is limited under dense shade; however, occasional seedlings can be found. Roy (1957) indicated that initial seedling height growth is greater in *Lithocarpus* than in *Pseudotsuga,* giving the broadleaf tree further competitive advantage.

Arbutus menziesii. This species is not as abundant as *Lithocarpus* within northern California forests. Rated as intermediate in shade tolerance (Terrant 1958), it grows well in open stands, once established. *Arbutus* sprouts grow faster than either *Pseudotsuga* seedlings or *Lithocarpus* sprouts (Roy 1957), giving it local success in competition as a common secondary associate.

Interactions. Competitive interactions among these species are the major factors determining existing forest patterns. Vegetation in early stages of succession after logging or burning is characteristically very heterogeneous because the relative densities of *Pseudotsuga* and *Lithocarpus* are dependent on the type and degree of disturbance, available seed sources, density of sprouts, and environmental conditions.

After logging or burning, a very dense, nearly pure even-aged stand of either *Pseudotsuga* or *Lithocarpus* can form. In other situations a mixture will result. As an example, we may start with a mature modal *Pseudotsuga*–hardwood community with *Pseudotsuga* forming an upper canopy and *Lithocarpus* a lower, secondary tree layer. In most stands, seedlings of both species will be found. If a very severe fire burned through this area and killed aerial portions of all trees, *Lithocarpus* would be the major regenerating species since *Pseudotsuga* does not sprout. *Lithocarpus* sprouts would grow from tree and shrub bases and completely dominate. *Pseudotsuga* seed would have to come from distant sources, slowing its rate of reestablishment. With time, though, the dense even-aged stand of *Lithocarpus* would mature and senesce, and *Pseudotsuga* would gradually invade the opening stand. Under these conditions, seedlings of both species would grow slowly, taking advantage of deaths of local trees, which produce canopy openings. All evidence appears to indicate that with increased age invading *Pseudotsuga* seedlings would outgrow *Lithocarpus* seedlings under these generally reduced light conditions. The potentially longer life span and greater height growth of *Pseudotsuga* ensure eventual stand dominance. However, *Lithocarpus* will continue to regenerate and maintain itself in a lower canopy. This successional sequence will occur at a faster rate on more productive soils to reestablish a mature mixed forest.

After a moderate burn, an adequate seed source of *Pseudotsuga* usually remains in the area. Consequently both *Pseudotsuga* and *Lithocarpus* regenerate. *Lithocarpus* initially surpasses *Pseudotsuga* in height growth and dominates. However, in most situations *Pseudotsuga* continues to slowly increase in height until it dominates *Lithocarpus* (Fig. 10-2). Alternatively, *Pseudotsuga* may form a dense canopy to the complete exclusion of *Lithocarpus* and other understory vegetation. Reestablishment of the *Lithocarpus* lower canopy takes place slowly as individual *Pseudotsuga* mortality begins to open up the canopy at 100–150 yr.

In other situations, mixed stands may be relatively open, allowing both *Pseudotsuga* and *Lithocarpus* to continually regenerate on the forest floor. All successional sequences, though, lead to the same mixed composition. Examples of the modal

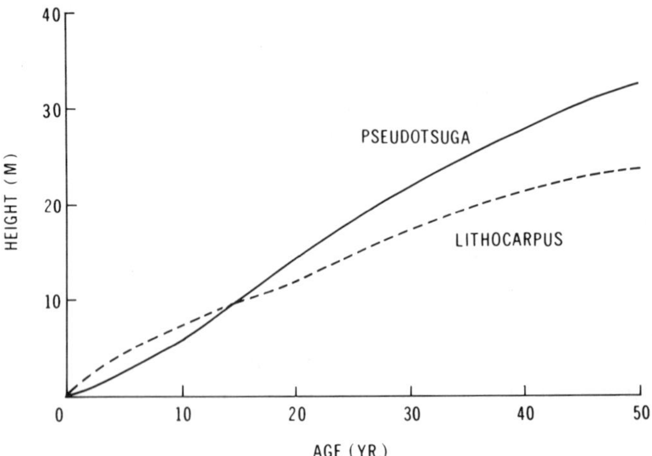

Figure 10-2. Height growth of *Pseudotsuga menziesii* seedlings and *Lithocarpus densiflora* sprouts after logging or burning of a modal *Pseudotsuga*–hardwood forest.

community are rare because of frequent fire. Consequently, most forests are presently composed of even-aged stands of either *Pseudotsuga* or *Lithocarpus*, or are composed of two or three different age classes, resulting from several disturbances. *Pseudotsuga* stands, *Lithocarpus* stands, or various mixtures have traditionally been misinterpreted as environmentally differentiated, whereas these differences are actually dynamic in nature. The pure Douglas fir forests (Munz and Keck 1959; Küchler 1964) of lower elevations in the Klamath region and north coastal mountains are not southern extensions of Pacific Northwest coniferous forests on more favorable habitats; instead they are part of the vegetation mosaic of the mixed evergreen *Pseudotsuga*-hardwood forests. In Pacific Northwest forests *Pseudotsuga* is an early successional species; here it plays a very different role as part of the climax pattern as well.

MIXED HARDWOOD FORESTS

Mixed hardwood forests occupy the southern portion of the mixed evergreen forest region, where *Pseudotsuga menziesii* makes only a minor contribution to the mixture or is absent. Pure stands of evergreen hardwoods are locally present throughout the mixed evergreen forest region, but in the North Coast Ranges and Klamath region, pure hardwood stands are often fragmented transitions between *Pseudotsuga*-hardwood forest and northern oak woodland or coastal prairie. From Napa, Sonoma, and Marin Cos. southward, mixed hardwood stands are recognizable over a large portion of the forested landscape. Although the mixed hardwood forest concept has not been commonly applied to Sierra Nevada situations, local areas in the lower Sierran montane forest from Yuba to Eldorado Cos. support *Lithocarpus densiflora–Arbutus menziesii–Quercus chrysolepis* stands similar to those of Coast Range forests. Even *Chrysolepis chrysophylla* is present (Griffin and Critchfield 1972).

In terms of physiognomical–ecological classes (Mueller-Dombois and Ellenberg 1974) portions of the mixed hardwood forest are winter rain evergreen broad-leaved sclerophyllous forest. But often the tree crowns are not literally interlocking over sizable areas, and mainly "evergreen broad-leaved woodland" is a more appropriate description. More open stands of mixed hardwood forest can be separated only arbitrarily from densest oak woodland stands (Chapter 11) in terms of tree identity and density. However, the mixed hardwood understory tends to have many forest shrubs and perennial herbs and few grassland annuals, so typical of oak woodland.

Mixed hardwood forests are difficult to recognize in the U.S. Forest Service records. In the Vegetation Type Map (VTM) Survey (Wieslander 1935), mixed hardwood stands usually can be identified by species symbols on the regional maps, but statistical summaries merge them either into generalized noncommercial woodlands or some adjacent commercial conifer type. The Cooperative Soil–Vegetation (S–V) Survey (State Cooperative Soil–Vegetation Survey 1969) did not map any significnt portion of mixed hardwood forest.

Quercus agrifolia–Arbutus Forest

The lowest-elevation forms of mixed hardwood forest are dominated by *Quercus agrifolia,* but away from the immediate coast *Arbutus menziesii* and other hardwoods may be common. This phase approaches Cooper's (1922) *Q. agrifolia–Arbutus* association. Cooper recognized a *Q. agrifolia* consociation in southern California, but pure *Q. agrifolia* stands south of the Santa Ynez Mts. are treated here as part of the southern oak woodland (Chapter 11). The *Q. agrifolia–Arbutus* phase corresponds to coastal portions of the oak–madrone forest type (Society of American Foresters 1954).

Along the central California coast, pure stands of *Q. agrifolia* forest extend inland up to 20 km on rolling hills of loosely consolidated sandy sediments, as in the Aromas area of Monterey Co., the Nipomo Mesa of San Luis Obispo Co., or the Orcutt Hills of Santa Barbara Co. The seaward portions of these live oak groves are severely wind pruned, and the trees are relatively short, even shrubby, as on the most exposed hills of Ft. Ord, Monterey Co. (NA 270671). Wells (1962) discussed two San Luis Obispo Co. examples of this type on Pleistocene sand. Both had *Q. agrifolia* as the only tree, a variety of understory shrubs including the widespread *Heteromeles arbutifolia* and *Rhamnus californica,* and only minor amounts of intervening grassland. In this type of *Q. agrifolia* community the interspersed chaparral patches have a diverse shrub flora with many endemic forms of *Arctostaphylos* and *Ceanothus*. The scrubby live oak forests have a floristic affinity with the closed-cone pine forest (Chapter 9), and they surround or adjoin many closed-cone pine stands.

On steeper topography, particularly on shale parent materials, *Q. agrifolia* communities may be more open and show considerable grassland between the oak groves. These stands become difficult to separate from foothill woodland (Chapter 11). McBride (1971) described some pure stands of *Q. agrifolia* in southwestern Santa Barbara Co. These stands were about 15 m tall, varied in density from savanna to forest, and had a shrub understory of *Heteromeles arbutifolia* and *Rhus diversiloba*. McBride viewed this forest as climax and well able to maintain itself in the absence of heavy grazing.

Wells (1962) outlined species composition and life-form classes for many examples of this open type in coastal San Luis Obispo Co. on a variety of parent materials. Although found on all aspects and soils, mixed hardwood forest was more common on north slopes and was uncommon on deep clay soils. *Quercus agrifolia* was by far the most important tree; *Heteromeles arbutifolia* was the most widespread understory shrub. Wells discussed the interaction of soils and fires in producing the moasic of vegetation types within the basically "forest" climate region.

On the steeper coastal front of the Santa Cruz and Santa Lucia Mts., *Q. agrifolia* forests form a zone between a complex of coastal sage scrub, grassland, and chaparral on the lower slopes and *Sequoia sempervirens* on the higher slopes or canyons. *Quercus agrifolia* may be locally replaced by a coastal form of *Q. wislizenii*. In the Santa Lucias, *Q. wislizenii* forests often occur in canyon bottoms and north slopes below 700 m, while *Q. agrifolia* communities occupy ridges up to 1000 m (Talley 1974). The upper part of these forests merges into *Lithocarpus-Arbutus-Quercus* forests.

On interior foothills, the *Q. agrifolia-Arbutus* forest has a more mixed composition. Deciduous *Q. kelloggii* and *Q. lobata* trees may be scattered along with colonies of *Aesculus californica*. These forests or, more often, actually dense woodlands are interspersed in a foothill woodland–grassland–chaparral complex below the *Q. chrysolepis–Pinus coulteri* forest. Near the Hastings Reservation (NA 270812) in the northeastern Santa Lucias, *Q. agrifolia* forms pure stands on shady alluvial terraces. One old-growth grove in a canyon bottom had 76 trees per hectare, totaling a 35.0 m^2 ha^{-1} basal area, with an average height of 25 m. These bottomland stands appear rooted down to the water table and have low xylem sap tensions throughout the dry season (Griffin 1973). Nearby north slopes have *Q. agrifolia* communities in which *Arbutus* and *Q. kelloggii* are common (Griffin 1974). Minimum xylem sap tensions rise much higher during the summer drought in these upland stands, with the driest *Q. agrifolia* stands approaching the moisture stress of some foothill woodland stands (Griffin 1973). *Holodiscus discolor* and *Symphoricarpos mollis* are common understory shrubs in the denser parts of these stands. In this region *Q. agrifolia* drops out of the forest above 1000 m, being replaced by *Q. chrysolepis*.

Jasper Ridge (NA 431010), on the eastern flank of the Santa Cruz Mts., has good examples of the interior *Q. agrifolia–Arbutus* forests (Cooper 1922; Porter 1962). Pelton (1962) studied *Arbutus* regeneration in this region, and his *Arbutus-Quercus* plot had 600 per hectare, trees totaling a 47.3 m^2 ha^{-1} basal area. The large number of stems must reflect sprouting after some old disturbance. The vegetation type map of the San Mateo quadrangle (Wieslander 1932) shows the general distribution of this phase in the northern Santa Cruz Mts.

Umbellularia californica is scattered throughout the mixed hardwood forest, but locally it achieves total dominance. Cooper (1922) listed an *Umbellularia* consociation with the *Q. agrifolia–Arbutus* forest; however, pure stands of *Umbellularia* also occur within other mixed hardwood forests. Unsicker (1974) gathered phytosociological data in 20 *Umbellularia* stands in the Santa Cruz Mts. These stands had conspicuously few understory species, and she explored the role of allelopathy and low light intensity in restricting understory growth under these dense stands.

McBride (1974) studied land use patterns and plant succession in the Berkeley Hills, describing 11 vegetation types including "oak woodland" dominated by *Q. agrifolia* and "bay woodland" dominated by *Umbellularia*. He demonstrated that *Q. agrifolia* and *Umbellularia* were invading *Baccharis pilularis* brushland, and that much of the brushland would be replaced by woodland in time. In woodland plots, age–class distributions suggested that *Q. agrifolia* was being replaced by *Umbellularia*. Fire and deer browsing seemed to play critical roles in determining the rate of this brushland–oak woodland–bay woodland succession.

Lithocarpus–Arbutus-Quercus Forest

Cooper (1922) recognized a *Pasania* (*Lithocarpus*)–*Quercus–Arbutus* association at lower elevations in the Coast Ranges. This community resembles the black oak–madrone forest that Mason (1947) characterized as a redwood border forest. This phase is also part of the oak–madrone forest type (Society of American Foresters 1954).

Although Cooper (1922) discussed North Coast Range examples of the *Lithocarpus–Arbutus-Quercus* community, much of the northern form of this variable type might be better treated as just a marginal fringe of the *Pseudotsuga*–hardwood forest. Here only the southern form is considered. The term "redwood border forest" is still appropriate in the south, for the best expression of hardwood forest with abundant *Lithocarpus* and minimal *Q. chrysolepis* is just interior to the *Sequoia* or *Sequoia–Pseudotsuga* forests. Howell (1970) referred to this type of redwood border forest in Marin Co., and it is well developed in the Santa Cruz Mts. (Cooper 1922; Thomas 1961). In the Santa Lucia Range, north-facing slopes of larger canyons adjacent to *Sequoia* forest support nearly pure stands of *Lithocarpus* (Shreve 1927; Talley 1974).

Few quantitative data are available for this phase. Jensen (1939) summarized VTM Survey and Forest Survey observations in the Santa Cruz Mts. which relate to *Lithocarpus–Arbutus-Quercus* forests and in part to the *Q. agrifolia–Arbutus* communities. He noted that the forest had a closed canopy 4–15 m tall with a range in diameter at breast height. The forest was indicative of deep soils, in contrast to adjacent chaparral and foothill woodland communities. Part of the forest is clearly successional, and Jensen estimated that one-half of the 40,000 ha related to these phases in the Santa Cruz Mts. was derived from *Sequoia–Pseudotsuga* forests by destructive logging and fire.

Quercus chrysolepis–Pinus coulteri Forest

Quercus chrysolepis is the most widely distributed oak in California (Griffin and Critchfield 1972), and it appears in a large number of vegetation types. This oak is prominent in canyon bottom riparian forests, steep rocky slope woodlands and forests, and rocky summit scrub throughout most forested regions (Myatt 1975). However, only the southern communities where *Q. chrysolepis* is less transitional and has more of a regional vegetation identity are considered here.

This forest includes the central Coast Range and southern California portions of Cooper's (1922) *Q. chrysolepis–kelloggii* association and the canyon live oak forest type (Society of American Foresters 1954). As the influence of *Pseudotsuga menziesii* declines in these southern mountains, *Q. chrysolepis* becomes a convenient species by which to characterize montane mixed hardwood forest over a wide range of slope and soil conditions. *Pinus coulteri* may be locally common in these forests, along with disjunct stands of montane conifers such as *Pinus ponderosa* and *P. lambertiana*.

Sprout thickets of short *Q. chrysolepis* trees appear on many Central Coast Range summits. Bowerman (1944) discussed a scrubby form of *Q. chrysolepis* association on the north slope of Mt. Diablo which was 8 m tall with a closed canopy. *Umbellularia* and *Q. wislizenii* were local subdominants. Common shrubs were *Holodiscus discolor, Rhus diversiloba, Ribes menziesii,* and *Symphoricarpos mollis*. *Pinus coulteri* is present on the peak but not in the *Q. chrysolepis* community. Cooper (1922) mentioned a related situation on the north slope of Mt. Tamalpais with thickets of *Q. chrysolepis* and *Q. wislinzenii* 3–5 m tall and occasional full-sized *Q. chrysolepis* trees. Depauperate forms of this phase were mapped in the Mt. Hamilton and Diablo ranges by the VTM survey, but no detailed descriptions of these scrubby interior forests are available.

This phase is not well developed in the Santz Cruz Mts., where *Q. chrysolepis* is seldom dominant and *P. coulteri* is absent. Some of the best examples of this phase are in the Santa Lucia Range above 1000 m (Fig. 10-3). These montane stands were

Figure 10-3. Mixed evergreen forest, mixed hardwood phase. *Quercus chrysolepis, Lithocarpus, Arbutus, Umbellularia,* and *Pinus coulteri* are present in foreground and on slopes beyond the fog. Near Cone Peak, Santa Lucia Range, Monterey Co.

the heart of Shreve's (1927) mesophytic forest. Talley (1974) viewed the *Q. chrysolepis* phase as the most extensive form of mixed hardwood forest in the Santa Lucias. Sheltered slopes and ravines have *Q. chrysolepis* mixed with *Lithocarpus, Arbutus,* and *Umbellularia. Acer macrophyllum* is common in the draws. More exposed slopes may have nearly pure stands of *Q. chrysolepis; Pinus coulteri* is widely scattered. In general this phase has a poor herb flora, and large patches of forest floor are nearly bare. Many slopes are so steep that not even much litter accumulates. On upper slopes swept by crown fires *Q. chrysolepis, Lithocarpus,* and *Q. wislizenii* form low thickets which resemble the Mt. Diablo scrub forest. It is difficult to tell how large these forests would be if fires were less frequent. In tiny, protected spots that have not had serious fires for centuries there are massive hardwoods with trunk diameters ranging from 1 to 2 m.

On summits where *Q. chrysolepis–P. coulteri* forests surround patches of *Q. lobata* savanna (Chapter 11), *P. coulteri* has been invading the savanna in recent decades (Griffin 1975a) *Quercus chrysolepis, Q. wislizenii,* and *Lithocarpus* seedlings and saplings are now established under young pines. Low fire frequency seems to be encouraging this conversion of deciduous oak savanna into mixed hardwood forest.

Although scrubby forms of the *Q. chrysolepis–P. coulteri* forest are present on most of the higher ranges in southern California, few items about these forests appear in the literature. Horton (1960) outlined vegetation types in the San Bernardino Mts., largely from VTM Survey data. He recognized a live oak woodland which frequently occurred on north-facing slopes within a woodland–chaparral zone. *Quercus chrysolepis* dominated the type with a canopy 6–15 m tall.

Minor Forest Types

***Abies bracteata* forest.** *Abies bracteata* (Santa Lucia fir, bristlecone fir) grows only in the Santa Lucia Range with a total distribution less than 90 km long and 20 km wide (Griffin and Critchfield 1972). Talley (1974) made a detailed study of the ecology of this narrow endemic and found that mature firs were largely confined to the *Quercus chrysolepis* phase of the mixed hardwood forest. There were no floristic differences between mixed hardwood areas supporting *A. bracteata* and comparable areas without the fir.

The fir groves are concentrated on steep, rough terrain that will not carry a severe fire. Within these fire-resistant areas, Talley described three physiographical situations supporting *Abies:* summit, transitional slope, and ravine. On summits, tree and shrub cover combined seldom exceed 25%, and fir regeneration is limited because of soil drought during the growing season. Many rock outcrop species grow between the firs and stunted hardwood on the summits, including *Arabis breweri, Cheilanthes intertexta, Erigeron petrophilus, Eriogonum saxatile, Penstemon corymbosus,* and *Streptanthus tortuosus* (Griffin 1975b; Talley 1974). The slopes have enough soil to support *Q. chrysolepis–A. bracteata* stands with up to 40% tree cover. Fir regeneration is scattered. The herb cover seldom exceeds 5% and is just a selection of species from the mixed hardwood flora. In the ravine sites, the fir grows with a greater diversity of hardwood trees, and the average tree cover is 65%. Fir regeneration is abundant in the ravines, but it may ultimately be limited by low light intensity under dense *Lithocarpus* stands.

In comparing fir and nonfir sites, Talley found that sites with mature firs had steeper, rockier soils that were unlikely to carry a fire because of topographical configuration and low fuel accumulations. During seasons with favorable seed supply and climatic conditions, the stable *A. bracteata* groves can extend out into mixed hardwood forest with deeper soils and smoother topography. But as soon as the next hot fire occurs, the fire-sensitive fir is again eliminated from the hardwood sites.

***Pseudotsuga macrocarpa* forest.** In southern California, *Pseudotsuga macrocarpa* (big-cone spruce, big-cone Douglas fir) geographically replaces *P. menziesii* (Griffin and Critchfield 1972). Clements (1920) suggested that *P. macrocarpa* replaced *P. menziesii* ecologically in the southern version of the coniferous montane forest, but this does not appear to be correct. Although the highest-elevation stands of *P. macrocarpa* do associate with montane species such as *Abies concolor* and *Pinus lambertiana,* the bulk of the *Pseudotsuga macrocarpa* distribution is within a mixed hardwood forest and chaparral matrix (Bolton and Vogl 1969; Gause 1966; Horton 1960; Knapp 1961; Littrell and McDonald 1974).

In the San Bernardino Mts., Horton (1960) recognized a big-cone Douglas fir forest type on north aspects below the Sierran montane forest. The *P. macrocarpa* stands, which ranged in height from 15 to 24 m, had a *Q. chrysolepis* understory and a very sparse herb layer. Horton suggested that repeated fires would convert the *P. macrocarpa* forest into live oak woodland or live oak chaparral.

Gause (1966) summarized the ecological features and common associates of *P. macrocarpa* in three physiographical zones: northern aspects and canyon bottoms at low elevation (down to 300 m), many aspects at middle elevation, and west to south aspects at higher elevation (up to 2400 m). Gause discussed the uncommon ability of this conifer to recover from moderate fires by sprouting from the trunk, but he nevertheless suggested that its range had been reduced at lower elevation by repeated fires. A good example of a low-elevation *P. macrocarpa* population is present in the Fern Canyon Research Natural Area (NA 190630) in the San Gabriel Mts.

Bolton and Vogl (1969) studied *P. macrocarpa* throughout the Santa Ana Mts. It was the most widely distributed tree in this predominantly chaparral-covered region. Although *P. macrocarpa* occurred on all exposures, the majority of stands were on mesic north aspects. The largest continuous stand occurred on a steep north slope that also contained a relic *Arbutus* colony. The density of *P. macrocarpa* ranged from 188 trees per hectare on level plots to 81 per hectare on steep plots. On some burned areas seedling densities up to 25,000 per hectare were observed, but seedlings and saplings are very susceptible to fire. Only larger trees have the ability to sprout and possess bark thick enough to resist hot fires. Bolton and Vogl concluded that the *P. macrocarpa* forest was a comparatively stable community on mesic, relatively fire-resistant habitats in the Santa Ana Mts. Littrell and McDonald (1974) surveyed size–class structure in 16 stands over the whole of the species. They also concluded that *P. macrocarpa* forests with a *Quercus chrysolepis* understory were relatively stable within their present range.

Summary

Mixed hardwood forest is dominated by one or more sclerophyllous tree species that sprout vigorously when burned. *Quercus agrifolia* lends conspicuous identity to the evergreen forest and woodland at low elevations south of Sonoma Co. On cool, foggy coastal hills, it may be the only tree over large areas. There are few conifers in the *Q. agrifolia* forest, although it is present under closed-cone pine forests and adjoins the lower margin of *Sequoia–Pseudotsuga* forests. On warmer, interior hills, *Q. agrifolia* occupies mesic areas within the oak woodland region. *Arbutus menziesii* and other hardwoods may be mixed in these stands. Stands with grassland species in the openings form a transition with oak woodland. At higher elevation on coastal slopes, *Lithocarpus densiflora* becomes conspicuous, particularly above canyons with *Sequoia* forest; *Umbellularia californica* may be locally dominant.

In montane areas, *Q. chrysolepis* replaces *Q. agrifolia* as the most widespread sclerophyll tree, forming pure stands on drier ridges and mixed stands with *Lithocarpus, Arbutus,* and other hardwood on mesic slopes. In southern California, the hardwoods associated with *Q. chrysolepis* are greatly reduced. *Pinus coulteri* is widely scattered in the *Q. chrysolepis* region, but it is only a minor part of mature stands. Very old forests are difficult to find in this fire-prone landscape; on frequently burned slopes, the hardwood forest is reduced to low thickets.

Quercus chrysolepis forests form the understory in relic stands of montane coniferous forest in the Central Coast Ranges. The nonsprouting conifers lose ground to the hardwoods after major fires. In the Santa Lucia Range, *Q. chrysolepis* forest also associates with *Abies bracteata* on the most fire-resistant topography. In southern California, *Q. chrysolepis* is associated with *Pseudotsuga macrocarpa* forest on mesic relatively fire-resistant areas within a basically chaparral region.

LITERATURE CITED

Black, P. E. 1964. A guide to the soil-vegetation maps of Humboldt County, Califronia, Mimeo rept., School of Nat. Resources, Humboldt State Univ., Arcata, Calif. 16 p.

Bolton, R. B. Jr., and R. J. Vogl. 1969. Ecological requirements of Pseudotsuga macrocarpa in the Santa Ana Mountains, California. J. For. 67:112-116.

Bowerman, M. L. 1944. The flowering plants and ferns of Mount Diablo, California. Gillick Press, Berkeley. 290 p.

Clark, H. W. 1937. Association types in North Coast Ranges of California. Ecology 18:214-230.

Clements, F. E. 1920. Plant indicators: the relation of plant communities to process and practice. Carnegie Inst. Wash. Pub. 290. 388 p.

Cooper, D. W., and R. Krohn. 1968. Forest soils field tour, Oct. 23, 1968. Univ. Calif. Agric. Ext. Serv., Eureka. 43 p.

Cooper, P. V. 1972. Soil moisture relations of selected soils in mixed evergreen type of northern California. M.S. thesis, Humboldt State Univ., Arcata, Calif. 48 p.

Cooper, W. S. 1922. The broad-sclerophyll vegetation of California. Carnegie Inst. Wash. Pub. 319. 124 p.

DeLapp, J. 1960. A preliminary guide to the woodland soils of Humboldt County. Mimeo rept., School of Forestry, Univ. Calif., Berkeley. 10 p.

Emmingham, W. H. 1973. Lower Illinois River ecosystem study. Unpub. rept. to Siskiyou National Forest. 74 p.

Franklin, J. F., and C. T. Dyrness. 1973. Natural vegetation of Oregon and Washington. USDA Forest Serv. Gen. Tech. Rept. PNW-8. 417 p.

Gause, G. W. 1966. Silvical characteristics of bigcone Douglas-fir (Pseudotsuga macrocarpa [Vasey] Mayr). USDA Forest Serv. Res. Paper PSW-31. 10 p.

Griffin, J. R. 1973. Xylem sap tension in three woodland oaks of central California. Ecology 54:152-159.

---. 1974. Notes on environment, vegetation, and flora, Hastings Natural History Reservation. Mimeo rept. on file, Hastings Reservation, Carmel Valley, Calif. 90 p.

---. 1976. Regeneration in Quercus lobata savannas, Santa Lucia Mountains, California. Amer. Midl. Natur. 95:422-435.

---. 1975. Plants of the highest Santa Lucia and Diablo Range peaks, California. USDA Forest Serv. Res. Paper PSW-110. 50 p.

Griffin, J. R., and W. B. Critchfield. 1972. The distribution of forest trees in California. USDA Forest Serv. Res. Paper PSW-82. 114 p.

Harshberger, J. W. 1911. Phytogeographic survey of North America. In Engler and Drude (eds.) Die Vegetation der Erde. Vol. 13. Stechert, New York. 790 p.

Horton, J. S. 1960. Vegetation types of the San Bernardino Mountains. USDA Forest Serv., Pac. SW Forest and Range Exp. Sta., Tech. Paper 44. 29 p.

Howell, J. T. 1970. Marin flora. 2nd ed. with supplement. Univ. Calif. Press, Berkeley. 366 p.

Irwin, W. P. 1960. Geologic reconnaissance of the northern Coast Ranges and Klamath Mountains, California. Calif. Div. Mines Bull. 179. 80 p.

Jensen, H. A. 1939. Vegetation types and forest conditions of the Santa Cruz Mountains unit of California. USDA Forest Serv., Calif. Forest and Range Exp. Sta., Forest Survey Release 1. 55 p.

Knapp, R. 1961. Uber die immergrune Hartlaub-Vegetation in Kalifornien. Schr. Naturw. ver. Schleswig-Holstein 32:3-20.

Knapp, R. 1965. Die Vegetation von Nord- und Mittelamerika und der Hawaii-Inseln. Gustav Fischer-Verlag, Stuttgart. 373 p.

Kruckeberg, A. R. 1954. The ecology of serpentine soils. III. Plant species in relation to serpentine soils. Ecology 35:267-274.

---. 1969a. Plant life on serpentine and other ferromagnesian rocks of northwestern North America. Syesis 2:15-114.

---. 1969b. Soil diversity and the distribution of plants, with examples from western North America. Madroño 20:129-154.

Küchler, A. W. 1964. Manual to accompany the map of potential natural vegetation of the conterminous United States. Amer. Geogr. Soc. Spec. Pub. 36. 116 p.

Littrell, E. E., and P. M. McDonald. 1974. Within-stand dynamics of bigcone Douglas-fir in southern California. Unpub. rept., Pac. SW Forest and Range Exp. Sta.

Mason, H. L. 1947. Evolution of certain floristic associations in western North America. Ecol. Monogr. 17:201-219.

McBride, J. 1971. Vegetation on the Jalama Ranch. Unpub. consultants, rept. on file, Forestry Library, Univ. Calif., Berkeley. 21 p.

---. 1974. Plant succession in the Berkeley Hills, California. Madroño 22:317-329.

Mize, C. W. 1973. Vegetation types of lower elevation forests in the Klamath Region, California. M.S. thesis, Humboldt State Univ., Arcata, Calif. 48 p.

Mueller-Dombois, D., and H. Ellenberg. 1974. Aims and methods of vegetation ecology. John Wiley, New York. 547 p.

Munz, P. A., and D. D. Keck. 1949. California plant communities. Aliso 2:87-105.

---. 1950. California plant communities-supplement. Aliso 2:199-202.

---. 1959. A California flora. Univ. Calif. Press, Berkeley. 1681 p.

Myatt, R. G. 1975. Geographical and ecological variation in Quercus chrysolepis Liebm. Ph.D. dissertation, Univ. Calif., Davis. 220 p.

Pelton, J. 1962. Factors influencing survival and growth of a seedling population of Arbutus menziesii in California. Madroño 16:237-256.

Porter, D. M. 1962. The vascular plants of the Jasper Ridge Biological Experimental Area of Stanford University. Dept. Biol. Sci. Stanford Univ., Res. Rept. 2. 28 p.

Rockey, R. E., K. E. Bradshaw, and A. G. Sherrell. 1966. Interim report, soil-vegetation survey, Orleans Ranger District, Six Rivers National Forest. 183 p.

Roy, D. F. 1957. Silvical characteristics of tanoak. USDA Forest Serv., Calif. Forest and Range Exp. Sta., Tech. paper 22. 21 p.

Shelford, V. E. 1926. Naturalists' guide to the Americas. Williams and Wilkins, Baltimore. 610 p.

Shreve, F. 1927. The vegetation of a coastal mountain range. Ecology 8:28-44.

Society of American Foresters. 1954. Forest cover types of North America. Washington, D.C. 67 p.

State Cooperative Soil-Vegetation Survey. 1969. Soil-vegetation surveys in California. Calif. Div. Forestry, Sacramento. 31 p.

Storie, R. E. and A. E. Wieslander. 1952. Dominant soils of the redwood-Douglas fir region of California. Soil Sci. Soc. Amer. Proc. 16:163-167.

Strothman, R. O. 1972. Douglas-fir in northern California: effects of shade on germination, survival, and growth. USDA Forest Serv. Res. Paper PSW-84. 10 p.

Talley, S. N. 1974. The ecology of Santa Lucia fir (*Abies bracteata*), a narrow endemic of California. Ph.D. dissertation, Duke Univ., Durham, N.C. 209 p.

Tarrant, R. F. 1958. Silvical characteristics of Pacific Madrone. USDA Forest Serv., Pac. NW Forest and Range Exp. Sta., Silvical Ser. 6. 10 p.

Thomas, J. H. 1961. Flora of the Santa Cruz Mountains of California. Stanford Univ. Press, Stanford. 434 p.

Unsicker, J. E. 1974. Synecology of the California bay tree, *Umbellularia californica* (H & A) Nutt. in the Santa Cruz Mountains. Ph.D. dissertation, Univ. Calif., Santa Cruz. 236 p.

Waring, R. H., and J. Major. 1964. Some vegetation of the California coastal redwood region in relation to gradients of moisture, nutrients, light, and temperature. Ecol. Monogr. 34:167-215.

Wells, P. V. 1962. Vegetation in relation to geological substratum and fire in the San Luis Obispo quadrangle, California. Ecol. Monogr. 32:79-103.

Whittaker, R. H. 1954. The ecology of serpentine soils. IV. The vegetational response to serpentine soils. Ecology 35:275-288.

---. 1960. Vegetation of the Siskiyou Mountains, Oregon and California. Ecol. Monogr. 30:279-338.

---. 1961. Vegetation history of the Pacific Coast States and the "central" significance of the Klamath Region. Madroño 16:5-23.

Wieslander, A. E. 1932. Vegetation types of California, San Mateo quadrangle. USDA Forest Serv.

---. 1935. A vegetation type map of California. Madroño 3:140-144.

CHAPTER 11

OAK WOODLAND

JAMES R. GRIFFIN
Hastings Natural History Reservation, Carmel Valley, California

Introduction 384
  Defining characters 384
  Synonyms and regional subdivisions 385
Vegetation types 387
  Foothill woodland 387
    Valley oak phase 388
      Central Valley 388
      Coast Ranges 389
    Blue oak phase 392
      Great Basin transition 392
      Coast Ranges 392
      Sierra Nevada–Cascades 397
      Central Valley 400
    Interior live oak phase 400
    North slope phase 401
  Southern oak woodland 402
    Engelmann oak phase 402
    Coast live oak phase 403
  Northern oak woodland 403
    Foothill woodland transition 403
    Bald hills 404
  Riparian forest 405
Moisture relations 405
Succession 407
Summary and areas for future research 408
Literature cited 411

INTRODUCTION

Defining Characters

The oak woodland has little floristic unity except the ubiquitous annuals in its ground cover. Species from adjacent grassland, chaparral, and forest communities associate with the "woodland" trees over a wide range of physiographical and climatic situations. Many California endemics grow in the oak woodland, but they either occupy only part of the woodland or else extend well beyond it into other vegetation types.

Physiognomical unity is also difficult to detect in vegetation with such variable combinations and densities of deciduous broad-leaved, evergreen broad-leaved, and evergreen needle-leaved trees. In the context of physiognomical–ecological classes (Mueller-Dombois and Ellenberg 1974) this oak woodland relates to "short grass savanna with isolated trees," "cold-deciduous woodland with evergreens," and "evergreen broad-leaved woodlands." But climate and fire frequency in California oak woodland do not entirely fit suggested conditions for these worldwide classes.

Open stands of deciduous "white oaks" characterize vast tracts of oak woodland, but evergreen "black oaks" are often present and sometimes dominant. Also, one or more species of pine may be scattered among the oaks. On the ground, the oak woodland has a significant grassland cover under and between the trees. The composition of the interspersed grassland may be similar whether the oaks are of savanna (isolated trees) or woodland (>30% cover) density. The combination of partial deciduous oak canopy and grassy ground cover distinguishes the typical form of this vegetation.

Oak woodland can be viewed as a group of variable communities geographically placed between grassland or scrub and montane forests. The xeric lower border of oak woodland is easily defined by the absence of oak trees—where the savanna form of oak woodland becomes true grassland or scrub. The mesic upper border, where the increasingly dense woodland becomes forest, is more difficult to establish. The presence of the appropriate woodland indicator oaks is as important as the actual tree density. Where the proportion of live oaks becomes great, the oak woodland can be only arbitrarily separated from interior forms of the mixed hardwood forest (Chapter 10). Often the upper woodland border is obscured by a chaparral zone between the typical woodland and the forest above.

In a geographical context the oak woodland is well known. Cattle ranches have occupied most of the community for over a century. The woodlands have been intensively prospected for minerals. Accessibility to the oak woodland has always been relatively easy, and floristic botanists have sampled a significant portion of the area. Probably few vascular plant species remain undescribed.

The Vegetation Type Map (VTM) Survey mapped the general distribution of oak woodland in central and southern California in the 1930s (Wieslander 1935), and large parts of the Cascade and North Coast Range woodlands have been covered by the State Cooperative Soil–Vegetation (S–V) Survey since the 1950s (State Cooperative Soil–Vegetation Survey 1969). Range managers have intensively studied the grassy cover between the woodland trees, and game managers know much about wildlife production in the heavily hunted woodland landscape. Detailed vegetational information for complete oak woodland community samples, however, is limited to a few localities. Because so few details exist, only dominant trees are emphasized in this chapter.

Synonyms and Regional Subdivisions

Clements (1920) included oak woodland within a "pine–oak woodland," stressing its relationship to the scrubby juniper–pine–oak communities of the southwestern United States and Mexico. In Cooper's (1922) often-quoted treatment of California woody vegetation, oak woodland was viewed as a derivative of "broad-sclerophyll" chaparral communities. He recognized no stable savanna type, but he interpreted scattered oaks and pines in grassland as occupying habitats where chaparral had been destroyed by fire.

In terms of the VTM and S–V Survey literature, oak woodland combines the "woodland," "woodland–grass," and "woodland–chaparral" vegetation types (Fig. 11-1*a,b,c*). Such woodlands cover at least 3.6 million ha in California (Wieslander and Gleason 1954). Küchler's (1964) map of the "California oakwoods" type seems to be derived from these U.S. Forest Service woodland types. Oak woodland includes the drier California portions of Knapp's (1965) "xeric oak and mixed-oak woodlands," which also embraced the mixed evergreen forest (Chapter 10) and extended into summer rainfall types in Arizona.

As used here, "oak woodland" approaches the "California oak woodland" of

Figure 11-1. Phases of oak woodland as categorized by the VTM and State Cooperative S–V Survey literature. (*a*) "Woodland" near Spenceville, Nevada Co., with *Quercus wislizenii* and *Pinus sabiniana* mixed in with *Q. douglasii* (VTM Survey photo 50-2520). (*b*) "Woodland–grass" in the same area, with *Q. douglasii* over an annual grassland (VTM Survey photo 50-2516). (*c*) "Woodland–chaparral" near Grass Valley, Nevada Co., with chaparral shrubs *Arctostaphylos viscida* and *Ceanothus cuneatus* between *P. sabiniana, Q. wislizenii,* and *Q. douglasii* (VTM Survey photo 50-2499).

Benson (1957) and the "woodland savanna" of Munz and Keck (1959). Munz subdivided woodland savanna into three regional units: northern oak woodland, foothill woodland, and southern oak woodland. The same names are used for similar units in this chapter. In each unit a different white oak species is locally dominant. Of these three woodlands, the foothill woodland has the best structural identity, covers the largest area, and has received the most study.

Oak woodland seldom forms a continuous cover over large areas. It is a major item in a mosaic including valley grassland (Chapter 14) and chaparral (Chapter 12) with strips of riparian forest. The islands of true grassland are not discussed here.

Figure 11-1 (Continued)

Although the trees influence the grassland matrix of the savanna, the grassy cover is considered in more detail in Chapter 14. Direct oak canopy effects on grassland are mentioned. The only chaparral items considered deal with scrubby oak thickets which bridge the gap between woodland and chaparral in some regions. The limited materials on riparian forests within the woodland are reviewed.

VEGETATION TYPES

Foothill Woodland

The term "foothill" is not ideal in this name since portions of the community grow on the level floor of the Central Valley, and other parts extend well into the mountains. But the name has a fairly consistent usage in California (Clark 1937; Griffin and Critchfield 1972; Hoover 1970; Klyver 1931; Munz and Keck 1959; Thomas 1961; White 1966), and the community does frequently grow on "foothill" terrain. Foothill woodland approaches the "Digger pine–oak forest type" of the Society of American Foresters (1954).

Geographically, the foothill woodland covers several million hectares around and partially in the Central Valley. This woodland is an exclusively Californian community. Three dominant trees are endemic to the community: *Quercus douglasii* (blue oak), *Q. lobata* (valley oak), and *Pinus sabiniana* (Digger pine). The combined distribution of these species (Griffin and Critchfield 1972) approximates the range of foothill woodland. Several other trees and many shrubs and herbs important in the foothill woodland are also California endemics.

The foothill woodland has two major physiographical subdivisions (Knapp 1965; Munz and Keck 1959; Society of American Foresters 1954). On deep soils, *Q. lobata* often forms nearly pure stands of large trees with no woody understory (valley oak phase). Such *Q. lobata* savannas appear similar in structure on valley bottoms and rolling slopes over a wide range of elevations. A few live oaks may be scattered in this phase: *Q. agrifolia* (coast live oak) in the Coast Range valleys and *Q. wislizenii* (interior live oak) in the Central Valley.

On shallower upland soils, a more mixed-species community develops. *Quercus douglasii* and *P. sabiniana* are the most characteristic trees in the hillside woodlands (blue oak phase). Live oaks are far more important in the uplands: *Q. agrifolia* in the South Coast Ranges and *Q. wislizenii* in the rest of the woodland. Throughout the blue oak phase *Q. lobata* may be present on deep soils. Nonsprouting chaparral shrubs are scattered about, such as *Ceanothus cuneatus*, various manzanitas related to *Arctostaphylos viscida* or *A. manzanita, Rhamnus californica* ssp. *tomentella, R. crocea* ssp. *ilicifolia,* and *Rhus diversiloba.* In all regions there is an elevational gradient with increased tree and shrub density of evergreens at higher elevations. Individual oaks are never as massive in the blue oak phase as in the valley oak phase, but dense stands of small trees on hills may have as much biomass as open stands of large trees in the valleys.

Foothill woodland canopy can become too dense to support a typical ground cover, as on steep north slopes, where thickets of scrubby evergreen trees are mixed with the deciduous *Aesculus californica* (north slope phase), or in semichaparral thickets of small *Q. wislizenii* trees (interior live oak phase).

Valley oak phase. Parklike *Quercus lobata* stands are best developed on most recent alluvial terraces of large valleys, but they may extend onto older terraces and low rolling hills. In the South Coast Ranges, *Q. lobata* appears in typical lowland situations, but similar-looking savannas also extend up to montane summits (Griffin 1976). Names for these parklike stands include "valley sonoran" (Jepson 1910), "savanna prairie" (Clark 1937), and "oak woodlands of Californian Central Valley" (Knapp 1965).

CENTRAL VALLEY. On appropriate soils of the Central Valley, the valley oak phase used to be extensive, particularly in the eastern portion of the valley (Jepson 1910). These savannas were associated with riparian forests along the active floodplains but were far simpler, often with only *Quercus lobata* trees towering above the grassland (Fig. 11-2).

In the Sacramento Valley, Thompson (1961) summarized early reports of the valley oak phase, which was common along the natural levees adjacent to riparian forests. He quoted from a valuable 1854 railroad survey near Cache Creek:

> This timber belt is composed of the most magnificant oaks I have ever seen. They are not crowded as in our forests, but grow scattered about in groups or singly, with open grass-covered glades between them. . . . There is no undergrowth beneath them, and as far as the eye can reach, when standing among them, an unending series of great trunks is seen rising from the lawn-like surface.

The only quantitative item in Thompson's compilation concerns a savanna of *Q. lobata* and *Q. wislizenii* near Willows, which had 49 "good trees" per hectare.

In the San Joaquin Valley, not even a review of qualitative remarks by early travelers and surveyors exists for the valley oak phase. The immense relics standing in parts of Fresno and Tulare Cos. suggest, however, that such savannas must have been impressive. One of Jepson's (1910) correspondents spoke of ". . . four hundred square miles of Valley Oaks . . ." on the Kaweah River alluvial soils.

Figure 11-2. Relatively dense valley oak phase savanna at Cone Ranch, Tehama Co., in 1929. The *Quercus lobata* trees were probably over 100 yr old then. Grazing and fire histories of the stand are not known. From Grinnell et al. (1930, photo MVZ 5890).

COAST RANGES. Less extensive stands of valley oak phase covered lowlands of the Coast Ranges, for example, in Napa, Santa Clara, and Santa Barbara Cos. Vancouver (1798) commented on the valley oak phase near Palo Alto before European disturbance:

> . . . we arrived at a very pleasant and enchanting lawn, situated amid a grove of trees. . . . For almost twenty miles it could be compared to a park which had originally been planted with the true old English oak; the underwood, that had probably attained its early growth, had the appearance of having been cleared away and had left the stately lords of the forest in complete possession of the soil. . . .

Cooper (1926) discussed the vegetation of the same region after agricultural disturbance but before complete urbanization. He emphasized soil and water table influences on woodland development and viewed the valley oak phase as successional to a mixed hardwood forest.

Some of the best examples of parklike valley oak phase still extant are in the Santa Lucia Range, Monterey Co. (Fieblekorn 1972; 1976). Size-class distributions for two of the least disturbed stands in this area are given in Table 11-1. In these lower-elevation stands, seedlings and saplings are virtually absent (Griffin 1976). Of the 24 0.1 ha plots sampled in the two stands, the plot with greatest biomass had 80 trees per hectare and a 61.3 $m^2\ ha^{-1}$ basal area. Probably few valley oak "savannas" ever exceeded this basal area over a very large region.

Table 11-1 also describes a montane ridgetop form of the valley oak phase. In this case, *Pinus coulteri* and several live oak species are regenerating under the overmature savanna oaks. Such conversion of savannas into pine–hardwood forest is occurring locally in widely scattered ridges in the South Coast Ranges. Causes of the conversion are not clear, but low fire frequency in recent decades is encouraging the pine invasion. The Gabilan Range *P. coulteri* plots are related to this situation (see Table 11-4).

TABLE 11-1. Tree density and basal area in two gentle lower slopes and one montane summit valley oak phase savannas in the Santa Lucia Range, Monterey Co. (Griffin 1976)

Diameter at Breast Height (cm)	Nacimiento Valley 7 (lower slope) *Quercus lobata*	San Antonio Valley 5 (lower slope) *Q. Lobata*	Chews Ridge 1 (montane summit) *Q. lobata*	*Pinus coulteri*	*Quercus chrysolepis*	*Q. wislizenii*	*Lithocarpus densiflorus*
I. Tree density (number per hectare)							
0-2.5	0.8		2.5	91.7	85.0	24.2	23.3
2.6-9				50.0	5.8	2.5	
10-29			27.5	70.0	6.7	1.7	
30-49	0.8	1.7	21.7	22.5	3.3		
50-69	0.8	5.8	5.0	7.5	1.7		
70-89	4.2	6.7	5.8				
90-109	5.1	8.3	10.8				
110-129	0.8	4.2	4.2				
130-149	3.3	3.3	5.0				
150-169	1.7	4.2	1.7				
170-189	4.2		0.8				
190-210	0.8	0.8					
Total	21.7	35.0	85.0	241.7	102.5	28.4	23.3

II. Total basal area ($m^2\ ha^{-1}$) for dbh 2.6-210

28.5	32.6	33.3	6.2	1.0	.04

Blue oak phase. GREAT BASIN TRANSITION. Foothill woodland is not entirely isolated from the Great Basin Floristic Province, for southern stands of the blue oak phase merge with the western fringe of the piñon–juniper woodland (Chapter 23). In the Temblor–La Panza–Caliente–San Emigdio ranges, *Quercus douglasii* associates locally with *Juniperus californica* (Hoover 1970; Twisselmann 1956). In the Tehachapi and Piute ranges, *Q. douglasii* and *Pinus sabiniana* grow with *J. californica* and *P. monophylla* (Bauer 1930; Twisselmann 1967). A few Kern Co. stands of *Q. douglasii* have a shrub understory of *Artemisia tridentata.*

The driest stands in the foothill woodland probably occur in Kern Co. Average precipitation in the Temblor Range is probably less than 25 cm (Twisselmann 1967). The highest-elevation stands also occur here; the blue oak phase reaches 1800 m on Mt. Pinos and in the Tehachapis (Twisselmann 1967).

In the Central Coast Range–Tehachapi transition, *J. californica* is present in the blue oak phase as a minor associate (Table 11-2). Juniper relics continue northward as widely separated colonies through the Coast Ranges to southwestern Shasta Co. (Bowermann 1944; Hoover 1970; Mallory et al. 1966; Vasek 1966), and in the Sierra Nevada–Cascade foothills to south-central Shasta Co. (Grinnell et al. 1930; Mallory et al. 1965b). These junipers may occur at any elevation within the blue oak phase. They are often sprinkled in the lowest savannas as near Bitterwater, San Benito Co., or Fruto, Colusa Co., but on Mt. Diablo junipers are scattered up to the rocky summit. In the Pit River area east of the Cascades, foothill woodland occurs in part on nonzonal soils. In this case, disjunct *Q. douglasii* stands mix with such northern juniper woodland species as *J. occidentalis, Purshia tridentata,* and *Artemisia tridentata.*

In addition to the juniper, a desert oak is significant in the southern blue oak phase. *Quercus turbinella* (shrub live oak) not only locally dominates the chaparral but also has had a profound genetic influence on *Q. douglasii* in the woodland. There has been widespread hybridization between evergreen *Q. turbinella* shrubs and deciduous *Q. douglasii* trees (Benson et al. 1967; Tucker 1952). In some Diablo and Santa Lucia Range woodlands, intermediates between these species (*Q.* × *alvordiana*) form an important part of the woodland canopy (White 1966). The *Q. turbinella* influence goes much further north in the Coast Ranges than in the Sierra Nevada.

COAST RANGES. The western margin of the blue oak phase in the Coast Ranges is quite irregular, and local soil conditions often control woodland distribution. Pockets of blue oak phase often appear well within the mixed evergreen forest on sterile soils. Adjacent to chaparral, they occur on deeper rocky soils. Typical blue oak phase makes its closest approach to the coast in the Nacimiento Valley hills of southern Monterey Co. Here *Quercus douglasii* stops some 10 km inland, but *Q. lobata* and *Pinus sabiniana* cross the seaward ridge and grow almost to the shore (Griffin 1962). The location of the coastal *P. sabiniana* is in part related to serpentine soils.

In the North Coast Ranges, Gardner et al. (1964) summarized the S–V Survey soil data for Mendocino Co. Of seven "woodland grass" soils, by far the most important was the Laughlin series. Blue oak phase grows over many thousands of hectares of Laughlin soils in southeastern Mendocino Co. The Hopland Field Station (NA 230870) in this region has blue oak phase with an average density of 500 trees per hectare and average diameter at breast height of 20 cm (Murphy and

TABLE 11-2. Relative tree density, absolute tree density, and grass-forb ground cover in six different geographical regions of the foothill woodland from 536 plots of 0.2 acres of the VTM Survey

	Coast Ranges			Cascade-Sierra Nevada		
	Napa	Gabilan	Lompoc	Redding	Chico	Jackson
I. Relative tree density (%)						
Quercus douglasii	54	34	19	52	57	54
Pinus sabiniana	2	9	8	18	28	12
Quercus lobata	7	9	7	3	0.5	1
Q. kelloggii	8	6		9	1	2
Q. wislizenii	7			13	12	30
Aesculus californica	5	6		0.2	0.3	1
Quercus agrifolia	15	26	66			
Arbutus menziesii	0.1	0.3		0.1		
Umbellularia californica	0.3			0.5		
Quercus garryana	1					
Q. chrysolepis		0.5			0.3	0.5
Pinus ponderosa				2	0.1	0.2
Arctostaphylos mariposa						0.1
Pinus attenuata				1		
P. coulteri		4				
Juniperus californica		0.4	0.1			
Quercus x *alvordiana*		4				
Prunus ilicifolia		0.1				
II. Absolute tree density (number per hectare)						
All foothill woodland	202	214	175	151	222	183
Blue oak phase only	166	184	126	138	195	181
III. Grass-forb ground cover (%)						
All foothill woodland	92	73	75	51	68	71
Blue oak phase only	99	88	79	66	85	89

Crampton 1964). The Hopland soil survey (Gowans 1958) described Laughlin silt loam–gravelly silt loam as supporting good grass cover with scattered *Q. douglasii* and dense *Q. wislizenii,* and Sutherlin loam–clay loam as supporting good grass cover with scattered *Q. douglasii* and *Q. wislizenii.*

The VTM Survey took a total of 132 0.2 acre (810 m²) plots in the 30-min USGS Napa quadrangle, Napa and Solano Cos. Sixty-six plots were in foothill woodland (Table 11-2); about one-half of the woodland plots could be called blue oak phase. The vast majority of *Q. douglasii* trees on the Napa and other VTM plots had dbh values of less than 30 cm (Table 11-3). This blue oak phase condition contrasts with the valley oak phase, where trees less than 30 cm are uncommon or absent.

The blue oak phase is poorly developed in Marin Co.; *P. sabiniana* is entirely absent (Howell 1970). The blue oak phase is better represented on the interior slopes of the Santa Cruz Mts. (Jensen 1939; Porter 1962; Thomas 1961). Of the local floristic studies in Coast Range woodlands, Bowerman's (1944) work on Mt. Diablo is the most detailed. She recognized a woodland association with *Q. douglasii* dominant and *P. sabiniana* less common; *Q. lobata* was local at lower elevation, and *Q. wislizenii* was scattered in the association on north aspects.

The Gabilan Range in Monterey and San Benito Cos. (VTM 15-min quads 104 B,C,D; 105 A,D; 107 A,B) provides a convenient geographical unit with complete VTM Survey coverage. The VTM Survey took 97 plots in foothill woodland (Table 11-2). The plots reveal four dominance trends: abundant *Q. douglasii* with much grass (Fig. 11-3), abundant *Q. agrifolia* with little grass, abundant *Q. lobata,* and abundant young *Pinus coulteri* (Table 11-4). Brooks (1971) studied 10 stands in the blue oak phase of the Pinnacles National Monument (NA 351650) region of the Gabilan Range. Of the 655 trees and saplings on his plots, 88% were *Q. douglasii* and 6% were *P. sabiniana.* Brooks's dbh data (Table 11-5) follow the VTM trend of few large and many small trees (Table 11-3). In Brooks's data, a distinct bulge appeared in the 11–21 cm dbh class.

Figure 11-3. Typical savanna form of oak woodland with widely spaced oaks and grassy ground cover. In this case, blue oak (*Q. douglasii*) is over an annual grassland dominated by *Avena fatua.* Photograph taken by Keith White in Carmel Valley, Monterey Co.

TABLE 11-3. Relative tree density (%) of *Quercus douglasii* by dbh category in six different geographical regions of the foothill woodland from 536 plots of 0.2 acres of the VTM Survey

Diameter at Breast Height (cm)	Coast Ranges			Cascade-Sierra Nevada		
	Napa	Gabilan	Lompoc	Redding	Chico	Jackson
10-29	87.9	74.6	74.4	82.3	82.7	82.9
30-60	11.6	23.1	20.5	15.4	16.0	15.4
61-90	0.3	2.1	5.1	2.0	1.3	1.8
90+	0.2	0.2		0.3		

In the northern Santa Lucia Range, much ecological work has been done around the Hastings Natural History Reservation (NA 270812). Detailed floristic–faunal behavior and local climatic data are available (Griffin 1974). White (1966) studied 49 foothill woodland stands here, ranging in elevation from 450 to 1500 m. Of the 3872 trees on White's plots 89% were *Q. douglasii.* White grouped his stands into three topographical–tree density classes. The steeper north slope dense stands had 1381 saplings and trees per hectare with a basal area of 24.6 m^2 ha^{-1}. Various gentle slopes of typical density had 575 saplings and trees per hectare with a basal area of 19.2 m^2 ha^{-1}. Open ridgetop stands had only 176 saplings and trees per hectare with a basal area of 14.7 m^2 ha^{-1}. Griffin (1971a, 1973, 1976) studied oak regeneration and moisture relations in woodlands of different densities. Table 11-6 gives details of size classes for two of the relatively dense stands.

In the Santa Lucias, *Q. douglasii* has regenerated poorly during the past 50 years, but the oak reproduced well at an earlier period. White (1966) found a large proportion of the saplings or small trees near Hastings to be 60–100 yr old. These are the trees that make the Hastings "savannas" so dense (Table 11-6). Brooks (1971) suggested that the youngest important age class in the Gabilans was about 80 yr old, a situation similar to that in the Santa Lucias and probably much of the South Coast Ranges.

Holland (1974) made an intensive study of the effect *Q. douglasii* trees have on vegetation and soils under them at the Hastings Reservation. He showed that vegetation under an oak crown grows faster and taller, produces about twice as much biomass, and matures later than vegetation in the open. Litter and leachate from the oak crown contribute to higher nutrient status and more desirable physical soil properties under the trees. Floristic differences under the trees, for example, higher proportions of *Bromus diandrus* and *Hordeum leporinum,* result from

TABLE 11-4. Variation in tree species composition, tree density, and grass-forb ground cover within the foothill woodlands of the Gabilan Range, Monterey-San Benito Cos., from a total of 97 plots of 0.2 acres of the VTM Survey

Species	Phase (and Number of Plots)			
	Blue Oak (65)	North Slope (16)	Valley Oak (10)	Pine Forest (6)
I. Tree density (number per hectare)				
Pinus coulteri				152
Quercus lobata	4		142	23
Q. agrifolia	22	205	25	78
Aesculus californica	4	62		
Quercus kelloggii	8	46		
Pinus sabiniana	25	7	15	10
Quercus douglasii	108	2	4	
Arbutus menziesii		3		
Quercus chrysolepis				10
Q. x *alvordiana*	12			
Juniperus californica	1			
Total tree density	184	325	186	273
II. Grass-forb cover (%)				
	88	26	86	36

interspecific competition for nutrients and light, perhaps reinforced by allelochemical interactions on seedling development. This crown effect can last for decades after the oak crown has been removed (Holland 1974).

At the southern end of the Coast Ranges, the VTM Survey sampled 96 foothill woodland plots in the 30-min Lompoc quad. Although typical blue oak phase occurs there, *Q. agrifolia* dominates a variety of woodland habitats (Table 11-2). When *Q.*

douglasii is absent in this region, the foothill woodland can be separated from southern oak woodland only arbitrarily.

SIERRA NEVADA–CASCADES. The vast strip of blue oak phase along the foothills east of the Central Valley seems more homogeneous than the interrupted Coast Range woodlands. The more continuous boundary of montane forest above the woodland may contribute to the appearance of uniformity. Although the dominant trees remain essentially the same throughout the Sierra Nevada–Cascades (Table 11-2), there are many latitudinal replacements in understory species along the way.

In the southern Sierra Nevada foothills, Brooks (1971) and Vankat (1970) studied blue oak phase in and adjacent to Sequoia National Park in Tulare Co. Of the 1663 trees and saplings on Brooks' plots 87% were *Quercus douglasii.* Half of Brooks's stands were on grazed ranchland near the park. Ungrazed stands inside the park generally had steeper slopes and shallower soils. Stands in the park had more *Q. wislizenii* and *Aesculus californica* (Table 11-7), and *Q. kelloggii* entered the highest portion of the blue oak phase in the park. The mean tree density on grazed plots was 230 per hectare and on ungrazed plots 329 per hectare. Oak density in stands with no known fire record was 336 per hectare; in stands burned in the last decade oak density was only 114 per hectare. Diameter growth was similar to that in the Gabilan Range, and up to 50% of the *Q. douglasii* trees in the Sequoia National Park area may be about 80 yr old.

TABLE 11-5. Relative tree density (%) of Quercus douglasii by dbh category in Sequoia National Park and Pinnacles National Monument foothill woodland regions. Some Sequoia stands were ungrazed for 80 yr, Pinnacles stands for 64 yr (Brooks 1971)

Diameter at Breast Height (cm)	Region, Grazing Status (and Number of Stands)			
	Pinnacles		Sequoia	
	Grazed (6)	Ungrazed (4)	Grazed (12)	Ungrazed (6)
2.6-10	31	18	16	5
11-21	49	40	50	41
22-32	13	19	25	34
33-42	3	9	6	11
43-53	3	3	2	4
54-63	1	6	1	4
64+	1	5	1	1

TABLE 11-6. Density of species by dbh category and total basal area in two "dense" blue oak phase savannas on the Hastings Reservation, Monterey Co. (Griffin 1973)

Diameter at Breast Height (cm)	Savanna D-7		Savanna D-10	
	Q. douglasii	*Q. agrifolia*	*Q. douglasii*	*Q. agrifolia*
	I. Density (trees per hectare)			
0-2.5	15		20	30
2.6-9	255	5	690	5
10-19	630		540	10
20-29	140		35	
30-39			15	
40-49			5	
50-59			25	
60-69	10			
Total	1050	5	1330	45
	II. Total basal area (m^2 ha^{-1}) for dbh 2.6-69 cm			
	20.47	0.05	19.36	0.14

South of the park near Glenville, Kern Co., Lawrence (1966) observed the effect of a July control burn on blue oak phase and chaparral. The fire and subsequent insect loss reduced tree cover on study transects by 16%. Large oaks and pines were little affected, but *Pinus sabiniana* saplings were severely damaged. Lawrence's main concern was with fire–vertebrate interactions. In general, this burn increased "grassland" birds and rodents, while "woodland" birds remained stable.

Although *P. sabiniana* is locally missing from the blue oak phase of the Coast Ranges, one of the most interesting absences of the pine is in the Tulare and Fresno Co. woodland. *Pinus sabiniana* is missing from 100 km of woodland stretching across the Kaweah drainage (Dudley 1901; Graves 1932; Griffin 1962; Jepson 1910; Shinn 1911; Whitney 1865). *Quercus douglasii* occurs in typical form throughout the pine gap (Brooks 1971). No obvious climatic or edaphic features seem related to the pine gap, and no satisfactory explanation of the situation has been offered. Graves

(1932) suggested that the gap could be "explained entirely on the basis of fire", but he offered no evidence that the fire history of the gap woodland differed from that of the rest of the Sierra Nevada woodland, where the pine thrives despite fire.

The San Joaquin Experimental Range lies within a classic example of the blue oak phase in Madera Co. (NA 201912). Detailed vascular plant lists, vertebrate animal lists, and local climatic data are available (Buttery and Green 1958; Newman and Duncan 1973). Duncan and Westfall (1975) summarized 243 publications relating to grassland–livestock–wildlife interactions on the range, but stand analyses for the oaks are not available. Holland (1974) replicated his *Q. douglasii* canopy effect trials on the range with results similar to those at Hastings Reservation.

The VTM Survey of the Jackson 30 min quadrangle in Calaveras Co. gives an example of foothill woodland composition in the central Sierra Nevada (Table 11-1). Much of the woodland lore from this region stems from "range improvement" projects that thin or eliminate trees from the woodland (Dal Porto and Carlson 1965; Emrich and Leonard 1954; Johnson et al. 1959). Lewis and Burgy (1967) measured all trees in a 17 ha watershed in Placer Co. before it was cleared, but the stand data have not been summarized by size classes. The area was covered with dense woodland dominated by *Q. wislizenii* on north aspects and with open woodland–grass dominated by *Q. douglasii* on south aspects.

The VTM Survey covered little of the volcanic Cascade foothills. The only stand samples were 43 woodland plots in the 30 min Chico quadrangle, Butte Co., and 94 plots in the 30 min Redding quadrangle (Table 11-2). The S–V Survey, however, mapped most of this region.

Helpful "legends" accompanying the S–V maps for Shasta Co. comment on local soil–vegetation relationships. On the 7.5 min Manton quadrangle, open *Q. douglasii*

TABLE 11-7. Ecological importance values (relative density + relative cover + relative frequency) for tree species in 20 stands from the Sequoia National Park region (Brooks 1971)

Species	Livestock Grazing Status (and Elevational Range)	
	Ungrazed (427-1400 m)	Grazed (213-1067 m)
Quercus douglasii	221	282
Q. wislizenii	32	6
Aesculus californica	26	9
Quercus kelloggii	11	0

woodland with occasional *P. sabiniana* and *Juniperus californica* trees grows below 900 m elevation on Guenoc loam–cobbly loam, Toomes very rocky loam, and Supan loam–clay loam (Mallory et al. 1965b). This is the northern limit of the juniper. The Whitmore quadrangle has somewhat denser woodland with more *Arctostaphylos viscida* understory. Seven rocky soil series are involved in these woodlands, including Kilarc cobbly loam–clay with "strongly acid subsoil" (Mallory et al. 1965a). A few *Q. garryana* trees are present in the *Q. douglasii* stands.

The northern limit of blue oak phase occurs on the Montgomery Creek quadrangle on Kilarc soils at elevations up to 1100 m (Mallory et al. 1965c). The wettest portion of the blue oak phase is here, with an average precipitation of 152 cm in typical woodland at Montgomery Creek. *Quercus douglasii* and *P. sabiniana* stragglers in the Pit River Canyon receive over 190 cm of average seasonal precipitation.

One disjunct stand of blue oak phase (NA 451670) occurs east of the Cascade crest in eastern Shasta Co. (Benson 1962; Griffin and Critchfield 1972). A relatively small *Q. douglasii* population lies within a larger *P. sabiniana* area (Griffin 1962, 1966). The blue oak phase is surrounded by a mixture of montane forest and northern juniper woodland species in which *Q. garryana* is the dominant oak. This central Pit River drainage has colder winters than any other area supporting *Q. douglasii* and *P. sabiniana.* The mean January temperature is about 2°C, and lows of −25°C are not uncommon (Griffin 1971b).

The S–V Survey has mapped all the blue oak phase in the geologically diverse Klamath Mt. foothills in western Shasta Co. (Mallory et al. 1966, 1968, 1973). The map legends describe several soil–woodland associations; for example, in the Chanchelulla Peak quadrangle, Millsholm loam–loam often supports a 20% cover of *Q. douglasii* and 5% *P. sabiniana.* On the deeper Parrish gravelly loam–gravelly clay, *Q. kelloggii* is added to the open canopy, and *Arctostaphylos manzanita* shrubs may invade the annual grass cover between the trees.

CENTRAL VALLEY. On uplands around the Central Valley the lowest savanna fringe usually has *Quercus douglasii* as the only tree. In the southern San Joaquin Valley, the lower border may be 600 m in elevation (Twisselmann 1967); in the Sacramento Valley the lower border may be only 50 m. At the head of the Sacramento Valley in Shasta and Tehama Cos., well-developed blue oak phase crosses the valley on nonalluvial soils (Wieslander 1939a). Woodland is poorly developed on the Sutter Buttes, which rise within the lower Sacramento Valley.

In only one part of the Central Valley does the blue oak phase extend significantly onto valley floor alluvium. Agricultural clearing has almost obliterated this example between Dunnigan and Arbuckle, but a few relic *Q. douglasii* and *Pinus sabiniana* trees remain on the gravelly Arbuckle soil series both along creeks and in fields and orchards (Griffin 1962). *Quercus lobata* is uncommon on Arbuckle soils, although present in typical valley oak phase on finer-textured soils nearby. Relic *Adenostoma fasciculatum* shrubs also remain in the area. This valley floor occurrence of woodland–chaparral relics on nonzonal soils is of interest because Cooper (1922) used it to postulate the former dominance of chaparral over all the Sacramento Valley floor.

Interior live oak phase. In low-elevation foothill woodlands, *Quercus wislizenii* appears as widely spaced trees or sprout clumps. In the Sierra Nevada, such live oak clumps may be concentrated around rock outcrops within the blue oak phase. With increasing elevation, particularly on north aspects, *Q. wislizenii* becomes a more significant part of the canopy (Brooks 1971; Griffin 1962; Klyver 1931; Lewis and Burgy 1967). Finally, in an irregular zone below the montane forest, *Q. wislizenii* almost excludes *Q. douglasii* from the woodland. The interior live oak thickets are too dense to support a typical grassland ground cover and are considered here as a scrubby evergreen phase of the foothill woodland.

In the Sierra Nevada–Cascade foothills, the interior live oak phase has a close floristic tie to chaparral; but the oaks are usually arborescent, and tall *Pinus sabiniana* trees may be common. In Fresno Co., Klyver (1931) listed a "live oak–Digger pine" association within the chaparral.

In the Klamath and North Coast Ranges, the interior live oak phase merges into both chaparral and mixed evergreen forest. *Quercus wislizenii* is present in montane chaparral and mixed hardwood forest communities in the South Coast Ranges, but interior live oak thickets are not important within the blue oak phase.

Some well-developed examples of live oak phase occur near Inskip Hill, Tehama Co., and Shingletown, Shasta Co. In this volcanic region, Guenoc loam–cobbly clay loam and related soils support blue oak phase, whereas deeper Aiken loam–clay soils support montane forest. Thickets of *Q. wislizenii* and chaparral dominated by scrubby *Q. garryana* are concentrated on intermediate soils between the savanna and forest in this area (Mallory et al. 1965b).

Low-elevation examples of interior live oak phase occur near the Redding airport, Shasta Co., at 150 m elevation (Wieslander 1939a). Extensive thickets of *Q. wislizenii* grow on deeper, better-drained soils than does the surrounding blue oak phase.

Myatt (1975) studied vegetation containing *Q. chrysolepis* in the Yuba and Stanislaus River canyons of the Sierra Nevada. He considered two of his eight dominance types as foothill woodland. The lowest-elevation type was essentially blue oak phase with very few *Q. chrysolepis*. The next highest type was related to the interior live oak type with thickets of *Q. wislizenii* and *Q. chrysolepis* and a prominent understory of common oak woodland shrubs. Above this were several types that Myatt assigned to mixed evergreen forest; they are closely related to the Coast Range mixed hardwood forest (Chapter 10).

North slope phase. In the Central Coast Ranges and to a lesser extent in the Sierra Nevada, steep north slopes may have a dense canopy in which *Quercus douglasii* is uncommon or absent. The deciduous *Aesculus californica* is often conspicuous in these thickets along with evergreens like *Umbellularia californica, Prunus ilicifolia,* and *Heteromeles arbutifolia* (Benseler 1968). This type can even be traced into the inner North Coast Ranges. Very steep north slopes of Putah and Cache creeks have this community where unburned. Scattered *Pinus sabiniana* tower over the dense, tall *Aesculus–Heteromeles–Fraxinus dipetala.*

Quercus agrifolia is present in the north slope phase in the Coast Ranges; *Q. wislizenii,* in the Sierra Nevada. Klyver's (1931) "buckthorn–live oak–buckeye" and "live oak–buckeye" associations refer to this kind of vegetation. The scrubby woodland dominated by *Aesculus californica* in the Kern River Canyon is also related (Twisselmann 1967). In the San Luis Obispo Co. foothills, where *Aesculus* is rare, *Prunus* may be important. There are some good examples of *Aesculus–Prunus–Umbellularia* thickets in the Santa Lucias of Monterey Co. Such thickets at the base of a north slope may merge with a very tall *Q. turbinella–Q.* × *alvordiana* chaparral on the upper slope, with the whole slope surrounded by blue oak phase.

This ill-defined north slope phase could be treated as a mesic form of the chaparral, but one can walk under the "shrubs." In the Santa Lucias *Prunus ilicifolia* may be up to 14 m tall, and *Heteromeles arbutifolia* is sometimes of tree size and form (Griffin 1974). From the Santa Cruz Mts. northward it may be more helpful to consider these north slope thickets as xeric forms of mixed hardwood forest rather than as mesic deviations of the foothill woodland. To the north they become an exposed, rocky canyon bottom type below forest, rather than a rocky, shady type within woodland.

Southern Oak Woodland

From the Santa Ynez Mts. of Santa Barbara Co. southeastward, significant floristic changes occur in the oak woodland. Introduced species in the grass cover continue with little change, but native shrubs and herbs become increasingly "southern Californian" (Munz 1974). Trees endemic to southern California also become locally dominant.

The blue oak phase has no significant distribution in southern California, for both *Quercus douglasii* and *Pinus sabiniana* barely reach into northern Ventura and Los Angeles Cos.; *Aesculus californica* also quits in this vicinity (Benseler 1968; Griffin and Critchfield 1972). *Quercus lobata* continues southward into western Los Angeles Co. The VTM Survey mapped the southern extreme of typical valley oak phase near Russel Valley, Ventura Co. (Wieslander 1939b). Of the woody dominants in the foothill woodland, only *Q. agrifolia* is widespread in southern California.

Munz and Keck's (1959) "southern oak woodland" is a convient name for the variable woodlands in coastal southern California. The savanna portion of the southern oak woodland with *Q. engelmannii* (Engelmann oak) prominent is called "Engelmann oak phase" here. The denser, more widespread woodlands with *Q. agrifolia* always important and *Juglans californica* sometimes dominant are termed "coast live oak phase." Together they constitute the "live oak woodlands of the southern California mainland" of Knapp (1965). The dominant trees in both phases are cismontaine, with no desert disjunctions.

Engelmann oak phase. The dominant white oaks in the Engelmann and blue oak phases are closely related, and *Quercus engelmannii* could be viewed as a sort of semievergreen southern California version of *Q. douglasii.* Compared to *Q. douglasii, Q. engelmannii* has a small, total range.

Quercus engelmannii used to grow between Pasadena and Claremont in Los Angeles Co. (Wieslander 1934a,b), but no intact woodland is left in this area now. While discussing the genetic and ecological separation between *Q. engelmannii* and *Q. dumosa* in this area, Benson (1962) suggested that *Q. engelmannii* woodland was restricted to deep valley clay soils, whereas *Q. dumosa* chaparral occurred on shallow upland soils. *Quercus engelmannii* did not occur on recent alluvial soils.

Quercus engelmannii locally dominates woodlands on the Santa Rosa plateau in the Santa Ana Mts., Riverside Co. This is one of the few spots in the Engelmann oak phase with any descriptive study (Lathrop and Thorn 1968; Snow 1972; Wieslander 1938; Zuill 1967). Zuill (1967) recognized two forms of the Engelmann oak phase there, both composed of *Q. agrifolia* and *Q. engelmannii.* The savanna form, "grass oak woodland," had a tree density of 27 per hectare, of which 90% was *Q. engelmannii.* The "dense oak woodland" had a tree density of 56 per hectare with both species equally represented. Zuill (1967) suggested that the "dense oak woodland" was concentrated around rock outcrops, and Snow (1972) documented an increased density of *Q. agrifolia* near rocks. Snow concluded that regeneration conditions for *Q. agrifolia* were optimum in the middle and on the north sides of rock outcrops, whereas establishment of *Q. engelmannii* was independent of rocks. Increased ground squirrel activity and decreased grazing near outcrops probably favored *Q. agrifolia* regeneration.

Snow (1972) discussed the climatic features of the Santa Rosa Plateau Engelmann oak phase, they do not differ greatly from those of the more coastal portions of the blue oak phase. For example, the seasonal precipitation on the plateau is 53 cm, and 51 cm at the Hastings Reservation in blue oak. But the precipitation extremes are greater in the south ranging from 21 to 138 cm per season, while only 30 to 106 cm at Hastings. Monthly temperatures also average 1 or 2°C higher in the Santa Rosa area than at Hastings. Both areas receive some fog in the spring.

The main distribution of Engelmann oak phase is in north–central San Diego Co., but it is still a relatively minor vegetation type there. In the Ramona–Santa Ysabel region (Wieslander 1934c), only 16 out of 152 VTM plots contained *Q. engelmannii.* These plots averaged 147 trees per hectare with 63% *Q. engelmannii,* 25% *Q. kelloggii,* and 12% *Q. agrifolia.* The dbh distribution of *Q. engelmannii* was similar to that of *Q. douglasii* in the blue oak phase. The ground layer averaged 79% grass–forb cover.

Coast live oak phase. Very little descriptive study has been devoted to the extensive woodlands that fill the space between the blue oak phase and its small southern homologue, the Engelmann oak phase. *Quercus agrifolia* usually dominates these predominantly evergreen communities, and *Heteromeles arbutifolia* is a conspicuous subdominant (McBride 1971). The best-developed stands on north slopes with deep soils or on shady alluvial terraces are very similar to the depleted mixed hardwood forests of the Central Coast Ranges (Chapter 10). Some scrubby portions of the coast live oak phase can be viewed as a southern, more coastal version of the north slope phase of the foothill woodland.

Between Santa Barbara and Orange Cos., *Juglans californica* is locally dominant in the coast live oak phase, usually on north slopes (Swanson 1967). Jepson (1910) was impressed with the open groves of *J. californica* and *Q. agrifolia* in the Ojai Valley, Ventura Co. There used to be many walnut woodlands in the Puente Hills region of Los Angeles Co. (Wieslander 1934a,b), but these woodlands have been severely disturbed by recent urban growth. The best remaining stands of *J. californica* are in the San Jose Hills south and east of Covina. In some respects the deciduous *J. californica* seems to serve the same function in the coast live oak phase as the deciduous *Aesculus californica* does in the north slope phase of the foothill woodland.

Northern Oak Woodland

Foothill woodland transition. The northern unit is the least satisfactory regional subdivision of the oak woodland, and only the presence of *Quercus garryana* provides any unity to the northern oak woodland. In a few North Coast Range and

Klamath Mt. localities, it seems appropriate to consider northern oak woodland as a continuation of foothill woodland with merely a change in the dominant white oak. For example, in the Trinity River drainage near Lewiston, the *Q. garryana* savannas are very similar to the blue oak phase some 15 km to the east in the Sacramento Valley foothills. The grassy cover is similar; the *Arctostaphylos* and *Ceanothus* shrubs are the same; and *Pinus sabiniana* is common in both. However, in much of the Klamath region and North Coast Ranges, *Q. garryana* stands appear more like xeric segregates from Douglas fir–hardwood forests than like extensions of oak woodland.

An example of the ecological separation of *Q. douglasii* and *Q. garryana* occurs near Browns Creek, Trinity Co. Here the only *Q. douglasii* stands in the Trinity drainage grow on small disjunct patches of shale within a metamorphic rock region. Along the edge of the Sacramento Valley, this shale forms Lodo shaley clay loam, which supports blue oak phase (Mallory et al. 1966). The shale at Browns Creek forms a similar clay loam, which supports a similar blue oak phase. Surrounding soils on metamorphic parent materials support Douglas fir–hardwood forest with *Q. garryana* prominent in the more open rocky spots. There has been some hybridization between the two white oaks, but the *Q. douglasii* savanna is essentially isolated on a locally unique, unproductive soil within a forest region.

In Mendocino Co. there is a shift from blue oak phase to *Q. garryana* communities within the Laughlin soil series (Gardner et al. 1964). *Quercus garryana* occurs on the more northern or higher rainfall portions of the Laughlin soils. Gardner et al. suggest that the *Q. garryana* communities are ecologically separated from the blue oak phase.

Bald hills. Another northern oak woodland element involves *Quercus garryana* communities loosely referred to as "bald hills." (Clark 1937; Jepson 1910). The bald hills not only have *Q. garryana* instead of *Q. douglasii* but also have a different structure. Although there is a superficial resemblance, the savannas have a more balanced mixture of *Q. douglasii* trees and grass, whereas on the bald hills either grass or *Q. garryana* trees are dominant (Clark 1937).

In part, this grass–tree mosaic reflects a soil mosaic. In Mendocino Co., the grassy "balds" are often on Yorkville soils and the patches of oaks on Laughlin soils (Gardner et al. 1964). In Humboldt Co., Laughlin and Yorkville soils plus many related series typically support grass, while Tyson loam–st. clay loam often supports *Q. garryana* or mixed stands of *Q. garryana* and *Q. kelloggii* (De Lapp 1960). Shallower phases of Tyson series are the most common soil under oak "scrub." Cooper (1972) studied a bald hills area east of the redwood forest in Humboldt Co. and found oaks on Tyson soils, grassland on Yorkville soils, and adjacent forest on Hugo soils.

The best development of the bald hills type of woodland occurs on ridgetops in Mendocino and Humboldt Cos. up to the 1600 m elevation (Clark 1937; Jepson 1910). Clark described such stands as "open woodland of large trees widely spreading, and with no understory" *Quercus garryana* on these North Coast Range ridges seems ecologically similar to *Q. lobata* in montane savannas of the South Coast Ranges.

The bald hills structure does not apply to the eastern limit of *Q. garryana* woodland. At low elevations in the central Pit River and Klamath River systems in the

Cascades, *Q. garryana* associates with *Juniperus occidentalis* (Griffin 1967). The resulting savannas on very rocky volcanic soils are transitional with the northern juniper woodland (Chapter 23). *Ceanothus cuneatus,* which is so widespread in the foothill woodland, is common in these oak–juniper stands. *Juniperus occidentalis* has a disjunct population in the western Yolla Bolly Mts. of Trinity Co., and *Q. garryana* is prominent in this woodland (Keeler-Wolf and Keeler-Wolf 1973).

Quercus garryana stands in the Scott and Shasta valleys of Siskiyou Co. are very similar to the oak woodlands of the Rogue Valley of southern Oregon (Detling 1961), but it probably is not helpful to relate any form of the California northern oak woodland too closely to *Q. garryana* communities of the Willamette Valley (Franklin and Dyrness 1973). Succession in Willamette region savannas and woodlands moves quickly to *Q. garryana* forest and often to *Pseudotsuga* forest in the long term (Thilenius 1968). The northern oak woodland in California is more closely tied to stable grassland patches and rocky forest openings than are oak communities in Oregon; and when succession to forest is involved in the northern oak woodland, it is to a different type of *Pseudotsuga* forest from that in the Pacific Northwest (Chapter 10).

Riparian Forest

The Central Valley floodplain forests were largely destroyed by agricultural clearing and flood control activities before they could be described. The strips of forest still remaining along Coast Range valleys or foothill streams have great floristic variability and are subject to frequent natural and cultural disturbances.

Knapp (1965) referred to the valley riparian forests as "Californian sycamore bottomland woods" and to the higher-elevation stands as "Californian white alder bottomland woods." The valley stands were included as a sycamore–walnut–ash variant of the western United States "cottonwood–willow" forest type (Society of American Foresters 1954). Munz and Keck (1959) did not include streambank communities in their vegetation classification.

The most helpful riparian reference for the Sacramento Valley is an historical review by Thompson (1961), which gives an excellent summary of comments by early travelers and surveyors. Thompson's review includes not only the dense floodplain forests but also the adjacent valley oak phase of the foothill woodland. No similar historical compilation is available for the San Joaquin Valley.

Local floristic studies in the oak woodland always note many riparian species, but no community information is given. For example, Klyver's (1931) qualitative comments on a "willow–poplar association" and "sycamore consociation" within valley oak phase and a "willow–sycamore association" within blue oak phase give as much vegetation information as does any reference. No quantitative descriptions of oak woodland riparian communities seem to be available.

MOISTURE RELATIONS

The xerophytic appearance of lowland oak communities matches that of upland woodlands, and soil drought during the summer is severe in both situations for shallow-rooted herbs. But for deep-rooted trees, large differences in water availability can exist in different topographical positions, often within short distances.

A relationship between large *Quercus lobata* trees and bottomland soils with shallow water tables was assumed in the older literature. Jepson (1910) pictured the best *Q. lobata* habitat as a fertile loam with water at depths of 3–12 m. Cannon (1914) thought that this oak reached its best development with water tables 3–6 m deep. In the Palo Alto region, Cooper (1926) studied *Q. lobata* distribution and water table depths in wells. Oaks were absent on higher terraces with water tables 9–55 m deep, whereas lower terraces with water tables 0–35 m deep supported both *Q. lobata* and *Q. agrifolia.* He found no close correlation between water table depth and oak dominance, but on any given terrace "with increasing depth to water the oaks sooner or later give way to chaparral." Cooper thought that *Q. lobata* had the deeper root system of these two oaks.

Lewis and Burgy (1964) studied the rooting depths of several oaks in Placer Co. foothill woodland. On deeply fractured metamorphic rocks, they found *Q. douglasii* and *Q. wislizenii* utilizing water from depths of at least 21 m during the summer. They stressed the importance of water tables to all oaks in all California situations.

Water table or rooting information, however, is rarely available for woodlands. Estimating the available soil moisture, even down to 1 or 2 m, during the peak of the summer drought has been tried in only a few woodlands in California. Waring and Major (1964) found that the soils of northern oak woodland supplied much less water than the adjacent mixed evergreen forest and redwood forest soils in Humboldt Co. Cooper (1972) suggested that northern oak woodland plots on Tyson soil had more available moisture than grassland on Yorkville soil but less than forest on Hugo soil. Several foothill woodland plots in Shasta Co. had much less available moisture than adjacent mixed conifer forest (Griffin 1967).

Severe limitations on soil sampling throughout the rooting zone have made pressure chamber sampling of oak crowns far more helpful in estimating moisture stress than soil sampling under the trees (Griffin 1973; Snow 1972; Syvertsen 1974; Waring 1969). Pressure chamber studies in alluvial habitats at Hastings suggest that both *Q. lobata* and *Q. agrifolia* exploit the water table. Their predawn xylem sap tension only declines from −2 to −5 bar during the dry season, levels similar to those for riparian trees. In mid-August of a very dry season, alluvial *Q. lobata* trees had a predawn xylem sap tension of only −4 bar, whereas an adjacent upland stand of *Q. douglasii* trees had a tension of −27 bar (Griffin 1973). The blue oak phase has higher moisture stress during the summer than any other oak community at Hastings. In very dry years, *Q. douglasii* trees become drought deciduous in late August or September, dropping much of their foliage when the predawn xylem sap tension drops below −35 bar (Griffin 1973).

Stand density influences moisture stress on uplands where the total water supply is limited. At Hastings, dense *Q. douglasii* stands (Table 11-6) have greater moisture stress than stands with widely spaced trees. The drier the season, the greater is this density effect (Griffin 1973). Snow (1972) found a similar situation in the Engelmann oak phase. After a wet season, there was no difference between oaks in open and dense plots. After a dry season, however, the open stand had a predawn xylem sap tension of −9 bar, whereas the dense stand averaged −24 bar.

Syvertsen (1974) studied moisture stress (stem water potential) in coast live oak phase during a dry year and found it to vary with slope position. All species had lower stress at the bottom of the slope. *Juglans californica* and *Heteromeles arbutifolia* had lower stress in the open south-aspect plots than in the denser north slope stand.

SUCCESSION

Comments by the earliest observers suggest that much of the lower-elevation oak woodland, particularly the valley oak phase, had very little woody understory before settlement (Cooper 1926; Jepson 1910; Thompson 1961). Throughout the oak woodland, historical records indicate gradual increases in density of evergreen oaks and shrubs, although long-term permanent plots to quantify this trend are lacking.

Deciduous oaks appear to have reproduced poorly in the past 50 yr (Brooks 1971; Fieblekorn 1972; Griffin 1971a, 1976; Snow 1972; White 1966). Live oak seedlings may be more browse resistant than the deciduous oaks (Griffin 1971a), or pocket gophers may prefer the roots of deciduous oaks to those of live oaks. In any case, live oaks have been more successful in producing saplings in recent decades.

Heavy damage to acorns and seedlings by cattle, deer, rodents, and insects explains most of the failure of deciduous oak seedlings (Griffin 1971a, 1976; Snow 1972). It is easy to document the difficulties of getting deciduous oaks past the seedling stage now; what is needed is an explanation of how these oaks managed to form saplings in the past, for in several regions deciduous oak regeneration was locally abundant before 1900 (Bauer 1930; Brooks 1971; Griffin 1971a, 1976; Snow 1972; White 1966). Favorable combinations of acorn supply, spring precipitation, and low pressure from acorn and seedling predators must have been involved. White (1966) suggested that deer populations were low then, and disastrous droughts may have temporarily reduced cattle grazing. One very important seedling predator is the pocket gopher (Griffin 1971a, 1976), but we have no information about fluctuations of gopher populations in the past.

Cooper (1926) observed one oak grove near Menlo Park that had not been disturbed for "many years," and his comments illustrate the state of oak woodland succession literature. The grove had widely spaced mature oaks with *Quercus agrifolia* dominant. Young live oaks formed groups between the big trees, and there was a shrubby understory of *Heteromeles arbutifolia, Rhamnus californica, Rhus diversiloba,* etc. Cooper concluded that his Santa Clara Valley stand "demonstrates quite clearly that the openness of the forest, and especially the bareness of the ground, are not 'natural' characteristics, but due to disturbance of some sort." Scraps of relic savanna throughout the oak woodland support this generality that, in the absence of heavy grazing or frequent fires, live oaks and bird-dispersed shrubs like *Rhamnus* and *Rhus* will increase.

Undisturbed savannas, however, do not uniformly acquire an understory. At the San Joaquin Experimental Range, *Pinus sabiniana* trees and *Ceanothus cuneatus* shrubs increased in an area ungrazed since 1934 and unburned since 1929. An adjacent unburned area with light grazing, however, showed no change in woody plants (Woolfolk and Reppert 1963). At the Hastings Reservation, *Q. agrifolia* and evergreen shrubs have generally increased in the woodland since 1937, but some savannas ungrazed by livestock during the same period and unburned for well over 50 yr persist with no understory. Probably islands of grassland and many open savannas always exist on shallow soils of south aspect, which are unfavorable for woodland regeneration, with or without disturbance.

Indian burning is usually invoked as the "disturbance" maintaining the savanna nature of presettlement oak woodland. Jepson's (1910) comments that open stands were "undoubtedly" the result of repeated annual fires have been very influential. Cooper (1922) and many recent authors echo Jepson's views in support of widespread and frequent Indian burning in the oak woodland. Although Indians lived in much of the oak woodland for thousands of years and must have influenced its development, it is now impossible to determine the extent and frequency of Indian burning. The few first-hand accounts of Indians deliberately firing the savanna refer to coastal areas and parts of the Central Valley. In vast areas of oak woodland, reliable witnesses never recorded any details about Indian burning.

During the American settlement period, however, countless fires were set in the oak woodland. Early newspapers are filled with accounts of these fires, some of which burned for weeks or months. Part of the difficulty in evaluating the influence of aboriginal fires is that virtually the entire oak woodland is still recovering from the fires set by ranchers and miners before 1900.

Lightning fires are scattered in the oak woodland; they usually result from major storms coming northward from Mexico, not from local convectional disturbances. Decades may pass in any given savanna between storms that actually start fires. At the Hastings Reservation, lightning has started no fires since record-keeping began in 1937, and it is questionable whether lightning fires in adjacent regions would have burned into Hastings savannas even if they had been uncontrolled.

Oak woodland has been locally exterminated in urban areas, as the Engelmann oak phase in Los Angeles. Greater woodland destruction occurred in prime agricultural areas (e.g., in the valley oak phase in Tulare or Yolo Co.). In some lower-elevation woodlands, range improvement clearing has removed all evidence of the blue oak phase. For example, Hoover (1970) thought that some of the grassland in eastern San Luis Obispo Co. was formerly savanna. In general, however, the boundaries of oak woodland have remained relatively stable, although great changes in density have occurred.

One of the more questionable situations has been the upper woodland limit in the Sierra Nevada. The present boundary between foothill woodland and montane forest can be easily seen, but this boundary may have moved upward after destructive logging and severe burning of the lowest-elevation forests (Graves 1932; Klyver 1931; Watts 1959; Wieslander 1935). The VTM Survey often noted the presence of relic forest trees well within the present woodland zone, and the S–V Survey has found apparently eroded forest soils that are now supporting woodland.

SUMMARY AND AREAS FOR FUTURE RESEARCH

Oak woodland is an ill-defined zone of oak-dominated communities growing between open grassland and montane forest. The woodlands grow on a variety of noncalcic Brown soils under a wide range of precipitation conditions. Different oaks are involved regionally. The important deciduous species are *Quercus garryana* in the northern oak woodland, *Q. douglasii* and *Q. lobata* in the foothill or central woodland, and *Q. engelmannii* in the southern woodland. In the coastal foothill woodland and southern oak woodland, the live oak *Q. agrifolia* is significant, while *Q. wislizenii* is an important live oak in the rest of the woodland. Several pines may be scattered about, particularly *Pinus sabiniana* in the foothill woodland.

The oak woodland always has some grassland ground cover, now dominated by introduced annual grasses and forbs. The density of the woody cover varies greatly. Much of the woodland literature emphasizes the open, low-elevation form of the community. Stands with less than 50 trees per hectare meet the common definition of savannas, having less than 30% tree cover. But a greater proportion of the oak woodland has live oak and pines mixed into denser, truly woodland stands, with 150–300 trees per hectare. When the deciduous "indicator" oaks are absent, it is difficult to separate the mesic forms of woodland from the xeric forms of hardwood forest by either floristic or physiognomic characters.

Soil differences between open grassland and oak savanna may be minor, and grass–savanna boundaries may shift, depending on the type and degree of disturbance. In some areas, savannas are shrinking because of lack of oak regeneration or range improvement clearing. In areas with low fire frequency and minimal grazing, savannas are increasing in density. Live oaks and shrubs like *Rhamnus* spp. and *Rhus diversiloba* may invade the least disturbed savannas.

Throughout the oak woodland, chaparral forms a mosaic with grass–woodland communities. These chaparral patches are concentrated on less developed, shallower soils than is the woodland. There is probably a stronger edaphic basis for chaparral–woodland boundaries than Cooper (1922) assumed. He believed that without disturbance oak savannas would become chaparral. It is more likely that much of the woodland would increase in live oak and shrub density, and woodland on mesic slopes would approach mixed hardwood forest.

Indian burning has traditionally been considered an important factor in maintaining the openness of oak savannas; however, direct evidence regarding the frequency and extent of Indian burning in oak woodland is lacking. Mature oaks are well adapted to survive ground fires, and some degree of Indian burning and infrequent lightning fires could have been involved in the development of open woodlands. In the more montane savannas, lack of fires in recent decades has permitted a highly flammable understory of pines and live oaks to develop. In this situation, the mature deciduous oaks may not survive the next fire, but the live oaks will recover by sprouting.

Past studies on the oak woodland have been so restricted that topics for future research are almost unlimited. In terms of decreasing availability of study sites, high priority should be given to the valley oak phase, for intensive urban and agricultural development is truly endangering this type of woodland. Relatively undisturbed savannas of *Quercus lobata* have not been available on prime alluvial soils for over a century, and soon even disturbed stands on marginal valley lands will be rare. The Engelmann oak phase, which has always been geographically limited, should also receive attention while some recognizable examples still exist.

Throughout the oak woodland, details on floristic composition, stand structure, successional trends, and fire dynamics are sorely needed. Careful studies on age-class distributions would help to clarify the extent and time span involved in the last major wave of *Q. douglasii* regeneration in the foothill woodland. More details on water relations and photosynthetic efficiencies of evergreen versus deciduous oaks would aid in understanding site preferences and succussional problems. A wide range of animal interactions is poorly known. The physiological basis for the browse resistance of some oaks should be studied, and the specific role of small mammals in acorn and seedling predation needs much work. Pocket gophers influence soil formation, grass production, litter recycling, and oak regeneration; they should receive serious study.

LITERATURE CITED

Bauer, H. L. 1930. Vegetation of the Techachapi Mountains, California. Ecology 11:263-280.

Benseler, R. W. 1968. Studies in the reproductive biology of *Aesculus californica* (Spach) Nutt. Ph.D. dissertation, Univ. Calif., Berkeley. 155 p.

Benson, L. 1957. Plant classification. D. C. Heath, Boston. 688 p.

---. 1962. Plant taxonomy, methods and principles. Ronald Press, New York. 494 p.

Benson, L., E. A. Phillips, and P. A. Wilder. 1967. Evolutionary sorting of characters in a hybrid swarm. I. direction of slope. Amer. J. Bot. 54:1017-1026.

Bowerman, M. L. 1944. The flowering plants and ferns of Mount Diablo, California. Gillick Press, Berkeley. 290 p.

Brooks, W. H. 1971. A quantitative ecological study of the vegetation in selected stands of the grass-oak woodland in Sequoia National Park, California. Progr. rept. to Superintendent, Sequoia-Kings Canyon Natl. Park, Three Rivers, Calif. (plus misc. data summaries).

Buttery, R. F., and L. R. Green. 1958. A checklist of plants of the San Joaquin Experimental Range. USDA Forest Serv., Calif. Forest and Range Exp. Sta., Misc. Paper 23. 32 p.

Cannon, W. A. 1914. Tree distribution in central California. Popular Sci. Monthly 85:417-424.

Clark, H. W. 1937. Association types in the north Coast Ranges of California. Ecology 18:214-230.

Clements, F. E. 1920. Plant indicators: the relation of plant communities to process and practice. Carnegie Inst. Wash. Pub. 290. 388 p.

Cooper, P. V. 1972. Soil moisture relations of selected soils in the mixed evergreen type, northern California. M.A. thesis Humboldt State Univ., Arcata, Calif.

Cooper, W. S. 1922. The broad-sclerophyll vegetation of California. Carnegie Inst. Wash. Pub. 319. 124 p.

---. 1926. Vegetational development upon alluvial fans in the vicinity of Palo Alto, California. Ecology 7:1-30.

Dal Porto, N. J., and C. E. Carlson. 1965. Blue oak cordwood from foothill range land. Calif. Div. Forest., Range Improve. Stud. 15. 4 p.

De Lapp, J. 1960. A preliminary guide to the wildland soils of Humboldt County. Mimeo rept., School of Forestry, Univ. Calif. 10 p.

Detling, L. E. 1961. The chaparral formation of southwestern Oregon with considerations of its postglacial history. Ecology 42:348-357.

Dudley, W. R. 1901. Zonal distribution of trees and shrubs in the southern Sierra. Sierra Club Bull. 24:289-312.

Duncan, D. A., and S. E. Westfall. 1975. Publications from the San Joaquin Experimental Range, 1935-73. Unpub. rept. on file, Pac. SW Forest and Range Exp. Sta., Berkeley. 68 p.

Emrick, W. E., and O. A. Leonard. 1954. Delayed kill of interior live oak by fall treatment with 2,4-D and 2,4,5-T. J. Range Manage. 7:75-76.
Fieblekorn, C. 1972. Interim report of oak regeneration study. Unpub. rept. on file, Nat. Resources Conserv. Office, Hunter Liggett Military Reservation, Jolon. 23 p.
Franklin, J. F., and C. T. Dyrness. 1973. Natural vegetation of Oregon and Washington. USDA Forest Serv. Gen. Tech. Rept. PNW-8. 417 p.
Gardner, R. A. et al. 1964. Wildland soils and associated vegetation of Mendocino County, California. State Cooperative Soil-Vegetation Survey, Calif. Div. Forestry, Sacramento. 113 p.
Gowans, K. D. 1958. Soil survey of the Hopland Field Station. Univ. Calif. Agric. Exp. Sta. 34 p.
Graves, G. W. 1932. Ecological relationships of Pinus sabiniana. Bot. Gaz. 94:106-133.
Griffin, J. R. 1962. Intraspecific variation in Pinus sabiniana Dougl. Ph.D. dissertation, Univ. Calif., Berkeley. 274 p.
---. 1966. Notes on disjunct foothill species near Burney, California. Leaf. West. Bot. 10:296-298.
---. 1967. Soil moisture and vegetation patterns in northern California forests. USDA Forest Serv., Res. Paper PSW-46. 22 p.
---. 1971a. Oak regeneration in the upper Carmel Valley, California. Ecology 52:862-868.
---. 1971b. Variability of germination in Digger pine in California. USDA Forest Serv. Res. Note PSW-248. 5 p.
---. 1973. Xylem sap tension in three woodland oaks of central California. Ecology 54:152-159.
---. 1974. Notes on environment, vegetation, and flora, Hastings Natural History Reservation. Mimeo rept. on file, Hastings Reservation, Carmel Valley, Calif. 90 p.
---. 1976. Regeneration in Quercus lobata savannas, Santa Lucia Mountains, California. Amer. Midl. Natur. 95:422-435.
Griffin, J. R., and W. B. Critchfield. 1972. The distribution of forest trees in California. USDA Forest Serv. Res. Paper PSW-82. 114 p.
Grinnell, J., J. Dixon, and J. M. Linsdale. 1930. Vertebrate natural history of a section of northern California through the Lassen Peak region. Univ. Calif. Press, Berkeley. 594 p.
Holland, V. L. 1974. A study of soil and vegetation under Quercus douglasii H. & A. compared to open grassland. Ph.D. dissertation, Univ. Calif., Berkeley. 369 p.
Hoover, R. F. 1970. The vascular plants of San Luis Obispo County, California. Univ. Calif. Press, Berkeley. 350 p.
Howell, J. T. 1970. Marin flora. 2nd ed. with supplement. Univ. Calif. Press, Berkeley. 366 p.
Jensen, H. A. 1939. Vegetation types and forest conditions of the Santa Cruz Mountains unit of California. USDA Forest Serv., Calif. Forest and Range Exp. Sta., Forest Survey Release 1. 55 p.
Jepson, W. L. 1910. The silva of California. Univ. Calif. Mem., Vol. 2. 480 p.

Johnson, W. C., C. M. McKell, R. A. Evans, and L. J. Berry. 1959. Yield and quality of annual range forage following 2,4-D application on blue oak trees. J. Range Manage. 12:18-20.

Keeler-Wolf, T., and V. Keeler-Wolf. 1973. A contribution to the natural history of the Yolla Bolly Mountains of California. Senior thesis, Univ. Calif., Santa Cruz. 607 p.

Kylver, F. D. 1931. Major plant communities in a transect of the Sierra Nevada mountains of California. Ecology 12:1-17.

Knapp, R. 1965. Die Vegetation von Nord- und Mittelamerika und der Hawaii-Inseln. Gustav Fischer-Verlag, Stuttgart. 373 p.

Küchler, A. W. 1964. Manual to accompany the map of potential natural vegetation of the conterminous United States. Amer. Geogr. Soc. Spec. Pub. 36. 116 p.

Lathrop, E. W., and R. F. Thorne. 1968. Flora of the Santa Rosa plateau of the Santa Ana Mountains, California. Aliso 6(4):17-40.

Lawrence, G. E. 1966. Ecology of vertebrate animals in relation to chaparral fire in the Sierra Nevada foothills. Ecology 47:278-291.

Lewis, D. C., and R. H. Burgy. 1964. The relationship between oak tree roots and ground water in fractured rock as determined by tritium tracing. J. Geophys. Res. 69:2579-2588.

---. 1967. Water use by native vegetation and hydrologic studies. Univ. Calif., Davis, Dept. Water Sci. and Eng., Ann. Rept. 7. 54 p.

Mallory, J. I., E. B. Alexander, and E. N. Gladish. 1966. Soils and vegetation of the Ono quadrangle. State Cooperative Soil-Vegetation Survey. U.S. Forest Serv., San Francisco, Calif. 50 p.

Mallory, J. I., E. B. Alexander, B. F. Smith, and E. N. Gladish. 1965a. Soils and vegetation of the Whitmore quadrangle. State Cooperative Soil-Vegetation Survey. U.S. Forest Serv., San Francisco, Calif. 55 p.

Mallory, J. I., B. F. Smith, E. B. Alexander, and E. N. Gladish. 1965b. Soils and vegetation of the Manton quadrangle. State Cooperative Soil-Vegetation Survey. U.S. Forest Serv., San Francisco, Calif. 36 p.

Mallory, J. I., C. O. Stone, E. B. Alexander, and J. M. Crawford. 1965c. Soils and vegetation of the Montgomery Creek quadrangle. State Cooperative Soil-Vegetation Survey. U.S. Forest Serv., San Francisco, Calif. 47 p.

Mallory, J. I., E. B. Alexander, W. L. Colwell Jr., and W. R. Powell. 1968. Soils and vegetation of the Chanchelulla Peak quadrangle. State Cooperative Soil-Vegetation Survey. U.S. Forest Serv., San Francisco, Calif. 48 p.

Mallory, J. I., W. L. Cowell, Jr., and W. R. Powell. 1973. Soils and vegetation of the French Gulch quadrangle (24D-1,2,3,4), Trinity and Shasta Counties, California. USDA Forest Serv. Resources Bull PSW-12. 42 p.

McBride, J. 1971. Vegetation on the Jalama Ranch. Unpub. consultants' rept. on file in Forestry Library, Univ. Calif., Berkeley. 21 p.

Mueller-Dombois, D., and H. Ellenberg. 1974. Aims and methods of vegetation ecology. John Wiley, New York. 547 p.
Munz, P. A. 1974. A flora of southern California. Univ. Calif. Press, Berkeley. 1086 p.
Munz, P. A., and D. D. Keck. 1959. A California flora. Univ. Calif. Press, Berkeley. 1681 p.
Murphy, A. H., and B. Crampton. 1964. Quality and yield of forage as affected by chemical removal of blue oak (*Quercus douglasii*). J. Range Manage. 17:142-144.
Myatt, R. G. 1975. Geographical and ecological variation in *Quercus chrysolepis* Liebm. Ph.D. dissertation, Univ. Calif., Davis. 220 p.
Newman, T. F., and D. A. Duncan. 1973. Vertebrate fauna of the San Joaquin Experimental Range, California. USDA Forest Serv. Gen. Tech. Rept. PSW-6. 17 p.
Porter, D. M. 1962. The vascular plants of the Jasper Ridge Biological Experimental Area of Stanford University. Stanford Univ., Dept. Biol. Sci., Res. Rept. 2. 28 p.
Shinn, C. 1911. Economic possibilities of *Pinus sabiniana*. Proc. Soc. Amer. Forest. 6:68-78.
Snow, G. E. 1972. Some factors controlling the establishment and distribution of *Quercus agrifolia* Nee and *Q. engelmannii* Grenne in certain southern California oak woodlands. Ph.D. dissertation, Oregon State Univ., Corvallis. 105 p.
Society of American Foresters. 1954. Forest cover types of North America. Washington, D.C. 67 p.
State Cooperative Soil-Vegetation Survey. 1969. Soil-vegetation surveys in California. Calif. Div. Forestry, Sacramento. 31 p.
Swanson, C. J. 1967. The ecology and distribution of *Juglans californica* in southern California. M.A. thesis, Calif. State Univ., Los Anageles. 115 p.
Syvertsen, J. P. 1974. Relative stem water potentials of three woody perennials in southern oak woodland community. Bull. South. Calif. Acad. Sci. 73:108-113.
Thilenius, J. F. 1968. The *Quercus garryana* forests of the Willamette Valley, Oregon. Ecology 49:1124-1133.
Thomas, J. H. 1961. Flora of the Santa Cruz Mountains of California. Stanford Univ. Press, Stanford, Calif. 434 p.
Thompson, K. 1961. Riparian forests of the Sacramento Valley, California. Ann. Assoc. Amer. Geogr. 51:294-315.
Tucker, J. M. 1952. Evolution of the Californian oak *Quercus alvordiana*. Evolution 6:162-180.
Twisselmann, E. C. 1956. A flora of the Temblor Range and neighboring parts of the San Joaquin Valley. Wasmann J. Biol. 14:161-300.
---. 1967. A flora of Kern County, California. Wasmann J. Biol. 25:1-395.
Vancouver, G. 1798. A voyage of discovery to the north Pacific Ocean and round the world. 3 vols. Robinson and Edwards, London.
Vankat, J. L. 1970. Vegetation change in Sequoia National Park, California. Ph.D. dissertation, Univ. Calif., Davis. 197 p.
Vasek, F. C. 1966. The distribution and taxonomy of three western junipers. Brittonia 18:350-372.

Waring, R. H. 1969. Forest plants of the eastern Siskiyous: their environmental and vegetational distribution. Northwest Sci. 43:1-17.

Waring, R. H., and J. Major. 1964. Some vegetation of the California coastal redwood region in relation to gradients of moisture, nutrients, light, and temperature. Ecol. Monogr. 34:167-215.

Watts, D. 1959. Human occupance as a factor in the distribution of the California Digger pine. M.A. thesis, Univ. Calif., Berkeley. 216 p.

White, K. L. 1966. Structure and composition of foothill woodland in central coastal California. Ecology 47:229-237.

Whitney, J. D. 1865. Geological survey of California. Vol. 1. Geology, report of progress and synopsis of the field-work from 1860-1864. Claxton Press, Philadelphia. 484 p.

Wieslander, A. E. 1934a. Vegetation types of California, Pasadena quadrangle (162D). U.S. Forest Serv. (map).

---. 1934b. Vegetation types of California, Pomona quadrangle (163C). U.S. Forest Serv. (map).

---. 1934c. Vegetation types of California, Romona quadrangle (181). U.S. Forest Serv. (map).

---. 1935. A vegetation type map of California. Madroño 3:140-144.

---. 1938. Vegetation types of California, Elsinore quadrangle (176). U.S. Forest Serv. (map).

---. 1939a. Vegetation types of California, Redding quadrangle (23). U.S. Forest Serv. (map).

---. 1939b. Vegetation types of California, Triunfo Pass quadrangle (161C). U.S. Forest Serv. (map).

Wieslander, A. E., and C. H. Gleason. 1954. Major brushland areas of the Coast Ranges and Sierra-Cascade foothills in California. USDA Forest Serv., Calif. Forest and Range Exp. Sta., Misc. Paper 15. 9 p.

Woolfolk, E. J., and J. N. Reppert. 1963. Then and now: changes in California annual-type range vegetation. USDA Forest Serv., Res. Note PSW-24. 9 p.

Zuill, H. 1967. Structure of two cover types of southern oak woodland in California. M.S. thesis, Loma Linda Univ., Loma Linda, Calif. 43 p.

CHAPTER 12

CALIFORNIA CHAPARRAL

TED L. HANES
Department of Biological Science,
California State University, Fullerton

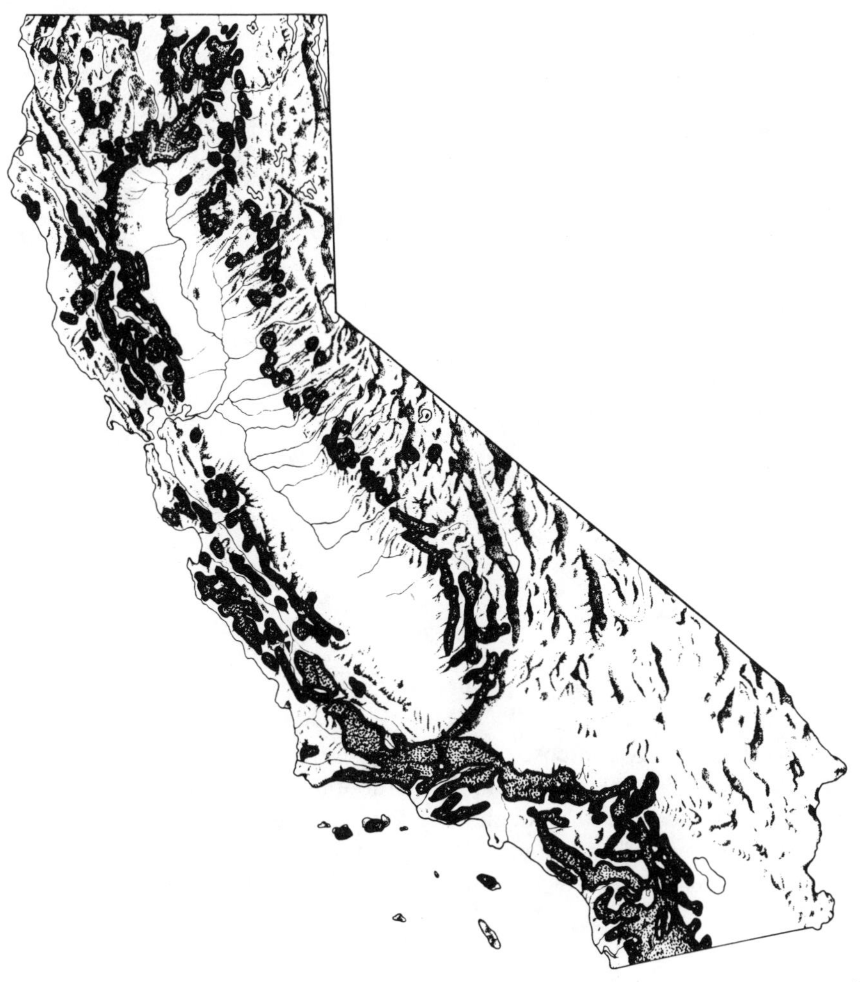

Introduction 418
 Defining characters 419
 Distribution 421
Chaparral types 422
 Chamise chaparral 422
 Ceanothus chaparral 425
 Scrub oak chaparral 426
 Manzanita chaparral 426
 Montane chaparral 427
 Red shanks chaparral 429
 Serpentine chaparral 429
 Desert chaparral 430
 Woodland chaparral 431
Chaparral succession 431
 Primary succession 431
 Secondary succession 432
  Role of fire 432
  Chaparral animals 433
  Herbland phase 433
  Shrub phase 434
  Montane chaparral succession 435
  Germination requirements 435
  Sprouting 446
  Chaparral climax 447
Autecology of chaparral 449
 Growth form 449
  Roots 450
  Leaves 450
 Water relations 451
 Photosynthesis 452
 Phenology 452
 Carbon balance 453
 Nutrition 454
Chaparral management 455
 Chaparral hydrology 455
 Chaparral fires 456
 Managing chaparral brushfields 456
 Replacing chaparral 456
Areas for future research 457
Literature cited 459

INTRODUCTION

California chaparral is composed mainly of evergreen woody shrubs, and it forms extensive shrublands that occupy most of the hills and lower mountain slopes of California. It is adapted to drought and fire, passing endlessly through cycles of burning and regrowth. Even though chaparral has no commercial value, it forms the most highly valued watershed cover of any vegetation in the state.

Defining Characters

"Chaparral" is a word of Spanish origin (*chaparro*) that originally denoted a thicket of shrubby evergreen oaks (Cronemiller 1942). In time the term was applied to dense brushlands in general, with descriptive names used to designate various chaparral types. Soft, hard, and montane chaparral have been described. For a discussion of soft chaparral, see Chapter 13. Island Chaparral is discussed in Chapter 26.

The term "chaparral" is also applied to Rocky Mt. foothill vegetation, but this is quite another type and should not be confused with California chaparral. Clements (1916) understood the difference, based on leaf life and seasonal activity patterns. He called the Rocky Mt. chaparral "Petran chaparral." It is summer active and winter deciduous, whereas the California chaparral is evergreen, winter active, and summer dormant. Cooper (1922) referred to the Petran chaparral as the "deciduous thicket scrub."

The physiognomy of chaparral is much the same throughout its range, yet there is a great amount of floristic diversity. Considerable floristic disparity is evident between northern and southern chaparral.

The leaf is the most distinctive diagnostic feature of chaparral shrubs, being small, thick, stiff, and evergreen. Schimper (1903) originated the term "sclerophyll" to apply to "hard-leaved," evergreen shrub communities adapted to Mediterranean type climates of the world. He was the first botanist to recognize the similarity of California chaparral to shrub formations in the Mediterranean region, the African Cape, central Chile, and southwestern Australia. Sclerophyll type vegetation is dense and nearly impenetrable, and occurs with or without an understory of shrub seedlings, annual herbs, or grasses (Naveh 1967; Specht 1969a; Mooney and Parsons 1973); it is adapted to periodic fire. The effects of fire are now being recognized as important in the evolution of sclerophyll shrub vegetation in Mediterranean-type climatic regions (Shantz 1947; Naveh 1967, 1975).

The climatic factors operative in chaparral lands have been abundantly described (Plummer 1911; Clements 1916; Cooper 1922; Shreve 1927b; Rowe and Colman 1951; Reimann 1959a,b; Reimann and Hamilton 1959; Hanes 1965). Generally these are mild temperature regions with limited winter rainfall (300–600 mm) occurring in a few intense storms, followed by hot summers accompanied by prolonged drought. Total rainfall and predictability of its arrival increase, and temperature and length of summer drought decrease, from lower to higher latitudes.

The overriding geographical factors that influence chaparral development are slope aspect, coastal–desert exposure, elevation, substrate, and fire (Hanes 1971; Vogl 1973). The characteristics of rainfall interception by chaparral vegetation have been studied (Hamilton and Rowe 1949), as has the influence of a chaparral canopy on soil thermal regime (Qashu and Zinke 1964). Kittredge (1939a,b) described the forest floor, annual accumulation and creep of litter, and other surface materials in the chaparral of the San Gabriel Mts. In comparison to agricultural soils, brushland soils are highly porous and rocky in texture and are notoriously low in essential plant nutrients (Vlamis et al. 1954; Crawford 1962). Some chaparral grows on serpentine soils, particularly in northern California (Whittaker 1954a; Walker 1954; Wells 1962). In mountainous terrain, chaparral stabilizes rocky slopes. After fire, large amounts of "dry creep" erode from denuded slopes (Anderson et al. 1959;

Krammes 1960), as well as surface erosion due to winter rains (Sinclair 1953). The presence of organic substances of plant origin causes coarse-textured chaparral soils to be nonwettable and is of interest to soil scientists, land managers, and plant ecologists (Krammes and DeBano 1965; Krammes and Osborn 1969). A large part of our knowledge of the physical milieu of chaparral has been gathered at the San Dimas Experimental Forest in the San Gabriel Mts., Los Angeles Co.

Axelrod (1958 and Chapter 5) has amassed a body of fossil evidence to support his claim that fire and summer drought were the major environmental agents that shaped California chaparral over the past 2+ million yr. Jepson (1930) suggested that fire played a leading role in chaparral speciation. The positive influence of fire in the evolution of chaparral has also been posed by Sweeney (1956). Vogl (1967) postulated that fire helped shape three-fourths of California's vegetation, including chaparral. Mooney and Dunn (1970) outlined the convergent evolutionary strategies of Mediterranean climate type sclerophyll shrubs and included fire as one of the major environmental stress factors. Fire caused by man has played a profound role in shaping our current chaparral landscape (Zivnuska 1968). The early role resulted from aboriginal burning. This was intensified by the practices of the Mexican rancheros, who followed the Spanish tradition of burning off the hills. The gold rush of 1849 opened up even more extensive burning and was followed by a whole wave of complex human impacts on the land, including deliberate burning.

The dominant woody genera of the California chaparral, such as *Adenostoma, Arctostophylos, Ceanothus, Heteromeles,* and *Rhus,* are absent from other regions having a Mediterranean type climate. Indeed, botanists have concluded that the entire California chaparral evolved in place (Axelrod 1958; Stebbins and Major 1965).

The Mediterranean type climate is the overriding environmental factor in the ecology of California chaparral. The winters are moderate and moist; the summers, hot and dry. From May to November, there is usually no rain other than a few local thundershowers in the mountains. During this long dry period, maximum temperatures may exceed 40°C and relative humidity may drop below 5% (Biswell 1963). The vegetation becomes extremely dry, yet remains alive, even the leaves. In late summer and early autumn hot, dry winds add to the severity of the drought stress to which the chaparral shrubs are exposed. In this desiccated state, conditions for fire in the chaparral are extraordinarily great.

California chaparral is broadly described in the literature. Major floras include those of Jepson (1925), Abrams (1910, 1951) Munz and Keck (1959), and McMinn (1964). Two small books have been written on California chaparral for the lay public (Fultz 1923; Head 1972). Ornduff (1974) included chaparral in his helpful guide to California vegetation. Southern California chaparral is described in two major floras (Munz 1935, 1974) and in a small book devoted to shrubs (Raven 1966). Collins (1972) presented an illustrated key to coastal and chaparral flowering plants of southern California. Several floristic and vegetation studies have dealt with chaparral in local sites, such as canyons (Howell 1929; Robinson 1953), plateaus (Lathrop and Thorne 1968), mountains (Parish 1917; Shreve 1927a, 1936; Bauer 1930; Klyver 1931; Bowerman 1944; Sharsmith 1945; Pequegnat 1951; Twisselmann 1956, 1967; Horton 1960; Whittaker 1960; Thomas 1961; Gillett et al. 1961; Cooke 1962; Raven and Thompson 1966; Vogl and Miller 1968; Griffin 1975), channel islands (Thorne 1969), counties (Howitt and Howell 1964; Hoover 1970), and Baja California (Brandegee 1889).

A few chaparral species have been singled out for systematic study: *Arctostaphylos* spp. (Wieslander and Schreiber 1939; Abrams 1940), *Arctostaphylos nissenana* and *A. viscida* (Schmid et al. 1968), *Cercocarpus* (Brayton and Mooney 1966; Searcy 1969), *Ceanothus* (Brandegee 1894; Van Renssalaer and McMinnn 1942; Nobs 1963), and *Rhus* (Young 1972, 1974).

Distribution

Chaparral is the most extensive vegetative type in California, covering about 3.5 million ha, or one-twentieth of the state (Weislander and Gleason 1954). It ranges from 250 km below the Baja California border of Mexico northward into southern Oregon, from coastal bluffs to high montane elevations, and down the interior flanks of mountains to desert margins, and is a dominant vegetation type on the coastal islands of southern and Baja California.

The extensive geographical range, varied habitats, semiarid climate, and recurrent fire have provided conditions conducive for the development of many highly adapted species. Munz and Keck (1959) listed several hundred species from diverse plant families which comprise California chaparral. Extensive variation and hybridization are rampant among certain genera of California chaparral (Cooper 1922). There is also a high degree of California endemic species. Stebbins and Major (1965) listed more than two dozen species endemic to California chaparral; some of these are considered rare or endangered by the California Native Plant Society (Powell 1974). Finally, there are several recognized community types that comprise California chaparral throughout its range.

Cooper (1922) considered southern California the center of distribution of chaparral since it reaches its maximum development there, where it ranges in elevation from about 300 to 3000 m. Epling and Lewis (1942) evaluated 27 chaparral species and subspecies and placed their centers of distribution in the San Diego–Baja California region. California chaparral is part of the flora of the Sierra Juarez and Sierra San Pedro Mártir in Baja California. These ranges and the Cuyamaca, Santa Rosa, San Jacinto, and Santa Ana Mts. constitute the Peninsular Ranges, in which chaparral is extensive in total area and diverse in species composition. Chaparral is well developed on the flanks of the San Bernardino, San Gabriel, Santa Monica, Santa Ynez, Sierra Madre, and La Panza Mts. in southern California. Many hilly coastal areas and lesser mountain ranges are vegetated with chaparral.

In northern California, chaparral is more widely scattered, circumscribing the Great Central Valley, and generally restricted to the drier slopes of the inner ranges. The Central Coast Ranges that include chaparral are the Santa Lucia, Cholame Hills, Gabilan, and Santa Cruz Mts., and the Diablo Range southeast of San Francisco Bay. The North Coast Ranges that have chaparral are the mountain ranges north of San Francisco Bay and west of the Great Central Valley. The fullest development of chaparral there occurs in the mountains and foothills bordering the western edge of the Sacramento Valley. Chaparral is found in large but scattered stands in the Klamath Range and Siskiyou Mts. of California and Rogue Valley in the Cascade Mts. of southwestern Oregon (Detling 1961). Chaparral forms a broken ribbon along the eastern side of the Great Central Valley in the foothills and higher regions of the Sierra Nevada. The Sierran chaparral merges with the chaparral of southern California at the southern extreme of the San Joaquin Valley through the Tehachapi Mts. The broad-scherophyll forests of the north are replaced by chaparral as rainfall decreases and length of drought increases (Morrow and Mooney 1974). Various types of woodlands bisect chaparral brushlands on clay foothills and along canyon bottoms. These present a striking physiognomical contrast to the widespread shrublands in the more xeric southland.

The fire–flood sequence of chaparral led to the creation of forest reserves (later national forests) in the late 1890s and to a fertile period of inquiry into the dynamics of California chaparral (Brandegee 1891; Leiberg 1899, 1900; Parish 1903; Sterling 1904; Miller 1906; Cannon 1914; Munns 1920; Jepson 1921; Calkins 1925).

CHAPARRAL TYPES

Table 12-1 summarizes the primary site factors influencing California chaparral development and maintenance. The overriding factors that affect different chaparral types are coastal and desert exposures, slope aspect (direction), and elevation. The type of chaparral that occurs on these various site conditions is also a function of the age of the stand. Table 12-2 presents ten most commonly encountered chaparral types in decreasing order of abundance and the dominant or indicator species of each.

Chamise Chaparral

Chamise chaparral is the dominant type of chaparral throughout California. It, in turn, in dominated by chamise (*Adenostoma fasciculatum*). Hanes (1971) found that this species occurred in over 70% of all chaparral stands sampled in southern California. The term "chamisal" is applied to pure stands of *A. fasciculatum* (80% or more of the stand is covered by the species). Chamise chaparral is associated with hot, xeric sites (south- and west-facing slopes and ridges), forming extensive stands. It occurs in the North Coast Ranges from Trinity, Shasta, and Mendocino Cos. south to San Francisco Bay; Central Coast Ranges from San Mateo, Alameda, and Contra Costa Cos. south to Kern, San Luis Obispo, and Santa Barbara Cos.; and Sierra Nevada foothills mainly in El Dorado, Sacramento, Tuolumne, Mariposa, and Tulare Cos.; it is the predominant chaparral type throughout the mountains of Ventura, Los Angeles, Orange, San Bernardino, Riverside, and San Diego Cos., and northern Baja California.

Chamise chaparral is a dense, interwoven vegetation 1–2(3)m high at maturity, without understory (Horton 1960; Hanes 1971) and with scanty litter (Fig. 12-1). Associated species of low frequency are *Arctostaphylos* spp., *Ceanothus* spp., *Elymus condensatus, Erigonum fasciculatum, Quercus dumosa, Rhus ovata, R.* (= *Malosma*) *laurina, Salvia apiana, S. mellifera,* and *Yucca whipplei*. None of these represents more than 10% by cover of a stand of chamise chaparral. The overall appearance of chamise chaparral, therefore, is a uniform sea of chamise with an occasional individual of an associated species.

Regrowth after fire is slow compared to other chaparral types because of poor site

TABLE 12-1. Primary site factors influencing California chaparral development and maintenance. Disturbance includes unstable slopes, trails and road construction, pipeline and powerline rights-of-way, logging, off-road vehicles, and bulldozing

Exposure	Slope Aspect	Elevation	Vegetation Type	Response After Fire or Other Disturbance
Coastal	North-facing	Above snow	Manzanita chaparral	Ceanothus chaparral
			Scrub oak chaparral	Ceanothus chaparral
			Montane chaparral	Ceanothus chaparral
			Yellow pine forest	Mixed chaparral
		Below snow	Scrub oak chaparral	Ceanothus chaparral
	South-facing	Above snow	Mixed chaparral	Ceanothus chaparral
		Below snow	Chamise chaparral	Coastal sage scrub (southern California)
			Serpentine chaparral	Grassland (northern California)
Desert	North-facing	Above snow	Yellow pine forest	Montane chaparral
		Below snow	Piñon-juniper woodland	Desert chaparral
			Desert chaparral	Grassland
	South-facing	Above snow	Mixed chaparral	Ceanothus chaparral
		Below snow	Chamise chaparral	Coastal sage scrub (southern California)
				Grassland (northern California)

TABLE 12-2. Ten California chaparral types, based on dominant or indicator species and geographical or topographical factors. Types are listed in decreasing order of abundance. Vegetation and composition cover values were arbitrarily derived

Chaparral Type	Dominant or Indicator Species	Percent of Ground Covered by Type	Percent of total Cover by Dominant Spp.	Understory Present
Chamise	*Adenostoma fasciculatum*	50–90	50–100	
Ceanothus	*Ceanothus* spp.	80–100	50–100	
Scrub oak	*Quercus dumosa*	90–100	20–50	X
Manzanita	*Arctostaphylos* spp.	80–100	50–100	
Montane	*Ceanothus* spp.	10–100	50–100	
	Arctostaphylos spp.			
	Castanopsis sempervirens			
Red shanks	*Adenostoma sparsifolium*	80–100	20–50	
Serpentine	Serpentine endemics	30–60	20–50	
Desert	*Adenostoma fasciculatum*	30–60	50–100	
	Ceanothus spp.			
	Quercus turbinella			
Woodland	*Quercus* spp.	20–100	20–100	X
Island	Island endemics	80–100	20–50	

Figure 12-1. Chamise chaparral, dominated by *Adenostoma fasciculatum*, San Bernardino Co.

conditions (Horton 1960). During the regrowth phase the stand is more open and lower in stature than mature stands. Ephermeral annuals and short-lived perennials occupy intershrub spaces during the first wet season after a fire, forming a vigorous herbaceaous carpet that persists for a few years (Horton and Kraebel 1955). Shrub cover in chamise chaparral may reach 50% in 10 yr old stands, 90% in 25 yr old stands, and 80% in 50 yr old stands.

Ceanothus Chaparral

Ceanothus chaparral is a successional form of chaparral in southern California and a climax form in northern California and southwestern Oregon, where it replaces chamise chaparral. There are several species of ceanothus, such as *Ceanothus crassifolius*, *C. oliganthus*, and *C. thysiflorus*, that may dominate as seedling shrubs in the early stages of chaparral development after fire. Such stands are commonly pure stands of a single species of *Ceanothus*.

Horton (1960) found that ceanothus chaparral is rare above 1200 m and that it occurs on more mesic sites than chamise chaparral. In northern California and southwestern Oregon, *Ceanothus cuneatus* is the dominant and is associated with *Adenostoma fasciculatum*. Cooper (1922) called this "climax chaparral." *Ceanothus* may also be codominant with *Arctostaphylos* spp., and it frequently occurs as the understory where deciduous oak woodland or ponderosa pine forest is clearly climax (Detling 1961).

Ceanothus chaparral tends to develop a more complete crown cover than chamise chaparral but is not as dense, since the branches do not intertwine. Ceanothus chaparral reaches 1–3 m at maturity without understory and with little litter. It is uniform in appearance for several decades. Some species are short lived and tend to die out as the stand reaches maturity in the second and third decades after establishment. Such stands develop a spotty appearance with erect, dead shrubs or gaps in the canopy where dead shrubs have fallen over.

Associated species in ceanothus chaparral are usually sprouters, such as *Adenostoma fasciculatum*, *Quercus dumosa*, *Heteromeles arbutifolia*, and *Rhus ovata*. In areas not occupied by *Ceanothus*, seedlings of coastal sage scrub elements may occur, such as *Eriogonum fasciculatum*, *Artemisia californica*, *Eriodictyon* spp., and a variety of forbs. Being on more mesic sites than chamise chaparral, ceanothus chaparral develops a crown cover of about 50% in 5 yr and 80–100% in 25 yr, but this may regress to 25% in 50 yr old stands.

Scrub Oak Chaparral

Scrub oak chaparral is a mesic type of chaparral composed of a great many large shrub species and woody vines. In southern California, it occupies north-facing slopes below 900 m and all slope aspects above 900 m (Hanes 1971). In northern California, it occurs above chamise chaparral on all slope aspects and ridges where soil is sufficiently deep, but is replaced by mixed evergreen forest on most shady slopes. In southern California, the dominant species is scrub oak (*Quercus dumosa*) but *Q. wislizenii* var. *frutescens* is also abundant in northern California.

Scrub oak chaparral is 2–4 m tall. The dense canopy nearly reaches the ground, which is covered by leaf litter but usually without an understory. If the canopy is open or broken, a scanty understory of forbs and ferns may develop. Leaf litter may be 20 cm deep, woody vines and ferns are apparent, and one can pass through what appears to be a low woodland of squat trees.

Generally, scrub oak chaparral shows the greatest richness of species. Some of these scrub oak associates are *Ceanothus leucodermis, Cercocarpus betuloides, Franxinus dipetela, Garrya* spp., *Heteromeles arbutifolia, Lonicera* spp., *Prunus ilicifolia, Rhamnus ilicifolia, R. californica, Rhus ovata, R.* (= *Toxicondendron*) *diversilobum, Ribes* spp., and *Sambucus* spp. Scrub oak chaparral has been described by Horton (1960), Patric and Hanes (1964), Hanes and Jones (1967), and Hanes (1971).

Scrub oak chaparral development is rapid because of the more mesic conditions where it occurs and the sprouting habit of most of its members. In southern California, there is no coastal sage scrub subclimax that occupies recently burned scrub oak chaparral lands. Seedlings of herbs and shrubs are common on burned scrub oak sites, but it is the vigorous sprouting shrubs that characterize the vegetation within a few years after fire. Scrub oak chaparral may have 50% shrub cover in 10 yr old stands, 80% cover in 25 yr old stands, and 100% cover in 50 yr old stands.

Manzanita Chaparral

Manzanita chaparral is not as extensive as chamise or ceanothus chaparral (Horton 1960) and generally occurs on deeper soils and at higher elevation (Fig. 12-2). Cooper (1922) compared slope and exposure differences, behavioral characteristics, humus accumulation, and evaporation rates of *Arctostaphylos* associations and other chaparral types. He cited 19 *Arctostaphylos* species as components of chaparral and classified manzanita site requirements as intermediate between those for chamise chaparral and those for coniferous forest. Wilson and Vogl (1965) considered manzanita chaparral "cold chaparral" since the major forms of precipitation it receives are fog drip, freezing moisture, and snow.

Several forms of manzanita chaparral are recognized, including manzanita–chamise, oak–manzanita, pine–manzanita, and pure manzanita (Horton 1960; Wilson and Vogl 1965; Gankin and Major 1964; Detling 1961).

Manzanita chaparral often forms stiff, almost impenetrable stands over uniform slopes. The appearance from a distance is that of a light green, velvety mantle. The

Figure 12-2. Manzanita chaparral, dominated by *Arctostaphylos glauca,* Los Angeles Co.

stature ranges from 1 to 2(4) m without understory but with a moderately well-developed litter layer. Manzanita chaparral develops from either abundant seedlings or resprouting species. About half of the chaparral manzanitas cannot resprout after fire and are therefore obligate seeders. Associated species are those listed above plus a variety of ephemerals and short-lived perennials that occur for a few years after fire. Manzanita chaparral develops at a moderate rate. Cover may be 50% in 10 yr old stands, 80% in 25 yr stands, and 100% in 50 yr stands. Whether seeders or sprouters, manzanitas are long lived. In old stands, may species of *Arctostaphylos* attain tree size.

Montane Chaparral

Montane chaparral (Fig. 12-3) occurs in higher altitudes under the same general climatic and soil conditions as the coniferous forest zone: cool to cold temperatures and plentiful precipitation, mainly as snow. Montane chaparral occurs in scattered to extensive thickets as dense, low to medium-high understory and interstory to the coniferous elements, or as a seral stage to the original coniferous forest after fire or logging (Show and Kotok 1924; Biswell 1952; Wilken 1967; Bock and Bock 1969). Montane chaparral occurs in the Cascade and Sierra Nevada Mts. from Siskiyou to

Figure 12-3. Montane chaparral, dominated by *Arctostaphylos parryana, Castanopsis* (= *Chrysolepis*) *sempervirens*, and *Ceanothus cordulatus.*

Kern Co., and in the North Coast Ranges and Klamath Mts. from Del Norte to Siskiyou, Trinity, and Shasta Cos.; it is found as a minor type from Tehama Co. south to Lake Co. Montane chaparral grows in southern California above 1500 m in the Transverse and Peninsular ranges of Los Angeles, San Bernardino, Riverside, and San Diego Cos. and in the high reaches of the Sierra San Pedro Mártir, Baja California.

Montane chaparral is a ground-hugging, compact form of chaparral, seldom exceeding 2 m in height and usually less than 1 m. Dominant shrub species are *Ceanothus* spp., *Arctostaphylos* spp., and *Castanopsis* (= *Chrysolepis*) *sempervirens*. Thickets of montane chaparral consisting of spinose *Ceanothus* species present formidable barriers to cross-country travel. The most commonly encountered species in montane chaparral are *Arctostaphylos nevadensis, A. parryana, A. patula, Castanopsis* (= *Chrysolepis*) *sempervirens, Ceanothus cordulatus, C. diversifolius, C. integerrimus, C. pinetorum, C. sanguineus, Cercocarpus ledifolius, Chrysothamnus nauseosus, Prunus emarginata, P. virginiana* var. *demissa,* and *Symphoricarpos parishii.* Montane chaparral has been called "yellow pine chaparral" (Clements 1920) and "timberland chaparral" (Horton 1960; Critchfield 1971).

A mixture of chaparral elements may invade the lower parts of ponderosa pine forest on exposed sites and become a form of montane chaparral. In northern California mountains, the most common species are *Arctostaphylos patula, A. acutifolia, A. mariposa, A. mewukka, A. viscida, A. canescens, Ceanothus cordulatus, C. fresnensis, C. parvifolia, C. pinetorum, C. prostratus, C. pumilus, C. tomentosus, C. velutina, Cercocarpus betuloides, Eriodictyon californicum, Garrya fremontii, G. buxifolia, G. flavescens* var. *pallida, Heteromeles arbutifolia, Rhamnus californica* vars. *crassifolia* and *cuspidata,* and *R. purshiana.* In southern California mountains, the most common species are *Arctostaphylos glandulosa, A. otayensis, A. pringlei* var. *drupacea, A. parryana, Ceanothus integerrimus, C. leucodermis, C. cordulatus, C. foliosus, Eriodictyon crassifolium, E. trichocalyx, Garrya fremontii, G. veatchii, Heteromeles arbutifolia, Prunus ilicifolia, Quercus wislizenii* var. *frutescens, Rhamnus californica,* and *R. californica* (= *ilicifolia*) var. *ilicifolia.*

Certain chaparral elements invade exposed transmontane slopes of the Jeffrey pine forest in southern California. Characteristic species of this montane chaparral are *Arctostaphylos glauca, A. pungens, Ceanothus cuneatus, C. greggii* vars. *vestitus* and *perplexans, Cercocarpus betuloides, Eriodictyon crassifolium, Fremontia californica* (= *Fremontodendron californicum*), *Garrya flavescens* var. *pallida, Quercus turbinella* spp. *californica,* and *Rhamnus californica* var. *cuspidata.*

The development of montane chaparral is slowed by cold temperatures, snow cover, and short growing season. However, its development is rapid on sites previously occupied by coniferous forest. The response of montane chaparral to fire, logging, and other environmental impacts is so variable that it is not possible to present a reliable general pattern.

Red Shanks Chaparral

Red shanks chaparral is found in only four locations of southern California and Baja California. It is dominated by *Adenostoma sparsifolium,* forming pure stands in some locations, and is associated with *A. fasciculatum, Ceanothus* spp., and *Rhus ovata* in most locations. Its center of distribution is in the San Jacinto and Santa Rosa Mts. and interior valleys of Riverside and San Diego Cos. Major stands occur in the south Laguna Mts. of San Diego Co. and the mountain plateaus of northern Baja California, the western Santa Monica Mts. in Los Angeles Co., and the western Cuyama Valley of Santa Barbara and San Luis Obispo Cos. It ranges in elevation from 600 to 1800 m, with both coastal and desert exposures on granitic soils. The distribution of *Adenostoma sparsifolium* has been considered by Marion (1943), Epling and Lewis (1942), and Hanes (1965).

Many people consider this to be the most attractive form of chaparral. The vegetation may be rather dense in some areas and in younger stands, but its typical physiognomy is an open one. *Adenostoma sparsifolium* itself is an open shrub or small tree with multiple branches from the base covered with rust-red, shaggy bark. The foliage is feather-like and chartreuse, and open enough to allow the major branches to show through in bold relief. Individual shrubs are 2–3(6) m tall, rising above the associated shrub species by 1–3 m.

Red shanks chaparral develops in much the same way as chamise chaparral, except that early in its development it has a sparse herbland phase which remains in the intershrub spaces even when the stand is mature. *Adenostoma sparsifolium* and most of its shrub associates are vigorous sprouters. The mature character of red shanks chaparral is attained by 25 yr, and in subsequent years the cover does not increase appreciably, although the size of individual crowns does.

Serpentine Chaparral

Serpentine chaparral is an open, low type associated with serpentine soils from San Luis Obispo Co. northward through the Coast Ranges and foothills of the northern Sierra Nevada. The shrubs are characterized by apparent "xeromorphism" (peinomorphism, serpentinomorphism) and dwarfed stature resulting from reduced productivity and growth (Whittaker 1954b). The dominant shrubs are *Adenostoma fasciculatum* and *Heteromeles arbutifolia,* but noteworthy are several localized endemic shrub species, *Arctostaphylos viscida* and *Ceanothus jepsonii.*

Cupressus sargentii, Garrya congdonii, and *Quercus durata* are unmistakable "indicator species" because of their typical restriction to, and numerical dominance on, serpentine soils (Kruckeberg 1954). Serpentine chaparral may be associated with foothill woodland (*Pinus sabiniana*) or montane coniferous forest (*P. jeffreyi, P. ponderosa, P. attenuata, Pseudotsuga mensiesii*) as an understory (Whittaker 1954b). The thousands of hectares of serpentine chaparral in the North Coast Ranges are easily distinguished from the oak–grasslands on hills of nonserpentine origin (Walker 1954). Wells (1962) has described serpentine chaparral in San Luis Obispo Co., as has Vogl (1973) for the Santa Ana Mts., Orange Co.

Serpentine chaparral is 0.5–2 m in height and, like desert chaparral, has considerable intershrub open space that may be covered by small grasses or annuals but little if any leaf litter. Often dark green serpentine rocks and rock outcrops characterize the intershrub spaces. Many of the shrubs are compact and close to the ground, similar to montane chaparral (*Cupressus sargentii* is an exception, being erect up to 2–3 m). The leaves of indicator species are typically reduced, curled, or thickened. Development of serpentine chaparral is slow because of the harsh conditions imposed by the substrate.

Desert Chaparral

Desert chaparral is an ill-defined type since it (1) is associated with desert scrub, Joshua tree, and pinon–juniper woodland communities; (2) contains species from coastal types of chaparral, as well as shrub species found only on desert exposures (Hanes 1971); and (3) contains coastal sage scrub and grassland elements. It occurs on the inner Coast Ranges bordering the San Joaquin Valley from San Benito to Kern Cos., west of Tejon Pass in Santa Barbara and Ventura Cos., on the lower slopes of the San Gabriel and San Bernardino Mts. facing the Mojave Desert, and on the lower slopes of the San Jacinto, Santa Rosa, and Laguna Mts. facing the California Sonoran Desert. Desert chaparral has been recognized by several workers (Clements 1920; Cooper 1922; Horton 1960; Critchfield 1971; Hanes 1971).

Desert chaparral (Fig. 12-4) is noticeably more open than all other chaparral

Figure 12-4. Desert chaparral, dominated by *Adenostoma fasciculatum* and *Dendromecon rigida*, Los Angeles Co.

types, except serpentine and montane chaparral. Vegetation cover is usually less than 50%, yet individual shrubs may attain heights comparable to those of coastal specimens (Hanes 1971). *Adenostoma fasciculatum, Arctostaphylos glauca, Ceanothus greggii* vars. *perplexans* and *vestitus, Cercocarpus betuloides, Dendromecon rigida, Ephedra* spp., *Eriodictyon trichocalyx, Eriogonum fasciculatum, Fremontia californica, Garrya flavescens* var. *pallida, Juniperus californica, Opuntia* spp., *Prunus fremontii, P. fasciculatum, Purshia tridentata, Quercus turbinella* var. *californica, Rhus trilobata,* and *Yucca whipplei* dominate desert chaparral. Lower-statured shrubs are associated with the skirts of larger shrubs. Various annual grasses and forbs occupy the intershrub spaces.

Desert chaparral does not burn as often as coastal forms since the presence of man is not as extensive and fire does not carry as well. Recovery rates are about the same as in chamise chaparral, but the herbland phase is not as apparent. Also, there is a noticeable variety of seedling species in all stages of desert chaparral development.

Woodland Chaparral

Woodland chaparral has two phases: (1) an assortment of large, woody shrubs associated with various overstory tree communities, and (2) a live oak chaparral dominated by *Quercus dumosa, Q. wislizenii,* and *Q. chrysolepis* associated with *Arctostaphylos* spp., *Cercocarpus betuloides,* and *Ceanothus* spp. (Horton 1960). This type tends to replace scrub oak chaparral on shady slopes above 900 m in the mountains of southern California.

Woodland chaparral, associated with foothill woodland, oak woodland, pinon–juniper woodland, and knobcone pine forest, generally forms localized patches of shrubs 2–4 m tall. Common woodland chaparral species are *Adenostoma fasciculatum, Ceanothus leucodermis, Pickeringia montana, Rhamnus californica,* and *Rhus trilobata.*

CHAPARRAL SUCCESSION

Primary Succession

Chaparral develops directly on local broken rock surfaces where weathering of parent rock, scaling, and downslope movement of the rock surface presents a perpetually disturbed area. The most important shrub pioneers are *Eriogonum fasciculatum, Yucca whipplei, Leptodactylon californicum, Opuntia* spp., *Eriodictyon* spp., and *Adenostoma fasciculatum.*

Chaparral develops on alluvial fans and washes adjacent to coastal sage scrub and riparian woodland. Periodic flooding and erosion create new surfaces for pioneer plants. Alluvial fans and continuous piedmont slopes occur throughout the state, but they are especially well developed in Los Angeles, Orange, Riverside, San Bernardino, and San Diego Cos., bordering high mountain ranges. The physiographical processes of periodic floods, erosion, and deposition continually change the courses of washes and drainages. Soils are azonal sands and gravels with various-sized rocks and boulders. Certain ancient fans now in the process of dissection show all stages of the physiographical cycle. Cooper (1922) described primary chaparral succession in San Timoteo Canyon near Redlands: the older stable areas were occupied by chamise chaparral, the unstable eroding slopes were inhabited by coastal sage scrub, and the dry washes contained *Baccharis viminea, Lepidospartum squamatum, Populus fremontii, Salix hindsiana,* and *S. lasiolepsis.*

Chaparral develops on coastal sand dunes after stabilization by dune succulents and half-shrubs (Cooper 1922); this is discussed in Chapter 7.

Secondary Succession

The overall fire ecology and the successional pattern of California chaparral are remarkably similar throughout its range (Hanes 1974). The great bulk of California chaparral is in some stage of secondary succession. What appears to be a stable community at any given time or place is in reality only a phase in a larger cycle of growth, maturity, removal, and regrowth that takes decades to complete. Fire is the main initiator of secondary succession in chaparral, but man and other natural agents of vegetation disturbance or removal also begin the cycle of chaparral development.

Role of fire. One cannot understand California chaparral apart from fire. It is a powerful force in the total ecology of California chaparral. Most of the fires in California natural areas are in chaparral, occurring with a frequency of once every 10–40 yr (Muller et al. 1968). The impact of fire is sudden and extensive. The entire complex of soil factors, microclimate, and vegetation that has been built up over many years of chaparral development is changed in moments. Temperatures as high as 700°C (1200°F) can be reached during chaparral fires (Sampson 1944; Bentley and Fenner 1958), killing nonsprouting species outright (Hadley 1961) and consuming all but the largest scaffold branches. In addition to the heat itself, fire leads to ash, charcoal, and gas formation, removal of duff, increased illumination, and removal of competition by living plants (Sampson 1944). Indications are that accumulated phytotoxic substances in the duff and litter are also removed and that both organic and inorganic nutrient levels are improved as a consequence of fire (Muller et al. 1968; Christensen 1973; Christensen and Muller 1975a,b).

The postfire environment has been described in definitive studies (Christensen 1973; Christensen and Muller 1975b). Bacteria and fungi are more abundant in burned than in unburned chaparral soils. Mineral levels increase dramatically, and a large store of readily available organic nutrients is added by the ash. However, removal of the shrubs by fire terminates the natural recycling of nutrients from foliar leaching. The moisture content of soils on burned sites is less than under mature chaparral because of increased insolation (McKell et al. 1968; Christensen and Muller 1975a,b). Nonwetting agents of plant origin in the soil may be concentrated by the heat from fire into an intense, water-repellent layer (DeBano 1966).

Fire serves as the major cause of secondary succession in California chaparral by creating the pioneer conditions necessary for seedling establishment (Craddock 1929). By virtue of its shrub density, summer dryness, and volitile substances, California chaparral is one of the most fire-susceptible vegetation types in the world (Lewis 1961). Yet fire actually maintains or assures the perpetuation of many chaparral species (Vogl 1970; Vogl and Schorr 1972).

Three lines of adaptive strategy to fire are evident in chaparral: (1) production of seeds at an early age (Biswell and Gilman 1961); (2) production of refractory seeds, that is, resistant seeds (Sweeney 1956; Christensen and Muller 1975a), that retain viability for decades, are resistant to fire (especially in a dry state; Sweeney 1956), and in some cases fail to germinate in the absence of fire (heat); and (3) sprouting from a lignotuber or root-crown burl (Jepson 1916).

Chaparral animals. Numerous rodents inhabit chaparral (Wirtz 1974), and their populations are greatly reduced by fire (Howard et al. 1959; Chew et al. 1959; Cook 1959; Lawrence 1966). Some brushland rodents avoid recently burned sites (Cook 1959; Wirtz 1974). Burrowing rodents and reptiles survive a chaparral fire if they are several centimeters below ground (Lawrence 1966). Birds and large mammals can move away from the fire unless trapped by the flames (Wirtz 1974). Mule deer exhibit a preference for more open foraging areas over relatively mature stands of chaparral. Unlike the eastern white-tailed deer, mule deer take much more grass and herbage in their diet and are therefore attracted to the herbage of chaparral in early stages of secondary succession (Wirtz 1974). Their foraging on shrub seedlings and sprouts may have a significant effect on the recovery rate and composition of chaparral (Biswell and Gilman 1961; Kinucan 1965; Davis 1967), but little is known of bird and rodent influence on the early stages of chaparral succession.

Herbland phase. The greatest germination on burned chaparral sites occurs during the first wet season after fire (Went et al. 1952; Horton and Kraebel 1955; Patric and Hanes 1964). The vast majority of plants occurring on burns germinate from seeds present in the duff and soil before the occurrence of fire (Sweeney 1968). These seeds are long lived, remaining viable for many decades (Quick 1959). The dispersal of seeds onto burns from unburned adjacent sites does not contribute materially to the herbaceous cover of chaparral burns (Sweeney 1956). Great quantities of ephemeral herbs and forbs may dominate the burn site as an herbland for several years after a fire (Sampson 1944; Horton and Kraebel 1955; Sweeney 1956). A distinctive feature of most of these herbaceous species on chaparral burns is that they are "pyrophyte endemics," that is, they are found on burn sites but not elsewhere. The herbland phase of chaparral succession is characterized by a dramatic increase in flowering and seed production (Stone 1951; Stocking 1966; Ammirati 1967). This has been attributed to increased light (Stone 1951) and nutrients, especially nitrogen compounds (Christensen 1973).

In northern California, chaparral burns are dominated by herbaceous forms for the first 3 yr. During the fourth and fifth years, forbs give way to grasses (Cooper 1922; Sampson 1944; Sweeney 1956). Further analysis shows that different species of forbs dominate the herbaceous cover during each of the first, second, third, and fourth years (Sweeney 1968). Light, phytotoxins, seedbed conditions, or herbivores become limiting about 7–9 yr after fire (Sampson 1944) when the brush overstory is fully developed; the herbaceous flora then becomes rare or absent (Sweeney 1968). The marked increase in flowering and subsequent seed production by herbs ensures an abundant seed cache for the continuation of these species after the next chaparral fire.

In Napa and Sonoma Cos., the fire type annuals are closely associated with an east–west moisture gradient and drop out of the herbaceous cover as moisture increases toward the west (Ammirati 1967). Many of the perennial herbs persist into the more mesic vegetation types of the middle and outer Coast Ranges in the north.

Table 12-3 lists the species of annual forbs and grasses and perennial forbs, grasses, and half-shrubs found in six major studies: Sampson (1944) and Sweeney (1956) represent northern California sites; McPherson and Muller (1969) and Christensen and Muller (1975a), Santa Barbara coastal sites; and Horton and Kraebel (1955) and Vogl and Schorr (1972), interior southern California mountains.

Shrub phase. Within the herbland seral stage of chaparral succession, seedlings of chaparral shrubs and sprouting root crowns are also present (Quick 1959; Hanes 1971; Fig. 12-5). For species that sprout from root-crown burls after fire, reproduction by seed is facultative, whereas for nonsprouting species reproduction by seed is obligatory. Both types of shrubs are heavy seed producers. Often shrub seedlings occur in great profusion, but attrition is high among them in the early years of succession. Summer drought is a major factor in seedling mortality, but other forms of competition are also factors (Horton 1950; Went et al. 1952). For example, Schultz et al. (1955) found poor shrub-seedling root development in dense herbaceous stands. Nitrogen-fixing species (*Ceanothus leucodermis* and *Lotus scoparius*) are better competitors than non-nitrogen-fixing shrub species (Schultz et al. 1955). As the shrub seedlings develop, there is a thinning of their numbers (Horton and Kraebel 1955; Hanes 1971).

Many shrub species are absent from mature chaparral stands but abundant on recently burned sites. Some of these pioneers are subshrubs (subligneous) with short life spans, such as *Artemisia californica, Baccharis pilularis, Chrysothamnus* spp., *Dendromecon rigida, Encelia californica, Eriodictyon* spp., *Eriogonum fasciculatum, Eriophyllum confertiflorum, Haplopappus* spp., *Helianthemum scoparium, Lotus scoparius, Lupinus* spp., *Malacothamnus fasciculatus, Marah macrocarpa,*

Figure 12-5. Emergence of shrubland phase of scrub oak chaparral. Eight year old seedlings of *Ceanothus leucodermis* and *C. greggii* var. *vestitus* in foreground; resprouts of *Quercus dumosa* in background. Lake Arrowhead, San Bernardino National Forest, San Bernardino Co.

Mimulus spp., *Nicotiana glauca, Penstemon* spp., *Senecio douglasii,* and *Trichostema* spp. The outstanding group of woody pioneer shrubs consists of *Ceanothus* and, to a lesser extent, *Pickeringia montana.* These shrubs may dominate burned sites several years after a fire, but die out gradually as the stand ages. It is significant that these woody pioneer shrub species support nitrogen-fixing bacteria on their roots.

Several species of *Ceanothus* regenerate from seed only. Their numbers are large the first growing season after fire but are reduced by attrition each year, and within 30–40 yr they may be absent from the chaparral community (Horton and Kraebel 1955; Hanes and Jones 1967; Hanes 1971; Keeley 1975). Seldom do any new species or individuals enter the vegetation as it develops (Patric and Hanes 1964).

Thus California chaparral does not conform to the usual relay floristics concept of community succession. The climax species are present the first growing season after fire, and there is a cyclic phenomenon with key species persisting throughout all phases of the cycle and others being present in postfire pulses. Odum (1971) referred to this type of succession as "pulse stable"; Hanes (1971) termed it "autosuccession."

Montane chaparral succession. Montane chaparral succession occurs in some of the coniferous forest regions of California. The successional sequence has more seral stages than are present for other types of chaparral (Biswell 1952; Wilken 1967). The sere is as follows: mixed conifer, fire, herbs, shrubs, pine, fir. A major difference between the herbaceous flora in burned mixed conifer types and that of the chaparral is that there are no fire type species in the mixed coniferous forest (Sweeney 1968); the herbaceous species on the burns are essentially like those in the surrounding forests. Also, there are fewer herbaceous species than on chaparral burns. Brush shows a marked increase in burned forest areas because of increased light and the heat scarification of shrub species in the understory (especially *Arctostaphylos patula, Ceanothus integerrimus,* and *C. velutinus*). When the conifers overtop the brushfields, the brush species diminish and may die out because of reduced light intensities in the forest understory (Sweeney 1968).

One may ask why it is possible for chaparral to extend into the montane zone, when the montane tree species do not extend down into the chaparral. Wright (1966) showed that in the San Bernardino Mts. summer depletion of soil moisture in the chaparral zone could prevent conifer seedling establishment, and that shallow-rooted montane *Pinus lambertiana* could not tolerate dry soils as well as deeper-rooted foothill *P. coulteri* and *P. attenuata.* He found seedlings of these three species within both recently burned and unburned areas and concluded that fire was not a restrictive factor to their lower elevational limits. *Pinus attenuata* is noteworthy in that it is a fire type pine found within extensive chaparral fields (Wright 1966; Vogl 1968, 1970, 1973; Chapter 9).

Germination requirements. Germination studies of many chaparral species have revealed a complex of factors that control germination. Many seeds of herb and shrub species are inhibited from germination and growth by plant-derived toxins. Some seeds require scarification by heat or mechanical means; others do not (Wright 1931; Quick 1935, 1947, 1959; Mirov 1936; Rogers 1949; Stone and Juhren 1951, 1953; Stone 1957; Went et al. 1952; Cronemiller 1959; Hadley 1961; Quick

TABLE 12-3. Herbland "chaparral" species found in successional studies in three regions of California. Northern coastal: Sam (Sampson 1944) and Swe (Sweeney 1956). South coastal: M&M (McPherson and Muller 1969) and C&M (Christensen and Muller 1975a). Southern interior: H&K (Horton and Kraebel 1955) and V&S (Vogl and Schorr 1972)

Annual Herbs and Grasses	North Coastal		South Coastal		Southern Interior	
	Sam	Swe	M&M	C&M	H&K	V&S
Aira caryophyllea	X					
Allophyllum gilioides	X				X	X
A. glutinosum			X	X		
Anagallis arvensis				X		
Antirrhinum cornutum	X					
Apiastrum angustifolium		X	X	X		X
Avena barbata	X					
A. fatua				X		
Brassica geniculata					X	X
B. nigra					X	
Briza minor	X					
Bromus madritensis	X					
B. mollis				X		
B. rubens	X	X	X	X	X	
B. tectorum	X				X	X
Calandrinia ciliata			X	X		

TABLE 12-3 [continued]

Species					
Calyptridium monandrum					X
Centaurea melitensis	X		X		
C. solstitialis	X				
Chaenactis glabriuscula var. *tenuifolia*					X
Chorizanthe californica					X
C. staticoides		X	X	X	
Clarkia concinna	X				
Collinsia concolor					X
Conyza canadensis	X	X			
Cryptantha affinis	X				
C. ambigua	X				
C. clevelandii		X	X		
C. intermedia			X	X	X
C. micromeres		X			
C. muricata var. *jonesii*					X
C. torreyana	X				
Cynosurus echinatus		X			
Daucus pusillus	X	X			
Delphinium californicum	X				
Elymus sp.					X
Emmenanthe penduliflora	X		X	X	X

TABLE 12-3 [continued]

Annual Herbs and Grasses	North Coastal		South Coastal		Southern Interior	
	Sam	Swe	M&M	C&M	H&K	V&S
Epilobium minutum	X					
E. paniculatum	X					
Eremocarpus setigerus	X					X
Eriastrum pluriflorum					X	
Eriogonum gracile	X				X	
E. thurberi				X		X
Erodium cicutarium					X	
Erysimum capitatum						X
Eucrypta chrysanthemifolia			X	X		
Festuca megalura	X				X	
F. myuros	X					
F. octoflora	X	X	X	X		
Filago californica	X	X	X	X		
Gilia capitata	X					
G. clivorum						X
Gnaphalium bicolor				X		
Hemizonia ramosissima			X	X		
Hordium leporinum	X					

Lactuca serriola				X		
Lessingia nemaclada var. *albiflora*					X	
Lolium multiflorum				X		
Lotus micranthus	X					
L. purshianus	X					
L. strigosus	X		X	X	X	X
L. subpinnatus	X					
Lupinus bicolor	X					X
L. concinnus						X
Madia exigua	X					
M. gracilis				X		
Mentzelia gracilenta				X		
Mimulus bolanderi	X					
M. fremontii						X
Montia perfoliata				X		X
Navarretia atractyloides				X		
N. leucocephala	X					
Oenothera micrantha	X	X	X	X	X	X
Orthocarpus attenuatus	X					
Papaver californicum				X		
Phacelia brachyloba					X	X

TABLE 12-3 [continued]

Annual Herbs and Grasses	North Coastal		South Coastal		Southern Interior	
	Sam	Swe	M&M	C&M	H&K	V&S
P. grandiflora			X	X	X	
P. imbricata ssp. *patula*						X
P. minor					X	
P. tanacetifolia						X
Pterostegia drymarioides					X	
Salvia columbariae					X	
Senecio vulgaris				X		
Silene multinervia			X			
Stephanomeria virgata					X	
Thelypodium lasiophyllum	X					
Thysanocarpus laciniatus				X		
Tillaea erecta			X			
Trifolium tridentatum	X					
Perennial Herbs, Grasses and Half-Shrubs						
Arabis holboellii var. *retrofracta*	X					
Bloomeria crocea				X		
Brodiaea laxa	X					

B. pulchella		X	X	X		
Calochortus albus				X		
C. weedii				X		
Chlorogalum pomeridianum	X	X	X	X		
Convolvulus cyclostegius			X	X		
C. occidentalis					X	
Corethrogyne filaginifolia var. *bernardina*						X
Daucus carota	X					
Delphinium menziesii	X					
D. nudicaule	X					
D. parryi						X
Dendromecon rigida						X
Dicentra chrysantha					X	
D. ochroleuca				X		
Dodecatheon clevelandii				X		
Elymus glaucus	X					
Epilobium angustifolium	X					
Erigeron divergens				X		
E. foliosus var. *stenophyllus*						X
Eriodictyon crassifolium						X
Eriogonum fasciculatum						X

TABLE 12-3 [continued]

Perennial Herbs, Grasses and Half-Shrubs	North Coastal		South Coastal		Southern Interior	
	Sam	Swe	M&M	C&M	H&K	V&S
Eriophyllum confertiflorum			X		X	X
E. lanatum var. *achillaeoides*	X					
Galium nuttallii	X			X		
Gnaphalium californicum	X		X			
G. microcephalum	X					
Haplopappus squarrosus					X	
H. squarrosus spp. *grindelioides*						X
Helianthemum scoparium				X	X	X
H. scoparium var. *vulgare*						X
Helianthus gracilentus					X	X
Hieracium albiflorum	X					
Hulsea heterochroma						X
Hypericum concinnum	X					
Lathyrus laetiflorus				X		X
Lomatium lucidum						X
L. utriculatum				X		
Lotus humistratus	X			X		
L. scoparius		X		X	X	

Lupinus leucophyllus	X					
L. longifolius					X	X
Malacothamnus fremontii	X					
Marah macrocarpus				X		X
Mimulus guttatus						X
Paeonia californica				X		X
Penstemon centranthifolius						X
P. heterophyllus	X					
P. spectabilis					X	X
Poa scabrella				X		
Polystichum munitum	X					
Rumex acetosella	X					
Salvia mellifera			X	X		
S. spathacea				X		
Sanicula aguta				X		
S. crassicaulis				X		
Scrophularia californica	X					
Scutellaria tuberosa		X	X			
Senecio douglasii					X	
Solanum xantii	X					
Stipa lepida				X		

TABLE 12-3 [continued]

Perennial Herbs, Grasses and Half-Shrubs	North Coastal		South Coastal		Southern Interior	
	Sam	Swe	M&M	C&M	H&K	V&S
Trichostema lanatum				X		X
Turricula parryi					X	
Wyethia ovata						X
Yucca whipplei				X		X
Zigadenus fremontii	X			X		

and Quick 1961; Gratkowski 1973). Quick and Quick (1961) found that montane *Ceanothus* species have two separate and essential germination requirements, plumping (hydration after scarification) followed by stratification. Better germination of species not requiring pretreatment seems to be related to removal of shading, herbivory, litter, and phytotoxins in the soil.

Quick (1935) showed that the germination behavior of *Ceanothus* species varies widely along ecological and taxonomic lines. Mirov (1936) summarized the results of a large number of germination tests on seeds of native California plants with respect to taxonomic position, altitudinal distribution, and growth form. Generally there was no consistent relation between germinative behavior and altitudinal distribution. There was a trend, however, toward germination failure due to seed coat dormancy among low-elevation species, and toward failure due to embryo dormancy among high-elevation species. He found that seeds of annuals were less dormant than seeds of herbaceous perennials, shrubs, and trees.

Some chaparral species have been shown to produce two types of seeds. Stone and Juhren (1953), working with *Adenostoma fasciculatum,* found that some seeds germinated readily from the shrub or soil duff, whereas most required scarification. This same seed dimorphism has been reported in the herbaceous species *Astragalus congdonii* and *Trifolium ciliolatum* (Stocking 1966), and in the shrubs and half-shrubs *Ceanothus crassifolius, Eriophyllum confertiflorum,* and *Lotus scoparius* (Christensen and Muller 1975b).

Germination in some species whose seeds lack endogenous dormancy mechanisms is inhibited by shrub-derived toxins (Christensen and Muller 1975a). Toxin removal by fire accounts for the abundance of such species after fire. Extensive studies at the University of California, Santa Barbara, led by C. H. Muller over the past 10 yr, have elucidated the presence and role of allelopathic substances in California chaparral (Muller 1966; McPherson and Muller 1967, 1969; Muller et al. 1968; Hanawalt 1971; Chou and Muller 1972; Muller and Chou 1972; Christensen and Muller 1975a,b).

Since chamise chaparral is the dominant chaparral type, and since *Adenostoma fasciculatum* was the first chaparral species suspected of producing allelopathic agents, major attention has been directed toward this species (Landers 1962; McPherson and Muller 1969; Muller and Chou 1972). It has been shown that leaves of *A. fasciculatum* accumulate water-soluble phenolic compounds as a result of normal metabolic processes during the summer drought. Roots and leaf litter are not involved. Fog drip and rain drip (as little as 5–10 mm precipitation) carry the toxins to the soil, where they inhibit the germination and growth of numerous species. The toxic factors are ephemeral, and herb suppression persists only as long as *A. fasciculatum* shrubs remain intact. Removal of above-ground parts of these shrubs, without soil disturbance, permits the same flush of herb growth in the next growing season as is observed after fire.

In spite of the herb-free understory of mature *Arctostaphylos* stands, the pattern of herb control by *Arctostaphylos* is different from *Adenostoma* control (Chou and Muller 1972). Herb control is a function of phenolic toxins which are more widely distributed in the plant: roots, fallen fruit, exfoliated bark, leaf litter, and, especially, freshly fallen leaves. Also, the phytotoxins produced by *Acrtostaphylos* are long lasting, extending beyond the second season's growth after shrub removal.

We now know that numerous extrinsic factors associated with the mature chaparral floor result in an extremely low probability of seedling establishment and survival. The presence of phytotoxins has been described. Low light intensities and nutrient levels in mature chaparral also limit seedling growth, as may nonwetting agents in the soil (Osborn et al. 1967). Animal grazing has drastic effects on seedling survival (Christensen and Muller 1975b); in particular, the wood rat has been described as a significant biotic factor in chaparral (Horton and Wright 1944). After fire, the negative effect of each factor is removed or reduced. These selective factors and fire have resulted in the evolution of very efficient seed dormancy mechanisms in several species which minimize germination during periods of low survival probability (Christensen and Muller 1975b).

Sprouting. About half of the species that comprise California chaparral are capable of sprouting after the top is killed, consumed by fire, or removed by mechanical means (Hanes 1971). This sprouting capability resides in a swollen root-crown burl (lignotuber) found just under the ground surface (its top may be a few centemeters above ground if the surface soil has been eroded away). The burl is evident in the seedling stage as a slight swelling at the root–stem juncture. In subsequent years, the burl enlarges to become a mass larger than the main stem or trunk of the shrub (Fig. 12-6). Top removal results in the activation of numerous burl buds which grow into vigorous sprouts and, in turn, recreate an entirely new shrub crown.

Sprouting response, in contrast to seed germination, is not dependent on the arrival of rains. Even during the summer months, sprouting may occur within a few weeks or months after fire (Plumb 1961, 1963). Since the sprouts are peripheral to

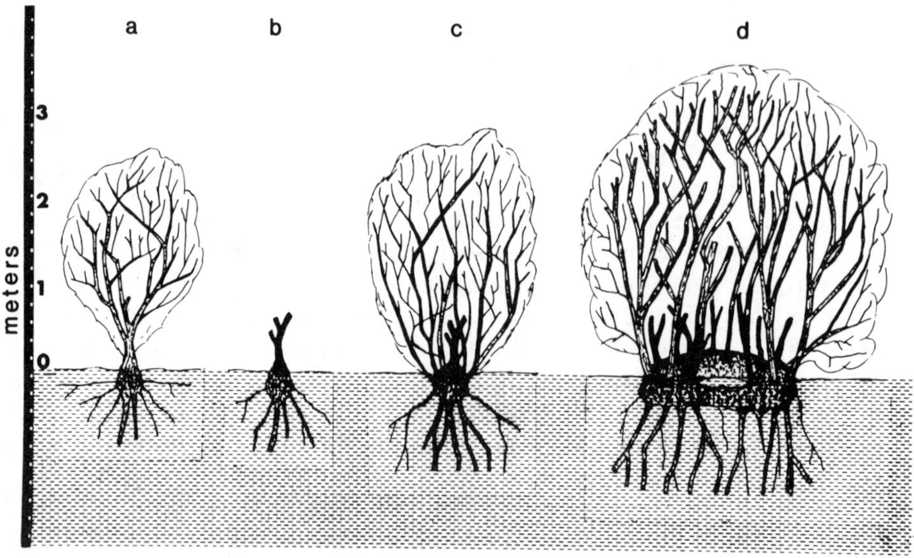

Figure 12-6. Generalized development of the root-crown burl in sprouting chaparral species: (*a*) appears as a swollen region in the first season of growth, (*b*) develops into a woody nodule below the ground surface of intact shrubs, (*c*) produces numerous sprouts after aerial portion of shrub is killed or removed, and (*d*) enlarges laterally as a consequence of repeated fire. In some species the root-crown burl develops into a large woody platform.

the previous stems or trunk, the girth of the new crown is larger than that of the original. Subsequent fires and regrowth result in a lateral expansion of the root-crown burl. In *Quercus dumosa, Arctostophylos* spp., and some other sprouting chaparral species, the repeated fire–regrowth sequence results in the development of subhypogeal woody platforms 1–4 m broad (Jepson 1930). These constitute a "fairy ring" type of root crown with the center rotted away, and in some cases there are gaps in the ring (Fig. 12-6). The age of chaparral burls can be 250+ yr (Hanes 1965).

The sprouting capacity of chaparral shrubs affords them several advantages. (1) They retain their already established position. (2) They can initiate growth immediately after fire by drawing on food and water stores in their fully developed root system. This can occur in the summer drought when competition is nonexistent, whereas dormant seeds must await winter rains in order to germinate and grow. (3) Their first-year growth is much more rapid than that of seedlings (Landers 1962). (4) They can be rejuvenated periodically without going through the rigors of the seedling stage. (5) Their new top can suppress seedlings by growing faster and overtopping them and by producing toxic substances.

Chaparral climax. In his classical study of the broad sclerophylls of California, Cooper (1922) considered *Adenostoma fasciculatum* to be the climax dominant species over the Central Coast Ranges. Cooper proposed that, along the mesophytic border of chaparral, fire had pushed back the coniferous forest, whereas along the xerophytic border the brush had been replaced by grasses. He also believed that coastal sage scrub was seral to chaparral.

Clements (1916, 1920) at first considered California chaparral a deflected or altered subclimax, based on the presence of so-called relics. Later Clements (Weaver and Clements 1929; Allred and Clements 1949) recognized some chaparral as true climax. He described mature chaparral as "a stable shrub community lying below a belt of subclimax chaparral due to forest fires." Jepson (1925) and Horton (1950) considered chaparral a transition vegetation which, if not for recurring fire, would eventually be replaced by oak woodland.

Other botanists (Bauer 1936; Munz and Keck 1959) have rejected the hypothesis that chaparral is a fire subclimax, contending that it is a true climax. This view is based on its widespread dominance, its evident stability, its occurrence on sites of diverse soil and topography, and the obvious adaptation of this xerophytic vegetation type to both semiarid climate and fire.

Part of the debate over the successional status of chaparral has resulted from an inadequate distinction between the northern and southern California environments as well as their vegetation. Grassland borders low-elevation chaparral in northern California, whereas coastal sage scrub borders low-elevation chaparral on coastal exposures in southern California. Insufficient attention has been paid to the distinctly different chaparral types which occur on north- and south-facing slopes. Also, no distinction has been made between coastal and desert chaparral types (Hanes 1971), and yet they are different in many ways.

There is a dynamic successional relationship between chamise chaparral and coastal sage scrub in southern California at elevations below 900 m, where the incidence of fire is high (Hanes 1971; Harrison *et al.* 1971). Disturbance of chamise chaparral often opens the way for coastal sage scrub species to invade the site

(Cooper 1922), and the resulting mix has been called coastal sage–chaparral subclimax (Hanes 1971). These scrub encroachments are usually low in species diversity and frequent components are *Eriogonum fasciculatum, Artemisia californica, Salvia apiana, S. mellifera, Eriophyllum confertiflorum, Gnaphalium* spp., and *Haplopappus* spp. (Fig. 12-7).

In isolated locations where fire is less frequent, coastal sage scrub elements are reduced gradually, in part by shading (McPherson and Muller 1967), and a chamise chaparral climax vegetation develops over a 20–30 yr period. This is a true climax since it is in harmony with its environment and does not give rise to another vegetation type. Grassland species are the main invaders of disturbed chaparral sites in northern California.

On coastal north-facing aspects at lower elevations in southern California, coastal sage scrub invasion is limited. There sprouts and seedlings of chaparral shrubs dominate after fire. This vegetation should be referred to as an immature scrub oak chaparral rather than a subclimax chaparral. If this chaparral is not burned for several decades, its height and cover increase and a subtle compositional change occurs (Patric and Hanes 1964). Certain species like *Ceanothus* die out, with a few other species such as *Prunus ilicifolia, Rhamnus ilicifolia,* and *Penstemon cordifolia* filling the void, and woody vines such as *Rhus diversiloba, Lonicera* spp., and *Clematis* spp. intermingle with the shrub canopy. The mature scrub oak chaparral develops over about a 50 yr period.

On south-facing slopes above 900 m in southern California, growth is more rapid than at lower elevations, and the near absence of coastal sage scrub elements in postfire vegetation lessens compositional changes. As the chaparral matures, it

Figure 12-7. Coastal sage scrub–chaparral subclimax in foreground resulting from removal of chamise chaparral 8 yr previously. *Eriogonum fasciculatum* and *Artemisia californica* dominate, but chaparral shrub seedlings and sprouts of *Adenostoma fasciculatum* and *Cercocarpus betuloides* are also present. Undistrubed chamise chaparral in background. Los Angeles Co.

resembles north-slope chaparral communities in species composition and appearance.

Most old, unburned chaparral stands become decadent In northern California this may occur in 20–25 yr (Sampson 1944). Chamise chaparral stands older than 60 yr are senile (Hedrick 1951; Hanes 1971); their species diversity is low, little annual growth occurs, and there is little if any herbaceous understory (McPherson and Miller 1969). Openings created by the death of individual shrubs may be filled with grasses (Sampson 1944), but they are not filled by new shrub seedlings. There is considerable standing and fallen dead material. Went *et al.* (1952) were the first to attribute this community stagnation to various biochemicals that accumulate in the soil of old stands. They suggested that the inferred biochemicals could interfere with normal decomposition, humification, and nitrification. Naveh (1960) and Landers (1962), working in northern California, indicated that the litter and roots of *Adenostoma fasciculatum* shrubs release substances which inhibit grass seed germination. Naveh (1960) speculated that lack of soil structure, severe erosion, and low nitrogen availability under some old chamise shrubs could be the result of such phytochemicals. The role of toxins in mature chamise chaparral is an important one, clearly established by a number of workers (McPherson and Muller 1969; Muller and Chou 1972; Christensen and Muller 1975a,b). The allelopathic agents may also inhibit the germination of *Adenostoma* itself. These substances, coupled with the build-up of water-repellent materials in soil, may account, in part, for the decadence of old chamise chaparral stands (Hanes 1971). In this context, fire may be a necessary environmental factor for chamise chaparral, keeping it young and productive.

Thus chamise chaparral is not a true climatic climax since it does not seem to have the ability to perpetuate itself indefinitely in the absence of fire.

AUTECOLOGY OF CHAPARRAL

Growth Form

The vast majority of California chaparral species are woody shrubs with oval crowns usually touching or greatly intertwined. They vary in height from 1 to 5 m, depending on species, site conditions, and age; most are 1.5–2.0 m. Some mesic species have the potential to become small trees, such as *Arctostaphylos elegans, A. glauca, A. manzanita, Heteromeles arbutifolia, Quercus wislizenii,* and *Q. chrysolesis.* In a single locality the height of individuals is strikingly uniform, creating a velvety appearance from a distance.

Chaparral physiognomy changes with geographical and elevational gradients (Shreve 1927a, 1936; Hanes 1971, 1974; Mooney and Harrison 1972) and moisture gradients (Griffin 1967). In Baja and southern California at the lowest elevations, there is a succulent sage scrub (see Chapter 13). At somewhat higher elevations, drought-deciduous coastal sage scrub predominates. Finally, evergreen chaparral dominates the highest shrub elevations. The stages in this sequence can be interpreted as adaptive responses to differences in moisture availability. It is significant that all chaparral shrubs are frost hardy, even the new growth which appears in the coldest months.

The shrub form of chaparral is ideal for plants exposed to fire, seasonal drought, and shading. Dominant chaparral genera (such as *Arctostophylos, Ceanothus,* and *Prunus*) exhibit strange stem morphologies (Jepson 1928; Adams 1934; Gankin and Major 1964; Patric and Hanes 1964; Waisel *et al.* 1972; Davis 1972, 1973; Keeley 1975). Conspicuous dead, decorticated patches and long strips, and the subsequent enlargement of the live stem portions, result in asymmetrical, contorted stems. Davis (1973) postulated that these dead portions are initiated by nonpathogenic environmental influences such as shade, fire, soil moisture stress, and extremes of temperature. As chaparral density increases, light becomes limiting to lower leaves, leaves and branches die, and they are aborted (McPherson and Muller 1967). Waisel *et al.* (1972) showed that bark striping is associated with dead branches and desiccated roots.

Whatever its mode of origin, bark striping represents a positive adaptation, because transpiration is reduced by the loss of all but the most healthy, vigorous and well-illuminated leaves, and excessive nonphotosynthetic tissue is eliminated by the inactivation and death of axial vascular tissues associated with dead leaves. Although growth continues at viable meristems, the volume of living shrub may increase little or not at all, since any increment of new tissues added by the meristems is apt to be canceled by the inactivation and death of older tissues. As the shrub ages, the proportion of functional stripes may be small compared to that of dead stripes (Davis 1973). Keeley (1975) described very old *Ceanothus* shrubs as having only a single live branch in the chaparral canopy connected by a single stripe of living tissue. The net result of this shrub growth pattern in mature stands is a sparse canopy held aloft by its own dead remains and sustained by stripes of live tissue. This growth form is reminiscent of that found in tree species exposed to harsh, subalpine environments, such as *Pinus aristata* and *P. flexilis.*

Roots. Cooper (1922) stated that one of the most important features of shrubs is their extensive root system in proportion to the size of the plant. Chaparral shrubs have a dual root system: an extensive lateral system suited to exploit surficial moisture supplies, and a deeply penetrating system suited to summer drought (Cannon 1911); Cooper 1922; Hellmers et al. 1955b). The primary roots of a shrub seedling develop quickly and deeply. Schultz et al. (1955) reported root depths of over 100 cm and lateral root growth of 60 cm for 3 mo old *Ceanothus* seedlings. Hanes (1965) found the root system of *Adenostoma sparsifolium* to be anomalous: both highly diffuse and succulent. Many dominant shrubs of chaparral lack root hairs but are abundantly supplied with mycorrhizal fungi (Cooper 1922).

Leaves. The leaves of California chaparral shrubs are moderately small (averaging 2–3 cm in length), simple, unlobed, elliptic, entire, and glabrous. Some are toothed, spinescent, revolute, or pubescent. A few species of *Ceanothus* and *Arctostaphylos* have glandular hairs. *Adenostoma fasciculatum* has very small, linear, terete leaves. The adaptive strategy of being evergreen requires that the leaves be small in order to avoid overheating during the summer (Mooney et al. 1974). Tannin is abundant in the leaves of most chaparral species. Sclerophylly promotes durability, water conservation, and protection from herbivores and is essential to gas-exchange capacity under water stress (Mooney and Dunn 1970; Dunn 1970; Morrow and Mooney 1974).

Cooper (1922) found that chaparral leaves are thick (~314 μ), about 3× as thick as those of deciduous tree species, and 2× as thick as those of oak woodland species. They have a greater mass per unit area than those of drought-deciduous coastal sage scrubs (Mooney and Dunn 1970). Chaparral leaves are xeromorphic, with an abundance of mechanical tissue and a very thick cuticle, and with stomates on the lower surface (Cooper 1922; Purer 1934; Nobs 1963), some in deep crypts. Although most species have leaves that lie in a horizontal position, some species have vertical leaves, and these exhibit stomates on both surfaces. Cooper (1922) demonstrated phenotypic plasticity in various leaf characteristics along a moisture gradient. There is an increased thickness of leaf and cuticle, an increased development of palisade tissue, and a decrease of spongy tissue with a decrease in soil moisture. Hanes (1965) showed that *Adenostoma fasciculatum* reduces its leaf surface during summer drought by shedding leaves only, whereas *A. sparsifolium* sheds entire leafy stems.

Mooney et al. (1974) related the incidence of vertically oriented leaves in some chaparral species to higher elevation. Vertical orientation reduces the radiant heat load on leaves in the middle of the day when the sun's rays are also vertical. The lower leaf temperatures provided by vertical orientation may be especially important when stomata close during summer drought stress periods and evaporative cooling is nonexistent. Since the upper leaves of such chaparral plants as *Heteromeles arbutifolia* are usually above photosaturation (Harrison 1971), the vertical orientation probably does not significantly reduce carbon gain. Furthermore, it increases light penetration to lower leaves.

Water Relations

Early attempts at determining drought resistance by measuring transpiration with crude photometers led Cooper (1922) to conclude that the number and size of chaparral leaves are more important than cuticular or stomatal effectiveness. Whitfield (1932a,b) found that transpiration rates are related to both edaphic and climatic factors, and that the osmotic concentrations of chaparral shrubs increase considerably during summer drought. Harrison et al. (1971) found that evergreen chaparral leaf transpiration rates are 0.25–0.33 those of drought-deciduous coastal sage scrub species.

Chaparral shrubs appear to use up all of the water available to them each year (Rowe and Reimann 1961; Pillsbury et al. 1963; Poole and Miller 1975). Their extensive root systems give them access to water stores beyond the reach of subshrubs and forbs. Morrow and Mooney (1974) attributed drought resistance in *Heteromeles arbutifolia* to its extensive rooting habit. Poole and Miller (1975) found differences in dawn water potentials to be species specific and related to rooting patterns. *Arctostaphylos glauca* and *Ceanothus greggii*, with major roots within the top 0.6 m of soil, had the most negative water potentials (<−65 bar) and showed stomata least responsive to water potential. *Rhus ovata*, *R. laurina*, and *R. integrifolia*, with roots penetrating rock outcrops, showed the least negative water potentials and most responsive stomata. *Adenostoma fasciculatum* and *Heteromeles arbutifolia*, with roots several meters deep, had minimum water potentials from −35 to −45 bar and showed stomatal closure at higher water potentials than did *Arctostaphylos* and *Ceanothus*.

Until recently it has been assumed that plants on south-facing slopes would be under greater water stress during the summer months than those on north-facing aspects. This is not the case. Syvertsen (1974) found that *Juglans* and *Heteromeles* on south-facing slopes were able to maintain higher diurnal stem water potentials than corresponding individuals on north-facing slopes. This apparent anomaly was attributed to possible xeric acclimation or selection pressure for drought resistance. The work of Poole and Miller (1975) on *Arctostaphylos, Ceanothus,* and *Adenostoma* revealed the same slope aspect anomaly. Their data also show that coastal sites are xeric for a longer period of time each year than inland sites, and that low winter soil moisture is a critical factor inhibiting evergreen shrubs along the coast.

Photosynthesis

Photosynthetic rates of chaparral species are low to moderate when compared with those of many other plants (Harrison et al. 1971). Peak photosynthesis occurs in the moist season, but there are minor variations on this theme. Hanes (1965) found that *Adenostoma fasciculatum* peaked in January and fell to low levels in July, whereas *A. sparsifolium* peaked in April and maintained a moderate photosynthetic rate in July. Both species fixed carbon more rapidly when soil moisture was below field capacity than above it. Morrow and Mooney (1974) found that *Heteromeles arbutifolia* is able to maintain stomata partly open throughout the year, thereby permitting continued CO_2 exchange and transpiration without water potentials dropping to critically low levels. Harrison (1971) and Morrow and Mooney (1974) showed that stomatal resistance to gas exchange is primarily a function of the amount of available soil moisture rather than of plant water potential. It has been shown that chaparral shrubs can carry on net photosynthesis in soils with water potentials below -15 bar (Hanes 1965; Dunn 1970). The sclerophyllous nature of the leaves prevents wilting and resulting mechanical damage (Harrison et al. 1971), and may contribute to the ability of chaparral plants to recover quickly from drought. Hanes (1965) showed that *Adenostoma fasciculatum* and *A. sparsifolium* under severe water stress could exhibit net photosynthesis 10 min after irrigation.

Phenology

Plummer (1911) early recognized that most of the growth in chaparral occurs in the cool season and little, if any, during summer. The seasonal growth cycle of *Adenostoma fasciculatum* is well known in the literature (Cooper 1922; Bauer 1936; Miller 1947; Hanes 1965) and illustrates the general pattern for California chaparral as a whole. Growth is initiated in late autumn or early winter, progresses greatly during late winter and spring, and terminates in June. Flower buds are initiated on the current season's growth as determinate inflorescences in late winter, and flowering occurs in late spring. Fruits mature in early summer. There are many variations on this cycle; for example, *Ceanothus crassifolius* flowers in early winter, and *Adenostoma sparsifolium* in August (Hanes 1965).

Studies of the controlling factors responsible for the cessation of growth in early summer and the resumption of growth during the cold season have established several points. Miller (1947) believed that some factor other than depletion of soil moisture ended the growing season, since plants on all aspects became dormant at the same time, although soil moisture was much greater on northerly exposures. Harvey (1962) related seasonal growth of chaparral shrubs to root types. In one study during an extremely dry year, no growth occurred (Harvey and Mooney 1964). Bauer (1936) and Miller (1947) showed that growth resumption of *Adenostoma* in January occurred with air temperatures lower by several degrees centrigrade than those in December. Also, mean monthly minimum air temperatures between March and May were as low as those in January, and yet this was the period of greatest growth activity. Cooper (1922) suggested that growth resumption was controlled by favorable soil temperatures, but Miller (1947) and Hanes (1965) showed that it was not dependent on soil temperatures. Jones and Laude (1960), working with northern California chaparral, concluded that *Adenostoma fasciculatum* failed to grow because of lowered starch reserves in the autumn and cold air temperatures in winter. Hanes (1965), working in the Santa Monica Mts., Los Angeles Co., found that reduced growth from December through February was related to lowered soil moisture, air relative humidity, and insolation. Apparently, once growth is initiated in early winter, general physiological activities continue to accelerate more or less independently of slight variations in the environmental factors that set them in motion. It is likely that an environmental factor may be limiting under one set of conditions, but has no controlling effect when conditions change (Hanes 1965).

Chaparral shrubs have been shown to exhibit positive thermoperiodism. The thermoperiod most suitable for top and root growth is warm days (23°C) and cool nights (4–10°C), although there is variability among species (Hellmers and Ashby 1958).

The seasonal cambial activity of a few chaparral species has been studied by Avila et al. (1975). *Rhus ovata* had an active cambium throughout the year, with peak activity in March. *Heteromeles arbutifolia* was active from December to mid-June, being most active between February and April, but was inactive through the drought period from late June to December. *Quercus dumosa* did not exhibit cambial activity until March, reached peak activity in May, and was inactive from late July through February. Guntle (1974) showed a significant correlation between seasonal rainfall totals and growth ring width in *Ceanothus crassifolius* and *Arctostaphylos glauca,* but with species differences.

Studies on reproductive strategies have been limited; far more needs to be done in this area. Hanes (1965) showed that *Adenostoma fasciculatum* adjusts its reproductive process to the seasonal conditions, ensuring maximum seed maturation during favorable years. *Adenostoma sparsifolium* possesses no such adjustment mechanism, and seed maturity in the autumn drought is uncertain. The different flowering periods of these two species constitute an effective barrier to hybridization. Both species have very low seed viability (0–4%; Horton 1949; Stone and Juhren 1953; Hanes 1965).

Carbon Balance

It has been shown that several chaparral shrubs fix carbon throughout the year (Hanes 1965; Harrison et al. 1971; Mooney and Dunn 1970; Mooney et al. 1975). Rates of fixation vary from year to year and with season, with both summer and winter depressions (Dunn 1970). During a dry year, summer production of *Rhus ovata* was considerably less than half of that during a wet year, but in both cases there was carbon fixation all year long (Mooney and Dunn 1970b). Morrow and Mooney (1974) found that *Heteromeles arbutifolia* exhibited the highest photosynthetic rates during February, when water stress was lowest, and the lowest rates in September, at the end of the drought period, when water stress was greatest. They found that the winter depressions were due primarily to the less intense photoperiod rather than to low temperatures. Plants from San Diego had photosynthetic thermal optima at a slightly higher temperature, and photosynthesized at a greater rate at comparable low water potentials, than plants from San Francisco. Mooney et al. (1975) suggested that evergreen *Heteromeles* is near its limits of adaptiveness in San Diego.

The differential partitioning of carbon compounds in *H. arbutifolia* varies seasonally (Mooney and Chu 1974). During the spring there is a priority allocation to the development of the canopy, apparently for competitive purposes, with little carbon going to reproduction, storage, and secondary compounds. During periods of little stem growth, carbon is channeled into root growth, fruit production, and toxic compounds. Dement and Mooney (1974) found that newly initiated leaves of *H. arbutifolia* exhibit high levels of tannins and cyanogenic glycosides, and they postulated that these high levels are possible because of favorable carbon and nutrient balances before leaf initiation. They found that the levels of nitrogen-containing cyanogenic glycosides in the leaves correlate positively with seasonal fluctuations in available nitrogen in this plant. Immature fruit exhibits extremely high tannin levels, as well as pulp cyanogenic glycosides. After a long maturation period with little fruit predation, the tannins decline and the glycosides are shifted from pulp to seed. At maturity the fruits are rapidly removed by birds.

Specht (1969b) provided data on biomass accumulation in a 37 yr old chamise chaparral stand in southern California. The total above-ground biomass was nearly 50,000 kg ha^{-1}. He estimated annual above-ground production to be 1000 kg ha^{-1}. Dunn (1970) calculated that the several chaparral species he studied fixed about 6.4 g CO_2 g^{-1} dry wt of leaves during 6 mo of winter and spring and 5.3 g during 6 mo of summer and autumn. He estimated that a stand of pure chamise chaparral in the San Dimas Experiment Station could fix about 50,000 kg CO_2 or produce about 30,000 kg dry matter during a wet year. Kittredge (1955) measured litter fall in chamise chaparral as 2800 kg ha^{-1}. Obviously, more measurements are needed to complete the analysis of chaparral productivity and carbon balance (Mooney and Parsons 1973).

Nutrition

California chaparral grows in shallow, coarse soils low in nutrients. Compared to agricultural soils, they are strikingly deficient in nitrogen and phosphorus, especially in the subsoil. Soil horizons can be separated only on the basis of color differences, since other profile distinctions are lacking (Jenny et al. 1950; Vlamis et al. 1954, 1959). The somewhat improved nutrient status in the upper soil reflects the leaching of nutrients from decomposing leaf litter. Hellmers et al. (1955a) showed that chaparral soils are so deficient in nitrogen that growth of natives is retarded.

Although there are no leguminous plants within the chaparral canopy (except for *Pickeringia montana* in isolated locations), some herbs and subshrub components of the herbland phase of chaparral are nitrogen fixers (*Lotus* spp. and *Lupinus* spp.), and species of *Ceanothus* and *Cercocarpus* have been shown to develop nitrogen-fixing nodules on their roots (Bottomley 1915; Quick 1944; Vlamis et al. 1958, 1964; Hellmers and Kelleher 1959; Zavitkovski and Newton 1968).

After fire, nitrate concentrations are comparatively high (Sampson 1944; Sweeney 1956; Christensen and Muller 1975a). Christensen (1973) showed that ash on chaparral burn sites not only contains NO_3^{2-} but may also provide a reservior of organic nitrogen. There is a rather striking increase in nitrate after the first autumn rains, accompanied by a decrease in soil ammonium, which may be due to the dependence of nitrifying bacteria on moisture. Throughout the remaining wet season, there is little nitrate output from the shrubs. The high mobility of nitrate ions in the soil results in low concentrations for the remainder of the annual cycle.

The high total nitrogen in the vegetation and low mineral nitrogen concentrations in the soil indicate inhibition of mineralization in unburned chaparral soils (Christensen 1973).

Serpentine chaparral presents a special case in chaparral nutrition. The fundamental reason for the intolerance exhibited by most species of plants for serpentine soils is low calcium levels (Mason 1946a,b; Kruckeberg 1951, 1954; Whittaker 1954b; Walker 1954). Walker (1948) proposed that serpentine endemics have the ability to obtain sufficient calcium, even at the low concentrations characteristic of serpentine soils. In addition, they must tolerate one or more of the following: high concentrations of chromium and nickel, high magnesium, low levels of major nutrients, low available molybdenum, and the drought and other undesirable aspects of shallow, stony soils (Walker et al. 1955).

Xeromorphism in plants on poor soils may best be interpreted as a pseudoxeromorphism (Whittaker 1954b) or peinomorphy (Daubenmire 1974). Walker (1954) suggested that plants which are widespread on serpentine soils possess heightened drought resistance as well as tolerance for serpentine soils.

CHAPARRAL MANAGEMENT

A considerable body of information has been amassed on the management of chaparral over the past 40 yr. Public concern over the loss of life and property related to California chaparral has led to the creation and support of public agencies designed to cope with chaparral-related environmental hazards. Research has followed four main lines of concern: chaparral hydrology, chaparral fires, managing chaparral brushfields, and replacing chaparral. No attempt is made here to cite the extensive literature on each of these areas.

Chaparral Hydrology

Research programs were initiated by state and federal forestry agencies during the 1930s to determine how watersheds covered with chaparral and other vegetation types function under various weather and seasonal conditions. This line of research established that chaparral is an extremely efficient watershed cover but that it also consumes most of the soil moisture. This finding led to field trials of the removal of native vegetation in order to conserve water. Current areas of research involve the problems of accelerated runoff and erosion during rainstorms and the occurrence of landslides that follow conversion of chaparral to grass cover on steep terrain.

Chaparral Fires

This line of research has studied the fire climate leading to chaparral fires, characteristics of the fuel (types and conditions of the vegetation) before and during a fire, fire storms, fire retardants, and fire-fighting equipment and techniques suitable to chaparral lands.

Managing Chaparral Brushfields

The extent of chaparral lands and their remoteness led to the trial and development of various management techniques. Roads and trails were the first means of access in suppressing fires in chaparral. By the 1920s firebreaks came into use. These consist of strips of cleared vegetation a few meters wide that encompass a given block of vegetation. Each year they are denuded to mineral soil mechanically or by hand.

During the late 1950s, the concept of fuelbreaks was devised. Fuelbreaks differ from firebreaks in that they are wide (65–200 m), extensive in scope, and clothed with an herbaceous cover. A fully developed fuelbreak system is designed to subdivide chaparral lands into relatively small, manageable blocks for better fire containment.

In recent years, consideration has been given to the design of fuelbreaks for wildlife habitat purposes. Undulating edges, leaving islands of brush vegetation within the fuelbreaks, and escape and migration routes are being assessed in relation to wildlife enhancement in chaparral lands.

Replacing Chaparral

During the 1950s, major research efforts were made to determine what species of grasses, herbs, and shrubs could be used to replace chaparral on fuelbreaks and on recently burned chaparral sites. Mustard (*Brassica* spp.) had been employed previously, but was rejected in favor of several annual and perennial grass species when it was demonstrated that it produced phytotoxic substances which inhibit shrub reestablishment. The current practice of seeding after fire is a questionable one, since germination and establishment are dependent on properly timed precipitation and storm intensity. A further complication is the suppression of native shrubs by the exotic grasses because of competition for light, moisture, and nutrients.

Another line of investigation initiated during the 1950s concerned the replacement of flammable chaparral species with less flammable exotic shrubs, such as rockroses (*Cistis* spp.). Three problems that have made this type of replacement impractical are the death of these exotic replacements after some years, their inability to reproduce, and their poor competitive abilities against chaparral.

Type conversion (changing chaparral stands to grassland) has become a common practice not only in the development of fuelbreaks, but also as a means to reduce fire hazard in urban areas and to expand rangeland. The increased demand for beef and sheep in the past two decades has caused many landholders in the foothill areas of the Central Valley to remove brushfields by various mechanical and chemical means and by controlled fire. The only chaparral terrain in southern California gentle enough for large-scale type conversion programs to be practiced is in San Diego Co. Introduced perennial grasses *Agropyron* spp., *Erharta calycina, Festuca arundinacea, Oryzopsis mileacca, Phalaris tuberosa* var. *stenoptera,* and *Poa ampla* mainly are used.

Where chaparral lands have been zoned residential, fuel modification has been employed. This type of chaparral management combines type conversion with fuel-break features by removing the bulk of the vegetation while retaining selected specimens for aesthetic and wildlife purposes. Such programs may also utilize high-moisture-content plants and sprinkler systems to keep the chaparral plants well hydrated during the summer drought.

In all these types of chaparral management, it should be apparent that man is working against nature. Replacing or modifying chaparral stands requires not only the initial effort, money, and energy but the ongoing maintenance obligations as well. Without maintenance subsidies for management plans, chaparral will reclaim its original territories. Left to itself, intact chaparral has no peer as a watershed cover. With the passage of time, however, it will eventually catch fire and burn. This natural course of events is intensified at the urban–chaparral interface and by the fire suppression policies of the past. The lay public, as well as the academic community, has shown an increasing awareness of and interest in the multifaceted nature of chaparral, as indicated by major conferences and symposia (Macey and Gilligan 1961; Rosenthal 1974).

AREAS FOR FUTURE RESEARCH

California chaparral has been the subject of many and varied investigations over the years. The pace has quickened in recent times, but the very complexity of chaparral leaves much to be explored. Moreover, Mooney and his group at Stanford University have demonstrated that similar shrub morphologies will evolve under similar climates regardless of the genetic histories of the floras involved (Mooney and Dunn 1970; Mooney et al. 1970; Parsons and Moldenke 1975).

Future explorations of the total ecology of California chaparral will certainly follow some of the inviting paths listed below.

1. Disjunct species. Local populations separated geographically from other members of the same species, and species at the limit of their range, could tell us much about their ecologies and their physiological capabilities and limitations. Griffin (1966) noted several disjunct species near Burney in the southern Cascade Mts.: *Ceanothus cuneatus, Lonicera interrupta, Rhamnus crocea* spp. *ilicifolia, Cercocarpus betuloides, Rhus diversiloba, Arctostaphylos viscida, A. manzanita, A. patula,* and *Chamaebatia foliolosa.* He pointed out that further study here may help in understanding the ecotone of floristic elements between the Sacramento Valley and the Modoc Plateau, as well as in understanding the northern and eastern limits of the California woodland and chaparral vegetation types. Axelrod (pers. commun.) has proposed comparative studies of Petran–Californian species-pairs.
2. Endemic species. The ecology of chaparral endemics needs to be understood in order to separate endemic species that are limited in distribution by narrow environmental tolerance from species limited by an environment composed of narrow

habitat factors (Stebbins and Major 1965). Such studies could be based on their physiology under varying environmental conditions, seed dispersal mechanisms, habitat alteration, and species response.

3. Reproductive ecology. The reproductive ecology of California chaparral species is little known. Young (1972), working with *Rhus integrifolia* and *R. ovata,* has opened up a new line of investigation that could be expanded to include important genera and species in many types of chaparral. Monogeneric families and monospecific genera such as *Cneoridium* (Rutaceae) and *Xylococcus* (Ericaceae) await imaginative studies. Pollination mechanisms, fruit structure (fleshy vs. dry), seed parasitism, predation and dispersal, reproductive success, and other aspects of population biology need study.
4. Vegetative reproduction. Obligatory resprouters are subject to heavy browsing by herbivores while their shoots are soft and their leaves are not yet lignified and thorny (Naveh 1975). Postfire grazing pressure has acted as an additional powerful selective agent, favoring species and biotypes which very soon develop hard, thorny, or distasteful leaves and twigs, and also those with the highest vegetative regeneration capacities to overcome defoliation stresses of both fire and grazing (Mooney and Dunn 1970). The effect of such biotic limiting and selective factors in the chaparral environment needs elucidation. Furthermore little is known of the significance, if any, of the ratio of sprouters to nonsprouters in the various chaparral types. What are the factors that maintain these ratios? What is the physiology of sprouting?
5. Adaptive strategies. We have only begun to view chaparral shrubs as expressions of adaptation through time. Viewed in this light, they "tell" us how they accommodate environmental factors, reduce stress, and enhance long-term success.
6. Serpentinomorphism. The relative roles of low moisture and low fertility in serpentine soils need to be clarified in serpentine chaparral.
7. Carbon balance. There is a wealth of basic ecological information to be gained by studies of the carbon economies of chaparral shrubs in natural environments. We do not know the efficiency of the chaparral ecosystem. Allocation and partitioning of resources constitute a new field of investigation awaiting imaginative minds. The role of secondary compounds has been firmly established by Muller's work, yet their full role at both the species and the community level awaits amplification. The dynamics of chaparral populations may hinge on these unseen as well as microbial agents.
8. Nutrition. Even though our understanding of the nutrition of chaparral is more than elementary, much more needs to be learned. The role of nitrogen-fixing species has been described but not quantified. The roles of micorrhizae and decomposing fungi in chaparral soils await elaboration.

LITERATURE CITED

Abrams, J. E. 1940. A systematic study of the genus Arctostaphylos. J. Elisha Mitchell Sci. Soc. 56:1-62.

Abrams, L. R. 1910. A phytogeographic and taxonomic study of the southern California trees and shrubs. Bull. N. Y. Bot. Gard. 6:300-485.

-----. 1951. Illustrated flora of the Pacific States. 3 vols. Stanford Univ. Press, Stanford, Calif. 865 p.

Adams, J. E. 1934. Some observations on two species of Arctostaphylos. Madroño 2:147-152.

Allred, B. W., and E. S. Clements (eds.). 1949. Dynamics of vegetation: selections from the writings of F. E. Clements. H. W. Wilson, New York. 296 p.

Ammirati, J. F. 1967. The occurrence of annual and perennial plants on chaparral burns. M.A. thesis, San Francisco State Coll., Calif. 144 p.

Anderson, H. W., G. B. Coleman, and P. J. Zinke. 1959. Summer slides and winter scour - dry-wet erosion in southern California mountains. USDA Forest Serv., Pac. SW Forest and Range Exp. Sta., Tech. Paper PSW-36. Berkeley, Calif. 12 p.

Avila, G., M. Lajaro, S. Araya, G. Montenegro, and J. Kummerow. 1975. The seasonal cambium activity of Chilean and Californian shrubs. Amer. J. Bot. 62:473-478.

Axelrod, D. I. 1958. Evolution of the Madro-Tertiary geoflora. Bot. Rev. 24:433-509.

Bauer, H. L. 1930. Vegetation of the Tehachapi Mountains. Ecology 11:263-280.

-----. 1936. Moisture relations in the chaparral of the Santa Monica Mountains, California. Ecol. Monogr. 6:409-454.

Bentley, J. R., and R. L. Fenner. 1958. Soil temperatures during burning related to postfire seedbeds on woodland range. J. For. 56:737-740.

Biswell, H. H. 1952. Factors affecting brush succession in the coast region and the Sierra Nevada foothills, pp. 63-97. In Proc. 4th ann. Calif. weed conf. San Luis Obispo, Calif.

-----. 1963. Research in wildland fire ecology in California, pp. 63-97. In Proc. Tall Timbers fire ecology conf. Tall Timbers Res. Sta., Tallahassee, Fla.

Biswell, H. H., and J. H. Gilman. 1961. Brush management in relation to fire and other environmental factors on the Tehama deer winter range. Calif. Fish and Game.

Bock, J. H., and C. E. Bock. 1969. Natural reforestation in the northern Sierra Nevada-Donner's Ridge burn, pp. 119-126. In Proc. Tall Timbers fire ecology conf. Tall Timbers Res. Sta., Tallahassee, Fla.

Bottomley, W. B. 1915. The root nodules of Ceanothus americanus. Ann. Bot. 29:605-610.

Bowerman, M. L. 1944. The flowering plants and ferns of Mount Diablo, California. Gillick Press, Berkeley, Calif. 290 p.

Brandegee, K. 1894. Studies in Ceanothus. Proc. Calif. Acad. Sci., Ser. 2, 4:173-222.

Brandegee, T. S. 1889. Plants from Baja California. Proc. Calif. Acad. Sci., Ser. 2, 2:117-216.
-----. 1891. The vegetation of "burns." Zoe 2:118-122.
Brayton, R., and H. A. Mooney. 1966. Population variability of *Cercocarpus* in the White Mountains of California as related to habitat. Evolution 20:383-391.
Calkins, H. C. 1925. In defense of brush. J. For. 23:30-33.
Cannon, W. A. 1911. The root habits of desert plants. Carnegie Inst. Wash. Pub. 131. 96 p.
-----. 1914. Specialization in vegetation and in environment in California. Plant World 17:223-237.
Chew, R. M., B. B. Butterworth, and R. Grecham. 1959. The effect of fire on the small mammal population of chaparral. J. Mammol. 40: 253.
Chou, C. H., and C. H. Muller. 1972. Allelopathic mechanisms of *Arctostaphylos glandulosa* var. *zacaensis*. Amer. Midl. Natur. 88:324-347.
Christensen, N. L. 1973. Fire and the nitrogen cycle in California chaparral. Science 181:66-68.
Christensen, N. L., and C. H. Muller. 1975a. Effects of fire on factors controlling plant growth in *Adenostoma* chaparral. Ecol. Monogr. 45:29-55.
-----. 1975b. Relative importance of factors controlling germination and seedling survival in *Adenostoma* chaparral. Amer. Midl. Natur. 93:71-78.
Clements, F. E. 1916. Plant succession - an analysis of the development of vegetation. Carnegie Inst. Wash. Pub. 242. 512 p.
-----. 1920. Plant indicators: the relation of plant communities to process and practice. Carnegie Inst. Wash. Pub. 290. 388 p.
Collins, B. J. 1972. Key to coastal and chaparral flowering plants of southern California. Calif. Lutheran Coll., Thousand Oaks, Calif. 249 p.
Cook, S. F., Jr. 1959. The effects of fire on a population of small rodents. Ecology 40:102-108.
Cooke, W. B. 1962. On the flora of the Cascade Mountains. Wasmann J. Biol. 20:1-67.
Cooper, W. S. 1922. The broad-scherophyll vegetation of California - and ecological study of the chaparral and its related communities. Carnegie Inst. Wash. Pub. 122 p.
Craddock, G. W. 1929. The successional influence of fire on the chaparral type. M.S. thesis, Univ. Calif., Berkeley.
Crawford, J. M. 1962. Soils of the San Dimas Experimental Forest. USDA Forest Serv., Pac. SW Forest and Range Exp. Sta., Misc. Paper PSW-76. Berkeley, Calif. 20 p.
Critchfield, W. B. 1971. Profiles of California vegetation. USDA Forest Serv., Pac. SW Forest and Range Exp. Sta., Res. Paper PSW-76. Berkeley, Calif. 54 p.
Cronemiller, F. P. 1942. Chaparral. Madroño 6:199.
-----. 1959. The life history of deerbrush - a fire type. J. Range Manage. 12:21-25.

Davis, C. B. 1972. Comparative ecology of six members of the Arctostaphylos andersonii complex. Ph.D. dissertation, Univ. Calif., Davis. 155 p.

-----. 1973. "Bark striping" in Arctostaphylos (Ericaceae). Madroño 22:145-149.

Davis, J. 1967. Some effects of deer browsing on chamise sprouts after fire. Amer. Midl. Natur. 77:234-238.

Daubenmire, R. F. 1974. Plants and environment. 3rd ed. John Wiley, New York. 442 p.

DeBano, L. F. 1966. Formation of non-wettable soils involves heat transfer mechanism. USDA Forest Serv., Pac. SW Forest and Range Exp. Sta., Res. Note PSW-132. Berkeley, Calif. 8 p.

Dement, W., and H. Mooney. 1974. Seasonal variation in the production of tannins and cyanogenic glucosides in the chaparral shrub, Heteromeles arbutifolia. Oecologia 15:65-76.

Detling, L. E. 1961. The chaparral formation of southeastern Oregon with consideration of its post-glacial history. Ecology 42:348-357.

Dunn, E. L. 1970. Seasonal patterns of carbon dioxide metabolism in evergreen sclerophylls in California and Chile. Ph.D. dissertation, Univ. Calif., Los Angeles.

Epling, C., and H. Lewis. 1942. The centers of distribution of the chaparral and coastal sage associations. Amer. Midl. Natur. 27: 445-462.

Fultz, F. M. 1923. The elfin-forest of California. Los Angeles Times Mirror Press, Los Angeles. 267 p.

Gankin, R., and J. Major. 1964. Arctostaphylos myrtifolia, its biology and relationship to the problem of endemism. Ecology 45:792-808.

Gillett, G. W., J. T. Howell, and H. Leschke. 1961. A flora of Lassen Volcanic National Park, California. Wasmann J. Biol. 19:1-185.

Gratkowski, H. J. 1973. Pregermination treatments for redstem ceanothus seeds. USDA Forest Serv., Pac. NW Forest and Range Exp. Sta., PNW-156. Portland, Ore. 10 p.

Griffin, J. R. 1966. Notes on disjunct foothill species near Burney, California. Leafl. West. Bot. 10:296-298.

-----. 1967. Soil moisture and vegetation patterns in northern California forests. USDA Forest Serv., Pac. SW Forest and Range Exp. Sta., Res. Paper PSW-46. Berkeley, Calif. 22 p.

-----. 1975. Plants of the highest Santa Lucia and Diablo Range Peaks, California. USDA Forest Serv., Pac. SW Forest and Range Exp. Sta., Res. Paper PSW-110. Berkeley, Calif. 50 p.

Guntle, G. R. 1974. Correlation of annual growth in Ceanothus crassifolius Torr. and Arctostaphylos glauca Lindl. to annual precipitation in the San Gabriel Mountains. M. A. thesis, Calif. State Poly. Univ., Pomona. 38 p.

Hadley, E. B. 1961. Influence of temperature and other factors on Ceanothus megacarpus seed germination. Madroño 16:132-138.

Hamilton, E. L., and P. B. Rowe. 1949. Rainfall interception by chaparral in California. USDA Forest Serv., Pac. SW Forest and Range Exp. Sta., Berkeley, Calif. 43 p.

Hanawalt, R. B. 1971. Inhibition of annual plants by Arctostaphylos, pp. 33-38. In Biochemical interactions among plants. Natl. Acad. Sci., Washington, D.C. 134 p.
Hanes, T. L. 1965. Ecological studies on two closely related chaparral shrubs in southern California. Ecol. Monogr. 35:213-235.
-----. 1971. Succession after fire in the chaparral of southern California. Ecol. Monogr. 41:27-52.
-----. 1974. The vegetation called chaparral, pp. 1-5. In Symp. on living with chaparral. Sierra Club, San Francisco.
Hanes, T. L., and H. Jones. 1967. Postfire chaparral succession in southern California. Ecology 48:259-264.
Harrison, A. T. 1971. Temperature related effects on photosynthesis in Heteromeles arbutifolia M. Roem. Ph.D. dissertation, Stanford Univ., Stanford, Calif.
Harrison, A. T., E. Small, and H. A. Mooney. 1971. Drought relationships and distribution of two mediterranean-climate California plant communities. Ecology 52:869-875.
Harvey, R. A. 1962. Seasonal growth of chaparral shrubs in relation to root types. M. A. thesis, Univ. Calif., Los Angeles.
Harvey, R. A., and H. A. Mooney, 1964. Extended dormancy of chaparral shrubs during severe drought. Madroño 17:161-163.
Head, W. S. 1972. The California chaparral - an elfin forest. Naturegraph, Healdsburg, Calif. 92 p.
Hedrick, D. W. 1951. Studies on the succession and manipulation of chamise brushlands in California. Ph.D. dissertation, Texas A & M Coll., College Station. 113 p.
Hellmers, H., and W. C. Ashby. 1958. Growth of native and exotic plants under controlled temperatures in the San Gabriel Mountains, California. Ecology 39:416-428.
Hellmers, H., J. F. Bonner, and J. M. Kelleher. 1955a. Soil fertility: a watershed management problem in the San Gabriel Mountains of southern California. Soil Sci. 80:189-197.
Hellmers, H., J. S. Horton, G. Juhren, and J. O'Keefe. 1955b. Root systems of some chaparral plants in southern California. Ecology 36:667-678.
Hellmers, H., and J. M. Kelleher. 1959. Ceanothus leucodermis and soil nitrogen in southern California mountains. Forest Sci. 5:275-278.
Hoover, R. F. 1970. Thevascular plants of San Luis Obispo County, California. Univ. Calif. Press, Berkeley. 350 p.
Horton, J. S. 1949. Trees and shrubs for erosion control of southern California mountains. USDA Forest Serv., Pac. SW Forest and Range Exp. Sta., Berkeley, Calif. 72 p.
-----. 1950. Effect of weed competition upon survival of planted pine and chaparral seedlings. USDA Forest Serv., Pac. SW Forest and Range Exp. Sta., Forest Res. Note PSW-72. Berkeley, Calif. 6 p.
-----. 1960. Vegetation types of the San Bernardino Mountains. USDA Forest Serv., Pac. SW Forest and Range Exp. Sta., Tech. Paper PSW-44. Berkeley, Calif. 29 p.
Horton, J. S., and C. J. Kraebel. 1955. Development of vegetation after fire in the chamise chaparral of southern California. Ecology 36:244-262.

-----. 1950. Effect of weed competition upon survival of planted pine and chaparral seedlings. USDA Forest Serv., Pac. SW Forest and Range Exp. Sta., Forest Res. Note PSW-72. Berkeley, Calif. 6 p.

-----. 1960. Vegetation types of the San Bernardino Mountains. USDA Forest Serv., Pac. SW Forest and Range Exp. Sta., Tech. Paper PSW-44. Berkeley, Calif. 29 p.

Horton, J. S., and C. J. Kraebel. 1955. Development of vegetation after fire in the chamise chaparral of southern California. Ecology 36:244-262.

Horton, J. S., and J. T- Wright. 1944. The wood rat as an ecological factor in southern California watersheds. Ecology 25:341-351.

Howard, W. E., R. L. Fenner, and H. E. Childs, Jr. 1959. Wildlife survival in brush burns. J. Range Manage. 12:230-234.

Howell, J. T. 1929. The flora of the Santa Ana Canyon region. Madroño 1:243-253.

Howitt, B. F., and J. T. Howell. 1964. The vascular plants of Monterey County, California. Wasmann J. Biol. 22:1-184.

Jenny, H., J. Vlamis, and W. E. Martin. 1950. Greenhouse assay of fertility of California soils. Hilgardia 20:1-8.

Jepson, W. L. 1916. Regeneration in manzanita. Madroño 1:3-11.

-----. 1921. The fire-type forest of the Sierra Nevada. Intercoll. For. Club Ann. 1:7-10.

-----. 1925. Manual of the flowering plants of California. Assoc. Students Store, Univ. Calif., Berkeley. 1238 p.

-----. 1928. Biological peculiarities of California flowering plants, Part I. Madroño 1:190-192.

-----. 1930. The role of fire in relation to the differentiation of species in the chaparral, pp. 114-116. In Proc. 5th Internat. Bot. Congr., Cambridge, England.

Jones, M. D., and H. M. Laude. 1960. Relationship between sprouting in chamise and physiological condition of the plant. J. Range Manage. 13:210-214.

Keeley, J. E. 1975. Longevity of nonsprouting ceanothus. Amer. Midl. Natur. 93:504-507.

Kinucan, E. S. 1965. Deer utilization of postfire chaparral shrubs and fire history of the San Gabriel Mountains. M. A. thesis, Calif. State Coll., Los Angeles. 61 p.

Kittredge, J., Jr. 1939a. The forest floor of the chaparral in San Gabriel Mountains, California. J. Agric. Res. 58:521-535.

-----. 1939b. The annual accumulation and creep of litter and other surface materials in the chaparral of the San Gabriel Mountains, California. J. Agric. Res. 58:537-541.

-----. 1955. Litter and forest floor of the chaparral in parts of San Dimas Experimental Forest, California. Hilgardia 23:563-596.

Klyver, F. D. 1931. Major plant communities in a transect of the Sierra Nevada Mountains of California. Ecology 12:18-20.

Krammes, J. S. 1960. Erosion from mountainside slopes after fire in southern California. USDA Forest Serv., Pac. SW Forest and Range Exp. Sta., Note PSW-171. Berkeley, Calif. 8 p.

Krammes, J. S., and L. F. DeBano. 1965. Soil wettability: a neglected factor in watershed management. Water Resour. Res. 1:283-286.

Krammes, J. S., and J. F. Osborn. 1969. Water-repellent soils and wetting agents as factors influencing erosion, p. 177. In L. F. DeBano and J. Letey (ed.). Water-repellent soils. Proc. symp. on water-repellent soils. Univ. Calif., Riverside.

Kruckeberg, A. R. 1951. Interspecific variability in the response of certain native plant species to serpentine soil. Amer. J. Bot. 38:408-419.

-----. 1954. The ecology of serpentine soils. III. Plant species in relation to serpentine soils. Ecology 35:267-274.

Landers, R. Q., Jr. 1962. The influence of chamise, Adenostoma fasciculatum, on vegetation and soil along chamise-grassland boundaries. Ph.D. dissertation, Univ. Calif., Berkeley. 194 p.

Lathrop, E. W., and R. F. Thorne. 1968. Flora of the Santa Rosa plateau of the Santa Ana Mountains, California. Aliso 6:17-40.

Lawrence, G. E. 1966. Ecology of vertebrate animals in relation to chaparral fire in the Sierra Nevada foothills. Ecology 47:278-291.

Leiberg, J. B. 1899. Forest reserves in southern California. U.S. Geol. Survey, 19th Ann. Rept., Part 5:351-371.

-----. 1900. Forest reserves in southern California. U.S. Geol. Survey, 20th Ann. Rept., Part 5:409-478.

Lewis, F. H. 1961. Chaparral lands of southern California. pp. 13-17. In A. Macey and J. Gilligan (eds.). Man, fire and chaparral - a conference on southern California wildland research problems. Univ. Calif. Agric. Exp. Sta., Berkeley.

Macey, A. and J. Gilligan (eds.). 1961. Man, fire and chaparral - a conference on southern California wildland research problems. Univ. Calif. Agric. Exp. Sta., Berkeley. 101 p.

Marion, L. H. 1943. The distribution of Adenostoma sparsifolium. Amer. Midl. Natur. 29:106-116.

Mason, H. L. 1946a. The edaphic factor in narrow endemism. I. The nature of environmental influences. Madroño 8:209-226.

-----. 1946b. The edaphic factor in narrow endemism. II. The geographic occurrence of plants of highly restricted patterns of distribution. Madroño 8:241-257.

McKell, C. M., J. R. Goodin, and C. C. Duncan. 1968. Chaparral fires change soil moisture depletion patterns. Calif. Agric. 22:15-16.

McMinn, H. E. 1964. An illustrated manual of California shrubs. Univ. Calif. Press. Berkeley. 663 p.

McPherson, J. K., and C. H. Muller. 1967. Light competition between Ceanothus and Salvia shrubs. Bull. Torrey Bot. Club 94:41-55.

-----. 1969. Allelopathic effects of Adenostoma fasciculatum, "chamise," in the California chaparral. Ecol. Monogr. 39:177-198.

Miller, E. H., Jr. 1947. Growth and environmental conditions in southern California chaparral. Amer. Midl. Natur. 37:379-420.

Miller, L. C. 1906. Chaparral as a watershed cover in southern California. Proc. Soc. Amer. Forest. 1:147-157.

Mirov, N. T. 1936. Germination behavior of some California plants. Ecology 17:667-672.

Mooney, H. A., and C. Chu. 1974. Seasonal carbon allocation in Heteromeles arbutifolia, a California evergreen shrub. Oecologia 14:295-306.

Mooney, H. A., and E. L. Dunn. 1970. Photosynthetic systems of mediterranean climate shrubs and trees in California and Chile. Amer. Natur. 104:447-453.

Mooney, H. A., E. L. Dunn, F. Shropshire, and L. Song, Jr. 1970. Vegetation comparisons between the Mediterranean climate areas of California and Chile. Flora 159:480-496.

Mooney, H. A., S. L. Gulmon, D. J. Parsons, and A. T. Harrison. 1974. Morphological changes within the chaparral vegetation type as related to elevational gradients. Madroño 22:281-316.

Mooney, H. A., and A. T. Harrison. 1972. The vegetational gradient on the lower slopes of the Sierra San Pedro Martir in northwest Baja California. Madroño 21:349-445.

Mooney, H. A., A. T. Harrison, and P. A. Morrow. 1975. Environmental limitations of photosynthesis on a California evergreen shrub. Oecologia 19:293-301.

Mooney, H. A., and D. J. Parsons. 1973. Structure and function in the California chaparral - an example from San Dimas, pp. 83-112. In F. di Castri and H. A. Mooney (eds.). Mediterranean type ecosystems, origin and -tructure. Springer-Verlag, Heidelberg.

Morrow, P. A., and H. A. Mooney. 1974. Drought adaptations in two California evergreen sclerophylls. Oecologia 15:205-222.

Muller, C. H. 1966. The role of chemical inhibition (allelopathy) in vegetational composition. Bull. Torrey Bot. Club 93:332-351.

-----. 1967. Relictual origins of insular endemics in Quercus, pp. 73-77. In R. N. Philbrick (ed.). Proc. symp. on biology of the California islands. Santa Barbara Bot. Gard., Santa Barbara, Calif.

Muller, C. H., and C. H. Chou. 1972. Phytotoxins: an ecological phase of phytochemistry, pp. 201-216. In J. B. Harborne (ed.). Phytochemical ecology. Proc. Phytochem. Soc., No. 8. Academic Press, New York.

Muller, C. H., R. B. Hanawalt, and J. K. McPherson. 1968. Allelopathic control of herb growth in the fire cycle of California chaparral. Bull. Torrey Bot. Club 95:225-231.

Munns, E. N., 1920. Chaparral cover, run-off, and erosion. J. For. 18: 806-814.

Munz, P. A. 1935. A manual of southern California botany. J. W. Stacey, San Francisco. 642 p.

-----. 1974. A flora of southern California. Univ. Calif. Press, Berkeley. 1086 p.

Munz, P. A., and D. D. Keck. 1959. A California flora. Univ. Calif. Press, Berkeley. 1681 p.

Mutch, R. W. 1970. Wildland fires and ecosystems - a hypothesis. Ecology 51:1046-1051.

Naveh, Z. 1960. The ecology of chamise (Adenostoma fasciculatum) as effected by its toxic leachates. Bull. Ecol. Soc. Amer. 41: 56-57 (abstr.).

-----. 1967. Mediterranean ecosystems and vegetation types in California and Israel. Ecology 48:445-459.

-----. 1975. The evolutionary significance of fire in the Mediterranean region. Vegetatio 29:199-208.

Nobs, M. A. 1963. Experimental studies on species relationships in Ceanothus. Carnegie Inst. Wash. Pub. 623. 94 p.

Odum, E. P. 1971. Fundamentals of ecology. 3rd ed. Saunders, Philadelphia. 574 p.

Ornduff, R. 1974. An introduction to California plant life. Univ. Calif. Press, Berkeley. 152 p.

Osborn, J., J. Letey, L. F. DeBano, and E. Terry. 1967. Seed germination and establishment as affected by non-wettable soils and wetting agents. Ecology 48:494-497.

Parish, S. B. 1903. A sketch of the flora of southern California. Bot. Gaz. 36:203-222, 259-279.

-----. 1917. An enumeration of the pteridophytes and spermatophytes of the San Gabriel Mountains, California. Plant World 20:163-178, 208-223, 245-259.

Parsons, D. J., and A. R. Moldenke. 1975. Convergence in vegetation structure along analogous climatic gradients in California and Chile. Ecology 56:950-957.

Patric, J. H., and T. L. Hanes. 1964. Chaparral succession in a San Gabriel Mountain area of California. Ecology 45:353-360.

Pequegnat, W. E. 1951. The biota of the Santa Ana Mountains, J. Entomol. Zool. 42:1-84.

Pillsbury, A. F., J F. Osborn, and R. E. Pelishek. 1963. Residual soil moisture below the root zone in southern California watersheds. J. Geophys. Res. 68:1089-1091.

Plumb, T. R. 1961. Sprouting of chaparral by December after a wildfire in July. USDA Forest Serv., Pac. SW Forest and Range Exp. Sta., Tech. Paper PSW-57. Berkeley, Calif. 12 p.

-----. 1963. Delayed sprouting of scrub oak after a fire. USDA Forest Serv., Pac. SW Forest and Range Exp. Sta., Res. Note PSW-1. Berkeley, Calif. 4 p.

Plummer, F. G. 1911. Chaparral - studies in the dwarf forests, or elfin-wood, or southern California. USDA Forest Serv. Bull. 85. Washington, D.C. 54 p.

Poole, D. K., and P. C. Miller. 1975. Water relations of selected species of chaparral and coastal sage communities. Ecology 56:1118-1128.

Powerll, W. R. 1974. Inventory of rare and endangered vascular plants of California. Calif. Native Plant Soc. Spec. Pub. 1. Berkeley, Calif.

Purer, E. A. 1934. Foliar differences in eight dune and chaparral species. Ecology 15:197-203.

Quashu, H. K., and P. J. Zinke. 1964. Th einfluence of vegetation on soil thermal regime at the San Dimas lysimeters. Soil Sci. Soc. Amer. Proc. 28:703-706.

Quick, C. R. 1935. Notes on the germination of Ceanothus seeds. Madroño 3:135-140.

-----. 1944. Effects of snowbush on the growth of Sierra gooseberry. J. For. 42:827-832.

-----. 1947. Germination of Phacelia seeds. Madroño 9:17-20.

-----. 1959. Ceanothus seeds and seedlings on burns. Madroño 15:79-81.

Quick, C. R., and A. S. Quick. 1961. Germination of Ceanothus seeds. Madroño 16:23-30.

Raven, P. H. 1966. Native shrubs of southern California. Univ. Calif. Press, Berkeley. 132 p.

-----. 1967. The floristics of the California islands, pp. 57-67. In R. N. Philbrick (ed.). Proc. symp. on biology of the California islands. Santa Barbara Bot. Gard., Santa Barbara, Calif.

Raven, P. H., and H. J. Thompson. 1966. Flora of the Santa Monica Mountains, California. Univ. Calif., Los Angeles. 189 p.

Reimann, L. F. 1959a. Mountain temperatures - data from the San Dimas Experimental Forest. USDA Forest Serv., Pac. SW Forest and Range Exp. Sta., Misc. Paper PSW-36. Berkeley, Calif. 33 p.

Robinson, J. T. 1953. A toxonomic and ecological study of the flora of the San Gabriel River Canyon. M. A. thesis, Whittier Coll., Whittier, Calif. 47 p.

Rogers, B. 1949. Effects of fire on germination of seeds of Arctostaphylos viscida. M. S. thesis, Univ. Calif., Berkeley. 38 p.

Rosenthal, M. (ed.). 1974. Proc. symp. on living with the chaparral. Sierra Club, San Francisco. 225 p.

Rowe, P. B., and E. A. Colman. 1951. Disposition of rainfall in two mountain areas of California. USDA Tech. Bull. 1048.

Rowe, P. B., and L. F. Reimann. 1961. Water use by brush, grass and grass-forb vegetation. J. For. 59:175-181.

Sampson, A. W. 1944. Plant succession on burned chaparral lands in northern California. Univ. Calif. Agr. Exp. Sta. Bull. 685. Berkeley, Calif. 144 p.

Schimper, A. F. W. 1903. Plant geography on a physiological basis (translated by W. R.Fisher). Clarendon Press, Oxford. 839 p.

Schmid, R., T. E. Mallory, and J. M. Tucker. 1968. Biosystematic evidence for hybridization between Arctostaphylos nissenana and A. viscida. Brittonia 20:34-43.

Schultz, A. M., J. L. Launchbauch, and H. H. Biswell. 1955. Relationship between grass density and brush seedling survival. Ecology 36:226-238.

Searcy, K. B. 1969. Variation in Cercocarpus in southern California. New Phytol. 68:829-839.

Shantz, H. L. 1947. The use of fire as a tool in the management of the ranges in California. Calif. State Board of Forestry. 156 p.

Sharsmith, H. K. 1945. Flora of the Mount Hamilton Range of California (a taxonomic study and floristic analysis of the vascular plants). Amer. Midl. Natur. 34:289-367.

Show, S. B. and E. I. Kotok. 1924. The role of fire in the California pine forests. USDA Bull. 1294. 80 p.

Shreve, F. 1927a. The vegetation of a coastal mountain range. Ecology 8:37-40.

-----. 1927b. The physical conditions of a coastal mountain range. Ecology 8:398-411.

-----. 1936. The transition from desert to chaparral in Baja California. Madroño 3:257-264.

Sinclair, J. D. 1953. Erosion in the San Gabriel Mountains of California. Trans. Amer. Geophys. Union 35:264-268.

Specht, R. L. 1969a. A comparison of the sclerophyllous vegetation characteristics of mediterranean type climates in France, California and Southern Australia. I. Structure, morphology, and succession. Austral. J. Bot. 17:277-292.

-----. 1969b. A comparison of the sclerophyllous vegetation characteristics of mediterranean type climates in France, California, and Southern Australia. II. Dry matter, energy, and nutrient accumulation. Austral. J. Bot. 17:293-308.

Stebbins, G. L., and J. Major. 1965. Endemism and speciation in the California flora. Ecol. Monogr. 35:1-35.

Sterling, E. A. 1904. Chaparral in northern California. For. Quart. 2:209-214.

Stocking, S. K. 1966. Influences of fire and sodium-calcium borate on chaparral vegetation. Madroño 18:193-203.

Stone, E. C. 1951. The stimulative effect of fire on the flowering of the golden brodiaea (Brodiaea ixioides Wats. var. lugens Jepson). Ecology 32:534-537.

-----. 1957. Dew as an ecological factor. II. The effect of artificial dew on the survival of Pinus ponderosa and associated species. Ecology 38:414-422.

Stone, E. C., and G. Juhren. 1951. The effect of fire on the germination of the seeds of Rhus ovata Wats. Amer. J. Bot. 38:368-372.

-----. 1953. Fire-stimulated germination. Calif. Agric. 7:13-14.

Sweeney, J. R. 1956. Responses of vegetation to fire: a study of the herbaceous vegetation following chaparral fires. Univ. Calif. Pub. Bot. 28:143-150.

-----. 1968. Ecology of some "fire type" vegetations in northern California, pp. 111-125. In Proc. Tall Timbers fire ecology conf. Tall Timbers Res. Sta., Tallahassee, Fla.

Syvertsen, J. P. 1974. Relative stem water potentials of three woody perennials in a southern oak woodland community. Bull. South. Calif. Acad. Sci. 73:108-113.

Thomas, J. H. 1961. Flora of the Santa Cruz Mountains of California. Stanford Univ. Press, Stanford, Calif. 434 p.

Thorne, R. F. 1969. A supplement to the floras of Santa Catalina and San Clemente Islands, Los Angeles County, California. Aliso 7:73-83.

Twisselmann, E. C. 1956. Flora of the Temblor Range. Wasmann J. Bot. 14:161-300.

-----. 1967. A flora of Kern County, California. Wasmann J.Biol. 25: 1-395.

Van Renssalaer, M., and H. E. McMinn. 1942. Ceanothus. Santa Barbara Bot. Gard., Santa Barbara, Calif. 308 p.

Vlamis, J., A. M. Schultz, and H. H. Biswell. 1958. Nitrogen-fixation by deerbrush. Calif. Agric. 12:12.

-----. 1959. Nutrient response of ponderosa pine and brush seedlings on forest and brush soils of California. Hilgardia 28:239-254.

-----. 1964. Nitrogen fixation by root nodules of western mountain mahogany. J. Range Manage. 17:73-74.

Vlamis, J., E. C. Stone, and C. L. Young. 1954. Nutrient status of brushland soils in southern California. Soil Sci. 78:51-55.

Vogl, R. J. 1967. Wood rat densities in southern California manzanita chaparral. South. Natur. 12:176-179.

-----. 1968. Fire adaptations of some southern California plants, pp. 79-110. In Proc. Tall Timbers fire ecology conf. Tall Timbers Res. Sta., Tallahassee, Fla.

-----. 1970. Fire and plant succession, pp. 65-75. In Symp. on the role of fire in the intermountain west. Intermountain Fire Council, Missoula, Mont.

Vogl, R. J., and B. C. Miller. 1968. The vegetational composition of the south slope of Mt. Pinos, California. Madroño 19:225-288.

Vogl, R. J., and P. K. Schorr. 1972. Fire and manzanita chaparral in the San Jacinto Mountains, California. Ecology 53:1179-1188.

Waisel, Y., N. Lipschitz, and Z. Kuller. 1972. Patterns of water movement in trees and shrubs. Ecology 53:520-523.

Walker, R. B. 1948. A study of serpentine soil infertility with special reference to edaphic endemism. Ph.D. dissertation, Univ. Calif., Berkeley.

-----. 1954. The ecology of serpentine soils. II. Factors affectint plant growth on serpentine soils. Ecology 35:259-266.

Walker, R. B., H. M. Walker, and P. R. Ashworth. 1955. Calcium-magnesium nutrition with special reference to serpentine soils. Plant Physiol. 30:214-221.

Warming, E. 1909. Oecology of plants (translated by P. Groom and T. T. Balfour). Clarendon Press, Oxford. 422 p.

Weaver, J. E. and F. E. Clements. 1929. Plant ecology. McGraw-Hill, New York. 601 p.

Wells, P. V. 1962. Vegetation in relation to geological substratum and fire in the San Luis Obispo Quadrangle, California. Ecol. Monogr. 32:79-103.

Went, F. W., G. Juhren, and M. C. Juhren. 1952. Fire and biotic factors affecting germination. Ecology 33:351-364.

Whitefield, C. J. 1932a. Osmotic concentration of chaparral, coastal sagebrush and dune species of southern California. Ecology 13: 279-285.

-----. 1932b. Ecological aspects of transpiration. II. Pikes Peak and Santa Barbara regions: edaphic and climatic aspects. Bot. Gaz. 94:183-196.

Whittaker, R. H. 1954a. The ecology of serpentine soils. I. Introduction. Ecology 35:258-259.

-----. 1954b. The ecology of serpentine soils. IV. The vegetational response to serpentine soils. Ecology 35:275-288.

-----. 1960. Vegetation of the Siskiyou Mountains, Oregon and California. Ecol. Monogr. 30:279-338.

Wieslander, A. E., and B. O. Schrieber. 1939. Notes on the genus Arctostaphylos. Madroño 5:38-47.

Wieslander, A. E., and C. H. Gleason. 1954. Major brushland areas of the coast ranges and Sierra-Cascade foothills of California. USDA Forest Serv., Pac. SW Forest and Range Exp. Sta., Misc. Paper PSW-15. Berkeley, Calif. 8 p.

Wilken, G. C. 1967. History and fire record of a timberland brush field in the Sierra Nevada of California. Ecology 48:302-304.

Wilson, R. C., and R. J. Vogl. 1965. Manzanita chaparral in the Santa Ana Mountains, California. Madroño 18:47-62.

Wirtz, W. O. 1974. Chaparral wildlife and fire ecology, pp. 7-18. In Symp. on living with the chaparral. Sierra Club, San Francisco.

Wright, E. 1931. The effect of high temperature on seed germination. J. For. 29:679-687.

Wright, R. D. 1966. Lower elevational limits of montane trees. I. Vegetational and environmental survey in the San Bernardino Mountains of California. Bot. Gaz. 127:184-193.

Young, D. A. 1972. The reproductive biology of Rhus integrifolia and Rhus ovata (Anacardiaceae). Evolution 26:406-414.

-----. 1974. Introgressive hybridization in two southern California species of Rhus (Anacardiaceae). Brittonia 26:241-255.

Zavitkovski, J., and M.Newton. 1968. Ecological importance of snowbrush, Ceanothus velutinus, in the Oregon Cascades. Ecology 49: 1134-1145.

Zivnuska, J. A. 1968. Some thoughts on the role of fire in California, pp. 1-3. In Proc. Tall Timbers fire ecology conf. Tall Timbers Res. Sta., Tallahassee, Fla.

CHAPTER

13

SOUTHERN COASTAL SCRUB

HAROLD A. MOONEY
Department of Biological Sciences, Stanford University, Stanford, California

Introduction 472
Distribution and composition 472
  General 472
  Local patterning 476
  The sage–grassland ecotone 478
Autecology 480
  Shrub structure 480
  Phenology 481
  Carbon gain and water balance 482
  Nutrient content 486
  Terpenes 486
Summary 487
Acknowledgments 487
Literature cited 488

INTRODUCTION

Even though coastal scrub is extensively distributed throughout California, and many universities and colleges are located within this type, remarkably little information is available on the characteristics and dynamics of this distinctive vegetation. This is particularly unfortunate, since many coastal scrub stands are disappearing because of the activities of man.

Coastal scrub consists of three phases: northern coastal scrub, coastal sage scrub, and coastal sage succulent scrub. Of these three phases, most is known about coastal sage scrub, and this phase serves as the central focus of this discussion. Northern coastal scrub is treated in more detail in Chapter 21.

In order to interpret the adaptive characteristics of the components of coastal scrub, comparisons will be made where possible with vegetation tyoes occupying slightly different environments, principally chaparral.

DISTRIBUTION AND COMPOSITION

General

Munz and Keck (1959) recognized a northern coastal scrub extending, often interrupted, along a narrow coastal strip from southern Oregon to Pt. Sur. Characteristic species of this type are *Baccharis pilularis, Mimulus aurantiacus, Castilleja latifolia, Rubus vitifolius, Lupinus variicolor, Heracleum lanatum, Eriophyllum staechadifolium, Gaultheria shallon, Anaphalis margaritacea, Artemisia suksdorfii,* and *Erigeron glaucus.*

To the south and continuing to Baja California, also principally along the coast and at elevations lower than those for chaparral, Munz and Keck recognized coastal sage scrub, characterized by *Artemisia californica, Salvia apiana, S. mellifera, S. leucophylla, Eriogonum fasciculatum, Rhus integrifolia, Encelia californica, Horkelia cuneata, Haplopappus squarrosus, H. venetus,* and *Eriophyllum confertiflorum* (Fig. 13-1). Chapter 7 pointed out the inclusion of some dune scrub dominants within this vegetation type as recognized by Munz and Keck.

Figure 13-1. Coastal sage scrub at Point Mugu, Ventura Co. A, aspect of hillsides; B, detail of above, showing *Artemisia californica*, *Salvia mellifera*, and *Encelia californica* as dominants.

To the south of San Diego, extending to El Rosario, again at elevations below those for chaparral, is another distinctive coastal scrub type, which can be termed coastal sage succulent scrub (Fig. 13-2). In addition to some of the above species, such as *Artemisia californica, Eriogonum fasciculatum, Rhus integrifolia, Encelia californica,* and *Eriophyllum confertiflorum,* it contains such distinctive and characteristic species as *Aesculus parryi, Adolphia californica, Bergerocactus emoryi, Agave shawii, Rosa minutifolia, Viguiera laciniata,* and *Salvia munzii.* This vegetation has been studied to only a limited degree (Shreve 1936; Epling and Lewis 1942; Mooney and Harrison 1972).

Limited quantitative analyses of coastal sage scrub and coastal sage succulent scrub have been made. Table 13-1 presents data for stands at 100 and 400 m in the San Pedro Mártir and near sea level at Camp Pendleton, California. The low-elevation San Pedro Mártir stand is rich in succulents and in desert-related elements such as *Simmondsia* and *Franseria.* Growth form diversity is high and plant cover low. At the somewhat moister locality higher in the San Pedro Mártir, the vegetation grows denser, loses the succulents, and becomes more similar in physiognomy and composition to the coastal sage scrub of southern California. Here *Salvia munzii,* a vicariad of the more northerly *S. mellifera,* is codominant with the relatively narrowly distributed *Viguiera laciniata* as well as *Lotus scoparius.* The Camp Pendleton stand is more or less characteristic of the vegetation of much of the coastal regions of southern California. Here total cover is high. *Salvia mellifera* and *Artemisia californica* are the dominant drought-deciduous species, along with the evergreen *Rhus laurina* and *Eriogonum fasciculatum.* At Camp Pendleton, in contrast to the more southerly stands, evergreens make up a proportionately greater part of the cover. This trend toward increasing evergreenness with decreasing aridity would no doubt be even more pronounced if data were available for northern coastal scrub,

Figure 13-2. Coastal sage succulent scrub north of Ensenada, Baja California.

TABLE 13-1. Percent cover and leaf type of perennial plants encountered in transects of coastal sage succulent scrub (San Pedro Mártir, two elevations) and coastal sage scrub (Camp Pendleton). From Mooney and Harrison (1972) and Keeley (unpublished data). Numbers refer to percentage plant cover; P notes presence of species. Letters in parentheses refer to plant leaf types: E = evergreen; D = drought-deciduous; and S = stem succulent

	Coastal Sage Succulent Scrub		
	100 m	400 m	Coastal Sage Scrub
Agave shawii	7.90 (S)		
Machaerocereus gummosus	4.00 (S)		
Echinocereus maritimus	0.25 (S)		
Mammillaria dioica	0.33 (S)		
Bergerocactus emoryi	1.16 (S)		
Dudleya ingens	0.83 (S)		
Myrtillocactus cochal	P (S)		
Opuntia rosarica	P (S)		
Franseria chenopodifolia	17.63 (D)		
Euphorbia misera	0.41 (D)		
Harfordia macroptera	1.91 (D)		
Lycium californicum	P (D)		
Galvezia juncea	P (D)		
Rhus integrifolia	P (E)		5.13 (E)
Rosa minutifolia	15.73 (D)	2.91 (D)	
Viguiera laciniata	P (D)	15.46 (D)	
Simmondsia chinensis	7.25 (E)	P (E)	
Eriogonum fasciculatum	0.83 (D)	8.95 (D)	7.15 (E)
Ephedra californica	P (E)	1.45 (E)	
Rhus laurina	P (E)	P (E)	16.58 (E)
Acalypha californica		0.16 (D)	
Eriogonum sp.		0.21 (D)	
Artemisia californica		2.49 (D)	13.31 (D)
Encelia californica		P (D)	
Aesculus parryi		P (D)	
Salvia munzii		15.83 (D)	
Lotus scoparius		27.69 (D)	
Cneoridium dumosum		P (E)	2.51 (E)

TABLE 13-1 [continued]

	Coastal Sage Succulent Scrub 100 m	Coastal Sage Succulent Scrub 400 m	Coastal Sage Scrub
Opuntia occidentalis		0.03 (S)	
Salvia mellifera		53.43 (D)	
Quercus dumosa		0.94 (E)	
Yucca whipplei		0.31 (E)	
Galium nuttallii		0.08	
Dudleya farinosa		0.03 (S)	
Total plant cover (%)	58.23	75.15	99.50
Relative cover (%) by plant leaf types:			
Stem succulents	24.8	0	0.06
Drought-deciduous	62.7	98.1	67.1
Evergreen	12.5	1.9	32.8

where the evergreen *Baccharis pilularis* predominates. The carbon balance implications of these trends are discussed in a subsequent section.

Local Patterning

Harrison et al. (1971) discussed the patterning of coastal sage scrub in respect to chaparral on a statewide as well as a local basis. They found that coastal sage always occurred on sites with less seasonal moisture availability, because of either lower rainfall or such substrate or habitat characteristics as finely textured soils or slope face. A common pattern in southern California coastal mountains is a predominance of coastal sage on the lower slopes of the mountains facing the ocean, interrupted by chaparral on the higher, more mesic slopes, and then a reoccurrence of sage on the rain shadow lower slopes of the mountain interior (Fig. 13-3). The interior stands of sage may differ in composition from the coastal stands. In particular, *Salvia apiana* may replace *S. mellifera* in more interior sites.

Similar patterning due to substrate mosaics can be seen in certain regions where stands of coastal sage on shale are embedded in a matrix of chaparral on sandstone soils. There are many instances, however, where the presence of coastal sage cannot be so simply related to habitat aridity. This is due to the fact that coastal sage scrub not only is "preclimax" to chaparral but may also be successional to it (Cooper 1922). Thus it will temporarily occupy disturbed sites. The principal woody sage species crown-sprout after fire, as do most of the chaparral species. Because of the

Figure 13-3. Distribution of the coastal sage (black) and chaparral (gray) vegetation in the Point Dume region of the Santa Monica Mts., according to the U.S. Forest Survey of 1930–34. The sage is limited to the lower elevations both on the coastal (lower) and interior (upper) regions of the mountains. Many of the areas in white were agricultural in 1930–34 and probably represented an even greater extent of sage. Suburban development has subsequently occupied much of this agricultural area.

rapid regrowth of sage species and their small, wind-dispersed seeds, they are often fire successional to chaparral (Wells 1962).

The complex relationships that can exist between community type, substrate, and disturbance history, particularly fire, have been discussed in detail by Wells (1962).

In other situations, which deserve more study, islands of coastal sage occur within the chaparral where there are no obvious patterns of disturbance, substrate, or slope change (Bradbury 1974). There is documentation that these islands have persisted in precisely their same positions for over 40 yr (Fig. 13-4).

Because of the successional nature of coastal sage elements, they are generally increasing in abundance on southern California landscapes as a result of the increased activities of man (Bradbury 1974). At the same time, the potential "climax" habitats are disappearing, since they generally occur on the lowest slopes of the coastal mountains in the most favorable building sites.

The Sage–Grassland Ecotone

Coastal sage scrub often makes direct contact with the annual grassland, or in many cases islands of sage may be embedded in a grassland matrix. The ecotone between these physiognomically distinctive vegetation types has been of considerable interest

Figure 13-4. Sage and chaparral patterning on the Banner Grade of San Diego Co. The top photo was taken in 1931, and the bottom one in 1972. From Bradbury (1974).

to ecologists. The characteristic pattern of the ecotone was first described by C. Muller et al. (1964). They noted that between the scrub and the grassland there was a transition zone nearly 1 m wide with no vegetation (the "bare zone"), and then toward the grassland a further zone of stunted herbs extended up to 9 m. In this initial description they attributed this patterning to the inhibitory effects of volatile terpenes from the sage species *Salvia leucophylla, S. apiana,* and *Artemisia californica.* Furthermore, they reported some success in cold-trapping atmospheric terpenes which inhibited germination, and hence they proposed dew as a principal mode of transfer of the volatiles to the zone of inhibition.

C. Muller, W. Muller, and their collaborators subsequently reported a number of studies that identified the inhibitory compounds, traced their probable routes of environmental transfer, and determined their modes of plant inhibition. Muller and Muller (1964) found six terpenes in *S. mellifera, S. leucophylla,* and *S. apiana,* of which cineole and camphor were the most abundant and also the most toxic as determined by bioassay. C. Muller (1965) further identified camphor and cineole from atmosphere collections within, and as far as 30 m from, *S. leucophylla* and *S. mellifera* shrubs. However, because of the low solubility of terpenes in water, he proposed a direct transfer of the volatiles from the sage to cuticular lipids of germinating seedlings, thus effecting toxicity without a dew transfer. This hypothesis was based on the high solubility of terpenes in paraffin.

Still later, C. Muller and del Moral (1966) proposed yet another transfer hypothesis, which was based on the accumulative adsorption of terpenes on soil during periods of high volatilization from the shrubs when they are in full leaf and temperatures are high (spring and summer). By this hypothesis, subsequent inhibition of the annual herbs occurred during germination on these charged soils during the fall rains. In this paper, they first suggested that, although *Salvia* terpenes are mandatory in producing the ecotone patterning, small animal activity, soil type, and microclimate may also be significant contributors.

Further evidence that volatiles were primarily involved in the patterning was given by C. Muller in 1966. He concluded that the edaphic factor was not important, since shrub roots did not extend into the zone of inhibition, ruling out competition for water as a possible cause. Furthermore, no physical or mineral soil differences in adjacent zones could be found. Apparently, cattle manure deposits did not alter the "bare zone" phenomenon, although they did enhance growth in the grassland. Since inhibition zones were noted uphill from the shrub contact, Muller concluded that volatile rather than soluble toxins were involved. He also noted that animal grazing, although occurring with greater preference near the shrubs, was rarely responsible for seedling mortality. Thus he suggested that grazing could augment the pattern but could not initiate or maintain it.

W. Muller detailed the mode of action of the volatiles through a series of papers. In 1965 he found that the inhibitory effect of volatiles from *Salvia leucophylla* was greatest during germination of assay plants. Both cell division and elongation were adversely affected. Subsequently, he and others (W. Muller et al. 1968, 1969) found that cineole, one of the *S. leucophylla* terpenes, inhibited respiration and root growth of herb seedlings. They proposed that such inhibited seedlings would then be susceptible to drought mortality.

The hypothesis that volatiles play the primary role in the maintenance of the "bare zone" has been questioned by several workers (Wells 1964; Bartholomew

1970, 1972; Halligan 1973, 1974). Bartholomew (1970) presented evidence of concentrated vertebrate feeding in the bare zones of *S. leucophylla* and *Baccharis pilularis* and showed by the use of exclosures that the "bare zone" phenomenon disappeared, that is, herbs grew. Halligan (1974) obtained similar results working with *Artemisia californica*. Both Bartholomew (1972) and Halligan (1974) indicated the complex composition of the ecotone of the shrub species they studied with the grassland. Both noted that along the shrub perimeter and within the "bare zone" there was a distinctive flora composed of apparently unpalatable herbs such as *Navarretia, Chorizanthe, Croton, Satureja* (Halligan 1974), *Centaurea,* and *Anagallis* (Bartholomew 1972). The ecotone may be even more complex, since Muller (1966) indicated that shrub seedlings of *Artemisia californica* also become established in the "bare zone" and area of inhibition.

To answer the criticism by Muller and del Moral (1971) that small enclosures alter the microclimate so that herbs can successfully grow within them in the "bare zone," Bartholomew (1972) designed large U-shaped fences which abutted directly on *A. californica* and *Baccharis pilularis* shrub–grassland ecotones. They had a 3 m long shrub vegetation contact with parallel 3 m long arms extending out into the grassland. The rationale for the design was that grazers would either have to go over the 0.6 m high mesh fences or travel out into the grassland and back into the "U" to graze next to the shrubs—a potentially highly precarious trip away from the protective cover of the shrubs. This design would discriminate particularly against grazing by small mammals. The "bare zone" that previously existed next to the shrubs was eliminated after a growing season.

From all of these studies it is clear that there is a unique ecotone between the coastal sage scrub and the grassland. All workers are in agreement that the causes of the "bare zone" are complex and at least involve interactions between climate, plant secondary chemicals, and vertebrates. It may be that the exact characteristics of the ecotone are quite dependent on the sage and vertebrate species locally predominating.

AUTECOLOGY

To understand the basis for the distribution of coastal sage scrub, it is necessary first to understand the ecology of the component species and to relate this not only to the environment but also to the environmental responses of their competitors.

Shrub Structure

The average physical characteristics of several coastal sage species contrast with those of chaparral shrubs (Table 13-2). The sage species have somewhat smaller volumes and considerably lower biomass densities. Even though the densities of wood and leaves (g shrub biomass per volume) are lower in the sage species, the proportions of leaves versus stems are similar.

The leaves of the sage species have an average lower specific weight (mg dry wt cm^{-2}) than the chaparral shrubs. The sage species, however, have a leaf life span of less than 1 yr, whereas leaves on the chaparral species may last 2–3 yr.

The sage species have a leaf area index (m^2 of total shrub leaf single surface per

TABLE 13-2. Mean structural characteristics of chaparral and coastal sage shrubs. Chaparral averages include data for five representatives of each of eight species: Rhus ovata, Ceanothus leucodermis, Heteromeles arbutifolia, Arctostaphylos glauca, Adenostoma fasciculatum, Ceanothus greggii, Quercus dumosa, and Q. agrifolia (shrub size). The shrubs were on a site in San Diego Co. burned 23 yr previously. Coastal sage data are based on five individuals each of Artemisia californica, Salvia mellifera, and Encelia californica harvested in coastal San Diego Co. Unpublished data from Kummerow, Mooney, and Giliberto

Characteristic	Chaparral	Coastal Sage
Height (m)	1.67	1.12
Diameter (m)	1.35	1.11
Projected area (m^2)	1.51	1.01
Total shoot weight (g)	4384.8	802.6
Total leaf weight (g)	787.2	135.5
Stem weight ($g\ m^{-2}$)	2332.4	569.6
Leaf weight ($g\ m^{-2}$)	503.4	113.3
Shoot weight ($g\ m^{-2}$)	2835.8	682.9
Percent stems	82.0	82.8
Percent leaves	18.0	17.2
Leaf area index ($m^2\ m^{-2}$)	2.65	1.31
Specific leaf weight ($mg\ cm^{-2}$)	19.4	8.3

m^2 of maximum shrub ground surface projection) only about one-half that of chaparral shrubs.

The root systems of several coastal sage species were examined by Hellmers et al. (1955) and compared with those of chaparral shrubs. Sage species, on the average, had roots that penetrated, at the maximum, only half as deeply as those of chaparral shrubs.

Phenology

The phenology of coastal sage species differs substantially from that of chaparral shrubs. The evergreen shrubs of chaparral produce new stem growth principally during the spring (Mooney et al. 1974). In contrast, the sage species initiate new stem growth soon after the commencement of the fall rains, during the coldest parts of the year. This is illustrated by the development of plants at Camp Pendleton during 1973–74 (Fig. 13-5). There the sage species *Artemisia californica, Salvia mellifera, Eriogonum fasciculatum,* and *Mimulus puniceus* all initiated new stem growth in November and December after the first rains of the season in October. In contrast, the evergreen chaparral type elements, such as *Heteromeles arbutifolia* and *Rhus integrifolia*, did not start growth until late March. The fact that the anomalous *Rhus laurina* has active stem growth year round may in part explain its noted frost sensitivity.

At the Camp Pendleton sage scrub site, at least one shrub species is flowering at any given time of the year. As has been shown for chaparral, the reproductive period of the community is more extensive than the vegetative growth period (Mooney et al. 1974).

The difference in canopy growth period between the sage subshrubs and the chaparral evergreen shrubs is no doubt due in part to differences in their root systems. Chaparral shrubs generally tap deeper soil water reserves and thus do not start growth until fall rains have penetrated to some depth. This was noted by Harvey and Mooney (1964) during the severe drought year of 1960–61. They found that, although the small amount of precipitation that fell was sufficient to initiate growth in the shallow-rooted sage species *Salvia apiana,* none of the chaparral shrubs at the same site produced stem growth that year.

Carbon Gain and Water Balance

The relationships between carbon-gaining capacity and water balance of evergreen chaparral shrubs and drought-deciduous coastal sage species have been discussed from several viewpoints by Mooney and Dunn (1970), Harrison et al. (1971), and Miller and Mooney (1974). The essence of these discussions is that the evergreen species are adapted to withstand the annual drought period, whereas the sage species evade it. In comparison to the drought-deciduous sage species, chaparral evergreens have lower photosynthetic rates and higher cuticular and stomatal resistances to water transfer. Thus the evergreen species have a long period of low gas exchange, and the sage species have a short period of very high gas exchange activity.

This is shown, in part, for the co-occurring shrubs *Heteromeles arbutifolia* and *Salvia mellifera,* an evergreen chaparral and a drought-deciduous sage species, respectively (Table 13-3). During the periods of lowest water stress in the winter, the sage species had photosynthetic rates about twice the value for the evergreen shrub. At this time, leaf resistances to water transfer were less than half that of the evergreen. During the height of the drought, the sage species had lost most of its leaves, and the few terminal ones left did not even have a positive photosynthetic rate. These shallow-rooted plants were under severe water stress with midday xylem water potentials of −64 bar. During the same time, the deeper-rooted *Heteromeles,* although in full leaf, was under less water stress and, furthermore, had photosynthetic rates reduced to only one-half those found during the optimal season. Not indicated in Table 13-3, however, is the fact that these relatively high drought photosynthetic rates of *Heteromeles* are maintained only in the morning. By midday, stomata close for the remainder of the day (Mooney et al. 1975).

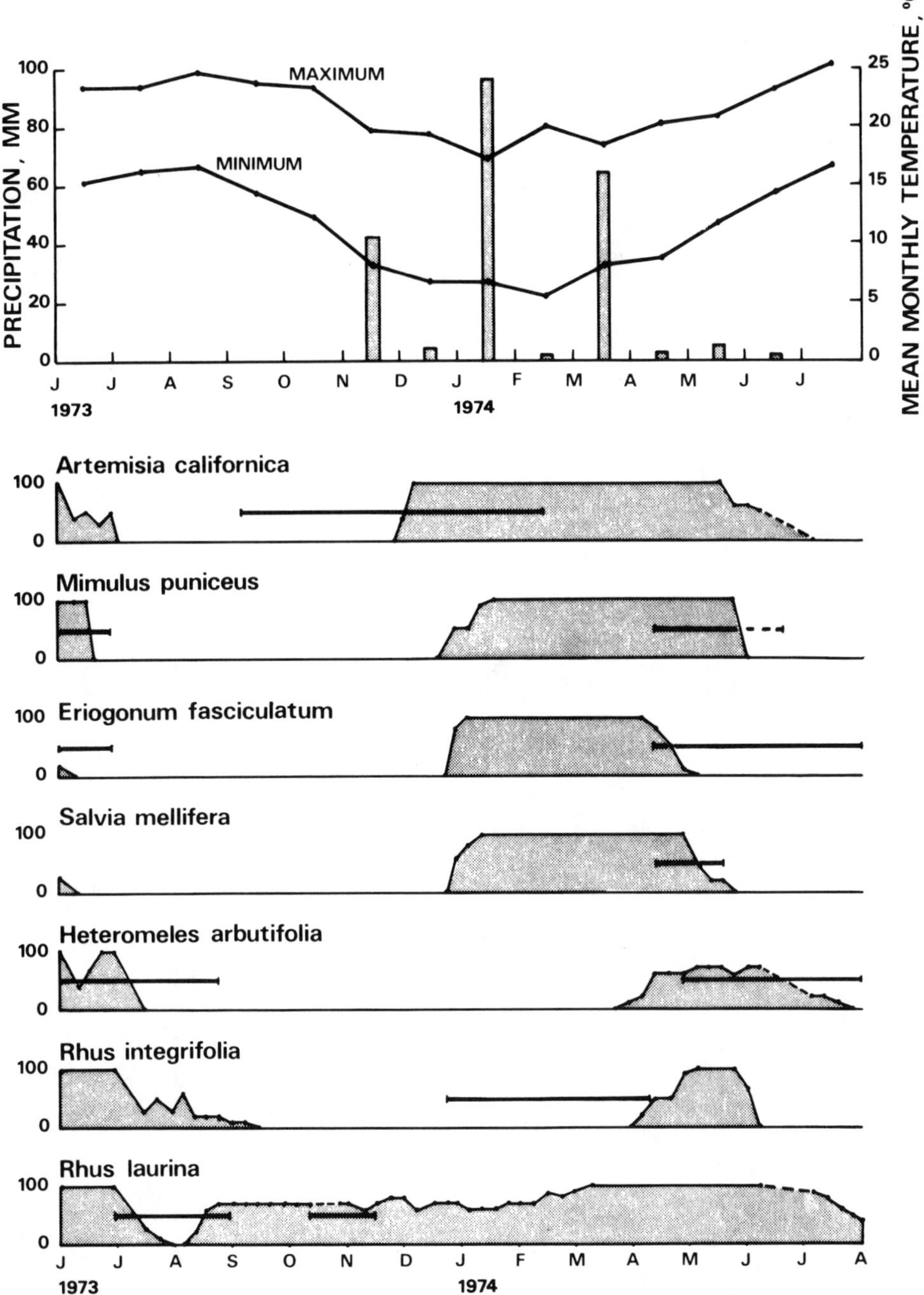

Figure 13-5. Phenological development of plants in a coastal sage community at Camp Pendleton, California. The gray areas represent the percentage of 10 plants that were elongating stems on a given date. The bars represent the period of flowering. The climate data are from nearby Oceanside.

TABLE 13-3. Seasonal changes in the maximum observed field photosynthetic rates of co-occurring evergreen chaparral (*Heteromeles arbutifolia*) and drought-deciduous sage species (*Salvia mellifera*) at Mira Mar Mesa, San Diego Co. Unpublished data from Mooney, Harrison, and Morrow

	Heteromeles arbutifolia			*Salvia mellifera*	
Date	Midday Xylem Water Potential (bar)	Maximum Net Photosynthetic Rate (mg CO_2 dm^{-2} hr^{-1})	Date	Midday Xylem Water Potential (bar)	Maximum Net Photosynthetic Rate (mg CO_2 dm^{-2} hr^{-1})
Feb. 8. 1970	-19	7.8	Feb. 10	-19	19.7
June 7, 1970	-23	8.4	June 9	-54	4.0
Aug. 7, 1970	-34	7.8	Aug. 18	-64	-0.2
Jan. 25, 1971	-22	13.4	Jan. 26	-12	23.0

TABLE 13-4. Leaf and photosynthetic characteristics of the dominants of plants of the chaparral, coastal sage scrub, and coastal sage succulent scrub. Data from Mooney et al. (1974)

	Echo Valley, San Diego Co.	Camp Pendleton, San Diego Co.	San Telmo, Baja California
Vegetation type	Chaparral	Coastal sage scrub	Coastal sage succulent scrub
Latitude	$32^{o}50'$	$33^{o}15'$	31°
Estimated annual precipitation (mm)	450	200	160
Relative cover (%) by leaf type:			
Evergreen	98.58	32.78	12.45
Drought-deciduous	1.41	67.08	62.70
Stem chlorophyllous	0.00	0.00	0.00
Succulent	0.00	0.06	24.85
Unclassified	0.01	0.08	0.00
Relative cover (%) by photosynthetic type:			
C_3	99.31	99.55	75.17
C_4	0.00	0.00	0.00
CAM	0.39	0.37	24.85
Unclassified	0.30	0.08	0.00

It has been shown by a cost (leaf production)/benefit (carbon gain) model that the evergreen species are favored in habitats of shorter drought duration than those characteristic of the sage species (Miller and Mooney 1974). This corresponds to their respective climatic distribution centers (Table 13-4). As the habitat becomes drier, evergreens become less abundant, and drought-deciduous species increase. In all cases, though, plants with the C_3 pathway predominate. In the driest sites, as the communities become more open and desert-like, the evergreens become even less important, and succulents, which have very low photosynthetic rates but the capacity to fix carbon during periods of low evaporative demand, become prevalent, along with the drought avoiders. In moister, and hence more closed, habitats, the slow-growing succulents are evidently noncompetitive. Thus the arrangement of the principal growth forms (evergreen shrubs, drought-deciduous shrubs, and succulents) along an aridity gradient is related to their gas exchange characteristics.

Nutrient Content

Little information is available on the nutrient balance of sage communities; however, there are indications that member species have high leaf contents of nitrogen and phosphorus in comparison to chaparral plants. Mature chaparral leaves ($n = 8$ species) averaged about 1% N and only 0.06% P, whereas averages for leaves of *Salvia mellifera, Encelia californica,* and *Artemisia californica* were 3.1% N and 0.25% P, according to an unpublished study by Mooney and Chu. High leaf nitrogen content (hence potentially high content of the carboxylating enzyme) in sage species may explain the fact that their capacity to fix carbon is higher than that of chaparral shrubs. Faster turnover time, lower biomass, and lower tissue density could all contribute to the high nutrient of the sage species in comparison to the chaparral shrubs. Furthermore, the shallow-rooted sage species "explore" a more nutrient-rich soil than do the chaparral shrubs.

Terpenes

One of the most distinctive features of many of the shrubs of coastal sage scrub is their highly aromatic nature due to the presence of monoterpenes. In species of *Salvia,* at least, the terpenes are produced in glandular leaf trichomes (Tyson et al. 1974) and are passively volatilized from the leaf at a rate in direct proportion to temperature.

The fact that these compounds can be present in relatively high concentrations in the leaves (3.5% leaf dry wt in *S. mellifera*) and are relatively costly to produce (5.2 g of CO_2 to produce 1 g of camphor: Tyson et al. 1974) would indicate an adaptive function. This view is further supported by the fact that in a comparable climate type in the Mediterranean region the garigue vegetation is constituted of a number of taxa such as *Rosmarinus, Thymus, Salvia,* and *Teucrium,* which are also distinctively terpenaceous.

The adaptive role of terpenes has been examined in varying degrees. As discussed earlier, the role of terpenes in allelopathy has been studied intensively, and in fact serves as the classic example of the phenomenon in textbooks. Little work, however, has centered on the possible role of terpenes as antiherbivore substances or in leaf–water relationships (Wellburn et al. 1974), both promising lines of research.

SUMMARY

Coastal scrub vegetation is restricted to coastal plateaus and the lower slopes of the coastal ranges of California. It changes in character from north (northern coastal scrub) to south (coastal sage scrub and coastal sage succulent scrub), with the principal trend being a decrease in evergreenness and a progressive increase in drought-deciduous and succulent species. In comparison to chaparral, the lower-growing, often more open coastal sage scrub occupies drier sites and is composed of dominants whose principal adaptive mode is exploitation of soil moisture in upper soil horizons during the cool winter season. Most sage dominants are winter active and avoid the summer drought by shedding their leaves. They are competitive with chaparral species only where drought is of sufficient length to make evergreenness a carbon balance liability.

The drought-avoiding features of the sage species, their fast growth rate, low investment in carbon per volume biomass, and lightweight seeds contribute to further their adoption of a seral role to chaparral species within habitats that support a chaparral climax.

Virtually no quantitative studies have been made of coastal sage scrub. This is especially unfortunate, because it often occupies choice development sites and is being destroyed over large areas of the state.

ACKNOWLEDGMENTS

Support of the National Science Foundation (GB 27151) in collecting certain data presented in this chapter is gratefully acknowledged. David Bradbury, Michael Barbour, C. H. Muller, and Richard Vogl provided helpful comments on a preliminary draft.

LITERATURE CITED

Bartholomew, B. 1970. Bare zone between California shrub and grassland communities, the role of animals. Science 170:1210-1212.

---. 1972. The role of animals in the interaction between California shrub and grassland communities. Ph.D. dissertation, Stanford Univ., Stanford, Calif. 65 p.

Bradbury, D. 1974. Vegetation history of the Ramona Quadrangle, San Diego County, California. Ph.D. dissertation, Univ. Calif., Los Angeles. 201 p.

Cooper, W. S. 1922. The broad-sclerophyll vegetation of California. Carnegie Inst. Wash. Pub. 319. 124 p.

Dement, W., B. Tyson, and H. Mooney. 1975. Mechanism of monterpene volatilization in Salvia mellifera. Phytochemistry 14 (in press).

Epling, C., and H. Lewis. 1942. The centers of distribution of the chaparral and coastal sage associations. Amer. Midl. Natur. 27:445-462.

Halligan, J. 1973. Bare areas associated with shrub stands in grassland: the case of Artemisia californica. BioScience 23:429-432.

---. 1974. Relationship between animal activity and bare areas associated with California sagebrush in annual grassland. J. Range Manage. 27:358-362.

---. 1976. Toxicity of Artemisia californica to four associated herb species. Amer. Midl. Natur. 95:406-421.

Harrison, A., E. Small, and H. Mooney. 1971. Drought relationships and distribution of two Mediterranean climate Californian plant communities. Ecology 52:869-875.

Harvey, R., and H. Mooney. 1964. Extended dormancy of chaparral shrubs during severe drought. Madroño 17:161-165.

Hellmers, H., J. Horton, G. Juhren, and J. O'Keefe. 1955. Root systems of some chaparral plants in southern California. Ecology 36:667-678.

Miller, P., and H. Mooney. 1974. The origin and structure of American arid-zone ecosystems. The producers: interactions between environment, form, and function, pp. 201-209. In Proc. First Internat. Congr. on Ecology. The Hague, Netherlands.

Mooney, H., and E. Dunn. 1970. Photosynthetic systems of mediterranean climate shrubs and trees of California and Chile. Amer. Natur. 104:447-453.

Mooney, H., and A. Harrison. 1972. The vegetational gradient on the lower slopes of the Sierra Pedro Martir in northwest Baja California. Madroño 21:439-445.

Mooney, H., A. T. Harrison, and P. Morrow. 1975. Environmental limitations of photosynthesis on a California evergreen shrub. Oecologia 19:293-301.

Mooney, H., D. Parsons, and J. Kummerow. 1974. Plant development in mediterranean climates, pp. 255-267. In H. Lieth (ed.). Phenology and seasonality modeling. Springer-Verlag, New York.

Mooney, H., J. Troughton, and J. Berry. 1974. Arid climates and photosynthetic systems. Carnegie Inst. Yearbook 73:793-805.

Muller, C. 1965. Inhibitory terpenes volatilized from *Salvia* shrubs. Bull. Torrey Bot. Club 92:38-45.
Muller, C., W. Muller, and B. Haines. 1964. Volatile growth inhibitors produced by aromatic shrubs. Science 143:471-473.
Muller, W. 1965. Volatile materials produced by *Salvia leucophylla*: effects on seedling growth and soil bacteria. Bot. Gaz. 126:195-200.
Muller, W., P. Lorber, and B. Haley. 1968. Volatile growth inhibitors produced by *Salvia leucophylla*: effect on seedling growth and respiration. Bull. Torrey Bot. Club 95:415-422.
---. 1969. Volatile growth inhibitors produced by *Salvia leucophylla*: effect on oxygen uptake by mitochondrial suspensions. Bull. Torrey Bot. Club 96:89-95.
Muller, W., and C. Muller. 1964. Volatile growth inhibitors produced by *Salvia* species. Bull. Torrey Bot. Club 91:327-330.
Munz, P., and D. D. Keck. 1959. A California flora. Univ. Calif. Press, Berkeley. 1681 p.
Shreve, F. 1936. The transition from desert to chaparral in Baja California. Madroño 3:257-264.
Tyson, B., W. Dement, and H. Mooney. 1974. Volatilisation of terpenes from *Salvia mellifera*. Nature 252:119-120.
Wellburn, A., A. Ogunkanmi, R. Fenton, and T. Mansfield. 1974. All-*trans*-farnesol, a naturally occurring antitranspirant? Planta 120:255-263.
Wells, P. 1962. Vegetation in relation to geological substratum and fire in the San Luis Obispo Quadrangle, California. Ecol. Monogr. 32:79-103.
---. 1964. Antibiosis as a factor in vegetation patterns. Science 144:889.

CHAPTER 14

VALLEY GRASSLAND

HAROLD F. HEADY
College of Natural Resources, University of California, Berkeley

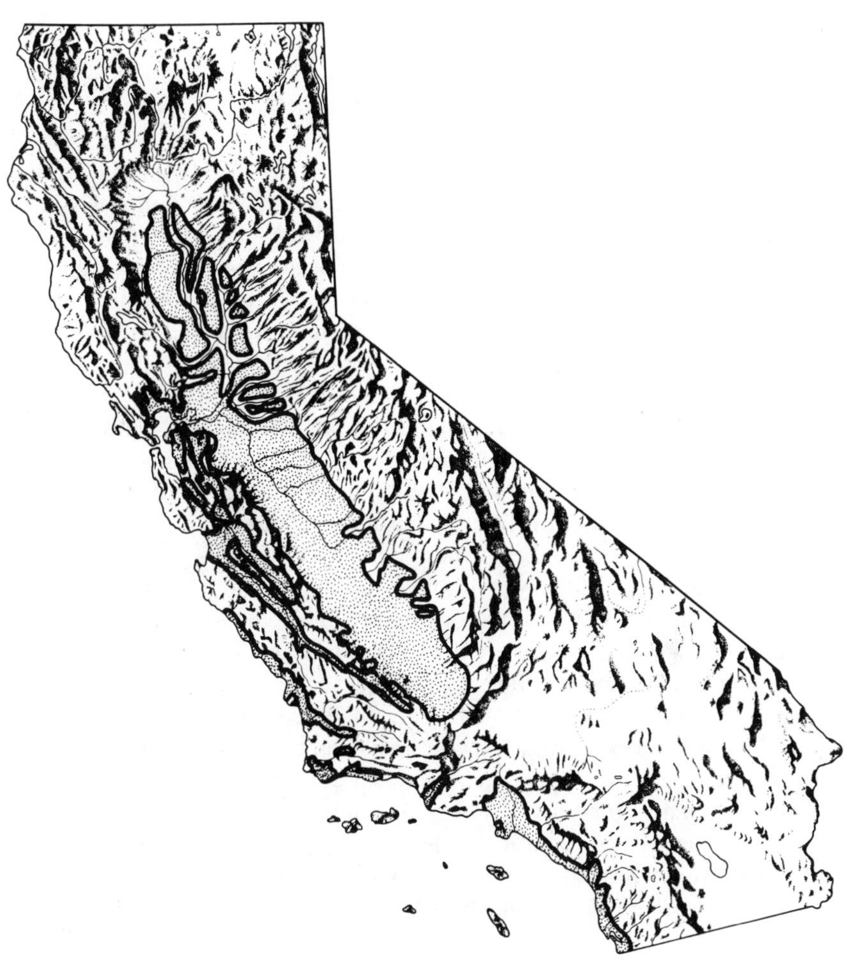

The California prairie **492**
  Relation to other vegetational types **493**
  Distribution **495**
  Composition of the perennial grassland **495**
Replacement of the pristine grassland **497**
  Invasion by alien plants **497**
  Changes in kinds of animals **498**
  Cultivation **498**
  Fire in the pristine grassland **499**
  Summary comments **499**
The annual grassland **501**
  Native versus introduced species **502**
  Dominants on different sites **502**
  Vegetational changes **503**
    Phenology **503**
    Successional changes in the annual grassland **506**
    Genetic changes in the annual species **506**
  Annual grassland responses to grazing **508**
  Annual grassland responses to fire **509**
Subjects for research **510**
Literature cited **511**

THE CALIFORNIA PRAIRIE

Several names commonly apply to the low-elevation grasslands that lie west of the Sierra–Cascade crest and in southern California. Shantz and Zon (1924) mapped this grassland as the *Stipa–Poa* bunch grass type and classified it as part of the Pacific grassland. Clements (1934) called single-species stands along rights-of-way in the Central Valley the "*Stipa pulchra* consociation of the California bunch grass prairie." Numerous later authors and speakers refer to the area simply as the "California prairie" or the "California annual grassland" (Fig. 14-1). The first name emphasizes the perennial nature of the pristine grassland species, and the latter draws attention to the current dominance of annual plants; neither, however, attempts to divide the area into two or more parts.

Burcham (1957) and Munz and Keck (1959) described the California prairie as a valley grassland with southern affinities and a north coastal prairie with northern relationships. Küchler (1964) labeled these parts as California steppe and *Festuca–Danthonia* grassland, respectively. Although the names differ, general descriptions leave little doubt as to their synonymous nature. This chapter concentrates on the valley grassland; Chapter 21 discusses the coastal prairie.

The first accounts of the California prairie described it simply as excellent pasture. Perhaps the earliest vegetational study of consequence was made by Burtt-Davy (1902), who described grasslands in the north coast mountains. Reports before 1900 by explorers, survey parties, and botanical collectors provided some information about the kinds of plants in the grasslands, but these limited vegetational descriptions give us grist for continuing deductive arguments. For example, on April 1, 1868, John Muir wrote of Santa Clara Valley, "the hills were so covered with flowers that they seemed to be painted" (Adams and Muir 1948). On the same trip,

Figure 14-1. Grazed valley grassland with *Aesculus californica* in the foreground and *Quercus* behind.

Muir referred to the Central Valley as a "garden of yellow compositae." But that was early April, when *Baeria chrysostoma, Amsinckia intermedia, Eschscholtzia californica,* and others make their brief show on restricted habitats. Six weeks later the descriptions could well have suggested a sea of mixed grasses gently moving in the morning breeze. One could argue from these observations that the vegetation was either annual or perennial, or a mixture of both.

History has not recorded the vegetational dynamics of the pristine California prairie. After 25 yr of studying this grassland, I believe that plant succession tended toward perennial bunch grass dominants on nearly all well-drained upland sites, that numerous annual species were present, and that they dominated intermediate and low successional stages, just as they do in many other grasslands. Also, I believe that introduced annual plants prevent many perennial grasses from attaining their dominance, that annuals are now a large part of the climax on many sites (if not all of it), and that alien species should be considered as new and permanent members of the grassland rather than as aliens. Their elimination from the California prairie is inconceivable. These hypotheses will be examined in detail throughout this chapter.

Relation to Other Vegetational Types

The Palouse prairie, centered in eastern Washington and Oregon, is related to the California prairie because of closely similar floras and growth habitats of the dominants. Species such as *Festuca idahoensis, Koeleria cristata, Poa scabrella,* and *Sitanion hystrix* occupy dominant positions in both grasslands. These species gradually lose their importance from north to south and from the coast inland, thus forming a continuum between the Palouse prairie and the California prairie. The Palouse dominant, *Agropyron spicatum,* and the California dominant, *Stipa pulchra,* are not shared (Beetle 1947). In addition, Stebbins and Major (1965) claimed from genetic and paleobotanical evidence that many endemic species of the California prairie originated from northern ancestors.

A transition between the California prairie and the herbaceous vegetation in the desert scrub types of southeastern California occurs over a short distance on the eastern slopes of the Tehachapi and other southern California mountains. *Stipa speciosa* replaces the *Stipa* species of the California prairie, and species in the grass genera *Aristida, Bouteloua,* and *Hilaria* of the desert grassland occur sparingly except on north-facing slopes with coarse-textured soils, where they may form relatively dense stands. *Bromus rubens, Erodium cicutarium, Salsola kali,* and *Schismus arabicus* dominate the annual layer.

The valley grassland extends into oak woodland and chaparral with little change in its herbaceous characteristics. Large specimens of *Quercus lobata* may be found in the Central Valley and smaller valleys in the Coast Ranges. The gentle beginning slopes of the mountains bordering the Central Valley are covered with California prairie, but *Q. douglasii* gives the grassland a savanna aspect. Burcham (1957) suggested that the savanna appearance covered more area in pristine times than it does today.

The present-day valley grassland occurs as a ring around the Central Valley. Toward the center it irregularly borders cultivated land (mostly irrigated) and marshes or remnants of the once-extensive tule swamps, such as along the San Joaquin River in Merced and Fresno Cos. (Fig. 14-2). Poorly drained alkali soils occur most abundantly in the southern and western sides of the San Joaquin Valley, which lacks oceanic drainage. Bordering bare soil within the sinks, alkali-tolerant species such as *Allenrolfea occidentalis, Kochia californica, Salicornia subterminalis,* and *Suaeda fruticosa* occur. Grasses, such as *Distichlis spicata, Hordeum depressum,* and *Sporobolus airoides,* also characterize the type. With somewhat better drainage and on foothill areas, *Atriplex polycarpa* dominates and the annual

Figure 14-2. Marshland along the San Joanquin River in Fresno Co. This type has been reduced in area by drainage control and cultivation.

grassland species appear as an understory (Twisselmann 1967). A small area of *Artemisia tridentata* is also present. Inner coastal hills in San Benito Co. support *Atriplex polycarpa, Ephedra californica,* and *Haplopappus racemosus*. Some years in late March, *Monolopia lanceolata* turns the landscape to a bright yellow, whereas in other years *Erodium* and the annual grasses cause greens to dominate.

In marked contrast to the alkali flats, "hog wallows" on the eastern side of the Central Valley are small depressions that fill with fresh water during the rainy season. An endemic vernal flora has evolved around these pools.

Distribution

The pristine California prairie appears to have been little different in distribution from the present-day grasslands,except the areas taken for cultivation. Boundaries with woodland and chaparral may be less or more distinct because of man's activities, but it is doubtful that their relative locations have changed.

The original valley grassland covered well-drained areas from sea level to approximately 1200 m in all the mountains surrounding the Central Valley. The valley grassland occurs in scattered patches throughout the Coast Ranges, with the possible exception of the narrow fringe of coastal facing slopes from Santa Cruz northward. Küchler's (1964) map shows that 5.35 million ha support this grassland, and an additional 3.87 million ha have an oak overstory (area determined by planimeter).

Such a large area has many habitats, and few accurate generalities can be made as to typical conditions under which the grasses grow. Most grassland soils are noncalcic Browns, medium to heavy in texture, granular in structure, moderate in organic matter content, and often about 0.5 m in depth. The valley grassland occurs on a wide variety of soils, sometimes closely associated in a complex mosaic (Fig. 14-3). Barry (1972) listed 195 soils series on which the grassland might be found. Other vegetational types, especially oak woodland, also occur on most of them.

Average annual rainfall for the valley grassland ranges from about 12 cm in the southwestern San Joaquin Valley to 200 cm in northwestern California. Regardless of the annual amount, soil water deficits characterize the grassland habitat for 4–8 summer months every year. Although germination of seed and breaking of dry season dormancy for many perennials in the grassland follow fall rains, winters are cool, allowing little growth. Only in spring do warm temperatures, beginning about March, stimulate rapid growth, flowering, and maturity. Cool-season species mature from April to June, and a few warm-season annuals reach their peak growth in summer. Annual variations in weather greatly influence species composition and biomass production.

Composition of the Perennial Grassland

Stipa pulchra, beyond all doubt, dominated the valley grassland. Toward the southern end of the grassland, *S. cernua* increased in importance. Perennial grasses associated with *Stipa* were *Aristida hamulosa, Elymus glaucus, E. triticoides, Festuca idahoensis, Koeleria cristata, Melica californica, M. imperfecta,* and *Poa scabrella.* Annual grasses included *Aristida oligantha, Deschampsia danthonioides, Festuca megalura, F. pacifica,* and *Orcuttia* spp. Broad-leaved herbs included many

Figure 14-3. Grassland variations in stages of vegetational drying that are associated with soil differences.

perennials, especially plants with bulbs, and annuals in the Caryophyllaceae, Compositae, Cruciferae, Labiatea, Leguminosae, and Umbelliferae (Stebbins 1965).

Quantitative descriptions based on plot sampling of the relict perennial grasslands have not appeared. White (1967) gave indices of commonness for species in 17 stands of *Stipa pulchra* on Hastings Reservation in Carmel Valley. Only one perennial grass (*Stipa*) was found in those stands. Barry (1972) described 19 examples of the valley grassland where *Stipa* dominates and mentioned 20 other sites that he did not examine. None of his descriptions included data more quantitative than species lists. Burcham's excellent review (1957) of historical accounts of California rangelands contained many references to abundant water and excellent native pasture. Botanical collections began about 1830, long after grazing by domestic livestock had become extensive along the California coast, but no authors recorded the relative importance of the perennial grassland species.

Although the perennial grasses may be difficult to find, they have by no means been eliminated. Fence corners, roadsides, rights-of-way, and brush plants protect sites which frequently support *Stipa* and other former dominants of the pristine grassland. *Aristida hamulosa* may be the least common of these species. Barry (1972) recorded it abundantly only where Salt Canyon enters the Sacramento Valley in western Colusa County (NA 061915).

Originally the valley grassland probably appeared as a bunch grass prairie, similar to the Palouse prairie, with the exception that native annual species filled the interspaces between the large bunches of *Stipa*. In one exclosure on the Hopland Field Station, a canopy of *Stipa pulchra* covers the soil, but many annual plants still remain. White (1967) mentioned 68 of these subdominant species in *Stipa* stands with 11% basal cover on Hastings Reservation. Of the 68, 71% were native species. On shallow soils, a thin or scattered perennial stand permits numerous annuals to increase in abundance. Annuals may dominate completely on certain sites, for example, the hog wallow communities at Dozier (Lin 1970; Luckenbach 1973).

REPLACEMENT OF THE PRISTINE GRASSLAND

Permanent alterations in the pristine grassland began when Europeans first reached the Americas. These changes resulted from a combination of (1) invasion by alien plant species, (2) changes in the kinds of animals and their grazing patterns, (3) cultivation, and (4) fire. Many authors suggest the relative importance of these factors by the order of listing above, but others believe that overgrazing was the principal destructive factor. That hypothesis has not been proved, nor can it be until something similar to the pristine grassland can be attained through separation and manipulation of each of the four factors. The following discussion takes them separately for reasons of simplicity, although their interacting impacts unquestionably changed the grassland.

Invasion by Alien Plants

Hendry (1931) suggested that *Erodium cicutarium, Rumex crispus,* and *Sonchus asper* may have preceded Europeans to California. Adobe bricks used to build the earliest Spanish missions contained these as well as *Hordeum leporinum, Lolium multiflorum,* and *Poa annua,* all introduced species. The earliest European arrivals to the Americas (Columbus, Cortes, Magellan, Coronado, and others) did not settle in California. However, seeds in packing and hay from Spain and in the debris associated with livestock on their sailing vessels undoubtedly were the sources of the alien plants. A brief land exploration, especially one with horses, from the first sailing vessel to reach the California shores probably left new plants. Many of the early ships carried a few live animals for meat; thus seeds in the manure thrown overboard could have resulted in alien plants reaching shore even though the sailors did not. Once a plant species produced seeds anywhere in the continent, birds could have carried them to other locations. California lies in the path of many bird migrations, especially north–south routes with one terminus in Mexico.

Burcham (1956, 1957) and Robbins (1940) presented evidence that suggests major replacement of the perennial grasses with introduced annuals in stages beginning in the 1850s. At that time, there appeared to be overgrazing in the coastal areas, and miners from all over the world had traveled to the gold fields, bringing seeds, bulbs, and cuttings of many plant species. Eleven years of drought caused much barren land. Perkins wrote in 1864, "Less than ten years ago, the traveller would ride for days through wild oats tall enough to tie across his saddle, now dwindled down to a stunted growth of six or ten inches, with wide reaches of utterly barren land," (as quoted in Burcham 1957). This statement suggests that *Avena fatua* and probably *Brassica nigra* invaded the Central Valley and became dominant before livestock had overgrazed the area. Similar botanical composition may be seen almost every year in the spring along the roadsides in the San Francisco Bay Area.

The second stage, about 1870, was dominated by *Bromus* spp., *Erodium* spp., *Gastridium ventricosum,* and *Hordeum leporinum.* Introduced annuals in the third stage, the 1880s, included *Aira caryophyllea, Bromus rubens, Centaurea melitensis,*

Hordeum hystrix, and *Madia sativa.* A fourth group, *Aegilops triuncialis, Brachypodium distachyon, Chondrilla juncea,* and *Taeniatherum asperum,* has increased recently. Species not in California, such as *Cryptostemma calendula,* or cape weed in Australia, dominate Mediterrean-type grassland in other parts of the world. They have the potential to become part of the California annual grassland. Today *Bromus mollis,* which arrived about 1860, is the matrix species through much of the grassland, but it may not continue in that role. Plant succession within the annuals will be described later.

Changes in Kinds of Animals

The well-drained soils covered with pristine valley grassland supported large numbers of pronghorn antelope (*Antilocapra americana*), deer (*Odocoileus hemionus*), jackrabbit (*Lepus californicus*), Beechey ground squirrel (*Spermophilus beecheyi*), kangaroo rat (*Dipodomys heermanni*), and pocket gopher (*Thomomys bottae*). The tule elk (*Cervus elophus nannodes*) used both the well-drained uplands and the marshland along the rivers. Antelope occurred in herds of 2–3,000, and elk in the hundreds. Several of these grazing animals developed into strains found only in the Central Valley, suggesting a lengthy association between the vegetation and the grazing.

As European man and his domestic animals rapidly increased in numbers in the 1850s, the larger wild animals diminished. Rodents remain numerous. Fitch and Bentley reported in 1949 that 25 pocket gophers, 4 ground squirrels, and 12 kangaroo rats on 1 ha of land would eliminate about a third of the forage. Batzli and Pitelka (1970) found that the meadow mouse (*Microtus californica*) significantly reduced plant height, ground cover, and standing crop in both the coastal prairie and the annual grassland. Bartholomew (1970) and Halligan (1973) reported that small mammals were responsible, at least in part, for creation of "bare zones" in grassland around *Artemisia* shrubs in southern California. The maximum number of livestock probably occurred about 1862 for cattle and 1876 for sheep. Therefore the maximum impacts of livestock grazing on the grassland came after many introduced species had arrived and made permanent places in the grassland vegetation. Although large numbers of livestock may have accumulated around the missions and on Spanish land grants before 1850, grazing in the Central Valley appears to have been relatively light until the gold rush period after 1850. The Spanish numbered about 10,000 in California by the early 1800's, but they did not settle extensively in the Central Valley (Künzel 1848). Only a part of the wild herbivorous fauna has been replaced. The introduction of livestock shifted the timing of grazing. Native ruminants may have been seasonal residents of the Central Valley (McCullough 1971, Sampson and Jespersen 1963), but they repeated that residence year after year. Nevertheless, some animal exclosures do not show a recovery by perennial grasses.

Cultivation

The largest acreage was cultivated during the 1880s, preceding extensive irrigation and depending on dry-land farming procedures. Grains and forage were the principal crops. They are still grown on some of the nonirrigated foothills. As was often the case in "boom" times, many more acres were plowed for a few crops than could be sustained in crop production. Many foothill hectares that now show little evidence of plowing, and that support annual grassland, once grew a wide variety of crops and, before that, supported perennial grasslands. Cultivation has contributed to the replacement of the pristine grassland with annuals.

Fire in the Pristine Grassland

The frequency of fires set by lightning today very likely approximates past occurrences. Although burning in pristine grassland cannot be quantified as to intensity and frequency, fire must have spread through the abundant, dry fuel, probably to a greater extent than it does today. The California Division of Forestry reported 312 lightning fires per year in its protection area, which is 43% woodland–grass. The evidence is deductive, but the conclusion is generally accepted that lightning-caused fires have been part of the entire evolutionary history of the grassland (Heady 1972).

Permanent settlements of Indians occurred along the streams and California coast during the last 8000 yr (Heizer and Whipple 1971) and perhaps longer. Early narrative reports about California mention fires set by Indians, many of which probably burned larger areas than intended (Sampson 1944). Burning by mankind came late in the evolution of the grassland species, however, and probably exerted only minor influences on the grassland.

Summary Comments

No single factor caused the demise of the pristine perennial grassland. Except for hectares in cultivation, roads, housing, and so on, the grassland boundaries probably are much the same today as they were 200 yr ago. Brush plants may have invaded in some places, but these probably balance the size of area where grassland increased after brush and woodland were removed by fire and herbicides. More likely, increase in brush density and cover after effective fire control resulted in thickened stands of woody plants rather than their invasion into new areas.

The pristine grassland received considerable but unknown impact from native grazing animals. Livestock replaced that pressure to a greater degree than it added a new impact. Rodents still remain in the grassland and perhaps occur as abundantly as they did during prelivestock times (Fitch 1948; Fitch and Bentley 1949). Droughts reduce the vigor of perennial grasses and the quantity of herbage. That, in turn, causes grazing pressures to be severe. This cyclic pressure of overgrazing in dry times and undergrazing in wet periods occurred with the wild animals before domestic animals arrived. It continues today; deer, for example, periodically overgraze their savanna type winter range on the foothills surrounding the Central Valley. Many herds of wild animals around the world fluctuate in numbers in concert with abundant or scanty forage resulting from weather cycles. I see no reason to doubt that these impacts occurred on the pristine valley grassland, that the perennial grassland occasionally received heavy use, and that livestock altered and increased grazing but did not add a completely new factor.

Into this situation came the introduced annuals, which were widely adapted to the Mediterranean climate and to the local soil, which were resistant to grazing, and which offered high competition to the perennials (Evans and Young 1972). Their passage through the dry summers in the seed stage enhanced their advantage over the perennials, especially during dry years. Evidence exists that some of the grass-

Figure 14-4. Perennial grassland dominated by *Stipa pulchra* behind and annuals in the foreground. La Jolla Valley, near Oxnard.

land became dominated by *Avena* before heavy grazing by domestic livestock occurred. Cultivation and other types of soil disturbance completely removed the perennials from many areas. The introduced annuals returned quickly after the abandonment of cultivation and held the land, preventing return of the perennials. Many factors contributed to the replacement of the perennials with introduced annuals, not the least being the competitive ability of the annuals under varying conditions of weather and grazing (Harris 1967). Replacement of the perennial grassland resulted, not from a single cause or a sequence, but from several causes operating together.

Complete absence of livestock is about the only situation that sometimes permits the reinvasion of perennial grasses. No livestock means no grazing by any large animals, as herds of elk and pronghorn antelope no longer exist. Removal of all livestock now may be as disruptive to pristine conditions as the presence of too many livestock (Heady 1968). Whether the perennial grasses will return on *all* ungrazed sites remain a question. Other than *Poa scabrella,* few perennial grass species can be found after 40 yr without grazing in the livestock-free area on the San Joaquin Experimental Range. Scattered plants and small stands of several perennial species developed in 10–15 yr without livestock grazing on the Hopland Field Station. *Stipa pulchra* cannot be found in areas with abundant annual grasses on Hastings Reservation, although it does occur in oak woodland. The stands of *S. pulchra* in La Jolla Valley near Oxnard remained after many decades of various kinds of use and have invaded the former cultivated fields and pastures in the valley center (Fig. 14-4; Barry 1972). *Stipa pulchra* abundantly appeared in one pasture on the Hopland Field Station after restriction of sheep grazing to the winter season. It decreased in the same pasture after spring, summer, and fall grazing a few years later.

Determination of the characteristics of the pristine valley grassland must not overshadow concern for the present and the future. Grassland ecologists should recognize that species labeled as "introduced" and "alien" cannot be removed and perhaps not even reduced from their present state. For example, *Bromus mollis* and *Avena fatua* usually increase when heavy grazing is reduced. They dominate on numerous soil types and over large areas. Regardless of whether managerial concerns consider them desirable or undesirable, they cannot be eliminated under any known rangeland management practice. Cultivation may remove them from the center of the field, but they remain at the edges. Ecologists and others should recognize these as "new natives".

The rest of this chapter takes that view. In accordance with these ideas, the present valley grassland system of organisms and abiotic factors will be described within the theory that the annual grassland exhibits its own location variations, vegetational changes, and responses to abiotic factors, and that it becomes relatively stable or climax if given a chance. The rationale for these views has been developed elsewhere (Heady 1975).

THE ANNUAL GRASSLAND

The term "California annual type" describes the present valley grassland. The boundaries show little change from those of the pristine perennial grassland. In other words, annual grasslands exist below 1200 m in elevation west of the Sierra Nevada–Cascade Mts. and in coastal southern California; they are best developed as a treeless ring that borders the cultivated Central Valley; they provide an understory to a *Quercus douglasii* savanna (Fig. 14-5) and in the oak woodland. Removal of the woodland and much of the chaparral results in a new or thickened stand of the herbaceous annuals until the woody plants return. The type has many variants, including climax characteristics in some areas and successional status to woody vegetation in others.

The annual vegetation contains many species. Janes (1969) encountered 124 species during the sampling of 20 sites along a transect between Kern and Humboldt

Figure 14-5. *Quercus douglasii*–annual grassland savanna.

Cos. About 50 species were recorded year after year in sampling a 625 m^2 plot on the Hopland Field Station by Heady (1956a). White (1966) found 55 species in North Field, Hastings Reservation.

Native versus Introduced Species

Robbins (1940) reported 526 species of alien plants growing without cultivation in California. Most of them are associated with cultivated and pasture lands and hence with the annual grassland. Of the species on Robbins' list, 111 were Gramineae and 96 Compositae. About 72% of the total number originated in Europe and western Asia. Of the 478 species of grasses in California, 175 were introduced and 156 are annual (Crampton 1974).

The number of species on a statewide basis tells little about the grassland. The proportion of native species in lists for individual stands varies from 71% at Hastings Reservation (White 1967) to less than 20% at Hopland (Heady 1956a) and perhaps more. Talbot et al. (1939) reported that annuals constituted 94% of the herbaceous cover in grassland, 98% in woodland, and 93% in chaparral. The introduced species constituting that cover accounted for 63%, 66%, and 54%, respectively, of the total cover. A number of studies have shown that the 6–10 most important species (various sampling methods) in the annual grassland are new natives (White 1966; Heady 1958; Talbot et al. 1939; Bentley and Talbot 1951). Bentley and Talbot (1948) reported that native annuals on the San Joaquin Experimental Range composed 20–60% of the cover, depending on rainfall pattern. The California annual grassland is a relatively new mixture of annual grasses and broad-leaved plants. Although many annual species occurred in the grassland of 200 yr ago, they are not the annuals that dominate today.

Dominants on Different Sites

Obviously, the annual grassland differs from place to place. McNaughton (1968) showed that the structure of the annual grassland varied greatly within small distances. Janes (1969) selected 20 sites at about 80 km intervals along a transect from the southern end of the San Joaquin Valley to southern Humboldt Co. Criteria for site selection included no evidence of grazing and fire for 3 yr, a minimum soil depth of 45 cm, a southerly aspect, and slope no greater than 35%. The gradient of average rainfall along the transect was 13–204 cm, as measured at the station nearest each site. In terms of foliage cover, *Bromus mollis, B. rigidus,* and *Erodium botrys,* in that order, were the principal species occurring with more than 20 cm of rainfall. *B. rigidus* reached highest proportions near 50 cm of rainfall. On 8 of the 20 sites *B. mollis* dominated the vegetation. *Bromus rubens* and *Erodium cicutarium* dominated on sites with less than 19 cm of rainfall and were found in measurable quantities to approximately 30 cm. *Avena fatua* occurred at midrange, 20–63 cm, and *A. barbata* peaked at 87–100 cm. Plots in the middle rainfall range, where *A. fatua* occured, had the highest total foliage cover and biomass per unit area. Although much variation occurs locally and from year to year, these data on botanical composition appear to represent reasonable generalitites about the distribution of the important annual species in relation to climate. Soils at the sites varied in texture between clay and gravely loams.

The California annual grassland includes a wide mixture of species, growing on many different habitats, and divisible into subtypes, depending on the purposes of the investigation. Subtypes may be designated according to soil types, as listed by Barry (1972); mapped, as in the State Cooperative Soil–Vegetation Survey (Zinke 1962); described as range sites (Bentley and Talbot 1951); or separated as objects of study for managerial purposes (Evans et al. 1975). The annual grassland responds to site variation where local differences may be controlled by soil nutrients (McGown and Williams 1968), broad north–south differences in temperature and moisture, or in allelopathic antagonisms. The leachate from straw of *Avena fatua* may reduce stands of *Centaurea melitensis* and *Silybum marianum,* thereby increasing the dominance of *A. fatua* (Tinnin and Muller 1971, 1972). Detailed analyses substantiating these general conditions exist for only a few locations. However, site-by-site data on the species composition of the grassland have been gathered for range condition evaluations by the Soil Conservation Service and for basic information by the State Cooperative Soil–Vegetation Survey (U.S. Forest Service 1954; University of California 1959).

Vegetational Changes

Phenology. The annual plants begin to germinate in the fall with the first rains exceeding about 15 mm, grow slowly through the winter, grow rapidly in the spring, and mature between late April and June. The proportion of different species varies to such an extent that grasses may dominate in some years and *Erodium* in others, and legumes may or may not be conspicuous. Several types of changes in the annual grassland have been described and quantified on the Hopland Field Station (Heady 1961) and the San Joaquin Experimental Range (Bentley and Talbot 1951; Biswell 1956). For detailed information, the reader should consult the review by McKell (1974) of autecological studies on germination, growth, flowering, and maturation of several annual grassland species.

The period of rapid spring growth brings on a progression of different dominant species. *Agoseris heterophylla, Baeria chrysostoma, Hypochoeris glabra, Lotus* spp., *Orthocarpus* spp., *Trifolium* spp., etc., flower and mature early, at about the same time. Among the grasses, *Aira caryophyllea* and *Briza minor* set seed early, *Festuca* spp. later, and *Bromus mollis* and *B. rigidus* still later. *Avena barbata* and *Erodium* spp. flower and seed throughout the spring. Early- and late-summer species include *Aristida oligantha, Eremocarpus setigerus, Gastridium ventricosum, Madia* spp., and *Taeniatherum asperum* (= *Elymus caput-medusae*).

Few descriptions of the annual grassland mention seed numbers and seed characteristics, yet most of the vegetation lives through the dry season in the seed stage. The soil seed bank makes an excellent starting point for a study of seed population dynamics. The range of numbers of seeds appears to be from about 300 (Sumner and Love 1961) to nearly 150,000 per square meter (Heady 1956b). Four other papers give numbers between 7500 and 60,000 per square meter (Batzli and Pitelka 1970; Heady and Torell 1959; Major et al. 1960; Major and Pyott 1966). Major et al. (1960) found that seeds of *Taeniatherum asperum* could be as numerous as 60,000 per square meter.

Buried viable seeds may live for several years. Major and Pyott (1966) found a poor correspondence between seed numbers and the vegetation above the seed bank.

Obviously, not all the species that can occur in the vegetation over a period of years will be evident at any one time. After extensive review of the literature on field germination in annual grassland, Bartolome (1976) concluded that the relationship between soil seed available (single species or *in toto*) at the beginning of the season and later vegetational patterns has not been quantitatively established. Numerous laboratory studies have been made on the characteristics of seed germination and seedling establishment of species in the annual grassland. Data concerning differential seed and seedling mortality in the field no not exist for the first few days or weeks of the growing season.

A 2 yr study of seedling establishment and plant numbers on the Hopland Field Station by Bartolome (1976) showed that near-maximum seedling densities for most species occurred before the second week of the growing season. However, seedlings of *Aira caryophyllea, Briza minor,* and *Hypochoeris glabra* appeared in the winter and spring. *Eremocarpus setigerus,* a summer-active species, germinated in the spring. Plant densities differed between the 2 yr, but total standing crop (biomass) did not. Soil seed numbers far exceeded seedling numbers in most species, suggesting that environmental factors exert more influence on plant densities than does seed supply.

Mulch, or above-ground plant residue, influences germination and seedling establishment through modification of several microenvironmental factors. Most seeds germinate on the soil surface or at depths to 1 cm below it. Seedlings must become established in a critical zone, which is the top few centimeters of soil and the lower 3–5 cm of atmosphere. Mulch on the soil surface results in physical conditions that favor microorganisms, seedlings, water infiltration, rapid decomposition, and other factors. More precise understanding of soil surface processes and relationships is needed. The lack of field data on germination, seedling establishment, dynamics of seed numbers, dynamics of seedling establishment, seed production, and seed dormancy is surprising in view of a widespread belief that the botanical composition of the annual grassland stems from patterns of seed numbers, germination, and seeding establishment.

By clipping a new set of replicates every 2 wk between October 15 and May 30, the normal growth patterns of eight annual species became clear (Savelle and Heady 1970). All plants were grown in the same soil and under the same conditions of watering and ambient temperatures. The period of maximum growth rate ranged from early March to early May (Fig. 14-6). Flower initiation began in early February in *Erodium botrys* but did not start until April in *Festuca megalura.* No two of the species studied were alike in growth pattern. Under drying spring weather in field conditions, rapid growth rates may show an abrupt ending.

Significant changes occur in numbers of plants and biomass per unit area as the season progresses. Biswell and Graham (1956) reported more than 20 seedlings per square centimeter after germination, but it is more likely that the average approximates 3–5 per square centimeter (Heady 1958). Both papers reported 10–90% mortality during the growing season, depending on species. The relative densities of *Bromus mollis* and *B. rigidus* tended to increase during the growing season, whereas those of shorter plants decreased. A study near Berkeley by Ratliff and Heady (1962) determined that several species reached their individual maximum biomass on the following dates in 1960: *Bromus mollis* on April 24, *Erodium cicutarium* on April 30, *Medicago hispida* on May 8, *Bromus rigidus* on May 21, *Avena fatua* on

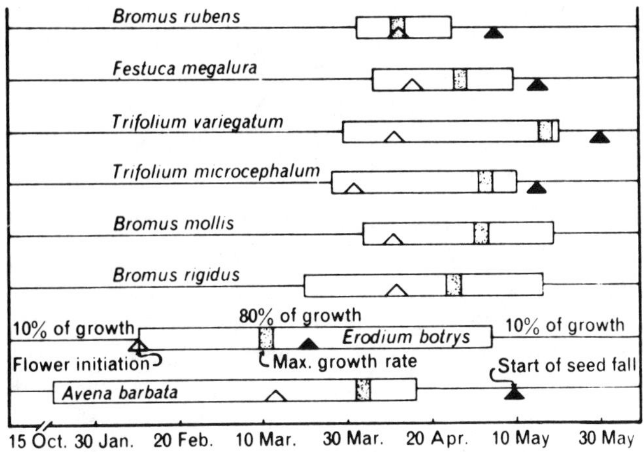

Figure 14-6. Time periods and approximate dates of growth, flower initiation, and seed fall for eight common annual species grown in pots under uniform soil, regular watering, and ambient temperatures. (Savelle and Heady 1970).

May 21, and *Lolium multiflorum* on May 28. The biomass for the whole community peaked on May 21. Two weeks later, 23% of the above-ground weight had been lost through shattering and seed fall. Broad-leaved species tended to disappear rapidly after plant maturity. In 1 wk, 79% of the *Erodium* biomass disappeared. A 19 yr study at Hopland by Heady (see Pitt 1975) showed that the March 1 biomass averaged 90 g m^{-2}, and the June 1 biomass 309 g m^{-2}. Year-to-year variation halved and doubled these averages.

Three sets of data on biomass, productivity, botanical composition, and weather have been analyzed by Murphy (1970) for the purposes of developing predictive relationships on the Hopland Field Station. He found that precipitation in November correlated (r^2 = 0.7) with forage yield several months later. Duncan and Woodmansee (1975) used 24 yr of data on the San Joaquin Experimental Range to find that peak forage yields were only poorly correlated with precipitation in any particular month or with annual precipitation. Pitt (1975), using Heady's data from Hopland Field Station, also obtained poor relationships between yield and precipitation alone. The inclusion of temperatures and periods without precipitation improved the regressions and explained more of the variation. If germination occurs early in the fall, production will be high, but periodic spring precipitation without dry soil between showers until May is also needed for greatest biomass yield (Hooper and Heady 1970).

The weight of herbage biomass at the end of the growing season differs severalfold among years. Duncan and Reppert (1960) reported a production range of 775–2,900 kg ha^{-1} over a 25 yr period on the San Joaquin Experimental Range. During the winter until about the first of March, green herbage might be scarce in some years and abundant in others. Above-ground standing crop on March 1 varied between 24 and 218 g m^{-2}, but high March weights were not always followed by high June weights (Pitt 1975).

Overseeding of a set of plots on the Hopland Field Station resulted in one plot having a high proportion of *Medicago hispida* in 1957 and the adjacent plot being relatively high in *Bromus mollis*. The following year, with no treatments and no grazing, both plots had high *Bromus* composition and no *Medicago*. These differences show the great and overriding impact of annual weather patterns on the annual grassland. An early fall rain followed by a long dry period results in large amounts of *Erodium botrys*. Grasses require continuous rainfall, or at least dry periods no longer than 2–3 wk.

Successional changes in the annual grassland. In 1951, Sampson et al. described successional trends in the annual grassland due to differences in grazing pressure. Under reasonable grazing management, tall species such as *Avena fatua* and *Bromus mollis* dominate, but with heavy grazing shorter species prevail. Close grazing delays fall growth and reduces winter growth (Talbot and Biswell 1942; Bentley and Talbot 1951). A study of the effects of natural plant residue on the composition and production of annual vegetation suggests that increasing the degree of herbage removal results in less biomass production and fewer floristic changes. With no mulch at the time of germination, *Aira carophyllea, Baeria chrysostoma, Hypochoeris glabra,* and *Orthocarpus erianthus* dominated the vegetation on plots in Mendocino Co. (Heady 1956a). Different but similarly short, broad-leaved species occupied plots with no mulch in other parts of the annual grassland (Heady, unpub. data). Ratliff et al. (1972) found that *Eschscholzia californica* var. *peninsularis, Festuca megalura, Orthocarpus purpurascens,* and *Trifolium* spp. had high frequencies in stubble fields. *Bromus rigidus* and *Erodium botrys* dominated in old fields. When plant residue was left on the ground in a plot at Hopland, *Bromus mollis* increased from 0.9% to 37.3% of the cover in 3 yr. When all the mulch was removed each year, the proportion of *B. mollis* continued at less than 2% (Heady 1965). Many other authors have recognized the importance of natural mulch in the maintenance of the annual grassland (Talbot and Biswell 1942; Hedrick 1948; Hervey 1949; Biswell 1956). Apparently mulch alters botanical composition through influence on physical conditions at the soil surface and on the bulk density of the surface soil horizon. Mulch has little influence on pH and phosphorus and nitrogen contents in the grassland system (Table 14-1).

General observations suggest that fields of forbs, especially in early to middle spring, indicate early successional stages, while dominance by grasses may be climax annual grassland (Fig. 14-7). With due regard for vegetational alterations caused by site, weather pattern, and other factors, a number of species are placed in successional position in Table 14-2. This list must be interpreted with care and altered to fit local areas. For example, *Bromus rubens* and *Erodium cicutarium* are middle successional or even lower on areas with annual precipitation greater than 25 cm or after removal of chaparral. *Erodium botrys* may characterize all the successional stages and fluctuate in abundance more in concert with seasonal rainfall patterns than with successional factors. The alphabetical arrangement in the list in Table 14-2 implies that relative importance cannot be assigned throughout the annual grassland. However, this list reflects a widely held view that succession exists among the wide spectrum of annual species. Native perennial grass species become part of the climax, but they will not completely replace the annuals.

Genetic changes in the annual species. Quantitative studies in the interplay of immigration, migration, and natural selection are insufficient to permit evaluation

TABLE 14-1. Average herbage production for 1954-60, and average total nitrogen, total phosphorus, and bulk density in the top 7.6 cm of soil in 1960 after 8 yr of mulch manipulation. Treatment 1 had all mulch removed, Treatment 8 had none removed, and the other treatments represented an intermediate scale of mulch removal (Heady 1965)

Mulch Treatment	Herbage ($g\ m^{-2}$)	Total Nitrogen ($kg\ ha^{-1}$)	Total Phosphorus ($kg\ ha^{-1}$)	Soil Bulk Density ($g\ cc^{-1}$)
1	115	1242	224	1.36
2	168	1547	245	1.30
3	196	1425	225	1.25
4	199	1478	226	1.21
5	204	1410	229	1.26
6	225	1623	249	1.26
7	232	1555	231	1.23
8	254	1524	232	1.14

of genetic and evolutionary changes in the annual grassland. Presumably the relatively small gene pool within a new arrival and the wide range of habitats available to the immigrants could have stimulated speciation. Jain et al. (1970) found that *Bromus mollis* was genetically variable but had not shown as much geographical differentiation as had other species. Knowles (1943) did distinguish a coastal and an inland strain of *B. mollis,* based on time from germination to flowering. Marshall and Jain (1969) showed that the genetic systems in *Avena barbata* and *A. fatua* may be important determinants of the relative population sizes in mixed stands. Stebbins (1965) described many of the annuals as weeds with wide adaptive ranges, so a few migrant individuals could quickly adapt to many habitats. McKell et al. (1962) claimed that probably a single introduction of *Taeniatherum asperum* evolved over 70 yr into a number of genotypes differing in morphological and physiological characteristics. These and other authors have used species in the annual grassland to study breeding systems of species, groups of species, and genetic principles, but little is known about the impact of genetic changes and speciation on the grassland vegetation.

Figure 14-7. Natural grassland on 2.5 cm thick strips of soil illustrating plant successional stages. Top: climax dominated by *Bromus rigidus* and *Bromus mollis*. Center: Intermediate stage dominated by *Medicago hispida* and *Erodium botrys*. Bottom: Early successional stage with abundant *Baeria chrysostoma*.

Annual Grassland Responses to Grazing

With reasonable intensities of grazing by livestock, no soil erodes, productivity of biomass ramains high, and floristic composition of the vegetation shows little change. In fact, annual grassland withstands remarkably heavy livestock use (Bentley and Talbot 1951; Heady 1961). Stocking rates 2–3× normal for a year or more do more economic damage to the livestock operation than permanent biological damage to the landscape. However, moderate season-long or year-long grazing is recommended.

Animals spread over the range take the most palatable forage and leave the coarse materials as plant residue. The residue serves to protect new seedlings and promote water infiltration into the soil, and it fosters a low soil bulk density. Heady (1965) has shown that approximately 500 kg ha^{-1} of above-ground plant residue at the beginning of the rainy season indicates proper intensity of grazing use. This standard applies at approximately 75–100 cm of rainfall. Criteria for areas with more and less rainfall have not been determined. Light grazing results in abundant tall grasses, and heavy grazing sets the succession back to low-growing plants, but unless continued indefinitely, overgrazing has short-term effects.

Annual Grassland Responses to Fire

Burning in the grassland probably diminished after the discovery of gold because increased plowing and overgrazing reduced the amounts of fuel. However, settlers, stockmen, and lumbermen burned the forest and chaparral types to convert them to grasslands.

Fires in the grassland have little permanent effect. Few seeds are destroyed by grassland fires (Bentley and Fenner 1958). Species composition sometimes changes toward more broad-leaved plants and fewer annual grasses (Hervey 1949). Burning apparently does not alter moisture content, temperature, and organic matter in soil where grazing has already removed most of the above-ground biomass. Cook (1959) documented a reduction in rodent numbers due to fire, but all species survived.

TABLE 14-2. Plants usually found in climax, middle, and low successional stages. N = native species

Climax	Middle in Succession	Low in Succession
Avena barbata	*Daucus pusillus* (N)	*Aira caryophyllea*
A. fatua	*Erodium botrys*	*Briza minor*
Bromus mollis	*Festuca dertonensis*	*Eremocarpus setigerus* (N)
B. rigidus	*F. megalura*	*Hordeum hystrix*
B. rubens	*F. myuros*	*Madia* spp.
Erodium cicutarium	*Gastridium ventricosum*	*Lupinus bicolor* (N)
Taeniatherum asperum	*Medicago hispida*	*Trifolium* spp. (N)

Many of the effects of burning duplicate those of mulch removal by hand (Talbot et al. 1939; Heady 1956a) and none of the changes has been detected beyond the third year after burning. Charred organic materials, in contrast to the white ash that follows intensive fires, leave the surface black and provide a soil cover.

Attempts to control *Taeniatherum asperum*, a late-maturing species, with prescribed fire have been abandoned. A fire in dry grass when the *Taeniatherum* seed is immature destroys many seeds, but enough escape to produce the next crop.

SUBJECTS FOR RESEARCH

A marjority of the research papers on the ecology of the valley grassland originated from studies at the San Joaquin Experimental Range, the Hopland Field Station, and the campuses of the University of California at Berkeley, Davis, and Santa Barbara. These papers place more emphasis on management than on basic ecological processes. Therefore subjects with minimal immediate practical application dominate my list of research suggestions.

1. With further study, the spatial variation of grassland species in relation to climate would be better understood.
2. Very little is known about the rates of organic matter accumulation and decomposition in grassland.
3. The taxonomy and population dynamics of decomposers need attention.
4. Experiments into the competition between annual and perennial herbaceous plants would help to define the new climax valley grassland.
5. The relative impacts of many kinds of grazing animals remain almost unknown.
6. Seed population dynamics in relation to grazing and climate needs more work.
7. The tule marshes and saltbush variations of this type have not been mapped, described, or studied in either their original or their present conditions.
8. On the practical side, measurement of pollution impacts on valley grassland would help to define the future of much land in the Central Valley of California.

LITERATURE CITED

Adams, A., and J. Muir. 1948. Yosemite and the Sierra Nevada. Houghton Mifflin, Boston. 130 p.

Barry, W. J. 1972. The Central Valley prairie. Calif. Dept. Parks and Recreation, Sacramento. 82 p.

Bartholomew, B. 1970. Bare zone between California shrub and grassland communities: the role of animals. Science 170:1210-1212.

Bartolome, J. W. 1976. Germination and establishment of plants in California annual grassland. Ph.D. dissertation, Univ. Calif., Berkeley.

Batzli, G. O., and F. A. Pitelka. 1970. Influence of meadow mouse populations on California grassland. Ecology 51:1027-1039.

Beetle, A. A. 1947. Distribution of the native grasses of California. Hilgardia 17:309-357.

Bentley, J. R., and R. L. Fenner. 1958. Soil temperatures during burning related to postfire seedbeds on woodland range. J. For. 56:737-740.

Bentley, J. R., and M. W. Talbot. 1948. Annual-plant vegetation of the California foothills as related to range management. Ecology 29:72-79.

-----. 1951. Efficient use of annual plants on cattle ranges in the California foothills. USDA Circ. 870. 52 p.

Biswell, H. H. 1956. Ecology of California grassland. J. Range Manage. 9:19-24.

Biswell, H. H., and C. A. Graham. 1956. Plant counts and seed production on California annual-type ranges. J. Range Mange. 9:116-118.

Burcham, L. T. 1956. Historical backgrounds of range land use in California. J. Range Manage. 9:81-86.

-----. 1957. California range land. Calif. Forestry, Sacramento. 261 p.

Burtt-Davy, J. 1902. Stock ranges of northwestern California. USDA Bur. Plant Ind. Bull. 12. 81 p.

Clements, F. E. 1934. The relict method in dynamic ecology. J. Ecol. 22:39-68.

Cook, S. F., Jr. 1959. The effect of fire on a population of small rodents. Ecology 40:102-108.

Crampton, B. 1974. Grasses in California. Calif. Nat. Hist. Guide 33. Univ. Calif. Press, Berkeley, Calif.

Duncan, D. A., and J. N. Reppert. 1960. A record drought in the foothills. USDA Forest Serv., Pac. SW Forest and Range Exp. Sta. Misc. Paper 46. 6 p.

Duncan, D. A., and R. G. Woodmansee. 1975. Forecasting forage yield from precipitation in California's annual rangeland. J. Range Manage. 28:327-329.

Evans, R. A., B. L. Kay, and J. A. Young. 1975. Microenvironment of a dynamic annual community in relation to range improvement. Hilgardia 43:79-102.

Evans, R. A., and J. A. Young. 1972. Competition within the grass community, pp. 230-246. In V. B. Younger and C. M. McKell (eds.). The biology and utilization of grasses. Academic Press, New York.

Fitch, H. S. 1948. Ecology of the California ground squirrel on grazing lands. Amer. Midl. Natur. 39:513-596.

Fitch, H. S., and J. R. Bentley. 1949. Use of California annual-plant forage by range rodents. Ecology 30:306-321.

Halligan, J. P. 1973. Bare areas associated with shrub stands in grassland: the case of Artemisia californica. BioScience 23:429-432.

Harris, G. 1967. Some competitive relationships between Agropyron spicatum and Bromus tectorum. Ecol. Monogr. 37:89-111.

Heady, H. F. 1956a. Changes in a California annual plant community induced by manipulation of natural mulch. Ecology 37:798-812.

-----. 1956b. Evaluation and measurement of the California annual type. J. Range Manage. 9:25-27.

-----. 1958. Vegetational changes in the California annual type. Ecology 39:402-416.

-----. 1961. Continuous vs. specialized grazing systems: a review and application to the California annual type. J. Range Manage. 14:182-193.

-----. 1965. The influence of mulch on herbage production in an annual grassland. Proc. 9th Internat. Grassland Congr. 1:391-394.

-----. 1968. Grassland response to changing animal species. J. Soil Water Conserv. 23:173-176.

-----. 1972. Burning and the grasslands in California. Proc. Ann. Tall Timbers Fire Ecol. Conf. 12:97-107.

-----. 1975. Structure and function of climax, pp. 73-80. In Proc. Third Workshop of the U.S./Austral. Rangelands Panel. Soc. for Range Manage.

Heady, H. F. and D. T. Torell. 1959. Forage preferences exhibited by sheep with esophageal fistulas. J. Range Manage. 12:28-34.

Hedrick, D. W. 1948. The mulch layer of California annual ranges. J. Range Manage. 1:22-25.

Heizer, R. F., and M. A. Whipple. 1971. The California Indians: a source book. Univ. Calif. Press, Berkeley. 619 p.

Hendry, G. W. 1931. The adobe brick as a historical source. Agric. Hist. 5:110-127.

Hervey, D. F. 1949. Reaction of a California annual-plant community to fire. J. Range Manage. 2:116-121.

Hooper, J. F., and H. F. Heady. 1970. An economic analysis of optimum rates of grazing in the California annual-type grassland. J. Range Manage. 23:307-311.

Jain, S. K., D. R. Marshall, and K. Wu. 1970. Genetic variability in natural populations of soft chess (Bromus mollis L.). Evolution 24:649-659.

Janes, E. B. 1969. Botanical composition and productivity in the California annual grassland in relation to rainfall. M.S. thesis, Univ. Calif., Berkeley. 47 p.

Knowles, P. F. 1943. Improving an annual brome grass, Bromus mollis L., for range purposes. J. Amer. Soc. Agron. 35:584-594.

Küchler, A. W. 1964. Potential natural vegetation of the conterminous United States. Amer. Geogr. Soc. Spec. Pub. 36.

Kunzel, H. 1848. Upper California (transl. from German by Anthony and Max Knight). Book Club of Calif., San Francisco, 1967.

Lin, J. W. Y. 1970. Floristics and plant succession in vernal pools. M.S. thesis, San Francisco State Univ. 99 p.

Luckenbach, R. 1973. Pogogyne, polliwogs, and puddles - the ecology of California's vernal pools. Fremontia 1(1):9-13.

Major, J., C. M. McKell, and L. J. Berry. 1960. Improvement of Medusahead infested rangelands. Calif. Agric. Exp. Sta. Leafl. 123. 3 p.

Major, J., and W. T. Pyott. 1966. Buried viable seeds in two California bunchgrass sites and their bearing on definition of a flora. Vegetatio 13:253-282.

Marshall, D. R., and S. K. Jain. 1969. Interference in pure and mixed populations of Avena fatua and A. barbata. J. Ecol. 57: 251-270.

McCullough, D. 1971. The tule elk. Univ. Calif. Press, Berkeley. 209 p.

McGown, R. L., and W. A. Williams. 1968. Competition for nutrients and light between the annual grassland species Bromus mollis and Erodium botrys. Ecology 49:981-990.

McKell, C. M. 1974. Morphogenesis and management of annual range plants in the United States, pp. 111-123. In K. W. Kreitlow and R. H. Hart (eds.). Plant morphogenesis as the basis for scientific management of range resources. USDA Misc. Pub. 1271.

McKell, C. M., J. P. Robinson, and J. Major. 1962. Ecotypic variation in medusahead, an introduced annual grass. Ecology 43:686-698.

McNaughton, S. J. 1968. Structure and function in California grasslands. Ecology 49:962-972.

Munz, P. A., and D. D. Keck. 1959. A California flora. Univ. Calif. Press, Berkeley. 1681 p.

Murphy, A. H. 1970. Predicted forage yield based on fall precipitation in California annual grasslands. J. Range Manage. 23:363-365.

Perkins, J. E. 1864. Sheep husbandry in California, pp. 134-145. In Trans. Calif. State Agric. Soc. 1863.

Pitt, M. D. 1975. The effects of site, season, weather patterns, grazing, and brush conversion on annual vegetation, Watershed II, Hopland Field Station. Ph.D. dissertation, Univ. Calif., Berkeley. 281 p.

Ratliff, R. D., and H. F. Heady. 1962. Seasonal changes in herbage weight in an annual grass community. J. Range Manage. 15:146-169.

Ratliff, R. D., S. E. Westfall, and R. W. Roberts. 1972. More California poppy in stubble field than in old field. USDA Forest Serv. Res. Note PSW-271. 4 p.

Robbins, W. W. 1940. Alien plants growing without cultivation in California. Calif. Agric. Exp. Sta. Bull. 637. 128 p.

Sampson, A. W. 1944. Plant succession on burned chaparral lands in northern California. Calif. Agr. Exp. Sta. Bull. 685. 144 p.

Sampson, A. W., A. Chase, and D. W. Hedrick. 1951. California grasslands and range forage grasses. Calif. Agric. Exp. Sta. Bull. 724. 130 p.

Sampson, A. W., and B. S. Jesperson. 1963. California range brushlands and browse plants. Univ. Calif. Agric. Exp. Sta. Manual 33. 162 p.

Savelle, G. D., and H. F. Heady. 1970. Mediterranean annual species: their response to defoliation, pp. 548-551. In Proc. 11th Internat. Grassland Cong.

Shantz, H. L., and R. Zon. 1924. Atlas of American agriculture. Section E: Natural vegetation. USDA Bur. Agric. Econ. 29 p.

Stebbins, G. L. 1965. Colonizing species of the native California flora, pp. 173-191. In H. G. Baker and G. L. Stebbins (eds.). The genetics of colonizing species. Academic Press, New York.

Stebbins, G. L., and J. Major. 1965. Endemism and speciation in the California flora. Ecol. Monogr. 35:1-35.

Sumner, D. C., and R. M. Love. 1961. Resident range cover often cause of seeding failures. Calif. Agric. 15(2):6.

Talbot, M. W., and H. H. Biswell. 1942. The forage crop and its management, pp. 13-49. In C. B. Hutchison and E. I. Kotok (eds.). The San Joaquin Experimental Range Calif. Agric. Exp. Sta. Bull. 663.

Talbot, M. W., H. H. Biswell, and A. L. Hormay. 1939. Fluctuations in the annual vegetation of California. Ecology 20:394-402.

Tinnin, R. O., and C. H. Muller. 1971. The allelopathic potential of _Avena fatua_: influence on herb distribution. Bull. Torrey Bot. Club 98:243-250.

-----. 1972. The allelopathic influence of _Avenu fatua_: the allelopathic mechanism. Bull. Torrey Bot. Club 99:287-292.

Twisselmann, E. C. 1967. A flora of Kern County, Calif. Wasmann J. Biol. 25(1-2):1-395.

U. S. Forest Service, Pacific Southwest Forest and Range Experiment Station. 1954. Field manual, soil-vegetation surveys in California. 95 p.

University of California Department of Agronomy. 1959. Field manual, grassland sampling. Supplement No. 1 to Field manual, Soil-Vegetation Survey. 14 p.

White, K. L. 1966. Old-field succession on Hastings Reservation, California. Ecology 47:865-868.

-----. 1967. Native bunchgrass (_Stipa pulchra_) on Hastings Reservation, California. Ecology 48:949-955.

Zinke, P. J. 1962. The soil-vegetation survey as a means of classifying land for multiple-use forestry. Proc. 5th World For. Congr. 1:542-546.

CHAPTER 15 VERNAL POOLS

ROBERT HOLLAND
SUBODH JAIN
Department of Agronomy and Range Science, University of California, Davis

Introduction 516
Edaphic and physiographical features 518
Vegetation 519
Species diversity among pools within sites 520
Zonation: mound-to-pool gradients 521
Seasonal patterns and autecological studies 527
Biosystematic and evolutionary studies 528
Areas for future research 529
Literature cited 532

INTRODUCTION

Generally speaking, a vernal pool or hogwallow is a small, hardpan-floored depression in a valley grassland mosaic that fills with water during the winter. As it dries up in the spring, various annual plant species flower, often in conspicuous concentric rings of showy colors. In certain areas of the Great Valley where such depressions are locally abundant, there is a characteristic semirolling landscape, with hummocks of annual grassland vegetation distinctly circumscribing the pools as islands with a different annual, herbaceous flora. It is clear that pool vegetation is azonal, with edaphic factors more important than the regional climate which affects the surrounding vegetation. Vernal pool vegetation is unique.

In a pioneering study of endemism in the California flora, Hoover (1937) recognized eight regions of rich endemism, including the "hogwallows" and alkali sinks. In his flora of Kern Co., Twisselmann (1967) identified two distinctive plant associations within the vernal pools, namely, those of the Tehachapi and Glennville regions versus those of the Temblor and San Emigdio ranges, the latter approximating alkali sink vegetation. In San Luis Obispo Co., Hoover (1970) described vernal pools, with such typical plants as *Downingia cuspidata, Eryngium vaseyi, Boisduvalia glabella, Callitriche marginata, Pilularia americana,* and *Allocarya stipitata,* as distinct from saline plains with *Distichlis, Frankenia, Cressa,* and *Lasthenia ferrisiae.* Stone (1957) concluded that "of the many factors which may be collectively responsible for vernal pools, low precipitation, poor drainage, and high evaporation are factors which are equally responsible in the genesis of alkali soils." Thus soil moisture and salinity together may define another special subclass of vernal pool vegetation. Lin's (1970) study of vernal pools near Dozier Station (south of Dixon), Crampton's (1959) studies on *Orcuttia* and *Neostapfia* in Alkali Lake in the same area, and our own studies in Sacramento Co. provide evidence of contrasts between saline and nonsaline vernal pool floristics.

Theories on the origin of vernal pool topography are numerous and mostly concern formation of the mounds between the pools rather than the pools themselves. Mima type topography has been attributed to fracture patterns in the underlying hardpan (near Lincoln in Placer Co.: Begg, pers. commun.), activities of fossorial rodents, hydrostatic groundwater pressure, the former presence of a shrubland which allowed localized accumulations of windblown soil beneath individual shrubs, and the alternate expansion and contraction upon wetting and drying of expanding-lattice clay minerals. Each of these theories can be supported by field evidence in some locations, although none is substantiated in all, or even most, locations. The time required for the development of a pool is unknown. Pools apparently have been a part of the California landscape for at least tens of thousands of years, as judged from the wide and, in many cases, disjunct distribution and large number of species which are restricted to the habitat.

Vernal pool vegetation attains its highest development on terrace soils bordering the east side of the Central Valley at the base of the Sierran foothills (Hoover 1937). Apparently in pristine times these pools were a common feature in most areas of the valley and were most numerous on the younger terrace soils such as the San Joaquin series, but agricultural expansion, mineral extraction, and urban sprawl have eliminated most of these areas. The best remaining pools are now found on higher, older terraces.

Vernal pools are not limited to terrace soils, however, or even to the Central Valley. Well-developed pools occur on soils derived from volcanic mudflows on the Vina Plains in Tehama Co. and on lava-capped mesas such as Table Mt. near Oroville in Butte Co. or at Mesa de Colorado in Riverside Co. (Lathrop and Thorne 1968; Thorne and Lathrop 1969; Griggs 1974). Single pools may also be found in the North Coast Ranges, as at Boggs lake in Lake Co., and at 1200 m in the Temblor Range and eastern San Luis Obispo Co. (Twisslemann 1967). They also occur in mountain valleys in the Sierran foothills (e.g., Lynns Valley, Kern Co.). They are

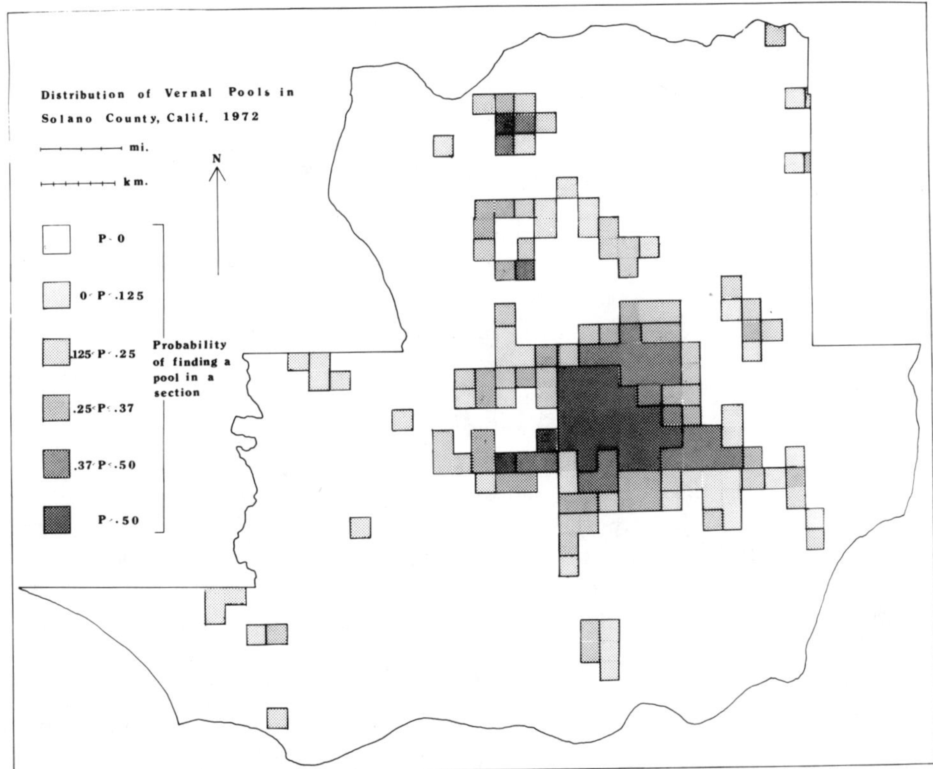

Figure 15-1. Map showing relative densities of vernal pools in Solano Co. Shades represent different probability classes per square mile.

most common and conspicuous in open grassland, but occur also under a variety of tree canopies ranging from *Quercus douglasii* savannah to *Pinus ponderosa* forests, in highly localized, understory herbaceous cover in the Coast Ranges, Pennisula Ranges, and Transverse Ranges, and as vernal ponds in dune slacks (Chapter 7).

The total area under vernal pool vegetation is perhaps a smaller fraction than might be suggested by the geographical range. It is rapidly diminishing in the Great Valley, with grazing, urbanization, and other land use developments. In Sacramento and Solano Cos., where vernal pools were presumably much more common in pristine times, we mapped the distribution of vernal pools from aerial photographs available in the county offices. Figure 15-1 shows the map for Solano Co. Three features should be noted: (1) there are variations in the density of pools per unit area so that the distribution is highly patchy and fragmented; (2) the pool area is continually shrinking and becoming further fragmented; and (3) the total area today of vernal pool vegetation is probably less than 1% (statewide it would be an even smaller figure).

Vernal pools offer several vegetational features of great interest to biogeographers, systematists, and ecologists. They are associations of unique native plants in "island" habitats that are well defined spatially, have a high proportion of endemics, exhibit an interplay between edaphic and other physical factors, and contain several genera with evidence of recent adaptive radiation. Since vegetation studies are inadequate in number, several excellent monographs on the biosystematics of individual genera and excellent floras will be our primary source for this chapter.

EDAPHIC AND PHYSIOGRAPHICAL FEATURES

Soil texture, moisture, salinity, and microrelief are important determinants of vernal pool vegetation. The terrace soils include some of the state's oldest; some have been isotopically dated at 600,000 yr (±3%; Arkley 1964). Such prolonged profile development has led to the formation of dense, impervious hardpans or claypans which inhibit subsurface drainage of rain water and result in water tables perched near the surface. The water in these pools evaporates during spring months, resulting in a concentration of ions in the pool water and soil.

Soil reactions may vary along the pool-to-mound gradient. Linhart (1972) reported soil pH data for three Solano Co. soils in October, March, and April. On Capay soils the pH difference between pool bottom and mound top in October was −0.3, while in March in the same pools the pH increased along the same gradient (ΔpH = +0.5). In the center, the pH was 6.4–6.5 in October, 8.3 in March, and 6.1 in April. Purer (1939) observed a similar seasonal increase in pH of the water standing in 25 pools in San Diego Co. Lin (1970) measured the soil pH of several Solano Co. pools and obtained values of 6.60–6.85, with smaller pools tending to have lower pH values. Linhart's data suggest that in Solano Co. the soil pH is usually, though not always, higher in the pool center than at the periphery, whereas in other areas (e.g., at Rancho Seco in Sacramento Co.) the pool soil may be slightly more acid than surrounding soils (Holland, unpub. data).

Thus hogwallows are in some respects analogous to playas in desert areas: conductivity measurements of soil-distilled water pastes (3:1) indicate about a threefold increase in ion concentration in pool bottom soils when compared with soils on adjacent mounds. The osmotic and qualitative effects of these ions on plant success have not been documented (see Walter 1973 for a good discussion).

Particle size distribution in pool surface soils is usually different from that in the surrounding areas, but no rules may be drawn. On the lower terraces bearing San Joaquin soils, pool soils (which are mapped as a separate soil series, the Alamo series, in the largest pools) contain considerably more clay than upland soils (Cole et al. 1954), whereas near Dozier Station in Solano Co. pool surface soils are more sandy than soils in surrounding areas (Carpenter and Cosby 1931).

Purer (1939) reported monthly precipitation data for 4 yr (1934–38) in relation to the duration of inundation in the pools. The total annual rainfall figures were 388, 216, 401, and 247 mm, respectively, with standing water only in February 1936 and February–April 1938. Thus year-to-year fluctuations in precipitation do provide a highly variable water regime.

Purer (1939) described the physiographical features of vernal pools in San Diego Co. in some detail. The pools are more or less oval in shape, with the long axis usually in a north–south direction, and they vary in size from nearly 700 to less than 1 m^2. Pool depth in the center varies in the range of 25–60 cm (also see Linhart 1972).

VEGETATION

We accumulated a list of 101 plant species known to occur typically in hogwallows. More than 70% were native annuals; introduced annuals comprised less than 7% of the species. All of the 3 ferns and fern allies were native. Two of the 19 perennials were introductions, but none of the 3 geophytes was introduced. Fifty-five of the species had ranges limited entirely within California, 14 barely crept over the state boundaries, and 18 were limited to western North America. The remaining species were either cosmopolitan (e.g., *Juncus bufonius*) or introduced from other continents. Surprisingly, South American species were more numerous than European, Mediterranean, and Eurasian taxa combined. No trees, shrubs, or stem succulents are known to occur in vernal pools.

Pools may be classified physiographically into three general types: valley pools, terrace pools, and pools of volcanic areas.

Pools on valley floors occur in basin areas and on low plains. They are usually saline and/or alkaline, often have lime-carbonate nodules in the lower parts of the soil (Carpenter and Cosby 1931, Arkley 1964), and usually are underlain by an impervious claypan. These pools support such alkali- or salt-associated plants as *Distichlis spicata, Downingia bella, Lepidium latipes, Arenaria californica, Astragalus tener, Cressa truxillensis* var. *vallicola, Grindelia camporum, Allocarya leptocladus,* and *Trifolium depauperatum.* Examples of such pools occur in the vicinity of Dozier Station in Solano Co., in Stanislaus Co. within a narrow belt 3–7 km east of the San Joaquin River, and at the Nature Conservancy's Pixley Reserve in Tulare Co.

Terrace pools comprise most of the pools remaining in the Central Valley. These old weathered soils are typically neutral or slightly acid, and there can be an iron–silica cemented hardpan (Cole et al. 1954). Most of the characteristic vernal pool taxa can be found here: *Alopecurus howellii, Blennosperma nanum, Boisduvalia glabella, Brodiaea hyacinthina, Deschampsia danthonoides, Evax caulescens, Gratiola ebracteata, Isoetes orcuttii, Juncus bufonius, J. uncialis, Limnanthes douglasii* var. *rosea, Lilaea scilloides, Allocarya stipitata, Lythrum hyssopifolia, Navarretia intertexta, Psilocarphus brevissimus,* and several species of *Downingia, Eryngium,* and *Lasthenia,* as well as most species in the unusual grass genus *Orcuttia.* Examples of such pools occur near Sloughhouse in Sacramento Co., in the vicinity of Planada (Merced Co.) and LaGrange (Stanislaus Co.), and near Woodlake in Tulare Co.

Pools in volcanic areas such as the Vina Plains in Tehama Co. and on the Mesa de Colorado in Riverside Co. owe their existence to a shallow R horizon. Although they are readily distinguished physiographically, they appear floristically similar to the terrace pools.

Thus pools occupy intermediate positions along several environmental gradients in California. They are intermediate in the duration of inundation between freshwater marsh and grasslands. They are also intermediate in soil ion concentration between grassland and the alkali sink communities (Mason 1957).

Species Diversity among Pools within Sites

In an unpublished study of aerial photographs of vernal pools in Sacramento Co., we determined that pools are not randomly mingled in grassland, as judged by the use of Pielou's (1964) patches-and-gaps test. Construction of a variance:block size diagram (Kershaw 1974) indicated a pattern on two scales: one at about 3500–4000 m^2, which corresponds roughly to the size of the larger individual pools in the area, and another at about 10 ha, corresponding roughly to clusters, or archipelagos, of pools.

The range of sizes of vernal pools spans many orders of magnitude; pools as small as 1–2 m^2 are common, whereas Alkali Lake, an enormous "pool" in Solano Co., covers nearly 60 ha. Some genera, notably *Orcuttia* and *Neostapfia,* are apparently limited to the largest of these pools (Crampton 1959).

If one compiles species lists for several individual pools within a single site, two striking points emerge: the diversity within individual pools (α-diversity) is usually quite low, whereas the diversity among the several pools (β-diversity) is rather high. In conducting such an analysis on 10 pools at Rancho Seco in Sacramento Co., we found 20.5 species per pool (range 16–25), yet of the 60 species encountered only 3 were common to all pools and 7 occurred in single pools only. The 3 ubiquitous species were *Navarretia leucocephala, Juncus bufonius* (actually a "strand species"), and *Eryngium vaseyi* var. *vallicola.* They had mean covers (point frame) over the 10 pools of 15, 4, and 5%, respectively. The 3 species together comprised only 35% of the total plant cover when all 10 pools were considered. Pairwise comparisons of the 10 pools indicate that 30–50% of the species may be common in any pair of pools. These results suggest that an insular biogeographical approach might be applied to vernal pool vegetation in an effort to understand the observed patterns of species diversity.

We performed a stepwise regression analysis of species diversity determinants (Johnson et al. 1968) on the 10 Rancho Seco pools, and concluded that species number is best predicted by a curvilinear multivariate expression involving pool area, pool depth, and the amount of bare ground. In every analysis using species number as the dependent variable, either area or the natural logarithm of area was "chosen" first by the regression algorithm, but in no case was the regression significant at even the 10% level unless at least one additional parameter, namely, some function of pool depth, was added.

The analysis was repeated using the Shannon–Weaver index, H' ($= -\Sigma\, p_i \log p_i$, where p_i is some measure of relative abundance of the ith species in the pool). The index was best predicted by a simple linear expression of the amount of bare ground ($H' = -2.41 + 0.02$ BG, $p < .10$, $R^2 = 0.34$). Incorporation of additional parameters in every case resulted in a decrease in the value of F. Logarithmic transformation of the index produced similar results, and simple expressions of bare ground gave the best predictions of diversity.

Table 15-1 gives a summary of Purer's (1939) data on 10 selected pools in San Diego Co. Pools were described as wet or dry at the time of sampling and as clayey or stony, based on broad soil texture differences. Actual plant census data are presented here in terms of relative frequencies, which are used to compute H' and J' ($J' = H'/\log$ total number of species) values as measures of α-diversity. Note that one or two species are dominants within any pool, and therefore the values of H' or J' tend to be low.

Zonation: Mound-to-Pool Gradients

The ephemeral nature of standing water in a pool results in many physical–chemical gradients which correspond to the topographical gradient from pool bottom to mound top. Prolonged standing water is apparently detrimental to the establishment of many annual grassland species. Sheep ranchers in the Dozier area report better establishment of grasses in low areas, and hence better forage, in years of slightly below normal precipitation. As a result of this flooding, a sharp ecotone between grassland and pool floras is usually immediately evident. Whereas in surrounding grassland plant cover may exceed 100%, most pools have a characteristically low total cover, frequently less than 15–30%.

One of the most conspicuous features of vernal pool vegetation is the concentric arrangement of species in relation to topography. At some times of the year, topographical maps of individual pools may be prepared on the basis of zonation of various plants. These maps are as accurate as maps prepared with transit and stadia. Specialization by pool taxa for portions of the topographical gradient is apparent. Lin (1970) established three zones to describe this concentric pattern.

Zone 1 corresponds roughly to the pool bottom and is characterized by such taxa as *Evax caulescens, Isoetes howellii, Marsilea mucronata, Boisduvalia glabella, Mimulus tricolor, Myosurus minimus* (several varieties), *Navarretia intertexta, N. leucocephala, Neostapfia colusana, Pilularia americana, Allocarya humistrata, A. hystricula, A. stipitata* var. *micrantha,* and the genera *Downingia, Eryngium, Psilocarphus,* and *Orcuttia.*

Zone 2 occurs around the margins of the pools and includes *Alopecurus howellii, Blennosperma nanum, Deschampsia danthonoides, Gratiola ebracteata, Juncus*

TABLE 15-1. A summary of Purer's data: relative densities of selected taxa in San Diego Co. Asterisk indicates underestimate due to groupings of many grasses and liverworts

Some Common Genera	Pools Damp; Center Samples					Pools Dry; Edge			Pools Dry and Stony	
	Pool: 1	2	5	12	13	14	15	18	6	8
Crassula	0.267	0.413	0.533	0.010	0.114	0.055	0.078	0.161	0.324	...
Psilocarphus	.251	.184	.029	.108	.161	.062	.263	.153	...	...
Downingia	.153	.150	.003	.360	.034	...	.184	.041	.312	0.368
Pilularia	.149	.155	.403	...	...	...		...	...	...
Isoetes	.133	.087	.002	.384	.331	.241	...	.557	.353	...
Callitriche	.036	.002	.013	...	.331	.482	.009	...	...	...
Pogogyne	...	...	...	...	...	...	...	...	...	.614
Deschampsia	...	.004	.001	.026	...	...	...	.021	.002	.013
Juncus	...	...	...	...	...	...	.263	...	...	...
Eleocharis	...	...	...	.096	.010	.002	.009	.008	...	...
Abs. density (Total number of plants per square meter)	12,340	20,545	8432	4161	2718	1451	5702	7253	3390	1385
Total number of species (S)	7*	7*	11*	9*	8*	8*	9*	11*	5*	6*
H'	1.69	1.49	0.95	1.40	1.51	1.39	1.72	1.38	1.14	0.82
J'	0.87	0.76	0.39	0.63	0.72	0.66	0.78	0.57	0.71	0.46

bufonius, J. capitatus, Layia chrysanthemoides, Lilaea scilloides, Limnanthes douglasii var. *rosea, Machaerocarpus californicus, Orthocarpus capestris, Allocarya acanthocarpus, A. distantiforus, A. stipitata, Pogogyne zizyphoroides, Trifolium barbigerum, T. cayathiferum, T. depauperatum, T. fucatum, T. variegatum, Veronica peregrina* spp. *xalapensis,* and most species of the genus *Lasthenia.*

Zone 3 occurs on the higher, better-drained mounds and supports California annual type grassland with such genera as *Bromus, Avena, Erodium, Festuca, Lomatium,* and *Hordeum,* and in some areas the native perennial *Stipa pulchra.*

Kopecko and Lathrop (1975) established five zones (Table 15-2) along the topographical gradient in their studies in Riverside Co. (1) A dry grassland zone dominated by *Bromus mollis, Avena barbata,* and *Erodium cicutarium.* (2) A "vernally moist zone" dominated by *Bromus mollis, Festuca megalura,* and *Hemizonia paniculata.* (3) A "muddy margin zone" dominated by *Tillaea aquatica, Allocarya undulata, Callitriche marginata, Eleocharis acicularis, Pilularia americana, Juncus bufonius, Elatine californica,* and *Downingia cuspidata.* In this zone, dominance is lower than in any other zone. (4) A zone characterized by vernally standing water and dominated by *Callitriche longipedunculata, Elatine californica, E. chilense, Eleocharis macrostachya, Isoetes orcuttii, Downingia cuspidata, Allocarya undulata,* and *Lilaea scilloides.* (5) A zone temporally as well as spatially defined as being the dried marsh bed in July. Here dominance was shared by *Navarretia prostrata, Psilocarphus brevissimus, Eryngium aristulatum* var. *parishii, Lythrum hyssopifolia, Eleocharis macrostachya,* and *Hordeum californicum.* Note from the values of H' and J' (Table 15-2) that diversity, as measured by total number of species (S), is higher than in Purer's site, and so are values of H' and J'.

We conducted a gradient analysis of 19 point frame transects in 10 pools at Rancho Seco in Sacramento Co. The transects were divided into six equal, relative segments. Species richness seems to vary in relation to the topographical gradient (Fig. 15-2); it is highest at the pool margin, slightly lower in the grassland, and considerably lower in the pool. Apparently, low total cover is correlated with species richness [species number = 3.18 + 0.61 BG (percent bare ground); $R^2 = 0.43$]. This may be a response to decreased competition (Daubenmire 1968) or to reduced dominance (McNaughton and Wolf 1970).

The pool habitat has resisted invasion by introductions. Unlike marshes, which have accumulated many exotic species, primarily rice culture weeds (Mason 1957), very few introduced species have been successful in the pool environment. In the grassland, 38% of the species are introduced, mostly from Europe and the Mediterranean region, whereas the pool flora (as noted earlier and see Fig. 15-2) contains only 5–10% introduced species. Introduced species in the pools come primarily from South America and Africa.

Species whose cover is greatest in the first segment of the gradient are typical annual grassland species (Fig. 15-3). Of 10 such species, only 3 occur at all in the actual pools. These plants, *Erodium botrys, Briza minor,* and *Festuca dertonensis,* were all observed as seedlings on still-moist soil late in the season. They probably represent seeds which matured in the adjacent grassland earlier in the same growing season.

Total plant cover is lowest and number of species is highest in the second segment of the gradient. Only two species reach their highest cover here, *Juncus bufonius* and *Pogogyne zizyphoroides,* the latter attaining nearly 4% cover.

TABLE 15-2. Zonation and measures of diversity. Data of Kopecko and Lathrop (1975)

Species	Relative Cover of a Few Selected Species				
	Zone				
	Hummock A	B	C	D	Center E
Avena barbata	20.3	12.7	0	0	0
Bromus mollis	18.1	19.8	0	0	3.0
B. rubens	5.6	0	0	0	0
Callitriche longipedunculata	0	0	0	11.4	0
C. marginata	0	0	2.5	1.4	0
Downingia cuspidata	0	0	4.9	5.7	5.0
Eleocharis macrostachya	0	4.2	8.6	11.4	7.4
E. acicularis	0	0	8.6	5.7	3.6
E. parishii	0	0	4.9	2.8	5.5
Elatine californica	0	0	4.9	4.3	0
E. chilensis	0	0	1.2	7.1	0
Eryngium aristulatum parishii	0	0	0	0	12.4
Isoetes howellii	0	0	1.2	2.9	4.7
I. orcuttii	0	0	1.2	18.6	4.1
Lasthenia chrysostoma	0.7	1.4	0	0	0
Lythrum hyssopifolia	3.7	0	4.9	0	3.6
Marsilea vestita	0	0	4.9	1.4	3.0
Navarretia prostrata	0	0	0	0	15.4
Allocarya nothofulvus	3.7	0	0	0	0
A. undulatam	0	1.6	9.9	15.7	0
Trifolium amplectens truncatum	2.07	3.1	3.7	0	0
Total number of species (S)	20	16	23	15	20
H'	2.40	2.48	2.89	2.44	2.80
J'	0.80	0.89	0.92	0.90	0.93

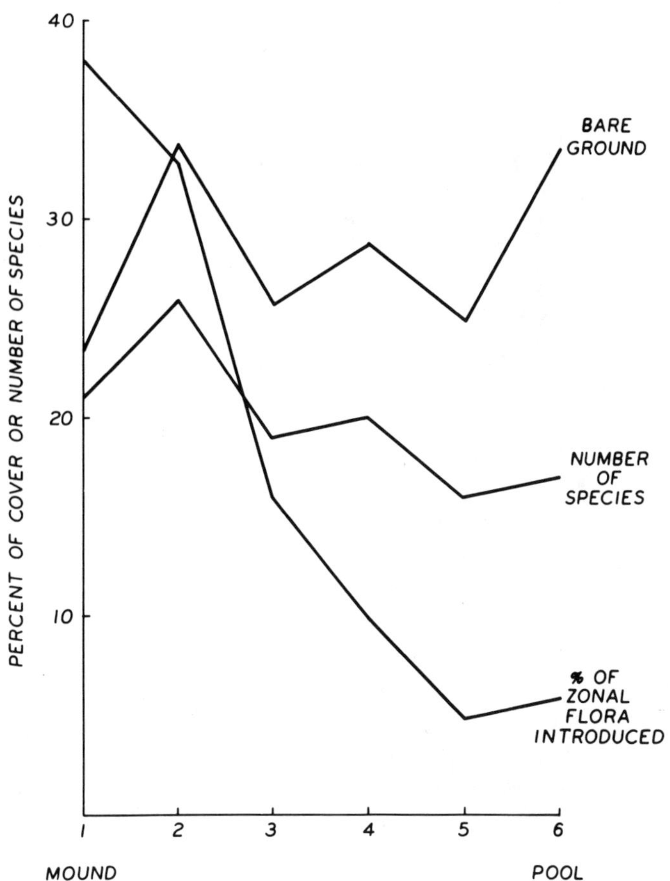

Figure 15-2. Some properties of vegetation along the mound-to-pool gradient (zones 1 to 6) at Rancho Seco site, Sacramento Co.

Not a single species attains high cover in the third segment of the gradient. This represents a discrete change in community composition, so that the species discussed above may be considered as grassland or strand species, and the species common in the central three segments as pool species.

Navarretia leucocephala, Crassula erecta, and *Brodiaea hyacinthina* attain highest cover in the fourth segment of the gradient, with the first species contributing nearly 33% of the total plant cover and dominating the vegetation.

Eryngium vaseyii var. *globosum, Allocarya stipitata* var. *micrantha,* and *Cressa truxellensis* var. *vallicola* peak in the fifth segment (Fig. 15-4). Together they contribute nearly 45% of the total plant cover, with the first two species having nearly equal abundance, and the third being much less important.

The last segment (Fig. 15-5) is characterized by *Lasthenia chrysostoma, Downingia cuspidata,* and *Psilocarphus oregonus,* with *Lasthenia glaberrima, Downingia bella,* and *Allocarya stipitata* also abundant here, but of considerably less importance.

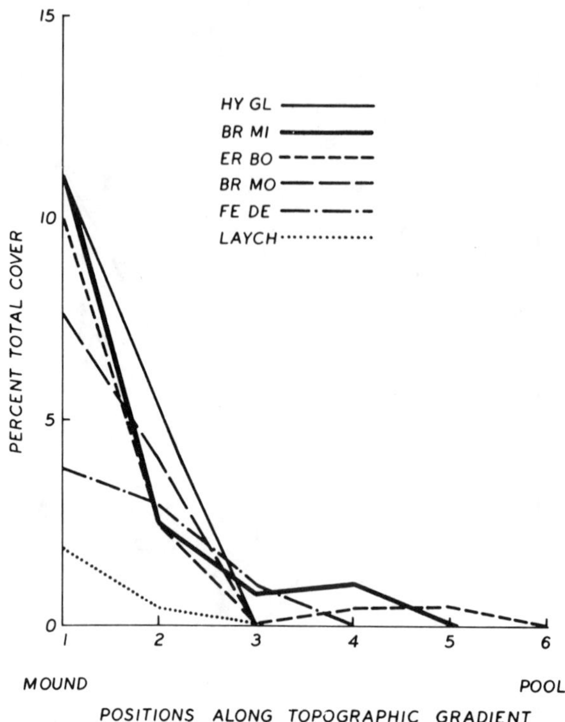

Figure 15-3. Cover of some hillock species at Rancho Seco. Species: HYGL = *Hypochoeris glabra*, BRMI = *Briza minor*, ERBO = *Erodium botrys*, BRMO = *Bromus mollis*, FEDE = *Festuca dertonensis*, LAYCH = *Layia chrysanthemoides*.

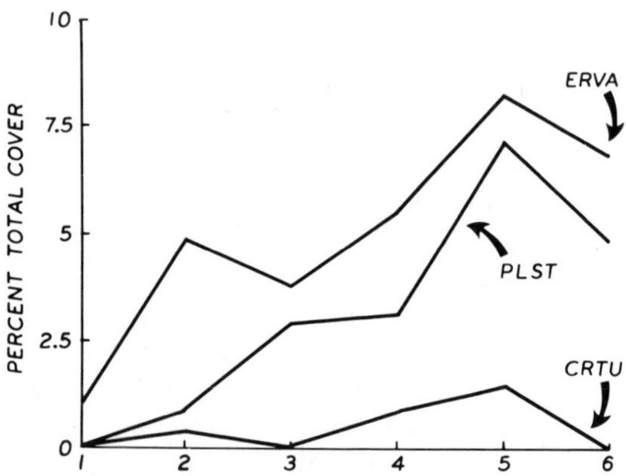

Figure 15-4. Cover of some pool species at Rancho Seco. Species: ERVA = *Eryngium vaseyii* var. *globosum*, PLST = *Plagiobothrys* (= *Allocarya*) *stipitata* var. *micranthus*, CRTU = *Cressa truxillensis*.

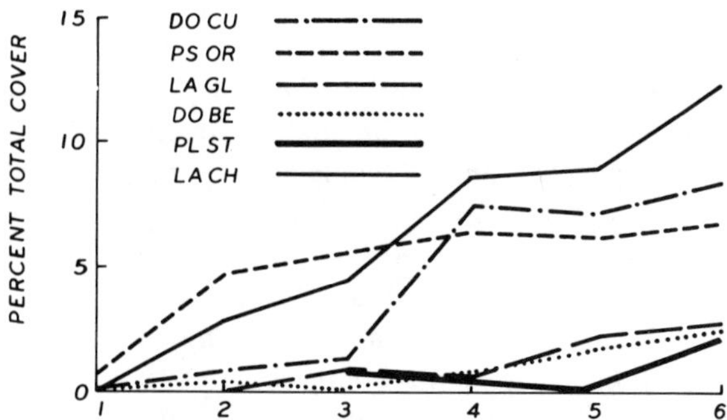

Figure 15-5. Cover of some pool species at Rancho Seco. DOCU = *Downingia cuspidata*, PSOR = *Psilocarphus oregonus*, LAGL = *Lasthenia glaberrima*, DOBE = *Downingia bella*, PLST = *Plagiobothrys stipitata*, LACH = *Lasthenia chrysostoma*.

In his study of saline pools near Byron, with *Myosurus* species relatively common, Stone (1957) noted distinct zonation of species in relation to the levels of salt concentration in different "rings": some species were largely restricted to one ring, whereas others could occur in two or more disjunct "rings." This pattern offers an excellent opportunity for analyzing the relative niche widths of different coexisting species in relation to soil moisture and salinity.

Seasonal Patterns and Autecological Studies

In his detailed phenological study, Lin (1970) scored the timing of germination, flowering, and maturity of numerous vernal pool species at weekly intervals. A sample of his data is given in Fig. 15-6. Differences in length of life cycle, "seasonal succession" in flowering, and congeneric overlap are interesting in relation to the

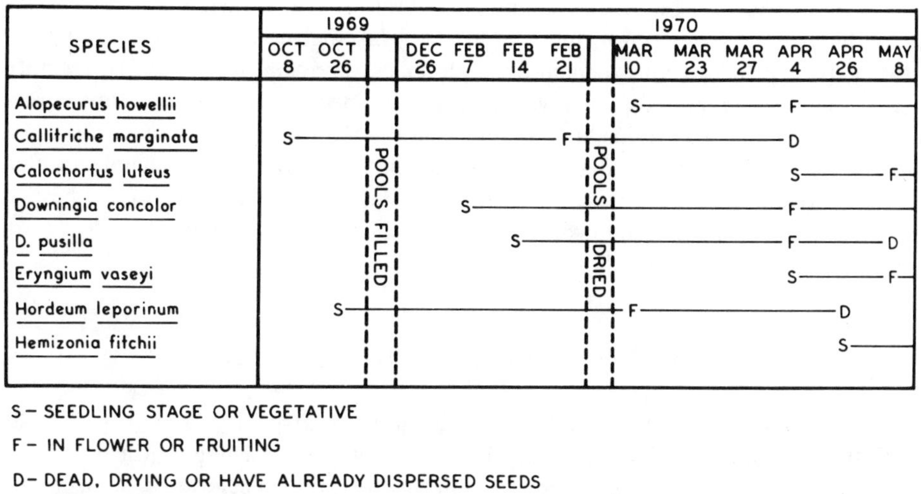

SPECIES	1969								1970					
	OCT 8	OCT 26		DEC 26	FEB 7	FEB 14	FEB 21		MAR 10	MAR 23	MAR 27	APR 4	APR 26	MAY 8
Alopecurus howellii			POOLS FILLED					POOLS DRIED	S			F		
Callitriche marginata	S						F					D		
Calochortus luteus												S		F
Downingia concolor					S							F		
D. pusilla						S						F		D
Eryngium vaseyi												S		F
Hordeum leporinum		S							F				D	
Hemizonia fitchii													S	

S – SEEDLING STAGE OR VEGETATIVE
F – IN FLOWER OR FRUITING
D – DEAD, DRYING OR HAVE ALREADY DISPERSED SEEDS

Figure 15-6. Phenology of selected species in vernal pools south of Dozier, Solano Co. After Lin (1970).

seasonality of temperature, moisture, photoperiod, and other environmental variables. Several autecological studies have paid special attention to the biology of seed germination: McLaughlin (1974) for *Callitriche*, Hauptli (unpub. data) for *Limnanthes*, Griggs (1974) for *Orcuttia*, and comments by Crampton (1954) for *Navarretia*. Dormancy varies among congeners, perhaps in relation to altitude, moisture regime, and summer temperature (Weiler 1962; Ornduff 1966; Cruden 1972). Lin (1970) experimentally studied the submergence factor by placing quadrats of soils under water for varying periods and recording germination sequences. Monocotyledons showed rapid germination with moisture but not submergence, whereas *Callitriche longipedunculata* seeds required a period of complete submergence. *Crassula aquatica*, on the other hand, germinated at several depths and with surface moisture only. Lin (1970) noted the potential interactions between aeration, light, and temperature variables and suggested the need for more accurate data. *Limnanthes* species showed a diversity of response to light quality, temperature, and storage time at high temperatures before germination. Another significant line of study in *Limnanthes* involved the role of seed surface micromorphology in flotation, safe-site requirements, and possibly zonation within pools (Webster, Jain, and Hauptli, unpub. data). Observations in controlled conditions need to be related to habitat differences; for example, *L. douglasii* var. *nivea* occurs in larger pools or floodplain areas in pastures, whereas *L. douglasii* var. *rosea* is found in well-defined smaller pools with conspicuous radial zonation around them. The role of fatty acid component in seed oils in *Limnanthes* may also have significance in adaptation to aquatic habitats. Electrophoretic assays of several enzymatic activities have also proved valuable, especially for alcohol dehydrogenase, known to be significant for flood tolerance.

Baker's (1972) survey of seed weights showed smaller seeds in aquatic and semiaquatic species, presumed to be an adaptation for better dispersal and faster root development. In *Limnanthes* species, a range of nutlet sizes seems to be related to other factors such as altitude, inbreeding, and colonizing ability, but further analysis is needed.

Comparisons of the ecology and morphology of floating or submerged, amphibious, and terrestrial groups might show fascinating differences. Crampton's (1954) study provided an excellent example in *Navarretia*, Griggs (1974) has noted several interesting patterns in revising *Orcuttia* species relationships, and Brown (unpub. data) has new extensions of numerical taxonomy of *Limnanthes alba* and *L. floccosa* that, with genetic data on variation patterns, would be valuable in niche analyses. In summary, vernal pool vegetation offers numerous areas for syntheses of traditional physiological and evolutionary approaches to the problems of species distribution, abundance, and adaptability.

BIOSYSTEMATIC AND EVOLUTIONARY STUDIES

Several biosystematic monographs on individual vernal pool genera (e.g., Ornduff 1966 on *Lasthenia*; Weiler 1962 and McVaugh 1941 on *Downingia*; Mason 1952 on *Limnanthes*) have noted that various conditions for rapid speciation, such as annual growth habit, entomophily and inbreeding, habitat subdivision, low dispersal, and karyotypic changes, are present in the vernal pool flora. New species were recently described in *Limnanthes, Lasthenia,* and *Blennosperma.* Mason (1957) commented on the varietal and ecotypic diversification in several genera. Recent systematic treatments of *Orcuttia, Myosurus,* and *Downingia* considered some of these factors of rapid speciation. It is significant to note that many genera with one or two species are confined to vernal pools, whereas others, like *Lasthenia* and *Limnanthes,* are undergoing adaptive radiation and are able to invade the habitats around the pools. The role of polyploidy and chromosome structural changes is perhaps relevant to larger species complexes in several genera, like *Downingia.* High rates of speciation might thus be accompanied by conditions favoring high rates of extinction. Stebbins (1974) analyzed the opportunities for adaptive radiation in terms of the average number of species per genus (S/G). For mesic, open country, which presumably included the vernal pool flora, the S/G figure is 9.7, in contrast to 3.0 for California chaparral and 5.4 for all California vegetation types.

Ornduff (1966) noted in *Lasthenia* that members of the same taxonomic section rarely grow together, although they may occupy similar habitats in one area. Sympatry was almost exclusively among members of different sections. Likewise, Weiler's (1962) study showed that sympatry within individual pools is low but is higher in terms of overlapping regional distributions. In *Limnanthes, Lasthenia,* and *Downingia,* with several species per genus and several regions with overlapping ranges of two or more species, "true" sympatry was found in less than 10% of all the pools reported. Studies on pollination biology of related species (e.g., Cruden 1967, 1972; Thorp, pers. commun.) are relevant to the origin of seasonal isolation in cases of sympatry, but in *Downingia* Weiler (1962) observed several cases of interspecific pollination by insects. Grant's (1971) model of speciation and origin of inbreeding under competition for pollinators deserves careful study. Ornduff (1966) further noted that lowered fertility of intraspecific hybrids in some species may be related to continued genetic divergence of spatially isolated populations (see also Wood 1961).

Several features of the population structure of vernal pool species would be of great interest for population biology research, namely, subdivided populations with low effective sizes, low dispersability, competition at the zonal boundaries with other species vis-à-vis intraspecific competition, low species α-diversity, and variations in the breeding system (see Stone 1957; Carlquist 1974; Jain 1976; MacArthur and Wilson 1967). A study of variation between central and peripheral populations of *Veronica peregrina* is summarized in Table 15-3, showing microdifferentiation over small distances. Brown and Jain (rept. in prep.) found almost no electrophoretic variation in populations of *Limnanthes floccosa,* an inbreeder, and rather low amounts in *L. alba,* an outbreeding species. Although *L. floccosa* occurs over a wide variety of habitats located peripherally north of the *L. alba* range, different subspecies represent adaptive radiation. Ornduff (1966), on the other hand, noted that marginal populations of several *Lasthenia* species had the highest concentration of morphological or cytological variants within a species.

AREAS FOR FUTURE RESEARCH

At least four research areas emerge as high priorities: (1) analysis of the ecology of vernal pool vegetation, with refined definitions of the types of pools and detailed surveys of species diversity patterns and their ordination along some environmental

TABLE 15-3. Microgeographical differentiation in *Veronica peregrina*. Adapted from Linhart (1972). Items marked with asterisk are from greenhouse cultures

Observation	Center	Periphery
Dominant species	*V. peregrina*	*Agrostis alba*, *Avena barbata*, *Lolium temulentum*
V. peregrina density	1.5 cm^2	0-0.02 cm^2
Period soil inundated	1-2 mo	1-2 days, periods infrequent
Relative predictability of soil moisture	High	Low
Minimum and maximum	0-16°C	-8 to 18°C
Seed germination*	Rapid and uniform	Variable
Seed output*	Fewer and larger seeds	Many, smaller seeds
Total biomass output*	Lower	Higher
Seed/vegetative biomass ratio*	Higher	Lower
Phenotypic variability for various metric traits*	Lower	Higher

variables; (2) genecology of selected large and small genera for parallel evolutionary studies; (3) applications of the theory of island biogeography to the development of conservation strategies; and (4) autecological and adaptation studies on the biology of seed germination, survivorship, and reproductive patterns.

A statewide survey of species composition in vernal pools of varying size, depth, association with other communities, and latitudinal position (north–south gradient) would be useful for a better analysis of the patterns of species diversity and gradient responses.

Several biosystematic studies on individual genera place us in a better position than most biogeographers of other vegetation types in the ability to define several evolutionary patterns and hypotheses. The applications of island models of biogeography should be obvious. We need population estimates of relative net reproductive rates, seed dispersal, rates of turnover in species numbers, and the role of genetic variation. This would allow developing models for defining conservation strategies in terms of the number, size, and relative location of vernal pool reserves (see Simberloff and Abele 1976; Jain 1975; and several papers in Duffey and Watt 1971 for discussions of these approaches).

Finally, in close relationship with the population ecological studies, several questions on adaptive strategies arise at the autecological level. We see certain features of seed biology, heterophylly, response to submergence versus amphibious environments, response to environmental stresses, and overall phenology giving an epiharmonious flora. This offers excellent opportunities for joint work on the genetic, physiological, and anatomical bases of adaptation. Walter (1973) has advanced compelling arguments for ecophysiological research in relation to the analysis of vegetation.

LITERATURE CITED

Arkley, R. J. 1964. Soil survey of the eastern Stanislaus area, California. USDA Soil Conserv. Serv. Ser. 1957(20):1-160.

Baker, H. G. 1972. Seed weight in relation to environmental conditions in California. Ecology 53:997-1010.

Carlquist, S. 1974. Island biology. Columbia Univ. Press, New York. 660 p.

Carpenter, E. J., and S. W. Cosby. 1931. Soil survey of the Suisun area, California. USDA Bur. Chem. Soils 18:1-60.

Cole, R. C., L. K. Stromberg, O. F. Bartholomew, and J. L. Retzer. 1954. Soil survey of the Sacramento area, California. USDA Soil Conserv. Serv. Ser. 1941(11):1-101.

Crampton, B. 1954. Morphological and ecological considerations in the classification of Navarretia (Polemoniaceae). Madroño 12:225-238.

---. 1959. The grass genera Orcuttia and Neostapfia: a study in habitat and morphological specializatin. Madroño 15:97-110.

Cruden, R. W. 1967. Genecological studies in Nemophila menziesii (Hydrophyllaceae). Ph.D. dissertation, Univ. Calif., Berkeley.

---. 1972. Pollination biology of Nemophila menziesii (Hydrophyllaceae) with comments on the evolution of obligolectic bees. Evolution 26:373-389.

Daubenmire, R. 1968. Plant communities. Harper & Row, New York. 300 p.

Duffey, E., and A. S. Watt (eds.). 1971. The scientific management of animal and plant communities. Blackwell, Oxford. 652 p.

Grant, V. 1971. Plant speciation. Columbia Univ. Press, New York. 435 p.

Griggs, F. T. 1974. Systematics and ecology of the genus Orcuttia (Gramineae). M.S. thesis, Chico State Univ., Chico, Calif. 70 p.

Hoover, R. F. 1937. Endemism in the flora of the Great Valley of California. Ph.D. dissertation, Univ. Calif., Berkeley.

---. 1970. The vascular plants of San Luis Obispo County, California. Univ. Calif. Press, Berkeley. 350 p.

Jain, S. K. 1975. Genetic reserves, pp. 379-396. In O. Frankel and J. G. Hawkes (eds.). Crop genetic resources for today and tomorrow. Cambridge Univ. Press.

---. 1976. The evolution of inbreeding in plants. Ann. Rev. Ecol. Syst. (in press).

Johnson, M. P., L. G. Mason, and P. H. Raven. 1968. Ecological parameters and plant species diversity. Amer. Natur. 102:297-306.

Kershaw, K. A. 1974. Quantitative and dynamic ecology. American Elsevier, New York. 183 pp.

Kopecko, K. J. P., and E. W. Lathrop. 1975. Vegetation zonation in a vernal marsh on the Santa Rosa Plateau of Riverside County, California. Alsio 8:281-288.

Lathrop, E. W., and R. F. Thorne. 1968. Flora of the Santa Rosa Plateau of the Santa Ana Mountains, California. Aliso 6:17-40.

Lin, J. 1970. The floristic and plant succession in vernal pools vegetation. M.A. dissertation, San Francisco State Coll., San Francisco. 99 p.

Linhart, Y. B. 1972. Intrapopulation differentiation in annual plants. I. Veronica peregrina L. raised under non-competitive conditions. Evolution 28:232-243.
MacArthur, R. H., and E. O. Wilson. 1967. The theory of island biogeography. Princeton Univ. Press, Princeton, N.J. 203 pp.
Mason, C. T. 1952. A systematic study of the genus Limnanthes. Univ. Calif. Pub. Bot. 25:455-512.
Mason, H. L. 1957. A flora of the marshes of California. Univ. Calif. Press, Berkeley. 878 p.
McLaughlin, E. G. 1974. Autecological studies of three species of Callitriche native in California. Ecol. Monogr. 44:1-16.
McNaughton, S. J., and L. Wolf. 1970. Dominance and the niche in ecological systems. Science 167:131-139.
McVaugh, R. 1941. A monograph on the genus Downingia. Mem. Torrey Bot. Club 19:1-57.
Ornduff, R. 1966. A biosystematic survey of the goldfield genus Lasthenia. Univ. Calif. Pub. Bot. 40:1-92.
Pielou, E. C. 1964. The spatial pattern of two-phase patchworks of vegetation. Biometrics 20:156-174.
Purer, E. A. 1939. Ecological study of vernal pools, San Diego County, California. Ecology 20:217-229.
Simberloff, D. S., and L. G. Abele. 1976. Island biogeography theory and conservation practice. Science 191:285-286.
Stebbins, G. L. 1974. Flowering plants: evolution above the species level. Harvard Univ. Press, Cambridge, Mass. 399 p.
Stone, D. 1957. A unique balanced breeding system in the vernal pool mousetails. Evolution 13:151-174.
Thorne, R. F., and E. W. Lathrop. 1969. A vernal marsh on the Santa Rosa Plateau of Riverside County, Calif. Aliso 7:85-95.
Twisselmann, E. C. 1967. A flora of Kern County, California. Wassmann J. Biol. 25:1-395.
Walter, H. 1973. Vegetation of the earth. Springer-Verlag, New York. 237 p.
Weiler, J. H. 1962. A systematic study of the genus Downingia. Ph.D. dissertation, Univ. Calif., Berkeley.
Wood, C. E., Jr. 1961. A study of hybridization in Downingia (Campanulaceae). J. Arnold Arbor. 42:219-262.

THE SIERRAN FLORISTIC PROVINCE

CHAPTER

16

MONTANE AND SUBALPINE FORESTS OF THE TRANSVERSE AND PENINSULAR RANGES

ROBERT F. THORNE

Rancho Santa Ana Botanic Garden, Claremont, California

Introduction **538**
Lower montane coniferous forests **539**
Coulter pine forest **539**
Yellow pine forests **540**
Ponderosa pine forest **541**
Jeffrey pine forest **542**
Piñon pine **544**
Rocky flats **544**
Upper montane coniferous forests **544**
White fir–sugar pine forest **545**
Mountain juniper woodland **546**
Subalpine coniferous forests **547**
Lodgepole pine forest **547**
Limber pine forest **549**
Other plant communities of higher elevations **551**
Aquatic communities **551**
Mountain meadow **552**
Aspen woodland **553**
Mountain talus and other rupicolous plants **554**
Status of knowledge **554**
Acknowledgments **555**
Literature cited **556**

INTRODUCTION

The Transverse Ranges of California, as usually defined, include the ranges oriented largely east and west which lie south of the Sierra Nevada and the Coast ranges and the linking Tehachapi Mts. They include, from west to east, the Santa Ynez, Santa Monica, Liebre, Sawmill, Pellona, San Gabriel, San Bernardino, Little San Bernardino, Eagle, and Pinto ranges. Mount Pinos, possibly best considered the southernmost and highest mountain of the Tehachapi Mts., will be mentioned occasionally with the Transverse Ranges because of its height (2760 m) and its transitional status between the northern and the Transverse ranges. The remaining high mountains of southwestern California—the San Jacinto, Santa Rosa, and Santa Ana Mts. on south into San Diego Co., and down the length of Baja California—have a north–south orientation and are generally called the Peninsular Ranges. Since their forests are rather similar to those of the Transverse Ranges, they too are included in this chapter.

Natural populations of conifers are missing from the Santa Monica Mts. (Raven and Thompson 1966), the highest elevation being only 890 m (National Geographic Society 1974). The Santa Ynez Mts., which rise in one peak to only 1520 m, possess stands of just two conifers, *Pinus coulteri* and *Pseudotsuga macrocarpa* (Smith 1952). Hence these and the relatively low Liebre (1620 m), Sawmill (1810 m at Burnt Peak), and Pellona (1510 m) ranges hold little interest for us here. The Little San Bernardino, Eagle, and Pinto Mts. are desert ranges too dry and too low to support the vegetation discussed below. Good altitudinal zonation in the coniferous forests, however, can be found on Mt. Pinos and the San Gabriel, San Bernardino, and San Jacinto Mts., and these will therefore receive the most attention.

LOWER MONTANE CONIFEROUS FORESTS

Above the southern California grassland and mixed chaparral of the Californian Floristic Province in the higher mountains are well-zoned conifer and oak forests and woodlands that belong mostly to the Sierran Floristic Province. They are essentially islands of Sierran vegetation surrounded on all sides by either Californian or Warm Desert Floristic Provinces.

Coulter Pine Forest

Pinus coulteri forms an open forest or woodland that is essentially a transitional community. Since this pine ranges from the San Francisco Bay Area down the western half of the state through the Coast, Transverse, and Peninsular ranges into northern Baja California and is absent from the Sierra Nevada, the Coulter pine forest has to be considered a component of the Pacific, rather than the Sierran, coniferous forest. In elevation it ranges as low as 230 m on Mt. Diabolo in the Bay Area, 460 or 610 m in the Santa Lucia Mts., 1000 m in the Santa Ynez Mts., and usually not below 1070 but exceptionally down to 550 m in the Transverse and Peninsular ranges. Its most extensive occurrence in the San Gabriel Mts. is in the area of Mt. Gleason, Chilao and Charlton Flats to Mill Creek Summit, and Alder Saddle east to Mt. Waterman, mostly at elevations between 1220 and 1830 m on dry, rocky slopes and ridges. Isolated stands in the San Gabriel Mts. are also known from the southwestern slope of Mt. Islip near Snow Spring and in Coldwater Canyon on the south slope of Mt. San Antonio (A. Lewis, pers. commun.). On cismontane slopes in the San Bernardino Mts., Coulter pines reach as high as 2135 m, and in the San Jacinto Mts., to 2290 m (Hall 1902).

The trees are often so widely spaced in Coulter pine forest that it can, with more accuracy, be called a woodland. In the northern part of its range, Coulter pine associates with *Pinus sabiniana* of the foothill woodland and with *P. ponderosa* of the ponderosa pine forest; in the Santa Lucia Mts., it associates with chaparral and with *Abies bracteata* of the northern mixed evergreen forest. In southern California it associates with chaparral, with *Pseudotsuga macrocarpa,* and with *Pinus ponderosa, P. jeffreyi, Quercus kelloggii, Q. chrysolepis,* and other plants of the yellow pine forest. In the areas of the San Gabriel, San Bernardino, and San Jacinto Mts. where it is the most abundant or the only pine, it forms the lowest zone of the montane pine forests, generally with an understory of chaparral shrubs or pineland annual and perennial herbs in very open forest or woodland. It thus comprises a rather broad ecotone community between mixed chapparal and yellow pine forest.

Between 1375 and 1830 m in the San Bernardino Mts. (Minnich 1976) are depauperate forests of *P. coulteri* and *Quercus kelloggii* or *Q. chrysolepis.* Farther south in the Peninsular Ranges, *P. coulteri* is widespread to the north end of the Sierra San Pedro Mártir. The Coulter pine groves are often found in the chaparral belt; and even when they form the lowermost forest belt in the lower montane zone, the sparse shrub understory is composed of such chaparral species as *Arctostaphylos glandulosa, A. pringlei* ssp. *drupacea, A. pungens, Ceanothus integerrimus,* and *Cercocarpus betuloides* ssp. *betuloides.*

In the Coulter pine flats of the Chilao–Charlton Flat–Alder Saddle–Mt.

Waterman area mentioned above, there are, among the wider-ranging plants of chaparral and pine forest, some species absent from or rare in other pine forests of the San Gabriel Mts. Representative of these distinctive species are *Deschampsia danthonioides, Frasera neglecta, Garrya flavescens* ssp. *flavescens, Heterocodon rariflorum, Horkelia bolanderi* ssp. *parryi, Nemacladus sigmoideus, Plagiobothrys californica, Rigiopappus leptocladus, Scutellaria austinae, Triteleia dudleyi, Viola douglasii,* and *Wyethia ovata.* That a number of these species are also found at lower elevations on the Mojave Desert slopes of the San Gabriels is not too surprising in view of the presence of scattered trees of the usually transmontaine *Pinus monophylla* at Pinyon Flats and the Mt. Waterman area.

Another conifer sometimes found in the chaparral belt is the closed-cone *P. attenuata,* rather rare in southern California. It is known from Sugarloaf Peak at about 1160 m in the Santa Ana Mts. and from the south face of the San Bernardino Mts. in a belt 1 km wide and about 400 ha in area from City Creek to Government Canyon (Minnich 1976), at an elevation of 1070–1220 m, in both ranges in the upper chaparral zone. Not indigenous in the San Gabriels, it has been planted by the U.S. Forest Service in a number of places and seems to be thriving (e.g., with *P. coulteri* in the old Charlton Flat burn at 1615 m).

Two other closed-cone conifers, *Cupressus forbesii* and *C. arizonica* ssp. *stephansonii,* similarly form small groves in chaparral, the former from 460 to 1525 m in the Santa Ana Mts. of Orange Co. and on Otay Mt. and Mt. Tecate of San Diego Co., and the latter in the headwaters of King Creek on the southwestern slope of Cuyamaca Peak from 915 to 1375 m in San Diego Co. (Munz 1974). The three closed-cone conifers can be regarded as representing small islands of inland closed-cone coniferous woodland in the mixed chaparral of southern California. One other cypress, *C. montana,* is, however, an element of the Jeffrey pine forest in the Sierra San Pedro Mártir in Baja California, at elevations of about 2300–2750 m. The closed-cone woodlands are discussed in more detail in Chapter 9.

Yellow Pine Forests

Most of the lower montane coniferous forests are dominated by the two yellow pines, *Pinus ponderosa* and *P. jeffreyi,* the latter being much more abundant in southern California. *Pinus ponderosa* in the Transverse and Peninsular ranges is the common pine on lower, more mesic cismontane slopes, usually between 1375 and 2135 m. *Pinus jeffreyi,* on the other hand, is dominant at higher levels between 2135 and 2750 m on both cismontane and transmontane slopes, and indeed is usually the only yellow pine on the transmontane slopes of southern California, there occurring as low as 1220 m.

The two yellow pines are surely closely related and, as demonstrated by Parratt (1965, 1967), appear to intergrade rather completely in a hybrid zone between 1985 and 2135 m, at least in San Antonio Canyon. Earlier, Haller (1962) studied hybridization between the two species, and found a relatively constant but low rate of hybridization in all localities where the two species grew together. The open, parklike ponderosa and Jeffrey pine forests (Fig. 16-1) are rather similar in appearance, and they have many associated species in common. Nevertheless, there are considerable floristic differences between the two yellow pine forests, perhaps

Figure 16-1. Ponderosa pine forest around Lake Fulmor in the San Jacinto Mts., San Bernardino National Forest, Riverside Co., elevation 1615 m. View looking south across Lake Fulmor to north-facing slope covered with open forest of *Pinus ponderosa, P. coulteri, P. lambertiana, Abies concolor, Calocedrus decurrens,* and *Quercus kelloggii.*

due to the different elevational and ecological zones they occupy. Hence the two forests are treated here as separate communities.

Ponderosa pine forest. In the Transverse and Peninsular ranges, *Pinus ponderosa* forms an open, parklike forest on rocky cismontaine slopes above the mixed chaparral and Coulter pine communities, generally at elevations between 1375 and 2015 m. It forms only relict colonies on Mt. Pinos and in the Tehachapi Mts. (Twisselmann 1967). Vogl and Miller (1968) identified no *P. ponderosa* in their ecological study of the coniferous forest on the south slope of Mt. Pinos. In the Cuyamaca Mts. of San Diego Co., ponderosa pine reaches its southern limit near Cuyamaca Reservoir at 1400 m. Rarely is it found on the transmontane slopes of the Transverse Ranges, since it apparently is less tolerant of drought and low temperatures than is *P. jeffreyi* (Haller 1959, 1962). Also, its susceptibility to smog damage has caused widespread destruction in some portions of the San Gabriel and San Bernardino Mts., with resultant clear-felling of dying stands in the local national forests, as in the San Sevaine and Arrowhead Lake areas.

In the San Gabriel Mts., the lowest stands of *P. ponderosa* reported are along the Angeles Crest Highway above 1370 m (Lewis and Gause 1966) and near Lower San Sevaine Flats at about 1430 m. Other stands examined were along the Angeles Crest Highway at 1525 m, on the summits of Mt. Wilson and Mt. Gleason at 1740 and 1890 m, respectively, and at Crystal Lake at 1830 m. Stands studied by Parratt (1965, 1967) were at Charlton Flats between 1615 and 1650 m, Barley Flats between 1675 and 1710 m, and San Antonio Canyon between 1710 and 2015 m. In San Antonio Canyon, between 2015 and 2135 m, a hybrid zone of pine forest exhibits varying combinations of characteristics of *P. ponderosa* and *p. jeffreyi.*

Associated with *P. ponderosa* at lower elevations, especially in ecotonal areas with Coulter pine forest, are *P. coulteri* and *Quercus kelloggii.* Under more mesic conditions, as on upper north- or east-facing slopes of canyons, associated trees are *Pseudotsuga macrocarpa, Calocedrus decurrens, Quercus chrysolepis,* and *Cornus nuttallii* of the southern mixed evergreen forest. At higher elevations on more mesic slopes, ponderosa pine mixes with *Abies concolor* and *Pinus lambertiana.* Shrubby undergrowth in the open forests is rather limited, with generally only scattered specimens of *Arctostaphylos glandulosa, A. pringlei* ssp. *drupacea, Ceanothus integerrimus, Eriodictyon trichocalyx, Garrya flavescens, Lupinus excubitus, L. formosus, Rhamnus californica* ssp. *cuspidata, Ribes roezlii,* and *Solanum xanti,* among others.

Perennial and annual herbs in the ponderosa pine forest are represented by numerous species, often presenting an attractive display at different times in the growing season. Species common here but not usually found in the Jeffrey pine forest are *Calystegia occidentalis* ssp. *fulcrata* (= *C. fulcrata*), *Carex multicaulis, Clarkia rhomboidea, Collinsia childii, Cordylanthus filifolius, Eriastrum densifolium* ssp. *austromontanum, Gilia splendens* ssp. *grantii, Iris hartwegii* ssp. *australis, Linanthus ciliatus, Phacelia imbricata* ssp. *imbricata, Silene lemmonii, Streptanthus bernardinus,* and *Viola purpurea* ssp. *purpurea.* Common native perennial grasses of the open forest are *Bromus breviaristatus, B. orcuttianus* ssp. *hallii, Koeleria macrantha, Melica imperfecta,* and *Poa scabrella.* Many of the most representative pineland species, however, appear equally at home in both yellow pine forests, as *Arceuthobium campylopodum, Asclepias eriocarpa, Castilleja martinii, Chaenactis santolinoides, Chimaphila menziesii, Galium johnstonii, Gayophytum heterozygum, Lotus davidsonii, Mimulus johnstonii, Penstemon grinnellii, P. labrosus, Sarcodes sanguinea,* and *Sitanion hystrix.*

Jeffrey pine forest. *Pinus jeffreyi* ranges as far south as the Sierra San Pedro Mártir in Baja California. It appears to be more tolerant of serpentine soils, drought, low temperatures, and smog than *P. ponderosa,* and equally tolerant of soil moisture and warm temperatures. Thus in southern California it predominates in the yellow pine forests at higher (Fig. 16-2) and more exposed sites, mostly to the exclusion of *P. ponderosa.* South of Lake Cuyamaca, *P. ponderosa* is entirely absent.

Pinus jeffreyi overlaps with *P. ponderosa* in a relatively narrow band on cismontane slopes, as in the hybrid zone described by Parratt (1967) in San Antonio Canyon. This hybrid zone can be considered an ecotone between the two yellow pine forests, much like the often rather wide bands occurring between other adjoining plant communities. On Mojave Desert slopes of the San Gabriel Mts. *P. jeffreyi* commonly is associated with *P. monophylla,* and on desert slopes of the San Bernardino Mts. it is even more abundantly mixed with *P. monophylla* and with *Juniperus occidentalis* ssp. *australis.* On pine flats and slopes with transmontane exposure in the Alder Saddle area, it associates with *P. coulteri,* also not far distant from stands of *P. monophylla* and *P. ponderosa.* On some of the higher, more mesic north- or east-facing slopes, *P. jeffreyi* forms a mixed forest with *Abies concolor, Pinus lambertiana,* and *Calocedrus decurrens.* Although the upper limits of the Jeffrey pine forest in the Transverse Ranges is usually 2590–2750 m, depending on exposure, individual trees have been observed to 2900 m on Mt. San Antonio and

Figure 16-2. Open, parklike Jeffrey pine forest at Icehouse Saddle, elevation 2320 m, on south slope of Timber Mt., Angeles National Forest—San Bernardino National Forest boundary, San Gabriel Mts. Forest floor largely bare except for pine-needle duff and colonies of *Arctostaphylos patula* ssp. *platyphylla.*

even to 3050 m in the San Bernardino Mts., usually in krummholz form, mingled with *P. contorta* ssp. *murrayana* and *P. flexilis.*

In the open Jeffrey pine forests of dry, rocky cismontane slopes in the San Gabriel Mts., a number of low, dense shrubs are associated with the conifers mentioned above. Most conspicuous are *Arctostaphylos patula* ssp. *platyphylla* (often heavily parasitized by *Boschniackia strobilacea*), *Chrysolepis sempervirens, Eriogonum wrightii* ssp. *subscaposum, Lupinus elatus,* and *Symphoricarpos parishii.* Most of these species also occur on the more arid ridges and transmontane slopes, but they are joined there by such shrubs rare or absent on cismontane slopes as *Artemisia tridentata, Ceanothus cordulatus, Chrysothamnus nauseosus, Prunus virginiana* ssp. *demissa, P. emarginata,* and *Tetradymia canescens.*

Quercus kelloggii is associated with *Pinus ponderosa* on cismontane slopes in the San Sevaine area, but it is an abundant small tree in lower parts of the *P. jeffreyi* forest on transmontane slopes in Prairie Fork, Swarthout Valley, and the Big Pines to Jackson Lake areas between 1830 and 2135 m and Big Pines to Inspiration Pt. (about 2257 m). *Quercus kelloggii* becomes more abundant in the San Bernardino and San Jacinto ranges between 1525 and 2135 m, reaching at least to 2225 m in the former range. In some areas, such as the Rim of the World, *Q. kelloggii* forms pure stands (Minnich 1976), probably because of the removal of associated conifers by crown fires or intensive logging. *Quercus chrysolepis,* normally a component of southern mixed evergreen forest, is an important subdominant of yellow pine forests on steep, south-facing slopes in the San Bernardino Mts. (Minnich 1976). Its wide elevational range varies from 760 m in all the canyons up to 2440 m on the ridges.

Numerous herbs characterize the open Jeffrey pine forest on the dry, rocky slopes and more level flats. Species largely limited to Jeffrey pine or higher forests are

Arabis repanda, Cordylanthus nevinii, Eriogonum parishii, Fritillaria pinetorum, Gayophytum diffusum ssp, *parviflorum, Ivesia santolinoides, Lupinus peirsonii, Melica stricta, Penstemon bridgesii, P. caesius*, and *Stipa parishii.*

PIÑON PINE. Although the piñon woodlands of the transmontane slopes of the Transverse Ranges are discussed in Chapter 24, the occasional occurrence of *Pinus monophylla* in yellow pine and higher forests in the ranges (sometimes on cismontane slopes) should be mentioned here. Stands have been examined by A. B. Cossey (pers. commun.) in Cow Canyon, Pinyon Gulch, and Cattle Canyon in the drainage of the East Fork of the San Gabriel River. In Cattle Canyon, from about 2045 m, the piñon ranges over about 325 ha, associated primarily with *Pinus jeffreyi, Cercocarpus ledifolius, C. betuloides* ssp. *betuloides, Quercus chrysolepis, Abies concolor*, and *Yucca whipplei.* Piñon is also adjacent to southern mixed evergreen forest in the shaded gullies. Stands of piñon have also been reported from the Mt. Waterman area, in the North Fork of Lytle Creek at about 1985 m, and on Pinyon Ridge east of Crystal Lake at about 1770 m. In the San Bernardino Mts., piñon is a frequent component—in some places one of the dominants—of Jeffrey pine and mountain juniper forests and woodlands, but usually on the northern transmontane slopes above 2135 m and up to 2440 m, where the influence of the desert climate is quite obvious. Associated species there, in addition to the three conifers, are *Cercocarpus ledifolius* and *Quercus chrysolepis.*

ROCKY FLATS. A plant community of open, gravelly flats, benches, ridges, or mesas, mostly within Jeffrey pine or Jeffrey pine–piñon forests from about 1985 to 2285 m on transmontane slopes, deserves mention here because of the rather special habitat, as yet hardly studied, which harbors an unusual number of local endemics and other plants of great interest. It would appear that these gravelly or rocky pavements are probably rather moist in the spring because of inadequate drainage but dry out quite thoroughly during the summer and fall, thus being somewhat reminiscent of the much wetter vernal pools of lower elevations in cismontane areas (Thorne and Lathrop 1969, 1970; Kopecko and Lathrop 1975). Referred to especially are the flat top of Table Mt. at about 2225 m near Big Pines in the San Gabriel Mts., and flats about Holcomb and Bear valleys, particularly near Sawmill Canyon at 2195 m, in the San Bernardino Mts. Among the more characteristic species of this habitat are *Androsace septentrionalis* ssp. *subumbellata, Antennaria dimorpha, Arabis holboellii* ssp. *retrofracta, *A. parishii, *Arenaria ursina, Astragalus bicristatus, A. leucolobus, A. purshii* ssp. *lectulus, Bouteloua gracilis, Carex douglasii, *Castilleja cinerea, *C. martinii* ssp. *ewanii, Draba douglasii* ssp. *crockeri, *Eriogonum kennedyi* ssp. *austromontanum, *Ivesia argyrocoma, I. santolinoides, Lewisia rediviva* ssp. *minor, Lomatium nevadense* ssp. *parishii, Lupinus breweri* ssp. *grandiflorus, Machaeranthera canescens, Phlox austromontana, Plantago purshii* var. *oblonga, Selaginella asprella*, and *Viola purpurea* ssp. *mohavensis.* The six taxa with asterisks are thought to be restricted to the San Bernardino Mts.

UPPER MONTANE CONIFEROUS FORESTS

The decision as to where one should treat the various coniferous communities has to be rather arbitrary because of the wide elevational ranges that many of the dominant conifers occupy. Because of the altitudes at which the white fir–sugar pine forest and the mountain juniper woodland are usually found, I prefer to treat them as upper montane forests, in contrast to the yellow pine forests of the lower montane zone. At the same time, I am well aware that Jeffrey pine forest on dry, rocky, exposed slopes can reach as high as 2900 m in the Transverse Ranges, with individual trees occasionally joining lodgepole and limber pines in the krummholz just below the alpine zone.

White Fir–Sugar Pine Forest

Abies concolor ssp. *concolor* and *Pinus lambertiana* are regular components on more mesic slopes in the yellow pine forests; however, they also form a community of their own on moister, steep, north- and east-facing slopes (Fig. 16-3) and ridges at higher elevations. This forest, found between 1675 and 2590 m, often includes *Calocedrus decurrens*. At about 2380 m in the San Gabriel Mts. (rarely as low as 1985 m, according to Lewis and Gause 1966), *Pinus contorta* ssp. *murrayana* becomes a component of the forest until it assumes dominance as lodgepole pine forest above 2440 m, as on the north slope of Mt. Baden-Powell. Both *Pinus lambertiana* and *Abies concolor* ssp. *concolor* (Hamrick and Libby 1972) have wide elevational ranges in southern California, ranging through the ponderosa and Jeffrey pine forests into the lodgepole pine forest. Some of the latitudinal extremes for *P. lambertiana* are 1525–2750 m in the San Jacinto Mts. (Hall 1902) and 1465–2750 m in

Figure 16-3. White fir–sugar pine forest on north-facing slopes (2000–2600 m) along West Fork of Lytle Creek, north of Mt. San Antonio, San Bernardino National Forest, San Gabriel Mts. The trees in the canyon bottom and on the south slopes are mostly *Pinus jeffreyi*, forming the usual very open Jeffrey pine forest.

the San Gabriel Mts., and for *A. concolor* ssp. *concolor,* 1280–2960 m in the San Jacinto Mts. and 1675–3035 m in the San Gabriel Mts.

Associated with *Abies concolor* ssp. *concolor, Pinus lambertiana,* and *Calocedrus decurrens* in the white fir–sugar pine forest are their parasites: *Phoradendron bolleanum* ssp. *pauciflorum* and *Arceuthobium abietinum* on fir, *A. californicum* on sugar pine, and *P. juniperinum* ssp. *libocedri* on incense-cedar. Also present are various shrubs like *Ribes nevadense, R. roezlii, Rubus parviflorus, Salix coulteri,* and *Sambucus caerulea.*

Mountain Juniper Woodland

The stands of *Juniperus occidentalis* ssp. *australis* in the San Gabriel and San Bernardino Mts. are widely disjunct from the subspecific center of distribution at high elevations in the Sierra Nevada (Vasek 1966). This gnarled, picturesque tree in the San Gabriel Mts. occupies dry, rocky, exposed ridges and slopes from about 2440 m on Mt. San Antonio's Gold Ridge and upper Cattle Canyon to 2750 m on the Devil's Backbone, 2865 m on Dawson's Peak, and 2925 m on the summit and north-trending ridge of Pine Mt. The lowest record is a remarkable tree at 2365 m in Icehouse Canyon, which, A. B. Cossey (pers. commun.) has suggested, is perhaps the largest tree in southern California. It is 14.5 m in circumference, 22.2 m tall, and possibly 3000 yr old.

The mountain juniper is usually scattered and seldom dominant in the San Gabriel Mts. With it are equally scattered and often much gnarled or warped specimens of *Pinus jeffreyi, P. contorta* ssp. *murrayana,* and *Abies concolor* ssp. *concolor.* Subdominants are *Cercocarpus ledifolius,* sometimes a sizable tree, and such shrubs as *Arctostaphylos patula* ssp. *platyphylla, Chrysolepis sempervirens, Holodiscus microphyllus, Prunus virginiana* ssp. *demissa,* and *Ribes roezlii.* The rather specialized rupicolous herbaceous flora of the juniper stand on the rocky ridge of Devil's Backbone includes *Allium burlewii, Caulanthus amplexicaulis, Chaenactis santolinoides, Cycladenia humilis* ssp. *venusta* (on steep talus), *Eriogonum davidsonii, E. saxatile, E. umbellatum* ssp. *minus, E. wrightii* ssp. *subscaposum, Gayophytum diffusum* ssp. *parviflorum, Hulsea vestita* ssp. *parryi, Mentzelia laevicaulis, Mimulus johnstonii, Oreonana vestita, Pellaea compacta, Penstemon bridgesii, P. grinnellii, Phacelia austromontana, Silene parishii* var. *latifolia,* and *Zauschneria californica* ssp. *californica.*

In the San Bernardino Mts., the mountain juniper is much more abundant and reaches lower elevations. Although essentially a high-elevation tree, it ranges down on the northern, or desert, side of the range from 2750 to about 2045 m with stragglers reaching even lower in the upper piñon woodland (Vasek 1966). On the dry flats and rocky slopes of Bear and Holcomb valleys it is one of the dominant trees, irregular in shape and usually 6–15 m tall, often growing intermixed with *Pinus jeffreyi, P. monophylla, Quercus kelloggii, Q. chrysolepis,* and *Cercocarpus ledifolius.* Other conspicuous associates are *Amelanchier pallida, Artemisia tridentata, Ceanothus greggii, Eriogonum wrightii* ssp. *subscaposum,* and *Fremontodendron californicum,* most of which emphasize the desert aspect of this community, and two parasites, *Phoradendron juniperinum juniperinum* on the juniper and *Arceuthobium divaricatum* on the piñon.

The other juniper woodlands of the Transverse and Peninsular ranges (discussed in Chapters 23 and 24) are not closely related to the mountain juniper woodland and are found chiefly below the piñon woodland at 1525 m or below.

SUBALPINE CONIFEROUS FORESTS

Some subalpine coniferous forests are present on the highest mountains of southern California, as on Mt. Pinos and the highest peaks of the San Gabriel, San Bernardino, San Jacinto, and Santa Rosa ranges. Their floras are, however, rather xerophytic and depauperate as compared with the extensive, often rather wet, subalpine forests of the Sierra Nevada.

Lodgepole Pine Forest

The lodgepole pine has a remarkably wide range, from the Black Hills of South Dakota and the Yukon to the Sierra San Pedro Mártir of Baja California, and from sea level to timberline. In southern California, *Pinus contorta* is represented in the San Gabriel, San Bernardino, and San Jacinto Mts. by the subspecies *murrayana* (Critchfield 1957), which forms extensive forests on the higher mountains from 2440 m on north slopes or 2600 m on south slopes to the tops of the mountains, about 3050 m on Mt. San Antonio and 3234 m on Mt. San Jacinto, to 3508 m (as krummholz on the peak) on Mt. San Gorgonio. On moist, shaded, north-facing slopes, *P. contorta* ssp. *murrayana* reaches as low as 1985 m in the San Gabriel Mts. (Lewis and Gause 1966), 2015 m in Bear Valley, and 2440 m in the San Jacinto Mts. It is absent from the Tehachapi Mts. and from Mt. Pinos.

A tall, straight tree to 18 or 21 m, it forms denser forests than the other pines. There is an especially beautiful forest of *P. contorta* ssp. *murrayana* on the north slope of Mt. Baden-Powell, where it is dominant from about 2440 to 2750 m (Fig. 16-4). *Pinus flexilis* joins it there at about 2590 m and is dominant from 2750 m to the summit at about 2867 m, where both species are still present. Underbrush in the lodgepole pine forest is usually not dense, but consists largely of *Ceanothus cordulatus* and *Chrysolepis sempervirens* in the openings. In denser stands, mycotropic plants like *Corallorhiza maculata, Pterospora andromedea, Sarcodes sanguinea,* and various species of *Pyrola* are among the few herbs present in the rather depauperate flora of this forest. Snowmelt gullies, described later in the "Mountain Meadow" section of this chapter, are much richer in herbs, and are found in lodgepole pine forests.

On Mt. San Antonio (Mt. Baldy), lodgepole pine appears above Baldy Notch and is relatively dominant in the forest from about 2590 m to near the summit, where it is the only "tree," present in low, matted, twisted krummholz form. On even the lower slopes it forms a very open forest with rather low, stunted trees scattered amid much montane chaparral of *Arctostaphylos patula* ssp. *platyphylla, Ceanothus cordulatus,* and *Chrysolepis sempervirens,* and accompanied here and there by equally stunted or warped trees of *Pinus jeffreyi, Abies concolor* ssp. *concolor, Juniperus occidentalis* ssp. *australis,* and *Cercocarpus ledifolius.*

Because of its wind-swept nature, the peak of Mt. San Antonio (3070 m) is relatively barren, without even krummholz, even though it is not above timberline. The community from the summit to 30 m below has dwarfed specimens of *P. contorta* ssp.

Figure 16-4. Lodgepole pine forest on the north slope of Mt. Baden-Powell, 2440 m, Angeles National Forest, San Gabriel Mts. The stand of *Pinus contorta ssp. murrayana* ia almost pure but more open than usual, allowing large colonies of *Chrysolepis sempervirens* and some *Ceanothus cordulatus* to occupy the openings as a low understory.

murrayana, Allium monticola, Arabis platysperma, Calyptridium parryi, Carex rossii, C. subfusca, Chrysolepis sempervirens, Collinsia torreyi ssp. *wrightii, Draba corrugata, Eriogonum saxatile, E. umbellatum* ssp. *minus, Galium parishii, Heuchera abramsii, Holodiscus microphyllus, Leptodactylon pungens* ssp. *pulchriflorum, Oreonana vestita, Ribes cereum, R. montigenum, Sitanion hystrix,* and *Viola purpurea* ssp. *xerophyta.* Mount Baldy is the type locality for several of these plants. A few other species found not far down slope are *Arenaria nuttallii* spp. *gracilis, Calochortus invenustus, Erigeron breweri* ssp. *jacinteus, Monardella cinerea,* and *Silene verecunda* ssp. *platyota.* Many of the same species occur with *Pinus contorta* ssp. *murrayana* on the peaks of other high mountains in the San Gabriel, San Bernardino, and San Jacinto ranges.

The lodgepole pine forest is far more extensive in the San Bernardino Mts., occupying some 8.7 km^2, including islands of limber pine forest (Grinnell 1908). Parish (1917) treated lodgepole pine as an indicator of the Canadian Zone, beginning usually about 2440 m on north slopes and rising to 2750 or 3050 m, where it merges into the Hudsonian Zone, as indicated by *Pinus flexilis.* "Canadian" islands of lodgepole pine forest occur as low as 2285 m around Bluff Lake and as low as 2044 m in Bear Valley near the dam (Minnich 1976). Other species rather characteristic of Parish's Canadian Zone in the San Bernardino Mts., which have not been reported from the San Gabriel Mts., are the ferns *Cryptogramma crispa* ssp. *acrostichoides, Polypodium hesperium, Woodsia oregana,* and *W. scopulina,* and the flowering plants *Antennaria rosea, Arabis breweri* ssp. *pecunearia, Chimaphila umbellata* ssp. *occidentalis, Phyllodoce breweri, Sambucus microbotrys, Sedum niveum,* and *Sibbaldia procumbens.* Some of these species are also found in the lodgepole pine forest on Mt. San Jacinto.

Limber Pine Forest

Of the various timberline pines characteristic of the high ranges farther north in California, only limber pine, *Pinus flexilis,* reaches southern California to dominate the upper subalpine forests of the desert-fronting mountains from Mt. Pinos through the San Gabriel and San Bernardino Mts. to the San Jacinto and Santa Rosa Mts. The peak of Mt. Pinos (2694 m) supports an open woodland of *P. flexilis* mixed with *P. jeffreyi* (Twisselmann 1967). In the San Gabriel Mts., only the peaks along the desert-fronting ridge, Mt. Hawkins, Throop Peak, Mt. Burnham, to Mt. Baden-Powell, support a forest of *P. flexilis,* generally intermixed with *P. contorta* ssp. *murrayana* and often with specimens of *P. jeffreyi* and *Abies concolor* ssp. *concolor,* at elevations of about 2590–2867 m. An isolated tree also has been found along the Blue Ridge at 2590 m. Only on Mt. Baden-Powell is the limber pine forest of any considerable extent (Fig. 16-5). It appears as scattered trees at about 2590 m on the north slope of the peak and gradually gains dominance by about 2750 m. The handsome, gnarled trees have an appearance of great age.

Figure 16-5. Limber pine forest on Mt. Baden-Powell, elevation 2590 m, Angeles National Forest, San Gabriel Mts. Between the two large, gnarled, ancient trees of *Pinus flexilis* can be seen the Angeles Crest Highway and the Mojave Desert in the distance.

On Mt. Pinos and along the Mt. Hawkins–Mt. Baden-Powell ridge the most characteristic associate of *P. flexilis* is the cushion-forming *Eriogonum kennedyi* ssp. *alpigenum*. The two species are also found together near the top of Mt. San Gorgonio (Old Grayback). The dwarf mistletoe parasitic on *P. flexilis, Arceuthobium cyanocarpum,* though present on the limber pines of Mt. Pinos and the San Bernardino Mts., has not been found in the San Gabriel Mts. Other associates along the ridge, none really restricted to limber pine forest, are *Arctostaphylos patula* ssp. *platyphylla, Cercocarpus ledifolius, Chrysolepis sempervirens, Chrysothamnus nauseosus, Erigeron breweri* ssp. *jacinteus, Eriogonum umbellatum* ssp. *minus, E. wrightii* ssp. *subscaposum, Heuchera elegans, Holodiscus microphyllus, Monardella cinerea, Oreonana vestita, Penstemon caesius, P. grinnellii, Ribes cereum, R. roezlii, Sarcodes sanguinea, Senecio ionophyllus, Silene verecunda* ssp. *platyota, Sitanion hystrix,* and *Stipa parishii,* an assemblage much like that on the Devil's Backbone of Mt. San Antonio, where *Pinus flexilis* has not been found.

In the San Bernardino Mts., the higher peaks of Sugarloaf, San Bernardino, and San Gorgonio support limber pine forest. Limber pine is found from 2590 to 2900 m on Sugarloaf and from perhaps 2900 to 3355 m on San Gorgonio, then almost to the peak (3508 m) as prostrate krummholz (Fig. 16-6). In Fish Hatchery Creek it is found as low as 2515 m and in Pipes Canyon and Little Morongo Creek as low as 2135 m (Minnich 1976). In another "Hudsonian" island in Snow Canyon it is reported between 1985 and 2135 m (Parish 1917). Few species, other than *Arceuthobium cyanocarpum,* are restricted to the limber pine forest. Plants like *Phyllodoce breweri* and *Eriogonum kennedyi* ssp. *alpigenum,* often found associated with limber pine, usually occur in the lodgepole pine forest below (or, on San Gorgonio, above) the forest in gravelly and bouldery fell-fields.

Pinus flexilis in the San Jacinto Mts. is the dominant tree on the ridges and high slopes of Marion and San Jacinto peaks, and it is common on Tahquitz Peak (Hall

Figure 16-6. Elfin woodland, or krummholz, of prostrate, shrubby *Pinus flexilis* above timberline and near summit of Mt. San Gorgonio (3508 m), San Bernardino National Forest, San Bernardino Mts.

1902). Hall reported it as low as 2500 m on the ridge between Tahquitz and Marion peaks, thus giving it a total range of 2500–3050 m. On Toro Peak in the Santa Rosa Mts., it has been collected between 2440 m and the summit (2658 m).

OTHER PLANT COMMUNITIES OF HIGHER ELEVATIONS

Several other well-marked plant communities do not fit into the altitudinal zonation of the communities described above. Their distributions are primarily controlled by edaphic factors, which are only in part affected by altitude. Aquatic communities, mountain meadow, and usually aspen woodland all depend on adequate supplies of standing or running water or saturated soil. Mountain talus and various other rupicolous communities must obtain their footholds in loose scree or rock crevices, and can occur at any level in the mountains.

Aquatic Communities

The freshwater communities of emersed, floating, and submersed aquatic plants are most effectively treated separately as emersed freshwater marsh, floating or submersed aquatics of lakes, ponds, and slow-flowing streams, and semiaquatics of fluctuating reservoir margins.

Freshwater marsh is a community of emersed plants that is separable from mountain meadows only with some difficulty, for in both communities the substratum is permanently saturated and consists of nutrient-rich, mineral soils. A mountain meadow generally has the water table just below the surface of the soil, whereas a marsh has the water table at or above the surface of the soil. The distinction, however, is rather tenuous, and there is much overlap in the dominant genera and even in many of the species. In the forested zone of the San Gabriel Mts., lakes and ponds are very few, so that marshy situations are mostly restricted to the margins of the permanent streams draining the larger canyons. Lakes, ponds, and nonfluctuating reservoirs are much more common in the San Bernardino Mts., and permanent streams issue from many of the canyons. Rather numerous species of *Carex, Cyperus, Eleocharis, Juncus, Polypogon, Rorippa, Rumex, Salix, Scirpus, Typha,* and *Veronica* make up the bulk of these fringing marshes. A few other genera are represented in freshwater marsh in the Transverse and Peninsular ranges by such species as *Alopecurus aequalis, Arundo donax, Asclepias fascicularis, Berula erecta, Glyceria elata, Lobelia cardinalis* ssp. *graminea, Nasturtium officinale* (= *Rorippa nasturtium-aquaticum*), *Sagittaria cuneata, Sparganium emersum,* and *Triglochin maritima.*

Lakes, ponds, and reservoirs in which water levels do not fluctuate severely and in which the water is not too deep, turbid, polluted, or alkaline, usually support a specialized flora of floating and submersed aquatics. Because of terrestrial limitations in this book, these communities will not be discussed further. Reservoirs with frequent and severe fluctuations of water level produce a rather different flora from that of lakes and ponds with reasonably constant water levels. The alternate flooding and emersion eliminates most of the true aquatics, leaving the desiccated mud banks to semiaquatic species that can tolerate or require fluctuations of water level. Frequently these are species that require submersion for germination but need a lowered water level to flower and set seed. Some of the species best adapted to these conditions are *Ammannia coccinea, Crassula aquatica, Crypsis aculeata, Elatine* spp., *Gnaphalium palustre, Juncus bufonius, Lythrum californicum, Sida leprosa* ssp. *hederacea,* and *Verbena bracteata.*

Mountain Meadow

As mentioned above, mountain meadows are not sharply differentiated from freshwater marshes. Generally the water table is just below the surface of the soil, and the most abundant species seem to thrive better in wet but not constantly flooded soil. Usually meadows are better drained than marshes. Meadows can occur at any level where there is adequate moisture; they are narrow along streams but often quite extensive where the terrain is sufficiently level, as in the San Bernardino and San Jacinto Mts. Even the steep, dry San Gabriel Mts. have a few meadows (locally called cienegas) and snowmelt gullies with a Sierran flora of grasses, sedges, and other herbs or shrubs. In southern California, the alpine areas on top of San Gorgonio and San Jacinto peaks are so limited and so steep and rocky that there are no alpine meadows. The montane and subalpine meadows are not differentiated here, since there is a great overlap in the numerous species characterizing them, although a few species do show some altitudinal zonation.

The flora of mountain meadows, even in relatively arid southern California, is very rich, with great numbers of species and individuals of grasses, sedges, and rushes, especially of the genera *Agrostis, Carex, Deschampsia, Juncus, Muhlenbergia, Poa,* and *Scirpus.* Conspicuous for size are the species of *Salix* and also, for color, are the large perennials or biennials, as *Barbarea orthoceras, Carex schottii, C. senta, Castilleja miniata, Dodecatheon redolens, Epilobium angustifolium* ssp. *circumvagum, Habenaria sparsiflora, Helenium bigelovii, Heracleum sphondylium* ssp. *montanum, Iris missouriensis, Lilium parryi, Lotus oblongifolius, Lupinus latifolius* ssp. *parishii, Oenothera hookeri* ssp. *venusta, Perideridia parishii* ssp. *latifolia, Psoralea orbicularis, Pycnanthemum californicum, Senecio triangularis, Sisyrinchium bellum, Smilacina stellata, Sphenosciadium capitellatum,* and *Veratrum californicum.* Many other species are represented in such genera as *Agropyron, Arnica, Aster, Epilobium, Equisetum, Erigeron, Frasera, Galium, Gentiana, Geum, Mentha, Mimulus, Polemonium, Polygonum, Potentilla, Prunella, Ranunculus, Sidalcea, Solidago, Spiranthes, Stellaria, Taraxacum, Thalictrum,* and *Trifolium.*

A rather special kind of upper montane to subalpine mountain meadow is the snowmelt gully (Fig. 16-7), as found in white fir–sugar pine and lodgepole pine forests on north-facing slopes of the Mt. Islip to Mt. Baden-Powell ridge of the San Gabriel Mts. On these high, protected slopes snowfields melt slowly, allowing more or less permanent cold-water seepage or drainage down the slopes and gullies during much of the growing season. These cold-water cienegas produce a meadow flora that is rich and essentially Sierran. Among the more abundant or characteristic species are *Agrostis idahoensis, Carex hassei, C. heteroneura, C. jonesii, Cystopteris fragilis, Dodecatheon redolens, Draba stenoloba* var. *nana, Eilobium brevistylum, E. glaberrimum, Galium bifolium, Habenaria dilatatum* var. *leucostachys, H. sparsiflora, Helenium bigelovii, Hypericum anagalloides, Juncus covillei, J. mertensianus* ssp. *duranii, Lewisia nevadensis, Lilium parryi, Lithophragma tenellum, Luzula*

Figure 16-7. Snowmelt gully on north slope of Throop Peak near Lodgepole picnic area southeast of Dawson Saddle along the Angeles Crest Highway (2335 m) at headwaters of Dorr Canyon branch of Big Rock Creek, Angeles National Forest, San Gabriel Mts. Visible are *Dodecatheon redolens* in capsule, *Aquilegia formosa* in flower, shrubs of *Ribes*, and small plants of *Abies concolor* in mesic forest of *A. concolor* ssp. *concolor*, *Pinus lambertiana*, and *P. contorta* ssp. *murrayana*.

comosa, *Mimulus moschatus*, *M. suksdorfii*, *Potentilla glandulosa* ssp. *ewanii*, *Ribes cereum*, *R. montigenum*, *R. nevadense*, *R. roezlii*, *Rubus leucodermis* var. *bernardinus*, *Salix lasiolepis*, *Sisyrinchium eastwoodiae*, *Solidago californica*, *Trifolium monanthum* var. *grantianum*, *Trisetum spicatum*, *Veratrum californicum*, and *Viola blanda* ssp. *mackloskeyi* (*V. macloskeyi*). Some of these species have not previously been reported from the San Gabriel Mts.

Aspen Woodland

Populus tremuloides is one of the most wide-ranging trees in North America. In the Sierra Nevada, it is common at high elevations on the margins of meadows or streams, especially among red fir and lodgepole pine forests. The community it dominates is basically a high-level riparian woodland, although the aspen also occurs occasionally and more scattered as a small tree or shrub on drier slopes and even at submontane elevations. In the Sierra Nevada, this woodland includes various species of *Salix*, sometimes other species of *Populus*, and generally some conifers, usually *Pinus contorta* ssp. *murrayana*, *P. jeffreyi*, and *Abies magnifica*.

Aspen woodland almost skips southern California. Between the Sierra Nevada and White Mts. to the north, and the Sierra San Pedro Mártir to the south in Baja California, where it is quite abundant, it has been reported only from Fish Creek in the San Gorgonio Wilderness of the San Bernardino Mts. There it forms groves from about 2135 to 2320 m on the canyon floor and on slopes along the creek on decomposed granite in yellow pine forest. I have not yet visited the groves and therefore cannot list the associated species. It surely remains the least known plant community in the southern California mountains.

Mountain Talus and Other Rupicolous Plants

The youthful nature of the Transverse and Peninsular ranges results in many rather sharp peaks and ridges, often with very steep, precipitous slopes. When these slopes are covered with scree or talus, a sparse but characteristic community of plants can be found inhabiting the deeper, gravelly debris of pebbles and rock flakes. Because of the instability of this habitat, most of the species in the community are perennials with heavily developed root systems, often with much more plant material below the ground level than above it. Expecially characteristic are *Allium monticola, Arabis platysperma, Calochortus invenustus, Draba corrugata, Eriogonum saxatile, E. umbellatum* ssp. *minus, Monardella cinerea, Oreonana vestita, Oxytheca parishii* (a lone annual), *Silene parishii,* and *Tauschia parishii.* Some of these highly specialized plants are quite restricted in their ranges, for example, *Monardella cinerea* and *Eriogonum umbellatum* ssp. *minus* to a few peaks in the San Gabriel Mts. and *Draba corrugata* from the San Gabriel to the Santa Rose Mts. On the other hand, the rare *Crepis nana* ssp. *ramosa,* known in southern California only from San Antonio and Pine Mts. in the San Gabriel Mts., has a wide western American range.

Other rupicolous plants have been discussed, for the most part, with the communities in which they are usually found, yet plants largely restricted to rock crevices probably deserve treatment as a special rupicolous community. In the Transverse and Peninsular ranges, the more characteristic of these rock-crevice plants are the spore-bearing *Cheilanthes covillei, Pellaea compacta, Selaginella asprella,* and *S. bigelovii,* the succulents *Dudleya cymosa, Sedum niveum,* and *S. spathulifolium* ssp. *anomalum,* and the perennials *Brickellia microphylla, Crepis acuminata, Diplacus calycinus* (= *Mimulus longiflorus* ssp. *calycinus*), *D. longiflorus* (= *Mimulus longiflorus*), *Eriophyllum confertiflorum, Galium angustifolium* ssp. *gabrielense, G. angustifolium* ssp. *nudicaule, Haplopappus cuneatus, Heuchera elegans, Silene lemmonii, Stephanomeria cichoriacea,* and *Zauschneria californica* ssp. *californica.* Other species often found with them, though ranging in addition over dry, rocky slopes, are *Brickella californica, Cryptantha muricata, Erigeron foliosus* var. *stenophyllus, E. parishii* Gray, *Eriogonum davidsonii, E. parishii* Wats., *Euphorbia albomarginata, Lotus davidsonii, Mentzelia laevicaulis, Poa scabrella, Turricula parryi,* and *Corethrogyne filaginifolia.*

STATUS OF KNOWLEDGE

Published information on the vegetation of the forests and other mountain communities of the Transverse and Peninsular ranges is indeed meager in comparison to our knowledge of their floras. Only the chaparral communities of the lower cismontane slopes seem to have received much attention from the ecologists. The rather taxonomic approach to mountain vegetation in this chapter results from this dearth of plant ecological and plant sociological publications. I have developed the pragmatic classification of California plant communities (Thorne 1976) followed here from my own fieldwork throughout the state, starting from the base developed by Munz and Keck (1949, 1950, 1959, 1973). More popular accounts of the plant communities of southern California can be found in handbooks by Jaeger and Smith (1971) and Ornduff (1974).

The only recent and detailed studies of the vegetation of any of the mountain ranges discussed here are those by Horton (1960) and Minnich (1976) on the San Bernardino Mts. The latter is especially informative and presents a map of community distribution based on modern techniques of surveying vegetation. Similar studies are needed for all the other ranges. Vogl and Miller (1968) described the vegetation of the south slope of Mt. Pinos, and Twisselmann (1967) included part of Mt. Pinos and surrounding peaks in his good discussion of the vegetation of Kern Co. Grinnell (1908) early gave a general account of the vegetation of the San Jacinto Mts. Parish (1917), in his treatment of the flora of the San Bernardino Mts., and Johnston (1919) in his similar approach to the flora of the San Antonio Mts., devoted minimal attention to the plant communities of the two ranges.

ACKNOWLEDGMENTS

I am indebted to many people who assisted me with the fieldwork that made possible this account of the vegetation of the higher mountains of southern California. Among the many botanical associates at the Rancho Santa Ana Botanic Garden who often accompanied me in the field, making the work both more instructive and more pleasant, are Mae Z. Thorne, C. W. Tilforth, John Dourley, A. C. Gibson, Larry DeBuhr, and J. D. Olmsted, who first introduced me to the San Gabriel and Santa Ana Mts. Others who supplied information or specimens from the southern ranges are J. Ewan, L. C. Wheeler, Reid V. Moran, R. Mitchell Beauchamp, Gary Wallace, Anselmo Lewis, Robert Frampton, A. B. Cossey, and a number of my former students, who made detailed surveys of several of the lower cismontane canyons of the San Gabriel Mts. as part of their field training. The District Foresters of the Angeles and San Bernardino National Forests have always been most cordial and cooperative in supplying permits, gate keys, and information, thus making our field efforts in these ranges most enjoyable and botanically profitable.

LITERATURE CITED

Critchfield, W. B. 1957. Geographic variation in Pinus contorta. Maria Moors Cabot Found. Pub. 3:1-118. Harvard Univ., Cambridge, Mass.

Grinnell, J. 1908. The biota of the San Bernardino Mountains. Univ. Calif. Pub. Zoo. 5:1-170.

Hall, H. M. 1902. A botanical survey of the San Jacinto Mountain. Univ. Calif. Pub. Bot. 1:1-140.

Haller, J. R. 1959. Factors affecting the distribution of ponderosa and Jeffrey pines in California. Madroño 15:65-71.

-----. 1962. Variation and hybridization in ponderosa and Jeffrey pines. Univ. Calif. Pub. Bot. 34:123-165.

Hamrick, J. L., and W. J. Libby. 1972. Variation and selection in western U.S. montane species. I. White fir. Silvae Genet. 21:29-35.

Horton, J. S. 1960. Vegetation types of the San Bernardino Mountains. USDA Forest Serv., Pac. SW Forest and Range Exp. Sta., Tech. Paper 44, Berkeley, Calif. 29p.

Jaeger, E. C., and A. C. Smith. 1971. Introduction to the natural history of southern California. Univ. Calif. Press, Berkeley. 104 p.

Johnston, I. M. 1919. The flora of the pine belt of the San Antonio Mountains of southern California. Plant World 22:71-90, 105-122.

Kopecko, K. J. P., and E. W. Lathrop. 1975. Vegetation zonation in a vernal marsh on the Santa Rosa Plateau of Riverside County, California. Aliso 8:281-288.

Lewis, A., and G. W. Gause. 1966. Trees of the Angeles National Forest. USDA Forest Serv., Covina, Calif. 18 p.

Minnich, R. A. 1976. Vegetation of the San Bernardino Mountains, pp. 99-124. In J. Latting (ed.). Plant communities of southern California. Calif. Native Plant Soc. Spec. Pub. 2.

Munz, P. A. 1974. A flora of southern California. Univ. Calif. Press, Berkeley. 1986 p.

Munz, P. A. and D. D. Keck. 1949. California plant communities. Aliso 2:87-105.

-----. 1950. California plant communities, supplement. Aliso 2: 199-202.

-----. 1959. A California flora. Univ. Calif. Press, Berkeley. 1681 p.

-----. 1973. A California flora, with supplement by P. A. Munz. Univ. Calif. Press, Berkeley. 1681 and 224 p.

National Georgraphic Society, Cartographic Division. Close-up: U.S.A.: California and Nevada. Washington, D. C. Map.

Ornduff, R. 1974. An introduction to California plant life. Univ. Calif. Press, Berkeley. 152 p.

Parish, S. B. 1917. An enumeration of the pteridophytes and spermatophytes of the San Bernardino Mountains, California. Plant World 20:163-178, 208-223, 245-259.

Parratt, M. W. 1965. A discussion of the hybridization and variation of the ponderosa and Jeffrey pines on Mt. San Antonio. M. S. thesis, Claremont Graduate School, Claremont, Calif. 46 p.

-----. Hybridization and variation of ponderosa and Jeffrey pines on Mt. San Antonio California. Aliso 6:79-96.

Raven, P. H., and H. J. Thompson. 1966. Flora of the Santa Monica Mountains, California. Univ. Calif. Press, Berkeley. 185 p.

Smith, C. F. 1952. A flora of Santa Barbara. Santa Barbara Bot. Gard., Santa Barbara, Calif. 100 p.

Thorne, R. F. 1976. The vascular plant communities of California, pp. 1-31. In J. Latting (ed.). Plant communities of southern California. Calif. Native Plant Soc. Spec. Pub. 2.

Thorne, R. F., and E. W. Lathrop. 1969. A vernal marsh on the Santa Rosa Plateau of Riverside County, California. Aliso 7:85-95.

-----. 1970. Pilularia americana on the Santa Rosa Plateau, Riverside County, California. Aliso 7:149-155.

Twisselmann, E. D. 1967. A flora of Kern County, California. Wasmann J. Biol. 25:1-395.

Vasek, F. C. 1966. The distribution and taxonomy of three western junipers. Brittonia 18:350-372.

Vogl, R. J., and B. C. Miller. 1968. The vegetational composition of the south slope of Mt. Pinos, California. Madroño 19:225-234.

CHAPTER 17

MONTANE AND SUBALPINE VEGETATION OF THE SIERRA NEVADA AND CASCADE RANGES

PHILIP W. RUNDEL

Department of Ecology and Evolutionary Biology, University of California, Irvine

DAVID J. PARSONS

Sequoia and Kings Canyon National Parks, Three Rivers, California

DONALD T. GORDON

Pacific Southwest Forest and Range Experiment Station, Redding, California

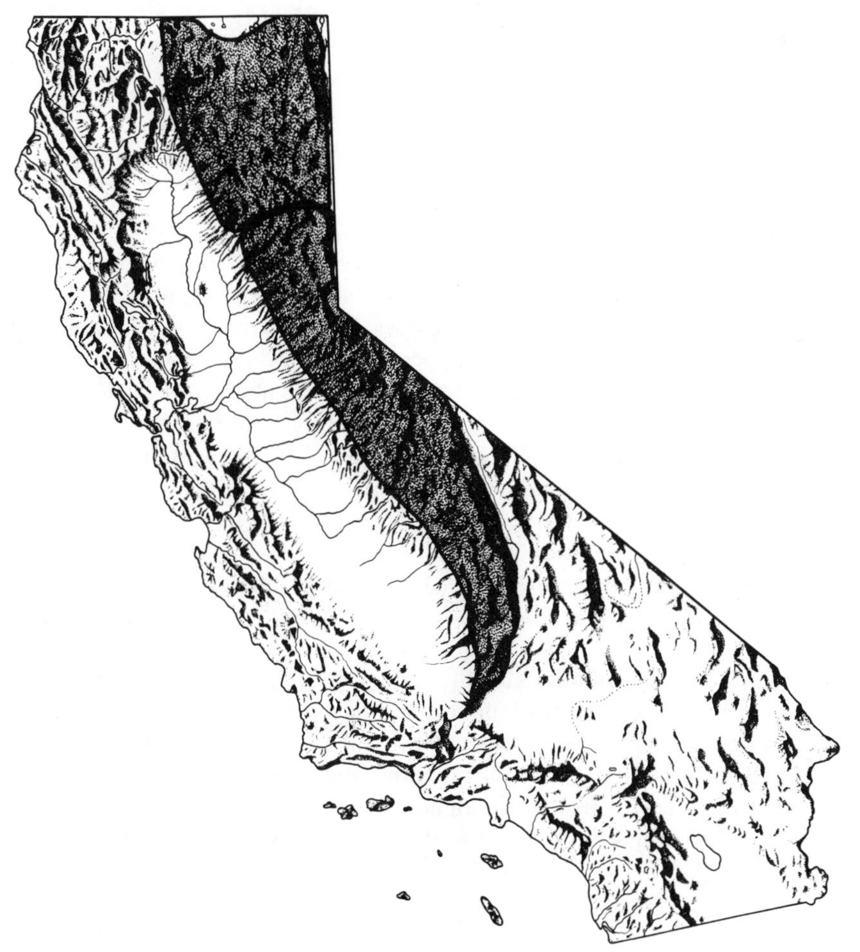

The Sierra Nevada **560**
Introduction **560**
Lower montane forests **561**
Ponderosa pine **561**
White fir–mixed conifer **563**
Giant sequoia **569**
Upper montane forests **571**
Red fir **571**
Jeffrey pine **574**
Lodgepole pine **577**
Subalpine forests **579**
Mountain hemlock **579**
Western white pine **579**
Whitebark pine **580**
Foxtail pine **580**
Limber pine **582**
Nonconiferous vegetation **583**
Deciduous forest **583**
Meadows **584**
Granite outcrops **585**
The ecological role of fire **586**
The Cascade Range **587**
Successional studies **590**
The Modoc Plateau **591**
The Warner Mountains **592**
Areas for future research **593**
Literature cited **595**

THE SIERRA NEVADA

Introduction

Montane and subalpine coniferous forests of the Sierra Nevada comprise one of the largest and most economically important vegetation regions in California. This region includes most of the area of both slopes of the Sierra from 600–1500 m at its lower margin to 3000–3500 m at its upper limit. The lower montane zone of the Sierran coniferous forests is composed of ponderosa pine (*Pinus ponderosa*) forest on more xeric sites and white fir (*Abies concolor*) forest on more mesic sites, including special areas with giant sequoia groves. Above this zone, forming a transition to the higher subalpine forests, are the upper montane red fir (*Abies magnifica*), Jeffrey pine (*Pinus jeffreyi*), and lodgepole pine (*Pinus contorta* ssp. *murrayana*) forests. The subalpine zone includes several geographically restricted types dominated by the mountain hemlock (*Tsuga mertensiana*), western white pine (*Pinus monticola*), whitebark pine (*P. albicaulis*), foxtail pine (*P. balfouriana*), and limber pine (*P. flexilis*).

A diagram of the distributions of these major types in relation to elevation and moisture for the southern Sierra (about 37°N) is shown in Fig. 17-1, and comparative values of cover, density, and basal area for six types appear in Table 17-1. Cover drops with elevation, from 100+% in montane forests to less than 25% in subalpine foxtail pine forest. Tree densities are highest in white fir, ponderosa pine, and red fir forests. Red fir forest has the largest basal area of any Sierran forest type.

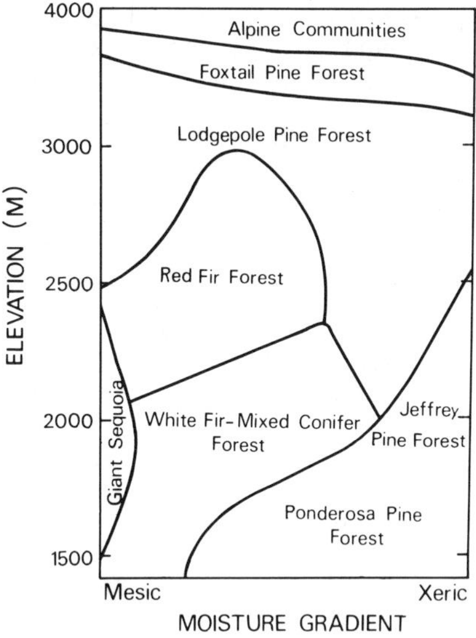

Figure 17-1. Diagrammatic relationships of major vegetation types in the southern Sierra (~ 37° N) in relation to elevation and moisture.

Less than half the ground surface within the montane and subalpine zones is dominated by forest vegetation (Richards 1959). The remaining area includes lakes, swamps, rock outcrops, talus slopes, meadows, and montane chaparral. Meadow and rock outcrop vegetation is discussed later in this chapter; montane chaparral was discussed in Chapter 12.

Lower Montane Forests

Ponderosa pine. Ponderosa pine forest dominates xeric sites in the lower montane zone. In the northern Sierra, this belt may occur from 300 to 1800 m (rarely higher, as around Lake Tahoe and northward), while to the south, at Sequoia and Kings Canyon National Parks, it forms a distinct zone at 1200–2100 m. On more mesic sites at middle and higher elevations, ponderosa pine forest is replaced by white fir forest or, in the northern Sierra, by Douglas fir (*Pseudotsuga menziesii*) on north slopes. At its lower margin, ponderosa pine forest intermingles with chaparral, and stands of *Quercus kelloggii* or individuals of *Pinus ponderosa* and *Calocedrus decurrens* may occur well down into chaparral and foothill woodland on favorable sites. At its upper margin (south of mid-Plumas Co.), *P. ponderosa* is replaced by *P. jeffreyi.* Within the elevation zone where ponderosa and Jeffrey overlap, hybrids are common (Haller 1962). Another high-elevation yellow pine, *P. washoensis,* is scattered in widely separate stands along the eastern slope of the northern Sierra Nevada and adjacent Great Basin ranges (Griffin and Critchfield 1972; Haller, pers. commun.). The literature on ponderosa pine has been summarized by Axelton (1967).

The characteristic structure and composition of ponderosa pine forest are determined primarily by two factors, soil moisture availability during the summer months and fire history. The relationship of stand composition to soil moisture availability was studied in Kings Canyon National Park by Sellers (1970), who correlated the importance values of five tree species with the height of each stand above

TABLE 17-1. Cover (%), total tree density per hectare, and total basal area (m^2ha^{-1}) for six forest types in Sequoia National Park. Data are averages for 8-16 sample sites for each type, and include all trees over 1 yr of age. From Vankat (1970)

Forest Type	Cover	Density	Basal Area
Ponderosa pine	134	6580	46
White fir-mixed conifer	110	5066	70
Red fir	74	6635	92
Jeffrey pine	55	1104	43
Lodgepole pine	56	6466	64
Foxtail pine	26	420	32

the Kings River. For both *Pinus ponderosa* and *Quercus kelloggii*, the importance values increased linearly from the most mesic sites along the river to the most xeric sites 35 m above the river. *Calocedrus decurrens* showed no strong statistical correlation with elevation, whereas *Abies concolor* and *Pinus lambertiana* had high importance values only within 8 m of the river and were virtually absent from the highest-elevation stands.

Pinus ponderosa and *Quercus kelloggii* are well adapted to light, regular ground fires. Seedlings of *P. ponderosa* are intolerant of shade in comparison to *Abies concolor* and *Calocedrus decurrens*, resulting in poor establishment of the former in dense stands. Where undergrowth of young trees is regularly thinned by ground fires, *P. ponderosa* becomes established and dominates. Typically, *P. ponderosa* increases in density as total stand density decreases, whereas the reverse relationship is typical for fire-intolerant *A. concolor, C. decurrens,* and *P. lambertiana* (Sellers 1970). Seedlings of *P. ponderosa* are intolerant of deep snow (Neal, unpub. data), and this factor may also be important in explaining its relative abundance.

Quercus kelloggii, adapted to fire by crown sprouting, decreases in importance in dense stands. Although this oak is moderately shade tolerant in its early stages of growth, it requires full sun for good growth when mature (McDonald 1969). It seems to have been virtually eliminated by shading from dense stands of *Calocedrus decurrens* in the Yosemite Valley (Gibbens and Heady 1964). The seedling abundance of *Q. kelloggii* is extremely high in relation to sapling and tree densities: Sellers (1970) counted 8800 seedlings per hectare in a stand with only 47 sapling and tree oaks per hectare. Seedling mortality is high, and growth during the first 25 yr (ca. 10 cm dbh) is slow, followed by a period of increase in diameter. Growth slows as trees reach maturity at about 90 yr, and few trees attain ages in excess of 350 yr (McDonald 1969). Although mature *Q. kelloggii* is intolerant of poor drainage or relatively mesic soils with poor aeration (McDonald 1969), it is extremely drought resistant and may grow on sites too poor for *P. ponderosa* (Sellers 1970).

Abundant evidence indicates that *Calocedrus decurrens* and *Abies concolor* have both increased their relative and absolute densities in ponderosa pine forest since the turn of the century, when fire control programs became widespread and effective in the Sierra (Vankat 1970). Seedlings of *C. decurrens* and *A. concolor* are highly tolerant of shade, readily become established in plots with light litter or brush cover (Fowells and Schubert 1951), and thus are favored in dense stands. Once released from shade, they grow rapidly in height (Mitchell 1918), although more slowly than associated coniferous trees (Dunning 1923). Sellers (1970) reported that trees 60–90 cm dbh were 155–240 yr old in Kings Canyon, but Sudworth (1908) gave ages of 360–546 yr for the same diameters. Once trees reach height maturity (45–60 m at 165–210 yr of age), they are subject to severe dry rot infections (Boyce 1920).

Mean ground cover, density, and basal area data for 12 stands of ponderosa pine forest in Sequoia National Park are shown in Table 17-2. Total tree cover ranged from 60 to 95% for the 12 stands, and basal area ranged from 26 to 172 m^2 ha^{-1}. Herb cover (not shown in the table) was slight, varying between 0.1 and 1.0%. Shrub cover and diversity were extremely variable among stands, but the characteristic species were *Chamaebatia foliolosa* and *Arctostaphylos viscida* (*A. patula* at higher elevations). These and other shrub species are favored by fire, and are eliminated by shading in dense older stands. Historical records indicate that shrub density and cover in ponderosa pine forest have decreased with fire suppression policies in the twentieth century (Vankat 1970).

The high densities of *Calocedrus decurrens,* particularly evident in Table 17-2, reflect recent increased establishment. Continuation of fire supression policies in the future will further shift the dominance in ponderosa pine forest to *Calocedrus* on more xeric sites and to *Abies concolor* on more mesic sites.

White fir–mixed conifer. White fir forest forms the dominant community on relatively mesic sites in the lower montane zone of the Sierra Nevada. Depending on latitude, this community occurs from 1250 to 2200 m. Although *Abies concolor* is the dominant species, up to six species of conifers may be present in individual stands—hence the common use of the name "mixed conifer" forest. White fir is the overwhelmingly dominant species, while *Pinus lambertiana* and *Calocedrus decurrens* are important associates, and *Sequoiadendron giganteum* may be dominant in terms of basal area in local groves (Rundel 1971). Community data for stands in Sequoia National Park are shown in Table 17-3. In the northern Sierra and southern Cascades, especially at 1200–1500 m, *A. concolor* is much less dominant, and *Pseudotsuga menziesii* is the prominent member of the community (Table 17-4, Fig. 17-2). At lower elevations and on drier sites at higher elevations, *A. concolor* may share dominance with *C. decurrens,* and *Pinus ponderosa* and *Quercus kelloggii* are important associates. Near the upper margin of the white fir forest, *A. concolor* mixes with *A. magnifica* in a transition to red fir forest communities.

The structure of individual stands within white fir forests is highly variable. This situation is due to the complex dynamics of stand mosaics, resulting from different fire histories and variation in rate of development on individual sites. Individual stands, varying from a few trees to a hectare or more in size, represent relatively

TABLE 17-2. Cover (%), density per hectare, and basal area ($m^2 ha^{-1}$) for ponderosa pine forest in Sequoia National Park. Data are averages for 12 100 m^2 samples, and include all trees over 1 yr of age. From Vankat (1970)

Taxon	Cover	Density	Basal Area
Abies concolor	19.6	760	8.56
Calocedrus decurrens	31.9	3570	9.85
Pinus lambertiana	8.7	480	2.16
P. ponderosa	30.0	370	11.96
Quercus chrysolepis	1.3	50	0.11
Q. kelloggii	42.9	1350	13.84
Arctostaphylos spp.	4.6		
Ceanothus integerrimus	0.3		
Chamaebatia foliolosa	28.8		
Rhamnus crocea	0.3		
Rhus diversiloba	<0.1		
Rosa spithamea	0.6		
Total	134.4	6580	46.49

even-aged stands in varying levels of successional development. Older trees are typically uniformly dispersed, but younger ones show contagious patterns of dispersion (Bonnicksen 1975). Giant sequoias clearly show a highly contagious pattern of dispersion throughout their lives.

Abies concolor commonly comprises 80% or more of the large trees in individual stands of white fir forests (Rundel 1971; Schumacher 1926). Mature trees of this species are commonly 50–63 m in height with diameters of 1–2 m or more, and probably reach ages of 300–400 yr. Although precipitation diminishes gradually from north to south in the Sierra Nevada, the population distribution of trees by diameter class is very similar in the central and southern Sierra (Fig. 17-3). Statistics for pure, even-aged stands of white fir are shown in Table 17-5. The literature on *A. concolor* is extensive and has recently been summarized by Bonnicksen (1972).

Pinus lambertiana, only rarely found in pure stands, is a typically important associate in white fir forests. With the exception of giant sequoia, *P. lambertiana* is

TABLE 17-3. Cover (%), density per hectare, and basal area (m^2ha^{-1}) for white fir-mixed conifer forest in Sequoia National Park. Data are averages for 16 100 m^2 samples, and include all trees over 1 yr of age. From Vankat (1970). Asterisk = data not available

Taxon	Cover	Density	Basal Area
Abies concolor	67.3	1410	55.05
A. magnifica	2.0	393	1.93
Calocedrus decurrens	10.5	2684	5.26
Pinus jeffreyi	*	*	0.76
P. labertiana	11.0	429	5.92
P. ponderosa	1.0	*	0.09
Quercus chrysolepis	*	20	*
Q. kelloggii	2.9	130	0.78
Total	109.7	5066	69.95

the largest member of the community, reaching heights of 78 m and diameters of 3.1 m. Seedlings germinate on both litter and bare mineral soil, but development is slow under shade conditions (Fowells 1965).

Understory trees and shrubs form an important element in white fir forest. The most significant aspects of this understory layer are the dense thickets of white fir and incense cedar saplings that have developed in this century in response to decreased fire frequency. Although total understory tree and shrub coverage is extremely variable, coverages of 5–10% are common, and values to 30% or more are not unusual. At lower elevations or on rocky sites, *Acer macrophyllum, Quercus kelloggii, Q. chrysolepis, Arctostaphylos patula, Chamaebatia foliolosa, Ceanothus integerrimus, C. cordulatus,* and *C. parvifolius* are common; many of these are favored by frequent fires. *Corylus cornuta* var. *californica* and *Cornus nuttallii* are restricted to relatively mesic sites and may be common in giant sequoia groves. Other shrub species occurring less commoningly are *Salix scouleriana, Castanopsis chrysolepis, Prunus emarginata,* and *Ribes* ssp.

Several understory trees typical of north coastal areas of California are present in the white fir forests of the northern Sierras but are absent to the south. These include *Arbutus menziesii, Chrysolepis chrysophylla,* and *Taxus brevifolia* (Griffin and Critchfield 1972; Rundel 1968). *Torreya californica,* a species with similar affinities, occurs rarely into the southern Sierra. *Acer glabrum,* typically a subalpine species elsewhere in its range, is rarely found in white fir forest communities.

Figure 17-2. (*a*) White fir–mixed conifer forest, west slope. (*b*) Same, with *Sequoiadendron,* Giant Forest, Sequoia National Park. Photograph courtesy of Bill Jones. The President's tree is in the foreground and is 7 m dbh.

TABLE 17-4. Importance values, average diameter at breast height (cm), and density per hectare of all conifer species in a virgin mixed conifer forest at Placer Co. Big Trees, 1700 m elevation. Data are means of about 80 points scattered over several hectares (point-centered quarter method) sampled in 1974 and 1975 by plant ecology classes under the direction of Michael Barbour. Importance values (IV: relative basal area + relative density + relative frequency) are for the size classes shown, but diameter at breast height and density are only for trees >3 cm dbh. Total basal area for trees >3 cm dbh was 49 m^2ha^{-1}

	IV for Three Size Classes (cm dbh)				
Taxon	<3	3-40	>40	Av. dbh	Density
Abies concolor	132	123	36	8.3	258
Libocedrus decurrens	62	35	27	15.0	67
Pinus lambertiana	59	38	69	23.1	87
P. ponderosa	10	47	89	16.6	114
Pseudotsuga menziesii	37	57	79	19.1	122

A special group of understory trees and shrubs forms a characteristic community in riparian habitats within white fir forests. These include *Alnus rhombifolia, A. tenuifolia, Acer macrophyllum, Cornus stolonifera, Rhododendron occidentale, Ribes nevadense, Rubus parviflorus, Fraxinus latifolia, Physocarpus capitatus,* and *Populus trichocarpa.* An herbaceous riparian community of *Carex senta, Peltiphyllum peltatum,* and *Aralia californica* is prominent in the northern and central Sierra.

Herbaceous cover is sparse in white fir forests (seldom > 5%), except in moist

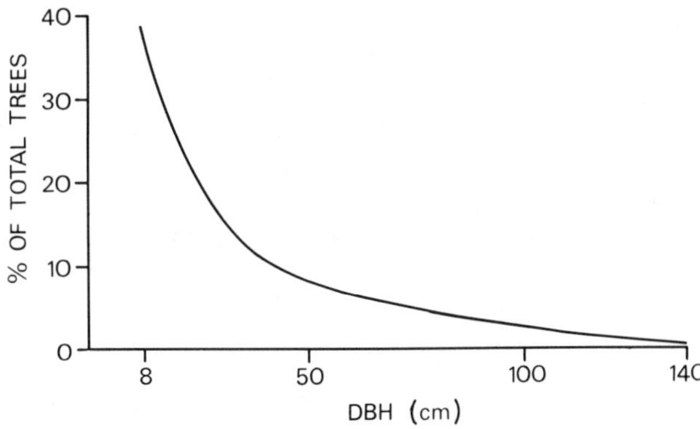

Figure 17-3. Size class distribution for *Abies concolor* throughout the Sierra Nevada, averaged for many sites.

TABLE 17-5. Statistics on pure, even-aged stands of *Abies concolor* in the Sierra Nevada greater than 10 cm dbh. Based on Schumacher (1926). Biomass and productivity figures are for above- and below-ground material. Wood density assumed to be 415 kg m^{-3}

Site Index	Age (yr)	Density per Hectare	Mean Height (m)	Basal area (m^2 ha^{-1})	Bole Wood Volume (m^3 ha^{-1})	Bole Wood Weight (kg m^{-3})	Biomass (kg m^{-2})	Increment Volume (m^3 ha^{-1})	Increment Weight (kg m^{-3})	Productivity (g m^{-2} hr^{-1})
27.4	50	1080	23	72	630	26.2	64	12.6	524	1870
	100	560	36	108	1370	56.8	140	8.4	349	1240
	150	410	40	108	1620	67.2	165	3.5	145	520
18.3	50	1870	15	61	370	15.3	38	7.4	307	1090
	100	970	25	91	800	33.2	82	4.9	204	730
	150	720	27	91	950	39.4	77	2.1	87	310
9.2	50	3930	8	38	150	6.2	14	2.8	116	410
	100	2010	12	57	320	13.3	33	2.1	87	310
	150	1480	14	57	380	15.8	39	.9	37	130

swales or drainage bottoms, where it may approach 100%. It may be completely absent over large areas with thick litter accumulation. Diversity, however, is commonly high: as many as 50–100 species may occur in relatively small areas if riparian habitats are included (Rundel 1971; Buchanan et al. 1966).

Below 1800 m, a number of species are present which do not occur at higher elevations in white fir communities. These include *Clintonia uniflora, Fragaria californica, Goodyera oblongifolia, Asarum hartwegii,* and *Iris hartwegii.* Little differentiation between herbaceous species in the higher elevations of white fir forest and the lower elevations of red fir forest is present, however. Two nitrogen fixers, *Lupinus latifolius* var. *columbianus* and *Ceanothus parvifolius,* commonly appear in great numbers of mesic sites after fires. Two bryophytes, *Funaria hygrometrica* and *Marchantia polymorpha,* may also cover large areas of bare soil (Rundel 1971).

In typical white fir communities where ground vegetation coverage is no more than 5%, dominant species include *Adenocaulon bicolor, Hieracium albiflorum, Galium sparsifolium, Disporum hookeri,* and *Osmorhiza chilensis.* In openings with more sun, *Pteridium aquilinum* var. *pubescens, Carex* spp., and *Viola lobata* may also be common.

In giant sequoia groves, ground vegetation coverage may be considerably greater. The frequency of 1 m^2 quadrats without herbaceous vegetation in the Muir Grove (Sequoia National Park) was only 16%, although samples from the less mesic Grant Grove, Giant Forest, and Lost Grove showed 50–70% of the quadrats bare (Rundel 1971).

In sites where a heavy carpet of litter coats the soil surface, herbaceous vegetation is restricted to scattered individuals of *Pyrola picta* and *Chimaphila menziesii.* The diversity of these and other mycorrhizal "saprophytes" (including *Allotropa virgata, Pyrola asarifolia, P. minor, P. secunda, Hemitomes congestum, Pityopus californicus, Sarcodes sanguinea, Corallorhiza striata, C. maculata,* and *Eburophyton austinae*) is remarkable in the Sierra Nevada. The heavy snow cover and moderate winter temperature extremes may provide conditions allowing critical fungal activity in the root zone throughout the year.

A final consideration here is the impact of human activities. Soil compaction and physical trampling eliminate many native species and often lead to the introduction of weedy invaders. Many studies have described the impact of man on white fir forest communities in giant sequoia groves (Rundel 1967; Hartesveldt 1962, 1967; Meinecke 1926). Beetham (1963) demonstrated that recreational and logging disturbance increased the relative densities of *Calocedrus decurrens* seedlings at the expense of *Abies concolor* and *Pinus lambertiana* seedlings. Recent documentation of oxidant-induced air pollution damage to mixed conifer forests points to an extremely serious development in California mountains (Miller 1973). *Abies concolor, P. ponderosa,* and *P. jeffreyi* are the species most sensitive to this type of oxidant damage.

Giant sequoia. Although the giant sequoia groves of the central and southern Sierra Nevada represent only a specific mesic segregate of typical white fir forest communities, these groves are often given special community recognition. Only *Sequoiadendron giganteum* is restricted to the groves.

Giant sequoia is restricted to 75 groves (Rundel 1972b) extending along the west slope of the Sierra Nevada from Placer Co. through Tulare Co. North of the Kings

River, giant sequoia is limited to 8 small disjunct groves, separated by as much as 90 km. South of the Kings River, the groves occur in a relatively continuous belt, never separated by more than 7 km. Groves range in size from the large Redwood Mt. and Giant Forest groves of Kings Canyon and Sequoia National Parks, averaging approximately 1000 ha and 20,000 giant sequoias each, to the tiny, isolated Placer Co. Grove with 6 living trees (and planted saplings). Typical densities of giant sequoias over 31 cm dbh in well-developed groves are 6–10 per hectare. Groves north of the Kings River occur primarily from 1400 to 2000 m, all on slopes with a general southern aspect. Southern groves reach to 2450 m, and a single tree occurs at 2700 m. Between the Kings River and the southern boundary of Sequoia National Park, groves occur on both north- and south-facing slopes with equal frequency. South of Sequoia National Park, 80% of the groves lie on slopes with a northern aspect.

Present data do not indicate that any significant enlargement or contraction of giant sequoia groves is taking place. With rare exceptions, mature giant sequoias mark the outer periphery of groves. Similarly, organic remains of giant sequoia are completely lacking outside the present grove boundaries, indicating a stability of such boundaries over a period of 500–1000+ yr (Rundel 1971).

Although reproduction is sparse, particularly in the last 50 yr because of grove management policies, high levels of reproduction are not necessary to maintain present population levels. The population distribution of giant sequoias greater than 31 cm dbh for six groves is shown in Fig. 17-4. Large mature groves, such as Giant Forest and Redwood Mt., show a relatively linear distribution of trees by diameter class. Smaller mature groves, like Muir and South Calaveras, typically lack young trees in the smaller diameter classes. The tiny Powderhorn Grove lacks trees in both of the smallest diameter classes, indicating a clearly senescent condition. An adjacent small grove in Sequoia National Forest at Long Meadow has abundant reproduction.

Few groves have sufficient young trees to maintain the present density of mature giant sequoias in the future; the great majority are undergoing a gradual decline in

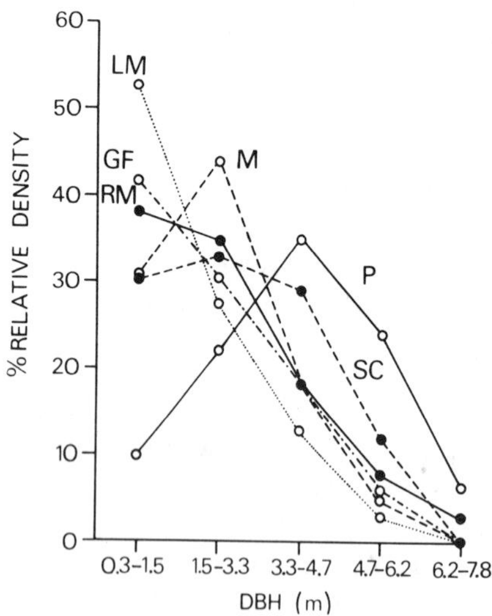

Figure 17-4. Size class distribution of giant sequoias in six groves of varying size. RM = Redwood Mt., LM = Long Meadow, SC = South Calaveras, GF = Giant Forest, P = Ponderosa, M = Muir. From Rundel (1971).

density—a decline that began long before the influence of western civilization. Under natural conditions, however, few if any groves face local extinction in the next millenium (Rundel 1971).

Soil moisture availability during the summer months is the most important single factor determining the maintenance of present grove boundaries. Groves appear to be restricted to sites with available soil moisture throughout the summer drought period (Rundel 1972a). Even with this availability of moisture, the upper height limits of giant sequoias may be determined by water stress (Rundel 1973). The disjunction of northern groves can be attributed to the harsh climatic regime of the Wisconsin glaciation, but the xeric conditions of the Altithermal Period were primarily responsible for the dissection of southern groves (Rundel 1972a).

A discussion of giant sequoia groves cannot ignore the importance of fire history. The groves represent a fire climax community whose stability is maintained by frequent fires (Hartesveldt and Harvey 1967; Hartesveldt et al. 1967; Biswell et al. 1966; Rundel 1971, 1973; Kilgore 1973a,b; Kilgore and Sando 1975). In the absence of regular ground fires, litter accumulates on the forest floor and limits germination and establishment. If these conditions are maintained in the future, the groves will become a long-standing seral community trending toward a mature white fir forest without giant sequoia (Bonnickson 1975).

Upper Montane Forests

Red fir. The red fir forest lies immediately above the montane mixed conifer forests, in an elevational band about 300 m wide. The lower and upper limits of this band are about 1800 and 2750 m, varying with latitude and local influences. The range extends from northern Lake Co. northward through the North Coast Ranges, and from the southern Sierra Nevada (Kern Co.) to the southern Cascade Range and northward into southwestern Oregon.

Snow is the characteristic form of precipitation, packs usually reaching depths of 2.5–4+ m. The top of the snow pack is marked by the lower limit of *Letharia vulpina* and *L. columbiana* on tree trunks. Annual precipitation is commonly 1000–1300 mm. There is usually some rain in fall and spring, but summers are dry except for infrequent, locally intense convectional (thunder) storms.

Abies magnifica is the overwhelmingly dominant species in red fir forests, often occurring in pure, dense stands. At its lower elevational limits, however, it is often equally mixed with *A. concolor,* but hybridization does not occur. On moist soils above 2500 m, red fir commonly shares dominance with *Pinus contorta* ssp. *murrayana.* Basal areas for 10 red fir communities in Sequoia National Park ranged from 74 to 112 m^2 ha^{-1} (Vankat 1970).

Studies of red fir forests in the southern Sierra Nevada are restricted to Kings Canyon National Park (Kilgore 1971) and Sequoia National Park (Vankat 1970; Hammon-Jensen-Wallen Forestry 1970). More detailed comparative data on community structure is available for red fir forests in the northern and central Sierra Nevada (Oosting and Billings 1943), and limited data for Yosemite National Park (Heath 1971). These data are shown in Table 17-6. *Pinus monticola* is a constant associate of *A. magnifica* in the northern and central Sierra, but becomes much less common in the southern Sierra. *Pinus lambertiana, P. jeffreyi* (on rocky sites), *Tsuga mertensiana,* and *Calocedrus decurrens* are occasional associates.

TABLE 17-6. Density per hectare and basal area (m^2 ha^{-1}) for red fir forests in the southern and northern Sierra Nevada. Southern data are means of 10 100 m^2 plots from Sequoia National Park (Vankat 1970); density includes all individuals over 1 yr old. Northern data are means of 75 225 m^2 plots scattered among 5 stands in Plumas, Sierra, Eldorado, Amador, and Alpine Cos. (Oosting and Billings 1943); density includes all individuals. Asterisk = data not available.

Taxon	Southern		Northern	
	Density	Basal Area	Density	Basal Area
Abies magnifica	5744	81.5	7152	99.8
A. concolor	456	3.3	22	0.2
Pinus contorta ssp. *murrayana*	189	3.8	86	0.6
P. lambertiana	*	1.2		
P. albicaulis	122	1.8		
P. jeffreyi	*	0.3		
P. monticola			159	1.4
Tsuga mertensiana			1	0.01
Total	6511	91.9	7420	102.0

The stand structure is commonly dominated by even-aged (established within a 20 yr period) groups of trees. Uniform stands may vary in size from a small fraction of a hectare to several hectares. Some stands which appear to be uneven aged are found upon close examination to be even aged by small groups, or to be two storied. The principal sources of quantitative information on even aged red fir stands are Schumacher (1928) and Dunning and Reineke (1933); summary statistics appear in Table 17-7.

Where *Abies magnifica* seedlings have an opportunity to become established, the most favorable microsites are filled by the first wave or two of seedlings from heavy seed crops. Additional seedlings may become established later for an indeterminate period, but most will eventually die. With time, the first wave of seedlings will become the dominant trees in a new stand. Snow suppression and competition for soil moisture constantly reduce the number of plants, and stand closure at the sapling stage eliminates nearly all the low vegetation.

Natural regeneration may occur in small openings where soil moisture becomes available after the death of one or a few old trees, or on larger areas created by wind damage, insect epidemics, fire, or logging. Seedlings become established best on mineral soil or litter less than 8 mm deep (Gordon 1970). Seedlings grow to a height

TABLE 17-7. Statistics on pure, even-aged stands of *Abies magnifica* in the Sierra Nevada greater than 10 cm dbh. Based on Schumacher (1928). Biomass and productivity figures are for above- and below-ground material. Wood density assumed to be 415 kg m^{-3}

Site Index	Age (yr)	Density per Hectare	Mean Height (m)	Basal area (m^2 ha^{-1})	Bole Wood		Biomass (kg m^{-2})	Increment		Productivity (g m^{-2} yr^{-1})
					Volume (m^3 ha^{-1})	Weight (kg m^{-3})		Volume (m^3 ha^{-1})	Weight (kg m^{-3})	
18.3	50	2360	12	68	350	14.5	36	12.6	523	1860
	100	690	31	110	1170	48.6	120	19.9	825	2940
	150	252	52	131	2230	92.4	228	13.3	551	1960
9.2	50	8890	6	54	171	7.1	17	5.9	246	880
	100	2590	14	87	570	23.6	58	9.4	392	1400
	150	954	24	104	1090	45.3	112	7.3	305	1090
6.1	50	15800	3	52	126	5.2	13	4.2	174	740
	100	4650	8	84	416	17.3	43	7.0	290	1030
	150	1704	13	100	794	32.9	81	5.2	217	770

of only 10–15 cm during the first 2–4 yr. If established where they remain badly suppressed by larger trees, they may grow only 3 cm a year for 60–80 yr or more (Gordon 1973). If established so as to have free growth from the beginning, after the 2–4 yr slow growth period they will attain a height growth rate of 20–30 cm yr^{-1} during the sampling stage.

Some shrub species remain until tree stems reach diameters of 15+ cm, but generally shrubs tend to dominate only areas that have been subject to hot fires. The most common brushfield species are *Ceanothus velutinus, C. cordulatus, Arctostaphylos patula, Ribes roezlii, Chrysolepis sempervirens, Prunus emarginata,* and *Salix scouleriana,* with *Quercus vaccinifolia* appearing on less fertile sites.

Abies magnifica often replaces lodgepole pine after fires. Large individuals of lodgepole pine that survive fires are highly susceptible to bark beetle damage. Under such conditions, the reproduction of *A. magnifica* is abundant, whereas lodgepole pine seedlings are restricted to openings and disturbed sites. The natural succession from pine to red fir is tremendously accelerated by bark beetle activity.

More commonly than any other species, lodgepole pine dominates scattered small areas within the red fir forest. Such areas often appear to be the result of small lightning-caused fires that killed fir trees and created a mineral seedbed at a time fortuitous to catch a seed crop from nearby pines. Such areas are transient, since fir regeneration is good under pine, and eventually firs (perhaps after 100–200 yr) overtop and kill the shorter and less tolerant pines. Other lodgepole-dominated areas are principally associated with shallow water tables near streams and meadows or on benches. In some of these areas, mixed stands have a visually grouped species distribution; in others, a more open and mixed appearance.

Understory vegetation is often slight in red fir forest because of the dense canopy which cuts out light and builds up a deep accumulation of litter. In many cases the understory flora closely resembles that of the lower white fir forest. Oosting and Billings (1943) found that *Ribes viscosissimum* was the most constant shrub in the northern and central Sierra, while *Symphoricarpos vaccinoides* was present in about half their stands. Densely shaded stands typically support an open herbaceous cover of less than 5%, dominated by *Hieracium albiflorum, Poa bolanderi, Corallorhiza maculata, Chimaphila umbellata, Pedicularis semibarbata, Chrysopsis breweri, Pyrola picta, P. secunda, Sarcodes sanguinea,* and *Pterospora andromeda.* Openings with coarse, gravelly soil typically support denser covers of *Arabis platysperma, Viola purpurea, Eriogonum nudum, Gayophytum nuttallii, Monardella odoratissima, Calyptridium umbellatum, Sitanion hystrix,* and *Wyethia mollis.*

Logged openings in red fir forest have a very different array of understory vegetation, partly because of large accumulations of debris and skidding activities. Studies from Siskyou to Calaveras Cos. of 32 cut stands found 58 species (D. T. Gordon and E. E. Bowen, unpub. data), the majority not included in the lists of Oosting and Billings (1943). *Gayophytum nuttallii, Phacelia mutabilis, Ribes roezlii, R. sanguineum,* and *Ceanothus cordulatus* were present in 40% or more of these samples.

Jeffrey pine. At higher elevations or drainage bottoms with cold temperatures, *Pinus jeffreyi* replaces *P. ponderosa* as the dominant pine. In the northern Sierra Nevada, Jeffrey pine forests occur commonly at 1520–1830 m, rising to 2130–2740 m in the south. On the west slope, Jeffrey pine typically occurs in mixed stands, although pure stands may be present on glaciated soils or granite outcrops. South of the Lake Tahoe Basin on the east slope of the Sierra Nevada, Jeffrey pine forms pure stands (Fig. 17-5), mixing only at higher elevations with *Abies magnifica* and *Pinus contorta* ssp. *murrayana,* and occasionally with *Abies concolor* on mesic sites.

Few quantitative stand data are available for Jeffrey pine forest. For the west slope of the Sierra Nevada, such data are limited to studies in Sequoia and Kings Canyon National Parks, where mixed stands predominate (Vankat 1970; Omi et al. 1975). *Pinus jeffreyi* is the dominant single species in terms of relative basal areas, but *Calocedrus decurrens* and *Abies concolor* may have greater densities (Table 17-8). Other commonly associated tree species are *Quercus kelloggii* and *Populus trichocarpa.* Scattered stands sampled by Sudworth (1900) included 60–210 *Pinus jeffreyi* per hectare and had a mean tree height of 27–40 m. Stands are open, with tree cover commonly varying from 40 to 65%. Total tree basal area ranges from 9.2 to 62.7 m^2 ha^{-1}. Shrub cover is highly variable, while herb cover averages around 20%. *Arctostaphylos patula* and *Chamaebatia foliolosa* are common shrub associates.

On the east slope of the Sierra Nevada, cold temperatures are assumed to restrict the growth and survival of *Pinus ponderosa,* and *P. jeffreyi* is the dominant yellow pine in montane forests. In the northern Sierra Nevada, east slope Jeffrey pine occurs in mixed stands, sharing dominance with *P. ponderosa, Calocedrus decurrens, Abies concolor,* and *Juniperus occidentalis* ssp. *australis.* Sample stands in Plumas National Forest (Table 17-8) have a mean tree cover of 64%, with a wide diversity of understory shrubs producing nearly 100% total plant coverage. Important understory shrubs include *Artemisia tridentata, Ceanothus prostratus, C. velutinus, Arctostaphylos patula, Purshia tridentata,* and *Haplopappus bloomeri.* The heavy dominance of *Ceanothus* and *Arctostaphylos,* as in other vegetation types, is indicative of succession after disturbance from fire or logging. Jeffrey pine ecosystems from adjacent parts of Nevada have been described in great detail by Stark (1973).

Figure 17-5. Jeffrey pine forest, east slope, Mono Co.

TABLE 17-8. Cover (%), density per hectare, and basal area ($m^2\ ha^{-1}$) for six stands of Jeffrey pine forest. Stands a-d are from Sequoia National Park and include all trees older than 1 yr (Vankat 1970); a: Crescent Meadow, b-d: Kern Canyon. Stands e-f are from the east slope and include all trees >2 m tall (Vasek 1976); e: Plumas National Forest, f: Tahoe and Toiyabe National Forest. Asterisk = data not available

Taxon	Cover				Density						Basal Area			
	a	b	c	d	a	b	c	d	e	f	a	b	c	d
Pinus jeffreyi	26	25	39	31	200	195	96	299	113	198	28	9	28	34
Calocedrus decurrens	*	10	19	11	*	405	400	598	36	5	3	22	9	16
Abies concolor		8	12	17		705	1008	403	35	39		*	*	12
Pinus contorta ssp. *murrayana*		5				105						*		
Quercus kelloggii		*				*						5		
Populus trichocarpa		*	16			105	96					2	5	
Pinus ponderosa									73					
Juniperus occidentalis ssp. *australis*									7	10				
Pinus labertiana										1				
Total	26	48	86	59	200	1515	1600	1300	264	253	31	38	42	62

In the Tahoe basin, *Pinus jeffreyi* dominates forest stands at elevations of 2200–2500 m, sometimes codominant with *Abies magnifica* (Smith 1973). On the adjacent eastern escarpment of the range in the Tahoe and Toiyabe National Forests, *P. jeffreyi* may share dominance with *Abies concolor* or *Juniperus occidentalis* (Table 17-8) with an open understory including *Ceanothus prostratus, Purshia tridentata, Arctostaphylos patula,* and *A. nevadensis.* Total tree cover is similar to stands to the north, with a mean value of 67% and a range of 50–100%. On xeric sites, *P. jeffreyi* commonly forms pure stands with an understory of *Purshia tridentata* and *Artemisia tridentata.*

In the northern Sierra Nevada and north into the Cascade Range, *Pinus washoensis* (a high-elevation form of *P. ponderosa*) may become locally important, associated with *P. jeffreyi* (Haller 1961). On Mt. Rose, from 1950 to 2560 m, *P. washoensis* reaches densities close to those of the dominant *P. jeffreyi* (Haller, pers. commun.). *Abies concolor* and *A. magnifica* are less important associates in these stands. *Ceanothus velutinus, Arctostaphylos patula, Cercocarpus ledifolius,* and *Purshia tridentata* are important understory shrubs. In the Bald Mt. Range, *P. washoensis* is commonly the dominant tree species above 2400 m, with *A. concolor, Juniperus occidentalis* ssp. *australis,* and *Populus tremuloides* var. *aurea* as important associates and *P. jeffreyi* and *A. magnifica* occurring only rarely (Haller, pers. commun.). Similarly, *P. washoensis* is the dominant species at about 2290 m in the Warner Mts., particularly on dry slopes. There it occurs with *A. concolor, Pinus contorta* ssp. *murrayana, P. albicaulis,* and typical understory associates of adjacent Jeffrey pine forests (Haller, pers. commun.).

Further south on the east slope of the central and southern Sierra Nevada, Jeffrey pine is present in relatively pure forest stands from 2590 to 2740 m, forming a distinct zone between higher red fir forests and lower sagebrush or piñon–juniper areas. Because of the low annual precipitation in this zone, typically 60–100 mm yr^{-1}, understory vegetation is dominated by Great Basin species. The most important of these are *Artemisia tridentata, Cercocorpus ledifolius, Leptodactylon pungens, Purshia tridentata,* and *Chrysothamnus parryi.* On steep slopes, however, shrub cover is virtually absent, and pure Jeffrey pine forests grade into dense red fir forests with admixtures of lodgepole and western white pine (Vasek 1976).

Although no quantitative data on herbaceous vegetation in Jeffrey pine forests are available, Vasek (1976) listed the conspicuous and common species for stands along the east slopes of the Sierra Nevada. *Sitanion hystrix, Elymus glaucus, Deschampsia elongata, Stipa occidentalis, Wyethia mollis,* and *Monardella odoratissima* occur in almost all stands. In relatively mesic sites, *Hordeum brachyanthemum, Bromus orcuttiana, Pyrola picta,* and *Wyethia angustifolia* are additionally common.

Lodgepole pine. Above the red fir zone, forest communities become more open and the trees are of shorter stature. Lodgepole pine (*Pinus contorta* ssp. *murrayana*) dominates the zone commonly found immediately above the red fir forest. In the northern Sierra, lodgepole pine occurs at 1830–2400 m but rises to 2440–3350 m in the south. The lodgepole forest is characterized by open stands with sparse litter accumulation and little shrub or herbaceous understory. Although it may intermingle with the red fir or mixed conifer forest below and subalpine forests above (occasionally even reaching up to tree line), it is most frequently found in extensive even-aged closed or open stands around meadows or moist areas. It is common on glacially scoured basins, where it colonizes joint planes on which soil has accumulated. With soil development, it may in time be replaced on these sites by *Abies magnifica*.

The lodgepole pine zone in the Sierra is characterized by an annual precipitation of 750–1500 mm, most of which falls as snow in the winter months. The growing season is short (2–3 mo), and temperatures are frequently cold. The stature and density of the trees are a function of the richness and depth of the soil and the protection of the habitat. With deep soil and a well-developed humus layer, lodgepole pine will grow in dense stands with individual trees reaching up to 30 m in height and 0.6–1.3 m dbh. On thin, poor soils of open granite slopes and exposed ridges, the tree occurs in widely scattered dwarfed forms. On the relatively thin granitic soils of much of the higher-elevation Sierra, it commonly attains a height of no more than 15–20 m. Although little ecological work has been done with lodgepole pine in the Sierra, considerable autecological and synecological research has been carried out with *Pinus contorta* ssp. *latifolia* in the Rocky Mts. (Baumgartner 1975).

In Sequoia National Park, Vankat (1970) sampled nine stands between 2600 and 3100 m. Lodgepole pine accounted for almost all of the ground cover. Total cover averaged 56%, lodgepole density averaged 3390 trees per hectare (including saplings), and lodgepole basal area averaged 58 m^2 ha^{-1}. An age analysis of a closed lodgepole forest in Kings Canyon National Park (Hammon-Jenson-Wallen Forestry 1970) revealed a bimodal distribution, with 50% of the trees in the seedling–sapling category and 20% in the 35–45 cm dbh class. Very few trees were over 65 cm dbh. Average tree height was 30+ m. Shrubs are generally of little important in lodgepole forests, but such species as *Arctostaphylos nevadensis, Ribes montigenum,* and *Phyllodoce breweri* are locally abundant.

Lodgepole pine is extremely successful in meadows and other moist areas, in addition to dominating open basins and ridges. Dense thickets of lodgepole commonly invade the peripheries of meadows. The invasion of many high Sierra meadows by lodgepole pine in recent years is a poorly understood phenomenon. One hypothesis is that lodgepole pine has only recently been able to grow freely in these areas, after years of effective suppression, first by natural and Indian-caused fires and more recently by the grazing of domestic sheep. The elimination of such disturbances during the present century has allowed for a high level of lodgepole reproduction in the meadows.

Physiological studies conducted on lodgepole and foxtail pines in the Rock Creek area of Sequoia National Park (Leonard et al. 1968) showed lodgepole pine to be a flood-tolerant species with remarkable control of its internal water balance over a wide range of soil moisture conditions. It is able to increase water uptake and transpiration rates much more rapidly than foxtail pine, and by several times as much, when placed under wet conditions. Sites that remain saturated throughout the growing season do not support lodgepole pine, however.

Vankat's (1970) study of vegetation change in Sequoia National Park showed a general increase in density and cover of lodgepole pine in the past half-century. There was extensive lodgepole reproduction sometime after 1900 in both mature stands and meadow peripheries. Periodic fires which served to keep lodgepole forest open before are no longer allowed to burn, preventing the thinning of young growth by fire and resulting in unnaturally heavy fuel accumulation. If continued, this trend may lead to a general degradation of the community or succession, leading to dominance by *Abies magnifica*.

In an effort to reintroduce fire in something approaching its natural role in these areas, the National Park Service has undertaken a program to allow lightning fires to run their course (Kilgore and Briggs 1972). Only areas where fuel accumulation has been low enough that fires can be safely allowed to burn are included in these Natural Fire Management Zones. Over 4000 ha of high-elevation forest in Yosemite, Sequoia, and Kings Canyon National Parks has burned under this program since 1973.

Subalpine Forests

Above the lodgepole forest is the subalpine forest, composed of widely spaced conifers from 25+ m in height to short, krummholz forms. Although occasionally intermingling with the lodgepole and red fir forests below, the subalpine forest associates are most commonly found between 2400 and 3050 m in the northern Sierra and between 2900 and 3660 m in the south. Characteristic species include *Tsuga mertensiana, Pinus albicaulis,* and *P. balfouriana,* and occasionally *P. monticola, P. flexilis,* and *Juniperus occidentalis* ssp. *australis* (Griffin and Critchfield 1972; Hudson 1933; Kylver 1931).

The subalpine forest zone is characterized by an annual precipitation of 750–1250 mm, which, except for an occasional summer thunderstorm, falls entirely as snow. The growing season is short, frequently no more than 7–9 weeks, and frosts can occur at any time of the year. Wind can be severe, especially on exposed ridges at higher elevations. Soils in this zone are relatively undeveloped, with little or no humus. The parent material is often simply decomposed granite, resulting in shallow, coarse-textured, quick-drying soils. All of the common tree species attain ages measurable in hundreds of years, and the foxtail pine is known to reach an age of nearly 2000 yr (Mastroguiseppe 1972). Litter and fuel accumulations are low; thus the impact of fire is minimal. Little shrubby vegetation is found in these communities, although *Salix* ssp., *Vaccinium occidentalis, V. nivictum,* and *Kalmia polifolia* are common around meadows and moist sites. Occasional shrubs include *Ribes cereum, Holodiscus microphyllus,* and *Artemisia tridentata.*

Mountain hemlock. *Tsuga mertensiana* is probably the most common representative of the subalpine forest in the northern Sierra. It grows as an erect tree, up to 30 m tall, in extensive groves southward to Yosemite. The *Tsuga* forest may be dense and dark, with a sparse ground flora.

South of Yosemite, *T. mertensiana* becomes increasingly limited to cold, moist slopes and high valleys at elevations above 2600 m (Parsons 1972), commonly in small stands at the heads of north- or east-facing canyons or in sheltered ravines where snow lingers well into the summer. These sites are characterized by shade and abundant soil moisture. These scattered southern stands, unlike the pure groves to the north, are commonly interspersed with such species as *Pinus contorta* ssp. *murrayana, P. monticola, P. balfouriana,* and *Abies magnifica.* At the southernmost known locality, below Silliman Lake in Tulare Co., a grove of about 60 trees contains individuals with heights as great as 24 m and diameters up to nearly 90 cm (Parson 1972), dimensions approaching those found in dense forests to the north. The stand is also characterized by a number of young individuals.

Western white pine. *Pinus monticola* is moderately abundant on the west slope of the Sierra, scattered in small groups and often interspersed with other species,

TABLE 17-9. Density per hectare of all trees by diameter class in a typical Pinus monticola subalpine forest at the headwaters of Copper Creek, Fresno Co., 3290 m elevation. Cumulative basal area for each species is also shown. Taken from Sudworth (1900). Lack of reproduction can be attributed to heavy sheep grazing around the turn of the century

Taxon	Diameter at Breast Height (cm)						Basal Area
	<15	15–30	31–45	46–60	61–100	>100	(m^2 ha^{-1})
Pinus monticola	0	0	20	69	104	10	103
Abies magnifica	0	0	0	0	20	30	59
Pinus contorta ssp. murrayana	0	10	10	0	10	10	25

throughout the subalpine forest zone. A common associate of red fir, lodgepole pine, and mountain hemlock forests in the central Sierra, *P. monticola* is more likely to be found in small groves of well-developed large trees on dry, exposed granite slopes to an elevation of 3400 m in the south. If given adequate shelter, it will grow as an erect tree nearly to tree line. Table 17-9 summarizes a typical stand of western white pine, 24–27 m tall, in what is now Kings Canyon National Park.

Whitebark pine. The most typical tree line conifer of the Sierra Nevada is probably *Pinus albicaulis,* though it is also frequent in pure or mixed stands with lodgepole or foxtail pines at lower elevations (see Table 17-10). Growing in areas where glacial scouring has eliminated most of the soil, the trees are generally small in stature (10–15 m), of candelabra form if erect, and often with gnarled or twisted branches, as on the southwest flanks of Mt. Tallac in the Lake Tahoe region (Smiley 1915).

In more exposed areas, whitebark pine forms stunted, multistemed trees. High on wind-swept slopes and ridges at tree line, they form low mats not more than 1 m high. Growth is very slow: a 12 cm long branchlet on *P. albicaulis* may be as much as 17 yr old (Storer and Usinger 1963), and a tree 43 cm dbh was found to be 800 yr old (Arno 1967).

Foxtail pine. *Pinus balfouriana* is found in the Klamath Mts. of northern California, but in the Sierra it is restricted to the region south of the South Fork of the San Joaquin River, at elevations of 2600–3660 m (Fig. 17-6). This disjunct distribution is unique among subalpine tree species. At its lower elevation limit, it occurs in extensive pure stands or mixed with other subalpine species. At higher elevations, the trees commonly grow as widely spaced individuals in pure stands. Intolerant of shade at all stages of growth, *P. balfouriana* does especially well on shallow, well-drained, decomposed granite soils on exposed slopes. It does not form a krummholz, retaining a single trunk throughout its range. A tree at 3250 m on Alta Peak in Sequoia National Park measures 2 m dbh and 24 m in height (Mas-

TABLE 17-10. Density per hectare of all trees by diameter class in a typical *Pinus albicaulis* subalpine forest at Rae Lakes in Kings Canyon National Park, 3230 m elevation. Taken from Hammon Jensen-Wallen Forestry (1970)

Taxon	Seedlings	Diameter at Breast Height (cm)					
		<8	8-15	16-30	31-45	46-60	>60
Pinus albicaulis	20	59	12	7	5	0	94
P. balfouriana	0	10	5	5	7	5	0
P. contorta ssp. *murrayana*	0	0	12	12	0	0	0

Figure 17-6. Foxtail pine forest, in foreground and on slopes in background. Photograph taken at Charlotte Lake, Kings Canyon National Park by S. DeBenedetti, courtesy of the National Park Service.

troguiseppe 1972). It has a deep, spreading root system from which it obtains melt-water in well-drained rocky sites (Arno 1973). The forest floor of a foxtail pine forest is characterized by bare earth and rocks, with only a scattering of litter (Fig. 17-6).

The open composition typical of foxtail pine "forest" was documented by Vankat (1970), who sampled eight stands in Sequoia National Park at elevations from 3170 to 3290 m. Average cover (almost all by foxtail) was 26%; average foxtail density was 418 per hectare; average foxtail basal area was 31 m^2 ha^{-1}. Much like its close relatives *Pinus longaeva* and *P. aristata, P. balfouriana* attains great age. A number of the larger trees in the southern Sierra are well over 1000 yr old (Mastroguiseppe 1972).

Limber pine. *Pinus flexilis* is fairly well represented on the east slope from Mono Pass south to 36°40′ N, where it apparently fills a niche provided by the absence of *P. albicaulis* (Arno 1967). It is basically a Rocky Mt. species, and only a small part of its North American range extends into the southern Sierra. Lepper (1974) investigated limber pine over its entire range, and 19 of her 54 sample sites were located in California. The material below is a summary of that work.

Limber pine typically occupies steep, eroded, rocky, coarse-textured, well-drained, nutrient-poor sites. When present in more mesic situations, its dominance is reduced and it may be viewed as a pioneer species. Lepper concluded that its distribution is limited by competition and that it persists longest on the driest and steepest sites. There was a remarkable degree of homogeneity in associated species over its entire range. For her California sites, average density of limber pine (older than seedlings) was 21 trees per hectare, and relative density was 35%. Importance value (relative basal area + relative density, maximum value 200) averaged 84. The oldest tree she found was on Mt. Baden-Powell, in excess of 1200 yr, but trees older than 250 yr were atypical. Seedlings were usually present on any site, but in low density (3–4 per hectare). She found that age did correlate with diameter at breast height, and Fig. 17-7 illustrates the extremely flat age–frequency curves of four California populations.

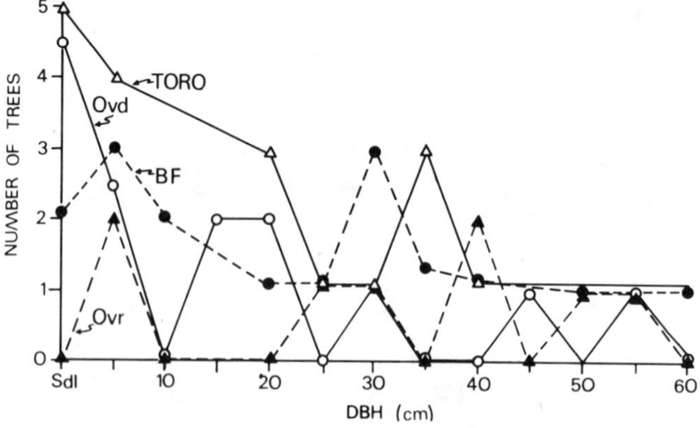

Figure 17-7. Size class distribution of *Pinus flexilis* in four Sierran stands. TORO = Santa Rosa Mts., Riverside Co.; Ovd and Ovr = Onion Valley, east slope of Sierra Nevada, Inyo Co.; BF = San Bernardino Mts., San Bernardino Co. From Lepper (1974).

Figure 17-8. Stand of *Populus tremuloides* var *aurea* in Murphy Meadows, Placer Co., 2400 m elevation.

Lepper collected seed from several sites and reported that optimum germination occurred at 13°C after stratification for 2 wk. Seedlings grew slowly, but optimally, at a 20/10°C regime, 15 hr photoperiod. Dormancy could be induced by lowering temperature to a constant 7°C and photoperiod to 8 hr. Limber pine seedlings were relatively drought tolerant, maximum net photosynthesis at a soil water potential of −20 atm being 29% of that at −0.3 atm. Wright (1970) has shown much poorer drought tolerance on the part of such montane pines as *P. attenuata, P. coulteri,* and *P. lambertiana.*

Nonconiferous Vegetation

Deciduous forest. A number of broad-leaved tree species are also found in the higher-elevation Sierra Nevada. On steep eastern slopes, *Betula occidentalis* occurs between about 1500 and 2750 m in the southern Sierra, but is lacking in the central and northern Sierra. It grows in groves of multistemmed trees 6–9 m tall along streams draining into the Owens Valley. The taller *Populus trichocarpa, P. tremuloides* var. *aurea,* and *Salix* spp. are also common in cool and moist canyons of both the east and west slopes.

Populus tremuloides var. *aurea,* which has one of the widest distributions of any tree in North America, extends throughout the Sierra Nevada, but it is especially abundant between 1800 and 3000 m in the southern Sierra. It has been monographed by Barry (1971). It occurs in pure groves fringing moist meadows, and in rock piles at the base of cliffs where there is an adequate supply of groundwater. Aspen groves are characterized by an openness that encourages a lush growth of grass and herbs (Fig. 17-8). While intolerant of shade, aspen groves typically expand and fill in the available habitat by sending up sprouts from an existing root system. This can result in a dense stand of trunks all of which belong to the same clone.

Barry (1971) showed that seedling mortality was strongly affected by soil characteristics, and young growth was optimal at pH 6 and salinity lower than 1000 ppm. He found aspen to occupy mesic sites with a high water table during the early part of the growing season; sites with a permanently high water table were occupied by *Salix* spp. The range of Sierran habitats for aspen includes Great Basin sagebrush scrub, Jeffrey pine woodland, northern juniper woodland, red fir forest, lodgepole pine forest, subalpine forest, mixed conifer forest, ponderosa pine forest, and (rarely) montane chaparral. Only in red fir forest did aspen groves appear to be successional, giving way to red fir in 200+ yr. Barry's dissertation contains a great number of environmental data across a gradient from *Wyethia*-dominated slopes through aspen, willow, and red fir communities in Murphy Meadows, Placer Co. (about 2400 m).

Meadows. Ranging in size from a few square meters to several hundred hectares, meadows are found scattered through virtually every forest type of the montane and subalpine Sierra Nevada. Although their total area is a small percentage of the mountainous terrain, meadows are very heavily used for both recreation and livestock grazing.

The origin of Sierran meadows has often been attributed to the filling in of glacial lakes or valleys (Storer and Usinger 1963). Although this explanation is valid for many meadows, it oversimplifies the intricacies and variety in meadow types and modes of formation. Wood (1975) has shown that many meadows in the 1500–2500 m range lie in stream valleys that were never glaciated, and stumps of mature trees are buried in the soil beneath. This implies that regressive succession may have occurred. Ratliff (1973) proposed that such species as *Kalmia polifolia* var. *microphylla, Vaccinium nivictum,* and several mosses become established around boulders at the edge of lakes. This vegetative mat then forms a foundation upon which a *Calamagrostis breweri* climax eventually gains hold. The single most important factor in explaining the distribution of meadows is the existence of a shallow water table which provides for a high soil moisture content the year around (Wood 1975).

Sierran meadows can be classified into three broad types: (1) wet meadow, (2) woodland meadow, and (3) short-hair sedge (Bennett 1965). The wet meadow type is generally found scattered above 1200 m in the north and 1800 m in the south. It is characterized as an open vegetation with a cover composed predominantly of perennial sedges, rushes, and grasses. The soil is especially dark in color because of large amounts of organic material. Dominant species reproduce chiefly through rhizomes. They store large amounts of carbohydrates below ground and burst forth with a quick flush of growth early in the spring (Mooney and Billings 1960).

Several subtypes of the wet meadow community have been identified by Bennett (1965) and Strand (1972). The subtype dominated by *Sphagnum* and species of fine-leaved sedges is characterized by an acidic, organic muck which is subject to oxidation if the meadow vegetation is disturbed. This subtype is highly susceptible to trampling. A subtype dominated by *Juncus* is more resistant to trampling because of the tough, fibrous roots of these species. This subtype is also noted for its low carbohydrate and protein availability, low palatability, and low tolerance to frost. A fine-leaved sedge and grass subtype, found on drier well-drained sandy loams, contains *Carex festivella, Heleocharis pauciflora, Calamagrostis breweri, Tricetum spicatum, Vaccinium nivictum,* and *Aster alpigenus* ssp. *andersonii.* It is the best developed and most common of the wet meadow types and is of significantly higher forage value. Smith (1973) noted that many of the meadows in the Lake Tahoe area show a shift in dominance of species from the wet to dry meadow type as the summer progresses.

The woodland meadow type, also found scattered above 1800 m, is typified by scattered grasses and forbs interspersed with *Pinus contorta* ssp. *murrayana, Salix* spp., *Populus tremuloides* var. *aurea,* and *P. trichocarpa.* Depending on the elevation and forests characteristic of the area, a great variety of herbaceous species occur within this type (Strand 1972). Common grasses include *Agropyron* spp., *Elymus glaucus,* and *Bromus carinatus. Lupinus* spp. are also common. These meadows hypothetically represent remnants of the wet meadow type that have dried out and been invaded by forests.

The short-hair sedge type is generally found at elevations above 2130–2440 m. This community is characterized by a tough *Carex exserta* sod which will withstand considerable disturbance. The foliage is extremely frost resistant and is high in carbohydrates and proteins (Bennett 1965). In a survey of the short-hair sedge meadows in Sequoia and Kings Canyon National Parks, Ratliff (1974) found average foliar cover values ~ 29% and oven-dry standing crop ~ 31 g m^{-2}, low values similar to those reported by Klikoff (1965a) for short-hair meadows of Yosemite National Park. Klikoff (1965b) also measured the net photosynthesis of three meadow species and found that their responses to moisture stress paralleled their natural distributions. As leaf water potential declined from about −2 to −15 atm, the net photosynthesis of *Carex exserta* from dry meadows did not significantly decline, but the rates of *Potentilla breweri* and *Calamagrostis breweri* from wet meadows fell sharply to 10% of the initial rates.

Short-hair sedge and its associates, *Antennaria spp., Lupinus breweri, Stipa occidentalis,* and *Trisetum spicatum,* do especially well on dry, gravelly soils. Growth begins early in the season, and plants turn dormant in the summer as moisture becomes limiting. Early growth, high nutrient content, and high palatability make the short-hair sedge meadow an extremely important source of early-season forage. Extensive sheep grazing in the higher Sierra around the turn of the century had detrimental effects on these meadows (as well as on all other types), for while the sod is extremely tough, once it is broken it becomes reestablished slowly and with great difficulty. The results of sheep grazing can still be seen today, 50 yr later (Sharsmith 1959). Coupled with the pack and saddle stock grazing of recent years, disturbance has changed the species composition and increased the percentage of barren ground and cover by nonpalatable species in many of the meadows. Further studies are needed on the long-term impacts of grazing and the feasibility of managing modified meadows so as to minimize deterioration and, if possible, return them to a more natural state.

Granite outcrops. Although open rock outcrops constitute a significant area within the coniferous zone of the Sierra Nevada, the colonization and development of vegetation on outcrops have rarely been discussed in the literature. The general pattern of succession on granite outcrops in the montane zone of the southern Sierra Nevada has been described in some detail by Rundel (1975), and his observations may have general applicability to other parts of the range. Successional development is primarily related to physiographical weathering, leading to the formation of fracture lines suitable for colonization by woody plants. Biological modification of the outcrop by primary colonization of cryptogams and herbaceous vascular plants is relatively unimportant in succession.

Two species of crustose lichens, *Rhizocarpon bolanderi* and *Lecanora gibbosa,* are the earliest colonizers on granite outcrops in the southern Sierra Nevada, although a variety of other species of lichens may also be present. Bryophytes are also early colonizers, particularly *Grimmia* species, which form mats along the margins of small drainage channels on the outcrop surfaces. Small discontinuous mats of *Polytrichum juniperinum* and *Ceratodon purpureus* are common along seepage zones and in areas of shallow soil accumulation. Despite this widespread primary colonization by lichens and bryophytes, only rarely do these species play a significant role in the establishment of woody perennials on the outcrops.

Significant colonization of perennial species first occurs on outcrops when physiographical succession proceeds from a bare granite surface to the formation of fracture lines with prominent crevices. These crevices, 2–5 cm wide at the surface, collect windblown sand and organic matter, thereby providing a substrate for vascular plant colonization. *Zauschneria californica* ssp. *latifolia* is the most widespread colonizer, but *Lomatium torreyi* may be more abundant locally. In moist, shaded crevices, *Pellaea brewerii, P. bridgesii, P. mucronata* var. *californica, Onychium densum, Cheilanthes gracillima,* and other ferns dominate.

Primary colonization of woody plants in crevices is the most important stage in succession leading to the development of broad soil mats. In moist years, seeds of a variety of woody shrubs and trees are able to germinate and become established in pockets of granitic sand and organic matter within crevices. The most important montane colonizers are *Pinus ponderosa, P. jeffreyi, Calocedrus decurrens,* and *Arctostaphylos patula.* Less commonly, *Quercus kelloggii, Q. chrysolepis, Abies concolor, Ceanothus cordulatus,* and *Arctostaphylos viscida* are present. In the subalpine zone, *Juniperus occidentalis* ssp. *australis, Arctostaphylos nevadensis, Pinus murrayana, P. balfouriana, P. albicaulis,* and *Quercus vaccinifolia* colonize outcrops in a similar manner. There is no evidence that the establishment of any of these species requires any previous biological succession on outcrops.

Alternatively, soil mats may evolve from weathered, broken outcrop surfaces. Shallow sands with little or no organic matter here support a modest diversity of herbaceous perennials, including *Calyptridium umbellatum, Eriogonum nudum, E. wrightii* var. *subscaposum, Lomatium torreyi, Streptanthus tortuosus,* and mats of *Polytrichum juniperinum* and *Ceratodon purpureus.* With only 2–5 cm of granitic soil, dense carpets of herbaceous annuals may cover outcrop surfaces, particularly *Linanthus montanus, Phacelia orogenes,* and *Eriogonum spergulinum.* As outcrop soils reach 5–15^{+} cm in depth with moderate amounts of organic matter, these species are replaced or overshadowed by a community of herbs characteristic of open areas within yellow pine forest communities. Woody trees and shrubs are able to colonize soils of this depth without reliance on crevice microhabitats.

The Ecological Role of Fire

Fire is a natural and characteristic factor in the environment of the coniferous forests of the Sierra Nevada. Hot, dry summers with occasional thunderstorms and severe lightning make coniferous forests of all types particularly susceptible to fire.

A great many studies have been made on the history of fire in the Sierra Nevada. Show and Kotok (1924), working in pine forests in the central Sierra, found the average interval between fires for a series of areas to be 3–11 yr with a mean of 8 yr. Burcham (1959) postulated less frequent fires for the same areas, and Reynolds (1959) gave a figure of 1.1 yr for fire interval in the Pinecrest area. Wagener (1961) believed such low intervals to be in error and reported a mean interval of 7–10 yr for the area from Plumas to Fresno Cos.

Studies of mixed conifer forests within and adjacent to Sequoia and Kings Canyon National Parks have established minimum average natural fire frequencies of 7–9 yr (range of 4–20 yr: Kilgore 1973b). Variations in local natural fire frequency appear to result from the quality and quantity of fuel accumulation and the statistical probabilities of ignition sources and favorable weather conditions for fire propagation (Van Wagtendonk 1972).

Intensive studies of the ecological role of fire have been made in *Sequoia*-mixed conifer forests (Kilgore 1973a,b; Kilgore and Sando 1975; Agee 1973; St. John and Rundel 1976), ponderosa pine forests (Weaver 1943, 1967; Biswell 1967), and red fir forests (Kilgore 1971). No fire studies of lodgepole pine and other subalpine forests in the Sierra Nevada have been made. In all of the forest types studied, natural fires, before the fire suppression policies of this century, were primarily restricted to mild-intensity ground fires of limited extent. Crown fires were probably rare or nonexistent in most Sierran forests before 1920 (Show and Kotok 1924; Kilgore and Sando 1975).

Frequent natural ground fires maintained ponderosa pine and *Sequoia*-mixed conifer forests in open, parklike stands by eliminating understory stands of shrubs and saplings. Fire suppression practices have resulted in a rapid accumulation of litter and understory vegetation, particularly dense stands of white fir saplings (Vankat 1970; see Table 17-4). The absence of regular ground fires has resulted in seedbed conditions unfavorable for many fire-adapted species (giant sequoia, ponderosa pine, Jeffrey pine, and *Ceanothus* spp.). Current management practices at Sequoia, Kings Canyon, and Yosemite National Parks are attempting to restore natural fire cycles to coniferous forests (Agee 1974; Van Wagtendonk 1974; Parsons 1975). The influence of man on fire frequencies and natural succession is less clear in red fir and subalpine forests than in mixed conifer forests, but similar patterns of change appear to have occurred, although to a much smaller extent (Vankat 1970; Kilgore 1971; Kilgore and Briggs 1972).

Overall, the ecological effects of fire on vegetation structure in the coniferous forests of the Sierra Nevada relate to four primary functions: (1) maintenance of fire-adapted species as community dominants, (2) production of uneven-age vegetation mosaics within communities (Weaver 1967; Kilgore 1973b; Bonnickson 1975), (3) cycling of nutrients (St. John and Rundel 1976), and (4) reduction of infestation centers (Wood 1972; Kilgore 1973b).

THE CASCADE RANGE

Low-elevation ponderosa pine forest, mixed conifer forest, and upper montane red fir forest typical of the Sierra Nevada continue northward into the Cascade Range. The distribution of coniferous communities in relation to elevation and moisture is shown in Fig. 17-9. On the higher volcanic peaks, particularly Mt. Shasta and

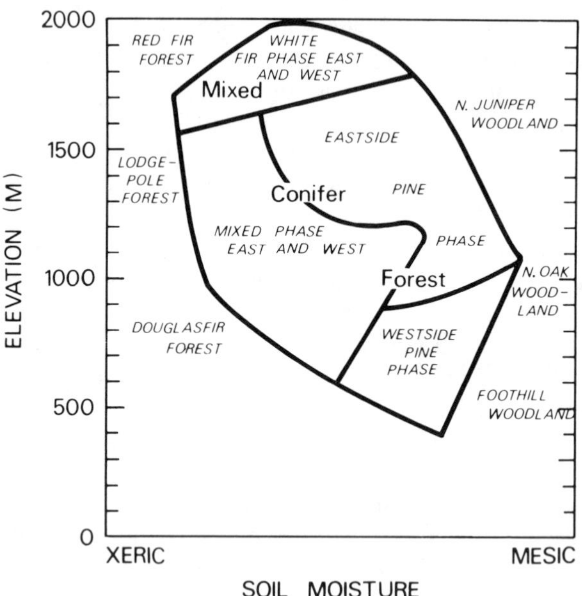

Figure 17-9. Diagrammatic relationships of major vegetation types in the southern Cascades (~ 41°N) in relation to elevation and moisture. Adapted from Griffin (1967).

Lassen Peak, good zonation of forest types is present. Ecological studies of the vegetation of these peaks have not been extensive, but descriptions for Mt. Shasta have been given by Merriam (1899), Dickson and Crocker (1953), and Cooke (1940, 1941, 1962), and for Lassen Peak by Gillett et al. (1961) and Bailey (1963). Much of the material below comes from Griffin (1967).

On the western slopes of the Cascade Range, ponderosa pine forest comprises a narrow belt through an elevational range of 500 m. These ponderosa pine stands are described by Griffin (1967) as often comprising little more than an ecotone between foothill woodland and mixed conifer forest. At higher elevations this mixed conifer forest on the western slopes includes mixed dominance of *Abies concolor, Pinus lambertiana, P. ponderosa, Calocedrus decurrens, Pseudotsuga menziesii,* and *Quercus kelloggii.* At yet higher elevations, upper montane forest communities of *Abies magnifica* and *Pinus contorta* ssp. *murrayana* dominate. On the east slope of the Cascades, montane communities are dominated by *P. ponderosa* and *P. jeffreyi,* occurring in both closed forest and open, parklike communities with shrub understories.

Community data for montane coniferous forests in the southern Cascade Range are given in Table 17-11. West slope ponderosa pine forests, described by Griffin (1967) as the west-side pine phase of the mixed conifer forest, resemble similar communities in the Sierra. *Pinus ponderosa* is the dominant species, with *Calocedrus decurrens* and *Quercus kelloggii* important associates. Important understory shrubs include *Arctostaphyos viscida, A. manzanita,* and *Rhus diversiloba.*

Truly mixed conifer forests at higher elevations include a diverse assemblage of species. Above 1500 m *Abies concolor* is clearly dominant, but from 800 to 1500 m any of the five conifer species previously mentioned may dominate individual stands

TABLE 17-11. Basal area (m^2 ha^{-1}) of conifers in 24 stands of the southern Cascade Range. Data are means of the numbers of stands shown in parentheses. From Shasta Co., 563-1798 m elevation (Griffin 1967). Plus = present outside the 0.2 ha plots

Taxon	West-Side Pine Phase (4)	East-Side Pine Phase (5)	West-Side Mixed Phase (8)	East-Side Mixed Phase (3)	West-Side White Fir Phase (4)
Quercus kelloggii	11.4	0.7	3.6	5.2	0.4
Q. garryana	2.7	1.1			
Pinus ponderosa	13.6	24.2	8.6	13.2	1.0
P. lambertiana	+	+	8.3	1.0	7.2
Pseudotsuga menziesii	+	1.5	8.2	2.4	5.8
Calocedrus decurrens	8.0	1.4	12.8	7.9	2.3
Abies concolor	0.8	5.0	12.4	0.8	28.4
A. magnifica					0.3
Pinus jeffreyi		1.7		0.2	1.7
P. monticola					+
Juniperus occidentalis ssp. *occidentalis*		0.4			
Total	36.5	36.0	53.9	30.7	47.1

(Table 17-11), with *Pinus lambertiana* and *Abies concolor* more abundant on mesic sites and *Calocedrus decurrens* and *P. ponderosa* more important on relatively xeric sites. Important understory shrubs include *Ceanothus velutinus, Chrysolepis sempervirens, Prunus emarginata, Symphoricarpos acutus, Rosa gymnocarpa,* and *Ribes roezlii.* On mesic sites, several understory trees may be present: *Cornus nuttallii, Taxus brevifolia,* and *Acer circinatum. Quercus chrysolepis* is occasionally present.

While west-side and east-side yellow pine forests are relatively isolated topographically along most of the Cascade Range, areas in Shasta Co. along the Pit River Canyon and lower portions of the Burney Basin provide corridors with intermediate environments and community structure. Eastern-slope pine forests are heavily dominated by *Pinus jeffreyi* and *P. ponderosa,* as shown in Table 17-11. Other mixed conifer species are occasional associates, but never reach major importance on xeric sites. Understory shrubs on the east slope reflect a strong Great Basin affinity: *Purshia tridentata, Artemisia tridentata, Arctostaphylos patula, Cercocarpus ledifolius,* and *C. betuloides* are all important.

Successional Studies

Major mudflows and avalanches have been a regular environmental feature on Lassen Peak and Mt. Shasta. These flows have killed large areas of coniferous forests, resulting in primary succession of conifers and shrub species on bare substrates. Sudden release of water on the south slope of Mt. Shasta during the summers of 1924–26 spread detrital material over several square kilometers to a maximum depth of 5+ m. These flows, studied by Beardsley and Cannon (1930) and Dickson and Crocker (1953), included not only large boulders but also considerable fine volcanic mud. The original mixed conifer forest community in the area of the flow was dominated by *Calocedrus decurrens, Abies concolor, Pinus ponderosa,* and *P. lambertiana,* with a scattering of *Quercus kelloggii.* Coniferous species showed early signs of injury in the mudflow area, and within a few months virtually all trees were dead in areas where mud reached over 1 m in depth. The direct effect of the mudflow on conifers was attributed to deprivation of oxygen from the roots. Willows in the area survived for several years, but eventually succumbed to desiccation of their substrate.

The recolonization of areas devastated by mudflows and avalanches on Lassen Peak was studied intensively by Bailey (1963) and later by Heath (1967), who described a successional series covering flows dated over the last 1500 yr. Heath's study areas included The Devastated Area on the eastern side of the peak (resulting from the 1914–15 eruption) and Chaos Jumbles 8 km to the north, both lying primarily at about 1800 m. Nine species of conifers were encountered in his successional study, eight of them (a mixture of montane and subalpine species) appearing in the early seral stages.

The early establishment of conifers is limited only by the availability of disseminules and the physical stresses of the environment, and some establishment occurs the first year. Thus in the early stages of recolonization high-elevation species such as *Pinus contorta* ssp. *murrayana, P. monticola,* and *Tsuga mertensiana* do well, but they are eventually replaced by *Abies concolor* and *Pinus jeffreyi,* which gradually increase in dominance. Successional trends in density and mean height of six major species are shown in Table 17-12.

TABLE 17-12. Density per hectare and mean height (m) of major conifers on dated mudflows and avalanches at Chaos Jumbles and The Devastated Area, Lassen National Park, 1800 m. From Heath (1967)

Taxon	Age of Stand (yr) 50	300	750	1500	"Climax"
Abies concolor					
Density	1377	753	895	602	2109
Height	0.5	0.6	0.6	3.8	5.5
Pinus jeffreyi					
Density	22	753	689	1570	65
Height	0.9	0.9	2.0	6.6	28.4
P. contorta ssp. *murrayana*					
Density	3593	215	1635		
Height	0.4	1.6	2.8		
P. monticola					
Density	344	398	551		
Height	1.2	2.2	2.3		
Abies magnifica					
Density	215	48	29		
Height	0.8	2.5	1.0		
Tsuga mertensiana					
Density	215	11			
Height	0.4	2.3			

THE MODOC PLATEAU

Within the Modoc Plateau, extensive areas of relatively level upland east of the Cascade–Sierran ridge are broken by the block-faulted Adin Range; these areas are high enough to support coniferous forest vegetation. Because of the rain shadow effect of the Cascades, precipitation in the conifer zone averages only about 600 mm yr^{-1}, most falling as winter snow.

The vegetation of the southern Modoc National Forest in the Adin Range has been described by Vasek (1976). This area of moderate relief includes elevations ranging from 1280 m at Adin to 1950 m at Hayden Hill. The vegetation cover is a mosaic of coniferous forest communities (39% of the area), open woodlands with heavy covery by shrubs (43%), and shrublands (17%). The woodlands are discussed in Chapter 23; scrub is treated in Chapter 22. In closed forest stands, *Abies concolor* and *Calocedrus decurrens* are most important, with *Quercus kelloggii* present in low numbers. Individual species of conifers sort out along environmental gradients in the Modoc Plateau in patterns similar to those described earlier for the Cascades and Sierra Nevada. Stands dominated by *Juniperus occidentalis* (mostly ssp. *australis*) are typically associated with dry, rocky substrates, although they may occur also on heavy soils on south-facing slopes. On the other end of the gradient, *Abies concolor* and *Calocedrus decurrens* dominate relatively mesic sites with deeper, less rocky soils. *Pinus ponderosa* occupies deep soils with little surface rock, manifesting a strong preference for north-facing slopes. *Pinus jeffreyi* is intermediate in all aspects. See also Chapter 23.

Jeffrey pine forests described by Vasek (1976) have an average tree cover of 49%, with a total vegetation cover of 61%. This total cover and the densities and basal areas of trees are similar to values for this community type in the Sierra Nevada. Although *Pinus jeffreyi* is the clear dominant, *P. ponderosa* and *Juniperus occidentalis* are important associates, primarily on north- and south-facing slopes, respectively. *Cercocarpus ledifolius* is the most important shrub, with *Ceanothus prostratus*, *Artemisia tridentata*, and *Purshia tridentata* associated.

On the western slopes of the Cascades, where annual precipitation is higher, the Jeffrey pine community is replaced by ponderosa pine forest (lower elevations) or red fir forest (higher elevations). To the east, with decreasing precipitation, Jeffrey pine forest grades into Great Basin sagebrush or western juniper woodland. *Pinus jeffreyi* reaches its northern limit as an important forest species in the Modoc Plateau area, although the species reaches southern Oregon on serpentine soils (Whittaker 1960). The elevational range of *P. jeffreyi* in northeastern California is relatively narrower than that to the south in the Sierra Nevada and Transverse ranges (Haller 1959, see also Chapter 16). At its northwestern extent, excessive moisture favors *P. ponderosa* (Haller 1959) and may limit Jeffrey pine. To the northeast, the cold, dry conditions transitional to the Great Basin appear limiting (Milligan 1969).

THE WARNER MOUNTAINS

The Warner Mts., running north–south along the Nevada border in northeastern California, reach a crest elevation of 2130–2440 m, with a maximum elevation of 3013 m at Eagle Peak. Few studies of the vegetation have been made. With the exception of rich volcanic soils on the western tilt-slope, much of the surface is too rocky to support more than scattered strands of conifers (Pease 1965). The vegetation and floristics of the Eagle Peak area have been described by Milligan (1969), and ponderosa and Jeffrey pine forests have been discussed briefly by Haller (1961) and Oliver (1972). Two typical Sierran conifers, *Abies magnifica* and *Pinus lambertiana*, are absent (Griffin and Critchfield 1972). Patches of Jeffrey pine forest were considered to be depauperate Sierran forests by Critchfield and Ashbaugh (1969).

Pinus washoensis, a high-elevation form of ponderosa pine, is an important member of upper montane communities in the Warner Mts. On volcanic ridges south of the Warner Wilderness area, *P. washoensis* comprises 50–75% of the mature trees on gentle west- and southwest-facing slopes at 2260 m (Haller 1961 and pers. commun.). Associated tree species are *Abies concolor, Pinus contorta* ssp. *murrayana, Juniperus occidentalis* ssp. *australis Populus tremuloides* var. *aurea*, and *Pinus albicaulis*. Understory species are those typical of Jeffrey pine communities: *Cercocarpus ledifolius, Artemisia tridentata,* and *Wyethia mollis*. At lower elevations, *P. washoensis* stands merge into *P. ponderosa* communities.

AREAS FOR FUTURE RESEARCH

There are several major areas of deficit in our knowledge of Sierran coniferous forest vegetation. Foremost of these, in the general sense, is the remarkable lack of quantitative data for most of the forest communities. Extensive sampling of communities has been carried out only in the southern Sierra in Sequoia National Park. Quantitative data on community structure and composition elsewhere are largely lacking, even at the most rudimentary level. Perhaps the greatest lack of knowledge concerns the subalpine forest communities. Lodgepole pine forests have never been studied in detail anywhere in the Sierra, and our assumptions of the ecological relationships in this community are based on data collected on *Pinus contorta* var. *contorta* in the Rocky Mts. (Baumgartner 1975). With the exception of *Pinus flexilis,* other, less extensive subalpine forest communities are equally poorly known.

In the montane zone, extensive data are available for giant sequoia groves within the white fir community, because of popular interest in *Sequoiadendron*. Quantitative data for other white fir communities and for ponderosa pine and Jeffrey pine forests, however, are fragmentary.

At the biogeographical level, unusual species disjunctions are common, yet there has been little intensive study on the origin of these disjunctions. The disjunctions of *Pinus balfouriana, Betula occidentalis,* and other species in the southern Sierra have never been adequately explained. Likewise, the origin of Great Basin species (e.g., *Artemisia tridentata* and *Pinus monophylla*) on the west slope of the Sierra remains conjectural.

Although considerable attention has been given in recent years to the ecological role of fire in montane mixed conifer forests, more effort needs to be fixed on the role of fires in other coniferous forest communities. Red fir forests have received only superficial attention, and Jeffrey pine and lodgepole pine communities have not been investigated. With increasing interest in the management implications of controlled burning, we would except a rapid expansion of research into these areas.

Other areas of research with management implications also need increased attention in the future. Limited research has evaluated the bio-geo-ecological consequences of logging practices in the Sierra Nevada, and these studies have been local in scope. Management of mountain meadows also requires increased attention. Grazing and recreational uses have greatly modified the structure of Sierran meadows; thus we know little about the pristine conditions that management efforts are striving to restore. The dynamics of meadow communities under natural successional cycles is also poorly understood, particularly with respect to the invasion of meadows by lodgepole pine. Further studies are also needed on the long-term impacts of grazing and the feasibility of managing modified meadow ecosystems in a steady-state condition.

The Cascade Range in northern California forms an important vegetational and floristic link between the Sierra Nevada to the south and the Oregon Cascades to the north, yet little research attention has been given to this region and many areas remain unstudied. Quantitative data for upper montane red fir and subalpine communities are almost totally lacking. As in the Sierra Nevada, studies of human impact on coniferous forests in northeastern California are badly needed. We know very little about the effects of logging and recreational activities on the structure and stability of natural forest ecosystems.

LITERATURE CITED

Agee, J. K. 1973. Prescribed fire effects on physical and hydrologic properties of mixed conifer forest floor and soil. Water Res. Center, Univ. Calif. Contr. Rept. 143.

-----. 1974. Fire management in the national parks. West. Wildlands 1(3):27-33.

Arno, S. F. 1967. Interpreting the timberline. M.S. thesis, Univ. Montana, Missoula, 206 p.

-----. 1973. Discovering Sierra trees. Yosemite and Sequoia Nat. Hist. Assoc. 87 p.

Axelton, E. A. 1967. Ponderosa pine bibliography through 1965. USDA Forest Ser. Res. Paper INT-40.

Bailey, W. H. 1963. Revegetation in the 1914-15 devasted area of Lassen Volcanic National Park. Ph.D. dissertation, Oregon State Univ., Corvallis. 195 p.

Barry, W. J. 1971. The ecology of Populus tremuloides, a monographic approach. Ph.D. dissertation, Univ. Calif., Davis. 730 p.

Baskerville, G. L. 1965. Estimation of dry weight of tree components and total standing crop in conifer stands. Ecology 46:867-869.

Baumgartner, D. M. 1975. Management of lodgepole pine ecosystems. Symp. proc., Wash. State Univ. Coop. Ext. Serv.

Beardsley, G. F., and W. A. Cannon. 1930. Note on the effects of a mudflow at Mt. Shasta on the vegetation. Ecology 11:326-337.

Beetham, N. M. 1963. The ecological tolerance range of the seedling stage of Sequoia gigantea. Ph.D. dissertation, Duke Univ., Durham, N. C. 135 p.

Bennett, P. S. 1965. An investigation of the impact of grazing on ten meadows in Sequoia and Kings Canyon National Parks. M.A. thesis, San Jose State Coll., San Jose, Calif. 164 p.

Biswell, H. H. 1967. The use of fire in wildland management in California, pp. 71-87. In Natural resources: quality and quantity. Univ. Calif. Press, Berkeley.

Biswell, H. H., H. Buchanan, and R. P. Gibbens. 1966. Ecology of the vegetation of a second-growth big tree forest. Ecology 46:630-634.

Bonnicksen, T. M. 1972. White fir [Abies concolor (Gord. & Glend.) Lindl.] - a bibliography with abstracts. School Forestry Conservation Univ. Calif., Berkeley. 98 p.

-----. 1975. Spatial pattern and succession within a mixed conifer-giant sequoia forest ecosystems. M.S. thesis, Univ. Calif., Berkeley. 239 p.

Boyce, J. S. 1920. Dry-rot in incense cedar. USDA Bull. 871. 58 p.

Buchanan, H., R. P. Gibbens, and H. H. Biswell. 1966. Checklist of higher plants of Whitaker's Forest, Tulare County, California. Weber State Coll. Printing Dept., Ogden, Utah. 38 p.

Burcham, L. T. 1959. Planning burning as a management practice for California wildlands. Calif. Dept. Nat. Resources Div. Forestry, Sacramento. 21 p.

Chabot, B. F., and Billings, W. D. 1972. Origins and ecology of the Sierran alpine flora and vegetation. Ecol. Monogr. 42:163-199.

Cooke, W. B. 1940. Flora of Mount Shasta. Amer. Midl. Natur. 23: 497-572.

-----. 1941. The problem of life zones on Mt. Shasta, California. Madroño 6:49-56.

-----. 1962. On the flora of the Cascade Mountains. Wasmann J. Biol. 20:1-68.

Critchfield, W. B., and G. A. Allenbaugh. 1969. The distribution of Pinaceae in and near northern Nevada. Madroño 20:12-26.

Dickson, B. A., and R. L. Crocker. 1953. A chronosequence of soils and vegetation near Mount Shasta. J. Soil Sci. 4:124-154.

Dunning, D. 1923. Some results of cutting in the Sierra forests of California. USDA Bull. 1176. 27 p.

Dunning, D., and L. H. Reineke. 1933. Preliminary yield tables for second growth stands in the California pine region. USDA Forest Serv. Tech. Bull. 354. 23 p.

Fowells, H. A. 1965. Silvics of Forest trees of the United States. USDA Forest Serv. Agric. Handbook 271. 762 p.

Fowells, H. A., and Schubert, G. H. 1951. Natural production in certain cutover pine-fir stands of California. J. For. 41:192-196.

Gibbens, R. P., and H. F. Heady. 1964. The influence of modern man on the vegetation of Yosemite Valley. Univ. Calif. Agric. Exp. Sta. Manual 36. 44 p.

Gillett, G. W., J. T. Howell, and H. Leschke. 1961. A flora of Lassen Volcanic National Park, California. Wasmann J. Biol. 19:1-185.

Gordon, D. T. 1970. Natural regeneration of white and red fir-influence of several factors. USDA Forest Serv. Res. Paper PSW-58.

-----. 1973. Released advance reproduction of white and red fir--growth, damage, mortality. USDA Forest Serv. Res. Paper PSW-95.

Griffin, J. R. 1967. Soil moisture and vegetation patterns in northern California forests. USDA Forest Serv., Pac. SW For. and Range Exp. Sta. 27 p.

Griffin, J. R., and W. B. Critchfield. 1972. The distribution of forest trees in California. USDA Forest Serv. Res. Paper PSW-82 114 p.

Haller, J. R. 1959. Factors affecting the distribution of ponderosa and Jeffrey pines in California. Madroño 15:65-96.

-----. 1961. Some recent observations on ponderosa, Jeffrey, and Washoe pine in northeastern California. Madroño 10:126-132.

-----. 1962. Variation and hybridization in ponderosa and Jeffrey pines. Univ. Calif. Pub. Bot. 34:123-166.

Hammon-Jensen-Wallen Forestry. 1970. Kings Canyon National Park, California; type acreages and sample plots. Rept. in Kings Canyon Natl. Park files, Three Rivers, Calif.

Hartesveldt, R. J. 1962. The effects of human impact upon *Sequoia gigantea* and its environment in the Mariposa Grove, Yosemite National Park, California. Ph.D. dissertation, Univ. Mich., Ann Arbor. 310 p.

-----. 1967. The ecology of human impact upon Sequoia groves. Bull. Ecol. Soc. Amer. 48:86.

Hartesveldt, R. J., and H. T. Harvey. 1967. The fire ecology of sequoia regeneration. Proc. Tall Timbers Fire Ecology Conf. 7:64-77.

Hartesveldt, R. J., H. T. Harvey, and H. S. Shellhammer. 1967. Giant sequoia ecology: final report on sequoia fire relationships. Rept. in Sequoia and Kings Canyon Natl. Park files, Three Rivers, Calif. 55 p.

Heath, J. P. 1967. Primary conifer succession, Lassen Volcanic National Park. Ecology 48:270-275.

-----. 1971. Changes in thirty-one years in a Sierra Nevada ecotone. Ecology 52:1090-1091.

Hudson, J. P. 1933. Zonal distribution of trees and shrubs in Yosemite National Park. M.A. thesis, Stanford Univ., Stanford, Calif. 155 p.

Kilgore, B. M. 1971. The role of fire in managing red fir forests. Trans. North Amer. Wild. Nat. Res. Conf. 36:405-416.

-----. 1973a. Impact of prescribed burning on a sequoia-mixed conifer forest. Proc. Tall Timbers Fire Ecology Conf. 12:345-375.

-----. 1973b. The ecological role of fire in Sierran conifer forests: its application to national park management. J. Quaternary Res. 3:396-513.

Kilgore, B. M., and G. F. Briggs. 1972. Restoring fire to high elevation forests in California. J. For. 70:226-271.

Kilgore, B. M., and R. W. Sando. 1972. Crown-fire potential in a sequoia forest after prescribed burning. For. Sci. 21:83-87.

Klikoff, L. L. 1965a. Microenvironmental influence on vegetational pattern near timberline in the central Sierra Nevada. Ecol. Monogr. 35:188-211.

-----. 1965b. Photosynthetic response to temperature and moisture stress of three timberline meadow species. Ecology 46:516-517.

Kylver, F. D. 1931. Major plant communities in a transect of the Sierra Nevada Mountains of California. Ecology 12:1-17.

Leonard, R., C. M. Johnson, P. Zinke, and A. Schultz. 1968. Ecological study of meadows in lower Rock Creek, Sequoia National Park. Progr. rept. in Sequoia Natl. Park files, Three Rivers, Calif. 67 p.

Lepper, M. G. 1974. Pinus flexilis James and its environmental relationships. Ph.D. dissertation, Univ. Calif., Davis. 197 p.

Mallory, J. I. and E. B. Alexander. 1965. Soils and vegetation of the Lassen Park quadrangle. USDA Forest Serv., Pac. SW Forest and Range Exp. Sta. 52 p.

-----. 1965b. Soils and vegetation of the Monton quadrangle. USDA Forest Serv., Pac. SW Forest and Range Exp. Sta. 36 p.

Mastroguiseppe, R. J. 1972. Geographic variation in foxtail pine, Pinus balfouriana Grev. & Balf. M.S. thesis, Calif. State Univ., Humboldt. 103 p.

McDonald, P. M. 1969. Silvical characteristics of California black oak (Quercus kelloggii Mewb.). USDA Forest Serv. Res. Paper PSW-53. 20 p.

Meinecke, E. P. 1926. Memorandum on the effects of tourist traffic on plant life, particularly Big Trees, Sequoia National Park, California. Rept. in Sequoia and Kings Canyon Natl. Park files. 19 p.

Merriam, C. H. 1899. Results of a biological survey of Mount Shasta, California. U. S. Biol. Survey, North Amer. Fauna 16.

Miller, P. C. 1973. Oxidant-induced community change in a mixed conifer forest, pp. 101-117. In J. A. Naegele (ed.). Air pollution damage to vegetation. Adv. in Chem., Ser. 122. Amer. Chem. Soc., Washington, D. C.

Milligan, M. T. 1969. Transect flora of Eagle Peak, Warner Mountains, Modoc County. M.S. thesis, Humboldt State Coll., Arcata, Calif.

Mitchell, J. A. 1918. Incense cedar. U.S. Dept. Agr. Bull 604. 40 p.

Mooney, H. A., and W. D. Billings. 1960. The annual carbohydrate cycle of alpine plants as related to growth. Amer. J. Bot. 47: 594-598.

Oliver, W. W. 1972. Growth after thinning ponderosa and Jeffrey pine pole stands in northeastern California. USDA Forest Serv., Pac. SW Forest and Range Exp. Sta. Res. Paper PSW-85.

Omi, P. N., et al. 1975. The effects of naturally ignited fires on components of a southern Sierra forest system. Rept. in Sequoia and Kings Canyon Natl. Park files, Three Rivers, Calif. 193 p.

Oostings, H. J., and W. D. Billings. 1943. The red fir forest of the Sierra Nevada: Abietum magnificae. Ecol. Monogr. 13:259-274.

Parsons, D. J. 1972. The southern extensions of Tsuga mertensiana (mountain hemlock) in the Sierra Nevada. Madroño 21:536-539.
-----. 1975. Fire in Sequoia and Kings Canyon National Parks. Fremontia 3(1):13-14.
Pease, R. W. 1965. Modoc County: a geographic time continuum on the California volcanic tableland. Univ. Calif. Pub. Geogr. 17:1-304.
Ratliff, R. D. 1973. Short-hair meadows in the Sierra Nevada...an hypothesis for their development. USDA Forest Serv. Res. Note PSW-281. 4 p.
-----. 1974. Short-hair sedge...its condition in the high Sierra Nevada of California. USDA Forest Serv. Res. Note PSW-293. 5 p.
Reynolds, R. 1959. Effect upon the forest of natural fire and aboriginal burning in the Sierra Nevada. M.A. thesis, Univ. Calif., Berkeley. 262 p.
Richards, L. G. 1959. Forest densities, ground cover and slopes in the snow zone of the Sierra Nevada west-side. USDA Forest Serv. Pac. SW Forest and Range Exp. Sta., Tech. Paper 40. 21 p.
Rundel, P. W. 1967. The influence of man and fire on the vegetation of the Calaveras Groves of Sequoiadendron giganteum. M.A. thesis, Duke Univ., Durham, N.C. 81 p.
-----. 1968. The southern limit of Taxus brevifolia in the Sierra Nevada, California. Madroño 19:300.
-----. 1971. Community structure and stability in the giant sequoia groves of the Sierra Nevada, California. Amer. Midl. Natur. 85: 478-492.
-----. 1972a. Habitat restriction in giant sequoia: the environmental control of grove boundaries. Amer. Midl. Natur. 87:81-99.
-----. 1972b. An annotated check list of the groves of Sequoiadendron giganteum in the Sierra Nevada, California. Madroño 21:319-328.
-----. 1973. The relationship between basal fire scars and crown damage in giant sequoia. Ecology 54:210-213.
-----. 1975. Primary succession on granite outcrops in the montane southern Sierra Nevada. Madroño 23:209-220.
St. John, T. V. and P. W. Rundel. 1976. The role of fire as a mineralizing agent in a Sierran coniferous forest. Ecology 25:35-45.
Schumacher, F. X. 1926. Yield, stand, and volume tables for white fir in the California pine region. Univ. Calif. Agric. Exp. Sta. Bull. 407. 26 p.
-----. 1928. Yield, stand, and volume tables for red fir in California. Univ. Calif. Agric. Exp. Sta. Bull. 456. 29 p.
Sellers, J. A. 1970. Mixed conifer forest ecology: a qualitative study in Kings Canyon National Park, Fresno County, California. M.A. thesis, Fresno State Coll. 65 p.
Sharsmith, C. W. 1959. A report on the status, changes, and ecology of the back-country meadows in Sequoia and Kings Canyon National Parks. Rept. in U.S. Natl. Park Service files, San Francisco, Calif. 122 p.
Show, S. B., and E. I. Kotok. 1924. The role of fire in the California pine forests. USDA Bull. 1294:1-80.
Smiley, F. J. 1915. The alpine and subalpine vegetation of the Lake Tahoe region. Bot. Gaz. 2:265-286.
Smith, G. L. 1973. A flora of the Tahoe Basin and neighborning areas. Wasmann J. Biol. 31:1-231.

Stark, N. 1973. Nutrient cycling in a Jeffrey pine ecosystem. Montana For. and Conserv. Exp. Sta., Missoula. 385 p.

Storer, T. I., and R. L. Usinger. 1963. Sierra Nevada natural history. Univ. Calif. Press, Berkeley. 374 p.

Strand, S. 1972. An investigation of the relationship of pack stock to some aspects of meadow ecology for seven meadows in Kings Canyon National Park. M.A. thesis, Calif. State Univ., San Jose. 125 p.

Sudworth, G. B. 1900. Notes on regions in the Sierra Forest Reserve. Unpub. rept. in files of Univ. Calif., Berkeley, For Library. 145 p.

-----. 1908. Forest trees of the Pacific Slope. USDA Forest Ser. 441 p.

Vankat, J. L. 1970. Vegetation change in Sequoia National Park, California. Ph.D. dissertation, Univ. Calif., Davis. 197 p.

Van Wagtendonk, J. W. 1972. Fire and fuel relationships in mixed conifer ecosystems of Yosemite National Park. Ph.D. dissertation, Univ. Calif., Berkeley.

-----. 1974. Refined burning prescriptions for Yosemite National Park. U.S. Natl. Park Serv. Occas. Paper 2. 21 p.

Vasek, F. C. 1976. Jeffrey pine, and vegetation of the southern Modoc National Forest. Madroño (in press).

Wagener, W. W. 1961. Past fire incidence in Sierra Nevada forests. J. For. 51:739-748.

Weaver, H. 1943. Fire as an ecological and sivicultural factor in the ponderosa pine region of the Pacific slope. J. For. 41:7-15.

-----. 1967. Fire and its relationship to pondersoa pine. Proc. Tall Timbers Fire Ecology Conf. 7:127-149.

Whittaker, R. H. 1960. Vegetation of the Siskiyou Mountains, Oregon, and California. Ecol. Monogr. 30:279-338.

Wood, D. L. 1972. Selection and colonization of ponderosa pine by bark beetles, pp. 101-117. In H.F. van Emden (ed.). Insect-plant relationshps. Sixth Symp. Royal Ent. Soc., Blackwell London.

Wright, R. D. 1970. CO_2 exchange of seedling pines in the laboratory as related to lower elevational limits. Amer. Midl. Natur. 83:291-300.

CHAPTER 18 ALPINE

JACK MAJOR
DEAN W. TAYLOR
Department of Botany, University of California,
Davis

Introduction 602
Floristic relationships of Californian alpine vegetation 608
Timberline 615
The vegetation 618
  Klamath Mountains 621
  Cascades 622
  Sierra Nevada 624
  Southern California 652
  Great Basin 654
    White Mountains 654
    Sweetwater Mountains 661
    Warner Mountains 664
Literature cited 666

INTRODUCTION

The alpine belt is above timberline, therefore includes the krummholz, and is below the snow line. Hermes (1955) gives timberline levels as >3500 m in southern California (San Gorgonio Peak 3510 m), 3300 in the Mt. Whitney (4415 m) region at 36½°N, dropping to 3200 m in Yosemite (Mt. Lyell 4000 m) at 38°N, to 3000–4200 m (snow line) at Donner Pass (39°N), where Castle Peak is 2770 m, 2800–3800 m at Mt. Lassen (3185 m) at 40½°N, 2700–3600 m at Mt. Shasta (4310 m), but 2500–3200 m westward in the less continental Klamath Mts. with Thompson Peak 2740 m at 41°N and Preston Peak 2223 m at 42°N. In the Great Basin to the east, in the rain shadow of the Sierra Nevada, the White Mts. have a timberline at 3600 m (Mooney et al. 1962) at 37½°N and the Sweetwater Mts. (38½°N) at 3350 m.

California has a nival belt (Reisigl and Pitschmann 1958, 1959; Ellenberg 1963:585 ff; Walter 1968:552), then, only on Mt. Shasta, but its vegetation has not been described. Other California mountains do not extend above the snow line. Nival habitats, however, occur down into the alpine belt. The vertical span of the alpine belt as the difference between timberline and snow line varies from 1300–1400 m in the southern Sierra Nevada to 1200 m in the north, to 900 m on Mt. Shasta, and to only 700 m in more coastal regions. Northern hemisphere extremes are 200–500 m in the Olympic Mts. and Coast Ranges of southern British Columbia to 1500–2500 m in the continental Himalaya and bordering ranges (Hermes 1955).

Vegetation has been described by using physiognomy, ecological factors, succession, geographical distribution, dominance, and floristics. We concentrate on the last since it is the most objective, is exhaustive, is repeatable, and pertains to vegetation as single-species chorology on any currently used scale cannot do by the definition of "vegetation." The taxonomic names of plants alone give data on these other necessary viewpoints. Floristic description as developed by Braun-Blanquet (1964), Ellenberg (1963), Walter (1968, 1973a), and others, paying adequate attention to the individuality of plant taxa, is sufficient. It has clear priority in studies of vegetation.

We can dismiss for lack of knowledge most of the ecological factors that should determine the alpine vegetation of California, its regional variants, and its local mosaics (Jenny 1941; Major 1951; Crocker 1952; Major and Bamberg 1967a). Instead we will discuss relevant ecological and ecophysiological (Walter 1960, 1973b) data where they pertain to particular floristic units or ecological sequences.

Regionally we hypothesize, following Walter (1973a:32, 225), Stanyukovich

(1973), Golubets (1974), and Grebenshchikov (1974), that mountain vegetation is usually most closely related floristically to adjacent lowland vegetation. Thus the alpine vegetations of the disjunct, island-like alpine areas of the southern California mountains, the Sierra Nevada, the Great Basin, the Cascades, and the Klamath Mts. are probably different.

We explicitly correct the view that the floras and vegetation of mid-latitude mountain ranges are extensions of arctic floras and vegetation. This relationship does increase in proportion as there is a reasonable connection in time or space to the arctic, provided reasonable climatic similarity exists (Major and Bamberg 1967a). Thus Bliss (1963:694–5) points out that Mt. Washington in New Hampshire has a much more arctic or circumpolar flora than has the Medicine Bows in southeastern Wyoming. His later work adds the Olympics of northwestern Washington. The compilation of U.S. alpine plants and their distributions by Zwinger and Willard (1972) is instructive. Some of the most interesting relationships of the flora of the Sierra Nevada exist across the Great Basin, not north–south (Major and Bamberg 1967b; Chabot and Billings 1972). Much of this directional relationship must be due to the north-south continuity of alpine topography, leading northward in the first instance to maritime areas of the Pacific Northwest, not directly to the arctic. In California the floras of such areas are related actually only to the maritime Klamath Mts. In glacial times California's interior climates were not maritime but even more continental than at present, and as a result the Sierra Nevadan flora is related to more continental regions, to the eastward-lying Great Basin. There may have been more topographical continuity of alpine-cold steppe (Walter 1975) habitats eastward across the Great Basin than there is or was topographical continuity to maritime Pacific Northwest high mountains.

A general classification of alpine landscapes was suggested by Tolmachev (1948; cf. Gorchakovsky 1966:15, 1975:13): (1) Alpine in the classical sense, as in the Alps, Caucasus, etc. (2) Goltsy, or mountain tundra, bordering the continental arctic and similar to arctic tundra in species composition, lack of deep snow cover, and widespread presence of permafrost. (3) Xerophytic or Mediterranean climates. (4) Paramo of northern South America and including tropical Africa. (5) High mountain steppe. (6) High mountain deserts as an extreme development of high-altitude steppes where precipitation decreases so much that even bunch grasses are excluded in favor of *Ceratoides* (*Eurotia*)–*Artemisia* vegetation, as in the eastern Pamir. The last two are in the most continental climates of temperate latitudes. California has only type 1 in the Klamath Mts. and Cascades–Sierra Nevada, type 3 in the Sierra Nevada and southern California mountains, and type 5 in the eastern Sierra Nevada and Great Basin ranges.

In Table 18-1 we suggest contrasts between these last three landscape types as they occur in California. Walter (1975), Ellenberg (1963:514), and Went (1948, 1953) discuss some of the relationships and similarities between steppe and continental alpine conditions.

Obviously these particular types of landscapes intermingle in most of our mountain regions. The types represent extremes of trends. The alpine and Mediterranean xerophytic types show oceanic to continental variation; severity of climate increases with increasing continentality. Tolmachev makes the point that there is a tendency for nival landscapes to be much the same in all the northern hemisphere types.

Much Mediterranean xerophytic alpine vegetation similar to California's in ecology is placed in its context of mountain belts by Grebenshchikov (1974) in a

TABLE 18-1. Three types of alpine landscapes in California (Tolmachev 1948; Gorchakovsky 1966:15, 1975:12)

Characteristic	Type of Alpine Ecosystem		
	Alpine	Xerophytic	Steppe
Climate			
Radiation as affected by atmospheric density and clarity	Moderate	High in summer	High all year
Summer cloudiness	Much	Little	Very little
Precipitation			
Yearly amount	High	High	Low
Summer amount	High to medium	Very low	Low in amount but high proportion
Distribution	<PotE only in late summer	<PotE in summer to early fall	<PotE except in winter
Summer intensity	Moderate	High	High
Water balance (precipitation + soil storage)	Larger than PotE in all months	Less than PotE in summer and fall	Less than PotE in spring, summer, fall
Water supply	Very high from snowmelt water plus summer precipitation	Low after snowmelt but temperatures then high	Very low, snowmelt rapid and in early spring when temperatures are low
Temperature			
Summer	Cool	Warm	Warm to hot
Increase soil over air temperature	Moderate	High	Very high
Winter temperature	Cold	Very cold	Very cold
Snow			
Depth	Deep to very deep	Deep	Shallow to none
Fresh density	High	High to medium	Low
Pack density	High	High	Low
Creep	Effective on plants	Effective on plants	Negligible effects

Metamorphosis	Equitemperature	Equitemperature	Temperature gradient
Avalanches	Wet, in spring, in couloirs, to ground, common	Loose, soft slab, large, uncommon	Slab, small, frequent
Wind redistribution	Not extensive except in continental areas	Moderate	Extensive
Insulation of vegetation and winter-active rodents by snow	High	High	Low
Activity of subnival rodents	High	High	Low
Length of vegetative period	Long normally, but short where snow accumulates	Short, limited by drought	Very short, limited by drought
Variability of snow cover	Low	High	Extreme
Variability of length of growing season	Low	High	Extreme
<u>Timberline and snowline</u>			
Timberline	Low	High	Highest
Snowline	Low, peaks above in many regions	High, above present highest summits	Highest, above present highest summits
Difference	Less than 900 m	Greater than 900 m	Greater than 1500 m
<u>Topography</u>			
Glaciation	Heavier in past, heavy snow	Heavy in past, light at present	Moderate in past, minimal now
Slope steepness	Extremely steep to flat	Steep to flat	Often gentle
Slope exposure contrasts	Great	Moderate	Extreme
<u>Soils</u>			
Types	Humid: podsols, rankers	Humid, skeletal	Arid, skeletal
Humus	High content	Moderate	Very low
Drainage	Moderate	High	Very high

TABLE 18-1 [continued]

Characteristic	Type of Alpine Ecosystem		
	Alpine	Xerophytic	Steppe
Root system development	Abundant, shallow	Deep, abundant only near surface	Very sparse and shallow
Sod cover	Complete	Incomplete	Sparse
Soil freezing in winter	None, typically	On ridges, thaws rapidly	Extensive, thaws and re-freezes in spring
Talus	Usually vegetated, bare in continental	Usually only scattered plants, extensive	Small areas, bare
Scree	Usually vegetated	Often bare, extensive	Desert pavements
Frost-structured soils			
Occurrence	Heavily modified by sod	Moderate development	Low development
Processes	Solifluction and frost creep	Solifluction with spring thaw, frost creep	Frost creep
Forms on slopes (Benedict 1970)	Turf-banked terraces and lobes, stone-banked terraces and lobes, sorted and unsorted stripes	Same	Stone-banked terraces, sorted and unsorted stripes
Forms on the level	Vegetation-centered rock polygons	Large, bare-centered rock polygons	Small, bare-centered rock polygons
Floristic composition			
Rocky Mt. elements	Common	Restricted to hydric habitats	Rare
Circumpolar elements	Common	Restricted to hydric habitats	Rare
Great Basin elements	Extremely rare	Common in xeric habitats	Common, including spp. normally found in the desert valleys
Endemic elements	Few	Common	Common but few in number

Vegetation			
Physiognomy	Meadows, tall, rhizomatous, many forbs	Open, cushion plants, tufted graminoids	Bunch grasses, low shrubs, cushion plants
Xerophily	All spp. mesic	Mesic to xeric	Xeric, mesic spp. rare
Biomass	High	Moderate	Low
Productivity	High	Moderate	Very low
Snowmelt geophytes	Abundant	Moderate	Uncommon
Ericaceous chamaephyte	Abundant	Uncommon	Absent
Therophytes	Rare	Abundant	Rare
California examples	Klamath Mts., Subalpine Cascades, and northern Sierra Nevada	Alpine Cascade ridges and subalpine Sierra Nevada, southern California mountains	Great Basin ranges (White, Sweetwater, Warner), eastern slope of and southern Sierra Nevada

transect of more than 6500 km over 80° of longitude at 35–40°N latitude on 14 mountain groups from southern Spain to Tadzhikistan.

Tolmachev (1948:173) points out a most interesting fact which should have bearing on the origin, the derivation, of the taxa now found in the high mountains. The meadows of the alpine type of high-mountain landscape can be derived from subalpine formations. The plants of high-altitude steppes and deserts have close relationships to nearby low-altitude steppes and deserts. The goltsy plants of northern latitudes have relationships to arctic tundra of both graminoid–moss-lichen types and dwarf shrubs. However, the Mediterranean xerophytic type of high-mountain vegetation is not closely related to the lower-altitude macchie or chaparral. We suggest a relationship and derivation from low-altitude, Mediterranean ecosystem chasmophytes. Such plants are numerous and important, and have been consistently present at least in Mediterranean (Ehrendorfer 1962), Colchic (Sokhadze 1974:208), and Californian floras.

A more specific scheme that suggests what kinds of alpine plant communities occur in what kinds of habitats in California can be extracted by analogy of alpine California with the French Alps, using Guinochet and Vilmorin's classification of French vegetation (1973). The Alps border a Mediterranean climate, and their vetetation has been well studied. It would violate ecologists' fixed faith in the correlation, even causation, between plant function and form, which was drawn from the experience of the earliest plant geographers such as Humboldt and Drude, to suggest that identical ecological conditions in Mediterranean France and California should not produce at least physiognomically similar vegetation. If France exhibits a certain kind of vegetation, similar ecological conditions in California should produce something similar. Thus the key tabulation presented by Guinochet and Vilmorin gives an ecological answer to the question, "What kinds of alpine plant communities are probably found in California?" It ignores the very real and great floristic differences between the Alps and California. The units named are classes, more or less equivalent to formations. Since the French flora differs from the Californian, the floristic names do not fit our conditions, but the associated ecological descriptions are suggestive: (1) Floating in quiet, clear water (Potametea). (2) Similar but submerged part of the year (Littorelletea). (3) Similar but water flowing (Montio-Cardaminetea). (4) Acid bogs (Oxycocco-Sphagnetea). (5) Fens (Scheuchzerio-Caricetea fuscae). (6) Cliffs (Asplenietea rupestris). (7) Talus and screes (Thlaspietea rotundifolii). (8) Nitrophile, animal congregation areas (Onopordetea). (9) Nitrophile, disturbed, plants of low stature (Chenopodietea). (10) Temporary pools (Isoeto-Nanojuncetea). (11) Rock surfaces (Sedo-Scleranthetea). (12) Snow patches, acid- or calcicole (Salicetea herbaceae). (13) Nutrient-rich alpine swards (Elyno-Seslerietea). (14) Nutrient-poor alpine swards (Caricetea curvulae). (15) Tall herbs (Betulo-Adenostyletea). (16) Wet meadows (Molinio-Juncetea). (17) Dry grassland (Festuco-Brometea). (18) Coniferous krummholz and ericads (Vaccinio-Piceetalia). (19) Willow thickets (Salicetea purpureae).

FLORISTIC RELATIONSHIPS OF CALIFORNIAN ALPINE VEGETATION

Floristic composition is a partial determinant of vegetation. The flora of an area is defined as the sum of its plant taxa, and vegetation is molded from this pool of taxa by the environment. Alpine habitats in California support vegetation that is distinct floristically as well as physiognomically from lowland vegetation. In consequence of the rigorous environment, alpine floras are depauperate. Few plants are able to live and compete in this environment; on the other hand, some plants can exist only in alpine environments.

In considering the floristic composition of the alpine vegetation of the Sierra Nevada, two distinct definitions of the alpine flora can be made. Any plants which can be found growing under alpine conditions can be considered to compose the alpine flora. On this basis, Munz and Keck (1959) list about 600 plants comprising the alpine flora of the Sierra (Major and Bamberg 1967b). On the other hand, some of these species can also grow in other vegetation types, where they may even be most common. Omitting these, Sharsmith (1940) recognized nearly 200 species confined to alpine habitats and therefore comprising the Sierran alpine flora. Although Sharsmith considered alpine species to be those which are restricted to the alpine zone, it is clear from a careful examination of his list that many species he considered to be exclusively alpine do, in fact, occur below timberline. Within the subalpine forest of the Sierra, typically alpine species can be encountered at the periphery of large summer snowbanks, in wet, acid bogs or *Carex*-dominated meadows and fens where cold air drainage is often evident, about shaded or moist cliff faces, in locally xeric sites with shallow soils or no snow cover in midwinter, and within avalanche tracks. These kinds of sites present conditions similar to those of above-timberline localities and are often vegetated in part by alpine species.

Alpine and subalpine habitats interdigitate at timberline. At a given locality, alpine vegetation may extend elevationally below the upper limit of Subalpine Forest for several hundred meters. Thus the alpine habitat and its flora can be defined only on an ecological and not simply on an elevational basis.

Sharsmith (1940) examined the phytogeographic affinities of species limited to the alpine belt in the Sierra. He recognized elements putatively on the basis of their centers of origin, but delimited groups by following the equiformal progressive areas of Hultén (1937). Sharsmith recognized three main elements in the Sierran alpine flora: Boreal, Arctic-Alpine, and Alpine (Table 18-2). His Boreal Element includes such widespread plants as *Phleum alpinum, Trisetum spicatum, Deschampsia caespitosa,* and *Cystopteris fragilis.* Sharsmith recognized that these species are not limited in their distribution to the alpine, but nevertheless included them in his analysis, whereas taxa such as *Botrychium lunaria* var. *minganense, Koeleria cristata, Epilobium angustifolium,* and *Potentilla fruticosa* were excluded even though they occur commonly in alpine situations. Taxa of this small boreal element are widespread throughout the Holarctic region.

Widespread taxa such as *Oxyria digyna, Carex capitata, Festuca brachyphylla, Sibbaldia procumbens, Stellaria umbellata, Anemone drummondii, Epilobium latifolium,* and *Crepis nana* comprise a portion of Sharsmith's Arctic-Alpine Element. Many important Sierran alpine species are derived from progenitors with an Arctic-Alpine distribution; among them are *Salix anglorum* (*S. artica* Pall.), *S. nivalis* (*S. reticulata* L.), *Arenaria capillaris* var. *compacta, Sedum rosea* ssp. *integrifolium, Saxifraga aprica,* and *Aster alpigenus* var. *andersonii.* Species characteristic of the Arctic-Alpine Element are widely distributed in Holarctic regions at high latitudes, and extend to low latitudes along the Cordilleran and Alpine orogenies in the New and Old World, respectively.

TABLE 18-2. Phytogeographical affinities of the Sierran alpine flora according to Sharsmith (1940)

Floristic Element	Percent of Flora
I. Boreal Element	3.8
II. Arctic-Alpine Element	14.2
III. Alpine Element	82.0
A. Cordilleran Group	64.0
1. Beringian Cordilleran Type	7.3
2. Arctic and Subarctic Type	8.4
3. Central Cordilleran Type	48.1
a. Rocky Mountain Species	17.3
b. Pacific Cordilleran Species	15.7
c. Sierra Nevada-Cordilleran Species	14.8
B. Western Group	17.9
1. Widespread Western Type	7.3
2. Great Basin Type	10.6

Sharsmith's Alpine Element is the largest and most diverse phytogeographical group in the alpine flora of the Sierra, being subdivided into a number of types (Table 18-2). Species with distributions limited to the Cordilleran mountain system comprise nearly half of the Sierran alpine flora. Within this type, there are three groups of species, each with characteristic relationships. Nearly a sixth of the flora is endemic, and another sixth is limited to the Cascade–Sierran axis, while a fifth of the species occur also in the Rocky Mts. Familiar species characteristic of these elements are listed in Table 18-3.

Sharsmith related the floristic composition of the Sierran alpine flora to past migrational events. He paid particular attention to the possibility that alpine plants were able to persist in ice-free areas (nunataks) during the glacial intervals of the Pleistocene and thus readily recolonize newly available habitats after ice retreat. However, his conclusions were not definitive in this regard, and the nunatak hypothesis is still moot (Ives 1974). Sharsmith contributed greatly to our knowledge of the Sierran alpine flora, but we still lack even simple dot maps giving the distribution of alpine taxa. Merxmuller's (1952–54; Merxmuller and Poelt 1954) conclusions for the Alps and Scandinavian work (Gjaerevoll 1963) illustrate the value of such maps.

Floristic plant geography has traditionally been a nonquantitative field. The delimitation of floristic elements based on the coincidental distribution of individualistic species is not without a subjective component. Quantitative floristic comparisons are possible, however. Taylor (1976b, unpub. data) has compared all available plant checklists for alpine areas of the Cordillera on the basis of shared species. When the Sierran alpine flora is considered as an extension of the Cascade–Sierran axis, several trends emerge. Figure 18-1 shows a plexus diagram depicting the floristic similarity of 22 localities along this mountain chain from Alaska and British Columbia to southern California. Three major floristic units are apparent: a north Cascade, a south Cascade, and a Sierra Nevadan unit. Of the California

TABLE 18-3. Taxa of the Central Cordilleran type of Sharsmith (1940) which are dominants or major components in the alpine vegetation of the Sierra Nevada

---

Rocky Mt. species

Antennaria alpina var. *media*, *A. umbrinella*, *Arabis lyallii*, *A. lemmonii*, *Arenaria nuttallii*, *Astragalus kentrophyta*, *Carex haydeniana*, *C. phaeocephala*, *C. pseudoscirpoidea*, *Chaenactis alpina*, *Erigeron vagus*, *Eriogonum ovaliforium*, *Gentiana calycosa*, *Haplopappus macronema*, *Linum lewisii*, *Mimullus tilingii*, *Phlox caespitosa* ssp. *pulvinata*, *Poa rupicola*, *Ribes cereum*, *R. montigenum*, *Stipa occidentalis*, *S. pinetorum*

Pacific Cordilleran species

Arabis platysperma var. *howellii*, *Astragulus whitneyi*, *Calamagrostis*, *breweri*, *Carex breweri*, *C. preslii*, *Dodecatheon alpinum*, *Draba breweri*, *Epilobium obcordatum*, *Eriogonum lobbii*, *Gentiana newberryi*, *Lupinus lobbii*, *L. lyallii*, *Mimulus primuloides* var. *pilosellus*, *Penstemon davidsonii*, *Phoenicaulis eurycarpa*, *Phyllodoce breweri*, *Polygonum shastense*, *Potentilla breweri*, *P. drummondii*, *P. flabellifolia*, *Ranunculus alismaefolius* var. *alismellus*, *Senecio fremontii*, *Silene watsonii*

Sierra Nevadan species (endemics)

Aquilegia pubescens, *Astragulus purshii* var. *lectulus*, *Carex incurviformis* var. *danaensis*, *C. helleri*, *C. subnigricans*, *Claytonia nevadensis*, *Draba cruciata* var. *integrifolia*, *D. lemmonii*, *D. sierrae*, *Erigeron pygmaeus*, *Hulsea algida*, *Lupinus brewerii*, *Primula suffrutescens*, *Raillardella argentea*

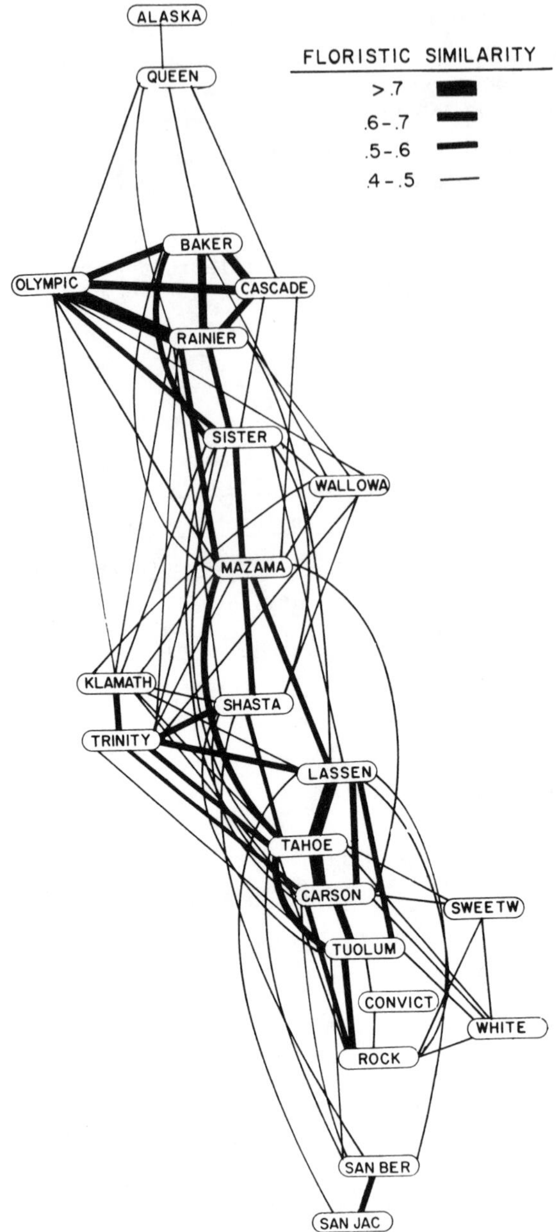

Figure 18-1. Plexus diagram of floristic similarities of 22 selected localities along the Cascade–Sierran axis. In addition to the 17 localities analyzed in Taylor (1976b) the following areas are included: Wallowa Mts., Oregon; Sweetwater Mts. (Taylor, unpub. data), Convict Creek (Major and Bamberg 1963 and unpub. data), White Mts. (Lloyd and Mitchell 1973), California. Floristic similarity was estimated using Sorenson's coefficient (high values correspond to high similarity) and inclusion criteria as given by Taylor (1976b).

localities, the alpine flora of the high peaks of the Klamath Mts. is more similar to that of the northern Cascades than to that of either of the southern Cascade volcanoes, Mt. Shasta and Mt. Lassen, or to the Sierra Nevada. Several important Pacific-Cordilleran species, like *Saxifraga caespitosa* and *Haplopappus lyallii*, are known to occur in California only in the Marble Mts. and the Trinity Alps. Mount Shasta and Mt. Lassen are most similar floristically to the other inland southern Cascade volcanoes, Mt. Mazama (Crater Lake) and the Three Sisters. The Sierran alpine flora is a distinctive unit, largely because of the high proportion of endemic species (cf. Table 18-3). The mountains of southern California have a depauperate alpine flora essentially similar to that of the Sierra Nevada. A large number of montane Rocky Mt. taxa, such as *Gentiana fremontii, Carex occidentalis,* and *Lewisia brachycalyx,* occur in the San Bernardino Mts. These disjunctions represent floristic exchange with the southern Rocky Mts. (Stebbins and Major 1965), which probably took place at intervals throughout the Pleistocene (Munz 1935:xvi). The Sweetwater and White Mts., lying east of and in the rain shadow of the Sierra Nevada, show a high degree of similarity to other alpine ranges of the Great Basin, such as the Toiyabe Mts., Ruby–East Humboldt Mts. and Wheeler Peak in Nevada, and the Deep Creek Mts. in Utah. Table 18-4 shows the similarity of these stations to other selected Cordilleran localities.

Chabot and Billings (1972) note that a distinction can be made between taxa of the alpine flora of the Sierra which have had long histories of association with arctic and alpine environments and those which have arisen, presumably quite recently, from lowland progenitors. This dualism between allochthonous and autochthonous origin of northern hemisphere mountain floras is a universal phenomenon. The former are abundant in direct proportion to the temporal and spatial proximity to the arctic (Major and Bamberg 1967a). Chabot and Billings attribute the present composition of the alpine flora of the Sierra to six predominant variables (1972:174): (1) recent uplift of the range, (2) isolation of the range from preexisting sources of alpine species, (3) restriction of species migrations by Pleistocene glaciations and (4) by post-Pleistocene climatic shifts, (5) mixing and telescoping of alpine and desert floras during Pleistocene migrations, and (6) the present summer aridity of the Sierran alpine climate.

Stebbins and Major (1965:23) noted that the number of Rocky Mt. taxa in the Sierra is greater in the southern portion of the range. This is correlated with a decrease in boreal species southward and with the more extensive alpine areas in the southern Sierra. Or it may, as Chabot and Billings (1972) suggest, indicate a migration of species from the Rocky Mts. south along the Cascade axis, rather than across the Great Basin as proposed by Harshberger (1911:255–6) and Major and Bamberg (1967a). Chabot and Billings point out that the importance of Rocky Mt. taxa in the alpine floras of the Great Basin decreases westward, and thus it would be unlikely for major floristic exchange to have taken place by this route. Major and Bamberg suggest by analogy to present-day Pamir habitats that several Rocky Mt. taxa—*Pedicularis crenulata, Draba nivalis* var. *elongata* (*D. lonchocarpa* Rydb.), *Arctostaphylos uva-ursi, Salix brachycarpa, Kobresia myosuroides,* and *Scirpus pumilis* [*Trichophorum pumilum* (M. Vahl) Schinz and Thell.]—migrated across the Great Basin via lowland bogs along the major watercourses. Many arctic-alpine taxa extend their range into the high-altitude deserts of the Pamir in central Asia in this manner today. The Pleistocene climate for the Great Basin suggested by Major

TABLE 18-4. Comparison of the floristic similarities of selected Californian alpine areas, (Taylor 1976b) with other alpine localities in the Western Cordillera. Similarities were calculated using Sørensen's coefficient

California Locality	Olympic Mts., Washington (Jones 1936)	Colorado Front Range (Weber 1972)	Ruby Mts., Nevada (Lewis 1971; Loope 1969)	Charleston Mts., Nevada (Clokey 1951)
Trinity Alps	0.53	0.28	0.40	0.22
Klamath Mts.	.46	.25	.46	.15
Mt. Shasta	.43	.26	.36	.19
Mt. Lassen	.45	.33	.43	.19
Lake Tahoe	.43	.36	.50	.20
Carson Pass	.42	.35	.51	.22
Sweetwater Mts.	.24	.21	.36	.25
White Mts.	.28	.33	.48	.31
Tuolumne Meadows	.42	.38	.47	.23
Rock Creek	.40	.36	.48	.18
San Bernardino Mts.	.32	.25	.34	.29
San Jacinto Mts.	.23	.17	.31	.30

and Bamberg could potentially have supported permafrost along the major watercourses. Bogs occurring over this permafrost and the greatly expanded alpine-steppe habitats of the numerous mountain ranges of the Great Basin make this an attractive migrational hypothesis. As Major and Bamberg (1967a:184) point out, many of our more maritime alpine plants, such as *Cassiope mertensiana, Stellaria umbellata, Luzula spicata,* and *Sibbaldia procumbens,* probably migrated along the Cascade axis to reach the Sierra Nevada.

TIMBERLINE

Physiognomically timberline can be a forest limit, an isolated tree limit, or an altitude limit for the tree species growing as a shrub. The second and third are most common in California.

Ecological kinds of limits to forest are climatic, related to temperature or wind; edaphic as lack of soil or solifluction; related to snow accumulation, snow avalanches, glaciers, fire, springs, and bogs; and anthropogene (Schröter 1926:42 ff; Gorchakovsky 1966:200, 1975:206; Shiyatov 1970). Most of these situations are obvious, at least in the extreme cases, and thermal timberlines are the zonal norm.

Avalanche timberlines deserve some explication. Snow avalanches that follow couloirs at some point in their paths and that penetrate deep into the forest as wet, spring snowslides to leave masses of broken trees and deep deposits of compact snow lasting well into summer have quite evident effects. Slab avalances at high altitudes may not. The latter occur in winter, are often loose, or at least blocky, even in their runout zones, and may not obviously damage trees, either because they run very infrequently or on too small a scale of depth and momentum or because they run so very frequently that trees do not exist at all on such slopes. Or such avalanches in steep terrain may become airborne and leave trees standing but partially stripped of branches and perhaps topped. In the Sierra Nevada, immediately after the Saddlebag Lake road turns off the Tioga Pass road it passes through an area where an airborne avalanche off the southwestern slopes of Tioga Peak in the heavy snow winter of 1968–69 had considerable thinning and topping effect on the *Pinus albicaulis* forest without altogether destroying it. Many cords of stove-sized firewood were scattered in June on the old snow over the road and across the creek on the opposite slope. At Mineral King the avalanche slopes on the southwest-facing side of Empire Mt. drop almost 1150 m in a 2.5 km horizontal distance to what was once proposed as a parking area. These slopes evidently run both as powder-slab and wet, loose, ground slides of very large proportions. A fine example of an area where slab avalanches have run and still run very frequently, but which was not originally suspected of being quite such an avalanche slope, is the Rendezvous Bowl at Jackson Hole, Wyoming. The area is the ski route down from the top of the aerial tram and is just below timberline, some 400 m wide, all of which may slide, dropping 220 m in a horizontal distance of 550 m.

The lower limit of the climatic alpine belt is marked by two structural modifications of the tree cover in addition to the forest-, individual tree-, krummholz-altitudinal sequence mentioned above. (1) The closed forest breaks up or thins out, forming tree islands. This is analogous to the forest tundra of arctic landscapes. Between these islands are meadows or steppes with alpine plants. (2) The upright, orthotropic trees become matlike, plagiotrophic krummholz. Meadow plants of the forest belt are carried upward under this elfin forest (Schröter 1926:41). We call both structures subalpine.

The ecological factors determining these two general structural kinds of timberline are different. The tree island–meadow type has been best explained by Brink (1959), Knapp (1957, 1960, 1965), Franklin and Dyrness (1973), Michaelis (1934 III:339, V:430), and Walter (1971–72, 1973a:217) as due to excessive accumulation and persistence of snow between the islands. The islands are often elevated areas. How does the snow affect the trees? It shortens the growing season, necessarily affecting reproduction most. It favors the fungus *Herpotrichia juniperi* (Duby) Petr. (= *H. nigra* Hartung), which kills foliage on lower branches (Gäumann et al. 1934). Snowmelt water cools the soil and maintains soil water above field capacity. Snow strips off branches as it settles. Deep snow favors meadow vegetation with taprooted forbs and winter gopher activity; see the discussion of *Polygonum daviseae* vegetation at Carson Pass. While such tree island timberlines can be observed in the Sierra Nevada, Cascades, and Klamath Mts., the reverse phenomenon of trees occurring only where snow persists longest is evident in early summer on the drier east face of the Sierra (north side of McGee Peak, e.g.).

The second, krummholz type of timberline has been studied at Carson Pass by Taylor (1976a). Here the density of *Pinus albicaulis* decreases rapidly and at an increasing rate along an altitude–exposure sequence. Associated changes in the trees occur at different rates relative to decrease in tree density: the mean height of individual trees decreases very abruptly, total tree cover declines more gradually, and the mean number of trunks per individual increases rapidly.

In California no relatives of ordinarily upright trees are genetically krummholz as is *Pinus mugo* Turra in the Alps (really the mountains of central Europe to the Balkans), related to upright *P. uncinata* Miller ex Mirbel, and *P. pumila* (Pall.) Rgl. of eastern Siberia to northern Japan, which, in turn, is related to the normally upright *P. sibirica* DuTour-*cembra* L.-*albicaulis*. However, Clausen (1965) did conclude, without uniform garden tests, that *P. albicaulis* of Slate Creek near Tioga Pass included both upright, tree-form and prostrate, krummholz ecotypes.

The formation of krummholz has been attributed to winter desiccation of needles projecting above the snow since Fries' and Kihlman's observations pre-1900 (Larcher 1972:317). Physiological data have not been collected in the Sierra. Our understanding of the physiological limits imposed at timberline, as the result of investigations by Michaelis (1932–34), Larcher (1957, 1963a, 1972), Tranquillini (1957, 1967), Wardle (1965, 1968, 1971, 1975), Schwarz (1971), and other work reviewed by Bauer et al. (1975), was about as follows. Michaelis (1934:191) found that *Picea abies* (L.) Karsten and *Pinus mugo* had closed stomata in winter. The Innsbruck ecophysiologists also measured transpiration (Pisek 1971) and found it drastically reduced in winter. The stomata close and stay closed even when temperature rises, and transpiration is reduced to cuticular levels. Recent experimental data show that, after the subfreezing temperatures of autumn, evergreen conifers drastically reduce their photosynthetic rates even under normally favorable conditions of temperature and insolation. The assumption is usually made that this is due to restriction of gas exchange by stomatal closure. The role of "residual" mesophyll resistance (Johnson and Caldwell 1975) has not been explicitly considered. Certainly cold soils, even if not frozen under an insulating snow mantle of 0.5 m or more, hinder the uptake of water (Larcher 1972:320), but they also reduce transpiration and photosynthesis (Havranek 1972). Finally, cuticular resistance increases from summer to winter (Larcher 1972:323).

Now Baig et al. (1974) have shown that twigs of *Pinus cembra,* the timberline tree of the continental Alps, has a very, very low cuticular transpiration with resistances of (1960)2400–3400 sec cm^{-1}. The least cuticular resistance was in trees exposed to wind, where in fact trees do have least success. The data for *Picea abies* showed resistances of (440)460-860 sec cm^{-1} at timberline, 1200 in the valley for 1 yr old twigs and (670)1000-1200 for 2 yr old twigs. In these cases also the sites actually least favorable for species growth had the lowest cuticular resistances, and there was a regular increase to the more favorable sites.*

Tranquillini (1974) has shown on 3 yr old seedlings that the shortened growing season at timberline on the Patscherkofel (1950 m) correlates well with the trend for cuticular transpiration of first-year twigs of *Picea abies* to be greater the higher the habitat of the tree. Cuticular resistances can be calculated from the data given. Twigs which developed over the long growing season at the valley elevation (700 m) had cuticular resistances in February of 690 sec cm^{-1}; those whose growing seasons were shortened had 340 sec cm^{-1}. Furthermore, the small cuticular resistances resulted in desiccation of the twigs, whereas the large ones did not. Thus the short growing season and wind exposure at timberline produce needles which lose more water than normal, including valley sites. Cuticular water loss in winter therefore produces the "frost drought" that limits the altitude at which trees can grow. A snow cover protects krummholz. Taylor's observations at Carson Pass show that the mean depth of midwinter snow cover is directly correlated with the height of *Pinus albicaulis* krummholz. Branches extending above this height are subjected to severe desiccation, with water potentials of -40 bar, intense bombardment by blowing snow and ice crystals, and very intense radiation which is not only large directly but also greatly augmented by the high albedo of the snow.

Snow cover also produces the flag forms of trees with a mat of branches at the base, and a bare trunk above with living branches only at the top. Again, the mat is protected by the snow, while immediately above the snow surface the additional heat from snow reflection produces increased water loss. Wind, then, is singularly ineffective except as it distributes a deep or shallow snow cover (Turner 1968).

In the White Mts. Schulze et al. (1967) have shown that *Pinus longaeva* ceases to photosynthesize in winter, as do the European conifers. Its stomata probably close in winter. Cuticular transpiration therefore must be decisive in the relative success of *P. longaeva.* Work on *Picea engelmannii* and *Abies lasiocarpa* in the Medicine Bows (Lindsay 1971) and Wasatch Mts. (Hansen and Klikoff 1972) showed very large tensions in the shoots of these plants in winter, to -35 and -90 bar.

Both structural kinds of timberline, then, are related to temperature. At least, both move as temperature changes.

Timberline evidently represents such a powerful concatenation of numerous kinds of ecological site variations and physiological responses that the tree life form sud-

* To calculate cuticular resistances from Baig et al.'s data we used for twigs of *Pinus cembra* a specific leaf area ratio of 1.7 dm^2 g^{-1} (Larcher 1963b, Table 5). Tranquillini (1974:182) gives area/weight ratios for the spruce of 0.98 in the valley, and at timberline 1.12 for fully developed twigs to 1.44 dm^2 g^{-1} for first-year twigs with a shortened vegetative period. We assumed that the wind speeds in the cuvette were sufficient to prevent leaf temperatures from rising above ambient.

denly fails across a very narrow ecotone. This does not apply to any particular kind of tree. In many areas scale or needle-leaved conifers form timberline, but the genera may be *Pinus, Picea, Abies, Tsuga, Juniperus, Chamaecyparis*, or *Cedrus*. In Siberia and the northern Rockies the life form is a deciduous, needle-leaved conifer, *Larix*. In northern oceanic climates and in the Himalaya *Betula*, and in middle European latitudes *Fagus*, even *Quercus*, are broad-leaved deciduous trees at timberline. In California *Pinus albicaulis, P. flexilis, P. balfouriana, P. longaeva, P. contorta* ssp. *murrayana, Tsuga mertensiana, Picea breweriana*, and *Chamaecyparis nootkatensis* are represented at timberline. Do these species have any physiological properties or function in common beyond the most general in the plant kingdom?

THE VEGETATION

We believe that use of the terms "tundra" and "fell-field" in relation to California's alpine vegetation not only adds nothing to our understanding but also implies false analogies (Table 18-1). We will characterize the types of vegetation as far as possible by the kinds of plants which comprise them.

Alpine vegetation is an altitudinal belt. Plants differ in their relationships to altitude, to the changes in ecological, including competitive, conditions that accompany high altitudes, and to the changes with elevation within the more or less broad alpine belt itself. Thus the alpine flora can be ranked by the upper and the lower limits of its species. Similarities of rankings between regions may allow conclusions on similarities of habitats. Dissimilarities suggest ecotypic differences or differences in habitats for example, a change in the floras available for competition. Aberg (1952:309) made such comparisons between Lule Lappmark in northernmost Sweden and Graubünden in Switzerland:

> On the whole there is a strong and positive correlation between the two series, a fact which is taken as an indication and an estimate of the ecological significance of the species concept. The altitudinal limit of a certain species in Lule Lappmark will thus give an approximate measure of its hardiness also in Graubünden, and vice versa. On the other hand there are naturally exceptions, caused for example by an unequal distribution of the alpine ecotype, and the lines of regression give the best basis for their detection and specification.

Gribbon (1968a,b) showed that in Greenland the altitudinal limits of species correlate with their general distributional restrictions to climatically differentiated areas, and the number of species of favorable habitats decreased with altitude about five times as fast as did the species of rigorous habitats. We have for Californian taxa only the gross altitudinal generalizations in floras. For future work we refer to Kuvaev's (1972) classification of eight types of changes of altitude limits with latitude in the prepolar Urals. He names species that we also have. His eight types can be generalized as sections of an altitude–latitude spindle-shaped figure that is wide in the middle latitudes and narrow at the northern and southern limits. A plant near its southern limit will show an upper height limit increasing in altitude with latitude; one near its northern limit will show an upper limit increasing in altitude toward the south.

Work on alpine vegetation elsewhere suggests ways of investigating California's alpine vegetation. The relationship of our floras to others is still imperfectly known. Taylor's work (1976b) is the first in America to combine field experience with published local floras and modern, computer-aided analyses of relationships between biological entities such as local floras by means of matrix analysis (Ceska and Roemer 1971; Spatz and Siegmund 1973), clustering, and ordination. (Taylor did not write this sentence.) We lack local floras with the kinds of precise species annotations that are possible only for local floras. Zoller et al.'s (1964) flora of the Swiss National Park, and its predecessors, can serve as models. We lack knowledge of the exact distributions of species. In this respect Cadbury et al.'s (1971) county flora of Warwickshire is a model. Malyshev's (1965), Shmidt's (1975), and Yurtsev's (1968) analyses of floras should be imitated (improved on?) in western North America. The elevational limits of our plants need study, as noted above. Backhuys (1968) has some fascinating suggestions on where and why alpine plants reach low, azonal altitudes. His and Van Steenis' explanation for low-altitude stations being found only on mountains above a certain minimal height is that only from such areas can a continual diaspore rain occur. But do not the plants at lowest altitudes produce sufficient viable diaspores themselves to maintain their populations? Are their populations, in fact, maintained? What is the role of changes in climate in eliminating the optimal-site populations from low mountains? Our kinds of timberlines need description (Shiyatov 1970; Gorchakovsky 1975:208 ff). Thus far, autecological data explaining timberline are limited to European species, as noted above. The autecological work so necessary for synecological studies of our alpine vegetation has barely begun, indispensable as it is (Mooney 1966; Mooney and Billings 1961; Billings and Mooney 1968; Billings 1975; Bliss 1971; Klikoff 1965b; Chabot and Billings 1972; Schulze et al. 1967; Mooney et al. 1964, 1965; Johnson and Caldwell 1975; Ehleringer and Miller 1975; Moser 1971; Larcher et al. 1973). Our alpine plants have root systems, but we have no information on them from the Sierra Nevada (Daubenmire 1941; Holch et al. 1941; Shaver and Billings 1975).

We can make some generalizations on the kinds of vegetation found in the Sierra Nevada on the basis of the French alpine analogues suggested in the Introduction to this chapter. At all altitudes, including the highest, isolated plants occur in Felsenmeer, cliffs, talus, screes. *Polemonium eximium, Draba lemmonii, D. Breweri, Hulsea algida, Calyptridium umbellatum,* and *Poa suksdorfii* are examples at high altitudes (Raven 1958). More stable surfaces, including peneplain remnants, were studied by Pemble (1970), and their treatment constitutes much of this chapter. The availability of water patterns most of the Sierran vegetation. Stream banks, edges of lakes and tarns, snow patches, springs, meadows—each has its own indicator species, its own plant community. There is little immediate evidence now of man's disturbance, but formerly nitrophilous vegetation must have surrounded sheep bed grounds. Although deer grazing can be heavy on the willows of some high basins, the mountain sheep that now produce nitrophilous vegetation spots on lookout points in the Yukon are only an ineffective remnant in the Sierra. Klikoff (1965a) ascribed considerable influence to gophers and ground squirrels on mountain meadow vegetation, but such effects are not always evident and natural erosion of swards probably takes place in a cyclical way. Ruderals are carried into wilderness areas by packstock and man, who make suitable habitats for these mostly annual exotics (Howell 1952, Went 1953). *Chenopodium album, Collomia linearis, Descurainia pinnata, Draba stenoloba,* and *Polygonum douglasii* are nitrophile examples. The Sierra does have bare flats of disintegrated granite which are saturated for considerable periods by inflow of snowmelt water and harbour annuals: *Collinsia parviflora, Deschampsia danthonoides, Epilobium minutum, Gayophytum racemosum, Juncus bufonius, Linanthus harknessii, Muhlenbergia filiformis, Plagiobothrys torreyi, Polygonum minimum, P. kelloggii, Saxifraga bryophora,* and so forth.

The lower elevations in the alpine and the treeless areas in the subalpine have taller vegetation: tall forb meadows, willow thickets [*Salix orestera, S. eastwoodiae, S. pseudocordata* (= *S. novae-angliae* Anderss., according to Argus 1973:140), *S. planifolia, S. drummondiana*].

Wetland vegetation at subalpine Grass Lake on Luther Pass has been described and mapped at 1:10,000 by Beguin and Major (1975). A remarkable series of communities is present, from eutrophic Caricetum nebraskensis meadows to Mimulo-Caricetum limosae fen. Accessory communities include an acidophile Kalmio-Pinetum, a bordering Salicetum rigidae (includes *S. mackenziana,* according to Argus 1973:124), a largely therophytic Sagino-Gnaphalietum, an aquatic Drepanoclado-Utricularietum with abundant *Nuphar polysepalum* and *Comarum palustre* L. (= *Potentilla palustris*), and a Torreyochloetum pauciflorae. Munz and Keck (1959:1480) include the freshwater species of *Torreyochloa* Church with the saline-alkaline genus *Puccinellia.*

Many of the plants at Grass Lake are the hygric, bog-fen, circumboreal element so prominent in western Europe and so uncommon in the Sierra Nevada, for example, *Sparganium minimum, Carex limosa, Comarum palustre, Drosera rotundifolia,* and *Eriophorum gracile.* Bog *Sphagnum* becomes less common southward in the Sierra. It may occur at high altitudes, as at Carson Pass (Taylor 1976a). In the southern subalpine Sierra it occurs at Charlotte Lake (Bennett 1965; Ratliff 1973) at the headwaters of Bubbs Creek. A small spot is also on the north shore of Third Lake at the very head of the Kern River with *Calamagrostis breweri, Carex kelloggii, C. ormantha* (?), and *Kalmia microphylla*—similar to the association described by Ratliff (1973).

Vegetation similar to that at Grass Lake, but richer in uncommon species, existed at Osgood swamp at the eastern base of Echo Summit behind a tiny morainal loop. Here occurred *Dulichium arundinaceum, Brasenia schreberi, Carex limosa,* etc. As part of "the destruction of California" the area was drained. Fortunately Adam (1967) obtained and analyzed a pollen record before the peat layer was entirely destroyed.

Many of the Grass Lake plants also occur in the Inglenook fen so well described by Baker (1972) from lowland coastal California near Ft. Bragg. Rae (1970) has studied the vegetation, ecology, and growth (3 cm yr^{-1}) of a sloping, neutral, montane non-*Sphagnum* bog with *Cratoneuron filicinum* and *Drepanocladus aduncus* at Sagehen Creek north of Lake Tahoe, particularly with reference to the autecology of *Drosera rotundifolia* and *D. anglica.* Savage (1973) has published the latest floristic checklist for the area.

Beguin and Major (1975) refer to European plant communities which aid interpretation of the ecology of the plant communities at Grass Lake. It is evident from the listing of the classes of these in the Introduction to this chapter that much study remains to be done of these fascinating, useful, and informative kinds of wetland vegetation in California. However, the bulldozers are coming!

Klamath Mountains

Alpine areas in the Klamath Mts. occur above 2680 m on Mt. Eddy (2750 m) at the eastern edge, above 2440–2590 m in the Trinity Alps on granite (Thompson Peak 2743 m), above only 2000 m in the Marble Mts. (Black Mt. 2285 m), where alpines go very low in the karst terrain developed on marble just as in the Colchis (eastern Black Sea) (Sokhadze 1974:207), and above 1600–1850 m on Preston Peak (2230 m) to the northwest. These areas are scattered as shown on the map in the pocket of this book. Here the subalpine is marked by the presence of *Pinus albicaulis, P. balfouriana, Abies lasiocarpa, Chamaecyparis nootkatensis* (Ground 1972), and the upper limits of *Tsuga mertensiana* and *Picea breweriana* (Griffin and Critchfield 1972). Ferlatte's flora (1974) for the Trinity Alps, Howell's account (1944), and theses by Muth (1967) and Ground (1972), plus Major's notes for the Marble Mts., suggest the following sketch.

This is an area of high floristic endemism. It also combines a strong element from the Pacific Northwest floristic region (Taylor 1976b) with southern Cascade, Sierran, and a few Great Basin (Taylor 1976c) taxa. Its vegetation is therefore unique.

The relationship to Pacific Northwest alpine and subalpine vegetation (Franklin and Dyrness 1973) is clearly evident from the occurrence of the mesic to hydric alpine species *Leutkea pectinata, Phyllodoce empetriformis, Cassiope mertensiana, Sedum rosea, Vahlodea* (= *Deschampsia*) *atropurpurea Fries, Xerophyllum tenax* (Ground 1972:83), *Gaultheria humifusa, Vaccinium membranaceum, V. scoparium, Veronica cusickii, V. alpina, Saxifraga ferruginea, S. tolmiei, S. fragosa* (= *S. integrifolia* Hook.), *Pinguicula macroceras,* and *Romanzoffia sitchensis* plus the tall meadow, mostly subalpine plants *Veratrum viride, Valeriana sitchensis, Polygonum phytolaccaefolium, Angelica arguta, A. lucida* (Ground 1972:168), *Carex mertensii, Epilobium angustifolium, Alnus sinuata* [cf. *A. viridis* (Chaix) DC–*A. crispa* (Ait.) Pursh], *Mimulus lewisii,* and *Bromus marginatus.* A snow course in subalpine forest in the Marble Mts. showed a spring average of 1300 mm water in 3.0 m snow (Muth 1967:27). This alone is enough to recharge the soils to field capacity some 10× over when the water is released by melting. Alpines of Pacific Northwest relationship but of rocky areas, overlapping the wet-site alpines in the habitats above, include the circumpolar *Saxifraga caespitosa, Polystichum lonchitis, Cardamine bellidifolia, Oxyria digyna, Campanula rotundifolia,* and *Artemisia campestris* ssp. *borealis.* Northwestern American plants of similar habitats include *Hieracium gracile, Senecio fremontii, Carex straminiformis, Lesquerella occidentalis* (Rollins and Shaw 1973:218 ff), *Penstemon deustus, Orthocarpus luteus, Anemone drummondii, Haplopappus lyallii, Thlaspi montanum* L. var. *montanum* P. Holmgren (1971), *Eriogonum compositum,* and *E. elatum.* Many of these are also northern Rocky Mt. taxa. *Erythronium grandiflorum* following immediately on snowmelt in meadows has this northern Rocky Mt.–Pacific Northwest distribution. Cascade-Sierran to Klamath Mt. plants include *Castilleja pruinosa, Polygonum davisiae,* and *Orthocarpus copelandii.* In only the Cascades and Klamath Mts. are *Schoenoliron album, Penstemon rupicola* (southward to the Yolla Bolly Mts.), *Silene grayi, Astragalus whitneyi* var. *siskiyouensis,* and *Arnica viscosa* (Ground 1972:870). In the Klamath Mts. plus the northern California Coast Ranges are *Haplopappus whitneyi* ssp. *discoidea* and *Machaeranthera shastensis* var *eradiata.* Klamath Mts. endemics are *Lewisia cotyledon, Draba howellii, Saxifraga fragarioides, Heuchera pringlei,*

Epilobium obcordatum (the species in the northwestern American group above) ssp. *siskiyouense,* and *Castilleja arachnoidea.* Other alpines such as *Eriogonum alpinum* are limited to Mt. Eddy and Scott Mt., *Raillardella pringlei* to subalpine boggy areas in the Trinity Alps and on Mt. Eddy, *Penstemon tracyi* to two stations in the Trinity Alps, *Draba pterosperma* to the bare marble of the Marble Mts. These strikingly bare, white, evidently arid bedrock areas with their more mesic sinkholes carry low-statured alpine plant communities well down into the subalpine belt.

In the subalpine are rich, tall herb meadows with the species mentioned above. Here past overgrazing is evident. Meadows are gullied; *Spergularia rubra* occupies bare soil where the tall meadow plants are coming back in the spotty way that marks colonization of once-bared ground. Several of these evident invaders are unpalatable to livestock—*Polygonum phytolaccaefolium, Veratrum viride, Bromus carinatus,* and mass invasions of low, shrubby *Erigonum umbellatum.*

Cascades

Alpine areas are scattered on the highest peaks: Mts. Shasta, Magee, and Lassen and fragmentarily south to the Sierra Nevada in the Lake Tahoe region. A series of equiformal progressive areas results in depauperization of the circumpolar flora southward, but Mt. Lassen has more boreal species than Mt. Shasta. At the same time endemics and derivatives from the Great Basin and Mediterranean California help to make up the alpine vegetation of the Cascades.

Cooke's work on Mt. Shasta (1940, 1941, 1949, 1955, 1963) enables us to sketch in the available alpine habitats and their plants. The undeveloped nature of the volcanic soils evidently produces a wide subalpine belt with *Pinus albicaulis, Tsuga mertensiana* (to 2440 m: Matthes 1942), and *Abies magnifica* var. *shastensis* as timberline trees. Much of the timberline is edaphic and locally advancing (Matthes 1942). *Pinus albicaulis* climbs highest as krummholz. Most alpines occur in various kinds of unstable, gravelly, or rocky habitats which may be wind-swept ridgetops or flats with clumps or colonies of the following species (asterisk indicates also on Mt. Lassen in similar habitats): *Carex phaeocephala* and *Poa epilis* (both to 3660 m), **P. pringlei,* **Eriogonum pyrolaefolium,* **E. marifolium,* **Polygonum shastense, P. newberryi,* **P. davisiae,* **Calyptridium umbellatum, Lupinus lyallii* (to 3050 m), *Phacelia frigida, Silene grayi, S. sargentii,* **Anemone occidentalis,* **Draba breweri* (reported to 3970 m), **Streptanthus tortuosus* var. *orbiculatus,* **Dicentra uniflora, Potentilla glandulosa* ssp. *pseudorupestris,* **Phlox diffusa, Polemonium pulcherrimum* (reported to 3660 m), **Monardella odoratissima, Penstemon gracilentus* (lower on Mt. Lassen), *Castilleja arachnoidea,* **Erigeron compositus,* **Chaenactis nevadensis, Hulsea nana,* **Achillea lanulosa* ssp. *alpicola, Arnica viscosa,* **Agoseris glauca, Machaeranthera shastensis* var. *montana* (lower on Mt. Lassen), and even an isolated **Fritillaria atropurpurea.* According to Matthes (1942), snow lies long enough in many places to shorten the growing season too much for trees to survive, just as happens in the Pacific Northwest (cf. the section entitled "Timberline" above). Steep slopes are unstable. Water percolates quickly downward in the loose volcanic material and is lost to plants.

Snow patches produce colonies of **Juncus parryi,* **Saxifraga tolmiei,* and **Penstemon davidsonii.* Around the edge of rocks in protected and more moist habitats, scoured of snow in winter (?), are **Cryptogramma crispa* ssp. *acrosti-*

choides and **Cardamine bellidifolia* [also on Preston Peak (Ground 1972:74), a circumpolar plant not southward into the Sierra]. Gravel flats with late snow-lie have **Arabis platysperma* and **Carex breweri.*

Cliffs and rock outcrops have **Ribes cereum, Woodsia scopulina,* **Penstemon newberri,* and **Hieracium horridum.* Seepage areas may be marked by *Alnus crispa* at subalpine altitudes, talus plus snow by **Athyrium alpestre* (= *A. distentifolium* Tausch ex Opiz), wet rocks by **Oxyria digyna* to 3350 m. Dry ridges may even have *Juniperus communis* or *Cercocarpus ledifolius* (rarer and lower on Mt. Lassen) and **Pteryxia terebinthina* with **Pinus albicaulis.* Dry openings in the subalpine forest have *Sitanion hystrix,* **Stipa occidentalis,* or **Haplopappus bloomeri.*

In heather meadows resulting from snow accumulation and springs near the treeline are *Phyllodoce empetriformis,* **Kalmia microphylla, Vaccinium nivictum, Danthonia intermedia, Allium validum* (rare and lower at Mt. Lassen), **Sagina saginoides,* and **Veronica alpina* (Olmsted 1976). Other snow-accumulation areas have *Leutkea pectinata* or **Juncus balticus* var. *montanus.*

In springs (moist meadows on Mt. Lassen) are **Phleum alpinum,* **Trisetum spicatum,* **Agrostis variabilis, A. thurberiana* [= *Podagrostis th.* (Hitchc.) Hulten], **Carex nigricans,* **C. spectabilis, Juncus mertensianus* (rare and at lower elevation on Mt. Lassen), **Luzula subcongesta, Saxifraga nidifica,* **Stellaria crispa,* **Sibbaldia procumbens,* **Potentilla flabellifolia, Hypericum anagalloides* (lower and rarer on Mt. Lassen), *Epilobium clavatum,* **Mimulus primuloides, Campanula wilkinsiana,* **Hieracium gracile,* and *Antennaria umbrinella.*

Situated almost equidistant between Mt. Shasta and Mt. Lassen is Magee Peak (2641 m). Its north-facing cirques support limited patches of alpine vegetation. Here several Cascade species reach their southern limit, including *Campanula scabrella, Phacelia frigida* ssp. *frigida, Polemonium pulcherimum* var. *pilosum,* and *Lutkea pectinata.* The last forms mats in snow accumulation areas between tree islands of *Tsuga mertensiana. Collomia larsenii* is here and on Mt. Lassen, disjunct from the Cascades in Oregon. Here *Arenaria nuttallii* ssp. *gracilis, Haplopappus apargioides,* and *Phyllodoce breweri* reach their northern limit along the Cascade–Sierran axis. *Phyllodoce* is dominant in several flats where snowmelt water accumulates in spring, exhibiting patterned distribution because of vegetative propagation and lack of seedling establishment. Olmsted (1976) has studied the distribution of *P. breweri* and its nothern Cordilleran relative, *P. empetriformis,* through the Cascade–Sierran axis. *Phyllodoce empetriformis* extends south to Mt. Shasta, while *P. breweri* occurs from Magee Peak south to the San Bernardino Mts. Olmsted's investigations demonstrate that *P. breweri* is more xeric in its moisture tolerance than *P. empetriformis,* and this correlates nicely with their respective distributions.

Again, for Mt. Lassen we have little information on the vegetation as distinct from the flora. The two alpine floras differ considerably. Mt. Lassen is much richer, and the plants are in more familiar habitats. Here we list by habitats, a most subjective and unsatisfying procedure, the plants on Mt. Lassen which are additions to the Mt. Shasta lists above or apparently change ecologies, and so forth.

Timberline is formed by *Pinus albicaulis,* to 2940 m with *Tsuga mertensiana* below. We neglect the subalpine meadows, but they have *Polygonum phytolaccaefolium, Sphenosciadeum capitellatum, Lewisia triphylla,* and *L. nevadensis*—a mixture of Pacific Northwestern and Sierra Nevadan species.

Open, rocky, gravelly flats have, in addition to the species on Mt. Shasta (indicated above by asterisk), *Poa nervosa, Sitanion hystrix, Trisetum spicatum* (also in dry meadows; see below), *Carex helleri, C. straminiformis, C. brainerdii, C. rossii, Eriogonum ochrocephalum* var. *agnellum, E. ovalifolium* var. *nivale, E. ursinum, Arenaria kingii* var. *glabrescens, A. nuttallii* (summit), *Silene douglasii, S. grayi* (summit), *Anemone drummondii, Aquilegia formosa, Arabis holboellii, A. lemmonii* (summit), *A. lyallii* (rare), *Draba aureola* (summit, disjunct to Oregon), *Erysimum capitatum, Lesquerella occidentalis* (summit), *Smelowskia ovalis* var. *congesta* (summit, the variety endemic, the species disjunct from the Three Sisters in Oregon), *Ribes montigenum, Holodiscus microphyllus* (with *Orobanche pinorum*), *Lupinus obtusilobus* (also in snow flats like *Carex breweri*), *Viola purpurea* ssp. *dimorpha, Collomia larsenii* (summit, disjunct to Oregon), *Ipomopsis congesta* (summit), *Leptodactylon pungens* ssp. *pulchriflorum, Phlox bryoides* (only California locality for this Great Basin and Rocky Mt. plant), *Polemonium pulcherrimum* (summit), *Phacelia frigida* var. *dasyphylla* (summit), *Cryptantha subretusa, Castilleja applegatei, C. payneae, Pedicularis attollens, Penstemon deustus, Galium grayanum, Erigeron barbellatus, Haplopappus greenei, Antennaria alpina* var. *media, Chaenactis douglasii* var. *achilleaefolia, Hulsea nana* (summit, southern limit), *Arnica nevadensis, Raillardella argentea, Senecio canus, S. fremontii* (summit), *Cirsium foliosum*, and *Stephanomeria lactucina.*

Wet, gravelly areas have *Claytonia nevadensis, Epilobium obcordatum, Cassiope mertensiana, Phyllodoce breweri,* and *Valeriana capitata.*

Species in moist meadows include *Carex vernacula, C. gymnoclada, Salix anglorum, Potentilla drummondii Dodecatheon alpinum, Hydrophyllum occidentale, Castilleja lassensis, Aster alpigenus* ssp. *andersonii, Erigeron peregrinus* ssp. *callianthemus,* and *Arnica longifolia.*

Included in very wet areas are *Mimulus tilingii* (at lower elevations on Mt. Shasta), *M. lewisii, M. moschatus,* and *Pedicularis groenlandica.*

Dry meadows have *Stipa californica* and *Carex preslii*—see the Shasta list.

Only in the subalpine forest are such alpines as *Carex phaeocephala, Lupinus lyallii,* and *Gayophytum diffusum.*

Sierra Nevada

Pemble (1970) used 188 stand surveys containing 175 species from the Slate Creek, Dana Plateau, and Convict Creek areas to describe the vegetation and relate it to two sequences of ecological factors: (1) soil parent material, namely, granitics, noncarbonate metamorphics, and marble; and (2) topography, particularly in relation to water availability. Since these two sequences cross, no linear arrangement of the entire matrix of species and stands is possible. Pemble therefore presented the same topographical sequence in three matrices, one for each soil parent material. His matrices are splendid documentations of one of the 29 kinds of Californian vegetation used by Munz and Keck (1949, 1950, 1959) to describe the habitats of members of the California flora, namely, "alpine fell-field." Pemble correlated his vegetation descriptions with those by Klikoff (1965a) from the Gaylor Lake basin just to the west of Slate Creek. Since we cannot reproduce his 311 pages of text and 8800 cm² of species-stand matrices here, we shall discuss his classification and its ecological meaning.

The classification is an extremely precise ecological tool. The data that enter it are also objective and documented. It regards vegetation as a botanist should, as associations of individualistic species that can be arranged to reflect the mosaic of habitat conditions in which those species exist. A more complete exposition of the method, the logic and history behind it, and some ecological applications can be found in Braun-Blanquet's textbook (1964). Later additions include computer use to lighten the load of arranging and rearranging the species-stand matrix and to test and document some working hypotheses on ecological relationships (Ceska and Roemer 1971; Moore et al. 1970; Ivimey-Cook and Proctor 1966; Spatz and Siegmund 1973; Frenkel and Harrison 1974; Westhoff and Van der Maarel 1973).

Pemble split his alpine vegetation into three hierarchical levels. The first level comprised an *Antennaria alpina–Carex vernacula* meadow, a sod–graminoid, moist-site type, and a *Sitanion hystrix–Phlox covillei* open, bunch grass–cushion plant, dry type. The former has, in addition to the *Antennaria* and *Carex,* the differential species, *Poa epilis, Potentilla diversifolia,* and *Sibbaldia procumbens. Thalictrum alpinum* differentiates the marble variation; *Calamagrostis breweri, Juncus mertensianus,* and *Draba crassifolia* differentiate the granite-metamorphic. The second type has the differential species shown in Table 18-5.

The meadow vegetation is split into that characterized by *Salix anglorum* var. *antiplasta* (= *S. arctica* Pall.) on the wettest sites, with *Epilobium anagallidifolium* as an additional differential species, and *Saxifraga aprica–Gentiana newberryi* vegetation on moist, seepage, but well-drained sites, with sod broken and plants spaced apart. The *Salix anglorum* vegetation is differentiated by *Luzula orestera, Castilleja culbertsonii,* and *Scirpus clementis* on granite, while *Thalictrum alpinum* differentiates the marble. This *Salix anglorum* vegetation has four variants, one of which is not on granite and is characterized by *Salix nivalis. Carex haydeniana* is a differential species on metamorphics. The three other types of *Salix anglorum* vegetation are characterized as (1) *Carex nigricans–Pedicularis attollens,* (2) *Trifolium monanthum–Phleum alpinum,* and (3) *Cassiope mertensiana–Phyllodoce breweri.*

The *Carex nigricans–Pedicularis attollens* vegetation is also differentiated by the presence of *Mimulus primuloides* with *Dodecatheon alpinum* not on marble and *Deschampsia caespitosa* only on granite. The marble vegetation of this type is positively differentiated by *Gentiana amarella, Kobresia myosuroides, Epilobium latifolium, Salix brachycarpa, Calamagrostis canadensis, Carex hassei,* and *Scirpus pumilus* [= *Trichophorum pumilum* (M. Vahl) Schinz & Thell.]. This is a rare type of Sierra Nevadan vegetation. It is found in depressions which are well supplied in early summer with water, which is often standing, and have a long-persisting snow cover and a peaty sod, with hummocks or solifluction.

The *Trifolium monanthum–Phleum alpinum* vegetation has the differential species shown in Table 18-5. Sites with this vegetation are well drained, convex, sloping, and cold, with running water, often below long-persisting snowfields.

The *Cassiope mertensiana–Phyllodoce breweri* vegetation does not occur on marble. *Carex nigricans, Vaccinium nivictum, Kalmia microphylla, Potentilla flabellifolia, Castilleja lemmonii,* and *Saxifraga bryophora* differentiate it on granite. It occurs in topographical lows at the base of slopes, among boulders, on glaciated terrain, on seepage sites with long-lasting snow.

The *Sitanion hystrix–Phlox covillei* vegetation was divided into three kinds: (1) *Carex breweri–Calyptridium umbellatum* not on marble, (2) *Juncus balticus* only on

TABLE 18-5. Differential species of 3 kinds of vegetation characterized by Pemble (1970:104, 75, 130) according to occurrence in variants on different soil parent materials

A. *Sitanion hystrix*-*Phlox covillei* vegetation

Community	Granite and Metamorphics	Marble
Sitanion hystrix	*Eriogonum ovalifolium* var. *nivale*	*Cymopterus cinerarius*
Phlox covillei		*Eriogonum microthecum*
Castilleja nana	*Selaginella watsonii*	*Erigeron clokeyi*
Erigeron pygmaeus	*Erigeron compositus*	*Machaeranthera shastensis* var. *montana*
Arabis lyallii var. *nubigena*	*Draba breweri*	
A. lemmoni var. *depauperata*	*Carex phaeocephala*	*Stipa occidentalis*
	Silene sargentii	*Linum lewisii*
	Penstemon davidsonii	
	Poa nervosa	

B. *Trifolium monanthus*-*Phleum alpinum* community

Association	Granite and Metamorphics	Metamorphics	Marble
Trifolium monanthum	*Carex spectabilis*	*Claytonia nevadensis*	*Juncus balticus*
Phleum alpinum	*Juncus drummondii*		
Mimulus tilingii	*Veronica alpina*		
Gentiana tenella	*Stellaria umbellata*		
	Mimulus moschatus		
	Epilobium pringleanum		
	Saxifraga nidifica		

C. *Arenaria kingii-Senecio werneriaefolius* community

Association	Granite and Metamorphics	Metamorphics and Marble	Only Metamorphics
Arenaria kingii	*Ivesia lycopodioiodes* ssp. *typica*	*Eriogonum ochrocephalum* ssp. *agnellum*	*Chrysothamnus parryi* ssp. *monocephalus*
Senecio werneriaefolius	*Lupinus lyallii*	*Koeleria cristata*	*Potentilla pseudosericea*
Antennaria rosea	*Epilobium minutum*	*Astragalus purshii* var. *lectulus*	
A. corymbosa		*A. lentiginosus* var. *ineptus*	
		Erysimum perenne	
		Cirsium tioganum	

marble, and (3) *Arenaria kingii–Senecio werneriaefolius*. The first has *Ivesia muirii, Hulsea algida,* and *Eriogonum lobbii* as differential species on granite. Variants are differentiated as *Carex helleri–Poa suksdorfii,* with *Luzula divaricata* a differential species on granite, and *Arabis platysperma–Penstemon heterodoxus* associations. These two associations occur on well-drained sites, sloping, without surface water, with snow accumulation on lee slopes and therefore a shortened growing season. Soils are poorly developed, raw but acid (pH 4.8–6.5), with sand or gravel covering the surface. Plant cover is low, and the communities are very open. The *Carex helleri–Poa suksdorfii* may include members of the meadow communities which respond positively to snow accumulation. The *Arabis–Penstemon* association has *Eriogonum marifolium* var. *icanum* and *Raillardella argentea* at lower elevations as differential species. It has more cover, of more forbs.

The *Juncus balticus* association considered here is a dry-site type. It includes the species of Table 18-5 listed as differential for the *Sitanion hystrix-Phlox covillei* vegetation on marble. The *Juncus* also occurs with high constancy in the *Carex nigricans–Pedicularis attollens* form of *Salix anglorum* vegetation on marble (see above). Does the *Juncus* behave in this way because its competitive relationships are disturbed by the presence of the rare species on marble? Or is *Juncus balticus* a poor species, ecologically considered? Many taxonomists have run into our difficulty, as is evident from the lengthy synonymy of the taxon (Munz and Keck 1959:1405; Hitchcock et al. 1969:187 ff) and the long string of different chromosome numbers associated with the name (Bolkhovshikh et al. 1969:359). Our marble stands are apparently dry, and the associated species on marble substantiate this impression. However, the sites are depressions and have snow accumulation in winter; although it melts quickly in summer, the constant supply of washed-in raw marble sand provides an efficient mulch to conserve soil moisture. Actually the soil is usually damp just below the dry surface.

The third major division of the dry-site alpine vegetation (*Sitanion hystrix–Phlox covillei*), namely, *Arenaria kingii–Senecio werneriaefolius,* has the differential species shown in Table 18-5. It is unique that no positive differential species separate this kind of vegetation on marble. Vegetation cover is always less than 50%, and the separate communities are diverse floristically. This vegetation was split by Pemble into three kinds: (1) *Draba oligosperma–Poa rupicola,* (2) *Muhlenbergia richardsonis–Stipa pinetorum,* and (3) *Haplopappus macronema–Phacelia frigida.*

The first has the differential species shown in Table 18-6. "This is the most common association in the Sierra alpine" (Pemble 1970:135). It occurs on exposed, level sites with a long snow-free season. Cover is less than 25% Soils are gravel-covered, not soddy, and of relatively high pH (5.2–6.1 on granites, 5.5-7.0 on metamorphics, 7.3-8.1 on marble). A third of the species occur also only in the Rocky Mts., except on marble where the proportion is 24%, and Great Basin species make up a third.

The differential species of the *Muhlenbergia richardsonis–Stipa pinetorum* association are also listed in Table 18-6. *Ribes cereum* is an additional association differential species. This vegetation is especially common on metamorphics, fairly common on marble, and only occasional on granitics. It avoids depressions, occupies stable slopes, windswept, snow-free, which therefore have a long growing season. Soils are rocky, dry, and slightly less acid than in the preceding association. Vegetation is sparse with only 20% cover. On marble this association has many Great Basin species (45%).

Differential species of the *Haplopappus macronema–Phacelia frigida* association are also listed in Table 18-6. *Pellaea breweri* and *Polemonium eximium* (at 3500–3800 m) help to differentiate the association on all substrates. It is found in scree between boulders, on sites with the rawest soils. Cover is least (15% on substrates with 50% rock and 34% gravel. No species on the marble are arctic, and 36% are confined to the Great Basin.

Pemble's arrangement of the stands and species he studied is a hypothesis. One test is its predictive value in the field.

Another way to arrange these same stands and species is to take those on each rock type, in each geographical area as a unit. At Convict Creek on marble 79 species in 57 stands produced 8 groups of species forming 6 communities (Table 18-7). Group 1 is found on very open, surficially dry to wet scree and rock crevices and is of limited occurrence. Group 2 is most typical of the open, disintegrated marble "soil" with some sod formation and is widespread on the marble. Group 3 is open, taller, shrubbier, rockier, with marble bedrock. Note the numerous taxa of low presence but high diagnostic value: shrubs, xerophytes of open, warm, rocky slopes, rock crevice plants. Mutually exclusive species groups are *Eriogonum microthecum, Erigeron clokeyi, Jamesia americana* versus *Phacelia frigida, Haplopappus macronema, Machaeranthera shastensis* var. *montana, Cirsium andersonii.* Group 4 is in the driest areas, often moving scree on slopes. *Castilleja breweri* is associated with the *Phacelia frigida* group noted above. Group 5 forms sod and is surficially dry but damp below the surface mulch of sand. Group 6 is wet, but the marble is at the surface. Cover is closed. Group 7 is transitional to Group 8, the wet end of this kind of vegetation.

The communities these eight groups form can again be divided into a *Sitanion hystrix–Phlox covillei* open, dry type and a *Carex subnigricans–Pedicularis attollens* wet, closed, turf type.

Extremely species-poor and occurring on open, raw areas, heterogeneous in moisture supply and of the scree to rock crevice to soil type, is the *Draba oligosperma–Arabis inyoensis* community. Three of the most characteristic species are *Oxyria digyna, Crepis nana,* and *Draba lonchocarpa.* The drier transition to the *Leptodactylon pungens–Oryzopsis hymenoides* type emphasizes the steppe to desert nature of this latter community. Rocky sites are occupied by group 3; raw soil has the bunch grasses, sod-formers, cushion and scree plants of group 2. *Stipa occidentalis* and *Juncus balticus* var. *montanus* are the best characteristic species of the calcareous sod type dominated by group 5. The most desertic group, 4, fades out in this community. Group 5 is both more abundant and more constant here than in the drier *Leptodactylon pungens–Oryzopsis hymenoides* community. The *Haplopappus apargioides* vegetation is related to the *Draba oligosperma–Arabis inyoensis* vegetation. It lacks the desertic group 4, and the rocky site group 3 also fades out. Note how nicely transitions between communities are shown in Table 18-7 by species presence, coherence of a group, and abundance. A full species-stand matrix is even more informative than the summary given in Table 18-7.

Salix nivalis vegetation is characterized by that species and also has *Festuca brachyphylla.* It is almost closed vegetation but has place for some plants of groups 1 and 2. It has the low, more mesic group 7 but lacks the taller group 5. The wet-site *Carex subnigricans–Pedicularis attollens* vegetation is as unique for California, with *Carex pseudoscirpoidea, Thalictrum alpinum, Salix brachycarpa, Kobresia*

TABLE 18-6. Differential species of variants of the *Arenaria kingii*-*Senecio werneriaefolius* vegetation as related to soil parent materials. From Pemble (1970:132, 141, 142)

Nonmarble	Granite	Metamorphics	Marble
A. *Draba oligosperma*-*Poa rupicola*			
Podistera nevadensis	*Ivesia shockleyi*	None	*Arabis inyoensis*
Poa nevadensis			*Crepis nana*
Arenaria rubella	*Androsace septentrionalis* ssp. *subumbellata*		*Draba lonchocarpa*
A. rossii			
Draba densifolia			
Astragalus kentrophyta var. *danaus*			
Carex tahoensis			
B. *Muhlenbergia richardsonis*-*Stipa pinetorum*			
Leptodactylon pungens ssp. *pulchriflorum*	*Ivesia muirii*	*Carex douglasii*	*Monardella odoratissima* ssp. *parviflora*
	Hulsea algida	*Astragalus whitneyi*	
	Eriogonum lobbii	*Polemonium pulcherrimum*	*Galium hypotrichium*
Carex rossii	*Draba sierrae*		*Oryzopsis hymenoides*
Pinus albicaulis			*Cryptantha hoffmanii*
			Stipa comata

			Oenothera caespitosa var. *crinita*
			Penstemon speciosus
			Frasera puberulenta
			Haplopappus suffruticosus
			Linanthus nuttallii
			Jamesia americana
C. *Haplopappus macronema-Phacelia frigida*			
Leptodactylon pungens ssp. *pulchriflorum*	*Hieracium horridum*	*Arabis holboellii*	*Monardella odoratissima* ssp. *parvifolia*
Calamagrostis purpurascens	*Cystopteris fragilis*	*Potentilla glandulosa* ssp. *nevadensis*	*Galium hypotrichium*
Pinus albicaulis	*Polemonium pulcherrimum*	*Streptanthus tortuosus* var. *orbiculatus*	*Oryzopsis hymenoides*
Aquilegia pubescens	*Carex rossii*	*Carex leporinella*	*Cryptantha hoffmanii*
	Draba sierrae	*Sedum stenopetalum* ssp. *radiatum*	*Castilleja breweri*
	Stipa californica		*Cirsium andersonii*

TABLE 18-7. Summary constancy-cover table for alpine vegetation on marble parent material in the Sierran Convict Creek basin. Constructed from data of Pemble (1970). Six community types are recognized: Drol-Arin = *Draba oligosperma*-*Arabis inyoensis*; Lepu-Orhy = *Leptodactylon pungens* ssp. *pulchriflorum*-*Oryzopsis hymenoides*; Stco-Juba = *Stipa comata*-*Juncus balticus*; Haap = *Haplopappus apargioides*; Sani = *Salix nivalis*; Casu-Peat = *Carex subnigricans*-*Pedicularis attolens*. Values entered for each taxon's occurrence in the six community types are constancy (Roman numeral, after Braun-Blanquet 1964) and modal cover (expressed on a 10-division Domin scale, after Pemble 1970)

		Plant Community and Number of Stands					
Group	Species	Drol-Arin 9	Lepu-Orhy 16	Stco-Juba 13	Haap 6	Sani 6	Casu-Peat 7
1.	*Draba oligosperma*	V.2	I.2				
	Arabis inyoensis	III.2			II.1		
	Aster alpigenus ssp. *andersonii*	III.2					
	Koeleria cristata	III.2	I.2	II.2	I.2	II.+	
	Castilleja nana	III.2			III.2	III.+	
	Crepis nana	II.+					
	Oxyria digyna	II.3					
	Draba lonchocarpa	I.2					
2.	*Phlox covillei*	IV.4	IV.2	III.2	IV.4	IV.1	I.1
	Sitanion hystrix	IV.1	V.3	IV.1	V.2	I.1	II.1
	Eriogonum rosense	V.3	V.2	IV.3	III.1	I.2	
	Astragalus purshii var. *lectulus*	III.3	IV.2	IV.3	III.1		
	Cymopterus cinerarius	II.1	IV.1	IV.1	III.1	I.1	
	Muhlenbergia richardsonis	II.2	III.2	IV.4	I.4	I.1	III.1
3.	*Eriogonum microthecum*	I.3	III.3	IV.3	III.2		
	Eriogeron clokeyi		IV.2	II.2	II.2		
	Jamesia americana		I.2				
	Phacelia frigida		III.2	III.1	III.2		

	Haplopappus macronema		III.2	III.2	I.1		
	Machaeranthera shastensis var. montana		II.2	IV.2	II.1		
	Cirsium andersonii		I.1				I.1
	Cryptantha nubigena	II.2	I.2	IV.1	II.+	I.1	
	Ribes cereum		I.3				
	Haplopappus suffruticosus		I.1	I.3			
	Cryptantha hoffmanii		I.+				
	Linanthus nuttallii		I.2				
	Heuchera rubescens		I.3				
4.	Leptodactylon pungens ssp. pulchriflorum	I.+	V.2	II.+			
	Monardella odoratissima ssp. parvifolia		III.2	I.+	I.1		
	Galium hypotrichium	I.2	III.2	I.+			
	Castilleja breweri		III.1				
	Penstemon speciosus		II.+				
	Oenothera caespitosa var. crinita	I.1	III.1	I.1			
	Frasera puberulenta	I.1	I.2	I.1			
	Oryzopsis hymenoides		III.2	I.2			
	Stipa pinetorum		II.3				
	S. comata	I.4	I.5	I.5			
5.	Achillea lanulosa ssp. alpicola		II.1	I.2	IV.2	I.+	II.2
	Stipa occidentalis		II.1	IV.4	I.3		
	Juncus balticus var. montanus		I.2	III.5	IV.5		V.4
6.	Salix nivalis					V.5	
	S. anglorum var. antiplasta					III.3	II.5
	Festuca brachyphylla	II.+				III.1	
7.	Solidago multiradiata	I.1	I.+		III.3	V.2	V.3
	Trisetum spicatum	I.+	I.1		V.3	III.2	V.2
	Haplopappus apargioides	I.3	I.1	I.3	V.5	V.2	II.2

TABLE 18-7 [continued]

Group	Species	Plant Community and Number of Stands					
		Drol-Arin 9	Lepu-Orhy 16	Stco-Juba 13	Haap 6	Sani 6	Casu-Peat 7
8.	Phleum alpinum					I.2	IV.1
	Carex pseudoscirpoidea						IV.3
	Thalictrum alpinum					II.1	IV.4
	Carex subnigricans					I.6	IV.3
	Salix brachycarpa					I.R	III.3
	Gentiana amarella						III.1
	Potentilla fruticosa						III.3
	Trifolium monanthum				I.1		III.2
	Pedicularis attollens						III.2
	Kobresia myosuroides						III.5
	Danthonia intermedia						III.3
	Epilobium latifolium					I.1	I.1
	Trichophorum pumilum						II.5
	*Carex exserta		I.+	I.5	I.4		II.2

myosuroides, Scirpus pumilus (= *Trichophorum pumilum*), and *Epilobium* (= *Chamaenerion*) *latifolium* (Fig. 18-2), as is the dry-site *Leptodactylon pungens–Oryzopsis hymenoides* for the alpine anywhere. Sod species of group 2, groups 7 and 5, and rare but exclusive species also help to characterize it.

Carex exserta characterizes none of our vegetation units. This is a common plant in the subalpine to alpine Sierra Nevada, and it should be studied (Ratliff 1973, 1974).

On metamorphics at Slate Creek and the Dana Plateau 50 species in 51 stands produced eight phytosociological–ecological groups upon analysis, and their combinations four plant communities (Table 18-8). Group 1 is dominated by the mat-forming *Podistera nevadensis,* which is evidently adapted to resist wind erosion, including snow corrasion, if such is really common in the alpine (Turner 1968). Group 2 is windblown also, but rocky, a mixture of very low and taller plants. Group 3 is similar but lacks the taller plants. Group 4 consists of many widespread low Sierran plants, of open ground, often in gravel or sand. Group 5 has plants of open, but long snow-covered, areas. Group 6 contains sod-forming plants, more mesic than all the preceding ones. Group 7 has wet-site plants, often of snow patches or running water, sod-formers. Group 8 is similar but occurs on even tighter sod.

The communities again are split between three types with *Sitanion hystrix,* open, spaced, with scattered rocks or bare ground, cushions, bunch grasses, mats and the wet-site sod *Antennaria alpina–Carex vernacula.* The *Carex breweri–Poa nervosa* does not form sod, has much open ground with few large rocks, but does have a long-lasting snow cover, at least in part. The *Calamagrostis purpurascens–Ribes cereum* is exposed to wind, has large rocks, lacks groups 1, 5, and 7-8. The *Podistera nevadensis–Erigeron pygmaeus* vegetation is the dry-site, open, rocky, windblown, metamorphics variant of the Sierran dry-site type.

Figure 18-2. Upper part of Convict Creek with marble on the left and metamorphic Red Slate Peak (4010 m) on the right. The calcareous subalpine meadow at Lake Mildred (3000 m) contains *Salix brachycarpa, S. orestera, Kobresia myosuroides, Trichophorum pumilum,* etc. The conifers are *Pinus albicualis.* August 24, 1962.

TABLE 18-8. Summary constancy-cover table for Sierran alpine vegetation on metamorphic parent materials. Constructed from data of Pemble (1970). Four community types are recognized: Poner-Erpy = *Podistera nevadensis-Erigeron pygmaeus*; Capu-Rice = *Calamagrostis purpurascens-Ribes cereum*; Cabr-Poner = *Carex brewerii-Poa nervosa*; Anal-Cave = *Antennaria alpina* var. *media-Carex vernacula*. Values entered for each taxon's occurrence in the four communities are constancy (Roman numeral after Braun-Blanquet 1964) and modal cover (expressed on a 10-division Domin scale, after Pemble 1970)

		Plant Community and Number of Stands			
		Poner-Erpy	Capu-Rice	Cabr-Poner	Anal-Cave
Group	Species	16	15	6	14
1.	*Podistera nevadensis*	V.4			
	Poa rupicola	II.2	I.2		
2.	*Erigeron pygmaeus*	IV.2	II.2		
	Astragalus kentrophyta var. *danaus*	IV.3	II.3		
	Draba densifolia	III.1	II.1		
	Carex tahoensis	II.2	I.1		
	Senecio werneriaefolius	II.1	III.2		
	Leptodactylon pungens ssp. *pulchriflorum*	II.1	V.2		
	Polemonium pulcherrimum	I.1	II.2		
	Ribes cereum	I.2	III.2		
	Penstemon davidsonii	I.1	II.1		I.+
	Calamagrostis purpurascens		IV.3		
3.	*Eriogonum ovalifolium* var. *nivale*	V.3	V.3	III.3	
	Sitanion hystrix	V.3	V.2	II.2	
	Phlox covillei	V.2	I.4	II.1	I.+
	Erigeron compositus vars.	III.1	V.1	II.1	
	Arenaria kingii vars.	III.3	I.+	III.1	I.2
4.	*Carex phaeocephala*	III.2	II.2	IV.2	
	Draba breweri	III.1	II.R	II.+	
	Arabis lemmonii	II.+	II.1	I.2	
	Castilleja nana	II.1	III.2	II.1	I.+
	Arabis lyallii	II.+	II.+	III.2	I.2
	Selaginella watsoni	I.2	III.2	I.2	I.3
	Antennaria umbrinella	II.1	I.1	II.2	
	Calyptridium umbellatum	II.+	I.+	IV.+	
	Raillardella argentea	I.1	I.2	II.2	
	Arenaria nuttallii ssp. *gracilis*	I.1	I.+	II.1	
	Phacelia frigida ssp. *dasyphylla*	I.2	I.1	I.R	
5.	*Carex breweri*			III.3	
	Poa nervosa		I.R	IV.2	

TABLE 18-8 [continued]

		Plant Community and Number of Stands			
Group	Species	Poner-Erpy 16	Capu-Rice 15	Cabr-Poner 6	Anal-Cave 14
6.	*Solidago multiradiata*		IV.2	IV.2	II.1
	Lupinus lyallii vars.		III.2	IV.4	II.2
	Potentilla drummondii		II.1	I.3	II.2
	Carex exserta		II.1	II.5	I.7
	Silene sargentii	I.3	II.1	III.1	I.3
	Juncus parryi	I.1	II.1	II.2	II.1
	Carex helleri	I.1		II.1	I.2
	Draba lemmonii		I.2	II.+	I.1
	Rumex paucifolius ssp. *gracilescens*		I.1	II.1	II.+
	Penstemon heterodoxus		I.1	I.2	I.+
7.	*Erigeron petiolaris*		I.1	III.2	IV.2
	Luzula spicata		I.1	I.R	III.1
	Carex vernacula				III.4
	Poa epilis				III.3
	Carex subnigricans		I.+		II.4
	C. albonigra		I.		II.3
	Saxifraga aprica				II.1
	Potentilla breweri		I.1		II.2
	Sedum rosea ssp. *integrifolium*				II.1
	Juncus mertensianus				I.2
	Lewisia sierrae			I.2	I.1
	Sibbaldia procumbens		I.2	I.2	I.+
8.	*Antennaria alpina* var. *media*				IV.4
	Calamagrostis breweri				IV.4
	Lewisia pygmaea vars.			I.3	IV.2

The 91 species on granites in 61 stands formed 13 groups (Table 18-9). Dry, open *Sitanion hystrix-Phlox covillei* through group 6 versus moist, sod *Antennaria alpina-Carex vernacula* with groups 9–13 can be recognized. Group 1 is exposed to wind in rocky sites, and is composed of cushion plants. Group 2 is similar but low in stature. Group 3 replaces group 2. Why? Group 4 is a shrub group in rocks with *Carex rossii* occupying the soil between the projecting rocks. Group 5 occurs on still rockier sites with plants within the crannies, not projecting out of them as in group 5. Group 6 is ubiquitous on dry, open habitats, although its species have low cover and are not constant. Group 7 is more site heterogeneous and rockier than group 6, at higher altitudes. Cover and constancy in group 7 are even lower than in group 6. In group 7 wet rocks, windblown areas with cushion plants, snow patches, rock crevices, and scree are all evident from the species found there. We have separated out *Carex exserta,* which often seems to dominate large areas but is singularly

TABLE 18-9. Summary constancy-cover table for Sierran alpine vegetation on granitic parent materials. Constructed from data of Pemble (1970). Nine community types are recognized: Ponev-Arki = *Podistera nevadensis*-*Arenaria kingii*; Drde-Aske = *Draba densifolia*-*Astragalus kentrophyta* var. *danaus*; Erin-Raar = *Eriogonum incanum*-*Raillardella argentea*; Lepu-Hama = *Leptodactylon pungens* ssp. *pulchriflorum*-*Haplopappus macronema*; Sewa-Erni = *Selaginella watsonii*-*Eriogonum ovalifolium* var. *nivale*; Cahe-Jupa = *Carex helleri*-*Juncus parryi*; Podr-Pobr = *Potentilla drummondii*-*P. breweri*; Sipr-Jume = *Sibbaldia procumbens*-*Juncus mertensianus*; Vani-Cani = *Vaccinium nivictum*-*Carex nigricans*. Values entered for each taxon's occurrence in the nine communities are constancy (Roman numeral, after Braun-Blanquet 1964) and modal cover (expressed on a 10-division Domin scale, after Pemble 1970)

		Plant Community and Number of Stands								
Group	Species	Ponev-Arki 7	Drde-Aske 5	Erin-Raar 5	Lepu-Hama 7	Sewa-Erni 7	Cahe-Jupa 3	Podr-Pobr 12	Sipr-Jume 6	Vani-Cani 9
1.	*Podistera nevadensis*	V.2	III.2	I.2						
	Arenaria kingii	III.2	III.2			I.+				
	Senecio werneriaefolius	III.2	III.2	I.1	II.2					
	Poa rupicola	II.3	II.1							
	Calamagrostis purpurascens	II.2	II.2		III.2					
2.	*Draba densifolia*		V.1	I.1						
	Astragalus kentrophyta var. *danaus*		V.2							
	Ivesia shockleyi		IV.2							
	Carex tahoensis		III.1							
	Androsace septentrionalis		III.1							
3.	*Raillardella argentea*			V.2						
	Eriogonum incanum			IV.3			I.R			
	Arenaria nuttallii ssp. *gracilis*			V.2	II.1					
	Eriogonum lobbii	I.R		III.2	II.1					
4.	*Leptodactylon pungens* ssp. *pulchriflorum*	I.2			IV.3					
	Haplopappus macronema				IV.3					

TABLE 18-9 [continued]

	Carex rossii				III.2				
	Ribes cereum				III.2				
5.	Polemonium pulcherrimum	I.2			III.2				
	Phacelia frigida			I.2	II.2				
	Cryptantha nubigena				II.1				
	Hieracium horridum				I.+				
6.	Selaginella watsoni	V.3	II.1	I.1	I.3	V.3			
	Eriogonum ovalifolium var. nivale	IV.2	V.3	IV.2	V.3	III.3			
	Antennaria umbrinella	IV.2	III.1	II.1	I.+	III.2			
	Castilleja nana	IV.5	IV.1	III.1	I.2	II.1			
	Sitanion hystrix	III.2	III.1	IV.2	V.2	I.+			
	Phlox covillei	III.3	V.3	III.2		II.1			
	Erigeron pygmaeus	II.2	IV.2	III.1	IV.2			I.1	
	Calyptridium umbellatum	I.2	I.+	III.2	II.R	II.2	IV.2		
7.	Draba breweri	III.1	II.1	I.+	I.+	I.+			
	Oxyria digyna	III.1		I.1	II.1	III.3		I.1	
	Ivesia muirii	II.2		II.3	I.3				
	Hulsea algida	II.1		I.+	II.2	II.2	I.+		
	Poa nervosa	II.+			III.3	II.1		I.1	I.2
	Arenaria obtusiloba	II.2	I.1			III.+	I.2		
	Haplopappus apargioides	II.2	I.2	I.1		II.2			
	Carex phaeocephala	II.2		I.3	II.1	III.3		I.3	
	Arabis lyallii	II.+	III.+	III.+		I.2			
	A. lemmonii	I.+	I.1		I.2	III.1			
	Erigeron compositus	I.2			I.+	II.1			
	Penstemon heterodoxus		III.1	III.2		III.1	IV.1		
	Silene sargentii	I.1	I.+	II.+		II.1	I.1		
	Poa "scabrella"	I.+	I.2	I.2		II.1	I.3	I.3	
	Aquilegia pubescens	I.1			II.1	I.+			
	Carex breweri			II.6	I.2	III.4	IV.5		

TABLE 18-9 [continued]

Group	Species	Ponev-Arki 7	Drde-Aske 5	Erin-Raar 5	Lepu-Hama 7	Sewa-Erni 7	Cahe-Jupa 3	Podr-Pobr 12	Sipr-Jume 6	Vani-Cani 9
		Plant Community and Number of Stands								
	Penstemon davidsonii	I.1	I.2	II.1	IV.2					
	Carex exserta	II.4	III.4	II.3		II.5		I.4		
8.	Carex helleri				I.2	I.3	IV.1			
	Juncus parryi			I.2		I.2	IV.3	I.1		I.1
	Draba lemmonii				I.1		IV.+			
	Poa suksdorfii			I.2	I.2		I.1			
	Arabis platysperma			I.3	II.1	I.+	I.1			
9.	Antennaria alpina var. media	I.2				I.2	V.4	V.4	III.4	V.3
	Carex vernacula						V.5	III.4	V.4	IV.5
	Calamagrostis breweri							V.5	V.5	IV.5
	Salix anglorum					I.4		III.4	V.5	IV.4
	Aster alpigenus ssp. andersonii							III.1		III.1
10.	Lewisia pygmaea	III.+		I.1		I.+	I.3	IV.1	I.R	III.1
	Luzula spicata	II.1		I.1	I.1	II.1	I.2	IV.1	I.2	II.1
	Potentilla drummondii			I.2		II.2	IV.2	III.2	II.1	I.1
	Luzula orestera						I.2	II.2	II.2	II.2
	Potentilla breweri	I.1						II.2	I.1	
	Erigeron petiolaris			I.+	I.+	I.1	I.2	III.1	I.2	I.R
	Carex pseudoscirpoidea							I.2		
	Poa epilis			I.3			I.3	III.3	I.1	III.2
	Phyllodoce breweri							I.+	II.+	IV.1
	Saxifraga aprica	I.+				I.+		II.1	I.3	I.1
	Gentiana newberryi							II.1		II.2
	Lewisia sierrae					I.1	I.1	I.1	I.3	I.2
	Juncus drummondii							I.1	II.1	IV.2
	Mimulus primuloides							I.3	II.3	I.2

11.	*Carex subnigricans*						I.4	II.3	II.5	
	Pedicularis attollens							I.1	II.1	I.1
	Deschampsia caespitosa							III.4	III.3	I.3
	Dodecatheon alpinum							II.2	IV.3	II.1
	Sibbaldia procumbens	III		I.2					IV.2	II.1
	Juncus mertensianus							II.3	III.3	II.+
	Carex spectabilis							I.2	I.4	III.3
12.	*Festuca brachyphylla*	V.3	V.2	II.2	IV.2	I_I.2	IV.1	II.1	I.+	
	Solidago multiradiata	III.3		II.+	I.2	III.2		III.3		I.1
	Trisetum spicatum	III.2		II.1	III.1	IV.2	I.3	III.3		III.2
	Ivesia lycopodioides	V.3	II.2	I.+		II.4	IV.2	IV.2		
	Rumex paucifolius		I.2	I.1		I.R		II.1		
	Sedum rosea	I.3		I.R				II.2		I.R
	Carex albonigra	III.2				I.3	I.1	III.3	I.2	
	Lupinus lyallii	IV.3	II.4	IV.3	II.3	II.4	I.4	III.3	I.1	II.2
13.	*Vaccinium nivictum*									III.2
	Kalmia microphylla									III.2
	Carex nigricans									III.4
	Potentilla flabellifolia									II.3

undiagnostic for the stands we have available. This kind of vegetation needs study and documentation (Ratliff 1973, 1974). Group 8 has long-lasting snow cover, often on talus so its sites are very moist, evidently followed by drought; it has a short growing season. Group 9 shifts to sod on wet sites. Group 10 is heterogeneous, wet but open, with dry spots. Group 11 separates out as very-wet-site, snow-patch plants that thrive where plants of group 10 are missing, of low cover or constancy, or both. Usually these sites dry out as snowmelt water disappears. Group 12 is found almost throughout the vegetation, in dry or wet habitats, but it does not occur in snowbank sites. Evidently its plants require good drainage. Group 13 is always wet, acid, on the peaty banks of lakes, tarns, or streams.

The nine plant communities we have extracted from our species-stand matrix are clearly insufficient in number to adequately describe all the variations in the Sierra Nevadan alpine vegetation on granite. They are, however, a start. The *Podistera nevadensis–Arenaria kingii* vegetation is species rich, with many mat plants, occupying niches with long-lasting snow or snowmelt water, not exposed to winter winds but protected by a snow cover. The *Draba densifolia–Eriogonum nivale* vegetation, on the other hand, is exposed to winter winds, is less heterogeneous, and is species poorer. The *Eriogonum incanum–Raillardella argentea* vegetation again picks up the snow accumulation of the *Podistera–Arenaria kingii* vegetation but still has open ground. It is wet in early summer, dry later, and species rich. The *Leptodactylon pungens–Haplopappus macronema* vegetation is a shrub type among projecting rocks, with forbs of group 5 in the rock crannies. More snow cover and less exposure—talus, in a word—is the habitat of the *Carex helleri–Juncus parryi* vegetation, which combines species of the *Leptodactylon-Haplopappus macronema* community with group 8 plus a selection of dry- and wet-site species (groups 7, 9, 10, and 12). The *Selaginella watsoni–Eriogonum nivale* vegetation takes us back to the dry-site division. This is species poor and site heterogeneous enough to include many plants of the wet-site groups 9, 10, and 12, plus transitional group 8 and dry-site groups 6 and 7. The *Potentilla drummondii–P. breweri* meadow vegetation is on the moist side, with a sod, mostly well drained, low in stature. Where snow accumulates or is long lasting, this vegetation is modified, usually only in constancy and abundance, to produce the *Sibbaldia procumbens–Juncus mertensianus* community, although some species of group 11 and even of group 7 drop out. The *Vaccinium nivictum–Carex nigricans* vegetation is well characterized positively by group 13 and negatively by the dropouts of group 7. It has a very distinctive habitat, although it never occupies areas larger than a very few square meters.

Carson Pass is the northern limit for many alpine plants in the Sierra, including *Epilobium latifolium, Carex pseudoscirpoidea, Phoenicaulis eurycarpa,* and *Lewisia sierrae.* The alpine vegetation is necessarily different from that of the southern Sierra studied by Pemble (1970) as a result of the differences in flora.

At Carson Pass alpine vegetation is mostly restricted to north exposures underlain by granodiorite; alpine steppe is common on all exposures where volcanic substrates predominate. Taylor (1976a) sampled 145 stands containing 297 species to characterize the timberline vegetation at Carson Pass. He related the vegetation to several complex sequences of ecological factors: (1) altitude; (2) wind exposure and snow distribution; (3) soil parent material–granodiorite, sedimentaries, andesitic volcanics, and metamorphics; (4) soil moisture availability; and (5) length of growing season. The relationships of some of the more important species in the vegetation to maximum soil water tensions and to snow depth are shown in Fig. 18-3. Table 18-10 presents a summary abstract of the alpine vegetation units. The association tables were arranged using input from species-stand matrix analyses (Ceska and Roemer 1971), reciprocal averaging, polar ordination, cluster analysis, and ecological data on the plant species and their habitats. The association tables document the vegetation. The data are objective, continua are shown, and the plant cover of the landscape is made intelligible.

In the moisture sequence, groups 1–5 and 8–10 are dry, 6 and 7 are intermediate, and 11–17 are moist to wet. Group 1 has fairly deep (<1 m) snow, although it melts quickly. Snow pack density is low because of the shrubs, which prevent normal settling. Group 2 appears very early in spring from beneath the snow. Rocks abound. Snow depth for group 3 is <0.5 m, the sites are windblown, the shrubs are low and wind trimmed, and the snow is compacted. Group 4 is still more windblown, and the snow is often scoured down to the ground. Soils are protected from wind erosion by a stone armor, a desert pavement, but in places fines are at the surface, in bits of sod formed by *Festuca brachyphylla* and *Carex rossii.* Group 5 has many dry-site, lowland sagebrush zone species with a very large range of altitudinal adaptation. The species range to the Rocky Mts. Why should these properties throw a group of species together in a plant association? The sites are rocky but have deep soil. Group

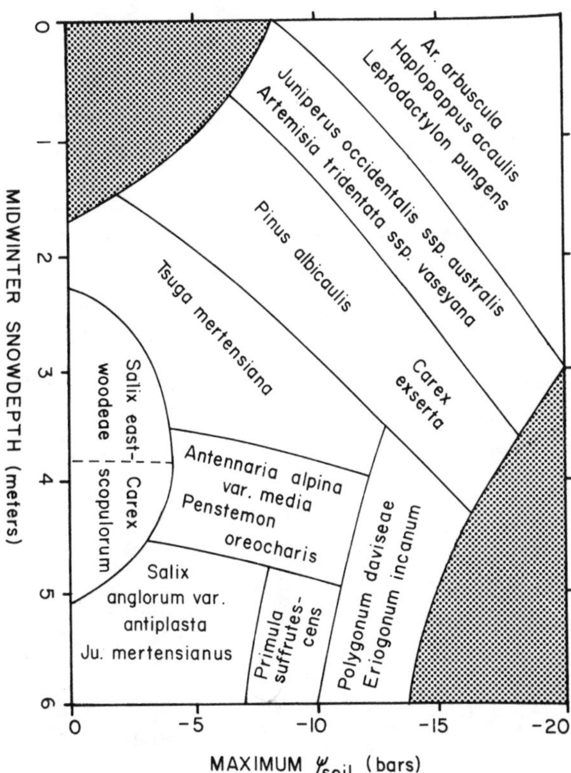

Figure 18-3. Quantitative relationships of some characteristic subalpine plant species to snow depth and matric ψ_{soil} (bar) at Carson Pass. "Maximum" here means the most negative ψ. From Taylor (1976a). Stippled areas do not exist in the landscape.

TABLE 18-10. Summary constancy-cover table. Taken from data of Taylor (1976a) for alpine vegetation at Carson Pass, Sierra Nevada. Twelve community types are recognized: Haac-Lepu = *Haplopappus acaulis*-*Leptodactylon pungens* ssp. *pulchriflorum*; Haac-Teca = *H. acaulis*-*Tetradymia canescens*; Haac-Geca = *H. acaulis*-*Geum canescens*; Caex = *Carex exserta*; Poda-Erin = *Polygonum daviseae*-*Eriogonum incanum*; Anal-Pehe = *Antennaria alpina* var. *media*-*Penstemon heterodoxus*; Prsu-Sisa = *Primula suffrutescens*-*Silene sargentii*; Pofr-Pobr = *Potentilla fruticosa*-*P. breweri*; Saea-Setr = *Salix eastwoodeae*-*Senecio triangularis*; Vani-Cabr = *Vaccinium nivictum*-*Calamagrostis breweri*; Casc = *Carex scopolurum*; Saan-Jume = *Salix anglorum* var. *antiplasta*-*Juncus mertensianus*. Values entered for each taxon's occurrence in the 12 communities are constancy (after Braun-Blanquet 1964) and modal cover (expressed on a 10-division Domin scale, after Taylor 1976a). Grp. = Group

		Community and Number of Stands											
		Haac-Lepu 15	Haac-Teca 10	Haac-Geca 3	Caex 5	Poda-Erin 9	Anal-Pehe 13	Prsu-Sisa 9	Pofr-Pobr 4	Saea-Setr 3	Vani-Cabr 3	Casc 3	Saan-Jume 7
	Taxa per Stand ($\bar{x}$):	16.5	20.4	25.0	4.7	18.1	20.2	21.4	28.3	22.3	26.3	19	19.7
	Taxa per Stand (max):	24	28	33	5	26	27	26	33	27	29	19	25
Grp.	Species												
1.	*Sitanion hystrix*	V.1	V.1	IV.1	I.R	II.+							
	Eriogonum umbellatum	III.1	V.1	V.2									
	Artemisia tridentata												
	ssp. *vaseyana*	II.+	III.1	IV.1	I.R								
	Phlox diffusa	I.+	II.+	IV.+	I.R	III.R							
	Symphoricarpos vaccinioides	I.+	V.1	II.+									
	Stipa californica	I.+	II.1	II.1	II.+	IV.+							
	Ipomopsis aggregata		I.+	II.+		III.+							
	Erysimum perenne	II.1	II.1										
	Ribes cereum	II.1	II.2										
2.	*Haplopappus acaulis*	V.2	V.2	IV.2									
	Leptodactylon pungens												
	ssp. *pulchriflorum*	V.2	V.2	IV.1									
	Penstemon speciosus	IV.1	V.1	IV.1									

	Arabis holboelii							
	var. retrofracta	IV.1	V.+	IV.+				
3.	Amelanchier utahensis		V.1					
	Tetradymia canescens		V.1					
	Purshia tridentata		III.1					
	Silene douglasii		IV.1	V.1				
	Eriogonum ovalifolium							
	var. nivale		III.+	IV.+	I.+		II.+	I.+
	Linum lewisii		III.+					
	Crepis acuminata		IV.+	III.+				
4.	Artemisia arbuscula	III.2	IV.2	V.2				
	Astragulus whitneyi	IV.1	IV.1	V.1				
	Senecio canus	II.1	V.1	I.+				
	Phoenicaulis cheiranthioides	IV.+	III.1	I.+				
	Festuca brachyphylla	III.1	V.1					
	Astragulus purshii							
	var. lectulus	III.1	V.1					
	Sedum lanceolatum	III.+	IV.1	II.+				
	Draba densifolia	III.1	IV.1					
	Carex rossii	III.1	II.+	IV.+				
	Eriogonum rosense	IV.1	IV.1					
	Ipomopsis congesta	III.2	III.1					
5.	Allium parvum	IV.+	III.R	II.R				
	Crepis modocensis							
	ssp. subacaulis	II.1	II.1	I.R				
	Penstemon deustus	I.+	II.1					
	Cryptantha humilis	II.R	III.R					
	Poa juncifolia	II.1						
	Lygodesmia spinosa	I.+	III.1					
	Poa cusickii	II.+						
	Zigadenus paniculatus	II.+	III.1					

TABLE 18-10 [continued]

Grp.	Species	Haac-Lepu 15	Haac-Teca 10	Haac-Geca 3	Caex 5	Poda-Erin 9	Anal-Pehe 13	Prsu-Sisa 9	Pofr-Pobr 4	Saea-Setr 3	Vani-Cabr 3	Casc 3	Saan-Jume 7
		Community and Number of Stands											
	Taxa per Stand ($\bar{x}$):	16.5	20.4	25.0	4.7	18.1	20.2	21.4	28.3	22.3	26.3	19	19.7
	Taxa per Stand (max):	24	28	33	5	26	27	26	33	27	29	19	25
	Haplopappus macronema	II.+	I.+										
	Lupinus caudatus	I.+	I.+										
6.	*Trisetum spicatum*			V.1									
	Agoseris glauca												
	var. *monticola*			IV.+	IV.+	IV.1	II.1	III.1					
	Erigeron barbellulatus			IV.1									
	Viola purpurea												
	ssp. *xerophyta*			IV.+									
	Geum canescens			V.1									
7.	*Carex exserta*				V.5								
	Lewisia nevadensis				II.+								
	Saxifraga aprica				III.+								
8.	*Polygonum daviseae*				II.R	V.3	IV.1	IV.+			I.R		
	Arabis platysperma												
	var. *howellii*					V.+	IV.1	IV.1					
	Eriogonum incanum					V.1	III.+	III.+					
	Poa nervosa				I.R	II.R	III.2	IV.1	I.+				
	Eriophyllum lanatum												
	var. *monoense*					III.1	II.1	I.1					
9.	*Pinus albicaulis*					IV.R		III.R	I.R				
	Haplopappus suffruticosus					IV.1							

	Castilleja nana	V.1		I.R					
	Arenaria kingii								
	var. glabrescens	III.+							
	A. nuttallii								
	ssp. gracilis		II.+						
	Raillardella argentea	III.+	II.+						
	Polygonum shastense	I.+	II.1						
	Eriogonum lobbii	I.R							
10.	Lupinus lobbii	III.1	IV.1						
	Poa rupicola		III.1						
	Carex abrupta	I.+	IV.+						
	Agropyron pringlei	II.+	II.+						
	Phacelia frigida								
	ssp. dasyphylla		II.+						
	Ivesia gordonii		II.+						
11.	Silene sargentii		I.+	IV.1					
	Primula suffrutescens		I.+	IV.3					
	Erigeron petiolaris			IV.1					
12.	Antennaria alpina								
	var. media		V.2	IV.2	V.1		V.1		V.2
	Juncus parryi		IV.1	III.1	II.1		III.1		IV.1
	Penstemon heterodoxus		IV.2	IV.1					
13.	Tsuga mertensiana		II.R	I.R			I.R		
	Pinus monticola			I.R					
	Ribes montigenum		II.1	I.R		II.1			
	Chrysopsis breweri			I.R					
	Abies magnifica		I.R	I.R					
14.	Ligusticum grayi		I.1	I.+					
	Carex spectabilis		III.2		IV.3		IV.1		II.1
	Erigeron peregrinus								
	ssp. callianthemus		II.1	I.R					

TABLE 18-10 [continued]

Grp.	Spcs.	Haac-Lepu 15	Haac-Teca 10	Haac-Geca 3	Caex 5	Poda-Erin 9	Anal-Pehe 13	Prsu-Sisa 9	Pofr-Pobe 4	Saea-Setr 3	Vani-Cabr 3	Casc 3	Saan-Jume 7
		Community and Number of Stands											
	Taxa per Stand ($\bar{x}$):	16.5	20.4	25.0	4.7	18.1	20.2	21.4	28.3	22.3	26.3	19	19.7
	Taxa per Stand (max):	24	28	33	5	26	27	26	33	27	29	19	25
15.	Potentilla breweri								V.2				
	Achillea lanulosa								III.+	II.+			
	Potentilla fruticosa								V.1				
	Gentiana amarella								III.+	I.R			
	Potentilla drummondii								II.1	II.+	II.1		
	Sedum rosea ssp. integrifolium								IV.1				
	Solidago multiradiata								IV.1				
16.	Salix eastwoodeae									V.4		II.R	IV.1
	Veratrum californicum								II.+	V.2	V.R		
	Castilleja miniata								I.+	III.1	I.+		
	Senecio triangularis									IV.2	II.1		
	Thalictrium fendleri								I.1	IV.1			
	Lupinus polyphyllus ssp. superbus									IV.2			
	Epilobium angustifolium									III.1			
17.	Carex scopulorum											V.3	
	Luzula comosa											IV.1	
	Scirpus clementis											IV.1	
	Arnica amplexicaulis											IV.+	
	Scirpus criniger											IV.1	

18.	Vaccinium nivictum					V.3		
	Hieracium gracile	III.+				V.1		
	Agrostis variabilis				I.+	IV.1		
	Botrychium simplex							
	ssp. compositum					II.+		
	Calamagrostis breweri					V.3		
19.	Veronica cusickii	I.+				V.1		
	Juncus mertensianus		I.+			IV.1	V.2	V.2
	Mimulus tilingii	I.+	I.R	II.1	II.+	I.+	IV.+	V.2
	Aster alpigenus							
	ssp. andersonii			II.1	I.1	IV.1	V.2	V.3
	Phleum alpinum			IV.1	IV.1			III.1
	Salix anglorum							
	var. antiplasta			III.1				IV.4
	Mimulus primuloides							
	var. pilosellus					IV.2	V.2	IV.1
	Potentilla flabelliformis			IV.1		IV.1	III.+	III.1
	Carex nigricans	II.+	I.+				IV.2	IV.1
	Caltha howellii			I.+	II.+	II.+	II.2	III.2
	Pedicularis attollens					II.+	I.+	III.+
	Dodecatheon alpinum							
	ssp. majus						II.+	II.+
20.	Kalmia polifolia							
	ssp. microphylla						III.2	III.1
	Pedicularis groenlandica			IV.1		II.1		III.1
	Phyllodoce breweri		I.1	I.1		III.2		III.2
	Cassiope mertensiana			I.2		III.2		II.1
	Sibbaldia procumbens	IV.1	III.1	V.2				
	Artemisia norvegica							
	ssp. saxatilis	II.1	II.1		I.1			

6 has "dry meadow" sites, with early-flowering, quickly drying plants. Group 7 is dominated by *Carex exserta,* with occasional *Saxifraga aprica* and *Lewisia nevadensis* in the sod. Sites are early snow free, often irrigated with snowmelt water for some time, but dry by late summer. Group 8 has deep snow, much winter gopher activity, considerable soil mixing, excessive drainage, and dry surface by summer, although the surrounding trees or exposure (north slopes) shade the sites much of the time. The areas may get cold air drainage. Group 9 is in sites also dry in summer, although shaded similarly to those for group 8, and the snowbanks piled up by *Pinus albicaulis* clumps protect taller shrubs like *Haplopappus suffruticosus* and shorten the growing season for them. At the same time, some areas occupied by parts of this group have shallow snow cover with *Polygonum shastense.* Group 10 sites are very rocky, dry, with shallow snow cover or at least warm exposure, no shade, and free air movement. Group 11 is very well shaded by its topographical position, wet from melting snow, very rocky; this is open vegetation. Group 12 consists of low plants in a meadow, often gravelly at the surface. Group 13 is forest clumps with fringing *Ribes montigenum.* Group 14 occupies a tall-herb snow patch with some bare soil. Group 15 has a moist, eutrophic meadow site, dry enough for *Potentilla fruticosa* but wet enough for *Sedum rosea.* Group 16 is a tall-herb, willow thicket combined with rivulets of more or less permanent water. Group 17 is sopping wet most of the year, in a bog, peaty, with plants of medium height (2–4 dm). Group 18 consists of low (<1 dm) acidicole vegetation over peat. Group 19 is in a meadow with abundant snowmelt water in early summer, drying later in places, heterogeneous. Group 20 has taller shrubs (1-2 dm), rock hugging, acidicole, on rankers, with the snow-patch plants *Pedicularis groenlandica* and *Sibbaldia procumbens.*

The above species groups combine in various ways to form 12 recognizable plant communities. The latter can be characterized by the plants that occur in them, their relative presence and abundance, and the ecological factors associated with the plant species. In Table 18-10 *Haplopappus acaulis* characterizes the first three communities. It is mat forming, of sagebrush and alpine belts, northern Great Basin to northern Rocky Mts. in distribution. With *Leptodactylon pungens* ssp. *pulchriflorum,* the lowered presence of most species of groups 1, 4, and 5, and the absence of group 3, *H. acaulis* forms a community limited to the driest, most wind-exposed sites (Fig. 18-4). Soils are rocky, derived chiefly from andesitic volcanics, and shallow. Plants are low-statured chamaephytes and mostly not evergreen. Snow cover is only intermittent in winter, and soils are subject to thaw on sunny days between storms. Drought is intense in summer but of short duration. On deeper soils, or sites where moderate snow depth (0.5–1.0 m) accumulates during most of the winter, the shrubby group 3 is added, forming the *Haplopappus acaulis–Tetradymia canescens* community. Finally, where snow accumulates to moderate depths (< 1.5 m), *Trisetum spicatum* and other dry-meadow plants of group 6 occur and characterize the *H. acaulis–Geum canescens* community. They replace the sod-formers of group 4, the shrubs of group 3, and the plants of widest altitudinal range of group 5, but this community has the highest presence of both *Artemisia tridentata* ssp. *vaseyana* and *A. arbuscula.* These two species are often mutually exclusive, on deep and shallow soils, respectively, and a measure of the heterogeneity of the *H. acaulis–G. canescens* sites is that both sagebrushes have high presence in the overall mosaic. All these communities are alpine steppe. Many disjuncts from the Great Basin are prominent (Taylor 1976c).

Figure 18-4. Windswept site at Carson Pass which is typical of many ridgetop localities in the Sierra Nevada. Snow cover is intermittent in winter, compared to less wind-exposed sites (background). The vegatation is alpine steppe of the *Haplopappus acaulis-Leptodactylon pungens* ssp. *pulchriflorum* type (Table 18-10). May 19, 1973; ca. 2900 m.

Carex exserta vegetation occurs as almost monospecific swards with a thick sod, surficial humus, very distinctively species poor. Only three rosette hemicryptophytes have substantial presence in the sod. The vegetation is patterned. Unfortunately we have no data on age changes in this *Carex* species (Rabotnov 1950, 1969). Small, open, gravelly areas have plants from nearby communities, but such areas are infrequent and have low cover. Soil water is utilized rapidly—*C. exserta* shows low stomatal resistance even at midday—until the vegetation dries.

Polygonum daviseae-Eriogonum incanum is unique in its combination of deep snow with high midsummer soil water stress (Fig. 18-2). This association also occurs in the subalpine forest in openings where snow piles in as it blows off the surrounding trees. The snow melts early, partly because the dark, surrounding trees reradiate much heat (Miller 1955). The gravelly soil drains quickly, partly because winter gopher activity is heavy in response to consistent snow cover and the abundance of tap-rooted forbs. The two species named differ in phenology. The *Polygonum* desiccates early whereas the cushions of *Eriogonum* apparently change little once they recover from their squashed and dirt-covered (from the melting snow) appearance just after snowmelt.

Antennaria alpina-Penstemon heterodoxus occupies sites where snow is often 5 m deep in spring, but the soils are too rocky to hold much available moisture, *Juncus parryi* occurs as scattered, small clumps. *Sibbaldia procumbens* of group 20 and the tall herb group, 14, are indicative of the deep snow.

Primula suffrutescens-Silene sargentii is also covered by deep snow and is typically not completely snow free until midsummer. Its soils are very rocky, with poor water-holding capacity, and become quite dry during the short growing season. Phenologically the life cycle of *Primula* is compressed into a short period of about 45 days from bud expansion to capsule dehiscence. Vegetative reproduction forms large (to 2 m diameter) clones of glowing magenta *Primula.* Adventitious roots are produced even while the plants are snow covered, while the soil temperature is very close to 0°C from percolating snowmelt water. The soil warms very late because of the persisting snowbanks and the topographically shaded sites. Chapin (1974) describes root growth in similarly cold soils at Barrow, Alaska.

Potentilla fruticosa–P. breweri forms small (a few square meters) communities composed chiefly of group 15 species but also containing plants of late-snow, tall forbs, even the clonal ericads of group 20. This association of plants is typically present on the lip of rock cliffs or steps where subsurface water appears at the surface. Evaporites are evident in late summer. *Potentilla fruticosa* [= *Dasiphora fruticosa* (L.) Rydb.] is elsewhere a calcicole (Gorchakovsky 1960; Pigott 1956; Elkington and Woodell 1963).

Salix eastwoodeae–Senecio triangularis with group 16 forms extensive thickets where seepage occurs all summer. Snow cover is moderate, and thus the growing season is relatively long. Other, similar species of willow are often present (*Salix orestera, S. lemmonii, S. planifolia*) in similar vegetation with tall herbs both within and outside the thicket, the last species in wetter places. Productivity is high, with the ample subirrigating water supply carrying mineral nutrients. Sites are warm and wind free.

Vaccinium nivictum–Calamagrostis breweri typifies mesic meadows and lake margins (Ratliff 1973), where snow melts by early July and moisture is not limiting. Soils are acid, with a thick peaty humus and an iron-eluviated A horizon. The heaths, *Phyllodoce breweri* and *Cassiope mertensiana* of group 20, are prominent here, often as dense clones excluding other species.

Carex scopulorum is confined to the wettest portions of meadow basins. Seepage is evident, and the water is cold. The soil is a thick, organic, mineral-rich muck.

In the *Salix anglorum* var. *antiplasta* (= *S. arctica* Pall.)–*Juncus mertensianus* vegetation (Fig. 18-5) the dry-site species of mineral-rich substrates of the preceding communities have dropped out. The plants of wet sites have increased in presence or abundance or both. This community occupies the areas with deepest snow, with least development of summer soil water stress, with no soil freezing, and with the shortest growing season. Its high cover seems to exclude *Sibbaldia procumbens,* although a closed-sward indicator of long-lasting snow, namely, the tall *Carex spectabilis,* is present.

Southern California

South of the Sierra Nevada only the three peaks in the San Bernardino and San Gabriel Mts. with elevations of 3050–3510 m (Horton 1960), plus Mt. San Jacinto (3290 m), have a few alpine plants (Hall 1902; Parish 1903, 1917; Munz 1974). Timberline trees are *Pinus flexilis* and *P. contorta* ssp. *murrayana.* Not only is the alpine flora poor, but also the alpine vegetation is fragmentary. Alpine, prostrate *Salix, Saxifraga,* and *Castilleja* species are lacking. Mount San Jacinto and the San Gabriels are the poorer floristically, corresponding to their lower altitudes.

Alpines of xeric habitats in the San Bernardino Mts. [asterisk indicates occurrence also on Mt. San Jacinto; dagger, occurrence on San Antonio Peak in the San Gabriel Mts. (Hanes 1976:75)] include **Selaginella watsonii, *Eriogonum umbellatum, *†E. saxatile, † E. kennedyi* ssp. *alpigenum, *Arenaria nuttallii* ssp. *gracilis, *Silene parishii, *Calyptridium monospermum* (Hinton 1975), **Ribes*

Figure 18-5. Moist meadow at Carson Pass typical of the *Salix anglorum* var. *antiplasta–Juncus mertensianus* type. *Carex nigricans* is an important species in this community, often comprising over 50% of the cover. *Antennaria alpina* var. *media-Penstemon oreocharis* vegetation occurs on the higher ground within the morainal rubble surrounding this meadow. July 19, 1973; 2931 m.

cereum, *†*Holodiscus microphyllus*, *†*Oreonana vestita*, **Monardella odoratissima*, **Carex mariposana* (which evidently includes Hall's record of *C. preslii* on Mt. San Jacinto and perhaps is *C. pachystachya*), **Sitanion hystrix*, **Stipa occidentalis*, *Arenaria rubella*, *Oxytropis oreophila*, *Androsace septentrionalis*, *Linanthus nuttallii*, *Leptodactylon pungens* ssp. *pulchriflorum*, *Phlox covillei*, *P. diffusa*, †*Hulsea vestita* ssp. *pygmaea*, *Raillardella argentea*, and *Festuca brachyphylla*.

The alpines of mesic to even hydric habitats are almost as numerous, although such habitats are uncommon: **Cryptogramma crispa* ssp. *acrostichoides*, **Oxyria digyna*, **Ranunculus eschscholtzii* ssp. *oxynotus*, **Aquilegia formosa*, **Ribes montigenum*, **Dodecatheon alpinum*, **Mimulus primuloides*, **M. tilingii*, **Muhlenbergia filiformis*, **Trisetum spicatum*, *Lewisia pygmaea*, *L. nevadensis*, *Draba corrugata*, *Heuchera alpestris*, *Potentilla wheeleri*, *Sibbaldia procumbens*, *Phyllodoce breweri*, *Gentiana amarella*, *Antennaria alpina* var. *media*, *Juncus parryi*. Many of the mesic species are subalpine, at lower altitudes than alpine, especially on Mt. San Jacinto.

Two plants of rocky but moist sites are found only on Mt. San Jacinto: *Draba corrugata* var. *saxosa* and *Heuchera hirsutissima*. That the summit of Mt. San Jacinto is alpine only in such spots as below long-lasting snowbanks is shown by the presence there of *Chrysolepis sempervirens*, *Pinus flexilis*, *P. contorta* ssp. *murrayana*, and *Pedicularis semibarbata* (Hall 1902:33). The summit rocks evidently provide the shade in which *P. semibarbata* usually grows in its forested habitats.

Hanes (1976) describes "weakly developed" alpine vegetation on Mt. San Antonio (3060 m) in the San Gabriel Mts. above the 2900 m timberline. He lists, in addition to the xerophytes noted above, *Lomatium nevadense* var. *parishii* and *Monardella cinerea*, both typically desert species (Munz 1974).

Most of the alpines and subalpines mentioned above are also of Rocky Mt. or even circumpolar distribution. The *Oxytropis* shows this in the extreme, being disjunct from Utah and northern Arizona. Others are southern Sierran: *Oreonana, Hulsea vestita* ssp. *pygmaea, Potentilla wheeleri,* and *Arenaria nuttallii* ssp. *gracilis.* Several are endemic: *Silene parishii, Eriogonum kennedyi* ssp. *alpigenum, Draba corrugata, Heuchera alpestris,* and *H. hirsutissima.* The *Calyptridum* is interesting in that *C. monospermum* is the cismontane California, lower-altitude taxon, whereas *C. umbellatum* is the Rocky Mt.–Great Basin–Sierran–Cascade, high-altitude taxon (Hinton 1975).

Again, any mountain flora is a unique combination of species, and its vegetation is therefore multidimensionally related to the vegetation of other mountainous areas.

Great Basin

Although the high mountains of the physiographical Great Basin have arctic–alpine circumpolar species and vegetation (Loope 1969; Lewis 1971), these occur mostly in the wetter sites. The zonal vegetation at alpine levels is cold steppe, its species are drawn from adjacent lowland steppe, and ecological analogies or even homologies of these alpine steppes are to be found in the mountains of middle and central Asia and Siberia (Walter 1974). Walter's discussion (1968:542 ff) of continental northern Greenland vegetation with <100 mm precipitation a year effectively points out ecological similarities with steppe vegetation (Walter 1975). Yurtsev has documented "tundra steppe" vegetation of the Chukotsk peninsula (1974).

The mountain ranges of California's part of the Great Basin are widely isolated both from each other and from the Sierra Nevada–Cascades. They differ in floras, climates, soil parent materials, relief, extent of glaciation, past livestock grazing use, extent of mining and lumbering, and so forth. Hence they are best treated separately.

Two of the westernmost of the Great Basin ranges are the White and Sweetwater Mts. in Mono Co., California. Located in the rain shadow of the Sierra, they rise to 4342 and 3557 m, respectively. Both have physiognomically similar vegetation: steppe. The alpine belt in the Sweetwater Mts. is mostly underlain by rhyolite and lacks the broad expanses of calcareous and granodiorite substrates of the White Mts. Thus calcicole vegetation occurs only in the White Mts. Dry but zonal sites in the two ranges are dominated in part by two closely related species of pulvinate *Phlox,* and they have several other taxa in common as well (Tables 18-11 and 18-12), but the composition of their vegetation differs markedly. Wet-site vegetation is characterized by *Carex subnigricans* in both ranges. These differences in vegetation must necessarily be the result of substrate, size, and elevational differences, but the proximity of the ranges to the Sierra Nevada is also a factor of importance, as it has been their main source of diaspores.

White Mountains. The White Mts. constitute a unique, high (to 4342 m), geologically diverse (Marchand 1973; LaMarche 1973) range in the rain shadow of the Sierra Nevada (Chapter 3).

Terjung et al. (1969:248) indicate that solar radiation here is "greater in intensity for this latitude than anywhere else on this planet." They collected data on the energy balance for the cloudless daytime hours of July 17, 1968, at 3580 m in the alpine. According to the results from the Oetztal (Turner 1958), radiation intensities might be greatest in early summer before snow has melted and with cumulus clouds present. Unfortunately neither the vegetation nor the soil parent material was described, but the albedo of the sod was 18–20%, of the bare soil 22%. Diffuse radiation was low, 12% of total at noon. Net radiation averaged 0.94 cal cm^{-2} min^{-1}, and its total of 550 cal cm^{-2} day^{-1} (daylight only) was partitioned into 330 cal to actual evapotranspiration (ActE) (5.5 mm water), 160 cal to heating the air (actually immediately radiated to the cold sky), and 60 cal to heating the soil. Terjung et al. describe numerous regressions between easily measured parameters such as temperatures at various levels, solar radiation, and elements of the energy balance. Using linear regressions between R_n and sod and air temperatures, one can derive $T_{sod} = 5.5T_{air} - 41.1$ in degrees centigrade. Thus, when the sod surface is freezing, the air temperature is +7.5°C, the two temperatures are equal at 9.1°, the sod heats to 40.0° when the air is 14.7°, and (beyond the limits of the data) air temperature at freezing means a sod temperature of −41.1°!.Terjung et al. emphasize and illustrate that the energy balance is a most useful conceptual scheme in describing environments.

LeDrew (1975) studied the surface energy balance at 3500 m in the alpine on Niwot Ridge on the eastern slope of the Front Range in Colorado during 2 summer months of 1973. Since his figures are for 24 hr days, it is impossible to compare his average terms of the energy balance except for ActE. Only this term may not be reversed in sign at night in his climate. The R_n of 298 cal cm^{-2} day^{-1} was apportioned as follows: to sensible heat 148, to soil flux 37, and to ActE 113 cal (1.9 mm water). He concluded that advection of cold dry air was significant. Terjung et al. assumed that daytime advection was negligible. LeDrew's values of ActE reached only some 40% of the White Mt. figure, and they are comparable to arctic tundra measurements at Barrow, Alaska.

Soviet work amplifies these climatological studies and incidentally disposes of the arctic–alpine analogy. Shikhlinskii (1966:136) gives monthly values of R_n for some Caucasian and Middle Asiatic stations. On a daily basis these can be above 300 cal cm^{-2} for high-altitude areas (Aragatz, 3229 m, 319 in July; Tien Shan Observatory, 3672 m, 296 in August; Fedchenko Glacier in the Pamir, 4169 m, 342 in August). Yearly values of R_n are about 30 kcal cm^{-2} at timberline. When only 75(60%) of this is used zonally for evapotranspiration, this amounts to less than 380 mm of water use. A summer snowbank will yield about half its depth in water. The opportunities for topoclimatic differences based on redistribution of snow water are obviously very great. Some interesting relationships between R_n, precipitation, potential evapotranspiration, actual evapotranspiration, July mean temperature, and other factors are calculated by Shikhinski. They deserve testing.

A similar picture of extreme environmental heterogeneity in the high mountains is evident from Aizenshtat's summary (1966; cf. Flohn 1974:58, who adds some additional data on heat and water balances in high mountains). Several Pamir localities have R_n values of 294–496 cal cm^{-2} day^{-1} In these alpine deserts, sensible heat may use from 83 to 54% of R_n. In one case, advected heat almost doubles the R_n amount used, for evaporation and melting of ice, to 934 cal cm^{-2} day^{-1}. In another case, steep north and south slopes in Turkestan at 3150 m have R_n's of 145 and 426 cal cm^{-2} day^{-1} with actual evapotranspiration of 19 and 47 cal cm^{-2} day^{-1} (0.03 and 0.08 mm of water a day, or 30–80 g m^{-2} day^{-1}).

A very useful plant geographical discussion of the White Mt. flora is given by Mitchell (1973).

An ecological survey of the alpine vegetation by Mooney et al. (1962) and extensive observational, experimental, and ordinational autecological work by Marchand (1973) point out the striking differences in species composition on dolomite, granitic, and sandstone substrates. Table 18-11, calculated from Mooney et al.'s data, lists the nine species with an average density greater than one plant per square meter above 3510 m elevation, as derived from a number of 25 m transects on which density was recorded by separate meters. *Eriogonum gracilipes* is exclusively on dolomite, whereas *Artemisia arbuscula, Carex eleocharis* (= *C. duriuscula* CAM according to Egorova 1966:126 ff), *Trifolium monoense,* and *Eriogonum ovalifolium* do not occur on it. The density of these nine more abundant species is greatest on granite, least on dolomite. In an example of a 25 m transect laid out across a contact between dolomite and sandstone, 16 species occurred with average numbers of 4.2 and 4.6 species per square meter (U indicates a local ubiquist; D, exclusively on dolomite; D^-, preferentially on dolomite; S, exclusively on sandstone; S^-, preferentially on sandstone; asterisk, low density and frequency): U *Arenaria kingii* ssp. *glabrescens, Koeleria cristata, Oxytropis parryi;* D *Phlox covillei, Eriogonum gracilipes, Haplopappus acaulis,* **H. apargioides,* **Castilleja nana;* D^- *Poa rupicola* (occurs around springs on granite) and *Erigeron clokeyi;* S *Carex eleocharis, Sitanion hystrix, Trifolium monoense,* **Stipa pinetorum,* **Leptodactylon pungens;* S^- *Artemisia arbuscula.*

From Marchand's work (1973) the very complex relationships between actual occurences of members of this rich flora on particular substrates, their altitudinal occurrences, their competitive relationships, their chemical compositions, and their relationships to measurable soil factors such as moisture constants, water-soluble and exchangeable cations, and pH are at once evident.

It is obvious from the available data that the exclusiveness of a taxon to a particular substrate is usually a function of the kinds of plants with which it is associated. That ecological requirements (temperature) and relative reduction of growth under competition are also effective is shown by Mooney's study (1966) of the substitution of the exclusively alpine *Erigeron pygmaeus* for the steppe to alpine *E. clokeyi* at higher altitudes in the White Mts., but more so on dolomite than on sandstone.

In the White Mts. 61 taxa occurred in 48 stands of alpine vegetation studied by Taylor. Of these, 44 species formed nine ecological–phytosociological groups (Table 18-12). Here, as elsewhere in the White Mts., the primary division of vegetation is by soil parent material: dolomite with groups 5, 6, and part of 1; granodiorite with 7, 8, 9, and part of 1; and sandstone with 2. In general, the far-ranging group 1 is held together by the occurrence of its members in the widespread *Carex albonigra–Koeleria cristata* community, but *Stipa pinetorum* and *Leptodactylon pungens* occur only on dolomitic substrates, whereas *Selaginella watsonii, Calyptridium umbellatum, Androsace septentrionalis,* and *Carex helleri* are found only on the granodiorite. Groups 3 and 4 are also widespread. The only wet-site group is 8.

The communities are easier to interpret. The *Phlox covillei-Linum lewisii* is open, with mostly cushion plants, a few bunch grasses. *Linum lewisii* was characterized by Marchand (1973) as a strict calcicole. It and *Hymenoxys cooperi* var. *canescens* are characteristic species. The *Phlox covillei–Eriogonum gracilipes* includes the calcicoles from group 1 and the more abundant of the cushion plants. The *Artemisia arbuscula* (Figure 18-6) community on sandstone has group 2 as characteristic

TABLE 18-11. Density (plants per square meter) of nine plants occurring in the White Mts. above 3510 m elevation on various substrates at various altitudinal intervals. From data of Mooney et al. (1962:264)

	Substrate and Substrate Elevation (m)							
	Granite				Sandstone		Dolomite	
Species	3510-3660	3660-3810	3810-3970	3970-4120	3510-3660	3660-3810	3510-3660	3660-3810
Eriogonum ovalifolium	0.3	1.0	1.5	1.5	0.3	0.1	0	0
Trifolium monoense	1.4	3.1	2.4	2.6	2.2	0.8	0	0
Carex eleocharis	5.2	2.8	4.0	10.3	0	3.1	0	0
Koeleria cristata	1.6	1.7	3.0	0	1.1	0.4	0.1	0
Artemisia arbuscula	0.5	0	0	0	0.4	0.7	0	0
Arenaria kingii var. *glabrescens*	0.3	0.1	0	0	0.4	0.2	0.2	0.1
Phlox covillei	0.3	0.3	0.3	0.2	1.2	0.2	1.6	2.2
Poa rupicola	0.1	0.4	0.1	0	0.2	0.3	1.0	0.2
Eriogonum gracilipes	0	0	0	0	0	0	1.2	0.9
Total	9.7	9.4	11.3	14.6	5.8	5.8	4.1	3.4

TABLE 18-12. Summary constancy-cover table for alpine vegetation in the White Mts., Mono Co., California. From unpublished work of D. W. Taylor. Elevations of stands sampled were 3470-4189 m. Seven community types are recognized: Phco-Lile = *Phlox covillei-Linum lewisii*; Phco-Ergr = *Phlox covillei-Eriogonum gracilipes*; Arab = *Artemisia arbuscula*; Caal-Kocr = *Carex albonigra-Koeleria cristata*; Ivly = *Ivesia lycopodioides* ssp. *scandalaris*; Casu-Deca = *Carex subnigricans-Deschampsia caespitosa*; Poch-Erva = *Polemonium chartaceum-Erigeron vagus*. Values entered for each taxon's occurrence in the seven communities are constancy and modal cover (after Braun-Blanquet 1964)

Group	Species	Community and Number of Stands Sampled						
		Phco-Lile 3	Phco-Ergr 4	Arab 4	Caal-Kocr 15	Ivly 8	Casu-Deca 8	Poch-Erva 6
	Taxa per Stand (mean)	12.6	11.2	13.5	13.6	13.6	10.8	4.3
	Taxa per Stand (max.)	14	13	15	17	16	16	5
1.	*Koeleria cristata*	V.+	V.1	IV.2	IV.2	IV.1	IV.+	
	Phlox covillei	V.2	V.2	IV.2	V.+	IV.1	III.+	
	Draba sierrae	II.+	V.+	II.+	V.+	IV.+	III.R	IV.+
	Penstemon heterodoxus				III.+	IV.+		
	Selaginella watsonii				III.1	II.2	I.R	
	Calyptridium umbellatum				II.+	I.+		III.R
	Androsace septentrionalis				II.+	I.+	I.+	
	Carex helleri				I.+	II.1	I.+	
	Linanthus nuttallii			I.+	I.+			
	Haplopappus suffruticosus		I.R	I.+	II.+			
	Stipa pinetorum		III.2	II.1				
	Leptodactylon pungens ssp. *pulchriflorum*		II.2	II.+				
2.	*Artemisia arbuscula*			V.3				
	Lupinus caespitosus			III.+				
	Agoseria glauca var. *monticola*			II.+				

3.	Castilleja nana	V.+	V.+		III.+	IV.+	II.+	
	Erigeron pygmaeus	V.+	V.1		IV.+	I.R	II.+	
	Potentilla pennsylvanica	IV.+	IV.+	III.+	IV.+	II.R	I.R	
4.	Haplopappus apargioides	II.+	III.+	V.+	V.+	IV.+	IV.+	
	Trifolium monoense			V.3	V.2	IV.2	III.1	II.R
	Lewisia pygmaea				IV.+	V.+	IV.1	
	Eriogonum ovalifolium var. nivale			IV.+	V.1	IV.1		
	Rumex paucifloliius ssp. gracilescens			IV.+	III.+	V.+	IV.+	
	Carex albonigra			V.1	V.3	III.1		
	Ivesia lycopodioides ssp. scandalaris			II.+	V.2	V.2		
5.	Arenaria kingii var. glabrescens	V.+	V.+	V.+				
	Eriogonum gracilipes	V.+	V.1	II.R				
	Leucopoa kingii	II.+	III.+	IV.1				
	Astragalus calycosus	II.1	IV.+					
6.	Cryptantha nana	V.+	II.R					
	Linum lewisii	V.1						
	Hymenoxys cooperi var. canescens	III.+						
7.	Carex subnigricans					V.4		
	Draba brewerii				II.+	IV.+	III.+	
	Antennaria alpina var. media		I.R		I.R	IV.2	IV.2	
	Festuca brachyphylla				I.+	IV.1	III.+	III.+
	Poa rupicola				I.+	IV.+	III.+	
	Carex phaeocephala					IV.+		
8.	Deschampsia caespitosa					IV.3		
	Luzula spicata					III.+		
	Juncus mertensianus					II.+		
9.	Sitanion hystrix				II.+	I.R		IV.+
	Polemonium chartaceum							V.+
	Erigeron vagus							V.+

Figure 18-6. *Artemisia arbuscula* dominates alpine vegetation on sandstone (foreground) in the White Mts. Sandstone soils are deeply weathered, often with a large quantity of cobbles on the surface. Dolomite in the background is vegetated by *Phlox covillei–Eriogonum gracilipes.* July 17, 1975; 3890 m.

species and lacks *Castilleja nana, Erigeron pygmaeus,* and *Astragalus calycosus* from the two *Phlox* communities and all of group 6. It has as differential species against the two *Phlox* communities *Linanthus nuttallii, Trifolium monoense, Eriogonum ovalifolium* var. *nivale, Rumex paucifolius* ssp. *gracilescens, Carex albonigra,* and *Ivesia lycopodioides* ssp. *scandalaris.* The *Carex albonigra–Koeleria cristata* has been characterized above and in addition lacks groups 5 and 6. Several species have considerable cover. The *Ivesia lycopodoides* ssp. *scandalaris* community lacks all the calcicoles and has only *Carex phaeocephala* as a differential species. *Carex subnigricans–Deschampsia caespitosa* is a hygrophile community with the *Carex* and group 8 unique to it. *Polemonium chartaceum–Erigeron vagus* has these as characteristic species but is otherwise species poor. It occurs on metasedimentary parent materials of the summit block of White Mt. Peak above 4000 m, where talus movement and solifluction are intense.

The active frost features of Pellisier Flats, located about 16 km north of the summit block, were described by Mitchell et al. (1966). Here intense frost action and solifluction have produced small, often bare-centered polygons. The authors describe the vegetation in terms of physiographic features and their associated plants. The flora is restricted to 49 vascular species, of which 23 are found only below 3960 m. Near Mt. Barcroft *Koeleria cristata, Carex albonigra,* and occasional *C. eleocharis* are the only perennial taxa which regularly vegetate centers of the most active polygons on granodiorite. *Ivesia lycopodioides* ssp. *scandalaris* often occurs on the snow-accumulation area in lee of these polygon rims. Old, inactive polygons on Mt. Barcroft and vicinity are vegetated by *Phlox covillei* and large mats of the pulvinate forb *Trifolium monoense.* Cyclic succession undoubtedly occurs (Billings and Mooney 1959). Functionally it is related to the age and activity of a given polygon. The rhizomatous *Carices* invade, and the caespitose-pulvinate species establish with time.

Sweetwater Mountains. The five groups that segregated from the matrix of 76 species by 51 stands can be characterized as follows (Table 18-13). The first contains taxa widespread in dry, wind-exposed alpine vegetation of the Sierra Nevada—the undefined "fell-field" of many western North American descriptions. The next (2) is found on unstable, alpine screes, wet below the surface and weathered from white rhyolite. Group 3 has more winter snow cover than group 1 and occurs on rockier sites, with large boulders covering much of the surface. Group 4 occupies extreme sites similar to those of groups 1 and 3 but with less plant cover. Soils here are wet in spring, and the growing season is short because of the considerable depth of snow. Drainage is good. Group 5 is unique in containing all the mesophytic plants in the entire table. This is a meadow group, where deep snow cover combines with a topographical position favoring the accumulation of soil water. The species composition is, however, heterogeneous as to ecology, with *Carex phaeocephala* and *Calamagrostis purpurescens* usually occurring on windblown, rocky, topographical eminences, and the other taxa in the wetter areas. *Sibbaldia procumbens* is a circumpolar, arctic–alpine, markedly snow-patch plant. This heterogeneity reflects the zonation surrounding these more mesic local sites, where *Carex subnigricans* dominates.

The communities that these groups characterize can be described as follows. The *Lupinus montigenus* vegetation is distinctively species poor, of low cover, open, the result of intense snowmelt-related solifluction of a sandy soil. The *Phlox pulvinata–L. montigenus* community is physiognomically dominated by cushion plants with forbs in the loose soil-scree between the cushions. The *P. pulvinata-Ivesia gordonii* association is similar but with less scree and less frost-disturbed soil, the characteristic

Figure 18-7. Alpine vegetation on dry sites in the Sweetwater Mts. Four community types are visible: (*A*) *Lupinus montigenus* here dominates a site with severe congeliturbation where wind erosion is active. underlain by andesite; (*B*) *Haplopappus suffruticosus–Ipomopsis congesta* type, underlain by andesite, with large cobbles and boulders on the surface; (*C*) *Phlox pulvinata-Ivesia gordonii* type, underlain here by granite; and (*D*) *Phlox pulvinata-Lupinus montigenus* type on slopes, underlain by chalky rhyolite. Ice axe in foreground is graduated in decimeters. July 27, 1975; 3575 m.

TABLE 18-13. Summary constancy-cover table for alpine vegetation in the Sweetwater Mts., Mono Co., California. From unpublished work of D. W. Taylor. Seven community types are recognizes: Lumo = *Lupinus montigenus*; Phpu-Lumo = *Phlox pulvinata-Lupinus montigenus*; Phpu-Ivgo = *Phlox pulvinata-Ivesia gordonii*; Phpu-Febr = *Phlox pulvinata-Festuca brachyphylla*; Phpu-Aneu = *Phlox pulvinata-Anelsonia eurycarpa*; Phpu-Hasu-Ipco = *Phlox pulvinata-Haplopappus suffruticosus-Ipomopsis congesta*; Casu-Analm = *Carex subnigricans-Antennaria alpina* var. *media*. Values entered for each taxon's occurrence in the seven communities are constancy and modal cover (after Braun-Blanquet 1964)

Group	Species (Total 43)	Community and Number of Stands Sampled						
		Lumo 7	Phpu-Lumo 9	Phpu-Ivgo 4	Phpu-Febr 7	Phpu-Aneu 5	Phpu-Hasu-Ipco 15	Casu-Analm 4
	Taxa per Stand (mean)	4.8	10.1	17.2	10.5	11.0	14.6	10.0
	Taxa per Stand (max.)	7	16	23	14	14	17	15
1.	*Sitanion hystrix*	I.+	IV.1	V.+	V.1	V.1	V.2	
	Phlox pulvinata		V.2	V.3	V.3	IV.1	III.2	
	Erysimum perenne	II.+	IV.+	IV.+	III.+	III.+	IV.+	
	Eriogonum ovalifolium ssp. *nivale*	II.R	IV.+	I.1	III.1	II.+	IV.1	
	Draba densifolia		II.+	III.+	V.1	II.+	III.+	
	Astragulus purshii var. *lectulus*		I.R	I.R	IV.+	I.R	II.+	
	Cymopteris cinerarius		I.R		II.+	III.+	II.+	
	Arabis inyoensis		I.+	IV.+	I.R	I.R	II.R	
	Agoseris glauca ssp. *monticola*	II.+	I.+	IV.+		I.+	I.+	
	Senecio werneriaefolius	I.+	I.+		I.R		II.+	
	Astragulus kentrophyta		I.1	I.+	III.+	III.+	I.+	
	Eriogonum rosense	I.+			II.+		I.+	
	Arenaria nuttallii ssp. *gracilis*		I.R	I.+			I.+	
	Ribes cereum	II.+	I.R				II.+	
	Lupinus montigenus	V.3	V.3	V.2	III.3	I.R	I.R	

	Species							
	Draba lemmonii var. incrassata	II.+	I.R	I.+				
	Calyptridium umbellatum	IV.1	III.1	IV.1	III.+			
	Senecio pattersonianus	III.2	II.+		I.+			
2.	Crepis nana					V.+		
	Anelsonia eurycarpa					V.+		
	Claytonia umbellata					IV.+		
3.	Ipomopsis congesta		III.+				IV.+	
	Haplopappus suffruticosus						IV.3	
	Leptodactylon pungens ssp. pulchriflorum						IV.1	
	Hymenoxys cooperi						IV.+	
	Erigeron pygmaeus			III.+	III.+		III.+	
	Festuca brachyphylla				III.1		III.+	
	Castilleja nana			V.1	II.+		II.+	
	Arenaria kingii ssp. glabrescens			IV.1	II.+		II.+	
4.	Ivesia gordonii			V.2				
	Androsace septentrionalis var. umbellata			IV.+				
	Eriophyllum lanatum ssp. monoense			IV.1				
	Silene sargentii			IV.1				
5.	Draba stenoloba var. nana			I.+				V.+
	Antennaria alpina var. media			I.+				V.2
	Carex subnigricans							V.5
	Rumex paucifolius			I.+				IV.+
	Carex phaeocephala			I.+				III.+
	Luzula spicata							IV.+
	Sibbaldia procumbens							IV.2
	Calamagrostis purpurascens							III.2
	Lewisia pygmaea							III.+
	Taraxacum officinale							III.+

species in the group, with *I. gordonii*, giving a unique aspect (Fig. 18-7). The *P. pulvinata–Festuca brachyphylla* has such good differential species as *Astragalus purshii* var. *lectulus, A. kentrophyta, Eriogonum rosense,* and the *Festuca* itself. The next *P. pulvinata* variant has *Anelsonia eurycarpa* and its group of unstable scree species as characteristic. The last has several half-shrubs of group 3 plus *Ribes cereum* as characteristic or differential species. This fits the designation steppe. Total plant cover is high, and many of the open-soil plants of group 1 are missing from this association. The *Carex subnigricans–Antennaria alpina* var. *media* association is unique as mentioned under group 5. It occupies snow-patch sites, depressions, and water-accumulating areas, but in the Sweetwaters its dryer margins also have bunch graminoids of more xeric sites.

The 43 plants used to characterize this Sweetwater Mt. vegetation, the "active" plants in Yurtsev's terminology (1968:154 ff), have these geographical distributions: 5 are circumpolar arctic–alpine, 17 are western North American Cordilleran, 3 are widespread in California, 2 in the Sierra-Cascades, 4 in the Sierra Nevada, 2 in the Great Basin, 6 in the western Great Basin, and 2 are endemic. These are Cordilleran–Great Basin vegetation types plus a few relationships to the immediately adjacent Californian flora.

Warner Mountains. The crest (just 3000 m on Eagle Peak) rises only to timberline on the gentle, west-facing slopes, according to U.S. Geological Survey 1:24,000 topographical maps (Warren and Eagle Peaks, 1963). Large areas above 27000 m, however, are treeless, bands of trees evidently follow the contour of exposure of particular flat-lying volcanic strata, and in east-facing cirques the alpines recorded for the range are mostly to be found.

Milligan (1968) has described the flora and vegetation on a transect across the range through Eagle Peak. He mentions the subalpine forest from 2600 m to the summit. It includes such alpines as *Pinus albicaulis, Arabis lemmonii, Calyptridium umbellatum,* and *Sitanion hystrix* and such subalpines as the pine, *Descurainia richardsonii, Oenothera heteranthera, Penstemon gracilentus, Phleum alpinum, Poa nevadensis, Ranunculus occidentalis* var. *dissectus,* and *Trisetum wolfii.* Other plants in nonforested, high-altitude vegetation in Milligan's flora include *Arenaria congesta* var. *subcongesta, Silene douglasii, Artemisia tridentata* (ssp.?), *Wyethia mollis, Erysimum perenne, Streptanthus cordatus, Linum lewisii* (= *L. perenne* ssp. *lewisii*), *Phlox diffusa, Eriogonum heracleoides, E. spergulinum* var. *reddingianum, E. umbellatum* var. *covillei, Geum ciliatum, Ivesia gordonii, Sibbaldia procumbens,* and *Trisetum spicatum.* Only the last two are mesic and circumpolar in distribution. The rest are part of the widespread, upper-elevation *Artemisia tridentata–Wyethia mollis* vegetation occuring in nonforested, rocky or gravelly, probably once heavily overgrazed, xeric sites (with snow mostly blown off in winter?) Note the abundance of typically Great Basin to Rocky Mt., lowland sagebrush belt plants in this list. This is an alpine steppe.

Using Munz and Keck's flora (1959), we could glean other alpines. The first 20% of the book gave this selection: *Ranunculus eschscholtzii, Thlaspi montanum* L. var. *montanum* P. Holmgren (1971), *Draba stenoloba, D. densifolia, Arabis lyallii, Sagina saginoides, Lewisia nevadensis, Claytonia bellidifolia* (= *C. megarhiza* (Gray) Parry var. *b.* (Rydb.) Hitchc. These are without exception Rocky Mt. plants—a clear indication of the floristic and therefore probably also the vegetational relationships of the alpine areas of the Warner Mts.

A difficulty with the last procedure can be illustrated by *Oxyria.* It probably occurs in the Warner Mts., but it is so widespread in the alpine areas of California, even occurring on Preston Peak (cf. Klamath Mts.) (Ground 1972:88), that Munz and Keck's flora does not list it specifically for Modoc Co. Our methods of describing the distribution of common plants are unsatisfactory.

When the vegetation of the Warner Mts. is studied, Faegri's paper (1966) on the nearby Steens Mts. of Oregon will provide a stimulus.

LITERATURE CITED

Åberg, B. 1952. Kärlväxternas höjdgränser i Lule Lappmark och i Graubünden, en jämförelse. Svensk Bot. Tidskr. 46:286-312.

Adam, D. P. 1967. Late Pleistocene and recent palynology in the Central Sierra Nevada, California, pp. 275-301. E. J. Cushing and H. E. Wright (eds.). Proc. 7th Internat. Quat. Assoc. Vol. 7: Quaternary paleoecology. Yale Univ. Press, New Haven, Conn.

Aizenshtat, B. A. 1966. [Investigations of the heat balance in Middle Asia], pp. 94-129. In M. I. Budyko (ed.) [Contemporary problems of climatology.] Gidromet. Izd., Leningrad.

Argus, G. W. 1973. The genus Salix in Alaska and the Yukon. Nat. Mus. Canada Pub. Bot. 2. 279 p.

Backhuys, W. 1968. Der Elevations-Effekt bei einigen Alpenpflanzen der Schweiz. Blumea 16:273-320.

Baig, M. N., W. Tranquillini, and W. M. Havranek. 1974. Cuticuläre Transpiration von Picea-abies- und Pinus-cembra-Zweigen aus verschiedener Seehöhe und ihre Bedeutung für die winterliche Austrocknung der Bäume an der alpinen Waldgrenze. Centralbl. Ges. Forstwesen 91:195-211.

Baker, H. G. 1972. A fen on the northern California Coast. Madroño 21:405-416.

Bauer, H., W. Larcher, and R. B. Walker. 1975. Influence of temperature stress on CO_2-gas exchange, pp. 557-586. In J. P. Cooper (ed.) Photosynthesis and productivity in different environments. Cambridge Univ. Press.

Beguin, C., and J. Major. 1975. Contribution a l'etude phytosociologique et ecologique des marais de la Sierra Nevada (Californie). Phytocoenologia 2:349-367.

Benedict, J. B. 1970. Downslope soil movement in a Colorado alpine region: rates, processes, and climatic significance. Arctic Alpine Res. 2:165-226.

Bennett, P. S. 1965. An investigation of the impact of grazing on 10 meadows in Sequoia and Kings Canyon National Parks. M.A. thesis, San Jose State Univ., San Jose, Calif. 164 p.

Billings, W. D. 1975. Arctic and alpine vegetation: plant adaptations to cold s-mmer climates, pp. 403-443. In J. D. Ives and R. G. Barry (eds.). Arctic and alpine environments. Methuen, London.

Billings, W. D., and H. A. Mooney. 1959. An apparent frost hummock-sorted polygon cycle in the alpine tundra of Wyoming. Ecology 40:16-20.

-----. 1968. The ecology of arctic and alpine plants. Biol. Rev. 43:481-529.

Bliss, L. C. 1963. Alpine plant communities of the Presidential Range, New Hampshire. Ecology 44:678-697.

-----. 1971. Arctic and alpine plant life cycles. Ann. Rev. Ecol. Syst. 2:405-438.

Bolkhovskikh, Z. V., V. G. Grif, O. I. Zakharieva, and T. S. Matveeva. 1969. [Chromosome numbers of flowering plants.] Nauka, Leningrad. 928 p.

Braun-Blanquet, J. 1964. Pflanzensoziologie. 3rd ed. Springer, Berlin. 868 p.

Brink, V. C. 1959. A directional change in the subalpine forest-heath ecotone in Garibaldi Park, British Columbia. Ecology 40: 10-16.

Cadbury, D. A., J. G. Hawkes, and R. C. Readett. 1971. A computer-mapped flora: A study of the county of Warwickshire. Academic Press, London. 768 p.

Ceska, A., and H. Roemer. 1971. A computer program for identifying species-relevé groups in vegetation studies. Vegetatio 23: 255-277.

Chabot, B. F., and W. D. Billings. 1972. Origins and ecology of the Sierran alpine flora and vegetation. Ecol. Monogr. 42:163-199.

Chapin, F. S., III. 1974. Morphological and physiological mechanisms of temperature compensation in phosphate absorption along a latitudinal gradient. Ecology 55:1180-1198.

Clausen, J. 1965. Population studies of alpine and subalpine races of conifers and willows in the California high Sierra Nevada. Evolution 19:52-68.

Clokey, I. W. 1951. Flora of the Charleston Mountains, Clark County, Nevada. Univ. Calif. Publ. Bot. 24:1-274.

Cooke, W. B. 1940. Flora of Mount Shasta. Amer. Midl. Natur. 23: 497-572. 1941. First su-plement to the flora of Mt. Shasta. Amer. Midl. Natur. 26:74-84. 1949. Second supplement to the flora of Mt. Shasta. Amer. Midl. Natur. 41:174-183. 1963. Third supplement to the flora of Mt. Shasta. Amer. Midl. Natur. 70:386-395.

-----. 1955. Fungi of Mount Shasta (1936-1951). Sydowia 9:94-215.

Crocker, R. L. 1952. Soil genesis and the pedogenic factors. Quart. Rev. Biol. 27:139-168.

Daubenmire, R. F. 1941. Some ecologic features of the subterranean organs of alpine plants. Ecology 22:370-378.

Egorova, T. V. 1966. [Sedges of the U.S.S.R., species of the subgenus Vignea.] Nauka, Leningrad. 266 p.

Ehleringer, J. R., and P. C. Miller. 1975. Water relations of selected plant species in the alpine tundra. Ecology 56:370-380.

Ehrendorfer, F. 1962. Cytotzxonomische Beiträge zur Genese der mitteleuropäischen Flora und Vegetation. -Ber. Deutsch. Bot. Ges. 85:137-152.

Elkington, T. T., and S. R. J. Woodell. 1963. Biological flora of the British Isles. *Potentilla fruticosa* L. J. Ecol. 51:769-781.

Ellenberg, H. 1963. Vegetation Mitteleuropas mit den Alpen. Vol. IV, Part 2; H. Walter's Einfuhrung in die Phytologie. Ulmer, Stuttgart. 945 p.

Faegri, K. 1966. A botanical excursion to Steens Mt., southeastern Oregon. Blyttia 24:177-181.

Ferlatte, W. J. 1974. A flora of the Trinity Alps of northern California. Univ. Calif. Press, Berkeley. 206 p.

Flohn, H. 1974. Contribution to a comparative meteorology of mountain areas, pp. 55-71. In J. D. Ives and R. G. Barry (eds.). Arctic and Alpine environments. Methuen, London.

Forstlichen Bundes-Versuchanstalt Mariabrunn, Mitteilungen 59 and 60. 1963. Oekologische Untersuchungen in der subalpinen Stufe. Parts I and II. 887 p. Forschungs. für Lawinenvorbeugung, Innsbruck.

Franklin, J. F., and C. T. Dyrness. 1973. Natural vegetation of Oregon and Washington. USDA Forest Ser., Pac. NW Forest and Range Exp. Sta., Gen. Tech. Rept. PNW-8. Portland, Ore. 417 p.

Frenkel, R. E., and C. M. Harrison. 1974. An assessment of the usefulness of phytosociological and numerical classification methods for the community biogeographer. J. Biogeogr. 1:27-56.

Gaumann, E., C. Roth, and J. Anliker. 1934. Ueber die Biologie der Herpotrichia nigra Hartig. Z. Pflanzenkr. 44:97-116.

Geiger, R. 1969. Topoclimates, pp. 105-138. In H. Flohn (ed.). World survey of climatology. Vol. 2: General climatology. Elsevier, Amsterdam.

Gillett, G. W., J. T. Howell, and H. Leschke. 1961. A flora of Lassen Volcanic National Park, California. Wasmann J. Biol. 19:1-185.

Gjaerevoll, O. 1963. Survival of plants on nunataks in Norway during the Pleistocene glaciation, pp. 261-283. In A. and D. Love (eds). North Atlantic biota and their history. Pergamon, London.

Golubets, M. A. 1974. [On the position of the high mountains in the system of regionalization of a mountainous country.] In A. I. Tolmachev (ed.) [Plant world of the high mountains and its mastery.] Probl. Bot. 12:123-127. Nauka, Leningrad.

Gorchakovsky, P. L. 1960. [On the distribution and growing conditions of shrubby cinquefoil (Dasiphora fruticosa (L.) Rydb.) in relation to the relict nature of its Ural stations.] Zap. Sverdlovsk. Otd. Vsesoyuz. Bot. Ob.-va 1:3-22.

-----. 1966. [Flora and vegetation of the high mountains of the Urals.] Ural. Filial, Akad. Nauk USSR, Trudy Inst. Biol. 48, Sverdloxk. 272 p.

-----. 1975. [The plant world of the high-mountain Urals.] Nauka, Moscow. 284 p.

Grebenshchikov, O. S. 1974. [On vegetation belts in the mountains of the Mediterranean region between 35 and 40°N latitude.] In A. I. Tolmachev (ed.) [Plant world of the high mountains and its mastery.] Probl. Bot. 12:128-134. Nauka, Leningrad.

Gribbon, P. W. F. 1968a. Altitudinal zonations in East Greenland. Bot. Tidsskr. 63(4):342-357.

-----. 1968b. Distributional-type spectra in Greenland montaine localities. Bot. Notiser 121:501-504.

Griffin, J. R., and W. B. Critchfield. 1972. The distribution of forest trees in California. USDA Forest Serv. Pac. SW Forest and Range Exp. Sta., Res. Paper PSW-82. 114 p.

Ground, C. A. 1972. A study of the flora of Preston Peak, Siskiyou County, California. M.A. thesis, Pacific Union Coll., Angwin, Calif. 180 p.

Guinochet, M., and R. de Vilmorin. 1973. Flore de France. Vol. 1. Centre Nat. Rech. Scien., Paris. 366 p.

Hall, H. M. 1902. A botanical survey of San Jacinto Mt. Univ. Calif. Publ Bot. 1:1-140.

Hanes, T. L. 1976. Vegetation types of the San Gabriel Mountains, pp. 65-76. In J. Latting (ed.) Plant communities of southern California. Calif. Native Plant Soc. Spec. Pub. 2.

Hansen, D. H., and L. G. Klifoff. 1972. Water stress in krummholz, Wasatch Mts., Utah. Bot. Gaz. 133(4):392-394.

Harshberger, J. W. 1911. Phytogeographic Survey of North America. W. Engelmann, Leipzig. 790 p.

Havranek, W. 1972. Ueber die Bedeutung der Bodentemperatur für Photosynthese und Transpiration junger Forstpflanzen und für die Stoffproduktion an der Waldgrenze. Angew. Botanik 46:101-116.

Hermes, K. 1955. Die Lage der oberen Waldgrenze in den Gebirgen der Erde und ihr Abstand zur Schneegrenze. Kolner Geogr. Arb. 5:1-277.

Hinton, W. F. 1975. Systematics of the Calyptridium umbellatum complex (Portulacaceae). Brittonia 27:197-208.

Hitchcock, C. L., A. Cronquist, and M. Ownbey. 1969. Vascular plants of the Pacific Northwest. Vol. 1: Vascular cryptogams, gymnosperms, and monocotyledons. Univ. Wash. Press, Seattle. 914 p.

Holch, A. E., E. W. Hertel, W. O. Oakes, and H. H. Whitwell. 1941. Root habits of certain plants in the foothill and alpine belts of Rocky Mtn. National Park. Ecol. Monogr. 11(2):327-345.

Holmgren, P. K. 1971. A biosystematic study of North American Thlaspi montanum and its allies. Mem. N. Y. Bot. Gard. 21(2): 1-106.

Horton, J. S. 1960. Vegetation types of the San Bernardino Mountains, California. USDA Forest Serv. Pac. SW forest and Range Exp. Sta., Tech. Paper 44:1-29.

Howell, J. T. 1944. Certain plants of the Marble Mts. in California with remarks on the boreal flora of the Klamath area. Wasmann Coll. 6:13-19.

-----. 1951. The arctic-alpine flora of three peaks in the Sierra Nevada. Leafl. West. Bot. 6(7):141-156.

-----. 1952. Mineral King and some of its plants. Leafl. West. Bot. 6(11):212-219.

Hultén, E. 1937. Outline of the history of arctic and boreal biota during the Quaternary period. Aktiebolaget Thule, Stockholm. 168 p.

Ivimey-Cook, R. B., and M. C. F. Proctor. 1966. The application of association-analysis to phytosociology. J. Ecol. 54(1):179-192.

Ives, J. D. 1974. Biological refugia and the nunatak hypothesis, pp. 605-636. In J. D. Ives and R. G. Barry (eds.) Arctic and alpine environments. Methuen, London.

Jenny, H. 1941. Factors of soil formation. McGraw-Hill, New York. 281 p.

Johnson, D. A. and M. M. Caldwell. 1975. Gas exchange of 4 arctic and alpine tundra plant species in relation to atmospheric and soil moisture stress. Oecologia 21:93-108.

Jones, G. N. 1936. A botanical survey of the Olympic Peninsula, Washington. Univ. Wash. Pub. Biol. 5:1-286.

Klikoff, L. G. 1965a. Microenvironmental influence on vegetational pattern near timberline in the central Sierra Nevada. Ecol. Monogr. 35:187-211.

-----. 1965b. Photosynthetic response to temperature and moisture stress of 3 timberline meadow species. Ecology 46(4):516-517.

Knapp, R. 1957. Probleme und besondere Eigenschaften der Hochgebrigs-Vegetation des westlichen Nordamerika. Ber. Oberhess. Ges. Natur- und Heilkunde N.F. Naturwiss. Abt. 28:128-132.

-----. 1960. Die Bedeutung der Dauer der Schneebedeckung fur die Vegetation in subalpinen Lagen. Ber. Bayer. Bot. Ges. 33:89-93.

-----. 1961. Bericht über Studienreisen in die Hochgebirge Nord-amerikas. Naturwiss. Ver. Darmstadt Ver. 1960:13-15.

-----. 1965. Die Vegetation von Nord- und Mittelamerika. Fischer, Stuttgart. 373 p.

Kuvaev, V. G. 1972. [Some regularities in the altitudinal distribution of plants.] Bot. Zhur. 57(9):1108-1115.

LaMarche, V. C. 1973. Guide to the geology, pp. 36-47. In R. M. Lloyd and R. S. Mitchell. A flora of the White Mountains, California and Nevada. Univ. Calif. Press, Berkeley.

Larcher, W., A. Cernusca, and L. Schmidt. 1973. Stoffproduktion und Energiebilanz in Zwergstrauchbeständen auf dem Patscherkofel bei Innsbruck, pp. 175-194. In H. Ellenberg (ed.). Oekosystemforschung. Springer, Berlin.

Larcher, W. 1957. Frosttrocknis an der Waldgrenze und in der alpinen Zwergstrauchheide. Veroff. Mus. Ferdinandeum Innsbruck 37:49-81

-----. 1963a. Zur spätwinterlichen Erschwerung der Wasserbilanz von Holzpflanzen an der Waldgrenze. Ber. naturwiss. -med. Ver. Innsbruck 53:125-137.

-----. 1963b. Die Leistungsfahigkeit der CO_2 -Assimilation höherer Pflanzen unter Laboratoriumsbedingungen und am natürlichen Standort. Mitt. Floristischsoziol. Arbeits. N.F. 10:20-33 + tables.

-----. 1972. Der Wasserhaushalt immergruner Pflanzen im Winter. Ber. Deutsch. Bot. Ges. 85(7/9):315-327.

LeDrew, E. 1975. The energy balance of a mid-latitude alpine site during the growing season, 1973. Arctic Alpine Res. 7(4):301-314.

Lewis, M. E. 1971. Flora and major plant communities of the Ruby-East Humboldt Mts., with special emphasis on Lamoille Canyon. USDA Forest Serv., Ogden, Utah. 62 p.

Lindsay, J. H. 1971. Annual cycle of leaf water potential in *Picea engelmannii* and *Abies lasiocarpa* at timberline in Wyoming. Arctic Alpine Res. 3(2):131-138.

Lloyd, R. M., and R. S. Mitchell. 1973. A flora of the White Mountains, California and Nevada. Univ. Calif. Press, Berkeley. 202 p.

Loope, L. L. 1969. Subalpine and alpine vegetation of northeastern Nevada. Ph.D. thesis, Duke University, Durham, N. C. 292 p.

Major, J. 1951. A functional, factorial approach to plant ecology. Ecology 32:392-412.

Major, J., and S. A. Bamberg. 1967a. Some Cordilleran plants disjunct in the Sierra Nevada of California and their bearing on Pleistocene ecological conditions, pp. 171-188. In H. E. Wright and W. H. Osburn (eds.) Arctic and alpine environments. Indiana Univ. Press, Bloomington.

-----. 1967b. Comparison of some North American and Eurasian alpine ecosystems, pp. 89-118. In H. E. Wright and W. H. Osburn (eds.) Arctic and alpine environments. Indiana Univ. Press, Bloomington.

Malyshev. L. I. 1965. [The high mountain flora of the eastern Sayan.] Akad, Nauk USSR, Sibir. Otd., Vostochno-Sibir. Biol. Inst. Nauka, Leningrad. 368 p.

Marchand, D. E. 1973. Edaphic control of plant distribution in the White Mountains, eastern California. Ecology 54(2):233-250.

Matthes, L. J. 1942. The plant communities of Mount Shasta and surrounding valleys. M.S. thesis, Univ. Minnesota, Minn. 153 p.

Merxmuller, H. 1952-54. Untersuchungen zur Sippengliederung und Arealbildung in den Alpen. Jahrb. Verein zum Schutz Alpenpflanzen und -Tiere 17:96-133; 18:135-158; 19:97-139.

Merxmuller, H., and J. Poelt. 1954. Beitrage zur florengeschichte der Alpen. Ber. Bayer. Bot. Ges. 30:91-101.

Michaelis, P. 1932. Oekologische Studien an der alpinen Baumgrenze. I. Das Klima und die Temperaturverhältnesse der Vegetationsorgane im Hochwinter. Ber. Deutsch. Bot. Ges. 50:31-42. 1934. II. Die Schichtung der Windgeschwindigkeit, Lufttemperatur and Evaporation über einer Schneefläche. Beih. Bot. Centralbl. 52B:310-332.

Michaelis, G., and P. Michaelis. 1934. Oekologische Studien an der alpinen Baumgrenze. III. Ueber die winterlichen Temperatur der pflanzlichen Organe, insbesondere der Fichte. Beih. Bot. Centralbl. 52B:333-377.

Michaelis, P. 1934. Oekologische Studien an der alpinen Baumgrenze. IV. Zur kenntnis des winterlichen Wasserhaushaltes. Jahrb. wiss. Bot. 80:169-247. V. Osmotischer Wert und Wassergehalt der Fichte während des Winters in den verschiedenen Höhenlagen. Jahrb. wiss. Bot. 80:337-362.

Miller, D. H. 1955. Snow cover and climate in the Sierra Nevada of California. Univ. Calif. Pub. Geogr. 11:218 p.

Milligan, M. T. 1969. Transect flora of the Eagle Peak region, Warner Mts. Modoc County. M.S. thesis, Humboldt State Univ., Arcata, Calif. 187 p.

Mitchell, R. S. 1973. Phytogeography and comparative floristics, pp. 17-36. In R. M. Lloyd and R. S. Mitchell. A flora of the Whtie Mountains, California and Nevada. Univ. Calif. Press, Berkeley.

Mitchell, R. S., V. C. LaMarche, and R. M. Lloyd. 1966. Alpine vegetation and active frost features of Pellisier Flats, White Mts., California. Amer. Midl. Natur. 75:516-525.

Mooney, H. A. 1966. Influence of soil type on the distribution of two closely related species of Erigeron. Ecology 47:950-958.

Mooney, H. A., and W. D. Billings. 1961. Comparative physiological ecology of arctic and alpine populations of Oxyria digyna. Ecol. Monogr. 31:1-29.

Mooney, H. A., W. D. Billings, and R. D. Hillier. 1965. Transpiration rates of alpine plants in the Sierra Nevada of California. Amer. Midl. Natur. 74:374-386.

Mooney, H. A., G. St. Andre, and R. D. Wright. 1962. Alpine and subalpine vegetation patterns in the White Mts. of California. Amer. Midl. Natur. 68:257-273.

Mooney, H. A., R. D. Wright, and B. R. Strain. 1964. The gas exchange capacity of plants in relation to vegetation zonation in the White Mountains of California. Amer. Midl. Natur. 72:281-297.

Moore, J. J., S. J. Fitsimmons, E. Lambe, and J. White. 1970. A comparison and evaluation of some phytosociological techniques. Vegetatio 20:1-20.

Moser, W. 1973. Licht, Temperatur und Photosynthese an der Station "Hoher Nebelkogel" (3184 m), pp. 203-223. In H. Ellenberg (ed.). Oekosystemforschung. Springer, Berlin.

Munz, P. A. 1935. A manual of southern California botany. Claremont Coll., Claremont, Calif. 642 p.

-----. 1974. A flora of southern California. Univ. Calif. Press, Berkeley. 1086 p.

Munz, P. A., and D. D. Keck. 1949. California plant communities. Aliso 2:87-105. 1950. Suppl. Aliso 2:199-202.

Muth, G. I. 1967. A flora of the Marble Mountain Wilderness area, Siskiyou County, California. M.A. thesis, Pacific Union Coll., Angwin, Calif.

Olmsted, I. C. 1976. Environmental factors influencing the distribution of two species of Phyllodoce in the mountains of western North America. Ph.D. thesis, Duke University, Durham, N. C. 176 p.

Parish, S. G. 1903. A sketch of the flora of southern California. Bot. Gaz. 36:203-222, 259-279.

-----. 1917. An enumeration of the pteriodophytes and spermatophytes of the San Bernardino Mts., California. Plant World 20:163-178, 208-223, 245-259.

Pemble, R. H. 1970. Alpine vegetation in the Sierra Nevada of California as lithosequences and in relation to local site factors. Ph.D. thesis, Univ. California, Davis. 311 p.

Pigott, C. D. 1956. The vegetation of Upper Teesdale in the North Pennines. J. Ecol. 44(2)545-586.

Pisek, A. 1971. Zur Geschichte der experimentellen Oekologie (besonders des in Innsbruck hierzu geleisteten Beitrages.) Ber. Deutsch. Bot. Ges. 84(7/8):365-379.

Rabotnov, T. A. 1950. [The life cycle of perennial herbs in meadow coenoses.] Trudy Bot. Inst. V.L. Komarov, Akad. Nauk USSR, Ser. 3, Geobotanika 6:7-204.

-----. 1969. Coenopopulations of perennial herbaceous plants in natural coenoses. Vegetatio 19(1-6):87-95.

Rae, S. P. 1970. Studies in the ecology of Mason Bog, Sagehen Creek, California. M.S. thesis, Univ. Calif., Davis. 106 p.

Ratliff, R. D. 1973. Short-hair meadows in the high Sierra Nevada. USDA Forest Serv. Pac. SW Forest and Range Exp. Sta. Res. Note PSW-281. 4 p.

-----. 1974. Short-hair sedge...its condition in the high Sierra Nevada of California. USDA Forest Serv. Pac. SW Forest and Range Exp. Sta. Res. Note PSW-293. 5 p.

Raven, P. H. 1958. Plants from Mt. Whitney, California. Leafl. West. Bot. 8(10):248.

Reisigl, H., and H. Pitschmann. 1958. Obere Grenzen von Flora und Vegetation in der Nivalstufe der zentralen Oetztaler Alpen (Tirol). Vegetatio 8:93-128.

-----. 1959. Zur Abgrenzung der Nivalstufe. Phyton 8:219-224.

Rollins, R. C., and E. A. Shaw. 1973. The genus Lesquerella (Cruciferae) in North America. Harvard Univ. Press, Cambridge, Mass. 287 p.

Savage, W. 1973. Annotated check-list of vascular plants of Sagehen Creek drainage basin, Nevada County, California. Madroño 22(3): 115-139.

Schröter, C. 1926. Das Pflanzenleben der Alpen. 2nd ed. Raustein, Zurich. 1288 p.

Schulze, E. D., H. A. Mooney, and E. L. Dunn. 1967. Wintertime photosynthesis of bristlecone pine (Pinus aristata) in the White Mountains of California. Ecology 48(6):1044-1047.

Schwarz, W. 1971. Das Photosynthesevermögen einiger Immergrüner während des Winters und seine Reaktivierungsgeschwindigkeit nach scharfen Frosten. Ber. Deutsch. Bot. Ges. 84(10):585-594.

Sharsmith, C. W. 1940. A contribution to the history of the alpine flora of the Sierra Nevada. Ph.D. thesis, Univ. Calif., Berkeley. 273 p.

Shaver, G. R., and W. D. Billings. 1975. Root production and root turnover in a wet tundra ecosystem, Barrow, Alaska. Ecology 56 (2):401-409.

Shikhlinskii, E. M. 1966. [On the heat balance of the Caucasus], pp. 130-146. In M. I. Budyko (ed.) [Contemporary problems of climatology.] Gidromet. Izd., Leningrad.

Shiyatov, S. G. 1970. [On the types of upper limits of the forest and their dynamics in the polar Urals], pp. 73-81. In B. A. Tikhomirov (ed.). [Biological basis of use of northern nature.] Komi Kniga Izd., Syktyvkar.

Schmidt, U. M. 1975. [The comparison of systematic structures of floras of Hokkaido, Sakhalin, Kamchatka, Kurile, Commander, and Aleutian Islands.] Bot. Zhur. 60(9):1225-1236.

Smiley, F. J. 1921. A report on the boreal flora of the Sierra Nevada of California. Univ. Calif. Pub. Bot. 9:1-425.

-----. 1915. The alpine and subalpine vegetation of the Lake Tahoe region. Bot. Gaz. 59:265-286.

Sokhadze, E. V. 1974. [Botanical geographical comparison of the high mountain karst of the Colchis with its analogues.] In A. I. Tolmachev (ed.) [Plant world of the high mountains and its mastery.] Probl. Bot. 12:205-213. Nauka, Leningrad.

Spatz, G., and J. Siegmund. 1973. Eine Methode zur tabellarischen Ordination, Klassifikation und ökologischen Auswertung von pflanzensoziogischen Bestandsaufnahmen. Vegetatio 28:1-17.

Stanyukovich, K. V. 1973. [Vegetation of the mountains of the USSR.] Donish, Dushanbe. 415 p.
Stebbins, G. L., and J. Major. 1965. Endemism and speciation in the California flora. Ecol. Monogr. 35:1-35.
Taylor, D. W. 1976a. Ecology of the timberline vegetation at Carson Pass, Alpine County, California. Ph.D. thesis, Univ. Calif., Davis.
-----. 1976b. Floristic relationships along the Cascade-Sierran axis. Amer. Midl. Natur. (in press).
-----. 1976c. Disjunction of Great Basin plants in the northern Sierra Nevada. Madroño 23(6):301-310.
Terjung, W. H., R. N. Kiekert, G. L. Potter, and S. W. Swarts. 1969. Energy and moisture balance of an alpine tundra in mid-July. Arctic Alpine Res. 1(4)247-266.
Tolmachev, A. I. 1948. [Basic paths of formation of the vegetation of the high mountain landscapes of the northern hemisphere.] Bot. Zhur. 33(2):161-180.
Tranquillini, W. 1957. Standortsklima, Wasserbilanz und CO_2-Gaswechsel junger Zirben (*Pinus cembra* L.) an der alpinen Waldgrenze. Planta 49(6):612-661.
-----. 1963. Climate and water relation of plants in the sub-alpine region, pp. 153-167. In The water relations of plants. Symp. British Ecol. Soc. London, April 5-8, 1961. Blackwell, London.
-----. 1967. Ueber die physiologischen Ursachen der Wald- und Baumgrenze. Mitt. Forstl. Versuch. Mariabrunn, Wien 75:457-487.
-----. 1974. Der Einfluss von Seehöhe und Länge der Vegetationszeit auf das cuticulare Transpirationsvermögen von Fichtensämlingen im Winter. Ber. Deutsch. Bot. Ges. 87(1):175-184.
Turner, H. 1958. Ueber das Licht- und Strahlungsklima einer Hanglage der Oetztaler Alpen bei Obergurgl und seine Auswirkung auf das Mikroklima und auf die Vegetation. Arch. Met. Geoph. Biokl. Ser. B, 8:273-325.
Turner, H. 1968. Ueber "Schneeschliff" in den Alpen. Das Problem der Wirkungen des winterlichen Schnee- und Eisgebläse auf dem forstlichen Jungwuches an der oberen Waldgrenze. Wetter und Leben 20(11):192-200.
Walter, H. 1960. Standortslehre. 2nd ed. Ulmer, Stuttgart. 566 p.
-----. 1968. Die Vegetation der Erde in ökophysiologischer Betrachtung. Vol. 2; Die gemässigten und arktischen Zonen. Gustav Fischer, Stuttgart. 1001 p.
-----. 1971-72. Oekologische Verhältnisse und Vegetationstypen in der Intermontanen Region des westlichen Nordamerikas. Verh. Zool. -Bot. Ges. in Wien 110/111:111-123.
-----. 1973a. Vegetation of the earth. Springer, New York. 237 p.
-----. 1973b. Allgemeine Geobotanik. Uni-Taschdnbücher 284. Ulmer, Stuttgart. 256 p.
-----. 1974. Die Vegetation Osteuropas, Nord- und Zentralasiens. Gustav Fischer, Stuttgart. 452 p.
-----. 1975. Ueber ökologische Beziehungen zwischen Steppenpflanzen und alpinen Elementen. Flora 164:339-346.
Wardle, P. 1965. A comparison of alpine timber lines in New Zealand and North America. New Zealand J. Bot. 3:113-135.

-----. 1968. Engelmann spruce (*Picea engelmannii* Engel.) at its upper limits on the Front Range, Colorado. Ecology 49:483-495.
-----. 1971. An explanation for alpine timberline. New Zealand J. Bot. 9:371-402.
-----. 1974. Alpine timberlines, pp. 371-402. In J. D. Ives and R. G. Barry (eds.). Arctic and alpine environments. Methuen, London.
Weber, W. A. 1972. Rocky Mountain flora. Colo. Assoc. Univ. Press, Boulder. 438 p.
Went, F. W. 1948. Some parallels between desert and alpine floras in California. Madroño 9:241-249.
-----. 1953. Annual plants at high altitudes in the Sierra Nevada, California. Madroño 12:109-114.
-----. 1964. Growing conditions of alpine plants. Israel J. Bot. 13: 82-92.
Westhoff, V., and E. Van der Maarel. 1973. The Braun-Blanquet approach, pp. 617-726. In R. H. Whittaker (ed.) Ordination and classification of communities. Part V: Handbook of Vegetation Science (R. Tuxen, ed.). Dr. W. Juuk, The Hague.
Yurtsev, B. A. 1968. [Flora of the Suntar-Khaiata Mts. Problems of the history of a high mountain landscape of northeastern Siberia. In B. A. Tikhomirov (ed.), Vegetation of the Far North of the USSR and its utilization. Vol. 9.] Nauka, Leningrad. 235 p. Rev. Ecol. 51(2)353-354, 1970.
-----. 1974. [Steppe communities of the Chukotsk tundra and Pleistocene tundra-steppe.] Bot. Zhur. 59(4):484-501.
Zoller, H., J. Braun-Blanquet, and P. Müller-Schneider. 1964. Flora des schweizerischen Nationalparks und seiner Umgebung. Ergebn. Wiss. Unters. schweiz. Nationalpark 9(51):1-408.
Zwinger, A. H., and B. E. Willard. 1972. Land above the trees: A guide to American alpine tundra. Harper and Row, New York. 489 p.

THE PACIFIC NORTHWEST FLORISTIC PROVINCE

CHAPTER

19 THE REDWOOD FOREST AND ASSOCIATED NORTH COAST FORESTS

PAUL J. ZINKE
Department of Forestry and Conservation,
University of California, Berkeley

Introduction 680
Vegetation 680
  Transects across north coast forests 683
  Successional relationships 684
    Grand fir–Sitka spruce–Douglas fir 685
    Redwood–grand fir 688
    Conifer–hardwood 689
  Climatic relationships 690
  Topographical relationships 691
  Geological relationships 692
Dynamics of redwood flat communities 693
Areas for future research 695
Literature cited 697

INTRODUCTION

Although the coniferous forests of coastal California from the Oregon border south to San Luis Obispo Co. are characterized by a California endemic (coast redwood, *Sequoia sempervirens*), they are mainly a southern extension of the great forests of Washington and Oregon. The coast redwoods make this forest outstanding; these magnificent trees are the world's tallest (112 m), growing at rates near world maximum (42 $m^3/ha^{-1}/yr^{-1}$), with accumulations of wood mass unequaled in any other place (Fritz 1945; Roy 1966; Fig. 19-1).

There are many references in the literature over the past 100 yr describing these forests (e.g., the Whitney Survey of 1865; Jepson 1910; the Cooperative Vegetation and Soil–Vegetation Surveys of the State of California and the U.S. Forest Service; and others documented in the bibliographic work of Fritz 1957). Recently, the National Park Service, the California Department of Parks and Recreation, and the University of California have carried out ecological studies. Cooper (1965) published an account of redwood ecology in which he concentrated on the forests in commercial timber production. Becking (1968) described the ecological conditions in old redwood groves, and Stone and Vasey (1968) discussed how these can be maintained.

A major portion of the data in this chapter is derived from vegetation and soil maps and study plots of the State Cooperative Soil–Vegetation Survey (see Chapter 6) which has covered the entire north coast forest area. First the vegetation of the area will be examined, then the environmental factors that seem to be related to the array of vegetation types will be discussed, and finally certain anomalies will be considered.

VEGETATION

Vegetation will be considered in terms of the most prevalent species of conifer and hardwood trees (Table 19-1). Common names will be emphasized in the text. Some of the conifers, such as Douglas fir, grand fir, hemlock, Sitka spruce, western red cedar, and Pt. Orford cedar, extend from the moist climates of Alaska, British Columbia, Washington, and Oregon, as noted by Pavari (1958). Other conifers, such

Figure 19-1. (*a*) Redwood forest near Prairie Creek, Humboldt Co. Note figure at base of tree in bottom middle foreground, for size scale. (*b*) Redwood forest on alluvial soil in Bull Creek Flat, Humboldt Co. Note the absence of trunk flaring at ground level, indicating deposition of sediment.

TABLE 19-1. Tree species of the redwood and north coastal forests and abbreviations used in text and figures

Category	Species		
	Common Name	Binomial Name	Symbol Abbreviation
Conifer trees	Redwood	*Sequoia sempervirens*	R
	Douglas fir	*Pseudotsuga menziesii*	D
	Grand fir	*Abies grandis*	G
	Coast hemlock	*Tsuga heterophylla*	H
	Sitka spruce	*Picea sitchensis*	S´
	Western red cedar	*Thuja plicata*	$\underline{C}$
	Pt. Orford cedar	*Chamaecyparis lawsoniana*	O
	Jeffrey pine	*Pinus jeffreyi*	J
	Incense cedar	*Libocedrus decurrens*	I
	Sugar pine	*Pinus lambertiana*	S
	Nutmeg	*Torreya californica*	N
	Yew	*Taxus brevifolia*	U
	Western white pine	*Pinus monticola*	W´
	Shore pine	*P. contorta*	L_C
	Bolander pine	*P. contorta* ssp. *bolanderi*	L_B
Hardwood trees	Tan oak	*Lithocarpus densiflora*	T
	Madrone	*Arbutus menziesii*	M
	Garry oak	*Quercus garryana*	G
	Black oak	*Q. kelloggii*	B
	Canyon oak	*Q. chrysolepis*	C
	Coast live oak	*Q. agrifolia*	A
	Interior live oak	*Q. wislizenii*	W
	Red alder	*Alnus oregona*	$\underline{R}$
	California bay	*Umbellularia californica*	L
	Big-leaf maple	*Acer macrophyllum*	$\underline{M}$
	Oregon ash	*Fraxinus latifolia*	$\overline{O}$

as the pines, represent extensions northward from more arid areas to the south and east. Redwood is the main species indigenous to the area, and the one that distinguishes the north coastal forests of California.

In traversing these forests from moist to dry locations (as along the coast from Crescent City to Ukiah), one progresses from Sitka spruce–grand fir–hemlock in moist areas, to redwood mixed with other conifers, to redwood mixed with hardwoods, to Douglas fir–hardwoods, and finally to grassland–oak woodland mosaics in the driest situations. The zonation of forest types is complex, following both latitudinal and inland gradients. To quantify the relationships, I have compiled a series of transects by referring to soil–vegetation type maps, recording types at 1 mile (1.6 km) intervals. Figure 19-2 shows the main tree species and their occurrences along each transect.

Transects Across North Coast Forests

The northernmost transect, inland from Big Lagoon at 41°11′10″N, shows an initial very narrow coastal strip with a forest of Sitka spruce and grand fir. Redwood appears 1.6 km inland. Sitka spruce drops out within 3–4 km of the coast, but grand fir continues further inland. A tan oak–hardwood component begins 5 km inland. Redwood remains the most abundant species up to 16 km inland, but thereafter Douglas fir becomes the most abundant conifer. Hardwood forest is dominant beyond 16 km inland, with tan oak, Garry oak, and madrone predominating.

In the transect at 40°32′54″N, a Sitka spruce–grand fir–Douglas fir forest extends inland across the Wildcat Hills for nearly 25 km to the Eel River valley. Here, red alder appears as a seral stage after logging of spruce and fir. This spruce–fir forest is apparently related to a strong northwest wind exposure in the Wildcat Hills south of Ferndale. Beyond 24 km inland, redwood first appears on this transect in a wind-sheltered ravine near Rhonerville and continues as the dominant, associated with Douglas fir, to 32 km inland. Beyond this, Douglas fir is most

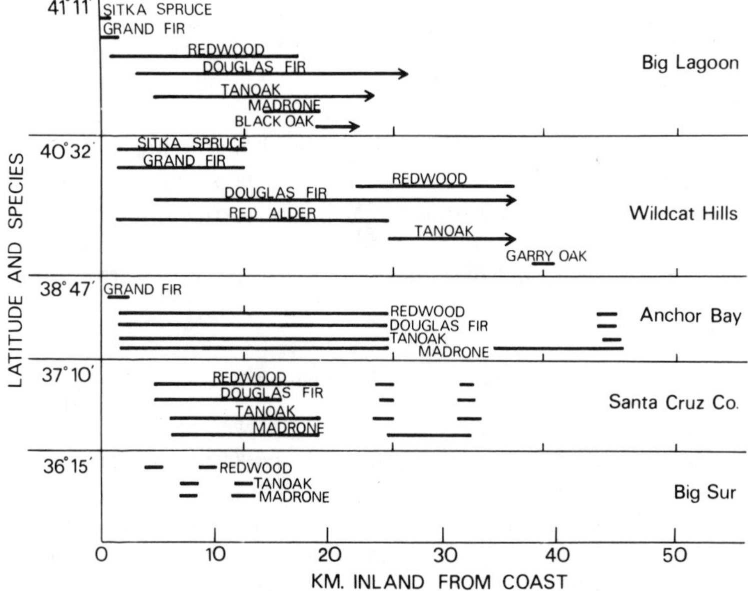

Figure 19-2. The occurrence of major tree species of north coastal forests on transects inland from the ocean at various latitudes, arranged from north to south. Arrows indicate that the species extend further inland.

abundant. There is an abrupt transition to Garry oak woodland at 39 km inland, with no redwood thereafter.

Further south, progressing inland from a point south of Anchor Bay at 38°47′36″ N, there is a narrow coastal bench with grassland and bishop pine (*Pinus muricata*), followed by a narrow 1–2 km wide strip of Douglas fir, redwood, and grand fir, with hardwoods, tan oak and madrone. Redwood continues as the major conifer until 12 km inland, after which Douglas fir predominates. Sugar pine enters the forest between 13 and 16 km from the coast. Black oak, madrone, and tan oak dominate beyond 19 km, with Douglas fir the only conifer except for an isolated stand of redwood in the Russian River canyon north of Cloverdale 37 km inland. This is the farthest inland that redwood extends in this area. Live oaks (*Quercus agrifolia* and *Q. wislizenii*) appear beyond 35 km. However, the absolutely farthest inland redwood as noted by Jepson (1910) is east of the Napa Valley near Angwin, California, in Pope Valley where, due east of Fort Ross on the coast, it would be 71.6 km (44.5 miles inland), or, at closest distance to the east of Duncan Mills, 60 km inland (36.8 miles).

Along a transect in Santa Cruz Co. at 37°10′0″ N, the first vegetation is a narrow coastal fringe of cultivated terrace land and hilly coastal scrub (*Artemisia, Baccharis*; see Chapter 21). Redwood is the main conifer from 5 to 19 km inland, with Douglas fir the only associated conifer and the hardwoods tan oak and madrone. A patch of this forest extends 32 km inland to Los Gatos Creek, but beyond 19 km it is interrupted by *Quercus agrifolia* woodland and *Ademostoma fasiculatum* chaparral, and chaparral becomes dominant beyond 32 km inland.

On a transect from Cooper Pt., Monterey Co., 36°15″0″ N, the north coastal forest is reduced to stringers of redwood in very moist canyon bottoms, such as along the Big Sur River. Hard and soft chaparral, coast live oak, tan oak, and California bay predominate. Finally, at Salmon Creek, 35°48′0″N, redwood reaches its southern limit as a small windswept clump on the south side of a ridge a few hundred meters from the coast, surrounded by coastal sage scrub (St. John and Raymond 1930).

In summary, the redwood belt is usually only about 16 km wide, and it is not always near the coast except as one approaches its southern limit. Hardwoods (except red alder) increase with distance inland. The narrow distribution of redwood is typified in Humboldt Co. There redwood forest is 16–24 km wide, portions of it trend northwest–southeast, and there are places where the western edge is very far inland.

Successional Relationships

The progression of climax vegetation types, from wet to dry locations, is summarized in Tables 19-2 and 19-3. The grasslands, however, occur in two situations, summer moist coastal grasslands and summer dry interior grasslands. Thus the complete progression of vegetation, by broad physiognomical types, is from coastal grassland to conifer forests, to conifer–hardwood forests, to oak woodlands, and to interior grasslands. Within this is an array of shrub types, usually on shallow, eroded soils (coastal scrub, *Arctostaphylos* chaparral, *Adenostoma* chaparral), and some miscellaneous types often on unusual parent material such as serpentine rock, or

TABLE 19-2. Vegetation type groups in the north coastal forests of California

Physiognomy	Type	Dominant Species
Conifer forest	Fir-spruce	Ĝ, S´, H
	Redwood-fir	R, Ĝ
	Redwood-Douglas fir	R, D
	Redwood	R
	Douglas fir	D
Conifer-hardwood forest	Redwood, Douglas fir, tan oak, madrone	R, D, T, M
	Douglas fir-hardwood	D, T, M, D, G
Hardwood forest	Oak woodlands	G, B. T
Shrub	Baccharis-coastal scrub	*Baccharis*, *Garrya*, *Ceanothus*
	Manzanita-chamise	*Adenostoma*, *Ceanothus*
Grasslands and fern prairie	Acid coastal grasslands	*Aira*, *Elymus*, *Danthonia*, *Rumex*, *Pteridium*
	Interior grasslands	*Bromus*, *Elymus*
Miscellaneous	Pine-cedar-cypress pygmy forest	J, I, *Cupressus sargentii*, *Arctostaphylos viscida*
	Redwood, sugar pine, hardwood	R, D, S, T, M

acid coastal plain deposits (*Pinus jeffreyi, P. lambertiana, Libocedrus decurrens, Cupressus* species, pygmy forest).

In addition to this gradient of climax types, mosaics of seral stages follow logging, fire, windthrow, landslide, sand encroachment, and other disturbances. The following are examples of some seres, based on evidence from vegetation types mapped in the field.

Grand fir–Sitka spruce–Douglas fir. Fir-spruce forests are found in greatest extent in the Wildcat Hills between the Bear River ridge and the Eel River, south of Ferndale. The vegetation was mapped by Colwell et al. (1959a,b). In the southwest corner of Township 2N, Range 2W (Humboldt Base and Meridian), in the Guthrie Creek watershed, almost all of the seral stages following disturbance of fir-spruce forest can be found. The apparent sequence is summarized in Fig. 19-3 and below.

Logging of grand fir (Ĝ), Sitka spruce (S^1), and Douglas fir (D), with some western red cedar (*C*), results in a pioneer stage dominated by bare ground (Ba),

TABLE 19-3. Vegetation in the overstory and understory of forests representing various vegetation types in the north coastal forest area. From selected plots and plot data of the State Cooperative soil-vegetation survey

Vegetation Type	Quadrangle and Plot	Mean Annual Rainfall (mm)	Mean Temperature (°C)			Species (for trees, see Table 19-1)	Overstory		Understory	
			Ann.	Jan.	July		Cover %	Max. Height (m)	Cover %	Max. Height (m)
Sitka spruce-Fir (cutover)	9A2 #1	2032	11.5	7.7	15.3	S	20	30.5		
						Ĝ	< 5	0.6		
						Rubus vitifolius	15	1.2		
						R. spectabilis	15	1.8		
						Polystichum munitum	15	0.9		
						Epilobium angustifolium	15	1.2		
						R	10	13.7		
						Pteridium aquilinum	< 5	0.6		
						Misc. herbs	< 5	0.6		
Redwood-Fir	9A4 #2	2083	11.7	7.8	15.6	R	70	76.2	5	4.6
						Polystichium munitum	...	...	70	0.9
						Vaccinium ovatum	...	...	5	0.9
						V. parvifolium	...	...	5	1.8
						D	5	38.1	...	...
						Ĝ	5	38.1	...	...
						H	5	38.1	5	22.9
Redwood-Douglas fir-Hardwood	61C4 #3	1321	11.1	9.4	13.9	R	50	36.6	...	7.6
						M	20	15.2	...	...
						T	20	15.2	5	2.4
						Barren	...	...	60	...
						Myrica californica	...	...	5	1.8
						Vaccinium ovatum	...	...	5	0.9
						Gaultheria shallon	...	...	5	0.3
						Polystichum munitum	...	...	5	0.6

						Pteridium	...	...	5	0.6
						Rhus diversiloba	...	...	5	0.6
						D	5	36.6	...	...
Douglas fir-hardwood	26C2 #4	1575	12.2	5.0	20.0	D	50	61.0	...	...
						T	20	24.4	50	9.1
						M	20	21.3	...	...
						Berberis nervosa	...	...	10	0.6
						M	...	...	10	12.2
						Vaccinium ovatum	...	...	10	0.9
						Polystichum munitum	...	...	10	0.6
						L´	10	18.3	...	...
Hardwood (Garry oak woodland)	26B1 #16	1829	11.7	3.9	20.6	G	100	15.2	10	9.1
						D	...	...	10	1.5
						Holodiscus discolor	...	...	10	0.9
						Rosa spp.	...	...	5	0.6
						Prunus emarginata	...	...	5	0.6
						Polystichum munitum	...	...	5	0.3

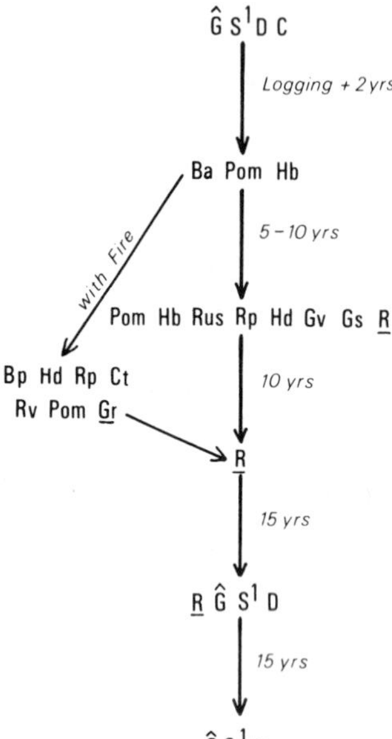

Figure 19-3. The seral stages in the grand fir–Sitka spruce forests after timber harvest in the Wildcat Hills southwest of Ferndale, Humboldt Co., at latitude 40°32′N.

Polystichum munitum (Pom), and various herbs (Hb). These herbs are usually composites with light seeds such as *Erechtites arguta* and *E. prenanthoides*. After 5–10 yr, *Polystichum* and herbs are still abundant, but a large number of shrubs have entered: *Rubus spectabilis* (Rus), *R. parviflorus* (Rp), *Holodiscus discolor* (Hd), *Garrya veatchii* (Gv), and *Gaultheria shallon* (Gs). Details of the species present immediately after logging have been presented by Roy (1966). Some red alder seedlings (*R*) begin overtopping the brush, and by another 10 yr (15–20 yr after the disturbance) red alder dominates the site but conifer seedlings appear beneath it. Over the course of the next 30 yr, these conifers (fir, spruce, Douglas fir) overtop the red alder and replace it, resulting in the climax community. If the pioneer community is burned, as it would be in preparation for a pasture, a slightly different sere results. *Baccharis pilularis* (BP), *Holodiscus discolor* (Hd), *Ceanothus thrysiflorus* (Ct), *Rubus vitifolius* (Rv), sword fern (Pom), and various grasses occur. Repeated burning ultimately produces a pasture with clumps of sword fern, but when burning ceases, red alder comes to dominate and conifers reinvade. More intense fires result in dominance by *C. thrysiflorus*.

Redwood–grand fir. Redwood–grand fir forest progresses through a similar sere after logging or windthrow. Examples of all stages can be found just east of Big Lagoon, where the vegetation was mapped by DeLapp et al. (1961) in quadrangle 10D1. The sere is shown in Figure 19-4 and is summarized below.

A pioneer community of bare ground (Ba), herbs (Hb), and sword fern (Pom) is followed within 10 yr by bracken fern (Pta), herbs (Hb), sword fern (Pom), redwood sprouts (R), *Baccharis pilularis* (Bp), and *Rubus* spp. (*Rx*). About 30 yr later, redwood again dominates the site and grand fir has overtopped the brush, but a few *Rubus spectabilis* and *R. parviflorus* shrubs remain. By 60–70 yr from the time of disturbance, the redwood–grand fir forest has returned. Redwood returns first

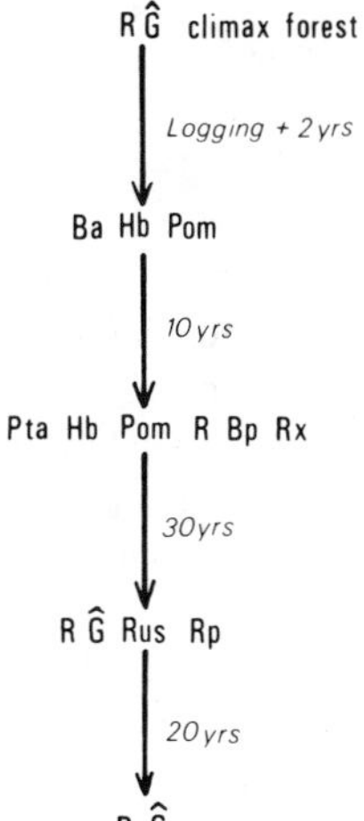

Figure 19-4. The seral stages in vegetation composition in the redwood–grand fir forest after timber harvest in an area immediately east of Big Lagoon, Humboldt Co., at latitude 41°10′N.

because of stump sprouting, whereas the other conifers depend on seed and return more slowly. Sitka spruce and western red cedar may also appear in this type. Stone and Vasey (1968) believed that redwood itself represents past disturbance such as fire, and that if fire is kept out a potential climax species is western red cedar.

Conifer–hardwood. The conifer–hardwood forests progress through a similar sere, as can be described from the vegetation mapped by Cuff et al. (1949) in the southeast quarter of the Branscomb quadrangle (44D4). The sere is shown in Fig. 19-5 and is summarized below.

After the climax forest, with redwood (R), Douglas fir (D), tan oak (T), and madrone (M) is logged, the pioneer community is dominated by *Erechtites* spp., *Whipplea modesta,* and sprouts of tan oak, madrone, and redwood. Seedlings of Douglas fir soon appear. The sprouting hardwoods dominate within 4–8 yr, but in another 15 yr a fairly tall stand of hardwoods and conifers results. By 70–85 yr after disturbance, redwood and Douglas fir have overtopped the hardwoods and the

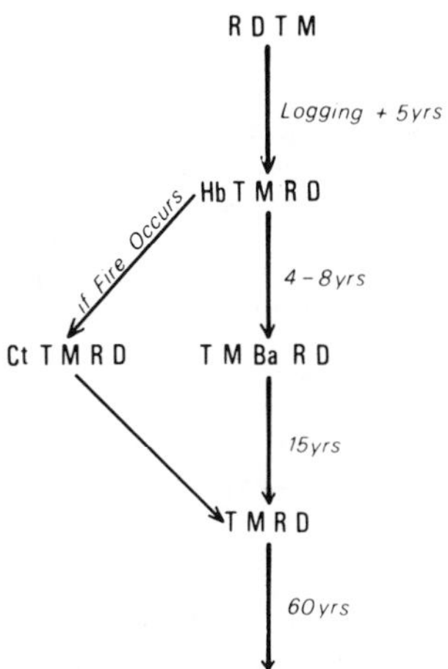

Figure 19-5. The seral stages in the redwood–Douglas fir-tan oak-madrone forest after timber harvest disturbance in an area along the south fork of the Ten Mile River, Mendocino Co., at latitude 39°33′N.

original community has returned. If fire occurs at any time, the sere is shunted to a *Ceanothus*-dominated stage (*C. thrysiflorus* or *C. velutinus*).

In addition to these seres which come after disturbance, there is a general increase in young stands of Douglas fir in the region, with age classes from 50 to 75 yr. These are invading either grassland areas near the coast or Garry oak forests in the interior. An example of an area with extensive invasion of grasslands is just south of the mouth of the Mattole River. In the sequence which occurs on grassland soils of the Wilder soil series in the first few kilometers east of Punta Gorda, the grassland is invaded by bracken fern, and this in turn by even-aged Douglas fir regeneration, which then takes over the grassland as a uniform stand of Douglas fir (Colwell et al. 1958). Similar Douglas fir invasions into grass openings have been observed throughout the northern coastal forest area.

The invasion of interior oak forests is typified in the area near Bridgeville, where Garry oak–black oak forests on Tyson soil series are gradually invaded by an understory of Douglas fir. In time these become oak–Douglas fir forests, with the Douglas fir gradually crowding the oaks to make pure young Douglas fir stands with an understory of oaks which gradually die out. This was noted on the areas at the head of Larabee Creek and near Bridgeville by Colwell et al. (1958). It has also been observed near Schoolhouse Peak in the Redwood Creek drainage and in the area north of Round Valley in Mendocino Co. The factors bringing this about may be cessation or lessened use of fire to keep grass openings and woodlands open, or a climatic change, or the occurrence of particular years favorable for Douglas fir regeneration. Many of these stands have been harvested since the area was mapped.

Climatic Relationships

The climatic variables influencing redwood forest have long been thought to be related to summer fog and a maritime setting. Figure 19-6 summarizes temperature along transects running from the coast inland. Temperature extremes generally

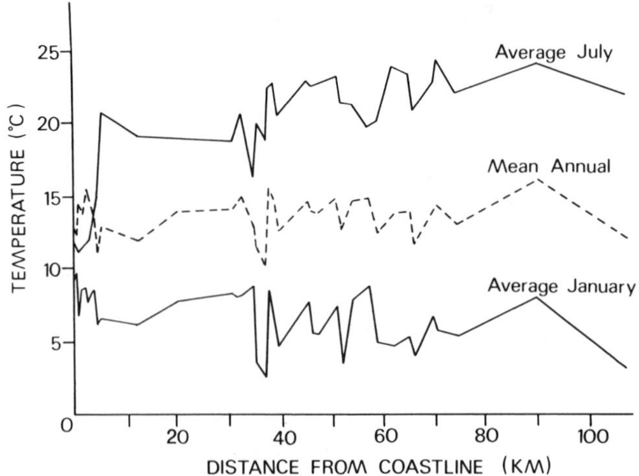

Figure 19-6. The mean annual temperatures, mean July and January temperatures, and distances inland from the coastline, based on data from 33 climate stations in the north coastal area of California. From data of W. L. Colwell, California State Cooperative Soil–Vegetation Survey.

increase with distance inland because of increasing elevation. However, topographical factors add to this, with interior valleys having broad ranges of temperature, from high maxima in summer to low minima in winter. It is apparent that position relative to the coast controls the gradient of increasing maximum temperatures and, to a lesser degree, minimum temperatures, resulting in the wider range of temperatures as one proceeds inland. In growth chamber studies, Hellmers and Sundahl (1959) found that redwood has very little thermoperiodism and thus does well in a stable temperature regime. Also they reported an optimum temperature for redwood seedling growth of 18.9°C. This is supported by Kuser (1976), who found optimum site quality or productivity of redwood to occur at a mean summer temperature of 17.8°C.

The wind patterns in the area and the influx of marine air, particularly during the summer, are partly the result of this heating of the inland areas during the summer, and the influx of marine air, as into a furnace. There is a pattern of such inflow of marine air up the Eel River canyon, as indicated by the windflagging of trees at various places in the redwood forest and adjacent forest types in the area. The influx of marine air seems to be related to the distribution of the redwood forest. Immediately along the coast, where the incoming wind is high in salt spray aerosols, there is a grassland strip, or forest area with trees that are tolerant to salt spray damage, such as Sitka spruce and Douglas fir. It is only inland of this effect that one begins to see a redwood forest, which extends inland until the marine air influence is overcome by inland heating of the land. The anomaly of the large gap in the redwood forest that occurs in the Mattole River basin seems to be related to a large back eddy in the summertime wind pattern. As the prevailing wind across the ocean (NW 18°) sweeps past the Kings Peak Range, there is actually an offshore wind pattern across the Mattole River basin, resulting in very dry downdrafts along the face of the Kings Peak Range. Related to this is a broad area of grassland and oak woodland in the Mattole River valley, despite the fact that the overall annual precipitation is the highest in the redwood region.

The generalization that one can make regarding wind and the distribution of vegetation types along the north coast is that, where the influx of marine air is excessive, there will be a coastal grassland, as along much of the coast from Eureka south, and inland in wind gaps such as those between Bodega Bay and Petaluma, the Golden Gate, and the Salinas Valley. Associated with the marine air influx is the occurrence of summer fog, which results in fog drip. Azevedo and Morgan (1974) have shown amounts of fog drip ranging from 18.4 to 30.4 cm. Of course, these effects will be interwoven with other factors which tend to favor or to mitigate against the presence of redwood forest or other forest types.

Topographical Relationships

Topography across the north coastal forest area shows an increasing elevation inland, dissected by deep valleys. This increase in elevation may be very abrupt, as in the Kings Peak Range at 40°10′N, where within 5 km of the ocean the elevation is 1246 m. Instead, the rise may be gradual, as near Ft. Bragg at 39°25′N, where the coast mountains crest at 1034 m nearly 40 km inland of Pine Ridge west of Ukiah. The rivers of the area have cut into these mountain ranges along fault alignments which tend to parallel the coastline and the San Andreas Fault. Thus the interior valleys trend from southeast to northwest. This tends to accentuate the inland climatic aspects of some of the interior valleys in their headwaters areas, isolating them from the coastal climate although they are only a short distance inland. As a result, one finds gaps in the range of species such as redwood which depend on the more equable marine climate. On the other hand, where the valleys open toward the ocean so as to reinforce the summer marine air indraft with its prevailing flow from the northwest, as at the mouth of the Eel River, the redwood belt extends further inland.

As one progresses inland, the climate becomes more severe because of extremes of summer heat and low humidity. The topographical effect of north slopes and of long slopes as sources of seepage water in their lower reaches becomes of more importance. Thus, in the upper south fork of the Eel River near Leggett Valley, the main extent of redwood forest ends abruptly with a general change from northerly to southerly slopes. Waring and Major (1964) reported details of the effects of altitude on the distribution of vegetation types on Grasshopper Peak in Humboldt Redwood State Park. Grasshopper Peak carries a range of vegetation types from nearly pure redwood forest at its base in Bull Creek at an elevation of 37 m to grasslands and Douglas fir–hardwood stands on Grasshopper Peak at 1070 m. Generally, they attributed this gradient of vegetation to the increasing dryness of the sites with increasing elevation.

Geological Relationships

The geology of the north coastal forest area in terms of the major rock types and the soils derived from them plays an important role in determining vegetation and the abrupt boundaries to some vegetation types.

The geological types trend in a manner similar to the topography of the area. Thus the topographical, geological, and soil controls on vegetation distribution reinforce each other. The rocks of the area are predominantly sedimentary. Younger, less consolidated sedimentary rocks near the coast give rise to deeper soils with greater water-holding capacities than those farther inland from older and harder sedimentary rocks. Coastal terraces adjacent to the coastline often have very old surfaces with old, infertile soils, and on these occurs the depauperate vegetation of pygmy forest types (*Pinus contorta* ssp. *bolanderi, Cupressus pygmaea*), as described by Gardner and Bradshaw (1954); see also Chapter 9). There are abrupt boundaries of these pygmy forest areas where erosion has cut into underlying rock types to expose younger, more fertile soils, on which the north coastal redwood and Douglas fir forests do better (Jenny et al. 1969).

Intrusions of serpentine and peridotite rock trend across the north coastal area from southwest to northwest, forming soil anomalies that are high in magnesium and low in potassium, calcium, and phosphorus. Frequently in contact with these are metamorphic rocks such as glaucophane schists weathering to heavy clay soils with high magnesium contents supporting grassland vegetation. The serpentine and peridotite areas themselves have anomalous vegetation for the area. An example of such is Red Mt. in northern Mendocino Co. The forest on the serpentine area consists of conifers only, with *Pinus jeffreyi, P. lambertiana, P. ponderosa, P. attenuata, Cupressus sargentii, C. macnabiana,* and *Calocedrus decurrens*. The lack of hardwoods seems attributable to the low soil fertility. This coniferous forest has an abrupt boundary with the more typical Douglas fir–hardwood forest on adjacent soils derived from sedimentary rocks that are more typical of the area. Thus one could conclude that the "climax" vegetation of this north coastal forest would be entirely different if by chance the geological factor were different; one would probably find a mixed conifer forest with very few hardwoods if the country rock had been all peridotite instead of the present sedimentary rock.

The boundary of the main redwood belt often ends abruptly at a change in rock type. An example of this occurs at Jedediah Smith Redwoods State Park on soil–vegetation quadrangle 9A4, classified and mapped by DeLapp and Smith (1976). The conifers on the soils derived from sedimentary rock are *Sequoia sempervirens, Pseudotsuga menziesii,* and *Tsuga heterophylla,* with infrequent *Chamaecyparis lawsoniana.* However, on the adjacent soils derived from peridotite rock, there is an abrupt change in vegetation within a few hundred meters. On these intrusive rocks, there is no redwood and the coniferous species are Douglas fir, Port Orford cedar, Jeffrey pine, knobcone pine, and western white pine. Some species such as Port Orford cedar, which is usually a serpentine endemic in northern California, cross this boundary, but others (e.g., western white pine and Jeffrey pine) stay strictly on the serpentine and peridotite rock. Exceptions occur where alluvium from the intrusive rocks is deposited downstream along benches on the Smith River; it supports species of both the redwood forest and the more interior mixed conifer forests.

The glades or grassland openings that often occur in the north coastal forest area are frequently vegetation anomalies related to soils derived from rocks richer in basic elements. These soils often are higher in clay content and pH than the adjacent soils from the less basic rocks that form the basis for the regional forest climax. It was noticed in mapping vegetation that ecotones of such natural grass openings with adjacent redwood, Douglas fir, or hardwood forests often resembled the sequence of vegetation types found east of the main redwood forest. The grass openings are surrounded by a margin of Douglas fir–hardwood, or of oak-hardwood that grades into Douglas fir, redwood-hardwood, and then the redwood-Douglas fir-tan oak-madrone forest typical of the area. In some areas, the inland side of the redwood forest terminates abruptly with a geological boundary that produces high-pH heavy clay grassland soils, which the redwood does not invade. Such is the case with the eastern boundary of the redwood forest on the slopes alongside the Van Duzen River just west of Bridgeville. Here, as with serpentine intrusions, the redwood ends abruptly.

Another local exception in the vegetation composition related in part to geology is the superlative redwood groves that occur on deposits of recent alluvium along the major rivers in the north coastal forest area. These form the redwood forests which laymen think of as typical redwood forests, although, in fact, they are only a small proportion of the area of the region. However, their uniqueness requires a thorough examination of their vegetation dynamics.

DYNAMICS OF REDWOOD FLAT COMMUNITIES

The life history of the superlative redwood groves along the rivers in the north coastal forest is tied into the dynamic regime of river flooding, sediment deposition, and soil buildup.

In Bull Creek Flat there is a stand of trees slightly more than 92 m tall growing on such a soil. In excavating this soil to more than 9 m depth, I found the soil profile shown in Fig. 19-7. A series of dark, organic-matter-rich layers at various depths indicated that soils were buried by subsequent sediment deposits. Below the lowest layer was a bed of stream gravels. The top layer, designated as I, was deposited during the flood of 1955, and it is evident that each of 15 layers below this represented at least 15 flood deposits. Charcoal deposits were associated with each of the layers, indicating that a fire–flood sequence may be associated with these groves. Dating of the charcoal from the lowest layer indicated an age of approximately 1000 yr. A tree that fell from an undercut bank nearby was, as determined from ring counts, 958 yr old, indicating that the forest itself was roughly the same age as the total sediment deposit. In addition, the flooding affected the ring growth in such a way as to allow a dating of the major floods that had occurred during the life of the tree. These floods occurred at intervals of 30–60 yr.

During work at Stephens Grove in Humboldt Redwoods State Park, I observed that each new deposit of sediment created ideal conditions for seed germination and seedling survival, and that the forest came in as even-aged waves of seedlings, dating the time of sediment deposition. These seedlings may end up as suppressed dwarfs because of lack of light if they are in the interior of a dense grove, or they may grow very rapidly if at the edge of a grove. Thus each of the sediment deposits may have an age class of trees associated with it, and at the margin of the grove with good light conditions these trees may be already 61 m tall, but where light conditions are poor within the stand, they may be less than 10 m in height. Many of the superlative groves have a margin of young growth seedlings now more than 100 yr old, dating the 1861 flood.

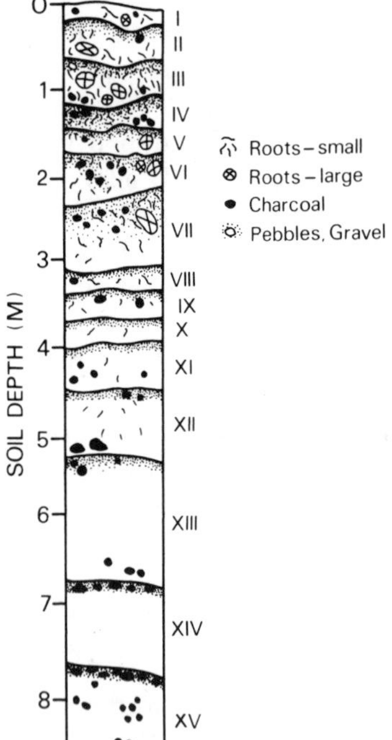

Figure 19-7. Soil under old redwood grove in Rockefeller Forest on Bull Creek Flat, showing sediment deposition layers from past floods. Layer I is from the 1955 flood, and layer XV is apparently the initial sediment layer on which the forest became established about 1000 yr ago, based on radiocarbon dating of coarcoal in the layer, but also coincident with tree ring ages for the forest.

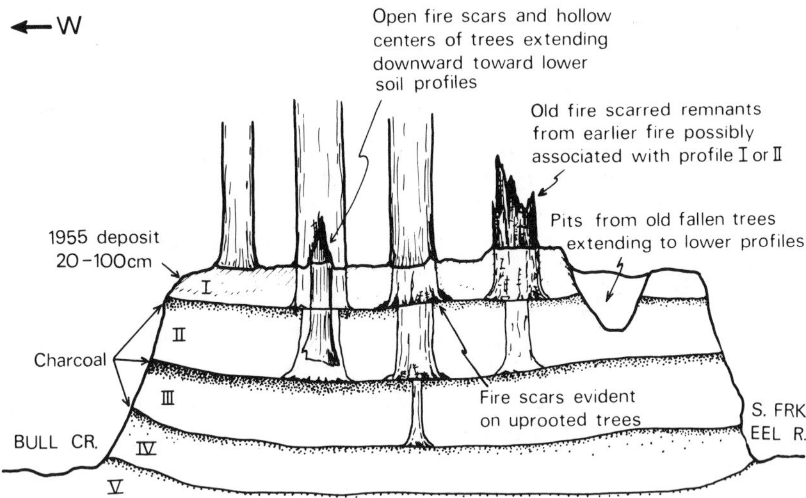

Figure 19-8. An old stand of redwood trees on Rockefeller Forest, Humboldt Redwoods State Park, in relation to various sediment deposition layers. Note multiple-layer root systems and their relation to buried soil layers, open fire scars, tree hollows extending to lower layers, consistent with old soil surface at time of fire, and pits where trees were burned out to lower soil layers.

This dynamic situation of continuous flooding and sediment deposition means that the understory species have to adapt to the same problems of new and rising soil surfaces. Stone and Vasey (1968) found that understory herbs such as *Oxalis oregona,* which typify a redwood grove understory after a long period of no disturbance, recover from deep sediment deposition by growing vertical shoots from their submerged parts, and gradually recolonize the surface. In addition, they found that the redwood itself grew roots into the new soil deposit.

Thus, as documented by Fritz (1933) at Richardson Grove, each time a new layer of soil is deposited on the alluvial flat, the tree grows a new set of surface roots corresponding to the new soil surface, giving the tree a multistoried root system corresponding to these buried layers. In addition, the broad outward taper at the lower base of the tree is buried, and the tree plunges directly into the soil. The new soil may be nutritionally important to continued good redwood growth (Florence 1965). Older trees that date back to deeply buried sediment layers may be burned out by fire, producing fire scars that open the interior of the tree trunk to the buried soil layers below; when a tree falls or burns out completely, a pit may develop down to lower soil layers. Thus one finds in a very old grove a scene as sketched in Figure 19-8. Each tree or the wreckage of past trees extends down to the flood deposit in which its major root system occurs. Research at Redwood National Park is further documenting the importance of flood and fire (see, e.g., Veirs 1975).

AREAS FOR FUTURE RESEARCH

Despite the fact that there is now a complete record of the distributions of the various vegetation types and dominant species in the north coastal area of California, there are many gaps in knowledge. Much remains to be learned about the history of these vegetation types and the changes presently occurring in them. There needs to be more study of the autecological requirements of the various species that comprise this forest vegetation. For example, most of the coniferous forest species are at the southern limit of their ranges in this portion of California, representing the southern limit of the great coniferous forest of the Pacific Northwest. Presumably this is related to the greater aridity and warmer temperatures that occur at this transition to oak woodlands and grasslands to the south and to the interior. Are these limits due to occasional extremes of drought, or to gradients of increasingly limiting average moisture? At what points in the life cycles of the species concerned are the factors limiting: is it seedling survival that is critical, or overall growth in relation to competitive advantage?

It is obvious in the case of some species that they can grow vegetatively far outside the limits of this forest area, as, for example, in the case of redwood trees planted at Stanley Park in Vancouver, B.C. Why does redwood have such abrupt boundaries in its range now, and why are these sometimes limited by geological and soil factors, and sometimes apparently by climatic factors? How do these factors interact in the limits of the range of any particular species? Does the present ecotone of the redwood forest indicate that it is extending its range again, or is it still retreating? What is the role of past fire in the present distribution of these species and the forest types that their mixtures create? What is the effect of changes brought about by human activity in these forests, as by use of fire or its control, or in the intermittent harvest of forests for timber production?

There is, in addition, an imposing conservation problem that will continue into the future. Redwood forest is estimated to have covered 800,000 ha 150 yr ago (Stone et al. 1972; Leydet 1969). Today less than 5% of this virgin redwood forest is left, and only half of that is in protected park land. Moreover, protected stands which occupy only the lower slopes of watersheds can be endangered by the activities of private developers above them, so the degree of protection is relative at present. In addition, it is possible that redwood stands will not maintain their population sizes without natural or managed fire and flood cycles. It may be necessary to develop a 500–1000 yr management plan for these parks if they are to remain well into the twenty-first century.

LITERATURE CITED

Azevedo, J., and D. L. Morgan. 1974. Fog precipitation in coastal California forests. Ecology 55:1135-1141.

Becking, R. W. 1968. The ecology of the coast redwood forest. Final Rept., Natl. Sci. Found. Grant 4690. 187 p.

Colwell, W., J. DeLapp, and E. Gladish. 1958. Soil-vegetation map, legend and interpretation, SE 1/4 Iagua Buttes quadrangle 26C4. Calif. Cooperative Soil-Vegetation Survey. Pac. SW Forest and Range Exp. Sta., Berkeley, Calif.

Colwell, W., J. DeLapp, E. Gladish, and J. Mallory. 1959a. Soil-vegetation map, legend and interpretation, SE 1/4 Ferndale quadrangle 27C4. Calif. Cooperative Soil-Vegetation Survey. Pac. SW Forest and Range Exp. Sta., Berkeley, Calif.

-----. 1959b. Soil-vegetation map, legend and interpretation, SW 1/4 Fortuna quadrangle 27D3. Calif. Cooperative Soil-Vegetation Survey, Pac. SW Forest and Range Exp. Sta., Berkeley, Calif.

Colwell, W., J. DeLapp, and P. Zinke. 1958a. Soil-vegetation map, legend and interpretation, SW 1/4 Iagua Buttes quadrangle 26C3. Calif. Cooperative Soil-Vegetation Survey, Pac. SW Forest and Range Exp. Sta., Berkeley, Calif.

Colwell, W., J. Mallory, and P. Zinke. 1958b. Soil and vegetation map, legends and interpretation, NE portion of Weott quadrangel. Calif. Cooperative Soil-Vegetation Survey, Pac. SW Forest and Range Exp. Sta., Berkeley, Calif.

Cooper, D. W. 1965. Coast redwood (Sequoia sempervirens) and its ecology. Humboldt County Agric. Ext. Serv., PO Box 1009, Eureka, Calif. 21 p.

Cuff, K., W. Hoffman, and P. Zinke. 1949. Soil-vegetation map, legend andinterpretation, SE 1/4 Branscomb quadrangle 44D4. Calif. Cooperative Soil-Vegetation Survey, Pac. SW Forest and Range Exp. Sta., Berkeley, Calif.

DeLapp, J., E. Gladish, and R. Volman. 1961. Soil-vegetation map, legend and interpretation, NE 1/4 Trinidad quadrangle 10D1. Calif. Cooperative Soil-Vegetation Survey, Pac. SW Forest and Range Exp. Sta., Berkeley, Calif.

DeLapp, J., and B. S. Smith. 1976. Soil-vegetation map, legend and interpretation, south half Crescent City quadrangel 9A3-4. Calif. Cooperative Soil-Vegetation Survey, Pac. SW Forest and Range Exp. Sta., Berkeley, Calif.

Florence, R. G. 1965. Decline of old-growth redwood forests in relation to some soil microbiological processes. Ecology 46:52-64.

Fritz, E. 1933. The story told by a fallen redwood. Save-the-Redwoods League, San Francisco. 8 p.

-----. 1945. Twenty years' growth on a redwood sample plot. J. For. 43:30-36.

-----. 1957. California coast redwood, an annotated bibliography of 2003 references. Found. Amer. Resources Manage., Recorder Sunset Press, San Francisco. 267 p.

Gardner, R. A., and K. E. Bradshaw. 1954. Characteristics and vegetation relationships of some podzolic soils near the coast of northern California. Soil Sci. Soc. Amer. Proc. 18:320-325.

Hellmers, H., and W. P. Sundahl. 1959. Response of Sequoia sempervirens (D. Don) Endl. and Pseudotsuga menziesii (Mirb.) Franc. seedlings to temperature. Nature 184:1247-1248.

Jenny, H., R. Arkley, and A. M. Schultz. 1969. The pygmy forest podsol ecosystem and its dune associates of the Mendocino coast. Madroño 20:60-74.

Jepson, W. L. 1910. The **silvae** of California. Mem. Univ. Calif. 2: 16-18, 128-139.

-----. 1936. Trees, shrubs, and flowers of the redwood region. **Save-the-Redwoods League, San Francisco. 15 p.**

Kuser, J. 1976. The site quality of redwood. M.S. thesis, Rutgers Univ., New Brunswick, N.J. 58 p.

Leydet, F. 1969. The last redwoods. Sierra Club, Ballantine Books, San Francisco. 160 p.

Pavari, A. 1958. Il genere Pseudostuga in America. Extracts from Monti e Boschi 1958 (7 and 8). 57 p.

Roy, D. F. 1966. Silvical characteristics of redwood. USDA Forest Serv. Res. Paper PSW-28. 20 p.

St. John, R., and F. M. Raymond. 1930. Vegetation type map, Jamesburg quadrangle 106A. Pac. SW Forest and Range Exp. Sta., Berkeley, Calif.

Stone, E. C., and R. B. Vasey. 1968. Preservation of coast redwood on alluvial flats. Science 159:157-161.

Stone, E. C., R. Grah, and P. Zinke. 1972. Preservation of the primeval redwoods in the Redwood National Park, parts I and II. Amer. For. 78(4):50-55; 78(5):48-56.

Veirs, S. D., Jr. 1975. Redwood vegetation dynamics. Bull. Ecol. Soc. Amer. 56(2):34-35 (abstr.).

Waring, R. M., and J. Major. 1964. Some vegetation of the California coastal redwood region in relation to gradients of moisture, nutrients, light, and temperature. Ecol. Monogr. 34:167-215.

Whitney, J. D. 1865. Geological Survey. Vol. 1: Geology of the Coast Ranges in California. Paxton Press, Philadelphia.

CHAPTER

20 MONTANE AND SUBALPINE VEGETATION OF THE KLAMATH MOUNTAINS

JOHN O. SAWYER
DALE A. THORNBURGH
Humboldt State University, Arcata, California

Introduction **700**
Montane vegetation patterns **701**
Vegetation on granitic and metamorphic parent materials **702**
***Abies concolor* zone** **702**
***Abies magnifica* zone** **708**
***Tsuga mertensiana* zone** **715**
***Pinus albicaulis* zone** **720**
Vegetation on ultrabasic parent materials **721**
Biology of Klamath montane trees **722**
Ecological dynamics of regional conifers **722**
Relict conifers **727**
Abies amabilis **727**
Abies lasiocarpa **727**
Chamaecyparis nootkatensis **729**
Picea breweriana **729**
Picea engelmannii **730**
Pinus balfouriana **730**
Areas for future research **730**
Acknowledgments **730**
Literature cited **731**

INTRODUCTION

The Klamath montane forests form a series of more or less discrete, island-like patches within a matrix of low-elevation forests and woodlands in northwestern California and southwestern Oregon. The Klamath Mts. represent a geological province of Paleozoic and Mesozoic formations distinct from surrounding younger rocks of the Coast Ranges, Cascades, and Sacramento Valley. The Klamath Mts. support distinctive conifer-dominated forests which are only beginning to be studied and understood.

Klamath montane forests grow mostly above low-elevation coniferous forests rather than chaparral, woodlands, or grasslands. Dominant species, such as *Pseudotsuga menziesii, Pinus ponderosa,* and *P. lambertiana,* are typical of low as well as montane elevations. The lower elevational limit of the montane zone in many cases is therefore difficult to establish, or is at best arbitrary. We have chosen to recognize the occurrence of *Abies concolor* on mesic habitats as the montane indicator. With this definition, the lower elevational limit varies from about 600 m in the western portion to as high as 1300 m in the southeastern portion of the region. Parent material, topography, and exposure locally adjust the boundary within this range. *Abies concolor* is regionally ubiquitous at middle elevations. Other dominant trees of middle elevations (*Pseudotsuga, Pinus* ssp.) are not appropriate indicators because they also occur in phases of mixed evergreen forest (Chapter 10).

Over much of their extent, montane forests occur above Klamath mixed evergreen *Pseudotsuga*-hardwood forests reminiscent of west-side pine on the Cascades (Griffin 1967; see also Chapter 17). Along the eastern boundary, open forests of *Pinus ponderosa* and *Quercus kelloggii* typically form the next lower zone, grading into oak juniper woodland (Chapter 23) in the Scott and Shasta valleys. In the Sacramento and Pitt River drainages, *P. ponderosa* and *Q. kelloggii* dominate at lower elevations where Paleozoic metasediments and ultrabasics give way to Cenozoic volcanics of the Cascades. In the upper Sacramento Valley, Klamath montane forests are bounded to some extent by chaparral (Chapter 12), but more typically by a band of *P. ponderosa* and *Q. kelloggii* forests on steep hillsides in western Shasta and Tehama Cos. The southern limit of the Klamath geological province occurs in the area of North Yolla Bolly, where contact is made with younger Franciscan Formation rocks at elevations well within the zone. Montane forests continue south through the Coast Range to around Snow Mt.

The general north coast mountain pattern is like that of the Klamath region: *Pseudotsuga*-hardwood forests to the west blending with elevational rise into montane forests, and on the eastern slopes montane forests giving way to *P. ponderosa*–*Q. kelloggii* forests or woodlands, and then to oak woodland (Chapter 11) or chaparral at still lower elevations.

The Klamath montane forests have attracted interest mainly because of their refugial nature, including many endemics and relicts (Chapter 4; Stebbins and Major 1965), their high floristic diversity, and their complex vegetation patterns (Whittaker 1960). Paleobotanical studies suggest that these mountains maintain forests most nearly equivalent to the western North American temperate Tertiary forests (Whittaker 1961; Wolfe 1969; Chapter 5).

The present floristic pattern reflects a comingling of Northwest and California-centered species. Trees regionally common elsewhere (*Abies amabilis* and *Chamaecyparis nootkatensis* in the Pacific Northwest, *Abies lasiocarpa* and *Picea engelmannii* in the western interior, and *Pinus balfouriana* in the southern Sierra Nevada) find their range limits within the Klamath area as scattered refugial stands. *Picea breweriana,* once widespread in Columbia Plateau–Cascade and Nevada Miocene floras (Wolfe 1969), is now endemic.

Existing vegetation patterns are controlled primarily by differences in parent material, and secondarily by elevation and by soil moisture as it relates to topography (Whittaker 1960; Waring 1969). Whittaker's work in southwestern Oregon documented different montane forest patterns on diorite, gabbro, and serpentine. In California, similar patterns are readily observed in eastern Del Norte Co. To the south and east, differences become greater as climate and rock patterns change.

The lithic pattern in California is a series of north–south-tending arcuate belts of distinctive rock assemblages of similar age and character. Geologists (Davis 1966; Irwin 1960, 1966) recognize four subprovinces (eastern Paleozoic, central metamorphic, western Paleozoic and Triassic, and western Jurassic). The composition of sedimentary and volcanic rock matrix and the degree of metamorphism vary. Granitic intrusions are mainly dioritic, and diverse ultrabasic intrusions, principally of peridotite and dunite which may be altered to serpentine, are common and may be extensive. In the montane zone, metamorphic, granitic, and ultrabasic parent materials differentially affect vegetation patterns. In comparing forest patterns of these classes, those on ultrabasics are most distinctive.

MONTANE VEGETATION PATTERNS

Montane forest patterns are reasonably predictable if mature, climax forests growing on deep soils are considered. These forests form a series of elevationally zoned belts dominated by one to several tree species. Four elevational zones can be conveniently recognized according to the dominant climax trees of the mesic habitats. These are, with increasing elevation, *Abies concolor, A. magnifica* var. *shastensis,*

Tsuga mertensiana, and *Pinus albicaulis*. These zones can then be divided into a series of vegetation types reflective of distinctive habitats. Within each zone, forests rich in herbs and shrubs along streams give way to simple ones barren of understories on dry slopes. The pattern is similar to that of other western mountain areas (Daubenmire and Daubenmire 1968; Franklin and Dyrness 1973), but here it is less obvious because equivalent environments are not continuous. In addition, variable fire history has created a confusing variety of seral stages. For this reason understory composition can be more indicative of habitat potential.

Klamath region forest patterns show decreasing geographical differentiation with increasing elevation (Waring 1969; Sawyer and Thornburgh 1969). Low and mid-elevation forests of the Siskiyou Mts., Marble Mts., western Salmon Mts., and western Trinity Alps are distinctive from those of the Trinity Mts., eastern Salmons, and eastern Trinity Alps. For this reason it is necessary to subdivide the Klamath region into western and eastern subregions.

In the western subregion, low-elevation *Pseudotsuga*-hardwood forests give way to *Abies concolor–Pseudotsuga menziesii* forests, then to *Abies magnifica–A. concolor* forests, and to *Tsuga mertensiana–A. magnifica* forests with increasing elevation. These forests grow in relatively cooler and more moist climatic areas which receive greater maritime influence because of location, as in the Siskiyous, or are of overall westerly aspect, as in the Marbles, western Salmons, and western Trinity Alps.

East and south of the highest ridges and peaks of the Salmons and Trinity Alps the climate is not only drier but also more continental in character. Here, in the eastern subregion, the forest zonation pattern is one of *Pinus ponderosa*-dominated forests at low elevations, giving way to *Abies concolor–Pinus* spp. forests, then to *Abies magnifica–A. concolor* forests, and to *Tsuga mertensiana–A. magnifica* forests with increasing elevation.

For these reasons we have made some distinctions between forest types of the eastern and western subregions in the *Abies concolor* zone. Subregional differences are not as great in the *Abies magnifica* zone, although some consistent variation is described. The highly discontinuous forests of the *Tsuga mertensiana* and *Pinus albicaulis* zones are regionally uniform.

Vegetation studies done in the montane zone in California are restricted to forest patterns on granitic parent materials (Sawyer and Thornburgh 1969, 1970, 1971, 1974), or are mainly floristic (Ferlatte 1974; Howell 1944; Muth 1967; Ottenger 1975). Soil-vegetation map information exists for some montane areas of the Orleans District, Six Rivers National Forest, Humboldt Co. (Rockey et al. 1966), the Chanchelulla Peak quadrangle (Mallory et al. 1968), and the French Gulch quadrangle (Mallory et al. 1973), Trinity and Shasta Cos. The Society of American Forester's cover types (1954) poorly define Klamath montane forests. Our work, done in the central Klamath region forests in Trinity, Siskiyou, Del Norte, and Humboldt Cos., forms the basis for the forest type descriptions discussed in this chapter.

Vegetation on Granitic and Metamorphic Parent Materials

Abies concolor zone. Five forest types may be recognized at midelevations (600–1600 m). Three are western subregion forests dominated by *Pseudotsuga* and *Abies* *concolor*. Two are in the eastern subregion, where *Pinus ponderosa* and *P. lambertiana* play more important roles, producing forests similar to Sierra Nevada mixed conifer (Chapter 17). Each forest type is characterized by overstory and understory taxa. We shall employ a slash to separate dominants of different strata. Western subregion forest types are characterized by *Trillium, Vicia,* and *Chimaphila*; eastern types, by *Berberis* and *Ceanothus*.

Forests dominated by *Pseudotsuga* with an *Abies* understory as mapped by Rockey et al. (1966) for the western Salmon Mts. can be readily categorized into the western subregion types. Ferlatte's (1974) mixed conifer zone is equivalent to eastern *Abies concolor/Berberis nervosa* and *Abies concolor/Ceanothus prostratus* types. In Shasta and Trinity Cos., Mallory et al. (1968, 1973) mapped several conifer–hardwood types for the *A. concolor* zone. These types (Chawanakee–Chiax–Holland/ponderosa pine–Douglas fir–California black oak, Sheetiron–Marapa–Josephine/ponderosa pine–Douglas fir–California black oak, Boomer–Sheetiron–Josephine/ponderosa pine–Douglas fir) are generally applicable to our *Abies concolor/Ceanothus prostratus*. The Corbett/shrub tan oak–mixed conifer appears as a still drier phase in these steep, interior mountains.

Abies concolor/Trillium ovatum. *Pseudotsuga menziesii* and *Abies concolor* dominate this tall, dense forest typical of draws, streamside terraces, seeps, and lower slopes with deep soils in the western subregion (Fig. 20-1). Forests approaching this type may also occur on some upper slopes. Dominance of *Pseudotsuga* in the canopy and *A. concolor* in the understory is typical, although *Pseudotsuga* consistenty reproduces in small numbers. *Pinus lambertiana* and *Calocedrus decurrens* are common secondary conifers.

Chrysolepis chrysophylla and *Taxus brevifolia* are typical understory trees above well-developed shrub and ground layers. The shrub layer is diverse, including up to 15 common regional species in a single stand (Table 20-1). The ground layer is rich in species and covers most of the soil.

This forest type commonly grades into *Abies concolor/Vicia americana* of mid-slopes, but can be differentiated from it by the extensive shrub layer. At higher elevations it grades into *Abies magnifica/Leucothoe davisiae* and *A. magnifica/Linnaea borealis*.

Abies concolor/Vicia americana. *Pseudotsuga menziesii* and *Abies concolor* dominate this dense western subregion forest type on mesic slopes. *Chrysolepis chrysophylla* and *Quercus chrysolepis* are common understory trees. The shrub layer is diverse, but of low coverage. The ground layer is diverse and typically well developed (Table 20-1).

This type is intermediate in character between the more moist *Abies concolor/Trillium ovatum* and the more xeric *Abies concolor/Chimaphila umbellata*. It is distinctive, though, with a sparse shrub layer and a well-developed ground layer.

Abies concolor/Chimaphila umbellata. This type is dominated by *Abies concolor*, with *Pseudotsuga menziesii, Pinus lambertiana, P. ponderosa,* and *Calocedrus decurrens* as less important associates. It grows on drier slopes and ridges with deep soils within the western subregion. This open mature forest allows reproduction of all five conifers. *Quercus chrysolepis* is a component at lower elevations. The depauperate shrub and ground layers distinguish the type (Table 20-1).

Immature forests are typically dense stands of *A. concolor*, a trait which contrasts with the more open character of *Abies concolor/Ceanothus prostratus* seres that

Figure 20-1. *Abies concolor/Trillium ovatum* forest in the *Abies concolor* zone of the Marble Mts., about 1400 m. Trees include *A. concolor, Pseudotsuga menziesii, Pinus ponderosa, P. lambertiana, Crysolepis chrysophylla,* and *Taxus brevifolia.*

allow maintenance of a chaparral component. This forest grades into *Abies concolor/Vicia americana* and the equivalent *Abies magnifica/Chimaphila umbellata* of the *Abies magnifica* zone.

Abies concolor/Berberis nervosa. In this eastern subregion type, *Abies concolor* dominates a typical assemblage of midelevation conifers, including *Pseudotsuga menziesii, Calocedrus decurrens, Pinus lambertiana,* and *P. ponderosa,* which form moderate to dense forests on mesic habitats with deep soils. The seedling and sapling layers are heavily dominated by *A. concolor* and *Calocedrus. Cornus nuttallii, Quercus chrysolepis, Q. kelloggii, Acer macrophyllum,* and *Taxus brevifolia* are common understory associates. Moderately developed shrub and ground layers are present (Table 20-1).

A mixture of canopy species, rather than *Abies concolor* and *Pseudotsuga* dominance, a mingling of mesophytic and xerophytic shrubs, and a less developed ground layer distinguish this type from *Abies concolor/Trillium ovatum* and *Abies concolor/Vicia americana* forests found on comparable habitats in the western

TABLE 20-1. Presence (Prs, %) and modal cover/abundance (C/A)[a] for shrubs with greater than 10% presence and herbs with greater than 20% presence in the *Abies concolor* zone types. Average relevé size for shrubs and herbs, 50 m^2. From Sawyer and Thornburgh (1974)

	Abies concolor				
Forest type:	*Trillium ovatum*	*Vicia americana*	*Chimaphila umbellata*	*Berberis nervosa*	*Ceanothus prostratus*
Number of relevés:	20	17	17	16	26
Species	Prs•C/A	Prs•C/A	Prs•C/A	Prs•C/A	Prs•C/A
Shrubs					
Vaccinium membranacium	15.3				
Quercus sadleriana	35.4				
Arctostaphylos nevadensis				19.2	
Rubus ursinus	20.2			6.2	
Berberis dictyota	15.2			6.2	
Ribes lacustre	15.3			13.3	
Holodiscus discolor[b]	15.3	6.3			
Cornus stolonifera	15.3	16.2		6.2	
Ribes lobbii	35.3	29.2		6.2	
Rubus parviflorus	75.4	39.2		19.2	
Ribes sanguineum	10.2	12.3		25.3	
Salix scouleriana	10.3	12.2		25.3	11.3
Amelanchier pallida	5.3	16.3		31.2	7.3
Cornus nuttallii	30.3	24.3		63.2	4.3
Berberis nervosa	45.3	41.3	13.2	88.3	
Corylus cornuta var. *californica*	65.4	59.3	38.3	25.4	
Rosa gymnocarpa	95.3	76.3	25.3	81.3	27.2
Symphoricarpos hesperius	45.3	24.2	31.2	50.3	23.2
Paxistema myrsinites	55.3	12.2	6.3	44.3	11.3
Ceanothus velutinus	25.2	6.2	25.2	6.3	11.3
Symphoricarpos mollis	30.3	24.3		6.3	23.3
Chrysolepis sempervirens	5.4	6.2	19.2	44.3	50.5

TABLE 20-1 [continued]

	Abies concolor				
Forest type:	Trillium ovatum	Vicia americana	Chimaphila umbellata	Berberis nervosa	Ceanothus prostratus
Number of relevés:	20	17	17	16	26
Species	Prs•C/A	Prs•C/A	Prs•C/A	Prs•C/A	Prs•C/A
Quercus vaccinifolia	15.3	6.2	6.3	25.3	46.4
Ceanothus prostratus			19.2	13.3	65.4
Berberis repens					15.3
Herbs					
Trillium ovatum	60.3				
Asarum caudatum	24.2				
Achlys triphylla	60.4	29.3			
Trientalis latifolia	55.3	29.3			
Penstemon anguineus	35.3	24.3			
Viola glabella	35.3	24.3			
Vancouveria hexandra	25.3	16.3			
Osmorhiza chilensis	25.3	24.2			
Smilacina racemosa					
var. amplexicaulis	20.3	24.3			
S. stellata[c]	30.3				11.3
Lupinus andersonii		29.3			
Adenocaulon bicolor	50.4	53.3		40.3	
Galium triflorum	50.3	41.3		27.2	
Frageria californica	35.3	12.2		27.2	
Clintonia uniflora	50.4	12.3		20.3	
Anemone deltoidea	30.3	16.3		20.2	
Hieracium albiflorum	25.3	47.3		20.3	
Pyrola secunda	25.3	12.2		27.2	

Linnaea borealis					
ssp. *longifolia*	35.3	12.2		33.3	
Arnica cordifolia[d]	10.2	12.2		20.2	
Calypso bulbosa		12.2		20.2	
Vicia americana var. *oregona*		59.3		20.2	
Disporum hookeri					
var. *trachyandrum*	45.3	35.3	19.2	40.2	
Corallorhiza maculata	25.2	29.2	31.2	12.2	
Apocynum pumila	30.2	24.3	19.2	33.2	
Campanula prenanthoides	10.2	29.3	12.2	12.2	
Stellaria jamesiana	10.2	24.3	19.2		
Goodyera oblongifolia	40.3	41.3	25.2	20.2	11.2
Pteridium aquilinum					
var. *lanuginosum*	45.3	35.3	12.2	53.3	34.3
Chimaphila umbellata					
var. *occidentalis*	70.3	53.3	56.3	67.3	69.3
Pyrola picta[e]	50.3	29.2	31.2	67.2	46.2
Chimaphila menziesii	20.3	29.3	31.2	27.2	15.2
Bromus marginata		12.3	19.2	20.2	11.2
Festuca occidentalis			12.2	27.2	

[a] 1 = one individual, 2 = rare, 3 = <10%, 4 = 10–25%, 5 = 25–50%, 6 = 50–75%, 7 = >75%.

[b] Including *H. discolor* var. *franciscanus*.

[c] Including *S. stellata* var. *sessifolia* in *Ceanothus prostratus* type.

[d] Including *A. cordifolia* var. *alata* and var. *pumila*.

[e] Including *P. picta* forma *aphylla* and ssp. *dentata*.

subregion. On drier slopes the type merges with *Abies concolor/Ceanothus prostratus* or *Abies concolor/Chimaphila umbellata* forests.

Abies concolor/Ceanothus prostratus. This is a generally open to moderately dense eastern subregion forest type on middle to upper slopes and ridges with thin soils. It is dominated by *Pinus ponderosa. Abies concolor* and *Pseudotsuga menziesii* are common secondary dominants, along with *Pinus lambertiana* and *Calocedrus decurrens.* A secondary layer of *Quercus chrysolepis, Q. kelloggii,* and *Chrysolopis chrysophylla* forms a lower canopy. The seedling and sapling layers are generally dominated by *Pseudotsuga* and *Abies.*

The understory is mainly of sclerophyllous shrubs, often forming dense patches under canopy openings, or of nonsclerophylls such as *Rosa gymnocarpa* and *Symphoricarpos hesperius.* Herbs are uncommon except for *Chimaphila umbellata, Pyrola picta,* and *Pteridium aquilinum.*

Successional trends show a mix of conifers gradually growing through and overtopping montane chaparral. The open, mixed character of the mature forest allows the light-demanding species to continue to reproduce well. The lack of a dense shade stage, as found in *Abies concolor/Chimaphila umbellata,* accounts for the understory differences between these closely related types which have mature forests of similar canopy composition. In addition, *Abies concolor/Ceanothus prostratus* does not show the more nearly complete *Abies* dominance of relatively more mesic *Abies concolor/Chimaphila umbellata.*

***Abies magnifica* zone.** A series of forest and meadow types may be recognized between 1400 and 1950 m where *Abies magnifica* dominates successional trends. There are both open- and closed-canopied climax forests at these elevations, creating greater variety than is found in the *Abies concolor* zone. Subregional differences are not as evident, although consistent differences occur. In the western Siskiyou Mts., *Abies procera* plays an ecologically equivalent role to *A. magnifica* elsewhere; however, field distinction between the two is difficult without rarely available mature cones, so *A. procera* data are included with those of *A. magnifica.* The taxonomic status of the two taxa, especially in the Klamath region, is still unclear (Franklin and Dyrness 1973; Griffin and Critchfield 1972).

Meadows are extensive within this zone, especially on metasediments of the western subregion mountains. Streamside terraces and broad valleys support lush, herb-rich communities. (*Actaea rubra, Agastache urticifolia, Angelica arguta, Carex* spp., *Delphinium glaucum, Heracleum lanatum, Hydrophyllum fendleri, Glyceria elata, Ligusticum californicum, Polygonum phytolaccaefolium, Solidago triangularis, Thalictrum polycarpum, Trifolium longipes, Valeriana sitchensis, Veratrum californicum, Vicia americana, Viola glabella,* and others).

Moist meadows grade into open to moderately dense *Abies magnifica* or *A. procera* forests with herb-rich understories of up to 80% cover (Fig. 20-2). These undescribed forests grow on various metasediments and are similar to herb-rich associations described by Franklin and Dyrness (1973) in southwestern Oregon.

Gravelly, internally drained upper slopes and broad ridges support extensive open meadows throughout the zone. Similar habitats within the *Tsuga mertensiana* zone also support these dry meadows, of which many are being invaded by *Abies magnifica.* Scattered *Haplopappus greenei* are interspersed with a variety of herbs and grasses. Grazing apparently increases herb dominance on these dry habitats. *Artemisia tridentata* is associated with these meadows in the eastern subregion.

Figure 20-2. *Abies concolor/Vicia americana* forest covering the slopes of the Marble Mts. in the *Abies concolor* zone. Major trees are *A. concolor* and *Pseudotsuga menziesii.* Picture was taken from about 1900 m in the *Abies magnifica* zone, and *Abies magnifica* is present on the foreground slopes.

Alder thickets are also common in the moist rocky talus and avalanche chutes centered in the *Abies magnifica* zone, but extending into the *A. concolor* zone, where they are less well developed. *Alnus sinuata, A. tenuifolia, Cornus stolonifera, Rhamnus purshiana, Ribes lacustre,* and *Sambucus melanocarpa* form impenetrable thickets throughout the region. The herbaceous understory is varied, including typical streamside and moist-habitat species.

As with the *Abies concolor* zone forest types, we shall employ a slash to separate characteristic taxa belonging to different strata. Four of the types are dominated by *A. magnifica* and differ in understory dominants. The fifth type is dominated by shrubs. Table 20-2 summarizes the five types.

Abies magnifica/Leucothoe davisiae. This dense forest type on streamside terraces, around meadow margins, and along seeps is dominated in the canopy, sapling, and seedling layers by *Abies magnifica* and *A. concolor. Pinus contorta, P. monticola,* and *Tsuga mertensiana* are common regional associates. Occasionally *Picea breweriana* is found at higher elevations, as are *Pinus lambertiana, P. ponderosa,* and *Calocedrus decurrens* at lower elevations. *Picea engelmannii* and *Abies lasiocarpa* are found in the Russian Peak area, and *Chamaecyparis nootkatensis* in the Siskiyou Mts.

The rich shrub layer is dominated in many areas by *Leucothoe davisiae. Ribes lacustre, Alnus tenuifolia, Sorbus californica,* and *Salix scouleriana* border the streams. *Taxus brevifolia* covers many alluvial flats. The ground layer is varied.

This type most commonly grades into *Abies magnifica/Linnaea borealis* forests. At higher elevations, it is associated with *Tsuga mertensiana/Phyllodoce empetriformis* and may be found on mesic slopes. At lower elevations, it is restricted to streamside locations and associates to some extent with *Abies concolor/Berberis nervosa,* but more commonly it grades into *Abies concolor/Trillium ovatum.*

Abies magnifica/Linnaea borealis. *Abies concolor* and *A. magnifica* dominate this

TABLE 20-2. Presence (Prs, %) and modal cover/abundance (C/A)[a] for shrubs with greater than 10% presence and herbs with greater than 20% presence in the *Abies magnifica* zone types. Average relevé size for shrubs and herbs, 50 m^2. From Sawyer and Thornburgh (1974)

	Abies magnifica				Quercus vaccinifolia–Arctostaphylos patula
Forest type:	Leucothoe davisae	Linnaea borealis	Chimaphila umbellata	Quercus vaccinifolia	
Number of relevés:	27	24	21	23	27
Species	Prs•C/A	Prs•C/A	Prs•C/A	Prs•C/A	Prs•C/A
Shrubs					
Kalmia polifolia var. *microphylla*	11.3				
Ledum glandulosum var. *californicum*	37.3				
Rhamnus californica		12.2			
Ribes lacustre	59.3	8.2			
Alnus tenuifolia	59.4	8.2			
Vaccinium arbuscula	30.3	17.3			
Alnus sinuata	26.3	17.3			
Berberis nervosa	11.3	25.2			
Vaccinium membranacium	7.4	17.3	10.3		
Leucothoe davisiae	85.7	8.3		4.3	
Salix scouleriana	37.3	29.3		13.3	
Spiraea douglasii	15.3	8.2		4.3	
Rosa gymnocarpa	33.2	67.3	29.3	26.3	
Symphoricarpos hesperius	37.3	67.3		17.3	7.4
Sorbus californica	22.2	12.3	10.2	4.1	7.2
Rubus californica	11.3	37.2	37.2	4.3	
Quercus sadleriana		37.4	33.3	4.5	22.3
Paxistema myrsinites		17.2		13.3	
Ribes lobbii	7.2	33.2	19.2	13.3	4.2

Lonicera conjugialis	15.2	4.3	10.2	17.2	7.3
Ribes viscosissimum	15.2	4.2	19.2	4.3	15.3
Chrysolepis sempervirens	22.3	25.3	19.2	76.4	22.5
Acer glabrum var. *torreyi*	7.4	33.3	10.3	17.3	7.2
Amelanchier pallida	19.3	33.3	5.2	30.3	59.4
Arctostaphylos nevadensis	4.3	17.2	14.3	69.4	44.4
A. patula	4.5	4.2		74.3	100.5
Ceanothus velutinus	4.2	4.3	5.3	26.4	70.4
Quercus vaccinifolia	4.2	25.3	5.3	74.5	70.5
Holodiscus microphyllus	7.2	17.3		22.3	48.3
Prunus emarginata		4.2	5.2	26.3	33.4
Garrya fremontii				9.3	37.3
Juniperus communis var. *saxatilus*					19.3
Haploppus greenii					15.2
Quercus garryana var. *breweri*					22.4
Herbs					
Listera convallarioides	26.2				
Mitella pentandra	30.3				
Streptopus amplexifolius var. *denticulatus*	22.3				
Senecio triangularis	48.3				
Athyrium filix-femina var. *californicum*	26.3				
Clintonia uniflora	63.3	42.3			
Trientalis latifolia	55.3	29.3			
Trillium ovatum[b]	48.2	29.2			
Linnaea borealis ssp. *longiflora*	11.4	29.3			
Anemone quinquefolia[c]	22.2	17.3			
Viola glabella	33.3	12.2			
Smilacina racemosa var. *amplexicaulis*	11.2	29.3			
Osmorhiza chilensis	22.3	25.3			
Smilacina stellata[d]	48.3	12.3			
Osmorhiza occidentalis		21.2			
Phacelia hastata		25.3			

TABLE 20-2 [continued]

Species	Leucothoe davisae	Linnaea borealis	Chimaphila umbellata	Quercus vaccinifolia	Quercus vaccinifolia-Arctostaphylos patula
Forest type:	Abies magnifica				
Number of relevés:	27	24	21	23	27
	Prs•C/A	Prs•C/A	Prs•C/A	Prs•C/A	Prs•C/A
Vicia americana var. *oregona*		21.3			
Anemone deltoidea	48.3	58.3	10.3		
Pyrola secunda	11.3	54.2	48.3		
Arnica cordifolia[e]	26.3	29.4	14.3		
Adenocaulon bicolor	11.3	37.3	10.2		
Goodyera oblongifolia	33.2	25.3	14.2		
Galium triflorum	15.3	46.3	10.2		
Penstemon anguineus		17.3	24.2		
Disporum hookeri var. *trachyandrum*		21.3	14.3		
Pteridium aquilinum var. *lanuginosum*	52.4	54.3	33.3	22.3	
Pyrola picta[f]	37.2	42.2	33.2	35.2	
Chimaphila umbellata var. *occidentalis*	70.3	71.3	57.3	65.3	
C. menziesii	22.3	37.2	52.2	35.2	
Hieracium albiflorum	30.2	33.3	29.3	17.2	
H. gracile	11.3				26.3
Apocynum pumila		21.3	19.2	61.3	22.2
Kelloggia galioides		12.3		35.3	22.3
Monardella odoratissima var. *glauca*			10.2	48.3	37.3
Senecio integerrimus var. *major*				17.3	26.3
Arenaria congesta[g]				13.4	26.3
Castilleja appelgatei				17.3	52.3
Penstemon newberryi ssp. *berryi*				35.3	63.3
Sedum obtusatum ssp. *boreale*				17.3	41.3

Phlox diffusa	13.3	41.3
Cheilanthes gracillima	17.2	56.3
Achillea lanulosa		22.3
Penstemon procerus ssp. *brachyanthus*		37.3
Onychium densum		22.3
Eriogonum umbellatum[h]		37.4
Smilacina racemonsa[i]		41.3
Calyptridium umbellatum		33.2

[a]1 = one individual, 2 = rare, 3 = <10%, 4 = 10–25%, 5 = 25–50%, 6 = 50–75%, 7 = >75%.

[b]Including *T. oratum* ssp. *ottingeri*, which replaces the species at higher elevations.

[c]*A. quinquefolia* var. *minor* and var. *oregona*.

[d]Including *S. stellata* var. *sessifolia*.

[e]Including *A. cordifolia* var. *alata* and var. *pumila*.

[f]Including *P. picta* forma *aphylla* and ssp. *dentata*.

[g]Including *A. congesta* var. *suffrutescens*.

[h]Including *E. umbellatum* var. *polyanthum* and var. *stellatum*.

[i]*S. racemosa* var. *glauca* in *Quercus vaccinifolia–Arctostaphylos patula* type.

forest type, with a moderate to dense canopy and an understory rich in shrubs and herbs. It occurs on lower slopes with uniformly deep, moist soils. *Pseudotsuga menziesii, Pinus lambertiana, P. monticola,* and *Calocedrus decurrens* are also common. *Abies* dominates the seedling and sapling layers, although all species are reproducing in mature forests.

Characteristically the shrub layer is dominated by *Quercus sadleriana,* but several other species have high cover/abundance values. The ground layer is well developed, with *Linnaea borealis* among others. This type grades into *Abies magnifica/Leucothoe davisiae* or the open forests of the *A. magnifica* zone, as well as *Tsuga mertensiana/Quercus vaccinifolia* at higher elevations.

On rocky, moist moraines, generally open forests of this type are very diverse, being characterized by a total of 16 conifers, with as many as 10 in a single stand. Shrub and herb composition is typical of the *Abies magnifica/Linnaea borealis* type. The canopy diversity results from midelevation taxa being enriched by high-elevation species. The dominants are *Pseudotsuga, Abies concolor,* and *A. magnifica*; the other species in order of prominence are *Pinus monticola, Picea breweriana, Tsuga, Pinus lambertiana, P. ponderosa, Taxus, Calocedrus, Picea engelmannii, Pinus contorta, Abies lasiocarpa, A. amabilis, Chamaecyparis nootkatensis,* and *C. lawsoniana.* All species reproduce in these forests, although the midelevation species gradually decline in importance with elevation. Forests like these were first found in the Russian Peak area (Sawyer et al. 1970, NA 471891), and described as enriched mixed conifer. Since then, other examples have been found in the eastern Trinity Alps, eastern Marble Mts., and central Siskiyous (Bear Basin Butte Botanical Area, NA 080231). Enriched stands like these are the exception for the type and occur in moist areas of spotty soil development.

Abies magnifica/Chimaphila umbellata. On drier middle to upper slopes and ridges with deep soils, a dense *Abies concolor–A. magnifica* forest type is common. *Pseudotsuga menziesii, Pinus lambertiana, P. monticola,* and *P. contorta* are less important than in more mesic forests at these elevations. Shrub and ground layers exhibit lower coverage and less variety than do the more mesic forests (Table 20-2). Common shrubs include *Rubus parviflorus, Quercus sadleriana,* and *Rosa gymnocarpa.* The ground layer is distinctive in its poor development rather than its composition, with *Pyrola* spp., *Pteridium aquilinum,* and *Chimaphila* spp. most common.

This forest type is ecologically equivalent to the lower *Abies concolor/Chimaphila umbellata,* but distinguished by the major role played by *Abies magnifica. Abies concolor*-dominated forests commonly found in the *A. magnifica* zone are seral, with *A. magnifica* heavily represented in the understory. At higher elevations, *A. concolor* is rare in *Abies magnifica/Pyrola picta* or comparable habitats in the *Tsuga mertensiana* zone. A midseral phase with dense *A. concolor* canopy may account for the lack of early seral shrubs continuing into more open, mature forests. This type also grades into mesic forests with *Abies* dominance.

Abies magnifica/Quercus vaccinifolia. This forest type, dominated by *Abies magnifica, A. concolor,* and *Pseudotsuga menziesii,* with *Pinus lambertiana, P. contorta, P. monticola,* and *Picea breweriana* as common associates, occurs on dry talus and rocky moraines. The canopy is composed of scattered groups of trees among a shrub matrix. Tree reproduction indicates a steady state for forest composition.

Shrubs cover the open, rocky slopes, with *Arctostaphylos patula, A. nevadensis, Quercus vaccinifolia,* and *Chrysolepis sempervirens* dominant. Other sclerophylls play a less dominant role than in the drier *Quercus vaccinifolia-Arctostaphylos patula*. Herbs associated with the mesic forests are common (Table 20-2).

This type gradually changes with elevation into *Tsuga mertensiana/Quercus vaccinifolia* and, on drier habitats of the *Abies magnifica* zone, into *Quercus vaccinifolia–Arctostaphylos patula*.

Quercus vaccinifolia–Arctostaphylos patula. Patches of generally dense sclerophyllous shrubs about 1 m tall cover dry slopes throughout the region. *Pinus monticola, Abies magnifica,* and *A. concolor* are scattered among the shrubs. *Pinus contorta, P. lambertiana, P. jeffreyi,* and *Calocedrus decurrens* are less common, although any tree may be locally abundant. *Amelanchier pallida, Arctostaphylos patula, A. nevadensis, Ceanothus velutinus, Holodiscus microphyllus, Prunus emarginata,* and *Quercus vaccinifolia* dominate the chaparral in varying degrees. Herbs are found scattered among the shrubs (Table 20-2).

Climax montane chaparral is developed best in the *Abies magnifica* zone and to a lesser extent in the *Abies concolor* and *Tsuga mertensiana* zones. It grades into *Pinus monticola/Holodiscus microphyllus* at higher elevations, into *Abies magnifica/Quercus vaccinifolia* within the *A. magnifica* zone, and into *Abies concolor/Chimaphila umbellata* and *Abies concolor/Ceanothus prostratus* at lower elevations where chaparral of similar composition is also seral.

This type is quite extensive in the central Trinity Alps, where it is described by Ferlatte (1974) as montane chaparral. On north-facing slopes and ridges of Paradise Peak, Shoemaker Bally, and Shasta Bally, Mallory et al. (1973) mapped scattered groves of *Abies concolor* and *A. magnifica* growing among montane chaparral dominated by *Lithocarpus densiflora* var. *echinoides, Arctostaphylos patula, Ceanothus prostratus* var. *laxus, Chrysolepis chrysophylla,* and *Quercus vaccinifolia* (Corbett/true fir, Corbett/ridgetop chaparral). Such groves appear to be similar to what we have described as *Quercus vaccinifolia–Arctostaphylos patula.*

***Tsuga mertensiana* zone.** A general lack of midelevation conifers characterizes five *Tsuga mertensiana*-dominated, high-elevation (1900–2200 m) forests. In contrast to the lower zones, these forests show little regional differentiation, although the distinction between open- and closed-canopied forests is convenient here as well. In most parts, meadows rather than forests characterize this zone. Glaciated valleys of the Trinity Alps and Salmon and Marble Mts. have soils too thin to even support open forests. The following brief descriptions, taken from observations in the Trinity Alps and Salmon Mts., suggest the nature of these unstudied communities.

Soils accumulated in depressions and among rocks support a rich array of plants watered by melting snow. A short season of quick growth and bloom occurs before these slopes dry in late July or August. *Carex* spp., including *C. interior* and *C. rostrata,* form patches in the larger depressions, along with *Angelica arguta, Athyrium alpestre, Castilleja miniata, Dodecatheon jeffreyi, Luzula parviflora, Schoenolirion album,* and *Veratrum viride.*

Wetter meadows are formed in larger valley floor depressions, on lake margins, and on slopes watered until late in the summer or fall. The herbs grow among *Carex* spp., *Scirpus microcarpus, Juncus mertensianus,* and patches of shrubs such as *Kalmia polifolia, Phyllodoce empetriformis,* and *Vaccinium arbuscula.*

Two of the five forest types in this zone are characterized mainly by *Tsuga mertensiana* in the overstory; *Abies concolor, A. magnifica,* or *Pinus monticola* play more important roles in the other three. Table 20-3 summarizes these types.

TABLE 20-3. Presence (Prs. %) and modal cover/abundance (C/A)[a] for shrubs with greater than 10% presence and herbs with greater than 20% presence in the *Tsuga mertensiana* and *Pinus albicaulis* zone types. Average relevé size for shrubs and herbs, 50 m^2. From Sawyer and Thornburgh (1974)

	Tsuga mertensiana/			Abies magnifica/	Pinus monticola/	Pinus albicaulis/
Forest type:	Phyllodoce empetriformis	Quercus vaccinifolia	Pyrola picta	Pyrola picta	Holodiscus micropyllus	Holodiscus microphyllus
Number of relevés:	29	18	21	18	23	21
Species	Prs•C/A	Prs•C/A	Prs•C/A	Prs•C/A	Prs•C/A	Prs•C/A
Shrubs						
Cassiope mertensiana	14.3					
Ledum glandulosum var. *californicum*	17.3					
Kalmia polifolia var. *macrophylla*	24.3					
Sambucus melanocarpa	14.3	6.3				
Alnus sinuata	28.4	6.5				
Phyllodoce empetriformis	58.4	6.3				
Sorbus californica	34.2	17.2				
Salix commutata	24.3	17.3				
Vaccinium membranacium	10.2	11.3				
Ribes lacustre	31.3	17.4				
Acer glabrum var. *torreyi*	7.4	22.4				
Vaccinium scoparium	34.4	11.3	11.3			
Leucothoe davisiae	37.3	6.1	6.3			
Vaccinium arbuscula	86.4	22.3			4.3	
Quercus vaccinifolia	14.2	56.4	6.2		65.4	5.4
Ribes viscosissimum	21.2	33.3	6.3	14.3	13.3	10.2
Arctostaphylos nevadensis	24.3	67.4	28.3	24.3	91.5	52.4
Chrysolepis sempervirens	7.3	44.4	11.2	14.2	22.2	43.3
Arctostaphylos patula	7.2	39.3	17.2	19.2	74.3	62.4
Quercus sadleriana	3.5	1.4		10.2	25.3	
Holodiscus microphyllus	10.2	17.4		5.2	61.3	86.3

Prunus emarginata	7.4	22.3		5.2	26.3	19.3
Amelanchier pallida	24.2	33.3			30.3	10.3
Lonicera conjugialis	38.3	11.3			13.3	5.2
Juniperus communis var. *saxatilus*	10.3				13.3	14.3
Haplopappus greenii					13.3	29.4
Cercocarpus ledifolius						14.4
Herbs						
Ligusticum grayi	24.3					
Senecio triangularis	38.3					
Veratrum californicum	58.3					
Potentilla flabellifolia	28.3					
Epilobium angustifolium	24.3	22.3				
Cystopteris fragilis	24.2	11.3				
Dodecatheon jeffreyi	65.3	17.2				
Arnica latifolia	31.2	28.3				
Clintonia uniflora		22.3				
Penstemon duestus		28.3				
Montia parvifolia		22.3				
Castilleja appelgatei		22.3				
Aster occidentalis		22.3				
Collinsia torreyi[b]		22.3				
Apocynum pumila		28.2				
Pteridium aquilinum var. *lanuginosum*		22.5				
Hieracium albiflorum		28.3		14.2	13.3	
Pyrola picta	14.3	11.3	28.2	38.2		
Chimaphila umbellata var. *occidentalis*	14.2	33.3	33.2	19.2	13.3	
Polygonum davisiae	31.3		17.2	17.2	30.3	29.3
Penstemon newberryi ssp. *berryi*	14.2	39.3	44.2	14.2	82.3	67.3
Monardella odoratissima var. *glauca*	17.2	28.3		14.2		14.3
Antennaria rosea	14.3	22.3			17.3	
Saxifraga ferruginea	24.3	11.2			11.2	

TABLE 20-3 [continued]

Species	Tsuga mertensiana/			Abies magnifica/	Pinus monticola/	Pinus albicaulis/
Forest types:	Phyllodoce empetriformis	Quercus vaccinifolia	Pyrola picta	Pyrola picta	Holodiscus micropyllus	Holodiscus microphyllus
Number of relevés:	29	18	21	18	23	21
	Prs•C/A	Prs•C/A	Prs•C/A	Prs•C/A	Prs•C/A	Prs•C/A
Sedum obtusatum ssp. *boreale*	31.3	44.3			69.3	
Cryptogramma arostichoides	28.2	22.3			35.2	
Senecio integerrimus var. *major*		28.3			17.3	
Penstemon procerus var. *brachyanthus*		11.2			26.3	
Selaginella densa var. *scopulorum*					22.3	
Lewisia leana					22.2	
Hieracium gracile					22.3	
Boschniakia strobilacea					22.2	
Lewisia cotyledon					43.3	
Juncus parryi					48.3	
Cheilanthes gracillima	14.3	17.3			56.2	38.2
Arenaria congesta	14.2	28.3			62.3	24.3
Phlox diffusa	21.2	28.3			43.3	67.3
Eriogonum umbellatum[c]		11.2			17.3	33.3
Achillea lanulosa		28.3			22.3	29.3
Calyptridium umbellatum					39.3	14.3
Aster ledifolius					17.3	19.3
Angelica arguta	31.3	33.3				19.2
Draba howellii						20.2
Polemonium pulcherrimum						33.2

[a] 1 = one individual, 2 = rare, 3 = <10%, 4 = 10-25%, 5 = 25-50%, 6 = 50-75%, 7 = >75%.

[b] Including *C. torreyi* var. *latifolia* and var. *wrightii*.

[c] Including *E. umbellatum* var. *polyanthum* and var. *stellatum*.

Tsuga mertensiana/Phyllodoce empetriformis. This forest type, growing along streams and lake margins and in local seeps, is regionally dominated by *Tsuga mertensiana* and *Abies magnifica,* along with *A. concolor, Pinus monticola,* and *P. contorta. Abies lasiocarpa* and *Picea engelmannii* are restricted to the Russian Peak stands. Reproduction is heavy for *Tsuga,* with others less abundant although all species are actively invading many wet meadows.

Phyllodoce empetriformis, among others, dominates the shrub layer, with *Ledum glandulosum* and *Kalmia polifolia* common at lake margins. The ground layer is composed of wet meadow herbs mingling with such forest associates as *Pyrola picta* and *Chimaphila umbellata,* under trees. *Sedum obtusatum, Phlox diffusa,* and *Cryptogramma acrostichoides* grow in xeric microhabitats which are especially characteristic of the seepage phase (Table 20-3).

This type grades into *Abies magnifica/Leucothoe davisiae* at lower elevations, but more commonly into *Tsuga mertensiana/Quercus vaccinifolia* which surrounds it in many areas. The seep phase is generally higher in elevation, forming a mosaic with *Pinus monticola/Holodiscus microphyllus.*

Tsuga mertensiana–Abies magnifica/Pyrola picta is a complex of types found on deeper, high-elevation soils. It is a moderate to dense forest of *Tsuga mertensiana* and *Abies magnifica. Pinus monticola* is a constant companion of minor importance. On drier habitats, *Abies* increases in dominance and may be the only tree in the stand (*Abies magnifica/Pyrola picta*). Such forests are typical of south- to southwest-facing slopes and ridges. On the cooler aspects *Tsuga* may dominate or be found without *Abies* (*Tsuga mertensiana/Pyrola picta*). Such forests are typical of upper slopes with east to northeast aspect. Since intermediate slopes maintain both species, the two types do not conveniently segregate except arbitrarily by canopy dominance. Occasionally other trees such as *Pinus albicaulis, P. contorta, Abies amabilis, A. lasiocarpa,* and *Picea breweriana* are found. Forests with these species usually tend towards the more open-canopied *Tsuga mertensiana/Quercus vaccinifolia.*

Shrub and herb coverage is extremely light, and in many cases nonexistent. Depauperate individuals of *Arctostaphylos patula, A. nevadensis, Chrysolepis sempervirens, Holodiscus microphyllus, Quercus vaccinifolia,* and *Ribes viscosissimum* are scattered and restricted to canopy openings. Wet areas support small stands of *Vaccinium scoparium* or *Leucothoe davisiae.* The ground layer is equally depauperate in both types. *Pyrola picta* is the most consistently associated, but only in small quantities. A few individuals of *Pyrola secunda, Penstemon newberryi, Chimaphila umbellata,* and *Polygonum davisiae* may be found (Table 20-3).

This complex of types merges into *Tsuga mertensiana/Quercus vaccinifolia* and *Pinus albicaulis/Holodiscus microphyllus.*

Tsuga mertensiana/Quercus vaccinifolia. This forest type grows on moraines, talus, and cool upper slopes with soils too thin to support closed stands. *Pinus monticola, Abies magnifica,* and *Tsuga mertensiana* dominate. *Picea breweriana* is commonly found in this type. *Pinus contorta* and, to a lesser extent, *Abies amabilis* and *A. concolor* are also associates. These trees are scattered around boulders and shrubs.

A well-developed shrub layer includes *Vaccinium arbuscula* in seepage areas. The ground layer is a mixture of mesophytic and xerophytic species (Table 20-3). The microhabitat variability is greater than that of *Abies magnifica/Quercus vaccinifolia.* In this respect *Tsuga mertensiana/Quercus vaccinifolia* is a high-elevation extension of the open phases of *Abies magnifica/Linnaea borealis* and *Abies magnifica/Leucothoe davisiae,* as well as an extension of *Abies magnifica/Quercus vaccinifolia.* Additionally this type grades into the closed *Tsuga–Abies magnifica* forest complex and into *Pinus monticola/Holodiscus microphyllus.*

Pinus monticola/Holodiscus microphyllus. This woodland type is most generally found in the Trinity Alps. It is represented by groves of *Abies magnifica, Pinus monticola, P. contorta,* and *P. jeffreyi* scattered along drainage courses and in small areas of soil accumulation as cracks, crevices, or concave surfaces on glacially polished bedrock slopes. Individual trees, especially *P. contorta,* root in small fissures.

A discontinuous, low shrub layer is scattered among the patches of herbs and bare rock. The chaparral component is dominated by *Arctostaphylos patula, A. nevadensis, Holodiscus microphyllus,* and *Amelanchier pallida. Vaccinium arbuscula* is found in localized seepage areas not sufficiently developed for *Tsuga mertensiana/Phyllodoce empetriformis.* Herbs are xerophytic (Table 20-3).

This type grades into *Quercus vaccinifolia–Arctostaphylos patula* at lower elevations, but differs in that *A. nevadensis* and *Holodiscus microphyllus* dominate, and *Q. vaccinifolia* is only a minor component. It also merges into *Tsuga mertensiana/Quercus vaccinifolia* or into meadow communities within the zone, or into *Pinus albicualis/Holodiscus microphyllus* on the higher ridges and summits.

***Pinus albicaulis* zone.** *Pinus albicaulis/Holodiscus microphyllus.* Ridge crests, summits, and upper southwesterly slopes above 2200 m support a dwarfed, open woodland where trees exceed 6 m only among the shelter of rocks. *Pinus albicaulis* dominates the canopy, with *Tsuga mertensiana* and *Abies magnifica* as common associates. *Pinus monticola, P. jeffreyi, P. balfouriana,* and *P. contorta* grow in some stands. Young trees suggest a steady state for this composition.

Shrubs, typically wind trained and pruned to less than 1 m tall, grow on gravel slopes or among rocks. *Holodiscus microphyllus* and *Haplopappus greenei* are common. *Juniperus communis,* shrubby *Populus tremuloides,* and *Cercocarpus ledifolius* show local dominance. Herbs are scattered among the rocks. Shrub and herb composition is similar to that of *Pinus monticola/Holodiscus microphyllus,* although *Ceanothus velutinus, Chrysolepis sempervirens,* and *Haplopappus greenei* are more common here. *Polemonium pulcherrimum* and *Draba howellii* are restricted to this habitat (Table 20-3).

This type most commonly grades into *Pinus monticola/Holodiscus microphyllus* and closed *Tsuga–Abies* forests of the *Tsuga mertensiana* zone, or into a series of meadow communities within the zone.

The *Pinus albicaulis* woodland is best developed on high, nonglaciated, south-facing slopes in the central Trinity Alps (Ferlatte 1974). Here these dwarfed, open woodlands grow on slopes above alpine-like meadows containing plants characteristic of western subalpine and alpine areas. Ferlatte recognized an alpine fell-field type with *Saxifraga tolmiei, Sibbaldia procumbens, Ranunculus eschscholtzii, Primula suffrutescens,* and *Oxyria digyna.* Although these species are of alpine distribution, a zone cannot be recognized since trees grow to the top of the highest peaks in the region. The alpine character of the Trinity Alps is due to the lack of soil development in the recently glaciated valleys (Sharp 1960), which provide alpine-like habitats within the *Abies magnifica, Tsuga mertensiana,* and *Pinus albicaulis* zones. Muth's (1967) alpine fell-fields are clearly dry meadows of the *A. magnifica* zone.

Vegetation on Ultrabasic Parent Materials

Open, mixed stands over a continuous layer of sclerophyllous shrubs typify middle-elevation mesic sites on peridotite, dunite, and serpentine. On more xeric sites and at higher elevations, open woodlands, lacking the continuous shrub understory, become common. Zonation is less apparent in these forests and woodlands, and stand variability is greater than in surrounding closed forests. As in the Pacific Northwest (Kruckeberg 1969), typical high-elevation species extend to lower elevations, and low-elevation species to higher elevations in these noncompetitive habitats, provided that species can tolerate dry soils, nutrient imbalances, and high levels of heavy metals (White 1971).

Western Klamath region montane forests on ultrabasics are mainly open stands of pines over a continuous shrub layer. Whittaker's (1960) descriptions for southwest Oregon apply to northwestern California as well. *Chamaecyparis–Pinus monticola–Pseudotsuga* forest, a forest-shrub complex, and *Pinus jeffreyi* woodlands continue well into the montane zone from lower elevations.

In the Onion Mt. area of eastern Del Norte and Humboldt Cos., *Pinus monticola/Xerophyllum tenax* communities dominate the mesic sites above 1200 m. These open climax forests of *Pinus monticola, P. lambertiana, P. jeffreyi, Calocedrus decurrens, Chamaecyparis lawsoniana, Pseudotsuga menziesii,* and *Abies concolor* grade into *Pinus jeffreyi*-dominated stands of xeric slopes and ridges. Shrubs include *Quercus vaccinifolia, Vaccinium parvifolium, Arctostaphylos nevadensis, A. canescens, Rhamnus californica* var. *occidentalis, Ceanothus pumilus, Amelanchier pallida, Juniperus communis* var. *jackii,* and *Berberis pumila. Xerophyllum tenax* is the single dominant herb. At these elevations, successional forests are largely dominated by *Pinus contorta* and *P. attenuata,* along with postfire-sprouting shrubs which rapidly dominate burned areas.

In the Orleans District of Six Rivers National Forest (Rockey et al. 1966), shallow, well-drained soils of the Dubakella, Weitchpec, and Ishi Pishi series, on very steep slopes, support open stands of *Pinus lambertiana, P. jeffreyi, P. monticola, P. attenuata,* and *Abies concolor* with understories of *Quercus vaccinifolia, Arctostaphylos patula, A. canescens,* and *A. nevadensis.* On more mesic, moderately deep soils of steep slopes, trees dominate, with *P. monticola* and *P. lambertiana* more common.

In the eastern Klamath region, Mallory et al. (1973) mapped small areas of Huse–Dubakella/Jeffrey pine–leather oak woodlands in western Shasta Co. These open stands generally fit Jeffrey pine woodland descriptions with *Pinus jeffreyi, Calocedrus,* bunch grasses, and scattered shrubs. *Quercus durata,* found here, is not regionally common. On nearby Bully Choop at 1800 m, scattered *P. monticola, P. jeffreyi,* and *Calocedrus* grow over a more typical shrub layer of *Quercus vaccinifolia, Arctostaphylos nevadensis, A. patula,* and *Ceanothus prostratus* var. *laxus.*

In Trinity and southern Siskiyou Cos., ultrabasic parent materials, especially peridotite, are common in the *Tsuga mertensiana* and *Pinus albicaulis* zones. Mount

Eddy, Scott Mt., Eagle Peak, Red Rock Mt., and Gibson Peak support forests similar in composition to those described for the western subregion, except that shrub layers tend to be discontinuous and herb diversity is greater. *Pinus balfouriana* becomes a common dominant in open woodlands with *P. monticola*. *Pinus jeffreyi* and *Abies magnifica* are occasional. The upper slopes of Red Rock Mt. at 2300 m exhibit typical woodland. Here *P. balfouriana* forms extensive stands with an understory of scattered *Arctostaphylos nevadensis* and occasional *Quercus vaccinifolia*. The ground layer includes *Castilleja arachnoides, C. applegatei, Aster ledophyllus, Sitanion hystrix, Crepis pleurocarpa, Lupinus croceus, Phlox diffusa, Arenaria nuttallii* ssp. *gregaria*, and *Lomatium macrocarpum*. Such woodlands continue to 2746 m on Mt. Eddy, where *P. albicaulis* and *Artemesia tridentata* associate with *P. balfouriana*.

BIOLOGY OF KLAMATH MONTANE TREES

Ecological Dynamics of Regional Conifers

The habitat requirements, competitive ability, fire resistance, and colonizing ability of individual conifer species have determined their ecological positions in elevational zones and habitats throughout the montane forests of the Klamath region (Table 20-4). Competitive ability is defined as the capability of individuals of a species to grow and achieve dominance in competition for light, water, nutrients, and space with individuals of other species.

With elevational change, each species assumes a different ecological role in the forests by virtue of its response to altered environmental conditions. The resulting pattern is one of changing forest composition with increasing elevation. From this pattern it is possible to infer a systematic sequence of zones, defined by the more competitive species at different elevations. These species should dominate the mature, climax forests if there has been sufficient time for them to outcompete others which are able to grow at these elevations. In addition, the relative habitat requirements and competitive abilities of the regional trees can also be considered as two-dimensional patterns within each zone (Figs. 20-3 to 20-6). Such analysis may lead to an understanding of potential and existing forest patterns.

In the western subregion of the *Abies concolor* zone (Fig. 20-3), *A. concolor* is the most competitive species on all habitats. Therefore this species, along with *Pseudotsuga menziesii*, dominates mature forests. In the drier eastern subregion (Fig. 20-4), *P. menziesii* plays a less competitive role, and others such as *Pinus ponderosa* are less environmentally restricted. The resulting mature forests are of more mixed character than those on comparable habitats in the western subregion.

In the *Abies magnifica* zone (Fig. 20-5), *A. magnifica* dominates mature, closed-canopied forests over continuous, well-developed soils. Moraines, talus, and colluvium are common within the zone, and create habitats where rocks separate pockets of soil. Diverse, open-canopied stands with lower *Abies* dominance grow over such sites (e.g., some *Abies magnifica/Linnaea borealis* stands). These habitats may be mosaics of xeric–mesic microhabitats, thus accounting for the broad mixture of species which they support, but the lessened degree of competitive interaction

TABLE 20-4. Successional status of trees in Klamath montane forests on granitic and metamorphic parent materials. C = major forest component throughout the sere, becoming a dominant climax species; c = minor forest component throughout the sere; S = major forest component in early and midseral stages, not present in climax forests; Sc = major forest component in early and midseral stages, and minor component in climax forests, s = minor forest component in early and midseral stages, not present in climax forests; x = occasional individuals; - = nonoccurrence.

	Species																							
Zone	*Abies amabilis*	*A. concolor*	*A. lasiocarpa*	*A. magnifica*	*Acer macrophyllum*	*Chrysolepis chrysophylla*	*Calocedrus decurrens*	*Chamaecyparis lawsoniana*	*C. nootkatensis*	*Picea breweriana*	*P. engelmannii*	*Pinus albicaulis*	*P. attenuata*	*P. balfouriana*	*P. contorta* var. *murrayana*	*P. jeffreyi*	*P. lambertiana*	*P. monticola*	*P. ponderosa*	*Pseudotsuga menziesii*	*Quercus chrysolepis*	*Q. kelloggii*	*Taxus brevifolia*	*Tsuga mertensiana*
White fir																								
Western subregion																								
Closed forests																								
Streamsides	-	C	-	x	c	c	c	c	-	-	-	-	-	-	-	-	s	-	-	Sc	-	-	c	-
Mesic slopes	-	C	-	-	-	c	s	c	-	-	-	-	-	-	-	-	s	-	-	Sc	c	-	-	-
Xeric slopes	-	C	-	-	-	-	c	-	-	x	-	-	s	-	-	-	c	-	s	c	c	s	-	-
Eastern subregion																								
Closed forests																								
Mesic slopes	-	C	-	x	c	-	C	-	-	-	x	-	s	-	-	-	Sc	-	Sc	Sc	x	s	c	x
Xeric slopes	-	C	-	-	-	-	Sc	-	-	-	-	-	s	-	-	-	C	-	C	C	c	c	-	-
Shasta fir																								
Closed forests																								
Streamsides	-	C	c	C	-	-	c	x	-	c	c	-	-	-	s	-	-	s	s	-	-	-	c	c
Mesic slopes	c	C	-	C	-	-	c	x	-	c	-	-	-	-	s	-	s	s	s	s	-	-	-	-
Xeric slopes	-	C	-	C	-	-	-	-	-	-	-	-	-	-	s	-	s	s	-	s	-	-	-	-
Open forests																								
Mesic slopes	c	c	c	C	-	-	c	x	c	c	c	-	-	-	c	-	c	c	c	c	-	-	c	c
Xeric slopes	-	c	-	c	-	-	c	-	c	c	-	-	-	-	c	c	c	c	-	-	-	-	-	-
Mountain hemlock																								
Closed forests																								
Streamsides	-	c	c	c	-	-	-	-	-	c	c	-	-	-	s	-	-	s	-	-	-	-	c	C
Mesic slopes	c	-	c	c	-	-	-	-	-	-	-	-	-	-	s	-	-	c	-	-	-	-	-	C
Xeric slopes	-	-	-	C	-	-	-	-	-	-	-	-	-	-	-	-	-	c	-	-	-	-	-	c
Open forests																								
Mesic slopes	c	x	-	c	-	-	x	-	c	c	-	c	-	-	c	-	-	c	-	-	-	-	-	c
Xeric slopes	-	x	-	c	-	-	-	-	c	c	-	c	-	c	C	c	-	C	-	x	-	-	-	x

TABLE 20-4 [continued]

Zone	Abies amabilis	A. concolor	A. lasiocarpa	A. magnifica	Acer macrophyllum	Chrysolepis chrysophylla	Calocedrus decurrens	Chamaecyparis lawsoniana	C. nootkatensis	Picea breweriana	P. engelmannii	Pinus albicaulis	P. attenuata	P. balfouriana	P. contorta var. murrayana	P. jeffreyi	P. lambertiana	P. monticola	P. ponderosa	Pseudotsuga menziesii	Quercus chrysolepis	Q. kelloggii	Taxus brevifolia	Tsuga mertensiana
	Species																							
Whitebark pine																								
Open forests																								
Xeric slopes	-	-	-	c	-	-	-	-	-	-	-	C	-	c	c	c	-		-	-	-	-	-	c

[a]Klamath populations are Abies magnifica var. shastensis; A. procera included here.

[b]Including poorly developed streamside phases.

[c]Including meadow margins, seeps.

[d]Including ridges.

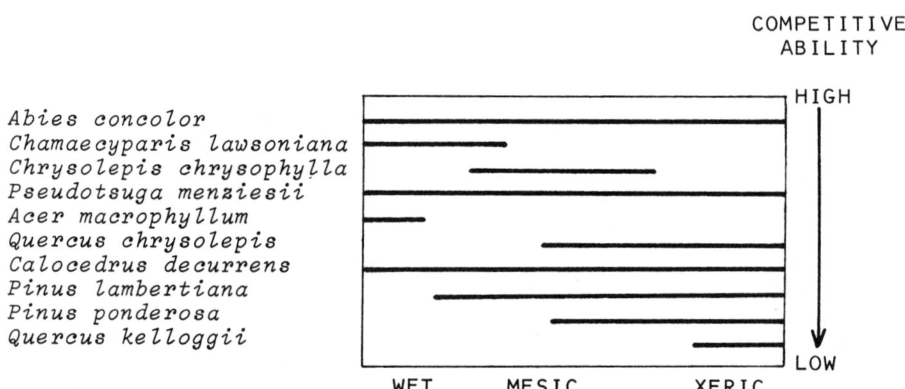

Figure 20-3. Habitat characteristics and competitive abilities of trees in western subregion *Abies concolor* zone forests on granitic and metamorphic parent materials.

Figure 20-4. Habitat characteristics and competitive abilities of trees in eastern subregion *Abies concolor* zone forests on granitic and metamorphic parent materials.

among the individual trees may be of significance in the maintenance of high species diversity.

In the *Tsuga mertensiana* zone (Fig. 20-6), *T. mertensiana* replaces *Abies magnifica* as the competitive dominant on all but dry, south-facing slopes. As in lower zones, closed forests tend to be dominated by a single species, whereas moraines, talus, glaciated slopes, and rocky ridges support a wide variety of species. The upper range limits of midelevation trees are found on such sites.

When considering only closed forests, the expected sequence of elevational zones is not always encountered because of disturbance. Historical events, such as fire, have continually altered these patterns by creating an array of successional stages

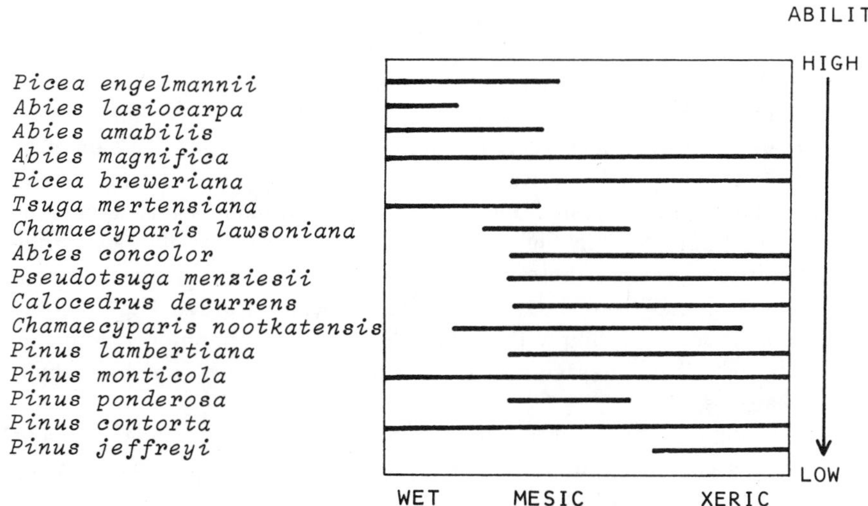

Figure 20-5. Habitat characteristics and competitive abilities of trees in *Abies magnifica* zone forests on granitic and metamorphic parent materials.

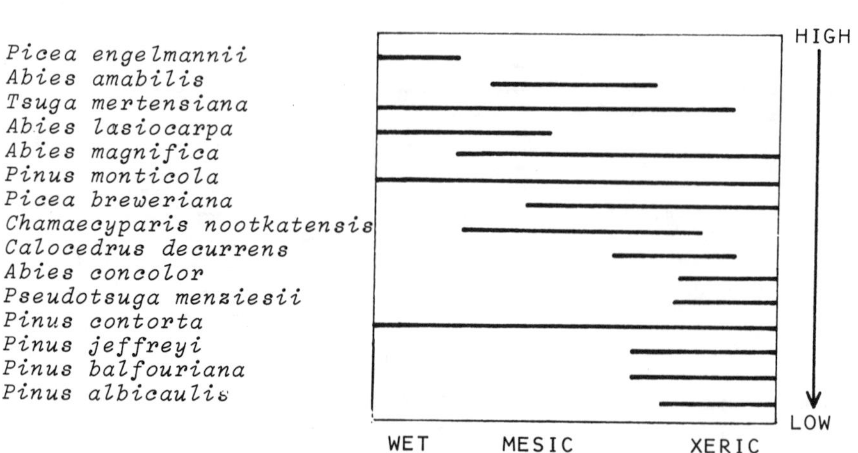

Figure 20-6. Habitat characteristics and competitive abilities of trees in *Tsuga mertensiana* and *Pinus albicaulis* zones on granitic and metamorphic parent materials.

colonized by a wide variety of species which are able to grow within the zone provided that competition is reduced or eliminated. In time, though, it is expected that these "pioneers" will be competitively eliminated from maturing forests.

Fire has played an important role in determining Klamath montane forest patterns. Fires, although common and burning large areas, are neither as destructive as those of the Pacific Northwest nor as frequent as those of the Sierra Nevada. Fires not only create diverse successional communities, but may also significantly reduce or eliminate populations of fire-sensitive species from areas of suitable habitat. Low fire resistance may also reduce colonizing ability because only small populations are maintained. Low fire resistance may be countered by other genetic adaptations, such as the serotinous cones of *Pinus attenuata.* Table 20-5 shows relative resistances to fire based on observations of Klamath populations.

After fire has completely or partially eliminated the trees from an area, some species will quickly recolonize. The colonizing ability of a species is a function of the amount of seed produced each year, the seed dissemination, and the capability to successfully germinate and survive as young seedlings. The relative colonizing abilities of conifers in the Klamath montane forests are listed in Table 20-6. It might be expected that species with high colonizing and competitive abilities would be regional dominants, whereas those with high colonizing ability but low competitive ability would be found throughout the region in localized areas. High colonizing ability does not necessarily imply that a species will be common or widespread. If the species is highly habitat specific, high colonizing ability may be nullified by lack of suitable habitats.

Regionally common trees have broad habitat requirements, high competitive abilities, or high colonizing abilities. For example, in the western subregion of the *Abies concolor* zone, *Pseudotsuga menziesii* is able to maintain significant stand importance in competition with *A. concolor* because of wide ecological tolerances, longevity, and high colonizing ability. Within the *Abies magnifica* zone forests,

Pinus lambertiana and *P. ponderosa* are minor successional species in dense forests dominated by shade-tolerant *Abies,* although they can maintain populations in open forests within the zone. In the *Tsuga mertensiana* zone, the wide tolerance range, high colonizing ability, and low competitive ability of *Pinus contorta* can account for its occurrence in a variety of habitats and its wide geographical range within the Klamath region in comparison to the more environmentally restricted *P. jeffreyi.*

Relict Conifers

Unlike the common regional trees, relicts such as *Abies amabilis, A. lasiocarpa, Chamaecyapris nootkatensis, Picea breweriana, P. engelmannii,* and *Pinus balfouriana* do not necessarily show predictable, environmentally controlled patterns. Some have narrow ecological amplitudes to account for their restriction, but others appear to be primarily controlled by historical events.

Abies amabilis. *Abies amabilis* grows in western subregion *Abies magnifica* and *Tusga mertensiana* zone forests in only two localities. A stand of *A. amabilis* and *A. procera,* approximately 3.5 km in length, is on steep, well-drained, north-facing slopes above Joe Creek and Dutch Creek in the Siskiyou Mts. Another is located in the Hancock–Abbott Lake area of the Marble Mts. Here *A. amabilis* grows with *Tsuga* and an occasional *Pinus monticola, Abies magnifica, A. concolor,* and *Picea breweriana.* In both areas *A. amabilis* is the dominant reproducing species under a dense canopy of *Abies* and *Tsuga.* The cool, moist, north-facing habitats of both areas seem comparable. There is little evidence of *A. amabilis* invading surrounding mesic sites. It appears that *A. amabilis* is restricted to a habitat which is highly scattered within the region. Inability to recolonize other suitable locations may account for its rarity.

Abies lasiocarpa. *Abies lasiocarpa* is known from only two California locations. The most extensive stands lie in four drainages in the Russian Peak area of the

TABLE 20-5. Relative fire resistance of Klamath montane conifers

Fire Resistance		
High	Medium	Low
Pinus ponderosa	Pinus lambertiana	Tsuga mertensiana
P. jefferyi	Calocedrus decurrens	Pinus balfouriana
Pseudotsuga menziesii	Chamaecyparis lawsoniana	P. monticola
	Abies concolor	P. albicaulis
	A. magnifica/A. procera	P. contorta
		P. attenuata
		Chamaecyparis nootkatensis
		Picea engelmannii
		P. breweriana
		Abies amabilis
		A. lasiocarpa

TABLE 20-6. Relative colonizing ability of Klamath montane conifers

Colonizing Ability			
High	Medium	Low	Very Low
Tsuga mertensiana	_Abies concolor_	_Chamaecyparis lawsoniana_	_Pinus ponderosa_
Pseudotsuga menziesii	_A. magnifica/A. procera_	_Calocedrus decurrens_	_P. jeffreyi_
	Picea breweriana	_Abies amabilis_	_P. lambertiana_
	P. engelmannii	_Pinus contorta_	_P. monticola_
		P. attenuata	_P. balfouriana_
			P. albicaulis
			Abies lasiocarpa
			Chamaecyparis nootkatensis

Salmon Mts. (Sawyer et al. 1970), where it is a dominant tree along with *Tsuga*, *Abies magnifica*, *Pinus monticola*, and *P. contorta*. It also occurs as populations of one to several trees in the northern Marble Mts. near Sky High and Deep lakes. The Russian Peak stands grow around meadows in deeply glaciated, northeast-tending valleys within the *Tsuga mertensiana* zone. The species can also be found as krummholz in seepage areas on middle to upper slopes of these same drainages. Individual shrubs are found near ridge crests. Existing stands are enlarging as young trees invade meadows and rather mesic slopes. High shade tolerance and ability to reproduce by layering make *A. lasiocarpa* a potentially aggressive competitor, but because of infrequent cone crops, heavy insect infestations of cones, and limited seed dissemination it is a weak colonizer. Apparently, *A. lasiocarpa* persisted above montane glaciers as krummholz and has since invaded deglaciated adjacent lower valley habitats. It has been unable, though, to recolonize nonadjacent valleys with similar habitats.

Chamaecyparis nootkatensis. *Chamaecyparis nootkatensis* is located in numerous isolated stands (Little Grayback, Preston Peak, Cyclone Gap, Devil's Punch-bowl, Bear Lake, Elk Hole, and others) between 1500 and 2200 m in the Siskiyou Mts. These stands usually involve only a small number of trees or shrubs growing on a wide variety of open habitats, such as lake margins, wet meadows, seepage areas, open ridges, and rocky slopes, and on a wide diversity of parent materials. Apparently in this region *C. nootkatensis* has a very low colonization ability and persists on a wide variety of sites because of its ability to reproduce vegetatively.

Picea breweriana. *Picea breweriana* is not restricted to any zone, habitat, or community. In the eastern subregion it generally occurs above 1500 m, but in the western subregion it has been found at lower elevations in the mixed evergreen-*Pseudotsuga*-hardwood forests. Although growing throughout the region, its range is one of local, disjunct populations of varying size. It grows on various parent materials, including ultrabasics. In the Siskiyou Mts. the species occurs in small, dense stands on north-, east-, and west-facing slopes, as single individuals invading seral pine stands and montane chaparral, and as scattered individuals in closed *Abies magnifica* zone forests. In the Indian Creek drainage, near Happy Camp, *P. brewerians* occurs as the dominant in several stands in association with *Abies concolor*, *A. procera*, and *Quercus sadleriana*. The species is found throughout the Marbles, Salmons, and Trinity Alps in scattered small stands, reproducing in a mixture of conditions. The most extensive stands occur in the Russian Peak area, where *Picea breweriana* is a major component of the *Abies magnifica/Linnaea borealis* and *Abies magnifica/Quercus vaccinifolia* types, as well as a minor component in others.

This species has a very broad ecological amplitude that should enable it to obtain a wider and more contiguous distribution. Waring (pers. commun.) reported that it can withstand moisture stresses below −20 bar and can outgrow *Pseudotsuga menziesii* at cool temperatures. The only habitat restriction appears to be a high water table. Past fires have probably played a major role in restricting populations to local areas. The thin bark and drooping nature of the branches make this species most sensitive. Thus areas of predictable *P. breweriana* occurrence are limited to fire-resistant open forests on north-facing slopes or rocky ridges.

Picea engelmannii. Klamath region populations of *Picea engelmannii* are located in seven adjacent drainages in the Russian Peak area. Here the species is typically found on streamside terraces and moist lower slopes of the *Abies magnifica/ Leucothoe davisiae* type. It is producing abundant seed, and reproducing well on fresh alluvium, rotten logs, and thick humus. *Picea engelmannii* has the ecological capability of growing throughout at least the eastern subregion on streamside terraces of the *A. magnifica* zone. It was restricted to the present drainages with extinction of most regional stands by montane glacial activity. Its reproduction cycle and fire resistance are like those of *P. breweriana,* but its ecological amplitude is narrow. Actual colonization ability is therefore low. The maintenance of only these few stands may be due to fire sensitivity as well as habitat specificity.

Pinus balfouriana. *Pinus balfouriana* occurs in numerous locations throughout the region: Lake Mt., Russian Peak, Scott Mt., Mt. Eddy, Eagle Peak, Red Rock Mt., Packer's Peak, Gibson Peak, North Yolla Bolly, and others (Mastrogiuseppe 1972). Dry ridges and south-facing slopes at high elevations characterize typical sites for the species. Although this tree is found on all parent material types, it is especially common on ultrabasics, where it forms open stands with *Pinus albicaulis, P. contorta, P. jeffreyi,* and *P. monticola.*

AREAS FOR FUTURE RESEARCH

Klamath region vegetation is only beginning to be studied and understood. The area awaits detailed analysis of the forest and woodland patterns on metasedimentary and ultrabasic parent materials. Studies of meadow communities should prove especially interesting. The vegetation of the southern Klamath region and north coastal mountains is essentially unknown. The ecological requirements of even the dominant regional species are poorly known. From the few studies done and from general observation, it is readily apparent that, although the Klamath Mts. maintain many species in common with other western mountain areas, local populations are not ecologically equivalent to those of the Sierra Nevada or Cascades. Additionally, the roles played by some environmental factors may be unique. For example, the fire ecology of the Klamath region seems to be intermediate between that of the Pacific Northwest and the Sierra Nevada, and therefore distinctive.

On the positive side, the region is still comparatively undisturbed by logging and other management practices, and extensive areas remain where natural patterns prevail. There is still much need for the field ecologist to prepare the way for analytical studies in a most interesting part of northern California.

ACKNOWLEDGMENTS

This paper summarizes research supported by McIntire–Stennis grants to the Department of Forestry, Humboldt State University; the Pacific Southwest Forest and Range Experiment Station; Klamath National Forest; and Humboldt State University Foundation.

LITERATURE CITED

Daubenmire, R. and J. Daubenmire. 1968. Forest vegetation of eastern Washington and northern Idaho. Wash. Agric. Exp. Sta. Tech. Bull. 60. 104 p.

Daubenmire, R. 1969. Structure and ecology of coniferous forests of the Rocky Mountains, pp. 25-41. In Center for Natural Resources (ed.). Coniferous forests of the northern Rocky Mountains. Univ. Mont. Found., Missoula.

Davis, G. A. 1966. Metamorphic and granitic history of the Klamath Mountains, pp. 39-50. In E. H. Bailey (ed.). Geology of northern California. Calif. Div. Mines Geol. Bull. 190.

Ferlatte, W. J. 1974. A flora of the Trinity Alps of northern California. Univ. Calif. Press, Berkeley. 206 p.

Fonda, R. W., and L. C. Bliss. 1969. Forest vegetation of the montane and subalpine zones, Olympic Mountains, Washington. Ecol. Monogr. 39:271-301.

Franklin, J. F., and C. T. Dyrness. 1973. Natural vegetation of Oregon and Washington. USDA Forest Serv. Gen. Tech. Rept. PNW-8. 417 p.

Griffin, J. R. 1967. Soil moisture and vegetation patterns in northern California forests. USDA Forest Serv. Res. Paper PSW-46. 22 p.

Griffin, J. R., and W. B. Critchfield. 1972. The distribution of forest trees in California. USDA Forest Serv. Res. Paper PSW-82. 114 p.

Howell, J. T. 1944. Certain plants of the Marble Mountains in California, with remarks on the boreal flora of the Klamath area. Wasmann Coll. 6:13-19.

Irwin, W. P. 1960. Geologic reconnaissance of North Coast Ranges and Klamath Mountains, California, with a summary of the mineral resources. Calif. Div. Mines Geol. Bull. 179. 80 p.

-----. 1966. Geology of the Klamath Mountains Province, pp. 19-28. In E. H. Bailey (ed.). Geology of northern California. Calif. Div. Mines Geol. Bull. 190.

Kruckeberg, A. R. 1969. Plant life on serpentinite and other ferromagnesian rocks in northwestern North America. Syesis 2:15-114.

Mallory, J. I., E. B. Alexander, W. L. Colwell, and W. B. Powell. 1968. Soils and vegetation of the Chanchelulla Peak quadrangle, Shasta and Tinity Counties, California. State Cooperative Soil-Vegetation Survey. 48 p. 1 map.

Mallory, J. I., W. L. Colwell, and W. R. Powell. 1973. Soils and vegetation of the French Gulch quadrangle, Trinity and Shasta Counties, California. USDA Forest Serv., Pac. SW Forest and Range Exp. Sta., Resources Bull. PSW-12. 42 p. 4 maps.

Mastrogiuseppe, R. J. 1972. Geographic variation in foxtail pine (*Pinus balfouriana* Grev. & Balf.). M.S. thesis, Humboldt State Univ., Arcata, Calif. 103 p.

Mueller-Dombois, D., and H. Ellenberg. 1974. Aims and methods of vegetation ecology. John Wiley, New York. 547 p.

Muth, G. J. 1967. A flora of Marble Valley, Siskiyou County, California. M.S. thesis, Pacific Union Coll., Angwin, Calif. 174 p.

Ottenger, F. 1975. Vascular plants of the High Lake Basin of English Peak, Siskiyou County, California. M.S. thesis, Claremont Grad. School, Claremont, Calif.

Rockey, R. E., K. E. Bradshaw, and A. G. Sherrell. 1966. Interim report, soil-vegetation survey, Orleans Ranger District, Six River National Forest. 183 p.

Sawyer, J. O., and D. A. Thornburgh. 1969. Ecological reconnaissance of relic conifers in the Klamath Region. Unpub. rept. to Pac. SW Forest and Range Exp. Sta. 97 p.

-----. 1970. Ecological studies in the Russian Peak area, Salmon Mountains, California. Unpub. rept. to Klamath Natl. Forest. 17 p.

-----. 1971. Vegetation types on granodiorite in the Klamath Mountains, California. Unpub. rept. to Pac. SW Forest and Range Exp. Sta. 47 p.

-----. 1974. Subalpine and montane forests on granodiorite in the central Klamath Mountains of California. Unpub. rept. to Pac. SW Forest and Range Exp. Sta. 53 p.

Sawyer, J. O., D. A. Thornburgh, and W. F. Bowman. 1970. Extension of the range of Abies lasiocarpa into California. Madroño 20: 413-415.

Sharp, R. P. 1960. Pleistocene glaciation in the Trinity Alps of northern California. Amer. J. Sci. 258:305-340.

Society of American Foresters. 1954. Forest cover types of North America. Washington, D.C. 67 p.

Stebbins, G. L., and J. Major. 1965. Endemism and speciation in the California flora. Ecol. Monogr. 35:1-35.

Waring, R. H. 1969. Forest plants of the eastern Siskiyous: their environmental and vegetational distribution. Northwest Sci. 43: 1-17.

White, C. D. 1971. Vegetation-soil chemistry correlations in serpentine ecosystems. Ph.D. dissertation, Univ. Oregon, Eugene. 274 p.

Whittaker, R. H. 1960. Vegetation of the Siskiyou Mountains, Oregon and California. Ecol. Monogr. 30:279-338.

-----. 1961. Vegetation history of the Pacific Coast States and the "central" significance of the Klamath Region. Madroño 16:5-23.

Wolfe, J. A. 1969. Neogene floristic and vegetational history of the Pacific Northwest. Madroño 20:83-111.

CHAPTER 21

COASTAL PRAIRIE AND NORTHERN COASTAL SCRUB

HAROLD F. HEADY
College of Natural Resources, University of California, Berkeley

THEODORE C. FOIN
MARY M. HEKTNER
Division of Environmental Studies and The Institute of Ecology, University of California, Davis

DEAN W. TAYLOR
MICHAEL G. BARBOUR
Botany Department, University of California, Davis

W. JAMES BARRY
California State Department of Parks and Recreation, Sacramento, California

Coastal prairie 734
Overview 734
Historical aspects 736
Sea Ranch: a case in point 739
Distribution of vegetation types 739
Composition of vegetation types 740
Succession 745
Northern coastal scrub 745
Overview 745
Baccharis scrub 747
North–south gradients 748
Lupinus scrub 754
Areas for future research 756
Acknowledgments 757
Literature cited 758

COASTAL PRAIRIE

Overview

The *Festuca–Danthonia* grassland, or coastal prairie, occurs along the California coast from Santa Cruz northward. Küchler (1964) mapped the type on 355,614 ha, as determined by planimetering his vegetational map of the United States. The type occupies the cool, wet northwestern area of the California prairie. Its relationships to the valley grassland were discussed in Chapter 14. The boundary between coastal prairie and valley grassland is unclear, but it probably lies near the inland side of the coastal redwood type. The prairie is a discontinuous grassland below 1000 m in elevation and seldom more than 100 km from the coast. Typically it occurs on the ridges and south-facing slopes, alternating with forest and scrub in the valleys and on north-facing slopes (Fig. 21-1). Most early descriptions refer to this type as "fine bunch grass pasturage" (Burcham 1957). Seven year average herbage production was 1700–13,000 kg ha^{-1} among nine soil series in Humboldt Co. (Cooper and Heady 1964). The type continues to produce excellent forage after 125 yr of use (Heady et al. 1963).

The dominant perennial grasses in this type are *Festuca idahoensis, F. rubra,* and *Danthonia californica* (Fig. 21-1). Beetle (1947) showed that many northern grass species reach their southern limits in the coastal prairie, including *Deschampsia caespitosa* spp. *holciformis, Trisetum cernuum, Melica geyeri,* and *Calamagrostis nutkaensis. Koeleria cristata, Poa pratensis,* and *Agrostis exarata* frequently occur (Burtt–Davy 1902). Several species of *Bromus, Poa, Melica, Agrostis,* and *Glyceria* extend widely but not abundantly throughout the type. Sandy sites support a number of perennial grass endemics in the genera *Poa, Elymus,* and *Agrostis,* as well as the ubiquitous *Phragmites, Distichlis,* and *Ammophila* (Beetle 1947). Two montane species, *Festuca ovina* and *Phleum alpinum,* occur rarely on coastal bluffs. Characteristic broad-leaved species are *Iris douglasiana, Ranunculus californicus,* and *Sysyrinchium bellum. Carex tumulicola* frequently is found with the perennial grasses.

Dominant species of the coastal prairie vary from north to south and with distance inland from the ocean. In Del Norte and Humboldt Cos., *Agrostis alba,*

Figure 21-1. A stand of *Festuca idahoensis* and *Danthonia californica* as an island of coastal prairie within a scrub–grassland mosaic in Humboldt Co.

Danthonia californica, Elymus mollis, Lasthenia chrysostoma, Poa pratensis, and *Pteridium aquilinum* var. *pubescens* are common close to dunes and bluffs. Further inland, in the prairie balds surrounded by forest, *Agrostis alba, Aira caryophyllea, Anthoxanthum odoratum, Bellis perennis, Bromus mollis, Carex obnupta, Dactylis glomerata, Danthonia californica, Festuca idahoensis, F. rubra, Iris douglasiana, Pteridium,* and *Trifolium pratense* are common in a mosaic of relatively xeric and mesic sites.

In Mendocino and Sonoma Cos., *Aira caryophyllea, Bromus mollis, Festuca megalura, Poa douglasii* ssp. *macrantha,* and *P. unilateralis* are common in sandy soils. Coastal terraces are commonly dominated by *Agrostis californica, Aira caryophyllea, Bromus marginatus, Carex obnupta, Danthonia californica, Deschampsia caespitosa* ssp. *holciformis, Festuca occidentalis, Holcus lanatus, Horkelia marinensis, Lolium multiflorum, Montia perfoliata,* and *Pteridium* (Barbour et al. 1973; Barry et al. 1974; Crampton 1974; Schlinger et al. 1976). Where coastal prairie meets closed-cone pine forest, *Avena barbata, Calamagrostis nutkaensis, Eryngium armatum, Festuca californica, F. idahoensis, Holcus lanatus, Hordeum brachyantherum, Pteridium,* and *Stipa pulchra* may dominate.

On Pt. Reyes peninsula, Marin Co, Elliott and Wehausen (1974) reported that coastal prairie was dominated by *Bromus carinatus, Deschampsia caespitosa* spp. *holciformis,* and *Rumex acetosella.* On Franciscan sandstone of the Tiburon Peninsula, dominants include *Aira caryophyllea, Avena barbata, A. fatua, Bromus carinatus, B. marginatus, B. diandrus, B. mollis, Luzula comosa,* and *Pteridium.* Serpentine soils, in contrast, support *Calamagrostis ophitidis, Cheilanthes carlottahalliae, Danthonia californica* var. *americana, Festuca dertonensis, F. pacifica, Hordeum californicum, Melica californica, Pellaea andromedifolia, Stipa lepida,* and *S. pulchra* (Peñalosa 1963). A similar community occurs at Mt. Tamalpais State Park, where coastal prairie meets the Central Valley prairie (Barry 1972). *Stipa pulchra* may occur in nearly pure stands on serpentine ridges. Associates include *Avena fatua, Bromus diandrus, B. mollis,* and *Plantago hirtella* var. *galeottiana.*

The southern limit of coastal prairie is at Mound Meadows, Pt. Lobos State Reserve in Monterey Co. Coastal terrace grassland occurs on Carmelo sandstone marked by poor drainage and a curious mima-mound topography (NA 271665). Dominants include *Aira caryophyllea, Brodiaea coronaria* var. *macropodia, B. lutea, Bromus mollis, Danthonia californica, Deschampsia caespitosa* ssp. *holciformis, Eryngium armatum, Grindelia latifolia* ssp. *platyphylla, Juncus bufonis, J. tenuis* var. *congestus, Luzula comosa, Perideridia gairdneri,* and *Rumex pulcher.*

Historical Aspects

The extent and the composition of pristine coastal prairie are not well recorded. The records of early explorers, settlers, and botanists give us only tantalizing, incomplete glimpses of the vegetation. An early description of the Russian River valley near the Sonoma Co. coast mentions "a great wide area of waving grasses higher than a man's head with deer, bear, and other big game everywhere . . ." (Barrett 1952). What grasses were these? According to Cronise (1868), *Deschampsia danthonoides* was common in the Russian River valley, and *Pleuropogon californicus* in wet meadows from Ukiah to Oakland. Other candidates, still present in the area, include *Agrostis diegoensis, Bromus carinatus, B. laevipes, B. orcuttianus, Danthonia californica* var. *americana, Deschampsia caespitosa* ssp. *holciformis, Elymus triticoides, Festuca californica, Melica californica, M. hardfordii, M. subulata, M. torreyana, Phalaris californica, Stipa pulchra,* and *Trisetum canescens.* Many of these species are also found in nearby Redwood Valley, in the upper Russian River watershed—a valley left in a natural state until relatively recently (Kniffen 1939). On an 1824 trip from Sonoma to Bodega Bay and Ft. Ross, Eschscholtz and Kotzebue noted a "luxuriant growth of grasses" around Ft. Ross, and they made special reference to a wild rye, probably *Elymus glaucus,* although today it rarely grows in pure stands (Howell 1970).

A more complete picture is available for Sherwood Valley, about 30 km north of Ukiah, Mendocine Co., well into the Coast Ranges. Burtt-Davy (1902) interviewed the first white settlers, who had lived there since 1853. He concluded that *Danthonia californica,* now scarce, was originally abundant on hillsides and valley floors. The only native grasses which Burtt-Davy encountered were *Danthonia californica, Deschampsia caespitosa, D. elongata, Hordeum brachyantherum,* and *Koeleria macrantha* in meadows; *Agrostis microphylla* in vernal pools; *Festuca californica* at the prairie–woodland ecotone; *Phalaris californica* at the prairie–redwood ecotone; and *Beckmannia syzigachne, Glyceria occidentalis,* and *Pleuropogon californicus* in wet areas.

Four major factors contributed to drastic changes in the distribution and composition of pristine coastal prairie: the introduction of highly competitive, exotic species, an increase in grazing pressures, the elimination of annual fires, and cultivation.

The use of fire, for various objectives, has been historically documented for almost every California Indian group, including those that inhabited the coastal prairie (Tolowa, Yurok, Wiote, Bear River, Mattole, Sinkone, Yuki, Pomo, Coast Miwok, Costanowan). It may be that coastal prairie species evolved under a relatively light grazing pressure but an intense fire-frequency regime.

Grazing use of the northern California coastal areas developed later than such use in southern California because the Spanish missions did not extend north of Sonoma Co. The Russian settlements at Ft. Ross and northward maintained only a few cattle and sheep. Burtt-Davy's (1902) extensive studies during the summers of 1899 and 1900 showed that *Danthonia californica* was the dominant grass in 1853, but that the "feed changed several times between 1870 and 1900."

Avena spp. were the first annuals to become abundant, and they were followed by introduced species of *Bromus, Festuca, Hordeum, Erodium, Centaurea, Hypochoeris,* and *Medicago,* just as described for the valley grassland, as discussed in Chapter 14. However, the "new native" perennials *Anthoxanthum odoratum, Danthonia pilosa,* and *Holcus lanatus* and the annuals *Cynosurus cristatus, C. echinatus, Festuca dertonensis,* and *Lolium multiflorum* contribute more to coastal prairie than they do to valley grassland. Another difference is that the coastal grassland is still dominated by perennials in many areas and the perennials will return on other areas under proper grazing practices. Batzli and Pitelka (1970) described relationships between fluctuating mouse populations and grassland productivity near San Francisco Bay. The coastal prairie supports large numbers of rodents belonging to several species.

Successional relationships, such as those related to grazing of domestic animals, may have resulted from heavy grazing by Roosevelt elk (*Cervus canadensis*) before European man arrived (Mason 1970). When heavy year-long grazing by cattle was replaced by moderate seasonal use in Humboldt Co., perennial grasses, mostly *Danthonia californica, Sitanion hystrix, Stipa* spp., *Poa* spp., and *Festuca idahoensis,* replaced annual vegetation, which was dominated by *Taeniatherum asperum* (= *Elymus caput-medusae*: Cooper 1960). Elliott and Wehausen (1974) inferred three stages of plant succession in the coastal grassland on Pt. Reyes Peninsula, Marin Co. (Table 21-1). After no grazing for 6 yr the vegetation was taller and had fewer annual species and a greater proportion of native species than in areas where grazing was heavy.

Hypericum perforatum, a large, perennial, broad-leaved plant, became established

TABLE 21-1. Relative heights and total cover of vegetation under three grazing situations on Pt. Reyes Peninsula, California. From Elliott and Wehausen (1974)

Height and Cover	Heavily Grazed	Moderately Grazed	Ungrazed for 6 yr
Average plant height (cm)	5.0	11.0	16.5
Relative cover of major species (%)			
Perennial and biennial	81.5	87.5	94.0
Annual	15.0	10.5	1.5
Native	65.0	63.5	78.5
Introduced	31.5	34.5	17.0

in the coastal prairie and adjacent areas about 1900. By 1945, it completely dominated the vegetation on about 1 million ha, largely in northwestern California. After release of the Klamath weed beetle (*Chrysolina quadrigemina*) in 1946, *Hypericum* rapidly declined, thus opening land area for plant succession. In 1949, Huffaker and Holloway reported increases in annual grasses and broad-leaved species. Huffaker (1951) later found that perennial grasses were increasing, but Murphy (1955) reported dominance of annuals on other sites 3–4 yr after *Hypericum* was reduced to a minor part of the vegetation. Data from remeasurement of the same plots over a 10 yr period (Huffaker and Kennett 1959) showed replacement of *Hypericum* by perennial grasses (Table 21-2, Fig. 21-2).

In some places, coastal prairie may be seral to *Lupinus* or *Baccharis* coastal scrub. At Tilden Park, in the Berkeley Hills, scrub components *Baccharis pilularis* spp. *consanguinea* and *Rhus diversiloba* invaded the grassland at the approximate rate of 40 cm yr^{-1} for a 20 yr period after grazing was eliminated (McBride and Heady 1968). More will be said about this relationship in a later section of the chapter.

Certainly, introduced plants will continue to be abundant on many hectares of coastal prairie, but succession will apparently move rapidly toward dominance of perennial grasses if land management practices are suitable. Whether or not forest will replace the grassland remains unclear. The coastal prairie has many typically grassland soils and endemic species, both suggesting that the type has existed for hundreds of years. Grazing in the coastal prairie has significantly modified the species composition of the coastal grassland. Floristic censuses at a number of coastal sites (Barbour 1970, 1972; Barbour et al. 1973; Batzli and Pitelka 1970; Elliott and Wehausen 1974; Hardham and True 1972; Howell 1970; Huffaker 1951; Huffaker and Holloway 1949; Huffaker and Kennett 1959; Peñalosa 1963) illustrate the phenomena of patchiness of species distributions, replacement of perennial grasses by forbs and annual grasses under grazing pressure, and recovery of perennial grasses with the cessation of grazing. There has been little other ecological research beyond this.

TABLE 21-2. Cover (%) of several species at Blocksburg, Humboldt Co., for 10 yr after release of the Klamath weed beetle. From Huffaker and Kennett (1959)

	1947	1949	1953	1957
Hypericum perforatum	57.6	0.0	0.0	0.0
Danthonia californica (perennial)	9.2	22.7	30.3	45.0
Bromus spp. (annual)	4.9	18.9	8.5	8.6
Aira caryophyllea (annual)	5.9	23.6	14.6	7.9
Others (mostly annual)	19.0	26.0	46.3	38.4
Bare ground	3.4	8.8	0.3	0.1

Figure 21-2. Perennial species of *Elymus, Festuca, Agrostis,* and *Danthonia* invading an annual grassland after reduction of *Hypericum perforatum* by the Klamath weed beetle in Humboldt Co.

Sea Ranch: A Case in Point

In 1963 the Del Mar Ranch, a 16 km segment of the northern Sonoma Co. coastline from the county line to just north of Stewart's Pt., was sold for development of what is now known as Sea Ranch. By 1965 the sheep that had grazed the landscape had been totally removed, and the grasslands began a recovery.

In this section we report the results of two seasons of study of the grasslands a decade after the recovery began. Our objective is to characterize the structure of a small area of coastal grassland in order to compare the relation of this area to the larger aspects of all California coastal grassland. Most of the results in the discussion that follows are based on 490 samples of the vegetation taken in August and September of 1974. Species cover and density were estimated using quadrats varying in size (50.25–0.25 m^2) and placed randomly along parallel transects.

After a thorough reconnaissance of Sea Ranch grasslands, five representative vegetation types were chosen for intensive sampling. The number of samples taken in each area varied with the size of the area, which affected both the number and the length of the transects. Assistance in species identification was given by Professor Beecher Crampton and Ms. June McCaskill of the University of California, Davis, and life history classifications are based on Reed et al. (1963).

Distribution of vegetation types. The classification of vegetation types was first based on color infrared aerial photographs taken by the National Aeronautics and Space Administration in 1972. Differences in infrared color were used to establish tentative vegetation boundaries that subsequently were verified by ground survey. During the ground surveys, arbitrary criteria of convenience were used to classify vegetation types; hence these classifications do not necessarily correspond to phytosociological units.

Headlands were recognized as areas of very low vegetative height, usually 20 cm of less, with abundant forbs. Grasslands were recognized as mixed when neither annual nor perennial grasses were estimated to reach 50% relative cover, and perennial grasslands were those having 50% relative cover of perennial grass species. Lupine areas were those where lupine cover (*Lupinus arboreus*) extended through more than two or three contiguous adult individuals; the patch of bushes defined the area. Other map units included *Calamagrostis* areas and annual grass stands.

If we draw a series of hypothetical transect lines perpendicular to the coast and randomly sample a number of them at fixed distances inland, we find significant differences in the distribution of types. Headlands dominate for the first 100 m back, mixed grassland peaks about 400 m back, and perennial grassland peaks 800–1600 m back; forest abrubtly increases beyond 1600 m. The distribution of mixed grassland may be more closely related to recent disturbance than to environmental gradients.

Composition of vegetation types. It is important to emphasize that the classification of vegetation types was made on the basis of dominant life history patterns which were readily visible on high-quality infrared aerial photographs. Thus dominance of perennial grasses in the perennial grassland type is by definition, although the composition of species in the type is not. Table 21-3 presents relative cover values of the species of the perennial type, listed by life history and in descending order of relative cover. Only four species have relative percent cover values exceeding 5%: three perennial grasses (*Deschampsia caespitosa* ssp. *holciformis, Anthoxanthum odoratum,* and *Holcus lanatus*) and one vine (*Rubus* spp.). Most of the remaining species are rare to moderately rare, and consequently the dominance-diversity curve is comparable to the log series model of Fisher et al. (1943) or to the log-linear model (Whittaker 1965). Further analysis of species composition may be found in Hektner and Foin (1976a), and the complete species list has also been reported elsewhere (Hektner and Foin 1976b).

The cover values of 18 selected species classified by vegetation type are shown in Table 21-4. Nine of these species are native, an equal number are introduced, and there is no clear trend of replacement of natives by introduced species or vice versa. Dominance shifts increasingly away from annual forbs to perennial grasses along the coast-inland gradient, and there are species composition changes as well. Correlation analysis (unpublished data) suggests that annual forbs are replaced by annual grasses and perennial forbs, which in turn lead to certain perennial grasses (*Danthonia pilosa, Holcus lanatus*) and finally to others (*Anthoxanthum ordoratum, Deschampsia caespitosa* ssp. *holciformis*). The reed grass, *Calamagrostis nutkaensis,* is restricted to areas of continuously high soil moisture, where its dominance is very high.

When all species are considered together (classified by life history and vegetation unit type), the same patterns may be seen (Table 21-5). Annual forbs, annual grasses, perennial forbs, and prostrate shrubs (principally *Lupinus variicolor*) dominate in headlands; in mixed grasslands, annual forbs and *L. variicolor* drop out in favor of perennial grasses; and in perennial grasslands, perennial grass dominance is increased to the near exclusion of everything else. *Calamagrostis nutkaensis* represents a special case of environmental conditions in which the size and bulk of *C. nutkaensis* are overwhelmingly dominant.

The dominance of *Holcus lanatus, Anthoxanthum odoratum,* and *Deschampsia caespitosa* ssp. *holciformis* at Sea Ranch is relatively unique when compared to descriptions of other coastal grasslands. Part of the difference must be due to local environmental patchiness (soil, slope, weather, climate, nutrients); another part, to

TABLE 21-3. Relative cover (%) and frequency for all species found in the perennial grassland type for 181 plots at Sea Ranch, August-September 1974. t = <0.01% relative cover. For conversion to absolute cover, multiply by 0.96. Two taxa of unknown longevity, *Galium* and *Lotus* species, are not included in the table

Taxon	Relative Cover	Frequency in Samples
Annual grasses		
Aira caryophyllea	1.24	61
Bromus mollis	0.83	51
Aira praecox	0.39	13
Lagurus ovatus	0.31	12
Festuca dertonensis	0.29	40
Briza minor	0.12	42
Cynosurus echinatus	0.06	22
Avena barbata	0.01	3
Bromus diandrus	t	6
Annual sedges and rushes		
Juncus bufonius	t	1
Annual forbs		
Sherardia arvensis	0.02	8
Trifolium dubium	0.01	3
Silene gallica	t	13
Plantago hookeriana var. *californica*	t	3
Anagallis arvensis	t	9
Pogogyne sp.	t	1
Hemizonia multicaulis	t	7
Lotus angustissimus	t	2
Galium aparine	t	2
Orthocarpus castillejoides	t	2
Daucus pusillus	t	2
Clarkia davyi	t	1
Lotus micranthus	t	2
Madia capitata	t	1
Trifolium sp.	t	1
Vicia benghalensis	t	1
Geranium dissectum	t	6
Biennial grasses		
Bromus carinatus	t	3
Biennial forbs		
Erechtites prenanthoides	0.09	2
Gnaphalium purpureum	0.01	6
Erechtites arguta	t	1
Annual or biennial forbs		
Linum bienne	0.01	19

TABLE 21-3 [continued]

Taxon	Relative Cover	Frequency in Samples
Perennial grasses		
Deschampsia caespitosa ssp. *holciformis*	28.21	118
Anthoxanthum odoratum	26.91	144
Holcus lanatus	13.18	111
Calamagrostis nutkaensis	0.55	1
Hordeum californicum	0.09	14
Danthonia pilosa	0.07	22
Elymus glaucus	t	10
Danthonia californica	t	5
Lolium perenne	t	4
Festuca arundinacea	t	1
Perennial sedges and rushes		
Juncus effusus var. *brunneus*	1.59	12
Carex spp.	0.48	13
Carex obnupta	0.29	4
Juncus effusus var. *pacificus*	0.10	11
Perennial forbs		
Plantago lanceolata	4.92	91
Hypochoeris radicata	1.89	67
Iris douglasiana	1.59	37
Rumex acetosella	0.28	26
Fragaria chilensis	0.24	8
Aster chilensis	0.09	1
Cirsium quercetorum	0.09	1
Eryngium armatum	0.08	7
Acaena californica	0.04	1
Oxalis pilosa	0.01	5
Sisyrinchium bellum	0.01	3
Convolvulus occidentalis var. *saxicola*	t	2
Horkelia californica	t	2
Lotus corniculatus	t	2
Cardionema ramosissimum	t	1
Scrophularia californica	t	1
Woody vines, shrubs and small trees		
Rubus ursinus-R. vitifolius	5.11	65
Lupinus arboreus	2.15	28
Myrica californica	0.55	1
Lupinus variicolor	0.02	2
Ferns		
Pteridium aquilinum var. *pubescens*	2.48	36

TABLE 21-4. Relative cover (%) of selected species by vegetation type at Sea Ranch. t = <0.01% relative cover; - = absent; I = introduced species; N = native; A = annual; P = perennial; G = grass; F = forb; S = shrub or vine; Fn = fern

		Vegetation Type				
Species	Life History Pattern	Perennial Grasses	Mixed Grasses	Headlands	*Calamagrostis*	*Lupinus*
Aira caryophyllea	IAG	1.24	3.45	-	-	0.35
A. praecox	IAG	0.39	0.55	34.87	-	-
Anthoxanthum odoratum	IPG	26.91	13.24	-	0.08	16.90
Briza minor	AG	0.12	0.27	-	-	t
Bromus mollis	IAG	0.83	1.48	0.25	-	0.62
Calamagrostis nutkaensis	NPG	0.55	-	-	55.15	-
Cynosurus echinatus	IAG	0.06	15.27	-	-	0.03
Danthonia pilosa	IPG	0.07	11.64	-	-	t
Deschampsia caespitosa holciformis	NPG	28.21	0.13	0.77	t	0.33
Festuca dertonensis	IAG	0.29	0.29	0.20	-	0.13
Holcus lanatus	IPG	13.18	0.96	-	3.14	16.33
Hypochoeris radicata	NPF	1.89	1.24	26.47	t	0.61
Lasthenia chrysostoma	NAF	-	0.24	12.34	-	-
Lupinus arboreus	NPS	2.15	0.04	-	-	28.37
L. variicolor	NPS	0.02	-	18.39	-	-
Plantago lanceolata	NPF	4.92	23.71	1.02	-	0.94
Pteridium aquilinum pubescens	NPFn	2.48	1.56	-	0.28	0.83
Rubus ursinus-R. vitifolius	NPS	5.11	4.67	-	11.03	2.13

TABLE 21-5. Relative cover (%) of all species, cross tablulated by vegetation type and life form at Sea Ranch.

	Vegetation Type				
Life History Pattern	Perennial Grasses	Mixed Grasses	Headlands	*Calamagrostis*	*Lupinus*
Annuals					
Grasses	3.25	21.99	35.32	-	2.72
Sedges and rushes	t	0.02	-	-	0.12
Forbs	0.03	1.07	12.67	t	3.27
Biennials					
Grasses	t	0.23	0.01	-	-
Forbs	0.10	-	0.95	2.24	-
Annual or biennial					
Forbs	0.01	0.48	-	t	0.20
Perennials					
Grasses	69.01	26.74	1.42	61.85	33.59
Sedges and rushes	2.46	0.38	-	6.26	3.48
Forbs	9.20	31.85	28.69	11.38	4.04
Shrubs	7.83	5.56	18.39	17.54	30.72
Ferns	2.48	1.56	-	0.28	0.83
Bare ground	5.93	10.13	2.44	0.45	21.48

the effects of grazing (Elliott and Wehausen 1974) and subsequent recovery favoring the dominance of perennial grasses. Environmental heterogeneity and recent grazing favor certain species over others [e.g., *Danthonia californica* as an early perennial dominant after grazing, *Festuca idahoensis* in drier areas of the coastal ranges (Crampton 1974)].

Measures of diversity provide an important contrast to cover and dominance measures when assessing the structure of a plant community. Table 21-6 presents three diversity measures: species richness, $\hat{S}$; the Shannon–Weaver index of diversity, H' (see Pielou 1969); and Pielou's measure of evenness, J'. Absence of bias in the estimate of J' depends on an accurate estimate of $\hat{S}$ and thus on a complete census; when rare species are missed, J' is overestimated. In these data, however, the census was relatively complete, and bias in J' should be minimal.

If diversity is considered only across vegetation types, species richness is highest in the perennial vegetation type and lowest in headlands, but H' (which includes the equitability component of species diversity) is highest for mixed grasslands and lowest for the *Calamagrostis* type. Trends in J' follow those in H' closely; J' is highest for the mixed grassland and lowest for *Calamagrostis*. If life history patterns within vegetation types are considered, annual forbs are most diverse by all three measures in the perennial grassland type, completely inverse to their cover values (Table 21-5); perennial forbs and annual grasses are most diverse in the mixed grassland. On the basis of species richness, perennial grasses are most diverse in the perennial type, but H' for perennial grasses is highest in the *Calamagrostis* type, and J' in the mixed type.

Succession. The grassland of Sea Ranch shows the same trend of succession toward dominance of perennial grasses as elsewhere, especially at Pt. Reyes (Elliott and Wehausen 1974), although the species composition is different. Correlation analysis of relative cover by species implies that *Deschampsia caespitosa* ssp. *holciformis* and *Anthoxanthum odoratum* will eventually dominate most of the landscape in association with *Rubus* spp. and a few others. Forbs and annual grasses will be reduced to disturbed areas and the coastal fringe. The evidence is strictly inferential, however, and evidence of succession at Sea Ranch must still be obtained.

One might expect that the coastal scrub type would eventually invade and dominate the grassland at Sea Ranch, as elsewhere (McBride and Heady 1968). *Lupinus arboreus* is present but has been planted here north of its natural range limit; thus it is uncertain whether the type is spreading, stable, or declining. Davidson (1975) has argued that bush lupine at Bodega Head, 50 km to the south, has stablized because of a combination of seedling mortality in competition with grasses and adult mortality from insect activity. However, stands at Bodega Head reach 100% cover by lupine, whereas Sea Ranch stands exhibit only 28% lupine cover (see Tables 21-4 and 21-5). Consequently, we may expect some increase in lupine density within the patches, but perhaps no extension of scrub borders.

NORTHERN COASTAL SCRUB

Overview

According to Munz and Keck (1959), this community extends in a narrow coastal strip from southern Oregon to Pt. Sur, Monterey Co. They described it as

TABLE 21-6. Diversity indices cross tabulated by vegetation type and life form at Sea Ranch. $\hat{S}$, H', and J' are defined in text

Life History Pattern	Vegetation Type														
	Calamagrostis			Perennial Grasses			Mixed Grasses			*Lupinus*			Headlands		
	$\hat{S}$	H'	J'	$\hat{S}$	H'	J'	$\hat{S}$	H'	J'	$\hat{S}$	H'	J'	$\hat{S}$	H'	J'
Annual forbs	1	0.000	0.000	15	2.587	0.955	12	1.983	0.798	5	1.261	0.783	6	1.050	0.586
Perennial forbs	13	0.344	0.134	14	2.259	0.862	16	2.471	0.891	11	2.035	0.849	5	1.130	0.702
Annual grasses	0	-	-	9	1.746	0.795	10	1.954	0.848	9	1.852	0.843	4	0.391	0.282
Perennial grasses	5	0.178	0.110	10	0.801	0.348	9	1.048	0.477	7	0.296	0.152	2	0.141	0.203
Overall	31	0.635	0.185	69	3.271	0.773	67	3.535	0.841	50	2.994	0.765	24	1.649	0.519

dominated mainly by evergreen shrubs less than 2 m tall, but with an additional herbaceous element to the extent that the scrub is interrupted by patches of coastal prairie. Important shrubs include *Baccharis pilularis* var. *consanguinea, Eriophyllum staechadifolium, Gaultheria shallon, Lupinus variicolor, Diplacus aurantiacus* (= *Mimulus a.*), and *Rubus vitifolius*. Perennial herbs such as *Anaphalis margaritacea, Castilleja latifolia,* and *Heracleum lanatum,* and grasses such as *Danthonia californica, Calamagrostis nutkaensis,* and *Holcus lanatus,* were also listed as prominent.

More recently, Ornduff (1974) mapped the distribution and briefly discussed the composition of northern coastal scrub. His map clearly shows its discontinuous nature within the 650 km range limits mentioned by Munz and Keck. South of San Francisco, it may alternate with coastal sage scrub, which there reaches its northern limit. Ornduff's list of important species does not significantly differ from that of Munz and Keck. Hoover (1970) has described an anomalous type of "northern coastal scrub" in northern San Luis Obispo Co. with a low, spreading form of coyote brush, *Baccharis pilularis* var. pilularis.

It appears to us that there are two distinct phases of northern coastal scrub, one dominated by *Baccharis*, the other by *Lupinus*. We shall discuss them separately in the sections below.

Baccharis Scrub

Bakker (1972) viewed northern coastal scrub as a northern version of "soft chaparral" (coastal sage scrub) because of superficial similarities in shrub and leaf morphology. However, we believe that major differences between the two, such as absence of drought-deciduous shrubs in the north, and absence of a two-layered, herb-rich community in the south, make such a view misleading. Bakker pointed out that northern coastal scrub shared some species with chaparral, such as *Baccharis pilularis* var. *consanguinea, Eriodictyon californicum,* and *Toxicodendron diversilobum* (= *Rhus diversiloba*), but that it was characterized by its own collection of semiwoody subshrubs, typically shorter than 1.6 m: *Ceanothus thyrsiflorus, Gaultheria shallon, Lupinus* spp., *Diplacus aurantiacus, Rubus parviflorus, R. spectabilis,* and *R. ursinus*. She also stressed the presence of some perennial herbs, such as *Castilleja* and *Heracleum lanatum*.

Howell (1970) described in his *Marin flora* a "coastal brush" which "includes the low wind-pruned plants of maritime bluffs and mesas as well as the dense rank thickets or 'soft chaparral' of interior hills and canyons." His extensive list of characteristic shrubs and herbs indicates, however, a collection of perhaps several diverse habitats, and we believe that his concept was too broad from a community standpoint. He correctly emphasized, on the other hand, the complex nature of ecotones between the scrub and chaparral, mixed evergreen forest, and coastal prairie. He believed that coastal prairie was seral to northern coastal scrub, *Baccharis* being the primary shrub colonizer.

The successional nature of northern coastal scrub has also been emphasized by McBride (1974) and McBride and Heady (1968). These authors described a "*Baccharis* brushland" in a study of the 90–580 m elevation belt of the Berkeley Hills in Contra Costa Co. The brushland covered 21% of their area, mainly alternating with coastal prairie and mixed evergreen forest. Overstory dominants were *Baccharis* and

Toxicodendron, with *Rhamnus californica* and *Rubus vitifolius* as associates. The understory ranged from essentially coastal prairie in very young, open brush to as few as 1 plant per 10 m² beneath mature, closed canopy scrub. In the latter case, the typical herbs were *Satureja douglasii* and *Scrophularia californica*. By resampling permanent line transects, it was concluded that not only did northern coastal scrub invade grassland, but also it in turn was invaded by mixed evergreen forest. Transition from scrub to forest was estimated to take 50 yr. Periodic fire, however, could maintain the scrub as a climax. We have no hard evidence to show that this grass–scrub–forest sere is widespread. In a much shorter-term study at Pt. Reyes, Elliott and Wehausen (1974) found no significant trend toward establishment of scrub indicators in coastal prairie protected from grazing for 6 yr. Another study at Pt. Reyes by Grams et al. (1977) also failed to show strong evidence for forest invasion of scrub.

North–south gradients. It should be expected that the composition of northern coastal scrub exhibits latitudinal gradients within its north–south limits. As an initial attempt to document such gradients, two of us (Barbour and Taylor) sampled northern coastal scrub at eight sites between Little Sur River in Monterey Co. and Stuarts Pt. in Sonoma Co., a distance of nearly 300 km. Sampling, done by relevé and/or line transect with the help of Robert Robichaux, was accomplished in November 1975; hence herbaceous growth was not at a maximum. The data are summarized in Table 21-7.

The most southerly two sites are clearly transition from coastal sage to northern coastal scrub, *Baccharis* being codominate with or subdominant to *Artemisia californica* and *Salvia mellifera* (Fig. 21-3). Herbaceous cover was low, about 1%, whereas bare ground (including litter and dead upright stems) was high, about 24%. Cover by drought-deciduous shrubs was 57%, close to the 67% value for coastal sage scrub at Camp Pendelton, San Diego Co. (see Chapter 13).

The next three sites to the north, in Monterey, Santa Cruz, and San Mateo Cos., still showed *Artemisia* at 10% or more cover, but *Salvia* dropped out and a number of northern-ranging species joined the community (*Pteridium aquilinum* var. *pubescens*, *Rubus ursinus*, *R. vitifolus*, *Sanicula bipinnata*, *Satureja douglasii*, *Scrophularia californica*, *Stachys rigida*). Herbaceous cover rose to 34%, while bare ground fell to 17%, and cover by drought-deciduous shrubs to 0%. Chandler (1955) sampled a *Baccharis* stand similar to these in Alameda Co. and reported aboveground biomass of 3200 kg ha^{-1}.

The three northern sites, especially the one high on Mt. Vision (Fig. 21-4), exhibited a distinct northern physiognomy and composition. Personal observations, not documented in Table 21-7, indicate a similar flora in Mendocino Co. scrub, but with more cover by *Rubus*, *Gaultheria*, *Stachys*, *Polystichum*, and possibly *Myrica* and *Lupinus* species. In the Marin and Sonoma Co. sites, the importance of *Toxicodendron* and *Diplacus* declined, but diversity in a stratum beneath the overstory shrubs increased overall: more cover by low shrubs, such as *Gaultheria shallon* and *Rubus ursinus*; more cover by ferns, such as *Pteridium* and *Polystichum*; and more cover by perennial herbs. Herbaceous cover on Mt. Vision rose to 43%, and bare ground was less than 1%; some 24 species were present, double the number in the southernmost sites. Scrub at Point Reyes has recently been sampled more intensively (Grams et al. 1977); these data show important differences between north- and south-facing slopes, with south-facing slopes having closer affinity to coastal sage scrub.

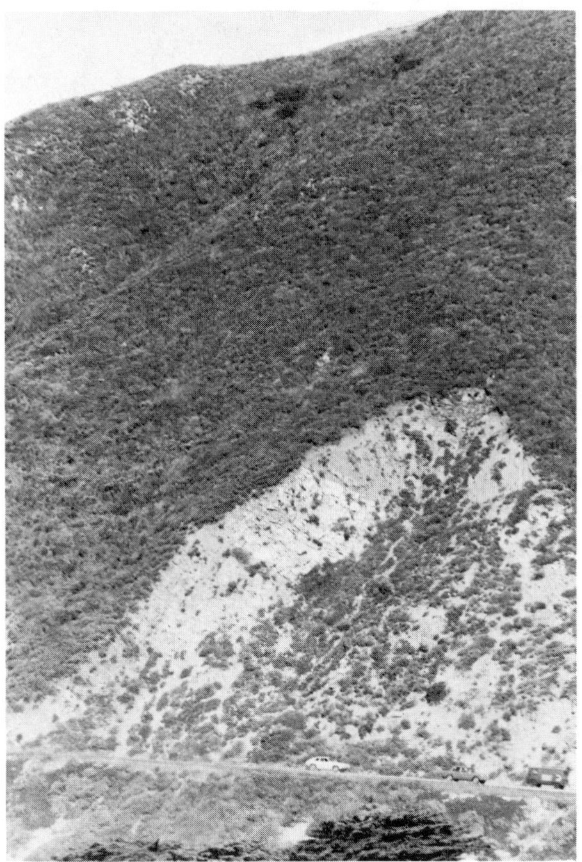

Figure 21-3. Northern coastal scrub (*Artemisia, Salvia, Baccharis, Toxicodendron*) at its southern limit on steep, generally west-facing slopes near Little Sur River, Monterey Co.

Our tentative conclusions are as follows. The components of *Baccharis* northern coastal scrub are found in (and perhaps derived from) adjacent vegetation types such as coastal sage scrub, chaparral, coastal prairie, closed-cone pine forest, and north coastal forest, depending on the latitude at which the scrub is examined. Community structure shifts from essentially a one-layered, somewhat open scrub about 1.3 m tall in the south to a two-layered community in the north. The upper synusium, about 2 m tall, is dominated by *Baccharis*, and its canopy may not be closed. Beneath is a closed canopy about 0.3 m tall, made up of ferns and low woody and herbaceous perennials. *Baccharis* appears to be the only dominant taxon continuous through the range of northern coastal scrub. It is one of the few species that do not join the understory of adjacent forest in the north; it may be an obligate heliophyte (Wright 1928).

We wish to reemphasize the superficial nature of this survey. The complexity of scrub vegetation at any one locale is not revealed in Table 21-7. Community patterns that could relate to slope, fire history, grazing intensity, and landslides were apparent to us, but were not sampled. As described by others, cited earlier in this

TABLE 21-7. Relevé and line transect data for eight northern coastal scrub (*Baccharis* phase) sites. Relevé numbers are cover-abundance values: r = rare, + = <1% cover, 1 = 1-10%, 2 = 10-25%, 3 = 25-50%, 4 = 50-75%, 5 = >75%. Line transect numbers (in parentheses) are in absolute percent cover; *t* = 0.01%. Occasionally, cover-abundance values do not match transect cover data. Total transect length was 80 m at Little Sur River, 20 m at Pescadero, and 30 m at Mt. Vision. Relevé areas were not usually demarcated and measured. Based on unpublished data by Taylor, Robichaux, and Barbour

	Site Location							
	2 km N of Little Sur River	20 km S of Monterey	12 km E of Monterey	15 km NW of Santa Cruz	Pescadero	Mt. Vision	2 km N of Jenner	2 km S of Stuarts Pt.
County:	Monterey	Monterey	Monterey	Santa Cruz	San Mateo	Marin	Sonoma	Sonoma
Lat. (N):	36°20'	36°25'	36°35'	37°00'	37°15'	38°05'	38°25'	38°40'
Slope and aspect:	27°WSW	20°SW	30°NE	15°WSW	15°S	10°NNW	19°NNW	10° all aspects
Elevation (m):	100	100	200	100	30	375	30	30
Overstory height (m):	1.3	1.5	2.0	2.0	1.0	2.0	2.0	1.8
Number of taxa:	12	10	10	12	12	24	14	23
Taxon								
Baccharis pilularis var. *consanguinea*	2(23.5)	3	4	3	4(57.0)	5(67.1)	5	4
Artemisia californica	3(12.1)	2	3	2	2(9.7)			
Salvia mellifera	4(56.6)	3						
Toxicodendron diversiloba	1(3.0)	+	1	2		+(0.3)	+	r
Rhamnus californica	1(0)	2		2		1(1.6)		1
Diplacus aurantiacus	1(4.5)	+	1		1(10.7)			I

Rhamnus crocea			+					
Rubus ursinus			r	+		2(4.6)	2	
Astragalus lentiginosus	+(0.3)	r						
Elymus condensatus	+(0.4)		+		+(0)			
Dryopteris arguta	+(0)							
Satureja douglasii			1	+	1(2.0)	1(2.8)		+
Pityrogramma triangularis			r		+(2.0)			
Ribes speciosum		r	+					
Anaphalis margaritacea	+(0.1)					+(1.0)	+	
Achillea borealis	+(0.6)	+					+	
Aster subspicatus				+	+(7.7)			
Pteridium aquilinum var. *pubescens*				+		1(4.6)	+	+
Eriophyllum staechadifolium		+		+				
Poa sp.					+(10.2)			
Calystegia occidentalis				+				
Sanicula bipinnata				+	+(0.7)		+	+
Stachys rigida				+	+(1.7)		+	+
Eriogonum arboreum					+(0)			
Scrophularia californica					+(1.0)		+	+
Lonicera involucrata							+	
Polystichum munitum						1(4.0)	+	+

TABLE 21-7 [continued]

	Site Location							
	2 km N of Little Sur River	20 km S of Monterey	12 km E of Monterey	15 km NW of Santa Cruz	Pescadero	Mt. Vision	2 km N of Jenner	2 km S of Stuarts Pt.
County:	Monterey	Monterey	Monterey	Santa Cruz	San Mateo	Marin	Sonoma	Sonoma
Lat. (N):	36°20'	36°25'	36°35'	37°00'	37°15'	38°05'	38°25'	38°40'
Slope and aspect:	27°WSW	20°SW	30°NE	15°WSW	15°S	10°NNW	19°NNW	10° all aspects
Elevation (m):	100	100	200	100	30	375	30	30
Overstory height (m):	1.3	1.5	2.0	2.0	1.0	2.0	2.0	1.8
Number of taxa:	12	10	10	12	12	24	14	23
Taxon								
Heracleum lanatum						1(4.6)	+	
Ribes sanguineum							+	
Iris douglasii						+(0.8)	+	
Erechtites arguta	+(0.1)							+
Potentilla glandulosa	r(0.1)							+
Poa douglasii						+(0)		
Gaultheria shallon						4(70.8)		1
Senecio arnicoides						+(0.8)		
Elymus glaucus						2(18.0)		
Galium californica						1(1.1)		+

Symphoricarpos alba	1(1.0)	
Lathyrus bolanderi	+(0.5)	
Fragaria californica	+(1.8)	+
Corylus californica	2(7.0)	
Viola adunca	+(0.1)	
Myrica californica	+(t)	
Pinus muricata	+(0)	
Polypodium californicum	+(t)	+
Rumex acetosella	+(0)	
Picris echeioides	+(t)	+
Umbellularia californica		3
Angelica arguta		+
Anagallis arvensis		+
Phacelia californica		+
Zigadenus fremontii		+
Pseudotsuga menziesii		2

Figure 21-4. (*a*) Northern coastal scrub dominated by *Baccharis* at 375 m elevation on Mt. Vision in Pt. Reyes National Seashore, Marin Co. (*b*) The understory, diverse in species and high in cover.

section, northern coastal scrub may be a seral community on certain sites, and this adds another element of complexity. Clearly, the status of knowledge regarding this vegetation type is unsatisfactory and remains at a very basic level.

Lupinus Scrub

Pockets of lupine-dominated coastal scrub, within a grassland matrix, appear on level terraces within about 200 m of ocean-facing bluffs. This scrub phase appears to be best developed between Santa Cruz and Sonoma Cos. In the "Sea Ranch" section of this chapter, this phase was identified as both "headlands" and "lupine areas."

On the most exposed sites, closest to the lip of ocean-facing bluffs, lupine scrub is dominated by the nearly prostrate *Lupinus variicolor.* Apart from a number of annual grasses and forbs, common associates include the perennials *Armeria maritima* var. *californica, Artemisia pycnocephala, Dudleya farinosa, Erigeron glaucus, Eriogonum latifolium, Eriophyllum staechadifolium, Grindelia stricta* ssp. *venulosa,* and *Mesembryanthemum chilense.* Because of the low, open nature of this community and its extremely narrow distribution, it does not strike a casual observer as "scrub" at all. Nevertheless, it is dominated by subshrubs, and its floristic composition is markedly different from that of grassland only 50 m inland (see Drysdale 1971; Pitelka 1974; Barbour et al. 1973).

From 50 to 200 m further inland, patches of a different lupine scrub may occur within a grassland matrix. This scrub is dominated by *Lupinus arboreus* almost to the exclusion of all other shrubs and herbs. Mature bush lupines reach 2 m in height and achieve a closed canopy (Fig. 21-5). Associated perennials are few but may include *Baccharis pilularis* ssp. *consanguinea, Eriophyllum staechadifolium, Myrica californica, Pteridium aquilinum* var. *lanuginosum,* and *Rubus* spp. At Bolsa Pt., San Mateo Co., the two lupine scrubs share a Sorensen coefficient of community of only 24.

Pitelka (1974) compared the microenvironments and adaptive strategies of *Lupinus variicolor* and *L. arboreus* at Bolsa Pt. He found the *L. variicolor* habitat at the lip to exhibit higher summer air and soil temperatures, greater wind speed even at 5 cm above the ground, higher levels of salt spray deposition, higher relative humidity, a finer soil texture, and higher amounts of soil moisture, compared to a *L. arboreus* habitat only 50 m inland. Similar gradients have been reported for Bodega Head, Sonoma Co., by Barbour et al. (1973 and unpub. data) and Drysdale (1971) with one exception: *L. variicolor* soils are often drier than those of *L. arboreus* at Bodega.

Transplant gardens at Bolsa Pt. revealed that *L. variicolor* could not survive summer conditions in *L. arboreus* sites, but that *L. arboreus* grew equally well during its

Figure 21-5. *Lupinus arboreus* scrub on coastal terrace at Tomales Point, Marin Co.

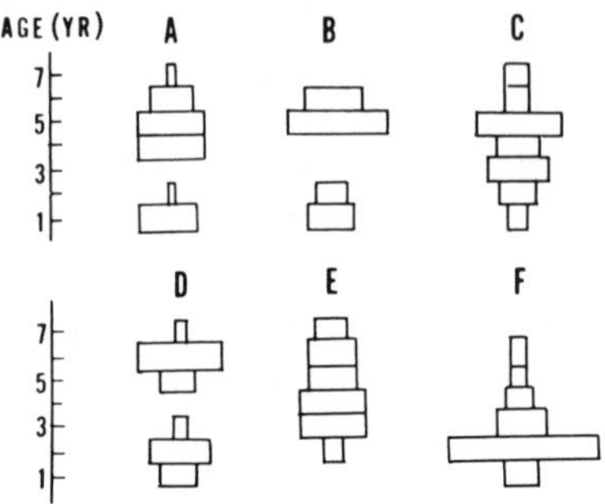

Figure 21-6. Age structure of six stands of *Lupinus arboreus* (*A–F*) at Bodega Head, Sonoma County. Horizontal extent of bars indicates relative number of shrubs (from Davidson 1975, and Davidson and Barbour 1977).

first year in either lip or inland sites. At Bodega, Drysdale (1971) transplanted an associate of *L. variicolor, Armeria maritima* var. *californica,* into a series of inland garden sites and similarly found that it grew and reproduced less successfully at inland sites.

Pitelka (1974) went on to report different strategies of carbon allocation in the two lupine species. *Lupinus arboreus* exhibited a 50% higher rate of net photosynthesis (dry weight basis), developed thinner leaves, and invested more energy in the above-ground portion of the plant (13% leaf, 48% stem, 20% root after 1 yr vs. 18% leaf, 23% stem, 42% root for *L. variicolor*). Optimum temperature for net photosynthesis was equal for both species, 20°C.

Davidson (1975) conducted a demographic study of *L. arboreus* at Bodega Head and concluded that maximum lupine age may be 7 yr. Major causes of mortality are drought and mammalian herbivore activity during the first year of life, and insect damage later. Principal insect herbivores are larvae of the swift moth (*Hepialus behrensi*), which burrow through the wood, and larvae of the California tussock moth (*Hemerocampa vetusta*) and the salt marsh moth caterpillar (*Estigmene acraea*), which eat the foliage. All three appear to be episodic in their population densities, and can reach such epidemic proportions that entire patches of lupine scrub are killed or denuded in one growing season. Figure 21-6 shows the age structure of six stands of *L. arboreus* at Bodega Head. On the basis of field and laboratory studies and the examination of aerial photographs for the past 40 yr, Davidson concluded that bush lupine had reached a point of dynamic equilibrium within the grassland and that it was therefore unlikely the grassland as a whole was successional to lupine scrub. Allelopathy did not appear to be a factor in this balance. He attributed the near absence of herbs beneath lupine canopy to activity by high rodent populations.

AREAS FOR FUTURE RESEARCH

The complex grassland–scrub–forest pattern along the central and northern coast appears to have a successional relationship, but studies have been so few that regional conclusions cannot yet be drawn. Even more basic work remains to be done: quantitative synecological descriptions of grassland and scrub in Del Norte, Humboldt, and Mendocino Cos. need to be accumulated. There does not seem to have been a detailed examination of the autecology of *Baccharis* scrub components for the past 50 yr—certainly nothing comparable to the recent demographic and carbon allocation studies of *Lupinus* scrub components. In short, our level of information and our level of interpretation for the vegetation of this part of California are not very sophisticated.

ACKNOWLEDGMENTS

Two of us (Foin and Hektner) gratefully acknowledge research support by the Rockefeller Foundation and the University of California Council for Advanced Studies of the Environment.

LITERATURE CITED

Bakker, E. 1972. An island called California. Univ. Calif., Press, Berkeley. 361 p.

Barbour, M. G. 1970. The flora and plant communities of Bodega Head, California. Madroño 20:239-313.

-----. 1972. Additions and corrections to the flora of Bodega Head, California. Madroño 21:446-448.

Barbour, M. G., R. B. Craig, F. R. Drysdale, and M. T. Ghiselin. 1973. Coastal ecology: Bodega Head. Univ. Calif. Press, Berkeley, 338 p.

Barrett, S. A. 1952. Material aspects of Pomo culture, Part I. Bull. Public Mus. Milwaukee 20:1-260.

Barry, W. J. 1972. California prairie ecosystems. Vol. I: The central valley prairie. State of Calif., Resources Agency, Dept. Parks and Recreation, Sacramento. 82 p.

Barry, W. J., D. M. Burns, and J. Chatfield. 1974. Pygmy forest ecological staircase feasibility study. State of Calif., Resources Agency, Dept. Parks and Recreation and Dept. Conservation, Sacramento. 32 p.

Batzlie, G. O., and F. A. Pitelka. 1970. Influences of meadow mouse populations on California grassland. Ecology 51:1027-1039.

Bettle, A. A. 1947. Distribution of the native grasses of California. Hilgardia 17:309-357.

Burcham, L. T. 1957. California grangeland. Calif. Div. Forestry, Sacramento. 261 p.

Burtt-Davy, J. 1902. Stock ranges of northwestern California. USDA Bur. Plant Ind. Bull. 12. 81 p.

Chandler, C. C. 1955. The classification of forest fuels for wildland fire control purposes. M.S. thesis, Univ. Calif., Berkeley. 101 p.

Cooper, D. W. 1960. Fort Baker ranges returned to champagne grasses. J. Range Manage. 13:203-205.

Cooper, D. W., H. F. Heady. 1964. Soil analysis aids grazing management in Humboldt County. Calif. Agric. 18(6):4-5.

Crampton, B. 1974. Grasses in California. Univ. Calif. Press, Berkeley. 178 p.

Cronise, T. F. 1868. The natural wealth of California. H. H. Bancroft, San Francisco. 696 p.

Davidson, E. D. 1975. Demography of _Lupinus arborues_ at Bodega Head, California. Ph.D. diss., Univ. Calif., Davis, 114 p. [see also p. 760].

Davidson, E. D., and M. G. Barbour. 1977. Germination, establishment, and demography of coastal bush lupine (_Lupinus arboreus_) at Bodega Head, California. Ecology (in press).

Drysdale, F. R. 1971. Studies in the biology of _Armeria maritima_ (Mill.) Willd. with emphasis upon the variety californica as it occurs at Bodega Head, Sonoma County, California. Ph.D. dissertation, Univ. Calif., Davis. 141 p.

Elliott, H. W., and J. D. Wehausen. 1974. Vegetational succession on coastal rangeland of Point Reyes Peninsula. Madroño 22:231-238.

Fisher, R. A., A. S. Corbet, and C. B. Williams. 1943. The relation between the number of species and the number of individuals in a random sample of an animal population. J. An. Ecol. 12:42-58.

Grams, H. J., K. R. McPherson, V. V. King, S. A. MacLeod, and M. G. Barbour. 1977. Northern coastal scrub on Point Reyes Peninsula, California. Madroño 24:18-24.

Hardham, C. B., and G. H. True. 1972. A floristic study of Point Arena, Mendocino County, California. Madroño 21:499-504.

Heady, H. F., D. W. Cooper, J. W. Rible, and J. E. Hooper. 1963. Comparative forage values of California oatgrass and soft chess. J. Range Manage. 16:51-54.

Hektner, M. M. and T. C. Foin. 1976a. Vegetation analysis of a Northern California coastal prairie: Sea Ranch, Sonoma County, California. Accepted by Madroño.

-----. 1976b. A flora of the coastal terraces of Sea Ranch, Sonoma County, California. To be submitted to Wasmann J. Biol.

Hoover, R. F. 1970. The vascular plants of San Luis Obispo County, California. Univ. Calif. Press, Berkeley. 350 p.

Howell, J. T. 1970. Marin flora. 2nd ed. Univ. Calif. Press, Berkeley. 366 p.

Huffaker, J. T. 1951. The return of native perennial bunchgrass following the removal of Klamath weed (Hypericum perforatum L.) by imported bettles. Ecology 32:443-458.

Huffaker, C. B., and J. K. Holloway. 1949. Changes in range plant population structure associated with feeding of imported enemies of Klamath weed (Hypericum perforatum L.). Ecology 30:167-175.

Huffaker, C. B., and C. E. Kennett. 1959. A ten-year study of vegetational changes associated with biological control of Klamath weed. J. Range Manage. 12:69-82.

Kniffen, F. B. 1939. Pomo geography. Univ. Calif. Pub. Amer. Archaeol. Ethol. 36:353-400.

Küchler, A. W. 1964. Potential natural vegetation of the conterminous United States. Amer. Geogr. Soc. Spec. Pub. 36.

Mason, J. 1970. Point Reyes -- the solemn land. DeWolf Printing, Pt. Reyes Station, Calif.

McBride, J. 1974. Plant succession in the Berkeley Hills, California. Madroño 22:317-329.

McBride, J., and H. F. Heady. 1968. Invasion of grassland by Baccharis pilularis D.C. J. Range Manage. 21:106-108.

Munz, P. A., and D. D. Keck. 1959. A California flora. Univ. Calif. Press, Berkeley. 1681 p.

Murphy, A. H. 1955. Vegetational changes during biological control of Klamath weed. J. Range Manage. 8:76-79.

Ornduff, R. 1974. Introduction to California plant life. Univ. Calif. Press, Berkeley. 152 p.

Peñalosa, J. 1963. A flora of the Tiburon Peninsula, Marin County, California. Wasmann J. Biol. 21:1-74.

Pielow, E. C. 1969. An introduction to mathematical ecology. John Wiley, New York. 286 p.

Pitelka, L. F. 1974. Energy allocation in annual and perennial lupines (Lupinus Leguminosae). Ph.D. dissertation, Stanford Univ., Stanford, Calif. 126 p.

Reed, M. J., W. R. Powell, and B. S. Bal. 1963. Electronic data processing codes for California wildland plants. USDA Forest Serv. Res. Note PSW-N20. 314 pp.

Schlinger, E. I., W. J. Barry, D. C. Erman, H. G. Baker, and R. Kawin. 1976. Biotic communities. In W. J. Barry and E. I. Schliner (eds.). Inglenook Fen, a study and plan. State of Calif., Resources Agency, Dept. Parks and Recreation, Sacramento (in press).

Whittaker, R. H. 1965. Dominance and diversity in land plant communities. Science 147:250-260.

Wright, A. D. 1928. An ecological study of *Baccharis pilularis*. M.S. thesis, Univ. Calif., Berkeley. 45 p.

THE GREAT BASIN FLORISTIC PROVINCE

CHAPTER

22 SAGEBRUSH STEPPE

JAMES A. YOUNG
RAYMOND A. EVANS
U.S. Department of Agriculture, Agricultural Research Service, Reno, Nevada

JACK MAJOR
Botany Department, University of California, Davis

Introduction 764
Forgotten ecological province 766
Characteristics of the vegetation 766
Artemisia 766
Diversity 766
Environmental relations 767
Relation to other vegetation types 767
Soils and topography 767
East slope of the Sierra Nevada 767
Modoc Plateau 767
Microenvironment 768
Succession 769
Pristine succession 769
Modern succession 770
Native animals 770
Plant communities 771
Artemisia communities 771
Artemisia tridentata communities 772
Agropyron spicatum as an understory dominant 772
Stipa species as understory dominants 773
Elymus cinereus as an understory dominant 779
Oryzopsis hymenoides as an understory dominant 779
Bromus tectorum as an understory dominant 780
Artemisia arbuscula communities 780
Artemisia nova communities 781
Artemisia cana communities 781
Other communities 784
Mountain brush communities 784
Cercocarpus ledifolius communities 785
Purshia tridentata communities 785
Pluvial basin communities 785
Atriplex confertifolia communities 788
Other communities 792
Future research 793
Literature cited 796

INTRODUCTION

The sagebrush steppe consists of a series of generally treeless, shrub-dominated communities located along the eastern and northeastern boundary of California. Species of *Artemisia* are the dominant shrubs, with perennial bunch grasses characterizing the understory.

The California sagebrush steppe is an elongated, discontinuous strip of land from the upper Owens River valley north along the eastern flanks of the Sierra Nevada (trans-Sierra) to Honey Lake near the 40th parallel (Fig. 22-1). In eastern California the type is a transitional series of communities between the coniferous forest of the Sierra Nevada Mts. and the extensive *Artemisia* steppe of the Great Basin. Where spur ranges of parallel mountain ranges extend to the state boundary, the sagebrush steppe is largely replaced by confierous forest. With topoedaphic exposure to winter winds, sagebrush communities may extend to the west slope of the Sierra Nevadas.

The most extensive areas of sagebrush steppe in California are north of the 40th

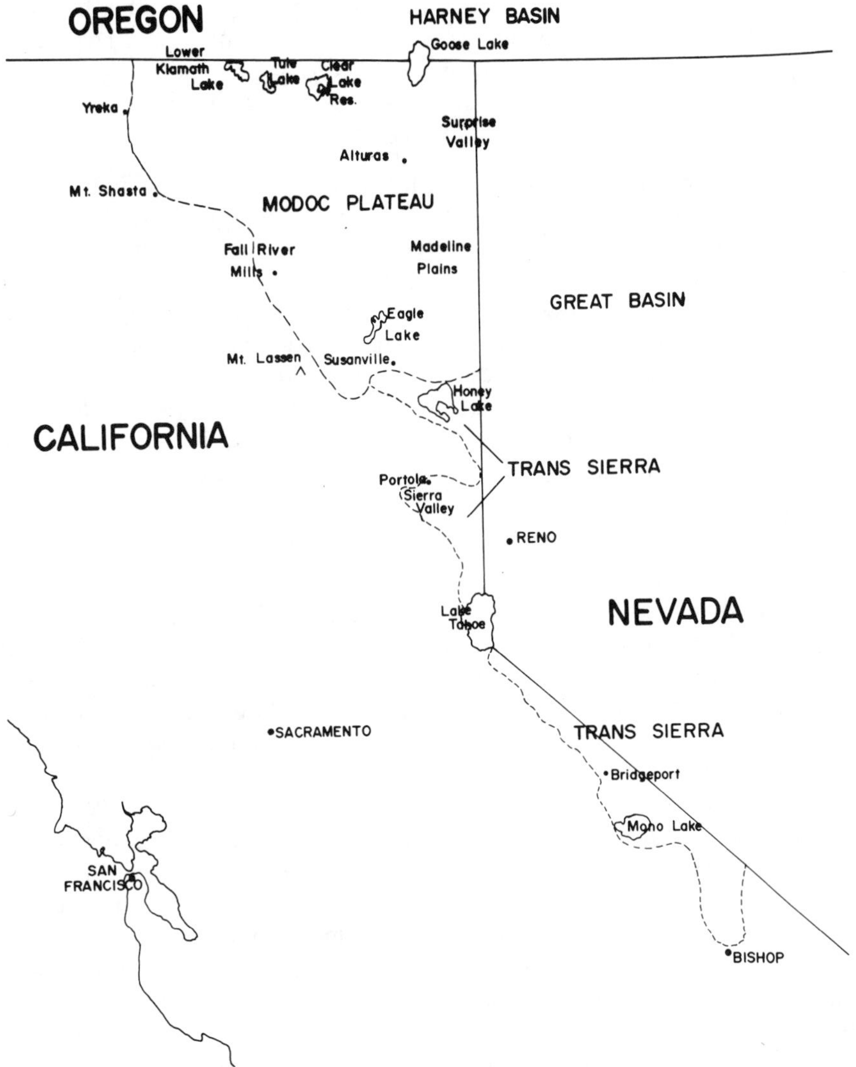

Figure 22-1. Location of sagebrush steppe vegetation in California. The sagebrush steppe of the Modoc Plateau is interspersed among various types of coniferous forest communities. South of Honey Lake the sagebrush steppe forms a narrow band along the trans-Sierra, the eastern slope of the Sierra Nevada Mts.

parallel, especially north of the Honey Lake valley on the Modoc Plateau (Fig. 22-1). North of the Sierra Nevada on the Modoc Plateau, the hydrological basin of the Great Basin extends as a westward bulge into California. The sagebrush steppe extends much farther westward than does the hydrological Great Basin. *Artemisia* communities occur west of the southern extension of the Cascade Mts. in Siskiyou Co.

The Modoc Plateau of northeastern California is a highland region capped by vast late Tertiary and Quaternary basalt plains and numerous volcanic shields (Macdonald and Gay 1966). It is a southern extension of the Pacific Northwest Volcanic

Provinces (King 1959). In topography, soils, and vegetation, this area has more in common with the high desert of southeastern Oregon (Franklin and Dyrness 1973) than with the Lahontan and Bonneville basins of Nevada and Utah.

Forgotten Ecological Province

The sagebrush steppe in California is much better known geologically than synecologically (e.g., see Russell 1885; Blackwelder 1931; Morrison 1964). The area does not contain a university, and most centers of ecological research are physically removed from it.

Considerable ecological research has been done on the phytosociology of the province at Squaw Butte Experimental Range and at the University of Nevada, Reno, in the Lahontan Basin. Forest or range scientists have done most of this research for purposes of environmental manipulation. The Devils Garden Natural Area (NA 250438) is located on the Modoc Plateau. This is apparently the only natural area set aside for study in the sagebrush steppe of California.

Characteristics of the Vegetation

Artemisia. The sagebrush steppe is obviously characterized by species of *Artemisia.* With one exception, *A. spinescens,* we are dealing with species of the section Tridentatae (Rydberg 1916). Species of this section occur discontinuously as dominants or partial dominants over one-third of the part of the contiguous United States west of 102°W longitude (Beetle 1960). Section Tridentatae is noted for both intraspecific and interspecific morphological variation (Hall and Clements 1923; Ward 1953; Beetle 1960).

The taxonomic treatments of Beetle (1960) and Beetle and Young (1965) and the chromatographic studies of Hanks et al. (1973) provide an ecologically significant breakdown of the section. For these reasons a departure from the standard taxonomic reference (i.e., Munz and Keck 1959) for this volume is worthwhile.

These modern treatments lead us to the following terminology: *A. tridentata* ssp. *tridentata* with at least three forms; *A. tridentata* ssp. *wyomingensis*; and *A. tridentata* ssp. *vaseyana* with at least three forms. The ecologically distinct *A. nova* is elevated from a subspecies of *A. arbuscula* to independent status. Combined ecological and taxonomic research has shown that the subspecies of *A. tridentata* generally reflect specific environments. Refinement of synecological technique may show specific community relations for these subspecies.

Diversity. The proximity of much of the sagebrush steppe in California to coniferous woodlands and forest enhances the diversity of many communities with subdominant species of *Prunus, Purshia, Ribes,* and *Symphoricarpos.* Most *Artemisia*-dominated communities contain root-sprouting species of *Chrysothamnus* and *Tetradymia* that become seral dominants with disturbance.

To complete our overview of the sagebrush steppe, we must include *Atriplex* and *Sarcobatus* communities of the Honey Lake and Surprise valleys. Both of these basins contained Lahontan-age pluvial lakes: Honey an embayment of Lahontan, Surprise an independent basin. These areas contain fine-textured, saline soils, and their vegetation is an extension of that of the northern Carson Desert (Billings 1949).

ENVIRONMENTAL RELATIONS

Relation to Other Vegetation Types

The proximity of much of the sagebrush steppe in California to coniferous woodlands or forest makes it difficult to separate piñon–juniper, three-needle pine, or juniper-woodland types. Sagebrush steppe communities may be relatively long-term seral dominants in the succession of these woodlands (Billings 1951). Certainly the ecotone boundaries between these types are highly mobile with climatic change (e.g., see Mehringer 1967). Further complicating this relation is the invasion of the sagebrush steppe type by coniferous woodlands in apparent response to the grazing of domestic livestock and to fire suppression (Burkhardt and Tisdale 1969; Blackburn and Tueller 1970).

The sagebrush steppe vegetation type in California, and a basic series of communities within it such as *Artemisia tridentata/Agropyron spicatum* and *Artemisi tridentata/Festuca idahoensis,* are related to the high desert of southeastern Oregon and to the Columbia Basin Province (Daubenmire 1970).

Soils and Topography

East slope of the Sierra Nevada. Much of the eastern slope of the Sierra Nevada from the California boundary to the crest of the mountains is an extremely abrupt physical and biological gradient. Annual precipitation can range from 80 mm at the bottom of an eastern basin to 1250 mm along the mountain crest (Nevada River Basin Survey Staff 1969). Elevations change from 1200 to 3000 m in a relatively short distance. Youth, or absence of soil profile development, is a predominant characteristic. Few landforms support soils older than late Pleistocene (Nevada River Basin Survey Staff 1969).

The Susan, Truckee, Carson, and Walker rivers of the Lahontan Basin and the streams of the Mono Basin bisect the sagebrush steppe east of the Sierra Nevada. Tills of four glaciations and downstream outwash terrace deposits are recognizable in these drainages (Birkeland 1968). These outwash terraces and tills can provide a 10^4–10^6 yr time disjunction between soils and parent material. These rivers flow west to east down the east slope of the Sierra Nevada; the tills and outwash reflect the same gradient. The older outwash deposits may support soils of the order of Mollisols (Soil Survey Staff 1960) at higher elevations, but deposits of the same age may have developed Aridisols (desert soils) further down the environmental gradient.

Apart from incised river gorges, the landscape is characterized by piedmonts of detritus from the steep mountain slopes, which frequently coalesce to form bajadas. Above the bajadas, the pediment surfaces often continue the sagebrush steppe environment to the transition into the conifer forest.

Modoc Plateau. The very fluid basalt flows that characterize much of the Modoc Plateau are often grouped as Warner basalt and are probably late Pliocene to Pleistocene in age (Macdonald and Gay 1966). Rocks of Pliocene age include both volcanic and lake deposits. The lake sediments are tuffaceous siltstones, ashy sandstones, and thick, extensive deposits of diatomite.

Lava rims, undulating upland plateaus, pahoehoe flows, lava tubes, pluvial lake beds, and large-volume springs issuing suddenly from porous formation characterize the Lassen-Modoc landscape. Upon this geological template, the sagebrush steppe is superimposed. Young and Evans (1970) provide a diagrammatic breakdown of how the physical and biological environments mesh on the Modoc Plateau (Fig. 22-2). Rocky ridges and rimrocks often reflect environments that support climax communities of *Juniperus occidentalis* (see Burkhardt and Tisdale 1969 for similar situations in Idaho). Relatively deep, well-drained soils, often highly modified by subaerially deposited volcanic ash, characterize *Artemisia tridentata* communities. The Modoc Plateau has extensive areas of soils that are relatively shallow above a restrictive layer of clay or bedrock. These are characteristic *A. arbuscula* environments (Fosberg and Hironaka 1964). both *Artemisia* types may support *Juniperus occidentalis* communities.

The landscape of the Modoc Plateau is dotted with pluvial lakebeds. These range from small depressions to the bed of pluvial Lake Madeline (Fig. 22-1), which is several thousand hectares in area (Hobbs and Miller 1948). *Artemisia cana* ssp. *bolanderi* communities characterize the vegetation of the old lakebeds. They generally occur as a ring surrounding seasonal ponds (Dealy 1971).

Microclimate

The sagebrush steppe environment is characterized by cold, harsh winters and dry summers (see Chapter 2). Available moisture is out of phase with temperatures suitable for growth. Evans et al. (1970) provide details of moisture and temperature parameters of the long, cold springs, when temperatures are too low for germination or growth. Most herbaceous growth takes place during a relatively short period in late spring and is terminated by the exhaustion of soil moisture in early summer.

Seasonal sequences in phenology are marked, with mosses and annuals developing first and larger grasses and forbs usually flowering in June. *Prunus andersonii* flowers in the spring, but most shrubs flower from early July (*Tetradymia canescens*) to October (*Artemisia tridentata*).

Shrub phenology may be typified by that of *Chrysothamnus viscidiflorus* (Young and Evans 1974a). Buds burst in late winter, followed by a long period of minimum growth. As much as 60% of the annual twig elongation may occur in a period of 2 wk in May. Deep-rooted perennials extend the period of stem elongation, but the depth of wetting in most soils is usually not more than 1 m, and indurate pans often limit rooting depth. Shrub growth ends with the onset of moisture stress. Daubenmire (1970) indicates that shrub growth in the Columbia Basin continues throughout the summer. In contrast, most sagebrush steppe species in California cease growth by midsummer. Flowering may or may not occur in late summer, depending on moisture stress (Young and Evans 1974a).

Variability among years is great. There may be no available soil moisture when temperatures warm in the spring, or rains may extend the growth period through June. Rains early enough in the fall to begin germination before the occurrence of cold temperature are an important variable in this environment.

Most herbaceous species flower near the peak of the active growth period, and their seeds lie dormant until the fall or spring growth period. The seedbed and seedling requirements of native herbaceous or shrubby perennials are virtually unknown.

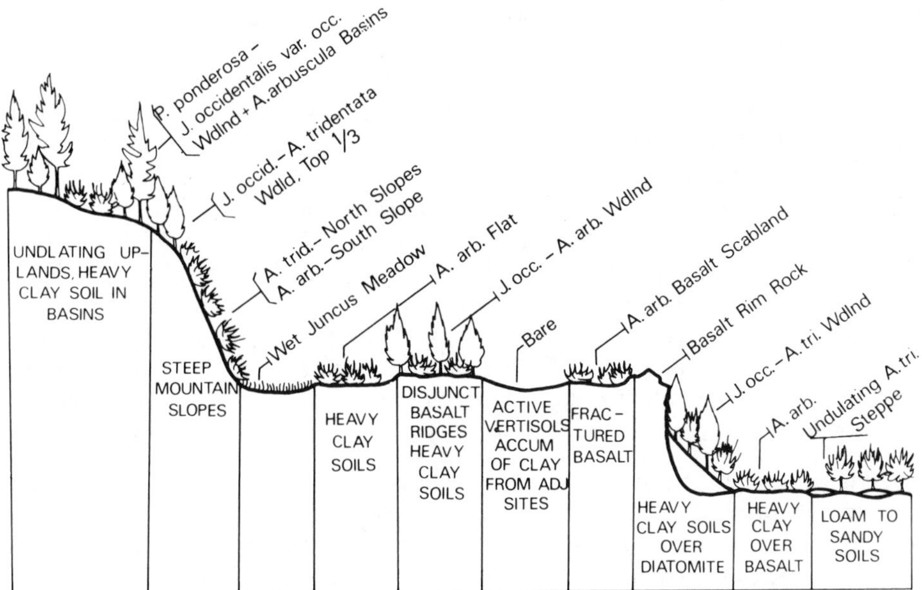

Figure 22-2. Diagrammatic presentation of the physiography and vegetation on the Modoc Plateau.

Possibly, an occasional summer with relatively abundant precipitation from convectional storms is important in the establishment of perennial grasses (Harris 1967).

Succession

In the past century, big sagebrush communities have known three conditions: (*a*) pristine, (*b*) unlimited exploitation by grazing and fire, and (*c*) attempted complete suppression of fire and attrition by grazing as stand renewal processes (Young and Evans 1970).

Pristine succession. Wildfire, grazing by large herbivores, and defoliation by larvae of the moth *Aroga websterii* undoubtedly contributed to stand renewal in the pristine sagebrush steppe of California (Young et al. 1975).

The incidence of wildfires under pristine condition is hard to assess. Relatively pristine *Artemisia*/bunch grass communities are quite fire-resistant compared with communities degraded by alien annual grasses, but they are not fireproof. We know from fire scars that wildfires have been frequent in adjacent pine woodlands (e.g., see Antevs 1938).

The impact of wildfire on sagebrush steppe is well known (Blaisdell 1953). *Artemisia tridentata* does not sprout after burning, but species of *Chrysothamnus, Tetradymia, Prunus, Ribes,* and some *Purshia* sprout from crowns or roots and dominate seral communities for periods up to 20 yr after wildfires (Young and Evans 1974b). *Artemisia arbuscula* communities have a reduced incidence of burning because of size and spacing of shrubs and probably because of herbaceous species composition. The perennial grass portions of pristine sagebrush steppe communities probably benefited from the reduced competition after wildfires. Burning may be necessary to make the sink of nutrients in *Artemisia* litter available for the competing herbaceous vegetation.

In contrast to much of the Lahonton Basin, the Modoc Plateau apparently supported small herds of American bison (*Bison bison*) (Merriam 1926). No great concentration of large herbivores consistently influenced succession, but occasionally animals drifted in from the north.

The dominant herbaceous seral species in *Artemisia*/grass communities are short-lived perennial grasses such as *Sitanion hystrix* or *Poa sandbergii*. These species have shorter growing seasons than do the climax perennial grasses, but both can be overgrazed by concentrations of large herbivores.

As we have previously noted, larvae of the native moth *Aroga websterii* occasionally defoliate large expanses of *Artemisia*-dominated communities. These outbreaks may be followed by fire storms, because the accumulation of insect webs and dead leaves greatly increases the flammability of the *Artemisia* communities.

Modern succession. Succession in the sagebrush steppe has been revolutionized by the introduction of domestic livestock (Young et al. 1972). Perennial grasses have been reduced or eliminated. The shrubs not preferred by herbivores, including the dominant species of *Artemisia*, have increased in vigor and density. The introduction of alien annuals has completely changed successional patterns. *Salsola iberica* (= *S. kali* var. *tenuifolia*), *Sisymbrium altissimum*, and *Bromus tectorum* form a seral continuum that closes many degraded *Artemisia* communities to the establishment of perennial grass seedlings (Robertson and Pearse 1945; Young et al. 1969; Young and Evans 1973). *Taeniatherum asperum* (= *Elymus caput-medusae*) has extended the seral continuum by replacing *B. tectorum* on degraded *Artemisia arbuscula* sites on the Modoc Plateau (Young and Evans 1970). The current role of alien annual plants in the succession of *Artemisia* communities was first recognized by Piemeisel (1951).

The range condition (the relation of existing plant communities to pristine potential) of sagebrush steppe ranges has probably improved since the turn of the century (e.g., Kennedy and Doten 1901). We must remember that plant communities today are the product of yesterday's stand renewal processes, and tomorrow's seral communities will be a product of current grazing attrition rather than past destructive grazing. For a well-documented presentation of the interaction of the environment of the Modoc Plateau and human development, see Pease (1965).

Native Animals

The native ruminants mule deer (*Odocoileus hemoinus*), pronghorn (*Antilocapra americana*), and desert bighorn (*Ovis canadensis*) are largely facultative browsers. Generally, these animals prefer sagebrush that grows in areas of high water potential, especially communities that are transitional to coniferous woodlands (Young et al. 1975).

Native browsers selectively prefer some forms of *Artemisia tridentata* (Hanks et al. 1973). *Artemisia tridentata* ssp. *wyomingensis* and *vaseyana* are generally more highly preferred by domestic or wild browsers than is *A. tridentata* ssp. *tridentata*. Generally, the herbage of big sagebrush contains essential oils that inhibit microbial action in rumens (Nagy et al. 1964). *Artemisia tridentata* ssp. *tridentata* is usually the characteristic species of low elevation. Because they are generally not browsed, the landscape-characterizing species of *Artemisia* increase under heavy browsing pressure. *Purshia tridentata* is the most important browse species of the sagebrush steppe (Hormay 1943; Nord 1965).

PLANT COMMUNITIES

We have listed and discussed the potential plant communities that comprise the natural vegetation of the sagebrush steppe of California. These communities are identified by the dominant shrub and herbaceous species. The potential or pristine plant communities are examples of habitat types (Daubenmire 1970). We also provide examples of seral plant communities that currently characterize much of sagebrush steppe landscape in California.

Ideally, quantitative cover and density data should be presented for each community mentioned. Unfortunately most of these communities have not been described quantitatively in California. Moreover, where quantitative descriptions have been published, the methodologies of various investigators have been so divergent that comparisons are meaningless. Therefore we have been forced to present comparative lists of the most important species in each community.

Artemisia Communities

Each of the four major *Artemisia* species characterizes a particular habitat.

Artemisia tridentata is usually found on the deeper soils. Characteristic communities are (dominant shrub/herbaceous species):

1. *A. tridentata–Purshia tridentata/Festuca idahoensis*
2. *A. tridentata/Festuca idahoensis*
3. *A. tridentate–Purshia tridentata/Agropyron spicatum*
4. *A. tridentata/Agropyron spicatum*
5. *A. tridentata/Bromus marginatus*
6. *A. tridentata/Stipa lettermanii*
7. *A. tridentata–Purshia tridentata/Stipa occidentalis*
8. *A. tridentata/Stipa occidentalis*
9. *A. tridentata/Stipa thurberiana*
10. *A. tridentata–Purshia tridentata/Stipa comata*
11. *A. tridentata/Stipa comata*
12. *A. tridentata/Stipa speciosa*
13. *A. tridentata/Elymus cinereus*
14. *A. tridentata/Oryzopsis hymenoides*
15. *A. tridentata/Hilaria jamesii*

Artemisia arbuscula grows on shallow, stony soils. Characteristic communities are:

1. *A. arbuscula/Festuca idahoensis*
2. *A. arbuscula–Purshia tridentata/Agropyron spicatum*

3. *A. arbuscula/Agropyron spicatum*
4. *A. arbuscula–Purshia tridentata/Stipa thurberiana*
5. *A. arbuscula/Stipa thurberiana*
6. *A. arbuscula/Koeleria cristata*

Artemisia nova grows on shallow soils with large quantities of gravel and mineral carbonates. Characteristic communities are:

1. *A. nova/Agropyron spicatum*
2. *A. nova/Stipa comata*
3. *A. nova/Oryzopsis hymenoides*

Artemisia cana grows on clay soils surrounding seasonal lakes. A characteristic community is: 1. *A. cana*

Artemisia tridentata communities

AGROPYRON SPICATUM AS AN UNDERSTORY DOMINANT. On the Modoc Plateau, the *Artemisia tridentata*-dominated communities are very similar to those described in adjacent Oregon (Franklin and Dyrness 1973). The landscape-characterizing community is *Artemisia tridentata/Agropyron spicatum* (Fig. 22-3 and Table 22-1). *Festuca idahoensis* becomes the dominant grass in more mesic situations, such as those on steep, north-facing slopes. Quantitative data for relatively pristine *Artemisia tridentate/Agropyron spicatum* communities in California are scarce, because most of the research has been done on disturbed range sites.

There are extensive areas north and east of Mt. Shasta where *Purshia tridentata* shares dominance or exceeds *Artemisia tridentata* in abundance (Table 22-1).

Soils associated with the *A. tridentata* communities are Haploxerolls (arid Brown great soil group). They usually have well-developed, clayey B horizons and are

Figure 22-3. *Artemisia tridentata/Agropyron spicatum* community, shown in relatively pristine condition. Clear Lake Hills, Modoc Co. Meter range pole marked in decimeter segments.

moderately deep (60–80 cm to cemented layers or bedrock). Many of the *Purshia–Artemisia* communities are associated with soils highly influenced by pumice and volcanic ash.

Artemisia tridentata/Agropyron spicatum becomes very restricted south of Honey Lake. In the Long Valley area of southern Lassen Co., *Agropyron spicatum* is restricted to sites on Pliocene volcanic ranges, such as Seven Lakes Mt., and the landscape is characterized by *Artemisia tridentata/Stipa communities.*

STIPA SPECIES AS UNDERSTORY DOMINANTS. Because the southeastern portion of the Modoc Plateau is relatively arid compared with the area of Medicine Lake and the Lassen–Warner uplands, the importance of *Stipa* species in the *Agropyron*-dominated communities increases. However, the character of the landscape changes beyond the plateau and across the Honey Lake valley.

Artemisia tridentata/Stipa lettermanii, A. tridentata/S. occidentalis, and *A. tridentata/S. thurberiana* characterize the environment from the pluvial lake shores to the margins of the coniferous woodlands south of Honey Lake (Table 22-2). *Stipa thurberiana* is a landscape-characterizing community south of Honey Lake, in the same way that *Agropyron spicatum* is the modal community of the high desert of Oregon and the Modoc Plateau. On the more mesic fringe of the *Stipa thurberiana* zone, we find *S. occidentalis*; low, sandy areas support *S. comata* (Table 22-1).

Artemisia tridentata/Stipa thurberiana is the most extensive community. The soils of much of the *S. thurberiana* zone are Aridisols, in contrast to the Mollisols of the Modoc Plateau.

Artemisia tridentata/Stipa occidentalis communities generally occur above the extensive *A. tridentata/S. thurberiana* zone (Fig. 22-4). *Artemisia tridentata–Purshia tridentata/S. occidentalis* communities occur extensively on sandy pumice soils of the Modoc Plateau. *Artemisia tridentata/S. lettermanii* extend the *Stipa*-dominated communities into high elevations.

In mesic situations at high elevations, *Bromus marginatus* replaces *Stipa let-*

Figure 22-4. *Artemisia tridentata* ssp. *vaseyana/Stipa occidentalis* community in relatively pristine condition, near Truckee, California. Meter range pole marked in decimeter segments.

TABLE 22-1. Characteristic species of A. tridentata communities from the Big sagebrush of California. a,b,c) and Northeastern California (Young and Evans 1970)

Community	Species Composition
Artemisia tridentata/ Bromus marginatus	Artemisia tridentata Symphoricarpos vaccinoides Ribes velutinum Tetradymia canescens Chrysothamnus viscidiflorus C. nauseosus Bromus marginatus Sitanion hystrix Stipa occidentalis Poa sandbergii Balsamorhiza sagittata Crepis acuminata Castilleja spp.
Artemisia tridentata- Purshia tridentata/ Festuca idahoensis	Artemisia tridentata Purshia tridentata Chrysothamnus viscidiflorus C. nauseosus Tetradymia canescens Sitanion hystrix Festuca idahoensis Carex rossii Stipa occidentalis S. thurberiana Agropyron spicatum Poa sandbergii Koeleria cristata Bromus tectorum Astragalus ssp. Lupinus ssp. Eriogonum ssp. Chaenactis douglasii Collinsia parviflora
Artemisia tridentata/ Elymus cinereus	Artemisia tridentata Poa sandbergii Elymus cinereus Agropyron spicatum Bromus tectorum Penstemon speciosus P. cusickii Thlaspi arvense Eriogonum umbellatum
Artemisia tridentata/ Agropyron spicatum	Artemisia tridentata Agropyron spicatum Poa sandbergii Phlox diffusa Aster scopulorum A. canescens Chaenactis douglasii Collinsia parviflora Phlox gracilis Lappula redowskii Gayophytum ramosissimum
Artemisia tridentata/ Festuca idahoensis	Artemisia tridentata Chrysothamnus viscidiflorus Symphoricarpos rotundifolius Ribes cereum Festuca idahoensis Agropyron spicatum Poa sandbergii Koeleria cristata Phlox diffusa Antennaria corymbosa Calochortus nitidus
Artemisia tridentata/ Stipa lettermanii	Artemisia tridentata Chrysothamnus viscidiflorus Symphoricarpos longiflorus Stipa lettermanii S. columbiana Agropyron trachycaulum Melica bulbosa Festuca idahoensis Poa sandbergii Lupinus caudatus Balsamorhiza sagittata Arabis Holboellii Phlox longifolia Crepis acuminata
Artemisia tridentata/ Stipa speciosa	Artemisia tridentata Chrysothamnus viscidiflorus Prunus andersonii Grayia spinosa Ephedra viridis Leptodactylon pungens Eriogonum umbellatum Stipa speciosa S. thurberiana S. occidentalis Oryzopsis hymenoides Bromus tectorum Lygodesmia spinosa Crepis accuminata Phlox hoodii
Artemisia tridentata/ Hilaria Jamesii	Artemisia tridentata Chrysothamnus viscidiflorus Grayia spinosa Atriplex confertifolia A. canescens Artemisia spinescens Eurotia lanata Ephedra nevadensis Hilaria jamesii Oryzopsis hymenoides Stipa comata Sporobolus cryptandrus Eriastrum diffusum Eriogonum deflexum Sphaeralcea ambigua Gilia latifolia

Examples are taken from the Oregon high desert (Dealy 1971), Great Basin (Blackburn et al. 1968

Community	Species Composition	Community	Species Composition
Artemisia tridentata/ Stipa thurberiana	Artemisia tridentata Chrysothamnus viscidiflorus Tetradymia canescens Ribes velutinum Prunus andersonii Ephedra viridus Leptodactylon pungens Eriogonum umbellatum Stipa thurberiana S. comata S. occidentalis Oryzopsis hymenoides Poa sandbergii Elymus cinereus Bromus tectorum Phlox hoodii P. longifolia Collinsia parviflora Lupinus Astragalus purshii Crepis	Artemisia tridentata/ Oryzopsis hymenoides	Artemisia tridentata Ephedra nevadensis Grayia spinosa Atriplex canescens Chrysothamnus nauseosus Prunus andersonii Tetradymia glabrata Oryzopsis hymenoides Sitanion hystrix Bromus tectorum Mentzelia multiflora Eriastrum diffusum Gayophytum ramosissimum Eriogonum thomasii
Artemisia tridentata/ Stipa occidentalis	Artemisia tridentata Chrysothamnus viscidiflorus C. nauseosus Prunus andersonii Purshia tridentata Tetradymia glabrata Ribes velutinum Ephedra viridis Eriogonum umbellatum Stipa occidentalis S. thurberiana S. speciosa Oryzopsis hymenoides O. bloomeri Bromus tectorum Carex douglasii Festuca octoflora Balsamorhiza sagittata Lupinus ssp. Astragalus filipes Wyethia mollis Collinsia parviflora Mimulus breweri Crepis accuminata Castilleja chromosa		
Artemisia tridentata/ Stipa comata	Artemisia tridentata Chrysothamnus viscidiflorus Leptodactylon pungens Stipa comata Oryzopsis hymenoides Bromus tectorum Sitanion hystrix Chaenactis douglasii Arabis holboellii Sphaeralcea coccinea		

TABLE 22-2. Characteristic species of *Artemisia arbuscula* communities found in the sagebrush steppe of California. Examples are taken from the Oregon high desert (Eckert 1957; Dealy 1971), Lahontan Basin B (Zamora and Tueller 1973), and northeastern California (Young and Evans 1973)

Community	Species Composition	Community	Species Composition
Artemisia arbuscula/ *Festuca idahoensis*	*Juniperus occidentalis*	*Artemisia arbuscula*/ *Purshia tridentata*/ *Stipa thurberiana*	*Artemisia arbuscula* *Purshia tridentata* *Chrysothamnus viscidiflorus* *C. nauseosus*
	Artemisia arbuscula		*Sitanion hystrix* *Poa sandbergii* *Koeleria cristata* *Agropyron spicatum* *Bromus tectorum*
	Festuca idahoensis *Agropyron spicatum* *Poa sandbergii*		*Lomatium triternatum* *Astragalus* ssp. *Erigeron* ssp.
	Phlox diffusa *P. hoodii* *P. longifolia* *Microseris troximoides* *Antennaria dimorpha* *Astragalus stenophyllus* *Lupinus saxosus* *Trifolium gymnocarpon* *T. macrocephalum*	*Artemisia arbuscula*/ *Stipa thurberiana*	*Artemisia arbuscula* *Chrysothamnus viscidiflorus* *Eriogonum microthecum* *Leptodactylon pungens*
			Stipa thurberiana *Poa sandbergii* *Agropyron spicatum* *Sitanion hystrix* *Koeleria cristata* *Festuca idahoensis* *Bromus tectorum*
Artemisia arbuscula/ *Purshia tridentata*/ *Agropyron spicatum*	*Artemisia arbuscula* *Chrysothamnus viscidiflorus* *Tetradymia canescens* *Eriogonum heracleoides* *E. microthecum*		

	Symphoricarpos longiflorus Artemisia tridentata Purshia tridentata Agropyron spicatum Poa sandbergii P. nevadensis Sitanion hystrix Stipa lettermanii Bromus tectorum		Collinsia parviflora Phlox longifolia P. hoodii Antennaria dimorpha Erigeron bloomeri Agoseris glauca Astragalus purshii Eriogonum ovalifolium Aster scopulorum
	Collinsia parviflora Crepis acuminata Balsamorhiza sagittata Phlox longifolia Calochortus nuttallii	Artemisia arbuscula/ Koeleria cristata	Artemisia arbuscula Poa sandbergii Koeleria cristata Sitanion hystrix Carex rossii Juncus ssp.
Artemisia arbuscula/ Agropyron spicatum	Juniperus occidentalis		
	Artemisia arbuscula Chrysothamnus viscidiflorus Tetradymia canescens Agropyron spicatum Poa sandbergii Sitanion hystrix Stipa thurberiana S. columbiana Taeniatherum asperum Bromus tectorum		Antennaria dimorpha Lomatium triternatum Erigeron bloomeri Lomatium nudicaule Geum ciliatum Polygonum douglasii Navarretia tagetina

TABLE 22-2 [continued]

Community	Species Composition	Community	Species Composition
	Epilobium paniculatum		
	Grindelia nana		
	Penstemon laetus		
	Eriogonum sphaerocephalum		
	E. umbellatum		
	Lomatium nudicaule		

termanii. Arctostaphylos patula may also occur, indicating a possible relationship to seral communities in the *Pinus jeffreyi* forest of the trans-Sierra.

Stipa comata communities often occur on deep sands, ranging in area from a few square meters to extensive upland areas (Blackburn et al. 1971). *Stipa comata* apparently can effectively utilize the low-nutrient level of the sands, rather than competing on the basis of reduced soil moisture (Daubenmire 1970).

Artemisia tridentata–Purshia tridentata/Stipa comata communities are important mule deer wintering areas on pluvial lakelain sands in the Honey Lake basin (Fig. 22-5).

The limited examples of *Artemisia tridentata/Stipa speciosa* that occur along the trans-Sierra are northern extensions of the same extensive community in the southern Great Basin. *Stipa speciosa* usually occurs on alluvial fans from granitic mountain escarpments with no profile development in the soils (Entisols). *Artemisia tridentata/Hilaria jamesii* is another southern Great Basin community sparingly represented in the Mono Co. valleys of the southern trans-Sierra.

ELYMUS CINEREUS AS AN UNDERSTORY DOMINANT. This community occurs in southeastern Oregon on moist alluvial bottomlands (Eckert 1957) and in upland situations (Culver 1964; Fig. 22-6). Both situations occur on the Modoc Plateau and along the east side of the Sierra Nevada. The presence of this community often reflects accumulations of erosional deposits derived from volcanic ash.

Artemisia tridentata/Elymus cinereus communities have been vital as food producers for Great Basin Indians, and livestock grazing in the Great Basin has probably altered this environment more than any other one mentioned in this chapter (Young et al. 1975).

ORYZOPSIS HYMENOIDES AS AN UNDERSTORY DOMINANT. This perennial grass occurs in many *Artemisia tridentata* communities, but below the *Stipa thurberiana* zone it is the dominant understory species over extensive areas. In this zone Aridisols with sand- or silt-texured surface soils have a topography shaped by wind.

Figure 22-5. *Purshia tridentata/Stipa comata* community. Old-growth community on pluvial-lake-deposited sand ridges. Herbaceous understory greatly reduced with abundance of *Tetradymia canescens* and *Chrysothamnus nauseosus*. Doyle, California. Meter range pole marked in decimeter segments.

Figure 22-6. *Artemisia tridentata/Elymus cinereus* community protected from grazing on margin of wet meadow, Grasshopper Valley, California. Meter range pole marked in decimeter segments.

BROMUS TECTORUM AS AN UNDERSTORY DOMINANT. Throughout the sagebrush steppe of California, the most abundant community is *Artemisia tridentata/Bromus tectorum*. *Bromus tectorum* invaded all of the potential communities we enumerate in this chapter and changed succession in an entire vegetation type. With overgrazing, *A. tridentata* has greatly increased in density. Stark *Artemisia/B. tectorum* communities often have vesicular-crusted soils between the shrubs and exhibit little diversity of species. Recurrent burning has eliminated *Artemisia tridentata* and resulted in dominance by *Chrysothamnus/B. tectorum* over extensive areas.

***Artemisia arbuscula* communities.** *Artemisia arbuscula* communities dominate a large portion of the sagebrush steppe in California (Fig. 22-7). This series of plant communities parallels the more mesic examples noted for *A. tridentata* (Table 22-2).

These communities occur as edaphic climaxes on shallow, stony phases of the Haploxerolls. On the Modoc Plateau, *A. arbuscula* communities are common as openings within the coniferous forest.

The *A. arbuscula* scab flats of the Modoc Plateau are virtually lakes in the spring when snow melts. The clayey soils of limited permeability yield much of this snowmelt as runoff, making the communities extremely important watershed areas. Poor aeration in the rooting zone due to perched water tables during the spring may be a major factor in the ecology of *A. arbuscula* (Zamora and Tueller 1973).

Examples of the *Artemisia arbuscula-Purshia tridentata/Agropyron spicatum* community occur in Honey Lake valley between Janesville and Susanville. These communities may reflect mosaics of shallow and deep, well-drained soils, with the *Purshia* on the deeper areas (Dealy 1971), or they may cover extensive areas with the two shrubs growing sympatrically (Zamora and Tueller 1973).

Artemisia arbuscula/Stipa thurberiana communities continue the parallel development of *Artemisia tridentata* in a series of communities toward more arid environments.

On the Modoc Plateau, many *Artemisia arbuscula* communities have been invaded by *Taeniatherum asperum*. *Artemisia arbuscula* communities are generally less susceptible to wildfires than are comparable *Artemisia tridentata* communities. When overgrazed, they exist as stark shrub communities with a great deal of bare ground between the low shrub canopies. *Bromus tectorum* invades these communities, but usually its densities are low in comparison to *A. tridentata* sites. *Taeniatherum asperum* invasion greatly increases the susceptibility to wildfires. Once the nonsprouting shrub is destroyed, dominance by the alien grass is assured (Young and Evans 1970).

Artemisia nova **communities.** *Artemisia nova* also continues the parallel evolution of *Artemisia* habitats into more arid portions of land (Table 22-3). The three species of *Artemisia—tridentata, arbuscula,* and *nova*—occupy different edaphic sites. However, the different shrubs have the same series of understory dominants.

Shallowness, high volumes of gravel, and the presence of large quantities of mineral carbonates are the most commonly occurring features of the soils of *Artemisia nova* communities. These soils are often the youngest land surfaces in the *Artemisia* environment. *Artemisia nova* is often confused with *A. arbuscula,* the species that occupies the oldest land surfaces in the *Artemisia* environment.

Artemisia nova is highly preferred by domestic sheep as winter browse. Long alluvial fans with widely spaced, perfectly hedged shrubs and a stark, empty understory characterize many *A. nova* communities. These communities provide insight into what the entire sagebrush steppe might look like if *A. tridentata* was generally preferred by domestic livestock!

The prevalence of this species on limestone areas is to be expected (St. Andre et al. 1965; Hironaka 1963), because it prefers soils high in calcium carbonate.

Artemisia cana **communities.** *Artemisia cana* communities are characteristic of pluvial lakebeds in the uplands of the sagebrush steppe. The soil surface horizons are saturated by spring accumulations of snowmelt runoff, which forms a perched water

Figure 22-7. Large areas in northeastern California that were dominated by *Artemisia arbuscula* (foreground) have been converted by grazing and recurrent fire to areas dominated by *Taeniatherum asperum* annual grasslands (background slopes).

TABLE 22-3. Characteristic species of *Artemisia nova* communities found in the sagebrush steppe of California. Examples are taken from Lahontan Basin (Blackburn et al. 1969 a,b,c; Zamora and Tueller 1973)

Community	Species Composition	Community	Species Composition
Artemisia nova/ *Agropyron spicatum*	*Juniperus osteosperma* *Pinus monophylla* *Artemisia nova* *Chrysothamnus viscidiflorus* *Eurotia lanata* *Ephedra viridis* *Agropyron spicatum* *Poa sandbergii* *Sitanion hystrix* *Calochortus nuttallii* *Phlox longifolia* *Arabis holboellii* *Haplopappus acaulis* *Lomatium macdougallii* *Erigeron pumilis*	*Artemisia nova*/ *Oryzopsis hymenoides*	*Artemisia nova* *Chrysothamnus viscidiflorus* *Eurotia lanata* *Ephedra viridis* *Atriplex confertifolia* *Eriogonum microthecum* *Oryzopsis hymenoides* *Poa sandbergii* *Sitanion hystrix* *Stipa comata* *Hilaria jamesii* *Phlox hoodii* *Eriogonum ovalifolium* *Haplopappus acaulis* *Caulanthus crassicaulis* *Erigeron pumilis*
Artemisia nova/ *Stipa comata*	*Juniperus osteosperma* *Pinus monophylla* *Artemisia nova* *Chrysothamnus viscidiflorus* *Eurotia lanata* *Ephedra viridis* *Stipa comata* *Poa sandbergii* *Sitanion hystrix*		

	Calochortus nuttallii *Phlox longifolia* *Arabis holboellii* *Astragalus purshii* *Gilia clokeyi*
Artemisia nova/ *Agropyron inerme*	*Artemisia nova* *Chrysothamnus viscidiflorus* *Eurotia lanata* *Ephedra viridis*
	Agropyron inerme *A. spicatum* *Poa sandbergii* *Sitanion hystrix* *Stipa comata*
	Calochortus nuttallii *Phlox longifolia* *Arabis holboellii* *Haplopappus acaulis* *Astragalus purshii* *Cryptantha humilis*

table on slowly permeable argillic horizons (formerly Solodized Solonetz). The only community that has been described is in the Oregon high desert, and it featured a *Muhlenbergia richardsonis* understory (Dealy 1971). The community we photographed in Lassen Co., California, had an herbaceous understory dominated by *Sitanion hystrix* (Fig. 22-8).

Other communities. Throughout the sagebrush steppe, small areas of Vertisols (self-churning soils) have unique plant communities. In dry years these areas may be devoid of vegetation, and in wet years they are covered by shallow lakes. Species of *Helianthus* form dense communities on the flats. These areas may be accumulations of clay-sized erosional products such as the extensive Sunflower Flat in eastern Lassen Co. or those with surface soil horizons eroded to expose argillic layers (Summerfield 1969; Young and Evans 1970).

Throughout the uplands of the Modoc Plateau, talus slopes below basalt rims support patches of *Prunus subcordata.* South along the trans-Sierra, erosion-resistant aplite dikes often form escarpments in granitic ranges. Talus along these dikes support a very restricted *Holodiscus discolor* community. We have already noted the widespread influence of volcanic ash, an erosional product from tuff formations, in numerous environments. Ash beds are highly erodable, specialized environments that often support *Salvia carnosa* communities.

The Clear Lake hills of western Modoc Co. support *Agropyron spicatum–Poa sandbergii* communities with limited shrub components. *Festuca idahoensis–Balsamorhiza sagittata* stands also grow on northern slopes. The occurrence of these communities with limited shrub representation is probably related to recurring fire, but it may also relate to the steppe communities of the Columbia Basin (Daubenmire 1970).

MOUNTAIN BRUSH COMMUNITIES

The term "mountain brush" has been applied in the Great Basin to plant communities that occur at high elevations and are composed of numerous species of shrubs (Table 22-4). The most characteristic species are *Cercocarpus ledifolius, Purshia tridentata, Amelanchier pallida,* and *Symphoricarpos vaccinoides.* Major plant communities are:

1. *Purshia tridentata/Festuca idahoensis,* at the margin of *Pinus ponderosa* woodlands on the Modoc Plateau.
2. *Cercocarpus ledifolius/Festuca idahoensis–Agropyron spicatum,* also at the margin of *Pinus ponderosa* woodlands on the Modoc Plateau.
3. *Cercocarpus ledifolius/Stipa lettermanii,* above the *Juniperus osteosperma/Pinus monophylla* zone on trans-Sierra ranges.
4. *Amelanchier pallida–Artemisia tridentata/Festuca idahoensis* mountain brush communities of trans-Sierra ranges.
5. *Symphoricarpos vaccinoides–Artemisia tridentata/Bromus marginatus* mountain brush communities of trans-Sierra ranges.

In the sagebrush steppe of California, these communities are most often found in

Figure 22-8. *Artemisia cana/Sitanion hystrix* community growing on Vertisol in area of seasonal flooding. Background vegetation is woodland of *Pinus ponderosa–Juniperus occidentalis*, with an *Artemisia arbuscula/Agropyron spicatum* understory. Willow Creek Valley, Lassen Co. Meter range pole marked in decimeter segments.

the ecotone between the coniferous forest and the sagebrush steppe. Only a few of the communities are discussed below.

Cercocarpus ledifolius Communities

On the Modoc Plateau, *Cercocarpus ledifolius* communities cover extensive areas, with the mature shrubs highlined (browsed to maximum height of deer) and with reproduction rarely apparent. Elsewhere it occurs on rimrocks.

Along the trans-Sierra on isolated mountain ranges such as Ft. Sage, the mountain brush communities are often found above *Juniperus osteosperma* woodlands. In the Walker River basin the communities occur with *Juniperus osteosperma/Pinus monophylla* woodlands.

Purshia tridentata Communities

Purshia tridentata–Artemisia tridentata/Stipa occidentalis communities on granitic alluvial fans between the Jackson Mt. escarpment and Honey Lake are remarkable for the stature of the large *Purshia* plants. They also are unusual in that the intermingling woodland is dominated by *Quercus kelloggii* and *Pinus jeffreyi*. The oak woodland offers a glimpse of the appearance of past Great Basin floras (Axelrod 1950).

Another *Purshia* community for which we lack an adequate description is *Purshia glandulosa–Artemisia tridentata*. After burning in wildfires, such communities are vivid with flowering *Castilleja, Monardella,* and *Lupinus*.

PLUVIAL BASIN COMMUNITIES

The landforms below the maximum shorelines of the pluvial lake basins are occupied by plant communities in which species of *Artemisia* of the section

TABLE 22-4. Characteristic species of mountain brush and transition communities in the sagebrush steppe of California. Examples are taken from the Oregon high desert (Dealy 1971) and Lahontan Basin (Blackburn et al. 1969 a,b,c)

Community	Species Composition	Community	Species Composition
Purshia tridentata/ Festuca idahoensis	Juniperus occidentalis Cerococarpus ledifolius Purshia tridentata Artemisia tridentata A. arbuscula Chrysothamnus nauseosus C. viscidiflorus Festuca idahoensis Poa sandbergii Agropyron spicatum Stipa thurberiana S. occidentalis Antennaria rosea Astragalus purshii	Cercocarpus ledifolius/ Stipa lettermanii	Cerocarpus ledifolius Artemisia tridentata Amelanchier alnifolia Symphoricarpos vaccinoides Ribes aureum Stipa lettermanii Melica stricta Sitanion hystrix Poa sandbergii Castilleja chromosa Crepis acuminata Monardella odoratissima Lupinus caudatus
Amelanchier pallida/ Artemisia tridentata/ Festuca idahoensis	Amelanchier pallida Artemisia tridentata Symphoricarpos longiflorus Ribes aureum Rosa woodsii Chrysothamnus nauseosus C. viscidiflorus Festuca idahoensis Bromus marginatus Poa sandbergii Machaeranthera canescens	Symphoricarpos vaccinoides/ Artemisia tridentata/ Bromus marginatus	Symphoricarpos vaccinoides Artemisia tridentata Ribes aureum Tetradymia canescens Amelanchier alnifolia Chrysothamnus viscidiflorus C. nauseosus Bromus marginatus Stipa lettermanii Lupinus caudatus Castilleja chromosa Lappula redowskii Crepis acuminata

	Amsinckia menziesii Phlox diffusa
Cercocarpus ledifolius/ Festuca idahoensis/ Agropyron spicatum	Cercocarpus ledifolius Juniperus occidentalis Pinus ponderosa
	Artemisia tridentata Chrysothamnus viscidiflorus C. nauseosus Ribes cereum Purshia tridentata
	Festuca idahoensis Agropyron spicatum Sitanion hystrix Poa sandbergii Koeleria cristata Bromus carinatus B. tectorum
	Balsamorhiza sagittata Microsteris gracilis Astragalus stenophyllus Lupinus caudatus Collinsia parviflora

Tridentatae are reduced in dominance or are absent. This environment is represented within the sagebrush steppe of California by Surprise Valley (Russell 1927), Honey Lake basin (Billings 1945), and the valleys east and southeast of Mono Lake (Russell 1885).

The major plant communities are (see Table 22-5):

1. *Atriplex confertifolia/Oryzopsis hymenoides*
2. *Sarcobatus baileyi/Oryzopsis hymenoides*
3. *Dalea polyadenia/Oryzopsis hymenoides*
4. *Grayia spinosa–Artemisia spinescens/Oryzopsis hymenoides*
5. *Eurotia lanata/Oryzopsis hymenoides*
6. *Sarcobatus vermiculatus/Elymus cinereus*
7. *Sarcobatus vermiculatus/Distichlis stricta*

Atriplex confertifolia Communities

The environments of the lower pluvial lake basins are often referred to as "salt deserts" (e.g., Franklin and Dyrness 1973), because they are internal drainage basins that accumulate soluble salts. Billings (1945) pointed out that *Atriplex confertifolia/Oryzopsis hymenoides*, instead of *Artemisia tridentata* communities exist in the northern Carson Desert because of drought tolerance rather than osmotic tolerance (Fig. 22-9). Many of the soils found under *Atriplex confertifolia* are relatively low in soluble salts, but rainfall is extremely limited in this environment.

Figure 22-9. Important plant communities of the pluvial lake basins. (*a*) *Atriplex confertifolia/Oryzopsis hymenoides* community. (*b*) *Oryzopsis hymenoides* dominance on sand dune area. (*c*) *Eurotia lanata* community.

(b)

(c)

Figure 22-9 (continued)

TABLE 22-5. Characteristic species of plant communities of pluvial lake basins of the sagebrush steppe of California. Examples are taken from the Great Basin (Blackburn et al. 1969 c) and Carson Desert (Billings 1949)

Community	Species Composition	Community	Species Composition
Atriplex confertifolia/ *Oryzopsis hymenoides*	*Atriplex confertifolia* *Artemisia spinescens* *Ephedra nevadensis* *Tetradymia glabrata* *Sarcobatus baileyi*	*Grayia spinosa*/ *Artemisia spinescens*/ *Oryzopsis hymenoides*	*Grayia spinosa* *Artemisia spinescens* *Atriplex confertifolia* *A. nuttallii*
	Oryzopsis hymenoides *Muhlenbergia richardsonis* *Sitanion hystrix* *Bromus tectorum*		*Oryzopsis hymenoides* *Hilaria jamesii* *Sitanion hystrix*
	Eriogonum densum *Eriastrum diffusum* *Amsinckia menziesii* *Lappula redowskii* *Sphaeralcea coccinea*		*Salsola paulsenii* *Halogeton glomeratus* *Chenopodium leptophyllum*
		Eurotia lanata/ *Oryzopsis hymenoides*	*Eurotia lanata* *Grayia spinosa*
Sarcobatus baileyi/ *Oryzopsis hymenoides*	*Sarcobatus baileyi* *Atriplex confertifolia* *Chrysothamnus viscidiflorus* *Artemisia spinescens* *Grayia spinosa*		*Oryzopsis hymenoides*
			Sphaeralcea coccinea *Eriogonum thomasii* *Salsola iberica* *Navarretia* spp.
	Oryzopsis hymenoides *Sitanion hystrix* *Bromus tectorum*	*Sarcobatus vermiculatus*/ *Elymus cinereus*	*Sarcobacus vermiculatus* *Artemisia tridentata* *Chrysothamnus nauseosus* *C. viscidiflorus*
	Navarretia spp. *Gayophytum ramosissimum* *Amsinckia menziesii* *Oenothera contorta*		*Elymus cinereus* *Distichlis stricta* *Poa sandbergii*

Dalea polyadenia/	*Dalea polyadenia*		*Iva axillaris*
Oryzopsis hymenoides	*Artemisia tridentata*		
	Chrysothamnus viscidiflorus	*Sarcobatus vermiculatus*/	*Sarcobatus vermiculatus*
	Ephedra nevadensis	*Distichlis stricta*	*Atriplex confertifolia*
	Eurotia lanata		
	Grayia spinosa		*Bromus tectorum*
			Distichlis stricta
	Oryzopsis hymenoides		
	Hilaria jamesii		*Iva axillaris*
	Sitanion hystrix		*Descurainia sophia*
	Eriastrum diffusum		
	Eriogonum densum		
	Mentzelia multiflora		

Other Communities

Grayia spinosa is one of the few species with sufficient ecological amplitude to occur in both *Artemisia*- and *Atriplex*-dominated communities. Species of *Chrysothamnus* seem to have the amplitude to extend from the coniferous forest to the saline–alkaline communities, but, in fact, ecologically distinct subspecies exist for each environmental subdivision.

Artemisia spinescens is a representative of a different section (Dracunalus) of the genus. This spring-blooming shrub is preferred by domestic sheep, and its communities have often been overgrazed.

Below the *Atriplex* zone, there are extensive *Sarcobatus baileyi/Oryzopsis hymenoides* communities. These communities may occur on soils with desert pavement, where very widely spaced shrubs are found on minimounds that rise 5–15 cm above the pavement surface. Since the depth of calcium carbonate cementation in the soil profile does not rise under the mound height, the mounds double the effective rooting depth.

Basins that contain abundant active sand fields are habitats for the *Dalea polydenia/Oryzopsis hymenoides* community. These basins have sparse herbaceous cover and a diversity of shrubs (Table 22-5).

Eurotia lanata/Oryzopsis hymenoides communities constitute a striking and probably dynamic soil alternation in the desert environment (Fig. 22-9). Much has been written about the patchy distribution of this economically important browse species (Gates et al. 1956; Mitchell et al. 1966).

Sarcobatus vermiculatus is the only green shrub in the pluvial desert. It dominates a community found in areas of available ground water, which is often of very low osmotic potential (Fig. 22-10).

Sarcobatus vermiculatus communities also occur in upland valleys near Alturas and in the Klamath Basin on the margins of Tule Lake and Lower Klamath sinks. In these situations they are not surrounded by a zone of *Atriplex,* as are the *Sarcobatus/Oryzopsis* communities.

Figure 22-10. *Sarcobatus vermiculatus* growing on sand dunes in Nevada.

The *Sarcobatus vermiculatus/Elymus cinereus* saline–alkaline meadows are an important, although much abused, forage resource.

Other than *Elymus cinereus* and *Distichlis stricta* on rather restricted sites, only *Oryzopsis hymenoides* grows as a perennial grass over much of the pluvial basins.

The origin and the evolution of these communities in relatively recent time since the pluvial basins were flooded have never been completely resolved (e.g., see Axelrod 1950).

FUTURE RESEARCH

Obviously, our understanding of the synecology of the California sagebrush steppe needs much refinement. The lack of quantitative data is most apparent. The many examples of *Artemisia tridentata/Agropyron spicatum* communities presented here preclude any argument over the existence of such a synecology. All experienced synecologists with whom we discussed the sagebrush steppe insisted on the validity of *Purshia tridentata–Artemisia tridentata/Agropyron spicatum* communities, but no documented descriptions in the literature are available.

We believe that modern taxonomic treatments of *Artemisia* have enumerated ecologically significant subspecies and forms. Unfortunately, we cannot yet verify the validity of these ecological separations.

A similar sequence of ecologically distinct subspecies and forms occurs in *Chrysothamnus* and probably in the woody species of *Eriogonum*. *Chrysothamnus viscidiflorus* is found in numerous communities (Sandford 1967), but a trend exists for ecologically distinct subspecies to occur in diverse communities. The synecology of the sagebrush steppe will be much more meaningful when these ecologically distinct entities can be identified as such taxonomically.

LITERATURE CITED

Antevs, E. 1938. Rainfall and tree growth in the Great Basin. Carnegie Inst. Wash. Pub. 469. 97 p.

Axelrod, D. I. 1950. Evolution of desert vegetation in western North American. Carnegie Inst. Wash. Pub. 590:215-260.

Beetle, A. A. 1960. A study of sagebrush, the section Tridentatae of Artemisia. Univ. Wyo. Agric. Exp. Sta. Bull. 368. 88 p.

Beetle, A. A., and A. Young. 1965. A third subspecies in the Artemisia tridentata complex. Rhodora 25:205-208.

Billings, W. D. 1945. The plant associations of the Carson Desert region, western Nevada. Butler Univ. Bot. Stud. 7:89-123.

-----. 1949. The shadescale vegetation zone of Nevada and eastern California in relation to climate and soil. Amer. Midl. Natur. 42:87-109.

-----. 1951. Vegetational zonation in the Great Basin of western North America. In Les bases ecologigues de la regeneration de la vegetation des zones arides. Internat. Colloq. Internat. Union Biol. Sci., Ser. B, 2:101-122.

Birkeland, P. W. 1968. Correlation of Quaternary stratigraphy of the Sierra Nevada with that of the Lake Lahontan Area, pp. 469-500. In R. B. Morrison and H. E. Wright, Jr. (eds.). Means of correlation of Quaternary successions. Proc. VII Congr. Internat. Assoc. Quaternary Res. Univ. Utah Press, Salt Lake City.

Blackburn, W. H., R. E. Eckert, Jr., and P. T. Tueller. 1968a. Vegetation and soils of the Duckwater watershed. Coll. Agric., Univ. Nevada, Bull. R40. 58 p.

-----. 1969a. Vegetation and soils of the Cow Creek watershed. Coll. Agric., Univ. Nevada, Bull. R49. 77 p.

-----. 1969b. Vegetation and soils of the Coils Creek watershed. Coll. Agric., Univ. Nevada, Bull. R48. 82 p.

-----. 1969c. Vegetation and soils of the Crane Springs watershed. Coll. Agric., Univ. Nevada, Bull. R55. 63 p.

-----. 1971. Vegetation and soils of the Rock Springs watershed. Coll. Agric., Univ. Nevada, Bull. R83. 115 p.

Blackburn, W. H., and P. T. Tueller. 1970. Pinyon and juniper in black sagebrush communities in east-central Nevada. Ecology 51: 841-848.

Blackburn, W. H., P. T. Tueller, and R. E. Eckert, Jr. 1968b. Vegetation and soils of Crowley Creek watershed. Coll. Agric., Univ. Nevada, Bull. R42. 58 p.

-----. 1968c. Vegetation and soils of the Mill Creek watershed. Coll. Agric., Univ. Nevada, Bull. R43. 72 p.

-----. 1969d. Vegetation and soils of the Churchill Canyon watershed. Coll. Agric., Univ. Nevada, Bull. R45. 157 p.

Blackwelder, E. B. 1931. Pleistocene glaciation in the Sierra Nevada and Basin ranges. Geol. Soc. Amer. Bull. 42:865-922.

Blaisdell, J. P. 1953. Ecological effects of planned burning of sagebrush-grass on the Upper River Plains. USDA Tech. Bull. 1075. 39 p.

Burkhardt, J. W., and E. W. Tisdale. 1969. Nature and successional status of western juniper vegetation in Idaho. J. Range Manage. 22:264-270.

Culver, R. N. 1964. An ecological reconnaissance of the Artemisia steppe on the east-central Owyhee uplands of Oregon. M.S. thesis, Oregon State Univ., Corvallis. 99 p.

Daubenmire, R. 1970. Steppe vegetation of Washington. Wash. Agric. Exp. Sta. Tech. Bull. 62. 131 p.

Dealy, J. E. 1971. Habitat characteristics of the Silver Lake mule deer range. USDA Forest Serv. Pac. NW Forest and Range Exp. Sta., Res. Paper PNW-125. Portland, Ore. 99 p.

Eckert, R. E., Jr. 1957. Vegetation-soil relationships in some _Artemisia_ types in northern Harney and Lake counties, Oregon. Ph.D. dissertation, Oregon State Univ., Corvallis. 208 p.

Evans, R. A., H. R. Holbo, R. E. Eckert, Jr., and J. A. Young. 1970. Influences of weed control and seedling methods on the functional environment of rangelands. Weed Sci. 18:154-162.

Fosberg, M. A., and M. Hironaka. 1964. Soil properties affecting the distribution of big and low sagebrush communities in southern Idaho. Amer. Agron. Spec. Pub. 5:230-236.

Franklin, J. F., and C. T. Dyrness. 1973. Natural vegetation of Oregon and Washington. USDA Forest Serv., Pac. NW Forest and Range Exp. Sta., Tech. Rept. PNW-8. p.

Gates, D. H., L. A. Stoddart, and C. W. Cook. 1956. Soil as a factor influencing plant distribution in salt deserts of Utah. Ecol. Monogr. 26:155-175.

Hall, H. M., and F. E. Clements. 1923. The phylogenetic method in taxonomy. The North American species of _Artemisia_, _Chrysothamnus_, and _Atriplex_. Carnegie Inst. Wash. Pub. 326. 355 p.

Hanks, D. L., E. D. McArthur, R. Stevens, and A. P. Plummer. 1973. Chromatographic characteristics and phylogenetic relationships of _Artemisia_, section Tridentatae. USDA Forest Serv. Res. Paper INT-141. 24 p.

Harris, G. A. 1967. Some competitive relationships between _Agropyron spicatum_ and _Bromus tectorum_. Ecol. Monogr. 37:89-111.

Hironaka, M. 1963. Plant-environment relations of major species in the sagebrush-grass vegetation of southern Idaho. Ph.D. dissertation, Univ. Wisc., Madison. 280 p.

Hobbs, C. L., and R. R. Miller. 1948. The zoological evidence in the Great Basin. Univ. Utah Bull. 38:17-167.

Hormay, A. L. 1943. Bitterbrush in California. USDA Forest Serv., Calif. Forest and Range Exp. Sta., Res. Note 34. 12 p.

Kennedy, P. B., and S. B. Doten. 1901. A preliminary report on the summer ranges of western Nevada sheep. Coll. Agric., Univ. Nevada, Bull. 51. 54 p.

King, P. B. 1959. The evolution of North America. Princeton Univ. Press, Princeton, N. J. 189 p.

Macdonald, G. A., and T. E. Gay, Jr. 1966. Geology of the Southern Cascade Range, Modoc Plateau, and Great Basin areas in northeastern California, pp. 43-48. In Mineral resources of California. Calif. Div. Mines Geol. Bull. 191. 450 p.

Mehringer, R. B. 1967. Pollen analysis of the Tule Springs area, Nevada, Part 3. Nevada State Mus. Anthropol. Paper 13. Carson City, Nev.

Merriam, C. H. 1926. The buffalo in northeastern California. J. Mammol. 7:211-214.

Mitchell, J. E., N. E. West, and R. W. Miller. 1966. Soil physical properties in relation to plant community patterns in the shadscale zone of northwestern Utah. Ecology 47:627-630.

Morrison, R. B. 1964. Lake Lahontan: geology of the southern Carson Desert, Nevada. U. S. Geol. Serv. Prof. Paper 401. 156 p.

Munz, P. A., and D. D. Keck. 1959. A California flora. Univ. Calif. Press, Berkeley. 1681 p.

Nagy, J. G., H. W. Steinoff, and G. M. Ward. 1964. Effect of essential oils of sagebrush on deer rumen microbial function. J. Wildlife Manage. 28:785-790.
Nevada River Basin Survey Staff. 1969. Water and related land resources of Central Lahontan Basin-Walker River Subbasin. USDA River Basin Survey, Carson City, Nev. 232 p.
Nord, E. C. 1965. Autecology of bitterbrush in California. Ecol. Monogr. 35:307-334.
Pease, R. W. 1965. Modoc County-a Geographic time continuum on the California Volcanic tableland. Univ. Calif. Pub. Geogr. 17:1-199.
Piemeisel, R. L. 1951. Causes affecting change and rate of change in vegetation of annuals in Idaho. Ecology 32:53-72.
Robertson, J. H., and C. K. Pearse. 1945. Artificial reseeding and the closed community. Northwest Sci. 19:58-66.
Russell, I. C. 1885. Geological history of Lake Lahontan, a Quaternary lake of northwestern Nevada. U. S. Geol. Survey Monogr. 11. 288 p.
Russell, R. J. 1927. The land forms of Surprise Valley, northwestern Great Basin. Univ. Calif. Pub. Geogr. 2:323-358.
Rydberg, P. A. 1916. Artemisia and Artemisiastrum. N. Amer. Flora 34:244-285.
Sandford, G. R. 1967. Changes in plant species dominance associated with succession near Eagle Lake Biological station. M. A. thesis, Chico State Coll., Chico, Calif. 95 p.
Soil Survey Staff, U. S. Department of Agriculture, Soil Conservation Service. 1960. Soil classification, a comprehensive system. 7th Approximation. U. S. Govt. Printing Office, Washington, D. C. 265 p.
St. Andre, G., H. A. Mooney, and R. C. Wright. 1965. The pinyon woodland zone in the White Mountains of California. Amer. Midl. Natur. 73:225-239.
Summerfield, H. B., Jr. 1969. Pedological factors related to the occurrence of big and low sagebrush in northern Washoe County, Nevada. M. S. thesis, Univ. Nevada, Reno. 94 p.
Ward, G. H. 1953. Artemisia, section Seriphidium, in North America: a cytotaxonomic study. Contri. Dudley Herb. 4:155-205.
Young, J. A., and R. A. Evans. 1970. Invasion of medusahead into the Great Basin. Weed Sci. 18:89-97.
_____. 1973. Downybrome--intruder in the plant succession of big sagebrush communities in the Great Basin. J. Range Manage. 26: 410-415.
-----. 1974a. Phenology of Chrysothamnus viscidiflorus subsp. viscidiflorus (Hook) Nutt. Weed Sci. 22:469-475.
-----. 1974b. Population dynamics of green rabbitbrush in disturbed big sagebrush communities. J. Range Manage. 27:127-132.
Young, J. A., R. A. Evans, and R. E. Eckert, Jr. 1969. Population dynamics of downy brome. Weed Sci. 17:20-26.
Young, J. A., R. A. Evans, and J. Major. 1972. Alien plants in the Great Basin. J. Range Manage. 25:194-201.
Young, J. A., R. A. Evans, and P. T. Tueller. 1975. Great Basin plant communities--pristine and grazed, pp. 186-215. In R. Elston (ed.). Holocene climates in the Great Basin. Occas. Paper, Nevada Archeol. Survey, Reno.
Zamora, B., and P. T. Tueller. 1973. Artemisia arbuscula, A. longiloba, and A. Nova habitat types in Northern Nevada. Great Basin Natur. 33:225-242.

CHAPTER

23 TRANSMONTANE CONIFEROUS VEGETATION

FRANK C. VASEK
Department of Biology, University of California, Riverside

ROBERT F. THORNE
Rancho Santa Ana Botanic Garden, Claremont, California

Introduction 798
Northern juniper woodlands 798
Western juniper woodland 799
Mountain juniper woodland 803
Scattered juniper occurrences 808
Piñon and juniper woodlands 808
Sierra Nevada piñon woodland 809
Utah juniper and one-leaf piñon woodlands 810
California juniper and one-leaf piñon woodlands 814
California juniper and four-leaf piñon woodlands 819
Scattered juniper occurrences 819
Two-leaf piñon woodland 821
Desert montane forests and woodlands 822
Montane white fir forest 822
Subalpine woodland 823
Limber pine woodland 823
Great Basin bristlecone pine woodland 825
Status of knowledge 826
Literature cited 829

INTRODUCTION

The transmontane region of California traditionally includes the portions of the state lying east of the main crests of the Cascade–Sierra axis and of the southern ranges forming the divide between coastal and desert drainages (Munz and Keck 1959). For this chapter, however, we restrict the region to interior areas of the state east of, and at elevations below, the mixed conifer forests of the major mountain systems.

Three broad categories of coniferous vegetation occur primarily in the transmontane region of California: northern juniper woodlands, piñon and juniper woodlands, and montane coniferous forests. Montane forests of interior ranges in northern California (e.g., the Warner Mts.) are more conveniently considered with other mixed conifer forests (Chapter 17). Therefore we include only forests in desert mountain ranges from the White Mts. southward.

NORTHERN JUNIPER WOODLANDS

The northern juniper woodland described by Munz and Keck (1949) is here interpreted to include two phases: a western juniper woodland in open, rolling country and a mountain juniper woodland on ridges and mountain slopes. For convenience, we adopt a common name convention whereby "western juniper" refers to *Juniperus occidentalis* ssp. *occidentalis* and "mountain juniper" applies to *J. occidentalis* ssp. *australis*.

Inclusion of mountain juniper woodland in transmontane vegetation is somewhat arbitrary in view of its strong subalpine aspect in the Sierra Nevada and its high montane aspect in the Transverse Ranges. However, the very close ecological relationship of mountain juniper woodland to western juniper woodland in northern California, and the close taxonomic relationship of the junipers dominant in those woodlands, suggest that our treatment is reasonable.

Few studies have been made regarding the composition or ecology of northern juniper woodlands; consequently the following account relies extensively on personal observation.

Western Juniper Woodland

The western juniper woodland (Fig. 23-1) is characterized by open stands or scattered trees of western juniper (Vasek 1966).

In California, western juniper woodlands may have a grassy understory, particularly where trees are close together, or they may have a shrub understory in more open stands. Understory shrubs, or interspersed stands of low shrubs, are primarily Great Basin sagebrush, *Artemisia tridentata,* on deep soils or well-drained slopes, and black sagebrush, *Artemisia arbuscula,* on heavy soils and rocky substrates. Other shrubs, such as *Cercocarpus ledifolius, Amelanchier utahensis* (or *A. pallida*?), *Ribes velutinum,* and *Purshia tridentata,* also occur with moderate to low frequencies.

The conspicuous species associated with western juniper all range into other plant communities. Consequently the range of western juniper woodland and its transitional variations is essentially defined by the range of western juniper (Griffin and Critchfield 1972; Vasek 1966).

Western juniper woodlands reach a northern limit in southern Washington and southwestern Idaho (Vasek 1966; Burkhardt and Tisdale 1969). They are extensive in Oregon east of the Cascades (Franklin and Dyrness 1969), where the detailed studies of Driscoll (1964a,b) provide a comprehensive description of plant associations and general ecology, as well as a series of photographs. A study of juniper invasion of sagebrush by Burkhardt and Tisdale (1969, 1976) provides a view of western juniper woodland ecology in Idaho. Detailed studies of western juniper woodlands have not been made in California.

In California, western juniper generally ranges south from the Oregon state line

Figure 23-1. Western juniper woodland on the volcanic plateau north of Alturas, Modoc Co.: *Juniperus occidentalis occidentalis* with an understory primarily of *Artemisia arbuscula.*

and west from the Nevada state line through much of Modoc and Lassen Cos. to the base of the Sierra Nevada and to the mountain complex (Scott Valley region) of west–central Siskiyou Co. (Griffin and Critchfield 1972). The range includes the northeast corner of Lassen National Park (Gillett et al. 1961), a portion of northeastern Shasta Co. (see photo in Love 1970), and the northeastern tip of Sierra Co.

An extensive stand of mature western juniper woodland occurs in northern Modoc Co. An increase in elevation toward the north on the volcanic plateau north of Alturas is accompanied by increasing density and size of western junipers. Within that elevational gradient (1370–1770 m), local topography controls woodland distribution. Toward the south, at low elevations and in depressions and valleys on the plateau, stands of sagebrush, varying in size up to many hundreds of hectares, are scattered intermittently through the woodland. Junipers extend with decreasing frequency down into the sagebrush lowlands, and sagebrush, both *Artemisia arbuscula* and *A. tridentata,* depending on substrate, extends up into the woodland understory.

With increasing elevation, the density of juniper trees increases and the density of *Artemisia* shrubs decreases. At still higher elevations and in locally mesic sites, ponderosa pines (*Pinus ponderosa*) occasionally occur in the juniper woodland. As the gradient continues, pines increase and junipers decrease in frequency and density until a ponderosa pine–western juniper woodland and finally a ponderosa pine forest are the predominant vegetation types.

The ponderosa pine forest of northern Modoc Co. is a particular example of mixed conifer forest in which *Pinus ponderosa* is by far the predominant species. *Abies concolor* and *Populus tremuloides* are only locally conspicuous. Within the ponderosa pine forest, western juniper occurs in small groups or as individual trees on local dry sites such as "rock balds" (see Billings and Mark 1957). These are openings in the forest clearly mediated by edaphic conditions, namely, the occurrence of volcanic rocks, slabs, and outcrops, rather than deep soil, on flats or sometimes ridges. The balds are sparsely inhabited, if at all, mostly by *Artemisia arbuscula* and sometimes western juniper and by various herbs (see Chapter 22).

Herbaceous plants associated with western juniper woodlands have not been detailed, but a list of all plants in the Lava Beds National Monument was compiled by Applegate (1938). For Modoc Co., a brief discussion and a photograph of *Leucocrinum montanum* were published by Rickabaugh (1975), along with a photograph of its sparse western juniper woodland habitat in the Devils Garden (NA 250438). In addition, spring flower displays include such interesting woodland plants as *Dodecatheon conjugens, Hesperochiron californicus, H. pumilus, Hydrophyllum capitatum, Phlox longifolia,* and *Trifolium macrocephalum.*

In Modoc Co., the western juniper woodland is utilized for grazing (cattle), hunting habitat (deer, pronghorn antelope), and wood cutting (posts, firewood). Three felled trees had attained ages of 450–500 yr, based on estimates of growth rings in stumps about 120 cm in diameter. The largest and oldest junipers generally occur on rocky substrates, often where rock ridges break the soil surface, or on flats of extensive rock slabs. In contrast, the junipers on more open, less rocky terrain seem generally smaller and hence younger. (Examples of older and younger trees are shown by photographs in Burkhardt and Tisdale 1969.) The intuitive interpretation is that rock ridges may be long-term refugia possibly offering protection from fire. Perhaps rocky ridges offer only the long-term adversity associated with extreme longevity (Schulman 1954), but Eckert (1957) found greater soil water supply on rocky outcrops and slopes occupied by junipers than the maximum required by nearby *Artemisia* (cited from Driscoll 1964a).

Toward the west, in Siskiyou Co., the western juniper woodland grades into a ponderosa pine forest similar to that of northern Modoc Co. (Keeler–Wolf and Keeler–Wolf 1973). A general description of vegetation in Lava Beds (NA 471220) National Monument (Applegate 1938) indicates that *Juniperus occidentalis* is the most common tree, especially in rocky areas such as recent volcanics. *Pinus ponderosa* also is a prominent tree among the 10 species listed, and *Purshia tridentata* the most common shrub. The decrease in relative juniper density and the increase in forest tree density at the woodland–forest transition are illustrated in Table 23-1. A

TABLE 23-1. Relative density of trees and shrubs along three transects in a woodland-forest transition area just south of Lava Beds National Monument, on sections 2, 3 and 4, R4E, T44N, Siskiyou Co., California. Sampled by F. C. Vasek on August 5, 1972, using point quarter transects with 20 points at intervals of 36 m. Transects A, B, and C are 1.6, 4.8 and 6.4 km from the main road on Tichener Cave Road; at transect C, point-to-shrub average distance = 1.29 m, point-to-tree average distance = 5.01 m. An asterisk indicates an estimation

Species	A	B	C
Trees and emergents			
Juniperus occidentalis ssp. occidentalis	53.8	5.0*	5.6
Pinus ponderosa	28.2	95.0*	62.5
Cercocarpus ledifolius	17.9		
Abies concolor			23.6
Calocedrus decurrens			6.9
Pinus lambertiana			1.4
Shrubs			
Purshia tridentata	58.8	95.0*	83.1
Artemisia tridentata	17.5		
Chrysothamnus viscidiflorus	12.5		8.5
C. nauseosus	8.8	3.0*	2.8
Ribes velutinum	2.5	2.0*	
Arctostaphylos patula			4.2
Rhamnus californica			1.4

parallel shift in shrub composition occurs from *Artemisia* and *Chrysothamnus* to *Purshia* and *Arctostaphylos*.

In western Siskiyou Co., large stands of western juniper, with an understory and intervening patches of *Purshia tridentata* or *Chrysothamnus nauseosus*, occur in the region north of Mt. Shasta. At the west edge of Shasta Valley, stands of western juniper occur with *Pinus jeffreyi, Quercus garryana, Ribes velutinum,* and *Amelanchier pallida* on rock breaks and lava outcrops (e.g., south of Gazelle) surrounded by open grassy areas or stands of *C. nauseosus*. The lower eastern slopes of the Scott Mts. have stands of western juniper on dry, warm exposures and ponderosa pine forest on more mesic exposures.

West of Gazelle, several vegetational phases are distributed according to slope and exposure. Large groups of western juniper occur on dry slopes dominated by *Purshia tridentata* and also inhabited by *Chrysothamnus nauseosus* and various grasses. A denser woodland phase dominated by western juniper includes scattered ponderosa and Jeffrey pines and small patches of *Quercus garryana*. The woodland grades into a ponderosa pine forest which includes *Pinus ponderosa* as an overdominant and sometimes *Q. garryana* as a local dominant.

At Gazelle Summit, a mixture of Great Basin and interior species on dry eastern exposures is abruptly replaced by the ponderosa pine forest. In terms of local distribution, western juniper, *Pinus jeffreyi, Purshia tridentata,* and *Cercocarpus ledifolius* reach a western limit at the summit.

West of the Scott Mts. the ponderosa pine forest becomes open and spotty in the drier conditions of Scott Valley. There *Ceanothus cuneatus* is a common understory plant, and *Quercus garryana* becomes locally abundant. In local mesic sites, a few *Pinus lambertiana* and *Calocedrus decurrens* (e.g., neat Etna) or a few *Pseudotsuga menziesii* (e.g., near Ft. Jones) occur in the ponderosa pine forest. Western juniper is found mostly as scattered trees within open phases of the ponderosa pine forest or as small groups at the lower edge and on rocky exposures within it. A few Jeffrey pines also enter on dry slopes with *Ceanothus cuneatus*. At the western limits of the western juniper woodland and of the western juniper itself, a vegetational transition to a ponderosa pine forest is apparent.

Toward the northeast, the western juniper woodland covers the lower western flanks of the Warner Mts. and extends around the mountains to Oregon and Nevada. At higher elevations in the Warner Mts., *Juniperus occidentalis* joins *Pinus jeffreyi* and *P. washoensis* in a forest–woodland complex (Thorne 1976; Milligan 1969). This marks the northernmost occurrence of mountain juniper and hence the extreme northern attenuation of the mountain juniper woodland (see the next section).

In Lassen Co., the western juniper woodland consists generally of smaller, younger trees than does the woodland described for northern Modoc Co. The southernmost stand of western juniper woodland occurs in a long, narrow strip along the California–Nevada state line from Honey Lake valley (Lassen Co.) to the south end of the Antelope Mts. in the northeastern corner of Sierra Co., California, and adjacent western Washoe Co., Nevada. The population also branches and extends southward along the immediate east flank of the Sierra Nevada to the Beckwourth Pass Road (State Highway 70).

The relationships eastward toward the Utah juniper and piñon woodlands of Nevada have not been studied.Probably, the western Nevada junipers east of Reno are *Juniperus osteosperma*, and those north of Reno are *J. osteosperma* with an introgressive influence from Western junipers along the state line (Vasek 1966). The transition to Utah juniper woodland is accompanied by a change in biotypes of *Leucocrinum montanum* (Ornduff and Cave 1975).

Mountain Juniper Woodland

The mountain juniper woodland is characterized by scattered trees of *Juniperus occidentalis* ssp. *australis*, commonly in association with *Pinus jeffreyi, Cercocarpus ledifolius, Purshia tridentata,* and *Artemisia tridentata.* It occurs on volcanic or granitic ridges, or on shallow rocky soils, in mountains with mixed conifer forests. In Modoc and Lassen Cos., mountain juniper woodland is transitional between a Jeffrey pine forest or mixed ponderosa–Jeffrey pine forest at higher elevations and the western juniper woodland and sagebrush at lower elevations.

The common pattern in this region places mountain junipers in forested uplands and western junipers in sagebrush lowlands. The transition between mountain juniper woodland and western juniper woodland is essentially clinal (Vasek 1966), but the pattern of transition is complicated by mountainous topography. The mountain–western woodland transition probably extends north to the Warner Mts. and south to about Susanville (Lassen Co.).

Mountain juniper woodland usually includes *Pinus jeffreyi* at the transition to sagebrush, whereas western juniper woodland usually does not.

In Lassen Co., a volcanic ridge north of Grasshopper Valley has an open forest–woodland dominated by *P. jeffreyi,* but *Cercocarpus ledifolius* and mountain juniper also contribute significant overstory cover. *Ribes* and *Wyethia* constitute most of the sparse understory (Table 23-2). On the lower slopes of the same ridge a more open woodland is dominated by an understory of sagebrush. The area between sagebrush plants is almost fully covered by perennial grasses and other herbs: *Festuca idahoensis, Agropyron spicatum, Arenaria congesta, Antennaria rosea, Arnica sororia,* etc. The open sagebrush scrub of Grasshopper Valley occurs on the same slope at still lower elevations.

South of Susanville, the mountain juniper woodland occurs on ridges, balds, and warm, dry slopes, usually within the range of the Sierran red fir forest or the mixed conifer forest. The north-trending ridge east of Sierra Valley (Sierraville) carries a fairly large mountain juniper woodland on its lower slopes below a Jeffrey pine forest. The Sierra Valley junipers were interpreted as *Juniperus osteosperma* by Griffin and Critchfield (1972). Well-developed but mostly small and scattered stands of mountain juniper occur in the Carson Pass region (with *Pinus murrayana*), and an analysis of vegetation on two adjacent substrates by Taylor (1976) indicated an edaphic control of mountain juniper woodland distribution here.

The broad occurrence of mountain juniper woodland in the Sierra Nevada is well illustrated by a transect over Ebbetts Pass. Scattered stands of mountain juniper occur from the ridge overlooking the Nevada line at Topaz Lake some 43 km east of Ebbetts Pass to the Hells Kitchen overlook some 35 km west of Ebbetts Pass. At the east end of this transect, common associates of mountain juniper are *Pinus monophylla, Cercocarpus ledifolius,* and *Artemisia tridentata.* In the Monitor Pass region, extensive high-elevation (2200–2500 m) brushfields of *Artemisia tridentata, Purshia tridentata, Prunus virginiana* var. *demissa, Ceanothus velutinus,* and *Symphoricarpos vaccinoides* include stray individuals or small groups of trees of *Abies*

TABLE 23-2. Forest-woodland vegetation on a volcanic ridge 5 km north of Grasshopper Valley (Termo Road junction) along State Highway 139, Lassen Co., California. Sampled by F. C. Vasek on August 4, 1972, using a 50 m line intercept on the ridge and a 20 point quarter method on the slope

Species	Ridge Top		Slope
	Number of Plants	Cover (%)	Relative[a] Density
Juniperus occidentalis ssp. _australis_	1	14.0[b]	1.2
Pinus jeffreyi	2	40.0	11.3
Cerocarpus ledifolius	4	13.4[b]	13.8
Understory			
Amelanchier pallida	1	0.5	
Artemisia tridentata			56.3
Chrysothamnus nauseosus			1.2
Haplopappus bloomeri			6.3
Lupinus caudatus			3.7
Ribes viscosissimum	1	2.4	3.8
Symphoricarpos acutus			1.2
S. vaccinoides	3	0.9	
Wyethia mollis	8	1.2	

Perennial grasses	
Stipa, *Sitanion*	15-20[c]

[a]Approximately 472 trees per hectare.
[b]5% overlap.
[c]Estimated.

concolor, mountain juniper, *Pinus jeffreyi, P. monticola, P. monophylla* and *Populus tremuloides.*

Patches of woodland, which occur mostly on rocky tors, ridges, and outcrops, increase in size and frequency at lower elevations toward Silver Creek and Markleeville (1678 m). Westward from Markleeville toward Ebbetts Pass, the incidence of *Pinus monophylla* decreases, whereas that of *P. jeffreyi* and then of *Abies concolor* increases. At middle elevations, an extensive forest is dominated first by *P. jeffreyi* plus woodland mixtures of *Cercocarpus ledifolius* and *P. monophylla* and then by forest mixtures of *A. concolor* and *P. jeffreyi.*

At higher elevations, a subalpine forest has several phases dominated by various mixtures of *Pinus murrayana, P. albicaulis, Abies magnifica, Tsuga mertensiana, Arctostaphylos nevadensis, Chrysolepis sempervirens,* etc. Patches of mountain juniper woodland or individuals and small groups of mountain juniper occur throughout the subalpine zone on exposed rocky substrates.

A large stand of mountain juniper occurs 0.5 km west of Ebbetts Pass (2660 m) with *Artemisia tridentata* ssp. *vaseyana, Arctostaphylos nevadensis,* and other subalpine shrubs. At slightly lower elevations, mountain juniper woodland merges with *Pinus murrayana* and *P. jeffreyi* in an open forest-woodland. Juniper woodlands continue with decreasing size and frequency through the red fir forest to the mixed conifer forest (as at Hells Kitchen overlook).

Mountain juniper grows on dry, sunny slopes of gravel moraine, weathered lava, or bare granite ridges in Alpine and Tuolumne Cos., where large trees approach ages of 2900–3250 yr (Glock 1937).

An example of the intricate relationships between *Pinus jeffreyi,* mountain juniper, and some of their associates occurs near Sonora Pass in Mono Co. At low elevations ($\pm$2200 m), Great Basin sagebrush, sometimes with a few scattered trees, is the common vegetation on gentle slopes or steep slopes on which the soil is not broken by rock outcrops. Steep slopes, especially those with conspicuous rock outcrops, support a forest vegetation. At Sonora bridge, a mixture of *P. jeffreyi,* mountain juniper, *Cercocarpus ledifolius,* and *Artemisia tridentata* dominates a forest vegetation mosaic, and any of these species may dominate a local patch. In addition, the quaking aspen, *Populus tremuloides,* enters this vegetation in wet places, and a few saplings of piñon, *Pinus monophylla,* also occur. Minor understory elements include *Purshia tridentata, Ribes velutinum, Symphoricarpos vaccinoides, Holodiscus microphyllous, Amelanchier pallida,* and *Chrysothamnus viscidiflorus.*

To the west, the forest increases in areal extent as *Abies concolor, Pinus murrayana,* and *Ceanothus velutinus* join in at higher elevations. From Leavitt Meadows to Leavitt Falls, an extensive open forest or woodland is dominated by Jeffrey pine and mountain juniper. Several phases occur in which mountain juniper, usually with *Cercocarpus ledifolius* and sometimes stray trees of *P. monophylla,* dominates on rocky exposures, Jeffrey pine dominates on moderate slopes, and a mixture of *P. jeffreyi, A. concolor,* and *Populus tremuloides* occupies relatively mesic areas. A narrow transition zone in which mountain junipers extend farther up into a lodgepole (*Pinus murrayana*) forest suggests that mountain juniper may be slightly more cold tolerant than Jeffrey pine. At still higher elevations, *P. murrayana* and *P. albicaulis* dominate an open subalpine forest that continues to Sonora Pass (2935 m).

At Devils Gate Pass (2290 m), a mountain juniper woodland occurs on cliffs and rock outcrops, usually with *Cercocarpus ledifolius*. Groves of aspen (*Populus tremuloides*) dominate some of the gentle slopes above the junipers, and the general forest is dominated by *Pinus jeffreyi* but includes a few *Abies concolor* and mountain junipers. Understory plants include *Ceanothus velutinus, Holodiscus microphyllous, Leptodactylon pungens, Ribes viscosissimum,* and *Symphoricarpos vaccinoides* in the more densely forested areas and *Purshia tridentata* and *Artemisia tridentata* in the more open areas. *Cercocarpus ledifolius* also forms groves in sagebrush or in Jeffrey pine forests or between them.

In the June Lake area (Mono Co.), a mountain juniper woodland is interspersed with stands of aspen and sagebrush (*Artemisia, Purshia*). The juniper woodland includes a variety of phases in which dominance shifts among *Cercocarpus ledifolius, Pinus jeffreyi,* mountain juniper, *Artemisia tridentata, Purshia tridentata, Abies concolor,* or *Prunus virginiana* var. *demissa.*

Near Crowley Lake, a small patch of mountain junipers occurs near the north end of McGee Mt., which is largely brush covered but supports a few scattered Jeffrey pines near the top. Southwest of Crowley Lake a few groups of mountain juniper occur with groves of aspen.

Other distributional details of mountain juniper woodland vegetation in the Sierra Nevada are obscure. Personal observations note the presence of mountain juniper on the Tioga Pass Road (Yosemite National Park), above Mist Falls in Kings Canyon National Park, and at Kaiser Pass (very large trees) in Fresno Co. The presence of mountain juniper was also reported on Owens Peak by Twisselmann (1967) in Kern Co.

Mountain juniper woodland recurs on ridges (Fig. 23-2) and scattered peaks in the San Gabriel Mts., for example, Devil's Backbone (NA 360433), and in a large area of the San Bernardino Mts. in southern California (Horton 1960; Minnich 1976; Vasek 1966). In the San Bernardino Mts., a coastal-facing mixed conifer forest of *Pinus ponderosa, P. lambertiana, Calocedrus decurrens, Abies concolor,* and

Figure 23-2. Wind-shaped mountain juniper on an exposed ridge near Mt. San Antonio, San Gabriel Mts., San Bernardino Co.

Quercus kelloggii gives way in the rain shadow of Mt. San Gorgonio to an open forest of *Pinus jeffreyi*, mountain juniper, *Cercocarpus ledifolius*, and *Artemisia tridentata*. Toward the desert, *P. jeffreyi* decreases as the forest merges with *P. monophylla* to form a mountain juniper–piñon woodland.

Scattered Juniper Occurrences

Mountain juniper occurs in small populations in the Panamint and Inyo Mts. between the piñon woodland and the bristlecone pine forest, and as a few relict individuals in the White Mts. (Vasek 1966). *Juniperus occidentalis* in the Yolla Bolly Mts. was interpreted as a disjunct stand of ssp. *australis* by Vasek (1966). However, Keeler-Wolf and Keeler-Wolf (1973) observed more stands and interpreted them as disjunct stands of ssp. *occidentalis*. Regardless of the subspecies, the Yolla Bolly disjunct populations suggest an incursion of interior plant species from the northeast, probably by way of the Pit River and doubtless during an arid time such as the Xerothermic.

PIÑON AND JUNIPER WOODLANDS

Piñon-juniper woodlands in California are here interpreted as vegetation types having one or more of the following species as a conspicuous emergent above a shrubby or herbaceous understory: *Pinus monophylla* (one-leaf piñon), *Juniperus osteosperma* (Utah juniper), *J. californica* (California juniper), *P. quadrifolia* (four-leaf piñon), *P. edulis* (two-leaf piñon). Collectively, piñon–juniper woodlands occur on mountain ranges, or other landforms within suitable elevational limits, east of the Sierra Nevada and southward from Alpine Co. They also continue along the transmontane slopes of the Sierra Nevada, Transverse, and Peninsular ranges to northern Baja California.

Generally, the California piñon–juniper woodland represents the westernmost expression of widespread vegetation types occurring in the Great Basin and Colorado Plateau regions. An extensive literature has been accumulating regarding the distribution, management, and ecology of piñon–juniper woodlands in the Great Basin and especially the Colorado Plateau regions, but study of piñon–juniper in California has lagged. Bibliographic reviews have been accumulated by West et al. (1973) and Aldon and Springfield (1973). A good general, illustrated description of distribution, vegetation, and management for piñon–juniper in Arizona was presented by Arnold et al. (1964). Other Arizona studies have been done by Herman (1953), Arnold (1964), and Woodbury (1947). The piñon–juniper of the Great Basin was described by Beeson (1974), who also included a detailed distribution map.

Much of the information about piñon–juniper woodland, especially that from Nevada and Arizona, is applicable also to the woodlands of California. More particularly, the zonation of vegetation observed in the Great Basin (Billings 1951) applies directly to California vegetation. A piñon–juniper zone commonly occurs above a brush zone in valleys, but a partial separation occurs with junipers on lower slopes, piñon on upper (steeper) slopes, and the two overlapping to some degree near midslope (Billings 1954; Reveal 1944). Similar partial separation occurs widely in

California, leading to a preference here for discussing piñon *and* juniper woodlands rather than a piñon–juniper woodland.

Sierra Nevada Piñon Woodland

Extensive woodlands of *Juniperus osteosperma* and *Pinus monophylla* in western Nevada (see the next section) extend into Alpine Co., California, and range southward over the mountainous area east of the Sierra Nevada. Some elements, especially *P. monophylla* but not *J. osteosperma,* range into the Sierra Nevada, where they form a distinctive piñon woodland on the steep east-facing mountain slopes between the sagebrush below and coniferous forests and woodlands above. In Alpine Co., *P. monophylla* merges with mountain juniper woodland. To the south, the piñon woodland is essentially continuous from Topaz Lake at the Nevada state line to Kern Co.

In the north, the Sierra Nevada piñon woodland is composed primarily of *Pinus monophylla* with an understory of *Artemisia tridentata* and sometimes *Chrysothamnus nauseosus, Purshia tridentata, Prunus andersonii, Haplopappus* spp., etc. In northern Mono Co., many piñon woodlands have an understory of *Purshia tridentata* (Nord 1965). Removal of piñon trees by hand cutting resulted, after a 10 yr lag, in a marked increase of *Purshia,* an invasion by *Artemisia tridentata,* and a reinvasion by *Pinus monophylla.* The predisturbance woodland was interpreted as an example of shrub–tree interaction unfavorable to *Purshia* (Nord 1965). Farther south, trees such as *Quercus chrysolepis* and shrubs such as *Ceanothus cuneatus* occur in this woodland (e.g., Tulare Co.), but vegetational analyses are lacking.

At low elevations, a common pattern is the occurrence of dense stands of large sagebrush plants in the valleys (deep soil) and an abrupt shift to open stands of smaller sagebrush plants on mountain slopes (thin soil). Piñon trees occur with sagebrush on the lower slopes and increase in size and density with elevation. At higher elevations or in more mesic areas, the Sierra Nevada piñon woodland blends with the mountain juniper woodland or with a Jeffrey pine forest.

At Mammoth Junction, piñon occurs on a rocky ridge with a few *Pinus jeffreyi.* The understory and treeless flats include *Purshia tridentata, Artemisia tridentata,* and *Chrysothamnus nauseosus.*

Glass Mountain Ridge northeast of Crowley Lake represents a piñon woodland–forest transition displaced eastward. It exhibits a forest mosaic in which the several vegetational phases are dominated either by *Pinus jeffreyi, P. monophylla,* or *Artemisia tridentata.* Mountain juniper occurs as an important but secondary element with both pines. Piñon are invading down-slope into open sagebrush. All the elements merge in various mixtures in the canyons, where *Populus tremuloides* and *Betula occidentalis* occupy the wetter areas. Understory shrubs include *Purshia tridentata* and *Chrysothamnus nauseosus,* as well as *Artemisia tridentata.* The annuals observed among piñon and Jeffrey pines are somewhat surprising in that several of them also occur in desert scrub vegetation far to the south, for example, *Anisocoma acaulis, Camissonia clavaeformis* ssp. *purpurascens, Plagiobothrys kingii,* and *Layia glandulosa.*

A few small stands of *Pinus monophylla* have been reported on the west side of the Sierra Nevada in the Tuolumne, San Joaquin, and Kings River basins (Griffin and Critchfield 1972).

Utah Juniper and One-Leaf Piñon Woodlands

Juniperus osteosperma is common in western Nevada and occurs also in Alpine Co., California, but only at the Piñon–Juniper Natural Area (NA 020150). *Juniperus osteosperma* does not occur on the Sierra Nevada west of Bridgeport (Mono Co.) and probably not on the southern Sweetwater Mts. north of Bridgeport, but the species is present as scattered individuals on the lower slopes of the piñon-covered mountains east of Bridgeport and east of Bridgeport Reservoir. That mixed woodland extends some 10 km south of Bridgeport and about 10 km eastward toward Bodie. Piñon are relatively more abundant toward the west and essentially merge with the Sierra Nevada piñon woodland. Utah junipers are relatively more common eastward, but groups of junipers decline in size and frequency as sagebrush increases in dominance at high elevations. The region south and west of Bodie is one of a very few areas where Utah junipers occur at the *upper* elevational margin of a piñon woodland.

About 4 km south of Bodie, a few Utah junipers occur in Upper Bridgeport Canyon, which cuts through an east–west-trending ridge at the north edge of Mono Basin. A piñon woodland occupies the lower 4 km of Bridgeport Canyon and extends along the ridge both east and west, forming a continuous woodland from the Nevada state line nearly to Conway Summit. Regional associates of *Pinus monophylla* are *Artemisia tridentata, Ephedra viridis, Purshia tridentata, Grayia spinosa* (upper elevations), and *Chrysothamnus nauseosus*.

A Utah juniper woodland occurs in Mono Basin on an alluvial terrace above and north of Mono Lake, and below and south of the piñon-covered ridge discussed above. This Utah juniper woodland extends about 27 km eastward in a nearly continuous band from a point south of Conway Summit grade to the Nevada state line. It spreads southward over much of the basin northeast of Mono Lake, where very large old plants occur. Regional associates of Utah juniper are *Artemisia tridentata, Chrysothamnus nauseosus, C. viscidiflorus, Purshia tridentata, Sarcobatus vermiculatus, Oryzopsis hymenoides,* and *Leptodactylon pungens*.

The piñon woodland on the ridge to the north and the Utah juniper woodland in Mono Basin are spatially separated for most of their local ranges. Toward the west, a few piñon drift down into the basin and meet the westernmost portions of the juniper woodland. As the topographical differences in their habitats diminish toward the east, the two woodlands merge near the state line and continue into Nevada, where these vegetation types are well developed.

Utah juniper woodlands range southeastward to the Adobe Hills and to an isolated little range of hills at the west edge of Adobe Valley. In these areas, piñon trees generally outnumber junipers by a factor of 5–10. Locally, however, Utah junipers may dominate, especially on lava rock breaks, rim rock, canyon sides, and so on. Associated species are *Artemisia tridentata, Purshia tridentata, Chrysothamnus viscidiflorus, C. nauseosus, Stipa thurberiana, S. comata,* and *Ephedra viridis*.

Piñon woodlands, without junipers, are very common on the remaining mountain ranges between Mono Lake and the White Mts. Utah juniper woodlands also occur at the north base of the White Mts. in Mono Co. and southward through the White–

Inyo Mts. to the higher mountains of the eastern Mojave Desert: the Panamint, Grapevine, Kingston, Clark, New York, and Providence ranges. In addition, disjunct populations occur in the San Bernardino Mts. and north of the San Gabriel Mts. (Vasek 1966) at the southwest margin of the Mojave Desert, but Utah juniper does not approach the Sierra Nevada south of Mono Lake. In the eastern desert areas, the common pattern of distribution finds stands of Utah juniper on lower slopes, alluvial fans, and occasionally mountain flats (Fig. 23-3) or gently sloping upland sites, such as Westgaard Pass and Whippoorwill Flat (NA 142345) in the Inyo Mts. The steeper slopes are usually occupied by piñon. In the White Mts., Utah juniper is rather sparse, occurring mainly in alluvial areas (St. Andre et al. 1965), so that the woodland should properly be called a piñon woodland. Shrub cover frequently exceeds tree cover with *Artemisia tridentata* and *Purshia glandulosa* being most important. At low elevations (1980 m), shrubs become of even greater importance in a transition to a desert scrub vegetation of *Artemisia tridentata, Chrysothamnus nauseosus, Menodora spinescens, Ephedra nevadensis,* etc. Shrubs also increase in importance toward high elevations, where an open sagebrush zone occurs between the piñon woodland and the bristlecone forest (Mooney 1973). Comparable zones, or "seams," between piñon woodland and bristlecone pine forest in the Inyo and Panamint Mts. are occupied by small stands of mountain juniper (Vasek 1966).

The White and Inyo Mts. are western Great Basin ranges with vegetational relationships to that region. Their low-elevation transition from piñon or juniper woodland is toward sagebrush scrub, as noted above for the White Mts. and as studied in some detail for several locations in Nevada (Blackburn and Tueller 1970; Zamora and Tueller 1973).

The White and Panamint Mts. were included in a study by Beeson (1974) of 53 Great Basin mountain ranges classified on the basis of major floristic groupings. In this classification, the Panamint Mts. were included in a group characterized by dry-

Figure 23-3. Utah juniper woodland at Westgaard Pass, Inyo Co.: *Juniperus osteosperma* and an understory of *Artemisia tridentata*.

warm climatic conditions and dominated in the piñon–juniper zone (Fig. 23-4) by *Pinus monophylla, Juniperus osteosperma, Opuntia* spp., *Sitanion hystrix, Oryzopsis hymenoides, Chrysothamnus viscidiflorus, Ephedra viridis, Artemisia nova, Cowania mexicana* var. *stansburiana*, and *Gutierrezia sarothrae*. The White Mts. have moist-cool conditions and include all the major plants of the Panamints except that *Ephedra, Artemisia nova,* and *Cowania* are replaced by *Artemisia tridentata* and *Poa secunda*. Vegetation analyses have not been detailed for the Panamint or Inyo Mts. to our knowledge, but both piñon woodland and bristlecone pine forest are apparently more extensive in the Inyo than in the Panamint Mts.

Little is known of the vegetation in the mountains east of Death Valley, but Miller (1946) described a piñon woodland at about 2100 m in the Grapevine Mts. and then a piñon "forest" of large trees in high density at 2650 m. The piñon forest assumes dimensions, physical aspects, and an associated avifauna ordinarily found in montane coniferous forests.

Vegetation analyses have not been made for the Kingston Mts. (NA 361140). However, the piñon woodland there appears extensive, dense, and widespread. The pattern of piñon on steeper slopes and Utah juniper on alluvium or gentle slopes continues through the Kingston, Clark, New York, and Providence Mts. (NA 361675) and the Mid Hills (NA 361325).

Compositions of the piñon and juniper woodlands for five samples in Caruthers Canyon, New York Mts., and one sample north of Clark Mt. are shown in Table 23-3. The abundance of *Artemisia, Coleogyne,* and *Thamnosma* on Clark Mt. and their absence from the New York Mts., and the converse occurrence of *Quercus turbinella*, are noteworthy. Variation in needle number for the piñon trees in the New York Mts. is interpreted by Lanner (1974b) as variation within *Pinus monophylla* rather than as evidence for the occurrence of *P. edulis* in that area. However, the two-

Figure 23-4. Piñon woodland on the slopes of upper Wildrose Canyon in the Panamint Mts., Inyo Co., looking down-canyon toward the beehive kilns below Mahogany Flat. The open woodland of *Pinus monophylla* and *Juniperus osteosperma* was cut to supply wood for the charcoal kilns during mining days.

TABLE 23-3. Relative density of perennial vegetation of Clark Mt. (north side) and in the New York Mts. (Caruthers Canyon), based on point quarter transects of 20 and 100 points, respectively, at intervals of about 18 m. Sampled by F. C. Vasek and H. B. Johnson in July 1972

Species	Clark	New York	Species	Clark	New York
Arctostaphylos pungens	...	2.75	*Haplopappus cooperi*	...	0.50
Artemisia tridentata	25.00	...	*H. cuneatus*	...	0.25
Atriplex canescens	2.50	...	*H. linearifolius*	...	0.75
Baccharis sergilloides	...	3.75	*Juniperus osteosperma*	3.75	3.75
Berberis haematocarpa	...	0.25	*Lycium* sp.	3.75	...
Brickellia californica	...	0.25	*Menodora scoparia*	2.50	...
Ceanothus greggii vestitus	...	2.00	*Opuntia erinacea*	2.50	1.50
Chrysothamnus depressus	...	1.00	*O. mojavensis*	3.75	1.25
Coleogyne ramosissima	23.75	...	*Pinus monophylla*	6.25	14.50
Cowania mexicana stansburiana	...	1.00	*Purshia glandulosa*	...	4.50
Ephedra viridis	3.75	...	*Quercus chrysolepis*	...	1.25
Eriodictyon angustifolia	...	1.50	*Q. turbinella*	...	28.50
Eriogonum sp.	1.25	...	*Rhamnus californica*	...	0.75
E. wrightii wrightii	...	1.25	*R. crocea*	...	0.50
Fallugia paradoxa	1.25	8.00	*Rhus trilobata anisophylla*	...	2.00
Forestieria neomexicana	...	1.50	*Salvia dorrii*	2.50	...
Garrya flavescens flavescens	1.25	9.25	*Thamnosma montana*	8.75	...
Gutierrezia sarothrae	2.50	3.25	*Yucca baccata*	5.00	4.25

needle piñon trees in the New York Mts. may be *P. edulis*, as they occur in a more or less distinctive two-needle piñon woodland (see the section on p. 822). The two pinyon species evidentally also occur together elsewhere (Lanner 1947b; Lanner and Hutchinson 1972).

Piñon and juniper woodlands on the Sheep, Spring, and McCullough ranges in adjacent Nevada have been discussed by Bradley and Deacon (1967). The McCullough Range is singled out by Beeson (1974) as the one range of 53 studied that does not include a single species of *Artemisia* among its dominant woodland plants. A few Utah juniper plants are isolated on top of Hackberry Mt., San Bernardino Co. Utah juniper woodlands then skip from the Providence Mts. to the San Bernardino Mts. of southern California, where large populations are present on the lower desert-facing slopes (Vasek 1966). *Pinus monophylla* occurs on the higher, steeper slopes, grading into a mountain juniper woodland toward higher elevations and grading into chaparral toward the coastal influence of Cajon Pass (Vasek and Clovis 1976). In this area, piñon woodland averages about 30% ground cover and includes *Pinus monophylla, Quercus turbinella, Juniperus osteosperma, Haplopappus linearifolius, Purshia glandulosa,* and *Eriogonum fasciculatum* ssp. *polifolium* among its dominants, along with a desert ecotype of *Arctostaphylos glauca*.

Finally, the southwesternmost stand of *Juniperus osteosperma* occurs on alluvium at the north base of the San Gabriel Mts. in San Bernardino and Los Angeles Cos. (Vasek 1966). This stand is surrounded by a large population of *J. californica* and borders a creosote bush scrub community.

California Juniper and One-Leaf Piñon Woodlands

Extensive stands of California juniper occur on alluvial fans and lower slopes at the desert edge on the north side of the San Gabriel Mts. Piñon are scattered on the steeper desert slopes, commonly intermixed with chaparral type shrubs. Surrounding vegetation includes creosote bush scrub at lower elevations, and a very open forest of Jeffrey pine, white fir, and montane chaparral at higher elevations. Joshua tree (*Yucca brevifolia*) woodlands occur at about the same elevation and associate with junipers or occur intermittently between patches of junipers along with Great Basin elements such as *Artemisia tridentata, Chrysothamnus nauseosus,* and *Tetradymia axillaris*. However, quantitative data are not available.

From the San Gabriel Mts., California juniper woodlands spread out in three main directions. Toward the east, they extend continuously across the Cajon Pass region and along the lower north slopes and fans of the San Bernardino Mts. approximately to Cushenbury. In this region, California juniper occurs with *Yucca brevifolia, Coleogyne ramosissima, Quercus turbinella, Salvia dorii,* and *Eriogonum fasciculatum* in an open shrub woodland, just above, and partly overlapping, the creosote bush scrub, and just below a woodland of *Pinus monophylla, Juniperus osteosperma, Arctostaphylos glauca,* and *Q. turbinella,* as described in the preceding section. With intermittent gaps, the California juniper woodland continues eastward at the base of the San Bernardino Mts. to the Little San Bernardino Mts., where the familiar pattern prevails of juniper on gentle slopes and piñon on higher, steeper slopes. The composition of the piñon woodland near the desert scrub transition north of Yucca Valley at the Burns Piñon Ridge Reserve (NA 360290) is relatively rich, with over 40 perennial species (Table 23-4). Piñon are conspicuous but sparse and infrequent.

Quantitative vegetational analyses for the Little San Bernardino Mts. have not been made; however, a checklist of plants for Joshua Tree National Monument (NA 331070) was compiled by Adams (1957). In addition, major plant belts in the Joshua Tree National Monument were mapped by Miller and Stebbins (1964), who also included brief descriptions and photographs of important habitats. The piñon woodland here is dominated by *Pinus monophylla, Quercus turbinella, Purshia glandulosa,* and *Juniperus californica.*

Still farther east, the Coxcomb Mts. (NA 360370) support a tiny woodland consisting of a handful of piñon trees on top of the range and two California junipers on small, level stream terraces within a steep, north-facing wash. Elsewhere in the Mojave Desert, California juniper woodland occurs at its northern extreme at Granite Pass (NA 360780) with a mixture of cold desert shrubs, such as *Coleogyne ramosissima* and *Ephedra nevadensis.* Immediately above, the steeper slopes of the Granite Mts. (south of the Kelso Dunes, NA 361130) support a piñon woodland.

South and east, California juniper woodlands occupy lower slopes of the Old Woman Mts. (NA 361560) marginally above a creosote bush scrub (Vasek 1966). Piñon occur higher in the Old Woman Mts. (Griffin and Critchfield 1972), but quantitative data do not exist.

The second major arm of California juniper distribution extends north and west from the San Gabriel Mts. Juniper woodlands occur along the margins of Antelope Valley and continue intermittently westward to Santa Barbara Co. and northward to Tehachapi Pass, Walker Pass, and the Kernville–Lake Isabella region. Junipers occur marginally above the creosote scrub vegetation and below piñon or oak woodlands. Fine woodlands of juniper and Joshua trees, with understory shrubs such as *Artemisia tridentata,* formerly occupied extensive acreage in Antelope Valley, but they have been cleared for agriculture and development. Extant stands of these woodlands continue toward the Mt. Pinos region, where the piñon element gradually replaces Joshua tree and then juniper as the elevation increases.

In Kern Co., piñon woodland occurs on the east side of the Sierra Nevada from the Tulare Co. line to south of Walker Pass. California junipers may occur at the lower borders. Piñon woodland is scattered along the east base of the Tehachapi Mts. and then forms a continuous belt around Mt. Pinos and in the San Emigdio Range to the upper Cuyama Valley (Twisselmann 1967).

On the south slope of Mt. Pinos, sagebrush and then piñon merge with a forest of Jeffrey pine (Vogl and Miller 1968). The distribution of piñon appears solid and extensive from Antelope Valley to western Ventura and eastern Santa Barbara Cos. (Griffin and Critchfield 1972).

Personal observations in the Lockwood Valley region indicate essentially continuous distribution of piñon woodland at various levels of density. In the valley bottom, sparse piñon woodlands are dominated by *Chrysothamnus nauseosus* and *Artemisia tridentata.* On uplands, ridges, and north slopes, *Pinus monophylla* dominates increasingly dense piñon woodlands, sometimes sharing dominance with *Quercus turbinella. Juniperus californica* occurs on south slopes, and in the valley where *Eriogonum fasciculatum* usually dominates. *Arctostaphylos parryana* is an occasional associate, especially where *Q. turbinella* is a codominant.

TABLE 23-4. Density per hectare and cover (%) of perennials in piñon woodland at Burns Pinyon Ridge, San Bernardino Co. Estimates (by F. C. Vasek in October 1975) were based on five sample plots with a total area of 2500 m^2 for the four major species (*), and on 25 plots with a total area of 250 m^2 for all other species. P = present in area near large plots; NR = not recorded on small plots; t = recorded cover less than 0.005%

Species	Density	Cover
Trees and shrubs		
Pinus monophylla	80	12.95*
Quercus turbinella	144	7.80*
Juniperus californica	88	4.19*
Arctostaphylos glauca	28	1.08*
Coleogyne ramosissima	644	1.41
Cercocarpus betuloides	48	1.02
Eriogonum fasciculatum	768	0.83
Yucca schidigera	84	0.33
Crossosoma bigelovii	16	0.13
Ephedra nevadensis	16	NR
Purshia glandulosa	32	NR
Yucca brevifolia	8	NR
Opuntia basilaris	8	NR
Prunus fasciculata	4	NR
Echinocereus engelmannii	12	NR
Nolina parryi	P	NR
Acacia greggii	P	NR
Brickellia desertorum	P	NR
Ephedra californica	P	NR
Rhamnus crocea	P	NR
Soft shrubs and suffrutescents		
Haplopappus linearifolius	1388	2.76
Chrysothamnus teretifolius	160	0.32
Eriophyllum confertiflorum	340	0.11
Salazaria mexicana	88	0.09
Galium sp	156	0.04
Eriogonum elongatum	16	0.02
E. ovalifolium	60	0.01

TABLE 23-4 [continued]

Species	Density	Cover
Erigeron parishii	16	0.01
Arabis sp.	20	t
Haplopappus cuneatus	52	t
Monardella linoides	36	NR
Sphaeralcea rosea	12	NR
Krameria grayi	4	NR
Mimulus longiflorus	4	NR
Encelia virginensis	P	NR
Salvia dorrii	P	NR
Viguiera deltoidea	P	NR
Lotus rigidus	P	NR
Mirabilis bigelovii	P	NR
Perennial grasses		
Stipa speciosa	1280	0.38
S. coronata depauperata	120	0.03
Poa scabrella	280	0.02
Hilaria rigida	P	NR
Total	6012	33.53

Piñon woodland continues to Pine Mt. summit in western Ventura Co., where dense stands of large *Pinus monophylla* trees are associated with *Cercocarpus betuloides*, *Fremontodendron californicum*, *Garrya flavescens*, and *Arctostaphylos glauca* (Table 23-5). The Great Basin character of piñon woodland essentially stops at Pine Mt. summit. *Quercus turbinella* occurs in denser stands of piñon, but *Q. dumosa* is fairly common in adjacent, more open, shrub-dominated stands. *Adenostoma fasciculatum*, *Ceanothus*, and other chaparral shrubs occur in the more open phases of the shrubby woodland, and chaparral becomes dominant at lower elevations a few kilometers south and west of Pine Mt. summit.

From the western Transverse Ranges of southern California, California juniper, but not piñon (except for one plant on Caliente Mt., reported by Hoover 1970), continues northward. A characteristic juniper–oak woodland (Hoover 1970) of *Juniperus californica*, *Quercus douglasii*, *Eriogonum fasciculatum* ssp. *polifolium*, *Gutierrezia bracteata*, and *Haplopappus linearifolius* occurs north of the La Panza

TABLE 23-5. Density per hectare and cover (%) of perennials in piñon woodland at Pine Mt. Summit, Ventura Co. Estimates are based on measurements by F. C. Vasek and P. Rowlands in November 1975 in a rectangular plot of 250 m^2; P = present in adjacent area

Species	Density	Cover
Pinus monophylla		
trees	440	44.67
seedlings	240	0.02
Fremontodendron californicum	240	15.61
Garrya flavescens	120	11.38
Cercocarpus betuloides	160	6.41
Arctostaphylos glauca	P	...
Ephedra viridis	P	...
Quercus turbinella	P	...
Eriogonum fasciculatum	120	0.75
Stipa speciosa	80	0.02
Penstemon centranthifolius	200	0.04

Range and along the San Juan River to Caliente Mt. and the higher parts of the Temblor Range. In the Temblor Range (Twisselmann 1956), *J. californica* occurs in a distinct local association with *Eriophyllum confertiflorum, Ephedra californica, E. viridis,* and *Yucca whipplei.* The best development is in Cedar Canyon between an upper Sonoran subshrub association and a Douglas oak woodland.

Northward, junipers gradually lose identity as a major vegetation type and tend to occur in small patches at the lower elevational margin of oak woodlands or on locally dry edaphic features, as on rock outcrops on Mt. Diablo (Bowerman 1944) and the arid east side of the Mt. Hamilton Range (Sharsmith 1945). The northern limit of distribution in the Coast Ranges is near Beegum in Tehama Co. (Vasek 1966). In the Sierra Nevada foothills, California junipers are uncommon, but scattered individuals or small groups occur, often on serpentine, from near Shingletown (Shasta Co.) southward (Mallory et al. 1965).

A third major distributional arm of California juniper woodland occurs along the Peninsular Ranges, southward from the San Jacinto Mts. Stands of California juniper occur with creosote bush scrub on the lower desert slopes and extend upward, where they merge with a piñon woodland. In turn, the latter woodland grades into chaparral at higher elevations away from the desert. An example of piñon woodland on a desert-edge mountain peak at the interior limits of chaparral is shown in Table 23-6 for the Sheep Mt. Area of the Boyd Deep Canyon Desert Research Center (NA 331679). Over 40 perennial species occur. Although *Leptodactylon pungens, Stipa speciosa, S. coronaria,* and *Gutierrezia sarothrae* are dominant in terms of density, they provide little cover. Conspicuous plants such as *Pinus monophylla, Juniperus californica, Nolina parryi, Quercus turbinella,* and *Yucca schidigera* dominate the vegetation in terms of cover but are relatively few in number.

Woodlands of *Pinus monophylla* also occur on Martinez Mt. at the desert edge of the Santa Rosa Mts. Piñon woodlands are infrequent to the south, but fairly large stands of *P. monophylla* occur in the Vallecito Mts., the confluent Pinyon Mts., and probably some of the higher, nearby landforms (e.g., Pinyon Ridge). Otherwise, piñon distribution is irregular and sparse (Griffin and Critchfield 1972), but it continues southward well into Baja California (Lanner 1974a), at least to the south end of the Sierra San Pedro Mártir.

California Juniper and Four-Leaf Piñon Woodlands

The four-leaf piñon, *Pinus quadrifolia,* reaches its northern limit in the broad saddle between the San Jacinto and Santa Rosa Mts. in Riverside Co. It occurs in a chaparral overlooking the desert at the lowermost margin of of coniferous forest–woodland (Jeffrey pine) and usually well above California juniper woodlands. *Pinus quadrifolia* occurs mostly in small groups of individuals southward in San Diego Co. along the Laguna Mts. (Griffin and Critchfield 1972) to the Jacumba Mts. It extends into the Sierra Juarez of Baja California, where more extensive woodlands occur (see photograph in Lanner 1974a).

The peninsular ranges south from the San Jacinto–Santa Rosa Mts. also support a few stands of *Juniperus californica* on the lower, desert-edge slopes. Examples may readily be seen in San Diego Co.: at Culp Valley on the grade southwest of Borrego Valley; below the Banner grade at the north base of the Oriflamme Mts.; on the Vallecitos Mts.; at Box Canyon southeast of the Oriflamme Mts.; and on the Jacumba Mts. near Mountain Springs grade (Harvey 1951). Populations of California juniper extend southward into Baja California in similar habitats.

In the Peninsular Range Province, California juniper woodlands usually occur between creosote bush scrub and chaparral, overlapping both to some extent. In a few places they grade into a piñon woodland at higher elevations or into sparse stands of piñon intermixed with chaparral.

Scattered Juniper Occurrences

Small groves or isolated individual shrubs of *Juniperus californica* occur widely in coastal southern California as emergents in coastal sage scrub or sometimes in chaparral, especially in washes south of mountain canyons such as Lytle Creek, Cucamonga, San Antonio, and San Gabriel.

Junipers straggle down into the chaparral of Cajon Canyon nearly to San Bernardino Valley from the large population at Cajon Pass. A small population

TABLE 23-6. Density per hectare and cover (%) of perennials in piñon woodland on Sheep Mt. at the south edge of Boyd Deep Canyon Desert Research Center, Riverside Co. Density estimates were based on three sample plots with a total area of 1500 m^2 for trees, shrubs, and suffrutescent perennials, and on 15 sample plots with a total area of 150 m^2 for perennial grasses. Cover estimates were based on an area of 1500 m^2 for *Pinus monophylla* and on 150 m^2 for all other species. P = present in area near large plots; NR = not recorded on small plots. Sampled by H. B. Johnson and F. C. Vasek in September 1975; t = recorded cover less than 0.005%

Species	Density	Cover
Trees and shrubs		
Pinus monophylla	160.0	14.83
Quercus turbinella	100.0	8.38
Juniperus californica	93.3	6.57
Arctostaphylos glauca	66.7	5.50
Nolina parryi	66.7	3.56
Agave deserti	6.7	1.51
Yucca schidigera	113.3	0.99
Eriogonum fasciculatum	273.3	0.46
Opuntia erinacea	40.0	0.01
O. echinocarpa	73.3	NR
Thamnosma montana	53.3	NR
Rhus ovata	13.3	NR
Opuntia basilaris	20.0	NR
Bernardia incana	13.3	NR
Ephedra nevadensis	6.7	NR
Adenostoma sparsifolius	P	NR
Soft shrubs and suffrutescent perennials		
Leptodactylon pungens	1420.0	1.28
Viguiera deltoidea	80.0	0.41
Spheralcea rosea	566.7	0.32
Chrysothamnus teretifolius	166.7	0.26
Gutierrezia sarothrae	820.0	0.08
Salvia vaseyi	80.0	0.07
Eriogonum wrightii	273.3	0.02
Galium aff. *parishii*	53.3	0.01

TABLE 23-6 [continued]

Species	Density	Cover
Mirabilis froebelli	20.0	0.01
Verbena lasiostachys	13.3	0.01
Artemisia ludoviciana	26.7	t
Haplopappus linearifolius	120.0	NR
Keckiella antirrhinoides	60.0	NR
Rhus trilobata	6.7	NR
Dyssodia porophylloides	6.7	NR
Brickellia arguta	P	NR
Lotus rigidus	P	NR
Krameria parvifolius	P	NR
Perennial grasses		
Stipa speciosa	3533.3	0.71
S. coronaria	1600.0	0.14
Aristida parishii	266.7	0.01
Bouteloua curtipendula	133.3	0.01
Hilaria rigida	P	NR
Muhlenbergia sp.	P	NR
Erioneuron pulchellum	P	NR

occurs in the Santa Ana River wash, north of Redlands. Scattered individuals are found near Perris, and a population occurs in the Lakeview Mts. near Hemet. Isolated individuals are present in the Santa Monica Mts. (Raven and Thompson 1966) and on the Santa Rosa Plateau in the Santa Ana Mts.

Two-Leaf piñon Woodland

The two-leaf piñons of the New York Mts. and Mid Hills of the Mojave Desert in eastern San Bernardino Co. have been treated by various botanists (Wolf 1938; Griffin and Critchfield 1972; Munz 1974; Henrickson and Prigge 1975) as the westernmost populations of *Pinus edulis*. Favoring this interpretation is the presence in these eastern Mojave Desert ranges of 50 Great Basin and even more eastern species that are not reported from elsewhere in California. This "eastern" influence probably can be attributed to the summer rains, which are more regular here than in other California desert ranges.

On the other hand, Lanner (1974b), in his study of natural hybridization between *Pinus edulis* and *P. monophylla,* presented rather convincingly his argument that these Mojave two-leaf piñons are a result of back-mutation of the normal one-leaf fascicle of *P. monophylla* to its probable ancestral two-leaf condition. In favor of this interpretation is the fact that the nearest surely determined *P. edulis* stands are at least 230 km farther east. Even the trees with largely two-needle fascicles have the large number of needle resin canals, needle length, cone size, and seed coat thickness typical of *P. monophylla.*

Whatever the ultimate identification of the Mojave two-leaf piñons, they do form their own woodland on New York Mt., largely above 1800 m and extending nearly to the peak (2296 m). In this altitudinal range Utah junipers are relatively rare, and one-leaf piñons are scarce, if present at all. It is a very open woodland with scattered small piñons and oaks (*Quercus chrysolepis, Q. turbinella*). A grove of *Abies concolor* ssp. *concolor* (discussed below) occurs in one north-facing canyon. A good many shrubs are found on New York Peak with the two-leaf piñon. Several that occur predominantly, but perhaps not exclusively, at the higher elevations with the two-leaf piñon are *Cercocarpus intricatus, Forsellesia nevadensis, Fraxinus anomala, Holodiscus microphyllus, Petrophytum caespitosum, Philadelphus microphyllus* ssp. *stramineus, Ribes cereum, R. velutinum* ssp. *glanduliferum,* and *Salvia pachyphylla.* Their more common occurrence here may be largely a function of altitude or of limestone substrate, found only at the higher elevations. *Fallugia paradoxa, Haplopappus linearifolius, Opuntia erinacea,* and *Rhus trilobata* var. *anisophylla* are more common at lower elevations among the one-leaf piñons. Shrubs that range throughout the piñon woodlands without apparent preference are *Arctostaphylos pungens, Artemisia arbuscula* ssp. *nova, A. bigelovii, Ceanothus greggii* ssp. *vestitus, Echinocereus triglochidiatus* ssp. *mojavensis, Ephedra viridis, Eriogonum heermannii* ssp. *floccosum, E. umbellatum* ssp. *ferrissii, E. wrightii* ssp. *trachygonum, Garrya flavescens* ssp. *flavescens, Haplopappus cuneatus, Petradoria pumila,* and *Symphoricarpos longiflorus.*

The report by Munz (1974) of *Pinus edulis* from the Little San Bernardino Mts. has not been confirmed (Griffin and Critchfield 1972).

DESERT MONTANE FORESTS AND WOODLANDS

Montane White Fir Forest

The Rocky Mt. white fir, *Abies concolor* ssp. *concolor,* has been found in the eastern Mojave Desert on only three California ranges; the Kingston, Clark, and New York Mts. (Miller 1940; Mehringer and Ferguson 1969; Henrickson and Prigge 1975), as well as on the nearby Charleston Mts. (Clokey 1951) of Clark Co., Nevada. It forms open groves, often mixed with piñon and other trees, in steep, mesic, north-facing ravines and on slopes usually below the crests of the ranges. It occurs at elevations of 1905–2345 m on either granitic or limestone substrata. The groves are not extensive, consisting of perhaps 150 trees in about 12 ha in the Kingston Mts., about 30 trees in 0.8 ha in the New York Mts., and about 1000 trees over 65 ha in the Clark Mts. The largest trees were measured at 71–87 cm dbh and 17-20 m in height and range in age from 250 to 430 yr. In the Charleston Mts. of adjacent Nevada, this white fir is scattered through a zone between 2270 and 3300 m.

This montane white fir forest of the Great Basin coniferous woodlands should not be confused with the montane white fir–sugar pine forests of the Transverse and Peninsular ranges of southern California, although the white fir of these forests has also been identified (Hamrick and Libby 1972) as *Abies concolor* ssp. *concolor.* The white fir of the Sierran mixed conifer forest is the Sierran white fir, *Abies concolor* ssp. *lowiana,* and is lacking, like *Pinus lambertiana,* from the desert ranges. Furthermore, there is no white fir at all on the intervening Panamint, Inyo, White, and other desert ranges between the eastern Mojave ranges and the Sierra Nevada. On the basis of present distributions, the two montane fir forests, Sierran and Great Basin, may never have been contiguous across the Mojave Desert. The montane white fir forest of the eastern Mojave Desert ranges is the westernmost attenuation of the white fir–Douglas fir–blue spruce zone of the Wasatch series of Great Basin vegetation (Billings 1951). The dominants of the Wasatch forest gradually drop out westward until only *Abies concolor* ssp. *concolor* is left in the eastern Mojave ranges, where it is largely telescoped into the uppermost piñon zone (Henrickson and Prigge 1975). Several species associated with white fir in the Charleston Mts. remain associated in the New York Mts: *Pinus monophylla* (including or additional to a two-leaf piñon), *Juniperus osteosperma, Acer glabrum* var. *diffusum, Amelanchier utahensis* (incl. *A. covillei*), *Fraxinus anomala, Holodiscus microphyllus, Leptodactylon pungens* ssp. *hallii, Petrophytum caespitosum, Philadelphus microphyllus* ssp. *stramineus, Quercus chrysolepis, Q. turbinella, Ribes cereum, R. velutinum* ssp. *glanduliferum,* and *Sambucus mexicana.*

Although fir forest is not found in the Death Valley and Inyo–White ranges, some of the associated species listed above are present in mesic canyons at montane levels between the piñon woodlands and subalpine woodlands of limber pine and bristlecone pine. Even the more arid Panamint Mts. west of Death Valley retain in some of the more mesic montane canyons scattered populations of such white fir (and often pinyon) associates as *Acer glabrum* var. *diffusum, Amelanchier utahensis, Betula fontinalis, Chamaebatiaria millefolium, Fraxinus anomala, Holodiscus microphyllus, Jamesia americana, Petrophytum caespitosum, Physocarpus alternans, Philadelphus microphyllus* ssp. *stramineus, Ribes cereum, Rosa woodsii, Sambucus caerulea,* and *Symphoricarpos longiflorus.*

Subalpine Woodland

A woodland community of white pines, either *Pinus flexilis* or *P. longaeva* or the two mixed, occupies the uppermost wooded slopes on the highest desert mountains between 2900 and 3500 m. The trees are usually too short and too widely spaced for this community to merit the term "forest," sometimes applied to it. Above this woodland is alpine tundra, without tree growth other than scattered elfinwood in the fell-field cushion community described in Chapter 18.

Limber pine woodland. *Pinus flexilis* in California is a desert-oriented white pine distributed from the Sweetwater Mts. (Griffin and Critchfield 1972) south along the Sierran crest and on the highest desert peaks to the Santa Rosa Mts. of southern California. The limber pine forest of Mt. Pinos and the Transverse and Peninsular ranges is described in Chapter 16. On desert mountain tops, this pine rarely forms a forest and is often mixed with *P. longaeva,* the Great Basin bristlecone pine. In the Sweetwater Mts., limber pine and white fir occur in groves at the heads of east-facing canyons at about 3350 m, whereas *P. albicaulis* and *Tsuga mertensiana* occur at the heads of west-facing canyons (D. Taylor, verbal commun.).

In the Mojave Desert, only the Panamint, Inyo, and White Mts. support limber pine woodland. On Bennett and Telescope peaks of the Panamint Mts., *P. flexilis* appears at about 2950 m where it barely overlaps *P. monophylla,* joins *P. longaeva* at about 3140 m, and continues with the latter in a mixed woodland to the peak at 3370 m (Fig. 23-5). In the Panamint Mts., the principal woody associates of the two pines are the relatively uncommon mountain juniper and *Acer glabrum* var. *diffusum,* and the more abundant shrubby *Artemisia tridentata* ssp. *tridentata, Chamaebatiaria millefolium, Chrysothamnus viscidiflorus* ssp. *pumilus, Ribes cereum, R. montigenum,* and *Symphoricarpos longiflorus.* The topmost 25 m of the summit ridge of Telescope Peak has a limited flora that includes both pines, *Artemisia dracunculus,*

Figure 23-5. Great Basin bristlecone pine, *Pinus longaeva,* in an open woodland at about 3200 m on Telescope Peak, Panamint Mts., Inyo Co. On the ridge to the lower right is a limber pine woodland; on the lower slopes in the middle distance is a pinyon–juniper woodland.

Figure 23-6. Bristlecone pine woodland of *Pinus longaeva* on dolomite in the Schulman Grove at about 3260 m in the White Mts., Inyo Co. Pine Alpha in this grove is estimated to be 4300 yrs old.

Chamaebatiaria millefolium, Chrysothamnus viscidiflorus ssp. *pumilus, Erigeron clokeyi, Galium hypotrichium* ssp. *tomentellum, Heuchera rubescens* var. *pachypoda, Lupinus alpestris, Oenothera caespitosa* ssp. *crinita, Pellaea breweri, Ribes cereum, Sitanion hystrix,* and *Tanacetum canum.*

In the White Mts., the limber pine is most abundant on favorably moist, granitic soils at elevations between 2900 and 3355 m, but it has a total range from 2074 m (Lone Tree Creek) to 3477 m (Mooney 1973; Mooney et al. 1962; Lloyd and Mitchell 1973). At the lower levels it forms pure stands, but over most of its altitudinal range it is mixed with *Pinus longaeva.* From a mixed stand on granite at 3203 m Mooney (1973) listed as associated species *Arenaria kingii, Artemisia tridentata, Cymopteris cinerarius, Erigeron clokeyi, Eriogonum ovalifolium, Hymenoxys cooperi, Koeleria macrantha, Leptodactylon pungens, Linanthus nuttallii, Poa rupicola,* and *Stipa pinetorum.* The total plant cover of 25.8% included *Pinus longaeva* with 11.1% and *P. flexilis* with 7.3% The plants associated with limber and bristlecone pines in the Inyo Mts. are much the same as those in the Panamint and White Mts. (DeDecker 1973). *Pinus flexilis* is noted as reaching down to 2225 m and *P. longaeva* to 2210 m, both in the Narrows of Marble Canyon.

Great Basin bristlecone pine woodland. *Pinus longaeva* (Fig. 23-5) is the Great Basin equivalent of the Rocky Mt. bristlecone pine, *P. aristata* (Bailey 1970). It extends from Utah into the desert ranges of eastern California, where it is the true timberline tree of the high desert ranges. It can form pure open stands (Fig. 23-6) at its highest elevations, often on dolomite, but otherwise it is mixed with *P. flexilis,* as in the Inyo and White ranges and on Telescope Peak of the Panamint Mts. It is also found as scattered trees on the Last Chance Mts., where *P. flexilis* does not occur. Its range and occurrence on Telescope Peak were discussed in the preceding section.

The Great Basin bristlecone pine is of special interest because the gnarled, sculptured, slow-growing specimens in the White Mts. are believed to be the world's oldest living organisms. Methuselah, in Schulman Memorial Grove, is said to be more than 4,6000 yr old. One older tree, aged 4900 yr by count of annual growth rings (Schmid and Schmid 1975), grew on Wheeler Peak in eastern Nevada before it was felled. The trees do not usually grow taller than 18 m but may attain considerable breadth. The relatively young, multistemmed Patriarch, only 1500 yr old, has a circumference of about 11.25 m. The dense, resinous wood of these White Mt. trees is resistant to decay in the dry subalpine environment; hence trees often remain standing for many years after their death, and fallen wood may persist for thousands of years. From examination of weathered remains of trees once growing above the present timberline in the White Mts., it has been found that the treeline has retreated downward about 100 m in the past 1000 yr (Schmid and Schmid 1975).

In the White Mts., *P. longaeva* ranges from 2620 to 3540 m (Lloyd and Mitchell 1973). Its best development occurs on nutritionally deficient dolomites and limestones, especially in the southern part of the range, but it is also present in lesser numbers on granitic substrates. The oldest bristlecones are found on the poorest soils. The average annual precipitation in this woodland in the White Mts. is less than 40 cm, falling mostly as snow in winter (Schmid and Schmid 1975). In a woodland community on dolomite at 3416 m, bristlecone pine provides 29.1% cover with a total cover of 35.5%. Associated species listed by Mooney (1973) are *Antennaria rosea, Arenaria kingii, Astragalus kentrophyta, Chrysothamnus viscidiflorus, Cymopteris cinerarius, Erigeron pygmaeus, Festuca brachyphylla, Haplopappus acaulis, Koeleria macrantha, Leptodactylon pungens, Linanthus nuttallii, Muhlenbergia richardsonis, Oxytropis parryi, Phlox covillei, Poa rupicola, Ribes cereum,* and *Sitanion hystrix.*

Other subalpine woodlands in the White Mts. consist of lodgepole pine and aspen. *Pinus contorta* ssp. *murrayana* is distributed sporadically in some of the northern canyons between 2745 and 3200 m, with the best representation on Chiatovich Flats and the glacial moraines of Perry Aiken Creek (Lloyd and Mitchell 1973). *Populus tremuloides* forms groves on moist slopes and meadows on the east slope of the range, but it is dwarfed and shrubby on exposed, rocky sites at the highest elevations on the west slope (Lloyd and Mitchell 1973). Aspen woodland ranges between 1830 and 3355 m. At about 2900 m aspen has a total cover of 64.4% and is the most productive woodland of the White Mts. It includes some 14 herbaceous species (Mooney 1973), which contribute 44.2% cover.

STATUS OF KNOWLEDGE

The major distributional features are known for the main woody species of northern juniper woodlands. However, comparatively little is known concerning the natural history of most of these species, and even less about their physiology and ecology. The state of autecological knowledge for Great Basin piñon–juniper was reviewed by Tueller and Clark (1975), who concluded that the piñon–juniper (ecosystem) can easily sustain important research efforts for many years to come (see also West et al. 1975). The outstanding phytogeographical problem concerns the altitudinal and latitudinal gradients involving the transition in woodland vegetation and in juniper species from mountain juniper to western juniper, and the apparently correlated change from a Jeffrey pine transition to a cold-adapted (Haller 1961) ponderosa pine transition.

Several other related phytogeographical problems are also apparent. One concerns the relationship of the mountain juniper woodland and the California juniper woodland, which approach each other a few miles north of Walker Pass in Kern Co. Present distributions suggest a large-scale westward movement of Great Basin species across the southern Mojave Desert (Wells and Berger 1967). For instance, *Abies concolor* of southern California has a closer relationship to the populations in the Mojave Desert and the southern Rocky Mts. than to those in the Sierra Nevada (Hamrick 1966; Hamrich and Libby 1972). The relationships for piñon–juniper species need clarification; and if a sharp break across the Tehachapi region is verified, a strong genetic difference between *Pinus monophylla* populations along the Sierra east face and those to the south would be predicted. Another problem concerns the nature, relationships, and limits of the two-needle piñon woodland in the New York Mts.

Our knowledge of the ecology of the upper montane and subalpine transmontane coniferous woodlands other than that of bristlecone pine woodland is still rather meager. Except for the note by Henrickson and Prigge (1975) on white fir in the mountains of the eastern Mojave Desert of California and earlier notes by Miller (1940) and Mehringer and Ferguson (1969) on the white fir groves in the Clark Mts., no attention has been given to the ecology of this montane Mojavean plant community. It deserves further study, if merely to ascertain how it relates to the piñon–juniper woodland surrounding or overlapping it. The gradual attenuation of the desert montane white fir forest westward through the Great Basin is worthy of study because of the implications it has for past vegetational history in the Great Basin (see Chapter 6).

Subalpine woodland has received somewhat more attention than montane woodland, but most of it has been devoted to the stands of bristlecone pine. Ecological information about transmontane woodlands of *Pinus flexilis* is rather limited and must be gleaned from such sources as Billings (1951), Lewis (1971), and Mooney (1973). Useful sylvic and distributional data for this species and bristlecone pine, as well as for white fir, have been made available by Fowells (1965), Critchfield and Allenbaugh (1969), Griffin and Critchfield (1972), and Little (1975), and in floras by Clokey (1951), Lloyd and Mitchell (1973), and DeDecker (1973).

In a series of related papers, Mooney et al. (1964, 1966, 1968) and Schulze et al. (1967) compared photosynthesis and transpiration rates of trees in the piñon and subalpine woodland zones of the White Mts. to corresponding values for woody desert plants below them and alpine plants above them. They showed that *Juniperus osteosperma, Pinus flexilis, P. longaeva, and P. monophylla* exhibited lower transpiration rates and less seasonal fluctuation than desert and alpine species. Photosynthetic rates were also low; net photosynthesis for *P. longaeva* at 20°C was only 1.2 mg CO_2 mg CO_2 g^{-1} hr^{-1}, compared to rates 9–20 times as high for desert shrubs *Artemisia tridentata* and *Hymnoclea salsola* and the alpine shrub *Artemisia arbuscula.* On an annual basis, *P. longaeva* shows very low net productivity. A high winter respiration rate produces a deficit of 140 mg CO_2 g^{-1}. The low summer photosynthesis rate means that 117 hr of peak photosynthesis is required to make up that deficit. With this sort of fine annual photosynthesis–respiration balance, it is not surprising that leaves may be retained for up to 30 yr.

Literature on Great Basin bristlecone pine includes papers by LaMarche and Mooney (1972), Billings and Thompson (1957), Mooney et al. (1962), Wright and Mooney (1965), LaMarche (1969), Bailey (1970), Mooney (1973), and Schmid and Schmid (1975).

The senior author acknowledges research support from the Academic Senate, Committee on Research, University of California, Riverside.

LITERATURE CITED

Adams, C. F. 1957. Plants of the Joshua Tree National Monument. Southwest. Monuments Assoc., Globe, Ariz. 44 p.

Aldon, E. F., and H. W. Springfield. 1973. The southwestern pinyon-juniper ecosystem: a bibliography. USDA Forest Serv. Gen. Tech. Rept. RM - 4. 20 p.

Applegate, E. I. 1938. Plants of the Lava Beds National Monument. Amer. Midl. Natur. 19:334-368.

Arnold, J. F. 1964. Zonation of understory vegetation around a juniper tree. J. Range Manage. 17:41-42.

Arnold, J. F., D. A. Jameson, and E. H. Reid. 1964. The pinyon-juniper type of Arizona: effects of grazing, fire and tree control. USDA For. Serv. Prod. Res. Rept. 84.

Bailey, D. K. 1970. Phytogeography and taxonomy of Pinus subsection Balfourianae. Ann. Missouri Bot. Gard. 57:210-249.

Beeson, C. D. 1974. The distribution and synecology of Great Basin pinyon-juniper. M.S. thesis, Univ. Nevada, Reno. 95 p.

Billings, W. D. 1951. Vegetation zonation in the Great Basin of western North America. In Les bases' ecologiques de la regeneration de le vegetation des zone arides. Internat. Colloq. Internat. Union Biol. Sci., Ser. B, 9:101-122.

-----. 1954. Temperature inversions in the pinyon-juniper zone of a Nevada mountain range. Butler Univ. Bot. Stud. XI:112-118.

Billings, W. D., and A. F. Mark. 1957. Factors involved in the persistence of montane treeless balds. Ecology 38:140-142.

Billings, W. D., and J. H. Thompson. 1957. Composition of a stand of old bristlecone pine in the White Mountains of California. Ecology 38:158-160.

Blackburn, W. H., and P. T. Tueller. 1970. Pinyon and juniper invasion in black sagebrush communities in east central Nevada. Ecology 51:841-848.

Bowerman, M. L. 1944. The flowering plants and ferns of Mount Diablo, California, Gillick Press, Berkeley, Calif. 290 p.

Bradley, W. G., and J. E. Deacon. 1967. The biotic communities of southern Nevada. Nevada State Mus. Anthropol. Papers 13:201-295.

Bradshaw, K. E., and J. L. Reveal. 1943 . Tree classification for Pinus monophylla and Juniperus utahensis. J. For. 41:100-104.

Burkhardt, J. W., and E. W. Tisdale. 1969. Nature and successional status of western juniper vegetation in Idaho. J. Range Manage. 22:264-270.

Burkhardt, J. W. and E. W. Tisdale. 1976 Causes of Juniper invasion in southwestern Idaho. Ecology 57: 472-484.

-----. 1976. Causes of juniper invasion in southwestern Idaho. Ecology 57:472-484.

Clokey, I. W. 1951. Flora of the Charleston Mountains, Clark County, Nevada. Univ. Calif. Press, Berkeley. 274 p.

Critchfield, W. B., and G. L. Allenbaugh. 1969. The distribution of Pinaceae in and near northern Nevada. Madroño 20:12-26

DeDecker, M. 1973. Plants of the Inyo Mountains of Inyo County, California (Unpub. mimeo). 28 p.

Driscoll, R. S. 1964a. Vegetation-soil units in the central Oregon juniper zone. USDA Forest Serv. Pac. NW Forest and Range Exp. Sta., Res. Paper PNW-19. 60 p.

-----. 1964b. A relict area in the central Oregon juniper zone. Ecology 45:345-353.

Eckert, R. E. 1957. Vegetation-soil relationships in some Artemisia types in northern Harney and Lake Counties, Oregon. Ph.D. dissertation, Oregon State Univ., Corvallis. 208 p.

Fowells, H. A. 1965. Silvics of forest trees of the United States. USDA Forest Serv. Agric. Handbook 271. 762 p.

Franklin, J. F., and C. T. Dryness. 1969. Vegetation of Oregon and Washington. USDA Forest Serv. Res. Paper PNW-80. 216 p.

Gillett, G. W., J. T. Howell, and H. Leschke. 1961. A flora of Lassen Volcanic National Park, California. Wasmann J. Biol. 19:1-185.

Glock, W. S. 1937. Observations on the western juniper. Madrono 4: 21-28.

Griffin, J. R., and W. B. Critchfield. 1972. Thedistribution of forest trees in California. USDA Forest Serv. Res. Paper PSW-82/ 1972. 114 p.

Haller, J. R. 1961. Some recent observations on ponderosa, Jeffrey and Washoe pines in northeastern California. Madrono 16:126-132.

Hamrick, J. L. 1966. Geographic variation in white fir. M.S. theiss, Univ. Calif., Berkeley. 143 p.

Hamrick, J. L., and W. J. Libby. 1972. Variation and selection in western U.S. montane species. I. White fir. Silvae Genet. 21: 29-35.

Harvey, H. T. 1951. Distributionof Juniperus californica in San Diego County with interpretations based on ecological factors. M.A. thesis, San Diego State Coll., San Diego, Calif.

Henrickson, J., and B. Prigge. 1975. White fir in the mountains of eastern Mojave Desert of California. Madroño 23:164-168.

Herman, F. R. 1953. A growth record of Utah juniper in Arizona. J. For. 51:200-201.

Hoover, R. F. 1970. The vascular plants of San Luis Obispo County, California. Univ. Calif. Press, Berkeley. 350 p.

Horton, J. S. 1960. Vegetation types of the San Bernardino Mountains. USDA Forest Serv., Pac. SW Forest and Range Exp. Sta., Tech. Paper 44. 29 p.

Keeler-Wolf, T., and V. Keeler-Wolf. 1973. A contribution to the natural history of the Yolla Bolly Mountains. Senior thesis, Univ. Calif., Santa Cruz. 607 p.

LaMarche, V. C., Jr. 1969. Environment in relation to age of bristlecone pines. Ecology 50:53-59.

LaMarche, V. C., Jr., and H. A. Mooney. 1972. Recent climatic change and development of the bristlecone pine krummholz zone, Mt. Washington, Nevada. Arctic Alpine Res. 4:61-72.

Lanner, R. M. 1974a. A new pine from Baja California and the hybrid origin of Pinus quadrifolia. Southwest. Natur. 19:75-95.

-----. 1974b. Natural hybridization between Pinus edulis and Pinus monophylla in the American Southwest. Silvae Genet. 23:108-116.

Lanner, R. M., and E. R. Hutchinson. 1972. Relict stands of pinyon hybrids in northern Utah. Great Basin Natur. 32:171-175.

Lewis, M. E. 1971. Flora and major plant communities of the Ruby-East Humboldt Mountains with special emphasis on Lamoille Canyon. U.S. Forest Serv., Humboldt Natl. Forest, Region 4. 62 p.

Little, E. L., Jr. 1975. Rare and local conifers in the United States. USDA Forest Serv. Conserv. Res. Rept. 19:1-25.

Lloyd, R. M., and R. S. Mitchell. 1973. A flora of the White Mountains, California and Nevada. Univ. Calif. Press, Berkeley. 202 p.

Love, R. M. 1970. The rangelands of the Western U.S. Sci. Amer. 222: 88-96.

Mallory, J. I., B. F. Smith, E. B. Alexander, and E. N. Gladish. 1965. Soils and vegetation of the Manton quadrangle. Calif. Div. Forestry, State Cooperative Soil-Vegetation Survey. 36 p.

Mehringer, P. J., Jr., and C. W. Ferguson. 1969. Pluvial occurrence of bristlecone pine (Pinus aristata) in a Mojave Desert mountain range. J. Ariz. Acad. Sci. 5:284-292.

Miller, A. H. 1940. A transition island in the Mohave Desert. Condor 42:161-163.

-----. 1946. Vertebrate inhabitants of the pinyon association in the Death Valley region. Ecology 27:54-60.

Miller, A. H., and R. C. Stebbins. 1964. Thelives of desert animals in Joshua Tree National Monument. Univ. Calif. Press, Berkeley. 452 p.

Milligan, M. T. 1969. A transect flora of the Eagle Peak Region, Warner Mountains. M.A. thesis, Humboldt State Coll., Arcata, Calif. 187 p.

Minnich, R. A. 1976. Vegetation of the San Bernardino Mountains, pp. 99-124. In J. Latting (ed.). Plant communities of southern California. Calif. Native Plant Soc. Spec. Pub. 2, Berkeley.

Mooney, H. A. 1973. Plant communities and vegetation, pp. 7-17. In Mountains, California and Nevada. Univ. Calif. Press, Berkeley.

Mooney, H. A., R. Brayton, and M. West. 1968. Transpiration intensity as related to vegetation zonation in the White Mountains of California. Amer. Midl. Natur. 80:407-412.

Mooney, H. A., G. St. Andre, and R. D. Wright. 1962. Alpine and subalpine vegetation patterns in the White Mountains of California. Amer. Midl. Natur. 68:257-273.

Mooney, H. A., M. West and R. Brayton. 1966. Field measurements of the metabolic responses of bristlecone pine and big sagebrush in the White Mountains of California. Bot. Gaz. 127:105-113.

Mooney, H. A., R. D. Wright, and B. R. Strain. 1964. The gas exchange capacity of plants in relation to vegetation zonation in the White Mountains of California. Amer. Midl. Natur. 72:281-297.

Munz, P. A. 1974. A flora of southern California. Univ. Calif. Press, Berkeley. 1086 p.

Munz, P. A., and D. D. Keck. 1949. California plant communities. Aliso 2:87-105.

-----. 1959. A California flora. Univ. Calif. Press, Berkeley. 1681 p.

Nord, E. C. 1965. Autecology of bitter brush in California. Ecol. Nomogr. 35:307-334.

Ornduff, R. and M. S. Cave. 1975. Geography of pollen and chromosomal heteromorphism in Leucocrinum montanum. Madroño 23:65-67.

Raven, P. H., and H. J. Thompson. 1966. A flora of the Santa Monica Mountains, California. Univ. Calif., Los Angeles, Student Bookstore (mimeo.)

Reveal, J. L. 1944. Single leaf pinyon and Utah juniper woodlands of western Nevada. J. For. 42:276-278.

Rickabaugh, H. 1975. Two lilies of Modoc County. Fremontia 3(1):15-17.

Schmid, R., and M. J. Schmid. 1975. Living links with the past. Nat. Hist. 84(3):38-45.

Schulman, E. 1954. Longevity under diversity in conifers. Science 119-396-3-9.

Schulze, E. D., H. A. Mooney, and E. L. Dunn. 1967. Wintertime photosynthesis of bristlecone pine in the White Mountains of California. Ecology 48:1044-1047.

Sharsmith, H. K. 1945. Flora of the Mount Hamilton Range of California. Amer. Midl. Natur. 34:289-267.

St. Andre, G., H. A. Mooney, and R. D. Wright. 1965. The pinyon woodland zone in the White Mountains of California. Amer. Midl. Natur. 73:225-239.

Taylor, D. W. 1976. Ecology of the timberline vegetation at Carson Pass, Alpine County, California. Ph.D. Dissertation, Univ. Calif., Davis (in prep.).

Thorne, R. F. 1976. California plant communities, pp. 1-31. In J. Latting (ed.). Plant communities of southern California. Calif. Native Plant Soc. Spec. Pub. 2, Berkeley.

Tueller, P. T., and J. E. Clark. 1975. Autecology of pinyon-juniper species of the Great Basin and Colorado Plateau, pp. 27-40. In Proc. pinyon-juniper ecosystem: a symposium. Utah State Univ., Logan.

Twisselman, E. C. 1956. A flora of the Temblor Range and the neighboring parts of the San Joaquin Valley. Wasmann J. Biol. 14:161-300.

-----. 1967. A flora of Kern County, California. Wasmann J. Biol. 25:1-395.

Vasek, F. C. 1966. The distribution and taxonomy of three western junipers. Brittonia 18:350-372.

Vasek, F. C., and J. F. Clovis. 1976. Growth forms in *Arctostaphylos glauca*. Amer. J. Bot. 63:189-195.

Vogl, R. J., and B. C. Miller, 1968. The vegetational composition of the south slope of Mt. Pinos, California. Madroño 19:225-234.

Wells, P. V., and R. Berger. 1967. Late Pleistocene history of the coniferous woodland in the Mojave Desert. Science 155:1640-1647.

West, N. E., D. R. Cain, and G. F. Gifford. 1973. Biology, ecology, and renewable resource management of the pygmy conifer woodlands of western North America: a bibliography. Utah Agric. Exp. Sta. Res. Rept. 12. 36 p.

West, N. E., K. H. Rea, and R. J. Tausch. 1975. Basic synecological relationships in pinyon-juniper woodland, pp. 41-53. In Proc. pinyon-juniper ecosystem: a symposium. Utah State Univ., Logan.

Wolf, C. G. 1938. California plant notes, Part II. Rancho Santa Ana Bot. Gard. Occas. Paper, Ser. 1(2):44-90.

Woodbury, A. M. 1947. Distribution of pigmy conifers in Utah and northeastern Arizona. Ecology 28:113-126.

Wright, R. D., and H. A. Mooney. 1965. Substrate-oriented distribution of bristlecone pine in the White Mountains of California. Amer. Midl. Natur. 73:257-284.

Zamora, B., and P. T. Tueller. 1973. *Artemisia arbuscula*, *A. longiloba* and *A. nova* habitat types in northern Nevada. Great Basin Natur. 33:225-242.

THE HOT DESERT FLORISTIC PROVINCE

CHAPTER

24 MOJAVE DESERT SCRUB VEGETATION

FRANK C. VASEK
Department of Biology, University of California, Riverside, California

MICHAEL G. BARBOUR
Botany Department, University of California, Davis, California

Introduction **836**
Creosote bush scrub **837**
Overview **837**
Regional details **841**
West, south, and east **841**
North **846**
Autecology of *Larrea* **848**
Germination and survival **849**
Phenology **849**
Water relations **849**
Rate of photosynthesis **849**
Other studies **850**
Saltbush scrub **850**
Synecology **850**
Autecology of saltbush scrub species **853**
Shadscale scrub **853**
Blackbush scrub **854**
Joshua tree woodland **857**
Annual vegetation **858**
Areas for future research **862**
Literature cited **863**

INTRODUCTION

The Mojave Desert Province is defined geologically in its western portion by the Garlock and San Andreas faults (Sharp 1972) and the associated mountain ranges, the Tehachapi and El Paso Mts. and the Transverse Ranges, respectively. Lines projected eastward from these two major faults are sometimes used to delimit the Mojave Desert in its eastern portions, but such definition is unsatisfactory from both geological and botanical considerations.

The geological characteristics of the Great Basin continue south through the Mojave Desert region, especially in the eastern portions (Oakeshot 1971). In addition, the vegetational features of the Mojave Desert extend northward to Big Pine in the Owens Valley and to Eureka Valley, thereby including Saline and Death valleys. We adopt the northern Mojave Desert boundary proposed by Shreve (1942), and accepted as the southern Great Basin boundary by Cronquist et al. (1972).

To the south, the San Bernardino, Little San Bernardino, Cottonwood, and Eagle Mts. form a fairly definite southern boundary. However, east of the Eagle Mts. a boundary can only be arbitrary. Especially near the Colorado River, many Sonoran species extend northward, for example, *Fouquieria splendens* and *Cercidium floridum* (Hastings et al. 1972), and the northern boundary of the Sonoran Desert is ill defined close to the Colorado River in the vicinity of Needles, California (Shreve and Wiggins 1964). Parish (1930) considered the Mojave Desert to extend south to the Chuckawalla Mts., but Shreve (1942) described typical Mojave Desert vegetation as extending south approximately to the northern Riverside Co. line. We adopt a zone, approximately following the Riverside–San Bernardino Co. line eastward from the Coxcomb Mts., as a southern boundary of convenience for the Mojave Desert.

Mojave Desert scrub vegetation types include all the shrub and woodland vegetation at elevations below coniferous woodland communities in the Mojave Desert. These types include saltbush scrub and creosote bush scrub at lower elevations and blackbush (blackbrush) scrub, shadscale scrub, and Joshua tree woodland at slightly higher or more northern locations (Munz and Keck 1949; Parish 1930; Shreve 1942; Shreve and Wiggins 1964). General descriptions of the creosote bush zone and the shadscale zone are given in Billings (1951).

Compared to the Sonoran Desert, the Mojave Desert exhibits a low diversity of perennial plants. Frequently, creosote bush (*Larrea tridentata*), alone or with a single associate, dominates Mojave Desert communities. *Larrea* and *Ambrosia dumosa* form the most characteristic association, dominating 70% of the Mojave Desert area, according to Shreve (1942). In the northern Mojave, Beatley (1969c) recognized six *Larrea*-dominated associations, the most widespread being *Larrea–Lycium andersonii–Grayia.*

CREOSOTE BUSH SCRUB

Overview

Creosote bush scrub occurs on well-drained sandy flats, bajadas, and upland slopes throughout much of the Mojave Desert. It is found below sea level in Death Valley, and up to elevations of 1500 m elsewhere. On washes, outcrops, and steep slopes, *Larrea* is present but not dominant, and it is excluded from dense soils in and around playas or other areas of high salt concentration (Wallace and Romney 1972). Lunt et al. (1973) showed that *Larrea* is relatively demanding of soil oxygen and may be excluded from fine-textured basin soils on this basis, rather than because of salinity. Alternatively, Beatley (1975) proposed cold air drainage as the most important playa feature. The northern limit of creosote bush scrub occurs in a broad zone of California, southern Nevada, and Utah, where basin elevations rise from 600–1000 m in the Mojave to 1500–1800 m in the Great Basin proper. Severity of winter frost is usually cited as the prime limiting factor here, but Beatley (1974b) has theorized that excessive rainfall during the winter period may be of key importance.

Although superficially the cresote bush scrub seems monotonously uniform throughout the Mojave Desert, there are significant local and regional differences in its composition. Diversity increases with topographical diversity and is influenced by relative community age (Johnson et al. 1975; Vasek et al. 1975a); it is also affected by the rockiness of the soil and annual precipitation. The contribution of *Larrea* to a number of Mojave sites is summarized in Table 24-1. Woodell et al. (1969) have shown that *Larrea* density generally increases with increasing annual precipitation, but Table 24-1 reveals that the relationship is not a close one. For example, Mercury Valley, Nevada, Rock Valley, Nevada, and Randsburg, California, all experience nearly the same rainfall, yet *Larrea* density ranges from 178 to 1003 shrubs per hectare. Note also the biomass differences between Mercury and Rock valleys. Clearly, other environmental factors are involved in determining shrub density.

In stable, old communities, creosote bushes or clones may attain ages of several thousand years (Vasek et al. 1975a) as a consequence of vegetative segmentation (Wallace and Romney 1972; Barbour 1969; Sternberg 1976). The 172 *Larrea* shrubs

TABLE 24-1. Density, cover, and biomass of *Larrea* and other perennial species for selected Mojave Desert sites. Authors of the data are indicated in footnotes. Some data are means of more than one sample. From Barbour et al. (1977)

Site	Annual Precipitation (mm)	*Larrea* Density (shrubs per hectare)	Other Shrub Density	*Larrea* Cover (%)	Total Cover (%)	(kg ha^{-1})
Death Valley[a]	40-60	285	1611	1.8	5.1	60
Death Valley[b]	40-60	393		6.2	14.0	
Stovepipe Wells[c]	40	60	47			
Inyokern[c]	91	262	7439			
Cantil[c]	100	207	744			
Emigrant Spring[c]	111	262	763			
Mojave[c]	128	495	138			
Frenchman Flat[d]	132	614		7.8	18.0	
Jackass Flat[d]	138	797		7.3	18.7	
N-80[b]	140	1790		4.0	11.0	
Haiwee[c]	143	299	5450			
Randsburg[c]	150	178	2795			
Mercury Valley[a]	155	695	10,293	2.9	19.2	299
Rock Valley[a]	162	1003	7941	4.3	20.1	122
Yucca Flat[d]	166	398		4.6	27.0	
Wildrose Canyon[c]	207	538	10,223			
Morongo Valley[c]	218	488	574			

[a] Wallace and Romney (1972).
[b] Barbour (1967, 1969).
[c] Woodell et al. (1969).
[d] Beatley (1974a).

in Table 24-3 include 21 separate shrubs plus 45 clones (Fig. 24-1). Each clone includes 2–9 satellite shrubs, and each satellite includes a number of stem crowns. Thus identification of an "individual" creosote bush is difficult and perhaps even arbitrary in mature communities. On actively eroding surfaces (e.g., rocky washes) or in young communities, creosote clones are relatively small or scarcely apparent. The consequences of cloning and community age have largely been ignored in considerations of population structure but would appear significant.

Barbour (1969) found that, within small stands, *Larrea* shrub height correlated with ring count at stem base, and he prepared frequency diagrams for shrub heights that he claimed were indicative of stand age structure. Figure 24-2 shows diagrams for five Mojave sites at increasing elevations along a bajada in Death Valley, California, and one site about 30 km north of Pahrump, Nevada. The Death Valley

Figure 24-1. Creosote bush scrub. (*a*) Community aspect (near Needles), with bajadas in background (*b*) Oblique aerial view (near Lucerne Valley) showing numerous "clonal creosote rings"

populations typically showed two or more peaks of size (age) and very few shrubs smaller than 40 cm, implying that mass germination and establishment is a rare but recurring event. Although these populations lack large numbers of young, they may not be senescent because of *Larrea's* long reproductive life span.

Shrubs in the southern Mojave are often widely spaced and appear to be regularly patterned (Went 1955). But in the northern Mojave, for example in the Nevada Test Site region, most shrubs are so closely associated with other species or other members of the same species that they occur in clumps with overlapping canopies. Table 24-2 summarizes association data from Nevada, but similar communities and

Figure 24-1 (continued) (*c*) Ground view of a clonal ring (near Lucerne Valley) with a bare area six meters long (*d*) New growth arising around the periphery of a creosote bush with a draught kelley top (near Pinto Basin).

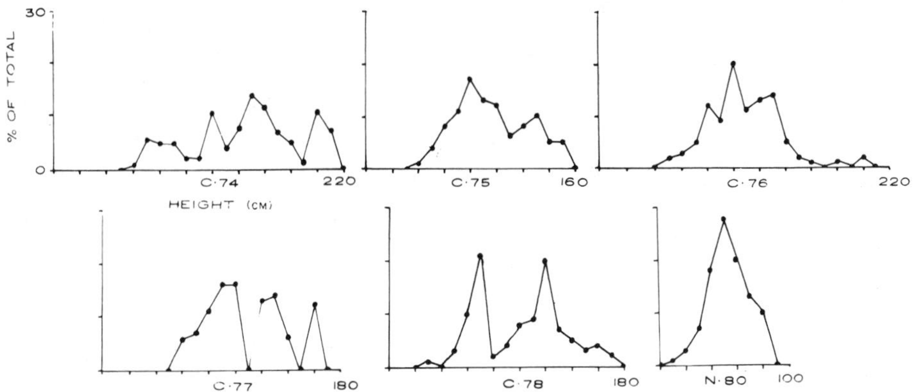

Figure 24-2. Height (age) frequency diagrams for Mojave Desert *Larrea* populations. One hundred shrubs, selected by a restricted random method, are summarized in each graph. From Barbour (1967) and Barbour et al. (1977).

patterns might be expected in California. *Larrea* does not appear to exhibit an unusual number of positive or negative associations, in comparison to other species. Even in more arid portions of the Mojave, *Larrea* shrubs may show a clumped or random, rather than a regular, pattern (Woodell et al. 1969; Anderson 1971; Barbour 1973, 1977; King and Woodell 1973).

The density and the pattern of *Larrea* have been attributed to differential seedling survival because of competition for moisture near established shrubs or alleopathic substances given off by established shrubs (Went 1955; Sheps 1973). In addition, Adams et al. (1970) have shown that a hydrophobic soil layer beneath *Larrea* shrubs may prevent seedling establishment.

Little has been written about plant succession in Mojave Desert vegetation. Shreve (1942) claimed that successions do not take place in desert vegetation. Nevertheless, destructive impacts on Mojave Desert vegetation, followed by community change, have been described by Beatley (1965, 1966), Davidson and Fox (1974), Stebbins (1974), Vasek et al. (1975a,b), and Wallace and Romney (1972). Sometimes winter annual species grow more abundantly and vigorously after shrub destruction (Beatley 1966). In other cases, short-lived perennials such as *Acamptopappus sphaerocephalus, Hymenoclea salsola, Encelia frutescens,* and *Stephanomeria pauciflora* increase rapidly in population size when long-lived species such as *Larrea tridentata, Ambrosia dumosa,* and *Ephedra nevadensis* are reduced in number by disturbance (Davidson and Fox 1974; Vasek et al. 1975a,b; Vasek, unpub. data).

Regional Details

West, south, and east. In the western Mojave Desert, *Larrea* attains a position of strong dominance on stable alluvial fans such as a sandy bajada 16 km northeast of Lucerne Valley, where erosion is minimal (Table 24-3). A dozen other perennial species were also recorded on the same sandy bajada, including long-lived perennials of the creosote bush scrub and other desert communities, and short-lived perennials

TABLE 24-2. Degree of association for all possible pairs of 28 perennial species at Mercury, Nevada. The nonsignificant interactions for each species are not tallied. From Wallace and Romney (1972)

Species	Number of Other Species Positively Associated with	Number of Other Species Negatively Associated with
Acamptopappus shockleyi	3	8
Atriplex confertifolia	8	1
Cacti (mixed spp.)	13	0
Coleogyne ramosissima	1	0
Dalea fremontii	0	2
Encelia virginensis	3	0
Ephedra funerea	14	0
E. nevadensis	14	0
Eurotia lanata	13	4
Franseria (*Ambrosia*) *dumosa*	17	2
Grayia spinosa	12	1
Hilaria rigida	3	2
Hymenoclea salsola	3	0
Krameria parvifolia	13	1
Larrea tridentata	12	1
Lepidium fremontii	6	0
Lycium andersonii	19	1
Machaeranthera tortifolia	13	0
Menodora spinescens	3	0
Oryzopsis hymenoides	10	0
Prunus fasciculata	3	0
Psilostrophe cooperi	1	0
Salazaria mexicana	11	0
Sphaeralcea ambigua	10	1
Stephanomeria pauciflora	4	0
Stipa speciosa	6	0
Thamnosma montana	1	0
Yucca schidigera	13	0

TABLE 24-3. Vegetational characteristics of an old creosote bush scrub community on a stable sandy bajada south of East Ord Mt., San Bernardino Co., California. Data by Johnson and Vasek, April 15, 1974, based on 20 circular plots of 100 m^2 each; density is based on 2000 m^2; IV = importance value; t = <0.01% cover

Taxon	Cover (%)	Density	Frequency	IV
Long-lived species				
Larrea tridentata	8.84	172	20	141.6
Ambrosia dumosa	0.85	107	19	55.6
Ephedra nevadensis	0.69	28	16	30.1
Thamnosma montana	0.38	20	10	19.0
Yucca schidigera	0.71	13	8	17.8
Salazaria mexicana	0.07	5	4	6.1
Stipa speciosa	0.03	5	4	5.7
Opuntia ramosissima	0.10	2	2	3.5
O. echinocarpa	t	1	1	1.4
Oryzopsis hymenoides	t	1	1	1.3
Short-lived species				
Hymenoclea salsola	0.08	9	5	8.3
Dyssodia cooperi	t	8	3	5.3
Acamptopappus sphaerocephalus	0.03	7	2	4.3
Total	11.78	378		

common on natural disturbances such as small washes; but their cover relative to that of *Larrea* was low.

North of the Lucerne Valley bajada and south of East Ord Mt., in more varied terrain, greater heterogeneity is apparent in the vegetation. Total perennial cover ranged from about 7 to 27% in 10 belt transects of 200 m^2 each taken by Vasek et al. (1975b). The proportion of that cover provided by *Larrea* ranged from 19 to 98%, and by *Ambrosia* from 0 to 18%. The total fraction of cover by all long-lived species was greatest on stable bajadas, alluvial slopes, and silty washes and was considerably less on more actively eroding gravelly and rocky washes. In addition to *Larrea* and *Ambrosia,* important long-lived species included *Atriplex canescens, A. polycarpa, Krameria parvifolia, Lycium andersonii, L. cooperi,* and *Opuntia ramosissima.* Minor additions to cover were provided by *Cassia armata, Encelia farinosa, Ephedra californica, Yucca schidigera,* etc. The primary short-lived pioneer, found mostly in small washes, was *Hymenoclea salsola.* Similar but slightly greater variation was observed along a 33 km transect with somewhat greater topographical diversity, north of Lucerne Valley (Vasek et al. 1975a).

In the southern Mojave Desert, creosote bush scrub at Indian Cove in Joshua Tree National Monument (NA 331070) includes *Hilaria rigida* as a dominant cover element among the conspicuous shrubs, such as *Larrea, Ambrosia dumosa,* and *Tetradymia axillaris* (Chew and Butterworth 1964). An annotated checklist of the plant species was compiled by Adams (1957), and a map of creosote bush scrub distribution in Joshua Tree National Monument was prepared by Miller and Stebbins (1964). The latter is accompanied by photographs and brief descriptions of the major plant zones. An open woodland of Joshua trees and California junipers occurs above the creosote bush scrub in what the authors call a "yucca zone." Slopes within the yucca zone support a mixture of shrubs such as *Purshia glandulosa, Coleogyne ramosissima,* and *Salazaria mexicana,* similar to the blackbush scrub vegetation described in a later section of this chapter. Shefi (1971) located some 15 sample plots along an elevational gradient from this *Coleogyne* zone, at about 850 m, all along bajada slopes down to Pinto Basin at 500 m. Absolute community cover decreased with declining elevation, from 7 to 1%, but *Larrea* consistently remained dominant, with *Ambrosia dumosa* and *Opuntia ramosissima* the most common associates. The average relative cover of *Larrea* was 33%, and the average *Larrea* shrub density was 226 per hectare.

In the southeastern Mojave Desert, vegetational heterogeneity is apparent in topographically diverse areas. For example, vegetation in the Bowman Canyon area of the Whipple Mts. (NA 362345; Table 24-4) was different on each of six transects. Although regional vegetation is creosote bush scrub, *Larrea* occurred on only four transects and had the highest importance value on only one transect. *Ambrosia dumosa* occurred on only two transects. *Acacia greggii* and *Cercidium floridum* assume major importance in washes, and *Opuntia bigelovii* apparently prefers level, stable, rocky sites. A steep (unstable?) boulder-strewn slope (transect II) had the greatest number of species. Most of the plants in the Whipple Mt. vegetation also occur widely in the Mojave Desert. However, the Whipple Mts. are also the northwestern outpost of some Sonoran Desert species, such as *Carnegiea gigantea* and *Cercidium microphyllum,* and hence are somewhat transitional between the two deserts (see Brum 1973).

TABLE 24-4. Importance values for perennial plants and total crown canopy cover for six areas in Bowman Caynon, Whipple Mts., San Bernardino Co., California. Data by Vasek, November 10, 1973, based on 50 m line intercepts; importance values are the sum of relative cover, relative density, and relative frequency, the latter based on 10 m subtransects

	Transect					
Taxon	I Rocky Slope	II Boulder Slope	III Level Terrace	IV Sandy Wash	V Rocky Terrace	VI Rocky Wash
Acacia greggii				110.6		114.4
Ambrosia dumosa	114.8	20.0				
Asclepias subulata		17.0				
Cercidium floridum						133.4
Encelia farinosa	121.7	161.2	176.0	51.9		
Ephedra viridis		19.4				
Hilaria rigida		15.3				
Krameria grayi		22.9				
Larrea tridentata	63.4		99.5	82.5	185.4	
Lycium fremontii				54.9		
Mirabilis bigelovii		18.2				
Opuntia acanthocarpa					59.2	52.2
O. bigelovii			24.5		55.4	
Peucephyllum schottii		26.4				
Total cover (%)	9.4	17.1	15.5	13.4	5.2	13.1

Another phase of the creosote bush scrub community is exemplified by a large stand of jumping cholla, *Opuntia bigelovii*, in the Sacramento Mts. (NA 361905) of the eastern Mojave Desert near Needles. This vegetation is dominated by *Ambrosia dumosa, O. bigelovii*, and *Larrea*, but some 18 other perennials also occur on the stable, flat, rocky benches and in the intermittent small, sandy drainageways or occasional larger washes. *Ambrosia dumosa* and *O. bigelovii* grow on stable, rocky substrates and also on deeply disturbed but level substrates. *Larrea, Krameria, Cassia, Salazaria*, and several cacti prefer undisturbed sites and usually decrease in relative abundance with soil disturbance. *Stephanomeria pauciflora, Encelia frutescens*, and *Acamptopappus sphaerocephalus* maintain low numbers in small natural disturbances within the generally mature community and increase rapidly on disturbed soils.

At Gold Valley, east of the Providence Mts., Garcia-Moya and McKell (1970) studied nitrogen economy, focusing on *Acacia greggii* and therefore on a desert wash phase of the creosote bush scrub community. The vegetation under study showed a total shrub cover of 19.7%, provided mostly by *Acacia, Cassia armata, Hymenoclea salsola, Larrea, Salazaria mexicana*, and *Thamnosma montana*. Lesser amounts of cover were provided by *Ephedra nevadensis, Eriogonum fasciculatum, Krameria* (two species), and *Opuntia echinocarpa*, and least by *Ambrosia dumosa* and *Brickellia incana*. The wash phase of the vegetation is emphasized by the presence of *Acacia, Brickellia*, and *Hymenoclea*, whereas the other species occur mostly in various other phases of creosote bush scrub. The distribution of creosote bush scrub in the general Providence Mt. area is shown on a map included in Johnson et al. (1948).

North. In the northern Mojave Desert, the general plant ecology of Death Valley has been described by Hunt (1966). This study concentrated on vegetation of the alluvial fans and bajadas and of salt flats, where a vegetational gradient is apparent. Several phreatophytes tolerant of salt concentration greater than 2% occur within a zone of available ground water (see section entitled "Saltbush Scrub"). A series of xerophytic communities occurs on the fans and slopes. A mixture of *Atriplex hymenolytra* or *A. polycarpa* with *Larrea tridentata* occurs just above the zone of available ground water. Next comes a zone of essentially pure *Larrea*, followed by *Larrea* mixed with *Tidestromia, Eriogonum*, and *Eucnide*. In turn, a mixture of *Larrea, Ambrosia dumosa*, and *Encelia farinosa* occurs high on the fans. *Larrea* extends up fans and slopes to about the lower snow line, which approximates the isoclimatic line for 6 consecutive days of freezing (Shreve 1942). *Larrea* is therefore probably limited by low temperatures (see also Beatley 1974b). At low elevations in Death Valley, it is probably limited by salt concentrations of about 2% or perhaps slightly less (Hunt 1966; Barbour 1969). Plant distributions in Death Valley are shown on a map prepared by Hunt.

In Death Valley, *Larrea* occurs in cobbly or rocky soil and evidently has a taproot. The latter observation suggests that the plants are young, as would be expected on active, unstable fans and bajadas. In contrast, the ancient creosote clonal rings (Vasek et al. 1975a) on stable sandy bajadas near Lucerne Valley or on stable rocky slopes in the Sacramento Mts. have spreading root systems. The taprooted seedlings probably grow, vegetatively reproduce, die, and decay in perhaps 500–1000 yr, leaving peripheral clones.

Still farther north and east, Mojave Desert vegetation and Great Basin Desert vegetation have a transitional zone in southern Nevada (Beatley 1974a,b, 1975; Rickard and Beatley 1968). *Larrea* comprises variable but usually major fractions of Mojave Desert communities on bajadas and fans in that region. Great Basin communities above, and sometimes below, creosote bush scrub are dominated by *Atriplex canescens, A. confertifolia, Artemisia tridentata,* or *A. arbuscula.* Transitional communities are dominated by (1) *Larrea* with low % cover, (2) *Coleogyne,* (3) *Larrea* and *Coleogyne,* (4) *Lycium,* (5) *Grayia* and *Lycium,* or (6) *Coleogyne* and *Artemisia.*

A similar transition area in Saline Valley, California, was surveyed by Randall (1972). Species presence and canopy cover were noted in circular quadrats at 46 locations along the north-facing bajada and in Grapevine Canyon at sites ranging in elevation from 600 to 1900 m. Randall attempted to ordinate community composition against several factors. Figure 24-3 shows a principal components ordination, using presence data alone. The horseshoe shape of the array correlates with elevation, which decreases in a clockwise direction from left to right. Most of the species did not respond monotonically to elevation. In general, the stands appear to segregate into three groups, with some stands intermediate.

Group I, at higher elevations, includes *Artemisia tridentata, Coleogyne ramosissima, Ephedra viridis, Haplopappus linearifolius,* and *Purshia glandulosa.* In addition, *Eriogonum fasciculatum* was present in all stands, even though not a dominant. *Artemisia, Ephedra,* and *Purshia* were exclusive to this group, as were the less abundant species *Chrysothamnus nauseosus, C. viscidiflorus, Ceanothus greggii, Eriogonum microthecum, E. umbellatum, Lupinus excubitus, Opuntia erinacea, Prunus fasciculata, Ribes velutinum,* and *Symphoricarpos longiflorus.*

Group II, at middle elevations, includes *Coleogyne ramosissima, Ephedra nevadensis, Grayia spinosa,* and *Salazaria mexicana* as dominants. In addition,

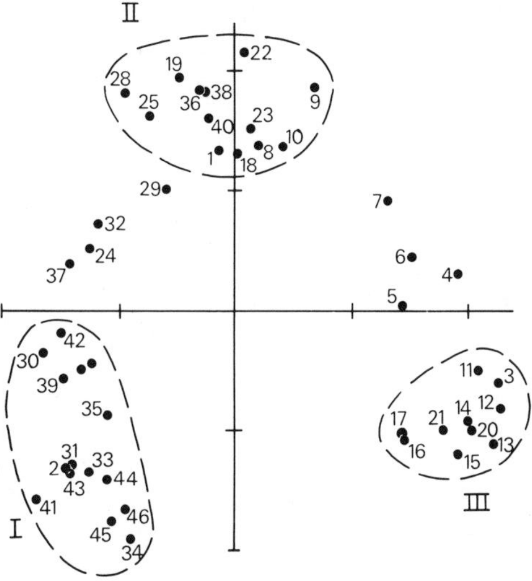

Figure 24-3. Principal components ordination of 46 stands in Saline Valley, California. Based on presence data only. From Randall (1972).

Eriogonum fasciculatum, Haplopappus cooperi, Hymenoclea salsola, Opuntia basilaris, and *Thamnosma montana* were present in 80% or more of the stands. *Haplopappus, Hymenoclea, Salazaria,* and *Eurotia lanata* were exclusive to this group.

Group III, at lower elevations, includes *Larrea tridentata* and *Ambrosia dumosa.* In addition, *Dalea fremontii, Amphipappus fremontii, Eriogonum inflatum,* and *Mirabilis bigelovii* var. *retrorsa* were present in 70% or more of the stands. *Amphipappus, Eriogonum, Larrea, Atriplex polycarpa,* and *Lycium pallidum* were exclusive to this group.

Randall found that total cover was correlated with elevation. The line in Fig. 24-4 is his estimation of the potential cover that could have been supported on the sites. He assumed carrying capacity to be reached when variance in cover from plot to plot in an elevational band was random (approached the Poisson distribution). Randall's concept that homogeneity of cover correlates with the habitat's carrying capacity being reached is a plausable, though unproven, hypothesis. Most sites are below their theoretical carrying capacity, especially those along the lower bajada. Randall did not believe that the reduced cover was due to herbivory, especially burro activity. It may instead reflect a catastrophic loss of cover in the recent past.

Contingency table analysis, using presence–absence, revealed that the species in any one segment of the transect were randomly distributed and did not show positive or negative associations. On the other hand, covariance analysis on cover data did in general reveal competitive interactions, "in the sense that an increase in the abundance of one species tends to be accompanied by decreases in others, and vice versa." This level of competition, Randall concluded, is nonspecific.

Randall attempted to display the stands along other gradients in addition to elevation, such as mean daily solar radiation, but he concluded that the principal components ordination was the most efficient method of summarizing the vegetation.

Autecology of *Larrea*

Because of the International Biological Program, many autecological studies of *Larrea* and other dominants have been carried out. Unfortunately for this book, that

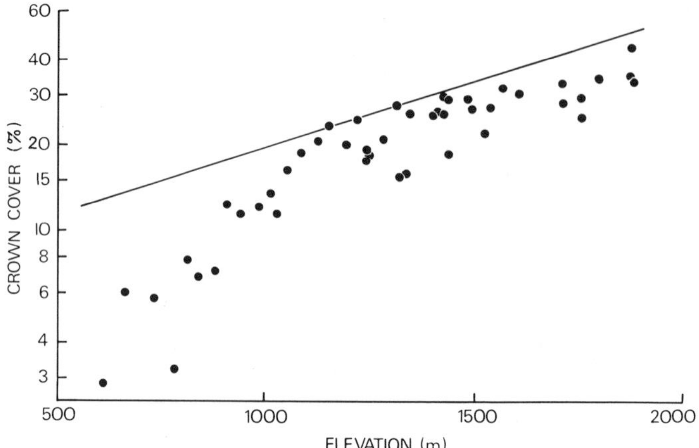

Figure 24-4. Plant cover as a function of elevation in Saline Valley, California. The line is an estimation of maximum potential cover (carrying capacity) along an elevational gradient. From Randall (1972).

work was largely done in southern Nevada, and it is not yet clear whether the lessons learned are applicable throughout the Mojave Desert. The autecological literature that centers on *Larrea* has been reviewed by Barbour et al. (1977), and this section borrows heavily from that review.

Germination and survival. Field studies of *Larrea* in the Mojave are not extensive. Went and Westergaard (1949) and Sheps (1973) both noted natural germination in Death Valley and correlated the event with temperature and precipitation. Sheps additionally sowed seeds and found survival to be lowest close to established *Larrea* shrubs. From laboratory studies, Barbour (1967, 1968) concluded that germination requirements for temperature, moisture, pH, salinity, and light were remarkably moderate and that germination in nature might be a rare event. Other Mojave studies (Beatley, pers. commun; Graves et al. 1975; Wallace and Romney 1972) show slightly different results, but none has shown allelopathy to be a significant factor in the laboratory, despite indications to the contrary from the field.

Phenology. A 6 yr phenological record has been compiled from Rock Valley, Nevada, by Ackerman and Bamberg (1974). Their data show that leaf and twig production is greatest between March and October, but otherwise production rate correlates poorly with rainfall and may be affected more by temperature. From their data and those of Bamberg et al. (1973) and Beatley (1974a), it appears that onset of leaf production and flowering in the spring requires 20-25 mm rain.

Flower buds develop in March and early April in Rock Valley, when soil moisture at rooting depth (15–30 cm) is at a maximum of about −10 bar. Flowers appear as soil moisture falls, and fruit begins to mature in May and June when soil moisture is −50 bar. A second flowering, in late summer, occurred at Rock Valley in 2 of the 6 yr, but it was not correlated with either summer or annual rainfall amounts. Fruit from both spring and late summer flowers reaches maturity from August through October. More arid sites may show a shorter growing season and an absence of any second flowering. Stark and Love (1969) reported that *Larrea* begins growth in Death Valley in March, flowers in April, and terminates growth in June.

Water relations. Transpiration rates of several species growing *in situ* in Rock Valley, Nevada, were measured throughout a year by Bamberg et al. (1973); *Larrea* clearly was the most efficient in restricting transpirational loss of water during dry periods. When sufficient moisture is available, however, *Larrea* transpires at rates comparable to those of mesophytes (Wallace and Romney 1972).

During May and June, in Rock Valley, *Larrea* water potential ranges between −51 and −65 bar, values similar to those for the evergreen *Krameria parvifolia* and for the deciduous *Lycium andersonii* and *L. pallidum* (Bamberg et al. 1973). Despite such low water potentials, *Larrea* showed positive net photosynthesis; the other three species were dormant. Water use efficiency varies between 0.3 and 1.5 mg CO_2 incorporated per milligram H_2O lost, depending on temperature and moisture status of the plant. These efficiencies are approximately equivalent to those of *Ambrosia dumosa, Lycium andersonii,* and *L. pallidum* when in leaf, but lower than those of another evergreen, *Krameria parvifolia.*

Rate of photosynthesis. Wallace et al. (1971) examined 14 perennials from the Mojave–Great Basin transition area for their ratio of C_3:C_4 photosynthetic enzymes.

They reported most taxa to be in the C_3 range, three (*Atriplex hymenelytra, Larrea tridentata, Ephedra nevadensis*) to be midway between C_3 and C_4, and three others (*Atriplex canescens, A. confertifolia, Salsola peetifera*) to be well into the C_4 range. A subsequent laboratory study by Barbour et al. (1974), however, failed to produce corroborating proof that *Larrea* was anything other than a typical C_3 species.

Although leaf temperatures of 42–46°C have been reported for Mojave *Larrea* (Wallace and Romney 1972), temperature optima for photosynthesis may be considerably lower, depending on soil moisture and preconditioning (Bamberg et al. 1973). Generally, net photosynthesis values for *Larrea* fit well into the range of values summarized by Mooney (1972a,b) for semiarid, evergreen, sclerophyllous shrubs.

Other studies. Strategies of gas exchange in C_3, C_4, and CAM plants of the southern California deserts were discussed in detail by Johnson (1975, 1976). Sixteen shrub species and winter annuals representative of 17 genera were found to have the C_3 type of anatomy. The C_4 type of anatomy was found in several shrubby species of *Atriplex* and in eight species of summer annuals, including species of *Pectis, Allionia, Amaranthus, Boerhaavia, Bouteloua,* and *Euphorbia.* All succulent perennials tested, including seven species of *Cactaceae,* plus *Agave deserti, Yucca brevifolia,* and *Y. schidigera,* were CAM plants. High carbon assimilation efficiencies and water use efficiencies are based on both physiological refinements and phenological adjustments. Physiological refinements are especially evident in C_4 and CAM pathways, where CO_2 assimilation approaches the theoretical maximum. CAM systems also recycle CO_2, thus reducing dependence on external CO_2, which can be acquired only at the expense of transpiration. The less physiologically efficient C_3 plants maintain favorable transpiration ratios by growing during the cooler part of the year. CAM plants adjust phenologically to the diurnal cycle by carrying on gas exchange during the cooler night.

Other aspects of photosynthetic rate and gas exchange were considered by Odening et al. (1974), Strain (1975), and Bamberg et al. (1975). Other aspects of general ecology that have been reported include nitrogen economy (Garcia-Moya and McKell 1970), shrub utilization (McKell 1975), and root biomass (Wallace et al. 1974).

SALTBUSH SCRUB

Synecology

Saltbush scrub vegetation is here interpreted as equivalent to the alkali sink community of Munz and Keck (1949). Saltbush scrub has a xerophytic phase on dry soils and a halophytic phase on soils with available ground water and usually with high concentrations of salt or alkali (Hunt 1966; Twisselmann 1967). The xerophytic saltbush scrub includes many of the species characteristic of the shadscale scrub, and the two communities probably have a close relationship which needs further study. However, shadscale scrub is considered separately in a later section of this chapter.

Important species of xerophytic saltbush scrub, such as *Atriplex polycarpa, A. confertifolia,* and *A. hymenolytra,* have limited salt tolerance. This vegetation usually occurs in basins and valleys throughout the Mojave Desert region but sometimes is found on slopes (e.g., west of Inyokern, south of Tecopa, and the Sacramento Mts. south of Camino) and also occurs in Lower Sonoran grassland (Twisselmann 1967) at the west edge of the San Joaquin Valley (from the Panoche Hills area southward). Extensive stands of *Atriplex confertifolia* occur with *A. polycarpa* and other species on the open rolling terrain south and west of Fremont Peak and toward Kramer Junction.

The halophytic saltbush scrub occurs on playas, in sinks, and near seeps with available surface or ground water high in mineral content. Its specialized habitats are topographically defined and scattered throughout the Mojave Desert region and in the San Joaquin Valley, as well as through the deserts of North America. Important species are succulent chenopods: *Allenrolfea occidentalis, Nitrophila occidentalis, Salicornia subterminalis, Suaeda* spp., and *Sarcobatus vermiculatis.* The halophytic saltbush scrub also bears many similarities to a nonshrubby alkali meadow community (Thorne 1976).

Gradients between halophytic and xerophytic scrub often occur around dry lakes. The salt pan or playa of a dry lake usually is salt encrusted and devoid of plants. Toward the edge of the playa, intermittent patches of halophytes occur on soil mounds raised a few centimeters or decimeters above playa level. Further away from the playa center, the halophyte patches increase in size and density, are joined by plants of xerophytic salt scrub species, and gradually are replaced by the latter. In turn, the xerophytic saltbush scrub is gradually replaced by creosote bush scrub or whatever community occurs in the region around the dry lake. The distribution of saltbush scrub communities in the Mojave Desert therefore depends on soil and microtopographical characteristics, as was also shown for saltbush communities in Utah (Mitchell et al. 1966; West and Ibrahim 1968).

In the western Mojave Desert, much of Lucerne Dry Lake has been converted to agriculture, but nearby Rabbit Dry Lake has an open salt pan surrounded by patches of halophytes (Fig. 24-5). *Kochia californica* commonly occurs on the playa

Figure 24-5. Halophytic saltbush scrub on Rabbit Dry Lake, San Bernardino Co., Calif. *Kochia americana, Suaeda torreyana, Atriplex lentiformis,* and *A. confertifolia* are most prominent.

or salt pan and also on small mounds of fine-grained soil a few centimeters above the level of the playa. *Suaeda fruticosa* and *S. torreyana* also occur on the salt pan but are more common on small to medium-sized mounds, on which also occur *Atriplex lentiformis* and *A. confertifolia*. Large mounds may reach about 1 m high and may cover irregular areas up to several dozen square meters, particularly when two or more mounds become confluent. Large mounds usually are occupied by *Atriplex polycarpa, A. canescens,* and *Haplopappus acradenius* ssp. *acradenius*. The correlation of particular species occurrences with mound size suggests a successional pattern in which mounds grow by the accumulation of material (organic debris, salt, fine soil), and composition of the plant cover shifts with changing soil conditions. However, detailed studies of mound development and concomitant plant succession have not been made to our knowledge.

Around both Rabbit and Lucerne dry lakes, the halophytic saltbush scrub grades into a xerophytic saltbush scrub of *Atriplex confertifolia, A. polycarpa, A. canescens,* and sometimes *Artemisia spinescens*. Still farther from the dry lakes, the saltbush scrub grades into a creosote bush scrub or, especially east of Rabbit Dry Lake, into a sparse Joshua tree woodland (Fig. 24-6).

A slightly different pattern exists at Soggy Dry Lake (a few kilometers east of Lucerne Valley), where nearly pure stands of *Atriplex polycarpa* extend down to the open salt pan. This pattern suggests rather mild salinity levels around the relatively small, dry lake.

In Death Valley, Hunt (1966) described a series of nine phreatophytes, with salt tolerances ranging sequentially from *Allenrolfea* (tolerant to 6% salt), through *Juncus cooperi, Distichlis spicata, Suaeda fruticosa, Tamarix* ssp., *Sporobilis airoides, Atriplex canescens,* and *Pluchea sericea,* to *Prosopis juliflora* (tolerant to 2% salt). Above the zone of available ground water, where salinity drops below 2%, a xerophytic saltbush scrub of *Atriplex hymenolytra* and *A. polycarpa* grades into a creosote bush scrub.

Figure 24-6. Xerophytic saltbush scrub at the east edge of Rabbit Dry Lake, San Bernardino Co., California. *Atriplex polycarpa, A. confertifolia,* and *A. canescens* are most abundant with an occasional Joshua tree (*Yucca brevifolia*) in evidence.

The vegetation around Saratoga Springs in southern Death Valley includes a salt marsh complex, four phreatophytic communities, and three xerophytic shrub communities (Bradley 1970). Water and salt concentrations clearly control relative community distributions. A sparse scrub vegetation of *Larrea* and *Atriplex polycarpa* occurs in dry soils with little salt. A slight increase in salinity is associated with a saltbush scrub of *Atriplex hymenolytra, A. parryi, Tidestromia oblongifolia,* and *Suaeda fruticosa.* With greater salinity, a xerophytic saltbush scrub develops with *Atriplex parryi, Tidestromia, Suaeda,* and *Distichlis.* Then four halophytic communities develop over soils with available ground water. These communities array over a range of fairly strong salt concentrations and include various combinations of the following species: *Suaeda fruticosa, Allenrolfea occidentalis, Distichlis spicata, Atriplex canescens, Nitrophila occidentalis,* and *Prosopis juliflora,* among others. The salt marsh complex includes a peripheral phase with *Distichlis, Nitrophila,* and *Juncus cooperi,* plus an aquatic bulrush phase with *Scirpus olneyi* and others.

Autecology of Saltbush Scrub Species

A comprehensive review of world literature for *Atriplex* has been compiled by Franclet and LeHouerou (1971), and another appears to be in preparation by Osmond (pers. commun.). More pertinent to the Mojave Desert, Bjorkman and his colleagues studied photosynthesis rates, temperature tolerance, and growth patterns of species of *Atriplex, Distichlis, Tidestromia,* and *Typha* from Death Valley. In particular, photosynthetic rates were studied in *Atriplex lentiformis* (Pearcy and Harrison 1974) and *A. hymenolytra* (Pearcy et al. 1974), and photosynthetic adaptation to high temperatures in *Tidestromia* were examined by Bjorkman et al. (1972). The latter also established a reciprocal transplant garden at Bodega Head, California. This work has been reviewed in the *Carnegie Institution Yearbook* (e.g., Mooney et al. 1974; Bjorkman et al. 1974). The physiologies of salt stress and salt tolerance were considered by Goodin and Mozafar (1972) and by Mozafar and Goodin (1970).

The germination requirements, salt tolerance, and water relations of *Atriplex polycarpa* have also been examined, but mainly from non-Mojave locations (Chatterton and McKell 1969; Chatterton et al. 1969, 1971; Sankary and Barbour 1972). Graves et al. (1975) investigated techniques to stimulate germination of Mojave *Atriplex canescens, A. polycarpa,* and *Hymenoclea salsola,* among others, for revegetation work. However, we still have a limited understanding of saltbush scrub autecology.

SHADSCALE SCRUB

Shadscale scrub vegetation was considered by Billings (1949) to form a distinct zone between sagebrush scrub and creosote bush scrub. It is a community of low, more or less spinescent, microphyllous shrubs of uniform physiognomy that occurs in western and southern Nevada and also in portions of eastern California. To the north, it is commonly found in broad desert valleys of the Great Basin region in distinctly alkaline soils. However, to the south, it occurs on steep mountain slopes, usually with heavy, rocky soils. Thus shadscale scrub vegetation occurs in drier areas than does sagebrush scrub, and the greater aridity is determined either by physiological–edaphic or by climatic factors.

In California, shadscale scrub is common in Owens Valley and continues north of Bishop in valleys and on low slopes. It also occurs within the Mojave Desert area on rocky slopes, in the Inyo Range and the Panamint Range (e.g., Townes Pass), and is very common in the Black and Funeral ranges east of Death Valley (e.g., the eastern reaches of Titus Canyon).

Shadscale scrub is dominated by *Atriplex confertifolia* and *Artemisia spinescens*, which occur in 97 and 83%, respectively, of all the stands examined by Billings (1949). In addition, *Sarcobatus baileyi* provides a significant fraction of the shrub cover in about two-thirds of the stands studied in detail by Billings (Table 24-5). Total shrub cover in those 12 stands ranged from 4.05 to 11.86% with an average of 6.95%. Density averaged 16.8 shrubs per 10 m^2 plot.

The species of shadscale scrub range geographically far beyond the area recognized by Billings as having a shadscale scrub vegetation. Nevertheless, Billings interpreted shadscale as a zonal vegetation on the basis that both floristic and physiognomical discontinuity between sagebrush scrub and shadscale scrub is far greater than any variation within either. Shadscale scrub adjoins Great Basin sagebrush scrub over much of Nevada. Shadscale continues eastward into Utah, where it has been studied by West and Ibrahim (1968) and by Ibraham et al. (1972). Along the southern border of the shadscale distribution discussed by Billings (1949), Joshua trees occasionally form open groves in the shadscale matrix. In other areas, especially in southeastern Nevada, shadscale vegetation seems to be replaced by a blackbush (*Coleogyne ramosissima*) community, which Billings also considered to be intermediate between sagebrush and creosote bush scrub.

BLACKBUSH SCRUB

In southern Nevada, a blackbush community (Great Basin authors use "blackbrush" as a common name for *Coleogyne ramosissima*, whereas California authors seem to prefer "blackbush") was considered by Bradley and Deacon (1967) to be a zonal desert scrub community. It is dominated by blackbush and interspersed with other desert shrubs such as *Yucca baccata*, *Thamnosma montana*, *Grayia spinosa*, *Eurotia lanata*, *Artemisia spinescens*, *Agave utahensis*, or species of *Ephedra*, *Dalea*, *Atriplex*, *Tetradymia*, *Eriogonum*, and *Lycium*. Sometimes Joshua trees form a conspicuous overstory. The blackbush community is widespread from elevations of about 1200 to over 1800 m on upper bajadas or old alluvium overlain with rocks. The climate is definitely colder than that in areas occupied by creosote bush, and snow is common for short periods during the winter. Some 185 species of vascular plants have been observed in the blackbush community, of which 68% also occur in creosote bush scrub and 46% occur in the "juniper–piñon" community. Only 9.2% of the species are restricted to the blackbush community. In contrast, 46.5% of the species in the creosote bush scrub of southern Nevada are limited to that community.

In the western Mojave Desert, vegetation above the creosote bush scrub commonly consists of either a California juniper woodland (see Chapter 23), a Joshua

TABLE 24-5. Relative shrub cover (%), total absolute shrub cover (%), and total shrub density per 10 m^2 in 12 shadscale stands in Nevada and eastern California. Adapted from Billings (1949). The high cover by "others" in stand 12 was due mostly to *Ambrosia dumosa* and *Kochia americana*

Taxon	Stand											
	1	2	3	4	5	6	7	8	9	10	11	12
Atriplex confertifolis	11.1	42.4	18.8	43.5	18.9	98.3	63.5	15.2	23.5	69.0	79.2	51.9
Artemisia spinescens	45.2	39.2	12.7	...	22.9	1.7	...	30.7	2.3	29.5	19.8	3.7
Sarcobatus baileyi	38.9	11.2	54.4	56.5	58.0	...	0.9	53.5	68.0	...	...	...
Lycium cooperi	2.4	...	12.1	...	0.2	...	27.8	...	...	...	1.0	...
Ephedra nevadensis	...	5.6	...	...	...	...	6.1	0.4	...	0.8	...	...
Eurotia lanata	1.6	0.8	...	...	...	...	...	0.2	3.9	...	...	...
Others	0.8	0.8	2.0	...	...	...	1.7	...	2.3	0.7	...	44.4
Total absolute cover (%)	6.30	6.25	5.96	7.64	11.86	8.90	5.75	10.02	5.13	6.45	5.05	4.05
Total density per 10 m^2	26.5	30.2	10.0	7.2	28.7	15.0	15.0	16.0	5.3	24.5	13.5	9.1

Figure 24-7. Blackbush scrub overlooking Lucerne Valley (San Bernardino Co., California), here intermixed with sparse California juniper woodland and an occasional Joshua tree.

tree woodland (see the next section), a vegetation of low dark shrubs, or some combination of these three (Fig. 24-7).

The vegetation of low dark shrubs bears a physiognomical resemblance to the blackbush community and therefore is considered a phase of blackbush scrub, even though *Coleogyne* may be absent. Considerable regional and local variation in composition is evident.

In the Covington Flat area of Joshua Tree National Monument, a blackbush scrub community includes *Coleogyne, Ephedra viridis, Eriogonum fasciculatum, Juniperus californica, Opuntia echinocarpa, Purshia glandulosa, Salazaria mexicana,* and some *Yucca brevifolia* and *Pinus monophylla* (Kornoelje 1973).

On East Ord Mt., north of Lucerne Valley, at about 1800 m, the vegetation conspicuously includes plants such as *Chrysothamnus teretifolius, Eurotia lanata, Ephedra nevadensis, E. viridis, Grayia spinosa,* and *Hilaria rigida.* On nearby Rodman Mt., at about the same elevation, blackbush occurs and is joined by *Ephedra viridis, Eriogonum fasciculatum, Haplopappus cooperi, Opuntia ramosissima, Salazaria mexicana,* and *Stipa speciosa.* A similar vegetation, dominated by *Eriogonum fasciculatum,* occurs on steep slopes of the southern Sierra Nevada between creosote bush scrub and piñon woodland (Twisselmann 1967). Considerable variation in composition is evident. Although Twisselmann listed about 20 shrubs as typical of this "arid shrub association," he stated that few of them are widespread within it. Nevertheless *Eurotia lanata, Coleogyne ramosissima, Salazaria mexicana, Salvia dorii, Tetradymia axillaris, Yucca brevifolia,* etc., occur sporadically with *Eriogonum fasciculatum* in a vegetation of distinctive appearance. In the southwestern corner of Inyo Co. (Nine-Mile Canyon), groves of Joshua trees seem edaphically separated from a "blackbush" scrub of *Eriogonum fasciculatum, Salazaria mexicana, Haplopappus cooperi,* and so on. With various shifts in composition, especially an increase in *Coleogyne,* this vegetation of low dark shrubs continues northward on steep slopes below the piñon woodland to Sherwin Grade (Mono Co.) north of Bishop, where it merges with shadscale scrub.

JOSHUA TREE WOODLAND

The Joshua tree woodland is a desert scrub vegetation consisting of a Joshua tree overstory (*Yucca brevifolia)* and an understory of various shrubs and perennial herbs. The Joshua tree is an extremely conspicuous plant and therefore lends a unique appearance to any vegetation in which it occurs. Limited to the Mojave Desert, it occurs at elevations ranging between the creosote bush scrub and piñon and juniper woodlands. The range of the Joshua tree pretty well outlines the extent of the Mojave Desert in California, southern Nevada, and northwestern Arizona (Jaeger 1957). However, Joshua trees do extend southward in western Arizona to southwestern Yavapai Co. almost to the vicinity of Wickenburg (Benson and Darrow 1954). The unique and sometimes bizarre growth forms of the Joshua tree have conferred upon it, along with the saguaro, a special status as a symbol of the desert. Within *Yucca brevifolia* a variety *jaegeriana* (McKelvy 1938) occurs in the eastern Mojave Desert (Fig. 24-8), and a form *herbertii,* characterized by rhizomatous clumps, is found along the western margin of the Mojave Desert and on the slopes of the Tehachapi Mts. and the southern Sierra Nevada (Webber 1953). The species continues north to Owens Valley and south and east to the San Bernardino Mts., where it occurs intermittently along the lower fringes of the piñon woodland and the upper fringes of the creosote bush scrub. It extends along the southern margin of the Mojave Desert to the Little San Bernardino Mts. In the eastern Mojave Desert extensive stands of *Yucca brevifolia* var. *jaegeriana* occur on Cima Dome and also east of

Figure 24-8. Joshua tree woodland near Cima, San Bernardino Co., showing *Yucca brevifolia* var. *jaegeriana* with an understory of *Y. baccata, Salazaria mexicana,* etc. Note the dense compact growth form of the tree in center.

the New York Mts. The general natural history, range, growth characteristics, and taxonomy of *Yucca*, including *Yucca brevifolia*, are presented in a monograph by Webber (1953).

Although the Joshua tree is considered well known (Jaeger 1957), we know of no quantitative stand data published for any Joshua tree woodland. A qualitative description of Joshua tree woodland in the Joshua Tree National Monument is provided by Miller and Stebbins (1964).

Relative to other desert vegetation Joshua tree woodland occurs in the same elevational zone as blackbush scrub and shadscale scrub. But the Joshua tree woodland usually occurs on loose soils and gentle substrates (Webber 1953), whereas blackbush scrub and shadscale scrub are often found on heavy or rocky soils. Nevertheless, some of the same species occur in the three communities. Gradations occur toward creosote bush scrub at low elevations, where *Yucca brevifolia* may be in codominance with *Larrea tridentata*. Gradations at higher elevations occur with piñon or juniper woodland, where *Y. brevifolia* may occur as a codominant with *Juniperus californica* or *J. osteosperma* and *Pinus monophylla*.

At intermediate elevations in the eastern Mojave, dense stands of *Yucca brevifolia* occur with a low shrub understory (Table 24-6). Despite the great conspicuousness of the Joshua tree, it has a rather low importance value as compared with *Hilaria rigida, Haplopappus cooperi, Salazaria mexicana, Hymenoclea salsola, Ephedra nevadensis, Muhlenbergia porteri, Opuntia ramosissima, Yucca baccata,* and so on. The estimated density of Joshua trees in the sample area for Table 24-6 was 125 trees per hectare, but an independent census of the entire plot showed 104 trees per hectare. Average Joshua tree height was 2.24 m, and average stem diameter 30 cm above the ground was 18.75 cm. On the basis of visual judgment, the density is somewhat greater slightly farther to the north near Kessler Spring, where Webber (1953) illustrated what he considered to be "undoubtedly the largest concentration of *Yucca* in the United States."

Joshua tree woodlands tend to occur on sandy, loamy, or fine gravelly soils, usually on fairly gentle slopes. The gentle terrain is also conducive to cattle-raising activities, and most Joshua tree woodlands that we have observed have been subjected to moderate or severe grazing pressure. In particular, clumps of *Stipa speciosa* and *Hilaria rigida* observed near Cima (Table 24-6) were heavily grazed, as were some of the shrubs. Many of the understory perennials attained large size as a consequence of cloning, and some underwent vegetative segmentation. Examples are *Ephedra nevadensis, Hilaria rigida, Menodora spinescens, Opuntia ramosissima, Salazaria mexicana,* and *Yucca baccata*. Some of the short-lived plants such as *Hymenoclea salsola* may have been present in fairly important numbers as a consequence of grazing activity.

Joshua tree woodland has also suffered from land clearing around new residential developments, such as California City in Kern Co.

ANNUAL VEGETATION

Considerable variation in winter annual vegetation is apparent regionally, locally, and seasonally. Patterns of distribution were described by Went (1942) in which certain species tended to occur under shrubs and other species tended to occur in intershrub areas. Regional differences in species association patterns were also

TABLE 24-6. Vegetational characteristics of a Joshua tree woodland community north of Cima, San Bernardino Co., California. Data by Vasek and Rowlands, September 14, 1975, based on 40 circular plots of 10 m^2 each. Density is based on plants per 400 m^2; cover is in percent, with t = <0.01%, cover; IV = importance value

Taxon	Density	Cover (%)	Frequency	IV
Coryphantha vivipara	1	t	1	0.59
Echinocereus engelmannii	5	0.01	5	3.05
Ephedra nevadensis	18	0.49	14	12.23
Eurotia lanata	23	0.43	13	11.99
Haplopappus cooperi	164	1.61	38	49.06
H. linearifolius	6	0.12	4	3.42
Hilaria rigida	488	2.72	40	96.92
Hymenoclea salsola	12	0.86	11	12.69
Lycium andersonii	5	0.56	5	6.94
L. cooperi	9	0.48	8	8.22
Menodora spinescens	12	0.68	8	10.00
Mirabilis bigelovii	1	0.01	1	0.64
Muhlenbergia porteri	28	0.10	18	12.57
Opuntia acanthocarpa	1	t	1	0.59
O. echinocarpa	1	t	1	0.59
O. ramosissima	9	1.08	8	12.48
Salazaria mexicana	21	2.33	15	26.15
Salvia dorii	2	0.07	1	1.21
Stipa speciosa	6	0.01	5	3.11
Thommosma montana	10	0.68	8	9.80
Yucca baccata	3	1.64	3	13.38
Y. brevifolia	5	0.20	5	4.36
Total	830	14.08		

explicitly indicated by Went. Some of the possible explanations for the occurrence of annuals under some shrubs but not under others have been explored by Muller (1953), Muller and Muller (1956), and Adams et al. (1970). Rickard and Beatley (1968) noted that sites dominated by *Larrea* had more annual species and greater annual plant cover than any of eight other shrub-dominated communities in southern Nevada.

The composition of the annual vegetation differs from year to year, depending on the time and amount of rainfall, as suggested by Shreve (1942) and inferred from the seed germination patterns described by Went (1948, 1949) and Tevis (1958a,b). Beatley catalogued differences in density of annual plants over a 5 yr period (1969a) and variations in annual plant biomass over 3 yr (1969b). Later she also correlated general phenology of annuals with climatic events (Beatley 1967, 1974a). The paragraph below is drawn from her work in southern Nevada.

Winter annuals germinate after autumn or winter rains in excess of 15 mm, and the life span extends to late May, when drought and high temperatures cause plant death. This 5–8 mo span certainly takes them out of the "ephemeral" category. The winter-hardy plants (at least down to -18°C) grow slowly until March or April. Then rapidly increasing soil and air temperatures lead to completion of vegetative growth and onset of flowering and fruiting by May. Depending mainly on rainfall, winter annual density may approach 1000 per square meter (but more typically less than 100 per square meter), a cover of 30% and a biomass of over 600 kg ha^{-1}. Herbivory accounts for some plant loss but is insignificant compared to mortality from drought stress in the top 25 cm of soil. Most winter annuals do not survive to maturity.

The occurrence of particular constellations of annual species on different substrates and surfaces can readily be observed at several scales. The preference of certain species for intershrub areas has already been mentioned. The preference of, for example, *Oenothera deltoides* for sandy areas and of *Chorizanthe rigida* for stony soils or mosaics (Munz 1974) is well known and may occur on a very small scale. On a larger, but still local scale, three different annual plant communities have been noted on three substrate types in the Whipple Mts. of eastern San Bernardino Co. (Table 24-7). A large, rocky slope, heterogeneous in surface conformation, supported 18 species with an average density of 55 plants per square meter. A more uniform, more open slope with stony soil had only 9 species, but small plants, such as *Plantago insularis,* were abundant and total plant density reached 190 per square meter. A sandy wash and adjacent stony soils supported 16 species at a total density of about 164 plants per square meter. The large rocky slope also supported a rich shrub (and suffrutescent perennial) vegetation of some 37 species, of which the most abundant (in order) were *Encelia farinosa, Ambrosia dumosa, Ephedra viridis, Opuntia bigelovii, Thamnosma montana, Dalea emoryi, Hibiscus denudatus, Larrea tridentata, Ferrococtus acanthodes,* and *Krameria grayi.*

A similar local difference in annual vegetation between sandy bajada and a rocky slope occurs in the western Mojave Desert (Johnson 1976). Furthermore, the spatial heterogeneity of annual communities seemed greater than that of perennial communities. In addition to the winter annuals discussed above, a few species of summer annuals germinate and grow in response to summer rain. Summer annuals grow in the intershrub areas, and all have a C_4 carbon pathway in contrast to the winter annuals, which have a C_3 pathway. Partial suppression of winter annuals was evident after growth of summer annuals (*Pectis papposa*), but the mechanism has yet to be explained (Johnson 1976). On a sandy bajada, the average density of winter annuals was 39.8 plants per square meter on a plot irrigated the previous summer and 52.1 plants per square meter on the adjacent nonirrigated area. In the Nevada Test Site, the summer annual flora consists of only seven species, though in the middle elevations they may occur in large numbers (84 plants per square meter, 8% cover: Beat-

TABLE 24-7. Relative density (*D*) and relative frequency (*F*) of annual species in three sample areas of the Whipple Mts. I, variable rocky slope northwest of Monument Peak; II, low, open, stony slope west of Monument Peak; III, sandy wash and gravelly flat in Gene Wash. Data by Vasek, May 1973, utilizing random circular plots of 0.1 m^2

	I		II		III	
	D	*F*	*D*	*F*	*D*	*F*
Atrichoseris platyphylla	0.5	0.9				
Calycoseris parryi					0.3	1.5
Camissonia brevipes			0.5	3.3	3.2	10.4
C. chamaenerioides					2.0	6.0
Chaenactis carphoclinia	6.4	6.5	6.3	23.3	1.2	4.5
Chorizanthe brevicornu	6.4	8.4	5.8	6.7	0.6	3.0
Cryptantha maritima	10.0	9.3	0.5	3.3		
Eriogonum trichopes	3.2	3.7	2.1	10.0	4.1	7.5
Gilia scopulorum	7.3	8.4			4.1	7.5
Langloisia punctata	2.7	0.9				
Lepidium lasiocarpum	0.5	0.9			0.3	1.5
Lotus salsuginosus	38.6	25.2	6.3	16.7	5.8	16.4
Lupinus arizonicus					1.2	3.0
Lygodesmia exigua	0.5	0.9				
Mentzelia involucrata	0.9	0.9				
Perityle emoryi	3.6	5.6			0.3	1.5
Phacelia crenulata	6.4	10.3				
P. distans			6.3	3.3	3.5	6.0
Plagiobathrys jonesii	4.5	8.4	0.5	3.3	0.3	1.5
Plantago insularis	5.5	2.8	71.6	30.0	69.8	16.4
Salvia columbariae					2.3	7.5
Schismus barbatus					0.6	3.0
Stylocline micropoides	0.9	1.9				
Trichoptilium incisum	0.5	0.9				
Unknown	1.8	3.7			0.6	3.0
Total number observed	220		190		344	
Number of sample plots	40		10		21	
Density per square meter	55		190		163.8	

ley 1974b). These germinate in August or September after heavy rains, remain small and inconspicuous, and flower and fruit until autumn frosts kill them. Unlike the winter annuals, their life span is measured in terms of weeks.

Very recently, Mulroy and Rundel (1977)[1] compiled a useful summary of 130 common species of winter and summer annuals of the Mojave and Sonoran Deserts. The summary lists type of leaf anatomy (Kranz or normal), flowering period, and leaf arrangement (rosette, cauline, prostrate, and combinations). Of 63 summer annuals, 42 (67%) exhibited Kranz anatomy and were judged to be C_4 taxa, whereas none of the 64 winter annuals had Kranz anatomy. Three species which were active in both winter and summer (*Aristida adscensionis, Euphorbia florida, E. gracillima*) were C_4 taxa. Most summer annuals exhibited entire, cauline leaves, but most winter annuals had dissected leaves which initially were restricted to basal rosette.

AREAS FOR FUTURE RESEARCH

The description of vegetation often ends with rough edges, left over vegetation units, and disagreement in interpretation. In the present discussion of Mojave Desert scrub vegetation, creosote bush scrub was broadly interpreted to include several special phases that sometimes are recognized as separate vegetational units. For example, Thorne (1976) lists desert rock plants, desert dune sand, stem-succulent scrub, semisucculent scrub, and desert microphyll woodland as separate recognizable communities. Several of those probably have greater currency in the Colorado Desert than in the Mojave.

Several major sand dune systems also occur in the Mojave Desert (Miller and Stebbins 1964). Perhaps a separate sand dune vegetation should be recognized, but we know of no quantitative analysis upon which to base a decision. For this chapter, therefore, we have interpreted sand dune vegetation as a special phase of creosote bush scrub.

Outstanding deficiencies in our knowledge of Mojave Desert vegetation occur in the area of basic quantitative description and evaluation. Quantitative assessment similar to that employed in Utah by Singh and West (1971) and Ibrahim et al. (1972), would permit rational consideration of several vegetational questions. The relationship of southern Mojave Desert xerophytic saltbush scrub to Great Basin shadscale scrub, the relationship of various transitional blackbush scrub vegetation types in and around the Mojave Desert, and the vegetational transition between the Mojave and Colorado (Sonoran) deserts could all come under critical review. With an understanding of current vegetation, consideration of the evolution of vegetation types, floras, and autecological strategies would have a sound basis.

The senior author acknowledges research support from the Academic Senate, Committee on Research, University of California, Riverside.

[1] This reference was seen when our chapter was already in proof, consequently the citation appears here: Mulroy, T. W., and P. W. Rundel. 1977. Annual plants: adaptations to desert environments. BioScience 27:109–114.

LITERATURE CITED

Ackerman, T. L., and S. A. Bamberg. 1974. Phenology studies in the Mojave Desert at Rock Valley Nevada Test Site, Nevada. Presented at Phenology and Seasonality Symp., AIBS, Minneapolis, Minn. 21 p. (mimeo).

Adams, C. F. 1957. Plants of the Joshua Tree National Monument. Southwestern Monuments Assoc., Globe, Ariz. 44 p.

Adams, S., B. R. Strain, and M. S. Adams. 1970. Water-repellent soils, fire and annual plant cover in a desert scrub community of southeastern California. Ecology 51:696-700.

Anderson, D. J. 1971. Pattern in desert perennials. J. Ecol. 59:555-560.

Bamberg, S. A., G. E. Kleinkopf, A. Wallace, and A. Vollmer. 1975. Comparative photosynthetic production of Mojave Desert shrubs. Ecology 56:732-736.

Bamberg, S., A. Wallace, G. Kelinkopf, and A. Vollmer. 1973. Plant productivity and nutrient interrelationships of perennials in the Mojave Desert, pp. 1-52. In Res. Memo. 73-10, Desert Biome Project, IBP, 2.3.1.3.

Barbour, M. G. 1967. Ecoclinal patterns in the physiological ecology of a desert shrub, _Larrea divaricata_. Ph.D. dissertation, Duke Univ., Durham, N. C. 242 p.

-----. 1968. Germination requirements of the desert shrub _Larrea divaricata_. Ecology 49:915-923.

-----. 1969. Age and space distribution of the desert shrub _Larrea divaricata_. Ecology 50:679-685.

-----. 1973. Desert dogma reexamined: Root/shoot production and plant spacing. Amer. Midl. Natur. 89:41-57.

-----. 1977. Plant-plant interactions, Vol. 4, Chapter 6. In R. A. Perry and D. Goodall (eds.). Structure, function, and management of arid land ecosystems. Academic Press, New York (in press).

Barbour, M. G., G. Cunningham, W. C. Oechel, and S. A. Bamberg. 1977. Growth and development, Chapter 4. In T. J. Mabry and J. H. Hunziker (eds.). _Larrea_ and its role in desert ecosystems. Dowden, Hutchison and Ross, Stroudsburg, Pa. (in press).

Barbour, M. G., D. V. Diaz, and R. W. Breidenbach. 1974. Contributions to the biology of _Larrea_ species. Ecology 55:1199-1215.

Beatley, J. C. 1965. Ecology of the Nevada Test Site. IV. Effects of the sedan detonations on desert shrub vegetation in northeastern Yucca Flat. UCLA 12-571. Univ. of Calif., Los Angeles, School of Med. 55 p

-----. 1966. Winter annual vegetation following a nuclear detonation in the northern Mojave Desert (Nevada test site). Radiation Bot. 6:69-82.

-----. 1967. Survival of winter annuals in the northern Mojave Desert. Ecology 48:745-750.

-----. 1969a. Dependence of desert rodents on winter annuals and precipitation. Ecology 50:721-724.

-----. 1969b. Biomass of desert winter annual plant populations in southern Nevada. Oikos 20:261-263.

-----. 1969c. Vascular plants of the Nevada Test Site, Nellis Air Force Range, Ash Meadows, UCLA 12-705. Univ. Calif., Los Angeles, Lab. Nucl. Med. Radiat. Biol. 122 p.

-----. 1974a. Phenological events and their environmental triggers in Mojave Desert ecosystems. Ecology 55:856-863.

-----. 1974b. Effects of rainfall and temperature on the distribution and behavior of Larrea tridentata (creosote bush) in the Mojave Desert of Nevada. Ecology 55:245-261.

-----. 1975. Climates and vegetation pattern across the Mojave/Great Basin Desert transition of southern Nevada. Amer. Midl. Natur. 93:53-70.

Benson, L., and R. A. Darrow. 1954. The trees and shrubs of the southwestern deserts. Univ. Ariz. Press, Tucson. 437 pp.

Billings, W. D. 1949. The shadscale vegetation zone of Nevada and eastern California in relation to climate and soils. Amer. Midl. Natur. 42:87-109.

-----. 1951. Vegetation zonation in the Great Basin of western North America. In: Les Bases ecologique de la régénération de la végétation des zones arides. Internat. Colloq. Internat. Union Biol. Sci., Ser. B, 9:101-122.

Bjorkman, O., M. Nobs, H. A. Mooney, J. Berry, F. Nicholson, and W. Ward. 1974. Growth responses of plants from habitats with contrasting thermal environments: transplant studies in the Death Valley and the Bodega Head Experimental Gardens. Carnegie Inst. Yearbook 73:748-757.

Bjorkman, O., R. W. Pearcy, A. T. Harrison, and H. A. Mooney. 1972. Photosynthetic adaptation to high temperatures: a field study in Death Valley, California. Science 172:786-789.

Bradley, W. G. 1970. The vegetation of Saratoga Springs, Death Valley National Monument. Southwest Natur. 15:111-129.

Bradley, W. G., and J. E. Deacon. 1967. The biotic communities of southern Nevada. Nevada State Mus. Anthropol. Paper 13, Part 4.

Brum, G. D. 1973. Ecology of the saguaro (Carnegiea gigantea): phenology and establishment in marginal populations. Madroño 22:195-204.

Chatterton, N. J., J. R. Goodin, and C. Duncan. 1971. Nitrogen metabolism in Atriplex polycarpa as affected by substrate nitrogen and NaCl salinity. Agron. J. 63:271-274.

Chatterton, N. J., and C. M. McKell. 1969. Atriplex polycarpa. I. Germination and growth as affected by sodium chloride in water cultures. Agron. J. 61:448-450.

Chatterton, N. J., C. M. McKell, J. R. Goodin and F. T. Bingham. 1969. Atriplex polycarpa. II. Germination and growth in water cultures containing high levels of boron. Agron. J. 61:451-453.

Chew, R. M., and B. B. Butterworth. 1964. Ecology of rodents in Indian Cove (Mojave Desert), Joshua Tree National Monument, California. J. Mammol. 45:203-226.

Cronquist, A., A. H. Holmgren, N. H. Holmgren, and J. L. Reveal. 1972. Intermountain flora: vascular plants of the intermountain west, U.S.A. Hafner, New York. 270 pp.

Davidson, E., and M. Fox. 1974. Effects of off-road motorcycle activity on Mojave Desert vegetation and soil. Madroño 22:381-390.

Franclet, A., and H. LeHouerou. 1971. The Atriplex in Tunisia and North Africa. FA: SF/Tun 11, Tech. Rept. 7. Food and Agriculture Organization of United Nations. 271 pp.

Garcia-Moya, E., and C. M. McKell. 1970. Contribution of shrubs to the nitrogen economy of a desert wash plant community. Ecology 51:81-88.

Goodin, J. R., and A. Mozafar. 1972. Physiology of salinity stress, pp. 255-259. In C. M. McKell, J. P. Blaisdell, and J. R. Goodin (eds.), Wildlife shrubs--their biology and utilization. USDA Forest Ser. Gen. Tech. Rept. Int.-1. Washington, D.C.

Graves, W. L., B. L. Kay, and W. A. Williams. 1975. Seed treatment of Mojave Desert shrubs. Agron. J. 67:773-777.

Hastings, J. R., R. M. Turner, and D. K. Warren. 1972. An atlas of some plant distributions in the Sonoran Desert. Univ. Ariz., Inst. Atmospheric Phys., Tech. Rept. 21. 255 pp.

Hunt, D. B. 1966. Plant ecology of Death Valley, California. Geol. Survey Prof. Paper 509. Washington, D.C.

Ibrahim, K. M., N. E. West, and D. L. Goodwin. 1972. Phytosociological characteristics of perennial Atriplex-dominated vegetation of southeastern Utah. Vegetatio 24:13-22.

Jaeger, E. C. 1957. The North American deserts. Stanford Univ. Press, Stanford, Calif. 308 pp.

Johnson, D. H., M. D. Bryant, and A. H. Miller. 1948. Vertebrate animals of the Providence Mountain area of California. Univ. Calif. Pub. Zool. 48:221-375.

Johnson, H. B. 1975. Gas-exchange strategies in desert plants, pp. 105-120. In D. M. Gates and R. B. Schmerl (eds.). Perspectives in biophysical ecology. Springer-Verlag, New York.

-----. 1976. Vegetation and plant communities of Southern California deserts--a functional view, pp. 125-164. In J. Latting (ed.). Symp. Proc. plant communities of southern California. Spec. Pub. No. 2, Calif. Native Plant Soc., Berkeley.

Johnson, H. B., F. C. Vasek, and T. Yonkers. 1975. Productivity, diversity and stability relationships in Mojave Desert roadside vegetation. Bull. Torrey Bot. Club 102:106-115.

King, T. J., and S. R. J. Woodell. 1973. The causes of regular pattern in desert perennials. J. Ecol. 61:761-765.

Kornoelje, T. A. 1973. Plant communities of the Covington area, Joshua Tree National Monument. M.A. thesis, Calif. State Univ., Long Beach.

Lunt, O. R., J. Lety, and S. B. Clark. 1973. Oxygen requirements for root growth in three species of desert shrubs. Ecology 54:1356-1362.

McKell, C. M. 1975. Shrubs--a neglected resource of arid lands. Science 187:803-809.

McKelvey, Susan D. 1938. Yuccas of the southwestern United States. Part I. Arnold Arboretum, Harvard Univ., Jamaica Plain, Mass. 150 p.

Miller, A. H., and R. C. Stebbins. 1964. The lives of desert animals in Joshua Tree National Monument. Univ. Calif. Press, Berkeley. 452 p.

Mitchell, J. E., N. E. West, and R. W. Miller. 1966. Soil physical properties in relation to plant community patterns in the shadscale zone of northwestern Utah. Ecology 47:627-630.

Mooney, H. A. 1972a. The carbon balance of plants. Ann. Rev. Ecol. System. 3:315-346.

-----. 1972b. Carbon dioxide exchange of plants in natural environments. Bot. Rev. 38:455-469.

Mooney, H. A., O. Bjorkman, and J. Troughton. 1974. Seasonal changes in the leaf characteristics of the desert shrub Atriplex hymenolytra. Carnegie Inst. Yearbook 73:846-852.

Mozafar, A., and J. R. Goodin. 1970. Vesiculated hairs: a mechanism for salt tolerance in Atriplex halimus L. Plant Phys. 45:62-65.

Muller, C. H. 1953. The association of desert annuals with shrubs. Amer. J. Bot. 40:53-60.

Muller, W. H., and C. H. Muller. 1956. Association patterns involving desert plants that contain toxic products. Amer. J. Bot. 43:354-361.

Munz, P. A. 1974. A flora of southern California. Univ. Calif. Press, Berkeley. 1086 pp.
Munz, P. A., and D. D. Keck. 1949. California plant communities. Aliso 2:87-105.
Oakeshot, G. B. 1971. California's changing landscapes: a guide to the geology of the state. McGraw-Hill, New York. 388 p.
Odening, W. R., B. R. Strain, and W. C. Oechel. 1974. The effect of decreasing water potential on net CO_2 exchange of intact desert shrubs. Ecology 55:1086-1095.
Parish, S. B. 1930. Vegetation of the Mojave and Colorado Deserts of southern California. Ecology 11:481-499.
Pearcy, R. W., and A. T. Harrison. 1974. Comparative photosynthetic and respiratory gas exchange characteristics of *Atriplex lentiformis* (Torry.) in coastal and desert habitats. Ecology 55:1104-1111.
Pearcy, R. W., A. T. Harrison, H. A. Mooney, and O. Bjorkman. 1974. Seasonal changes in net photosynthesis of *Atriplex hymenolytra* shrubs growing in Death Valley, California. Oecologia 17:111-124.
Randall, D. C. 1972. An analysis of some desert shrub vegetation of Saline Valley, California. Ph.D. dissertation, Univ. Calif., Davis. 186 pp.
Rickard, W. H., and J. C. Beatley. 1968. Canopy coverage of the desert shrub vegetation mosaic of the Nevada Test Site. Ecology 46:524-529.
Sankary, M. N., and M. G. Barbour. 1972. Autecology of *Atriplex polycarpa* from California. Ecology 53:1155-1162.
Sharp, R. P. 1972. Geology field guide to southern California. Wm. C. Brown, Dubuque, Iowa. 181 pp.
Shefi, R. M. 1971. Plant communities of the Pinto Basin, Joshua Tree National Monument. M.A. thesis, Calif. State Coll., Long Beach 133 p.
Sheps, L. O. 1973. Survival of *Larrea tridentata* S & M seedlings in Death Valley National Monument, California. Israel J. Bot. 22:8-17.
Shreve, F. 1942. The desert vegetation of North America. Bot. Rev. 8:195-246.
Shreve, F., and I. L. Wiggins. 1964. Vegetation and flora of the Sonoran Desert. 2 vols. Stanford Univ. Press, Stanford, Calif.
Singh, T., and N. E. West. 1971. Comparison of some multivariate analyses of perennial *Atriplex* vegetation in southeastern Utah. Vegetatio 23:289-313.
Stark, N., and L. D. Love. 1969. Water relations of three warm desert species. Israel J. Bot. 18:175-190.
Stebbins, R. C. 1974. Off-road vehicles and the fragile desert. Amer. Biol. Teacher 36:203-208, 294-304.
Sternberg, L. 1976. Growth forms of *Larrea tridentata*. Madroño 23:408-417.
Strain, B. R. 1975. Field measurements of carbon dioxide exchange in some woody perennials, pp. 145-158. *In* D. M. Gates and R. B. Schmerl (eds.). Perspectives in biophysical ecology. Springer-Verlag, New York.
Tevis, L., Jr. 1958a. Germination and growth of ephemerals induced by sprinkling a sandy desert. Ecology 39:681-688.
-----. 1958b. A population of desert ephemerals germinated by less than one inch of rain. Ecology 39:688-695.

Thorne, R. F. 1976. The vascular plant communities of California, pp. 1-31. In J. Latting (ed.). Proc. symposium on plant communities of southern California. Calif. Native Plant Soc. Spec. Pub. No. 2.

Twisselmann, E. C. 1967. A flora of Kern County, California. Wasmann J. Biol. 25:1-395.

Vasek, F. C., H. B. Johnson, and D. H. Eslinger. 1975a. Effects of pipeline construction on creosote bush scrub vegetation of the Mojave Desert. Madroño 23:1-13.

Vasek, F. C., H. B. Johnson, and G. D. Brum. 1975b. Effects of power transmission lines on vegetation of the Mojave Desert, Madroño 23:114-130.

Wallace, A., S. A. Bamberg, and J. W. Cha. 1974. Quantitative studies of roots of perennial plants in the Mojave Desert. Ecology 55: 1160-1162.

Wallace, A., V. Q. Hale, G. E. Kleinkopf, and R. C. Huffaker. 1971. Carboxydismutase and phosphoenolpyruvate carboxylase activities from leaves of some plant species from the northern Mojave and southern Great Basin Deserts. Ecology 52:1093-1095.

Wallace, A., and E. M. Romney. 1972. Radioecology and ecophysiology of desert plants at the Nevada Test Site. U.S. Atomic Energy Commission T.I.D. -25954. 439 p.

Webber, J. M. 1953. Yuccas of the southwest. USDA Agric. Monogr. 17:1-97.

Went, F. W. 1942. The dependence of certain annual plants on shrubs in southern California deserts. Bull. Torrey Bot. Club 69:100-114.

-----. 1948. Ecology of desert plants. I. Observations on germination in Joshua Tree National Monument. Ecology 29:242-253.

-----. 1949. Ecology of desert plants. II. The effect of rain and temperature on germination and growth. Ecology 30:1-13.

-----. 1955. Ecology of desert plants. Sci. Amer. 192-68-75.

Went, F. W., and M. Westergaard. 1949. Ecology of desert plants. III. Development of plants in the Death Valley National Monument, California. Ecology 30:26-38.

West, N. E., and K. I. Ibrahim. 1968. Soil-vegetation relationships in the shadscale zone of southeastern Utah. Ecology 49:445-455.

Woodell, S. R. J., H. A. Mooney, and A. J. Hill. 1969. The behaviour of *Larrea divaricata* (creosote bush) in response to rainfall in California. J. Ecol. 57:37-44.

CHAPTER 25

SONORAN DESERT

JACK H. BURK

Department of Biological Sciences, California State University, Fullerton

Introduction 870
Vegetation 871
Creosote bush scrub 873
Cactus scrub 876
Wash woodland 877
Palm oasis 877
Saltbush scrub 878
Alkali sink 880
Growth form and physiological response 880
Evergreen 880
Winter deciduous 882
Drought deciduous 883
Drought and winter deciduous 883
Succulent 884
Other 884
Concluding remarks 884
Literature cited 886

INTRODUCTION

The Colorado Desert of California represents the northwesternmost portion of the Sonoran Desert, which extends into Arizona, Baja California, and Sonora. In California, the Sonoran is delimited on the west by the Penninsular Ranges (Laguna, Santa Rosa, San Jacinto Mts.) and on the north by a gradual ecotone, around 34–35° N, into the Mojave Desert at an elevation of 350–500 m. The eastern and southern limits are represented by provincial boundaries: on the east by the California–Arizona state line, and on the south by the International Border. Physiographically, the area is composed of the Salton Basin and the plains and bajadas surrounding the lower Colorado River Valley, extending north along the river to near Needles, California. The Salton Basin was formed by alluvial deposits of the Colorado River, which separate it from the Gulf of California, and it is the only major undrained feature of the California Sonoran Desert.

Climatically, the subtropical Colorado Desert results from the descent of cold air which is heated by compression and arrives hot and dry at the earth's surface. Precipitation is frontal in nature during the winter and convectional in the summer. A bimodal rainfall pattern is a significant factor distinguishing the Sonoran from the more northern Mojave, in which summer precipitation is rare. The pattern of summer precipitation also varies within the Sonoran Desert (Mallery 1936). Shreve (1925) reported that summer rainfall increased from approximately 5% of the total in the western edge to 34% at the Colorado River and 50% at Tucson, Arizona. Tevis (1958) reported summer and winter rainfall data for an Indio station from 1940 to 1956; there was measurable summer precipitation in this far western portion of the desert each year, and it varied from 1 to 65% of the annual total. If rains of < 6 mm are excluded [Shreve and Wiggins (1964) reported that rains of less than 6 mm have little influence], there still was potentially significant rainfall during 10 of the 16 yr reported for Indio.

Reduced summer rainfall and high potential evapotranspiration make the Colorado Desert one of the most arid areas in North America (Walter 1971; Shantz and Piemeisel 1924). Summer temperatures frequently reach 45°C. The mean 24 hr maximum may exceed 38°C for several weeks (Shreve and Wiggins 1964). The effect of high temperature is moderated toward the east as summer rainfall becomes more frequent and predictable; however, the high-intensity summer precipitation results in proportionately less available water than an equivalent amount of winter or spring precipitation.

The geographical location and the sometimes inhospitable environment of the Colorado Desert may be responsible for a general lack of knowledge concerning the vegetation of the area. Since it is on the fringes of the Sonoran Desert (Dice 1939), and possibly because of provincialism, research has centered in Arizona. In the early part of this century the Desert Laboratory of the Carnegie Institution of Washington, at Tucson, carried out a significant research program concerning the vegetation of the Sonoran Desert. Much of the autecological research (e.g., Shreve 1922; Cannon 1905, 1911; Spalding 1909) has been assumed to be applicable to the western section of the desert. Field excursions supported by the Carnegie Institution into the Colorado Desert before 1920 (e.g., MacDougal 1914) are responsible for the greater part of the early literature on the desert. Many of these researchers did not consider areas outside the Salton Basin as part of the Colorado Desert; consequently, they did not discuss the large bajadas off the eastern slopes of the Chocolate Mts. and areas to the north. During the past 50 yr, physiognomical studies continued to be rare, and autecological studies on Colorado Desert material represent only isolated geographical areas and species.

The literature included in this chapter is limited almost exclusively to studies that deal directly with material from the Colorado Desert. The premise is that, until confirming data are available, unqualified extrapolation from local studies to all Colorado Desert populations and communities is risky. Nomenclature follows that of Munz (1974).

VEGETATION

The Colorado Desert can be considered a biotic reality only if the area under consideration includes the lower-elevation areas of Arizona which drain immediately into the Colorado River, the lower Gila River drainage in Arizona, northeastern Baja California, and northwestern Sonora, Mexico. This is a somewhat modified version of what Shreve and Wiggins (1964) referred to as the Lower Colorado Valley.

Table 25-1 shows the affinities of the more common perennial plants that occur in the Colorado Desert. A large number of Colorado Desert plants are also found in the Mojave and Arizona Sonoran Deserts. There are, however, several species that occur only in the low-elevation Colorado. Some are relictual (e.g., *Washingtonia filifera, Beleperone californica*) and occur only in isolated microhabitats. Others (e.g., *Dalea spinosa*) have a broader distribution but are restricted primarily to the low-elevation desert. There are a significant number of arboreal species in the Sonoran Desert that do not occur in the Mojave. Axelrod (1950) attributed this characteristic to cold winter temperatures and lack of summer rainfall in the Mojave. Arboreal species increase in size in areas of greater summer rainfall. Some arboreal species of the Arizona Sonoran are not found in the Colorado. *Cercidium*

TABLE 25-1. The affinities of Colorado Desert shrubs and trees for three southwestern deserts: Mojave, Colorado (including extreme southwestern Arizona), and Arizona Sonoran. Asterisks indicate taxa restricted to higher elevations in the Colorado Desert

Colorado	Colorado-Mojave	Colorado-Arizona Sonoran	Colorado-Mojave-Arizona Sonoran
Ambrosia ilicifolia	*_Acamptopappus sphaerocephalus_	_Agave deserti_	_Acacia greggii_
Asclepias subulata	_Atriplex hymenolytra_	_Baccharis sarothroides_	_Ambrosia dumosa_
Atriplex canescens var. _linearis_	_A. lentiformis_ ssp. _lentiformis_	_Bernardia incana_	_Atriplex canescens_ ssp. _canescens_
Beleperone californica	*_Cassia armata_	_Bursera microphylla_	_A. polycarpa_
Condaliopsis parryi	*_Coleogyne ramosissima_	_Calliandra eriophylla_	_Bebbia juncea_
Dalea californica	*_Encelia frutescens_	_Cercidium floridum_	_Chilopsis linearis_
D. schottii	_Ephedra californica_	_Condaliopsis lysioides_ var. _canescens_	_Echinocereus engelmanii_
D. spinosa	*_Happlopappus linearifolius_	_Crossosoma bigelovii_	_Encelia farinosa_
Hoffmannseggia microphylla	*_Keckiella antirrhinoides_ var. _microphylla_	_Ephedra trifurca_	_Ephedra nevadensis_
Prunus fremontii	*_Lycium cooperi_	_Fagonia laevis_	_Eriogonum fasiculatum_
Salvia eremostachya	*_Menodora scoparia_	_Fouquieria splendens_	_Fraxinus velutina_ var. _velutina_
S. vaseyi	*_Peucephyllum schottii_	_Hyptis emoryi_	_Hilaria ridgida_
Washingtonia filifera	*_Prunus fasiculata_	_Krameria grayi_	_Hymenoclea salsola_
	*_Thamnosma montana_	_Lycium fremontii_	_Krameria parvifolia_
	*_Yucca schidigera_	_Olneya tesota_	_Lycium torreyi_
		Simmondsia chinensis	_Physalis crassifolius_
			Pluchea sericea
			Prosopis glandulosa var. _torreyana_

microphyllum occurs in California only in isolated populations near the Colorado River (Hastings et al. 1972).

The vicinity of the Colorado River also marks the western extent of *Carnegiea gigantea.* Steenbergh and Lowe (1969) and Turner et al. (1966) discussed the requirements for establishment of *C. gigantea* in Arizona and showed that seedling mortality was reduced when young plants were shaded by "nurse plants or rocks". Later, Brum (1972) examined the reproductive potential and success of the largest California population of *Carnegiea* (Whipple Mts., San Bernardino Co.). Even though adequate pollinators were present, reproductive activity was significantly lower than in more easterly populations. A greater portion of the California population was associated with shade-producing objects. *Cercidium* species were the nurse plants at the California site, while there was a greater diversity of associated plants at Arizona sites. McDonough (1964) reported that *C. floridum* reduces incoming solar energy more effectively than do other potential nurse plants such as *Dalea schottii, Larrea tridentata,* and *Fouquieria splendens.* A lower density of potential nurse plants may also be a limiting factor to *Carnegiea* survival in California.

The Colorado River is an artificial boundary, in a biological sense, but represents closely the point at which summer rainfall is great enough to support species such as *Carnegiea* and *Cercidium microphyllum.* Shreve (1925) also noted that, within 32 km east of the Colorado River, *Larrea tridentata* becomes larger and more dense and trees associated with watercourses are 2–3 m taller. The species which reach a western limit in the area of the Colorado River and the smaller size of broadly distributed plants on the California side of the river attest to the importance of summer rainfall as a factor controlling the distribution and performance of many Sonoran Desert species.

A number of genera occur in the Colorado Desert only at higher elevations on the northern and western borders of the Salton Basin (Table 25-1). This is a highly diverse grouping of plants occurring in a habitat which is moderated considerably from the surrounding lower-elevation bajadas and may represent a habitat more like the Mojave than the adjacent Colorado Desert. Stebbins and Major (1965) pointed out that a relatively large number of mono- and ditypic genera and diploid endemic species which are related to more widespread polyploid species (patroendemics) occur in this habitat.

The colder winter temperatures of the Mojave may serve to limit Colorado Desert species. Turnage and Hinckley (1938) surveyed the damage following a severe freeze in the Sonoran Desert. They reported that frost damage to *Olneya tesota* was severe enough to be a potential limiting factor for this species. The northern distributional limit of *Olneya* falls in the vicinity of the critical isotherm (around 34°N) and approximates the boundary between the Sonoran and Mojave Deserts.

Colorado Desert vegetation can be categorized into six basic types. These vegetation types are arbitrary and are presented as nonoverlapping entities only for purposes of discussion. They exist on clines of soil salinity and water availability.

Creosote Bush Scrub

The most widespread vegetation type in the region is dominated by *Larrea tridentata* and *Ambrosia dumosa* (Fig. 25-1). Creosote bush scrub characterizes the vast intermountain bajadas, reaching greatest development on coarse, well-drained soil with a

Figure 25-1. Creosote bush scrub near Cabazon, Riverside Co.

total salinity of less than 0.02% (Shantz and Piemeisel 1924; Marks 1950). *Larrea* and *Ambrosia* become decreasingly important at higher elevations on the bajadas.

Highly diverse stands occur on lower-elevation (above *Larrea–Ambrosia*-dominated areas) rocky slopes of the Laguna, Santa Rosa, and San Jacinto Mts., and to a lesser extent in the mountains east and northeast of the Salton Basin (Johnson 1968). These stands include species which are (1) primarily cismontane (e.g., *Isomeris arborea, Eriogonum fasiculatum*), reflecting a moderated climate when compared with the lower-elevation creosote bush scrub (McHargue 1973), and (2) primarily Mojave (e.g., *Cassia armata, Peucephyllum schottii*). Species which reach their greatest development on these slopes include *Simmondsia chinensis, Mammillaria tetrancistra, Echinocereus englemannii, Thamnosma montana,* and *Salvia vaseyi* (McHargue 1973).

Where sand accumulates on the bajadas, *Coldenia palmeri, Croton californicus, Dalea schottii, D. emoryi, Hilaria ridgida,* and *Ephedra trifurca* gain importance (McHargue 1973; Rempel 1936; Shreve and Wiggins 1964). These species are adapted to areas of unstable soil, are relatively short lived, and are frequently found as pioneer species after disturbance in the Coachella Valley (McHargue 1973).

Creosote bush scrub occupies areas surrounding the Salton Basin between the higher, rocky hillsides and the desert saltbush community. The transition to desert saltbush occurs as the soil becomes heavier and the salt content increases to approximately 0.2% (Shantz and Piemeisel 1924). The factor that reduces the dominance of *Larrea* in the desert saltbush community is not clear. Lunt et al. (1973) reported that *Larrea* has a low tolerance for oxygen deficiency (oxygen diffusion rate of 0.3 mg cm^{-2} min^{-1} required for 50% of maximum growth rate). Barbour (1968) found that osmotic pressures of −20 bar (approximately equivalent to 0.2–0.3% total salts) reduced germination and root growth of *Larrea.* Inhibition by increasing salinity and reduced soil aeration must synergistically influence *Larrea* as dominance switches to *Atriplex polycarpa.*

The physiognomy of the creosote bush scrub community is simple because of low species diversity and the broad spacing of shrubs. The Sonoran ecotype (Barbour 1967; Yang 1967, 1970) of *Larrea* is taller and more slender and has a more open growth form than the other North American ecotypes. *Larrea* size varies greatly with respect to soil depth (Chew and Chew 1965; Burk and Dick-Peddie 1973; Shantz and Piemeisel 1924). I have observed apparently mature *Larrea* individuals in the Colorado Desert which range from approximately 90 to 300 cm tall, depending on conditions. The codominant *Ambrosia dumosa* is shorter (20–60 cm), and the size of mature individuals is less variable.

Shreve (1942) and Went (1955) commented on the regular spacing of *Larrea* individuals. Shreve assumed that spacing was due to competition for water, and Went speculated that water-soluble plant toxins might be responsible. There is no direct evidence to support the latter hypothesis. Woodell et al. (1969) and Barbour (1969) reported that a uniform distribution of *Larrea* individuals is more common in areas of lowest rainfall in the Colorado Desert. It appears that *Larrea* pattern is an expression of complex factors which have not yet been enumerated in detail. For example, Wright (1970) has examined the pattern resulting from asexual reproduction (i.e., root sprouting on buried branches and crown splitting; see also chapter 24). This type of reproduction results in a small-scale pattern of mounds. A small-scale pattern would also be caused by aggregations of seedlings in available microhabitats. These seedling aggregations would not be correlated with patterns due to other causes because reproduction by seed is a rare event, particularly in the most arid sections of the desert (Barbour 1968). The rate of asexual reproduction by crown splitting has not been compared in areas with different amounts of rainfall, nor do we know how asexual reproduction influences pattern after the original "mound" is no longer apparent. Plants that arise by crown splitting also do not necessarily have a symmetrical root system with respect to the shoot. This would result in a nonuniform distribution which might be maintained by intraspecific competition. The evidence indicates that *Larrea* spacing is modified by available water, soil depth, mode of reproduction, and symmetry of the root system. The relative importance of these factors at a particular site determines the pattern of *Larrea*. Woodell et al. (1969) showed that the *Larrea* pattern is apparently independent of the density of associated shrubs.

Desert shrubs exhibit differences in branching pattern which, in turn, modify their influence on the environment. Went (1942), Muller (1953), and Muller and Muller (1956) noted that certain shrubs (e.g., *Thamnosma montana, Ambrosia dumosa, Larrea tridentata*) characteristically form new branches from the crown, resulting in multiple stems arising from the ground. In contrast, *Encelia farinosa* emerges with a central stem branching above the soil surface. *Encelia farinosa* does not accumulate organic material and windblown soil beneath its crown, as do multiple-stemmed shrubs. These shrub mounds serve as an increased source of nutrients and water supply for the shrub in addition to providing a unique microhabitat for desert ephemerals.

Shrub mounds are typically occupied by *Chaenactis fremontii, Emmenanthe penduliflora, Phacelia distans, Malacothrix glabrata, Rafinesquia neomexicana,* and others, while many other species occur primarily between shrubs. Muller (1953) and Muller and Muller (1956) reported that *Encelia farinosa* does not harbor ephemerals, as do *Thamnosma montana* and *Ambrosia dumosa.* Toxic exudates

Figure 25-2. Cactus scrub community near Whitewater. Dominant succulent species are *Echinocactus acanthoides* and *Echinocercus engelmanii.*

were extracted from *E. farinosa* (Gray and Bonner 1948), *T. montana* (Bennet and Bonner 1953), and *A. dumosa* (Muller 1953). Muller and Muller (1956) concluded that these chemicals were rendered inactive by the presence of organic material accumulated by multiple-stemmed plants. Associated annuals may have higher moisture and nutrient requirements, be more susceptible to high soil temperatures in the intershrub areas, be more sensitive to wind damage, and so forth. Therefore conclusions that toxins are the controlling factor in spacing are premature until the microenvironmental requirements of the ephemerals are enumerated.

Adams et al. (1970) showed that the water repellency of soil under mound-forming plants also influences the distribution of annual plants. The occurrence of hydrophobic topsoil needs to be examined in more detail for other areas and beneath other species. This phenomenon represents an additional factor which may influence the pattern of perennial and ephemeral plants in the desert.

Cactus Scrub

Cactus scrub (Fig. 25-2) is of limited distribution, and it is usually surrounded by creosote bush scrub. The dominant species are *Opuntia bigelovii,* other *Opuntia* spp., *Echinocactus acanthoides,* and *Echinocereus engelmanii.* This is the only community of the Colorado Desert which is dominated by stem-succulent plants. It is commonly found on south-facing slopes with finer soil texture than in the surrounding *Larrea–Ambrosia* community (Marks 1950; McHargue 1973). These observations are consistant with those of Shreve and Wiggins (1964), who noted the importance of simultaneously warm temperatures and high soil moisture for growth and estabilshment of stem succulents. The south-facing aspect and finer-textured soils apparently provide a habitat similar to those further east where summer rainfall is more predictable. The soil type associated with cactus scrub has been reported as rocky (Benson 1969) with the implication of a coarse, undeveloped soil, but the implication is not always warranted. For example, McHargue (1973) reported that round stones are often present in a soil which he referred to as a sandy loam with almost no intermediate-sized particles.

Wash Woodland

The margins of arroyos in the Colorado Desert support a relatively dense growth of trees. This vegetation type is typical of higher-elevation bajadas as well as of arroyo margins in the Arizona Sonoran. However, lower precipitation to the west restricts these trees to areas of greater water availability. The common species include *Cercidium floridum, Olneya tesota, Dalea spinosa, Lycium andersonii, Prosopis glandulosa* var. *torreyana, Baccharis sarothroides, Condaliopsis lysiodes,* and *Chilopsis linearis*. Height varies from 1 to 5 m and is proportional to the size of the arroyo (Shreve and Wiggins 1964).

Certain topographical situations (e.g., the fans at the eastern base of the Chocolate Mts., Imperial Co.) cause the formation of a vegetation primarily associated with watercourses. Very flat areas between arroyos are devoid of perennial vegetation, and only a limited number of ephemerals (e.g., *Chorizanthe ridgida*) are able to survive on the black pebble beds. *Larrea* and *Ambrosia* are also associated with watercourses in this habitat. The vegetation appears as a continuous stand of riparian species because of the presence of large numbers of arroyos (Shreve and Wiggins 1964). The only quantitative data available on this vegetation type are general phytosociological data obtained by McHargue (1973) for the Coachella Valley (Table 25-2).

Palm Oasis

Washingtonia filifera is associated with *Sporobolus airoides, Juncus acutus, Pluchea sericea,* and *Happlopappus acradenius* on the northern and western margins of the Salton Basin (Fig. 25-3). Palm oases have been described qualitatively by

Figure 25-3. Palm oasis. Indio Hills, along the San Andreas Fault.

MacDougal (1908), Jepson (1925), Benson and Darrow (1954), Jaeger (1957), and Shreve and Wiggins (1964). The only detailed quantitative description is that of Vogl and McHargue (1966).

Washingtonia is a relict species (Axelrod 1950) now limited to sites with a permanent water supply in and around the Salton Basin and south into Sonora and Baja California, Mexico. Washes along the San Andreas Fault are the site of emergence of underground water and, therefore, the location of many oases. Other oases are present in washes and on hillsides where exposed strata or other geological structures produce permanent water.

Vogl and McHargue (1966) delimited hydric, oasis, and transition zones dominated by *Scirpus olneyi, Washingtonia,* and *Happlopappus acradenius,* respectively. The relative number of species increases away from the point of water emergence (10%) to the oasis (17%) and to the transition zone (73%). The oases are highly variable, with *Washingtonia* the only species present in all 24 study sites. Salinity is low in the root zone of the palms (0.16%: Shantz and Piemeisel 1924), but increases toward the surface, where evaporation leaves accumulations of salt. Successful reproduction is limited by surface salinity, water supply, rainfall, and fire.

Oases are the only community in the Colorado Desert in which fire is a significant abiotic factor. The palms are fire tolerant, whereas understory species are not. Fire opens the understory, allowing seedlings to establish. It also increases the water supply to the fire-tolerant palm by removal of competition. Therefore fire is an important factor in the reproduction and maintenance of the *Washingtonia* community (Vogl and McHargue 1966).

Saltbush Scrub

Saltbush scrub occupies habitats which are generally moist, with a sandy, loam soil, and a total salinity in the range of 0.2–0.7% (Shantz and Piemeisel 1924). *Atriplex polycarpa* is the dominant species. Physiognomically, the community is often composed of a nearly uniform stand of shrubs about 1 m tall forming a more complete cover than in the *Larrea–Ambrosia*-dominated area (Table 25-2). Species number is lower than in the higher-elevation communities. *Atriplex polycarpa* has a wide range of salinity tolerance and is a facultative phreatophyte when growing in coarse soil (Chatterton 1970). In silty loams more typical of saltbush scrub, however, development of a taproot is reduced in favor of a shallow, more fibrous root system.

Atriplex canescens var. *linearis* and *Prosopis glandulosa* var. *torreyana* are common associates (McHargue 1973). *Atriplex canescens* shows greater dominance in dryer, coarser soils and occurs throughout the saltbush scrub habitat. *Prosopis glandulosa* reaches greater development in lower-elevation areas with a shallow water table or capillary fringe. *Haplopappus acradenius* ssp. *eremophylus* is common in areas where *P. glandulosa* is dominant.

Shantz and Piemeisel (1924) noted a paucity of ephemerals in *Atriplex* stands in the Coachella Valley. These frequently flooded areas are covered with a layer of silt which dries and cracks, making establishment difficult.

Marks (1950) and Shantz and Piemeisel (1924) comment on the agricultural potential of the *Atriplex–Prosopis* habitats. Most of these areas are now under cultivation in the Coachella and Imperial valleys.

TABLE 25-2. Summary of existing phytosociological data on Colorado Desert plant communities

Location and Community Type	Annual Precipitation (mm)	Total Cover (%)	Shrub Density per Hectare: All Species	Shrub Density per Hectare: *Larrea*	Number of Species per Hectare	Source
Coachella Valley	90-168					
Creosote bush scrub						McHargue 1973
Rocky hillsides[a]		6.7	1436		73	
Rocky bajadas		7.8	895		77	
Gravel bajadas		6.8	684		80	
Sandy bajadas		7.8	727		59	
Saltbush scrub (level, moist)		14.9	998		27	
Alkali sink (level, wet)		50.8	679		33	
Wash woodland		18.4	466		79	
San Andreas Fault						Vogl and McHargue 1966
Oasis		34				
Cabazon						
Creosote bush scrub[a]	299		1500	239		Woodell et al. 1969
Yuma						
Creosote bush scrub[b]	91		110	108		Woodell et al. 1969
Niland						
Creosote bush scrub	48		460	86		Woodell et al. 1969
Southern Imperial Valley						Barbour 1969,
Creosote bush scrub	65	22		130		Barbour et al. 1976
Creosote bush scrub	65	22		830		

[a] High-elevation hillsides (higher diversity of plants with Mojave affinities).
[b] Sandy bajada.

Alkali Sink

Allenrolfea occidentalis and *Suaeda torreyana* characterize the vegetation of low-elevation, wet, high-salinity habitats along the margins of the Salton Sea and the Colorado River (Fig. 25-4). *Suaeda* occurs in dense stands where salinity ranges from 0.5 to 1%, and *Allenrolfea* is dominant in habitats with a totally saline profile (1.5–2%) and heavy soil (Marks 1950).

Tamarix chinensis and *Pluchea sericea* dominate areas frequently flushed of salt buildup along the Colorado River. Associated phreatophytes include *Prosopis pubescens, Baccharis glutinosa, Atriplex lentiformis, Aster spinosis, Salix gooddingii,* and *Populus fremontii.* The average salinity is on the order of 0.14% (Marks 1950), and the water table is seldom below 3 m (Horton and Campbell 1974).

Atriplex is a minor component of the sink vegetation and has the ability to survive as a nonphreatophyte (McDonald and Hughes 1964). McDonald and Hughes compared the water loss of the facultative phreatophyte *Atriplex* with that of an obligate phreatophyte *Pluchea* and found that *Atriplex* was more conservative with water than *Pluchea* when both were growing in areas with nearly optimal water table.

GROWTH FORM AND PHYSIOLOGICAL RESPONSE

Considerable interest has been shown in the physiological adaptations of Colorado Desert plants. This interest has included comparison of desert and cismontane populations of *Simmondsia chinensis, Isomeris arborea,* and *Atriplex lentiformis* (Al-Ani et al. 1972; Tobiessen 1970; Pearcy and Harrison 1974). Various adaptive strategies, such as evergreen, winter deciduous, summer deciduous, and various combinations, have been examined.

Evergreen

Larrea tridentata, Simmondsia chinensis, and *Atriplex* spp. are among those that maintain leaf tissue during all seasons. These evergreen species also have a partial seasonal leaf drop which coincides with the available moisture. However, if all leaves are shed, the plant will not recover (true at least for *Larrea*: Ashby 1932; Runyon 1934; Dalton 1962).

Though these species are evergreen, other phenological events (stem growth, flowering, germination) are seasonal. In a review, Barbour et al. (1977) concluded that most investigators view moisture as the most important factor in driving Sonoran plant phenology, but certainly the relative importance of moisture, temperature, and day length has not been agreed upon unanimously. In *Cercidium floridum,* for example, stem growth seems photoperiodically attuned, beginning several weeks before the onset of summer rains, and its flowering is also initiated by photoperiod (Turner 1963). All phenological events of *Larrea,* however, seem moisture regulated first, and temperature regulated second. Oechel et al. (1972a) reported for Deep Canyon that *Larrea* began active growth in January, flowered in April, bore mature fruit in June, and fell back to a low vegetative growth rate in June. For most of the California portion of the Sonoran Desert, there are no phenological data.

Positive photosynthesis under a wide range of environmental conditions seems

Figure 25-4. Alkali sink community along the Colorado River near Blythe. Dominant species are *Tamarix chinensis* and *Pluchea sericea.*

characteristic of evergreens. Strain and Chase (1966) compared the temperature response of *Larrea* net photosynthesis to the responses of *Chilopsis linearis, Encelia farinosa,* and *Hymenoclea salsola* after acclimation to day/night temperatures of 15/7, 30/15, and 40/25°C. *Larrea* showed the greatest acclimation potential of the species tested. Strain (1969) monitored respiration in *H. salsola, E. farinosa,* and *Acacia greggii.* Dark respiration decreased at temperatures greater than 25°C. *Larrea*, the only evergreen tested, showed the most dramatic seasonal fluctuations in respiration rates: 0.43 and 1.02 mg CO_2 g^{-1} dry wt hr^{-1} at 25°C in July and January, respectively. These rates are about half the values Strain and Chase reported for the drought-deciduous shrubs *Hymenoclea* and *Encelia,* but about the same as the rate for the winter-deciduous *Acacia,*

A similar reduction in respiration at high acclimation temperature has been reported by Chatterton et al. (1970) for *Atriplex polycarpa,* in which plants acclimated at 32–43°C showed no measurable respiration at 10°C. *Simmondsia chinensis* (Al-Ani et al. 1972) and *Atriplex lentiformis* (Pearcy and Harrison 1974) from desert habitats have lower rates of dark respiration at high temperatures than do coastal plants of the same species. An important mechanism increasing the survival of desert plants, particularly evergreens, is reduction of Q_{10} for respiration under stress situations, which also lower photosynthetic rates. This results in a positive 24 hr net CO_2 accumulation at 43°C in *Larrea* (Oechel et al. 1972b) and up to 47°C in *Simmondsia* (Al-Ani et al. 1972).

Runyon (1936) noted that *Larrea* leaves formed under stress were more xeromorphic than those formed under more mesic conditions. The leaf anatomy of *Larrea* is not classically xerophytic (Barbour et al. 1974, 1977), the strategy apparently being to tolerate rather than prevent desiccation (Ashby 1932). Odening et al. (unpub. data) have quantified desiccation tolerance and reported that *Larrea* may withstand extended periods of negative turgor pressure in the range of –10 to

−16 bar. Barbour et al. (1974) reported that net photosynthesis is independent of water potential between −8.5 and −26.0 bar, but Odening et al. (1974) reported a decrease in net photosynthesis over the same range. Further experimentation should clarify this discrepancy, which may be due to genetic variability, problems relating to acclimation, and inappropriate humidity control (Lange et al. 1971).

Odening et al. (1974) reported that stem elongation, growth, flowering, and so forth are initiated only when the plant water potential is above −24 to −28 bar. Oechel (1972a,b) and Barbour et al. (1974) also reported a rapid drop in net photosynthesis at water potentials below −26 bar. Oechel et al. (1972a) showed a reduction of diurnal water potential extremes at low water potentials apparently caused by stomatal closure. Reduction of evaporative cooling by stomatal closure may be compensated for by small leaf size, maintaining the leaf temperature near ambient via convection.

Larrea exhibits one of the lowest seasonal ranges of tissue water potentials among its associated perennials. At Deep Canyon, maximum (dawn) daily xylem water potential was −24 bar in February, −56 bar in September; minimum (afternoon) daily potential was −40 bar in February, −65 bar in September (Oechel et al. 1972a). A subsequent study at Deep Canyon (Odening et al., unpub. data) demonstrated that, during much of the year, tissue moisture levels are sufficiently low to place *Larrea* under conditions of negative turgor pressure. This may explain why cell formation can continue all year, but cell elongation is restricted to brief periods when tissue water potential and thus cell turgor pressure are above critical levels (Oechel et al. 1972a).

Strain (1970)reported a midday depression in *Larrea* net photosynthesis, but Pearcy and Harrison (1974) showed no such midday depression for *Atriplex lentiformis,* and no stomatal closure due to water potentials down to −30 bar. This behavior probably reflects the phreatophytic habit of *A. lentiformis.*

The evergreen growth habit is successful in the desert only when coupled with (1) the ability to maintain positive net photosynthesis at high temperatures (*Larrea, Simmondsia*) or to adopt the C_4 photosynthetic pathway (*Atriplex lentiformis, A. polycarpa*); and (2) the ability to maintain positive photosynthesis under low water potentials (*Larrea,* −70 bar; *Simmondsia,* −36 bar) or to grow as a phreatophyte (*A. lentiformis*).

Winter Deciduous

Prosopis glandulosa var. *torreyana, Acacia greggii,* and *Chilopsis linearis* are taxa which maintain leaves during the hot, dry season. Abscission occurs, presumably in response to photoperiod, in late fall. These species tend to be limited to watercourses and other areas of greater summer water supply. Water potential seldom drops below −20 bar in *C. linearis in situ. Chilopsis* is more conservative with water than are evergreen species and maintains higher water potentials, apparently by stomatal closure (Odening et al. 1974). Acclimation studies by Strain and Chase (1966) showed that *Chilopsis* had a broad acclimation potential at low (15°C) temperatures but that net photosynthesis was depressed under high (40°C) preconditioning temperatures. They concluded that, even though *Chilopsis* maintains photosynthetic material during the summer months, it is poorly adapted to high temperatures. High relative productivity during cooler periods and summer survival in the presence of a perennial water supply apparently compensate for a low acclimation potential. A modified dark respiration rate at high acclimation temperatures is also important in compensating for low CO_2 uptake in summer.

The annual pattern of net photosynthesis in *Acacia greggii* is similar to that of *Chilopsis*, except that the summer depression is less dramatic. The nonstructural carbohydrate pools are the reserve carbon resources available for respiration. Strain (1969) measured the seasonal variation in alcohol-soluble carbohydrates in leaves and stems of *A. greggii in situ* in Deep Canyon. He found the seasonal variation to be 12–19% of the alcohol-insoluble dry weight for stems, and 4–7% for leaves. This pattern was shared by evergreen *Larrea*, but variation was considerably greater for drought-deciduous *Hymenoclea* and *Encelia* (about 8–45% for leaves, 4-25% for stems). The size of the nonstructural carbon pool in *Acacia* and *Larrea* is low throughout the year.

Drought Deciduous

Hymenoclea salsola and *Encelia farinosa* shed leaves, not in response to photoperiod or low temperatures, but to the onset of drought stress. *Encelia farinosa* shows seasonal variation in leaf density/thickness. During times of available water, leaves expand more, are less pubescent, are less capable of reducing water loss, and have lower resistance to CO_2 flux. These characteristics reverse as soil water decreases and the more mesophytic leaves abscise (Cunningham and Strain 1969b).

Deciduous species show significant seasonal change in storage material and therefore must possess mechanisms (e.g., anatomical and morphological leaf changes) to increase productivity during favorable periods. *Encelia farinosa* has periods of extremely high productivity (up to 48 mg dm^{-2} hr^{-1} net CO_2 exchange), made possible because the photosynthetic system does not saturate even at nearly full sunlight. These high rates of CO_2 accumulation provide large carbohydrate reserves which apparently serve as respiratory substrate during periods of leaflessness (Cunningham and Strain 1969a,b).

Strain and Chase (1966) found that acclimation at high temperatures increased the photosynthetic rate at higher temperatures much more in *Encelia* than in the winter-deciduous *Chilopsis*, but less than in *Larrea*.

Drought and Winter Deciduous

Cercidium floridum is without leaves during much of the year because leaves lost in response to drought are not replaced until the following spring. Survival is possible because the mean annual net photosynthetic rate of green stem tissue is comparable to that of leaves on a unit area basis (Adams and Strain 1968, 1969; Adams et al. 1967). A summer reduction in net photosynthesis is probably the result of low water potentials rather than of a decrease in metabolic rate due to high temperatures. The highest photosynthetic rates were measured at 0600 hr when water potential was highest and leaf angle was nearly perpendicular to the sunlight.

Fouquieria splendens is also drought and winter deciduous but differs from *Cercidium* in that positive bark photosynthesis occurs only when plants are in a well-hydrated condition (Mooney and Strain 1964). Long periods when positive synthesis is not possible are apparently compensated for by the rapidity with which

Fouquieria can replenish photosynthetic tissues. Cannon (1905) timed leaf development after a rain and reported buds enlarged in 48 hr, leaves 1 cm long in 72 hr, and leaves matured in 5 days. The importance of bark photosynthesis may be in the period of leaf development, when increased respiratory substrate may increase the rate of leaf formation. Additional metabolic products for rapid leaf formation are present in special leaf axil tissues which store water, oil, and sugar (Scott 1932).

Succulent

Szarek et al. (1973) reported that *Opuntia basilaris* maintained photosynthetic activity during drought periods, presumably utilizing endogenously produced CO_2 recycled through dark CO_2 fixation, organic acid transformations, photosynthesis, and respiration. Immediately after precipitation, nighttime stomatal opening permitted atmospheric CO_2 assimilation and organic acid synthesis. This species maintains a moderate energy state which allows a rapid response to soil moisture. No other data are available on the physiological response of succulents from the Colorado Desert.

Other

There are no gas exchange data on annuals or herbaceous perennials for material originating in the Colorado Desert. Capon and Van Asdall (1967) reported that heat pretreatment (50°C) increased the germination percentage of Colorado Desert annuals (*Lepidium lasiocarpum, Sisymbrium altissimum, Streptanthus arizonicus*), whereas a 5 mo maturation period was necessary to achieve comparable germination percentages in seeds stored at 20°C. *Plantago insularis* germination was induced by leaching in running water and was not influenced by heat treatment.

CONCLUDING REMARKS

The distribution of major species in the Colorado Desert is determined primarily by variations in soil water availability and salt accumulation. McHargue (1973) characterized the major habitats of the Colorado Desert according to the geographical affinities of common genera. Most plants endemic to the Colorado Desert are found on sandy bajadas or in level, moist habitats. Genera centered in the Arizona Sonoran Desert but occurring also in the Colorado Desert are typically found in riparian situations. Plants common to all the warm deserts occur in environmentally severe habitats, whereas plants widely distributed over southwestern deserts are commonly halophytic. Physiological modifications such as maximum photosynthetic rates, dark respiration rates, and leaf xeromorphism make it possible for evergreen, drought-deciduous, winter-deciduous, and drought- and winter-deciduous growth forms to successfully coexist in the desert environment.

Figure 25-5 summarizes the relationship between growth form and maximum photosynthesis (Strain 1975). Winter-deciduous shrubs have the highest overall photosynthetic rate. Evergreen plants have the lowest rates, and drought-deciduous species are intermediate. The exception is *Isomeris,* which is evergreen in the Colorado Desert but which Tobiessen (1970) has experimentally shown to be drought

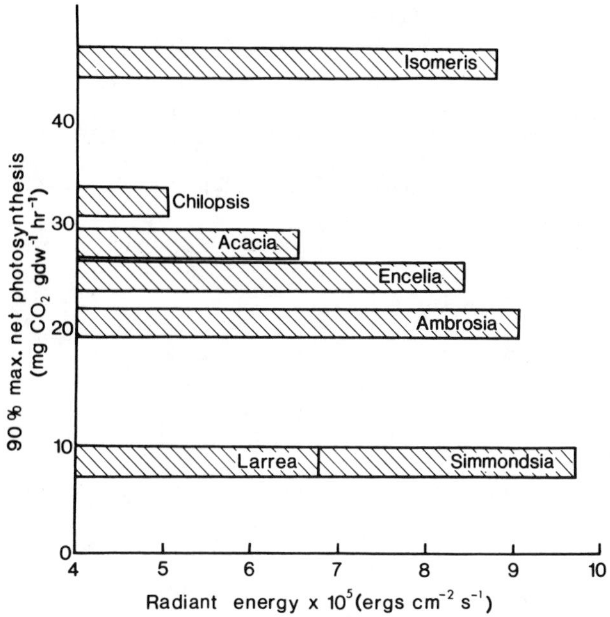

Figure 25-5. Radiant energy required to reach 90% of maximum net photosynthesis for greenhouse-grown plants of the Colorado Desert. Modified from Strain (1975).

deciduous. This species apparently grows only in areas with soil moisture adequate to maintain relatively mesophytic leaves year round. *Cercidium* and *Fouquieria* are unique because a combination of photosynthetic bark and deciduous leaves yields photosynthetic rates characteristic of both evergreen and deciduous growth forms.

Our knowledge of Colorado Desert plant ecology is sketchy at best, and a review of this type is particularly difficult in view of the absence of basic background data. The future approach should seek to extend basic, descriptive information on phytosociological characters and physiological responses of characteristic taxa of each type. It is essential that, as we pursue "modern" plant ecological research, an effort be made to improve the descriptive data base. The Colorado Desert includes the hottest and driest areas of the United States; thus future studies here may make a significant contribution to our understanding of desert adaptations.

LITERATURE CITED

Adams, M. S., and B. R. Strain. 1968. Photosynthesis in stems and leaves of Cercidium floridum: spring and summer diurnal field response and relation to temperature. Oecol. Plant. 3:285-297.

-----. 1969. Seasonal photosynthetic rates in stems of Cercidium floridum Benth. Photosynthetica 3:55-62.

Adams, S., B. R. Strain, and M. S. Adams. 1970. Water repellent soil, fire, and annual plant cover in a desert shrub community of southeastern California. Ecology 51:696-700.

Adams, M. S., B. R. Strain, and J. P. Ting. 1967. Photosynthesis in chlorophyllous stem tissue and leaves of Cercidium floridum: accumulation and distribution of ^{14}C from $^{14}CO_2$. Plant Physiol. 42:1797-1799.

Al-Ani, H. A., B. R. Strain, and H. A. Mooney. 1972. The physiological ecology of diverse populations of the desert shrub Simmondsia chinensis. J. Ecol. 60:41-57.

Ashby, E. 1932. Transpiratory organs of Larrea tridentata and their ecological significance. Ecology 13:182-188.

Axelrod, D. I. 1950. Studies in late Tertiary paleobotany. Carnegie Inst. Wash. Pub. 590. 323 p.

Barbour, M. G. 1967. Ecoclinal patterns in the physiological ecology of a desert shrub, Larrea divaricata. Ph.D. dissertation, Duke Univ., Durham, N. C. 242 p.

-----. 1968. Germination requirements of the desert shrub Larrea divaricata. Ecology 49:915-923.

-----. 1969. Age and space distribution of the desert shrub Larrea divaricata. Ecology 50:679-685.

Barbour, M. G., G. Cunnigham, W. C. Oechel, and S. A. Bamberg. 1977. Growth and development, Chapter 4. In T. Mabry and J. Hunziker (eds.). Larrea and its role in desert ecosystems. Dowden, Hutchinson, and Ross, Stroudsburg, Pa. (in press).

Barbour, M. G., D. V. Diaz, and R. W. Breindenbach. 1974. Contributions to the biology of Larrea species. Ecology 55:1199-1215.

Bennet, E. L, and J. Bonner. 1953. Isolation of plant growth inhibitors from Thamnosma montana. Amer. J. Bot. 40:29-33.

Benson, L. 1969. The native cacti of California. Stanford Univ. Press, Stanford, Calif. 243 p.

Benson, L., and R. A. Darrow. 1954. The trees and shrubs of the southwestern desert. Univ. New Mex. Press, Albuquerque. 437 p.

Brum, G. D., Jr. 1972. Ecology of the saguaro (Carnegiea gigantea): phenology and establishment in marginal populations. M.S. thesis, Univ. Calif., Riverside. 42 p.

Burk, J. H., and W. A. Dick-Peddie. 1973. Comparative production of Larrea divaricata Cav. on three geomorphic surfaces in southern New Mexico. Ecology 54:1094-1102.

Cannon, W. A. 1905. On the transpiration of Fouquieria splendens. Bull. Torrey Bot. Club 32:397-414.

-----. 1911. The root habits of desert plants. Carnegie Inst. Wash. Pub. 131. 96 p.

Capon, B., and W. Van Asdall. 1967. Heat pre-treatment as a means of increasing germination of desert annual seeds. Ecology 48:305-306.

Chatterton, N. J. 1970. Physiological ecology of Atriplex polycarpa: growth, salt tolerance, ion accumulation and soil-plant-water relations. Ph.D. dissertation, Univ. Calif., Riverside. 120 p.

Chatterton, N. J., C. M. McKell, and B. R. Strain. 1970. Intraspecific differences in temperature-induced respiratory acclimation of desert saltbush. Ecology 51:545-547.

Chew, R. M., and A. E. Chew. 1965. The primary productivity of a desert-shrub (Larrea tridentata) community. Ecol. Monogr. 33:355-375.

Cunningham, G. L., and B. R. Strain. 1969a. Irradiance and productivity in a desert shrub. Photosynthetica 3:69-71.

-----. 1969b. An ecological significance of seasonal leaf variability in a desert shrub. Ecology 50:400-408.

Dalton, P. D., Jr. 1962. Ecology of the creosotebush Larrea tridentata. Ph.D. dissertation, Univ. Ariz., Tucson. 162 p.

Dice, L. R. 1939. The sonoran biotic province. Ecology 20:118-129.

Gray, R., and J. Bonner. 1948. An inhibiter of plant growth from the leaves of Encelia farinosa. Amer. J. Bot. 35:52-57.

Hastings, J. R., R. M. Turner, and D. K. Warren. 1972. An atlas of some plant distributions in the Sonoran Desert. Meteor. Climatol. Arid Regions Tech. Rept. 21, Univ. of Ariz. Inst. Atmos. Physics, Tucson. 255 p.

Horton, J. S., and C. J. Campbell. 1974. Management of phreatophyte and riparian vegetation for maximum multiple use values. USDA Forest Ser. Res. Paper RM-117. 23 p.

Jaeger, E. C. 1957. The North American deserts. Stanford Univ. Press, Stanford, Calif. 308 p.

Jepson, W. L. 1925. A manual of the flowering plants of California. Univ. Calif. Press, Berkeley. 1238 p.

Johnson, A. W. 1968. The evolution of desert vegetation in western North America, pp. 590-606. In G. W. Brown, Jr. (ed.). Desert biology. Academic Press, New York.

Lange, O. L., R. Losch, E. D. Schulze, and L. Kappen. 1971. Responses of stomata to changes in humidity. Planta 100:76-86.

Lunt, O. R., J. Letey, and S. B. Clark. 1973. Oxygen requirements for root growth in three species of desert shrubs. Ecology 54:1356-1362.

MacDougal, D. T. 1908. Botanical features of North American deserts. Carnegie Inst. Wash. Pub. 99. 111 p.

-----. 1914. The Salton Sea. Carnegie Inst. Wash. Pub. 193. 182 p.

McDonald, C. D., and G. H. Hughes. 1964. Studies of consumptive use of water by phreatophytes and hydrophytes near Yuma, Arizona. Geol. Survey Prof. Paper 486-F. 24 p.

McDonough, W. T. 1964. Reduction in incident solar energy by desert shrub cover. Ohio J. Sci. 64:250-251.

McHargue, L. T. 1973. A vegetational analysis of the Coachella Valley, California. Ph.D. dissertation, Univ. Calif., Irvine. 340 p.

Mallery, T. D. 1936. Rainfall records for the Sonoran Desert. Ecology 17:110-121.

Marks, J. B. 1950. Vegetation and soil relations in the lower Colorado Desert. Ecology 31:176-193.

Mooney, H. A., and B. R. Strain. 1964. Bark photosynthesis in ocotillo. Madroño 17:230-233.

Muller, C. H. 1953. The association of desert annuals with shrubs. Amer. J. Bot. 40:53-60.

Muller, W. H., and C. H. Muller. 1956. Association patterns involving desert plants that contain toxic products. Amer. J. Bot. 43:354-361.

Munz, P. A. 1974. A flora of southern California. Univ. Calif. Press, Berkeley. 1086 p.

Odening, W. R., B. R. Strain, and W. C. Oechel. 1974. The effect of decreasing water potential on net CO_2 exchange of intact desert shrubs. Ecology 55:1086-1095.

Oechel, W. C., B. R. Strain, and W. R. Odening. 1972a. Tissue water potential, photosynthesis, ^{14}C-labeled photosynthetic utilization and growth in the desert shrub Larrea divaricata. Ecol. Monogr. 42:127-141.

-----. 1972b. Photosynthetic rates of a desert shrub, Larrea divaricata Cav., under field conditions. Photosynthetica 6:183-188.

Pearcy, R. W., and A. T. Harrison. 1974. Comparative photosynthetic and respiratory gas exchange characteristics of Atriplex lentiformis (Torr.) Wats. in coastal and desert habitats. Ecology 55:1104-1111.

Rempel, R. J. 1936. The crescentic dunes of the Salton Sea and their relations to the vegetation. Ecology 17:347-358.

Runyon, E. H. 1934. The organization of the creosote bush with respect to drought. Ecology 15:128-138.

-----. 1936. Ratio of water content to dry weight in leaves of the creosote bush. Bot. Gaz. 97:518-553.

Scott, F. M. 1932. Some features of the anatomy of Fouquieria splendens. Amer. J. Bot. 19:673-678.

Shantz, H. L., and R. L. Piemeisel. 1924. Indicator significance of the natural vegetation of the southwestern desert region. J. Agric. Res. 28:721-801.

Shreve, F. 1922. Conditions indirectly affecting vertical distribution on desert mountains. Ecology 3:269-274.

-----. 1925. Ecological aspects of the deserts of California. Ecology 6:93-103.

-----. 1942. The desert vegetation of North America. Bot. Rev. 8:195-246.

Shreve, F., and I. L. Wiggins. 1964. Vegetation and flora of the Sonoran Desert. Vol. I. Stanford Univ. Press., Stanford, Calif. 840 p.

Spalding, V. M. 1909. Distribution and movements of desert plants. Carnegie Inst. Wash. Pub. 113. 144p.

Stebbins, G. L., and J. Major. 1965. Endemism and speciation in the California flora. Ecol. Monogr. 35:1-35.

Steenbergh, W. F., and C. H. Lowe. 1969. Critical factors during the first years of life of the Saguaro (Cereus gigantea) at the Saguaro National Monument. Ecology 50:825-854.

Strain, B. R. 1969. Seasonal adaptations in photosynthesis and respiration in four desert shrubs growing in situ. Ecology 50:511-513.

-----. 1970. Field measurements of tissue water potential and carbon dioxide exchange in the desert shrubs Prosopis juliflora and Larrea divaricata. Photosynthetica 4:118-122.

-----. 1975. Field measurements of carbon dioxide exchange in some woody perennials, pp. 145-158. In D. M. Gates and R. B. Schmerl (eds.). Perspectives of biophysical ecology. Springer-Verlag, New York.

Strain, B. R., and V. C. Chase. 1966. Effect of past and prevailing temperatures on the carbon dioxide exchange capacities of some desert perennials. Ecology 47:1043-1045.

Szareck, S. R., H. B. Johnson, and J. P. Ting. 1973. Drought adaptation in Opuntia basilaris. Plant Physiol. 52:539-541.

Tevis, L., Jr. 1958. Germination and growth of ephemerals induced by sprinkling a sandy desert. Ecology 39:681-688.

Tobiessen, P. L. 1970. Temperature and drought stress adaptations of desert and coastal populations of *Isomeris arborea*. Ph.D. dissertation, Duke Univ., Durham, N. C. 141 p.

Turnage, W. V., and A. L. Hinckley. 1938. Freezing weather in relation to plant distribution in the Sonoran Desert. Ecol. Monogr. 8:529-550.

Turner, R. M. 1963. Growth in four species of Sonoran Desert trees. Ecology 44:760-765.

Turner, R. M., S. M. Alcorn, G. Olin, and J. A. Booth. 1966. The influence of shade, soil and water on saguaro seedling establishment. Bot. Gaz. 127:95-102.

Vogl, R. M., and L. T. McHargue. 1966. Vegetation of California fan palm oases on the San Andreas fault. Ecology 47:532-540.

Walter, H. 1971. Ecology of tropical and subtropical vegetation. Van Nostrand Reinhold, New York. 539 p.

Went, F. W. 1942. The dependence of certain annual plants on shrubs in Southern California deserts. Bull. Torrey Bot. Club 69:100-114.

-----. 1955. The ecology of desert plants. Sci. Amer. 192:68-75.

Woodell, S. R. J., H. A. Mooney, and A. J. Hill. 1969. The behavior of *Larrea divaricata* (creosote bush) in response to rainfall in California. J. Ecol. 57:37-44.

Wright, R. A. 1970. The distribution of *Larrea tridentata* (D. C.) Coville in the Avra Valley, Arizona. J. Ariz. Acad. Sci. 6:58-63.

Yang, T. W. 1967. Ecotypic variation in *Larrea divaricata*. Amer. J. Bot. 54:1041-1044.

-----. 1970. Major chromosome races of *Larrea divaricata* in North America. J. Ariz. Acad. Sci. 6:41-45.

THE SOUTHERN CALIFORNIA ISLANDS

CHAPTER

26 THE SOUTHERN CALIFORNIA ISLANDS

RALPH N. PHILBRICK
Santa Barbara Botanic Garden, Santa Barbara, California

J. ROBERT HALLER
Department of Biological Sciences, University of California, Santa Barbara

Introduction 894
Southern beach and dune 896
Coastal bluff 897
Coastal sage scrub 898
Maritime cactus scrub 899
Island chaparral 899
Valley and foothill grassland 900
Oak woodland 901
Southern riparian woodland 902
Coniferous forest 902
Coastal marsh 903
Discussion 903
Acknowledgment 905
Literature cited 906

INTRODUCTION

There are eight major islands off the coast of southern California between Pt. Conception and the Mexican border (Fig. 26-1). The more northern islands (San Miguel, Santa Rosa, Santa Cruz, and Anacapa) lie along an east–west line that roughly parallels the coast of Santa Barbara and Ventura Cos. The four more southern islands are scattered off the coast between Los Angeles and San Diego, with Santa Catalina about 32 km offshore and San Nicolas 98 km from the mainland; completing the southern group are tiny Santa Barbara Island and endemic-rich San Clemente.

The members of the northern group share many floristic affinities with each other and with the coastal areas north of their geographical position. The southern group forms a slightly less cohesive floristic unit with more xeric components.

The four largest islands (Santa Cruz, Santa Rosa, Santa Catalina, and San Clemente), with more varied topography, have the greatest diversity of vegetation.

With minor exception, the climate of the islands is highly maritime, characterized by mild temperatures with little fluctuation during the year. The mean annual temperature range is 3°C on San Miguel, and 6°C on Santa Catalina (Dunkle 1950). The former island is small, low, and fully exposed to the prevailing northwesterly wind, whereas the latter is large and high with its weather station located on the lee side. The documented mean July temperatures range from about 15°C on San Miguel to about 19°C on Santa Catalina (Dunkle 1950). The size of an island and its location in relation to the wind seem to be more important than its latitude in determining temperatures. The islands are essentially frost free, except for the interior central valley on Santa Cruz and possibly a few localities on the lee sides of the other larger islands.

Annual precipitation ranges from 52 cm on Santa Cruz (Santa Cruz Island Company weather records) to 27 cm on San Clemente (Dunkle 1950) and shows a fairly consistent decrease from the northern to the southern islands. This pattern is reflected by the vegetation, with relatively large areas of woodland and closed grassland on the northern islands and comparably large areas of open xeric scrub on San Clemente in the south. The summer drought that characterizes all of the islands is ameliorated somewhat by frequent fog or low clouds, especially on windward slopes.

Wind is an important factor in limiting the growth of vegetation on exposed sites.

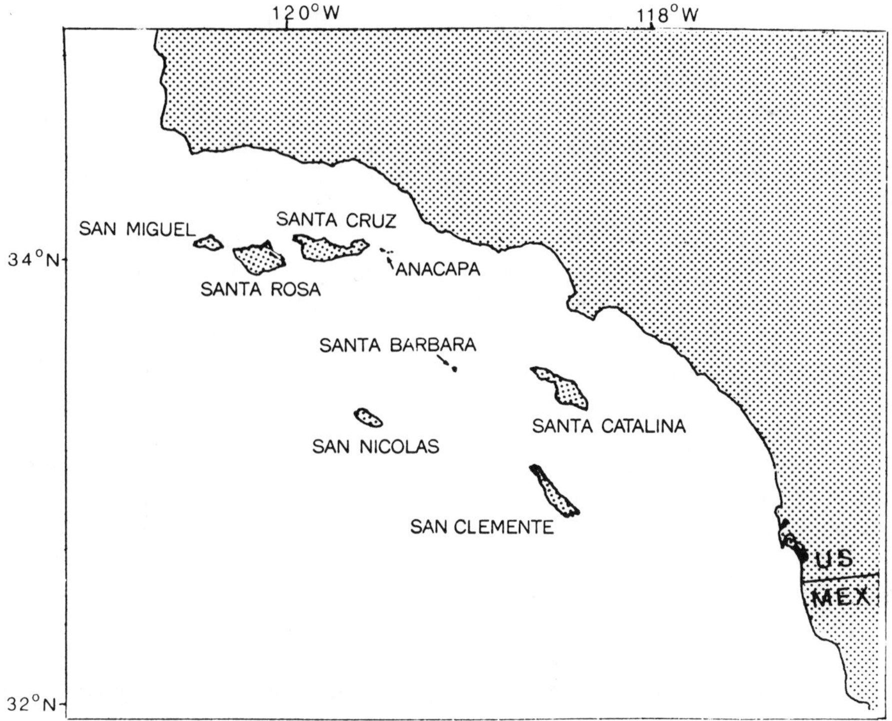

Figure 26-1. The southern California islands occur along the coast from Pt. Conception to San Diego. Their distance offshore ranges from 98 (San Nicolas) to 20 km (Anacapa), and their size is between 249 (Santa Cruz) and 2.6 km^2 (Santa Barbara).

The wind blows from the west or northwest most of the year, probably most strongly and steadily during the winter and spring. Wind may have a relatively direct effect on exposed vegetation by causing desiccation and dieback; but wind-borne salt from the ocean, rather than the wind itself, is probably the primary cause of dieback (Ogden 1975).

The community classification used in this chapter roughly follows that of Cheatham and Haller (1975). The vegetation is treated here under about a dozen categories. A case can easily be made for increasing or decreasing this number, depending primarily on the recognition given to minor associations.

There is an initial suspicion that the current concepts of California plant communities do not fit all insular associations. The island plant and community distributions have been altered by intensive grazing rather than the extensive bulldozing and construction of the mainland. On the mainland, therefore, entire coastal marshes and major areas of coastal and alluvial shrub vegetation have been physically removed, whereas modification on the islands is more likely to allow sea bluffs and interior cliffs to act as refuges. Another conspicuous feature of island communities is that many species, including insular endemics, are relatively wide-ranging and are not restricted to a particular community.

Although there is considerable taxonimic literature on insular endemics, there is no corresponding ecological literature. There are virtually no synecological data for the island communities and no autecological data for island-mainland species pairs.

SOUTHERN BEACH AND DUNE

The islands' beaches are relatively undisturbed by man. There are often broad sandy stretches without dunes, which support a low vegetation of *Abronia maritima, A. umbellata, Ambrosia chamissonis* (= *Franseria chamissonis*), *Cakile maritima, Camissonia cheiranthifolia* ssp. *cheiranthifolia* (= *Oenothera cheiranthifolia*), and *Distichlis spicata.*

The windiest sandy locations of San Miguel, Santa Rosa, Santa Cruz, San Nicolas, and San Clemente islands are areas of active dune-building (Fig. 26-2) and characteristic sites of *Abronia maritima, Ambrosia chamissonis, Cakile maritima,* and *Camissonia cheiranthifolia* ssp. *cheiranthifolia.* This occurrence of the northern subspecies of *Camissonia* suggests a slight insular modification of this southern unstabilized coastal sand dune association. Poorly developed dunes also occur on Santa Catalina.

The abundant sand and strong winds on San Miguel have caused the formation of more or less continuous dunes across large portions of the island. Several species of dune plants follow these dunes across the island, and therefore often occur farther inland than is usual for these species.

Abronia umbellata, Carpobrotus aequilaterus (= *Mesembryanthemum chilense*), and *Distichlis spicata* occur on the more protected dunes of most of the larger southern California islands and form a southern stabilized coastal sand dune association. *Astragalus miguelensis, Eschscholzia californica, Haplopappus venetus, Lotus argophyllus, L. scoparius,* and plants of the *Malacothrix incana* complex are also conspicuous in local areas.

A major portion of northwestern San Nicolas Island is covered by a *Lupinus albifrons* association on relatively stabilized sand (Foreman 1967). In a somewhat more consolidated sandy soil this same lupine is dominant northwest of Beecher's Bay on Santa Rosa.

Figure 26-2. Unstabilized coastal sand dune community on San Miguel Island. Sand extends beyond top of hill at right. James Tatum photograph.

Often the more stabilized sandy habitats are provided by Pleistocene dunes, which may occur some distance from the immediate coast. Such sites tend to allow the accumulation of more woody plants such as *Artemisia californica, Baccharis pilularis* ssp. *consanguinea, Haplopappus venetus, Lupinus albifrons,* and *Rhus integrifolia.* This association has similarities to the mainland dune scrub (see Chapter 7).

COASTAL BLUFF

The bluff communities of the islands are indistinct and variable. The more mesophytic sea cliffs are characteristic sites of a modified southern coastal bluff community, and this association often occurs on cliffs of similar substrate well away from the coast. This community varies from island to island along a north–south cline. Taken as a whole, it includes a relatively large number of island endemics.

Coreopsis gigantea, Eriogonum grande (= *E. latifolium* ssp. *grande*), *Malacothrix saxatilis,* and crustose and foliose lichens are nearly always present on the ocean bluffs of the northern islands from San Miguel to Anacapa; however, these plants are certainly not restricted to this community. The plants of a particular east-facing sea cliff on Santa Cruz, between Scorpion Rock and San Pedro Pt., include *Calystegia macrostegia* ssp. *macrostegia* (= *Convolvulus macrostegius*), *Erigeron glaucus, Eriophyllum staechadifolium,* and *Marah macrocarpus* in addition to the more widespread indicators mentioned above. *Eriogonum arobrescens* and *Haplopappus detonsus* are other frequent northern island associates. On the west end of East Anacapa, the bluff plants extend uninterrupted from bluff to bluff across the terrace of this narrow island.

Much of the northern island sea cliff community differs from the typical southern mainland bluffs by the insular presence of northern indicators, such as *Castilleja, Erigeron glaucus, Eriophyllum staechadifolium, Poa douglasii,* and *Polypodium* aff. *scouleri.*

Thorne's (1967) generalized description for Santa Catalina sea bluffs included *Calandrinia maritima, Crossosoma californicum, Diplacus puniceus* (= *Mimulus puniceus*), *Dudleya hassei, D. virens, Eriophyllum nevinii, Galium catalinense* ssp. *catalinense, Galvezia speciosa, Gilia nevinii, Hemizonia clementina,* and *Phacelia lyonii,* plus the widespread indicators. Several of these same species are also characteristic of the sea bluffs of San Clemente Island. On Santa Barbara the main coastal bluff plants are *Atriplex californica, Coreopsis gigantea, Eriogonum giganteum* ssp. *compactum, Eriophyllum nevinii,* and *Hemizonia clementina.* Many of the plants of this community have gray foliage.

The insular *Coreopsis gigantea* thickets, typically on the sandy deep soils of sea terraces, have an affinity to the coastal bluff communities (Fig. 26-3). This association is best preserved on San Miguel, Anacapa, Santa Barbara, and San Nicolas, although at least some of these stands are only a fraction of what they were 30 yr ago (Philbrick 1972). On Santa Barbara the usual *Coreopsis* associates are *Amblyopappus pusillus, Calystegia macrostegia* ssp. *macrostegia, Malacothrix foliosa, Marah macrocarpus,* and *Trifolium palmeri.*

Mainland *Coreopsis gigantea* is relatively confined in its distribution, and it occurs only at a few disjunct maritime sites. On the islands, however, *Coreopsis* may occur on sea terraces, on interior or coastal rocky cliffs, or in grasslands; many of the thickest stands are now confined to north-facing canyon slopes.

Figure 26-3. Giant coreopsis thicket in full bloom on San Miguel Island sea terrace.

The ice plant carpet, especially on San Miguel, western Santa Cruz, Santa Barbara, southeastern San Nicolas, and San Clemente islands, where it may extend for hectares, is an aggressive introduced community replacing the native plants of the coastal bluffs and the southern stabilized coastal sand dunes. *Mesembryanthemum crystallinum,* the chief ice plant for all except the most southern islands, excludes other vegetation by increasing the salt content of the soil above the tolerance of potential invaders (Vivrette 1973). However, in spite of the toxic effect of ice plant, aggressive native and introduced plants do exist in intermixed patches. On Santa Barbara, for example, *Amblyopappus pusillus, Amsinckia intermedia, Atriplex semibaccata, Hordeum glaucum* (= *H. stebbinsii*), and *Lasthenia chrysostoma* (= *Baeria chrysostoma*) are present within the ice plant carpet.

An association in which *Suaeda californica* and sea gulls predominate differs slightly from the ice plant carpet described above. Trampling by the gulls reduces much of the potential vegetation, and bird excrement alters the macronutrient content of the soil. Relatively few species endure under such conditions. The usual associates are *Atriplex semibaccata, Frankenia grandifolia, Mesembryanthemum crystallinum, M. nodiflorum,* and *Spergularia macrotheca* var. *macrotheca.*

COASTAL SAGE SCRUB

Artemisia californica, Encelia californica, Opuntia littoralis (= *O. occidentalis* var. *littoralis*), *O. oricola* (and hybrids between these two), *Rhus integrifolia,* and *Salvia mellifera* dominate much of the dry, rocky, south-facing slopes. This association differs from most of the mainland coastal sage by the consistent addition of the prickly pear opuntias. The insular coastal sage is closely related to the maritime cactus scrub; the latter is characterized primarily by its more southern distribution and its richness in cactus species.

Examples of insular coastal sage occur on the dry offshore slopes of San Miguel,

Santa Rosa, Santa Cruz, and Anacapa. This association is also characteristic of parts of San Nicolas Island; Foreman (1967) treated a mixture of dune, bluff, and sage vegetation as a "mixed-shrub community." The south-facing slopes near Little Harbor Beach, Santa Catalina, and the high southeastern end of San Clemente are other examples of the *Artemisia–Opuntia* association (Thorne 1967; Benedict, pers. commun.)

Baccharis pilularis ssp. *consanguinea* is the dominant plant on portions of the larger southern California islands, especially the northern ones. On Santa Cruz it is dominant in large areas inland from Christi Beach, near the western end of the island. On San Clemente it is primarily associated with dunes (Raven 1963). For San Nicolas, however, Foreman (1967) mapped much of the vegetation as a "baccharis community," which may have increased in density and extent after the removal of grazing. New reports of this *Baccharis* on Santa Barbara (Philbrick 1972) and on East Anacapa (1974 herbarium specimen at Santa Barbara Botanic Garden) may also indicate a spreading of this species. The prevalence of *Baccharis* on several islands, at least as a successional component of the vegetation, may relate these insular habitats to the central coastal scrub community on the mainland north of Pt. Conception (see Chapter 21). *Baccharis* is even more common in this community than on the islands and occurs both as an aggressive colonizer and as a component of the more persistent scrub vegetation.

MARITIME CACTUS SCRUB

Beginning on Santa Catalina and San Clemente and extending to the islands off northwestern Baja California is a community related to coastal sage scrub. This maritime cactus scrub supports various cacti in addition to those of the insular coastal sage and is equivalent to the "maritime desert shrub" of Dunkle (1950).

Although the cactus scrub community is more fully developed to the south, it was photographed by Thorne (1967) at Indian Head Pt., Santa Catalina, where *Bergerocactus emori* (= *Cereus emori*), *Lycium californicum*, *Opuntia littoralis*, and *O. prolifera* are dominants. In various sites *Brickellia californica*, *Mirabilis californica* (= *M. laevis*), and *Perityle emoryi* (all plants that occur abundantly in the coastal sage) supplement the cactus scrub community.

Opuntia prolifera occurs on all the islands from Santa Rosa to San Clemente. On Santa Barbara, for example, short plants of this cholla form dense patches on the south-facing slopes of the eastern part of the island, and associates include *Amblyopappus pusillus*, *Lycium californicum*, *Marah macrocarpus*, *Opuntia oricola*, and *Perityle emoryi*. Such populations are just beyond the northern limit of maritime cactus scrub and form a phase transitional to coastal sage scrub.

ISLAND CHAPARRAL

The dry, steep, rocky, north-facing slopes of the larger islands are frequently vegetated with chaparral that differs from that of the mainland by having uniquely insular plants such as *Crossosoma californicum*, *Eriodictyon traskiae* ssp. *traskiae*, *Prunus ilicifolia* ssp. *lyonii* (= *P. lyonii*), *Quercus* aff. *dumosa*, *Rhamnus pirifolia*

(= *R. crocea* ssp. *pirifolia*), and species of *Arctostaphylos, Ceanothus, Dendromecon,* and *Garrya.* The insular community also has more open space between the plants and more heterogeneous habits of growth, ranging from small to large plants and even grading into woodland. Windswept portions of Santa Rosa and Santa Cruz islands contain patches of low to prostrate *Adenostoma fasciculatum, Diplacus,* and *Quercus dumosa.* These *Adenostoma* plants appear to be genetically distinct in that they remain low growing when cultivated in wind-free situations at the Santa Barbara Botanic Garden. Nonendemic chaparral associates include *Cercocarpus betuloides, Heteromeles arbutifolia, Malosma laurina* (= *Rhus laurina*), *Rhus integrifolia, R. ovata, Toxicodendron diversilobum* (= *Rhus diversiloba*), and *Xylococcus bicolor.*Many of these plants occur together in canyons and on slopes of interior southeastern Santa Catalina.

An interesting group of plants is often associated with chaparral on steep, rocky cliffs, that are essentially free from grazing. On Santa Cruz the plants of such north-facing ledges usually include *Dudleya greenei, Selaginella bigelovii,* and *Silene laciniata.* Other species often present are *Haplopappus detonsus, Heuchera maxima, Polypodium californicum,* and *Stephanomeria cichoriacea* along with *Diplacus longiflorus* (= *Mimulus longiflorus*), *D. parviflorus* (= *Mimulus flemingii*), *Keckiella cordifolia* (= *Penstemon cordifolius*), *Toxicodendron diversilobum,* and small plants of *Prunus ilicifolia* ssp. *lyonii, Quercus dumosa,* and *Rhamnus pirifolia.*

VALLEY AND FOOTHILL GRASSLAND

In spite of grazing pressures, the four larger islands still contain fairly significant areas of at least depauperate native grasslands. This association occurs on seasonally dry but reasonably deep soils. On Santa Rosa (west of Beecher's Bay) and on Santa Cruz (south of China Harbor) the conspicuous nongrass associates include *Achillea borealis, Dichelostemma pulchella* (= *Brodiaea pulchella*), *Ranunculus californicus,* and *Sisyrinchium bellum,* an assemblage that suggests the more northern coastal prairie. *Stipa* is also present in this grassland, but it is usually not nearly as extensive or abundant as the introduced annual grasses.

Small areas of San Miguel and Anacapa also support native grasslands. The interior part of San Clemente, on the other hand, contains impressive areas of native grasslands with dominant *Stipa.*

Other grassland plants on the islands are *Achyrachaena mollis, Agoseris heterophylla, Allium praecox, Amsinckia, Aristida divaricata, Astragalus, Bloomeria crocea* var. *crocea, Calandrinia ciliata* var. *menziesii, Calochortus, Castilleja affinis, Clarkia, Cryptantha, Delphinium parryi, Dodecatheon clevelandii* ssp. *insulare, Eschscholzia californica, Gilia, Lasthenia chrysostoma, Layia platyglossa, Lotus, Lupinus, Melica imperfecta, Microseris, Orthocarpus, Poa scabrella, Sanicula arguta, Sidalcea malviflora, Trifolium,* and *Viola pedunculata.*

All of the islands have extensive areas dominated by introduced grasses, such as *Avena fatua, Bromus diandrus* (= *B. rigidus*), *B. rubens,* and *Hordeum glaucum.* Often the associated species of the native grasslands persist, but frequent weedy associates are *Erodium cicutarium, Malva parviflora,* and *Medicago polymorpha* (= *M. hispida*).

OAK WOODLAND

The better soils of bottomland and riparian borders, especially on Santa Rosa and Santa Cruz, are occupied by large trees of *Quercus agrifolia,* which form a southern coastal oak woodland. Occasionally a continuous stand with intermeshing crowns produces a coast live oak forest, as in upper Cañada del Puerto, Santa Cruz Island, where *Heteromeles arbutifolia, Marah macrocarpus,* and *Symphoricarpos mollis* are also present.

Often grading into island chaparral or southern coastal oak woodland is an island woodland dominated by such unique island trees and large shrubs as *Arctostaphylos insularis, A. tomentosa* ssp. *insulicola, Ceanothus arboreus, Lyonothamnus floribundus, Pinus* aff. *remorata, Quercus* ×*macdonaldii,* or *Q. tomentella* (Fig. 26-4). This woodland is often on dry, rocky, steep slopes at higher elevations. Good examples are found on the high, north-facing onshore slopes of Santa Cruz and (in more depauperate form) on the rocky sea cliffs of southeastern San Clemente. *Arbutus menziesii* (only two struggling, old populations of about four clones

Figure 26-4. Santa Cruz Island ironwood often forms small groves of island woodland without an associated understory.

on Santa Cruz), *Comarostaphylis diversifolia, Quercus chrysolepis,* and *Xylococcus bicolor* are widespread nonendemic associates of this island woodland.

Island woodland also occurs as relatively discrete groves (especially of *Lyonothamnus floribundus* or *Quercus tomentella*) in subriparian situations near stands of *Quercus* ×*macdonaldii* or *Q. agrifolia.* This is well illustrated in the upper reaches of canyons on the northeast slope of Black Mt. on Santa Rosa, in various locations in the interior of Santa Cruz, and in several of the major canyons on southeastern Santa Catalina. Island woodland may have characterized much larger areas of the islands in prior times.

SOUTHERN RIPARIAN WOODLAND

Riparian woodlands are confined to wet soils along major streams on Santa Rosa, Santa Cruz, and Santa Catalina and are usually quite depauperate, dominated by a few species of *Populus* and *Salix.* The expression of this community is probably restricted by limited rainfall, relatively small watersheds, a lack of consistent underground and surface water, and a lack of more mesic refuges from which to repopulate after extended drought.

Thorne (1967) described riparian communities in Middle Ranch and Cottonwood canyons, Santa Catalina, with *Populus fremontii, P. trichocarpa, Salix laevigata, S. lasiolepis,* and *Sambucus mexicana.* Similar riparian woodlands occur at several interior locations on.Santa Cruz.

A half-dozen small to medium-sized stands of *Acer macrophyllum* occur at relatively low elevations on the north side of Santa Cruz Island. This occurrence parallels that of *Arbutus menziesii* in the island woodland and suggests a remnant of an insular mixed evergreen forest.

CONIFEROUS FOREST

The closed-cone pines of the *Pinus muricata* – *P. remorata* complex dominate extensive, more or less north-facing slopes in four areas of Santa Cruz Island. Each grove, or series of groves, differs in cone morphology, and each exhibits a partially different assemblage of community associates. A single smaller grove (about 250 trees) occurs in the central part of Santa Rosa Island at elevations between 150 and 300 m.

The majority of the trees in most of the groves of insular bishop pine forest resemble *Pinus remorata,* although not completely. A smaller number of trees vary in the direction of *P. muricata.* The northernmost of the Santa Cruz Island groves is exceptional because there is no single, obvious morphological mode, and the variability seems unusually large.

The pines are relatively short-lived, reaching an age of about 65 yr (Linhart et al. 1967). With age they become increasingly susceptible to insect damage and diseases. Where the groves have been protected from fire for extended periods (as in Christi Canyon, western Santa Cruz Island), many dead and diseased trees are evident.

Diplacus parviflorus, Lilium humboldtii, Quercus parvula (= *Q. wislizenii* var. *frutescens*), and *Sanicula arguta* are a few of the more regular associates of the insular bishop pine forest. Other frequently occurring species are *Adenostoma fasciculatum, Arctostaphylos confertifolia* (= *A. subcordata* var. *confertifolia*), *A. tomentosa* ssp. *subcordata* (= *A. subcordata*), *Comarostaphylis diversifolia, Heteromeles arbutifolia, Lyonothamnus floribundus* ssp. *asplenifolius, Marah macrocarpus,* and *Vaccinium ovatum.*

The main island stand of *Pinus torreyana* consists of a grove which runs about 1 km along the north-facing slope of a ridge just south of Beecher's Bay on Santa Rosa. This Torrey pine forest begins near the base of the slope at an elevation of 60 m and extends up-slope about 0.4 km to an elevation of 175 m near the top of the ridge. Growing with these pines are several other insular endemics or near-endemics, such as *Castilleja hololeuca, Ceanothus arboreus, Coreopsis gigantea, Dendromecon rigida* ssp. *harfordii* (= *D. harfordii*), *Diplacus parviflorus, Dudleya greenei, Eriogonum grande, Galium catalinense* ssp. *buxifolium* (= *G. buxifolium*), *Lyonothamnus floribundus* ssp. *asplenifolius, Salvia brandegei,* and *Solanum wallacei* var. *clokeyi.* A second smaller grove occurs about 1 km southeast of the main grove.

The morphologically distinct Torrey pine of the mainland (Haller 1967) has its own group of associated species, including a number that do not occur on the island at all. Several, such as *Adenostoma fasciculatum, Heteromeles arbutifolia, Pityrogramma triangularis, Polypodium californicum,* and *Rhus integrifolia,* occur with both the mainland and island Torrey pines.

Much as in the case of the island bishop pines, few mesophytic herbs grow directly under mature Torrey pines on Santa Rosa. This lack of herbaceous understory is undoubtedly influenced by dense tree crowns and abundant needle litter.

COASTAL MARSH

The mouths of various canyons and streams, especially on Santa Rosa, Santa Cruz, San Nicolas, and Santa Catalina, are sites of freshwater, brackish, or saltwater marshes. Examples of coastal salt marsh occur on the central east coast of Santa Rosa and on Jehemy Beach at the southeastern end of San Nicolas, where *Distichlis spicata, Frankenia grandifolia,* and *Salicornia* predominate. Thorne (1967) described a small salina in Catalina Harbor, Santa Catalina Island, as containing *Atriplex leucophylla, A. watsonii, Distichlis spicata, Frankenia grandifolia, Jaumea carnosa, Monanthochloe littoralis, Salicornia subterminalis, S. virginica, Spergularia macrotheca* var. *macrotheca,* and *Suaeda californica.* The lagoon at the mouth of Cañada del Puerto, Santa Cruz Island, with *Scirpus californicus* and *Typha domingensis,* is nearer the freshwater end of the spectrum.

DISCUSSION

Paleoclimatic and fossil evidence suggests that the modern plant communities on the southern California islands are the result of successive waves of migration from different source areas (Axelrod 1967). The more mesic components of these communities, exemplified by *Pinus remorata,* were better represented in the late Pleistocene than at present. Maritime cactus scrub, the most xeric vegetation, presumably expanded on the islands during the more recent Xerothermic. The island chaparral

TABLE 26-1. A generalized summary of plant communities on the southern California islands. Presence of a community, regardless of its extent, is indicated by (+) and absence by (-). Isolated occurrences of individual species, apart from their characteristic communities, are not shown.

	Island and Area (km^2)								
Community	Santa Cruz (249)	Santa Rosa (217)	Santa Catalina (194)	San Clemente (145)	San Nicolas (58)	San Miguel (37)	Anacapa (2.9)	Santa Barbara (2.6)	Total
Southern coastal dune	+	+	+	+	+	+	-	-	6
Coastal bluff	+	+	+	+	-	+	+	+	7
Coastal sage scrub	+	+	+	+	+	+	+	-	7
Maritime cactus scrub	-	-	+	+	-	-	-	-	2
Island chaparral	+	+	+	-	-	-	-	-	3
Valley and foothill grassland	+	+	+	+	+	+	+	+	8
Southern coastal oak woodland	+	+	+	-	-	-	-	-	3
Island woodland	+	+	+	+	-	-	-	-	4
Southern riparian woodland	+	+	+	-	-	-	-	-	3
Bishop pine forest	+	+	-	-	-	-	-	-	2
Torrey pine forest	-	+	-	-	-	-	-	-	1
Coastal marsh	+	+	+	+	+	-	-	-	5
Total	10	11	10	7	4	4	3	2	

and island woodland may be the oldest island communities. Their modern taxonomic and vegetational relatives occur today in the temperate highlands of Mexico. Similar communities apparently occurred on the California mainland during the Pliocene. It seems likely that the present island chaparral and island woodland are Pliocene relicts, preserved by the equable insular climate. One factor in the divergence of the mainland vegetation from the older insular types may have been the more frequent occurrence of fire, which could have favored the development of dense chaparral of uniform height on the mainland.

The plant communities of the southern California islands now show a number of affinities with more northerly coastal associations. These northern affinities are correlated with, and probably made possible by, the more moderate temperature and higher humidity on the islands than on the adjacent southern California mainland. Examples of island plant associations with some degree of northern affinity are coastal sand dune, coastal bluff, native grassland, bishop pine forest, and the isolated patches of *Acer macrophyllum* and *Arbutus menziesii* reminiscent of mixed evergreen forest.

A half-dozen species are conspicuous in that they occur in three to four different insular communities. The wide-ranging *Coreopsis gigantea* occurs on sea terraces, grasslands, interior or coastal cliffs, and canyon slopes. *Adenostoma fasciculatum, Heteromeles arbutifolia, Lyonothamnus floribundus, Marah macrocarpus,* and *Rhus integrifolia* all occur in island chaparral and in at least two other communities. Most of these wide-ranging plants are also present in or adjacent to biship pine and Torrey pine forests.

Island chaparral, island woodland, and Torrey pine forest are all, as might be expected, rich in insular endemics. The coastal bluff association contains an equally impressive number of endemic plants and may have once been the major plant community of the islands, covering extensive areas in addition to the cliffs and few sea terraces to which it is presently restricted. Many of the coastal bluff plants are succulent, fragile, and particularly susceptible to the intensive grazing that has occurred on the islands; thus the present restriction of these plants is not surprising.

In summary, the coastal bluff, coastal sage, and grassland communities occur on all or nearly all of the southern California islands. Torrey pine forest, on the other hand, occurs on only a single island, Santa Rosa. Santa Rosa, Santa Cruz, and Santa Catalina are the only islands having chaparral, oak woodland, and riparian woodland. These three islands have more communities than the other islands, whereas Santa Barbara Island has the fewest, as summarized in Table 26-1.

ACKNOWLEDGMENT

Special appreciation is due Michael R. Benedict for his extensive criticism of an earlier draft of this chapter.

LITERATURE CITED

Axelrod, D. I. 1967. Geologic history of the Californian insular flora, pp. 267-315. In R. N. Philbrick (ed.). Proc. symp. on biology of California Islands. Santa Barbara Bot. Gard., Santa Barbara, Calif.

Cheatham, N. H., and J. R. Haller. [Unpubl.] Dec. 1975. An annotated list of California habitat types.

Dunkle, M. B. 1950. Plant ecology of the Channel Islands of California. Pub. Allan Hancock Pac. Exped. 13:247-386.

Foreman, R. E. 1967. Observations on the flora and ecology of San Nicolas Island. U.S. Naval Radiol. Defense Lab. TR-67-8. 79 pp.

Haller, J. R. 1967. A comparison of the mainland and island populations of Torrey pine, pp. 79-88. In R. N. Philbrick (ed.). Proc. symp. on biology of California Islands. Santa Barbara Bot. Gard., Santa Barbara, Calif.

Linhart, Y. B., B. Burr, and M. T. Conkle. 1967. The closed-cone pines of the Northern Channel Islands, pp. 151-177. In R. N. Philbrick (ed.). Proc. symp. on biology of California Islands. Santa Barbara Bot. Gard., Santa Barbara, Calif.

Ogden, G. L. 1975. Differential response of two oak species to far inland advection of sea-salt spray aerosol. Ph.D. dissertation, Univ. Calif., Santa Barbara. 230 pp.

Philbrick, R. N. 1972. The plants of Santa Barbara Island, California. Madroño 21:329-393.

Raven, P. H. 1963. A flora of San Clemente Island, California. Aliso 5:289-347.

Thorne, R. F. 1967. A flora of Santa Catalina Island, California. Aliso 6:1-77.

Vivrette, N. J. 1973. Mechanism of invasion and dominance of coastal grassland by Mesembryanthemum crystallinum. Ph.D. dissertation, Univ. Calif., Santa Barbara. 78 pp.

APPENDIX

THE MAP OF THE NATURAL VEGETATION OF CALIFORNIA

A. W. KÜCHLER
Department of Geography, University of Kansas, Lawrence

Introduction	909
Some aspects of vegetation	910
The map legend	911
Overview of the expanded legend	913
Acknowledgments	914
Literature cited	915
Description of the vegetation: the expanded legend	916

INTRODUCTION

The land boundaries of California follow mostly straight lines. They serve to establish the state's precise geographical location by latitude and longitude but this also means that, from an ecological point of view, the boundaries are entirely artificial. Therefore, California cannot at all be considered an ecological region. However, the geographical location of the area with a "Mediterranean" climate results in a region with a vegetation all its own. This more strictly Californian vegetation is surrounded by extensive deserts, and it is through these that the boundaries pass. The deserts are shared, therefore, with the neighboring states of Oregon, Nevada, and Arizona, forming a broad belt around the more specifically Californian core. This isolating belt comes to an end in the far northwest where the forests continue across the state line without interruption. California's geographical location, its long history, and its vigorous topography combine to produce a great number and variety of plant communities.

This is not the first attempt to map the vegetation of California. It is not necessary here to recount the history of vegetation mapping in this state; suffice it to say that available vegetation maps of small areas (e.g. quadrangles) are numerous whereas maps of the whole state are very few indeed. California is, of course, included on the vegetation maps of the entire United States by Shantz and Zon (1924) and by Küchler (1964, 1967b). Wieslander published a map of the state in 1945. But the entirely new and up-to-date treatment of the Californian vegetation by Barbour and Major and their colleagues necessarily calls for a correspondingly modern vegetation map of the entire state. The new vegetation map of California at the scale of 1 : 1,000,000 tries to fulfill this need.

SOME ASPECTS OF VEGETATION

Vegetation may be defined as the mosaic of plant communities in the landscape. Plant communities, or phytocenoses, consist of growth forms such as trees, shrubs, grasslike plants (graminoids), forbs (nongraminoid herbs), and bryoids (mosses and lichens). These various growth forms occur in different combinations and result in the structure of the phytocenose. Structure can therefore be defined as the spatial distribution pattern of growth forms in a phytocenose. One or a few growth forms usually dominate, giving the phytocenose its peculiar appearance or physiognomy.

In addition, the members of a plant community can be identified as taxa such as genera, species, or units of some other rank. The aggregate of these taxa is called the floristic composition of the phytocenose.

All phytocenoses consist of growth forms and of taxa, and the combination of structure and floristic composition identifies a plant community more precisely than anything else can. This statement applies to all phytocenoses anywhere on the globe, and there is no exception to the rule.

The vegetation is an integral part of the ecosystem. It is thus the best and most logical foundation on which to base our inquiry into the ecological conditions of the landscape because it alone expresses the intricate interrelations between all factors of the environment and the plants that survive in it. The whole organization of our living space is, in fact, based on these interrelations. *The vegetation is the tangible, integrated expression of the entire ecosystem, and mapping the vegetation is the only effective method to present the ecological order of our living space.* This is so because only on a map can the geographical location, extent and distribution of plant communities be shown objectively and in a really meaningful way (Küchler 1973, 1974).

An important distinction among the various plant communities is based on the degree to which they are affected by man. Plant communities range from undisturbed types such as some forests in an inaccessible landscape through disturbed types where grazing or selective logging may have resulted in basic changes. At the other end of the spectrum, man has removed the vegetation altogether and replaced it with different types such as wheat fields, vineyards, and the like. These types are also phytocenoses, composed of the crop plants and their associated weed communities. It is characteristic of man-made plant communities that they are short-lived, and are frequently changed into quite different communities.

A vegetation map that shows man-made phytocenoses quickly becomes an historical document as the vegetation follows economic exigencies. From an ecological point of view, however, a vegetation map gains in value if it is so organized as to retain its usefulness over a longer period, and if it portrays the vegetation types as indicators of environmental conditions, that is, above all the natural vegetation.

Vegetation that is not appreciably disturbed by man is termed natural. Natural phytocenoses have definite structures and floristic compositions as dictated by their evolution and by their respective environments. If man destroys the natural vegetation and then abandons the land, the natural vegetation will usually return. Therefore, the natural vegetation is actually there to be observed, or else it exists potentially. Be it actual or potential, it is essentially the same as the climax vegetation *sensu* Whittaker (1970). Much of the Californian natural vegetation is today potential rather than actual.

Some readers may wonder whether the potential natural vegetation of California is sufficiently well known to be shown on a map. The answer is twofold. First, in large parts of the state, the natural vegetation is still extant or has been disturbed so recently that satisfactory information on it is available. This permits mapping it with considerable accuracy. Second, there are parts of California where the potential natural vegetation as portrayed on the new map is indeed no more than an approximation. However, considering the small scale of the map and our present knowledge, this approximation is relatively close to the mark, close enough at any rate to be accepted and included here. In addition, areas of approximation need further investigation, and the new map will have fulfilled one of its major objectives if it stimulates research into the nature of the potential natural vegetation in regions of uncertainty.

Therefore, in the light of the philosophy outlined above, the new vegetation map of California portrays the natural vegetation. However, it is not practical to distinguish on the map between what is actual and what is potential.

A vegetation map showing potential natural vegetation should always bear a date. The date of the new map of California, 1977, implies that man's influences during the past centuries may have had a lasting effect on the potential natural vegetation of today. One such effect may be the accidental or deliberate introduction of Old World plant species. Some of these became acclimatized and managed to spread on their own, once man had helped them across the ocean. Included here is a host of weeds, but most of these quickly succumb to the competition of native species wherever the vegetation is free to develop toward a relatively stable equilibrium.

Some species, however, seem to be able to maintain themselves, as for instance some annual grasses from southern Europe which successfully invaded many Californian grassland communities. Complete acclimatization is entirely feasible but there is no definite certainty that the immigrants in California can actually maintain themselves in the face of competition from the natives where human interference is withdrawn and the vegetation is permitted to return to climax conditions. Known immigrant species have therefore not been listed in the legend of the new map (with one exception).

THE MAP LEGEND

One of the basic problems a vegetation mapper must face is to select the manner of portraying the vegetation. The units he must establish are the vegetation types he presents on his map. A number of classifications of vegetation permit the mapper to break down the vegetation of a region into a variety of categories of any rank and size. Unfortunately, however, no classification of vegetation has found worldwide approval (Küchler 1967a).

The ideal legend items of a vegetation map describe the individual plant communities in such a manner that the reader can visualize them. But on a map at the scale of 1:1,000,000 the rich vegetation of California will easily result in a legend much larger than the map itself. The names of all legend items are therefore kept as brief as is consistent with the needs of both laymen and scientists in California and elsewhere.

A uniform approach is maintained throughout the entire legend in order to keep all units on the same level and to keep them comparable. The names of the individual legend items consist of an English name, followed by the scientific name of the leading taxon or taxa. Local names, such as chaparral, have been used wherever they exist. The English names show no consistency whatever as some of them refer to environmental conditions (e.g. alpine communities) or to location (e.g. seashore communities). The scientific names of the dominant taxa have been translated into English where no local name is in use. The combination of English names with the scientific names of dominant taxa allows the identification of the vegetation types in a clear and brief form. This identification rests on the dominant genera or species. Species are stipulated wherever only one dominant is listed. Communities with two or more dominants are characterized by genera only, or by the name of a genus followed by "spp." if two or more dominants are species of the same genus (e.g. *Opuntia* spp.). The English names serve to distinguish between two phytocenoses if their scientific names are the same.

The legend items are numbered consecutively, and these numbers appear in the legend as well as on the face of the map. The numbers and colors tie map and legend together, thus facilitating the interpretation of the map. Every area on the map has a number unless it is too small. In such a case, the colors alone will guide the reader to the correct legend item. Even though some of the colors may resemble one another somewhat, the reader will nevertheless be able to distinguish them readily because the numbers reveal whether they are the same or not.

The taxa listed in the title of each legend item are joined by hyphens. This ties them together and indicates that they belong together forming a community of which each is an important part. But there are exceptions, such as the pine, cypress forests (*Pinus, Cupressus*), where the generic names do not form a single type and do not necessarily occur together. In order to maintain such a vegetation type on the map and at the same time indicate that the connection between the listed taxa is very loose, their names are separated by commas rather than joined by hyphens.

Both the growth forms and the taxa may have a very even distribution throughout a given phytocenose. More often, however, their distribution is quite irregular. The types of vegetation shown on the map are therefore not so uniform as the reader might expect at first sight. Frequently, a type may occur in its "pure" form, but often there are inclusions of other types, islands as it were, of subtypes or even altogether different types. Such variations and inclusions make a type more heterogeneous than appears on the map but they are too small to be shown at the given scale. For instance, there are numerous inclusions of chaparral in the area of the blue oak woodlands but only the larger ones can be shown separately.

Inclusions are largely a matter of scale. But there are also local variations of the vegetation, resulting from the varying prominence of individual taxa or their occasional absence. This may apply even to the dominants. Such variations are based on local environmental differences and cannot be recorded on a map at 1:1,000,000.

In addition, vegetation units extend horizontally in plains and vertically in mountains from one set of environmental conditions to another. Any given type of vegetation may therefore differ more or less, and sometimes quite markedly, at its opposite borders where it merges with adjacent categories.

The vegetation types on the new map are not units of some hierarchic classification. This places all legend items on the same level. The classless approach is not affected by the grouping of vegetation types into physiognomic and floristic units such as needle-leaved evergreen forests or creosote bush. Such categories may perhaps serve as nuclei for a classification, but as used here, they are simply a device to assist the reader in establishing and locating a type more readily.

All plant names are according to Munz and Keck (1973).

OVERVIEW OF THE EXPANDED LEGEND

The new vegetation map of California was prepared in conjunction with *Terrestrial Vegetation of California* edited by Barbour and Major (1977). The editors and many of their collaborators contributed to the map. However, the map is an independent creation with its own philosophy, organization, and method. In due time, the map will become available separately, accompanied by a commentary and an expanded legend, that is, a systematic "description of the vegetation." The names of the legend items are as brief as possible without impairing their usefulness. However, many a reader, on studying the map, will wish to explore individual vegetation types and seek further information on them. The description of the vegetation is to serve this purpose. The following paragraphs will allow the reader to appreciate just how this description is organized and what kind of information he may expect to find in it.

In any one phytocenose, there is a limited number of growth forms and of taxa, some of which may be more prominent than others. It is the presence and the particular proportion of each that give a plant community its unique and unmistakable character, that is, its structure and floristic composition. Because of their universal applicability, structure and floristic composition have been selected to serve as the exclusive criteria for establishing the units of vegetation on the new vegetation map of California. This permits a uniform approach to all phytocenoses throughout the state as well as meaningful comparisons of any individual plant community with any other one. The reader can visualize each vegetation type and at the same time determine what it is that distinguishes one type from another. The method has been successfully employed on the vegetation maps of the United States (Küchler 1964), northwestern Thailand (Küchler and Sawyer 1967), and Kansas (Küchler 1974).

The "Description of the Vegetation" is designed to supplement the map legend by greatly expanding the information on the legend items. There are five parts to each individual description of a vegetation type:

1. The title.
2. The structure.
3. The dominants.
4. Other characteristic components.
5. Location

The title consists of a number, an English name, and the scientific names of the dominant taxa. In the description, the number and the title of a legend item are repeated exactly as they appear on the map. This means the closest coordination of the map legend with the description. Therefore, a reader who studies the map and wishes more information on a given type of vegetation is enabled to find it instantly.

The structure is described briefly and in general terms. It should be understood that it may vary considerably within a given type of vegetation, especially where a type covers large areas or very uneven terrain.

The dominants are, of course, particularly important in characterizing a vegetation type. They may range in number from one to several but an effort has been made to keep the number to a minimum. The dominants are given on the species level, identified by both their vernacular and their scientific names. The former will be appreciated by the layman whereas the latter is essential for proper identification.

The other characteristic components of a given vegetation type assist in establishing the plant community. No effort has been made to present complete check lists of all species occurring in the community. This is neither necessary nor desirable for a vegetation map at 1:1,000,000. These additional species are listed by their scientific names only. This is not only adequate but necessary because many of them have no common names.

All taxa are listed in alphabetical order, and definitely not in the order of their prominence. The prominence varies too much within a vegetation type, and therefore cannot serve as the basis of a listing order as serious misinterpretation might result.

Sometimes, comments like "northern part" or "higher elevations" follow the name of a species. They refer only to the specific part of the area of the vegetation type in which this species is listed, not to the whole state of California. Thus red fir occurs at lower elevations in type 17 even though that may be over 3000 m above sea level.

The last part of the description does not refer to the character of a phytocenose but rather to its location, directing the reader in his endeavor to find it on the map. This direction is kept in very general terms.

Finally, there is a brief indication of the chapter of *Terrestrial Vegetation of California* where the reader may find additional details. Thus, B&M ch. 16 means that there is more information on the subject in Chapter 16 of Barbour and Major's book.

ACKNOWLEDGMENTS

It would have been a difficult and very time-consuming task to prepare a meaningful vegetation map of California with the desired detail, had it not been for the generous help and advice of many colleagues and other interested persons throughout the state. Many of the contributors to *Terrestrial Vegetation of California* helped me with valuable information, advice, and cartographic material in the areas of their specialties. I happily express my gratitude to all; their help has been invaluable. I cannot but mention a few to whom I feel particularly indebted. They include M. G. Barbour, J. H. Burk, D. T. Gordon, J. R. Griffin, J. R. Haller, T. L. Hanes, H. F. Heady, K. B. Macdonald, J. Major, H. A. Mooney, J. O. Sawyer, R. F. Thorne, F. C. Vasek, R. J. Vogl, J. A. Young, and P. Zinke.

The preparation of the vegetation map was therefore to a considerable degree a cooperative venture. As author, I alone am responsible for the map content and its organization, and only I can be blamed for avoidable errors and misinterpretations. On the other hand, where praise is due, I am happy to share it with all who helped to make this map a success. My heartfelt thanks go to them all.

Finally, I am most grateful to various agencies who contributed most generously to the cost of preparing the new vegetation map of California. They include the Ahmanson Foundation, the California Native Plant Society, the Sierra Club, the U.S. Bureau of Reclamation, the U.S. Forest Service, the University of Kansas, and John Wiley & Sons. Their support helped to make the new vegetation map a reality.

LITERATURE CITED

Barbour, M. G., and J. Major (eds.). 1977. Terrestrial Vegetation of California. Wiley, New York.

California Soil and Vegetation Survey. Continuing. Pacific Southwestern Forest and Range Experiment Station, Berkeley.

Küchler, A. W. 1964. Potential natural vegetation of the conterminous United States. American Geographical Society, New York, Special Publication No. 36. 154 pp.

-----. 1967a. Vegetation mapping. Ronald Press, New York. 472 pp.

-----. 1967b. Potential natural vegetation of the United States. U.S. Geological Survey, National Atlas, Washington, D.C.

-----. 1973. Problems in classifying and mapping vegetation for ecological regionalization. Ecology 54:512-523.

-----. 1974. A new vegetation map of Kansas. Ecology 55:586-604.

Küchler, A. W. and J. O. Sawyer. 1967. A study of the vegetation near Chiengmai, Thailand. Trans. Kansas Acad. Sci. 70:281-348. With map.

Latting, J. (ed.). 1976. Plant Communities of Southern California. California Native Plant Society, Berkeley, Special Publication No. 2. 164 pp.

Munz, P. A., and D. D. Keck. 1973. A California flora, with supplement. Univ. Calif. Press, Berkeley. 1816 pp.

Shantz, H. L., and R. Zon. 1924. The natural vegetation of the United States. In: O. E. Baker (ed.). Atlas of American Agriculture, section 7. Washington, D.C., U.S.D.A.

Whittaker, H. 1970. Communities and ecosystems. MacMillan Company, New York. 161 pp.

Wieslander, A. E., 1945. Vegetation types of California. U.S.D.A. Forest Service, Calif. Forest & Range Exp. Stat., Berkeley.

DESCRIPTION OF THE VEGETATION: THE EXPANDED LEGEND

1. GRAND FIR-SITKA SPRUCE FOREST (*Abies-Picea*)

STRUCTURE: Very tall, dense, needle-leaved evergreen forest. Broad-leaved deciduous trees very rare; very little undergrowth. Bryoids are sometimes extensive.

DOMINANTS: Grand fir (*Abies grandis*), Sitka spruce (*Picea sitchensis*)

OTHER CHARACTERISTIC COMPONENTS: *Alnus rubra, Epilobium augustifolium, Polystichum munitum, Pteridium aquilinum* var. *pubescens, Rubus spectabilis, R. vitifolius, Thuja plicata, Tsuga heterophylla.*

LOCATION: Along the coast in Humboldt and Del Norte counties.

B&M ch. 19

2. REDWOOD FOREST (*Pseudotsuga-Sequoia*)

STRUCTURE: Very tall, dense, needle-leaved evergreen forest. Broad-leaved evergreen, medium tall trees gradually increase eastward. Undergrowth is low and patchy with forbs mainly on alluvial sites, shrubs and low trees on the uplands.

DOMINANTS: Douglas fir (*Pseudotsuga menziesii*), redwood (*Sequoia sempervirens*)

OTHER CHARACTERISTIC COMPONENTS: *Alnus rubra* (northern part), *Arbutus menziesii, Chamaecyparis lawsoniana* (northern part), *Cornus nuttallii, Gaultheria shallon, Lithocarpus densiflora, Myrica californica, Oxalis oregona, Polystichum munitum, Pteridium aquilinum* var. *pubescens, Quercus garryana, Rhododendron californica, R. occidentalis, Rhus diversiloba, Thuja plicata* (northern part), *Tsuga heterophylla* (northern part), *Umbellularia californica, Vaccinium ovatum, V. parvifolium, Vancouveria planipetala.*

LOCATION: Mainly western side of Coast Ranges from Monterey county to Oregon.

B&M ch. 19

3. NORTHERN JEFFREY PINE FOREST (*Pinus jeffreyi*)

STRUCTURE: A tall, needle-leaved evergreen forest, moderately open (coverage about 65%). A rather open layer of broad-leaved shrubs, both evergreen and deciduous, may be dense locally, rarely exceeding 2 m in height. A low herbaceous layer varies much in density.

DOMINANTS: Jeffrey pine (*Pinus jeffreyi*)

OTHER CHARACTERISTIC COMPONENTS: *Abies concolor, A. magnifica, Arctostaphylos patula, Artemisia tridentata, Calocedrus decurrens, Ceanothus prostratus, Cercocarpus ledifolius, Deschampsia elongata,*

Elymus glaucus, Juniperus occidentalis, Monardella odoratissima, Pinus ponderosa, Purshia tridentata, Sitanion hystrix, Stipa occidentalis, Wyethia mollis.

LOCATION: East side of Sierra Nevada and Modoc Plateau

B&M ch. 17

4. SOUTHERN JEFFREY PINE FOREST (*Pinus jeffreyi*)

STRUCTURE: Open, tall or medium tall, needle-leaved evergreen forest. Some lower, broad-leaved evergreen and/or deciduous trees. Broad-leaved evergreen, sclerophyll shrubs may be dense. Graminoids and forbs are common in more open spaces.

DOMINANTS: Jeffrey pine (*Pinus jeffreyi*)

OTHER CHARACTERISTIC COMPONENTS: *Abies concolor* (higher, more mesic slopes), *Arabis repanda, Arctostaphylos patula, Artemisia tridentata, Calocedrus decurrens* (higher, more mesic slopes), *Ceanothus cordulatus, Chrysolepis sempervirens, Cordylanthus nevinii, Chrysothamnus nauseosus* (north and east facing slopes), *Eriogonum wrightii* ssp. *subscaposum, Ivesia santolinoides, Juniperus occidentalis* ssp. *australis* (north and east facing slopes of San Bernardino Mountains where it may codominate), *Lupinus elatus, Penstemon caesius, Pinus lambertiana* (higher, more mesic slopes), *P. monophylla* (north and east facing slopes where it may codominate), *Prunus emarginata* (north and east facing slopes), *P. virginiana* var. *demissa* (north and east facing slopes), *Stipa coronata* var. *depauperata, Symphoricarpos parishii, Tetradymia canescens* (north and east facing slopes).

LOCATION: Transverse and Peninsular Ranges of southern California, mainly at elevations from 2135-2590 m.

B&M ch. 16

5. NORTHERN YELLOW PINE FOREST (*Pinus ponderosa*)

STRUCTURE: Tall, rather open, needle-leaved evergreen forest with a much lower layer of broad-leaved deciduous trees. Mainly broad-leaved evergreen shrubs vary much in height and density, rarely exceeding 2 m in height.

DOMINANTS: Yellow pine (*Pinus ponderosa*)

OTHER CHARACTERISTIC COMPONENTS: *Abies concolor, Arctostaphylos manzanita, A. viscida, Calocedrus decurrens, Quercus garryana, Q. kelloggii, Rhus diversiloba.*

LOCATION: On the western slopes of the Cascade Mountains, usually at elevations of 500-1000 m.

B&M ch. 17

6. SOUTHERN YELLOW PINE FOREST (*Pinus ponderosa*)

STRUCTURE: Open, tall, needle-leaved evergreen forest. Broad-leaved evergreen and deciduous trees and shrubs usually form very open lower layers, interspersed with forbs.

DOMINANTS: Yellow pine (*Pinus ponderosa*)

OTHER CHARACTERISTIC COMPONENTS: *Abies concolor* (higher mesic slopes), *Arctostaphylos glandulosa, Calocedrus decurrens* (more mesic slopes), *Ceanothus integerrimus, Cornus nuttallii* (more mesic slopes), *Eriastrum densifolium* ssp. *austromontanum, Eriodictyon trichocalyx, Garrya flavescens, Gilia splendens* ssp. *grantii, Iris hartwegii* spp. *australis, Lupinus excubitus, L. formosus, Pinus coulteri, P. lambertiana* (higher mesic slopes), *Pseudotsuga macrocarpa* (more mesic slopes), *Quercus chrysolepis* (more mesic slopes), *Q. kelloggii* (more mesic slopes), *Rhamnus californica, Ribes roezlii, Solanum xanti, Streptanthus bernardinus.*

LOCATION: Mainly on rocky, south and west facing slopes in the San Gabriel, San Bernardino and San Jacinto Mountains, usually at elevations of 1375-2015 m.

B&M ch. 16

7. SIERRAN YELLOW PINE FOREST (*Pinus ponderosa*)

STRUCTURE: Tall, moderately dense, needle-leaved evergreen forest. Low, broad-leaved evergreen and/or deciduous trees occur occasionally. A broad-leaved evergreen shrub layer varies much in height and density, rarely exceeding 2 m in height. Numerous inclusions of chaparral (29), and many chaparral shrubs grow as an undergrowth under the pines.

DOMINANTS: Yellow pine (*Pinus ponderosa*)

OTHER CHARACTERISTIC COMPONENTS: *Abies concolor, Arctostaphylos mariposa, A. viscida, Calocedrus decurrens, Chamaebatia foliolosa, Pinus lambertiana, Quercus chrysolepis, Q. kelloggii.*

LOCATION: Western slopes of the Sierra Nevada, mainly at lower or intermediate elevations.

B&M ch. 17

8. YELLOW PINE-SHRUB FOREST (*Pinus-Purshia*)

STRUCTURE: A moderately open, needle-leaved evergreen forest, 20-35 m tall. A layer of shrubs can be dense and surpass 2 m in height but is usually open and less high. A low layer of graminoids is always open.

DOMINANTS: Yellow pine (*Pinus ponderosa*), Antelope bush (*Purshia tridentata*)

OTHER CHARACTERISTIC COMPONENTS: *Abies concolor, Agropyron spicatum, Arctostaphylos parryana* var. *pinetorum, A. patula, Artemisia tridentata, Balsamorrhiza sagittata, Calamagrostis rubescens, Calocedrus decurrens, Ceanothus velutinus, Cercocarpus betuloides, C. ledifolius, Festuca idahoensis, Holodiscus discolor, Juniperus occidentalis, Symphoricarpos vaccinioides.*

LOCATION: Mainly east of the Cascade Mountains.

B&M ch. 17

9. COASTAL CYPRESS AND PINE FORESTS (*Cupressus, Pinus*)

STRUCTURE: Fairly dense, needle-leaved evergreen forest, low and shrubby to medium tall, the height depending largely on the floristic composition and the soil.

DOMINANTS: No species dominates throughout the area of this type. Some species strongly dominate, but always in very small areas.

OTHER CHARACTERISTIC COMPONENTS: *Adenostoma fasciculatum* (southern part), *Cupressus goveniana, C. macrocarpa* (around Carmel Bay), *C. pygmaea* (mostly in Mendocino county), *Gaultheria shallon, Lonicera involucrata, Myrica californica, Pinus contorta* (Mendocino county and north), *P. contorta* var. *bolanderi* (Mendocino county), *P. muricata, P. radiata, P. remorata* (Santa Barbara county), *P. torreyana* (San Diego county), *Pteridium aquilinum* var. *pubescens, Quercus agrifolia* (locally), *Rhamnus californica, Rubus vitifolius, Vaccinium ovatum.*

LOCATION: A very narrow, much interrupted belt along the coast.

B&M ch. 9

10. SAN BENITO FOREST (*Calocedrus decurrens*)

STRUCTURE: Open, low to medium tall, needle-leaved evergreen forest.

DOMINANTS: Incense cedar (*Calocedrus decurrens*)

OTHER CHARACTERISTIC COMPONENTS: *Pinus coulteri, P. sabiniana, P. jeffreyi*

LOCATION: Mostly in southeastern San Benito county on serpentine.

B&M ch. 11

11. JUNIPER-PINYON WOODLAND (*Juniperus-Pinus*)

STRUCTURE: Open groves of needle-leaved evergreen shrubby trees, usually less than 10 m tall. Coverage rarely exceeds 50%. A dense to open layer of broad-leaved evergreen (and some deciduous) shrubs and dwarfshrubs reaches a height of 1.5 m. Low herbaceous plants may be present but remain inconspicuous.

DOMINANTS: California juniper (*Juniperus californica*), singleleaf pinyon (*Pinus monophylla*)

OTHER CHARACTERISTIC COMPONENTS: *Anisocoma acaulis, Artemisia tridentata* (may codominate), *Caulanthus amplexicaulis, Chrysothamnus nauseosus* (northern part), *Coleogyne ramosissima, Gutierrezia sarothrae* (southern part), *Haplopappus linearifolius, Juniperus osteosperma* (may dominate locally, e.g. northeast of Mono Lake), *Layia glandulosa, Leptodactylon pungens* (southern part), *Nolina parryi* (southern part), *Oreonana vestita, Pellaea compacta, Penstemon grinellii, Phacelia austromontana, Pinus quadrifolia* (Riverside and San Diego counties), *Plagiobothrys kingii, Purshia tridentata* (northern part), *Quercus dumosa, Q. turbinella, Silene parishii* var. *latifolia, Stipa coronaria* (southern part), *S. speciosa* (southern part), *Tetradymia axillaris, Yucca schidigera* (southern part).

LOCATION: Southeastern California

B&M ch. 23

12. KLAMATH MONTANE FOREST WITH DOUGLAS FIR (*Abies-Pseudotsuga-Tsuga*)

STRUCTURE: Tall, dense to moderately open, needle-leaved evergreen forest with patches of broad-leaved evergreen shrubs and islands of meadows.

DOMINANTS: White fir (*Abies concolor*), Shasta fir (*Abies magnifica* var. *shastensis*), Douglas fir (*Pseudotsuga menziesii*), mountain hemlock (*Tsuga mertensiana*)

OTHER CHARACTERISTIC COMPONENTS: *Abies procera, Corylus cornuta, Chimaphila umbellata, Dodecatheon jeffreyi, Leucothoe davisiae, Linnaea borealis, Monardella odoratissima, Phyllodoce empetriformis, Pyrola picta, Rosa gymnocarpa, Rubus parviflorus, Valeriana sitchensis, Veratrum californicum.*

Shrub community: *Arctostaphylos patula, Ceanothus velutinus, Prunus emarginata, Quercus sadleriana, Q. vaccinifolia*

Meadow communities: the discrete meadows are dense and up to 1 m tall. The role of forbs increases with soil moisture. Mesic meadows may be distinguished from drier ones.

Mesic meadows: *Angelica arguta, Carex* spp., *Glyceria elata, Heracleum lanatum, Phyllodoce empetriformis* (higher altitudes), *Solidago triangularis, Vaccinium arbuscula* (higher altitudes), *Valeriana sitchensis, Veratrum californicum.*

Drier meadows: *Bromus orcuttianus, Calamagrostis koelerioides, Calyptridium umbellatum, Eriogonum umbellatum, Festuca idahoensis, Haplopappus greenei, Monardella odoratissima, Polygonum davisiae, Sitanion hystrix.*

LOCATION: Western Klamath Mountains and adjacent areas.

B&M ch. 20

13. KLAMATH MONTANE FOREST WITH YELLOW PINE (*Abies-Pinus-Tsuga*)

STRUCTURE: Tall, dense to moderately open, needle-leaved evergreen forest with patches of broad-leaved evergreen and deciduous low trees and shrubs and islands of meadows.

DOMINANTS: White fir (*Abies concolor*), Shasta fir (*Abies magnifica* var. *shastensis*), yellow pine (*Pinus ponderosa*), mountain hemlock (*Tsuga mertensiana*)

OTHER CHARACTERISTIC COMPONENTS: *Berberis nervosa, Ceanothus prostratus, Chimaphila umbellata, Leucothoe davisiae, Linnaea borealis, Monardella odoratissima, Phyllodoce empetriformis, Pinus albicaulis, P. lambertiana, Pseudotsuga menziesii, Pyrola picta.*

Shrub community: *Amelanchier pallida, Arctostaphylos nevadensis, A. patula, Ceanothus velutinus, Holodiscus microphyllus, Prunus emarginata, Quercus vaccinifolia.*

Meadow communities: both in structure and in floristic composition, these meadows are essentially like those of Klamath montane forest with Douglas fir (12).

LOCATION: Eastern Klamath Mountains and adjacent areas.

B&M ch. 20

14. COAST RANGE MONTANE FOREST (*Abies-Pinus-Pseudotsuga*)

STRUCTURE: Tall, dense to moderately open, needle-leaved evergreen forest with occasional patches of broad-leaved evergreen shrubs.

DOMINANTS: White fir (*Abies concolor*), Red fir (*Abies magnifica*), Yellow pine (*Pinus ponderosa*), Douglas fir (*Pseudotsuga menziesii*)

OTHER CHARACTERISTIC COMPONENTS: *Arctostaphylos nevadensis, A. patula, Ceanothus cordulatus, Chrysolepis sempervirens, Pinus lambertiana, Quercus kelloggii, Q. vaccinifolia.*

LOCATION: Higher elevations of the northern Coast Ranges.

B&M ch. 20

15. SIERRAN MONTANE FOREST (*Abies-Pinus*)

STRUCTURE: Tall, dense, needle-leaved evergreen forest. A layer of low trees, rarely 10 m high, is very open (coverage less than 10%). A shrub layer, 1-2 m high, also covers less than 10%. A herbaceous layer covers less than 5%.

DOMINANTS: White fir (*Abies concolor*, mainly at higher elevations), sugar pine (*Pinus lambertiana*)

OTHER CHARACTERISTIC COMPONENTS: *Acer macrophyllum, Adenocaulon bicolor, Arbutus menziesii, Arctostaphylos patula, Calocedrus decurrens, Ceanothus integerrimus, C. parvifolius, C. velutinus, Chamaebatia foliolosa, Chimaphila menziesii, C. umbellata, Chrysolepis chrysophylla, C. sempervirens, Cornus nuttallii, Disporum hookeri, Galium sparsifolium, Hieracium albiflorum, Pinus ponderosa, Prunus emarginata, Pseudotsuga menziesii* (may dominate in the north, absent in the south), *Pyrola picta, Quercus chrysolepis, Q. kelloggii, Ribes roezlii, Rosa gymnocarpa, Sequoiadendron giganteum* (locally in the Sierra Nevada), *Symphoricarpos acutus, Taxus brevifolia, Viola glabella.*

LOCATION: Western slopes of Sierra Nevada and Cascade Mountains.

B&M ch. 17

16. MOJAVE MONTANE FOREST (*Abies concolor*)

STRUCTURE: Low, fairly dense to open, needle-leaved evergreen forest, up to 20 m tall. Needle-leaved and broad-leaved shrubs form a rather open lower layer.

DOMINANTS: White fir (*Abies concolor*)

OTHER CHARACTERISTIC COMPONENTS: *Acer glabrum* var. *diffusum, Amelanchier utahensis, Fraxinus anomala, Holodiscus microphyllus, Juniperus osteosperma, Pinus monophylla, Quercus chrysolepis, Q. turbinella, Ribes cereum, R. velutinum* var. *glanduliferum, Sambucus mexicana*

LOCATION: Eastern Mojave mountain ranges, usually above 1900 m.

B&M ch. 23

17. UPPER MONTANE-SUBAPLINE FORESTS (*Abies-Pinus*)

STRUCTURE: A tall, dense, needle-leaved evergreen forest. Height and density decrease with altitude: height about 40 m at lower elevations, 10-30 m at higher elevations. A very open shrub layer is usually 1-2 m tall. Herbaceous plants cover less than 5%; epiphytic lichens are prominent, and saprophytes are common where the forest is dense. Montane and subalpine meadows may be extensive (see below), and inclusions of sagebrush steppe (31) occur east of the crest of the Sierra Nevada.

DOMINANTS: Red Fir (*Abies magnifica*, mainly at lower elevations), lodgepole pine (*Pinus cortorta* ssp. *murrayana*, mainly at higher elevations)

OTHER CHARACTERISTIC COMPONENTS: *Arctostaphylos nevadensis, A. patula, Artemisia tridentata, Ceanothus cordulatus, Chimaphila umbellata, Chrysopsis breweri, Corallorhiza maculata, Hieracium albiflorum, Holodiscus microphyllus, Juniperus occidentalis* ssp. *australis, Kalmia polifolia* var. *microphylla, Letharia vulpinus, Pedicularis semibarbata, Phyllodoce breweri, Pinus albicaulis* (higher elevations), *P. balfouriana* (higher elevations), *P. flexilis* (higher elevations), *P. jeffreyi*, *P. lambertiana* (lower elevations), *P. monticola* (higher elevations), *P. washoensis* (Warner Mountains), *Poa bolanderi*, *Pyrola picta*, *P. secunda*, *Ribes cereum*, *R. viscosissimum*, *Salix eastwoodiae*, *S. jepsonii*, *S. lutea*, *Sarcodes sanguinea*, *Symphoricarpos vaccinioides*, *Tsuga mertensiana* (may dominate locally in the north), *Vaccinium nivictum*, *V. occidentalis*.

Montane and subalpine meadows (*Calamagrostis-Carex*)

STRUCTURE: Dense communities of graminoids and forbs, their height decreasing with altitude from 60 cm to 10 cm.

DOMINANTS: Reedgrass (*Calamagrostis breweri*), sedges (*Carex exserta*, *C. nigricans*, *C. subnigricans*)

OTHER CHARACTERISTIC COMPONENTS: *Antennaria alpina* var. *media*, *Aster alpigenus*, *Carex nebrascensis*, *C. rostrata*, *C. scopulorum*, *Deschampsia caespitosa*, *Gentiana heterophylla*, *Heleocharis pauciflora*, *Mimulus primuloides*, *Vaccinium nivictum*.

LOCATION: Higher altitudes of Sierra Nevada and Cascade Mountains, upward to timberline which is more or less at 3000 m in the north and 3600 m in the south.

B&M ch. 17

18. SOUTHERN MONTANE SUBALPINE FOREST (*Abies-Pinus*)

STRUCTURE: Low to medium tall, dense to open, needle-leaved evergreen forest with a more or less open layer of broad-leaved evergreen and some deciduous shrubs. Low broad-leaved evergreen and deciduous trees may occur, mainly at lower elevations.

DOMINANTS: White fir (*Abies concolor*), sugar pine (*Pinus lambertiana)*, lodgepole pine (*Pinus contorta* ssp. *murrayana*: at higher elevations to timberline and as krummholz beyond; absent on Mount Pinos and Santa Rosa Mountains)

OTHER CHARACTERISTIC COMPONENTS: *Allium monticola*, *Amelanchier pallida*, *Arctostaphylos patula*, *Calocedrus decurrens* (mainly on moister north and east facing slopes), *Ceanothus cordulatus*, *C. greggii*, *Cercocarpus ledifolius*, *Chrysolepis sempervirens*, *Draba corrugata*, *Eriogonum kennedyi* ssp. *alpigenum* (higher elevations), *E. umbellatum* var. *minus*, *E. wrightii* ssp. *subscaposum*, *Fremontodendron californicum*, *Holodiscus microphyllus*, *Juniperus occidentalis* ssp. *australis* (on exposed ridges), *Leptodactylon pungens* ssp. *pulchriflorum*, *Pinus flexilis* (higher elevations), *P. jeffreyi*, *P. monophylla* (on north and east facing slopes), *Prunus virginiana* var. *demissa*, *Quercus chrysolepis*, *Q. kelloggii*, *Ribes cereum*, *R. montigenus*, *R. nevadense*, *R. roezlii*, *Rubus parviflorus*, *Salix coulteri*, *Sambucus caerulea*.

LOCATION: Mainly in the San Gabriel, San Bernardino and San Jacinto Mountains, usually above 2400 m.

B&M ch. 16

19. GREAT BASIN SUBALPINE FOREST (*Pinus* spp.)

STRUCTURE: Low, very open, needle-leaved evergreen forest with an open layer of shrubs, dwarf shrubs and herbaceous plants.

DOMINANTS: Bristlecone pine (*Pinus aristata* = *P. longaeva*), limber pine (*P. flexilis*)

OTHER CHARACTERISTIC COMPONENTS: *Acer glabrum* var. *diffusum, Arenaria kingii, Artemisia tridentata, Chamaebatiaria millefolium, Chrysothamnus viscidiflorus* ssp. *pumilus, Cymopteris cinerarius, Erigeron clokeyi, E. pygmaeus, Galium hypotrichium* ssp. *tomentellum, Ipomopsis aggregata* ssp. *arizonica, Koeleria macrantha, Leptodactylon pungens, Linanthus nuttallii, Lupinus alpestris, Mimulus bigelovii* var. *panamintensis, Poa rupicola, Ribes cereum, R. montigenum, Senecio spartioides, Sitanion hystrix, Symphoricarpos longiflorus.*

LOCATION: Mainly in the White, Inyo and Panamint Mountains at elevations above 2620 m.

B&M ch. 23

20. OREGON OAK FOREST (*Quercus garryana*)

STRUCTURE: Medium tall, open to dense, broad-leaved deciduous forest with an admixture of usually much taller needle-leaved evergreen trees. Where the canopy is more open, broad-leaved evergreen, and some deciduous shrubs form an undergrowth of varying density. A ground cover of low to medium tall graminoids may be continuous.

DOMINANTS: Oregon oak (*Quercus garryana*)

OTHER CHARACTERISTIC COMPONENTS: *Arctostaphylos patula, Arbutus menziesii, Danthonia californica, Festuca californica, Holodiscus discolor, Juniperus occidentalis* (northern part), *Pinus jeffreyi* (northern part), *P. ponderosa* (northern part), *P. sabiniana, Polystichum munitum, Prunus emarginata, Pseudotsuga menziesii, Purshia tridentata, Quercus kelloggii, Rhus diversiloba.*

LOCATION: Mainly in northwestern California.

B&M ch. 11

21. MIXED EVERGREEN FOREST WITH CHINQUAPIN (*Arbutus-Chrysolepis-Lithocarpus-Pseudotsuga-Quercus*)

STRUCTURE: Dense, broad-leaved evergreen forest, about 35 m tall, overtowered by an open layer of very tall needle-leaved evergreen trees reaching a height of 65 m. A shrubby undergrowth (up to 2 m tall) and a lower herbaceous layer are quite open.

DOMINANTS: Madrone (*Arbutus menziesii*), chinquapin (*Chrysolepis chrysophylla*), tanbark oak (*Lithocarpus densiflora*), Douglas fir (*Pseudotsuga menziesii*), canyon oak (*Quercus chrysolepis*)

OTHER CHARACTERISTIC COMPONENTS: *Achlys triphylla, Berberis nervosa, Chamaecyparis lawsoniana, Chimaphila umbellata, Corylus cornuta, Disporum hookeri* var. *trachyandrum, Gaultheria shallon, Pinus lambertiana, P. ponderosa, Quercus kelloggii, Rhus diversiloba, Taxus brevifolia, Vancouveria planipetala, Whipplea modesta.*

LOCATION: Klamath Mountains and adjacent areas.

B&M ch. 10

22. MIXED EVERGREEN FOREST WITH RHODODENDRON (*Arbutus-Lithocarpus-Pseudotsuga-Quercus-Rhododendron*)

STRUCTURE: Tall, dense, needle-leaved and broad-leaved evergreen forest with an admixture of broad-leaved deciduous trees and shrubs.

DOMINANTS: Madrone (*Arbutus menziesii*), tanbark oak (*Lithocarpus densiflora*), Douglas fir (*Pseudotsuga menziesii*), canyon oak (*Quercus chrysolepis*), California rose-bay (*Rhododendron macrophyllum*)

OTHER CHARACTERISTIC COMPONENTS: *Acer macrophyllum, Corylus cornuta, Oxalis oregona, Pinus lambertiana, P. ponderosa, Polystichum munitum, Quercus agrifolia, Q. garryana, Q. kelloggii, Vaccinium ovatum, Whipplea modesta.*

LOCATION: Northern Coast Ranges

B&M ch. 10

23. MIXED HARDWOOD FOREST (*Arbutus-Quercus*)

STRUCTURE: Low to medium tall, broad-leaved evergreen forest with an admixture of broad-leaved deciduous and needle-leaved evergreen trees; the latter may be towering above the canopy. The forest is more or less dense, and in higher elevations, its height may be so low as to make the forest appear shrubby. Inclusions of chaparral (29) are common.

DOMINANTS: Madrone (*Arbutus menziesii*), coast live oak (*Quercus agrifolia*, lower elevations), canyon oak (*Quercus chrysolepis*, higher elevations)

OTHER CHARACTERISTIC COMPONENTS: *Abies bracteata* (Monterey county), *Aesculus californica, Baccharis pilularis* var. *consanguinea* (northern part), *Heteromeles arbutifolia, Lithocarpus densiflora, Pinus coulteri* (higher elevations), *P. lambertiana* (higher elevations, Monterey county), *P. ponderosa* (higher elevations), *Pseudotsuga macrocarpa* (southern part), *P. menziesii* (northern part), *Quercus kelloggii, Q. wislizenii, Rhamnus californica, Rhus diversiloba, Umbellularia californica.*

LOCATION: Central and southern Coast Ranges

B&M ch. 11

24. MIXED HARDWOOD AND REDWOOD FOREST (*Arbutus-Quercus-Sequoia*)

STRUCTURE: Same as No. 23 with groves of redwood along many (but not all) streams. This type includes long narrow bands of coastal sagebrush (32) along the coast.

DOMINANTS: Madrone (*Arbutus menziesii*), interior live oak (*Quercus wislizenii*), redwood (*Sequoia sempervirens*)

OTHER CHARACTERISTIC COMPONENTS: Same as No. 23

LOCATION: In the Coast Ranges, south of Monterey Bay.

B&M ch. 11

25. BLUE OAK-DIGGER PINE FOREST (*Pinus-Quercus*)

STRUCTURE: Medium tall, dense to open broad-leaved deciduous forest with an admixture of needle-leaved evergreen trees. Low broad-leaved evergreen trees and/or shrubs are common. Near the California Prairie (36) the canopy opens up, the prairie forms a continuous ground cover, and the forest changes into a savanna. Inclusions of chaparral (29) may be numerous.

DOMINANTS: Digger pine (*Pinus sabiniana*), blue oak (*Quercus douglasii*)

OTHER CHARACTERISTIC COMPONENTS: *Aesculus californica, Arctostaphylos manzanita, A. viscida, Ceanothus cuneatus, Heteromeles arbutifolia, Juniperus californica* (especially in the southwest), *Prunus ilicifolia, Quercus lobata, Q. wislizenii* (north-facing slopes and higher elevations), *Rhamnus californica* ssp. *ilicifolia, R. crocea* ssp. *ilicifolia, Rhus diversiloba.*

LOCATION: Distributed mainly in a belt of varying width around the Central Valley.

B&M ch. 11

26. COULTER PINE FOREST (*Pinus coulteri*)

STRUCTURE: Rather low, open groves of needle-leaved evergreen trees and some broad-leaved evergreen and/or deciduous trees. A lower layer of broad-leaved evergreen, sclerophyll shrubs varies much in density.

DOMINANTS: Coulter pine (*Pinus coulteri*)

OTHER CHARACTERISTIC COMPONENTS: *Arctostaphylos glandulosa, A. glauca, Ceanothus integerrimus, Cercocarpus betuloides, Deschampsia danthonioides, Frasera neglecta, Garrya flavescens, Horkelia bolanderi* ssp. *parryi, Pinus jeffreyi* (north and east facing slopes),

P. monophylla (north and east facing slopes), *P. ponderosa* (south and west facing slopes), *Pseudotsuga macrocarpa* (north facing slopes), *Quercus chrysolepis, Q. kelloggii, Scutellaria austinae, Viola douglasii, Wyethia ovata.*

LOCATION: Western San Gabriel Mountains, mainly at elevations of 1120-1830 m.

B&M ch. 16

27. SOUTHERN OAK FOREST (*Quercus agrifolia*)

STRUCTURE: Rather low, dense to open, broad-leaved evergreen forest with occasional broad-leaved deciduous trees and some shrubs. The California prairie (36) forms a ground layer where the canopy opens, and inclusions of chaparral (29) are common.

DOMINANTS: Coast live oak (*Quercus agrifolia*)

OTHER CHARACTERISTIC COMPONENTS: *Heteromeles arbutifolia, Juglans californica* (San Bernardino to southeastern Santa Barbara counties where it may dominate locally), *Quercus engelmannii* (Los Angeles to San Diego counties where it may dominate locally).

LOCATION: Hilly to mountainous terrain in the Peninsular, Transverse and southern Coast Ranges.

B&M ch. 11

28. RIPARIAN FOREST (*Populus fremontii*)

STRUCTURE: Medium tall to tall, broad-leaved deciduous forest with lianas. Islands of tule marsh (37) occur occasionally.

DOMINANTS: Cottonwood (*Populus fremontii*)

OTHER CHARACTERISTIC COMPONENTS: *Acer negundo* ssp. *californicum, Alnus rhombifolia, Baccharis viminea, Cephalanthus occidentalis* var. *californicus, Clematis ligusticifolia, Platanus racemosa, Quercus lobata, Rubus vitifolius, Salix gooddingii* var. *variabilis, S. laevigata, Urtica holosericea, Vitis californica.*

LOCATION: Mainly in the Central Valley and in areas of valley oak savanna (33).

B&M ch. 11

29. CHAPARRAL (*Adenostoma-Arctostaphylos-Ceanothus*)

STRUCTURE: Dense communities of needle-leaved and broad-leaved evergreen sclerophyll shrubs, varying in height from 1-3 m, rarely to 5 m. An understory is usually lacking. Inclusions of southern oak forest (27) are common in the coastal regions of Santa Barbara to San Diego counties in canyons and other drainage areas and on clay.

DOMINANTS: Chamise (*Adenostoma fasciculatum*), manzanita (*Arctostaphylos* spp.), California lilac (*Ceanothus* spp.)

OTHER CHARACTERISTIC COMPONENTS: *Adenostoma sparsifolium* (southern part where it may dominate), *Arctostaphylos elegans, A. glandulosa, A. glauca, A. manzanita, A. myrtifolia, A. parryana, A. stanfordiana, A. viscida, Artemisia californica, Ceanothus crassifolius* (southern part), *C. cuneatus, C. dentatus, C. impressus, C. jepsonii* (codominates on serpentine soils), *C. leucodermis, C. megacarpus, C. oliganthus, C. palmeri* (southern part), *C. parryi, C. ramulosus, C. sorediatus, C. spinosus, C. tomentosus, C. velutinus* (northern part), *C. verrucosus* (southern part), *Cercocarpus betuloides, Eriogonum fasciculatum, Fraxinus dipetala, Garrya elliptica, G. flavescens* var. *pallida, G. veatchii, Heteromeles arbutifolia* (codominant on serpentine soils), *Lonicera interrupta, L. subspicata, Prunus ilicifolia, Quercus dumosa* (codominates locally), *Q. durata* (may codominate on serpentine soils), *Q. wislizenii* var. *frutescens* (codominates locally), *Rhamnus ilicifolia, Rhus integrifolia, R. laurina, R. ovata, Ribes amarum, R. californicum, R. indecorum, R. malvaceum, R. viburnifolium, Salvia apiana, S. mellifera, S. spathacea.*

LOCATION: Lower elevations of most mountain ranges but absent in plains, deserts and high elevations.

B&M ch. 12

30. ISLAND CHAPARRAL (*Adenostoma-Ceanothus-Quercus*)

STRUCTURE: Dense vegetation of needle-leaved and broad-leaved evergreen, sclerophyll shrubs. The height varies from 1-3 m; there is no undergrowth. There are inclusions of southern oak forest (27).

DOMINANTS: Chamise (*Adenostoma fasciculatum*), California lilac (*Ceanothus insularis*), island oak (*Quercus tomentella*)

OTHER CHARACTERISTIC COMPONENTS: *Arctostaphylos insularis, Ceanothus arboreus, Cercocarpus betuloides* var. *blancheae, Heteromeles arbutifolia* var. *macrocarpa, Lyonothamnus floribundus, Prunus lyonii, Rhamnus crocea* ssp. *pirifolia, Rhus ovata.*

LOCATION: Islands off the shores of southern California.

B&M ch. 26

31. SAGEBRUSH STEPPE (*Agropyron-Artemisia*)

STRUCTURE: Low, open (15% coverage), broad-leaved evergreen shrub formation, up to 1 m tall, with an admixture of deciduous shrubs. An open (5-15% coverage) layer of graminoids and forbs is most common in the north.

DOMINANTS: Bluebunch wheatgrass (*Agropyron spicatum*), Basin sagebrush (*Artemisia tridentata*)

OTHER CHARACTERISTIC COMPONENTS: *Artemisia arbuscula* (dominates where snow melt water accumulates), *Castilleja chromosa, Chrysothamnus nauseosus, C. viscidiflorus, Collinsia parviflora, Crepis acuminata, Juniperus occidentalis, Lupinus caudatus, Lygodesmia spinosa, Phlox hoodii, Poa sandbergii, Purshia tridentata, Sitanion hystrix, Stipa thurberiana* (mainly eastern flanks of Sierra Nevada), *Tetradymia canescens*.

LOCATION: Northeastern California

B&M ch. 22

32. COASTAL SAGEBRUSH (*Artemisia-Eriogonum-Salvia*)

STRUCTURE: Moderately dense communities of drought-deciduous shrubs rarely more than 1.5 m tall. Forbs are numerous.

DOMINANTS: California sagebrush (*Artemisia californica*), California buckwheat (*Eriogonum fasciculatum*), white sage (*Salvia apiana*), black sage (*S. mellifera*)

OTHER CHARACTERISTIC COMPONENTS: *Baccharis pilularis* var. *consanguinea, Encelia californica, E. farinosa, Eriophyllum confertiflorum, Grindelia hirsutula, Haplopappus squarrosus, H. venetus* ssp. *vernonioides, Horkelia cuneata, Mesembryanthemum chilense* (coastal bluffs), *Opuntia occidentalis* var. *littoralis* (southern part), *Rhus integrifolia, R. laurina, Salvia leucophylla, Yucca whipplei* (southern part).

LOCATION: A much interrupted belt in the coastal regions of central and southern California.

B&M ch. 13

33. VALLEY OAK SAVANNA (*Quercus-Stipa*)

STRUCTURE: Savanna of tall, broad-leaved deciduous trees, widely spaced and stately. Woody undergrowth is insignificant. Broad-leaved evergreen and/or needle-leaved evergreen trees occur occasionally. The density increases where merging with riparian forest (28). The California prairie (36) covers the ground.

DOMINANTS: Valley oak (*Quercus lobata*)

OTHER CHARACTERISTIC COMPONENTS: *Pinus sabiniana, Quercus agrifolia* (in the Coast Ranges), *Q. douglasii, Q. wislizenii* (Central Valley)

LOCATION: Central Valley and adjacent low hills; valleys of the Coast Ranges.

B&M ch. 11

34. JUNIPER SAVANNA (*Juniperus-Stipa*)

STRUCTURE: Dense to somewhat open medium tall bunchgrass community with many forbs. Scattered throughout are low, rather shrubby needle-leaved evergreen trees. Low, broad-leaved deciduous trees occur occasionally.

DOMINANTS: California juniper (*Juniperus californica*), needlegrass (*Stipa cernua*), speargrass (*Stipa pulchra*)

OTHER CHARACTERISTIC COMPONENTS: *Ephedra californica, E. viridis, Eriogonum fasciculatum* ssp. *polifolium, Eriophyllum confertiflorum, Gutierrezia bracteata, Haplopappus linearifolius, Quercus douglasii, Yucca whipplei.*

In addition, there are numerous species which are essentially those of the California prairie (36).

LOCATION: Inner Coast Ranges, mainly in Fresno, Kern and San Luis Obispo counties.

B&M ch. 14

35. JUNIPER SHRUB SAVANNA (*Agropyron-Artemisia-Juniperus*)

STRUCTURE: Low, fairly dense to open, broad-leaved evergreen shrub community with a strong admixture of graminoids and forbs. Widely spaced, rather shrubby, needle-leaved evergreen trees occur throughout, rarely surpassing 15 m in height.

DOMINANTS: Bluebunch wheatgrass (*Agropyron spicatum*), basin sagebrush (*Artemisia tridentata*), western juniper (*Juniperus occidentalis*).

OTHER CHARACTERISTIC COMPONENTS: *Artemisia arbuscula, Balsamorrhiza sagittata, Dodecatheon conjugens, Festuca idahoensis, Hesperochiron californicus, H. pumilus, Hydrophyllum capitatum* var. *alpinum* (in moist places), *Leucocrinum montanum, Lithospermum ruderale, Lupinus sericeus, Phlox longifolia, Purshia tridentata, Sitanion hystrix, S. jubatum, Trifolium macrocephalum.*

LOCATION: Northeastern California.

B&M ch. 23

36. CALIFORNIA PRAIRIE (*Stipa* spp.)

STRUCTURE: Dense to somewhat open, medium tall bunchgrass community with many forbs. Height and seasonal aspects of this prairie can vary greatly.

DOMINANTS: Needlegrass (*Stipa cernua*), speargrass (*Stipa pulchra*)

OTHER CHARACTERISTIC COMPONENTS: *Aristida divaricata, Baeria chrysostoma, Elymus glaucus, E. triticoides, Eschscholtzia californica, Festuca megalura, Gilia clivorum, G. interior, G. minor, G. tricolor, Lupinus bicolor, L. luteolus, Orthocarpus attenuatus, O. campestris, O. erianthus, O. linearilobus, O. purpurascens, Plagiobothrys nothofulvus, Poa scabrella, Sisyrinchium bellum, Stipa coronata* (southern part), *S. lepida, Trifolium albopurpureum, T. amplectens, T. depauperatum, T. gracilentum, T. microdon, T. olivaceum.*

Many introduced species have become well established and are now likely to survive even under natural conditions. Among the most prominent are:

Avena barbata, A. fatua, Bromus mollis, B. rigidus, B. rubens, Erodium botrys, E. cicutarium, Festuca dertonensis, Medicago hispida.

LOCATION: Mainly in the Central Valley, often extending into the adjacent foothills of the surrounding mountains; coastal areas of southern California.

B&M ch. 14

37. TULE MARSH (*Scirpus-Typha*)

STRUCTURE: Tall, dense graminoid plant communities, occasionally interrupted by open water.

DOMINANTS: Common tule (*Scirpus acutus*), cattail (*Typha latifolia*)

OTHER CHARACTERISTIC COMPONENTS: *Carex obnupta, C. senta, Heleocharis palustris, Juncus balticus, J. effusus* var. *pacificus, Scirpus americanus, S. californicus, S. olneyi, S. robustus, S. validus, Typha domingensis.*

LOCATION: Along Sacramento and San Joaquin Rivers, extensive in the delta region.

B&M ch. 14

38. COASTAL SALTMARSH (*Salicornia-Spartina*)

STRUCTURE: Community of perennial graminoids and succulent forbs, the former 1 m tall or more, the latter usually less. Forbs usually dominate at higher elevations. Coverage 100% but may be less at lower elevations. Algae may colonize frequently flooded areas of bare ground.

DOMINANTS: Glasswort (*Salicornia virginica*), cordgrass (*Spartina foliosa*)

OTHER CHARACTERISTIC COMPONENTS: *Batis maritima* (southern part), *Distichlis spicata* var. *stolonifera, Frankenia grandifolia, Jaumea carnosa, Limonium californicum, Monanthochloe littoralis* (southern part), *Salicornia bigelovii* (southern part), *S. subterminalis* (southern part), *Scirpus californicus, Suaeda californica, Triglochin maritima.*

LOCATION: Around sheltered bays, estuaries and coastal lagoons, usually above mean high water level and inland from intertidal sand and mud flats.

B&M ch. 8

39. ALPINE COMMUNITIES (*Draba, Eriogonum, Festuca, Phlox, Poa, Sitanion*)

STRUCTURE: Low graminoid and forb communities with an admixture of dwarfshrubs (often cushion plants). Coverage may reach 100% at lower elevations but becomes increasingly open with growing altitude. On mesic sites, a continuous turf contrasts with bunchgrasses and cushion plants on drier sites. Numerous, more or less extensive inclusions of cold alpine desert (54) occur at higher elevations.

DOMINANTS: Draba (*Draba oligosperma*), wild buckwheat (*Eriogonum ovalifolium* var. *nivale*), fescue (*Festuca brachyphylla*), phlox (*Phlox covillei*), bluegrass (*Poa rupicola*), squirreltail (*Sitanion hystrix*).

OTHER CHARACTERISTIC COMPONENTS: *Astragalus kentrophyta* var. *implexus, Calyptridium umbellatum, Carex breweri, C. exserta, C. helleri, Cassiope mertensiana, Castilleja nana, Cryptantha nubigena, Draba breweri, D. densifolia, D. lemmonii, Epilobium anagallidifolium, E. obcordatum, Haplopappus macronema, Hulsea algida, Ivesia lycopodioides, Lupinus lyallii, Luzula spicata, Oxyria digyna, Penstemon davidsonii, Phoenicaulis eurycarpa, Phyllodoce breweri, Poa suksdorfii, Podistera nevadensis, Polemonium eximium, Potentilla diversifolia*.

LOCATION: High altitudes above timberline.

B&M ch. 18

40. JOSHUA TREE SCRUB (*Salazaria-Tetradymia-Yucca*)

STRUCTURE: Low, more or less dense community of broad-leaved, evergreen and/or deciduous shrubs with widely scattered Joshua trees, up to 12 m tall, rarely more.

DOMINANTS: Bladder sage (*Salazaria mexicana*), cottonthorn (*Tetradymia axillaris*), Joshua tree (*Yucca brevifolia*).

OTHER CHARACTERISTIC COMPONENTS: *Artemisia tridentata, Coleogyne ramosissima, Grayia spinosa, Juniperus californica, Larrea divaricata* (may codominate locally), *Lycium andersonii, L. cooperi, L. torreyi*.

LOCATION: Mainly in the Mojave Desert.

B&M ch. 24

41. MOJAVE CREOSOTE BUSH (*Larrea divaricata*)

STRUCTURE: Open community of broad-leaved evergreen, microphyll shrubs, rarely more than 2 m high; coverage usually less than 50%. Various smaller shrubs may form a very open lower layer.

DOMINANTS: Creosote bush (*Larrea divaricata*)

OTHER CHARACTERISTIC COMPONENTS: *Ambrosia dumosa, Baccharis sergiloides, Encelia farinosa, E. frutescens, Hymenochloa salsola, Lycium andersonii, Opuntia bigelovii* (locally), *O. basilaris, Sphaeralcea ambigua.*

LOCATION: Mojave Desert

B&M ch. 24

42. SONORAN CREOSOTE BUSH (*Ambrosia-Larrea*)

STRUCTURE: Open stands of broad-leaved evergreen and deciduous shrubs of varying height, rarely exceeding 3 m. Low, widely spaced graminoids occur occasionally. A denser and taller community of broad-leaved deciduous trees and shrubs occurs in usually dry stream beds or washes, the so-called wash-woodland (see below).

DOMINANTS: Bursage (*Ambrosia dumosa*), creosote bush (*Larrea divaricata*).

OTHER CHARACTERISTIC COMPONENTS: *Dalea californica, Echinocereus engelmannii* (on rocky slopes above *Larrea*), *Encelia farinosa, Fouquiera splendens, Hilaria rigida, Hymenochloa salsola, Isomeris arborea* (on rocky slopes above *Larrea*), *Opuntia bigelovii* (locally), *O. basilaris, Simmondsia chinensis, Thamnosma montana* (on rocky slopes above *Larrea*)

Wash-woodland (*Cercidium-Dalea-Olneya*)

DOMINANTS: Palo verde (*Cercidium floridum*), smoke tree (*Dalea spinosa*), desert ironwood (*Olneya tesota*).

OTHER CHARACTERISTIC COMPONENTS: *Cercidium microphyllum, Chilopsis linearis, Condaliopsis lycioides, Dalea schottii, Hymenochloa salsola, Lycium andersonii, L. brevipes, Prosopis juliflora* var. *torreyana.*

LOCATION: Colorado Desert

B&M ch. 25

43. BLACKBUSH SCRUB (*Coleogyne ramosissima*)

STRUCTURE: Dense to open community of low, mostly deciduous shrubs. The height can vary much but usually remains well below 2 m.

DOMINANTS: Blackbush (*Coleogyne ramosissima*)

OTHER CHARACTERISTIC COMPONENTS: *Ephedra nevadensis, Eriogonum fasciculatum* ssp. *polifolium, Grayia spinosa, Purshia glandulosa, Thamnosma montana.*

LOCATION: Deserts of southeastern California.

B&M ch. 24

44. CACTUS SCRUB (*Opuntia* spp.)

STRUCTURE: Low, open stands of stem succulents. The height remains well below 2 m.

DOMINANTS: Jumping cholla (*Opuntia bigelovii*), pencil cactus (*Opuntia ramosissima*)

OTHER CHARACTERISTIC COMPONENTS: *Echinocactus acanthodes, Echinocereus engelmannii, Opuntia echinocarpa, Yucca schidigera.*

LOCATION: Northern and western margins of Colorado Desert, mainly on south-facing slopes.

B&M ch. 25

45. DESERT SALTBUSH (*Atriplex-Grayia-Sarcobatus*)

STRUCTURE: Open formation of broad-leaved evergreen and broad-leaved drought-deciduous shrubs. The height varies from less than 1 m to over 2 m. Graminoids and forbs rarely exceed 40 cm in height, usually covering less than 5%.

DOMINANTS: Shadscale (*Atriplex confertifolia*), hopsage (*Grayia spinosa*), greasewood (*Sarcobatus vermiculatus*).

OTHER CHARACTERISTIC COMPONENTS: *Allenrolfea occidentalis, Artemisia spinescens, Atriplex canescens, A. nuttallii, A. polycarpa* (may dominate locally), *Chaenactis douglasii, Chenopodium leptophyllum, Cryptandra circumscissa, Dalea polyadenia, Distichlis spicata* var. *stricta* (may dominate locally), *Ephedra nevadensis, Eurotia lanata, Frankenia grandifolia* var. *campestris, Gilia leptomeria, Gutierrezia sarothrae, Hilaria jamesii, Kochia californica, Lycium cooperi, Menodora spinescens, Oryzopsis hymenoides, Sarcobatus baileyi, Sitanion hystrix, Suaeda depressa, S. fruticosa, Tetradymia glabrata.*

LOCATION: Arid regions of eastern California.

B&M ch. 24, 25

46. SALTON SEA SALTBUSH (*Atriplex polycarpa*)

STRUCTURE: Open stands of low, broad-leaved, mainly deciduous shrubs.

DOMINANTS: Saltbush (*Atriplex polycarpa*)

OTHER CHARACTERISTIC COMPONENTS: *Atriplex canescens* ssp. *linearis, Haplopappus acradenius* ssp. *eremophilus, Prosopis juliflora* var. *torreyana*

LOCATION: Around the Salton Sea.

B&M ch. 25

47. SAN JOAQUIN SALTBUSH (*Atriplex polycarpa*)

STRUCTURE: open, broad-leaved evergreen and/or deciduous shrub community. An undergrowth of herbaceous plants may vary from medium dense to absent. The structure can vary much within short distances.

DOMINANTS: Saltbush (*Atriplex polycarpa*)

OTHER CHARACTERISTIC COMPONENTS: *Allenrolfea occidentalis, Artemisia tridentata* (in well drained, non-saline soils), *Atriplex fruticulosa, A. lentiformis, A. phyllostegia, A. spinifera, A. tularensis, Distichlis spicata* var. *stricta, Ephedra californica, Haplopappus racemosus, Kochia californica, Lycium cooperi, Salicornia subterminalis, Sarcobatus vermiculatus, Sporobolus airoides, Suaeda torreyana.*

LOCATION: Mainly southern and western part of San Joaquin Valley.

B&M ch. 14

48. ALKALI SCRUB-WOODLAND (*Allenrolfea-Tamarix*)

STRUCTURE: Dense groves of low, shrubby trees or tall shrubs, tall broad-leaved deciduous trees (locally) and/or open stands of low shrubs.

DOMINANTS: Iodine bush (*Allenrolfea occidentalis*), tamarisk (*Tamarix chinensis* = introduced)

OTHER CHARACTERISTIC COMPONENTS: *Aster spinosus, Atriplex lentiformis, Baccharis glutinosa, Cercidium microphyllum, Pluchea sericea, Populus fremontii, Prosopis pubescens, Salix gooddingii, Suaeda torreyana.*

LOCATION: Along Colorado River and the south end of the Salton Sea.

B&M ch. 25

49. OASIS SCRUB-WOODLAND (*Washingtonia filifera*)

STRUCTURE: More or less dense, scrubby woodlands of palms and deciduous broad-leaved trees and shrubs and some graminoids.

DOMINANTS: Fan palm (*Washingtonia filifera*)

OTHER CHARACTERISTIC COMPONENTS: *Baccharis sergiloides, Fraxinus velutina, Haplopappus acradenius, Juncus acutus, Platanus racemosa, Pluchea sericea, Populus fremontii, Salix gooddingii, S. laevigata, Sporobolus airoides.*

LOCATION: At heads and/or in bottoms of canyons or where springs and seeps issue from desert mountains in southern California.

B&M ch. 25

50. NORTHERN SEASHORE COMMUNITIES (*Elymus, Baccharis*)

In this complex vegetation type, the northern beach community is distinguished from northern dune scrub.

Northern beach community (*Elymus mollis*)

STRUCTURE: Graminoid community, about 70 cm tall with a layer of evergreen, perennial, often succulent forbs, about 10 cm high. The graminoids lose their dominance in the south, becoming scattered. Coverage is 5-20% on the upper beach, 50-80% on the foredunes.

DOMINANTS: Silver beach weed (*Ambrosia chamissonis*), rye grass (*Elymus mollis*)

OTHER CHARACTERISTIC COMPONENTS: *Abronia latifolia, Ammophila arenaria* (introduced, now often dominant), *Atriplex leucophylla* (southern part), *Cakile maritima* (introduced), *Calystegia soldanella, Mesembryanthemum chinense* (southern part where it may dominate), *Oenothera cheiranthifolia*.

Northern dune scrub (*Baccharis-Haplopappus-Lupinus*)

STRUCTURE: Open community of broad-leaved evergreen shrubs, about 1 m high, and a lower, denser layer of evergreen subshrubs, about 50 cm high. Evergreen, prostrate, perennial (and some annual) forbs form a ground cover, about 10 cm high. Coverage about 70%.

DOMINANTS: Coyote brush (*Baccharis pilularis* ssp. *consanguinea*), goldenweed (*Haplopappus ericoides*), lupines (*Lupinus arboreus* and *L. chamissonis*).

OTHER CHARACTERISTIC COMPONENTS: *Abronia latifolia, Ambrosia chamissonis, Artemisia pycnocephala, A. californica* (southern part), *Carex obnupta, Eriogonum latifolium, Fragaria chiloensis, Juncus lesueurii,J. phaeocephalus, Lathyrus littoralis, Poa douglasii, Salix lasiolepis*.

LOCATION: Narrow, much interrupted belt along the coast, north of 36° northern latitude.

B&M ch. 7

51. SOUTHERN SEASHORE COMMUNITIES (*Abronia, Haplopappus*)

In this complex vegetation type, the southern beach community is distinguished from dune scrub.

Southern beach community (*Abronia maritima*)

STRUCTURE: Evergreen, perennial, often succulent forbs and subshrubs. Graminoids are largely absent. Height about 30 cm, coverage 5-20% on the upper beaches, 50-80% on the foredunes.

DOMINANTS: Sand verbena (*Abronia maritima*)

OTHER CHARACTERISTIC COMPONENTS: *Ambrosia chamissonis, Atriplex leucophylla, Cakile maritima* (introduced), *Calystegia soldanella, Mesembryanthemum chilense* (may dominate locally)

Dune scrub (*Haplopappus ericoides*)

Dune scrub covers up to 70% of mobile and stabilized dunes. Both structurally and floristically it is desirable to distinguish a central from a southern dune scrub.

Central dune scrub (*Artemisia-Haplopappus-Lotus*)

STRUCTURE: Scattered, broad-leaved evergreen shrubs, about 1 m tall, and a much denser layer of evergreen subshrubs, about 50 cm tall. A ground layer of evergreen, prostrate, perennial (and some annual) forbs and some succulents is about 10 cm high.

DOMINANTS: California sagebrush (*Artemisia californica*), goldenweed (*Haplopappus ericoides*), bird's foot trefoil (*Lotus scoparius*).

OTHER CHARACTERISTIC COMPONENTS: *Abronia umbellata, Corethrogyne filaginifolia, Croton californicus, Dudleya caespitosa, Lupinus chamissonis, Mesembryanthemum chilense.*

Southern dune scrub (*Ephedra-Haplopappus-Lycium*)

STRUCTURE: Dense, broad-leaved evergreen subshrubs, about 50 cm tall, and a lower layer of succulent, perennial forbs, about 10 cm tall.

DOMINANTS: Mormon tea (*Ephedra californica*), goldenweed (*Haplopappus ericoides*), boxthorn (*Lycium brevipes*)

OTHER CHARACTERISTIC COMPONENTS: *Abronia maritima, Atriplex canescens, Haplopappus venetus* (may dominate locally), *Lupinus chamissonis* (may dominate locally in northern part), *Mesembryanthemum chilense, M. crystallinum, Opuntia occidentalis, Rhus integrifolia* (may dominate locally), *Simmondsia chinensis.*

LOCATION: The southern seashore communities are located in a narrow much interrupted belt along the coast south of 36° northern latitude. The boundary between the central and southern dune scrub communities lies at approximately 34° northern latitude.

B&M ch. 7

52. COASTAL PRAIRIE-SCRUB MOSAIC (*Baccharis, Dantonia-Festuca*)

In this complex mosaic, coastal prairie is distinguished from coastal scrub. Farther inland, the scrub disappears and only the prairie prevails.

Coastal prairie (*Danthonia-Festuca*)

STRUCTURE: Dense graminoid community of perennial bunchgrasses, about 50 cm tall when in flower, with a lower layer of annual and perennial forbs, about 10 cm high. The coverage usually approaches 100%.

DOMINANTS: Oatgrass (*Danthonia californica*), red fescue (*Festuca rubra*).

OTHER CHARACTERISTIC COMPONENTS: *Bromus maritimus, Calamagrostis nutkaensis, Deschampsia caespitosa* var. *holciformis, Elymus pacificus, E. vancouverensis, Festuca idahoensis, Holcus lanatus* (introduced).

Coastal scrub (*Baccharis pilularis* ssp. *consanguinea*)

STRUCTURE: Open to dense, broad-leaved evergreen shrub community, about 1.5 m high. Evergreen and/or deciduous subshrubs, vines, perennial forbs and graminoids form a dense lower layer, about 30-50 cm high.

DOMINANTS: Coyote brush (*Baccharis pilularis* ssp. *consanguinea*)

OTHER CHARACTERISTIC COMPONENTS: *Artemisia californica* (southern part), *A. suksdorfii, Ceanothus thyrsiflorus* (northern part), *Elymus mollis, E. pacificus, E. vancouverensis, Erigeron glaucus, Gaultheria shallon* (northern part), *Heracleum lanatum* (northern part), *Mimulus aurantiacus, Polystichum munitum* (northern part), *Pteridium aquilinum* var. *pubescens, Rhamnus californica, Rhus diversiloba* (southern part), *Rubus parviflorus* var. *velutinus, R. vitifolius, Satureja douglasii, Scrophularia californica, Stachys rigida*.

LOCATION: The coastal prairie-scrub mosaic is located in a narrow, much interrupted belt along the northern part of the coast.

B&M ch. 21

53. HOT SANDY DESERT (Vegetation largely absent)

STRUCTURE: Growthforms largely absent. If any plants occur at all, they are members of the surrounding communities.

DOMINANTS: none

OTHER CHARACTERISTIC COMPONENTS: none

LOCATION: Southeastern Imperial county.

54. COLD ALPINE DESERT (Vegetation largely absent)

STRUCTURE: Growthforms largely absent. If any plants occur at all, they are members of the alpine communities (39).

DOMINANTS: none

OTHER CHARACTERISTIC COMPONENTS: none

LOCATION: Alpine fields of naked rock, snow and ice, usually above the alpine communities (39) of the Sierra Nevada and Cascade Range.

INDEX

This index includes subjects, place names, authors of cited literature, and taxa. Page numbers of authors refer only to literature cited sections. Only those common names of plants used in the text are listed here, and cross references to scientific names are given. Page references are shown only after the scientific names. Common names are listed by their first, not generic, name; lodgepole pine, for example, is indexed under l. With a few exceptions, taxa above the species level are not listed. Not all contributors were equally precise in the use of subspecific ranks, so references to such categories may not be complete. The index reflects and does not correct the confusing use of subspecies and varietal categories by botanists working with the western American flora. We have, however, tried to cross-list synonyms at the species and genus level. Some place names may have slipped by us in the indexing process, but the omission of counties was a deliberate choice. To save space, consecutive pages are combined (as 000-0), even when those pages do not deal exclusively with the item indexed. The sequence is strictly alphabetical, regardless of the number of words or initials. The reader can supplement the index with the very full chapter tables of contents.

Abele, L.G., 533
Aberg, B., 666
Abies
  amabilis, 154, 701, 714, 719, 723-28
  bractaeata, 151, 160, 376, 378, 539, 925
  concolor, 68, 104, 143, 145, 148-51, 153, 161, 216, 312, 364, 376, 541-2, 544-6, 549, 553, 560, 562-9, 571-7, 586-93, 700-5, 708-9, 714-5, 719, 721-9, 800-1, 806-7, 822, 827, 916-23
    ssp. concolor, 546-7, 822
    ssp. lowiana, 823
  grandis, 145, 147, 154, 244, 368, 680, 682-6, 688-9, 916
  lasiocarpa, 104, 154-5, 617, 621, 701, 709, 714, 719, 723-9
  longirostris, 146
  magnifica, 84, 151-4, 161, 312, 553, 560, 562-3, 565, 571-5, 578, 580, 587-9, 591-2, 647,
Abies (cont.)
    702, 704, 708-10, 714-6, 719-30, 806, 916, 922
    var. shastensis, 104, 148-9, 312, 622, 701, 920-21
  procera, 154, 708, 724, 727-29, 920
Abrams, J.E., 549
Abrams, L., 131, 351, 459
Abronia
  crux-maltae, 41
  latifolia, 121, 229, 230-31, 235-6, 240, 243-4, 246, 249, 254, 257, 261, 936
  maritima, 228-31, 235, 241, 246, 248, 250, 252, 254, 257, 261, 896, 936-7
  turbinata, 41
  umbellata, 230, 235-6, 248, 250, 254, 257, 896
Acacia greggii, 181, 185, 816, 844-6, 872, 881-3, 885
Acaena californica, 742
Acalypha californica, 181, 475
Acamptopappus
  shockleyi, 842

Acamptopappus (cont.)
  sphacrocephalus, 841, 843, 846, 872
Acclimation, 881, 883
Acer
  brachypterum, 153-4, 163
  circinatum, 362-3, 590
  glabrum, 153, 565, 711
    var. diffusum, 823-4, 922, 924
    var. torrey, 716
  grandidentatum, 153, 166
  macrophyllum, 141, 145, 151, 159-61, 308, 360-1, 363-4, 367, 375, 565, 567, 682, 704, 723-4, 902, 905, 922, 925
  negundo, 145, 150-1, 165
    ssp. californicum, 926
  neomexicanum, 146
  pictum, 145
  saccharinum, 145, 161
Achillea
  borealis, 254, 335, 344, 751, 900
    ssp. arenicola, 243
  lanulosa, 648, 713, 718
    ssp. alpicola, 622, 633
  millefolium, 246, 248, 254
Achlys triphylla, 363, 368, 706, 925
Ackerman, T.L., 863
Actaea rubra, 708
Actual evapotranspiration, See Evapotranspiration
Adam, D.F., 188
Adam, D.P., 666
Adams, A., 511
Adams, C.F., 829, 863
Adams, D.A., 289
Adams, J.E., 459
Adams, M.S., 863, 886
Adams, S., 863, 886
Adenocaulon bicolor, 363, 569, 706, 712, 922
Adenostoma
  fasciculatum, 93, 166, 178, 212, 246-8, 298, 301, 304, 308, 311-2, 322, 326, 329, 331, 400, 422-5, 429-31, 445, 447-53, 481, 684-5, 817, 900, 902, 905, 919, 927-8
Adenostoma (cont.)
  sparsifolium, 177, 424, 429, 450-3, 820, 928
Adiantum
  capillus-veneris, 122
  jordanii, 335
Adobe Hills, 810
Adolphia californica, 474
Advection, 15, 655
Aegilops triuncialis, 498
Aesculus
  californica, 372, 388, 393, 396-7, 399, 401-3, 493, 925-6
  parryi, 164, 474-5
Agastache urticifolia, 708
Agave
  deserti, 181, 820, 872
  shawii, 252, 474-5
  utahensis, 854
    var. nevadensis, 129
Agee, J.K., 595
Age structure of stands or plant species, 300, 303, 318-20, 330, 332, 340-2, 345, 369, 395, 409, 425-7, 448-9, 564, 567-8, 570, 572-3, 577-8, 580-2, 756, 800, 806, 822-3, 825-7, 837, 839-41, 846, 902, 910
Agoseris
  apargioides, 235, 249
  glauca, 622, 777
    var. monticola, 646, 658, 662
  heterophylla, 503, 900
Agropyron
  pringlei, 647
  spicatum, 40, 493, 767, 771-2, 774, 776, 780, 782-6, 793, 803, 919, 928, 930
  trachycaulum, 774
Agrostis
  alba, 530, 734-5
  californica, 735
  diegoensis, 344, 736
  exarata, 734
  hallii, 365
  idahoensis, 552
  microphylla, 736
  thurberiana, See Podagrostis thurberiana

Agrostis variabilis, 623, 649
Aiken soil, 401
Ailanthus altissima, 128, 145, 160, 162
Aira
  caryophyllea, 436, 497, 503-4, 506, 509, 685, 735-6, 738, 741, 743
  praecox, 254, 741, 743
Aizenshtat, B.A., 666
Alamitos Bay, 265
Al-Ani, H.A., 69, 886
Albion, 303
Alcorn, S.M., 889
Alder, See Alnus
Alder Saddle, 542
Aldon, E.F., 829
Alexander, E., 220, 413, 597, 731, 831
Alien plants, See Introduced Taxa
Alkali sink, 494, 516, 519-20, 850-3, 879-81, 935
Alkali soil, 853
Allard, R.W., 133
Allelopathy, 328, 373, 396, 445, 449, 454, 479, 486, 503, 756, 841, 849, 875-6
Allenbaugh, G.A., 595, 829
Allenrolfea occidentalis, 851, 853, 880, 934-5
Allium
  burlewii, 152
  monticola, 554, 923
  parvum, 645
  praecox, 900
  validum, 623
Allocarya
  acanthocarpus, 523
  distantiflorus, 523
  humistrata, 521
  hystricula, 521
  leptocladus, 519
  nothofulvus, 524
  stipitata, 516, 520, 523, 527
    var. micrantha, 521, 525-6
  undulata, 523-4
Allophyllum gilioides, 436
Allophyllum glutinosum, 436
Allotropa virgata, 569
Allred, B.W., 459
Alnus
  crispa, 623
Alnus (cont.)
  oregana, 239, 254, 682, 684, 688, 916
  rhombifolia, 145, 159, 161, 165, 567, 927
  rubra, See A. oregana
  sinuata, 145, 621, 709-10, 716
  tenuifolia, 145, 567, 709-10
Alopecurus
  aequalis, 551
  howellii, 520-1, 527
Alpine, 3, 5, 16, 82, 84, 552, 601-675
Altithermal, See Xerothermic
Alturas, 38
Amblyopappus pusillus, 897-9
Ambrosia
  chamissonis, 121, 228-31, 235-6, 240, 243-4, 246, 254, 257, 261, 896, 936-7
    ssp. bipinnatisecta, 231, 235, 241, 246, 248, 250, 254, 261
  dumosa, 837, 841-5, 848-9, 855, 860, 872, 885, 933
  ilicifolia, 872
  psilostachya, 254
Amelanchier
  alnifolia, 145, 159, 160-1, 786
  covillei, 823
  pallida, 312, 546, 705, 711, 715, 717, 720-1, 784, 786, 799, 802, 804, 806, 921, 923
  utahensis, 164, 170, 645, 799, 823, 922
American dune grass, See Elymus mollis
Ammania coccinea, 552
Ammirati, J.F., 459
Ammophila
  arenaria, 225-6, 229-31, 236, 254, 261, 734, 936
  breviligulata, 226, 254
Amorpha californica, 164
Amphipappus fremontii, 848
Amsinkia
  intermedia, 254, 493, 898
  menziesii, 787, 790
Anacapa Islands, 84, 894, 897, 899, 900, 904, See also East Anacapa Island

Anagallis arvensis, 436, 741, 753
Anaheim Bay, 265, 276
Anaphalis margaritaceae, 236, 254, 472, 747, 751
Anchor Bay, 305, 315, 684
Anderson, B.R., 131
Anderson, C.E., 289
Anderson, D.J., 863
Anderson, E., 131
Anderson, H.W., 459
Anderson Island, 87
Androsace
  septentrionalis, 637, 653, 656, 658
    ssp. subumbellata, 544, 630, 663
Anelsonia eurycarpa, See Phoenicaulis eurycarpa
Anemone
  deltoidea, 706, 712
  drummondii, 569, 621, 624
  occidentalis, 622
  quinquefolia, 711
Angeles National Forest, 543, 548-9, 553
Angelica
  arguta, 621, 708, 715, 718, 753
  hendersonii, 237
  lucida, 621
Angwin, 684
Anisocoma acaulis, 809, 920
Anliker, J., 668
Annuals, 554, 878, 884
  summer, 850, 860, 862
  winter, 841, 858-62, 875-7
Año Nuevo, 88, 92, 342
Antelope bush, See Purshia tridentata
Antelope Mountains, 802
Antelope, pronghorn, See Antilocapra americana
Antennaria
  alpina, 625, 635, 651
    var. media, 611, 624, 637, 640, 643, 647, 653, 659, 663-4, 923
  corymbosa, 627, 774
  dimorpha, 544, 776
  rosea, 548, 627, 717, 786, 803, 826
  umbrinella, 611, 623, 636, 639
Antevs, E., 794
Anthoxanthum odoratum, 737, 740 742-5
Antilocapra americana, 498, 500, 770, 800
Antirrhinum cornutum, 436
Apiastrum angustifolium, 436
Aplopappus, See Haplopappus
Apocynum pumila, 706, 712, 717
Apomixis, 126, 127
Applegate, E.I., 829
Aquatic vegetation, 551-2
Aquilegia
  formosa, 553, 624, 653
  pubescens, 611, 631, 639
Arabis
  breweri, 376, 548
  holboellii, 624, 631, 774-5, 782
    var. retrofracta, 440, 544, 645
  inyoensis, 629-30, 632, 662
  layallii, 6, 11, 636, 639, 644
    var. nubigena, 626
  lemmonii, 611, 624, 636, 639
    var. depauperata, 626
  parishii, 544
  platysperma, 574, 623, 628, 640
    var. howellii, 611, 646
  repanda, 544, 917
Aralia californica, 567
Araya, S., 459
Arbutus
  arizonica, 170
  glandulosa, 163
  menziesii, 147-8, 150-2, 159-63; 167, 171, 308, 333, 343, 360-4, 366-7, 369, 371-3 375, 377, 393, 565, 682-3, 685, 689, 693, 901-2, 905, 916, 925-6
  xalapensis, 162; 171
Arcata, 230, 236-40, 242-5, 275, 339
Arceuthobium
  abietinum, 546
  californicum, 546
  campylopodum, 542
  cyanocarpum, 550
  divaricatum, 546
Arctomecon, 116, 118

Arctostaphylos
acutifolia, 428
canescens, 308, 366, 428, 721
columbiana, 304-5
confertiflora, 903
crustacea, 334
elegans, 122, 449, 928
glandulosa, 161, 301, 304, 316, 322, 331, 428, 539, 542, 918, 926, 928
glauca, 165, 169, 170, 177, 308, 427, 429, 449, 451, 453, 481, 814, 816-8, 820, 926, 928
hookeri, 303-4
insularis, 901, 928
manzanita, 161, 165, 175, 388, 400, 449, 457, 588, 917, 926, 928
mariposa, 175, 216, 312, 393, 428, 918
mewukka, 428
morroensis, 178
myrtifolia, 122, 928
nevadensis, 149, 428, 577-8, 586, 705, 711, 714-6, 719-22, 806, 921-2
nissenana, 421
nummularia, 122, 304-5, 307-8, 340
otayensis, 300-1, 428
parryana, 428, 928
var. pinetorum, 919
patula, 312, 362, 428, 435, 457, 547, 563, 565, 574-5, 577, 586, 590, 710-11, 714-6, 719-21, 779, 801, 916-17, 919-24
ssp. platyphylla, 543, 546
pechoensis, 247
pringlei, 153, 539
ssp. drupacea, 428, 542
pumila, 178, 246
pungens, 164, 170-1, 174, 177, 308, 429, 539, 813, 822
rudis, 178, 247
silvicola, 121, 307-8
stanfordiana, 928
subcordata, See A. confertifolia, A. tomentosa ssp. subcordata
tomentosa, 304, 307-8, 344-5
ssp. insulicola, 901
Arctostaphylos
uva-ursi, 122, 156, 238, 244, 613
var. coactilis, 254
vestita, 246
viscida, 177, 308, 311-12, 365, 386, 388, 400, 421, 428-9, 457, 563, 586, 588, 685, 917-8, 926, 928
Arcto-Tertiary Geoflora, 112-3, 115, 119-21, 159, 162, 181, 332
Arenaria
californica, 519
capillaris var. compacta, 609
congesta, 712, 718, 803
var. subcongesta, 664
kingii, 627-8, 636, 638, 642, 825-6, 924
var. glabrescens, 623, 647, 656-7, 659, 663
nuttallii, 611, 623
ssp. gracilis, 623, 636, 638, 647, 652, 654, 662
ssp. gregaria, 722
obtusiloba, 639
rossii, 630
rubella, 630, 653
ursina, 544
Argus, G.W., 666
Arid-humid boundary, 55-6, 60-1, 65-66, 68
Aridiosols, 767, 773, 779
Aridity, radiative index, 18, 37, 40, 64
Aristida
divaricata, 900, 930
hamulosa, 495-6
oligantha, 503
parishii, 821
Arizona, 297, 385, 857, 870-2, 909
Arkley, R.J., 69, 133, 354, 532, 697
Armeria
maritima, 238
ssp. californica, 254, 755-6
Armstrong Redwoods, 88
Armstrong, W., 351
Arnica
amplexicaulis, 648
cordifolia, 707, 712
latifolia, 717

Arnica (cont.)
longifolia, 624
nevadensis, 624
sororia, 803
viscosa, 621, 622
Arnold, C.A., 188
Arnold, J.F., 829
Arno, S.F., 595
Aroga websterii, 769, 770
Arrowhead, Lake, 541
Arroyo, See Wash
Artemisia
arbuscula, 643, 645, 650, 656-8, 660, 766, 768-72, 776, 780-1, 785-6, 800 827, 847, 928, 930
bigelovii, 822
californica, 88, 247, 249, 301, 304, 323, 334, 341, 425, 434, 448, 472-5, 479-83, 748, 750, 897-8, 928-9, 936-8
campestris ssp. borealis, 621
cana, 772, 781, 785
ssp. bolanderi, 768
dracunculus, 335, 824
frigida, 180
ludoviciana, 821
norvegica ssp. saxatilis, 649
nova, 766, 772, 781-2, 799, 812, 822
pycnocephala, 235, 238, 244, 246, 249, 254, 755, 936
spinescens, 766, 774, 788, 790, 792, 852, 854-5, 934
suksdorfii, 472, 938
tridentata, 38, 42, 65, 79, 82-3, 152-3, 179-80, 201, 311-12, 392, 495, 543, 575, 577, 579, 590, 592-3, 664, 708, 767-9, 771-7, 779-81, 784, 786, 788, 790, 793, 799-801, 803-4, 806-15, 824-5, 827, 847, 916, 919-20, 924, 928, 930, 932, 935
ssp. tridentata, 766, 770
ssp. vaseyana, 643, 644, 650, 766, 770, 806
ssp. wyomingensis, 766, 770
vulgare, 254
Arthrocnemum, See Salicornia
Arundo donax, 551
Arvanitis, L.G., 219
Asarum
caudatum, 368, 706
hartwegii, 569
Asclepias
eriocarpa, 542
fascicularis, 551
subulata, 845, 872
Ash, See Fraxinus
Ashby, E., 886
Ashby, W.C., 462
Ashe, W.W.
Ash Mountain, 33, 63-4
Ashville, N.C., 29-30
Ashworth, P.R., 469
Asilomar State Park, 227-8, 246
Aspect, 55, 316, 318, 494, 502, 542-3, 545, 576, 725, 734, 815, 822, 824, 876, 897-903, 926
Aspen, See Populus
Aspen woodland, 551. 553
Association tables, 1, 236-9, 632-4, 636-41, 644-9, 658-9, 662-3, 705-7, 710-13, 716-8, 750-3
Aster
alpigenus, 923
var. andersonii, 585, 609, 624, 632, 640, 649
canescens, 774, See also Machaeranthera
chilensis, 742
ledophyllus, 718, 722
occidentalis, 717
scopulorum, 774, 777
spinosis, 880, 935
subspicatus, 751
Astragalus
bicristatus, 544
calycosus, 659-60
congdonii, 445
filipes, 775
kentrophyta, 611, 662, 664, 826
var. danaus, 630, 636, 638
var. implexus, 932
lentiginosus, 751
var. ineptus, 627
leucolobus, 544
leucopsis, 251
miguelensis, 896
porrectus, 41
purshii, 775, 777, 783, 786
var. lectulus, 544, 611, 627, 632, 645, 662, 664

Astragalus (cont.)
  stenophyllus, 776, 787
  tener, 519
  whitneyi, 611, 630, 645
    var. siskiyouensis, 621
Athens, Greece, 25, 26
Athyrium
  alpestre, 714, See also
    A. distentifolium
  distentifolium, 623
  filix-femina var. californicum,
    711
Atrichoseris platyphylla, 861
Atriplex
  californicum, 897
  canescens, 41, 185, 250, 774-5,
    813, 844, 847, 850, 852, 872,
    878, 934, 937
  confertifolia, 41, 64-5, 87,
    774, 788, 790, 842, 847,
    850-5, 934
  fruticulosa, 935
  hymenelytra, 846, 850-3, 872
  julacea, 251-2
  lentiformis, 851-3, 872, 880-2,
    935
  leucophylla, 230-1, 235, 241,
    253-4, 257, 261, 903, 936-7
  nuttallii, 790, 934
  parishii, 250
  parryi, 853
  patula, 273-6
  phyllostegia, 935
  polycarpa, 494-5, 844, 846,
    850-3, 872, 881-2, 934-5
  semibaccata, 274, 898
  spinifera, 935
  tularensis, 935
  watsonii, 279, 903
Atwater, B.F., 289
Auburn, 54, 56
Audubon, 96, 100
Avalanche, 605, 615, 709
Avalon, 37, 39
Avena
  barbata, 436, 502-3, 505, 509,
    523-4, 530, 735, 741, 931
  fatua, 394, 436, 501-5, 509,
    735-6, 900, 931
Avila, G., 459
Axelrod, D.I., 69, 136, 131, 188,
  190, 192, 351, 459, 794, 886,
  906
Axelton, E.A., 595
Azalea (area), 88
Azevedo, J., 69, 351, 697

Babcock, E.G., 131
Baccharis
  douglasii, 237
  emoryi, 185
  glutinosa, 880, 935
  pilularis, 237, 244, 246, 255,
    304, 373, 434, 472, 476, 480,
    685, 688, 748-9, 754
    ssp. consanguinea, 249, 255,
      304, 334, 738, 747, 750,
      897, 925, 929, 938
    ssp. pilularis, 747, 899, 936
  sarothroides, 181, 872, 877
  sergiloides, 178, 813, 932,
    935
  viminea, 432, 927
Backbone Creek, 82
Backhuys, W., 666
Badran, O.A., 351
Baeria, See Lasthenia
Bagdad, 43
Baig, M.N., 666
Bailey, D.K., 829
Bailey, H.P., 69, 189
Bailey, W.H., 595
Baja California, 250, 279, 298,
  300, 325, 332, 421, 474, 485,
  538-40, 542, 547, 553, 808,
  819, 870-1, 878, 899
Bajada, 667, 837-8, 840-1, 844,
  846-7, 854, 871, 873-7, 879,
  884
Baker, F.S., 9, 69
Baker, H.G., 131, 136, 532, 666,
  759
Bakersfield, 33-7, 64
Bakker, E.S., 9, 758
Bal, B.S., 759
Balds, 404, 800, 803
Ball, F.M., 132
Ball, W.S., 136
Balsamorhiza
  hookeri See B. platylepis
  platylepis, 152
  sagittata, 774-5, 777, 784,
    786, 919, 930
Bamberg, S.A., 69, 72, 134, 191,
  671, 863, 867, 886

Banner Grade, 478
Bannister, M.H., 351, 354
Barbarea orthoceras, 552
Barbour, M.G., 131, 258-9, 261, 289, 291, 758, 863, 866, 886
Barley Flats, 541
Barnes, I., 351
Barrett, S.A., 758
Barry, J.W., 351
Barry, W.J., 9, 258, 511, 595, 758-9
Bartholomew, B.L., 131, 488, 511
Bartholomew, O.F., 532
Bartolome, J.W., 511
Bartschot, R., 354
Basal area 364, 373, 398, 560, 562-5, 567-8, 571-3, 575-6, 578, 580, 589, 592
Basalt, 315, 337, 765, 767, 769
Bates, D.M., 131
Batiquitos Lagoon, 285
Batis maritima, 273, 276-9, 281-2, 284-5, 287, 931
Batzli, G.O., 511, 758
Bauer, H., 666
Bauer, H.L., 411, 459
Baum, B.R., 131
Baumgartner, D.M., 595
Bay, See Umbellularia
Bazilevich, N.I., 73
Beach, 87, 225, 228-41, 896-7
Beach pine, See Pinus contorta ssp. contorta
Beals, E., 289
Beardsley, G.F., 595
Bear Valley, 544, 546-7
Beatley, J.C., 863-4, 866
Bebbia juncea, 185, 872
Becker, H.F., 189
Becking, R.W., 697
Beckmannia syzigachne, 736
Bedford, H., 351
Beechers' Bay, 897, 900, 903
Beechey ground squirrel, See Spermophilus beecheyi
Beeks, R.M., 131
Beeson, C.D., 69 829
Beetham, N.M., 595
Beetle, A.A., 69, 511, 758, 794
Beguin, C., 666
Beleperone, 118
  californica, 871-2
Bellue, M.K., 136
Benedict, J.B., 666
Bennet, E.L., 886
Bennett, P.S., 595, 666
Benseler, R.W., 411
Benson, L., 131, 351, 411, 864
Bentley, J.R., 459, 511-12
Berberis
  dictyota, 705
  haematocarpa, 813
  nervosa, 333, 362-3, 365, 368, 687, 703-5, 709-10, 921, 925
  pumila, 721
  repens, 706
Bergerocactus emoryi, 474-5
Berger, R., 189, 192, 832
Berm, 228
Bernardia incana, 820, 872
Berry, J., 413, 488, 573, 864
Berula erecta, 541
Betula
  fontinalis, See B. occidentalis
  lenta, 145
  occidentalis, 583, 593, 809
  papyrifera, 145, 161
Bigcone spruce, See Pseudotsuga macrocarpa
Big Horn, See Ovis
Big Lagoon, 265, 275, 683, 688-9
Bigleaf maple, See Acer macrophyllum
Big Rock Creek, 553
Big Sur River, 684
Billings, W.D., 69-70, 132, 189, 595, 597, 666-7, 672-3, 794, 829, 864
Binet, P., 289
Bingham, F.T., 864
Biomass, 229, 481, 504-7, 509, 530, 568, 573, 607, 748, 838, 850, 860
Birkeland, P.W., 69, 794
Bishop, 38, 42-3, 64-6, 854, 857
Bishop pine, See Pinus muricata
Bishop Pine Preserve, 97
Bison, 770
Biswell, H.H., 459, 467-8, 511, 514, 595
Björkman, O., 864-6
Blackbrush scrub, 839, 844, 858, 862
Blackburn, W.H., 794, 829
Blackbush, See Coleogyne ramosissima

Blacklock soil series, 93
Black Mountain, 902
Black oak, See Quercus kelloggii
Blacksage, See Salvia mellifera
Black sagebrush, See Artemisia nova or A. arbuscula
Blacks Mountain, 83, 854
Blackwelder, E.B., 794
Bladdersage, See Salazaria mexicana
Blaisdell, J.P., 794
Blennosperma nanum, 520-1
Blight, M.M., 352
Bliss, L.C., 666, 731
Bloomeria crocea, 440
  var. crocea, 900
Bluebunch wheatgrass, See Agropyron spicatum
Blue Canyon, 54-5, 58
Blue oak, See Quercus douglasii
Blue Ridge, 549
Bluff Lake, 548
Bluthgen, J., 69
Blythe, 38, 181, 881
Boblaine, 100
Boca, 54, 56, 59
Bock, J.H. and C.E., 459
Bodega, 92, 226, 244, 246, 265, 275, 691, 736, 745, 755-6, 853
Bodfish, 83, 315
Bodie, 42, 810
Boerhaavia, 117
Bogel, R., 71
Boggs Lake, 517
Bog, sphagnum, 340, 620
Boisduvalia glabella, 516, 520-1
Bolander pine, See Pinus contorta ssp. bolanderi
Bolinas, 265, 275
Bolsa Bay, 265, 282
Bolsa chica, 91
Boltehovskitch, Z.V., 666
Bolton, R.B., 379
Bonner, J., 886-7
Bonner, J.F., 462
Bonneville Basin, 766
Bonnicksen, T.M., 595
Boomer soil series, 362, 703
Booth, J.A., 889
Borisov, A.A., 69
Borrego Valley, 819
Boschniakia strobilacea, 543, 718
Botrychium
  lunaria var. minganense, 609
  simplex ssp. compositum, 649
Bottomley, W.B., 459
Bouma, A.H., 239
Bouteloua
  curtipendula, 821
  gracilis, 544
Bowen, O.E., 351
Bowerman, M.L., 379, 411, 459, 829
Bowers, N.A., 351
Bowman Canyon, 844-5
Bowman, W.F., 732
Box Canyon, 819
Boxthorn, See Lycium brevipes
Boyce, J.S., 595
Boyde Deep Canyon, 92, 819-21
Brachypodium distachyon, 498
Bradbury, D., 488
Bradley, W.G., 829, 864
Bradshaw, D.S., 289, 293
Bradshaw, K.E., 219, 381, 697, 732, 829
Brandegee, K., 459-60
Brasenia schreberi, 620
Brassica
  campestris, 255
  geniculata, 323, 436
  nigra, 436, 497
Braun-Blanquet, J., 69, 666, 675
Brayton, R., 460, 831
Breckon, G.J., 131, 258
Breidenbach, R.W., 863, 886
Brickellia
  arguta, 821
  californica, 554, 813
  desertorum, 816
  incana, 847
  microphylla, 554
Bridgeport, 810
Briggs, G.F., 597
Brink, V.C., 667
Bristlecone pine, See Pinus longaeva
Briza minor, 436, 503-4, 509, 523, 526, 741, 743
Brockmann-Jerosch, H., 69
Brodiaea
  coronaria var. macropodia, 736
  hyacinthina, 520, 525
  laxa, 440

Brodiaea (cont.)
lutea, 736
pulchella, 441, 900
Bromus
breviaristatus, 365, 542
carinatus, 249, 344, 585, 622, 735, 736, 741
diandrus, 344, 395, 502-6, 508-9, 735-6, 741, 787, 900, 931
laevipes, 736
madratensis, 436
marginatus, 621, 707, 735, 771, 773-4, 784, 786
manitimus, 938
mollis, 436, 498, 501-6, 508-9, 523-4, 526, 735-6, 741, 743, 931
orcuttianus, 577, 736, 920
ssp. hallii, 542
rigidus, See B. diandrus
rubens, 41, 436, 494, 497, 502, 505-6, 509, 524, 900
tectorum, 436, 770, 774-6, 780-1, 787, 790
Brooks, W.H., 411
Broome, S.W., 294
Brown, D.E., 9
Browning, B.M., 289-90, 292
Brown, R.L., 258-9
Browsing, effect of, See Herbivory
Bruce, D., 220
Brum, G.D., 864, 867, 886
Bryant, M.D., 865
Buchanan, H., 595
Buckman, A.R., 290
Buckman, R.E., 107
Buckwheat, See Eriogonum
Budyko, M.I., 69-70
Buell, M.F., 9
Buena Vista Lagoon, 91, 97, 265
Bull Creek, 691, 694
Bumelia lanuginosa, 164, 170
Bumpy camp, 97
Bunchgrass, 492-7, 900
Burcham, L.T., 511, 595, 758
Burckhardt, J.W., 794, 829
Burgy, R.H., 413
Burk, J.H., 886
Burns, D.M., 758
Burns Piñon Ridge Reserve, 92 816-7
Burr, B., 354, 906
Bursage, See Ambrosia dumosa
Bursera microphylla, 118, 181, 872
Burtt-Davy, J., 511, 758
Butterfly Valley, 83
Butterworth, B.B., 460, 864
Buttery, R.F., 411
Byers, H.R., 70
Byler, J.W., 351
Bynum, H.H., 351

C_4 photosynthesis, 253, 485, 849-50, 860, 862, 882
Cabazon, 874, 878
Cache Creek, 388
Cactus scrub, desert, 876-7
Cactus scrub, maritine, 898-9, 903
Cadbury, D.A., 667
Cain, D.R., 832
Cain, S.E., 351
Cajon Pass, 814, 819
Cakile
edentula, 228, 230, 235, 240-1, 255, 261
maritima, 228-31, 235-6, 240-1, 246, 248, 251, 255, 257, 261, 895, 936-7
Calamagrestis
breweri, 584-5, 611, 625, 637, 649, 652, 923
canadensis, 625
ephitidis, 735
koeleroides, 365, 920
nutkaensis, 734-5, 740, 742-7, 938
purpurascens, 631, 635-6, 638, 661, 663
rubescens, 919
Calandrinia
ciliata var. menziesii, 400
maritima, 897
Calaveras Big Trees, 60
Caldwell, M.M., 669
Caliente Mountain, 817-8

California bay, See
  Umbellularia
California Department of Fish and
  Game, 86, 90, 91
California floristic provinces,
  6, 31, 111-18
California history, 497-9, 736-7
California Islands, See
  Southern California Islands
California juniper, See
  Juniperus californica
California Natural Areas
  Coordinating Council (CNACC),
  103-5
California natural features
  (map), 8, 51, 54
California place names (map),
  7, 32-3, 265, 765, 894
California rose bay, See
  Rhododendron macrophyllum
California sagebrush, See
  Artemisia californica
California State Board of
  Forestry, 219
California State Parks,
  Natural Preserves, State
  Reserves, Wildernesses,
  Seashores, Parks, 85-9
California steppe, See
  Grassland, valley
California tussock moth, See
  Hemerocampa
Calkins, H.C., 460
Callander Beach, 233
Calliandra eriophylla, 181, 872
Callitriche
  longipedunculata, 523-4, 528
  marginata, 516, 523-4, 527
Calocedrus decurrens, 88, 104,
  151, 160, 216, 309, 312, 326,
  364-6, 541-2, 545-6, 561-5,
  567, 569, 571, 575-6, 586,
  588-90, 592, 682, 685, 692,
  703-4, 708-9, 714-5, 721,
  723-8, 801-2, 807, 916-9,
  922-3
Calochortus
  albus, 441
  eurycarpus, 774
  invenustus, 548, 554
  luteus, 527
  nitidus, See C. eurycarpus
  nuttallii, 777, 782
Calochortus (cont.)
  splendens, 323
  weedii, 441
Caltha howellii, 649
Calycoseris parryi, 861
Calypso bulbosa, 707
Calyptridium
  monandrum, 437
  monospermum, 121, 652, 654
  parryi, 548
  umbellatum, 574, 586, 619,
    622, 625, 636, 639, 654, 656,
    658, 663-4, 713, 718, 932
Calystegia
  cyclostegium, 441
  macrostegia, 323-4, 897
  occidentalis, 441, 751
    ssp. fulcrata, 542
  soldanella, 229-31, 235-6, 241,
    246, 250, 255, 257, 261,
    936-7
CAM (crassulacean acid
  metabolism), 485, 850
Cambria, 342
Cameron, G.N., 289
Camino, 851
Camissonia
  brevipes, 861
  chamaenerioides, 861
  cheiranthifolia, 121, 230-1,
    235, 237, 241, 244, 246, 248,
    251, 255, 261, 896, 936
  clavaeformis
    ssp. purpurascens, 809
  hardamiae, 118
  intermedia, 118
  micrantha, 118
Campanula
  californica, 87-8
  prenanthoides, 707
  rotundifolia, 621
  scabrella, 623
  wilkinsiana, 623
Campbell, C.J., 887
Campbell, D.H., 132
Campbell, R.E., 70
Camp Pendleton, 474-5, 482-3, 485
Cañada de Puerto, 901, 903
Canadian Zone, 548
Canker, See Coryneum cardinale
Cannon, W.A., 411, 460, 595, 886
Canotia holocantha, 181
Cantil, 838

Canyon live oak, See Quercus chrysolepis
Capaetown, 25-6
Capon, B., 886
Carbon balance, 482, 486, 756, 883
Cardamine bellidifolia, 621, 623
Cardionema ramosissimum, 238, 249, 742
Careaga sandstone, 337
Carex
albonigra, 637, 641, 656, 659-60
brainerdii, 624
brewerii, 611, 623-5, 635-6, 639, 932
capitata, 609
douglasii, 544, 630, 775
duriuscula, 656
eleocharis, 656-7, 660
exserta, 584-5, 634-5, 637, 640, 643, 646, 650-1, 923, 932
festivella, 585
globosa, 344
gymnoclada, 624
hassei, 552, 625
haydeniana, 611, 625
helleri, 611, 624, 628, 637, 640, 656, 658, 932
heteroneura, 552
incurriformis var. danaensis, 611
interior, 715
jonesii, 552
kelloggii, 620
leporinella, 631
limosa, 620
macrocephala, 255
mariposana, 653
mertensii, 621
multicaulis, 542
nebraskensis, 923
nigricans, 623, 625, 628, 641-2, 649, 653, 923
obnupta, 239, 243, 255, 275, 647, 735, 742, 931, 936
occidentalis, 613
ormantha, 620
pachystachya, 653
pansa, 246
Carex (cont.)
phaeocephala, 611, 622, 624, 626, 636, 639, 654, 660-1, 663
preslii, 611, 624, 653
pseudoscirpoidea, 611, 629 634, 640, 642
rossii, 548, 624, 630-1, 637, 639, 643, 645, 774, 777
rostrata, 715, 923
schottii, 552
scopulorum, 643, 648, 652, 923
senta, 552, 567, 931
spectabilis, 623, 626, 641, 647, 652
straminiformis, 621, 624
subfusca, 548
subnigricans, 611, 629, 634, 637, 641, 654, 659-61, 663-4, 923
tahoensis, 630, 636, 638
tumicola, 734
vernacula, 624-5, 635, 637, 640
Carl Inn, 84
Carlquist, S., 132, 136, 351, 532
Carlson, C.E., 411
Carlton Flats, 539-40
Carmel, 265, 496
Carneggie, D.M., 219
Carnegia gigantea, 118, 844, 873
Carpelan, L.H., 289
Carpenter, E.J., 532
Carpenteria (area), 83, 265
Carpenteria californica, 82-3, 112, 276
Carpobrotus, See Mesembryanthemum
Carrying capacity, 848
Carson Desert, 790
Carson Pass, 60, 149, 614, 642-3, 803
Carter, D.G., 70
Caruthers Canyon, 812-3
Cascade Range, 53-64, 397-400, 587-91, 611-5, 622-4, 917. 919, 922-3
Caspar, 88
Cassia armata, 844, 846, 872, 874
Cassiope mertensiana, 615, 621, 624-5, 649, 652, 716, 932
Castanares, A.A., 289
Castanea americana, 163

Castanopsis, See Chrysolepis
Castilleja
affinis, 900
applegatei, 624, 712, 717, 722
arachnoidea, 622, 722
breweri, 629, 631, 633
chromosa, 775, 786
cinerea, 544
culbertsonii, 625
hololeuca, 903
lassensis, 624
latifolia, 472, 747
lemmonii, 625
lyngbyei, 275
martinii, 542
ssp. ewanii, 544
miniata, 152-2, 648, 715
nana, 626, 632, 636, 639, 647, 656, 659-60, 663, 932
payneae, 624
pruinosa, 621
wrightii, 88
Castle Peak, 54-6
Castle Rock, 84
Catalina Harbor, 903
Catalina Island, See Santa Catalina Island
Cattle Canyon, 544, 546
Caulanthus
amplexicaulis, 920
crassicaulis, 782
Cave, M.S., 831
Ceanothus
arboreus, 171, 903, 928
coeruleus, 171
cordulatus, 312, 428, 543, 547-8, 565, 574, 586, 917, 921-3
crassifolius, 164, 174, 301, 322, 329, 425, 445, 452-3 928
cuneatus, 149, 151, 161, 164-5, 170, 174-5, 177, 247, 307, 309, 311-2, 366, 386, 388, 405, 407, 425, 429, 457, 802, 809, 926, 928
dentatus, 246, 928
divaricatus, 177, 426-8, 431, 434
fresnensis, 428
foliosus, 301, 304, 428
gloriosus, 304
greggii, 125, 312, 429, 431, 434, 481, 546, 813, 822, 847, 923
Ceanothus (cont.)
impressus, 178, 247, 928
insularis, 164, 928
integerrimus, 148, 151, 312, 428, 435, 539, 542, 564-5, 918, 922, 926
jepsonii, 125, 311-2, 429, 928
lanuginosus, 125
lemmonii, 312
leucodermis, 301, 481, 928, See also C. divaricatus
martinii, 153, 178
megacarpus, 928
oliganthus, 177, 425, 928
otayensis, 300-1
palmeri, 928
papillosus, 331
parryi, 928
parvifolius, 428, 565, 569, 922
perplexens, 174
pinetorum, 428
prostratus, 312, 428, 575, 577, 592, 703, 705, 708, 715, 721, 916, 921
pumilus, 428, 721
purpureus, 122
ramulosus, 928
rigidus, 178, 246
sorediatus, 928
sanguineus, 428
spinosus, 166, 168, 177, 928
thyrsiflorus, 304, 344, 425, 688, 690, 747, 938
tomentosus, 174, 301, 321-2, 329, 428, 928
velutinus, 148, 150, 312, 362, 428, 435, 574-5, 577, 590, 705, 711, 715, 720, 803, 919-22, 928
verrucosus, 164, 928
Cedar, See Calocedrus, Chamaecyparis, Thuja, Juniperus
Cedar Canyon, 818
Cedros Island, 332, 347
Celtis reticulata, 165
Cenchrus pauciflora, 250
Centaurea
melitensis, 437, 497, 503
solstitialis, 437
Central Valley, 298, 387-402, 405, 421, 491-510, 516-31, 851, 926

Cephalanthus occidentalis
var. californicus, 927
Cerastium holosteoides, 255
Ceratodon purpureus, 586
Ceratoides lanata, 774, 782, 788, 790, 792, 842, 848, 854-6, 859
Cercidium
floridum, 181, 836, 844, 871-3, 877, 880, 883, 933
microphyllum, 844, 933, 935
Cercis occidentalis, 164, 312
Cercocarpus
betuloides, 151, 160, 164-5, 170, 174-5, 177, 301, 312, 329, 426, 428-9, 431, 457, 539, 544, 590, 816-9, 926, 928
blancheae, 168-9, 334, 928
breviflorus, 153, 164, 170, 717
intricatus, 122, 822
ledifolius, 153, 170, 174, 312, 428, 544, 546-7, 590, 592-3, 623, 720, 784, 785-6, 799, 801-4, 806-8, 916, 919
minutiflorus, 301, 341
mojadensis, 171
montanus, 147
paucidentatus, 146
traskiae, 171
Cernuska, A., 670
Cervus
canadensis, 737
elaphusnannodes, 89, 498, 500
Ceska, A., 667
Cha, J.W., 867
Chabot, B.F., 70, 132, 595, 667
Chaenactis
alpina, 611
carphoclinia, 861
douglasii, 774-5, 934
var. achilleaefolia, 624
fremontii, 875
glabriuscula, 437
nevadensis, 622
santolinoides, 542, 546
Chaetadelpha, 116
Chamaebatia
australis, 300-1
foliolosa, 216, 457, 563-5, 575, 918, 922
Chamaebatiaria millefolium, 146, 312, 824-5, 924
Chamaecyparis
lawsoniana, 83, 145, 147-8,
Chamaecyparis (cont.)
150, 154, 161, 362, 365-6, 368, 680, 682, 693, 714, 723-5, 727-8, 916, 925
nootkatensis, 618, 621, 701, 709, 714, 721, 723-9
Chamise, See Adenostoma fasciculatum
Chandler, C.C., 758
Chaney, R.W., 190, 351
Channel Islands, See Southern California Islands
Channel Islands National Monument, 84
Chaparral
general, hard, 3-4, 82, 92, 94, 97, 99-100, 123-4, 157, 173-8, 181, 184-6, 196, 198, 200, 202, 205, 216, 218, 247-8, 300, 307, 311, 316-7, 326, 328-9, 332, 341-2, 360, 373-4, 376-8, 385, 392, 401-2, 406, 409, 417-69, 476-8, 480-2, 485, 495, 501-2, 509, 539-40, 554, 606, 684, 701, 747, 749, 814, 817, 819, 918, 925-8
chamise, 422-5
desert, 423-4, 430-1, 819
insular, 899, 903-4
montane, See Montane chaparral
Petran, 419
serpentine, 428-30
soft, See Coastal sage scrub
Chapin, F.S., 289, 667
Chapman, V.J., 290
Charleston (Spring) Mountains, 614, 822-3
Chase, A., 513
Chase, V.C., 888
Chasmophyte, 606
Chatfield, J., 758
Chatterton, N.J., 864, 886-7
Cheatham, N.H., 906
Cheatham Reserve, 92
Cheilanthes
carlotta-halliae, 735
covillei, 554
feei, 122
gracillima, 586, 713, 718
intertexta, 376
jonesii, 122
sinuata
var. cochisensis, 122

Chenopodium
  album, 250, 619
  ambrosioides, 274
  leptophyllum, 790
Chew, A.E., 887
Chew, R.M., 460, 864, 887
Chews Ridge, 390
Chiatovich Flats, 826
Chilao, 539
Childs, H.E., 463
Chilopsis linearis, 118, 172, 872, 877, 881-3, 885, 932
Chimaphila
  menziesii, 542, 569, 707, 712, 922
  umbellata, 363, 365, 574, 703, 704-5, 708-9, 712, 714-5, 717, 719, 921-2, 925
    ssp. occidentalis, 548, 707
China Island, 700
Chinquapin, See Chrysolepis chrysophylla
Chlorogalum pomeridianum, 329, 441
Chocolate Mountains, 871, 877
Chondrilla juncea, 498
Chorizanthe
  brevicornu, 861
  californica, 437
  rigida, 860, 898
  staticoides, 437
Chou, C.H., 460, 465
Christensen, N.L., 352, 460
Christi Beach, 899
Christi Canyon, 902
Chrysolepis
  chrysophylla, 146, 150, 154, 160, 362, 364-6, 371, 565, 703-5, 708, 715, 723-4, 922, 925
    var. minor, 304, 309
  sempervirens, 313, 424, 428, 543, 546-8, 550, 574, 590, 653, 711, 714, 716, 719-20, 806, 917, 921-2
Chrysolina quadrigemina, 738
Chrysopsis breweri, 574, 647, 880, 922
Chrysothamnus
  depressus, 813
  nauseosus, 153, 428, 543, 550, 774-6, 779, 782, 786, 801-2, 804, 809-11, 814-5, 847, 917, 920, 928
Chrysothamnus (cont.)
  parryi, 577
    ssp. monocephalus, 627
  teretifolius, 816, 820, 856
  viscidiflorus, 313, 768, 774-6, 786, 790, 793, 801, 806, 810, 812, 826, 847, 928
    ssp. pumilus, 824, 924
Chuang, T.I., 132
Chu, C., 464
Chuckawalla Mountains, 836
Cienegas, See Meadow, mountain
Cimadon, 857-9
Cirsium
  andersonii, 629, 631, 633
  californicum, 255
  foliosum, 624
  occidentale, 335
  quercetorum, 742
  rhothophilum, 121
  tioganum, 627
Cisco, 58
Clarke, S.H., 192
Clark, H.W., 352, 379, 411
Clarkia
  biloba
    ssp. australis, 127
  conginna, 437
  davyi, 741
  exilis, 127
  franciscana, 126-7
  lingulata, 127
  rhomboidea, 542
  rubicunda, 126-7
  springvillensis, 127
  tembloriensis, 127
  unguiculata, 127
  xantiana, 127
Clark, J.E., 832
Clark, L.D., 290
Clark Mountains, 811-3, 822-3, 827
Clark, S.B., 865, 887
Clausen, J., 70, 107, 132, 667
Claytonia
  bellidifolia, 664
  lanceolata, 152
  megarhiza, 664
  nevadensis, 611, 624, 626
  perfoliata, See Montia perfoliata
  umbellata, 663
Clear Lake, 52

Clematis ligusticifolia, 927
Clements, F.E., 190, 259, 379, 411, 459-60, 469, 511, 795
Clethra lanata, 162
Climate, ancient, 139-187
Climate, continental, 13, 19, 40-1, 45, 53, 63, 603, 702
Climate, general, 11-74
Climate, maritime, 13, 178, 225, 268-271, 346, 690-1, 894
Climate, Mediterranean, 23, 34-8, 47, 176, 224-5, 316, 419-20, 499, 909
Climate, See also particular vegetation types and taxa
Clintonia uniflora, 568, 706, 711, 717
Clokey, I.W., 190, 667, 829
Closed-cone pine forest, 4, 171-2, 297, 325-50, 303, 305, 307, 325-30, 372, 540, 749, 902-5
Clovis, J.F., 832
Cneoridium dumosum, 124, 301, 358, 475
Coachella Valley, 878
Coal Oil Point, 92
Coastal bluff, 91-3, 278, 280, 282, 284, 339-42, 895, 896-9, 904-5
Coastal dune, 87, 92, 98, 224-5, 241-52, 278, 339-40, 905
Coastal dune scrub, See dune scrub
Coastal prairie, 5, 88, 278, 332-3, 360, 366, 478, 733-45, 749
Coastal sage scrub, 87, 89, 178-9, 244, 246, 278, 282, 303, 332, 334-8, 341-2, 426, 431-2, 447, 471-89, 684, 747, 749, 904-5, 929
Coastal sage succulent scrub, 472, 474-5, 485, 487, 899, 904
Coastal salt marsh, See salt marsh
Coastal scrub, northern, 4, 100, 303, 472, 487, 745-6, 899, 937
Coastal stand, See also coastal dune, beach
Coast hemlock, See Tsuga heterophylla
Coast live oak, See Quercus agrifolia
Cobb, F.W., 351
Cold air drainage, 837
Cold Desert, See Great Basin
Coldenia
  nuttallii, 41
  palmeri, 874
Coldwater Canyon, 539
Cole, K.L., 352
Coleman, G.A., 352
Coleman, G.B., 459
Coleogyne ramosissima, 42, 64, 813-4, 816, 844, 847, 854-7, 872, 920, 932-3
Cole, R.C., 532
Colfax, 54, 58
Collier, G., 291
Collins, B.J., 460
Collinsia
  childii, 542
  concolor, 437
  parviflora, 620, 774-5, 777, 787, 928
  torreyi, 717
    ssp. wrightii, 548
Collomia
  larsenii, 623-4
  linearis, 619
Colman, E.A., 467
Colorado Desert, See Sonoran Desert
Colorado River, 836, 870-7, 873, 880-1
Colubrina californica, 181
Columbia Basin, 767-8, 784
Colusa, 52
Colwell, R.N., 219
Colwell, W.L., 219-20, 355, 413, 697, 731
Comarostaphylis diversifolia, 169, 902
Comarum palustre, 620
Competition, 328, 486, 499, 510, 523, 688-90, 722-7, 848, 875
Condalia
  lycioides, 877, See Condaliopsis lycioides
  parryi, See Condaliopsis parryi
Condaliopsis
  lycioides, 181, 872, 933
  parryi, 181, 872
Condit, C., 190
Conkle, M.T., 352, 354, 906
Conrad, V., 70

Constance, L., 132
Contagne, M.A., 70
Contandriopoulos, J., 132
Continentality, See Climate, continental
Convict Creek, 122, 156, 624, 635
Convolvulus, See Calystegia occidentalis var. saxicola, 742
Conway Summit, 810
Cook, C.W., 795
Cooke, W.B., 460, 595, 667
Cook, S.F., 460, 511
Cooper, C.F., 107
Cooper, D.W., 379, 697, 758-9
Cooper, P.V., 379, 411
Cooper, W.S., 258, 352, 379, 411, 460, 488
Corallorhiza
  maculata, 547, 569, 574, 707, 922
  striata, 569
Corbet, A.S., 758
Cordgrass, See Spartina foliosa
Cordylanthus
  nevinii, 544, 917
  palmatus, 129
  maritimus, 273, 275
Core, E.L., 9
Coreopsis gigantea, 169, 897, 903, 905
Corethrogyne
  filaginifolia, 249, 335, 441, 554, 937
  leucophylla, 246, 248, 255
Cornus
  californica, 161
  nuttallii, 152, 364, 368, 542, 565, 590, 704-5, 916, 918, 922
  stolonifera, 567, 705, 709
Correll, D.S., 132
Corylus
  cornuta, 920, 925
    var. californica, 362-3, 368, 565, 705, 753
  rostratus, 150
Coryneum cardinale, 303
Coryphantha vivipara, 858
Cosby, S.W., 532
Costello, K., 355
Cottonthorn, See Tetradymia axillaris
Cottonwood Canyon, 902
Cottonwood Mountains, 836
Cottonwood, See Populus
Cotula coronopifolia, 239, 273-4, 478-9
Couch, E.B., 258
Coulombe, H.N., 107
Coulter pine, See Pinus coulteri
Covington Flat, 856
Cowan, B., 258
Cowania mexicana
  var. stansburiana, 812
Cow Canyon, 544
Cow Creek, 33
Cowles, F.H., 352
Coxcomb Mountains, 815, 836
Coyote brush, See Baccharis pilularis ssp. consanguinea
Craddock, G.W., 460
Craig, R.B., 258, 289, 758
Crampton, B., 414, 511, 532, 758
Crassula, See Tillaea
Crataegus hupehensis, 145
Cratoneuron filicinum, 620
Crawford, J.M., 413, 460
Creosote bush scrub, 5, 814-5, 819, 851-2, 856-7, 873-6
Creosote bush, See Larrea tridentata
Crepis
  acuminata, 554, 645, 742, 775, 777, 786, 929
  modocensis
    ssp. subacaulis, 645
  nana, 609, 629-30, 632, 663
    ssp. ramosa, 554
  pleurocarpa, 722
Crescent City, 19-21, 23-4, 26, 31, 38, 45, 48, 145
Cressa truxillensis var. vallicola, 274, 278, 516, 519, 525, 527
Critchfield, W.B., 71, 190-1, 219, 352-3, 379, 412, 460, 556, 595-6, 731, 829-30
Crocker, R.L., 70, 596, 667
Cronemiller, F.P., 460
Cronise, T.F., 758
Cronquist, A., 70, 132, 669, 864
Crossosoma
  bigelovii, 182, 816, 872
  californicum, 897, 899
Croton californicus, 246, 249, 255, 875, 937

Crotophytus wislizenii, 185
Crowley Lake, 807, 809
Crown-sprouting, See Sprouting
Cruden, R.W., 532
Crypsis aculeata, 552
Cryptantha
affinis, 437
ambigua, 437
clevelandii, 437
hoffmanii, 630-1, 633
humilis, 645, 783
intermedia, 437
leiocarpa, 238
maritima, 861
micromeres, 437
muricata, 437, 554
nana, 654
nubigena, 633, 639, 932
subretusa, 624
torreyana, 437
Cryptogramma crispa ssp. acrostichoides, 622, 653, 717, 719
Crystal Lake, 541
Cucamonga Canyon, 819
Cuesta Pass, 325
Cuesta Ridge, 83
Cuff, K., 697
Culp Valley, 819
Cultivation, farming, ranching, effect of, 497-9, 878, 910
Culver, R.N., 794
Cunningham, G.L., 260, 863, 886-7
Cupressus
abramsiana, 297, 299, 307-10, 315, 320, 326
arizonica, 164, 299, 540
bakeri, 297-8, 311-4, 320
ssp. matthewsii, 297, 299, 311-5
forbesii, 297, 299, 303, 315-7, 319-22, 540
funebris, 149
glabra, 299
goveniana, 89, 297, 299, 303-5, 315, 320-1, 324, 919
guadalupensis, 299
macnabiana, 82, 297-9, 303, 307-15, 692
macrocarpa, 89, 297-9, 303-5, 315, 317, 919
montana, 299
nevadensis, 83, 297, 299, 311-5, 317, 320-1
Cupressus (cont.)
pygmaea, 89, 93, 121, 297, 303-6, 315, 319, 321, 340, 691-2, 919
sargentii, 82-3, 123, 297, 300, 303, 307-11, 315, 320-1, 429-30, 685, 692
stephensonii, 297, 299-303, 311, 316-7, 319-20, 324
torulosa, 159
Cupressus, taxonomy of, 297-9
Curray, J.R., 258
Curtis, B.F., 190
Curtis, J.T., 9
Cuscuta salina, 273-6, 278, 284
Cuyamaca cypress, See Cupressus stephensonii
Cuyamaca Mountain, 300, 311, 318, 324, 540-1
Cuyamaca Rancho State Park, 317
Cuyama Valley, 815
Cymopterus cinerarius, 626, 632, 662, 826, 924
Cynosurus
cristatus, 737
echinatus, 437, 737, 741, 743
Cypress, 297-324, 347-50, See also Cupressus
Cypress Mountain, 307
Cystopteris fragilis, 552, 609, 631, 717

Dactylis glomerata, 735
Daiber, F.C., 290
Dale, R.F., 70
Dalea
Californica, 872, 933
emoryi, 182, 874
fremontii, 848
polydenia, 41, 788, 790, 792, 842, 934
schottii, 872-4, 933
spinosa, 181, 871-2, 877, 933
Dal Porto, K.J., 411
Dalton, P.D., 887
Dana Plateau, 62, 624, 635
Danthonia
californica, 365, 734-8, 742, 745, 757, 924, 938
intermedia, 623, 634
pilosa, 737, 740, 742-3
Darlingtonia californica, 83
Darrow, R.A., 864, 886

Dasiphora, See Potentilla fruticosa
Dasmann, R.F., 9
Daubenmire, R., 461, 532, 667, 731, 795
Daucus
carota, 441
pusillus, 437, 509, 741
Daus, S.J., 219
Davidson, E.D., 258, 758, 864
Davies, J.L., 258
Davis, 23-4, 26, 34-5
Davis, C.B., 461
Davis, D., 461
Davis, G.A., 731
Davis, W.S., 132
Dawson Los Monos Canyon Reserve, 92
Day, A., 136
Deacon, J.E., 829, 864
Dealy, J.E., 795
Death Valley, 43, 66, 117-8, 122, 812, 838, 840, 846, 852-4
De Bano, L.F., 461, 463, 465
Decker, H.F., 136
De Decker, M., 829
Deep Canyon, 882-3
Deer, See Odocoileus hemionus
De Gloria, S.D., 219
De Jong, T.M., 258-9, 261
De Lapp, J., 379, 411, 697
Del Mar, 341
Delphinium
californicum, 437
glaucum, 708
menziesii, 411
nudicaule, 441
parryi, 441, 900
Dement, W., 461, 488-9
Dempster, L.T., 132
Dendromecon
harfordii, 903
rigida, 301, 308, 329, 331, 341, 430-1, 434, 441
Dennis, L.J., 261
Dentaria californica, 335
Deschampsia
atropurpurea, See Vahlodea
caespitosa, 609, 625, 641, 659-60
ssp. holciformis, 734-6, 740, 742-3, 745, 938
Deschampsia (cont.)
danthonoides, 495, 520-1, 540, 620, 736, 926
elongata, 577, 736, 916
Descurainia
pinnata, 619
richardsonii, 664
Desert, 16, See also Mojave desert, Sonoran desert, Great Basin, and particular vegetation types
Desert dunes, 862
Desert ironwood, See Aneya fesota
Detling, L.E., 190, 411, 461
Devil's backbone, 807
Devil's bathtub, 546
Devil's Garden, 79, 82, 766, 800
Devil's gate, 806-7
Devil's Postpile National Monument, 209
Dew, 242, 479
Diatomaceous soil, 122, 334-8, 342, 767, 769
Diaz, D.V., 863, 886
Dice, L.R., 887
Dicentra
chrysantha, 324, 441
ochroleuca, 491
Dichelostema, See Brodiaea
Dick-Peddie, W.A., 886
Dickson, B.A., 596
Digger pine-oak woodland, See oak woodland
Digger Pine, See Pinus sabinana
Dillon Beach, 233
Dingman, R., 291
Diorite, 701
Diospyros virginiana, 161
Diplacus, See Mimulus
Dipodomys heermanni, 498
Disjuncts, 169, 178, 148, 187 580, 593, 650, 729
Disporum
hookeri, 368, 569, 921
var. trachyandrum, 363, 365, 707, 712, 925
Distichlis
spicata, 231, 273-6, 278-9, 282, 284, 494, 516, 519, 734, 790, 793, 852-3, 896, 903, 934-5
stricta, 788, 791
Disturbance, 809, 841, 874, 895

Dithyrea californica, 250
Diversity, See Species diversity
Dixon, J., 412
Doane Valley, 87
Dobzhansky, T., 132
Dodecatheon
  alpinum, 611, 624-5, 641, 653
    ssp. majus, 649
  clevelandii, 441
  conjugens, 800, 930
  jeffreyi, 715, 717, 920
  rodolens, 552
Dolan, R., 259
Dolomite, 656, 660, 825-6
Donner, 55, 57, 61, 63
Door Canyon, 553
Dorf, E., 190
Dorland Preserve, 97
Doten, S.B., 795
Doty, M.S., 290
Douglas fir forest, 305, 326, 333, 361
Douglas fir, See Pseudotsuga menziesii
Downingia
  bella, 519, 525, 527
  concolor, 527
  cuspidata, 516, 523-5, 527
  pusilla, 527
Dozier, 517, 519, 521, 527
Draba
  aureola, 624
  breweri, 611, 619, 622, 626, 636, 639, 659, 932
  corrugata, 653-4, 923
    var. saxosa, 548, 554, 653
  crassifolia, 625
  cruciata
    var. integrifolia, 611
  densifolia, 636, 638, 642, 645, 662, 664, 932
  douglasii, 630
    ssp. crockeri, 544
  howellii, 621, 718, 720
  lemmonii, 611, 619, 637, 640, 932
    var. incrassata, 663
  lonchocarpa, 122, 163, 629-30, 632
  nivalis var. elongata, See D. lonchocarpa
  oligosperma, 628-30, 632, 932
Draba (cont.)
  pterosperma, 622
  sierrae, 611, 630-1, 658
  stenoloba, 619
    var. nana, 552, 663-4
Drake's Estero, 265
Drepanocladus aduncus, 620
Driscoll, R.S., 829
Drosera
  anglica, 620
  rotundifolia, 87, 620
Drought, 27, 142, 144-5, 150, 497, 499, 541-2, 571, 583, 617, 650, 860
Drought tolerance, See Moisture stress
Dryopteris arguta, 331, 335, 344, 751
Drysdale, F.R., 258-9, 289, 758
Dubakella soil, 366, 721
Dudleya
  caespitosa, 246, 248, 255, 937
  cymosa, 554
  farinosa, 249, 304, 476, 755
  greenei, 900
  hassei, 897
  ingens, 475
  lanceolata, 248
  virens, 897
Dudleys, 62
Dudley, W.R., 411
Duffey, E., 532
Duffield, J.W., 352
Dulichium arundinaceum, 620
Duncan, C.C., 464, 846
Duncan, D.A., 411, 414, 571
Dune forest, 243-4, 246, 339-40
Dune scrub, 3, 244-52, 341, 472, 897, 936
Dune slack, 241
Dunes, 241-52, 339-40, See alo coastal dunes, desert dunes
Dunkle, M.B., 906
Dunn, E.L., 9, 135, 461, 464, 488, 673, 831
Dunning, D., 219, 352, 596
Durham, J.W., 354
Durrenberger, R.W., 259
Dryness, C.T., 9, 254, 353, 379, 412, 668, 731, 795, 830
Dyssodia porophylloides, 821, 843

Eagle peak, 307, 592, 664
Eagle Range, 538
East Anacapa Island, 877, 899
East Ord Mountain, 842, 844, 856
Eaton, L.C., 131
Ebbett's Pass, 803, 806
Eburophyton austinae, 569
Echinocactus
  acanthoides, 876, 934
  viridescens, 341
Echinocereus
  engelmanii, 816, 859, 872, 874, 876, 933-4
  maritimus, 475
Echo Valley, 485
Eckenwalder, J.E., 352
Eckert, R.E., 794-6, 830
Ecotypes, 253, 507, 529, 618, 814, 875
Ectosperma, See Swallenia
Eel River, 265, 275
Efimova, N.A., 70
Egorova, T.V., 664
Ehleringer, J.R., 667
Ehrendorfer, F., 667
Eilers, H.P., 290
Elatine
  californica, 523-4
  chilense, 523-4
Eleocharis
  acicularis, 244, 523-4
  macrostachya, 523-4
  palustris, 931
  parishii, 524
  pauciflora, 585, 923
Elk, See Cervus elephas nannodes
Elkhorn Slough, 97, 265
Elkington, T.T., 667
Ellenberg, H., 70, 380, 414, 667, 731
Ellery Lake, 59, 61-2, 65
Elliott, H.W., 758
Elliott Reserve, 93
El Rosario, 474
El Socorro, 251-2
Elymus
  caput-medusae, See Taeniatherum asperum
  cinereus, 771, 774-5, 779-80, 788, 790, 793
  codensatus, 334, 422, 751
Elymus (cont.)
  glaucus, 344, 365, 441, 495, 577, 585, 736, 742, 752, 917, 930
  mollis, 225-6, 229-31, 236, 240, 255, 257, 735, 936, 938
  pacificus, 246, 938
  triticoides, 329, 495, 736, 930
  vancouverinsis, 938
Emery, K.O., 259, 292-3
Emigrant Gap, 58
Emigrant Spring, 838
Emmenanthe penduliflora, 437, 875
Emmingham, W.H., 379
Emplectocladus andersonii, See Prunus andersonii
Emrick, W.E., 412
Encelia
  californica, 284, 434, 472-4, 481, 898, 929
  farinosa, 845, 872, 875, 881, 883, 885, 929, 933
  frutescens, 841, 846, 872, 933
  virginensis, 817, 842
Endangered taxa, 129
Endemism, 3, 110-3, 116, 118, 120-1, 129, 421, 424, 429, 433, 457, 495, 516, 606, 610, 621-2, 664, 701, 873, 894-5, 897, 903, 905
Energy balance, 15-6, 21, 225, 268, 654-5, 885
Engler, A., 190
Ensenada, 250
Entisols, 779
Ephedra
  californica, 250-2, 475, 495, 844, 872, 930, 935-6
  funerea, 842
  nevadensis, 41, 774-5, 790, 811, 815-6, 818, 820, 841-3, 846-7, 850, 855-6, 858-9, 872, 933
  trifurca, 181-2, 874
  viridis, 313, 775, 782, 810, 812-3, 822, 845, 856, 860, 930
Epilobium
  anagallidifolium, 621, 624, 648, 932

Epilobium (cont.)
  angustifolium, 441, 552, 609, 686, 717, 916
  brevistylum, 552
  clavatum, 623
  glaberrimum, 552
  latifolium, 609, 625, 634-5, 642
  minutum, 438, 620, 626
  obcordatum, 611, 624, 932
    ssp. siskiyouense, 622
  paniculatum, 438, 778
  pringleanum, 626
Epling, C., 132, 356, 461, 488
Equability, 13, 41, 127, 164, 167, 179, 186
Erechtites
  arguta, 236, 688, 741, 752
  prenanthoides, 236, 688, 741
Eremocarpus setigerus, 438, 503-4, 509
Erharta calycina, 456
Eriastrum
  brandegeae, 122
  densifolium, 249
    ssp. austromontanum, 542, 918
  diffusum, 774-5, 790-1
  pluriflorum, 438
Erigeron
  barbellulatus, 624, 646
  blochmanae, 250
  bloomeri, 777
  breweri
    ssp. jacinteus, 548, 550
  clokeyi, 122, 626, 629, 632, 656, 825, 924
  compositus, 622, 626, 636, 639
  divergens, 441
  foliosus var. stenophyllus, 441, 554
  glaucus, 235, 237, 250, 255, 304, 472, 755, 897, 938
  latifolius, 237
  parishii, 554, 817
  peregrinus ssp. callianthemus, 624, 647
  petiolaris, 637, 640, 647
  pumilus, 782
  pygmaeus, 611, 626, 635-6, 639, 656, 659-60, 663, 826, 924
  vagus, 611, 659-60
Eriodictyon
  angustifolium, 813
Eriodictyon (cont.)
  californicum, 177, 308, 311, 314, 428, 747
  crassifolium, 301, 322, 329, 428-9, 441
  traskiae ssp. traskiae, 899
  trichocalyx, 301, 428, 431, 542, 918
Eriogonum
  alpinum, 622
  anemophilum, 41
  apricum, 122
  arborescens, 897
  arboreum, 751
  compositum, 621
  davidsonii, 546, 554
  deflexum, 774
  densum, 790-1
  elatum, 621
  elongatum, 816
  fasciculatum, 301, 322, 329, 422, 425, 431, 434, 441, 448, 472, 474-5, 482, 488, 814-8, 820, 846-8, 872, 874, 928-9, 934
    ssp. polifolium, 814, 930
  gracile, 438
  gracilipes, 656-7, 659-60
  grande, 897, 903
  heermanii, 41
    ssp. floccosum, 822
  heracloides, 664, 776
  incanum, See E. marifolium
  inflatum, 848
  intrafractum, 117-8
  kennedyi ssp. alpigenum, 550, 652, 654, 923
  latifolium, 240, 246, 249, 255, 755, 936
  lobbii, 611, 626, 630, 638, 647
  marifolium, 152, 622
    var. incanum, 626, 638, 642-3, 646
  microthecum, 626, 628, 632, 776, 782, 847
  nivale, See E. ovalifolium var. nivale
  nudum, 574, 586
  ochrocephalum var. agnellum, 624, 626, 632, 645, 662, 664
  ovalifolium, 611, 656-7, 776, 782, 816, 825
    var. nivale, 624, 626, 636, 639, 642, 645, 659-60, 662, 930

Eriogonum (cont.)
parishii, 544, 554
parvifolium, 246, 249, 255, 335
pyrolaefolium, 622
rosense, See E. ochrocephalum var. agnellum
rubricaule, 41
saxatile, 376, 546, 548, 554, 652
spergulinum, 586
var. reddingianum, 664
sphaerocephalum, 778
thomasii, 775, 790
thurberii, 438
trichopes, 861
umbellatum, 622, 644, 652, 713, 718, 774-5, 778, 847, 920
var. covillei, 664
ssp. ferrissii, 822
ssp. minus, 546, 548, 550, 558, 923
ursinum, 624
wrightii
ssp. subscaposum, 343, 546, 550, 586, 917, 923
ssp. trachygonum, 822
ssp. wrightii, 813
Erionenon pulchellum, 821
Eriophorum gracile, 620
Eriophyllum
confertiflorum, 301, 305, 321-2, 324, 329, 334, 434, 442, 445, 448, 472, 474, 554, 816, 818, 929, 930
lanatum
var. achillaeioides, 442
var. monoense, 646, 663
nevinii, 897
stachaedifolium, 244, 472, 747, 751, 755, 897
Erman, D.C., 759
Erodium
botrys, 502, 505-6, 509, 523, 526, 931
cicutarium, 127, 438, 494, 497, 502, 504, 506, 509, 523, 900, 931
Eryngium
aristulatum
var. parishii, 523-24
armatum, 735-6, 742
vaseyi, 516, 520
var. globosum, 525-6
Erysimum
capitatum, 438, 624
concinnum, 250, 261
insulare, 248, 250, 255
perenne, 626, 644, 662, 664
suffrutescens, 248
teretifolium, 121
Erythronium grandiflorum, 621
Eschscholzia californica, 248, 250, 255, 493, 506, 896, 930
Eslinger, D.H., 867
Esterode Limantour, 84
Estigmene acraea, 756
Eucrypta chrysanthemifolia, 438
Euphorbia
albomarginata, 554
misera, 185, 252, 475
Eureka Valley, 836
European beach grass, See Ammophila arenaria
Eurotia, See Ceratoides
Evans, R.A., 219, 413, 511, 595-6
Evapotranspiration, actual, 19
Evapotranspiration, potential, 15, 18
Evax caulescens, 520-1
Everett, P.C., 352

Faegri, K., 667
Fagonia laevis, 872
Fairfield-Osborn Preserve, 97
Fallon, 54-6, 58-9, 179
Fallugia paradoxa, 117
Pan palm, See Washingtonia filifera
Favarger, C., 132
Feather, T.P., 260
Feldmeth, R., 291
Fen, 87, 608, 620
Fenner, R.L., 459, 463, 511
Fenton, R., 489
Ferguson, C.W., 831
Ferlatte, W.J., 667, 731
Fernald, M.L., 132
Fern Canyon, 82
Ferndale, 688
Ferocactus acanthodes, 860
Fescue, See Festuca
Fescue-oatgrass, See Coastal prairie
Festuca
arundinacea, 457, 742

Festuca (cont.)
  brachyphylla, 609, 629, 633, 641, 643, 645, 653, 659, 663-4, 930
  californica, 735-6, 924
  dertonensis, 509, 523, 526, 735, 737, 741, 743, 931
  idahoensis, 493, 495, 734-5, 737, 745, 767, 771-2, 774, 776, 784, 786, 803, 919, 920, 930, 938
  megalura, 439, 495, 505-6, 509, 523, 735, 930
  myuros, 439, 509
  occidentalis, 707, 735
  octoflora, 439, 775
  ovina, 365, 734
  pacifica, 499, 735
  rubra, 734-5, 938
Fieblekorn, C., 412
Filago californica, 439
Filice, F.P., 290
Fir, See Abies, Pseudotsuga
Fire
  adaptations to, See Sprouting, Germination, Age distributions
  effect of, 218, 297, 307, 316-7, 319-25, 328, 330, 332, 337, 340, 343, 347-50, 366, 368-70, 373-7, 384, 389, 399-400, 407-9, 418-9, 422, 425-7, 429, 431-5, 446-50, 455-6, 458, 476-8, 497, 499, 502, 509-10, 543, 561, 563, 565, 571-2, 574-5, 578-9, 586-7, 593, 688-90, 694-6, 721-2, 726, 729-30, 736, 748-9, 767, 769-70, 780-1, 878, 902, 905
  frequency, 316-17, 319-25, 332, 341, 587
Fish Creek, 553
Fisher, R.A., 758
Fish Hatchery Creek, 550
Fish Slough area, 91
Fitch, H.S., 511-2
Fitsimmons, S.J., 672
Flandrian sediments, 258
Flint, R.F., 259
Flohn, H., 667
Flooding, 306, 551-2, 693-5
Flora, California, 109-37
Florence, R.G., 697
Floristic provinces, 6, 31, 111-18
Fog, fog drip, 13, 19, 34, 303, 307, 316, 328, 332, 340, 342, 346, 426, 445, 690-1, 894
Foin, T.C., 759
Fonela, R.W., 731
Foothill woodland, See Oak woodland
Forde, M.B., 352
Ford-Robertson, F.C., 107
Fordyce Dam, 58
Foredune, 225, 241
Foreman, R.E., 906
Forest condition maps, 198-200
Forestieria neomexicana, 185, 813
Forest of Nicene Marks, 97
Forsellesia nevadensis, 822
Fort Bidwell, 32
Fort Bragg, 147, 303, 620, 692
Fosberg, M.A., 795
Fossil floras, 112-3, 139-86, 903
Fouquieria splendens, 118, 181, 836, 872-3, 883, 933
Fowells, H.A., 190, 259, 353, 596, 830
Fox, M., 864
Fragaria
  californica, 344, 569, 706, 753
  chiloensis, 231, 235, 243-4, 255, 742, 936
Franciscan formation, 361, 367
Franclet, A., 864
Frankenia grandifolia, 273, 275-6, 277-8, 284, 516, 898, 931, 934
Franklin, J.F., 9, 107, 259, 353, 379, 411, 468, 731, 795, 830
Franseria, See also Ambrosia chenopodifolia, 252, 475
Frasera
  neglecta, 540
  puberulenta, 631
Fraxinus
  anomala, 153, 175, 822-3, 922, 928
  dipetala, 313, 329, 401, 426
  latifolia, See F. oregona
  oregona, 150, 567, 682
  velutina, 153, 872, 935
Fremontia, See Fremontodendron

Fremontodendron
  californicum, 174, 313, 429, 431, 546, 817-8, 923
  mexicanum, 300-1
Fremont Peak, 851
Frenchman Flat, 838
Frenkel, R.E., 132, 668
Frenzel Creek, 82
Fresno, 34, 36-7
Fritillaria
  atropurpurea, 152, 622
  pinetorum, 544
Fritz, E., 697
Frost, 482, 837, 846, 873, 894
Frost, F.H., 190
Fultz, F.M., 461
Funaria hygrometrica, 569

Gabbro, 701
Gabilan Range, 389, 393, 395-6, 421
Gale, J., 70
Galium
  angustifolium, 329
    ssp. gabrielense, 554
    ssp. nudicaule, 554
  aparine, 344, 741
  bifolium, 552
  californicum, 344, 752
  grayanum, 624
  hypotrichium, 630-1, 633
    ssp. tomentellum, 825, 924
  johnstonii, 542
  nuttallii, 442, 476
  parishii, 548, 820
  sparsifolium, 569, 922
  triflorum, 706, 712
Gallery forest, See Riparian
Galloway Creek, 305
Galvezia juncea, 475
Gankin, R., 70, 132, 461
Garcia-Moya, E., 864
Gardner, R.A., 219, 412, 697
Garfield, 84
Garfunkel, Z., 190
Garlock Fault, 836
Garrya
  buxifolia, 309, 428
  congdoni, 309, 311, 313, 429
  elliptica, 171, 244, 928
  flavescens, 428-9, 431, 542, 817, 818, 822, 918, 926
Garrya (cont.)
    ssp. flavescens, 540, 813
    ssp. pallida, 301, 313, 928
  fremontii, 177, 313, 329, 428, 711
  ovata, 171
  veatchii, 428, 688, 928
Garry oak, See Quercus garryana
Gastridium ventricosum, 497, 503, 509
Gates, D.H., 795
Gaultheria
  humifusa, 621
  ovatifolia, 150
  shallon, 171, 244, 303-5, 362-3, 368, 472, 686, 688, 747, 748, 752, 916, 919, 925, 930
Gaumann, E., 668
Gause, G.W., 379, 556
Gaussen, H., 70
Gayophytum
  diffusum, 624
    ssp. parviflorum, 544, 546
  heterozygum, 542
  nuttallii, 574
  racemosum, 620, 774-5
  ramosissimum, 790
Gay, T.E., 795
Gazelle, 802
Geiger, R., 71, 668
Gentiana
  amarella, 625, 648, 653
  calycosa, 611
  fremontii, 613
  heterophylla, 923
  newberryi, 611, 624, 640
  tenella, 626
Geranium dissectum, 741
Gerbode Preserve, 97
Gerdes, G.L., 290
Germination, 435, 445-6, 503, 530, 587, 684-90, 694, 726, 849, 853, 859, 874, 884
Geum
  canescens, 646, 651
  ciliatum, 664, 777
  macrophyllum, 152
Ghiselin, M.T., 258, 289, 758
Gialdini, M.J., 219
Giant Forest, 33, 56, 59, 64
Gibbens, R.P., 595-6
Gibson Creek, 303

Giguere, P.E., 290
Gifford, G.F., 832
Gila River, 871
Gilia
  capitata, 438
  clivorum, 438, 930
  clokeyi, 783
  interior, 930
  latifolia, 774
  leptomeria, 934
  minor, 930
  scopulorum, 861
  splendens, 542
    ssp. grantii, 918
  tricolor, 930
Gillett, G.W., 353, 461, 596, 668, 830
Gilligan, J., 464
Gill, R., 290
Gilluly, J., 190
Gilman, J.H., 459
Gilmannia, 111, 117-8
Ginkgo biloba, 141
Gjaerevoll, O., 668
Gladish, E.N., 413, 697, 831
Glass Mountain, 809
Glasswort, See Salicornia
Glaucophane, 367, 692
Glaux maritima, 273-4
Gleason, C.H., 415, 469
Glehnia leiocarpa, 237, 255
Glock, W.S., 830
Glyceria
  elata, 551, 708, 920
  occidentalis, 736
Glyptopleura marginata, 41
Gnaphalium
  bicolor, 438
  californicum, 323, 442
  chilense, 236
  microcephalum, 336, 442
  purpureum, 237, 741
  ramosissimum, 237
Godfrey, P.J., 259
Goggans, J.F., 356
Gold Bluffs Beach, 233
Gold Run, 58
Gold Valley, 846
Goleta, 265
Goltsy, 603, 608
Golubets, M.A., 668
Goodin, J.R., 464, 864-5
Goodwin, D.L., 865
Goodwin, R.H., 107
Goodyera oblongifolia, 333, 363, 365, 368, 569, 707, 712
Goosenest Mountain, 311
Gopher, See Thomomys
Gorchakovsky, P.L., 668
Gordon, D.T., 596
Gosselink, J.G., 290
Gottlieb, L.D., 133
Gowans, K.D., 412
Graham, C.A., 511
Grah, R., 698
Grams, H.J., 758
Grand fir, See Abies grandis
Grand fir-sitka spruce forest, 916
Granite Creek, 84
Granite Pass, 815
Granitic parent material, 315-6, 327, 329, 339, 803, 806, 822, 825
Granodiorite, 315
Grant, K.A., 133
Grant, V., 133, 532
Grapevine Canyon, 847
Grapevine Mountains, 811-2
Grasshopper Valley, 805
Grass Lake, 620
Grassland
  coastal, See Coastal prairie
  general, 3, 278, 478, 539, 926, 930
  insular, 894, 897, 900, 904-5
  valley, 5, 491-510, 515-33, 900, 904-5
Gratiola ebracteata, 520-1
Gratkowski, H.J., 461
Graves, G.W., 412
Graves, W.L., 865
Gray, A., 133
Grayia spinosa, 774-5, 788, 790, 792, 810, 842, 847, 854, 856, 932, 934
Gray, J., 191
Gray, R., 887
Grazing, See Herbivory, Livestock, California history
Greasewood, See Sarcobatus vermiculatus
Great Basin
  general, 4, 64-6, 116, 140, 142, 147-8, 150, 152-3, 156-8, 164, 179-81, 392, 561, 592, 602-3, 613, 624, 654, 664, 808, 817, 823, 836, 847, 924

Great Basin (cont.)
sagebrush, 4, 41-2, 763-96, 853-4, 862, 922, 928
Grebenshchitcov, O.S., 668
Grecham, R., 460
Greco Island, 100
Green, H.A., 353
Greenhorn Mountain, 321
Green, J., 290
Green, L.R., 411
Greenstone, 315
Gribbon, P.W.F., 668
Griffin, J.R., 71, 133, 190, 353-4, 379, 412, 461, 596, 731, 830
Grif, V.G., 666
Griggs, F.T., 532
Grimmia unicolor
Grindelia
camporum, 519
hirsitula, 929
humilis, 273
latifolia
ssp. platyphylla, 736
nana, 778
paludosa, 273
stricta, 272, 755
Grinnell, J., 412, 556
Ground, C.A., 668
Ground squirrel, Beechey, See Spermophilus beecheyi
Groveland, 62
Guadalupe Island, 332
Guatay Mountain, 298, 300, 316-8
Guinochet, M., 668
Gulf of California, 870
Gulmon, S.L., 465
Guntle, G.R., 461
Gutierrezia
bracteata, 817, 930
californica, 185
sarothrae, 812-3, 819-20, 920, 934
Gymnocladus dioica, 161

Habenaria sparsiflora, 552
Hackberry Mountains, 814
Hadley, E.B., 461
Hafenrichter, A.L., 258
Hagen Canyon, 87
Haines, B., 489
Haiwee, 878
Hale, V.Q., 867
Haley, B., 489
Half Moon Bay, 233
Haller, J.R., 133, 353, 556, 596, 830, 906
Hall, H.M., 107, 556, 668, 795
Halligan, J., 488, 512
Halogeton glomeratus, 790
Galophyte, 850-3, 884
Hamilton, E.L., 71, 461
Hamilton Sanctuary, 98
Hamrick, J.L., 556, 830
Hanawalt, R.B., 462, 465
Hanes, T.L., 462, 466, 669
Hanks, D.L., 795
Hannon, N.J., 290
Hansen, D.H., 669
Hansen, H.P., 191
Haplopappus
acaulis, 643-4, 650-1, 656, 782, 826
acradenius, 852, 877, 934-5
ssp. eremophylus, 878
apargioides, 623, 629, 633, 639, 656, 639
arborescens, 313
bloomeri, 575, 623, 804
cooperi, 813, 848, 856, 858-9
cuneatus, 554, 813, 817
detonsus, 897, 900
eastwoodea, 303-4
ericoides, 244, 246, 248-50, 256, 309, 335, 936-7
greenei, 624, 708, 717, 720, 920
linearifolius, 813, 816-7, 821-2, 847, 859, 872, 920, 930
lyallii, 613, 621
macronema, 611, 628-9, 631, 633, 638, 642, 646, 932
pinifolia, 185
racemosa, 494, 935
squarrosa, 256, 301, 341, 442, 472, 929
ssp. grindelioides, 442
suffruticosus, 631, 633, 646, 650, 658, 661, 663
venetus, 250-2, 472, 896-7, 929, 937
whitneyi
ssp. discoidea, 621
Haploxerolls, 772, 780
Happy Camp, 31-2, 46, 48, 729
Hardham, C.B., 353, 759

Hardpan, 315, 517-8, 520
Hare, F.C., 171
Harfordia macroptera, 475
Harney Basin, 765
Harnickell, E., 74
Harris, G., 512, 795
Harrison, A.T., 72, 462, 465, 488, 864, 866, 888
Harrison, C.M., 668
Harshberger, J.W., 379, 669
Hartesveldt, R.J., 596
Harvey, H.T., 596, 830
Harvey Monroe Hall Research Natural Area, 79, 82
Harvey, R.A., 462, 488
Hassouna, M.G., 295
Hastings, J.R., 865, 887
Hastings Natural History Reservation, 33, 93, 395, 398, 403, 406-8, 496, 502
Hathaway, W., 358
Havranek, W.M., 666, 669
Hawkes, J.G., 667
Hawksworth, G.H., 353
Hays, R.L., 107
Head, W.S., 462
Heady, H.F., 512-4, 596, 758-9
Heat, See Energy balance
Heather Lake, 84
Heath, J.P., 596
Hecastocleia, 117-8
Heckard, L.R., 132-3
Hedel, C.W., 289
Hedrick, D.W., 462, 512-3
Heizer, R.F., 512
Hektner, M.M., 759
Helenium bigelovii, 552
Helen Mine, 49-50
Helianthemum
  scoparium, 301, 323-4, 331, 434, 442
  suffrutescens, 122
Helianthus
  gracilentus, 323, 442
  niveus, 251-2
Heliocharis, See Eleocharis
Hellmers, H., 462, 488, 697
Hell's Kitchen, 803, 806
Helm Creek, 59
Hemerocampa vetusta, 756
Hemet, 821
Hemitomes congestum, 569
Hemizonia
  arida, 87
Hemizonia (cont.)
  clementina, 897
  fitchia, 527
  multicaulis, 741
  paniculata, 523
  ramosissima, 438
Hemlock, See Tsuga
Hendry, G.W., 133, 512
Henrickson, J., 291, 830
Henry, A.J., 71
Hepialus behrensi, 756
Heracleum
  lanatum, 237, 472, 708, 747, 752, 920, 938
  sphondylium ssp montanum, 552
Herbivory, effect of, 347-8, 480, 493, 496-500, 502, 508-10, 585, 745, 748-9, 767, 770, 792, 800, 848, 858, 860, 895, 900, 905, 910
Hermann, F.J., 133
Herman, F.R., 830
Hermes, K., 669
Hermidium alipes, 41
Heron Rookery, 87
Herpotrichia juniperi, 616
Hertel, E.W., 669
Hervey, D.F., 512
Hesperelaea, 112
Hesperochiron
  californicus, 800, 930
  pumilus, 800, 930
Hesperocnide, 112
Hetch Hetchy, 62
Heterocodon rariflorum, 540
Heteromeles arbutifolia, 124, 147, 151, 160, 165, 168, 171, 175, 256, 301, 309, 323, 329, 331, 333, 335, 341, 372, 401-3, 407, 245-6, 428-9, 451-4, 481, 483-4, 900-1, 903, 905, 925-8
Heterotheca grandiflora, 256
Heuchera
  abramsii, 548
  alpestris, 653-4
  elegans, 550, 554
  hirsutissima, 653-4
  maxima, 900
  pringlei, 621
  rubescens, 633
    var. pachypoda, 825
Heusser, C.J., 191
Hibbard Reserve, 98
Hibiscus denudatus, 860

Hidden Palms, 91
Hieracium
albiflorum, 442, 569, 574, 706, 712, 717, 972
gracile, 621, 648, 712, 717
horridum, 623, 631, 639
Hiesey, W.M., 107, 132
Hilaria
jamesii, 41, 771, 774, 779, 782, 790, 934
rigida, 817, 821, 842, 844-5, 858-9, 872, 874, 933
Hill, A.J., 867, 889
Hillier, R.D., 672
Himmelberg, G., 351
Hi Mountain Condor Sanctuary, 98
Hinckley, A.L., 889
Hinde, H.P., 290
Hintikka, V., 71
Hinton, W.F., 133, 669
Hironaka, M., 795
History, recent, See California history
Hitchcock, C.L., 669
Hobbs, C.L., 795
Hobergs, 51-3
Hodges, J.D., 71
Hoffmanseggia microphylla, 118, 181, 872
Hoffman, W., 697
Hogwallows, See Vernal pool
Holbo, H.R., 795
Holch, A.E., 669
Holcomb Valley, 544, 546
Holcus lanatus, 735, 737, 740, 742-3, 747, 938
Holland soil, 703
Holland, V.L., 412
Holloway, J.K., 759
Holmgren, A.H., 70, 864
Holmgren, N.H., 70, 864
Holmgren, P.K., 669
Holmwood Reserve, 98
Holodiscus
discolor, 373-4, 687-8, 784, 919, 924
dumosus, 145
microphyllus, 153, 546, 548, 550, 579, 624, 653, 710, 715-6, 719-20, 806, 821, 823
Honey Lake, 764-6, 773, 779-80, 785, 788, 802
Hooper, J.F., 512, 759
Hoover, R.F., 259, 290, 353, 412, 462, 532, 759, 830, 921-3
Hopland, 50-3, 392, 496, 500, 502, 505-6, 510
Hopsage, See Grayia spinosa
Hordeum
brachyantherum, 577, 735-6
californicum, 523, 735, 742
depressum, 494
geniculatum, 498
glaucum, 898, 900
hystrix, See H. geniculatum
leporinum, 395, 438, 496, 527
Horkelia
bolanderi ssp. parryi, 540, 926
californica, 742
cuneata, 334, 472, 929
marinensis, 735
Hormay, A.L., 514, 795
Horton, J.S., 353, 379, 462-3, 488, 556, 669, 830, 887
Hot desert, See Mojave or Sonoran Desert
Howard, W.E., 463
Howell, J.T., 133, 191, 259, 290, 353, 379, 412, 461, 463, 596, 668-9, 731, 759, 830
Howitt, B.F., 353, 463
Hubbard, J.C.E., 290
Hübl, E., 71
Huckleberry Hill, 303, 324, 348
Hudsonian zone, 548, 550
Hudson, J.P., 596
Huffaker, C.B., 759
Huffaker, R.C., 867
Hughes, G.H., 887
Hulsea
algida, 611, 619, 626, 639, 932
heterochroma, 442
nana, 622, 624
vestita
ssp. parryi, 546
ssp. pygmaea, 653, 654
Hultén, E., 669
Humboldt, Bay, 265-7, 275
Hunt, C.B., 71
Hunt, D.B., 865
Huntington Lake, 62
Hurlbut, L.W., 73
Hutchinson, E.R., 830
Hybridization, 124-5, 540-2, 822, 898

Hydrangea aspera, 145
Hydrophyllum
capitatum, 800, 930
fendleri, 708
occidentale, 624
Hymenoclea
monogyra, 41
salsola, 41, 827, 842-4, 848, 853, 858-9, 872, 881, 883, 933
Hymenoxys cooperi, 825
var. canescens, 656, 659, 663
Hypericum
anagalloides, 552, 623
concinnum, 442
perforatum, 737-9
Hypochoeris
glabra, 503-4, 506, 526
radicata, 256, 742-3
Hyptis emoryi, 872

Ibrahim, K.M., 865, 865
Icehouse Canyon, 546
Icehouse saddle, 543
Idaho, 799
Iltis, H.H., 133
Iman, A.G., 133
Imperial Valley, 879
Importance value, 399, 561, 567, 582, 843, 845, 858-9
Incense cedar, See Calocedrus decurrens
Independence, 63
Indiana Summit, 82
Indian cover, 844
Indian Creek, 88
Indians, affect on vegetation, 266, 499
Indio, 33
Indio Hills, 877
Inglenook fen, 87
Insular woodland, 169, 904-5
Interior live oaks, See Quercus wislizenii
Intertidal zone, See Salt marsh
Introduced taxa, 127-9, 226-8, 497-9, 502, 519, 523, 525, 736-8, 770, 780, 898, 900, 911, 931
Inverness Ridge, 97
Inyokern, 838, 851
Inyo Mountains, 808, 811-2, 823-5, 854, 924
Inyo Region, 117-8
Iodine bush, See Allenrolfia occidentalis
Ione, 122
Ipomopsis
aggregata, 152, 644
ssp. arizonica, 924
congesta, 624, 645, 661, 663
Iris
douglasiana, 344, 734-5, 742, 752
hartwegii, 569
ssp. australis, 542, 918
missouriensis, 552
pseudocorus, 127
Ironwood, See Lyonothamnus, Aneya
Irwin, W.P., 380, 731
Ishi Pishi soil series, 721
Island oak, See Quercus tomentella
Islands, See Southern California Islands
Isoetes
howellii, 521, 524
orcuttii, 520, 523-4
Isomeris arborea, 185, 250, 874, 880, 885, 933
Iva
axillaris, 790
nevadensis, 41
Ivesia
argyrocoma, 544
gordonii, 647, 661, 663-4
lycopodioides, 641, 932
ssp. scandalaris, 658, 660
ssp. typica, 627
muirii, 628, 630, 639
santolinoides, 544, 917
shockleyi, 630, 637
Ives, J.D., 669
Ivimey-Cook, R.B., 669

Jackass Flatt, 838
Jackrabbit, See Lepus californicus
Jackson Lake, 543
Jacksonville, 62
Jaeger, E.C., 556, 865, 887
Jaeger Sanctuary, 98
Jain, S.K., 133, 512-3, 532
Jamesia americana, 146, 153, 629, 631-2, 823

Jameson, D.A., 829
Janes, E.B., 512
Jasper Ridge Nature Reserve. 95
Jaumea carnosa, 273-6, 278-9, 285, 903, 931
Jefferies, R.L., 291
Jefferson, C.A., 291
Jeffrey pine, See Pinus jeffreyi
Jeffrey pine forest, 540, 542-5, 574-7, 592-3, 803, 809, 815, 916
Jehemy Beach, 903
Jenkins, R.E., 107
Jenny, H., 71, 133, 354, 463, 669, 697
Jensen, H.A., 219-20, 354, 380, 412
Jepsonia, 112
Jepson, W.L., 133, 191, 354, 412, 463, 698, 887
Jesperson, B.Z., 513
John Little (area), 89
Johnson, A.W., 191, 258-9, 887
Johnson, C.M., 597
Johnson, D.A., 669
Johnson, D.H., 865
Johnson, H.B., 865, 867, 888
Johnson, J.W., 259
Johnson, M.P., 532
Johnson, T.H., 107
Johnson, W.C., 413
Johnston, I.M., 556
Johnston, M.C., 132, 134
Jones, H., 462
Jones, G.N., 669
Jones, M.D., 463
Jorgensen, C.D., 193
Josephine soil series, 362, 703
Joshua tree, See Yucca brevifolia
Joshua Tree National Monument, 815, 844, 856, 858
Joshua tree woodland, 814, 836, 852, 854-5, 857-9
Juglans
  californica, 167, 170, 172, 402-3, 407
  major, 164, 170
Juhren, G., 462, 468-9, 488
Juhren, M.C., 469
Jumping cholla, See Opuntia bigelovii
Juncus
  acutus, 246, 279, 284, 877, 935
  balticus, 275, 552, 625-6, 628, 931
    var. montanus, 623, 629, 633
  bufonius, 519-21, 523, 620, 736, 741
  capitatus, 523
  cooperi, 852-3
  covillei, 552
  drummondii, 626, 641
  effusus, 275
    var. brunneus, 742
    var. pacificus, 742, 931
  lesueurii, 239, 243-4, 256, 275, 936
  mertensianus, 623, 625, 637, 642-3, 649, 652-3, 658, 715
    ssp. duranii, 552
  mexicanus, 279
  parryi, 622, 627, 640, 642, 647, 651, 653, 718
  patens, 335
  phaeocephalus, 244, 936
  tenuis var. congestus, 736
  uncialis, 520
June Lake, 807
Juniper, See Juniperus
Juniperus
  californica, 164, 169-70, 185, 311, 313, 393, 396, 400, 431, 814-21, 856, 858, 919, 930, 932
  communis, 104, 622, 720
    var. jackiie, 721
    var. saxatilix, 711, 717
  occidentalis, 40, 79, 82-3, 153, 313, 392, 405, 767, 776, 785-6, 917, 920, 924, 929-30
    ssp. australis, 546-7, 575-6, 586, 592-3, 643, 798, 803-8, 917, 922-3
    ssp. occidentalis, 589, 769, 798-803
  osteosperma, 40, 164, 167, 782, 784-5, 803, 808-14, 827, 858, 920, 922
  sibirica, 152, 180

Kaiser Pass, 807
Kalmia
  microphylla, 579, 620, 623, 625, 641
  polifolia var. microphylla, 585, 649, 710, 715-6, 719, 922
Kangaroo rat, See Dipodomys heermanni
Kappen, L., 887
Karwinskia humboldtiana, 166, 169
Kasapligil, B., 291
Kaweah Basin, 84
Kawin, R., 759
Kay, B.L., 511, 865
Kearney, R.H., 258
Keck, D.D., 9, 107, 132, 135, 260, 355, 380, 414, 465, 489, 513, 556, 672, 759, 795, 831, 866, 915
Keckiella
  antirrhinoides, 301, 821, 872
  cordifolia, 301
Keeler-Wolf, T. and V., 413, 830
Keeley, J.E., 463
Keeleher, J.M., 462
Kelloggia galioides, 712
Kelso Dunes, 815
Kennedy, P.B., 795
Kennett, C.E., 759
Kentfield, 49
Kern River, 311
Kernville, 815
Kershaw, K.A., 532
Kesseli, J.E., 71
Kessler Spring, 858
Keteleeria davidiana, 163
Khromov, S.P., 71
Kiekert, R.N., 674
Kiepert, D.W., 72
Kilgore, B.M., 597
King Creek, 324, 540
King, P.B., 191, 795
King, R.C., 354
King's Canyon National Park, 84, 209, 807
King's River, 810
Kingston Mountains, 811-2, 822
King, T.J., 865
King, V.V., 758
Kinncan, E.S., 463
Kittredge, J., 463
Klamath Basin, 792
Klamath Mountains, 119, 140, 147, 152, 154-5, 361-2, 366, 371,
Klamath Mountains (cont.) 404, 421, 580, 602-3, 613-4, 616, 621-2, 665, 699-732, 920-1
Klamath River, 265
Klamath weed, See Hypericum perforatum
Klamath weed bettle, See Chrysolina
Kleinkopf, G.E., 863, 867
Klikoff, L.L., 597, 669-70
Klyver, F.D., 413, 463, 597
Knapp, R., 9, 354, 380, 670
Kneeland soil series, 367
Kniffen, F.B., 759
Knobcone pine, See Pinus attenuata
Knawkes, P.F., 512
Kobresia myosuroides, 122, 613, 625, 629, 634
Kochia
  americana, 855
  californica, 494, 851, 934-5
Koeberlinia, 118
Koeleria
  cristata, 65, 365, 493, 609, 627, 632, 656-7, 660, 734, 772, 774, 776, 787
  macrantha, 542, 736, 825-6, 924
Komarek, E.V., 354
Kopecko, K.J.P., 532, 556
Köppen, V., 18, 229
Kornoelje, T.A., 865
Kotok, E.I., 467, 598
Kraebel, C.J., 462-3
Krameria
  grayi, 817, 845, 860, 872
  parvifolia, 821, 842, 844, 844, 849, 872
Krammer, J.S., 463
Kramer, P.J., 191
Kranz syndrome, 861
Krohn, R., 379
Kruckeberg, A.R., 134, 354, 380, 464, 731
Krummholz, 543, 547, 550, 579-80, 602, 606, 615-7, 622-3
Kruse Rhododendron area, 89
Küchler, A.W., 9, 218-9, 380, 413, 512, 759, 915
Kuller, Z., 468
Kumler, M.L., 259
Kummerow, J., 459, 488
Kunzel, H., 512
Kuser, J., 698

Kuvaev, V.G., 670
Kyhos, D.W., 134

Lack Creek, 311
Lactuca serriola, 439
Laguna Mountains, 819, 870, 874
Lagurus ovatus, 741
Lajaro, M., 459
La Jolla, 87, 500
Lake Cuyamaca, 542
Lake Earl, 265, 275
Lake Eleanor, 62
Lake Fulmore, 541
Lake Isabella, 815
Lakeport, 52
Lake Spaulding, 54
Lake Tahoe, 614
Lake Talawa, 265, 275
Lakeview Mountain, 821
La Mareche, V.C., 351, 670-1, 830
Lambe, E., 672
Lamb, H.H., 71
Lancher, W., 71
Landers, R.Q., 463
Lane, R.S., 291
Langenheim, J.H., 354
Lange, O.L., 887
Langloisia punctata, 861
Lanner, R.M., 830
Lanphere-Christensen dunes, 8, 244
La Panza Range, 817
Lappula redowshii, 774, 786, 790
Lapse rate, evaporation, 44-6
Lapse rate, precipitation, 44-6, 55, 60-1, 63, 66, 68
Lapse rate, temperature, 53, 55, 60-1, 63, 65-6, 68
La Purisma Hills, 332, 334-8
Larcher, W., 666, 670
Larrea
  divaricata, See L. tridentata
  tridentata, 3, 5, 42, 87, 185, 837-50, 858, 860, 873-6, 880-2, 895, 932-3
Larsen, L.T., 354
Las Flores Marsh, 279
Lassen Volcanic National Park, 209, 800
Last Chance Creek, 825
Lasthenia
  chrysostoma, 493, 506, 508, 516, 524-5, 527, 734, 743, 898, 900, 930
Lasthenia (cont.)
  glaberrima, 525, 527
  glabrata, 274, 278
  minor ssp. maritima, 122
Las Vegas, 67
Lathrop, E.W., 413, 464, 532-3, 556-7
Lathyrus
  bolanderi, 753
  japonica, 261
  laetiflorus, 301, 441
  littoralis, 230-1, 235, 237, 240, 243-4, 256, 261, 936
  splendens, 300-1
  vestitus, 344
Latting, J., 10, 915
Laude, H.M., 463
Lauff, G.N., 290
Laughlin soil series, 367, 404
Launchbaugh, J.L., 467
Laurocerasus lyonii, See Prunus lyonii
Lautensach, 11, 71
Lava Beds National Monument, 84, 800-1
Lawrence, D.B., 107
Lawrence, G.E., 413, 464
Lawrence, L., 354, 358
Laycock, W.A., 107
Layia
  carnosa, 238, 261
  chrysanthemoides, 523, 526
  glandulosa, 809, 920
Leaf Area Index, 481
Leaf Morphology, 480-1, 485, 861, 881-2, 884
Least Tern (area), 87
Leavitt, Fa-ls, 806
Leavitt Meadows, 806
Lecanora gibbosa, 586
Le Conté, J., 191
Le Drew, E., 670
Ledum glandulosum, 304-5, 333, 719
    var. californicum, 710, 716
Lee, J.A., 294
Le Houerou, H., 864
Leiberg, J.B., 464
Leonard, O.A., 412
Leonard, R., 597
Leopold, L., 71
Lepechinia
  calycina, 309
  cardiophylla, 300-1, 323, 329, 331

Lepechinia (cont.)
ganderi, 300-1
Lepidium
fremontii, 41, 842
lasiocarpum, 861, 884
latipes, 519
Lepidospartum squamatum, 185, 432
Lepper, M.G., 597
Leptodactylon
californicum, 247, 431
pungens, 548, 577, 529, 635-6, 642-3, 656, 774-6, 807, 810, 819-20, 823, 825-6, 920, 924
ssp. pulchriflorum, 624, 630-1, 633, 638, 644, 650-1, 653, 658, 663, 923
Lepus californicus, 498
Leschke, H., 353, 461, 596, 668, 830
Leskinen, P.H., 191
Lesquerella occidentalis, 620, 624
Lessingia hemaclada, 439
Letey, J., 465, 865, 887
Letharia
columbiana, 571
vulpina, 571, 922
Leucocrinum montanum, 803, 930
Leucopoa kingii, 658
Leucothoe davisiae, 150, 703, 709-10, 714, 716, 719-20, 729, 920-1
Leutkea pectinata, 621, 623
Levering, C.A., 291
Lewis, A., 556
Lewis, D.C., 413
Lewis, F.H., 464
Lewis, H., 132, 134-5, 461, 488
Lewisia
brachycalyx, 153, 613
cotyledon, 621, 718
leana, 718
nevadensis, 552, 623, 646, 650, 653-4
pygmaea, 637, 640, 653, 658, 663
redivia ssp. minor, 544
sierrae, 640, 642
triphylla, 623
Lewis, M.E., 670, 830
Leydet, F., 698
Libby, W.F., 189
Libby, W.J., 354, 556, 830
Libocedrus decurrens, See Calocedrus decurrens
Lichens, 340, 897, 922
Liebra Range, 538
Lieth, H., 74
Ligusticum
californicum, 708
grayi, 647, 717
Lilaea scilloides, 520, 523
Lilium
humboldtii, 902
parryi, 552
Limber pine, See Pinus flexilis
Limber Pine Forest, 549-51, 823-5
Limestone, 117, 122, 822
Limnanthes
alba, 528-9
douglasii
var. nivea, 528
var. rosea, 520, 523, 528
floccosa, 528-9
Limonium californicum, 273-6, 278, 284, 931
Lin, J.W.Y., 512, 532
Linanthus
ciliatus, 542
harknessii, 620
montanus, 586
nuttallii, 631, 633, 653, 658, 660, 825-6, 924
parryae, 126
Lindguist, J., 219
Lindsay, A.D., 354
Lindsay, J.H., 670
Linhart, Y.B., 354, 358, 533, 906
Linnaea borealis, 152, 707, 709-10, 714, 720, 722, 729, 920-1
ssp. longiflora, 711
Linum lewisii, 611, 626, 645, 656, 658, 664
Lipschitz, N., 468
Listera convallarioides, 711
Lithocarpus densiflora, 45, 89, 145, 147-52, 154, 159-61, 167, 304, 333, 360-2, 364-75, 390, 682-5, 689, 693, 715, 916, 925
var. echinoides, 365, 708
Lithophragma tenellum, 552
Lithospermum ruderale, 930
Little, E.L., 191, 352, 354-5, 357, 830
Little Harbor Beach, 899

Little Lost Man Creek, 84
Little Morongo Creek, 550
Little Salmon Creek, 303
Little San Bernardino Mountains, 538, 815, 822, 836, 857
Little Sur River, 750-3
Littrell, E.E., 380
Livestock, 498, 500, 578, 580, 585, 619, 770, 781, 792, 858, See also California history, Herbivory
Lloyd, R.M., 134, 670-1, 830
Lobelia cardinalis ssp. graminea, 551
Lackwood Valley, 815
Lodgepole pine, See Pinus contorta ssp. murrayana
Lodgepole pine forest, 4, 545, 547-8, 553, 577-9, 588, 593
Logging, 348, 366, 369-70, 374, 574-5, 593-4, 730, 812, 910
Lolium
  multiflorum, 439, 497, 505, 735, 737
  perenne, 742
  temulentum, 530
Lomatium
  lucidum, 442
  macdougali, 782
  macrocarpum, 722
  nevadense ssp. parishii, 653
  nudicaule, 778-8
  ravenii, 116
  torreyi, 586
  triternatum, 776
  utriculatum, 442
Lompoc, 897
Lone Tree Creek, 825
Long Beach, 265
Longevity, See Age structure, Seed longevity
Lonicera
  conjugialis, 711, 717
  hispidula, 333
  interupta, 457, 928
  involucrata, 239, 244, 751, 919
  subspicata, 331, 928
Leope, L.L., 191, 670
Lorber, P., 489
Los Angeles, 241, 894
Losch, R., 887
Los Osos Oaks area, 89
Los Peñasquitos, 87, 265, 278-9
Lotus
  argophyllus, 896
  angustissimus, 741
  consinnus, 439
  corniculatus, 742
  crassifolius, 329, 331
  davidsonii, 542, 554
  humistratus, 442
  junceus, 248
  micranthus, 238, 439, 741
  oblongifolius, 552
  purshianus, 439
  rigidus, 817, 821
  salsuginosus, 861
  scoparius, 246, 248-9, 256, 301, 323, 334, 434, 442, 445, 474-5, 896, 937
  strigosus, 439
  subpinatus, 439
Love, L.D., 866
Love, R.M., 219, 514, 830
Lowe, C.H., 9, 888
Lower Tubbs Island, 98
Lucerne Dry Lake, 851-2
Lucerne Valley, 841, 844, 856
Luckenbach, R., 513
Lunt, O.R., 865, 887
Lupinus
  albifrons, 896-7
  alpestris, 825, 924
  andersonii, 706
  arboreus, 88, 228, 235, 243-4, 253, 256, 261, 740, 742-3, 745-6, 755-6. 936
  arizonicus, 861
  bicolor, 238, 439, 509, 930
  breweri, 585
    ssp. grandiflorus, 544
  caespitosus, 658
  caudatus, 645, 774, 786, 804, 929
  chamissonis, 235, 244, 246, 248-50, 256, 936-7
  croceus, 722
  elatus, 543, 917
  excubitus, 542, 847, 918
  formosus, 542, 918
  latifolius
    var. columbianus, 569
    ssp. parishii, 552
  leucophyllas, 443
  littoralis, 238, 256
  lobbii, 611, 647

Lupinus (cont.)
  longifolius, 443
  luteolus, 930
  lyallii, 611, 622, 624, 627, 637, 641, 932
  montigenus, 661-2
  obtusilobus, 624
  peirsonii, 544
  polyphyllus ssp. superbus, 648
  saxosus, 776
  sericeus, 930
  variicolor, 472, 740, 742-3, 747, 755-6
Luzula
  campestris, 365
  comosa, 552-3, 648, 735-6
  orestera, 625, 640
  parviflora, 715
  spicata, 615, 637, 640, 658, 663, 932
  subcongesta, 623
Lycium
  andersonii, 41, 837, 842, 844, 849, 859, 877, 932-3
  brevipes, 250-2, 933, 937
  californicum, 475, 899
  cooperi, 844, 855, 859, 872, 932, 934-5
  fremontii, 872
  pallidum, 182, 848-9
  torreyi, 181, 932
Lygodesmia
  exigua, 861
  spinosa, 645, 774
Lyonothamnus
  aspleniformis, 168, 170
  floribundus, 112, 165-6, 168, 901, 903, 905, 928
Lythrum
  californicum, 592
  hyssopifolia, 520, 523-4
Lytle Creek, 544-5, 819

MacArthur, E.D., 795
MacArthur, R.H., 533
MacDonald, G.A., 795
Macdonald, K.B., 259, 291
Macey, A., 464
MacGinitie, H.D., 191
Machaeranthera
  canescens, 544, 786
Machaeranthera (cont.)
  shastensis, 629
    var. eradiata, 621
    var. montana, 622, 626, 633
  tortifolia, 842
Machaerocarpus californicus, 523
Machaerocereus gummosus, 475
MacKerricher Beach, 233
MacLeod, S.A., 758
Madeline, 59, 768
Madia
  capitata, 741
  exigus, 439
  gracilis, 439
  nutans, 122
  sativa, 498
Madrean-Tethyan floristic links, 113, 115
Mad River, 368
Madrone, See Arbutus menziesii
Madro-Tertiary Geoflora, 112-3, 115, 120-1, 159, 181
Magalia Grove, 311
Mahogany Flat, 812
Mahonia
  aqufolium, 145
  fremontii, 153, 174
  haematocarpa, 175
  pinnata, 147
  repens, 153
Maillard Redwoods, 89
Maino, E., 355
Major, J., 9, 69-70, 72, 132, 134, 136, 191, 381, 415, 461, 467, 513-4, 666, 670-1, 674
Malacothamnus
  fasciculatus, 301, 434
  fremontii, 443
Malacothrix
  foliosa, 897
  glabrata, 875
  incana, 248, 256-7
  saxatilis, 897
Mallery, T.D., 887
Mallory, J.I., 220, 355, 413, 597, 697, 731, 831
Mallory, T.E., 465
Mall, R.E., 291
Malosma laurina, See Rhus laurina
Malva parviflora, 900
Malyshev, L.I., 671
Mammilaria
  dioica, 341, 475

Mammilaria (cont.)
  tetrancistra, 874
Mammoth Junction, 809
Management, reclamation, 226-8, 253, 287, 349-50, 455-7, 500-1, 508-10, 579, 585, 587, 593, 696, 730
Manchester, 233, 339
Manhattan, 248
Mansfield, T., 489
Manzanita, See _Arctostaphylos_
Mapes, B.J., 292
Mapping, See Vegetation mapping
Marah
  fabaceus, 244, 261, 335
  macrocarpus, 301, 329, 443, 897, 899, 901, 903, 905
Marble, 625, 628
Marble Canyon, 825
Marble Mountains, 613, 621-2, 702, 704, 714-5, 727, 729
Marchantia polymorpha, 569
Marion, L.H., 464
Marion Peak, 550-1
Maritime bluffs, See Coastal bluff
Maritime climate, See Climate, maritime
Mark, A.F., 829
Markleeville, 806
Marks, J.B., 887
Marsh, See Salt marsh, Fresh water marsh, Tule marsh
Marshall, D.R., 133, 572-3
Marshall, J.T., 292
Marsilia
  mucronata, 521
  vestita, 524
Martin, E.V., 259
Martinez Canyon, 98
Martinez Mountains, 819
Martin, P.S., 191
Martin, W.E., 463
Mason, C.T., 533
Mason, H.L., 134, 190-1, 351, 355, 380, 464
Mason, J., 759
Mason, L.G., 532
Mastroguiseppe, R.J., 597, 731
Mather, 32, 62
Mather, J.R., 70, 72-3
Mathews, W.H., 192
Mathews, W.C., 355
Matthes, L.J., 671
Mattole River, 690-1
Matveeva, T.S., 666
Maurandya petrophila, 122
McArdle, R.W., 220
McBride, J.R., 259, 355, 380, 413, 759
McCloud River, 98
McCullough, D., 513
McDonald, C.D., 887
McDonald, J.B., 355
McDonald, P.M., 380, 597
McDonough, W.T., 887
McDougall, D.T., 887
McGee Mountain, 807
McGown, R.L., 513
McGrath Beach 233
McHargue, L.T., 74, 887, 889
McIlwee, W.R., 292
McKell, C.M., 413, 464, 513, 864-6, 887
McKelvey, S.D., 865
McLaughlin, E.G., 533
McLaughlin, W.T., 259
McLoughlin, P.L., 292
McMillan, C., 134, 555
McMinn, H.E., 355, 464, 468
McNab cypress, See _Cupressus macnabiana_
McNaughton, S.J., 71, 513, 533
McPherson, K.R., 758
McPherson, J.K., 464-5
McVaugh, R., 533
Meadow mouse, See _Microtus californica_
Meadows, alpine, 5, 552, 601-75
Meadows, mountain, 547, 552-3, 561, 584-5, 593, 622, 624-5, 628, 708, 920-1, 923
Medicago hispida, See _M. polymorpha_
Medicago polymorpha, 279, 504-5, 508-9, 900, 931
Medicine Bow Mountains, 57
Mediterranean climate, See Climate
Mehringer, R.B., 795, 831
Meineke, E.P., 597
Melica
  bulbosa, 774
  californica, 495, 735-6
  geyeri, 365, 734
  harfordii, 736
  imperfecta, 344, 495, 542
  stricta, 544
  subulata, 736

Melica (cont.)
torreyana, 736
Melilotus indicus, 279
Mendocino, 306, 315, 332, 337
Mendocino cypress, See Cupressus pygmaea
Menodora
scoparia, 813, 872
spinescens, 811, 842, 858-9, 934
Mentzelia
gracilenta, 439
involucarata, 861
laevicaulis, 546, 554
multiflora, 775, 791
Menyanthes trifoliata, 87
Merced River, 84
Mercury, 842
Mercury Valley, 838
Merriam, C.H., 597, 795
Merxmüller, H., 671
Mesembryanthemum
chilense, 121, 228-9, 231, 235, 237, 241, 246, 248, 250, 256-8, 755, 896, 929, 936-7
crystallinum, 250-1, 256, 284, 898, 937
edule, 228, 250
nodiflorum, 279, 898
Metasequoia, 140-1, 143, 146, 162
Meusel, H., 134
Mexico, 870, 905
Meyer, W.H., 220
Michaelis, G., 671
Michaelis, P., 671
Microseris troximoides, 776
Microsteris gracilis, 787
Microtus californica, 498
Midden, 266
Middle Ranch Canyon, 902
Miller, A.H., 831, 865
Miller, B.C., 557, 832
Miller, C.H., 464
Miller, D.E., 351
Miller, D.H., 671
Miller, G.N., 135
Miller Gulch, 305
Miller, L.C., 464
Miller, P.C., 466, 488, 597, 667
Miller, P.R., 355
Miller, R.R., 795
Miller, R.W., 795, 865
Milligan, M.T., 597, 671, 831
Milliman, J.D., 292
Milliman, J.P., 259
Millsholm soil series, 400
Mima mound topography, 736
Mimulus
aurantiacus, 302, 304, 334, 344-6, 472, 747-8, 750, 938
bigelovii var. panamintensis, 924
bolanderi, 439
breweri, 775
clevelandii, 302
flemingii, 900
fremontii, 439
guttatus, 128, 244, 443
johnstonii, 542, 546
lewisii, 621, 624
longiflorus, 302, 323, 554, 817, 900
ssp. calycinus, 554
moschatus, 331, 553, 624, 626
parviflorus, 902-3
primuloides, 623, 625, 641, 653
var. pilosellus, 611
puniceus, 482-3, 897
pygmaeus, 116
suksdorfii, 553
tilingii, 611, 624, 649, 653
tricolor, 521
Minard, C.R., 260
Mineral King, 56
Mining, 348
Minnich, R.A., 355-6, 831
Mirabilis
californica, 250
bigelovii, 41, 817, 845, 848, 859
var. retrorsa
froebelii, 821
laevis, See M. californica
Mira Mar Masa, 484
Mirov, N.T., 355, 464
Mission Bay, 93, 265, 270, 276
Mistletoe, See Arceuthobium, Phoradendron
Mitchell caverns, 67-8, 87
Mitchell, J.A., 597
Mitchell, J.E., 795, 865
Mitchell, R.S., 134, 670-1, 830
Mitella pentandra, 711
Mixed conifer forest, 34, 83-4, 147-54, 300, 326, 406, 545-6, 563-9, 588, 593, 701-15, 722-4,

Mixed conifer forest (cont.)
800, 803, 806-7, 823
Mixed evergreen forest, 3, 4, 88, 158, 162, 307, 326, 333, 359-81, 401, 406, 539, 542, 700, 748, 902, 905, 924-5
Mize, C.W., 380
Mobberley, D.G., 292
Modesto, 62
Modoc cypress, See Cupressus bakeri
Modoc Plateau, 457, 591-2, 765-8, 770, 772-3, 779-80, 748-5, 917
Moir, W.H., 107
Moisture stress, 373, 405, 409, 450, 452, 454, 482-4, 617, 643, 651, 849, 860, 881-3
Mojave, 838
Mojave Desert, 37, 41-4, 66-8, 92, 117-8, 164, 166, 170, 179-85, 298, 539-40, 542, 549, 811, 815, 821-2, 824, 835-62, 871-4, 922, 932-3, 938
Moldenke, A.R., 460
Mollisols, 767
Monanthochloe littoralis, 273, 276, 278, 282, 284-5, 287, 931
Monardella
  antonia, 122
  cinerea, 550, 554, 653
  lanceolata, 331
  linoides, 817
  odoratissima, 152, 574, 577, 622, 653, 786, 916, 920-1
    var. glauca, 712, 717
    ssp. parvifolia, 630-1, 633
  undulata, 250
Monerma cyclindrica, 274
Mono Lake, 61-3, 94, 765, 788, 810, 920
Monolopia lanceolata, 495
Monroe, G.W., 292
Montane chaparral, 547, 561, 584, 715, 784-5
Montane forest, 537-47, 560-79, 588-9, 701-15, 812, 822-6, 921-2
Monteith, J.L., 72
Montenegro, G., 459
Monteny, B., 72
Monterey, 31, 235, 246-7, 315, 337, 342, 345, 750-3
Monterey cypress, See Cupressus macrocarpa
Monterey pine, See Pinus radiata
Montgomery woods, 89
Montia
  parvifolia, 717
  perfoliata, 439, 735
Monticello Dam, 13-4
Monumental, 31, 45
Monument Peak, 861
Mooney, H.A., 9, 69, 72, 294, 460-2, 464-5, 488-9, 597, 666, 671-3, 830-2, 864-5, 866-7, 886-7, 889
Moore, D.M., 135
Moore, J.J., 672
Mooring, J.S., 135
Morgan, D.L., 69, 351, 697
Mormon tea, See Ephedra
Morongo Canyon, 99, 838
Morrison, R.B., 795
Morro Bay, 87, 91, 241, 246-8
Morrow, P.H., 465, 488
Moser, W., 672
Moss, 768
Moulds, F.R., 355
Mountain hemlock, See Tsuga mertensiana
Mountain Pass, 67
Mount Baden-Powell, 545, 547-50, 552
Mount Baldy, 547-8
Mount Burnham, 549
Mount Diablo, 375, 392, 818
Mount Eddy, 621-2, 722, 730
Mount Gleason, 539, 541
Mount Hamilton, 13-5, 47
Mount Hawkins Range, 549-50, 590-1, 613-4, 622
Mount Islip, 539, 552
Mount Lassen, 623
Mount Palomar, 33, 50, 66, 68
Mount Pinos, 153, 164, 170, 392, 538, 541, 547, 549-50, 554, 815, 823, 923
Mount Ranier, 55
Mount Rose, 577
Mount Saint Helena, 49
Mount San Antonio, 539, 542, 545-7, 550, 552, 653
Mount San Gorgonio, 550, 552, 602, 808
Mount San Jacinto, 548, 653
Mount Shasta, 590, 602, 613-4, 622-4, 722, 802
Mount Tamalpais, 47, 49

Mount Vision, 748, 750-4
Mount Waterman, 539-40, 544
Mount Wilson, 50, 541
Mouse, meadow, See
  Microtus californica
Mozafar, A., 864-5
Mudie, P.J., 289, 292-3
Mueller-Dombois, D., 74, 380, 414, 731
Mugu Lagoon, 265, 275-6, 278
Muhlenbergia
  filiformis, 620, 653
  porteri, 858-9
  richardsonis, 628, 630, 632, 784, 790, 826
Muir, J., 192, 511
Mule deer, See Odocoileus hemionus
Muller, C.H., 352, 460, 464-5, 489, 514, 865, 887
Müller-Schneider, P., 675
Muller, W., 489, 865, 887
Mulroy, T.W., 862
Munns, E.N., 465
Munz, P.A., 9, 135, 192, 260, 292, 355, 380, 414, 465, 489, 513, 556, 672, 759, 795, 831, 866, 888, 915
Murphy, A.H., 414, 513, 759
Murphy Preserve, 99
Mutch, R.W., 465
Muth, G.J., 672, 731
Myatt, R.G., 380, 414
Myosurus minimus, 273-5, 521
Myrica
  californica, 171, 239, 256, 303-4, 686, 742, 748, 753, 755, 916, 919
  mexicana, 171
Myrtillocactus cochal, 475
Myrup, L.O., 72

Nagy, J.G., 796
Napa, 34, 36
Nash, A.J., 72
Nasturtium officinale, 551
Natural areas, nature reserves, 75-108, 349-350
Nature Conservancy, 85, 95, 96, 108
Navarretia
  atractyloides, 303-4, 307, 309, 439
  intertexta, 520, 521
  leucocephala, 439, 520-1, 525
  mellita, 307, 309
  prostrata, 523, 525
  squarrosa, 303-5
  tagetina, 777
Naveh, Z., 135, 465
Nectria fuckeliana, 338
Needlegrass, See Stipa
Needles, 67, 870
Nemacladus sigmoideus, 540
Neostapfia colusana, 520-1
Nesbit, D.M., 292
Net radiation, 15, 20, 655
Neuenschwander, L.F., 292
Neuns soil series, 362
Nevada, 800, 802, 809, 814, 826, 853-4, 857, 860, 909
Newcombe, C.L., 292
Newcombe, G.B., 356
Newman, T.F., 414
New Mexico, 297
Newport Bay, 265, 276, 284
Newton, M., 469
New York Mountains, 811-14, 821-23, 827, 858
Nichols, D.R., 292
Nichols, S.P., 219
Nicotiana glauca, 256, 435
Niklfeld, H., 71
Niland, 879
Nilsen, T.H., 192
Nine Mile Canyon, 856
Nissen, E.M., 72
Nitrogen fixation, 226, 434, 454, 458
Nitrophila occidentalis, 851, 853
Nitrophile, 608, 619
Nival belt, 602, 938
Nobs, M.A., 135, 465, 864
Noldecke, A.M., 136
Nolina
  biglovii, 181
  interrata, 129
  parryi, 300, 302, 321-2, 816, 819-20, 920
Noncalcic brown soils
Nord, E.C., 796, 831

Norland, E.D., 192
Norris, K.S., 107
North coast forest, 4, 145-7, 305, 326, 340, 679-93, 749
Northern California Coast Range Preserve, 99
Northern coastal scrub, See Coastal scrub
Northern oak woodland, See Oak woodland
Nowacki, E., 135
Nuphar polysepalum, 620
Nurse plants, 873
Nutmeg, See Torreya californica
Nutrients, See Soil nutrients, Carbon balance
Nyssa
  aquatica, 163
  sylvatica, 166

Oak, See Quercus, Lithocarpus
Oakdale, 62-3
Oakeshot, G.B., 866
Oakes, W.O., 669
Oak woodland, 3-4, 11, 162-169, 307, 311, 372, 383-415, 451, 494-5, 500, 502, 539, 683-4, 691, 696, 701, 815, 817-8, 901-2, 904, 926
  Northern, 403-4
  Southern, 402-3
Oasis, 99, 877-79, 935
Oatgrass, See Danthonia californica
Oberlander, G.T., 72
Oceanside, 483
Odening, W.R., 866, 888
Odocoileus hemionus, 348, 498-9 770, 800
Odontostomum, 112
Odum, E.P., 290, 465
Odum, W.E., 259
Oechel, W.C., 863, 866, 886, 888
Oenothera, See also Camissonia
Oenothera
  caespitosa, 633
    var. crinita, 631, 825
  contorta, 790
  cruciata, 336
  deltoides, 860
  heteranthera, 664
  hookeri ssp. venusta, 552
Oenothera (cont.)
  micrantha, 439
  nevadensis, 41
Offord, H.R., 356
Ogden, G.L., 906
Ogunkanmi, A., 489
Ohman, L.F., 107
Ohwi, J., 135
O'Keefe, J., 462, 488
Old Woman Mountains, 815
Oligomeris linifolia, 115
Olin, G., 889
Oliver, W.W., 597
Olmsted, I.C., 672
Olneya tesota, 118, 181, 872-3, 877, 933
Olsson-Seffer, P., 260
Olympic Mountains, Washington, 612, 614
Omi, P.N., 597
Onychium densum, 586, 713
Oosting, H.J., 597
Opuntia
  acanthocarpa, 845, 859
  basilaris, 816, 820, 848, 884, 933
  bigelovii, 844-5, 860, 876, 933-4
  echinocarpa, 820, 843, 846, 856, 859, 934
  erinacea, 813, 820, 822, 847
  mojavensis, 813
  occidentalis, 250, 476, 937
    var. littoralis, 898, 929
  oricola, 898, 899
  ramosissima, 843-4, 856, 858-9, 934
  rosarica, 475
Orcutt sand, 337
Ordination, 643, 847
Orebamjo, T.O., 294
Oregon, 297, 311, 325, 328, 339, 360-1, 364, 366, 425, 493, 624, 772-3, 775-6, 779, 784-5, 799, 802, 909
Oregon oak, See Quercus garryana
Oreonana vestita, 546, 550, 554, 653-4, 920
Oriflamme Mountains, 819
Orme, A.R., 260
Ornduff, R., 9, 135, 356, 465, 533, 556, 759, 831
Orobanche pinorum, 624

Orthocarpus
  attenuatus, 439, 930
  campestris, 523, 930
  castillejoides, 741
  copelandii, 621
  erianthus, 506, 930
  linearilobus, 930
  luteus, 621
  purpurascens, 238, 335, 506, 930
Oryctes nevadensis, 41
Oryzopsis
  bloomeri, 775
  hymenoides, 629-31, 771-2, 774-5, 779, 782, 788, 790, 792-3, 810, 812, 842-3, 934
  miliacea, 457
Osborne, J.F., 463, 466
Osborn, J., 465
Osmorhiza
  chilensis, 569, 706, 711
  occidentalis, 711
Otay Mountain, 208, 300, 316-8, 321, 549
Ottenger, F., 731
Ovis canadensis, 770
Owens Peak, 807
Owens River, 765
Owens Valley, 836, 854
Ownbey, M., 669
Oxalis
  oregona, 368, 695, 916, 925
  pilosa, 742
Oxnard, 500
Oxygen deficiency, See Flooding, Soil aeration
Oxyria digyna, 609, 621, 623, 629, 632, 639, 653, 665, 932
Oxystylis, 117-8
Oxytheca parishii, 554
Oxytropis
  oreophila, 654
  parryi, 656, 826

Pace, N., 72
Pachystima myrsinites, 153, 705, 710
Pacific Southwest Forest and Range Experiment Station, 198, 206, 209
Packard, E.L., 191
Paeonia californica, 443
Pahrump, 840
Paine Wildflower Memorial, 99
Palley, M., 219
Palm oasis, 877-9
Palm Springs, 23-4, 26
Palouse, 493, 496
Paloverde, See Cercidium floridum
Pamir, 613
Panamint Mountains, 808, 811-2, 823-5, 854, 924
Papaver californicum, 439
Parapholis incurva, 274, 279
Parish, S.B., 135, 356, 465-6, 556, 672, 866
Parker, J., 260
Parker, L., 107
Park, R.B., 291
Parnell, D.R., 293
Parratt, M.W., 556
Parsons, D.J., 465-6, 488, 593
Parsons, J.J., 72
Patric, J.H., 72, 466
Pattern, 839-42, 848, 858-9, 875
Patton, C.P., 72
Pavari, A., 698
Payne, W.W., 73
Pearcy, R.W., 72, 864, 866, 888
Pearse, C.K., 796
Pearson, G.A., 107
Pease, R.W., 598, 796
Peattie, D.C., 356
Peck, D., 192
Peck, M.E., 192
Pectis papposa, 860
Pectocarya setosa, 41
Pedicularis
  attollens, 628, 628, 634, 641, 649
  crenulata, 122, 613
  groenlandica, 624, 649-50
  semibarbata, 574, 653, 922
Peinomorphism, 455
Pelishek, R.E., 466
Pellaea
  breweri, 586, 629, 825
  bridgesii, 586
  compacta, 546, 554, 920
  mucronata var. californica, 586
Pellona Mountains, 538
Peltiphyllum peltatum, 567
Pelton, J., 380
Pemble, R.H., 672
Peñalosa, J., 759

Pencil cactus, See *Opuntia ramosissima*
Penck, A., 73
Peninsular Ranges, 537-55, 818, 823, 870, 917, 927
Penman, 20
Penstemon
  anguineus, 706, 712
  bridgesii, 544, 546
  caesius, 544, 550, 917
  centranthifolius, 443, 818
  cordifolius, 329, 334, 448
  corymbosus, 376
  cusickii, 774
  davidsonii, 611, 622, 626, 636, 640, 932
  deustus, 621, 624, 645, 717
  gracilentus, 622, 664
  grinnellii, 542, 546, 550, 920
  heterodoxus, 628, 637, 639, 647, 651, 658
  heterophyllus, 443
  labrosus, 542
  laetus, 778
  newberryi, 152, 623, 712, 717, 719
  oreocharis, 643, 653
  procerus ssp. *brachyanthus*, 713, 718
  rupicola, 621
  speciosus, 631, 633, 644, 774
  spectabilis, 443
  tracyi, 622
Pegnegnat, W.E., 356, 446
Peraphyllum ramosissimum 146, 164, 170
Perideridia
  gairdneri, 736
  parishii, 552
Peridermium harknessii, 338
Peridotite, 692-3, 721
Perityle emoryi, 861, 899
Perkins, E.J., 260
Perkins, J.B., 513
Perris, 821
Perry Akin Creek, 826
Persea
  borbonia, 161, 163
  podadenia, 165
Perth, 25-6
Pert, M., 356
Pescadero, 88, 233, 303, 749-53
Peterson, K.M., 73
Peterson, R.S., 356
Peter, W.G., 107
Petradoria pumila, 822
Petran, 156, 158, 419
Petrong, R., 293
Petrophytum caespitosum, 117, 122, 822-3
Peucyphyllum schottii, 845, 872, 874
Phacelia
  austromontana, 546, 920
  brachyloba, 321, 324, 439
  californica, 753
  crenulata, 861
  dalesiana, 118
  distans, 251, 861, 875
  douglasii, 248
  frigida, 622, 628-9, 631-2, 639
    ssp. *dasyphylla*, 624, 636, 647
    ssp. *frigida*, 623
  grandiflora, 440
  hastata, 711
  imbricata, 440
  lyonii, 897
  minor, 440
  mustelina, 122
  mutabilis, 574
  orogenes, 155, 586
  ramosissima, 256, 335
  tanacetifolia, 440
Phalaris
  californica, 736
  tuberosa, 457
Phenology, 247, 279-83, 452, 481-4, 495, 503-6, 527, 531, 768, 849-50, 860, 862, 880
Philbrick, R.N., 906
Phillips, E.A., 131
Phleger, F.S., 289, 293
Phleum alpinum, 609, 623, 625-6, 634, 649, 664, 734
Phlox
  austromontana, 544
  bryoides, 624
  caespitosa ssp. *pulvinata*, 611
  covillei, 625-6, 628-9, 632, 636-7, 639, 653, 656-8, 660, 826, 932
  diffusa, 152, 622, 644, 653, 664, 713, 718-9, 722, 774, 776, 787
  gracilis, 774
  hoodii, 774-6, 782, 929
  longifolia, 774-7, 782, 800, 930

Phlox (cont.)
  pulvinata, 661-2, 664
Phoenicaulis
  cheiranthioides, 645
  eurycarpa, 611, 642, 663-4, 932
Phoradendron
  bolleanum, 313
    ssp. pauciflorum, 546
  juniperinum ssp. libocedri, 546
Photinia, See Heteromeles
Photosynthesis, 45, 482, 484-6, 827, 849-50, 853, 881-5
Phragmites communis, 734
Phreatophyte, 846, 852-3, 878, 880, 882
Phyllodoce
  breweri, 548, 550, 578, 611, 621, 623-5, 640, 649, 652-3, 922, 932
  empetriformis, 623, 709, 715-6, 719-20, 920-1
Physalis crassifolius, 872
Physocarpus
  alternans, 823
  capitatus, 567
Picea
  albies, 616-7
  breweriana, 104, 145-6, 148-9, 151, 154, 161, 618, 621, 701, 709, 714, 719, 723-30
  engelmannii, 104, 145, 154-5, 617, 701, 709, 714, 719, 723-8, 730
  pungens, 146
  sitchensis, 34, 83, 88-9, 145, 243, 256, 304, 339, 360, 680, 682-3, 685-6, 689, 691, 916
Pickeringia montana, 124, 300, 302, 309, 314, 431, 435, 454
Picris echioides, 753
Pielou, E.C., 293, 533, 759
Piemeisel, R.L., 796, 888
Pierce, W.D., 260
Pigmy forest, 333
Pigott, C.D., 293, 672
Pillsbury, A.F., 466
Pine barrens, See Pygmy forest
Pine, closed-cone, See Closed cone pine forest
Pine mountain, 817-8
Pinguicula macroceras, 621
Pinnacles National Monument, 209, 394, 397
Piñon pine, See Pinyon pine
Pinto Basin, 844
Pinto Range, 538
Pinus
  albicaulis, 82, 104, 154-5, 560, 572, 577, 579, 581-2, 586, 593, 615-6, 618, 621-3, 630-1, 635, 643, 646, 650, 664, 702, 716, 719-24, 726-8, 730, 806, 824, 921-2
  aristata, See P. longaeva
  attenuardiata, 326
  attenuata, 149, 297, 307, 309, 311, 313, 319, 343, 365-6, 393, 430-1, 435, 540, 583, 692-3, 721, 723-8
  balfouriana, 104, 154-5, 560, 562, 578-82, 586, 593, 618, 621, 701, 720, 722-4, 726-8, 730, 923
  banksiana, 339
  cembra, 616-7
  contorta, 152-3, 243, 256, 341, 366
    ssp. bolanderi, 93, 121-2, 297, 304-5, 340-1, 682, 692, 919
    ssp. contorta, 297, 304, 338-40, 682
    ssp. latifolia, 339, 578
    ssp. murrayana, 82, 104, 313, 339, 543, 545-9, 553, 560, 562, 571-2, 575-8, 581, 585-6, 588, 590-1, 593, 618, 652-3, 709, 715, 719-29, 803, 806, 826, 922-3
  coulteri, 87, 300, 302, 309, 326, 361, 373-7, 389, 390, 393-4, 396, 435, 538-42, 583, 918-9, 925-6
  edulis, 167, 808, 814, 821-2, 827
  flexilis, 84, 149, 153-5, 450, 543, 547-51, 560, 579, 582, 593, 618, 652-3, 823-5, 827, 923-4
  gregii, 171
  jeffreyi, 79, 82, 104, 151-3, 166, 178, 313, 326, 365-6, 430, 539-47, 549, 560-2, 565, 569, 571-2, 575-7,

Pinus (cont.)
588-92, 682, 685, 692-3, 715, 720-1, 723-8, 730, 779, 785, 802-4, 806-9, 916, 919, 923-4, 926
lambertiana, 104, 149-51, 309, 313, 326, 362-6, 374, 376, 435, 541-2, 545-6, 553, 562-5, 562, 569, 571-2, 576, 583, 588-90, 592, 682, 684-5, 692, 700, 703-4, 708-9, 714, 721-5, 727-8, 801-2, 807, 823, 917-8, 921, 923, 925
longaeva, 65, 82-3, 122, 146, 149, 155, 450, 582, 618, 824-8, 924
monophylla, 40, 65, 153, 164, 170, 313, 392, 540, 542, 546, 593, 782, 784-5, 803, 806, 808-10, 812-24, 827, 856, 858, 917, 920, 922-3, 927
monticola, 104, 145, 148-9, 152-3, 161, 326, 365, 560, 571-2, 579-80, 589-91, 647, 682, 693, 709, 714-6, 719-30, 806, 923
mugo, 616
muricata, 97, 121, 168-9, 171-2, 244, 297-8, 303-5, 325, 328, 332-8, 340, 684, 753, 902, 919
var. borealis, 332
var. cedroensis, 332
murrayana, See P. contorta ssp. murrayana
oocarpa, 171
patula, 171
ponderosa, 42, 60, 82-3, 87, 89, 104, 143, 147-51, 153, 161, 166, 178, 307, 309, 311, 313, 320, 363-6, 368, 374, 393, 430, 518, 539-43, 545, 560-5, 569, 574-7, 586-90, 592-3, 692, 700-4, 708-9, 714, 722-5, 727-8, 768, 784-5, 787, 800-2, 807, 917-8, 921-2, 924-5, 927
ssp. scopulorum, 153
pringlei, 171
pumila, 616
quadrifolia, 808, 819-20
radiata, 89, 166, 168, 170-2, 246, 297, 303-4, 325-6, 328,

Pinus (cont.)
337, 342-4, 919
var. binata, 332
remorata, 169, 171-2, 297, 332-8, 901-3, 919
sabiniana, 11, 82, 167, 169, 173, 307, 309, 313, 326, 341, 386-8, 392-4, 396, 398, 400-2, 404, 407, 409, 430, 539, 919, 924, 926, 929
sibirica, 616
torreyana, 88-9, 169, 297, 341-2, 903, 919
uncinata, 616
washoensis, 561, 577, 592-3, 802, 923
Pinyon Flat, 540
Pinyon-juniper natural area, 810
Pinyon-juniper woodland, 4, 169-71, 311, 317, 544, 546, 767, 798, 808-22, 854, 856-8, 919-20
Pinyon mountains, 819
Pinyon pine, See Pinus monophylla, P. edulis, P. quadrifolia
Pit River, 808
Pisek, A., 672
Pismo dunes, 88
Pitelka, F.A., 511, 758
Pitelka, L.F., 260, 759
Pitschmann, H., 673
Pitt, M.D., 513
Pitts, W.D., 261
Pityopus californicus, 569
Pityrogramma triangularis, 336, 751, 903
Piute cypress, See Cupressus nevadensis
Piute Mountains, 315
Plagiobothrys, See also Allocarya
californica, 540
jonesii, 861
kingii, 920
nothofulvus, 930
torreyi, 620
Plantago
hirtella var. galeottiana, 736
hookeriana, 238
var. californica, 741
insularis, 860-1, 884
lancelata, 742-3
maritima, 273-4

Plantago (cont.)
  ovata, 115
  purshii, 544
Plant-animal interactions, 246, 347-8, 398-9, 407, 410, 433, 479-80, 497-8, 506, 508-9, 619, 736-8, 756, 770-1, 848, 873, 898, 902, See also Herbivory, Livestock, and particular taxa
Plantation, 305
Platanus racemosa, 100, 151, 161, 164-5, 927, 935
Platt, W.D., 351
Playa, 837, 851
Plectritis samolifolia, 238
Pleuropogon californicus, 736
Pluchea sericea, 185, 852, 872, 877, 880-1, 935
Plumas National Forest, 565-6
Plumb, T.R., 466
Plummer, A.P., 795
Plummer, F.G., 466
Poa
  ampla, 457
  annua, 497
  bolanderi, 574, 923
  confinis, 238, 256
  cusickii, 645
  douglasii, 231, 235, 238, 243-4, 246, 250, 257, 752, 897, 936
    ssp. macrantha, 735
  epilis, 622, 625, 637, 640
  juneifolia, 645
  macrantha, 257
  nervosa, 624, 626, 635-6, 639, 646
  nevadensis, 630, 664, 777
  pratensis, 734-5
  pringlei, 622
  rhizomata, 365
  rupicola, 611, 628, 630, 636, 638, 647, 656-7, 659, 825-6, 924, 932
  sandbergii, 770, 774-6, 782, 784, 786, 790, 812, 929
  scabrella, 118, 443, 493, 495, 554, 639, 817, 900
  secunda, See P. sandbergii
  suksdorfii, 619, 628, 640, 932
  tenerrima, 118
  unilateralis, 735
Pocket gophers, See Thomomys
Podagrostis thurberiana, 623
Podistera nevadensis, 630, 635-6, 638, 642, 932
Podsol soils, 121, 305, 315, 337, 340
Pogogyne zizyphoroides, 523
Point Arena, 50-2, 145, 244, 305
Point Conception, 229, 275, 899
Point Cypress, 303
Point Dume, 476
Point Lobos, 89, 303, 348, 736
Point Mugu, 473
Point Reyes, 31, 33, 47, 49, 84, 209, 229, 233, 244, 248-51, 735, 737, 745, 748, 754
Point Saint George, 230
Point Sur, 745
Polemonium
  chartaceum, 659-600
  eximium, 619, 629, 932
  pulcherrimum, 622-4, 630-1, 636, 639, 718, 720
Pollution, air, 348, 510, 541-2, 569
Polygonum
  bidwelliae, 122
  daviseae, 616, 621-2, 643, 646, 651, 717, 719-20
  douglasii, 619, 777
  minimum, 620
  newberryi, 622
  paronychia, 238, 243, 249, 257
  phytolaccaefolium, 621-3, 708
  shastense, 611, 622, 647, 650
Polyploidy, 126-7, 529, 873
Polypodium
  californicum, 238, 753, 900, 903
  hesperium, 548
  scouleri, 897
Polystichum
  lonchitis, 621
  munitum, 237, 333, 363, 368, 443, 686-8, 748, 751, 916, 924-5, 938
Polytrichum juniperinum, 586
Ponderosa pine forest, 541-2, 561-3, 588, 592, 801-2, See also Pinus ponderosa, Yellow pine forest
Pool, D., 260
Poole, D.K., 466
Pope, R.M., 290

Populus
- angustifolia, 145, 153
- balsamifera, 146
- brandegeei, 163
- deltoides, 165
- fremontii, 161, 163, 165, 432, 880, 902, 927, 936
- grandidentata, 145, 161-2
- heterophylla, 161
- mexicana, 169
- tremula, 180
- tremuloides, 147-50, 165, 551, 553, 720, 800, 806-7, 809, 826
  - var. aurea, 153, 577, 583, 585
- trichocarpa, 145, 149, 152, 164, 168, 567, 575-6, 583, 585, 902
- wislizenii, 165

Porcupine, 348
Porter, D.M., 380, 414
Port Orford Cedar, See Chamaecyparis lawsoniana
Posey, C.E., 356
Potbury, S.S., 192
Potentilla
- breweri, 585, 611, 637, 640, 642, 648, 652
- diversifolia, 625, 932
- drummondii, 611, 624, 637, 640, 642, 648
- egedii, 238, 257
- flabellifolia, 623, 625, 641, 717
- flabelliformis, 611, 649
- fruticosa, 609, 634, 648, 650, 652
- glandulosa, 752
  - ssp. ewanii, 553
  - ssp. nevadensis, 631
  - ssp. pseudorupestris, 622
- pacifica, 273-5
- palustris, See Comarum
- pennsylvanica, 659
- pseudosericea, 627
- wheeleri, 653-4

Potter, G.L., 674
Potter, J.R., 261
Powell, W.R., 135, 220, 355-6, 413, 466, 731, 759
Prairie Fork, 543
Preston Peak, 621, 623, 729
Pride, C.R., 292
Prigge, B., 830
Primbs, E.R.J., 290
Primula suffrutescens, 611, 643, 647, 720
Proctor, J., 135, 356
Proctor, M.C.F., 669
Productivity, 280-1, 504-5, 568, 573-4, 827
Pronghorn antelope, See Antilocapra americana
Prosopis
- glandulosa, 785
  - var. torreyana, 872, 877-8, 882
- juliflora, 181-2, 185, 852-3, 933-4
- pubescens, 880, 935

Providence Mountains, 811-2, 814
Prunus
- andersonii, 180, 768, 774-5, 809
- davidiana, 146
- demissa, 145, 159-60, 314, 546, 807, 917, 923
- emarginata, 149, 159, 313, 428, 543, 565, 574, 590, 687, 711, 715, 717, 917, 920-2, 924
- fasciculata, 431, 816, 842, 847, 872
- fremontii, 431, 872
- ilicifolia, 165, 170, 174-5, 177, 302, 393, 401-2, 426, 428, 448, 926, 928
  - ssp. lyonii, See P. lyonii
- lyonii, 162-3, 168-9, 171, 899-900, 928
- prionophylla, 171
- subcordata, 313, 784
- virginiana, 314, 428, 543, 803
  - var. demissa, See P. demissa

Pseudotsuga
- glauca, 146
- macrocarpa, 82, 87, 309, 376-8, 538-9, 542, 918, 925, 927
- menziesii, 84, 89, 99, 104, 145, 147-53, 161, 164, 169, 244, 305, 309, 314, 326, 333, 343, 360-71, 374, 376, 405, 430, 561, 563, 567, 589, 680, 682-3, 685-6, 688-9, 691-3, 700-4, 708-9, 714, 721-6, 728-9, 753, 802, 916, 920-2, 924-5

Psilocarphus
  brevissimus, 520, 523
  oregonus, 525, 527
Psilostrphe cooperi, 842
Psoralea orbicularis, 552
Ptelea crenulata, 309, 314
Pteridium aquilinum, 333, 339, 344, 365, 368, 685-90, 707-8, 712, 714, 717, 755
    var. pubescens, 569, 735, 742-3, 748, 751, 916, 919, 938
Pterospora andromedea, 547, 574
Pterostegia drymarioides, 440
Pteryxia terebinthina, 623
Puccinellia pauciflora, See Torreyochloa
Pumice, 773
Purer, E.A., 260, 293, 466, 533
Purshia
  glandulosa, 811, 813-6, 844, 847, 856, 933
  tridentata, 153, 165, 170, 180, 314, 392, 431, 575, 577, 590, 592, 645, 771-6, 779-80, 784, 786, 793, 799, 801-3, 806, 809-10, 917-8, 920, 924, 929
Putram, W.C., 192
Pychanthemum californicum, 552
Pyrola
  asarifolia, 569
  minor, 569
  picta, 363, 569, 574, 577, 707-8, 712, 714, 717, 719-23
  secunda, 569, 574, 706, 712, 719, 923
Pygmy forest, 89, 93, 121, 315, 692, 919
Pygmy pine, See Pinus contorta ssp. bolanderi
Pyott, W.T., 72, 513

Q_{10}, 881
Quashu, H.K., 466
Queen, W.H., 293
Quercus
  agrifolia, 89, 151, 163, 165-6, 246-7, 302, 333-4, 343, 346, 360-1, 371-3, 377, 388, 392-4, 396, 398, 401, 403, 406-8, 481, 682, 684, 901-2, 919, 925, 927, 929
  x alvordiana, 392-3, 396, 402

Quercus (cont.)
  arizonica, 162, 164, 170
  borealis, 163
  brandegeei, 162
  chrysolepis, 84, 145, 148-51, 159-64, 168, 170, 216, 302, 309, 360-7, 369, 371, 373-8, 390, 393, 396, 401, 431, 449, 539, 542-4, 546, 564-5, 586, 590, 682, 703-4, 708, 723-5, 809, 813, 822-3, 902, 918, 922-3, 925, 927
  douglasii, 82, 87, 151, 161, 165, 169, 314, 386-7, 392-404, 406, 408, 494, 501, 518, 817-8, 926, 929-30
  dumosa, 165, 175, 302, 307, 309, 329, 341, 402, 422, 424-6, 431, 434, 447, 453, 476, 481, 721, 817, 899-900, 920, 928
  dunnii, 170, 174, 177
  durata, 123, 307, 309, 311, 314, 429
  eduardi, 162
  emoryi, 162, 164, 168
  engelmannii, 93, 163, 165-7, 170, 172, 402-3, 406, 408-9, 927
  falcata, 163
  gambelii, 153
  garryana, 145, 367, 393, 400-1, 403-5, 408, 589, 682-4, 687, 690, 802, 916-7, 924-5
    var. breweri, 314, 365, 711
  kelloggii, 59, 145, 159, 216, 314, 361-2, 364-5, 367, 372-4, 393, 396-7, 399-400, 403-4, 539, 541-3, 546, 561-5, 575-6, 586, 588-9, 592, 682-4, 690, 700-1, 703-4, 708, 723-5, 785, 808, 917-8, 921-5, 927
  lobata, 149-51, 165, 170, 309, 372, 375, 387-91, 393, 396, 400, 402, 404, 406, 408-9, 494, 926-7
  x macdonaldii, 901-2
  montana, 145
  palmeri, See Q. dunnii
  parvula, See, Q. wislizenii var. frutescens
  potosiana, 162, 167-8
  pseudolyrata, 161
  reticulata, 161, 163

Quercus (cont.)
rugosa, See Q. reticulata
sadleriana, 314, 366, 705, 710, 714, 716, 729, 920
tomentella, 165-6, 168, 901-2, 928
turbinella, 164, 167, 174-5, 392, 402, 424, 429, 431, 813-20, 822-9, 920, 922
vaccinifolia, 152, 309, 362, 365-6, 574, 586, 706, 710-1, 714-6, 719-22, 729, 920-1
vaseyana, 162, 165
virginiana, 161-3, 167
wislizenii, 82, 147-51, 161-5, 168, 170, 216, 310, 314, 372, 374-5, 386, 388, 390, 393-4, 397, 399, 401, 406, 408, 431, 449, 682, 684, 925-6, 929
var. frutescens, 331, 426, 428, 902, 928
Quick, A.S., 466
Quick, C.R., 357, 466
Quintus, R.L., 107

Rabbit Dry Lake, 851-2
Rabotnov, T.A. 672
Radiation, See Net radiation, energy balance
Rae, S.P., 672
Rafinesquia neomexicana, 875
Raillardella
argentea, 611, 624, 628, 636, 638, 642, 647, 653
pringlei, 622
Rain shadow, 36, 40, 144, 808
Ramaley, F., 260
Ramalina reticulata, 303, 305
Rancho Las Cruces, 99
Rancho Seco, 521, 523, 526
Randall, O.C., 73, 866
Randsburg, 838
Ratliff, R.D., 513, 593, 672-3
Ranunculus
alismaefolius, 611
californicus, 734, 900
eschscholtzii, 664, 720
ssp. oxynotus, 653
occidentalis var. dissectus, 664
Ranwell, D., 260, 293
Raven, P.H., 136, 131, 134-5, 132-3, 192, 466, 532, 556, 673, 831, 906
Raymond, F.M., 698
Readett, R.C., 667
Rea, K.H., 832
Red alder, See Alnus oregona
Red Bluff, 32, 57, 60-1, 64
Red cliffs, 88
Redfield, A.C., 293
Red fir, See Abies magnifica
Red fir forest, 4, 553, 571-4, 588, 593-4, 708-15, 726-7, 729-30, 803, 806
Redlands, 821
Red Mountain, 307
Redshanks, See Adenostoma sparsifolium
Redwood, See Sequoia sempervirens
Redwood forest, 3-4, 305, 307, 333, 340, 679-98, 916, 926
Redwood National Park, 84, 209
Reedgrass, See Calamagrostis
Reed, M.J., 759
Reichert, T., 358
Reid, E.M., 829
Reimann, L.F., 466-7
Reimold, R.J., 293
Reineke, L.H., 596
Reisigl, H., 673
Relevé, See Association Table
Relict, 110, 112, 119-20, 147, 152, 154, 156, 179, 182, 701, 727
Remote Sensing, 209
Rempel, R.J., 888
Renney, K.M., 192
Reno, 38, 40, 54, 56, 59
Repenning, C.A., 192
Reppert, J.N., 415, 511
Research natural areas, See Natural areas, Nature reserves
Respiration, 881
Retzer, J.L., 532
Reveal, J.L., 70, 132, 829, 831, 864
Reynoldson, F., 260
Reynolds, R., 598
Rhamnus
betulafolii, 153
californica, 145, 161, 164-5, 174, 244, 246, 310, 314, 344, 372, 407, 426, 428-9, 431, 712, 748, 750, 801, 813, 918-9, 925-6, 938
ssp. cuspidata, 542

Rhamnus (cont.)
var. occidentalis, 365-6, 721
var. tomentella, 314, 388
crocea, 174, 177, 302, 331, 334, 564, 751, 813, 816
ssp. ilicifolia, See R. ilicifolia
ilicifolia, 175, 302, 388, 426, 448, 457, 926, 928
insula, 169
pirifolia, 899-900, 928
purshiana, 150, 159, 333, 428, 709
rubra, 314
Rhizocarpon bolanderi, 586
Rhododendron
californica, 916
macrophyllum, 303, 305, 333, 368, 925
occidentale, 88, 567, 916
Rhus
diversiloba, 89, 165, 246, 305, 329, 334, 344-6, 362, 365, 368, 372, 374, 388, 407, 409, 426, 448, 457, 564, 588, 687, 738, 747-8, 750, 900, 916-7, 924-6
glabra, 153
integrifolia, 171, 178, 250-1, 257, 341, 451, 458, 472, 474-5, 483, 898, 900, 903, 905, 928-9, 937
lanceolata, 165, 168, 170
laurina, 166, 168, 171, 178, 257, 302, 316, 322, 341, 422, 451, 474-5, 482-3, 900, 928-9
ovata, 164, 166, 168, 170, 174, 302, 329, 422, 425-6, 429, 451, 453, 458, 481-2, 820, 900, 928
trilobata, 314, 431, 820
var. anisophylla, 813, 822
Ribes
amarum, 928
aureum, 146, 553, 786
californicum, 928
cereum, 146, 170, 548, 550, 553, 579, 611, 628, 633, 635-6, 639, 644, 652, 662, 664, 774, 822-6, 922-4
indecorum, 928
lacustre, 705, 709-10, 716
lobbii, 705
malvaceum, 177, 928
Ribes (cont.)
menziesii, 374
montigenum, 548, 553, 578, 611, 624, 647, 650, 653-4, 923-4
nevadense, 546-7, 923
roezlii, 149, 546, 550, 553-4, 590, 918, 922-3
sanguineum, 574, 705, 752
speciosum, 177, 334, 751
velutinum, 774-5, 799, 801, 806, 847
ssp. glanduliferum, 822-3, 922
viburnifolia, 169, 928
viscosissimum, 574, 711, 716, 719, 804, 807, 923
Rible, J.W., 759
Richards, L.G., 598
Richardson Bay, 100
Richardsons Grove, 32
Rickabaugh, H., 831
Rickard, W.H., 866
Righter, F.I., 357
Rim of the World, 543
Riparian, 88, 92, 99-100, 148-9, 163, 279, 367, 374, 388, 405-6, 431, 553, 567, 569, 693, 884, 901-2, 904-5, 927
Ripley, S.D., 136
Robbins, W.W., 136, 513
Roberts, M.R., 134
Roberts, R.W., 513
Robertson, J.H., 796
Robichaux, R.H., 258
Robinia neomexicana, 164-5
Robinson, J.P., 513
Robinson, J.T., 467
Robinson, W., 356
Rock Creek, 614
Rockey, R.E., 381, 732
Rocky Mountains, 606, 610-11, 614, 827
Rock Valley, Nevada, 838, 849
Rodin, L.E., 73
Rodman, J.E., 258, 260
Rodman Mountain, 856
Roemer, H., 667
Rogers, B., 467
Rogue, 405, 421
Rollins, G.L., 293
Rollins, R.C., 673
Romanzoffia sitchensis, 621
Romancier, R.M., 107
Romneya tricocalyx, 112, 302

Romney, E.M., 867
Root systems, 406, 482, 606, 846, 850, 874-5, 878
Rorippa nasturtium aquatica, See Nasturtium officinale
Rosa
 californica, 302
 gymnocarpa, 314, 365, 368, 590, 705, 708, 710, 714, 920, 922
 minutifolia, 474-5
 spithamea, 564
 woodsii, 786, 823
Roseman, 305
Rosenthal, M., 467
Rosenzweig, M.L., 73
Roth, C., 668
Rouse, G.E., 192
Rouse, W.R., 73
Rowe, P.B., 461, 467
Roy, D.F., 356, 381, 698
Rubtzoff, P., 133
Rubus
 californicus, 710
 leucodermis var. bernardinus, 553
 parviflorus, 24, 546, 567, 688, 705, 714, 747, 920, 923, 938
 spectabilis, 686, 688, 747, 916
 ursinus, 344-5, 705, 742-3, 747-8, 751
 vitifolius, 239, 246, 365, 472, 686, 688, 742-3, 747-8, 916, 919, 927, 938
Ruby Mountains, Nevada, 614
Rumex
 acetosella, 443, 735, 742, 753
 crassus, 257
 crispus, 127, 279, 497
 paucifolius, 641, 663
  ssp. gracilescens, 637, 658, 660
 pulcher, 736
Rundel, P.W., 598, 862
Rune, O., 136
Runyon, E.H., 888
Rupicoles, 554
Russell, I.C., 796
Russell, R.I., 73, 260
Russian Peak, 104, 709, 714, 719, 727, 729-30
Russian River, 265, 736
Rust gall, See Peridermium harknessii
Ryan Oak Glen, 93
Rydberg, P.A., 796

Sabal uresana, 163
Sacramento, 34-5, 54-5, 58
Sacramento Mountains, 93, 846, 851
Sacramento Valley, See Central Valley
Sagebrush, See Artemisia tridentata and Artemisia californica
Sagebrush steppe, See Great Basin sagebrush
Sagina saginoides, 623, 664
Sagittaria cuneata, 551
Saint Helena, 34, 49, 147
Salazaria mexicana, 817, 842-4, 846-7, 856, 858-9, 932
Salicornia
 bigelovii, 273, 276-9, 282-3, 285, 931
 europaea, 273-4
 subterminalis, 273, 276, 278, 282-5, 287, 494, 851, 903, 931, 935
 virginica, 271, 273, 275-6, 278, 280-5, 287, 903, 931
Salinas River, 36, 235
Saline Valley, 66, 847-8
Salinity, See Soil salinity
Salisbury, E., 260
Salix
 anglorum, See S. arctica
 arctica, 609, 624-5, 628, 633, 640, 643, 649, 652-3
 brachycarpa, 122, 613, 625, 629, 634-5
 caudata, 145
 commutata, 716
 coulteri, 546, 923
 drummondiana, 620
 eastwoodiae, 620, 643, 648, 652, 923
 exigua, 164
 gooddingii, 164, 880, 927, 935
 hindsiana, 279, 432
 hookeriana, 145, 239, 243
 jepsonii, 923
 laevigata, 902, 927, 935
 lasiandra, 145, 164-5
 lasiolepis, 165, 244, 257, 279, 432, 553, 902, 936
 lemmonii, 652

Salix, (cont.)
lutea, 923
mackenziana, 620
melanopsis, 164
nivalis, See S. reticulata
orestera, 620, 635, 652
planifolia, 620, 652
pseudocordata, 620
reticulata, 609, 625, 629, 633
scouleriana, 145, 565, 574, 705, 709-10
Salmon Creek, 684
Salsola
iberica, 770, 790
kali, 250, 494, See also S. iberica
paulsenii, 790
pestifera, 850
Saltbush, See Atriplex polycarpa
Saltbush scrub, 5, 510, 837, 846, 850-3, 862, 874, 878-9
Salt Canyon, 496
Salt marsh, 3, 4, 91, 93, 97, 100, 263-94, 341, 895, 903-4, 931
low, 285-7
high, 285-7
Salt marsh moth, See Estigmene
Salton Basin, 870-1
Salton Sea, 880, 934-5
Salt spray, 225, 303, 337, 340, 342, 691, 755, 895
Salvia
apiana, 302, 341, 422, 448, 472, 476, 479, 482, 928, 929
brandegei, 903
carnosa, 784
clevelandii, 302
columbariae, 440, 861
dorrii, 41, 813-4, 817, 856, 859
eremostachya, 872
leucophylla, 171, 472, 479-80, 929
mellifera, 247-8, 302, 322, 329, 341, 422, 443, 448, 472, 474, 476, 479, 481, 483-4, 748, 750, 898, 928-9
munzii, 302, 474-5
pachyphylla, 822
sonomensis, 311, 314
spathacea, 335, 443, 928
vaseyi, 872, 874
Sambucus
coerulea, 546, 823, 923
melanocarpa, 709, 716
mexicana, 823, 902, 922
microbotrys, 548
Samoa, 235
Sampson, A.W., 467, 513
San Andreas Fault, 140, 159, 162-163, 836, 877-9
San Antonio Canyon, 541, 819
San Antonio Mountains, 554-5, 652
San Benito Forest, 919
San Bernardino Mountains, 50, 92, 325, 376, 538-9, 541, 543, 546-50, 553, 555, 582, 613-4, 623, 652, 807, 811, 814, 836, 857, 917-8, 924
San Bernardino National Forest, 541, 543, 545
San Bernardino Valley, 819
San Clemente Island, 185, 894, 896-901, 904
San Diego, 33, 37-8, 241, 265-7, 341, 894
San Diego Bay, 265, 276, 279-81, 474
San Dieguito Lagoon, 265
San Dimas, 420
Sandford, G.R., 796
Sando, R.W., 597
Sand Ridge Wildflower Preserve, 99
Sandstone soils, 315-6, 321, 476
Sand verbena, See Abronia
San Elijo Lagoon, 265
San Francisco, 25-6, 235, 244, 265-7, 271, 273-7
San Francisco Bay, 100, 265, 286
San Gabriel Mountains, 377, 419-20, 538-50, 552, 554, 652, 807, 811, 814-5, 924, 927
San Gabriel River, 544, 819
Sangamon interglacial, 224
San Gorgonio Wilderness, 553
Sanicula
arctopoides, 238
arguta, 443, 900, 902
bipinnata, 748, 751
crassicaulis, 334, 344, 443
laciniata, 344
San Jacinto Mountains, 93, 538-9, 541, 543, 545, 547-59, 552, 555, 614, 818-9, 870, 874, 918, 924
San Joaquin Experimental Range, 33, 63, 82, 399, 407, 500, 502, 505, 510

San Joaquin River, 810
San Joaquin soil series, 517
San Joaquin Valley, 93, 494, 851
San Jose Creek, 303
San Juan River, 818
Sankary, M.N., 866
San Luis Obispo, 328
San Mateo, 265, 279
San Miguel Island, 894, 896-8, 904
San Nicholas Island, 894, 896, 898-9, 903-4
San Pablo Bay, 275, 279, 281
San Pedro Mártir, 540, 542, 819
San Pedro Point, 897
San Rafael, 49
San Savaine Lake, 541, 543
San Telmo, 485
Santa Ana Mountains, 298, 300, 328, 331, 348, 538, 540, 821
Santa Ana River, 821
Santa Ana winds, 177, 316, 325
Santa Barbara Island, 894, 897-9, 904
Santa Catalina Island, 37, 39, 894, 897, 899, 902-5
Santa Catarina, 300
Santa Clara, 265, 492
Santa Cruz area, 91, 750-3
Santa Cruz cypress, See *Cupressus abramsiana*
Santa Cruz Island, 94, 337, 347, 402, 894, 896-905
Santa Cruz Mountains, 394
Santa Cruz pine, See *Pinus remorata*
Santa Lucia Mountains, 316, 360-1, 372, 374, 389-90, 392, 395, 401-2, 539
Santa Margarita, 265, 279
Santa Maria River, 265
Santa Monica Mountains, 93, 99, 372, 453, 477, 538, 821
Santa Rosa Island, 337, 341, 896, 899-905
Santa Rosa Mountains, 539, 547, 549, 551, 554, 582, 819, 823, 870, 874, 923
Santa Rosa Plateau, 402
Santa Ynez, 265, 538
San Vicento, 332, 429
Sapindus drummondii, 164-5, 169-70
Sap tension, See Moisture Stress
Saratoga Springs, 852
Sarcobatus
baileyi, 41, 788, 790-1, 854-5, 934
vermiculatus, 788, 790, 793, 810, 851, 934-5
Sarcodes sanguinea, 112, 542, 547, 550, 569, 574, 923
Sargent, C.S., 356
Sargent cypress, See *Cupressus sargentii*
Satureja douglasii, 344, 748, 751, 938
Savage, W., 673
Savanna, 375-6, 384-5, 387-91, 395, 404, 407-9, 494, 501, 518, 926-7, 929-30
Savelle, G.D., 514
Saver, R.H., 136
Save-the-Redwoods League, 85
Sawmill Mountains, 538
Sawyer, J.O., 356, 732, 915
Saxifraga
aprica, 609, 625, 637, 640, 646, 650
bryophora, 620, 625
caespitosa, 613, 621
ferruginea, 621, 717
fragorioides, 621
fragosa, 621
nidifica, 623, 626
tolmiei, 621-2, 720
Schimper, A.F.W., 467
Schismus
arabicus, 494
barbatus, 861
Schlinger, E.I., 759
Schmaltzia ovata, See *Rhus ovata*
Schmid, M.J., 831
Schmid, R., 467, 831
Schmidt, L., 670
Schmidt, U.M., 673
Schoenolirion album, 621, 715
Schonchin, 84
Schorr, P.K., 468
Schreiber, B.O., 469
Schröter, C., 673
Schubert, G.H., 596
Schulman, E., 831
Schulman Grove, 825-6
Schultz, A.M., 133, 354, 467-8, 597, 697
Schulze, E.D., 673, 831, 887

Schumacher, F.X., 593
Schwarz, W., 673
Scirpus
  acutus, 274-5, 931
  americanus, 275, 931
  californicus, 274, 279, 284, 903, 931
  clementis, 625, 648
  criniger, 648
  microcarpus, 715
  olneyi, 274, 279, 853, 878, 931
  pumilus, See Trichophorum pumilum
  robustus, 274-5, 931
  rollandii, See Trichophorum pumilum
  validus, 931
Sciurus griseus, 337-8
Scopulophila, 117-8
Scott, C.W., 356
Scott, D.B., 293
Scott, F.M., 888
Scott Mountains, 802
Scott Valley, 802
Scree slope, 554, 619
Scripps Shoreline Underwater Reserve, 93
Scrophularia californica, 237, 244, 335, 443, 742, 748, 751, 938
Scrub, See Dune scrub, Coastal scrub, Coastal sage scrub, etc.
Scutellaria
  austinae, 540, 927
  tuberosa, 443
Sea level, changes, 224
Sea Ranch, 739, 741-3
Searey, K.B., 467
Sedge, See Carex
Sedimentation, 268-9
Sedum
  lanceolatum, 645
  niveum, 548, 554
  obtusatum, 712, 718-9
  rosea, 621, 637, 641, 648, 650
    ssp. integrifolium, 609
  stenopetalum ssp. radiatum, 631
Seed dispersal, 458
Seedlings, See Germination
Seed longevity, 503-4
Seed numbers, 503
Seeman, E.L., 356
Seep, 851
Selaginella
  asprella, 544, 554
  bigelovii, 554, 900
  densa var. scopulorum, 718
  watsoni, 626, 636, 639, 642, 652, 656, 658
Sellers, J.A., 598
Sellers, W.D., 73
Semion, V.P., 355
Seneca, E.D., 293-4
Senecio
  arnicoides, 752
  blochmanae, 249
  canus, 624, 645
  douglasii, 435, 443
  fremontii, 611, 621, 624
  integerrimus var. major, 712, 718
  ionophyllus, 550
  pattersonianus, 663
  spartioides, 924
  triangularis, 552, 648, 652, 711, 717
  vulgaris, 440
  werneriaefolius, 627-8, 636, 638, 662
Sequoiadendron giganteum, 63-4, 112, 143, 148-52, 156, 161, 560, 563, 565-6, 569-70, 587, 593, 922
Sequoia National Park, 63, 84, 209, 397, 561, 563, 569, 570-1, 575-6, 578-80, 585
Sequoia sempervirens, 34, 83-4, 88-9, 92, 97, 99-100, 112, 118, 140, 145, 147, 150, 160-1, 164-5, 169, 172, 175, 213, 308, 310, 360, 367-8, 372, 374, 377, 404, 406, 679-8, 916, 926
Serpentine, serpentinite, 83, 89, 94, 122-3, 125, 152, 305, 307, 311, 315, 326-9, 392, 422, 424, 429-31, 455, 458, 532, 544, 592, 692-3, 701, 721, 735-6, 818, See also Ultrabasic soils
Settlers, effect on vegetation, See California history
Shade tolerance, 343, 364, 684-90, 726-7
Shadscale, See Atriplex confertifolia
Shadscale scrub, 837, 850, 853, 855, 858

Shale, 476
Shanholtzer, A.F., 293
Shantz, H.L., 467, 514, 888, 915
Sharp, R.P., 732, 866
Sharsmith, C.W., 598, 673
Sharsmith, H.K., 192, 467, 831
Shasta Dam, 32
Shasta fir, See Abies magnifica var. shastensis
Shasta Mudflow, 82
Shasta Valley, 802
Shaver, G.R., 673
Shaw, E.A., 673
Sheep, See Livestock
Sheep Canyon, 88
Sheep Mountains, 814, 819, 820-1
Sheetiron soil series, 362
Shefi, R.M., 866
Shelford, V.E., 381
Shellhammer, H.S., 596
Sheps, L.O., 866
Sherardia arvensis, 741
Sherrell, A.G., 381, 732
Sherwin Grade, 857
Shikhlinskii, E.M., 673
Shingletown, 818
Shinn, C., 414
Shiyatov, S.G., 673
Shreve, F., 136, 192, 381, 476, 489
Shropshire, F., 464
Shore pine, See Pinus contorta ssp. contorta
Show, S.B., 467, 598
Sibbaldia procumbens, 548, 609, 615, 623, 625, 632, 641-2, 649-53, 661, 663-4, 710
Sidalcea malviflora, 900
Sida leprosa ssp. hederacea, 552
Siegmund, J., 673
Sierra Juarez, 300, 819
Sierra Nevada Mountains, 53-64, 525, 327, 397-400, 427-8, 538-9, 546-7, 553, 559-87, 593, 609-15, 624-54, 764-5, 767, 784-5, 798, 800, 808-11, 815, 818, 856-7
Sierra Peak, 298, 300, 316-721, 324, 348
Sierra Valley, 765
Silene
  douglasii, 624-5, 664
  gallica, 741
  grayi, 621-2, 624
Silene (cont.)
  laciniata, 900
  lemmonii, 542, 554
  multinervis, 440
  parishii, 554, 652, 654, 920
    var. latifolia, 546
  sargentii, 622, 626-7, 639, 647, 651, 663
  verecunda, 548
  watsonii, 611
Silver Creek, 806
Silver Strand, 233
Silverwood, 100
Silybum marianum, 503
Simberloff, D.S., 533
Simmondsia chinensis, 124, 185, 250-2, 341, 475, 872, 874, 880-1, 885, 933, 937
Simpson, G.G., 192
Sinclair, J.D., 467
Singh, T., 866
Sink, See Alkali sink
Siskiyou cypress, See Cupressus bakerii ssp. matthewsii
Siskiyou Mountains, 311, 315, 325
Sisymbrium altissimum, 770, 884
Sisyrinchium
  bellum, 552, 734, 742, 900, 930
  eastwoodiae, 553
Sitanion
  hystrix, 493, 542, 548, 574, 577, 622, 624-6, 628-9, 632, 635-7, 639, 644, 653, 656, 659, 662, 664, 722, 737, 770, 774-6, 782, 784-5, 790, 812, 825-6, 917, 920, 924, 929-30, 932, 934
  jubatum, 365, 930
Site Index, 200, 217, 568, 573
Sitka spruce, See Picea sitchensis
Slate Creek, 624
Slope aspect, See Aspect
Small, E., 462, 488
Smelowskia ovalis var. congesta, 624
Smilacina
  racemosa var. amplexicaulis, 706, 711, 713
  stellata, 706, 711
Smiley, F.J., 598, 673
Smith, A.C., 556
Smith, B.F., 413, 831
Smith, B.S., 697
Smith, C.F., 260, 293, 557

Smithe Redwoods, 89
Smith, F.H., 261
Smith, G.L., 136, 598
Smith, H.V., 192
Smith River, 265, 275,
Smog, See Pollution
Smoke tree, See Dalea spinosa
Snajberg, K., 355
Snow, 12-3, 44, 56-7, 150-1, 158, 426-7, 562, 571-2, 578-9, 602, 604-5, 609, 615-7, 621-3, 625, 628, 635, 637, 642-3, 650-3, 661, 664, 715
Snow, G.E., 414
Snow Mountain, 701
Snow Spring, 539
Sochava, V., 9
Society of American Foresters, 381, 414, 732
Soda Springs, 54
Soderstrom, T.R., 136
Soft chaparral, See Coastal sage scrub
Soggy Dry Lake, 852
Soil
- aeration, 837, 874
- Conservation Society of America, 107
- development, 82
- field capacity, 19
- general, 305, 307, 311, 315-6, 327-8, 337, 346, 349, 860, 875, 897, 900-1
- mapping, 201-6, 210-8
- moisture, 367, 487, 496, 516, 530, 801, 841, 489-50, 876
- nutrient content, 245, 272, 315-6, 327-8, 337, 346, 395, 506-7, 826, 850, 875
- organic matter, 269, 495, 509-10
- pH, 316, 340, 518, 520, 849
- physical attributes, 268-9, 495, 502-3, 507, 510, 569, 799, 803, 809, 818, 837, 845-6
- salinity, 225, 271-2, 285, 837, 846, 850-3, 873-4, 878, 880, 884
- survey staff, 796
- texture, 853, 858, 861, 873-4, 876-8
- vegetation and timber stand maps, 205-6, 218
- vegetation survey, 201-6, 210-8

Sokhadze, E.V., 673
Solanum
- douglasii, 257,
- nigrum, 250
- nodiflorum, 237
- umbelliferum, 126, 335
- wallacei var. clokeyi, 903
- xantii, 443, 542, 918

Solar radiation, See Energy balance
Solidago
- californica, 553
- multiradiata, 633, 637, 641, 648
- spathulata, 237, 243, 257
- triangularis, 708, 920

Solodized solonetz, 784
Sonchus asper, 127, 497
Song, L., 464
Sonora, 62
Sonoran Desert, 37, 41, 43, 66-8, 92, 118, 120, 166, 179-85, 298, 430, 539, 862, 869-85, 933-4, 938
Sonora Pass, 806
Sonora (State of, Mexico), 871, 878
Sorbus
- californica, 709-10, 716
- scopulina, 148

South Dakota, 547
Southern California Islands, 4, 94, 298, 347, 893-905
Southern oak woodland, See Oak woodland
Spalding, V.M., 888
Sparganium
- emersum, 551
- minimum, 620

Spartina foliosa, 271, 273-5, 278-9, 284-5, 329, 931
Spatz, G., 673
Specht, R.L., 467, 523-4, 529
Speciation, 123, 126-7, 507, 529
Species richness, diversity, 520-1, 745-6, 748
Spergularia
- macrotheca, 51, 273
  - var. macrotheca, 898, 903
- marina, 273-4, 279
- rubra, 622

Spermophilus beecheyi, 498
Speth, J., 292

Sphaeralcea
ambigua, 774. 842, 933
coccinea, 775, 790, 817
rosea, 820
Sphagnum, 340
Sphenosciadeum capitellatum, 552, 623
Spiraea densiflora, 152
Sporobolus
airoides, 494, 852, 877, 935
cryptandrus, 774
Springfield, H.W., 829
Sprouting, 123-5, 175, 476, 769, 858, 875
Spruce, See Picea, Pseudotsuga macrocarpa
Squaw Butte, 766
Squaw Valley, 53
Squirrel, California Grey, See Sciurus griseus
Squirrel tail, See Sitanion hystrix
Stachys
bullata, 336, 344
chamissonis, 334
rigida, 748, 751, 938
Standing, J., 293
St. Andre, G., 672, 796, 831-2
Stanyukovich, K.V., 674
Stark, N., 599, 866
Star Ranch, 100
State Cooperative Soil-vegetation Survey, See Soil vegetation survey
Stebbins, G.L., 9, 103, 108, 131-2, 136, 356-7, 467, 514, 533, 674, 732, 888
Stebbins, R.C., 10, 831, 865-6
Steenbergh, W.F., 888
Steens Mountains, 665
Steinoff, H.W., 796
Stellaria
crispa, 623
jamesiana, 707
umbellata, 609, 615, 626
Stenotopsis linearifolius, 185
Stephanomeria
cichoriacea, 554, 900
lactucina, 624
pauciflora, 841-2, 846
virgata, 440
Stephenson, S.N., 293
Steppe, 3, 5, 604, 608, 615, 629, 650-1, 654, 656, 664, 763-96
Sterling, E.A., 467
Sternberg, L., 866
Stevens, R., 795
Stevenson, R.E., 293
Stewart, G.R., 294
Stipa
californica, 624, 631, 644
cernua, 495, 930
columbiana, 774, 777
comata, 630, 633, 771-5, 779, 782, 810
coronata, 817, 819, 821, 919-20, 930
lemmoni, 365, 779
lepida, 87, 344, 443, 735, 930
lettermanii, 771, 773-4, 777, 786
occidentalis, 577, 585, 611, 623, 626, 629, 633, 653, 771, 773-5, 785-6, 916
parishii, 544, 500
pinetorum, 611, 628, 630, 633, 656, 658, 825
pulchra, 87, 492-6, 500, 523 735-6, 930
speciosa, 41, 771, 774-5, 779, 817-9, 821, 842-3, 856, 858, 920
thurberiana, 771-6, 779-80, 786, 810, 929
St. John, R., 698
St. John, T.V., 598
Stocking, S.K., 467
Stockwell, P., 357
Stoddard, C.H., 357
Stoddart, L.A., 795
Stomatal resistance, 452, 616
Stone, C.O., 353, 357, 413
Stone, D., 533
Stone, E.C., 259, 468, 698
Stone Lagoon, 275
Storer, T.I., 599
Storie, R.E., 219-20, 381
Stovepipe wells, 838
Strain, B.R., 69, 672, 831, 863, 865-6, 886-8
Strand, See Beach, Coastal dune
Strand, S., 599
Streptanthus
bernardinus, 542, 918
cordatus, 664
tortuosus, 376, 586
var. orbiculatus, 622, 631

Streptopus amplexifolius
var. denticulatus, 711
Strogonov, B.P., 294
Stromberg, L.K., 532
Strotham, R.O., 381
Strother, J.L., 73
Stuart's Point, 739, 750-3
Stylocline micropoides, 861
Styrax
occidentalis, 175
officinalis, 172, 310
Suaeda
californica, 273-6, 278, 284, 898, 903, 931
depressa, 257, 934
fruticosa, 494, 852-3, 934
torreyana, 851, 858, 880. 935
Subalpine forest, 4, 154-6, 547-51, 560, 579-83, 593-4, 615-8, 715-21, 730, 798, 806, 823-7, 922
Succession, 242-4, 246-7, 324, 345, 368-9, 373-4, 407-9, 425, 427, 431-9, 476-8, 499, 501, 506, 508-9, 575, 585, 587, 590, 593, 602, 660, 684-5, 688-90, 708, 723, 737-8, 745, 747-8, 754, 756, 767, 769-70, 841, 852, 874
Succulent, 474, 485, 554, 850, 876, 884, 905
Sudworth, G.B., 357, 599
Sugarloaf Peak, 550
Sugarpine, See Pinus lambertiana
Sugar Pine Point, 88
Suisun, 266, 281
Summerfield, H.B., 796
Sumner, D.C., 574
Sundahl, W.P., 697
Sunset Bay, 265
Surprise Valley, 788
Susanville, 57, 60-1, 803
Sutherlin soil series, 394
Sutter Buttes, 400
Swallenia, 111, 117-8
Swanson, C.J., 414
Swarthout Valley, 543
Swarts, S.W., 674
Sweeney, J.R., 468
Sweetwater Mountains, 613-4, 654, 661-2, 664, 810, 823
Swift moth, See Hepialus
Symphoricarpos
acutus, 590, 804, 922
alba, 753
hesperius, 705, 708, 710
longiflorus, 774, 777, 786, 822-3, 847, 924
mollis, 302, 329, 331, 344, 346, 373-4, 705, 901
parishii, 428, 543, 917
racemosus, 145
rotundifolius, 774
vaccinioides, 314, 574, 644, 774, 784, 786, 803-4, 807, 919, 923
Syvertsen, J.P., 260, 414, 468
Szareck, S.R., 888

Table Mountain, 317, 544
Taeniatherum asperum, 498, 503, 507, 509-10, 737, 770, 777, 781
Tahachapi Mountains, 538, 547
Tahquitz Peak, 550
Taiga, 16, 180
Talbot, M.W., 511, 514
Talley, S.N., 381
Talus, 554, 561, 583, 619, 642, 725, 784
Tamarack, 56, 59-60
Tamarisk, See Tamarix
Tamarix chinensis, 880-1, 935
Tanacetum
camphoratum, 244
canum, 825
douglasii, 237, 257
Tan oak, Tanbark oak, See Lithocarpus densiflora
Taraxacum officinale, 663
Tarrant, R.F., 381
Tauschia
arguta, 331
parishii, 554
Tausch, R.J., 832
Taxus brevifolia, 104, 150, 152, 154, 363, 366, 565, 590, 682, 703-4, 709, 714, 723-4, 922, 925
Taylor, B., 357
Taylor, D.W., 108, 674, 832
Taylor, R.F., 357
Teal, J., 294
Teal, M., 294

Tecate cypress, See Cupressus forbesii
Tecate Peak, 298, 316-21, 324
Tehachapi Mountains, 494, 541, 815, 827, 836, 857
Telescope Peak, 824-5
Temblor Range, 516-7, 818
Ten Mile River, 233
Terjung, W.H., 674
Termo, 42
Terpenes, 486
Terry, E., 465
Tetracoccus ilicifolius, 117, 122
Tetradymia
  axillaris, 645, 814, 844, 856, 920, 932
  canescens, 543, 645, 650, 768, 774-6, 779, 786, 917, 924
  comosa, 185
  glabrata, 775, 790, 934
  tetrameres, 41
Tevis, L., 866, 889
Texas, 297
Thalictrum
  alpinum, 625, 629, 634
  fendleri, 152, 648
  polycarpum, 708
Thamnosma montana, 813, 820, 843, 846, 848, 854, 859-60, 872, 874-6, 933
Thellypobium lasiophyllum, 440
Thermoperiodism, 453, 691
Thilenius, J.F., 414
Thlaspi
  alpestre, 152, 621
  arvense, 774
  montanum var. montanum, 664
Thomas, J.H., 136, 357, 381, 414, 468
Thomas, M.C., 358
Thomas, R.W., 219
Thomomys, 407, 410, 498, 650-1
Thompson, H.J., 466, 556, 831
Thompson, J.H., 829
Thompson, K., 414
Thompson, W.W., 291
Thornburgh, D.A., 356, 732
Thorne, R.F., 10, 413, 464, 468, 532-3, 557, 832, 867, 906
Thorn scrub, 163-4, 176, 181-4
Thornthwaite, C.W., 73, 260
Throop Peak, 549, 553
Thuja plicata, 145, 368, 680, 682, 685, 689, 916
Thysanocarpus lacinactus, 440
Tidal marsh, See Salt marsh
Tides, 269-70, 278, 286
Tidestromia oblongifolia, 853
Tijuana River, 265, 285
Tikhomirov, B.A., 73
Tillaea
  aquatica, 523, 528, 552
  erecta, 441, 525
Timber croplands map, 201
Timbered Crater, 311
Timberline, 64, 547, 580, 602, 605, 609, 615-8, 622-3, 652, 923
Timberline Station, 62
Timber Mountain, 543
Timber stand-vegetation cover maps, 201-5
Ting, J.P., 887-8
Ting, W.S., 189
Tinnin, R.O., 514
Tisdale, E.W., 794, 829
Titus Canyon, 854
Titus, S., 219
Tobiessen, P.L., 889
Tolmachev, A.I., 674
Tomales, 265, 755
Tomaks Bay, 91, 265
Toomes soil series, 400
Topax Lake, 803, 809
Topography, 6
Toro Peak, 551
Torrell, D.T., 512
Torreya californica, 150, 152, 565, 682
Torreyochloa pauciflora, 620
Torrey Pine, See Pinus torreyana
Torrey Pines State Park, 88-9
Towle, 58
Towns Pass, 854
Toxicodendron, See Rhus diversiloba
Trancas Beach, 234
Tranguillini, W., 666, 674
Transmontane Floristic Province, 797-828
Transpiration, 45, 451, 616-7, 827, 849-50
Transverse Ranges, 537-55, 798, 808, 817, 823, 836, 919, 927
Trappe, J.M., 107
Trentepohlia aurea, 303, 305
Trewartha, G.T., 73

Trichophorum pumilum, 122, 613, 625, 634-5
Trichoptilium incisum, 861
Trichostoma
  lanatum, 302, 329, 444
  parishii, 302
Trientalis latifolia, 706, 711
Trifolium
  albopurpureum, 930
  amplectens, 930
    ssp. truncatum, 524
  barbigerum, 523
  ciliolatum, 445
  cyathiferum, 523
  depauperatum, 519, 523, 930
  dubium, 741
  fucatum, 523
  gracilentum, 930
  gymnocarpum, 776
  longipes, 708
  macrocephalum, 505, 776, 800, 930
  microdon, 930
  monanthum, 625-6
    var. grantianum, 553, 634
  monoense, 656-8, 660
  olivaceum, 930
  palmeri, 897
  pratense, 735
  tridentatum, 440
  variegatum, 505, 523
  wormskjoldii, 273
Triglochin
  concinna, 273-4
  maritima, 273-5, 277-9, 284, 551, 931
Trillium ovatum, 703-6, 709, 711
Trinidad head, 230, 236-40, 242-5
Trinity Alps, 93, 614, 702, 720-1
Trisetum
  canescens, 365, 736
  cerunum, 734
  spicatum, 553, 585, 609, 623-4, 633, 641, 645, 650, 653, 664
  wolfii, 664
Triteleia dudleyi, 540
Tropical rain forest, 17
Troughton, J., 488, 865
Truckee, 54, 58
True, G.H., 353, 357, 759
Tsuga
  heterophylla, 84, 145-6, 154,
Tsuga (cont.)
  305, 361, 368, 682, 693, 916
  mertensiana, 82, 104, 148, 153-4, 161, 560, 572, 579, 590-1, 618, 621-3, 643, 647, 702, 708-9, 714-6, 719-21, 723-9, 806, 824, 920-1, 923
Tucker, J.M., 136, 414, 467
Tucson, 871
Tueller, P.T., 794, 796, 829, 832
Tuff, 315
Tugai, 21
Tule elk, See Cervus elephus nannodes
Tule Elk area, 89
Tule Lake, 48, 792
Tule marsh, 5, 36, 494, 510, 931, See also Fresh water marsh, see also Scirpus
Tuolumne Meadows, 614
Tuolumne River, 810
Turnage, W.V., 889
Turner, H., 674
Turner, R.M., 865, 887, 889
Turricula parryi, 444, 554
Twenty-nine Palms, 33, 66
Twisselmann, E.C., 357, 414, 468, 514, 533, 557, 832, 867
Typha
  domingensis, 279, 903, 931
  latifolia, 274-5, 279, 931
Tyson, B., 488-9
Tyson soil series, 367, 404, 406, 690

Ukiah, 28-30, 36, 50, 53
Ulmus
  alata, 161
  americana, 161, 162
Ulrich, R., 69
Ultrabasics, 123, 152, 364-7, 700-1, 721, 729-30, See also Serpentine
Umbellularia, 172
  californica, 97, 112, 114, 147, 161-3, 165, 310, 360, 365-6, 373, 375, 377, 393, 401, 682, 684, 753, 916, 925
Ungar, I.A., 294
Ungnadia speciosa, 161
United States Bureau of Land Management, Natural Areas, 80-1

United States Bureau of Land Management, Vegetation mapping, 208
United States Federal Committee on Research Natural Areas, 108
United States Fish and Wildlife Service, Natural Areas, 81
United States Forest Service, Natural Areas, 79-80, 82-3
United States Forest Service, Vegetation mapping, 207-8
United States Park Service, Natural Areas, 81, 84
United States Park Service, Vegetation mapping, 208-9
United States Weather Bureau, 73
University of California, Natural Areas, NLWRS, 90, 92-5
University of California, Remote Sensing, 209-10
Unsicker, J.E., 381
Upper Lake, 52
Upson, J.E., 294
Urtica holosericea, 927
Utah, 825, 854, 862
Utah juniper, See *Juniperus osteosperma*
Uttinger, H., 74

Vaartaja, O., 357
Vaccinium
  arbuscula, 710, 715-6, 719-20
  confertum, 171
  membranaceum, 621, 705, 710, 716
  nivictum, 579, 584-5, 623, 625, 641-2, 649, 652, 923,
  occidentale, 579, 923
  ovatum, 171, 244, 257, 303, 305, 333, 335, 344-5, 362, 365, 368, 686-7, 903, 916, 919, 925
  parvifolium, 145, 150, 333, 686, 721, 916
  scoparium, 621, 716, 719
Vahlodea atropurpurea, 621
Valentine Eastern Sierra Reserve, 94
Valeriana
  capitata, 624
  sitchensis, 621, 708, 920
Vallecito Mountains, 819
Valiels, I., 294
Valley oak, See *Quercus lobata*
Valparaiso, 25-6
Van Asdall, 886
Vancouver, G., 414
Vancouveria
  hexandra, 368, 706
  planipetala, 363, 368, 916, 925
Vandenberg Beach, 233, 247-51
Van der Maarel, E., 675
Van Duzen Grove, 99
Vankat, J.L., 414, 599
Van Renssalaer, M., 468
Van Wagtendonk, J.W., 599
Vaquelinia, 182
Varney, B.M., 74
Vasek, F.C., 136, 414, 557, 599, 832, 865, 867
Vasey, R.B., 698
Vegetation mapping, 7, 195-220, 909-38
Vegetation, potential natural, 3, 910-11
Vegetation Type Map Survey, 197-201
Vegetation types, area of, 4-5
Veirs, S.D., 698
Ventura River, 265
Veratrum
  californicum, 152, 552-3, 648, 708, 717, 920
  viride, 621-2, 715
Verbena
  bracteata, 552
  lasiostachys, 335, 820
Vernal pool, lake, pond, 4, 81, 99, 241, 243, 278-9, 495-7, 515-54
Veronica
  alpina, 621, 623, 626
  cusickii, 621, 649
  peregrina, 529-30
    ssp. *xalapensis*, 523
Vertisols, 784-5
Vicia
  americana, 344, 704-5, 707-9, 712
  benghalensis, 741
Viguiera
  deltoidea, 817, 820
  laciniata, 747-5

Vilmorin, R. de, 668
Viola
  adunca, 753
  blanda ssp. mackloskeyi, 553
    (V. mackloskeyi)
  douglasii, 540, 926
  glabella, 706, 708, 711, 922
  lobata, 569
  pedunculata, 900
  purpurea, 574
    ssp. mohavensis, 544
    ssp. purpurea, 542, 624
    ssp. xerophyta, 548, 656
  sempervirens, 305, 368
Virgin Creek, 303
Visalia, 63-4
Vitis californica, 927
Vivrette, N.J., 261, 906
Vlamis, J., 463, 468
Vogl, R.J., 74, 294, 357-8, 379, 468-9, 557, 832, 889
Volcanic soils, 311, 800-4, 806, 810
Vollmer, A., 863
Volman, R., 697
Volpes macrotis, 185

Waginer, W.W., 357, 599
Waisel, Y., 294, 468
Walker, H.M., 469
Walker, Pass, 815, 827
Walker, R.B., 136, 357, 469, 666
Walkington, D.L., 131
Wallace, A., 863, 867
Walnut woodland, 167-8, 403, See also Juglans
Walter, H., 74, 533, 674, 889
Wana, C.W., 192
Ward, G.H., 796
Ward, G.M., 796
Wardle, P., 674-5
Wareing, P.F., 259
Waring, R.H., 357, 381, 415, 698, 732
Warm desert, See Mojave or Sonoran desert
Warme, J.E., 294
Warming, E., 469
Warner Mountains, 116, 118, 592, 664-5, 802-3, 923
Warren, D.K., 865, 887
Wasatch, 823
Wash, 815, 819, 837, 839, 844-6, 860-1, 877-9, 933
Washingtonia filifera, 44, 88, 172, 181, 871-2, 877-8, 935
Washington, state, 493, 799
Water balance, 18, 21, 23
Water potential, See Moisture stress
Water table, 551, 880
Watt, A.S., 532
Watts, D., 415
Weaver, H., 599
Weaver, J.E., 469
Webb, A., 136
Webber, I.E., 192
Webber, J.M., 867
Webb, K.L., 294
Weber, W.A., 675
Weeds, See Introduced taxa
Wehansen, J.D., 758
Weiler, J.H., 533
Weitchpee soil series, 366
Welburn, A., 489
Wells, P.V., 136, 192-3, 261, 357, 381, 469, 489, 832
Went, F.W., 469, 675, 867, 889
Westergaard, M., 867
Western juniper, See Juniperus occidentalis
Western red cedar, See Thuja plicata
Western white pine, See Pinus monticola
Westfall, S.E., 411
Westgaard pass, 812
Westhoff, V., 675
West, M., 831
Westman, W.E., 137, 357
West, N.E., 74, 795, 832, 865-7
Wheeler Peak, 826
Whipplea modesta, 363, 368, 689, 925
Whipple M.A., 512
Whipple Mountains, 844-5, 860-1, 873
Whippoorwill Mountain, 811
Whitebark pine, See Pinus albicaulis
White, C.D., 732
White fir, See Abies concolor
White, J., 672
White, K.L., 415, 514

White Mountains, 32, 42, 64-6, 82, 116-7, 553, 602, 613-4, 617, 654-8, 660, 798, 808, 810, 812, 823-5, 924
White sage, See Salvia apiana
Whitewater, 826, 876
Whitfield, C.H.
Whitney Creek, 84
Whitney, J.D., 415, 698
Whittaker, R.H., 137, 193, 357-8, 381, 469, 599, 732, 759, 915
Whitwell, H.H., 669
Wickenburg, 857
Wiedemann, A., 261, 358
Wiens, D., 353
Wieslander, A.E., 219-20, 381, 415, 469, 915
Wiggins, I., 132, 136, 193, 358, 866, 888
Wilder, P.A., 131
Wilder soil series, 690
Wilde, S.A., 358
Wild Rose Canyon, 812, 838
Wilken, G.C., 469
Willard, B.E., 675
Williams, 36, 51, 52
Williams, C.B., 758
Williams, J.W., 358
Williams, W.A., 513, 865
Williams, W.T., 261
Wilson, E.O., 533
Wilson, R.C., 75, 137, 358, 469
Wind, 340, 342, 605, 876, 894, 896, 900
Winters, 13-4, 27
Wirtz, W.O., 469
Wolf, C.B., 137, 358
Wolf, C.G., 832
Wolfe, J.A., 137, 193, 732
Wolf, L., 533
Wolfolk, J., 415
Wona, G., 294
Woodard, D.W., 261
Woodbury, A.M., 832
Wood, C.E., 137, 533
Wood, D.L., 599
Woodell, S.R.J., 135, 294, 356, 667, 865, 867, 889
Woodhouse, W.W., 294
Woodland, See Oak-, Walnut-, Insular-woodland, etc.
Woodmansee, R.G., 511
Woodsia
  oregana, 548
  scopulina, 548, 623
Woodson Bridge, 88
Wright, A.D., 759
Wright, D.T., 463
Wright, E., 469
Wright, N.A., 292
Wright, R.A., 889
Wright, R.D., 358, 469, 599, 672, 831-2
Wu, K., 512
Wyethia
  angustifolia, 577
  mollis, 574, 577, 584, 593, 644, 775, 804, 919
  ovata, 444, 540, 927

Xantusia vigilis, 185
Xerophyllum tenax, 303, 305, 310, 365, 368, 621, 721
Xerothermic, 150-2, 155-6, 166, 168-9, 171-2, 177, 182, 184-7, 808, 903
Xylococcus bicolor, 169, 302, 321, 341, 458, 900, 902

Yana, T.W., 889
Yeaton, R.I., 358
Yellow pine forest, 307, 311, 539, 540-44, 917, See also Ponderosa pine forest, Jeffrey pine forest
Yew, See Taxus brevifolia
Yollabolly Mts., 405, 621, 671,808
Yonkers, T., 865
Yorkville soil series, 367, 404, 406
Yosemite, 61-2, 64, 84-5, 209, 571, 579, 585
Young, A., 794
Young, C.L., 468
Young, D.A., 469
Young, D.H., 187
Young, J.A., 511, 795-6
Yreka, 48
Yucca
  baccata, 87, 854, 858-9
  brevifolia, 87-8, 430, 814, 850, 852, 856, 932

Yucca (cont.)
  brevifolia (cont.)
    var. herbertii, 88, 857
    var. jaegeriana, 857
  schidigera, 87, 185, 302, 341, 816, 819-20, 842-4, 850, 872, 934
  whipplei, 302, 310, 314, 322, 422, 431, 444, 476, 544, 818, 929-30
Yucca Flat, 838
Yukon, 547
Yuma, 23-4, 879
Yurok, 83
Yurtsev, B.A., 193, 675

Zaca Peak, 307
Zahner, R., 193
Zakharieva, 666
Zamora, B., 796, 832
Zauschneria californica
    ssp. californica, 546, 554
    ssp. latifolia, 586
Zavarin, E., 354-5, 358
Zavitkovski, J., 469
Zedler, J.B., 294
Zedler, P.H., 107
Zeiner, D., 107
Zelkova serrata, 161
Zigadenus
  fremontii, 444, 753
  paniculatus, 645
Zinke, P.J., 220, 459, 466, 514, 597, 697-8
Zivnuska, J.A., 469
Zmudowski, 234
Zoller, H., 675
Zon, R., 514, 915
Zuill, H., 415
Zwinger, A.H., 675

SUPPLEMENT

This reprint would not have been possible without financial help from the California Native Plant Society and the California Botanical Society. We also thank David Comstock for his important aid.

The Supplement follows the chapter sequence of Terrestrial Vegetation of California. Short updates were solicited from the original authors for the reprint of the book, which went out of print in 1986. Citations to the literature used in the 1977 text and included in the 1977 index are marked *with a page reference.

Knapp (1981 & 1982) produced a valuable bibliography of the literature on California's vegetation. We intend to keep this current.

Unfortunately our book is still "terrestrial." However the massive volume on "California riparian systems" (Warner & Hendrix 1984) covers in detail many of the problems associated with the almost total loss or drastic change of our riparian vegetation. There is much interest in riparian ecosystems.

1. Introduction

The uniqueness of California's vegetation is illustrated by 3 record properties: 1) The International Biological Program focused on vegetation productivity. A world record, high biomass occurs in the redwood (*Sequoia sempervirens*) groves of the northwest coast, $\leq$346.1 kg/m^2 according to Fujimori (1977). 2) The oldest living plant, older than even the 4000-year-old *Pinus longaeva* of the White Mts. of Calif. and the Snake Range of western Utah, is the clonal creosote bush (*Larrea tridentata*) of the Mojave Desert. Vasek (1980) has found a plant he believes to be 11,700 years old. 3) A third record is Gentry's (1986) for numbers of rare and endangered, and mostly endemic, plants in the various states of the U.S. He records 1518 for Calif., for Hawaii 883, Florida 385, Texas 379, Utah 169, Arizona 164, and all the midwestern and northeastern states <6. The history, relationships, and endemism of the Calif. flora has been treated *in extenso* by Raven and Axelrod (1978).

In *Bioscience* (1987) are 6 excellent plant physiological ecology review papers on present status of the subject, on carbon gain by plants, water balance, stress physiology, responses to multiple environmental factors and allocation of resources to reproduction and defense. The papers make a fine, 47-page review of the state of the subject. References number some 350+. Much of the work was done in Calif., on plants of our flora. In no case is the vegetation which provides the environment of a discussed plant itself defined or discussed. However, the authors do not go so far as to suggest plants can be adequately understood if moved into a garden environment (Thimann loc. cit. p. 5).

The books by Barbour et al. (1987) and Billings and Barbour (1987) consider California's plant ecology and its vegetation in a broader context.

The new edition of Takhtajan's book (1986) on Floristic regions of the world summarizes that most basic of Jenny's "factors of soil formation" (1941:*71, 1980) which can be considered ecosystem determining factors (Major 1951:*72, Crocker 1952:*70), namely the floristic factor (Jenny 1958). According to Takhtajan, California's flora is in the Holarctic Kingdom of which the Madrean subkingdom with one, Madrean, Region includes a Californian, a Great Basin, and Sonoran Provinces. The also Holarctic, Rocky Mt. Region's Vancouverian Province includes not only northwestern, mostly coastal Calif. but is extended south inland to include the Sierra Nevada. This extension is a mistake in our opinion, but detailed dot mapping of critical species will be needed to locate a boundary. Tentatively the southern limits of *Picea breweri, Alnus viridis (= Alnus sinuata), Luetkea pectinata, Phyllodoce empetriformis,* etc., can be used to separate the Californian Province's Sierra Nevada from the Vancouverian Mts. to the north.

Also needing research is the western limit of the Great Basin flora. How far and high has it penetrated the eastern slope of the Sierra Nevada (Taylor 1976, Raven & Axelrod 1978:58, 63)?

The migrational history of the Sierra Nevadan flora is still unclear. Taylor's (1977) Investigation of floras of the Cascade-Sierran axis is an excellent summary of current data (add *610-615). Granted there has been massive plant migration southward along the Cascade-Sierran axis, but such plants as *Asplenium viride, Tsuga mertensianus, Taxus brevifolia, Xerophyllum tenax, Lithocarpus densiflorus, Corydalis caseana, Leucothoe davisiae, Gaultheria ovatifolia, Acer circinatum, Arbutus menziesii, Rhamnus purshianus,* etc., which clearly illustrate this movement (Raven & Axelrod 1978:58) are exclusively plants of oceanic climates. The few Convict Creek calcicole disjuncts (Major & Bamberg 1963:*134) on the other hand are exclusively plants of continental climates, and their hypothesized Pleistocene migration route to the Sierra Nevada ecologically fits the Late Glacial very cold and dry climate of the Great Basin (Pewe 1983, Mears 1981, Malde 1964) of the uplands with cold bogs along the through-flowing rivers such as the Humboldt (Lavrenko 1956, Major & Bamberg 1967a:*671, Walter & Box in West 1983).

We complained in 1977 about the lack of species-stand tables to document and aid analysis of the various kinds of California's vegetation. Little progress has been made over the last decade. Such data often are buried in theses, but needed stand data are also effectively buried by current methods of ordination and classification. The results, only, of these methods alone are insufficient. Is our technology so limited that the original data for ordination and classification cannot be made available to the sceptical reader? Surely the best test of an ordination is an ordinated table of stands of species abundances. Surely the best test of a mathematically derived classification is a table of stands arranged as the computer suggests. Stand data can be analyzed from many points of view as Komarkova (1979) and many others have shown. They should be. They cannot be if they are not available, if the plant species that compose vegetation are not made known.

But we need a classification of all of California's vegetation right now. Holland (1986) suggests a physiognomic scheme of 375 vegetation units which improves Cheatham and Haller's (1975:*906) classification. The paper is a most valuable culmination of efforts to excavate from the literature descriptions of much (most? all?) of the vegetational diversity within Calif. The Natural Diversity Data Base is the logical place for a classification of California's plant communities to be developed. We believe the first hierarchial break should be floristic, following Takhtajan as indicated above unless someone can come up with a better floristic classification. Some communities will transgress Provincial floristic boundaries, but such azonal communities are few. Further floristic classification can only follow accumulation of floristic data, namely stand surveys (releves, Aufnahmen). The Data Base is where such data belong.

We have splendid models of ecologically centered vegetation classifications to follow in Oberdorfer (1979), Ellenberg (1978), Horvat et al. (1974), Komarkova (1979).

Classification of forest and range vegetation by the US Forest Svc. has recently burgeoned in all the Regions except Calif. Following the Daubenmires' lead (1968:*731) which had spilled over into the admirable volume by Franklin and Dyrness (1969:*830, 1973:*731) on the "Natural vegetation of Oregon and Washington," Pfister et al. (1977) produced "The forest habitat types of Montana." Mueggler and Stewart (1980) extended the idea to grass- and shrublands. Other forest habitat type volumes have followed for the National Forests in Wyoming (Alexander 1986), Colorado, New Mexico, Arizona, Utah, Idaho, the Black Hills (Alexander 1985). The forest vegetation of Calif. has been left out. An exception is Riegel's thesis on forest habitat types in the Warner Mts. (1982). Sawyer mentions other work in progress in northern Calif. (Chapter 20).

Originally only "climax" vegetation was sampled in this Forest Svc. research. However there are too many aspen (*Populus tremuloides*) and lodgepole pine (*Pinus contorta* ssp. *latifolia*) forests in the rockies to ignore these species. They are often clearly seral, but in other sites their communities often show no evidence of change with time (Despain 1983). See Chapter 17.

In this habitat type work much European experience with vegetation classification was ignored. Dominance dominated selection of sites, naming, the hierarchy of units. But the assumptions and methods do roughly parallel those associated with Braun-Blanquet (1928–1964:*666). The model treatment of Swiss forests by Ellenberg and Klötzli (1972) has been approached but still offers some easy, un-used advantages, e.g. maps locating the stands surveyed are now given but a simple chart showing the altitudinal distribution of stands is not. The Swiss also print a diagram for each plant community showing the field occupied by a forest type on axes of soil acidity and moisture (cf. Burke 1979, 1982). Pojar et al. (1982) have produced such diagrams for many of the forest species of British Columbia. We still lack agreement on a general, floristic vegetation classification such as the Europeans have (Oberdorfer 1979).

Unfortunately in the Forest Svc. work individual stand surveys are not always printed, but only constancy and average cover values for the habitat types as a whole. Species are tabulated by life form and then alphabetically but not by their contribution to the taxonomic hierarchy of vegetation units.

Homogeneity has evidently not been a prime criterion for stand selection. In a recent study in southern Utah by Youngblood and Mauk (1985) the plot size was 500 m^2, and this was evidently too large. Number of stands in a habitat type varied from 5 to 128. Number of single occurrences of species in a particular plant community varied from 23 to 5 (2), percentage of singles from 13 (3) to 61%. Roughly, when there are more than 4 singles in a habitat type, the percentage of singles increases at a rate 3.55 times the number of singles. Singles make cluttered tables.

2. California climate in relation to vegetation

The water balance at a site measures both the water and heat available to a plant or a stand of vegetation. Actual evapotranspiration is a measure of water availability; potential evapotranspiration is a measure of heat availability. The Hennings (1977a & b) have used Penman's approach to calculate potential evapotranspiration in North America. They discuss and map the net water balance as well as water surplus and water deficit. The Penman formula takes into account more of the climatic parameters which influence water loss than does Thornthwaite's formula. It is correspondingly more difficult to apply on a detailed scale. Müller's book (1982) gives a set of 16 climatic data for some 1045 stations more or less evenly distributed over the globe. Thornthwaite's method is used to calculate potential evapotranspiration. A program is available to plot the data into a Walter-style diagram.

In a climatic study of North American deserts the Hennings (1984) use the Bowen ratio of the fluxes of sensible (H) and latent (LE) heat, Budyko's Index of Aridity as the ratio of net radiation (R_n) to the heat needed to evaporate cm of precipitation (LP), and Transeau's ratio of precipitation and potential evapotranspiration. Data for 22 stations are listed. Note that $L = 2501 - 0.24T$ in J/g of water and T in °C, and on p. *19, line 4 from the top "inversely" should be stricken.

Arid and humid mountain systems are described based on potential evapotranspiration and precipitation (Henning, I. & D. 1981). Evapotranspiration amounts decrease slowly with altitude in the arid mountains of Afghanistan and Middle Asia, but they decrease precipitously in the humid Alps. Where does the Sierra Nevada fit? The Klamath Mts.? The Great Basin ranges?

Using the notation above, Lettau (1969) points out that Budyko's index is equal to the product of 1 minus the ratio of runoff to precipitation times 1 plus the Bowen ratio [$R_n/LP = (1-r/P)(1+H/LE)$]. Thus, the evaporative power of the air (Budyko's index) is inversely proportional to precipitation, approaching 0 in the wettest climates and ∞ in the driest. The evaporative power of the air is the product of 2 terms. The first is <1 but increases as a humid climate becomes drier to $r=0$, when it equals 1. The second term is >1 and approaches ∞ as the climate becomes drier. Storr (1972) suggests modifications necessary for a snowy climate where some energy is used to melt the snow.

Unfortunately precipitation as snow is undermeasured. Wendler (1979:17) describes a case on Barter Island off northern Alaska where a standard gauge registered 27.3 mm water equivalent whereas a nearby Wyoming gauge (Barry 1981:203) registered 91.9 mm. In the high mountains of southern Norway the ratio between measured runoff and measured precipitation is 1.0 at the coast and increases steadily to 2.0 at the highest, glaciated altitudes. Since the source of runoff is precipitation, of which some must be evapotranspired, such ratios are "clearly impossible" (Barry 1979:66-67). Measurements using a snow board are recommended. At Alta, Utah, at 2680 m elevation snow board measurements were made, and mean precipitation turns out to be, from the record through 1977, 1456 mm/yr, perhaps 50% higher than any other measured annual precipitation in Utah. Water content of snowpacks alone on 1 April is far greater than 1000 mm at 3000 m elevation (9840 ft) in most of the northern and western Great Basin ranges (1686 mm on the west flank of the Wasatch, 1000 on the east, 1598 mm in the Owyhee, 1254 in the Steens, 1193 in the Carson, 1100 in the Warner Mts., etc.) according to data assembled by Zielinski and McCoy (1987). How much precipitation as snow really does occur in the Sierra Nevada and other high mountains in California?

The arid-humid boundary in Budyko's terms is where net radiation equals the energy just necessary to evaporate the precipitation. It is the arid/humid, pedocal/pedalfer, steppe/forest boundary. In our terms it is where potential evapotranspiration just equals precipitation. Using lapse rates from Chapter 2 we generate the following table (Burke et al. 1982:25).

The White Mts. obviously exaggerate the conditions on the eastern slope of the Sierra Nevada. Is the more or less arid eastern slope of the Sierra Nevada really arid to the elevations indicated?

For a modern, global look at "Climate and plant distribution" see Woodward (1987). Unfortunately our descriptions of vegetation, our knowledge of the details of plant distribution and our mastery of pertinent physical climatology are inadequate. Simple scale gives us trouble (Jarvis & McNaughton 1986).

3. Research Natural Areas and related programs in California

***M. C. Roshovsky**, Land & Natural Areas Project, Calif. Dept. of Fish & Game*

A discussion of natural area preservation needs no excuse for inclusion in a book on vegetation. Although students of vegetation and its ecology require natural areas for research and comparison, such areas are rapidly being lost to development projects and land use changes. Thus, researchers have difficulty finding pristine examples of natural ecosystems to study. Rare communities which can provide valuable insights for ecological and evolutionary theory are disappearing.

Ecologists cannot resign themselves to such losses. They need to inform themselves about programs for protecting natural areas, and need to take advantage of these programs to protect California's unique diversity. Their professional expertise gives them an influential position in protecting natural areas.

Cochrane (1984) discusses a variety of programs for such protection. Six federal agencies collectively administer 32 of them. Among the 8 state agencies which have biological conservation projects, 4 (Calif. Water Resources Board, Dept. of Fish & Game, Parks & Recreation, and Forestry) officially designate and manage protected natural areas. In addition, 3 colleges or universities, 7 private conservation groups and over 40 local land trusts work toward protecting natural areas.

There are advantages in having a variety of conservation programs. However, by the late 1970s the disadvantages of a scattered approach became apparent. For example, while certain high profile species were receiving inordinate amounts of attention, other more critically endangered species or habitats were being virtually ignored. Conservation agencies had difficulty identifying the most critical areas for protection because no comprehensive inventory of the state's important

Location	Lat.N	Elevation range (m)	Elev. of humid/arid boundary & values of PotE = Ppt. West slope		East slope	
			Elev m	Value mm	Elev m	Value mm
Red Bluff– Susanville	40°15′	104–1262	750	780	1280	640
Sacramento–Reno	30° ± 20′	5–1340	375	800	1610	570
Stockton– Tamarack	38°38′ 1/2	7–2450	500	770	–	–
Modesto– Mono Lake	37°50′	28–1985	380	780	2400	500
Madera–Bishop	37°00′- 37°20′	82–1251	750	780	3100	450
Visalia–Grant Grove	36°20′	108–2004	1000	770	–	–
White Mts.	37°40′	1251–4342	3180	380	–	–

biological resources existed. Most important, little or no coordination existed between conservation programs. Most were individually pursuing their own goals, unaware of potential assistance from other organizations. The result was an inefficient allocation of funds and staff.

In response to this situation 16 conservation organizations organized a consortium in 1978 to propose development of a statewide natural area program. By 1979 the resultant program was administratively established in the Calif. Dept. of Fish and Game (CDFG). Two years later it received a firm legislative basis by passage of Assembly Bill 1039.

This legislation required the CDFG to meet five significant goals relating to biological diversity: 1) Develop and maintain a data management system for natural resources, 2) Identify the most significant natural areas of the state, 3) Ensure the recognition of these areas, 4) Seek the long-term protection of these areas, 5) Provide coordinating services for other public and private conservation organizations interested in protecting natural areas.

Since the passage of this legislation, CDFG's natural area program has evolved into two interactive programs within the department's Nongame Heritage Program: the Natural Diversity Data Base (NDDB) and the Lands and Natural Areas Project (LNAP). Both programs work cooperatively alongside staff from the Calif. Native Plant Society and the Nature Conservancy, a mutualistic relationship which benefits all parties involved.

NDDB's main purpose is to be the data management system for natural resources. It is now the most comprehensive inventory of California's rare species and natural communities. Based on methodology developed by the Nature Conservancy, this inventory is part of a nationwide network of 48 natural heritage programs. NDDB continually gathers and updates its information from museum and herbarium collections, biological inventory reports, and field data of knowledgeable biologists throughout the state. Since NDDB was established, over 16,000 location records on rare species and natural communities have been computerized.

Terrestrial vegetation is catalogued in NDDB according to Holland's (1986) plant community classification, a modification of Cheatham and Haller's (1975:*906) system, which recognizes a total of 262 terrestrial plant communities. Over half of these (137) are sufficiently uncommon to be given high inventory priority by NDDB.

LNAP is responsible for meeting the remainder of the legislation: identifying and protecting significant natural areas as well as helping to coordinate protection activities with other conservation organizations. Information from NDDB and other sources is used to identify Significant Natural Area sites throughout Calif. A computerized inventory tracks the condition and protection status of each Significant Natural Area.

LNAP ensures recognition of Significant Natural Areas by publishing and promoting the use of a regularly updated report which briefly describes each area (Roshovsky 1987). This report can be valuable in identifying sites for research projects, habitat acquisition proposals and mitigation projects. It may also help protect natural areas by providing early notice to project developers and resource managers. Such early notice allows greatest flexibility in adjusting project plans to minimize adverse impacts.

LNAP is assisted in protecting natural areas by other CDFG programs and personnel. Funds for land acquisition are available from many sources, including the Environmental License Plate Fund. Proposals for land acquisition are prepared by CDFG personnel, approved by Regional Managers, and acquired by the Wildlife Conservation Board. Income Tax Check-off funds are available for recovery and management of threatened or endangered species. CDFG endangered species biologists provide technical advice for natural area protection, and regional biologists and game wardens provide local field support throughout the state.

To recruit further assistance and to coordinate with other agencies, LNAP has organized the Interagency Natural Areas Coordinating Committee. This committee allows a variety of conservation organizations to regularly meet and work cooperatively toward their common goal of protecting natural areas. In addition to CDFG, the committee presently has representatives from the Nature Conservancy, the Calif. Dept. of Parks & Recreation, US Fish & Wildlife Svc., US Forest Svc., Natl. Park Svc., Bureau of Land Management and the University of California's Natural Reserves System. This cooperation reduces duplication of effort and allows the pooling of resources between agencies for protecting natural areas.

* * *

Progress in establishing new Forest Service RNAs has been slow. Their descriptions need to be accumulated and published so further use can be made of them. The Pacific Northwest region offers a model (Franklin et al. 1972, Greene et al. 1986).

The Nature Conservancy has become more active in Calif., and the University of California's Land and Water Preserve system has grown.

4. The California flora

Raven and Axelrod (1978) greatly expanded their treatment of the origin and relationships of the Calif. flora. Specific examples are cited to back up generalizations. Theirs is a masterly treatment. The projected new flora from the Jepson Herbarium at Berkeley will make possible the use by plant ecologists of current plant taxonomic progress.

We need dot maps of distributions of the species of the Calif. flora. The Swiss volumes by Welten and Sutter (1982) are a model.

Axelrod (1982) considers the Monterey endemic area anew and proposes that fog associated with the deep submarine Monterey canyon ameliorates the local climate enough to account for the presence of many of the endemics of the Monterey area. Local podzolic soils support his hypothesis.

5. Outline history of California vegetation

The geological history and therefore the origin of coastal sage (Chap. 13) and of Sonoran desert (Chap. 25) vegetation are fully and expertly elucidated by Axelrod (1978 & 1979). Both are modern assemblages of species resulting from selection over a long period of geological history, from various floras, growing in various climates.

Recently discovered Pleistocene occurrences of various woody plant species are used to interpret the history of the vegetation of coastal Calif. (Axelrod 1983a). In a worldwide survey of oak biogeography and history he (1983b) emphasizes both the uniqueness and relationships of California's flora.

The first use in Calif. of plant remains in pack rat middens to decipher Pleistocene and Neogene climatic, and therefore vegetational, changes is Cole's work (1983) in King's Canyon on the western slope of the Sierra Nevada. The site at present has *Quercus chrysolepis, Cercocarpus betuloides, Ceanothus integerrimus, Cercis occidentalis, Penstemon breviflorus,* etc. In Full Glacial time it had *Juniperus occidentalis, Pinus monophylla, P. ponderosa, Sequoiadendron* and other up-slope conifers—an odd combination of xeric and cold-mesic species. Cole does not believe the marble at his site produces the odd combination of xeric and mesic species.

In Death Valley Wells and Woodcock (1985) have documented full glacial vegetation at 425 m elevation with *Juniperus osteosperma, Atriplex confertifolia, Yucca brevifolia, Y. whipplei,* plus the more hot desert species *Chrysothamnus teretifolius* and *Opuntia basilaris.* Changes in elevations occupied at present are evidently 1200 to 1700 m. Although *Ambrosia dumosa* was at this site 10,230 years ago, *Larrea tridentata* was not. *Larrea* arrived much later, and the authors trace its history in the Southwest.

7. Beach and dune

M. G. Barbour, *Botany Dept., University of California, Davis*

Most of the recent literature on Calif. strand plants has centered around demographic or physiological autecological studies, using sophisticated equipment for field measurements of gas exchange and water stress. Pavlik (1977; 1982; 1983a,b,c; 1984; 1985) examined photosynthesis, transpiration, N use, growth and C allocation of two important foredune grasses, *Elymus mollis* (native) and *Ammophila arenaria* (introduced). The competitve advantage of European beach grass was due to its ultimate pattern of allocation, rather than to more proximate physiological traits. DeJong (1978, 1979, DeJong & Barbour 1979) followed the seasonal performance of *Atriplex leucophylla* populations for more than a year, comparing the gas exchange and water relations of this uniquely C_4 beach plant with adjacent C_3 species. North-south morphological ecotypes of this *Atriplex* occur. The effects of salt spray and soil salinity on growth, phenology and survival of a variety of beach taxa were reported by Barbour and DeJong (1977), Barbour (1979) and Boyd and Barbour (1986). In general, zonation of species back from tide line corresponded to relative salt tolerance, as measured in growth chamber experiments. Salt tolerance of *Cakile edentula* from marine and fresh water strands was shown to be different, genetically fixed and of adaptive value. Holton (1980) measured N inputs to, and losses from, the strand at Pt. Reyes. He then measured the combined effects of soil N and salinity on tissue chemistry, photosynthesis and growth of *Cakile maritima.*

Turning to more demographic studies, Pitts (1975, Pitts & Barbour 1979) was successful in demonstrating a spatial relationship between herbivore activity and plant distribution on the strand at Pt. Reyes. Boyd (1986) also examined the role of herbivores on population size in an elegant study of two *Cakile* species. In addition, he compared them in terms of germination requirements, life history traits and reproductive success. The competitive advantage of *C. maritima* over *C. edentula* lay in the former's plastic life span—its ability to survive into a second or even third reproductive season—in contrast to the more strictly annual habit of the latter.

Synecological studies in the past decade have emphasized spatial or temporal dynamics. Holton and Johnson (1979) described 4 associations of dune scrub at Pt. Reyes and related their distribution to clines of soil organic matter and texture and to salt spray. Barbour et al. (1981) compared dune scrub along two coasts with mediterranean climate—Calif. and Israel. Beyond a superficial similarity in vegetation, there were major differences in physiognomy, life form spectra, floristic affinities and leaf types. Johnson compiled detailed records of latitudinal and zonational distributions of dune plants along Baja Calif. (1977) and the adjacent Mexican coast (1982) from Atascadero (Calif., 35°N) to Los Corchos (Nayarit, 20°N). She later investigated 23 populations of *Abronia maritima* within this latitudinal range with regard to salt tolerance, growth, phenology, germination and response to burial. She reported ecotypic differences. Williams (1985, W. & W. 1985) has continued his exceptional, long-term study of dune vegetation dynamics at Morro Bay, begun in 1968. Such native species as *Haplopappus ericoides, Dudleya caespitosa* and *Croton californicum* show density-stable relationships, but the introduced iceplant *Carpobrotus chilense* exhibits a linear rate of increase, cumulative over many years. The continuing impact of off-road vehicles on substrate movement and vegetation at Nipomo Dunes has been dramatically described by Chipping and McCoy (1982) and Jones (1984). Finally, in a synthesis volume on North American vegetation Barbour et al. (1985) pointed out some unique features of Pacific Coast strand vegetation, particularly in terms of life form spectra.

William Cooper, author of a 2-volume work on the coastal dunes of Oregon-Washington and Calif., died in 1978. The executors of his estate have deposited with Barbour Cooper's raw material for those volumes: original ground-level photographs, aerial photographs and negatives dating back to 1922, miscellaneous field notes, USGS maps with site locations marked, tree ring cores, and a card file index of dune references. This material is available for examination by investigators by contacting Barbour.

Dunes

Andrea Pickart, *Lanphere-Christiansen Dunes Preserve, Arcata, California*

Dune restoration. A recent series of dune restoration projects has generated literature which contains useful data on the biology and management of dune species. The Buhne Point Shoreline Erosion Control Demonstration Project at Humboldt Bay produced reports documenting all phases of an experimental dune revegetation project. Pickart (1985) reviewed dune restoration projects throughout Calif. Pre-project investigations examined germination, survival and development of 9 native dune plants under experimental treatments and controls (Pickart 1986, 1987a). Clark (1986) monitored nutrient levels in experimental treatments and developed fertilizer recommendations. Methods and costs of seed collection and processing were well documented (Newton 1985a, 1986). Several planting techniques were used (Newton 1985b, Bioflora 1986) and evaluated (Pickart 1987b). The Calif. Dept. of Parks and Recreation initiated restoration projects at 7 state parks. Methods are detailed in individual project plans for McKerricher (Barry 1985), Sunset (Gray 1985a), Marina (Gray 1985b) and Asilomar (Moss 1987) State Parks and Beaches. Results have been assessed for Marina State Beach (Gray & Ferreira 1987) and Tijuana Estuary (Jorgenson 1987). Test plots were established at the Guadalupe Dunes to measure reinvasion of dune species in a disturbed pipeline corridor and to test revegetation methods (Bowland & Jordan 1987). Management problems encountered in a Monterey dune revegetation project are outlined by Guinon and Allen (1987).

Vegetation. Duebendorfer (1985) classified natural and degraded vegetation of semi-stabilized dunes on the Lower North Spit of Humboldt Bay, and he correlated the resultant types with population parameters of *Erysimum menziesii.* Pickart (1987) further explored this relationship but focused on undisturbed vegetation and increased the geographical scope to include the entire North Spit.

Management. Brown (1987) quantified the effects of human trampling on semi-stabilized dunes at Humboldt Bay. Van Hook (1983) tested 7 methods of controlling *Ammophila arenaria* and developed management recommendations for the Lanphere-Christensen Dunes Preserve in Arcata. Using historic photos, interviews, herbarium specimens and records of historic exploration Miller (1987) documented the introduction history of *Lupinus arboreus* north of San Francisco Bay.

8. Coastal salt marsh

K. B. Macdonald, *Phillips Brandt Reddick, San Diego, California*

References describing Calif. salt marsh vegetation have more than tripled since late 1975 when Chap. 8 was first prepared. This striking increase reflects the greater State and Federal protection now afforded wetland habitats. While basic research continues to expand, increasingly greater emphasis is being put on habitat protection and restoration issues. Indeed, some 40% of the 94 salt marsh references listed in this addendum concern habitat values (Nixon 1977, 1980; Onut et al. 1978; Dennis & Marcus 1984; Macdonald 1986), boundary delineation (State Lands Comm. 1976; Harvey, Kutilak & DiVittoria 1978; Calif. Coastal Comm. 1981; Eilers, Taylor & Sanville 1982; Zedler & Cox 1985; Environmental Lab. 1987; Sipple 1987), habitat restoration (Josselyn 1982; J. & Atwater 1982; Race & Christie 1982; Faber 1983; Metz & Zedler 1983; Williams & Harvey 1983; Zedler 1983, 1984, 1986, 1987a & b; Josselyn & Buchholz 1984; Harvey & Josselyn 1985; Macdonald & Williams 1985; Race 1985, 1986; J. Dobbin 1986; Natl. Audubon Soc. 1986; ABA 1987; Andrecht & Firle 1987; Williams & Swanson 1987), and/or restoration monitoring (Niesen & Josselyn 1981; Claycomb 1983; Springer & Sawyer 1984; Eliot 1985; Jacobson 1986).

Another key development since 1975 has been publication by the US Fish & Wildlife Svc. of an outstanding series of "Community profiles" and "Estuarine profiles." These profiles provide excellent regional data summaries describing southern Calif. marshes (Zedler 1982), Tijuana Estuary (Zedler 1986), Mugu Lagoon (Onuf 1987), San Francisco bay marshes (Josselyn 1983), and the marshes of the Pacific Northwest coast (Seliskar & Gallagher 1983). Site history, physical environment, the biotic community and management issues are all addressed in each profile. Detailed descriptions of salt marsh vegetation at several other Calif. sites are also now available: Santa Ana River mouth (Macdonald et al. 1985), Bollana Creek (Gustafson 1981), Carpinteria (Ferren 1985), Goleta Slough (Ferren & Rindlaub 1983), Elkhorn Slough (Mayer 1986, Oliver & Mayer 1987), San Francisco Bay (Atwater et al. 1979), Sacramento-San Joaquin Delta (Atwater 1980), and Humboldt Bay (Rogers 1981, Eicher 1987).

Additional salt marsh research topics covered by groups of papers include: Geological history and physical processes (Atwater et al. 1975; Lohmar et al. 1980; Mudie & Byrne 1980; Zedler & Onuf 1984; Collins et al. 1985, 1986; Macdonald & Williams 1985; M. & Bilharn 1987; Bilharn & Macdonald 1987; Williams & Swanson 1987); latitudinal ranges of *Spartina* spp. (Spicher 1984, S. & Josselyn 1985); aspects of the autecology of *Cordylanthus maritimus* ssp *maritimus* (Dunn

1981, Murphy et al. 1981, Vandervier 1983, V. & Newman 1984, Zedler 1984, Lincoln 1985), *Salicornia virginica* (McIntyre 1977), *Spartina foliosa* (Seneca 1974, S. & Blum 1984, Covin 1984, Zedler 1986) and *Typha domingensis* (Beare 1984); plant salinity relationships (Barbour & Davis 1970, Barbour 1978, Stephenson et al. 1980, Ustin et al. 1982, Zedler 1983, Pearcy & Ustin 1984, Zedler & Beare 1986) and primary production (Zedler et al. 1978, 1980; Z. 1980, 1982; Winfield 1980; Eilers 1981; Onuf 1981; Rogers 1981; Balling & Resh 1983).

Many of the above citations have additional references and most cover more than the single topics highlighted above. An overview suggests several key conclusions developed over the last decade.

1. Extensive studies of the tussocky (caespitose) form of *Spartina* recorded from intermediate marsh elevations at Humboldt Bay (Eicher 1987) suggest it is *S. densiflora* rather than an ecotype of *S. foliosa* as previously believed. *S. densiflora* is a native of Chile and was probably introduced into Humboldt Bay in the 1850s. Thus, the northern coastal limit of *S. foliosa* is Bodega Bay, just north of San Francisco Bay (Josselyn 1983, Spieker 1984, S. & J. 1985).

2. Physiological studies of *Spartina foliosa, Salicornia virginica* and *Scirpus robustus* in San Pablo Bay confirm that soil salinities control soil moisture availability as well as plant growth and photosynthetic rates. Different responses among the 3 species closely correlate with their field distributions and behavior (Ustin et al. 1982, Pearcy & U. 1984).

3. At Elkhorn Slough Oliver and Mayer (1987) confirm that certain salt marsh species mosaics reflect the effects of interspecific competition rather than responses to physical gradients. Perennial *Salicornia virginica* rapidly colonizes newly available intertidal habitat, shading out initially successful annuals such as *Atriplex patula* and *Spergularia marina*. Other perennials such as *Jaumea, Frankenia* and *Distichlis* are able to invade the dense *Salicornia* stands only if the canopy is opened up by strandline wrack deposits or parasitic dodder (*Cuscuta salina*) patches.

4. Detailed geomorphic studies at Petaluma Marsh, San Francisco Bay (vertical control ±2 mm, Collins et al. 1985) confirm that the distribution of elevations across a mature marsh surface is not random but rather follows a predictable, explainable pattern. Tidal datums such as MHW or MHHW vary several cms from the mouth to headwaters of individual tidal creeks. Thus, "remote" tidal data may be meaningless. Distribution patterns of some species, *Jaumea, Frankenia* and *Scirpus* for example, parallel microtopographic changes much more closely than previously realized. Again, are physical variables or interspecific competition controlling the development of species mosaics?

5. Zedler (1986) has proposed an alternative to the *Spartina* colonization ⟶ sediment accretion ⟶ *Salicornia* succession outlined in Chap. 8. Long-term monitoring at Tijuana Estuary suggests that extensive *Spartina* stands become established only at sites that experience continuous tidal flushing. Major flooding, lowered soil salinities and the availability of viable seeds promote clone establishment. Vegetative reproduction would accelerate following subsequent floods and reduced soil salinities (that result in increased plant density and height) until the clones merge into a continuous *Spartina* stand. The combined stresses of high soil salinity and low soil moisture, due to increased intertidal elevation and/or reduced tidal flushing, tip the competitive balance in favor of succulents such as *Salicornia*. Increases in *Spartina* density and biomass following removal of other species from the upper part of its range confirm that interspecifc competition also plays a role in plant distribution patterns.

Roots and rhizomes identified from subsurface marsh sediments in San Francisco Bay (Atwater et al. 1979) offer support to both salt marsh models. Palo Alto Baylands, Richardson Bay and China Camp locations each yielded *Spartina*-dominated sediments overlain by a *Salicornia*-dominated sequence, as expected in the Chap. 8 model. Sections at Mare Island and Southampton Bay, however, reveal a broad, *Salicornia*-dominated marsh plain, fronted by *Spartina,* but lacking any subsurface *Spartina*-dominated sediments. This more closely fits Zedler's (1986) model. Comparable paleoenvironmental studies are not yet available for southern Calif.

6. Some salt marsh plants exhibit seemingly different environmental responses at different latitudes. *Salicornia virginica* grows readily from seed in San Francisco Bay and Elkhorn Slough, yet rarely germinates at Tijuana Estuary (Williams & Harvey 1983, Oliver & Mayer 1987, Zedler 1986). *Spartina foliosa* occupies a consistent position in San Francisco Bay marshes, but it is only patchily distributed among and within southern Calif. salt marshes (Atwater et al. 1979, Zedler 1986).

Two concepts emphasized by Zedler (1986) in her extensive studies at Tijuana Estuary help to explain these apparent differences. First, germination and seedling survival may differ greatly from conditions under which an established species population persists. Zedler here refers to the "regeneration niche" concept of Grubb (1977). Second, extreme events such as major episodes of flooding (low-salinity gap) or prolonged drought (soil hypersalinity) can substantially alter salt marsh species competition and vegetation structure.

7. The increased regulatory protection of wetlands and urgent need to develop more effective salt marsh restoration programs have become a major driving force behind salt marsh research.

8. Regulatory requirements for wetlands identification and boundary determination have focused more attention on salt marsh-upland transition habitats. Single parameter approaches to identification and delineation have been increasingly replaced by multi-parameter methods that focus on wetlands hydrology and soils, as well

as vegetation (Environmental Lab 1987, Sipple 1987).

9. Manipulative field and laboratory experiments (reflooding of diked marshes, reciprocal transplants, canopy removal, controlled growth experiments, etc.) as well as monitoring of "natural experiments" such as the effects of major floods, droughts or ocean inlet closures have greatly increased understanding of complex plant/habitat interrelationships (Zedler 1986, Onuf 1987, Oliver & Mayer 1987).

10. Long-term field monitoring, as conducted by Zedler at Tijuana Estuary (1986), Oliver at Elkhorn Slough (O. & Mayer 1987) and Collins at Petaluma Marsh (C. et al. 1986), has yielded critical research results otherwise unavailable through intense but short-term studies.

11. It is now apparent that Calif. salt marsh vegetation is much more dynamic and responsive to environmental perturbations than was previously believed.

12. Calif. salt marshes are increasingly recognized as unique regional habitats whose preservation and management must depend on local research, rather than indiscriminate application of values and concepts developed elsewhere.

13. Careful comparisons among historic maps and early aerial photographs have revealed considerable information on former environmental conditions and vegetative development at coastal salt marsh sites (State Lands Comm. 1976, Lohmar et al. 1980, Ferren 1985, Macdonald & Williams 1985, Zedler 1986).

14. Despite some progress, knowledge of the historical development of Calif. salt marshes, as well as their tidal patterns, groundwater hydrology, and soils all remain woefully inadequate.

Macdonald gratefully acknowledges discussions of Calif. salt marsh research trends with B. Atwater, J. Buchholtz, J. Collins, T. Harvey, M. Josselyn, J. Oliver, W. Patrick, R. Pearcy, M. Race, M. Silberstein, P. Springer and J. Zedler.

9. The closed-cone pines and cypresses

The endemic *Cupressus forbesii* requires fire to open its cones for seed dissemination. It also requires 10+ years to reach cone-bearing age and 50+ years for maximum seed production. Fires on a too frequent schedule will eliminate the species (Zedler 1977).

10. Mixed evergreen forest

A great variety of presently non-commercial, low-elevation forests suffers under this rubric. Stand data are few. Wainwright and Barbour (1984) ordinated some stands in Sonoma Co. Clark (1984) investigated mineral nutrition of epiphytes.

Many Douglas fir forests fit here. John Sawyer assembled references. Old growth Douglas fir is controversial—how much acreage, if any, should be preserved (Barrows 1983, Old Growth Definition Task Group 1986)? Succession is a problem (Strothmann & Roy 1984, Thornburgh 1984).

Harrington et al. (1984) investigated leaf area and biomass of the hardwood sprouts and seedlings (Tappeiner et al. 1986) that follow logging of overstory *Pseudotsuga menziesii* (Tappeiner et al. 1984, Tappeiner & McDonald 1984). Minore (1986, 1987) further investigated hardwood sprout and duff effects on physical factors of the environment.

11. Oak woodland

James Griffin*, Hastings Natural History Reservation, Carmel Valley, California*

Over 70 post-1977 research papers relating to oak woodland were presented at 2 symposia at Claremont in 1979 and San Luis Obispo in 1986. Proceedings of both symposia were published by the US Forest Svc. (Plumb 1980 & in press). These publications provide an excellent review of recent oak woodland research. Another US Forest Svc. publication helpful in reviewing the oak woodland literature is an oak bibliography with detailed, topical outline (Griffin, McDonald & Muick in press).

Regeneration. A frequent theme in recent oak woodland reports is a lack of species recruitment, particularly in *Quercus douglasii.* Concern about poor oak regeneration and the advent of extensive harvesting of oaks for fuel stimulated activity by wildland regulatory agencies. One result was a statewide survey of oak regeneration which confirmed that *Q. douglasii* regeneration was not adequate to maintain current stand structure (Muick & Bartolome 1987). A significant regeneration study evolved on the Los Padres Natl. Forest in San Luis Obispo Co. where 2 sets of experimental exclosures confirmed the positive effects of acorn burial and the strong negative effects of rodent predation (Borchert et al. 1986). In 1986 a major program with 4 sets of regional exclosures and study plots in foothill woodland was established by J. Menke and associates, University of California, Davis.

Stand structure. One important problem in understanding the dynamics of oak woodland populations has been the lack of age-class data. The status of current regeneration cannot be properly understood without knowing more about when various portions of the current stands started. McLaren (1986) demonstrated that age-classes cannot be accurately estimated from size-classes. He also made age-class analyses of *Q. douglasii* stands in the northern and southern Sierra Nevada foothill woodland. V. L. Holland (in press) studied age-classes in 3 Coast Range *Q. douglasii* foothill woodland stands. Lathrop and Arct (in press) have started age-class analyses of *Q. engelmannii* stands in the southern oak woodland.

Physiology. Few physiological studies have been completed since 1977. Parker and Billow (in press) compared available N under evergreen and deciduous canopies. Matsuda and McBride (1986) compared root growth patterns in 3 oak species under field and laboratory conditions. They also compared germination and early seedling growth in 6 oak species (McBride & Matsuda, in press).

Miscellaneous studies. Recent work in the little-studied northern oak woodland has included interactions of fire frequency, heavy grazing and conifer invasion (Anderson & Pasquinelli 1984, Reed & Sugihara in press, Sugihara & Reed in press). Saenz and Sawyer (1986) looked at floristic composition under and outside northern oak woodland canopies.

New work in the southern oak woodland has been done by Griggs (in press), Lathrop and Wong (1986), Lathrop and Zuill (1984). Knudsen (1984) completed an interesting ecological study of *Q. lobata* in a riparian community in the Sacramento Valley. At another extreme the *Quercus garryana* woodlands of coastal northern Calif. in the Redwoods Natl. Park were studied by Sugihara et al. (1987). Using Twinspan 7 plant communities were recognized, and a summary table of the 135 species out of the 305 found in the vegetation is given.

12. Chaparral

***T. L. Hanes**, Dept. Biol. Sci., California State University, Fullerton*

How did chaparral originate? Its association with a mediterranean climate has been assumed to be causal. Axelrod (1986) however proposes a recent expansion during the hot-dry period of 4000-8000 years ago. Relict woodlands and potreros are evidence. Minnich (1986) agrees with Axelrod that fire is of recent origin, and that it has been the major selective force in the moulding of chaparral rather than the mediterranean climate. Minnich cites the many chaparral species that live in non-mediterranean climates and that all chaparral species grow well in response to summer rains.

Chaparral is variable both in Calif. and in Arizona as are its habitats (e.g. Griffin 1978). Holland (1986) lists 28 chaparral types in Calif., based on dominance.

Succession. Schlesinger and Gill (1980) follow the ecological development of pure stands of *Ceanothus megacarpus.* Annuals are cued to follow fire (Keeley & Keeley 1982). For some species charcoal influences germination more than does heat alone. Non-sprouting species are more likely to be self-pollinators than sprouters (Keeley & Keeley 1977, Carpenter & Recher 1979, Fulton & Carpenter 1979) which are mostly tetraploids (Wells 1986). Survival of seedlings is variable and dependent on rooting depth and tolerance of water stress (Davis 1986).

Fire reduces soil levels of organic N but increases available N (Rundel 1981). In the latter case N fixation is low (Roth 1982)—root nodule formation is limited for several years after fire. Fire seasonality and intensity strongly influence patterns of post-fire community development (Parker 1987, Florence 1987). Rundel and Parsons (1980) have followed the nutrient changes in 2 chaparral shrubs along an age gradient. Keeley (1984) has assembled a bibliography.

Root systems of chaparral shrubs have been described as deep compared to coastal sage scrub and grasses; now Kummerow et al. (1977, 1978) and Miller and Ng (1977) working on chaparral near San Diego provide modifying details. Where bedrock is near the surface (soil depth less than a meter) the root system conforms to the soil depth, with spacing between fine roots of ca. 2 cm. Fine root growth starts in early spring and continues to late summer. Under arid conditions more fine roots are produced than leaves. Fine roots are shed each year whereas leaves are evergreen. Root/shoot ratio is 3–5.

Physiological studies have demonstrated a great range of species differences. There are now data on phenology (Baker et al. 1980), water relations (Burke 1978, Barnes 1981), leaf age (Field 1983, F. & Mooney 1983), growth and water use efficiency (Mahall & Schlesinger 1982), resistance to insects (Mooney et al. 1983), photosynthesis of montane species (Conard & Radosevich 1981) and as related to elevation (Oechel & Mustafa 1982) and decomposition rates of shrub litter (Schlesinger & Hasey 1981).

Management of chaparral is necessary in forestry and with regard to fire (Green 1977, Mooney & Conrad 1977, DeBano et al. 1979, Conard et al. 1985, Conrad et al. 1986).

* * *

Several recent volumes have been devoted to chaparral, including diCastri & Mooney (1973), Mooney (1977), Mooney & Conrad (1977), Miller (1981), Conrad & Oechel (1982), Kruger et al. (1983). The short chapter by Mooney and Miller (1985) is an excellent summary for Calif. The periodical literature is rich, and only little of it is mentioned here.

Almost pure stands of fire-following, seed-originated, even-aged *Ceanothus megacarpus* of the Santa Ynez Mts. were found to have high productivity (850 g/m^{2}yr) and biomass of up to 6563 g/m^2 at 22 years following establishment (Schlesinger et al. 1982). Competition for water and light and nutrient cycling are also discussed. Monty-gierd-Loyaba and J. E. Keeley (1987) found a 55-year-old mixed species stand, and they conclude "chaparral senescence" is a concept in need of re-evaluation. Gray (1982) compared adjacent stands of the *Ceanothus* chaparral and coastal sage scrub. He concluded (1983) that where the two are in contact, fire results in replacement of the coastal sage by the chaparral. The opposite effect is predicted by Zedler et al. (1983) following fires in successive years. J. E. Keeley (1987, K. et al. 1985) found a variety of germination responses to light/dark, with or without powdered charcoal, heat and wetting in 36 woody, 28 herbaceous and 2 suffrutescent species found in these two plant formations. Further work on 57 species is reported in Keeley and Keeley (1987).

Chaparral species in Marin Co. which stump sprout transpired at half the rate of seed-regenerating species. *Adenostema fasciculatum* was an exception (V.T. Parker 1984). Further variations in the water relations of *Quercus durata, Heteromeles arbutifolia, Adenostema fasciculatum* and *Rhamnus californica* have been carefully studied by Davis and Mooney (1986).

Some chaparral species lose their leaves during drought. Physiological ecology studies of the early successional *Lotus scoparius* (Nilsen & Muller 1980, 1981; N. & Schlesinger 1981) and of the successional and rocky face adapted *Diplacus aurantiacus* (Gulmon & Chu 1981) may help to understand the role of such plants in chaparral.

Studies of chaparral have investigated few of the numerous plant species involved. The two "chaparral genera" *Ceanothus* and *Arctostaphylos* have at least 32 and 29 chaparral species, most essentially unstudied either autecologically or synecologically. With numerous shrubby "chaparral species" from such other genera as *Quercus, Rhus, Rhamnus, Diplacus, Penstemon (Keckiella), Eriogonum, Lepechinia, Dendromecon, Ribes* and single species of *Styrax, Umbellularia, Fremontia, Lithocarpus, Chrysolepis, Cercis, Pickeringia, Comarostaphylos, Xylococcus* plus exotics in *Sarothamnus, Cytisus, Ulex* there is ample opportunity for further investigations.

13. Southern coastal scrub

This vegetation adjoins chaparral, and the two vegetation types are frequently studied together as noted above. Westman (1981a & b) gives overviews of the type. He points out that the understory plants of coastal sage scrub are the invaders of burned evergreen chaparral (1979). This must be a southern Californian phenomenon judging from the known, exclusively southern distribution of coastal sage scrub. Why? In fact, the location of coastal scrub is often determined by the presence of a fine-grained geological substratum. Cole (1980) describes an example.

Biomass and related vegetation parameters were measured by Gray and Schlesinger (1981).

The vegetation of alluvial floodplains in southern Calif. is somewhat related to southern coastal scrub, but it has important differences in species composition and of course in ecology (Smith 1980).

14. Valley grassland

Most of the introduced species of the present Californian flora occur in "the Valley Grassland." Jackson (1985) found that at present the annual grasses which now dominate the valley grasslands of Calif. are ruderals in the coastal Mediterrean region from which many (most?) of them came. Is the conclusion proper that the present annual grasslands in Calif. are still highly unstable with respect to their flora, subject to drastic modification as other weedy species arrive? Is the recent history of invasion by *Taeniatherum asperum, Aegilops triuncialis, Brachypodium distachyon, Schismus barbatus,* etc.—each of which has a different, well-defined niche—a portent of the future? Should "the Mediterranean region" as a source area for future plant invasions include the area of mediterranean climate extending eastward, altitudinally above the deserts to Soviet Turkestan? There are already ecotypic differences between the French examples of some of our annual grasses and their putative Californian derivatives, and the 5 global "mediterranean" areas do not have the same rainfall patterns, floras, soils, etc., so plant behavior also is different (Jackson & Roy 1986).

The valley grassland lies below and to the west of the area of *Bromus tectorum*. From a phytosociological point of view it is inadequately defined, inadequately documented, inadequately subdivided into ecologically more or less homogeneous units, inadequately ordinated. Stand tables of species composition are lacking. In a fine paper on the water relations-phenology of "3 co-occurring, serpentine-grassland annuals," namely *Plantago erecta, Clarkia rubicunda,* and *Hemizonia luzulifolia,* the associated species which determine (?) the reactions of the named species are ignored (Gulmon et al. 1983). A penetrating field and laboratory study of ecological behavior of *Bromus mollis, Lolium multiflorum* and *Avena fatua* (Gulmon 1979) concluded (p. 415) that "the components of the grassland system are not necessarily stable over long time periods, but persist on the basis of the frequency of conditions at which they are competitive." What happens when other, different species are in the mix? Ewing and Menke (1983a & b) studied drought effects on reproductive potential and herbage production of *Bromus mollis* and *Avena barbata*. A wider variety of species' responses to variations in temperature and rainfall at Hopland were studied by Pitt and Heady (1978). Also at Hopland Bartolome (1979) investigated germination and seedling establishment. Germinable seed density in 2 years was 6.7 and 6.1 per cm^2. Within the one species, *Bromus mollis,* Jackson et al. (1987) studied the seasonal shift of photosynthetic activity from leaves to inflorescence. Young et al. (1981) exhaustively investigated quantitative aspects of seed availability for germination in 2 Sierran foothill sites.

Comparing populations of *Bromus mollis* and *B. rubens,* Wu and Jain (1979) found that for non-desert sites the former had a higher survival rate and the latter a higher reproductive rate in their respective habitats. Borchert and Jain (1978) documented the considerable effects populations of *Microtus californicus* and *Mus musculus* had on shifts in species composition in annual grassland. *Avena fatua* and *Hordeum leporinum* were most selected, and the other, associated species increased in size and reproductive output. In a serpentine annual grassland Hobbs and Mooney (1985) studied production, dispersal and storage of seed and germination, survivorship and growth in both the undisturbed grassland and on gopher mounds. Movement of seeds was short-range.

What were the original kinds of grassland vegetation in California? Most field botanists believe *Stipa pulchra* was prominent inland. However, there is no theoretical reason to believe that the drastic change in California's flora could not exclude a native like *Stipa pulchra* from the vegetation. But has it? *Stipa* is not part of present grassland succession at the north coastal Sea Ranch (Foin & Hektner 1986, see Chapter 21). Bartolome and Gemmill (1981) found that *Stipa* did not seed in on permanent, ungrazed-by-livestock transects at Hopland. However, at many sites the *Stipa* has increased with exclusion of grazing (Hull & Miller, Savelle 1977, personal observations). Heady's mulch plots at Hopland, the Hastings Reservation (Griffin personal communication, White 1966:*514, 1967:*514) and the present Jepson Prairie are also past and present examples, but in these cases the species did not have to invade as seedlings—a residual population of grazed-to-the-ground clumps had persisted. At the Jepson Prairie Bartolome et al. (1986) used grass-derived phytoliths to demonstrate replacement of dominant, perennial *Stipa pulchra* by the exotic annual grasses which now form the "Calif. grassland."

Grasses interact with several kinds of shrubs or trees. The short-lived shrub *Baccharis pilularis* invades and modifies annual grassland (Hobbs & Mooney 1986, 1987). Very interesting data on the life cycle of the *Baccharis* and on seed production, seed rain, and seed storage in the soil are presented. An experiment showed seedling *Bromus mollis* interferes with seedling *Baccharis*, especially under dry conditions (Da Silva & Bartolome 1984), but the shrub favors moister areas (Hobbs & Mooney 1987). Sharp boundaries between grassland and chaparral are usual. Davis and Mooney (1985) showed that early season and surface-soil water use by the grasses, their limited seed dispersal, and herbivory within and adjacent to the chaparral combined with the water relations of the chaparral species and the herbs within the adjacent bare zone could explain the observed, stable plant distributions.

Parker and Muller (1982) studied vegetation differences under and adjacent to *Quercus agrifolia* in southern Calif. grasslands. Whereas *Avena fatua* was prominent in the grassland, *Bromus diandrus* was prominent under the tree canopies along with the evidently allelopathic *Pholistima auritum*. Under and adjacent to *Quercus garryana* in northwestern Calif. occur 62 species of perennial (13) and annual (9) grasses plus 21 perennial and 19 annual forb species (Saenz & Sawyer 1986). Season-long grazing reduces the number of perennial grass and forb species and increases the number of annual grass species. *Cynosurus cristatus* has most cover under full season grazing whereas several perennial grasses have most cover under partial season grazing.

15. Vernal pools

R. Holland, *Natural Diversity Data Base, Non-game Heritage Program, California Dept. of Fish & Game*

Since 1977 two symposia on vernal pools have been held (Jain 1976, J. & Moyle 1984), and research has advanced as follows:

Species diversity/richness. The relationship between vernal pool area and species diversity has been investigated using the MacArthur-Wilson insular biogeographic model. Holland and Jain (1981) reported very small slopes for the log-log linear reagression of species richness on vernal pool area ($z=0.04$). Similar data have been reported by Ebert and Balko (1984, $z=0.17$), Rosario and Lathrop (1984, $z=0.13$–0.15) and Zedler (1987, $z=0.10$ in a re-analysis of Ebert and Balko's data). These values are as low as any reported for insular biotas anywhere. Zedler (1987) has recently reviewed these results.

Habitat loss. Holland (1978) mapped the distribution of vernal pools throughout the Great Valley, based on air photo interpretation. His survey of about 6×10^6 ha indicates that about 1/3 of the Great Valley had soil conditions appropriate for vernal pools in pre-agricultural Calif., but only 11% of this area still supported such habitats as of the date of the imagery (late 1960s to early 1970s). Zedler's generalization of Holland's map overstates vernal pool acreage. Habitat losses have continued since the early 1970s. Bauder (1986) found that 27% of the vernal pools extant in the San Diego area in 1979 had been lost to urbanization by 1986. Only 7% of the original area persists as present-day vernal pools.

Plant associations. Most ecological sutdies have examined how vernal pool vegetation is structured along the inundation gradient (Zedler 1987). Study areas have been limited as to number of pools and geographically. Almost no stand table data which might show recurrent groups of species are available. Classifications have been proposed by Holland (1976) and Holstein (1983).

MacDonald's (1976) may be the first stand table of vernal pool vegetation. He used data from 10 m^2 plots to identify 6 "association groups" of species and then described how these groups related to the microtopography at a site in northeastern Sacramento Co. His table is arranged to show the influence of standing water on the species composition, with species from the surrounding grassland on one side of the table, grading to species found in the deepest parts of the pools on the other side.

Holland's unpublished dissertation (1978) includes a stand table for 83 vernal pools sampled over a wide latitudinal range and over several soil types. He used whole pools as stands. His analysis identified several species groups. Common to virtually every vernal pool was a group of ubiquists: *Eryngium vaseyi, Deschampsia danthonoides, Allocarya stipitata micrantha, Lasthenia fremontii, Layia fremontii, Eleocaris macrostachya* and *Psilocarphus tenellus*. These plants were also frequently dominants.

Each of Holland's study sites had a unique suite of species, demonstrating strong endemism. San Diego pools were most conspicuous in this respect, being characterized by *Pogogyne abramsii, Eryngium aristulatum parishii, Anagallis arvense,* and *Crassula erecta*. The first 2 are endemic to the San Diego-Tijuana area. Both endemics and ubiquists were dominants.

All of the pools sampled from Redding or related soil series had *Brodiaea hyacinthina, Juncus bufonius, Downingia cuspidata* and (except for the San Diego area) *Pogogyne zizyphoroides*. Most of these species had low cover values.

Shallow pools on Redding soils in Shasta Co. were characterized by *Thysanocarpus curvipes, Allocarya austinae* and *Sidalcea calycosa* while deeper ones also included *Orthocarpus attenuatus, Marsilea mucronata, Juncus patens, Blennosper-*

ma nanum and *Isoetes howelii.* Of these species, only *Blennosperma* and *Sidalcea* ever attained high cover. These pools were quite similar to those described by MacDonald (1976).

Introduced taxa were present in low cover in most pools. Plants such as *Hypochoeris glabra, Bromus mollis* and *Erodium botrys* were observed late in the summer. They were interpreted as plants from seed produced earlier that year in the surrounding grassland. Curiously, Holland included with these weeds the native *Navarretia leucocephala,* a plant that accounted for over 20% of the cover in several pools.

Pools on Redding soils in eastern Stanislaus Co. often were dominated by *Eryngium vaseyi* and *Allocarya stipitata-micrantha,* but also had high relative covers of *Hordeum geniculatum, Allocarya stipitata stipitata* and *Limnanthes douglasii rosea.* In contrast Stanislaus Co. vernal pools on Miekle soils (recent, clayey soils formed in drainageways between surrounding, low hills—the drainageways being blocked by deposition of alluvium from surrounding major streams) had fewer of these species and less cover. These pools are known to support large populations of several species of *Orcuttia,* but low rainfall had precluded their growth.

Holland's last group characterized the alkali pools of Solano and Merced Cos. Their soils have lime-silica cemented hardpans (Fresno, Lewis, Landlow and related soils) or dense, very slowly permeable subsoils (Antioch, Pescadero and similar soils) rather than iron-silica hardpans such as are found in Redding, San Joaquin and related high terrace soils. These pools also failed to fill with water so they were not intensively sampled. Dominant plants here included *Lythrum hyssopifolia, Cressa truxillensis vallicola, Distichlis spicata, Eryngium aristulatum, Navarretia bakeri* and *Lepidium latipes.*

Holland and Jain (1978) present a species-association table for 58 pools sampled at 4 sites over 3 years from near Redding (Shasta Co.) to near Pixley (Tulare Co.). Their table has several similarities with the two described above: a group of ubiquists, several edaphically limited groups specific to shallow or deep pools on particular soils, groups found at only one site but recurring there over 3 years, and a group of grassland species—waifs.

All 3 of these tables were prepared using the same computer program (Ceska & Roemer 1971:*667). What is most remarkable about them is a uniform result—the algorithm arranged the pools (stands) monotonically by pool depth, even though pool depth data were not used by the algorithm, only species composition. Other features are: several species groups may be present at a given site, reflecting the zonation of taxa with respect to the inundation gradient; not all species groups are common to all sites, reflecting regional endemics and apparently edaphic specialization in the vegetation; and the number of species groups at a given site is not constant, reflecting regional variation in community complexity. Many of the differences between vernal pool soils reflect varying ages of soil profile development (Harden 1987). Presumably the observed differences in vernal pool vegetation in turn reflect these differences in age (Walker et al. 1983).

16. Montane and subalpine forests of the Transverse Ranges

Borchert and Hibberd (1984) measured tree basal areas and density in 126 sample plots located between 1675 and 2135 m elevation on the north exposure of Pine Mt. in Ventura Co. Relative values were combined to give importance values which were ordinated by detrended correspondence analysis and classified (Twinspan). Elevation and a topographic gradient of increasing aridity from canyon bottom through north to east to west to south exposures to exposed ridges provide the coordinates within which the tree species attain various importance values. Roughly, the low altitude species are less mesic in the order *Pseudotsuga macrocarpa*<*Calocedrus decurrens*<*Quercus chrysolepis*<*Pinus ponderosa* and those of higher altitudes *Pinus jeffreyi*<*P. lambertiana*<*Abies concolor.* This is a crude summary of a very interesting, if heavily massaged, data set.

Is ordination a helpful means to understanding vegetation-environment relationships? Wartenberg et al. (1987) conclude detrended correspondence analysis excessively simplifies the problem. The equation is not just 2- or 3-dimensional. Jenny (1941:*71) suggests a minimum of 5 groups of factors. It is easy to assemble 5 or more pertinent or effective aspects of climate alone. It is difficult to hold uninteresting factors constant. Smooth curves of overlapping tolerances (Whittaker 1960:*137, 1975:115, Barbour et al. 1987:159, 227) are probably also excessive simplifications (Major 1980). Certainly the overlapping curves of species measures against an environmental gradient need to be backed up by a tabulated matrix of species' abundance versus stands. If the stands can be arranged by some environmental measure, so much the better.

17. Montane and subalpine vegetation of the Sierra Nevada and Cascade Range

Robert J. Laacke, *Pacific Southwest Forest & Range Exp. Sta., US Forest Svc., Redding, California*

Although more is being learned about natural regeneration of the true firs in Calif., there is still much uncertainty as to the combination of conditions conducive to natural regeneration.

Establishment of natural regeneration of shasta red fir (*Abies magnifica* var *shastensis* Lemmon) in undisturbed forests of California's northern Coast Ranges correlated with sunfleck intensity, as did natural seedling establishment of Calif. red fir in the central Sierra Nevada (Ustin et al. 1984, Selter & Pitts 1986). On the other hand, establishment of natural regeneration on disturbed sites in northeastern Calif. suggests that the species has broad tolerances to levels of direct solar radiation. On these 4-acre shelterwood sites, where duff and litter were removed and a bare soil seedbed was created, natural regeneration after 8 years reached 48,000 seedlings and saplings per hectare.

On the most open of these sites (25 seed trees/ha) full sun is the rule with but transitory shade. The heaviest shelterwood level (75 trees/ha) resulted in an environment with little direct solar radiation (except in sunflecks) but significantly more indirect lighting than in a closed, undisturbed red fir forest. Seedling and sapling numbers were not different at the 3 levels of cover although growth of the seedlings, after the 4- to 5-year slow-growth period, was greatest on the areas with the least overstory shade (Laacke & Tomascheski 1986).

A possibly significant difference between these studies is the surface condition of the soil, natural litter and duff as opposed to disturbed bare mineral soil. The rapidly increasing temperature of the dark duff as sunfleck intensity and duration increased (Selter & Pitts 1986) could well be the critical difference.

Once established, both red fir and white fir in Calif. have demonstrated the capacity to increase growth following reduction in overstory cover, even after up to 90 years of severe suppression. The longer and more severe the period of suppression, the slower the growth response (Helms & Standiford 1985, Oliver 1985). This ability to resume normal growth after periods as an understory tree makes determination of age distributions difficult in undisturbed, naturally regenerated forests.

Understory vegetation present following severe disturbance (such as clearcutting) varies significantly from what was present before the disturbance. A study of logged sites in the central Sierra Nevada (Zieroth 1978) emphasizes The complexity of predicting community composition noted earlier. Vegetation on the clearcuts represented 9 genera (10 species) of grasses and sedges, 57 genera (80 species) of herbs, and 9 genera (12 species) of brush. Present on more than 80% of the clearcuts, but not on the uncut controls, were 1 genus of grass, 17 genera of herbs, and 6 genera of shrubs. However, there were 3 genera of grasses, 4 of herbs, and 4 of shrubs that were present on 90% or more of all sites including the uncut controls.

* * *

Vankat (1982) ordinated 108 stands from Sequoia Natl. Park into 15 vegetation types along axes of elevation and topographically determined moisture. Tabulated stand surveys are not given. Changes in the vegetation of Sequoia Natl. Park are documented and attributed to cessation of heavy grazing by domestic livestock, mostly sheep, and to effective wildfire control (Vankat & Major 1978). Parker (1982, 1984, 1986a) has applied ordination and Twinspan to forests of the southern Sierra Nevada (Yosemite region). Elevation and a topographic index separate different kinds of forests. Density differences between small- and large-diameter trees indicate stand dynamics. Diameter class patterns are different for white and red firs. Tabulated stand surveys for the sampled sites to illustrate the results are not given. Ordered species-stand tables would confirm and illustrate conclusions, suggest others, add to our autecological knowledge of species and lead to a hierarchial classification of forest vegetation. Tree species could advantageously be entered into the tables not only as species but by various other properties—by species' age and diameter distributions, site index, reproduction, form, biomass, leaf area, clumping, etc., within their populations.

Some lodgepole pine forests are not successional. This holds true for the Sierra Nevada (Parker 1986b) as well as Yellowstone Natl. Park (Despain 1983).

Westman's (1987) numbers for white and red fir biomasses (59–97 and 40–67 kg/m^2) and productivities (1630–2200 and 810–1030 g/m^{2}yr) in the southern Sierra Nevada need amplification and definition of just what kinds of forests were sampled. *Pinus lambertiana* is one of the loveliest and most valuable of our trees, and Yeaton (1984, 1983) describes its elevational distribution, growth, size class distribution and replacement of *Pinus ponderosa*.

White fir forests of the northern Sierra Nevada and their fire-related, chaparral-dominated successional stages are well described and documented by Conard and Radosevich (1982). Full presentation of standard tree measures for all stands and tabular comparison of species composition document the hypothesized succession, lend credence to the conclusions and provide a factual data base for further studies.

Not all forest types in Calif. have been well described by such standard measures as density of various age or size classes, basal area, height, locations, as well as associated vegetation. Shasta red fir forests are described by Barbour and Woodward (1985).

18. Alpine

Alpine vegetation can be paramo, mediterranean shrub, meadows, tundra, steppe or desert vegetation. Its lower boundary can be placed where the tree life form stops. Where steppe or desert alpine occurs, trees are confined to nonzonal water sources. In the Great Basin the highest altitude aspen are in snowpatch environments. Woody, "steppe" *Artemisia* species such as *A. arbuscula* and *A. nova* often occur above this altitude limit. We would therefore modify the definition of "alpine" by Hunter and Johnson (1983) and Bell and Johnson (1980) to include these *Artemisia* species and their associates when they occur above the last plants of tree life form, even though these trees are in azonal habitats.

The alpine flora of the Sierra Nevada makes up not only the vegetation above timberline but also much of the vegetation of openings in the subalpine forests below. These openings may be caused by excessive snow accumulation, wind exposure so snow cannot accumulate, snow avalanches, rockfalls, high water tables—either moving or stagnant, springs, soils of extraordinarily low water-holding capacity such as sand, gruss or gravel, talus blocks both sloping and flat-lying (Felsenmeer), bare rock outcrops—flat-lying or cliffs, including glacier-scoured bedrock, stream overflows, ice-shoved lakeshores, solifluction features such as polygons, stripes, mounds and hollows, etc.

The alpine flora is derived from 4 source floras according to Stebbins (1982). Of the total High Sierran flora of 876 species, 342 are old Cordilleran, 227 are Circumboreal, 168 are lowland cismontane Californian and 139 are Great Basin. Considering the 207 species occurring exclusively in the alpine zone, the figures are 103, 58, 15 and 31. The Sierra is a center of speciation for *Ivesia, Hackelia, Castilleja* and possibly other genera, but the Sierran species of genera such as *Carex* almost all also occur elsewhere—66 out of the 77 High Sierran species.

While most of the species are herbaceous perennials, a high proportion of woody species occurs in the Circumboreal group while annuals are from lowland Calif. or, to a lesser degree, from the Great Basin. As to habitats, only the Circumboreal group has an abundance of species of continuously moist habitats. Species of continuously dry habitats predominate in the other source groups, strikingly so in plants of the Great Basin.

The Sierra Nevada is unusual in its abundance of annual and spring-ephemeral perennial plant species at high altitudes. Jackson and Bliss (1982) describe the *Carex exserta* open sand and gravel communities where these species occur. Numbers of both annuals and ephemerals (9 and 11 respectively) decrease from the subalpine upward into the alpine zone.

There is an obvious conflict between the very evident invasion of many Sierran mountain meadows by *Pinus contorta* ssp. *murrayana* and a high degree of meadow stability concluded from palynological and stratigraphic studies (Benedict 1982, Benedict & Major 1981). Bruce Johnson's fine thesis (1986) concludes that large (>0.5 m Ø) open, bare spots are avoided by voles, etc., which normally consume tree seeds. Seedlings cannot persist on permanently wet and unshaded sites. Helms (1987) shows that lodgepole seed rain is abundant in Yosemite meadows, but establishment is not. Establishment does seem to be positively related to dry periods. Using plant species' cover values to ordinate plots along a moisture gradient. Helms and Ratliff (1987) show that sites which are wet early in the growing season but dry somewhat later are most invaded.

Aspects of the ecology of several other characteristic, high altitude species of the High Sierra have been studied. The species include *Mimulus primuloides* (Douglas 1981), *Carex exserta* (Ratliff 1973:*672, 1974:*674, 1985a & b), *C. nebrascensis* (Ratliff 1983, Steele et al. 1984), *Gentiana* ssp (Spira 1986), *Pinus albicaulis* (Tomback 1986, Mattes 1982).

We are still many stand surveys away from a conclusive classification, and thereby description, of Sierran meadow vegetation (Ratliff 1982, 1985). What should a classification be based on? Obviously on vegetation properties. The most evident properties of vegetation are qualitative, namely changes in species composition. But "quantitative methods" are current. And quantitative is defined as numerical when the best of our data are only ordinal. "Only" if we use the increasing precision of numbers from naming to ordinal to numerical to scalar. Halpern's (1986) study of montane meadows (at 1493 to 2390 m elevations and hence not alpine) in Sequoia Natl. Park essentially followed Braun-Blanquet's methods of sampling (Mueller-Dombois & Ellenberg 1974:*731). 71 plots were analyzed into 12 associations. Twinspan (Hill et al. 1975) aided analysis, and Decorana (Hill & Gauch 1980) was used to ordinate associations. Stand-species matrices to illustrate or test the classification and the ordination are not presented. Haase (1987) has shown how valuable a documented vegetation classification of a limited area (Komarkova 1979) can be in mapping the kinds of vegetation and their ecological relationships in geographically separated areas.

Alpine vegetation and its ecology in the Rae Lakes Basin has been described and analyzed by Burke (1979, 1982). 126 stands were grouped by their species composition into 12 associations of 5 alliances. 15 species groups characterize singly or in combination the associations and their ecological differences. Tables of stands versus species groups were used to characterize plant associations and to describe their ecology. The main site factors are elevation, site moisture, snow cover, slope and exposure and substrate characteristics.

Benedict (1981, 1983) investigated meadow vegetation in the Rock Creek area of Sequoia Natl. Park. Again, he recognized plant associations characterized by particular groups of species. The association tables are printed in full. They can be extended, tested, and revised by recording the species composition of additional stands in other sites. The 19 associations were ordinated. "Of the 16 environmental factors examined, soil moisture, depth to the water table, and depth of standing water were related to the x-axis of the ordination, and water temperature was related to the y-axis." Other site factors were correlated with specific associations.

19. The redwood forest and associated north coast forests

The extraordinary biomass of an old growth redwood stand (Fujimori 1977) was mentioned in the **Introduction.** In the new Redwoods Natl. Park work began (Viers 1982) and in the state parks continued on ecology of the vegetation, both old growth (Mathews 1986) and following the devastation of logging (Muldavin et al. 1981, Lenihan et al. 1982). Bingham (1984) estimated 35 g/m^2 of woody debris on the ground following logging, and he showed it was an important substrate for conifer seedling establishment. Stand structure in the Little Lost Man RNA (Lenihan et al. 1983) was analyzed by Combs (1984). Stuart described the long fire history of an old growth stand. Fire interval halved after settlement. John Sawyer supplied most of the information for this account.

20. Montane and subalpine vegetation of the Klamath Mountains

John Sawyer, *Dept. Biol. Sci., Humboldt State University, Arcata, California*

Floristic studies. Interest in the region's rare and endemic conifers continues. New locations for *Abies lasiocarpa* were found in the Marble Mts. (Sawyer & Cope 1982), and in 1986 Don Hemphill discovered new southern populations in the Scott Mts. Cope 1978, 1983) studied the monoterpene composition of the Russian Peak population. Knowledge of two endemic conifers was summarized: *Chamaecyparis lawsoniana* by Zobel et al. (1985) and *Picea breweriana* by Thornburgh (1987). Smith and Sawyer supplied keys (1981) and a well-documented checklist (1987a) for the vascular plants.

Floristic richness and phytogeographical relationships were interpreted in a series of local floras: Del Norte Co. (Barker 1978), The Lassics in Humboldt Co. (Nelson 1979), Mt. Eddy in Trinity Co. (Whipple 1981), Snow Mt. in Mendocino Co. (Heckard & Hickman 1978), Klamath Mts. (Quinn 1977), the region as a whole (Smith & Sawyer 1978b) and the region's rich serpentine flora (Baker 1984, Gofoth 1984, Sawyer 1984, Whipple 1984).

Vegetation studies. Intensive vegetation inventory and classification make possible the evaluation of rarity and uniqueness (Sawyer 1987). The Forest Service has surveyed many local areas as potential Research Natural Areas: Adoral (Sawyer 1981a), North Trinity Mt. (Sawyer 1981b), and Ruth (Thornburgh 1983) in Humboldt Co.; Bridge Creek (Keller-Wolf 1985), Indian Creek Brewer Spruce (Sawyer 1978), Perch Creek (Keller-Wolf 1987), Specimen Creek (Sawyer & Stillman 1977a), Williams Point (Sawyer & Stillman 1977b) and the Mt. Shasta mud flow area (Keeler-Wolf 1984c) in Siskiyou Co.; Preacher Meadows (Sawyer et al 1978), Mt. Eddy (Whipple & Cope 1979) and Smoky Creek (Taylor & Teare 1978) in Trinity Co.; Cedar Basin (Keller-Wolf 1982) on the Shasta-Trinity Natl. Forest, and Stone Corral-Josephine (Keeler-Wolf 1986) on the Six Rivers Natl. Forest.

The vegetation of the upper drainage of Sugar Creek in Siskiyou Co. was mapped for the Klamath Natl. Forest (Keller-Wolf 1984b). Forests of the Siskiyou Mts. in Del Norte Co. were described for the Six Rivers Natl. Forest (Thornburgh & White 1979). The vegetation of 3 areas in the South Fork area of the Shasta-Trinity Natl. Forest in Trinity Co. were compared by Keller-Wolf (1984a).

More thorough descriptions and analyses of the region's vegetation are in MS theses done at Humboldt State University. Analyses consist of tabular and cluster analyses and ordination of extensive data sets. Three forest studies were in the Yolla Bolly area (Laidlaw-Holmes 1981, Muldavin 1982—see also Becking et al. 1979, and Waddell 1982) and one in the southern Siskiyou Mts. (Simpson 1980). Palmer (1979) analyzed open vegetation on diorite at the Bear Lakes in Trinity Co. Stillman (1980) classified the meadows of the Marble Mts. in Siskiyou Co. Other theses include Dubendorfer's (1987) explanation for the 2-phase vegetation found on the ultramafics of Del Norte Co. (Whittaker 1960:*137), Regalia's (1978) demonstration that the Shasta red fir forests of the Marble Mts. are all-aged, Ross' (1983) finding a decreasing effect of parent material on species diversity with increasing altitude in the Trinity Mts. and Hunt's (1976) study of environmental restrictions on disjunct *Abies amabilis*.

In the late 1980s a comprehensive ecosystem-type vegetation classification was begun for Forest Service lands in Calif. (Kosco et al. 1985). Extensive data sets have been obtained for the region's forests on the Klamath, Six Rivers and Shasta-Trinity Forests and are being analyzed using a sequence of multivariate techniques. Jimerson and Imper (1986) prepared a classification of the red fir ecosystem types. Jimerson (1987) developed Douglas fir/hardwood, white fir, red fir and Port Orford cedar ecosystem types for the Gasquet District of the Six Rivers Natl. Forest and for the Ukonom and Happy Camp Districts (1988) of the Klamath Natl. Forest. Adamson (1988) is classifying the forests on ultramafic rocks of Trinity Co.

21. Coastal prairie and northern coastal scrub

In continuing studies at the real estate development at Sea Ranch, which was formerly heavily sheep-grazed, Foin and Hektner (1986) show increasing cover of the exotic *Anthoxanthum odoratum.* This short-lived perennial grass is characterized by Oberdorfer (1979:253) as a Magerkeitszeiger and by Ellenberg (1978:917) as oceanic and on somewhat acid soils. Snaydon (1970) found that at Rothamsted it had developed races adapted to both Ca-rich and Ca-poor soils. It can endure shade. Its invasion and relative abundance at Sea Ranch may be due to soil impoverishment in what was a secondary, sheep-induced annual (except for the weedy *Rytidosperma (Danthonia) pilosum*) vegetation under a heavily leaching precipitation regime. Of the precipitation of 1030 mm/yr at Sea Ranch, at least 550 mm must leach through the soil profile. There is an abrupt decrease in annual precipitation, and an equal decrease in water available for soil leaching, southward from Sea Ranch and Ft. Ross to Pt. Reyes where other perennial grasses dominate (Elliott & Wehausen *1974).

By looking closely at such life history behavior of the dominant species as seed production, dispersal, and seedling survival Peart and Foin (1985) were able to identify and analyze successfully the factors affecting population changes in the vegetation at Sea Ranch.

Foin and Jain (1977) earlier suggested approaches to community ecology based on experience at Sea Ranch and with grasslands of the Central Valley.

22. Sagebrush steppe

James A. Young, *Agricultural Research Service, USDA, Reno, Nevada*

Coordination of studies of sagebrush ecosystems has been facilitated by the publishing of *Temperate Deserts and Semi-deserts*, edited by Neil West (1983). West provides ecological data and an excellent perspective on the vegetation of the temperate deserts and semi-deserts of North America. In this same volume Walter and Box summarize some of the information available on the temperate deserts and desert-steppes of Asia.

During the last decade, a cadre of scientists interested in various aspects of the biology of shrubs native to our temperate deserts has evolved. Annual meetings sponsored by the Shrub Consortium, under the leadership of the Shrub Laboratory of the US Forest Service, have been held at Brigham Young University in Provo, Utah. Proceedings have been published by the Intermountain Research Station of the Forest Service located in Ogden, Utah (for example, Tiedemann et al. 1983). The proceedings of these meetings contain much synecological information on sagebrush and related plant communities.

The emphasis on autecological and cytogenetic studies of the various woody species of *Artemisia* has heightened awareness and importance of recognizing related species and subspecies of *Artemisia tridentata* Nutt. Synecological studies in sagebrush communities during the last decade have specified the subspecies of big sagebrush present on the study sites (for example, Hironaka et al. 1983).

Also during the last 10 years there has been extensive application of the habitat type concept in management of rangelands. This has produced a greater awareness of synecological concepts, and it has enlarged the data base available for analysis of vegetation dominated by sagebrush.

A series of studies by Bruce Roundy (for example, 1985) have provided information on plant and soil relationships in saline/alkaline sites located on pluvial lake plains. The information was used to help delineate and study the ecology of plant communities in an extensive salt desert ecosystem (Young et al. 1986).

Much of the interest during the last decade in sagebrush and salt desert plant communities in eastern Calif. has concerned the influence of withdrawing surface and/or ground water on wildland plant communities. Mono Lake is an example and Owens Lake a warning. This interest has heightened awareness of the influence of subaerially deposited sediments rich in soluble salts on soil development and distribution of plant communities (Young & Evans 1986).

Data from eco-physiological studies of *Atriplex confertifolia* and *Ceratoides lanata* are meaningfully summarized by Caldwell (1985). These kinds of data are needed for other species and over the range of ecosystems in which all the species occur.

* * *

Walter and Box (West 1983) discuss the makeup and ecology of *Artemisia* and *Ceratoides* steppes found at high elevations in Middle (within the USSR) and Central (outside the USSR) Asia. Similar communities evidently exist in California's White and Sweetwater Mts. as well as on the eastern slope of the Sierra Nevada. A common species is *Carex duriuscula* (*C. Eleocharis* of Mooney et al. 1962:*672). The shrubs' ecologies should be compared.

23. Transmontane coniferous vegetation

Pinyon-juniper is an important part of this vegetation as it extends eastward from the Sierran scarp across the Great Basin and Colorado Plateaus. A recent symposium (Everett 1987) considered it, and a Soviet note (Vasilchenko & Konnov 1986) mentions work on *Juniperus* forests in the USSR and reports a conference.

Young and Evans (1981) investigated the age, density and fire history of *Juniperus occidentalis* ssp *occidentalis* in *Artemisia tridentata* ssp *wyomingensis* and *A. arbuscula* stands in northeastern Calif. They found that the oldest tree in the former was established in 1855 and 90% of the trees dated to 1890–1920 while the dates for the latter site were 1600 and before 1900. "At the beginning of the 20th century, the western juniper population on big sagebrush sites was doubling in density every 3 years." This kind of settlement-related invasion is common across the Great Basin, involving *Juniperus osteosperma* (Tausch et al. 1981).

Biomass in an Oregon juniper stand was 21,160 kg/ha, primary production 1100 kg/ha/yr and foliage biomass 4315 kg/ha with a LAR of 2.0 ha/ha (Gholz 1980). Most soil nutrient cations, but including H^+ and Na^+, under *Pinus monophylla* in west-central Nevada growing on basalt-andesite were concentrated under the canopy (Everett et al. 1986), and Young et al. (1984) found that during the winter western juniper trees averaged 0.53 liter of water as stem flow per cm of precipitation.

24. Mojave desert scrub vegetation

Vegetation characterized by *Coleogyne ramosissima* is widespread at elevations and latitudes between sagebrush and salt desert shrub and lower and more southern *Larrea tridentata* communities. West (1983) discusses blackbrush occurrence only in Utah. As he points out, using K.F. Wells' biomass data from San Bernardino Co., root/shoot ratios are low in this community, 0.29 and 0.14 in the two stands investigated. Provensa and Urness (1981) have applied shoot diameter and length relationships to shoot weight to determine blackbrush grazing capacity. In the Panamint Mts. the long-lived *Coleogyne* occupies the very oldest geomorphic surfaces (Webb et al. 1987). Younger surfaces with younger, less solidly $CaCO_3$-cemented soils have more species-rich plant communities. On fields homesteaded, plowed and now used only for grazing in the Lanfair Valley southeast of the New York Mts. at elevations from 1100 to 1615 m Carpenter et al. (1986) studied plant succession. At low elevations *Larrea tridentata* has reached a density somewhat more than half the density in unplowed areas although cover is less than 1/3. *Yucca brevifolia* invaded at the medium elevations but very slowly. In the Alabama Hills Yoder et al. (1983) found unique combinations of arid, western American species. Their large (700 m^2) sample plots were distributed mechanically so the mixing of species may be a consequence of their sampling methods. Phytosociological stand surveys might be more informative.

Limestone frequently outcrops in the Mojave desert ranges. Just as in more humid climates (Major & Bamberg 1963:*134) limestone (and dolomite) often has a unique flora. Desert endemics on limestone include *Notholaena jonesii, Scopulophia rixfordii, Eriogonum esmeraldense, E. intrafractum, Tetracoccus spinulosa, Purpusia saxosa, Phacelia barnebyana, Ph. mustalina, Ph. perityloides, Erigeron uncialis, Haplopappus gilmanii, Hazardia brickelloides, Hecastocleis shockleyi, Lepidospartum latisquamum, Peucephyllum schottii*, etc. (DeDecker 1984).

Regional vegetation has long been regarded as that of playa, bajada and arroyo. Mountain washes have been neglected. This is the habitat of a spectacular, new-to-science species, *Dedeckera eurekensis* plus a range extension of the calcicole *Buddleja utahensis*. Others?

Water relations and photosynthesis of plants of the Eureka Valley sand dunes, with the endemic *Swallenia (Ectosperma) alexandrae*, have been studied by Pavlik (1980). He has also described the sand dune flora of 20 Mojave and mostly nearby Great Basin sites (1985).

Water availability is the first factor in desert ecology. Lane et al. (1984) have calculated water balances for desert conditions and show a good correlation between calculated water use and vegetation production. By carefully defining, and limiting, their ecosystem to shallow-rooting cacti on a sandy soil of known bulk and particle densities, water contents at field capacity, and water potentials Young and Nobel (1986) could calculate evaporation and water available to the cacti.

Heat loads on *Echinocereus engelmannii* and *E. triglochidiatus* are related to the form of the plants and thereby to their altitudinal (community) distribution (Yeaton 1982). Yeaton et al. (1985) studied differences between the ecology of *Yucca brevifolia* and *Y. schidigera*.

25. Sonoran desert vegetation

***J. H. Burk**, Dept. Biol. Sci., California State University, Fullerton*

Creosote bush scrub usually has 2 sources of water—precipitation and runoff. Schlesinger & Jones (1984) found that the 45-year-old diversion of sheet flow and stream channel runoff on bajadas along the base of the Coxcomb Mts. caused a significant decrease in the density and biomasses of *Larrea tridentata* and *Ambrosia dumosa*. Average plant size was not changed in *Ambrosia*, but in *Larrea* mortality was concentrated in the larger, older individuals.

Fires in desert vegetation have increased since the years of high precipitation in 1976–1984 (Brown & Minnich 1986), and fires favor the establishment of short-lived shrubs such as *Encelia farinosa*, native ephemerals, and annual aliens such as *Bromus rubens* and *Schismus barbatus* at the expense of *Larrea* and *Ambrosia dumosa*. The biomass of the exotics has continued to be high even during drier years. This increase in flammable biomass may encourage fire and cause a change in the character of some creosote scrub communities without changes in climate.

Studies in physiological ecology include N-fixation, water relations and photosynthetic rates. Record photosynthetic rates have been found in C_4 *Amaranthus palmeri* with 82 μmol $CO_2 m^{-2} s^{-1}$ (Ehleringer 1983) and in the C_4 *Hilaria rigida* with 67 (Nobel 1980). Higher rates than usual occur in C_3 annuals in Death Valley. Werk et al. (1983) associated high rates with high leaf conductances to water vapor and high N content of leaves. Ehleringer (1984) used changes in water relations, growth and reproduction to show competitive release in *Encelia farinosa* when intraspecific competitors were removed.

Nilsen et al. (1986) reported that *Prosopis glandulosa* var *torreyana* from Harpers Well had 2 leaf morphs with different abilities to acclimate depending on season and water status. The smaller size of dry-season leaves may be due to low turgor potential during development rather than an adaptation to tolerate a stressful environment. Nilsen et al. (1983) showed that *Prosopis glandulosa* is a phreatophyte which acclimates on a seasonal basis through adjustments in osmotic concentration and changing stomatal sensitivity to VPD. Nilsen et al. (1984) reported that winter deciduous *Olneya tesota, Prosopis glandulosa* and *Acacia greggii* tolerated summer drought by acclimatizing. *Chilopsis linearis* and *Hyptis emoryi* avoided drought by changes in leaf biomass and low leaf conductance. *Dalea (Psorothamnus) spinosa* avoided stress by maintaining a small evaporative surface. Sparks (1982) compared water relation parameters of *Cercidium floridum* and *Chilopsis linearis*. The more drought tolerant *Cercidium* had lower osmotic potentials, higher turgor potentials, lower points of zero turgor and less elastic cell walls throughout the season of an unusually dry year.

Shearer et al. (1983) found that N is fixed by members of the sub-family *Papilionoideae* and by *Prosopis glandulosa*. Rundel et al. (1982) reported that decoupling the *Prosopis* from limitations of water and N results in a significantly higher than expected productivity and accumulation of biomass.

Nobel has contributed extensively to our knowledge of the autecology of cacti (see below). Cladode orientation of platyopuntias is such that interception of PAR is maximized (Nobel 1982a). Plant size was used to estimate times of seedling establishment of *Ferocactus acanthodes* at various geographic locations (Jordan & Nobel 1982). Heat tolerance of *Opuntia bigelovii* was found to be as high as 59°C and subject to acclimation (Didden-Zopfy & Nobel 1982). Nobel (1982b) concluded that cold hardening increased cold temperature tolerance in cacti; *Opuntia bigelovii* and *O. ramosissima* are geographically limited by low cold tolerance.

Burk (1982) recorded patterns of germination and phenology of ephemerals. Survivorship curves of winter annuals varied with seasonality of precipitation. Clark and Burk (1980) compared patterns of structural and nonstructural C in *Plantago insularis* and *Camissonia boothii*. They found that *Camissonia* had a longer growing season through more allocation of C to vegetative tissues and storage, presumably as investment toward longer life and higher levels of reproduction.

* * *

Cacti and other succulents such as *Agave* are not restricted to the Sonoran desert, but they are common (>125 species of cacti according to Gibson & Nobel 1986:144) and wonderfully diverse there. The Cactus Primer by Gibson and Nobel (1986) is exhaustive, thorough, new, rigorous in its terminology and units, very readable and not a primer but a treasure. Nobel (1985) concisely supplements it.

Soil properties, including nitrate accumulation, under mesquite have been carefully studied by Virginia and Jarrell (1983).

Data on plant ages are necessary for both aut- and synecological studies. Goldberg and Turner (1986) summarize age data for Tumamoc Hill at the famous Carnegie Desert Laboratory near Tucson, Arizona. Records date from 1906. Many of the species are also found in Calif.

26. The southern California islands

Wallace (1985) provides a flora (621 species) and discusses endemism (137 species) and introductions (222 species). Westman (1983) hypothesizes that the correlation between island area and species richness on the Channel islands is related to increased habitat heterogeneity with increasing area rather than to extinction. A million-year-old chronosequence of soils on San Clemente Island has been studied by Muhs (1982). Unfortunately the kind of soil formation occurring on the island is not widespread.

Jack Major
Botany Dept., Univ. California, Davis

References

ABA Consultants. 1987. Elkhorn Slough wetlands management plan (draft report). Monterey Co. Plan. Dept. & State Coastal Conservancy, Oakland, CA. 128 p.

Adamson, B. 1988. Draft ecological types of the Trinity ultramafic sheet, Shasta-Trinity Natl. Forest. Unpubl. rpt. Redding, CA.

Alexander, R.R. 1985. Major habitat types, community types, and plant communities in the Rocky Mts. US For. Svc., Rocky Mt. For. & Range Exp. Sta. Gen. Tech. Rpt. RM-123.

Alexander, R.R. 1986. Classification of the forest vegetation of Wyoming. US For. Svc., Rocky Mt. For. & Range Exp. Sta. Research Note RM-466. 10 p.

Anderson, M.V. & R.L. Pasquinelli. 1984. Ecology and management of the northern oak woodland community, Sonoma Co., California. M.S. thesis, Sonoma State Univ., Rohnert Park, CA. 125 p.

Andrecht, K.L. & T.E. Firle. 1987. The Chula Vista wildlife reserve: Development of a coastal salt marsh in South San Diego Bay. Assoc. Envir. Professionals, Ann. Conf., San Diego, CA. 17 p.

Atwater, B.F. 1980. Distribution of vascular plant species in 6 remnants of intertidal wetland in the Sacramento-San Joaquin Delta, California. US Geol. Surv., Open File Rpt. 80-883. 46 p.

Atwater, B.F., et al. 1979. History, landforms and vegetation of the estuary's tidal marshes. *In* Conomas, T.J., ed. San Francisco Bay: the urbanized estuary. Pac. Div., Am. Assoc. Adv. Sci., San Francisco, CA. Pp. 347-400.

Axelrod, D.I. 1986. Age and origin of chaparral. Symposium on the California chaparral: Paradigms revisited. 7-8 Nov. Natur. Hist. Mus., Los Angeles. In press.

Axelrod, D.I. 1983a. New Pleistocene conifer records, coastal California. Univ. of Calif. Publ. Geol. Sci. 127:ix + 108.

Axelrod, D.I. 1983b. Biogeography of oaks in the Arcto-Tertiary Province. Ann. Missouri Bot. Garden 70(4):629-657.

Axelrod, D.I. 1982. Age and origin of the Monterey endemic area. Madroño 29(3):127-147.

Axelrod, D.I. 1980. The origin of coastal sage vegetation, Alta and Baja California. Amer. J. Bot. 65(10):1117-1131.

Axelrod, D.I. 1979. Age and origin of Sonoran desert vegetation. Occ. Papers Calif. Acad. Sci. 132:vi + 74.

Baker, G.A., P.W. Rundel & D.J. Parsons. 1982. Comparative phenology and growth in 3 chaparral shrubs. Bot. Gaz. 143:94-100.

Balling, S.S. & V.H. Resh. 1983. The influence of mosquito control recirculation ditches on plant biomass, production and composition in 2 San Francisco Bay salt marshes. Estuaries & Coastal Shelf Sci. 16:151-161.

Barbour, M.G., J.H. Burk & W.D. Pitts. 1987. Terrestrial plant ecology. 2nd ed. Benjamin/Cummings, Menlo Park, CA. xiii + 634.

Barbour, M.G. & R.A. Woodward. 1985. The Shasta red fir forests of California. Canad. J. For. Res. 15:570-576.

Barbour, M.G., T.M. DeJong & B.M. Pavlik. 1985. Marine beach and dune plant communities. *In* Chabot, B.F. & H.A. Mooney, eds. Physiological ecology of North American plant communities. Chapman & Hall, NY. Pp. 296-322.

Barbour, M.G., A. Shmida, A.F. Johnson & B. Holton, Jr. 1981. Comparison of coastal dune scrub in Israel and California. Israel J. Bot. 30:181-198.

Barbour, M.G. 1978a. Salt spray as a microenvironmental factor in the distribution of beach plants at Pt. Reyes, California. Oecologia 32:213-224.

Barbour, M.G. 1978b. The effect of competition and salinity on the growth of a salt marsh plant species. Oecologia 37:93-100.

Barbour, M.G. & T.M. DeJong. 1977. Response of west coast beach taxa to salt spray, seawater inundation, and soil salinity. Bull. Torrey Bot. Club 104:29-34.

Barbour, M.G. & C.B. Davis. 1970. Salt tolerance of 5 California salt marsh plants. Amer. Midl. Natur. 84:262-265.

Barker, L.M. 1978. A flora of the Old Gasquet Toll Road, Del Norte Co., California. M.S. thesis, Humboldt State Univ., Arcata, CA. 147 p.

Barker, L.M. 1984. Serpentine flora notes on prominent sites in California: Scott Mt./China Mt. crest zone. Fremontia 11:13-14.

Barnes, F.J. 1981. Water relations of 4 species of *Ceanothus*. *In* Proc. Intl. Bot. Congr. 13.

Barrows, K. 1983. Old-growth Douglas fir forest. Fremontia 11:20-23.

Barry, R.G. 1981. Mountain weather and climate. Methuen, London. xii + 313.

Barry, R.G. 1979. High altitude climates. *In* High altitude geocology, ed. P.J. Webber. AAAS Selected Symp. 12. Westview Press, Boulder, CO. Pp. 55-74.

Barry, W.J. 1985. MacKerricher state park Ten Mile Dunes restoration plan. Rpt. of Res. Agency, Calif. Dept. Parks & Rec., Natur. Heritage Sect., Sacramento. 16 p.

Bauder, E.T. 1986. San Diego vernal pools: recent and projected losses, their condition and threats to their existence. Calif. Dept. Fish & Game, Sacramento.

Bartolome, J.W., S.E. Klukkert & W.J. Barry. 1986. Opal phytoliths as evidence for displacement of native Californian grassland. Madroño 33(3):217-222.

Bartolome, J.W. & B.A. Gemmill. 1981. The ecological status of *Stipa pulchra (Poaceae)* in California. Madroño 28(3):172-184.

Bartolome, J.W. 1979. Germination and seedling establishment in California annual grassland. J. Ecology 67(1):273-281.

Beare, P.F.A. 1984. Salinity tolerance in cattails (*Typha domingensis* Pers.): explanations for invasion and persistence in a coastal salt marsh. M.S. thesis, Calif. State Univ., San Diego. 57 p.

Becking, R.W., J.M. Lenihan & E.H. Muldavin. 1979. The forest communities of the Big Butte/Shinbone region of northern California's Yolla Bolly Mts. Freestone Wilderness Inst., Covelo, CA. 47 p.

Bell, K.B. & R.E. Johnson. 1980. Alpine flora of the Wassuk Range, Mineral Co., Nevada. Madroño 27:25-35.

Benedict, N.B. 1983. Plant associations of subalpine meadows, Sequoia Natl. Park, California, USA. Arctic & Alpine Res. 15(3):383-396.

Benedict, N.B. 1982. Mountain meadows: stability and change. Madroño 29(3):148-153.

Benedict, N.B. & J. Major. 1981. A physiographic classification of subalpine meadows of the Sierra Nevada, California. Madroño 29(1):1-12.

Benedict, N.B. 1981. The vegetation and ecology of subalpine meadows of the southern Sierra Nevada, California. PhD thesis, Univ. of Calif., Davis. 128 p.

Bilhorn, T.W. & E.B. Macdonald. 1987. Hydrologic regime of an altered California coastal marsh. Abstract. Natl. Symp. Wetland Hydrology, Assoc. State Wetland Mgrs, Inc., Chicago.

Billings, W.D. & M.G. Barbour, eds. 1987. North American terrestrial vegetation. Cambridge Univ. Press, NY. In press.

Billow, C.R. 1986. A comparison of seasonal patterns of soil N beneath an evergreen and deciduous California oak. *In* Intl. Congr. Ecol. Prog. Syracuse, NY. Abstract.

Bingham, B.B. 1984. Decaying logs as a substrate for conifer regeneration in an upland, old-growth redwood forest. M.A. thesis, Humboldt State Univ., Arcata, CA. 56 p.

Bio-flora Research Inc. 1986. Buhne Point Shoreline Erosion Demonstration Project, Humboldt Bay, Humboldt Co., California. Phase 2, planting methods and costs. USA Corps of Eng., San Francisco Dist. 24 p.

Bioscience. 1987. Mooney, H.A., et al. Plant physiological ecology. 37(1):18-67.

Borchert, M., F.W. Davis & L.D. Oyler. 1986. Factors affecting acorn losses and seedling recruitment for blue oak (*Quercus douglasii* Hook & Arn.) in San Luis Obispo Co., California. *In* Intl. Congr. Ecol. Prog. Pp. 96-97. Abstract.

Borchert, M.I. & M. Hibbard, 1984. Gradient analysis of a north slope montane forest in the western Transverse ranges of southern California. Madroño 31(3):129-139.

Borchert, M.I. & S.K. Jain. 1978. The effect of rodent seed predation on 4 species of California annual grasses. Oecologia 33(1):115-127.

Bowland, J.L. & E.T. Jordan. 1987. Guadalupe Dunes revegetation program. *In* Proc. 5th Symp. on Coastal & Ocean Mgt., Seattle. Pp. 1704-1715.

Boyd, R.S. 1986. Comparative ecology of two west coast *Cakile* species at Pt. Reyes, California. PhD thesis, Univ. of Calif., Davis.

Boyd, R.S. & M.G. Barbour. 1986. Relative salt spray tolerance of *Cakile edentula* from lacustrine and marine beaches. Amer. J. Bot. 73:236-241.

Brown, D.E. & R.A. Minnich. 1986. Fire and changes in creosote bush scrub of the western Sonoran Desert. Amer. Midl. Natur. 116:411-422.

Brown, D.R. 1987. The effect of human trampling on dune-mat vegetation of the Lanphere-Christensen Dune Preserve. I. Prelim. rpt. Nature Conservancy, Calif. Field Off., San Francisco. 12 p.

Burk, J.H. 1982. Phenology, germination and survival of desert ephemerals in Deep Canyon, Riverside Co., California. Madroño 29:154-163.

Burk, J.H. 1978. Seasonal and diurnal water potentials in selected chaparral shrubs. Amer. Midl. Natur. 99(1):144-148.

Burke, M.T. 1982. The vegetation of the Rae Lakes Basin, southern Sierra Nevada. Madroño 29(3):154-176.

Burke, M.T., R. Curry, J. Major & D.W. Taylor. 1982. Natural Landmarks of the Sierra Nevada. Heritage Cons. & Rec. Svc., Natl. Park Svc. Bot. Dept., Univ. of Calif., Davis. 529 p.

Burke, M.T. 1979. The flora and vegetation of the Rae Lakes Basin, southern Sierra Nevada: an ecological overview. M.S. thesis, Univ. of Calif., Davis. 166 p. + tables, diagram.

Caldwell, M.M. 1985. Cold desert*in* Physiological ecology of North American plant communities, ed. B.F. Chabot & H.A. Mooney. Chapman & Hall, NY. Pp. 198-212.

California Coastal Comm. 1981. Statewide interpretative guidelines for wetlands and other wet, environmentally sensitive habitat areas. Calif. Coastal Comm., San Francisco.

Carpenter, D.E., M.G. Barbour & C.J. Bahre. 1986. Old field succession in Mojave desert scrub. Madroño 33(2):111-122.

Carpentera, F.L. & H.F. Recher. 1979. Pollination, reproduction and fire. Amer. Natur. 113:871-879.

Chipping, D.H. & R. McCoy. 1982. Coastal sand dune complexes, Pismo Beach and Monterey Bay. Calif. Geol. 315(1):7-12.

Clark, D.D. & J.H. Burk. 1980. Resource allocation patterns of two California Sonoran Desert ephemerals. Oecologia 46:84-91.

Clark, K.L. 1986. Soil nutrient analysis, beach and dune nutrient cycling, fertilizer use and long-term management recommendations, Buhne Point Shoreline Erosion Demonstration Project. USA Corps of Eng., San Francisco Dist. 35 p.

Clark, K. 1984. Nutrient analysis of throughfall and stemflow, and its implications for epiphytic mineral nutrition. M.A. thesis, Humboldt State Univ., Arcata, CA. 167 p.

Claycomb, D.W. 1983. Vegetational changes in a tidal marsh restoration project at Humboldt Bay, California. M.A. thesis, Humboldt State Univ., Arcata, CA. 81 p.

Cochrane, S.A. 1986. Programs for the preservation of natural diversity in California. Nongame Heritage Prog. Adm. Rpt. 84-2. Calif. Dept. Fish & Game, Sacramento. 230 p.

Cole, K. 1983. Late Pleistocene vegetation of King's Canyon, Sierra Nevada, California. Quaternary Res. 19:117-129.

Cole, K. 1980. Geologic control of vegetation in the Purisima Hills, California. Madroño 27(2):79-89.

Collins, J.N. & V.H. Resh. 1985. Utilization of natural and man-made habitats by the salt marsh song sparrow, *Melospiza melodia samuelis* (Baird). Calif. Fish & Game 71:40-52.

Collins, J.N. & L.M., & L.B. Leopold. 1986. The influence of mosquito control ditches on the geomorphology of tidal marshes in the San Francisco Bay area: evolution of salt marsh mosquito habitats. *In* Proc. 54th Annual Conf. Calif. Mosquito & Vector Control Assoc. Pp. 91-95.

Collins, L.M. & J.N., & L.B. Leopold. 1985. Geomorphic process of an estuarine marsh: Prelim. results and hypotheses. *In* Proc. 1st Intl. Conf. on Geomorphology. Manchester, England.

Combs, W.E. 1984. Stand structure and composition of the Little Lost Man Creek Research Natural Area, Redwood National Park. M.S. thesis, Humboldt State Univ., Arcata, CA. 93 p.
Conard, S.G., A.E. Jaramillo, K. Cromack & S. Rose. 1985. The role of the genus *Ceanothus* in western forest ecosystems. US For. Svc., Pac. NW For. & Range Exp. Sta. Gen. Tech. Rpt. PNW-182.
Conard, S.G. & S.R. Radosevich. 1982. Post-fire succession in white fir (*Abies concolor*) vegetation of the northern Sierra Nevada. Madroño 29(1):42-56.
Conard, S.G. & S.R. Radosevich. 1981. Photosynthesis, xylem pressure potential and leaf conductance of 3 montane chaparral species in California (*Abies concolor, Ceanothus velutinus* and *Arctostaphylos patula*). For. Sci. 27(4):627-639.
Conrad, C.E., G.A. Roby & S.C. Hunter. 1986. Chaparral and associated ecosystems management: A 5-year research and development program. US For. Svc., Pac. SW For. & Range Exp. Sta. 15 p.
Conrad, C.E. & W.C. Oechel, eds. 1982. Dynamics and management of mediterranean-type ecosystems. US For. Svc., Pac. SW For. & Range Exp. Sta. Gen. Tech. Rpt. PSW-58. 637 p.
Cope, E.A. 1983. Chemosystematic affinities of a California population of *Abies lasiocarpa.* Madroño 30:110-115.
Cope, E.A. 1978. Phytogeographic implications of monoterpene variation in *Abies lasiocarpa* (Hook.) Nutt. M.A. thesis, Humboldt State Univ., Arcata, CA. 34 p.
Covin, J. 1984. The role of inorganic N in the growth and distribution of *Spartina foliosa* at Tijuana Estuary, California. M.S. thesis, Calif. State Univ. San Diego. 60 p.
Cox, G.W. & J.B. Zedler. 1986. The influence of mima mounds on vegetation patterns in the Tijuana Estuary salt marsh, San Diego Co., California. Bull. So. Calif. Acad. Sci. 85:158-172.
DaSilva, P.G. & J.W. Bartolome. 1984. Interaction between a shrub, *Baccharis pilularis* ssp *consanguinea (Asteraceae),* and an annual grass, *Bromus mollis (Poaceae),* in coastal California. Madroño 31(2):93-101.
Davis, S. 1986. Physiology of resprouts and seedlings. Symp. on the California chaparral: Paradigms revisited. 7-8 Nov. Natur. Hist. Mus., Los Angeles. In press.
Davis, S.D. & H.A. Mooney. 1986. Tissue water relations of 4 co-occurring chaparral shrubs. Oecologia 70:527-535.
Davis, S.D. & H.A. Mooney. 1985. Comparative water relations of adjacent California shrub and grassland communities. Oecologia 66:522-529.
DeBano, L.F., R.M. Rice & C.E. Conrad. 1979. Soil heating in chaparral fires: effects on soil properties, plant nutrients, erosion and runoff. US For. Svc., Pac. SW For. & Range Exp. Sta. Res. Paper PSW-145. 21 p.
DeDecker, M. 1984. Flora of the northern Mojave desert, California. Calif. Nat. Plant Soc., Spec. Publ. 7. xv + 164.
DeJong, T.M. & M.G. Barbour. 1979. Contributions to the biology of *Atriplex leucophylla,* a C_4 California beach plant. Bull. Torrey Bot. Club 106:9-19.
DeJong, T.M. 1979. Water and salinity relations of California beach species. J. Ecology 67:647-663.
DeJong, T.M. 1978. Comparative gas exchange of 4 California beach taxa. Oecologia 34:343-351.
Dennis, N.B. & M.L. Marcus. 1984. Status and trends of California wetlands. ESA/Madrone, San Francisco, for Calif. Assembly Res. Subcomm. on Status & Trends, Sacramento. 135 p.
Despain, D.G. 1983. Nonpyrogenous climax lodgepole pine communities in Yellowstone Natl. Park. Ecology 64(2):231-234.
diCastri, F. & H.A. Mooney, eds. 1973. Mediterranean-type ecosystems—origin and structure. Ecological Studies 7. Springer Verlag, NY. 405 p.
Didden-Zopfy, B. & P.S. Nobel. 1982. High temperature tolerance and heat acclimation of *Opuntia bigelovii.* Oecologia 52:176-180.
Douglas, D.A. 1981. The balance between vegetative and sexual reproduction of *Mimulus primuloides (Scrophulariaceae).* J. Ecology 69(1):295-310.
Duebendorfer, T.E. 1987. Vegetation-soil relations in ultramafic parent materials, Pine Flat Mts., Del Norte Co., California. M.A. thesis, Humboldt State Univ., Arcata, CA. 62 p.
Duebendorfer, T.E. 1985. Habitat survey of *Erysimum menziesii* on the North Spit of Humboldt Bay. Humboldt Co. Pub. Works Dept., Natur. Res. Div., Eureka, CA. 40 p.
Dunn, P. 1981. Field observations of *Cordylanthus maritimus* Nutt. ssp *maritumus* at Tijuana Estuary, California. Final Rpt. USN, Naval Air Sta., North Island, San Diego. 18 p.
Ebert, T.A. & M.L. Balko. 1984. Vernal pools as islands in space and time. Pp. 90-101 *in* Jain & Moyle, eds.
Ehleringer, J.R. 1984. Intraspecific competitive effects on water relations, growth and reproduction in *Encelia farinosa.* Oecologia 63:153-158.
Ehleringer, J.R. 1983. Ecophysiology of *Amaranthus palmeri,* a Sonoran Desert summer ephemeral. Oecologia 57:107-112.
Eicher, A.L. 1987. Salt marsh vascular plant distribution in relation to tidal elevation, Humboldt Bay, California. M.S. thesis, Humboldt State Univ., Arcata, CA. 86 p.
Eilers, H.P. 1981. Production in coastal salt marshes of southern California. US Envir. Protec. Agency, Natl. Tech. Info. Svc., Washington, DC. Tech. Rpt. EPA-60013-81-023.
Eilers, H.P. 1980. Ecology of a coastal salt marsh after long-term absence of tidal fluctuation. Bull. So. Calif. Acad. Sci. 72:55-64.
Eliot, W.P. 1985. Implementing mitigation policies in San Francisco Bay: a critique. State Coastal Conservancy, Oakland, CA. 47 p.
Ellenberg, H. 1978. Vegetation Mitteleuropas mit den Alpen in ökologischer Sicht. 2nd ed. Ulmer, Stuttgart. 982 p.
Ellenberg, H. & F. Klötzli. 1972. Waldgesellschaften und Waldstandorte in der Schweiz. Mitt. Schweizerische Anstalt für das forstl. Versuchswesen 48(4):587-930 + tables. Rev. Ecology 59(5):1086-1088, 1978.
Environmental Laboratory. 1987. Corps of Engineers wetlands delineation manual. USA Corps of Eng. Waterways Exp. Sta., Vicksburg, MS.
Everett, R.L., compiler. 1987. Proceedings—Pinyon-juniper conference. Reno, Nevada, Jan. 13-16, 1986. US For. Svc., Intermount. Res. Sta. Gen. Tech. Rpt. INT-215. 581 p.
Everett, R., S. Sharrow & D. Thran. 1986. Soil nutrient distribution under and adjacent to singleleaf pinyon crowns. Soil Sci. Soc. Amer. J. 50:788-792.
Ewing, A.L. & J.W. Menke. 1983a. Reproductive potential of *Bromus mollis* and *Avena barbata* under drought conditions. Madroño 30(3):159-167.
Ewing, A.L. & J.W. Menke. 1983b. Responses of soft chess (*Bromus mollis*) and slender oat (*Avena barbata*) to simulated drought cycles. J. Range Mgt. 36(4):415-418.
Ferren, W.R., Jr. 1985. Carpinteria salt marsh: Environment, history and botanical resources of a southern California estuary. Herbarium, Dept. Biol. Sci., Univ. of Calif., Santa Barbara. Publ. 4. 300 p.
Ferren, W.R., Jr. & K.W. Rindlaub. 1983. Botanical resources of Goleta Slough. Chapt. 3 *in* Environmental Assessment/Impact Rpt. for Santa Barbara Munic. Airport & Goleta Slough. App. B, Biol. component. 106 p.
Field, C. 1983. Allocating leaf N to maximize C gain: Leaf age as a control on the allocation program. Oecologia 56:341-347.
Field, C. & H.A. Mooney. 1983. Leaf age and seasonal effects on light, water and N use efficiency in a California shrub. Oecologia 56:348-355.
Florence, M. & S. 1987. Prescribed burns of chaparral on BLM lands. Fremontia 15(2):7-10.
Foin, T.C. & M.M. Hektner. 1986. Secondary succession

and the fate of native species in a California coastal prairie community. Madroño 33(3):189-206.
Foin, T.C. & S.K. Jain. 1977. Ecosystems analysis and population biology: Lessons for the development of community ecology. Bioscience 27(8):532-538.
Franklin, J.F., F.C. Hall, C.T. Dyrness & C. Maser. 1972. Federal Research Natural Areas in Oregon and Washington. US For. Svc., Pac. NW For. & Range Exp. Sta., Portland, OR. Subsequent Supplements (17 in 1984).
Fujimori, T.J. 1977. Stem biomass and structure of a mature *Sequoia sempervirens* stand on the Pacific coast of northern California. J. Japan. For. Soc. 59(12):435-441.
Fulton, R.E. & F.L. Carpenter. 1979. Pollination, reproduction and fire in *Arctostaphylos*. Oecologia 38:147-157.
Gentry, A.H. 1986. Endemism in tropical versus temperate plant communities. *In* Conservation biology, the science of scarcity and diversity, ed. M.E. Soule. Sinauer Assoc., Sunderland, MA. Pp. 153-181.
Gholz, H.L. 1980. Structure and productivity of *Juniperus occidentalis* in central Oregon. Amer. Midl. Natur. 103(2):251-261.
Gibson, A.C. & P.S. Nobel. 1986. The cactus primer. Harvard Univ. Press, Cambridge, MA. ix + 286.
Goforth, D. 1984. Serpentine flora notes on prominent sites in California: Gasquet Mt. Fremontia 11:11-12.
Goldberg, D.E. & R.M. Turner. 1986. Vegetation change and plant demography in permanent plots in the Sonoran desert. Ecology 67(3):695-712.
Gray, J.T. 1983. Competition for light and a dynamic boundary between chaparral and coastal sage scrub. Madroño 30(1):43-49.
Gray, J.T. 1982. Community structure and productivity in *Ceanothus* chaparral and coastal sage scrub of southern California. Ecol. Monogr. 52(4):415-435.
Gray, J.T. & W.H. Schlesinger. 1981. Biomass, production and litterfall in the coastal sage scrub of southern California. Amer. J. Bot. 68:24-33.
Gray, K.L. & J.E. Perreira. 1987. Marina State Beach dune revegetation. *In* Proc. 2nd Nat. Plant Reveg. Symp., San Diego.
Gray, K.L. 1985a. Sunset State Beach dune revegetation plan. Res. Agency, Calif. Dept. Parks & Rec., Natur. Heritage Sect., Sacramento. 7 p.
Gray, K.L. 1985b. Marina State Beach dune revegetation plan. Res. Agency, Calif. Dept. Parks & Rec., Natur. Heritage Sect., Sacramento. 10 p.
Green, L.R. 1977. Fuelbreaks and other fuel modifications for wildland fire control. US For. Svc., Pac. SW For. & Range Exp. Sta., Agric. Handbook 499. 79 p.
Greene, S.E., T. Blinn & J.F. Franklin. 1986. Research Natural Areas in Oregon and Washington; past and current research and related literature. US For. Svc. Pac. NW Res. Sta., Gen. Tech. Rpt. PNW-197. 115 p.
Griffin, J.R., P.M. McDonald & P.C. Muick. In press. California oaks: a bibliography. US For. Svc., Gen. Tech. Rpt. PSW-.
Griffin, J.R. 1978. Maritime chaparral and endemic shrubs of the Monterey Bay region, California. Madroño 25(2):65-81.
Griggs, F.T. In press. The ecological setting for the natural regeneration of *Quercus engelmannii* in the Santa Rosa Plateau, Riverside Co., California. US For. Svc., Gen. Tech. Rpt. PSW-.
Grubb, P.J. 1977. The maintenance of species richness in plant communities: the importance of regeneration niche. Biol. Rev. 52:107-145guinon, m. 9 d. allen. 1987. the restoration of dune habitats at spanish bay. II. Prelim. results. *In* Proc. 2nd Nat. Plant Reveg. Symp., San Diego.
Gulmon, S.L. & C.C. Chu. 1981. The effects of light and nitrogen on photosynthesis, leaf characteristics, and dry matter allocation in the chaparral shrub, *Diplacus aurianticus*. Oecologia 49:207-212.
Gulmon, S.L., N.R. Chiariello, H.A. Mooney & C.C. Chu. 1983. Phenology and resource use in 3 co-occurring grassland annuals. Oecologia 58:33-42.
Gulmon, S.L. 1979. Competition and coexistence: Three annual grass species. Amer. Midl. Natur. 101(2):403-416.
Gustafson, R.J. 1981. The vegetation of Ballona. *In* Schreiber, R.W., ed. The biota of the Ballona region, Los Angeles Co. Natur. Hist. Mus. Foundation, Los Angeles. 39 p.
Haase, R. 1986. An alpine vegetation map of Caribou Lake Valley and Fourth of July Valley, Front Range, Colorado, USA. Arctic & Alpine Res. 19(1):1-10.
Halpern, C.B. 1986. Montane meadow plant association of Sequoia National Park, California. Madroño 33(1):1-23.
Harden, J.W. 1987. Soils developed on granitic alluvium near Merced, California. US Geol. Surv. Bull. 1590A. 65 p.
Harrington, T.B., J.C. Tappeiner & J.D. Walstad. 1984. Predicting leaf area and biomass of 1- to 6-year-old tanoak (*Lithocarpus densiflorus*) and Pacific madrone (*Arbutus menziesii*) sprout clumps in southwestern Oregon. Canad. J. For. 14:209-213.
Harvey, M.T. & M.N. Josselyn. 1985. Wetlands restoration and mitigation policies: comment in reply. Envir. Mgt. 10:569-570.
Heckard, L.R. & J.C. Hickman. 1984. The phytogeographical significance of Snow Mt., North Coast Ranges, California. Madroño 31:30-47.
Helms, J.A. & R.A. Ratliff. 1987. Germination and establishment of *Pinus contorta* var *murrayana (Pinaceae)* in mountain meadows at Yosemite National Park, California. Madroño 34(2):77-90.
Helms, J.A. 1987. Invasion of *Pinus contorta* var *murrayana (Pinaceae)* into mountain meadows of Yosemite National Park, California. Madroño 34(2):91-97.
Helms, J.A. & R.B. Sandiford. 1985. Predicting release of advance reproduction of mixed conifer species in California following overstory removal. For. Sci. 31(1):3-15.
Henning, I. & D. 1977. Klimatologische Wasserbilanz in Nordamerika. Arch. Met. Geoph. Biokl. B. 25:51-66.
Henning, I. & D. 1977. Zur klimatologischen Wasserbilanz von Nordamerika: mittlere Jahressummen von Wasserüberschuss und Wasserdefizit. Arch. Met. Geoph. Biokl. B. 25:117-125.
Henning, I. & D. 1981. Potential evapotranspiration in mountain geoecosystems of different altitudes and latitudes. Mountain Res. & Dev. 1(3/4):267-274.
Henning, I. & D. 1984. Zu Wämehaushalt und Hydroklima der nordamerikanischen Deserts. Arch. Met. Geoph. Biokl. B. 34:341-352.
Hill, M.O., R.G.H. Bunce & M.W. Shaw. 1975. Indicator species analysis, a divisive polythetic method of classification and its application to a survey of native pinewoods in Scotland. J. Ecol. 63:597-613.
Hill, M.O. & H.G. Gauch. 1980. Detrended correspondence analysis: an improved ordination technique. Vegetatio 42:47-58.
Hironaka, M., M.A. Fosberg & A.H. Winward. 1983. Sagebrush grass habitat types in southern Idaho. College of Forestry, Wildlife and Range Sci., Univ. of Idaho, Moscow. 35 p.
Hobbs, R.J. & H.A. Mooney. 1987. Leaf and shoot demography in *Baccharis* shrubs of different ages. Amer. J. Bot. 74(7):1111-1115.
Hobbs, R.J. & H.A. Mooney. 1986. Community changes following shrub invasion of grassland. Oecologia 70:508-513.
Hobbs, R.J. & H.A. Mooney. 1985. Community and population dynamics of serpentine grassland annuals in relation to gopher disturbance. Oecologia 67:342-351.
Holland, R.F. 1986. Preliminary descriptions of the terrestrial natural communities of California. Calif. Dept. of Fish & Game, Sacramento. 156 p., mimeo.
Holland, R.F. & S.K. Jain. 1984. Spatial and temporal variation in plant species diversity of vernal pools. Pp. 198-209. *In* S.K. Jain & P. Moyle, eds.
Holland, R.F. 1976. The vegetation of vernal pools: a survey. Pp. 11-15 *in* Jain, ed.
Holland, R.F. 1978a. The geographic and edaphic dis-

tribution of vernal pools in the Great Central Valley, California. Calif. Nat. Plant Soc. Spec. Publ. 4. 12 p. + 2 maps.

Holland, R.F. 1978b. Biogeography and ecology of vernal pools in California. PhD thesis, Univ. of Calif., Davis. 121 p.

Holland, V.L. In press. A study of blue oak regeneration in California. US For. Svc., Gen. Tech. Rpt. PSW-.

Holstein, G. 1983. List of California natural communities. Calif. Dept. of Fish & Game, Sacramento.

Holton, B., Jr. 1980. some aspects of the N cycle in a northern California coastal dune-beach ecosystem, with emphasis on *Cakile maritima*. PhD thesis, Univ. of Calif., Davis.

Holton, B., Jr. & A.F. Johnson. 1979. Dune scrub communities and their correlation with environmental factors at Pt. Reyes National Seashore, California. J. Biogeography 6:317-328.

Horvat, I., V. Glavac & H. Ellenberg. 1974. Vegetation Südösteuropas. Gustav Fischer, Jena. xxxii + 768.

Hoshovsky, M.C. 1987. Significant natural areas of California 1987. Annual Update. Calif. Dept. Fish & Game, Sacramento.

Hull, J.C. & C.H. Muller. 1977. The potential for dominance by *Stipa pulchra* in a California grassland. Amer. Midl. Natur. 97(1):147-175.

Hunt, J.L. 1976. Environmental restrictions of *Abies amabilis* in the English Peak area, Marble Mts., California. M.S. thesis, Humboldt State Univ., Arcata, CA. 44 p.

Hunter, K.B. & R.E. Johnson. 1983. Alpine flora of the Sweetwater Mts., Mono Co., California. Madroño 30(4):89-105.

Jackson, L.E., J.L.J. Houpis & M.W. Diemer. 1987. The role of leaf position in the ecophysiology of an annual grass during reproductive growth. Amer. Midl. Natur. 117(1):56-62.

Jackson, L.E. & J. Roy. 1986. Growth patterns of Mediterranean annual and perennial grasses under simulated rainfall regimes of southern France and California. Acta Oecologia, Oecol. Plant. 7[21](2):191-212.

Jackson, L.E. 1985. Ecological origin of California's Mediterranean grasses. J. Biogeography 12:349-361.

Jackson, L.E. & L.C. Bliss. 1982. Distribution of ephemeral herbaceous plants near timberline in the Sierra Nevada, California, USA. Arctic & Alpine Res. 14(1):33-42.

Jacobson, S.L. 1986. Vertebrate response to a tidal marsh restoration in Humboldt Bay, California. M.S. thesis, Humboldt State Univ., Arcata, CA. 126 pp.

Jain, S.K., ed. 1976. Vernal pools: their ecology and conservation. Inst. of Ecol., Univ. of Calif. Spec. Publ. 9. 93 p.

Jain, S.K. & P. Moyle. 1984. Vernal pools and intermittent streams. Inst. of Ecol., Univ. of Calif. Spec. Publ. 28. 280 p.

James Dobbin Assoc., Inc. 1986. Tijuana River National Estuarine Sanctuary management plan. James Dobbin Assoc., Alexandria, VA. 170 p.

Jarvis, P.G. & K.G. McNaughton. 1986. Stomatal control of transpiration: Scaling up from leaf to region. Advances in Ecol. Res. 15:1-49.

Jenny, H. 1958. Role of the plant factor in the pedogenic functions. Ecology 39(1):5-16.

Jenny, H. 1980. The soil resource. Origin and behavior. Ecological Studies 17. Springer-Verlag, NY. xx + 377.

Jimerson, T.M. In press. Draft ecological types of the Ukonom and Happy Camp Ranger Districts, Klamath National Forest. Unpubl. rpt., Klamath Natl. For., Yreka, CA.

Jimerson, T.M. 1987. Draft ecological types of the Gasquet Ranger District, Six Rivers National Forest. Unpubl. rpt., Six Rivers Natl. For., Eureka, CA.

Jimerson, T.M. & D.K. Imper. 1986. Draft of prelimary ecological types of the red fir zone, Six Rivers National Forest. Six Rivers Natl. For., Eureka, CA. 49 p.

Johnson, A.F. 1985. Ecologia de *Abronia maritima*, especie pionera de las dunas del oeste de Mexico. Biotica 10:19-34.

Johnson, A.F. 1982. Dune vegetation along the eastern shore of the Gulf of California. J. Biogeography 9:317-330.

Johnson, A.F. 1977. A survey of strand and dune vegetation along the Pacific and southern Gulf coasts of Baja California and Mexico. J. Biogeography 7:83-99.

Johnson, B.R. 1986. Influence of soil moisture, herbaceous competition, and small mammals on lodgepole pine (*Pinus contorta* var *murrayana* (Grey & Balf.) Engelm.) seedling establishment in Sierra Nevada meadows of California. PhD thesis, Univ. of Calif., Davis.

Jones, K.G. 1984. The Nipomo dunes. Fremontia 11(4):3-10.

Jordan, P.W. & P.S. Nobel. 1982. Height distribution of 2 species of cacti in relation to rainfall, seedling establishment and growth. Bot. Gaz. 143:511-517.

Jorgenson, P. 1987. Dune restoration at Tijuana Estuary. *In* Proc. 2nd Nat. Plant Reveg. Symp., San Diego.

Josselyn, M.N. & J.W. Buchholz. 1984. Marsh restoration in San Francisco Bay: A guide to design and planning. Tech. Rpt. 3. Center for Envir. Studies, San Francisco State Univ., Tiburon, CA. 103 p.

Josselyn, M.N. 1983. The ecology of San Francisco Bay tidal marshes: A community profile. US Fish & Wildlife Svc., Div. Biol. Svcs., Washington, DC. FWS/OBS-83/23. 102 p.

Josselyn, M.N., ed. 1982. Wetland restoration and enhancement in California. Calif. Sea Grant College Progress Rpt. T-CSGCP-007, La Jolla, CA. 110 p.

Josselyn, M.N. & B.F. Atwater. 1982. Physical and biological constraints on man's use of the shore zone of the San Francisco Bay estuary. *In* Kockelman, W.J., et al., eds. San Francisco Bay: Use and protection. Pac. Div., Am. Assoc. Adv. Sci., San Francisco. Pp. 57-84.

Keeley, J.E. 1987. Role of fire in seed germination of woody taxa in California chaparral. Ecology 68(2):434-443.

Keeley, J.E. & S.C. 1987. Role of fire in the germination of chaparral herbs and suffrutescents. Madroño 34(3):240-249.

Keeley, J.E., B.A. Morton, A. Pednosa & P. Trotter. 1985. Role of allelopathy, heat and charred wood in the germination of chaparral herbs and suffrutescents. J. Ecology 73(3):445-458.

Keeley, J.E. 1984. Bibliographies on chaparral and the fire ecology of other mediterranean systems. Calif. Water Res. Center, Univ. of Calif., Davis. Rpt. 58. 190 p.

Keeley, J.E. 1977. Seed production, seed populations in soil and seedling production after fire for 2 congeneric pairs of sprouting and nonsprouting chaparral shrubs. Ecology 58(4):820-829.

Keeley, J.E. & S.C. 1977. Energy allocation patterns of sprouting and non-sprouting species of *Arctostaphylos* in California chaparral. Amer. Midl. Natur. 98:1-10.

Keeley, S.C. & J.E. 1982. The role of allelopathy, heat and charred wood in the germination of chaparral herbs. Pp. 128-134. *In* Conrad, C.E. & W.C. Oechel, eds. Proc. Symp. on dynamics & mgt. of mediterranean type ecosystems. US For. Svc., Pac. SW For. & Range Exp. Sta. Gen. Tech. Rpt. PSW-58.

Keller-Wolf, T. 1987. Ecological survey of the proposed Perch Creek Research Natural Area, Klamath National Forest, Siskiyou Co., California. Unpubl. rpt. on file, US For. Svc., Pac. SW For. & Range Exp. Sta., Berkeley, CA. 25 p.

Keller-Wolf, T. 1986. Ecological survey of the proposed Stone Corral-Josephine peridotite Research Natural Area, Shasta-Trinity National Forest, California. Unpubl. rpt. on file, US For. Svc., Pac. SW For. & Range Exp. Sta., Berkeley, CA. 71 p.

Keller-Wolf, T. 1985. Ecological survey of the proposed Bridge Creek Research Natural Area, Klamath National Forest, Siskiyou Co., California. Unpubl. rpt. on file, US For. Svc., Pac. SW For. & Range Exp. Sta., Berkeley, CA. 57 p.

Keller-Wolf, T. 1984a. Ecological evaluation for the Rough Gulch drainage area with comparisons to the adjacent Chinquapin and Yolla Bolly candidate Research Natural Areas, Shasta Trinity National Forest.

Unpubl. rpt, Klamath Natl. Forest. 53 p.
Keller-Wolf, T. 1984b. Vegetation map of the upper Sugar Creek drainage, Siskiyou Co., California. Unpubl. rpt, Klamath Natl. For. 22 p.
Keller-Wolf, T. 1984c. Ecological survey of the proposed Mt. Shasta mud flow Research Natural Area, Shasta-Trinity National Forest, California. Unpubl. rpt. on file, US For. Svc., Pac. SW For. & Range Exp. Sta., Berkeley, CA. 72 p.
Keller-Wolf, T. 1982. Ecological survey of the proposed Cedar Basin Research Natural Area, Shasta-Trinity National Forest, California. Unpubl. rpt. on file, US For. Svc., Pac. SW For. & Range Exp. Sta., Berkeley, CA. 75 p.
Knapp, R. 1981 & 1982. Bibliographical review on the vegetation of California. Excerpta Botanica, Sectio B 21(2):121-153 & 22(3):175-188.
Knudsen, M.D. 1984. Life history aspects of *Quercus lobata* Nee in a riparian community, Bobelaine Audubon Sanctuary, Sutter Co., California. M.S. thesis, Calif. State Univ., Sacramento. 135 p.
Komarkova, V. 1979. Alpine vegetation of the Indian Peaks area, Front Range, Colorado Rocky Mts. Flora et Vegetatio Mundi 7, ed. R. Tüxen. J. Cramer, Vaduz. 2 vols. xiv + 591 + tables.
Kosco, B., D. Diaz & Regional Ecology Advisory Committee. 1985. Draft ecosystem classification handbook. US For. Svc., Reg. 5 Off., San Francisco. 53 p.
Kruger, F.J., D.T. Mitchell & J.U.M. Jarvis. 1983. Mediterranean-type ecosystems: the role of nutrients. Ecological Studies 43. Springer Verlag. 552 p.
Kummerow, J., D. Krause & W. Jow. 1977. Root systems of chaparral shrubs. Oecologia 29(2):163-177.
Kummerow, J., D. Krause & W. Jow. 1978. Seasonal changes in fine root density in the southern California chaparral. Oecologia 37(2):201-212
Laacke, R.J. & J.H. Tomascheski. 1986. Shelterwood regeneration of true firs: conclusions after 8 years. US For. Svc., Pac. SW For. & Range Exp. Sta. Res. Paper PSW-184. 7 p.
Laidlaw-Holmes, J.M. 1981. Forest habitat types on metasedimentary soil of the South Fork Mt. region of California. M.S. thesis, Humboldt State Univ., Arcata, CA. 46 p.
Lane, L.J., E.M. Romney & T.E. Hakonson. 1984. Water balance calculations and net production of perennial vegetation in the northern Mojave desert. J. Range Mgt. 37(1):12-18.
Lathrop, E.W. & M.J. Arct. In press. Age structure of Engelmann oak (*Quercus engelmannii* Greene) populations on the Santa Rosa Plateau, Riverside Co., California. Proc. Symp. on Multiple-use Mgt. of Calif. Hardwood Res. US For. Svc. Gen. Tech. Rpt. PSW-.
Lathrop, E.W. & B. Wong. 1986. Stand characteristics of southern oak woodland on the Santa Rosa Plateau, southern California. Crossosoma 12:1-7.
Lathrop, E.W. & H.A. Zuill. 1984. Southern oak woodlands of the Santa Rosa Plateau, Riverside Co., California. Aliso 10:603-611.
Lavrenko, E.M. 1956. [On central Asiatic mountain *Carex* bogs and on the Siberian-Mongolian element in the flora of the Caucasus.] *In* Lavrenko, E.M., S.Yu. Lipshits, V.B. Sochava & B.K. Shishkin, eds. Akademiku V.N. Sukachevu k 75-letiyu so Dnia Rozhdeniia. Leningrad, Akad. Nauk SSSR. Pp. 340-353.
Lenihan, J.M. 1983. The forest communities of the Little Lost Man Creek Research Natural Area, Redwood National Park, California. *In* Natl. Park Svc., Proc. 1st Biennial Conf. at Univ. of Calif., Davis. Pp. 139-152.
Lenihan, J.M., W.S. Lennox, E.H. Muldavin & S.D. Viers. 1982. A handbook for classifying early post-logging vegetation in the lower Redwood Creek Basin. Natl. Park Svc., Redwood Natl Park, Arcata, CA. Tech. Rpt. 7. 40 p.
Lettau, H. 1969. Evapotranspiration climatonomy. Monthly Weather Review 97:691-699.
Lincoln, P.A. 1985. Pollinator effectiveness and ecology of seed set in *Cordylanthus maritimus* ssp *maritimus* at Pt. Mugu, California. Endangered Species Off. US Fish & Wildlife Svc., Sacramento. No. 10181-9750. 48 p.
Lohmar, J.M., K.B. Macdonald & S.A. Janes. 1980. Late Pleistocene-Holocene sedimentary infilling and faunal change in a southern California coastal lagoon. *In* M.E. Field et al., eds. Quaternary depositional environments of the Pacific Coast. Pac. Coast Paleogeography Symp. 4. Soc. Econ. Paleontol. Mineral, Pac. Sect., Los Angeles. Pp. 231-240.
McBride, J.R. & K. Matsuda. In press. The pattern and timing of seed germination and early seedling development in California oaks. US For. Svc. Gen. Tech. Rpt.
MacClaran, M.P. 1986. Age structure of *Quercus douglasii* in relation to livestock grazing and fire. PhD thesis, Univ. of Calif., Berkeley. 119 p.
Macdonald, K.B. & T.W. Bilhorn. 1987. Soil oxygen, redox potential and ferrous iron testing in southern California coastal salt marshes. Abstract. 9th Biennial Intl. Estuarine Res. Conf., New Orleans.
Macdonald, K.B. ed. 1986. Pacific Coast wetlands: function and values. Symp. Proc., West. Soc. Naturalists, 63rd Ann. Meeting, Calif. State Univ., Long Beach. Dec. 1982. K.B. Macdonald & Assoc., San Diego. 84 p.
Macdonald, K.B. & P.B. Williams. 1985. The effects of hydrologic factors on light-footed clapper rail and salt marsh bird's beak habitat quality at Tijuana Estuary, San Diego, California: A conceptual management model. Rpt. to Endangered Species Off., US Fish & Wildlife Svc., Sacramento. No. 14-16-0001-84283 (NR). 104 p. + app.
Macdonald, K.B., et al. 1985 Santa Ana Marsh and adjacent lowlands: terrestrial resources report. Envir. Res. Branch, USA Corps of Eng., Los Angeles Dist. No. DACWOG-83-M-2581 (0800). 225 p. + app.
MacDonald, R. 1976. Vegetation of the Phoenix Park vernal pools on the American River bluffs, Sacramento Co., California. Pp. 69-75. *In* S.K. Jain, ed.
McIntire, M.B. 1977. The abiotic and physical effects of trampling salt marsh (*Salicornia virginica* L). M.S. thesis, Calif. State Univ., San Diego.
Mahall, B.E. & W.H. Schlesinger. 1982. Effects of irradiance on growth, photosynthesis and water use efficiency of seedlings of the chaparral shrub *Ceanothus megacarpus*. Oecologia 54:291-299.
Major, J. 1980. Overlapping tolerance curves of species. Phytocoenologia 7:26-34.
Malde, H.E. 1964. Patterned ground in the western Snake River plain, Idaho, and its possible cold-climate origin. Bull. Geol. Soc. Am. 75:191-208.
Mathews, S.C. 1986. Old-growth forest associations of the Bull Creek watershed, Humboldt Redwoods State Park, California. M.S. thesis, Humboldt State Univ., Arcata, CA. 93 p.
Mattes, H. 1982. Die Lebensgemeinschaft von Tannenhäher und Arve. Ber. Eidg. Anstalt forstl. Versuchswesen 241:3-74. *From* O. Wilmanns, 1985, On the significance of demographic processes in phytosociology, pp. 15-31 *in* J. White, ed. The population structure of vegetation, Dr. W. Junk, Dordrecht. Handbook of Veg. Sci. III.
Matsuda, K. & J.R. McBride. 1986. Differences in seedling growth morphology as a factor in the distribution of 3 oaks in central California. Madroño 33:207-216.
Mayer, M.A. 1986. Recruitment of plants into a newly established salt marsh in Elkhorn Slough, California. Final Rpt., Off. of Sanctuary Prog., NOAA, Washington, DC.
Mears, B. 1981. Periglacial wedges in late Pleistocene environments in Wyoming's intermontane basins. Quaternary Res. 15:171-198.
Metz, E. & J.B. Zedler. 1983. Who says science can't influence decision making? Coastal Zone '83, Amer. Soc. Civil Eng., NY. 1:584-600.
Miller, L. 1987. The introduction history of yellow bush lupine (*Lupinus arboreus* Sims.) on the North Spit of Humboldt Bay, California. Nature Conservancy, Calif. Field Off., San Francisco. 40 p.
Miller, P.C., ed. 1981. Resource use by chaparral and matorral: a comparison of vegetation function in two mediterranean-type ecosystems. Ecol. Studies 39. Springer, NY. 455 p.

Miller, P.C. & E. Ng. 1977. Root/shoot biomass ratios in shrubs in southern California and central Chile. Madroño 24(4):215-223.
Minnich, R.A. 1986. Evolutionary convergence or phenotypic plasticity? Phys. Geography. In press.
Minore, D. 1987. Madrone duff and the natural regeneration of Douglas fir. US For. Svc., Pac. NW For. & Range Exp. Sta. Res. Note PNW-456. 7 p.
Minore, D. 1986. Effects of madrone, chinkapin and tanoak sprouts on light intensity, soil moisture and soil temperature. Canad. J. For. Res. 16:645-658.
Montygierd-Loyba, T. & J.E. Keeley. 1987. Demographic structure of *Ceanothus megacarpus* chaparral in the long absence of fire. Ecology. 68(1):211-213.
Mooney, H.A. & P.C. Miller. 1985. Chaparral *in* Physiological ecology of North American plant communities, ed. B.F. Chabot & H.A. Mooney. Chapman & Hall, NY. Pp. 213-231.
Mooney, H.A., S.L. Gulmon & N. Johnson. 1983. Physiological constraints on plant chemical defenses. Pp. 21-36. *In* Hardin, D.A., ed. Plant resistance to insects. Amer. Chem. Soc. Symp. Ser. 208.
Mooney, H.A., ed. 1977. Convergent evolution in Chile and California mediterranean climate ecosystems. Dowden, Hutchinson & Ross, Stroudsberg, PA. 224 p.
Mooney, H.A. & C.E. Conrad, eds. 1977. Proceedings of the symposium on the environmental consequences of fire and fuel management in mediterranean ecosystems. US For. Svc. Gen. Tech. Rpt. WO-3. 498 p.
Moss, T.K. 1987. Asilomar State Beach dune restoration plan. Res. Agency, Calif. Dept. Parks & Rec., Natur. Heritage Sect., Sacramento. 41 p.
Mudie, P.J. & R. Byrne. 1980. Pollen evidence for historic sedimentation rates in California coastal marshes. Estuarine Coastal Mar. Sci. 10:305-316.
Mueggler, W.F. & W.L. Stewart. 1980. Grassland and shrubland habitat types of western Montana. US For. Svc., Intermount. For. & Range Exp. Sta. Gen. Tech. Rpt. INT-66.
Muhs, D.R. 1982. A soil chronosequence on Quaternary marine terraces, San Clemente Island, California. Geoderma 28:257-283.
Muick, P.C. & J.W. Bartolome. 1987. An assessment of natural regeneration of oaks in California. Calif. Dept. Forestry, For. & Rangeland Res. Prog. 100 p.
Muldavin, E.H. 1982. Forest communities of Shinbone Ridge in the Yolla Bolly Mts. of northern California. M.S. thesis, Humboldt State Univ., Arcata, CA. 82 p.
Muldavin, E.H., J.M. Lenihan, W.S. Lennox & S.D. Veirs. 1981. Vegetation succession in the first 10 years following logging of coast redwood forests. Natl. Park Svc., Redwood Natl. Park, Arcata, CA. Tech. Rpt 6. 69 p.
Müller, M.J. 1982. Selected climatic data for a global set of standard stations for vegetation science. Tasks for Veg. Sci. 5. Dr. W. Junk, The Hague. 306 p.
Murphy, T.G., J.C. Newman & R.J. Dow. 1981. Ecology and natural history of salt marsh bird's beak (*Cordylanthus maritimus* ssp *maritimus*) in Mugu Lagoon. *In* Dow, R.J., ed. Biennial Mugu Lagoon/San Nicolas Island Ecol. Research Symp., Pac. Missile Test Center, Naval Air Sta., Pt. Mugu, CA. Pp. 187-190.
National Audubon Society. 1986. Ballona wetland habitat management plan (draft). Natl. Audubon Soc. West. Reg. Off., Sacramento. 99 p. + 3 app.
Nelson, T.W. 1979. A flora of the Lassica, Humboldt and Trinity Cos., California. M.A. thesis, Humboldt State Univ., Arcata, CA. 127 p.
Newton, G.A. 1986. Phase 2 seed collection. Buhne Point Shoreline Erosion Demonstration Project. USA Corps of Eng., San Francisco Dist. 32 p.
Newton, G.A. 1985a. Phase One planting methods and cost analysis. Buhne Point Shoreline Erosion Demonstration Project. USA Corps of Eng., San Francisco Dist. 51 p.
Newton, G.A. 1985b. Seed collection. Buhne Point Shoreline Erosion Demonstration Project. USA Corps of Eng., San Francisco Dist. 71 p.
Niesen, T. & M. Josselyn, eds. 1981. The Hayward Regional Shoreline marsh restoration: biological succession during the first year following dike removal. Tech. Rpt. 1. Center for Envir. Studies, San Francisco State Univ., Tiburon, CA. 185 p.
Nilsen, E.T., M.R. Sharifi, P.W. Rundel & R.A. Virginia. 1986. Influence of microclimatic conditions and water relations on seasonal dimorphism of *Prosopis glandulosa* var *torreyana* in the Sonoran Desert, California. Oecologia 69:95-100.
Nilsen, E.T., M.R. Sharifi & P.W. Rundel. 1984. Comparative water relations of phreatophytes in the Sonoran Desert of California. Ecology 65:767-778.
Nilsen, E.T., M.R. Sharifi, P.W. Rundel, W.M. Jarrell & R.A. Virginia. 1983. Diurnal and seasonal water relations of the desert phreatophyte *Prosopis glandulosa* (honey mesquite) in the Sonoran Desert of California. Ecology 64:1381-1393.
Nilsen, L.T. & C.H. Muller. 1980. An evaluation of summer deciduousness in *Lotus scoparius* (Nutt. in T.&G.) Ottley. Amer. Midl. Natur. 103(1):88-95.
Nilsen, L.T. & C.H. Muller. 1981. Phenology of the drought deciduous shrub *Lotus scoparius*: climatic controls and adaptive significance. Ecol. Monogr. 51(3):323-341.
Nilsen, L.T. & W.H. Schlesinger. 1981. Phenology, productivity, and nutrient accumulation in the post-fire chaparral shrub *Lotus scoparius*. Oecologia 50:217-224.
Nixon, S. 1980. Between coastal marshes and coastal waters: A review of 20 years of speculation and research on the role of salt marshes in estuarine productivity and water chemistry. *In* Hamilton, P. & K.B. Macdonald, eds. Estuarine and wetland processes. Plenum Press, NY. Pp. 437-525.
Nixon, S. 1977. Variation and evaluation of coastal salt marshes. Envir. Mgt. 1:201-211.
Nobel, P.S. 1985. Desert succulents. *In* Chabot, B.F. & H.A. Mooney, eds. 1985. Physiological ecology of North American plant communities. Chapman & Hall, NY. Pp. 181-197.
Nobel, P.S. 1982a. Orientations of terminal cladodes of platyopuntias. Bot. Gaz. 143:219-224.
Nobel, P.S. 1982b. Low temperature tolerance and cold hardening of cacti. Ecology 63:1650-1656.
Nobel, P.S. 1980. Water vapor conductance and CO_2 uptake for leaves of a C_4 desert grass, *Hilaria rigida*. Ecology 61:252-258.
Oberdorfer, E. 1979. Pflanzensoziologische Exkursionsflora. 4th ed. Ulmer, Stuttgart. 979 p.
Oechel, W.C. & J. Mustafa. 1982. Photosynthetic variability along an elevational gradient in the chaparral. Pp. 417-433. *In* Quezel, P., ed. Definition and localization of terrestrial and marine mediterranean biota. Plenum, NY.
Old-growth Definition Task Group. 1986. Interim definitions of old-growth, Douglas-fir and mixed-conifer forests in the Pacific Northwest and California. US For. Svc., Pac. NW For. & Range Exp. Sta. Res. Note PNW-447. 7 p.
Oliver, W.W. & M.A. Mayer. 1987. Development of plant zonation and preemption of space during the colonization of a salt marsh. Moss Landing Marine Lab., CA. In review. 30 p. + tables + figs.
Oliver, W.W. 1985. Growth of California red fir advance regeneration after overstory removal and thinning. US For. Svc., Pac. SW For. & Range Exp. Sta. Res. Paper PSW-180. 6 p.
Olmsted, N. 1986. Vanished waters: A history of San Francisco's Mission Bay. Mission Creek Conservancy, San Francisco. 70 p.
Onuf, C.P. 1987. The ecology of Mugu Lagoon, California: an estuarine profile. US Fish & Wildlife Svc. Biol. Rpt. 85(7.15). 122 p.
Onuf, C.P. 1981. Demography and production of *Salicornia virginica* L. in a southern California salt marsh. Marine Sci. Inst., Univ. of Calif., Santa Barbara. 53 p.
Onuf, C.P. et al. 1978. An analysis of the values of central and southern California wetlands. *In* P.W. Greeson et al., eds. Wetlands function and values: the state of our understanding. Amer. Water Res. Assoc., Minneapolis.

Palmer, J.S. 1979. Vegetation on quartz diorite in the Bear Lakes area, Trinity Co., California. M.A. thesis, Humboldt State Univ., Arcata, CA. 81 p.

Parker, A.J. 1986a. Environmental and historical factors affecting red and white fir regeneration in ecotonal forests. For. Sci. 32(2):339-347.

Parker, A.J. 1986b. Persistence of lodgepole pine forests in the central Sierra Nevada. Ecology 67(6):1560-1567.

Parker, A.J. 1985. Mixed forests of red fir and white fir in Yosemite National Park, California. Amer. Midl. Natur. 112:15-33.

Parker, A.J. 1982. Environmental and compositional ordinations of conifer forests in Yosemite National Park, California. Madroño 29(2):109-118.

Parker, V.F. 1987. Can the native flora survive prescribed burns? Fremontia 15(2):3-6.

Parker, V.T. & C.R. Billow. In press. A comparative survey of soil N beneath evergreen and deciduous California oaks. US For. Svc., Gen. Tech. Rpt. PSW-.

Parker, V.T. 1984. Correlation of physiological divergence with reproductive mode in chaparral shrubs. Madroño 31(4):231-242.

Parker, V.T. & C.H. Muller. 1982. Vegetational and environmental changes beneath isolated live oak trees (*Quercus agrifolia*) in a California annual grassland. Amer. Midl. Natur. 107(1):69-81.

Parker, V.T. & C.H. Muller. 1979. Allelopathic dominance by a tree-associated herb in a California annual grassland. Oecologia 37:315-320.

Pavlik, B.M. 1985a. Sand dune flora of the Great Basin and Mojave deserts of California, Nevada and Oregon. Madroño 32(4):197-213.

Pavlik, B.M. 1985b. Water relations of the dune grasses *Ammophila arenaria* and *Elymus mollis* on the coast of Oregon, USA. Oikos 45:197-205.

Pavlik, B.M. 1984. Seasonal changes of osmotic pressure, symplasmic water content and tissue elasticity in the blades of dune grasses growing along the coast of Oregon. Plant, Cell & Envir. 7:531-539.

Pavlik, B.M. 1983. Nutrient and productivity relations of the dune grasses *Ammophila arenaria* and *Elymus mollis*. I. Blade photosynthesis and N use efficiency in the laboratory and field. Oecologia 57:227-232. II. Growth and patterns of dry matter and N allocation as influenced by N supply. Do. 57:233-238. III. Spatial aspects of clonal expansion with reference to rhizome growth and the dispersal of buds. Bull. Torrey Bot. Club 110:271-279.

Pavlik, B.M. 1982. Nutrient and productivity relations of the beach grasses *Ammophila arenaria* and *Elymus mollis*. PhD thesis, Univ. of Calif., Davis.

Pavlik, B.M. 1980. Patterns of water potential and photosynthesis of desert sand dune plants, Eureka Valley, California. Oecologia 46:147-154.

Pearcy, R.W. & S.L. Ustin. 1984. Effects of salinity on growth and photosynthesis of 3 California tidal marsh species. Oecologia 62:68-73.

Peart, D.R. & T.C. Foin. 1985. Analysis and prediction of population and community change: A grassland case study. *In* White, J., ed. The population structure of vegetation. Handbook of Veg. Sci. 3:313-339.

Pewe, T.L. 1983. Alpine permafrost in the contiguous United States: A review. Arctic & Alpine Res. 15(2):145-156.

Pfister, R.D., B.L. Kovalchik, S. Arno, R.C. Presby. 1977. Forest habitat types of Montana. US For. Svc., Intermount. For. & Range Exp. Sta. Gen. Tech. Rpt. INT-34. 174 p. Rev. Ecology 59(5):1086-1088, 1978.

Pickart, A.J. 1987a. Parameters controlling the success of dune revegetation at King Salmon, California. *In* Proc. 5th Symp. on Coastal & Ocean Mgt., Seattle. Pp. 1746-1758.

Pickart, A.J. 1987b. Qualitative evaluation of Phase 2 Planting. Buhne Point Shoreline Erosion Demonstration Project. USA Corps of Eng., San Francisco Dist. 23 p.

Pickart, A.J. 1987c. A classification of Northern Foredune and its relationship to Menzies wallflower on the North Spit of Humboldt Bay, California. Nature Conservancy, Calif. Field Off., San Francisco. 14 p.

Pickart, A.J. 1986. Phase 1 monitoring. Buhne Point Shoreline Erosion Demonstration Project. USA Corps of Eng., San Francisco Dist. 102 p.

Pickart, A.J. 1985. A review of California coastal dune restoration/revegetation projects. USA Corps of Eng., San Francisco Dist. 26 p.

Pitt, M.D. & H.F. Heady. 1978. Responses of annual vegetation to temperature and rainfall patterns in northern California. Ecology 59(2):336-350.

Pitts, W.D. & M.G. Barbour. 1979. The microdistribution and feeding preferences of *Peromyscus maniculatus* in the strand at Pt. Reyes National Seashore, California. Amer. Midl. Natur. 101:38-48.

Pitts, W.D. 1976. Plant/animal interactions in the beach and dunes of Pt. Reyes. PhD thesis, Univ. of Calif., Davis.

Plumb, T.R. In press. Proceedings of symposium on multiple-use management of California's hardwood resources. US For. Svc., Gen. Tech. Rpt. PSW-.

Plumb, T.R. 1980. Proceedings of the symposium on the ecology, management and utilization of California oaks. US For. Svc., Gen Tech. Rpt. PSW-44.

Pojar, J., R. Love, D. Meidinger & R. Scagel. 1982. Some common plants of the sub-boreal spruce zone. Brit. Columbia Ministry of Forests, Research Br. Land Mgt. Handbook 6. iv + 102.

Poth, M.A. 1982. Biological dinitrogen fixation in chaparral. Pp. 285-290. *In* Proc. Symp. Dynamics & Mgt. of Mediterranean-type Ecosystems. US For. Svc., Pac. SW For. & Range Exp. Sta. Gen. Tech. Rpt. PSW-58.

Provensa, F.D. & P.J. Urness. Diameter-length-weight relationships for blackbrush (*Coleogyne ramossissima*) branches. J. Range Mgt. 34:215-217.

Quinn, T.F. 1977. Flora of the Klamath Mts. M.S. thesis, Pacific Union College, Angwin, CA. 503 p.

Race, M.S. 1986. Wetlands restoration and mitigation policies: Reply. Envir. Mgt. 10:571-572.

Race, M.S. 1985. Critique of present wetlands mitigation policies in the US based on analysis of past restoration projects in San Francisco Bay. Envir. Mgt. 9:71-82.

Race, M.S. & D.R. Christie. 1982. Coastal zone development: mitigation, marsh creation and decision making. Envir. Mgt. 6:317-328.

Ratliff, R.D. 1985a. Nutrients in *Carex exserta* sod and gravel in Sequoia National Park, California. Great Basin Natur. 45(1):61-66.

Ratliff, R.D. 1985b. Rehabilitating gravel areas with short-hair sedge sod plugs and fertilizer. US For. Svc., Pac. SW For. & Range Exp. Sta. Res. Note PSW-371. 4 p.

Ratliff, R.D. 1985c. Meadows in the Sierra Nevada of California: state of knowledge. US For. Svc., Pac. SW For. & Range Exp. Sta. Gen. Tech. Rpt. PSW-84. 52 p.

Ratliff, R.D. 1983. Nebraska sedge (*Carex nebraskensis* Dewey). Observations on shoot life history and management. J. Range Mgt. 36(4):429-430.

Ratliff, R.D. 1982. A meadow site classification for the Sierra Nevada, California. US For. Svc., Pac. SW For. & Range Exp. Sta. Gen. Tech. Rpt. PSW-60. 16 p.

Raven, P.H. & D.I. Axelrod. 1978. Origin and relationships of the California flora. Univ of Calif. Publ. Bot. 72:viii + 134.

Reed, L.J. & N.G. Sugihara. In press. Northern oak woodlands—ecosystem in jeopardy or is it already too late? US For. Svc., Gen. Tech. Rpt. PSW-.

Regalia, M.R. 1987. The stand dynamics of Shasta fir in northwest California. M.S. thesis, Humboldt State Univ., Arcata, CA. 21 p.

Riegel, G.M. 1982. Forest habitat types of the southern Warner Mts., Modoc Co., northeastern California. M.S. thesis, Humboldt State Univ., Arcata, CA. 114 p.

Rogers, J.D. 1981. Net primary productivity of *Spartina foliosa, Salicornia virginica* and *Distichlis spicata* in salt marshes at Humboldt Bay, California. M.A. thesis, Humboldt State Univ., Arcata, CA. 122 p.

Rosario, J.A. & E.W. Lathrop. 1984. Distributional ecology of vegetation in the vernal pools of the Santa Rosa Plateau, Riverside Co., California. Pp. 210-217. *In* S.K. Jain & P. Moyle, eds.

Ross, R. 1983. Elevational effects on vegetation diversity of ultramafic and granitic substrata in the Trinity Alps of northern California. M.A. thesis, Humboldt State Univ., Arcata, CA. 76 p.

Roundy, B.A. 1985. Emergence and establishment of basin wildrye and tall wheatgrass in relation to moisture and salinity. J. Range Mgt. 38(2):126-131.

Rundel, P.W., E.T. Nilsen, M.R. Sharifi, R.A. Virginia, W.M. Jarrell, D.H. Kohl & G.B. Shearer. 1982. Seasonal dynamics of N cycling for a *Prosopis* woodland in the Sonoran Desert. Plant & Soil 67:343-353.

Rundel, P.W. 1981. Successional dynamics of chamise chaparral: the interface of basic research and management. *In* Abstracts Dynamics & Mgt. of mediterranean type ecosystems: an international symposium. Calif. State Univ., San Diego. 7 p.

Rundel, P.W. & D.J. Parsons. 1980. Nutrient changes in 2 chaparral shrubs along a fire induced age gradient. Amer. J. Bot. 67(1):51-58.

Saenz, L. & J.O. Sawyer, Jr. 1986. Grassland as compared to adjacent *Quercus garryana* woodland understories exposed to different grazing regimes. Madroño 33:40-46.

Savelle, G.D. 1977. Comparative structure and function in a California annual and native bunchgrass community. PhD thesis, Univ. of Calif., Berkeley. 276 p.

Sawyer, J.O. 1987. The problem of the Salmon Mt. conifers. *In* Conservation and management of rare and endangered plants, T. S. Elias, ed. Calif. Nat. Plant Soc., Sacramento.

Sawyer, J.O. 1984. Serpentine flora notes on prominent sites in California: The Lassics. Fremontia 11:15-16.

Sawyer, J.O. & E.M. Cope. 1982. Noteworthy collections, California: *Abies lasiocarpa* (Hook.) Nutt. Madroño 29:219.

Sawyer, J.O. 1981a. An ecological survey of the Adorni candidate Research Natural Area, Humboldt Co., California. Unpubl. rpt. on file, US For. Svc., Pac. SW For. & Range Exp. Sta., Berkeley. 22 p.

Sawyer, J.O. 1981b. An ecological survey of the North Trinity Mt. candidate Research Natural Area, Humboldt Co., California. Unpubl. rpt. on file, US For. Svc., Pac. SW For. & Range Exp. Sta., Berkeley. 27 p.

Sawyer, J.O. 1978. An ecological survey of the proposed Indian Creek Brewer Spruce Research Natural Area, Humboldt Co., California. Unpubl. rpt. on file, US For. Svc., Pac. SW For. & Range Exp. Sta., Berkeley. 21 p.

Sawyer, J.O., J.S. Palmer & E.A. Cope. 1978. An ecological survey of the proposed Preacher Meadows Research Natural Area, Trinity Co., California. Unpubl. rpt. on file, US For. Svc., Pac. SW For. & Range Exp. Sta., Berkeley. 24 p.

Sawyer, J.O. & K.T. Stillman. 1977a. An ecological survey of the proposed Specimen Creek Research Natural Area, Siskiyou Co., California. Unpubl. rpt. on file, US For. Svc., Pac. SW For. & Range Exp. Sta., Berkeley. 21 p.

Sawyer, J.O. & K.T. Stillman. 1977b. An ecological survey of the proposed Williams Pt. Research Natural Area, Humboldt Co., California. Unpubl. rpt. on file, US For. Svc., Pac. SW For. & Range Exp. Sta., Berkeley. 22 p.

Schlesinger, W.H. & C.S. Jones. 1984. The comparative importance of overland runoff and mean annual rainfall to shrub communities of the Mojave Desert. Bot. Gaz. 145:116-124.

Schlesinger, W.H., J.T. Gray, D.S. Gill & B.E. Mahall. 1982. *Ceanothus megacarpus* chaparral: a synthesis of ecosystem processes during development and annual growth. Bot. Rev. 48(1):71-117.

Schlesinger, W.H. & M.M. Hasey. 1981. Decomposition of chaparral shrub foliage: Losses of organic and inorganic constituents from deciduous and evergreen leaves. Ecology 62:762-774.

Schlesinger, W.H. & D.S. Gill. 1980. Biomass, production and changes in the availability of light, water and nutrients during the development of pure stands of the chaparral shrub *Ceanothus megacarpus* after fire. Ecology 61(4):181-189.

Seliskar, D.M. & J.L. Gallagher. 1983. The ecology of tidal marshes of the Pacific Northwest Coast: a community profile. US Fish & Wildlife Svc., Washington, DC. FWS/OBS-82/32. 65 p.

Selter, C.M. & W.D. Pitts. 1986. Site microenvironment and seedling survival of Shasta Red Fir. Amer. Midl. Natur. 115(2):288-300.

Seneca. E.D. & U. Blum. 1984. Response to photoperiod and temperature by *Spartina alterniflora (Poaceae)* from North Carolina and *Spartina foliosa* from California. Amer. J. Bot. 71:91-99.

Seneca, E.D. 1974. A preliminary germination study of *Spartina foliosa,* California cordgrass. Wasmann J. Biol. 33:215-219.

Shearer, G., D.M. Kohl, R.A. Virginia, B.A. Bryan, J.L. Skeeters, E.T. Nilsen, M.R. Sharifi & P.W. Rundel. 1983. Estimates of N_2-fixation from variation in the natural abundance of ^{15}N in Sonoran Desert ecosystems. Oecologia 56:365-373.

Simpson, L.G. 1980. Forest types on ultramafic parent material of the southern Siskiyou Mts. in the Klamath Region of California. M.S. thesis, Humboldt State Univ., Arcata, CA. 74 p.

Sipple, W.S. 1987. Wetland identification and delineation manual. Off. of Wetlands Protection, US EPA, Washington, DC. Interim Rpt. 2 vols.

Smith, J.P. & J.O. Sawyer. 1987a. A checklist of the vascular plants of northwest California. Herbarium Misc. Publ. 2. Humboldt State Univ., Arcata, CA. 78 p.

Smith, J.P. & J.O. Sawyer. 1987b. Endemic vascular plants of northwest California and southwest Oregon. Madroño, in press.

Smith, J.P. & J.O. Sawyer. 1981. Keys to the vascular plants of northwest California. 4th ed. Mad River Press, Eureka, CA. 202 p.

Smith, R.L. 1980. Alluvial scrub vegetation of the San Gabriel River floodplain, California. Madroño 27(3):126-138.

Snaydon, R.W. 1970. Rapid population differentiation in a mosaic environment. I. The response of *Anthoxanthum odoratum* populations to soils. Evolution 4:257-269.

Sparks, S.R. 1982. Seasonal water relations components of 2 California Sonoran Desert riparian trees. M.S. thesis, Calif. State. Univ., Fullerton.

Spicher, D.P. & M. Josselyn. 1985. *Spartina (Poaceae)* in northern California: distribution and taxonomic notes. Madroño 32:153-167.

Spicher, D.P. 1984. The ecology of a caespitose cordgrass (*Spartina* sp.) introduced to San Francisco Bay. M.A. thesis, Calif. State Univ., San Francisco. 83 p.

Spira, T.P. & O.D. Pollack. 1986. Comparative reproductive biology of alpine biennial and perennial gentians (*Gentiana: Gentianaceae*). Amer. J. Bot. 73(1):39-47.

Springer, P.F. & J.O. Sawyer. 1984. Response of vegetation and vertebrates other than fish to a tidal marsh restoration project in Humboldt Bay, California. Humboldt State Univ. Foundation, Arcata, CA. 184 p.

Stebbins, G.L. 1982. Floristic affinities of the High Sierra Nevada. Madroño 29(3):189-199.

Steele, J.M., R.D. Ratliff & G.L. Ritenour. 1984. Seasonal variation in total nonstructural carbohydrate levels in Nebraska sedge. J. Range Mgt. 37(5):465-467.

Stephenson, M. et al. 1980. The environmental requirements of aquatic plants. The use and potential of aquatic species for waste water treatment. Calif. State Water Res. Control Board, Sacramento. Publ. No. 65, App. A, 8-131-400-0. 655 p.

Stillman, K.T. 1980. Meadow vegetation on metasedimentary and metavolcanic parent materials in the north central Marble Mts., California. M.S. thesis, Humboldt State Univ., Arcata, CA. 26 p.

Storr, D. 1972. On the inclusion of a snowfall term in the relationship between the energy and water balances over land. J. Appl. Meteorology 11:1151-1152.

Strothmann, R.O. & D.F. Roy. 1984. Regeneration of Douglas-fir in the Klamath Mts. Region, California and Oregon. US For. Svc., Pac NW For. & Range Exp.

Sta., Gen. Tech. Rpt. PNW-81. 35 p.
Stuart, J.D. 1987. Fire history of old-growth forest of *Sequoia sempervirens (Taxodiaceae)* forest in Humboldt Redwoods State Park, California. Madroño 34(2):128-141.
Sugihara, N.G. & L.J. Reed. 1987. Vegetation of the Bald Hills oak woodlands, Redwood National Park, California. Madroño 34(3):193-208.
Sugihara, N.G. & L.J. Reed. In press. Prescribed fire for restoration and maintenance of Bald Hills oak woodland. US For. Svc., Gen. Tech. Rpt. PSW-.
Takhtajan, A.I. 1986. Floristic regions of the world. Transl. T.J. Crovello. Ed. A. Cronquist. Univ. of Calif. Press, Berkeley. xxii + 522. Rev. Quart. Rev. Biol. 62(2):193-194, and Ecology 61(2):439-440, 1980. (1st ed. 1978.)
Tappeiner, J.C., P.M. McDonald & T.F. Hughes. 1986. Survival of tanoak (*Lithocarpus densiflorus*) and Pacific madrone (*Arbutus menziesii*) seedlings in forests of southwestern Oregon. New Forests 1:43-55.
Tappeiner, J.C., T.B. Harrington & J.D. Worley. 1984. Predicting recovery of tanoak (*Lithocarpus densiflorus*) and Pacific madrone (*Arbutus menziesii*) after cutting and burning. Weed Sci. 32:413-417.
Tappeiner, J.C. & P.M. McDonald. 1984. Development of tanoak understory in conifer stands. Canad. J. For. Res. 14:271-277.
Tausch, H.J., N.E. West & A.A. Nabi. 1981. Tree age and dominance patterns in Great Basin pinyon-juniper woodlands. J. Range Mgt. 34(4):259-264.
Taylor, D.W. & K.A. Teare. 1978. An ecological survey of the proposed Smoky Creek Research Natural Area, Trinity Co., California. Unpubl. rpt. on file, US For. Svc., Pac. SW For. & Range Exp. Sta., Berkeley. 44 p.
Taylor, D.W. 1977. Floristic relationships along the Cascade-Sierran axis. Amer. Midl. Natur. 97(2):333-349.
Thimann, K.V. 1987. Foster habitats. Bioscience 37(1):5.
Thornburgh, D.A. 1987. *Picea breweriana. In* A silvics of forest trees of the United States. USDA Handbook, in press.
Thornburgh, D.A. 1984a. An ecological survey of the proposed Ruth Research Natural Area, Humboldt Co., California. Unpubl. rpt. on file, US For. Svc., Pac. SW For. & Range Exp. Sta., Berkeley. 11 p.
Thornburgh, D.A. 1984b. Succession in the mixed evergreen forests of northwest California. *In* Forest succession and stand development research in the Northwest, J.E. Means, ed. For. Res. Lab., Oregon State Univ., Corvallis. Pp. 87-91.
Thornburgh, D.A. & R.C. White. 1979. Vegetation habitat-types of montane Siskiyou Mts. Unpubl. rpt. on file, Six Rivers Natl. For., Eureka, CA. 42 p.
Tiedemann, A.R., E.D. McArthur, H.C. Stutz, R. Stevens & K.L. Johnson. 1983. Proc. Symp. on the biol. of *Atriplex* and related chenopods. US For. Svc., Intermount. For. & Range Exp. Sta., Ogden, UT. Gen. Tech. Rpt. INT-172.
Tomback, D.F. 1986. Post-fire regeneration of Krummholz whitebark pine: a consequence of nutcracker seed caching. Madroño 33(2):100-110.
Ustin, S.L., R.A. Woodward, M.G. Barbour & J.T. Hatfield. 1984. Relationships between sunfleck dynamics and red fir seedling distribution. Ecology 65(5):1420-1428.
Ustin, S.L., R.W. Pearcy & D.E. Bayer. 1982. Plant water relationships in a San Francisco Bay salt marsh. Bot. Gaz. 143:368-373.
Vandervier, J.M. & J.C. Newman. 1984. Observations of haustoria and host preference in *Cordylanthus maritimus* ssp *maritimus (Scrophulariaceae)* at Mugu Lagoon. Madroño 31:185-190.
Vandervier, J.M. 1983. A study of *Cordylanthus maritimus* ssp *maritimus* at Mugu Lagoon, Ventura Co., California. *In* Third Biennial Mugu Lagoon/San Nicolas Island Ecol. Res. Symp., Pac. Missile Test Center, Pt. Mugu, CA. Pp. 117-126.
Van Hook, S.S. 1983. A study of European beachgrass, *Ammophila arenaria* (L.) Link: Control methods and a management plan for the Lanphere-Christensen Dunes Preserve. Nature Conservancy, Calif. Field Off., San Francisco. 54 p.
Vankat, J.L. 1982. A gradient perspective on the vegetation of Sequoia National Park, California. Madroño 29(3):200-214.
Vankat, J.L. & J. Major. 1978. Vegetation changes in Sequoia National Park, California. J. Biogeography 5:377-402.
Vasek, F.C. 1980. Creosote bush: long-lived clones in the Mohave Desert. Amer. J. Bot. 67(2):245-255.
Vassilchenko, I.T. & A.A. Konnov. 1986. [Juniper forests of Middle Asia—immediate tasks of their study, preservation and rational use.] Bot. Zhurnal 71(4):554-556.
Viers, S.D. 1982. Coast redwood forest stand dynamics, successional status and the role of fire. *In* Forest succession and stand development research in the Pacific Northwest, ed. J.E. Means. For. Res. Lab., Oregon State Univ., Corvallis. Pp. 119-141.
Virginia, R.A. & W.M. Jarrell. 1983. Soil properties in a mesquite-dominated Sonoran Desert ecosystem. Soil Sci. Soc. Am. J. 47(1):138-144.
Waddel, D.R. 1982. Mountain forest vegetation-soil relationships in the Yolla Bolly Mts., northern California. M.S. thesis, Humboldt State Univ., Arcata, CA. 89 p.
Wainwright, T.C. & M.G. Barbour. 1984. Characteristics of mixed evergreen forest in the Sonoma Mts. of California. Madroño 31(4):219-230.
Walker, J., C.H. Thompson & W. Jehne. 1983. Soil weathering stage, vegetation succession and canopy dieback. Pac. Sci. 37:471-481.
Wallace, G.D. 1985. Vascular plants of the Channel Islands of southern California and Guadalupe Island, Baja California. Contrib. Sci. (Bot. Sect.), Natur. Hist. Mus., Los Angeles. 365:1-136.
Warner, R.E. & K.M. Hendrix. 1984. California riparian systems. Univ. of Calif. Press, Berkeley. xxix + 1035.
Wartenberg, D., S. Ferson & F.J. Rohlf. 1987. Putting things in order: A criticism of detrended correspondence analysis. Amer. Natur. 129(3):434-448.
Webb, R.H., J.W. Steiger & R.M. Turner. 1987. Dynamics of Mojave desert shrub assemblages in the Panamint Mts., California. Ecology 68(3):478-490.
Wells, K.F. 1967. Aspects of shrub-herb productivity in an arid environment. M.S. thesis, Univ. of Calif., Berkeley.
Wells, P.V. 1986. Ecological aspects of evolution in *Arctostaphylos*. Symp. on the California chaparral: Paradigms revisited. Natur. Hist. Mus., Los Angeles.
Wells, P.V. & D. Woodcock. 1985. Full-glacial vegetation of Death Valley, California. Juniper woodland opening to *Yucca* semidesert. Madroño 32(1):11-23.
Welten, M. & R. Sutter. 1982. Verbreitungsatlas der Farn- und Blütenpflanzen der Schweiz. 2 vols. Birkhäuser, Basel.
Wendler, G. 1979. Snow blowing and snowfall on the North Slope, Alaska. 47th Ann. Meeting, West. Snow Conf. Pp. 10-19.
Werk, K.B., J. Ehleringer, I.N. Forseth & C.S. Cook. 1983. Photosynthetic characteristics of Sonoran Desert winter annuals. Oecologia 59:101-105.
West, N.E., ed. 1983. Temperate deserts and semi-deserts. Ecosystems of the world 5. Elsevier Scientific, Amsterdam. 522 p.
Westman, W.E. 1987. Above-ground biomass, surface area and production relationships of red fiir (*Abies magnifica*) and white fiir (*Abies concolor*). Canad. J. For. Res. 17:311-319.
Westman, W.E. 1983. Island biogeographical studies on the xeric shrublands of the inner Channel Islands, California. J. Biogeogr. 10(2):97-118.
Westman, W.E. 1981a. Factors influencing the distribution of species of Californian coastal sage scrub. Ecology 62(2)L:439-455.
Westman, W.E. 1981b. Diversity relations and succession in Californian coastal sage scrub. Ecology 62(1):170-184.
Westman, W.E. 1979. A potential role of coastal sage scrub understories in the recovery of chaparral after fire. Madroño 26(2):64-68.

Whipple, J.J. 1984. Serpentine flora notes on prominent sites in California: Mt. Eddy. Fremontia 11:15-16.
Whipple, J.J. 1981. A flora of Mt. Eddy, Klamath Mts., California. M.A. thesis, Humboldt State Univ., Arcata, CA. 206 p.
Whipple, J.J. & E.A. Cope. 1979. An ecological survey of the proposed Mt. Eddy Research Natural Area, California. Unpubl. rpt. on file, US For. Svc., Pac. SW For. & Range Exp. Sta., Berkeley. 32 p.
Whittaker, R.H. 1975. Communities and ecosystems. 2nd ed. Macmillan, NY. xviii + 387.
Williams, P.B. & M.L. Swanson. 1987. Tijuana Estuary enhancement hydrologic analysis (draft). Philip Williams & Assoc., San Francisco, for San Diego State Univ. Foundation. 50 p.
Williams, P.B. & H.T. Harvey. 1983. California coastal salt marsh restoration design. Coastal Zone '83, Amer. Soc. Civil Eng., NY. Pp. 1445-1456.
Williams, W.T. 1985. The Morro Bay sand spit, a California treasure. Fremontia 13(3):11-16.
Williams, W.T. & J.A. 1985. Invasion rate of an introduced iceplant, *Carpobrotus chilense* in the coastal strand community at Morro Bay. Bull. Ecol. Soc. Amer. 66(2):295.
Winfield, T.P., Jr. 1980. Dynamics of C and N in a southern California salt marsh. PhD thesis, Univ. of Calif., Riverside & Calif. State Univ., San Diego. 76 p.
Woodward, F.I. 1987. Climate and plant distribution. Cambridge Univ. Press, Cambridge. xi + 174.
Yeaton, R.I. & R.Y., J.P. Waggoner & J.E. Horenstein. 1985. The ecology of *Yucca (Agavaceae)* over an environmental gradient in the Mohave Desert: Distribution and interspecific interactions. J. Arid Envir. 8(1):33-44.
Yeaton, R.I. 1984. Aspects of the population biology of sugar pine (*Pinus lambertiana*) on an elevational gradient in the Sierra Nevada of central California. Amer. Midl. Natur. 111:126-137.
Yeaton, R.I. 1983. The successional replacement of ponderosa pine by sugar pine in the Sierra Nevada. Bull. Torrey Bot. Club. 110(3):292-297.
Yeaton, R.I. 1982. Ecomorphology and habitat utilization of *Echinocereus engelmannii* and *E. triglochidiatus (Cactaceae)* in southeastern California. Great Basin Natur. 42(3):353-359.
Yoder, V., M.G. Barbour, R.S. Boyd & R.A. Woodward. 1983. Vegetation of the Alabama Hills region, Inyo Co., California. Madroño 30(2):118-126.
Young, D.R. & P.S. Nobel. 1986. Predictions of soil water potentials in the northwestern Sonoran Desert. J. Ecology 74(1):143-154.
Young, J.A. & R.A. Evans. 1986. Erosion and deposition of fine textured particles from playas. J. Arid Environments 10:103-118.
Young, J.A., R.A. Evans, B.A. Roundy & J.A. Brown. 1986. Dynamic landforms and plant communities in a pluvial lake basin. Great Basin Natur. 46:1-21.
Young, J.A., R.A. Evans & D.A. East. 1984. Stem flow in western juniper (*Juniperus occidentalis*) trees. Weed Sci. 32:320-327.
Young, J.A., R.A. Evans, C.A. Raguse & J.R. Larson. 1981. Germinable seeds and periodicity of germination in annual grasslands. Hilgardia 49(2):1-37.
Young, J.A. & R.A. Evans. 1981. Demography and fire history of a western juniper stand. J. Range Mgt. 34(6):501-506.
Youngblood, A.P. & R.L. Mauk. 1985. Coniferous forest types of central and southern Utah. US For. Svc., Intermount. Res. Sta. Gen. Tech. Rpt. INT-187. 90 p. + color photos.
Wu, K.K. & S.K. Jain. 1979. Population regulation in *Bromus rubens* and *B. mollis:* Life cycle components and competition. Oecologia 39:337-357.
Zedler, J.B. 1987. Salt marsh restoration: Lessons from California. *In* Cairns, J., ed. Management for rehabilitation and enhancement of ecosystems. CRC Press, Boca Raton, FL. In press.
Zedler, J.B. 1986a. Wetland restoration: Trials and errors or ecotechnology? *In* Wetland functions, rehabilitation and creation in the Pacific Northwest. Washington State Dept. Ecol., Olympia. Publ. 86-14:11-16.
Zedler, J.B. 1986b. Catastrophic flooding and distributional patterns of Pacific cordgrass (*Spartina foliosa* Trin.). Bull. So. Calif. Acad. Sci. 85:74-86.
Zedler, J.B. & P.A. Beare. 1986. Temporal variability of salt marsh vegetation: The role of low-salinity gaps and environmental stress. *In* Wolfe, D.A., ed. Estuarine variability. Academic Press, NY. Pp. 295-306.
Zedler, J.B. & C.S. Nordby. 1986. The ecology of Tijuana Estuary, California: an estuarine profile. US Fish & Wildlife Svc., Biol. Rpt. 85(7.5). 104 p.
Zedler, J.B. et al. 1986. Catastrophic events reveal the dynamic nature of salt marsh vegetation in southern California. Estuaries 9:75-80.
Zedler, J.B. 1984. Salt marsh restoration: A guidebook for southern California. Calif. Sea Grant College Prog. Rpt. T-CSGCP-009. 46 p.
Zedler, J.B. & C.P. Onuf. 1984. Biological and physical filtering in arid-region estuaries: seasonality, extreme events and effects of watershed modification. *In* V.S. Kennedy, ed. The estuary as a filter. Academic Press, NY. Pp. 415-432.
Zedler, J.B. 1983a. Salt marsh restoration: The experimental approach. Coastal Zone '83, Amer. Soc. Civil. Eng., NY. 3:2578-2586.
Zedler, J.B. 1983b. Freshwater impacts in normally hypersaline marshes. Estuaries 6:346-355.
Zedler, J.B. 1982a. The ecology of southern California coastal salt marshes: A community profile. US Fish & Wildlife Svc., Biol. Svc. Prog., Washington, DC. FWS/OBS-81/54. 110 p.
Zedler, J.B. 1982b. Salt marsh algal mat composition: Spatial and temporal comparisons. Bull. So. Calif. Acad. Sci. 8:41-50.
Zedler, J.B. 1980. Algal mat productivitiy: Comparisons in a salt marsh. Estuaries 3:122-131.
Zedler, J.B., T. Winfield & P. Williams. 1980. Salt marsh productivity with natural and altered tidal circulation. Oecologia 44:236-240.
Zedler, J.B., T. Winfield & D. Mauriello. 1978. Primary production in a southern California estuary. Coastal Zone '78. Amer. Soc. Civil Eng., NY. 1:649-662.
Zedler, P.H. 1987. The ecology of southern California vernal pools: a community profile. US Fish & Wildlife Svc. Biol. Rpt. 85(7.11). 136 p.
Zedler, P.H., C.R. Gautier & G.S. McMaster. 1983. Vegetation change in response to extreme events: the effect of a short interval between fires in California chaparral and coastal scrub. Ecology 64(4):809-818.
Zedler, P.H. 1977. Life history attributes of plants and the fire cycle: a case study in chaparral dominated by *Cupressus forbesii. In* H.A. Mooney & C.E. Conrad, eds., Proc. Symp. on the Envir. Consequences of Fire and Fuel Mgt. in Mediter. Ecosystems, 1-5 Aug. 1977, Palo Alto, CA. US For. Svc. Gen. Tech. Rpt. WO-3. Pp. 451-458.
Zielinski, G.A. & W.D. McCoy. 1987. Paleoclimatic implications of the relationship between modern snowpack and late Pleistocene equilibrium-line altitudes in the mountains of the Great Basin, western USA. Arctic & Alpine Res. 19(2):127-134.
Zieroth, Elaine. 1978. The vegetation and environment of red fir clearcuts in the central Sierra Nevada, California. M.A. thesis, Calif. State Univ., Fresno. 80 p.
Zobel, D.B., F. Roth & G.M. Hawk. 1985. Ecology, pathology and management of Port Orford cedar (*Chamaecyparis lawsoniana*). US For. Svc., Pac NW For. & Range Exp. Sta., Gen. Tech. Rpt. PNW-184. 161 p.